Journal of
Physical and Chemical Reference Data

Monograph No. 5

Atlas of Surface Structures: Volume 1A (1994) Based on the NIST Surface Structure Database (SSD)

P. R. Watson
Department of Chemistry
Oregon State University
Corvallis, Oregon 97331

M. A. Van Hove
Materials Sciences Division
Lawrence Berkeley Laboratory
University of California
Berkeley, California 94720

K. Hermann
Abteilung Theorie
Fritz-Haber-Institut
Faradayweg 4–6
D-14195 Berlin, Germany

Major Contributors:
S. Crampin, G. Darling, P. Pinkava, A. Barbieri, J. Batteas,
A. Kaduwela, N. Materer, U. Starke, W. Weiss

Consultants:
G. A. Somorjai and J. B. Pendry

Published by the **American Chemical Society**
and the **American Institute of Physics** for
the **National Institute of Standards and Technology**

International Standard Book Number
1-56396-413-9

American Institute of Physics

500 Sunnyside Boulevard

Woodbury, New York 11797-2999

Printed in the United States of America

Foreword

The *Journal of Physical and Chemical Reference Data* is published jointly by the American Institute of Physics and the American Chemical Society for the National Institute of Standards and Technology (NIST). Its objective is to provide critically evaluated physical and chemical property data, fully documented as to the original sources and the criteria used for evaluation. One of the principal sources of material for the journal is the NIST Standard Reference Data Program, a program promoting the compilation and critical evaluation of property data.

The regular issues of the *Journal of Physical and Chemical Reference Data* are published bimonthly and contain compilations and critical data reviews of moderate length. Longer works, volumes of collected tables, and other material unsuited to a periodical format have previously been published as *Supplements to the Journal*. Beginning in 1989 the generic title of these works has been changed to *Monograph*, which reflects their character as independent publications. This volume, ''Atlas of Surface Structures: Volume 1A (1994)'' by P.R. Watson, M.A. Van Hove, and K. Hermann is the first part of *Monograph No. 5* of the *Journal of Physical and Chemical Reference Data.*

Jean W. Gallagher, Editor
Journal of Physical and Chemical Reference Data

Atlas of Surface Structures: Volume 1A (1994)
Based on the NIST Surface Structure Database (SSD)

P.R. Watson
Department of Chemistry, Oregon State University, Corvallis, OR 97331

M.A. Van Hove
Materials Sciences Division, Lawrence Berkeley Laboratory, University of California, Berkeley, CA 94720

and

K. Hermann
Abteilung Theorie, Fritz-Haber-Institut, Faradayweg 4–6, D–14195 Berlin, Germany

Major Contributors:
S. Crampin, G. Darling, P. Pinkava, A. Barbieri, J. Batteas,
A. Kaduwela, N. Materer, U. Starke, W. Weiss

Consultants:
G. A. Somorjai and J. B. Pendry

These volumes contain listings of crystallographic information on almost 600 surface structures contained in the NIST Surface Structure Database (SSD). These are prefaced by explanatory text and are combined with high quality computer-generated views of the surface. The listings include full literature references, statements on experimental and theoretical methods used, bulk and surface unit cells, a complete list of atomic coordinates for overlayer, epilayer, interfacial and bulk atoms and selected bond lengths and angles.

Key words: electron diffraction; ion scattering; LEED; NEXAFS; photoelectron diffraction; SEXAFS, surface crystallography; surface structure; surface structure database; X-ray diffraction.

Contents

Volume 1A

1. Introduction 5
   1.1. Background 5
   1.2. Purpose of This Atlas 5
   1.3. How to use This Atlas 6
   1.4. Information Content of This Atlas 6
      1.4.1. General Information 7
      1.4.2. Surface Type 7
      1.4.3. Structure Type 9
      1.4.4. Comments 9
      1.4.5. Sample Preparation 9
      1.4.6. Data Collection 9
      1.4.7. Theory/Data Treatment 9
      1.4.8. Structures Examined 9
      1.4.9. Quality of Experiment-Theory Fit 9
      1.4.10. 2D Unit Cells 10
      1.4.11. Lattice Definitions 10
      1.4.12. 3D Coordinates 11
      1.4.13. Bond Distances and Angles 13
   1.5. Relation of this Atlas to Other Databases and Graphics Software 13
2. Catalog of Structures 14
   2.1. Alphanumerical Index of Structures 14
   2.2. Main Table of Structures 19
3. Acknowledgments 879
4. Bibliography and Related Software 879
   4.1. References 879
   4.2. Software 879
5. Appendices 880
   A. Entering New Structural Data into SSD 880
      A.1. Submitting New Structures for Inclusion in SSD 880
      A.2. Format of ASCII Files for SSD 880
      A.3. Generic ASCII File 882
      A.4. Sample ASCII Files 886
      A.5. Utility Program SSDCHK 894
   B. List of Symbols and Abbreviations 895
   C. Major Surface Structure Techniques 899
   D. List of Covalent Radii 900
   E. The 17 2D Space Groups 901
   F. Examples of Superlattice Cells and Notations 903
   G. Contacts 906
6. Index 907

Volume 1B

1. Introduction 5
2. Structural Figures 5
   2.1. Categorized Index of Figures 5
   2.2. Figures 11
3. Acknowledgments 343
4. Bibliography and Related Software 343
   4.1. References 343
   4.2. Software 343

1. Introduction

1.1. Background

Surface science has undergone explosive growth in the last decades. Of fundamental importance has been the knowledge of the atomic-scale geometrical surface structure. In close analogy with three-dimensional solids, the atomic-scale structure of surfaces is the basis for many surface-related properties of materials: electronic, chemical, magnetic, mechanical, biological, etc. Structural understanding is particularly critical for producing novel materials with new properties.

In the early 1980s, the need to gather all solved surface structures in a uniform database became obvious. A collaboration between groups in Berkeley in the USA and in Daresbury (and later at Imperial College) in the UK led to the publication in 1987 of the Surface Crystallographic Information Service (SCIS) Handbook and Software. It made available 256 surface structures in both printed and electronic forms.

The effort was subsequently reorganized as a collaboration between Berkeley, Corvallis (Oregon, USA) and Berlin (Germany). In Corvallis, P.R. Watson had independently started critical evaluations of surface structures, supported by NIST, while, in Berlin, K. Hermann was developing surface visualization software, in collaboration with M.A. Van Hove at Berkeley. By joining corresponding database software developed by Watson and visualization software written by Hermann to the data collection effort, now centralized in Berkeley, the Surface Structure Database (SSD) was created, with partial NIST support. It was published in 1993 in an electronic form for PCs and compatible computers. This atlas is a printed version of SSD and contains the same information (save for the correction of a few typographical errors).

The number of included structures has sharply risen from 256 for SCIS to 597 for SSD and this atlas, illustrating the vitality of surface crystallography in particular, and surface science in general. One can estimate the present rate of production of new structures to be one per week.

In a sense, this atlas succeeds the well-known (but now out-of-print) ''Atlas of Models of Crystal Surfaces'' by J.F. Nicholas, which described in considerable geometric detail clean, unrelaxed and unreconstructed surfaces obtained by ideal termination of the most common bulk lattices. When Nicholas' atlas was published, in 1965, no actual surface structures had been solved, while we can now encompass the very wide variety of real surface structures determined since then, whether relaxed or reconstructed, clean or covered with adsorbed atoms and molecules, and including adatom penetration or epitaxial growth. Nicholas' atlas used photographs of hand-made ball models of surfaces.

This new atlas illustrates the actual surface structures, with figures generated electronically from the experimentally-determined atomic coordinates. Standard numerical output of SSD was used, then postprocessed by the visualization tools PLOT3D and BALSAC (written by K. Hermann) and put into PostScript format for final printing.

1.2. Purpose of This Atlas

This atlas of surface structures is aimed at scientists and students in physics, chemistry and materials sciences who wish to know and compare the detailed atomic-scale structures of surfaces obtained from experiment.

This first edition attempts to cover all known surface structures since the inception of surface crystallography in the early 1970s through the end of 1991. It provides extensive structural information about surface structures determined from experiment. A unified format is used to allow convenient direct comparisons of related but different structures, or of results obtained with different techniques for the same structure. The criteria for inclusion of structures are:

a. publication in refereed journals or proceedings before 1992;
b. determination with proven techniques and analysis methods;
c. reasonable completeness of the structure determination;
d. complete description of the structure.

The source of the data is an extensive search of the refereed literature in the major journals and conference proceedings, including especially: Surface Science, Physical Review Letters, Physical Review B, Journal of Physics (London), Solid State Communications, and the Proceedings of the International Conferences on the Structure of Surfaces. A more complete list of included journals is provided in Appendix B.

By selecting only refereed publications, a first filtering of acceptable structures was performed. Where a structure determination was described in several publications, we have generally chosen the latest or most complete publication. A new analysis of a given structure by the same authors or group, using the same technique, generally implies exclusion of an earlier result from this atlas, except where the older analysis has historical value. Controversial structures have generally not been included, except again for historical value: solved structures are normally included only when a clear consensus appears. When serious doubts exist about any included structure, an editorial comment is added to that effect. This atlas therefore collects all atomic-scale surface structures known with a high degree of reliability.

A multitude of techniques has been developed since the early 1970s to determine surface structure from experiment, especially for crystalline surfaces. These techniques employ various probes: photons, electrons, ions, atoms, or combinations thereof. Thanks to the differing characteristics of each technique, independent studies of the same structure with different techniques provide a powerful check on the reliability of results. Many of the surface structures in this atlas have been verified in this manner, enhancing both the reliability of those particular structures and the reliability of the techniques used to determine other structures. Techniques that give only qualitative structural information are not included, such as vibrational spectroscopies, STM and FIM (unless coupled with theoretical simulations that give quantitative determinations).

Results obtained with a wide variety of surface structure determination techniques have been included in the data. The most prominent techniques and their basic principles are summarized in Appendix C):

To be accepted in this atlas, structures must have been determined with sufficient completeness. Atomic positions must be known with respect to the substrate lattice in three dimensions (e.g., interlayer spacing or bond length information alone is insufficient). However, hydrogen positions in molecules are generally unknown and omitted (they may nonetheless be included, using guessed coordinates, and are thus labeled). Also, the positions of impurities that stabilize certain metastable structures are usually unknown and therefore omitted. It is of course understood that, in all cases, the actual surface structure may be more complex and variable than the idealized and average structure determined from experiment.

Finally, structures included in this atlas must have been described completely and unambiguously in the corresponding publication (see also in this regard the last paragraph of this section).

The data in this atlas include commensurate structures (which case covers most solved structures), incommensurate structures (formed for instance with physisorption) and disordered structures of adsorbates and alloys, as well as solid/solid interfaces.

The data in this atlas are available as an electronic database, called Surface Structure Database (SSD), which runs on PCs and compatible computers. SSD contains the same information as does this atlas, and provides the following main advantages:

a. targeted searches for particular structures or classes of structures;
b. interactive color graphics, including rotations, dissections, on-screen bond length and angle measurements, etc.;
c. option of color or gray-tone publication-quality printing of surface views on PostScript laser printers, with many user-selected presentation options.

The Surface Structure Database can be obtained from NIST directly at the following address:

U.S. Department of Commerce
National Institute of Standards and Technology
Standard Reference Data Program
Building 221, Room A320
Gaithersburg, MD 20899
(301) 975–2204

This atlas, in contrast with the electronic database, offers the advantage of quick off-the-shelf access to specific information, without need for a computer. Thanks to carefully selected views of surface structures, the illustrations in Volume 1B of this atlas provide a valuable source of visual information to complement the extensive tabular data shown in Volume 1A.

Unintentional omissions of structures and errors in our tabulations are possible. The authors of affected structures are invited to submit their structures or corrections directly to one of the authors. Appendix A.1 describes in detail the submission of surface structures for possible inclusion in future editions. The ultimate decision on inclusion rests with the database authors, so prior consultation is advisable.

New and updated structures may also be submitted by their authors for inclusion in future editions of SSD and this atlas. See Appendix A for details.

We take this opportunity to address a plea to authors of surface structures: It is our experience that many published structural results are not described in a clear, accessible manner. Often it is necessary to piece together the structure from disparate sentences in various parts of a paper. It is sometimes even difficult to find a clear statement of the type of structure favored by the authors. We strongly suggest that structures be summarized completely in the conclusions section of a paper for easy access by others, e.g., as a complete set of atomic coordinates.

1.3. How to use This Atlas

This atlas, relative to the electronic database (SSD), provides the most rapid access to specific information about individual surface structures, via the alphabetical listing given in Sec. 2.1. The structural tables themselves are ordered alphanumerically in Sec. 2.2. The accompanying illustrations are provided in Volume 1B, using a logical categorization scheme described in Sec. 2.1 of that volume. In this way the reader can directly compare the illustration with the accompanying numerical data.

Each individual structure is given about a page of tabular information in Sec. 2.2. The meaning of the particular data is generally self-explanatory, but in any case explained in great detail in Sec. 1.4. Most structures refer to a particular illustration in Volume 1B.

Abbreviations used in the data are listed in Appendix B.

Appendix E illustrates the 17 two-dimensional space groups, whose notations are used in the data to characterize the surface symmetries. Likewise, Appendix F shows general examples of many common superlattice cells and notations.

Authors of structures who wish to contribute new results to the database should read Appendix A for instructions.

Finally, information on how to contact the database authors is listed in Appendix G.

1.4. Information Content of This Atlas

Any given structure in the database is accompanied by several groups of information:

GENERAL INFORMATION: identifies the structure and the bibliographic reference that describes its determination;

SURFACE TYPE: identifies the nature of the surface, e.g. substrate material and crystallographic plane, adsorbate nature and coverage, surface pattern and symmetry;

STRUCTURE TYPE: describes the kind of structure in words, e.g. unreconstructed or reconstructed substrate, overlayer or underlayer adsorption, adsorption site, molecular orientation;

COMMENTS: gives other pertinent information;

SAMPLE PREPARATION: indicates the experimental procedures;

DATA COLLECTION: describes measurement methods and experimental data base used for the structural determination;

THEORY/DATA TREATMENT: indicates analysis methods of the experimental data;

STRUCTURES EXAMINED: lists those structures or models tested against experiment;

QUALITY OF EXPERIMENT-THEORY FIT: estimates the reliability of the structure determination;

2D UNIT CELLS: gives unit cell information for substrate and surface or interface, including commensurate lattices, incommensurate lattices and disordered cases;

3D COORDINATES: tabulates full atomic coordinates, and a layer-by-layer description of the structure type, whenever helpful;

BOND DISTANCES AND ANGLES: lists selected interatomic distances and angles.

The detailed information below should be used in conjunction with the sample data sheet which is on the next page.

1.4.1. General Information

[s1] "Common name".

This is a commonly used name for the structure, put in uniform notation, e.g., Ni(100)-c(2×2)-CO. It may be non-unique, i.e. other names may exist for this structure (e.g., Ni(100)-($\sqrt{2}\times\sqrt{2}$)R45°-CO), or other structures may have the same common name (e.g., an overlayer and an underlayer structure could have the same common name).

Note: different inequivalent structures may coexist on the same sample at the same time; such different structures are treated as independent structures in SSD, with separate structure classification numbers [s2].

[s2] "Classification number".

Each structure determination in the database has a unique structure classification number for unambiguous identification.

The first parts of the classification number are atomic numbers that indicate all chemical elements present in the structure. Thus, the classification number 28.6.8.2b implies that this structure consists of 28 = Ni, 6 = C and 8 = O; the last number, like 2b, labels different studies of surfaces containing the same elements. The classification number is assigned by the database authors.

[s3] "Date entered into SSD".

This item of information is used by the authors to update the SSD Database and is not shown here.

[s4] "Technique".

Gives the main experimental technique used to determine the surface structure. (If other techniques were also used, these should appear in the "Comments" items [s28-s32]). See Appendix B for explanations of the abbreviations used for techniques.

[s5] "Authors".

Authors of the primary publication (if additional publications should be mentioned which describe the same analysis, they are included in the "Comments" items [s28-s32]).

[s6] "Journal".

Journal name of the primary publication. See Appendix B for abbreviations used for journal names.

[s7] "Volume".

Journal volume of the primary publication.

[s8] "Page".

Journal page of the primary publication.

[s9] "Year".

Journal year of the primary publication.

1.4.2. Surface Type

[s10] "Substrate".

Chemical formula of the substrate, e.g., Ni, GaAs, TiO_2.

[s11] "Bulk lattice".

Gives the 3D bulk substrate lattice: could be fcc, bcc, hcp, sc, diamond, zincblende, wurtzite, NaCl, CsCl, etc.

[s12] "Crystal face".

Gives the Miller (hkl) or Bravais-Miller (hkil) indices of the substrate surface. Commas are used to separate more complex indices, e.g., (12,1,1), (1,0,-1,0), but are omitted otherwise, e.g., (110).

[s13–14] "2D bulk symm" and "2D surf symm".

Gives the 2D symmetry of the ideally terminated substrate and of one domain of the actual surface, respectively. This can be any of the 17 2D space groups in the following list. (See the illustrations in Appendix E for definitions of these space groups). Only the short notation is used, e.g., p4m instead of p4mm.

p1
p2 (= p211)
pm (= p1m1)
pg (= p1g1)
cm (= c1m1)
pmm (= p2mm)
pmg (= p2mg)
pgg (= p2gg)
cmm (= c2mm)
p4
p4m (= p4mm)
p4g (= p4gm)
p3
p3m1
p31m
p6
p6m (= p6mm)
none (for disordered surface structures).

COMMON NAME : Ni(100)-c(2x2)-CO [s1]
CLASSIFICATION : 28.6.8.11 [s2]
TECHNIQUE : LEED [s4]
AUTHORS : K. Heinz, E. Lang and K. Mueller [s5]
REFERENCE : Surf. Sci., 87, 595 (1979) [s6-9]

ILLUSTRATION: 71

SURFACE TYPE
Substrate : Ni [s10] Adsorbate: CO [s15]
Crystal face: 100 [s12] Coverage : 1/2 CO/Ni [s16]
Temperature : 100 K [s22] Pattern : c(2x2) [s17]
Bulk lattice: fcc [s11] Matrix : (1.000, 1.000)
2D bulk symm: p4m [s13] (-1.000, 1.000)
2D surf symm: p4m [s14] [s18-21]

STRUCTURE TYPE [s23-27]
Molecular on-top adsorption, C bonding to Ni

SAMPLE PREPARATION (1 sample) [t1]
Treatment : CO exposure 1E-5Pa x 50sec [t2]
Crystallinity: LEED intensities match published data[t3]
Anal. methods: [t4]
Contamination: AES: very little C on clean surface [t5]

COMMENTS [s28-32]
Spectra were taken within 16 sec after termination of adsorption process

DATA COLLECTION
Technique: LEED [s4,t6]
Dataset : I-V spectra: (1,1) (1,0) (1/2,1/2) beams at normal incidence (0,0) beam at Θ=4°,φ=0°; E range 30-400 eV [t7-8]

THEORY/DATA TREATMENT [t9-10]
Comparison with earlier dynamical LEED I-V spectra by Pendry

STRUCTURES EXAMINED [t11-15]
Linearly bonded CO perpendicular to surface; variation of C-O bond length: 0.9-1.2Å in steps of 0.05Å; variation of Ni-C bond length: 1.7, 1.8, 1.9Å; bulk interlayer spacings assumed in metal

QUALITY OF EXPERIMENT-THEORY FIT
RZJ=0.19 [t16]

2D UNIT CELLS (1 domain observed) [2d6]

Cell	Ax (Å)	Ay (Å)	Bx (Å)	By (Å)	α (°)	Matrix	Pattern	Cell type
	[2d1]	[2d2]	[2d3]	[2d4]	[2d5]			
Bulk	2.489	0.000	0.000	2.489	90.0	(1.000, 0.000) (0.000, 1.000)	(1x1)	b: bulk lattice
	[2d9]	[2d10]	[2d11]	[2d12]	[2d13]	[2d15-18]	[2d14]	[2d8]
Surface 1	2.489	2.489	-2.489	2.489	90.0	(1.000, 1.000) (-1.000, 1.000)	c(2x2)	s1: commens. superlattice

3D COORDINATES

O1-C2: upright on-top molecule (C bonded to Ni3) [3d1-4]

Dx/Dy in Å, or as a fraction of layer's unit cell vectors; atom 0 at the origin. Epir/subr are bulk repeat vectors.

No. of atoms: 4 [3d7]

Bulk z = 1.760 Å [3d6]

Region	Chem el.	At. no.	Cell type	Site occ.	Rel. to	Dx ± εx		Dy ± εy		Dz ± εz		Dz/Bz(%) ± εz/Bz
[3d8]	[3d9]	[3d10]	[3d11]	[3d12]	[3d13]	[3d14-16]		[3d17-19]		[3d20,21]		[3d22,23]
epir		-2					f		f		Å	
subr		-1				1.245	Å	1.245	Å	1.760	Å	
ovrl	O	1	s1	.50	0	0.000	f	0.000	f	0.000	Å	0.0
ovrl	C	2	s1	.50	1	0.000	f	0.000	f	1.150 ± .100	Å	65.3 ± 5.7
intf	Ni	3	b	1.00	2	0.000	f	0.000	f	1.800 ± .100	Å	102.3 ± 5.7
subl	Ni	4	b	1.00	3	0.500	f	0.500	f	1.760	Å	100.0

BOND DISTANCES AND ANGLES

Bond distances and angles are derived from coordinates [b1]

No. of distances/angles: 4 [b2]

Interatomic dist. A-B (Å)	Atom A	Atom B	Atom C	Bond angle A-B-C (°)
[b3]	[b4]	[b5]	[b6]	[b7]
1.150	O1	C2	Ni3	180.0
1.800	C2	Ni3		
2.489	Ni3	Ni3(1,0)		
2.489	Ni3	Ni4		

[s15] ''Adsorbate''.

Identifies the adsorbed species after adsorption, i.e. not the gas-phase species used in order to adsorb. Coadsorbed species are separated by semicolons, e.g., ''C;O'' stands for coadsorbed C and O atoms, while ''CO'' stands for adsorbed molecular CO.

[s16] ''Coverage''.

The coverage is measured as the number of adsorbate units per (1×1) substrate unit cell. The adsorbate units may be specified for clarity. Thus, ''1/2 CO'' means one half of a molecular adsorbate unit per (1×1) unit cell, i.e. one adsorbate unit for every two (1×1) cells. In the case of coadsorption, the coadsorbates should be specified separately. For example, ''1/2(Na),1/4(S)'' would mean that the structure contains a half monolayer of Na and a quarter monolayer of S (relative to the (1×1) unit cell). The substrate unit may be included for clarity, e.g., 0.25 O/1×1. This reminder is helpful for stepped surfaces, where the (1×1) cell includes the full step-to-step width of one terrace.

[s17] ''Pattern''.

Gives the Wood or Bibérian (''rect'') notation for the surface superlattice, including the simplest case of (1×1) if there is no superlattice. Examples of common superlattices and their notations are given in Appendix F. If several surface superlattices coexist in different layers in the same structure (as shown under ''Surface 2D unit cells''), they are all listed (or explained in [s23–27] or [s28–32]).

For disordered lattices of type ndk (see definition of ndk below under [2d8]), we use an ordered lattice of available sites together with an occupation probability of these sites. For instance, a (1×1) lattice could be used for disorder on top of an unreconstructed clean substrate. This implies that one site per (1×1) lattice (in this example) will be occupied with a probability <1. (This probability is given by the site occupancy; see [3d12]).

[s18–21] ''Matrix notation''.

Describes the same surface lattice or superlattice as [s17]. ([s18–21] is left blank if more than one superlattice is present).

[s22] ''Temperature''.

Experimental temperature in K or °C units (can be *RT* for room temperature; *RT** is used when the temperature is not specified by the authors, but probably is room temperature).

1.4.3. Structure Type

[s23–27] ''Structure type''.

Describes the characteristic features of the structure in words. This information could, for instance, state whether a surface is relaxed (and, if so, to what depth) or reconstructed (and, if so, in what fashion), whether the surface has an overlayer or buried underlayer, which adsorption sites are occupied, etc.

1.4.4. Comments

[s28–32] ''Comments''.

Any pertinent information that does not fit elsewhere.

1.4.5. Sample Preparation

[t1] ''No. of samples''.

Gives the number of samples used (to check the experimental reproducibility).

[t2] ''Treatment''.

Describes the substrate preparation and/or the adsorption method.

[t3] ''Crystallinity''.

Any indication of the degree of crystalline perfection (e.g., sharp LEED spots).

[t4] ''Analytical methods''.

Lists any methods other than the primary structure determination technique used to evaluate the preparation, perfection, coverage, adsorbate state, etc.

[t5] ''Contamination''.

Any indication of the level of contamination by impurities (e.g., Auger estimates).

1.4.6. Data Collection

[t6] ''Technique''.

Describes the main measurement technique, apparatus, etc.

[t7–8] ''Data set''.

Gives the nature of the experimental database (e.g., I-V curves in LEED) and measure of its size (e.g., number of beams in LEED, cumulative energy range summed over all LEED beams, or momentum transfer range used in SEXAFS or ARPEFS).

1.4.7. Theory/Data Treatment

[t9–10] ''Theory/data treatment''.

Gives the method of the data treatment (averaging, smoothing, Fourier filtering, etc.) and the theoretical analysis (dynamical LEED, tensor LEED, quasidynamical LEED, Fourier transform, curved-wave, spherical-wave, multiple scattering cluster calculation, etc.).

1.4.8. Structures Examined

[t11–15] ''Structures examined''.

Lists the structures which have been tested in the analysis. This may specify which structures can be excluded based on this study.

1.4.9. Quality of Experiment-Theory Fit

[t16] ''Quality of experiment-theory fit''.

Gives a qualitative or, better, quantitative measure of the experiment-theory fit, such as R-factor values.

1.4.10. 2D Unit Cells

[2d1–5] "Bulk substrate unit cell".

Defines the 2D unit cell of the (1×1) lattice of the ideally terminated bulk substrate by its lattice vectors (Ax,Ay) and (Bx,By), in 2D Cartesian coordinates, in Angstroms, and the angle (in degrees) spanned by those two vectors.

[2d6] "Number of observed domain or site orientations".

When a superlattice has lower symmetry [s14] than the 2D bulk symmetry [s13], then several domain orientations are allowed and normally observed (all rotational axes and mirror planes of the substrate generate allowed domain orientations). The number of unique domain orientations is counted up and entered here. For instance, a (2×1) overlayer in hollow sites on fcc(100) has two domain orientations, the second one corresponding to the notation (1×2); a totally asymmetrical adsorption site in this case would give rise to 8 domain orientations.

The entry "many" is used for incommensurate superlattices with many orientations, perhaps continuous orientations.

In the case of disordered layers, [2d6] indicates the number of differently oriented but otherwise equivalent sites (e.g., two different bridge site orientations can occur on fcc(100), since they are symmetrically equivalent but differently oriented).

[2d7–18] These items (described below) list the one or more surface lattices that occur in the structure. Only one domain orientation is shown for any lattice. A minimum of one lattice is shown. (It could be the (1×1) lattice for simple cases like an unreconstructed clean surface).

[2d7] (not shown) "Number of surface lattices".

This is the number of different surface lattices present in different layers of the structure, which are individually defined under "Cell".

In SSD, most structures show only one 2D unit cell entered under "Cell": either the (1×1) unit cell if there is no superlattice, or a superlattice different from (1×1). Other entries under "Cell" are shown only when there is a second, third, etc. superlattice present in other layers. Thus, if an incommensurate layer lies on top of an overlayer with c(2×2) lattice, both the c(2×2) lattice and the incommensurate lattice would be shown under "Cell". (Note: if other superlattices exist on different patches of the surface, they are considered as independent structures, not as different layers of the same structure.)

The types of allowed lattices are defined next under [2d8].

[2d8] "Type of 2D surface lattice".

This item defines the type of each surface lattice present. The type is labeled s1, s2, i1, etc., and named explicitly as "commens. superlattice", "incommensurate lattice", etc., according to the definitions given below. The labels s1, s2, i1, etc. are used with the 3D coordinates in item [3d11] to assign a particular lattice to each atom of the surface, see the "Cell type" column under the 3D Coordinates.

The following 5 different types of surface lattices are allowed (with k=1, 2, 3, or 4, see below):

Note: the lattice types rdk and mk are not really randomly disordered in SSD. They use a periodic representation of a random lattice (see their exact definitions below).

If more than one superlattice of the same type exists in the same structure, these are distinguished by the suffix k, e.g., if both a c(2×2) layer and a p(2×2) layer coexist on top of each other, both of which are of the same commensurate type, one would need to specify s1 for c(2×2) and s2 for p(2×2).

A maximum of 4 surface lattices is allowed for a given structure (under "Cell"). As an example, a structure could thus contain layers with cells labeled s1, s2, nd1 and i1, but no more.

1.4.11. Lattice Definitions

sk (k=1, 2, 3 or 4): "commensurate superlattice".

A commensurate superlattice matches periodically with the substrate lattice. This type of superlattice can be written in the matrix notation with integer coefficients (the Wood notation may involve non-integer numbers, like $\sqrt{3}$). It includes the case of the (1×1) ideal 2D bulk substrate lattice (this then duplicates the items [2d1–5]), and any commensurate Wood or Bibérian (rect) superlattices, such as: (2×1), c(2×2), (2×2), ($\sqrt{3}\times\sqrt{3}$)R30°, ($2\sqrt{3}\times3$)rect. See Appendix F for illustrated examples of these and other commensurate lattices, with their notations.

ik (k=1, 2, 3 or 4): "incommensurate lattice".

An incommensurate superlattice in no way matches the substrate lattice, at least in one of the two surface dimensions. An incommensurate superlattice has a matrix notation with at least one irrational non-fractional number. (If a fractional coefficient appears, i.e. one which is the fraction of two integers, a larger unit cell should be chosen, so as to make all matrix coefficients integers). The corresponding Wood notation may also involve special numbers, typically irrational. Often many domain orientations coexist on the same sample for incommensurate lattices. Only one should be listed (and [2d6] should indicate the number of orientations present, e.g., "many" for continuous orientations).

It is assumed that an incommensurate layer has a well-defined unit cell that is not modulated (i.e. not "wavy") across the surface due to the variable local registry. For example, an incommensurate monolayer sheet of graphite on a substrate would be assumed to be perfectly planar and regular, unaffected by the substrate. A substrate-induced structural modulation, for instance a buckling, is in practice possible and expected, and cannot be described in SSD as an incommensurate lattice. It should be treated as a commensurate superlattice instead with a large unit cell.

ndk (k=1, 2, 3 or 4): "non-reconstructive lattice-gas disorder".

This type corresponds to single-site adsorption without relaxations being induced in the substrate (but rigid translations of (1×1) layers are allowed).

Symmetry may allow several site orientations. For instance, 2 equivalent bridge sites mutually rotated by 90° coexist on a surface with 4-fold rotational symmetry. In that case one has two options within SSD: 1) Only one site orientation is listed, while the number of site orientations is given by [2d6], namely 2 in our example; 2) Both orientations are included separately, but labeled with different values of k (e.g., nd1 and nd2 in our example), while [2d6] now has the value 1.

If several inequivalent sites or several inequivalent adsorbates are present simultaneously, a separate ndk-type lattice must be specified for each. Rigid clusters, like molecules, which are randomly positioned at lattice sites are denoted thus: the different atoms composing one such rigid cluster or molecule should be given the same value of k in ndk.

For example, assume that a CO molecule adsorbs randomly at bridge sites with two different orientations, or at two different sites (e.g., top and bridge). Then one would (with the 3D coordinates) assign nd1 to one C and nd1 to one O (for the first site), while assigning nd2 to a second C and nd2 to a second O (for the second site). Thereby, the first C and first O, labeled nd1, will be known to bond together and to randomize as a pair rather than being randomized separately. The second C and the second O will likewise be known to bond together and will be randomized separately from the first CO. If the same nd1 were specified for all of these 4 atoms, they would incorrectly form a rigid 4-atom cluster. If nd1, nd2, nd3 and nd4 were assigned to these 4 atoms, they would be considered to adsorb totally independently of each other, thereby not forming rigid molecules.

A set of lattice vectors [2d9–13] should be specified that defines the lattice of sites which will be occupied with a probability given by the occupancy [3d12], which should be <1. Thus, specifying lattice vectors that correspond to a (1×1) lattice will allow random adsorption on one site in each (1×1) cell. This is the most common situation. Specifying lattice vectors corresponding to a $c(2\times2)$ lattice will only allow random adsorption in every other unit cell, leaving one half of the unit cells systematically unoccupied. (In that case, an occupancy [3d12] of 0.5 will therefore yield a coverage of only 0.25, since only one half of the sites will be occupied with a probability of 0.5 each).

rdk (k=1, 2, 3 or 4): "reconstructive lattice-gas disorder".

This corresponds to disordered on-site adsorption or vacancies or other defects, that include induced local substrate relaxations, such as lateral relaxations and layer bucklings.

This case is simulated by an ordered commensurate superlattice giving a comparable coverage and a supercell which contains enough atoms to adequately represent the local atomic arrangement. For the 3D coordinates, the same ordered lattice is assumed.

For example, for a 1/5-monolayer disordered overlayer, a (2×2) superlattice might be chosen with one adsorbate per (2×2) cell (this gives a 1/4 coverage to approximate the actual 1/5 coverage), which would include relaxations in the substrate atoms nearest to the adsorbate.

mk (k=1, 2, 3 or 4): "randomly mixed layer".

This corresponds to random alloys and other cases of random mixed occupation of lattice points.

This case is simulated as an ordered commensurate superlattice with a suitable periodic supercell (in close analogy with reconstructive lattice-gas disorder, case rdk).

(Note: case rdk is equivalent to case mk when mixing one kind of atoms with vacancies: then rdk is the preferred representation.)

[2d9–13] "Surface unit cell vectors".

These give the unit cell vectors and angles for the individual cells. (They are similar to those [2d1–5] for the bulk substrate). For disordered lattices of type ndk, see the definition of ndk.

[2d14, 2d15–18] "Wood notation" and "Matrix notation".

These are the Wood and matrix notations for each surface lattice present. (Examples of common superlattices and their notations are given in Appendix F).

1.4.12. 3D Coordinates

[3d1–4] These notes are used to clarify in words the atomic relationships, e.g., to specify which atoms together form a particular molecule, which together form a buckled layer, which form an epilayer, which the periodic substrate, etc. It is also appropriate to include a note here if the coordinates are indirectly derived from interatomic distances and angles, as in SEXAFS, rather than being direct results from experiment.

[3d5] (not currently used).

[3d6] "Bulk substrate interlayer spacing".

This bulk substrate spacing is used as a reference to determine the relative interlayer spacings shown in items [3d22, 3d23]. Typically, this spacing is the layer-to-layer repeat distance for simple substrates.

More generally, this spacing can be the bulk periodicity perpendicular to the surface, or anything else that is meaningful. (This spacing need not be equal to the Dz-component of the bulk substrate repeat vector given in the table as "subr").

Examples: for fcc(100) one would choose a/2 (if a is the bulk cube edge); for hcp(0001) one would choose c/2 (if c is the hcp lattice constant perpendicular to the basal plane, such that c is twice the interlayer spacing in that direction, while in this hcp case "subr" would have c as z-component); for Si(111) one could choose the bilayer-to-bilayer repeat spacing, rather than the smaller interlayer spacings that exist between adjacent layers.

[3d7] "No. of atoms".

Gives the number of tabulated atoms (not counting the always present tabular entries labeled "epir" and "subr").

[3d8–23] These items specify the individual atomic positions and identities in the coordinate table. Each line in the table specifies one atom, together with its associated layer obtained by the 2-dimensional periodicity given by the "Cell type" [3d11]. We shall therefore use the terms "atom" and "layer" interchangeably in the following.

In the coordinate table, the first 2 lines contain not atoms, but 2 bulk repeat vectors [3d14, 3d17–18, 3d20–21] labeled ''epir'' and ''subr'' for the epilayer (if it exists) and the substrate, respectively. The components of ''epir'' are blank if the structure does not contain an epilayer. The vector ''subr'' is always defined, since SSD assumes that a substrate is present. These 2 vectors describe how the surface and interface region propagate into the bulk epilayer and substrate regions. The table contains corresponding atoms labeled ''epil'' and ''subl''. These constitute the units that repeat periodically away from the surface according to the vectors ''epir'' and ''subr'' to build the semi-infinite epilayer and substrate. (The ''epir'' and ''subr'' components are always given in Ångström units.)

In the table, each atom's position is referred to either a fixed origin at position (0,0,0) (called ''0'' in the ''Relative to'' column [3d13]) or another atom's position (defined higher up in the table), whichever is more convenient. The choice can be made individually for each atom. We often choose to refer an atom's position to that of the preceding atom in the table. Dx and Dy are coordinates parallel to the surface; Dz is perpendicular to the surface, relative to the origin or that other atom.

Dx and Dy can be given either in Cartesian coordinates (in Ångström) or in fractional cell coordinates (i.e. as fractions of the 2D cell vectors defined for the atom in question), whichever is more convenient. The choice can be made individually for each atom. When fractional cell coordinates are used, they refer to the ''Cell type'' [3d11] of the atom in question (not of the ''Relative to'' atom and not necessarily the (1×1) cell, for example). Dz must be given in Ångström.

[3d8] ''Region''.

Each atom belongs to a well-defined region, which can be any of the following 4 types:

a. ''subl'': substrate. The atoms labeled thus repeat indefinitely into the substrate according to the repeat vector labeled ''subr''. When two or more atoms are labeled thus, they are assumed to form a group of layers that will be repeated together into the substrate. The relative coordinates specified for these atoms are thus substrate-bulk-like.

   Therefore, for a substrate bulk with one atom per primitive 3D unit cell, only one layer labeled ''subl'' is needed, e.g., for fcc and bcc; for a substrate bulk with N atoms per 3D primitive unit cell, N layers labeled ''subl'' are needed: e.g., two for hcp, diamond, zincblende and wurtzite, and four for graphite.

b. ''intf'': interface. This typically represents surface-relaxed substrate atoms at the interface between bulk and vacuum. Atoms of this type can also form underlayers (in the case of compound formation), or be relaxed atoms at a substrate-epilayer interface.

c. ''ovrl'': overlayer. Such atoms are typically foreign adatoms or admolecules. (The distinction between interface and overlayer atoms may not always be clearcut, leaving a choice in labeling.) They may also be relaxed epilayer atoms at an interface.

d. ''epil'': epilayer. The atoms labeled thus repeat indefinitely into the epilayer, according to the repeat vector labeled ''epil''. When two or more atoms are labeled thus, they are assumed to form a group of layers that will be repeated together into the epilayer. The relative coordinates specified for these atoms are thus epilayer-bulk-like.

Caution: The z-components of both the vectors ''subr'' and ''epir'' point in the positive direction towards the substrate (not towards the epilayer).

[3d9] ''Chemical element''.

Specifies the chemical element for one 2D-periodic layer. Coplanar layers of the same element or other elements are treated as separate entries in this table.

[3d10] ''Atom number''.

Counts the tabulated atoms sequentially from 1 up. This number serves as label for the ''Relative to'' column [3d13]. The entries -2 and -1 labeling ''epir'' and ''subr'' are not used in [3d13].

[3d11] ''Cell type''.

Assigns a 2D periodicity to each tabulated atom, by referring to [2d1–5] and [2d8], the type of 2D surface lattice defined with the 2D structure. Examples are b (for the bulk substrate (1×1) lattice), s1, s2, i1, nd1, etc.

[3d12] ''Site occupancy''.

For ordered and incommensurate layers, this gives the coverage of a layer with respect to the substrate (1×1) cell. Thus, a (2×1) layer typically has a site occupancy of 0.5, while (2×2) gives 0.25. For a displacive reconstruction (which breaks the periodicity of a layer into a larger unit cell by displacements such as out-of-plane bucklings), this occupancy will be <1, e.g., 0.5 for each of the two atoms defining a c(2×2) buckled layer. (Note that the two atoms would be tabulated separately).

For disordered layers, the ''Site occupancy'' gives the probability of occupying each site of the lattice defined by [3d11], which refers back to the lattice vectors of [2d9–13]. Thus, specifying through [3d11] a set of (1×1) lattice vectors allows random adsorption on one site in each (1×1) cell. This is the most common situation. Specifying lattice vectors corresponding to a c(2×2) lattice will only allow random adsorption in every other unit cell, leaving one half of the unit cells systematically unoccupied. (In that case, an occupancy of 0.5 will therefore yield a coverage of only 0.25, since only one half of the sites will be occupied with a probability of 0.5 each).

For the lattice types rdk (reconstructive lattice-gas disorder) and mk (randomly mixed layer), described under [2d8], the ''Site occupancy'' is the experimentally determined or estimated coverage or fraction, rather than the value appropriate for the artificial periodic structure representing it in SSD. For instance, if a disordered overlayer of 0.2 coverage is modeled as a (2×2) overlayer of type rdk because it induces substrate relaxations, the site occupancy is given as 0.2 and not 0.25.

[3d13] ''Relative to''.

Gives the origin of the vector [3d14, 3d17, 3d20] that points to this atom's position. The origin is given as the number [3d10] of a previously listed atom or as the origin of coordinates 0=(0,0,0). The entries -2 and -1 labeling ''epir'' and ''subr'' should not be used in [3d13].

[3d14–15, 3d17–18] ''Parallel position coordinates''.

Dx [3d14] and Dy [3d17] are position coordinates parallel to the surface relative to a previously tabulated atom given by [3d13]. The Dx and Dy coordinates may be expressed in one of two ways: either in Cartesian coordinates (then A for Ångström is appended as [3d15] or [3d18]); or in fractional cell coordinates of the cell [3d11] defined for this atom's layer (then f is appended as [3d15] or [3d18]). A mix of absolute and fractional coordinates can be used for clarity: e.g., Dx=0.5f and Dy=0.21A may be used simultaneously for a given atom.

[3d20] ''Perpendicular position coordinate''.

Dz is the perpendicular position coordinate relative to a previously tabulated atom (the same reference atom [3d13] as for Dx and Dy). Dz is always given in Cartesian coordinates (in Ångström). Dz>0 points down towards the substrate, while Dz<0 points up away from substrate towards the overlying vacuum or epilayer. Dz=0 corresponds to coplanar atoms.

[3d16, 3d19, 3d21] ''Error bars''.

These give the experimentally determined uncertainties on the coordinates Dx, Dy and Dz, in the same units (A or f) as the corresponding coordinates. Error bars are shown only for coordinates that were fit to experiment; a blank entry is used when the corresponding coordinate value was assumed rather than fit.

[3d22] ''Relative interlayer spacing''.

Gives Dz/Bulk z ([3d20]/[3d6]) as a percent, i.e. the interlayer spacing divided by the bulk substrate spacing. This is most useful to exhibit near-surface interlayer spacing relaxations compared to the bulk substrate value Bulk z. Thus the last entry in this column [3d22] is typically 100% for simple structures (fcc and bcc), representing the unrelaxed bulk spacing. See [3d6] for more details on appropriate choices of Bulk z.

[3d23] ''Relative spacing error''.

Gives the relative uncertainty [3d21]/[3d6] on the ''Relative interlayer spacing'' [3d22], i.e. the error bar as a percent relative to Bulk z. Thus, [3d22] = 94 and [3d23] = 2 would indicate an interlayer spacing relaxed to 94±2% of the bulk spacing.

1.4.13. Bond Distances and Angles

[b1] ''Note''.

Gives any comments. This might mention whether the distances and angles are derived from the coordinates, rather than vice versa (the SEXAFS case).

[b2] ''No. of distances/angles''.

Gives the number of lines containing distance and/or angle information. This information is, in general, only a selection of distances and angles, i.e. only important distances and angles are included, rather than a complete set from which the surface structure could be derived.

[b3] ''Interatomic distance A-B''.

Gives an interatomic distance (in Ångström) between the next two atoms (A and B) on the same line.

[b4-b6] ''Atom A'', ''Atom B'', ''Atom C''.

Gives atoms between which interatomic distances [b3] and bond angles [b7] are tabulated on the same line. The atom labels, such as O1 and Ni3, refer to the 3D coordinates table. (The labels are made up of items [3d9] and [3d10].)

The information between parentheses, as in Ni3(1,0), indicates an atom that is shifted parallel to the surface to a neighboring 2D unit cell, relative to the atom tabulated with its 3D coordinates. Thus Ni3(1,0) is atom Ni3 shifted by 1 × (Ax,Ay) + 0 × (Bx,By) = (Ax,Ay), where (Ax,Ay) and (Bx,By) are the unit cell vectors assigned to atom Ni3 in the 3D coordinates table (i.e., the bulk 2D (1×1) cell, since ''b'' is specified for atom Ni3.)

[b7] ''Bond angle A-B-C''.

Gives the angle (in degrees) subtended by the apex atom B for the triplet of atoms A, B, C specified on the same line.

1.5. Relation of This Atlas to Other Databases and Graphics Software

In the following, it should be remembered that this atlas and SSD contain the same data, SSD being an electronic version of this atlas, with associated software to use the database and to visualize its data.

Relation to J.F. Nicholas' atlas (An Atlas of Models of Crystal Surfaces, Gordon and Breach, New York, 1965):

This atlas contains actual surface structures determined from experiment, while Nicholas' atlas is limited to ideal (clean, unrelaxed) terminations of simple bulk lattices.

Relation to SCIS handbook (Surface Crystallographic Information Service, A Handbook of Surface Structures; authors J.M. MacLaren, J.B. Pendry, P.J. Rous, D.K. Saldin, G.A. Somorjai, M.A. Van Hove and D.D. Vvedensky; see ch. 4):

This atlas contains a much revised, corrected and updated database compared to its forerunner SCIS, covering 597 rather than 256 structures.

Relation to SCIS software (Surface Crystallographic Information Service; author J.B. Pendry; distributed by D. Reidel; PC-based; see ch. 4):

SSD is a much expanded program providing many more options than the SCIS software. Because of the extensive enhancements of SSD, the database files of SCIS and SSD are incompatible.

Relation to SARCH software (Surface Architect, author M.A. Van Hove; bundled with LATUSE discussed below; PC-based; see ch. 4):

SARCH emphasizes the interactive construction of surface structures from the perspective of two-dimensional periodicity and disorder, and provides a wide variety of presentation modes, as well as analysis options. Further, SARCH allows one to analyze lattice diffraction patterns (kinematic LEED patterns) and can simulate surface topographic geometries for the interpretation of experimental STM data. A surface structure produced within SARCH (version 3.2 or higher) can be output to a SSD/ASCII file, which in turn can be inserted by the authors of SSD into the SSD database (the published SSD database files have a different non-ASCII Paradox format). You could thus use SARCH to produce a structure in a format that allows the authors of SSD to insert the structure into the SSD electronic database. SARCH reads and writes both SCIS and SSD/ASCII files, providing a bridge between SCIS and SSD.

Relation to LATUSE/PLOT3D software (Lattice Use and Plotting in 3D, bundled with SARCH discussed above; author K. Hermann; PC-based; see ch. 4):

LATUSE emphasizes the construction (within the program or through modification of an input file) and interactive display of periodic surface structures. It is especially convenient to choose and display any Miller-index terminations of bulk 3-dimensional lattices. LATUSE, like SARCH, provides a wide variety of presentation modes, as well as numerical and graphical analysis options. PLOT3D is similar in its menu handling to LATUSE, but is aimed at clusters of atoms, including non-periodic sections of surfaces. Transfers of structures between SARCH, LATUSE and PLOT3D are possible.

Relation to BALSAC (Build and Analyze Lattices, Surfaces, and Clusters; author K. Hermann; PC-based; see ch. 4):

The BALSAC program system ((C) Copyright K. Hermann 1992) combines all features of the two PC-based program systems LATUSE and PLOT3D; but it offers easier handling, more analysis features, and enhanced graphics capabilities. It is available for both PC's and Unix based workstations.

2. Catalog of Structures

2.1. Alphanumerical Index of Structures

The structures are listed alphanumerically by elements in the substrate and adsorbate (if any). Clean surfaces are listed first (in order of crystal face by increasing Miller indices) and then adsorbate structures, again ordered by crystal face. In addition the unique classification number assigned to a structure in the database is provided with the number of the appropriate figure in Volume 1B. A figure number of — indicates that no figure is provided.

2.1. Alphanumerical Index of Structures

Common name of structure	Class. no.	Figure	Page
Ag(100)-(1×1)	47.22	2	19
Ag(110)-(1×1)	47.15	4	20
Ag(110)-(1×1)	47.16	4	21
Ag(110)-(1×1)	47.17	4	22
Ag(110)-(1×1)	47.19	4	23
Ag(100)-C_2H_4 disordered	47.6.1.2	74	24
Ag(100)-c(2×2)-Cl	47.17.1	28,29	25
Ag(100)-c(2×2)-Cl	47.17.3a	28,29	26
Ag(100)-c(2×2)-Cl	47.17.4	28,29	27
Ag(100)-c(2×2)-Cl	47.17.8	28,29	28
Ag(100)-c(2×2)-Cl	47.17.9	28,29	29
Ag(111)-($\sqrt{3}\times\sqrt{3}$)R30°-Cl	47.17.5a	22,24	30
Ag(111)-($\sqrt{3}\times\sqrt{3}$)R30°-2Cl	47.17.5b	22,24	31
Ag(100)-(1×1)-3Co	47.27.1	83	32
Ag(110)-(1×2) Cs-induced	47.55.2	5	33
Ag(111)-Cs 0.15ML disordered	47.55.1a	22	34
Ag(111)-Cs 0.3ML disordered	47.55.1b	22	35
Ag(100)-(1×1)-Cu multilayer	47.29.3	83	36
Ag(100)-(1×1)-2Cu	47.29.2a	83	37
Ag(100)-(1×1)-5Fe (bcc)	47.26.0a	83	38
Ag(100)-(1×1)-Fe multilayer	47.26.1	83	39
Ag(111)-($\sqrt{3}\times\sqrt{3}$)R30°-I	47.53.1	22,24	40
Ag(111)-($\sqrt{3}\times\sqrt{3}$)R30°-I	47.53.4	22,24	41
Ag(110)-(2×1)-O	47.8.4	35	42
Ag(100)-c(2×2)-Se	47.34.1	28,29	43
Ag(111)-Xe incommensurate	47.54.1	82	44
Ag(111)-Xe incommensurate	47.54.2	82	45
AgBr(100)-(1×1)	47.35.2a	149	46
AgBr(111)-(2×1)	47.35.2b	151	47
Al(100)-(1×1)	13.15a	2	49
Al(100)-(1×1)	13.16a	2	50
Al(100)-(1×1)	13.26	2	51
Al(110)-(1×1)	13.16b	4	52
Al(110)-(1×1)	13.25	4	53
Al(110)-(1×1)	13.27	4	55
Al(111)-(1×1)	13.19	1	57
Al(111)-(1×1)	13.20a	1	58
Al(111)-(1×1)	13.21	1	59
Al(111)-(1×1)	13.21a	1	60
Al(111)-(1×1)	13.41	1	61
Al(210)-(1×1)	13.36	9	62
Al(311)-(1×1)	13.30	8	64
Al(331)-(1×1)	13.31	10	66
Al(100)-c(2×2)-Na	13.11.1	28,29	68
Al(100)-c(2×2)-Na	13.11.2	28,29	70
Al(111)-($\sqrt{3}\times\sqrt{3}$)R30°-Na	13.11.3	27	72
Al(111)-O	13.8.8	22,23	73
Al(111)-(1×1)-O	13.8.12	22,23	74
Al(111)-(1×1)-O	13.8.15	22,23	75
Al(111)-(1×1)-O	13.8.6a	22,23	76
Al(111)-(1×1)-O	13.8.6b	26	77
AlP(110)-(1×1)	13.15.2	116	78
Au(100)-(1×1)	79.8a	2	80
Au(100)-hex incommensurate	79.80	3	81
Au(110)-(1×2)	79.25	5	82
Au(110)-(1×2)	79.32	5	84
Au(110)-(1×2)	79.34	5	86
Au(110)-(1×2)	79.66a	5	87
Au(110)-(1×3)	79.55.2	7	89
Au(110)-c(2×2)-K	79.19.3	41	91
$AuCu_3$(100) disordered	29.79.5	135	93
C(0001)-(1×1) graphite	6.4	157	95
C(111)-(1×1) diamond	6.5	166	96
C(0001)-(2×2)-Cs	6.55.3	163	97

2.1. Alphanumerical Index of Structures — Continued

Common name of structure	Class no.	Figure	Page
C(0001)-(√3×√3)R30°-Cs	6.55.2	162	98
C(111)-(1×1)-H (diamond)	6.1.1a	166	100
C(0001)-1K disordered underlayer	6.19.2a	164	101
C(0001)-2K disordered underlayer	6.19.2b	165	103
CaO(100)-(1×1)	20.8.1	149	105
CdS(11-20)-(1×1)	48.16.1	125	106
CdSe(10-10)-(1×1)	48.34.2	124	108
CdSe(10-10)-(1×1)	48.34.4b	124	110
CdSe(11-20)-(1×1)	48.34.4a	125	112
CdTe(110)-(1×1)	48.52.2	116	114
CdTe(110)-(1×1)	48.52.6	116	116
Co(0001)-(1×1)	27.5a	19	118
Co(10-10)-(1×1)	27.9a	20	119
Co(10-10)-(1×1)	27.10	20	121
Co(11-20)-(1×1)	27.8	21	123
Co(100)-(1×1)	27.4	2	125
Co(111)-(1×1)	27.5b	1	126
Co(10-10)-c(2×2)-K	27.19.1	60	127
Co(100)-c(2×2)-O	27.8.1	28,29	129
Co(100)-c(2×2)-S	27.16.1	28,29	130
CoO(100)-(1×1)	27.8.3	149	131
CoO(111)-(1×1)	27.8.2	150	132
$CoSi_2$(111)-(1×1)	14.27.1	145	133
$CoSi_2$(111)-(1×1)	14.27.14	145	135
$CoSi_2$(111)-(1×1) Co-rich	14.27.11b	145	137
$CoSi_2$(111)-(1×1) Si-rich	14.27.11a	146	139
Cr(100)-(1×1)-N	24.7.1	48,49	141
Cr(100)-c(2×2)-S	24.16.1	48,50	143
Cu(100)-(1×1)	29.25a	2	145
Cu(100)-(1×1)	29.36a	2	146
Cu(100)-(1×1)	29.43	2	147
Cu(100)-(1×1)	29.59	2	148
Cu(110)-(1×1)	29.25b	4	149
Cu(110)-(1×1)	29.28	4	150
Cu(110)-(1×1)	29.29	4	151
Cu(110)-(1×1)	29.37	4	152
Cu(110)-(1×1)	29.38	4	153
Cu(110)-(1×1)	29.41a	4	154
Cu(110)-(1×1)	29.48	4	155
Cu(110)-(1×2)	29.52	5	156
Cu(110)-(1×2) (Li induced)	29.3.1	5	157
Cu(111)-(1×1)	29.31	1	158
Cu(311)-(1×1)	29.11	8	159
Cu(311)-(1×1)	29.46	8	160
Cu(100)-c(2×2)-Au	29.79.2	33	161
Cu(100)-CH_3O disordered	29.6.1.8.10	76	162
Cu(100)-C_2H_2 disordered	29.6.1.2a	73	163
Cu(100)-C_2H_4 disordered	29.6.1.2b	74	164
Cu(100)-C_2H_4 disordered	29.6.1.3	—	165
Cu(100)-c(2×2)-CO	29.6.8.0a	71	166
Cu(100)-c(2×2)-CO	29.6.8.1	71	167
Cu(100)-c(2×2)-Cl	29.17.12	28,29	168
Cu(100)-c(2×2)-Cl	29.17.13	28,29	169
Cu(100)-c(2×2)-Cl	29.17.5	28,29	171
Cu(100)-c(2×2)-Cl	29.17.7	28,29	173
Cu(111)-(√3×√3)R30°-Cl	29.17.8	22,24	174
Cu(111)-(√3×√3)R30°-Cl	29.17.9	22,24	175
Cu(100)-(1×1)-1Co	29.27.2a	83	176
Cu(100)-(1×1)-8Co	29.27.2b	83	177
Cu(100)-(1×1)-20Co	29.27.3	83	179
Cu(111)-(1×1)-1Co	29.27.1	81	181
Cu(111)-(2×2)-Cs	29.55.1	22	182
Cu(100)-(1×1)-1Fe	29.26.2a	83	183
Cu(100)-(1×1)-1Fe	29.26.3a	83	184
Cu(100)-(1×1)-2Fe	29.26.3b	83	185
Cu(100)-(1×1)-3Fe	29.26.3c	83	186
Cu(100)-(1×1)-4Fe	29.26.3d	83	187

2.1. Alphanumerical Index of Structures — Continued

Common name of structure	Class no.	Figure	Page
Cu(100)-(1×1)-5Fe	29.26.3e	83	188
Cu(100)-(1×1)-10Fe	29.26.2b	83	189
Cu(110)-(1×1)-Fe	29.26.11	85	191
Cu(111)-(1×1)-1Fe	29.26.4	81	193
Cu(100)-p(2×1)-Fe multilayer	29.26.13	83	194
Cu(110)-(1×1)-H 2L	29.1.2a	4	196
Cu(110)-(1×1)-H 10L	29.1.2b	4	197
Cu(110)-(1×1)-H 50L	29.1.2c	4	198
Cu(110)-(1×1)-H 200L	29.1.2d	4	199
Cu(110)-HCO_2 disordered	29.6.1.8.3	79	200
Cu(100)-HCO_2 disordered	29.6.1.8.8a	75	201
Cu(110)-HCO_2 disordered	29.6.1.8.8b	79	202
Cu(100)-(2×2)-I	29.53.2b	28,30	203
Cu(111)-(√3×√3)R30°-I	29.53.2a	22,24	204
Cu(100)-c(2×2)-N	29.7.3	28,29	205
Cu(100)-c(2×2)-N	29.7.4	28,29	207
Cu(100)-(1×1)-1Ni	29.28.2a	83	208
Cu(100)-(1×1)-2Ni	29.28.4a	83	209
Cu(100)-(1×1)-3Ni	29.28.2b	83	210
Cu(100)-(1×1)-4Ni	29.28.4b	83	211
Cu(111)-(1×1)-Ni	29.28.1	81	212
Cu(100)-c(2×2)-O	29.8.15	28,29	213
Cu(100)-c(2×2)-O	29.8.2	28,29	214
Cu(100)-c(2×2)-O	29.8.7	28,29	215
Cu(100)-(2√2×√2)R45°-O	29.8.39	34	216
Cu(110)-(1×2)-O	29.8.17	39	218
Cu(110)-(1×2)-O	29.8.18a	39	219
Cu(110)-(2×1)-O	29.8.30	39	220
Cu(110)-(2×1)-O	29.8.31	39	222
Cu(110)-(2×1)-O	29.8.51	39	224
Cu(410)-(1×1)-O	29.8.12a	42	226
Cu(410)-(1×1)-2O	29.8.12b	43	228
Cu(100)-c(2×2)-Pb	29.82.1a	28,29	230
Cu(100)-c(2×2)-Pb	29.82.2	28,29	231
Cu(100)-c(2×2)-Pd	29.46.2	33	232
Cu(100)-c(5√2×√2)R45°-3Pb	29.82.1b	31	233
Cu(100)-c(5√2×√2)R45°-3Pb	29.82.3	31	235
Cu(100)-p(2×2)-S	29.16.0a	28,30	236
Cu(100)-p(2×2)-S	29.16.1	28,30	237
Cu(100)-p(2×2)-S	29.16.10	28,30	238
Cu(100)-p(2×2)-S	29.16.2	28,30	239
Cu(100)-p(2×2)-S	29.16.3	28,30	240
Cu(100)-(2×2)-S	29.16.12	28,30	241
Cu(100)-(2×2)-S	29.16.13	28,30	243
Cu(100)-(2×2)-S	29.16.14	28,30	244
Cu(100)-(2×2)-Te	29.52.1	28,30	246
Cu(100)-(2×2)-Te	29.52.2a	28,30	247
Cu(111)-16%Al-(√3×√3)R30°	29.13.2	133	248
$Cu_{0.85}Pd_{0.15}$(110)-(2×1)	29.46.4	141	249
Fe(100)-(1×1)	26.4	12	251
Fe(100)-(1×1)	26.18	12	252
Fe(100)-(1×1)	26.18a	12	253
Fe(100)-(1×1) epitaxial on Cu(100)	29.26.7	83	254
Fe(100)-(1×1) epitaxial on Cu(100)	29.26.8	83	255
Fe(100)-(1×1) epitaxial on Ni(100)	28.26.2c	83	256
Fe(110)-(1×1)	26.8	11	257
Fe(111)-(1×1)	26.17	15	258
Fe(111)-(1×1)	26.9	15	260
Fe(210)-(1×1)	26.16	16	262
Fe(211)-(1×1)	26.14	17	264
Fe(310)-(1×1)	26.13	18	266
Fe(100)-c(2×2)-C+O disordered	26.6.8.1	48,50	268
Fe(100)-(1×1)-3Co	26.27.1	86	270
Fe(100)-(1×1)-Cu multilayer	26.29.1	86	271
Fe(110)-(2×1)-H	26.1.1a	44,45	272
Fe(110)-(3×1)-2H	26.1.1b	46	274
Fe(100)-c(2×2)-N	26.7.1	48,50	276

2.1. Alphanumerical Index of Structures — Continued

Common name of structure	Class no.	Figure	Page
Fe(100)-(1×1)-3Ni	26.28.0a	86	278
Fe(100)-(1×1)-O	26.8.10	48,49	279
Fe(100)-(1×1)-O	26.8.3	48,49	280
Fe(100)-(1×1)-O	26.8.7	48,49	282
Fe(100)-(1×1)-O	26.8.9	48,49	283
Fe(211)-(2×1)-O	26.8.8	54	285
Fe(100)-c(2×2)-S	26.16.1	48,50	287
Fe(100)-c(2×2)-S	26.16.6	48,50	288
Fe(110)-p(2×2)-S	26.16.4	47	289
GaAs(110)-(1×1)	31.33.26	116	291
GaAs(110)-(1×1)	31.33.27	116	293
GaAs(110)-(1×1)	31.33.29a	116	295
GaAs(110)-(1×1)	31.33.68	116	297
GaAs(111)-(2×2)	31.33.24	117	299
GaAs(311)-(1×1) As termination	31.33.53b	121	301
GaAs(311)-(1×1) Ga termination	31.33.53a	122	302
GaAs(110)-(1×1)-1Al (low coverage)	31.33.13.4a	127	304
GaAs(110)-(1×1)-2Al (med coverage)	31.33.13.4b	127	306
GaAs(110)-(1×1)-3Al (high coverage)	31.33.13.4c	127	308
GaAs(110)-(1×1)-2Bi	31.33.83.2	126	310
GaAs(110)-(1×1)-2Sb	31.33.51.2	126	312
GaAs(110)-(1×1)-2Sb	31.33.51.5	126	314
GaP(110)-(1×1)	31.15.4	116	316
GaP(111)-(2×2)	31.15.5	117	318
GaSb(110)-(1×1)	31.51.2	116	320
GaSb(110)-(1×1)	31.51.3	116	322
GaSb(110)-(1×1)	31.51.3a	116	324
GaSb(111)-(2×2)	31.51.4	117	326
Ge(100)-(2×1)	32.1	94	328
Ge(100)-(2×1)	32.5	94	330
Ge(111)-c(2×8)	32.21	92	332
Ge(111)-($\sqrt{3}$×$\sqrt{3}$)R30°-Bi	32.83.1	96,97	334
Ge(111)-(1×1)-Cl	32.17.1	96,99	336
Ge(111)-(1×1)-H	32.1.1	88	337
Ge(111)-(1×1)-I	32.53.2	96,99	339
Ge(111)-(1×1)-PH_x	32.15.1.1	110	340
Ge(111)-($\sqrt{3}$×$\sqrt{3}$)R30°-Pb (1/3ML)	32.82.6a	96,97	342
Ge(111)-($\sqrt{3}$×$\sqrt{3}$)R30°-4Pb (4/3ML)	32.82.6b	115	344
Ge(100)-(2×1)-S	32.16.2	104,106	346
Ge(111)-(2×2)-S	32.16.1	96,98	348
HfC(100)-(1×1)	72.6.2	149	350
InAs(110)-(1×1)	49.33.1	116	352
InAs(110)-(1×1)	49.33.2	116	354
InP(110)-(1×1)	49.15.2	116	356
InP(110)-(1×1)	49.15.3	116	358
InSb(110)-(1×1)	49.51.1	116	360
InSb(110)-(1×1)	49.51.6	116	362
Ir(100)-(1×1)	77.10	2	364
Ir(100)-(1×5)	77.11	3	365
Ir(100)-(1×5)	77.6	3	367
Ir(110)-(1×1)	77.3	4	369
Ir(110)-(1×2)	77.16	5	370
Ir(110)-(1×3)	77.26	7	372
Ir(111)-(1×1)	77.2	1	374
Ir(110)-c(2×2)-O	77.8.1	35	375
Ir(111)-(2×2)-O	77.8.2	22,25	377
Ir(110)-(2×2)-2S	77.16.2	40	378
Ir(111)-($\sqrt{3}$×$\sqrt{3}$)R30°-S	77.16.1	22,24	380
MgO(100)-(1×1)	12.8.4	149	381
MgO(100)-(1×1)	12.8.5	149	383
MgO(100)-(1×1)	12.8.8	149	384
MgO(100)-(1×1)	12.8.9	149	385
Mn(100)-(1×1) epitaxial on Pd(100)	46.25.3	83	386
Mo(100)-(1×1)	42.4	12	387
Mo(100)-(1×1) discrdered	42.10	14	388
Mo(110)-(1×1)	42.7	11	389
Mo(111)-(1×1)	42.8	15	390

2.1. Alphanumerical Index of Structures — Continued

Common name of structure	Class no.	Figure	Page
Mo(100)-c(2×2)-C	42.6.2	48,50	391
Mo(100)-(1×1)-2H (D)	42.1.1	12	393
Mo(100)-c(2×2)-N	42.7.1	48,50	394
Mo(100)-c(2×2)-S	42.16.10	48,50	395
Mo(100)-(1×1)-Si	42.14.1	48,49	397
MoS_2(0001)-(1×1)	42.16.4b	159	398
Na(0001)-(1×1)	11.2	19	400
Na(110)-(1×1)	11.1	11	401
Na(110)-(1×1)	11.1a	11	402
Na_2O(111)-(1×1)	11.8.1	152	403
Nb(110)-(1×1)	41.2	11	405
$NbSe_2$(0001)-(1×1)	42.16.4a	160	406
Ni(100)-(1×1)	28.16a	2	408
Ni(100)-(1×1)	28.29a	2	409
Ni(100)-(1×1)	28.4b	2	410
Ni(110)-(1×1)	28.11a	4	411
Ni(110)-(1×1)	28.15	4	412
Ni(110)-(1×1)	28.17	4	413
Ni(110)-(1×1)	28.18	4	415
Ni(110)-(1×1)	28.22	4	416
Ni(110)-(1×1)	28.23	4	417
Ni(110)-(1×1)	28.25	4	418
Ni(110)-(1×1)	28.26a	4	419
Ni(111)-(1×1)	28.11b	1	420
Ni(111)-(1×1)	28.4a	1	421
Ni(311)-(1×1)	28.12	8	422
Ni(311)-(1×1)	28.21	8	424
Ni(100)-Ag(111) multilayers	28.47.1	84	426
Ni(100)-c(2×2)-Bi	28.83.1	28,29	427
Ni(100)-p4g(2×2)-2C	28.6.16	32	428
Ni(100)-p4g(2×2)-2C	28.6.17	32	430
Ni(100)-p4g(2×2)-2C	28.6.2	32	432
Ni(110)-(2×1)-C	28.6.18	35	434
Ni(111)-(2×2)-C_2H_2	28.6.1.2	64	435
Ni(100)-C_6H_5S disordered	28.6.16.1	77	437
Ni(100)-c(2×2)-CO	28.6.8.11	71	438
Ni(100)-c(2×2)-CO	28.6.8.12b	71	439
Ni(100)-c(2×2)-CO	28.6.8.4	71	441
Ni(100)-c(2×2)-CO	28.6.8.6	71	442
Ni(100)-c(2×2)-CO	28.6.8.7	71	443
Ni(100)-c(2×2)-CO	28.6.8.8	71	444
Ni(110)-p(2×1)-2CO	28.6.8.20	78	445
Ni(111)-($\sqrt{3}$×$\sqrt{3}$)R30°-CO	28.6.8.12a	61	447
Ni(100)-c(2×2)-Cl	28.17.1	28,29	449
Ni(111)-($\sqrt{3}$×$\sqrt{3}$)R30°-Cl	28.17.2	22,24	450
Ni(111)-($\sqrt{3}$×$\sqrt{3}$)R30°-Cl	28.17.3	22,24	451
Ni(100)-(1×1)-3Co	28.27.1	83	452
Ni(100)-(1×1)-Cu	28.29.3	83	453
Ni(100)-(1×1)-3Cu	28.29.5a	83	454
Ni(100)-(1×1)-Cu multilayer	28.29.5	83	455
Ni(100)-(1×1)-1Fe	28.26.2a	83	456
Ni(100)-(1×1)-2Fe	28.26.2b	83	457
Ni(100)-H (D) disordered	28.1.17	28	458
Ni(110)-(2×1)-2H	28.1.23	35,37	459
Ni(111)-(2×2)-2H	28.1.6	22,25	461
Ni(111)-(2×2)-2H (D)	28.1.31	22,25	463
Ni(100)-c(2×2)-Hg	28.80.1	28	464
Ni(100)-c(2×2)-I	28.53.2	28,29	465
Ni(100)-p4g(2×2)-2N	28.7.2	32	466
Ni(100)-p4g(2×2)-2N	28.7.4	32	468
Ni(100)-c(2×2)-N+O disordered	28.7.8.1	28,29	470
Ni(100)-c(2×2)-Na	28.11.3	28,29	472
Ni(100)-c(2×2)-Na	28.11.4	28,29	473
Ni(100)-c(2×2)-Na+S	28.11.16.1a	28,29	474
Ni(100)-p(2×2)-Na+2S	28.11.16.1b	28,30	476
Ni(100)-p(2×2)-Na+S	28.11.16.1c	28,30	478
Ni(100)-c(2×2)-O	28.8.21	28,29	480

2.1. Alphanumerical Index of Structures — Continued

Common name of structure	Class no.	Figure	Page
Ni(100)-c(2×2)-O	28.8.30a	28,29	482
Ni(100)-c(2×2)-O	28.8.32a	28,29	483
Ni(100)-c(2×2)-O	28.8.35	28	485
Ni(100)-c(2×2)-O	28.8.36	28,29	486
Ni(100)-c(2×2)-O	28.8.37	28,29	488
Ni(100)-c(2×2)-O	28.8.42	28,29	490
Ni(100)-c(2×2)-O	28.8.48	28,29	492
Ni(100)-c(2×2)-O	28.8.59	28,29	494
Ni(100)-c(2×2)-O	28.8.61	28,29	495
Ni(100)-c(2×2)-O	28.8.7	28,29	496
Ni(100)-c(2×2)-O	28.8.71	28,29	498
Ni(100)-p(2×2)-O	28.8.30b	28,30	500
Ni(100)-p(2×2)-O	28.8.32b	28,30	501
Ni(100)-p(2×2)-O	28.8.5	28,30	503
Ni(100)-p(2×2)-O	28.8.72	28,30	504
Ni(100)-O disordered	28.8.82	28	506
Ni(110)-(2×1)-O	28.8.53	39	508
Ni(110)-(2×1)-O	28.8.58	—	510
Ni(110)-(2×1)-O	28.8.70	39	511
Ni(111)-p(2×2)-O	28.8.75a	22,25	513
Ni(111)-($\sqrt{3}\times\sqrt{3}$)R30°-O	28.8.27	22,24	514
Ni(111)-($\sqrt{3}\times\sqrt{3}$)R30°-O	28.8.85	22,24	515
Ni(100)-c(2×2)-S	28.16.10a	28,29	516
Ni(100)-c(2×2)-S	28.16.13	28,29	517
Ni(100)-c(2×2)-S	28.16.15	28,29	518
Ni(100)-c(2×2)-S	28.16.16	28,29	519
Ni(100)-c(2×2)-S	28.16.18	28,29	520
Ni(100)-c(2×2)-S	28.16.19	28,29	521
Ni(100)-c(2×2)-S	28.16.22	28,29	522
Ni(100)-c(2×2)-S	28.16.23a	28,29	523
Ni(100)-c(2×2)-S	28.16.28	28,29	524
Ni(100)-c(2×2)-S	28.16.35	28,29	525
Ni(100)-c(2×2)-S	28.16.4c	28,29	526
Ni(100)-c(2×2)-S	28.16.9	28,29	527
Ni(100)-p(2×2)-S	28.16.36a	28,30	528
Ni(100)-p(2×2)-S	28.16.4a	28,30	530
Ni(100)-S disordered	28.16.36b	28	531
Ni(110)-c(2×2)-S	28.16.17	35,36	533
Ni(110)-c(2×2)-S	28.16.20	35,36	535
Ni(110)-c(2×2)-S	28.16.23b	35,36	537
Ni(110)-c(2×2)-S	28.16.25	35,36	538
Ni(110)-c(2×2)-S	28.16.26	35,36	539
Ni(110)-c(2×2)-S	28.16.7	35,36	540
Ni(110)-p(2×2)-S	28.16.4d	35	541
Ni(111)-(2×2)-S	28.16.23c	22,25	542
Ni(111)-(2×2)-S	28.16.4b	22,25	543
Ni(111)-p(2×2)-S	28.16.33	22,25	544
Ni(100)-c(2×2)-Se	28.34.0a	28,29	545
Ni(100)-c(2×2)-Se	28.34.1	28,29	546
Ni(100)-p(2×2)-Se	28.34.0b	28,30	548
Ni(110)-c(2×2)-Se	28.34.0c	35,36	549
Ni(111)-p(2×2)-Se	28.34.0d	22,25	550
Ni(100)-c(2×2)-Te	28.52.1	28,29	551
Ni(100)-p(2×2)-Te	28.52.1a	28,30	552
Ni$_3$Al(100)-(1×1)	28.13.5	134	553
Ni$_3$Al(110)-(1×1)	28.13.13	137	555
Ni$_3$Al(111)-(1×1)	28.13.12	128	557
NiAl(100)-(1×1)	28.13.15c	143	559
NiAl(110)-(1×1)	28.13.11	142	560
NiAl(110)-(1×1)	28.13.15b	142	561
NiAl(110)-(1×1)	28.13.19	142	563
NiAl(110)-(1×1)	28.13.4	142	564
NiAl(111)-(1×1) Al-terminated	28.13.14b	144	565
NiAl(111)-(1×1) Ni-terminated	28.13.14a	144	566
NiO(100)-(1×1)	28.8.20	149	567
NiSi$_2$(100)-(1×1)	14.28.6	148	568
NiSi$_2$(100)-(1×1)	14.28.9	147	570

2.1. Alphanumerical Index of Structures — Continued

Common name of structure	Class no.	Figure	Page
NiSi$_2$(111)-(1×1)	14.28.1	145	571
NiSi$_2$(111)-(1×1)	14.28.13	145	573
Pb(100)-(1×1)	82.13	2	575
Pb(110)-(1×1)	82.1	4	576
Pb(110)-(1×1)	82.12	4	577
Pb(110)-(1×1)	82.7	4	579
Pb(111)-(1×1)	82.16	1	580
Pb(311)-(1×1)	82.18	8	581
Pd(100)-(1×1)	46.2	2	583
Pd(100)-(1×1)	46.8	2	584
Pd(110)-(1×1)	46.4a	4	585
Pd(110)-(1×1)	46.5b	4	586
Pd(110)-(1×2)	46.4b	5	587
Pd(110)-(1×2)	46.6	5	588
Pd(111)-(1×1)	46.3	1	590
Pd(111)-(1×1)	46.5a	1	591
Pd(111)-(1×1)-Au	46.79.1	81	592
Pd(111)-(3×3)-C$_6$H$_6$+2CO	46.6.1.8.2	70	593
Pd(111)-($\sqrt{3}\times\sqrt{3}$)R30°-CO	46.6.8.2	61	595
Pd(100)-($2\sqrt{2}\times\sqrt{2}$)R45°-2CO	46.6.8.0a	71,72	597
Pd(100)-(1×1)-Cu multilayer	46.29.1	83	598
Pd(100)-(1×1)-H (D)	46.1.13a	28	599
Pd(100)-c(2×2)-H (D)	46.1.13b	28,29	600
Pd(110)-(1×2)-H	46.1.11a	—	601
Pd(110)-(2×1)-2H	46.1.12	37	603
Pd(100)-(1×1)-12Fe	46.26.2a	83	605
Pd(100)-(1×1)-53Fe	46.26.2b	83	606
Pd(100)-(1×1)-200Fe	46.26.2c	83	607
Pd(100)-c(2×2)-Mn / Pd$_3$Mn(100)	46.25.5b	134	608
Pd(100)-c(2×2)-Mn mixed top layer	46.25.5a	33	610
Pd(100)-c(2×2)-S	46.16.1	28,29	612
Pd(111)-($\sqrt{3}\times\sqrt{3}$)R30°-S	46.16.2	22,24	613
Pt(100)-(1×1)	78.16	2	614
Pt(100)-(1×1)	78.16a	2	615
Pt(100)-(1×1)	78.6	2	616
Pt(110)-(1×2)	78.19	5	617
Pt(110)-(1×2)	78.32	5	618
Pt(110)-(1×2)	78.33a	5	620
Pt(110)-(1×2)	78.34	5	622
Pt(110)-(1×2)	78.41	5	624
Pt(110)-(1×2)	78.47	5	626
Pt(110)-(1×2)	78.49	5	628
Pt(110)-(1×3)	78.33b	7	629
Pt(110)-(1×3)	78.48	6	631
Pt(111)-(1×1)	78.12	1	633
Pt(111)-(1×1)	78.15	1	634
Pt(111)-(1×1)	78.20a	1	635
Pt(111)-(1×1)	78.8	1	636
Pt(210)-(1×1)	78.43	9	637
Pt(111)-(1×1)-H (D)	78.1.7	22,23	639
Pt(111)-(2×2)-C$_2$H$_3$	78.6.1.5	65	640
Pt(111)-C$_6$H$_6$ disordered	78.6.1.18	67	642
Pt(111)-($2\sqrt{3}\times4$)rect-2C$_6$H$_6$+4CO	78.6.1.8.1	68	644
Pt(111)+CO 1/3ML disordered	78.6.8.7	61	646
Pt(111)-c(4×2)-2CO	78.6.8.4	61,62	648
Pt(100)-(1×1)-Cu multilayer	78.29.2	83	649
Pt(111)-($\sqrt{3}\times\sqrt{3}$)R30°-S	78.16.1	22,24	650
Pt(111)-(2×2)-Sn	78.50.1a	27	651
Pt(111)-($\sqrt{3}\times\sqrt{3}$)R30°-Sn	78.50.1b	27	652
Pt$_{0.8}$Fe$_{0.2}$(110)-(1×2)	78.26.2	140	653
Pt$_{0.8}$Fe$_{0.2}$(111)-(1×1)	78.26.1	132	655
Pt$_{0.1}$Ni$_{0.9}$(100)-(1×1)	78.28.8	136	657
Pt$_{0.1}$Ni$_{0.9}$(110)-(1×1)	78.28.7	138	659
Pt$_{0.1}$Ni$_{0.9}$(111)-(1×1)	78.28.2	129	661
Pt$_{0.5}$Ni$_{0.5}$(110)-(1×1)	78.28.3	139	663
Pt$_{0.5}$Ni$_{0.5}$(111)-(1×1)	78.28.1b	130	665
Pt$_{0.5}$Ni$_{0.5}$(111)-(1×1)	78.28.9	130	667

2.1. Alphanumerical Index of Structures — Continued

Common name of structure	Class no.	Figure	Page
$Pt_{0.78}Ni_{0.22}$(111)-(1×1)	78.28.1a	131	669
Re(10-10)-(1×1)	75.2	20	671
Rh(100)-(1×1)	45.7b	2	673
Rh(110)-(1×1)	45.7c	4	674
Rh(110)-(1×1)	45.9a	4	675
Rh(100)-(1×1)	45.9b	2	676
Rh(111)-(1×1)	45.7a	1	677
Rh(111)-(1×1)	45.8	1	678
Rh(311)-(1×1)	45.11	8	679
Rh(111)-(2×2)-C_2H_3	45.6.1.11	65	680
Rh(111)-(2×2)-C_2H_3	45.6.1.3	65	682
Rh(111)-c(4×2)-C_2H_3+CO	45.6.1.8.4a	66	684
Rh(111)-c(4×2)-C_2H_3+NO	45.6.7.8.1.1	66	686
Rh(111)-(3×3)-C_6H_6+2CO	45.6.1.8.3	70	688
Rh(111)-c(2$\sqrt{3}$×4)rect-C_6H_6+CO	45.6.1.8.2	69	690
Rh(111)-(2×2)-3CO	45.6.8.4	63	692
Rh(111)-($\sqrt{3}$×$\sqrt{3}$)R30°-CO	45.6.8.1	61	694
Rh(100)-c(4×2)-Cs	45.55.1	28,30	695
Rh(110)-(1×1)-2H	45.1.10	37	696
Rh(110)-(1×1)-2H	45.1.4	37	698
Rh(110)-(1×2)-H	45.1.8	38	700
Rh(110)-(1×2)-3H	45.1.6	—	702
Rh(110)-(1×3)-H	45.1.5	38	704
Rh(111)-(2×2)-3NO	45.7.8.1	63	706
Rh(100)-(2×2)-O	45.8.2	28,30	708
Rh(111)-(2×2)-O	45.8.1	22,25	709
Rh(100)-(2×2)-S	45.16.1	28,30	710
Rh(110)-c(2×2)-S	45.16.2	35,36	711
Rh(111)-($\sqrt{3}$×$\sqrt{3}$)R30°-S	45.16.3	22,24	712
Ru(0001)-(1×1)	44.1	19	713
Ru(0001)-(1×1)-H	44.1.1	55,56	714
Ru(0001)-($\sqrt{3}$×$\sqrt{3}$)R30°-CO	44.6.8.1	80	715
Ru(0001)-CO disordered	44.6.8.2	80	717
Ru(0001)-(1×1)-1Fe	44.26.1a	87	718
Ru(0001)-p(2×1)-O	44.8.1	55,57	719
Ru(0001)-p(2×2)-O	44.8.2	55,57	721
Sc(0001)-(1×1)	21.2	19	723
Si(100)-(2×1)	14.170	94	725
Si(100)-(2×1)	14.182a	94	727
Si(100)-(2×1)	14.75	94	729
Si(100)-(2×1)	14.85	94	731
Si(100)-c(4×2)	14.182b	95	733
Si(111)-(2×1)	14.120	89	735
Si(111)-(2×1)	14.25	—	737
Si(111)-(2×1)	14.89	89	739
Si(111)-(2×1)	14.96	89	741
Si(111) laser annealed	14.108	88	743
Si(111)-(1×1) laser-annealed	14.99	88	745
Si(111)-(7×7)	14.132	91	746
Si(111)-($\sqrt{3}$×$\sqrt{3}$)R30°-Al	14.152	90	750
Si(111)-($\sqrt{3}$×$\sqrt{3}$)R30°-Al	14.13.12	96,97	752
Si(111)-(1×1)-As	14.33.10	102	754
Si(111)-(1×1)-As	14.33.7	102	755
Si(111)-(1×1)-As	14.33.8	102	756
Si(111)-($\sqrt{3}$×$\sqrt{3}$)R30°-B	14.5.6	101	757
Si(111)-($\sqrt{3}$×$\sqrt{3}$)R30°-Bi	14.83.2	103	759
Si(111)-($\sqrt{3}$×$\sqrt{3}$)R30°-Bi (1/3 ML)	14.83.3a	96,97	761
Si(111)-($\sqrt{3}$×$\sqrt{3}$)R30°-Bi (1 ML)	14.83.3b	100	763
Si(111)-Br 0.25ML	14.35.2	96,99	765
Si(111)-Br 0.67ML	14.35.1	96,99	766
Si(111)-(1×1)-Cl	14.17.4b	96,99	767
Si(111)-(7×7)-Cl	14.17.4a	96,99	768
Si(100)-Co 0.4ML	14.27.16	104,105	769
Si(111)-(1×1)-$CoSi_2$(111) interface	14.27.2	111	770
Si(111)-(1×1)-$CoSi_2$(111) interface	14.27.3	111	772
Si(111)-(1×1)-$CoSi_2$(111) interface	14.27.8	112	774
Si(111)-($\sqrt{3}$×$\sqrt{3}$)R30°-Ga	14.31.4	96,97	776
Si(111)-(7×7)-I	14.53.2	96,99	778
Si(100)-(2×1)-2K	14.19.9	108	779
Si(100)-(2×1)-Na	14.11.4	107	781
Si(111)-(1×1)-$NiSi_2$(111) interface	14.28.12a	114	783
Si(111)-(1×1)-$NiSi_2$(111) interface	14.28.12b	113	785
Si(111)-(1×1)-$NiSi_2$(111) interface	14.28.2	114	787
Si(111)-(1×1)-$NiSi_2$(111) interface	14.28.8	114	789
Si(111)-($\sqrt{3}$×$\sqrt{3}$)R30°-Pb	14.82.1	96	790
Si(111)-($\sqrt{3}$×$\sqrt{3}$)R30°-Pb (a phase)	14.82.2	96,97	791
Si(111)-($\sqrt{3}$×$\sqrt{3}$)R30°-Sn	14.50.2	96,97	793
Si(100)-(2×1)-2Sb	14.51.7	109	795
Si(111)-(7×7)-Te	14.52.1	96	796
SiC(100)-c(2×2) (C_2H_4 exposed)	14.6.7a	118	798
SiC(100)-c(2×2) (Si sublimation)	14.6.7b	119	800
SiC(100)-p(2×1)	14.6.8	120	802
$SrTiO_3$(100)-(1×1) O-Ti-O terminated)	38.22.8.1b	154	804
$SrTiO_3$(100)-(1×1) Sr-O terminated)	38.22.8.1a	155	806
Ta(100)-(1×1)	73.1	12	808
Ta(100)-(1×1)	73.4	12	809
Ta(100)-(1×3)-O	73.8.1	53	810
TaC(100)-(1×1)	73.6.2	149	812
TaC(100)-(1×1)	73.6.4	149	814
Tb(0001)-(1×1)	65.1	19	816
Te(10-10)-(1×1)	52.1	158	817
Ti(0001)-(1×1)	22.1	19	819
Ti(10-10)-(1×1)	22.3	20	821
Ti(0001)-(1×1)-2Cd	22.48.3a	87	823
Ti(0001)-(1×1)-4Cd	22.48.3b	87	824
Ti(0001)-(1×1)-Cd	22.48.2	87	825
Ti(0001)-(1×1)-N	22.7.2	58	827
TiC(111)-($\sqrt{3}$×$\sqrt{3}$)R30°-O	22.6.8.3	156	829
TiO_2(100)-(3×1)	22.8.1	153	831
$TiSe_2$(0001)-(1×1)	22.34.2	161	833
V(100)-(1×1)	23.4	12	835
V(110)-(1×1)	23.2	11	837
VN_0.89(100)-(1×1)	23.7.1	149	838
W(100)-(1×1)	74.1.21a	12	840
W(100)-(1×1)	74.21	12	841
W(100)-(1×1)	74.2a	12	842
W(100)-(1×1) disordered	74.47	14	843
W(110)-(1×1)	74.2b	11	844
W(110)-(1×1)	74.45	11	845
W(100)-c(2×2)	74.14	13	846
W(100)-c(2×2)	74.53	13	847
W(100)-c(2×2)	74.59	13	848
W(211)-(1×1)	74.55	17	850
W(310)-(1×1)	74.63	18	851
W(100)-(1×1)-2H	74.1.10	48,50	852
W(100)-c(2×2)-N	74.7.2	48,50	854
W(100)-O disordered	74.8.8	51	856
W(100)-p(2×1)-disordered O	74.8.12	52	858
W(110)-(2×1)-O	74.8.1	44,45	859
Zn(0001)-(1×1)	30.1	19	860
ZnO(0001)-(1×1)	30.8.2	123	861
ZnO(10-10)-(1×1)	30.8.2a	124	862
ZnO(11-20)-(1×1)	30.8.2b	125	864
ZnS(110)-(1×1)	30.16.2	116	866
ZnSe(110)-(1×1)	30.34.2a	116	868
ZnSe(110)-(1×1)	30.34.2b	116	870
ZnTe(110)-(1×1)	30.52.2	116	872
Zr(0001)-(1×1)	40.1	19	874
Zr(0001)-(1×1)-C	40.6.1	58	875
Zr(0001)-(1×1)-N	40.7.1	58	877
Zr(0001)-(2×2)-O	40.8.1	59	878

2.2. Main Table of Structures

COMMON NAME : Ag(100)-(1x1)   ILLUSTRATION: 2
CLASSIFICATION : 47.22
TECHNIQUE : LEED
AUTHORS : H. Li, J. Quinn, Y.S. Li, D. Tian, F. Jona and P.M. Marcus
REFERENCE : Phys. Rev., B43, 7305 (1991)

SURFACE TYPE
Substrate : Ag
Crystal face: 100
Temperature : RT*
Bulk lattice: fcc
2D bulk symm: p4m
2D surf symm: p4m

Adsorbate:
Coverage :
Pattern : (1x1)
Matrix : (1.000, 0.000)
(0.000, 1.000)

STRUCTURE TYPE
No multilayer relaxation

SAMPLE PREPARATION (1 sample)
Treatment : ion bombardment followed by annealing
Crystallinity: sharp LEED pattern
Anal. methods: AES
Contamination:

COMMENTS

DATA COLLECTION
Technique: LEED
Dataset : IV spectra for 5 non-equivalent beams at normal incidence 3 beams at 10° off-normal

THEORY/DATA TREATMENT
Dynamical LEED (CHANGE program)

STRUCTURES EXAMINED
Variation of 1st and 2nd interlayer spacings

QUALITY OF EXPERIMENT-THEORY FIT
PRE=0.39

2D UNIT CELLS (1 domain observed)

Cell	Ax (Å)	Ay (Å)	Bx (Å)	By (Å)	α (°)	Matrix	Pattern	Cell type
Bulk	2.889	0.000	0.000	2.889	90.0	(1.000, 0.000) (0.000, 1.000)	(1x1)	b: bulk lattice
Surface 1	2.889	0.000	0.000	2.889	90.0	(1.000, 0.000) (0.000, 1.000)	(1x1)	s1: commens. superlattice

3D COORDINATES

Dx/Dy in Å, or as a fraction of layer's unit cell vectors; atom 0 at the origin. Epir/subr are bulk repeat vectors.

No. of atoms: 3   Bulk z = 2.043 Å

Region	Chem el.	At. no.	Cell type	Site occ.	Rel. to	Dx ± εx		Dy ± εy		Dz ± εz		Dz/Bz(%) ± εz/Bz
epir		-2					f		f		Å	
subr		-1				1.445	Å	1.445	Å	2.043	Å	
intf	Ag	1	s1	1.00	0	0.000	Å	0.000	Å	0.000	Å	0.0
intf	Ag	2	s1	1.00	0	1.445	Å	1.445	Å	2.043 ± .030	Å	100.0 ± 1.5
subl	Ag	3	b	1.00	0	0.000	Å	0.000	Å	4.086 ± .030	Å	200.0 ± 1.5

BOND DISTANCES AND ANGLES

Bond distances and angles are derived from coordinates

No. of distances/angles: 2

Interatomic dist. A-B (Å)	Atom A	Atom B	Atom C	Bond angle A-B-C (°)
2.889	Ag1	Ag2		
2.889	Ag2	Ag3		

COMMON NAME : Ag(110)-(1x1) ILLUSTRATION: 4
CLASSIFICATION : 47.15
TECHNIQUE : HEIS
AUTHORS : Y. Kuk and L.C. Feldman
REFERENCE : Phys. Rev., B30, 5811 (1984)

SURFACE TYPE
Substrate : Ag
Crystal face: 110
Temperature : RT*
Bulk lattice: fcc
2D bulk symm: pmm
2D surf symm: pmm

Adsorbate:
Coverage :
Pattern : (1x1)
Matrix : (1.000, 0.000)
(0.000, 1.000)

STRUCTURE TYPE
Bulk termination with multilayer relaxation

SAMPLE PREPARATION (1 sample)
Treatment : electropolishing, etching, sputtering and annealing
Crystallinity:
Anal. methods:
Contamination: monitored by AES and ion scattering

COMMENTS

R-factor defined as:
R = 100*√(Σ((Ycalc-Yexpt)/Yexpt)²)/n, where n is the number of data points, Yexpt the experimental energy of the surface peak, and Ycalc the theoretical energy

DATA COLLECTION
Technique: HEIS
Dataset : angular and energy scans in the <101> and <100> directions

THEORY/DATA TREATMENT
High energy ion scattering with computer simulations;
ΘD=149 K(surf), 215K(bulk)

STRUCTURES EXAMINED
Top two layer spacings varied from -10% to +10% from bulk; ΘD varied: 105K, 149K, 215K

QUALITY OF EXPERIMENT-THEORY FIT
R=0.45 (see comments)

2D UNIT CELLS (1 domain observed)

Cell	Ax (Å)	Ay (Å)	Bx (Å)	By (Å)	α (°)	Matrix	Pattern	Cell type
Bulk	2.880	0.000	0.000	4.090	90.0	(1.000, 0.000) (0.000, 1.000)	(1x1)	b: bulk lattice
Surface 1	2.880	0.000	0.000	4.090	90.0	(1.000, 0.000) (0.000, 1.000)	(1x1)	s1: commens. superlattice

3D COORDINATES

Dx/Dy in Å, or as a fraction of layer's unit cell vectors; atom 0 at the origin. Epir/subr are bulk repeat vectors.

No. of atoms: 4 Bulk z = 1.440 Å

Region	Chem el.	At. no.	Cell type	Site occ.	Rel. to	Dx ± εx		Dy ± εy		Dz ± εz		Dz/Bz(%) ± εz/Bz
epir		-2					f		f		Å	
subr		-1				1.440	Å	2.045	Å	1.440	Å	
intf	Ag	1	b	1.00	0	0.000	f	0.000	f	0.000	Å	0.0
intf	Ag	2	b	1.00	1	0.500	f	0.500	f	1.330 ± .040	Å	92.4 ± 2.8
intf	Ag	3	b	1.00	2	-0.500	f	-0.500	f	1.500 ± .040	Å	104.2 ± 2.8
subl	Ag	4	b	1.00	3	0.500	f	0.500	f	1.440	Å	100.0

BOND DISTANCES AND ANGLES

Bond distances and angles are derived from coordinates

No. of distances/angles: 3

Interatomic dist. A-B (Å)	Atom A	Atom B	Atom C	Bond angle A-B-C (°)
2.880	Ag1	Ag1(1,0)	Ag2	59.5
2.833	Ag1	Ag2	Ag3	59.0
2.916	Ag2	Ag3	Ag4	60.9

COMMON NAME : Ag(110)-(1x1)
CLASSIFICATION : 47.16
TECHNIQUE : LEED
AUTHORS : J.R. Noonan and H.L. Davis
REFERENCE : Vacuum, 32, 107 (1982)

ILLUSTRATION: 4

SURFACE TYPE
Substrate : Ag
Crystal face: 110
Temperature : RT
Bulk lattice: fcc
2D bulk symm: pmm
2D surf symm: pmm

Adsorbate:
Coverage :
Pattern : (1x1)
Matrix : (1.000, 0.000)
(0.000, 1.000)

STRUCTURE TYPE
Bulk termination with 6.6% top spacing contraction

SAMPLE PREPARATION (1 sample)
Treatment : spark erosion, lapping, electropolish, sputter and anneal
Crystallinity:
Anal. methods:
Contamination: monitored by AES and LEED

COMMENTS
Averaging of symmetrically related beams and careful variations of non-structural parameters enables an accuracy of 0.02Å to be achieved

DATA COLLECTION
Technique: LEED
Dataset : I-V curves: 7 symm.-averaged beams (01) (10) (02) (11) (12) (21)

THEORY/DATA TREATMENT
Dynamical LEED (angular momentum basis): ΘD=190 K

STRUCTURES EXAMINED
Top layer spacing varied; various atomic potentials tried, as well as several smooth variations of Voi, with Vor constant

QUALITY OF EXPERIMENT-THEORY FIT
RZJ=0.098

2D UNIT CELLS (1 domain observed)

Cell	Ax (Å)	Ay (Å)	Bx (Å)	By (Å)	α (°)	Matrix	Pattern	Cell type
Bulk	2.890	0.000	0.000	4.087	90.0	(1.000, 0.000) (0.000, 1.000)	(1x1)	b: bulk lattice
Surface 1	2.890	0.000	0.000	4.087	90.0	(1.000, 0.000) (0.000, 1.000)	(1x1)	s1: commens. superlattice

3D COORDINATES

Dx/Dy in Å, or as a fraction of layer's unit cell vectors; atom 0 at the origin. Epir/subr are bulk repeat vectors.

No. of atoms: 3

Bulk z = 1.445 Å

Reg ion	Chem el.	At. no.	Cell type	Site occ.	Rel. to	Dx ± εx		Dy ± εy		Dz ± εz		Dz/Bz(%) ± εz/Bz
epir		-2					f		f		Å	
subr		-1				-1.445	Å	-2.044	Å	1.445	Å	
intf	Ag	1	b	1.00	0	0.000	f	0.000	f	0.000	Å	0.0
intf	Ag	2	b	1.00	1	0.500	f	0.500	f	1.349 ± .022	Å	93.4 ± 1.5
subl	Ag	3	b	1.00	2	-0.500	f	-0.500	f	1.445	Å	100.0

BOND DISTANCES AND ANGLES

Bond distances and angles are derived from coordinates

No. of distances/angles: 3

Interatomic dist. A-B (Å)	Atom A	Atom B	Atom C	Bond angle A-B-C (°)
2.890	Ag1	Ag1(1,0)	Ag2	59.5
2.843	Ag1	Ag2	Ag3	58.3
2.890	Ag2	Ag3		

COMMON NAME : Ag(110)-(1x1)
CLASSIFICATION : 47.17
TECHNIQUE : MEIS
AUTHORS : E. Holub-Krappe, K. Horn, J.W.M. Frenken, R.L. Krans and J.F. van der Veen
REFERENCE : Surf. Sci., 188, 335 (1987)

ILLUSTRATION: 4

SURFACE TYPE
Substrate : Ag
Crystal face: 110
Temperature : RT
Bulk lattice: fcc
2D bulk symm: pmm
2D surf symm: pmm

Adsorbate:
Coverage :
Pattern : (1x1)
Matrix : (1.000, 0.000)
(0.000, 1.000)

STRUCTURE TYPE
Relaxations in top two interlayer spacings

SAMPLE PREPARATION (1 sample)
Treatment : cycles of sputter/anneal with a base pressure of 5E-9 Pa
Crystallinity: sharp (1x1) LEED pattern
Anal. methods:
Contamination: no impurities detected by AES and RBS

COMMENTS

DATA COLLECTION
Technique: MEIS; 50.6 and 97.5keV proton beams
Dataset : blocking curves measured in (1,-1,1), (1,-1,0), and (0,0,1) scattering planes

THEORY/DATA TREATMENT
Monte Carlo analysis in the shadowing and blocking geometry; ΘD=215 K, enhanced 65% and 12% in layers 1,2

STRUCTURES EXAMINED
Top two interlayer spacings varied

2D UNIT CELLS (1 domain observed)

Cell	Ax (Å)	Ay (Å)	Bx (Å)	By (Å)	α (°)	Matrix	Pattern	Cell type
Bulk	4.090	0.000	0.000	2.892	90.0	(1.000, 0.000) (0.000, 1.000)	(1x1)	b: bulk lattice
Surface 1	4.090	0.000	0.000	2.892	90.0	(1.000, 0.000) (0.000, 1.000)	(1x1)	s1: commens. superlattice

3D COORDINATES

Dx/Dy in Å, or as a fraction of layer's unit cell vectors; atom 0 at the origin. Epir/subr are bulk repeat vectors.

No. of atoms: 4 Bulk z = 1.446 Å

Region	Chem el.	At. no.	Cell type	Site occ.	Rel. to	Dx ± εx		Dy ± εy		Dz ± εz		Dz/Bz(%) ± εz/Bz
epir		-2					f		f		Å	
subr		-1				2.045	Å	1.446	Å	1.446	Å	
intf	Ag	1	b	1.00	0	0.000	f	0.000	f	0.000	Å	0.0
intf	Ag	2	b	1.00	1	0.500	f	0.500	f	1.309 ± .030	Å	90.5 ± 2.1
intf	Ag	3	b	1.00	2	-0.500	f	-0.500	f	1.533 ± .040	Å	106.0 ± 2.8
subl	Ag	4	b	1.00	3	0.500	f	0.500	f	1.446	Å	100.0

BOND DISTANCES AND ANGLES

Bond distances and angles are derived from coordinates

No. of distances/angles: 4

Interatomic dist. A-B (Å)	Atom A	Atom B	Atom C	Bond angle A-B-C (°)
2.892	Ag1	Ag1(0,1)		
2.826	Ag1	Ag2	Ag3	59.1
2.826	Ag1	Ag2	Ag4	117.6
2.937	Ag2	Ag3	Ag4	61.5

COMMON NAME : Ag(110)-(1x1)
CLASSIFICATION : 47.19
TECHNIQUE : LEED
AUTHORS : M. Lindroos, C.J. Barnes, M. Valden and D.A. King
REFERENCE : Surf. Sci., 218, 269 (1989)

ILLUSTRATION: 4

SURFACE TYPE
Substrate : Ag
Crystal face: 110
Temperature : 100 K
Bulk lattice: fcc
2D bulk symm: pmm
2D surf symm: pmm

Adsorbate:
Coverage :
Pattern : (1x1)
Matrix : (1.000, 0.000)
(0.000, 1.000)

STRUCTURE TYPE
Multilayer relaxations of (-7, +1, -2, 0%) to the fourth layer, with bulk termination

SAMPLE PREPARATION (1 sample)
Treatment : Ar+ sputtering with 900 K annealing
Crystallinity: surface within 1° of the [110] plane
Anal. methods:
Contamination: no contamination by AES

COMMENTS

DATA COLLECTION
Technique: LEED
Dataset : I-V curves for 8 symmetrically inequivalent beams; E range 50-250 eV

THEORY/DATA TREATMENT
Dynamical LEED (Van Hove/Tong): 9 ph shs (Moruzzi et al); Vor=-15 eV(then fit),Voi=-3.5eV;ΘD(bulk)=215 K,(surf)=150K

STRUCTURES EXAMINED
Variation of top 4 interlayer spacings; 7 R-factors used

QUALITY OF EXPERIMENT-THEORY FIT
RPE=0.153

2D UNIT CELLS (1 domain observed)

Cell	Ax (Å)	Ay (Å)	Bx (Å)	By (Å)	α (°)	Matrix	Pattern	Cell type
Bulk	2.878	0.000	0.000	4.070	90.0	(1.000, 0.000) (0.000, 1.000)	(1x1)	b: bulk lattice
Surface 1	2.878	0.000	0.000	4.070	90.0	(1.000, 0.000) (0.000, 1.000)	(1x1)	s1: commens. superlattice

3D COORDINATES

Dx/Dy in Å, or as a fraction of layer's unit cell vectors; atom 0 at the origin. Epir/subr are bulk repeat vectors

No. of atoms: 5

Bulk z = 1.445 Å

Region	Chem el.	At. no.	Cell type	Site occ.	Rel. to	Dx ± εx		Dy ± εy		Dz ± εz		Dz/Bz(%) ± εz/Bz
epir		-2					f		f		Å	
subr		-1				1.439	Å	2.035	Å	1.439	Å	
intf	Ag	1	b	1.00	0	0.000	f	0.000	f	0.000	Å	0.0
intf	Ag	2	b	1.00	1	0.500	f	0.500	f	1.338 ± .029	Å	93.0 ± 2.0
intf	Ag	3	b	1.00	2	-0.500	f	-0.500	f	1.453 ± .029	Å	101.0 ± 2.0
intf	Ag	4	b	1.00	3	0.500	f	0.500	f	1.410 ± .029	Å	98.0 ± 2.0
subl	Ag	5	b	1.00	4	-0.500	f	-0.500	f	1.439 ± .029	Å	100.0 ± 2.0

BOND DISTANCES AND ANGLES

Bond distances and angles are derived from coordinates

No. of distances/angles: 5

Interatomic dist. A-B (Å)	Atom A	Atom B	Atom C	Bond angle A-B-C (°)
2.878	Ag1	Ag1(1,0)		
2.829	Ag1	Ag2		
2.791	Ag1	Ag3		
2.885	Ag2	Ag3		
2.863	Ag2	Ag4		

COMMON NAME : Ag(100)-C2H4 disordered
CLASSIFICATION : 47.6.1.2
TECHNIQUE : NEXAFS
AUTHORS : J. C. Tang, J. F. Shen and Y. B. Chen
REFERENCE : Surf. Sci., 244, L125 (1991)

ILLUSTRATION: 74

SURFACE TYPE
Substrate : Ag
Crystal face: 100
Temperature : RT*
Bulk lattice: fcc
2D bulk symm: p4m
2D surf symm: none

Adsorbate: C2H4 (ethylene)
Coverage : 0.1 ML
Pattern : disordered
Matrix : (1.000, 0.000)
(0.000, 1.000)

STRUCTURE TYPE
Intact molecular adsorption over 4-fold hollow site with C-C bond parallel to the [001] or [010] direction

SAMPLE PREPARATION (sample)
Treatment :
Crystallinity:
Anal. methods:
Contamination:

COMMENTS
Coverage assumed to be 0.1 ML for tabulation

DATA COLLECTION
Technique: NEXAFS
Dataset :

THEORY/DATA TREATMENT
Multiple-scattering cluster method of NEXAFS

STRUCTURES EXAMINED
Aligned-hollow, diagonal-hollow, aligned-top, diagonal-top aligned-bridge and perpendicular-bridge sites

QUALITY OF EXPERIMENT-THEORY FIT
Visual

2D UNIT CELLS (2 domains observed)

Cell	Ax (Å)	Ay (Å)	Bx (Å)	By (Å)	α (°)	Matrix	Pattern	Cell type
Bulk	2.889	0.000	0.000	2.889	90.0	(1.000, 0.000) (0.000, 1.000)	(1x1)	b: bulk lattice
Surface 1	2.889	0.000	0.000	2.889	90.0	(1.000, 0.000) (0.000, 1.000)	disordered	nd1: non-recon. lattice-gas dis

3D COORDINATES

C1-C2: ethylene over 4-fold hollow site with C-C bond parallel to the [001] or [010] direction

Dx/Dy in Å, or as a fraction of layer's unit cell vectors; atom 0 at the origin. Epir/subr are bulk repeat vectors.

No. of atoms: 4

Bulk z = 2.043 Å

Region	Chem el.	At. no.	Cell type	Site occ.	Rel. to	Dx ± εx		Dy ± εy		Dz ± εz		Dz/Bz(%) ± εz/Bz
epir		-2					f		f		Å	
subr		-1				1.445	Å	1.445	Å	2.043	Å	
ovrl	C	1	nd1	.10	0	0.495	Å	0.495	Å	0.000	Å	0.0
ovrl	C	2	nd1	.10	0	-0.495	Å	-0.495	Å	0.000	Å	0.0
subl	Ag	3	b	1.00	0	1.445	Å	1.445	Å	1.715 ± .100	Å	83.9 ± 4.9
subl	Ag	4	b	1.00	0	0.000	Å	0.000	Å	3.758	Å	183.9

BOND DISTANCES AND ANGLES

Bond distances and angles are derived from coordinates

No. of distances/angles: 3

Interatomic dist. A-B (Å)	Atom A	Atom B	Atom C	Bond angle A-B-C (°)
1.400	C1	C2		
1.960	C1	Ag3		
1.960	C2	Ag3		

COMMON NAME : Ag(100)-c(2x2)-Cl
CLASSIFICATION : 47.17.1
TECHNIQUE : LEED
AUTHORS : E. Zanazzi, F. Jona, D.W. Jepsen and P.M. Marcus
REFERENCE : Phys. Rev., B14, 432 (1976)

ILLUSTRATION: 28,29

SURFACE TYPE
Substrate : Ag
Crystal face: 100
Temperature : 300 K
Bulk lattice: fcc
2D bulk symm: p4m
2D surf symm: p4m

Adsorbate: Cl
Coverage : 0.5 Cl/Ag
Pattern : c(2x2)
Matrix : (1.000, 1.000)
(-1.000, 1.000)

STRUCTURE TYPE
Atomic adsorption in hollow sites

SAMPLE PREPARATION (1 sample)
Treatment : exposure to 100mtorr C2H4Cl2 at 423 K
Crystallinity:
Anal. methods:
Contamination: monitored by AES and LEED

COMMENTS
Cf. R-factor analysis of same data: class. no. 47.17.3a

DATA COLLECTION
Technique: LEED
Dataset : I-V spectra: 15 inequivalent beams from 3 inc. angles: Θ=0°; Θ=10°,φ=29.5°; Θ=20°,φ=29.5°; 20<E<150 eV

THEORY/DATA TREATMENT
Dynamical LEED (layer KKR): 58 beams, 8 phase shifts
Vor=-10 eV; ms ampl=0.024Å²(Cl), 0.012<(Ag)<0.048

STRUCTURES EXAMINED
Bridge and hollow sites; mixed buckled Cl-Ag overlayer; variable overlayer-Ag spacing in both cases; relaxations of top Ag-Ag spacing of 10%

QUALITY OF EXPERIMENT-THEORY FIT
Visual

2D UNIT CELLS (1 domain observed)

Cell	Ax (Å)	Ay (Å)	Bx (Å)	By (Å)	α (°)	Matrix	Pattern	Cell type
Bulk	2.889	0.000	0.000	2.889	90.0	(1.000, 0.000) (0.000, 1.000)	(1x1)	b: bulk lattice
Surface 1	2.889	2.889	-2.889	2.889	90.0	(1.000, 1.000) (-1.000, 1.000)	c(2x2)	s1: commens. superlattice

3D COORDINATES

Cl1: overlayer in hollow sites; 0.1Å error bars assumed for tabulation

Dx/Dy in Å, or as a fraction of layer's unit cell vectors; atom 0 at the origin. Epir/subr are bulk repeat vectors.

No. of atoms: 3

Bulk z = 2.043 Å

Reg ion	Chem el.	At. no.	Cell type	Site occ.	Rel. to	Dx ± εx		Dy ± εy		Dz ± εz		Dz/Bz(%) ± εz/Bz
epir		-2					f		f		Å	
subr		-1				-1.445	Å	-1.445	Å	2.043	Å	
ovrl	Cl	1	s1	.50	0	0.000	f	0.000	f	0.000	Å	0.0
intf	Ag	2	b	1.00	1	0.500	f	0.500	f	1.720 ± .100	Å	84.2 ± 4.9
subl	Ag	3	b	1.00	2	-0.500	f	-0.500	f	2.043 ± .100	Å	100.0 ± 4.9

BOND DISTANCES AND ANGLES

Bond distances and angles are derived from coordinates

No. of distances/angles: 3

Interatomic dist. A-B (Å)	Atom A	Atom B	Atom C	Bond angle A-B-C (°)
2.671	Cl1	Ag2	Ag2(1,0)	122.8
2.671	Cl1	Ag2	Ag3	85.1
2.889	Ag2	Ag3		

COMMON NAME : Ag(100)-c(2x2)-Cl
CLASSIFICATION : 47.17.3a
TECHNIQUE : LEED
AUTHORS : E. Zanazzi and F. Jona
REFERENCE : Surf. Sci., 62, 61 (1977)

ILLUSTRATION: 28,29

SURFACE TYPE
Substrate : Ag
Crystal face: 100
Temperature : 300 K
Bulk lattice: fcc
2D bulk symm: p4m
2D surf symm: p4m

Adsorbate: Cl
Coverage : 0.5 Cl/Ag
Pattern : c(2x2)
Matrix : (1.000, 1.000)
(-1.000, 1.000)

STRUCTURE TYPE
Atomic adsorption in hollow sites

SAMPLE PREPARATION (1 sample)
Treatment : exposure to 100mtorr C2H4Cl2 at 423 K
Crystallinity:
Anal. methods:
Contamination: monitored by AES and LEED

COMMENTS
Cf. visual analysis of same data: class. no. 47.17.1

DATA COLLECTION
Technique: LEED
Dataset : I-V spectra: 4 beams at Θ=0°, 9 beams at Θ=10°,φ=29.5° 7 beams at Θ=20°,φ=29.5°

THEORY/DATA TREATMENT
Dynamical LEED (layer KKR): 58 beams, 8 phase shifts; rms vibr=0.155Å

STRUCTURES EXAMINED
Hollow, bridge and top sites; mixed buckled Cl-Ag overlayer; variable overlayer-Ag spacing in both cases; relaxations of top Ag-Ag spacing

QUALITY OF EXPERIMENT-THEORY FIT
RZJ=0.14

2D UNIT CELLS (1 domain observed)

Cell	Ax (Å)	Ay (Å)	Bx (Å)	By (Å)	α (°)	Matrix	Pattern	Cell type
Bulk	2.889	0.000	0.000	2.889	90.0	(1.000, 0.000) (0.000, 1.000)	(1x1)	b: bulk lattice
Surface 1	2.889	2.889	-2.889	2.889	90.0	(1.000, 1.000) (-1.000, 1.000)	c(2x2)	s1: commens. superlattice

3D COORDINATES

Cl1: overlayer in hollow sites; 0.1Å error bars assumed for tabulation

Dx/Dy in Å, or as a fraction of layer's unit cell vectors; atom 0 at the origin. Epir/subr are bulk repeat vectors.

No. of atoms: 3 Bulk z = 2.043 Å

Region	Chem el.	At. no.	Cell type	Site occ.	Rel. to	Dx ± εx		Dy ± εy		Dz ± εz		Dz/Bz(%) ± εz/Bz
epir		-2					f		f		Å	
subr		-1				-1.445	Å	-1.445	Å	2.043	Å	
ovrl	Cl	1	s1	.50	0	0.000	f	0.000	f	0.000	Å	0.0
intf	Ag	2	b	1.00	1	0.500	f	0.500	f	1.670 ± .100	Å	81.7 ± 4.9
subl	Ag	3	b	1.00	2	-0.500	f	-0.500	f	2.043 ± .100	Å	100.0 ± 4.9

BOND DISTANCES AND ANGLES

Bond distances and angles are derived from coordinates

No. of distances/angles: 3

Interatomic dist. A-B (Å)	Atom A	Atom B	Atom C	Bond angle A-B-C (°)
2.639	Cl1	Ag2	Ag2(1,0)	123.2
2.639	Cl1	Ag2	Ag3	84.3
2.889	Ag2	Ag3		

COMMON NAME : Ag(100)-c(2x2)-Cl
CLASSIFICATION : 47.17.4
TECHNIQUE : atom diffraction
AUTHORS : M.J. Cardillo, G.E. Becker, D.R. Hamann, J.A. Serri, L. Whitman and L.F. Mattheiss
REFERENCE : Phys. Rev., B28, 494 (1983)

ILLUSTRATION: 28,29

SURFACE TYPE
Substrate : Ag
Crystal face: 100
Temperature : 250 K
Bulk lattice: fcc
2D bulk symm: p4m
2D surf symm: p4m

Adsorbate: Cl
Coverage : 0.5 Cl/Ag
Pattern : c(2x2)
Matrix : (1.000, 1.000)
(-1.000, 1.000)

STRUCTURE TYPE
Atomic adsorption in hollow sites

SAMPLE PREPARATION (1 sample)
Treatment : Cl2 exposure at 300 K at 3.0E-7 torr for 270L
Crystallinity:
Anal. methods:
Contamination: monitored by AES, LEED, and He diffr.

COMMENTS
Cl-Ag spacing derived from self-consistent electronic structure: for each model, surface electron charge density evaluated using SLAPW method and used to determine He-surface potential corrugation

DATA COLLECTION
Technique: atom diffraction
Dataset : plots of reduced scattered He beam intensity as function of polar scattering angle for ϕ=0 and 45° azimuths

THEORY/DATA TREATMENT
He atom diffraction: fit of specular intensity scans calculated from He-surface potential corrugation (see comments)

STRUCTURES EXAMINED
1. simple Cl overlayer on hollow sites (1Å corrugation); 2. mixed Cl/Ag coplanar layer (<0.1Å corrugation); positions of the rainbow maxima in specular scans indicate corrugation of about 1 Å

QUALITY OF EXPERIMENT-THEORY FIT
Visual

2D UNIT CELLS (1 domain observed)

Cell	Ax (Å)	Ay (Å)	Bx (Å)	By (Å)	α (°)	Matrix	Pattern	Cell type
Bulk	2.889	0.000	0.000	2.889	90.0	(1.000, 0.000) (0.000, 1.000)	(1x1)	b: bulk lattice
Surface 1	2.889	2.889	-2.889	2.889	90.0	(1.000, 1.000) (-1.000, 1.000)	c(2x2)	s1: commens. superlattice

3D COORDINATES

Cl1: overlayer in hollow sites; 0.2Å error bar assumed for tabulation

Dx/Dy in Å, or as a fraction of layer's unit cell vectors; atom 0 at the origin. Epir/subr are bulk repeat vectors.

No. of atoms: 3

Bulk z = 2.043 Å

Region	Chem el.	At. no.	Cell type	Site occ.	Rel. to	Dx ± εx		Dy ± εy		Dz ± εz		Dz/Bz(%) ± εz/Bz
epir		-2					f		f		Å	
subr		-1				-1.445	Å	-1.445	Å	2.043	Å	
ovrl	Cl	1	s1	.50	0	0.000	f	0.000	f	0.000	Å	0.0
intf	Ag	2	b	1.00	1	0.500	f	0.500	f	1.960 ± .200	Å	95.9 ± 9.8
subl	Ag	3	b	1.00	2	-0.500	f	-0.500	f	2.043	Å	100.0

BOND DISTANCES AND ANGLES

Bond distances and angles are derived from coordinates

No. of distances/angles: 3

Interatomic dist. A-B (Å)	Atom A	Atom B	Atom C	Bond angle A-B-C (°)
2.831	Cl1	Ag2	Ag2(1,0)	120.7
2.831	Cl1	Ag2	Ag3	88.8
2.889	Ag2	Ag3		

COMMON NAME : Ag(100)-c(2x2)-Cl
CLASSIFICATION : 47.17.8
TECHNIQUE : SEXAFS
AUTHORS : G.M. Lamble, R.S. Brooks, J.C. Campuzano, D.A. King and D. Norman
REFERENCE : Phys. Rev., B36, 1796 (1987)

ILLUSTRATION: 28,29

SURFACE TYPE
Substrate : Ag
Crystal face: 100
Temperature : RT
Bulk lattice: fcc
2D bulk symm: p4m
2D surf symm: p4m

Adsorbate: Cl
Coverage : 0.5 Cl/Ag
Pattern : c(2x2)
Matrix : (1.000, 1.000)
(-1.000, 1.000)

STRUCTURE TYPE
Atomic adsorption in 4-fold hollow sites on unrelaxed substrate

SAMPLE PREPARATION (1 sample)
Treatment : exposure to chlorine from an electrolytic source
Crystallinity: checked by LEED
Anal. methods:
Contamination: monitored by AES

COMMENTS

DATA COLLECTION
Technique: SEXAFS
Dataset : SEXAFS spectra: 2800<E<3200 eV (Cl edge); photons at normal incidence only

THEORY/DATA TREATMENT
Fourier transform: correction for phase shifts optimized by fitting to EXAFS spectra for AgCl

2D UNIT CELLS (1 domain observed)

Cell	Ax (Å)	Ay (Å)	Bx (Å)	By (Å)	α (°)	Matrix	Pattern	Cell type
Bulk	2.889	0.000	0.000	2.889	90.0	(1.000, 0.000) (0.000, 1.000)	(1x1)	b: bulk lattice
Surface 1	2.889	2.889	-2.889	2.889	90.0	(1.000, 1.000) (-1.000, 1.000)	c(2x2)	s1: commens. superlattice

3D COORDINATES

Cl1: overlayer in 4-fold hollow sites

Dx/Dy in Å, or as a fraction of layer's unit cell vectors; atom 0 at the origin. Epir/subr are bulk repeat vectors.

No. of atoms: 3

Bulk z = 2.043 Å

Reg ion	Chem el.	At. no.	Cell type	Site occ.	Rel. to	Dx ± εx		Dy ± εy		Dz ± εz		Dz/Bz(%) ± εz/Bz
epir		-2					f		f		Å	
subr		-1				-1.445	Å	-1.445	Å	2.043	Å	
ovrl	Cl	1	s1	.50	0	0.000	f	0.000	f	0.000	Å	0.0
intf	Ag	2	b	1.00	1	0.500	f	0.500	f	1.750 ± .050	Å	85.7 ± 2.5
subl	Ag	3	b	1.00	2	-0.500	f	-0.500	f	2.043	Å	100.0

BOND DISTANCES AND ANGLES

Bond distances and angles are derived from coordinates

No. of distances/angles: 3

Interatomic dist. A-B (Å)	Atom A	Atom B	Atom C	Bond angle A-B-C (°)
2.690	Cl1	Ag2	Cl1(1,0)	98.8
2.690	Cl1	Ag2	Ag3	85.6
2.889	Ag2	Ag2(1,0)		

COMMON NAME : Ag(100)-c(2x2)-Cl
CLASSIFICATION : 47.17.9
TECHNIQUE : SIMS
AUTHORS : Che-Chen Chang and N. Winograd
REFERENCE : Surf. Sci., 230, 27 (1990)

ILLUSTRATION: 28,29

SURFACE TYPE
Substrate : Ag
Crystal face: 100
Temperature : RT
Bulk lattice: fcc
2D bulk symm: p4m
2D surf symm: p4m

Adsorbate: Cl
Coverage : 0.5 ML
Pattern : c(2x2)
Matrix : (1.000, 1.000)
(-1.000, 1.000)

STRUCTURE TYPE
Atomic adsorption in 4-fold hollow site

SAMPLE PREPARATION (1 sample)
Treatment : sputtering and annealing; exposure to 5L Cl2 at RT
Crystallinity: perfect LEED pattern
Anal. methods: SIMS, LEED
Contamination:

COMMENTS

DATA COLLECTION
Technique: SIMS
Dataset : scans in the [001] and [011] directions
scans with varying incident angles

THEORY/DATA TREATMENT
Shadow-cone-enhanced desorption
channeling-blocking

QUALITY OF EXPERIMENT-THEORY FIT
Visual

2D UNIT CELLS (1 domain observed)

Cell	Ax (Å)	Ay (Å)	Bx (Å)	By (Å)	α (°)	Matrix	Pattern	Cell type
Bulk	2.889	0.000	0.000	2.889	90.0	(1.000, 0.000) (0.000, 1.000)	(1x1)	b: bulk lattice
Surface 1	2.889	2.889	-2.889	2.889	90.0	(1.000, 1.000) (-1.000, 1.000)	c(2x2)	s1: commens. superlattice

3D COORDINATES

Cl1: atomic overlayer in the 4-fold hollow site

Dx/Dy in Å, or as a fraction of layer's unit cell vectors; atom 0 at the origin. Epir/subr are bulk repeat vectors.

No. of atoms: 2 Bulk z = 2.043 Å

Region	Chem el.	At. no.	Cell type	Site occ.	Rel. to	Dx ± εx	Dy ± εy	Dz ± εz	Dz/Bz(%) ± εz/Bz
epir		-2				f	f	Å	
subr		-1				1.445 Å	1.445 Å	2.043 Å	
ovrl	Cl	1	s1	.50	0	0.000 Å	0.000 Å	0.000 Å	0.0
subl	Ag	2	b	1.00	0	1.445 Å	1.445 Å	1.608 ± .040 Å	78.7 ± 2.0

BOND DISTANCES AND ANGLES

Bond distances and angles are derived from coordinates

No. of distances/angles: 1

Interatomic dist. A-B (Å)	Atom A	Atom B	Atom C	Bond angle A-B-C (°)
2.600	Cl1	Ag2		

COMMON NAME : Ag(111)-($\sqrt{3}$x$\sqrt{3}$)R30°-Cl   ILLUSTRATION: 22,24
CLASSIFICATION : 47.17.5a
TECHNIQUE : SEXAFS
AUTHORS : G.M. Lamble, R.S. Brooks, S. Ferrer, D.A. King and D. Norman
REFERENCE : Phys. Rev., B34, 2975 (1986)

SURFACE TYPE
Substrate : Ag
Crystal face: 111
Temperature : 100 K
Bulk lattice: fcc
2D bulk symm: p3m1
2D surf symm: p31m

Adsorbate: Cl
Coverage : 1/3 Cl/Ag
Pattern : ($\sqrt{3}$x$\sqrt{3}$)R30°
Matrix : (1.000, 1.000)
(-2.000, 1.000)

STRUCTURE TYPE
Atomic adsorption in fcc hollow sites on unrelaxed substrate

SAMPLE PREPARATION (1 sample)
Treatment : dosed at room temperature with Cl from electrolytic source
Crystallinity:
Anal. methods:
Contamination: checked by AES and LEED

COMMENTS

DATA COLLECTION
Technique: SEXAFS
Dataset : spectra measured above Cl K edge at around 2820 eV; photon beam at normal incidence

THEORY/DATA TREATMENT
EXAFS: phase shifts for Ag and Cl calculated from muffin-tin potential; SEXAFS: full multishell calculation

STRUCTURES EXAMINED
Fcc hollow site: Cl-Ag spacing varied

2D UNIT CELLS (1 domain observed)

Cell	Ax (Å)	Ay (Å)	Bx (Å)	By (Å)	α (°)	Matrix	Pattern	Cell type
Bulk	2.893	0.000	1.446	2.505	60.0	(1.000, 0.000) (0.000, 1.000)	(1x1)	b: bulk lattice
Surface 1	4.339	2.505	-4.339	2.505	120.0	(1.000, 1.000) (-2.000, 1.000)	($\sqrt{3}$x$\sqrt{3}$)R30°	s1: commens. superlattice

3D COORDINATES

Cl1: overlayer in fcc hollow sites

Dx/Dy in Å, or as a fraction of layer's unit cell vectors; atom 0 at the origin. Epir/subr are bulk repeat vectors.

No. of atoms: 3   Bulk z = 2.360 Å

Region	Chem el.	At. no.	Cell type	Site occ.	Rel. to	Dx ± εx		Dy ± εy		Dz ± εz		Dz/Bz(%) ± εz/Bz
epir		-2					f		f		Å	
subr		-1				1.446	Å	0.835	Å	2.360	Å	
ovrl	Cl	1	s1	.33	0	0.000	f	0.000	f	0.000	Å	0.0
intf	Ag	2	b	1.00	1	0.333	f	0.333	f	2.120 ± .010	Å	89.8 ± .4
subl	Ag	3	b	1.00	2	0.333	f	0.333	f	2.360	Å	100.0

BOND DISTANCES AND ANGLES

No. of distances/angles: 3

Interatomic dist. A-B (Å)	Atom A	Atom B	Atom C	Bond angle A-B-C (°)
2.699	Cl1	Ag2	Ag3	177.1
2.893	Ag2	Ag2(1,0)		
2.891	Ag2	Ag3		

COMMON NAME : Ag(111)-(√3x√3)R30°-2Cl
CLASSIFICATION : 47.17.5b
TECHNIQUE : SEXAFS
AUTHORS : G.M. Lamble, R.S. Brooks, S. Ferrer, D.A. King and D. Norman
REFERENCE : Phys. Rev., B34, 2975 (1986)

ILLUSTRATION: 22,24

SURFACE TYPE
Substrate : Ag
Crystal face: 111
Temperature : 100 K
Bulk lattice: fcc
2D bulk symm: p3m1
2D surf symm: p31m

Adsorbate: Cl
Coverage : 2/3 Cl/Ag
Pattern : (√3x√3)R30°
Matrix : (1.000, 1.000)
(-2.000, 1.000)

SAMPLE PREPARATION (1 sample)
Treatment : dosed at room temperature with Cl from electrolytic source
Crystallinity:
Anal. methods:
Contamination: checked by AES and LEED

DATA COLLECTION
Technique: SEXAFS
Dataset : spectra measured above Cl K edge at around 2820 eV; photon beam at normal incidence

STRUCTURES EXAMINED
Fcc hollow sites (2 per cell): Cl-Ag spacing varied

STRUCTURE TYPE
Atomic adsorption in fcc hollow sites on unrelaxed substrate with 2 Cl per unit cell forming honeycomb lattice

COMMENTS

THEORY/DATA TREATMENT
EXAFS: phase shifts for Ag and Cl calculated from muffin-tin potential; SEXAFS: full multishell calculation

2D UNIT CELLS (1 domain observed)

Cell	Ax (Å)	Ay (Å)	Bx (Å)	By (Å)	α (°)	Matrix	Pattern	Cell type
Bulk	2.893	0.000	1.446	2.505	60.0	(1.000, 0.000) (0.000, 1.000)	(1x1)	b: bulk lattice
Surface 1	4.339	2.505	-4.339	2.505	120.0	(1.000, 1.000) (-2.000, 1.000)	(√3x√3)R30°	s1: commens. superlattice

3D COORDINATES

Cl1-Cl2: honeycomb overlayer in fcc hollow sites

Dx/Dy in Å, or as a fraction of layer's unit cell vectors; atom 0 at the origin. Epir/subr are bulk repeat vectors.

No. of atoms: 4

Bulk z = 2.360 Å

Reg ion	Chem el.	At. no.	Cell type	Site occ.	Rel. to	Dx ± εx	Dy ± εy	Dz ± εz	Dz/Bz(%) ± εz/Bz
epir		-2				f	f	Å	
subr		-1				1.446 Å	0.835 Å	2.360 Å	
ovrl	Cl	1	s1	.33	0	0.000 f	0.000 f	0.000 Å	0.0
ovrl	Cl	2	s1	.33	1	0.333 f	0.667 f	0.000 Å	0.0
intf	Ag	3	b	1.00	2	1.333 f	-0.667 f	2.120 ± .010 Å	89.8 ± .4
subl	Ag	4	b	1.00	3	0.333 f	0.333 f	2.360 Å	100.0

BOND DISTANCES AND ANGLES

No. of distances/angles: 3

Interatomic dist. A-B (Å)	Atom A	Atom B	Atom C	Bond angle A-B-C (°)
2.699	Cl1	Ag3	Cl2(0,-1)	64.8
2.893	Ag3	Ag3(1,0)		
2.891	Ag3	Ag4		

COMMON NAME : Ag(100)-(1x1)-3Co
CLASSIFICATION : 47.27.1
TECHNIQUE : ARXPS
AUTHORS : Hong Li and B.P. Tonner
REFERENCE : Phys. Rev., B40, 10241 (1989)

ILLUSTRATION: 83

SURFACE TYPE
Substrate : Ag
Crystal face: 100
Temperature : RT
Bulk lattice: fcc
2D bulk symm: p4m
2D surf symm: p4m

Adsorbate: Co
Coverage : 3 Co/(1x1)
Pattern : (1x1)
Matrix : (1.000, 0.000)
(0.000, 1.000)

STRUCTURE TYPE
Co grows in the bct structure, with its [110] parallel to the [100] of the fcc Ag(100) lattice; May have high long-range disorder; but we here show ordered layers

SAMPLE PREPARATION (1 sample)
Treatment : resistive heating of a 99.99-purity Fe wire
Crystallinity:
Anal. methods: LEED; XPS; AES; quartz microbalance for
Contamination: <2% of C and O

COMMENTS
ARXPS was used in the fingerprint mode to identify the overlayer crystallography: no attempt was made to optimize structural parameters

DATA COLLECTION
Technique: ARXPS
Dataset : Co 2p polar intensity scans at two azimuths

THEORY/DATA TREATMENT

2D UNIT CELLS (1 domain observed)

Cell	Ax (Å)	Ay (Å)	Bx (Å)	By (Å)	α (°)	Matrix	Pattern	Cell type
Bulk	2.892	0.000	0.000	2.892	90.0	(1.000, 0.000) (0.000, 1.000)	(1x1)	b: bulk lattice
Surface 1	2.892	0.000	0.000	2.892	90.0	(1.000, 0.000) (0.000, 1.000)	(1x1)	s1: commens. superlattice

3D COORDINATES

Co1-Co3: 3 bct layers of Fe

Dx/Dy in Å, or as a fraction of layer's unit cell vectors; atom 0 at the origin. Epir/subr are bulk repeat vectors.

No. of atoms: 4 Bulk z = 2.892 Å

Region	Chem el.	At. no.	Cell type	Site occ.	Rel. to	Dx ± εx		Dy ± εy		Dz ± εz		Dz/Bz(%) ± εz/Bz
epir		-2					f		f		Å	
subr		-1				1.446	Å	1.446	Å	2.045	Å	
ovrl	Co	1	s1	1.00	0	0.000	f	0.000	f	0.000	Å	0.0
ovrl	Co	2	s1	1.00	1	0.500	f	0.500	f	1.663	Å	81.3
ovrl	Co	3	s1	1.00	2	-0.500	f	-0.500	f	1.663	Å	81.3
subl	Ag	4	b	1.00	3	0.500	f	0.500	f	2.045	Å	100.0

BOND DISTANCES AND ANGLES

Bond distances and angles are derived from coordinates

No. of distances/angles: 3

Interatomic dist. A-B (Å)	Atom A	Atom B	Atom C	Bond angle A-B-C (°)
2.636	Co1	Co2	Co3	101.8
2.892	Co1	Co1(1,0)	Co1(1,1)	90.0
2.892	Co3	Ag1	Ag1(1,1)	45.0

COMMON NAME : Ag(110)-(1x2) Cs-induced
CLASSIFICATION : 47.55.2
TECHNIQUE : LEED
AUTHORS : C.J. Barnes, M. Lindroos, D.J. Holmes and D.A. King
REFERENCE : Surf. Sci., 219, 143 (1989)

ILLUSTRATION: 5

SURFACE TYPE
Substrate : Ag
Crystal face: 110
Temperature : 100 K
Bulk lattice: fcc
2D bulk symm: pmm
2D surf symm: pmm

Adsorbate: Cs
Coverage : 0.16 Cs/1x1
Pattern : (1x2)
Matrix : (1.000, 0.000)
(0.000, 2.000)

STRUCTURE TYPE
Mesured (1x2) reconstructive phase transition due to 0.16ML Cs adsorption; no Cs positions
optimum (1x1) to (1x2) from 0.16ML Cs, anneal 550-600K

SAMPLE PREPARATION (1 sample)
Treatment : Ar+ sputtering, anneal 800 K
Crystallinity: crystal <1° from (110) plane
Anal. methods:
Contamination:

COMMENTS

DATA COLLECTION
Technique: LEED
Dataset : E range 50-250 eV

THEORY/DATA TREATMENT
Dynamical LEED (layer doubling): 9 phase shifts from Moruzzi et al potential; Voi=-3.5 eV

STRUCTURES EXAMINED
Tested paired row and buckled models: pairing 0-0.6Å in 0.1Å steps; buckling 0-1.6Å in 0.1Å steps

QUALITY OF EXPERIMENT-THEORY FIT
RPE=0.29

2D UNIT CELLS (1 domain observed)

Cell	Ax (Å)	Ay (Å)	Bx (Å)	By (Å)	α (°)	Matrix	Pattern	Cell type
Bulk	2.889	0.000	0.000	4.086	90.0	(1.000, 0.000) (0.000, 1.000)	(1x1)	b: bulk lattice
Surface 1	2.889	0.000	0.000	8.172	90.0	(1.000, 0.000) (0.000, 2.000)	(1x2)	s1: commens. superlattice

3D COORDINATES

Dx/Dy in Å, or as a fraction of layer's unit cell vectors; atom 0 at the origin. Epir/subr are bulk repeat vectors.

No. of atoms: 6 Bulk z = 1.445 Å

Region	Chem el.	At. no.	Cell type	Site occ.	Rel. to	Dx ± εx		Dy ± εy		Dz ± εz		Dz/Bz(%) ± εz/Bz
epir		-2					f		f	1.445	Å	
subr		-1				1.445	Å	2.043	Å	1.445	Å	
intf	Ag	1	s1	.50	0	0.000	f	0.000	f	0.000	Å	0.0
intf	Ag	2	s1	.50	1	0.500	f	0.262 ± .012	f	1.280 ± .100	Å	88.6 ± 6.9
intf	Ag	3	s1	.50	2	0.000	f	-0.524 ± .012	f	0.000	Å	0.0
intf	Ag	4	s1	.50	3	-0.500	f	-0.238 ± .012	f	1.460 ± .100	Å	101.1 ± 6.9
intf	Ag	5	s1	.50	4	0.000	f	0.500	f	0.100 ± .100	Å	6.9 ± 6.9
subl	Ag	6	b	1.00	5	-0.500	f	-0.500	f	1.360 ± .100	Å	94.1 ± 6.9

BOND DISTANCES AND ANGLES

Bond distances and angles are derived from coordinates

No. of distances/angles: 5

Interatomic dist. A-B (Å)	Atom A	Atom B	Atom C	Bond angle A-B-C (°)
2.889	Ag1	Ag1(1,0)		
2.884	Ag1	Ag2		
2.840	Ag1	Ag5		
2.828	Ag2	Ag4		
3.019	Ag2	Ag5		

COMMON NAME : Ag(111)-Cs 0.15ML disordered
CLASSIFICATION : 47.55.1a
TECHNIQUE : SEXAFS
AUTHORS : G.M. Lamble, R.S. Brooks, D.A. King and D. Norman
REFERENCE : Phys. Rev. Lett., 61, 1112 (1988)

ILLUSTRATION: 22

SURFACE TYPE
Substrate : Ag
Crystal face: 111
Temperature : 120 K
Bulk lattice: fcc
2D bulk symm: p3m1
2D surf symm: none

Adsorbate: Cs
Coverage : 0.15 Cs/Ag
Pattern : disordered
Matrix : (1.000, 0.000)
(0.000, 1.000)

STRUCTURE TYPE
Atomic adsorption in 3-fold hollow site (fcc assumed here)

SAMPLE PREPARATION (1 sample)
Treatment : Cs evaporation at RT; coverage from AES and saturation
Crystallinity:
Anal. methods:
Contamination: monitored by AES and LEED

COMMENTS
Ag-Cs bond length coverage dependent (cf 0.3ML structure)

DATA COLLECTION
Technique: SEXAFS
Dataset : SEXAFS total electron yield spectra for Cs L3 edge

THEORY/DATA TREATMENT
Comparison of Fourier filtered data and multishell curved wave calc; ph shs fit to EXAFS ph shs from CsBr and AgCl

QUALITY OF EXPERIMENT-THEORY FIT
Visual

2D UNIT CELLS (1 domain observed)

Cell	Ax (Å)	Ay (Å)	Bx (Å)	By (Å)	α (°)	Matrix	Pattern	Cell type
Bulk	2.887	0.000	1.443	2.500	60.0	(1.000, 0.000) (0.000, 1.000)	(1x1)	b: bulk lattice
Surface 1	4.330	2.500	-4.330	2.500	120.0	(1.000, 0.000) (0.000, 1.000)	disordered	nd1: non-recon. lattice-gas dis

3D COORDINATES

Cs1: disordered overlayer in 3-fold hollow sites (here fcc hollow assumed)

Dx/Dy in Å, or as a fraction of layer's unit cell vectors; atom 0 at the origin. Epir/subr are bulk repeat vectors.

No. of atoms: 3

Bulk z = 2.360 Å

Region	Chem el.	At. no.	Cell type	Site occ.	Rel. to	Dx ± εx		Dy ± εy		Dz ± εz		Dz/Bz(%) ± εz/Bz
epir		-2					f		f		Å	
subr		-1				1.443	Å	0.833	Å	2.360	Å	
ovrl	Cs	1	nd1	.15	0	0.000	f	0.000	f	0.000	Å	0.0
intf	Ag	2	b	1.00	1	0.333	f	0.333	f	2.730 ± .030	Å	115.7 ± 1.3
subl	Ag	3	b	1.00	2	0.333	f	0.333	f	2.360	Å	100.0

BOND DISTANCES AND ANGLES

No. of distances/angles: 3

Interatomic dist. A-B (Å)	Atom A	Atom B	Atom C	Bond angle A-B-C (°)
3.199	Cs1	Ag2	Cs1(1,0)	82.1
3.199	Cs1	Ag2	Ag2(1,0)	116.8
2.887	Ag2	Ag2(1,0)		

COMMON NAME : Ag(111)-Cs 0.3ML disordered
CLASSIFICATION : 47.55.1b
TECHNIQUE : SEXAFS
AUTHORS : G.M. Lamble, R.S. Brooks, D.A. King and D. Norman
REFERENCE : Phys. Rev. Lett., 61, 1112 (1988)

ILLUSTRATION: 22

SURFACE TYPE
Substrate : Ag
Crystal face: 111
Temperature : 120 K
Bulk lattice: fcc
2D bulk symm: p3m1
2D surf symm: none

Adsorbate: Cs
Coverage : 0.3 Cs/Ag
Pattern : disordered
Matrix : (1.000, 0.000)
(0.000, 1.000)

STRUCTURE TYPE
Atomic adsorption in 3-fold hollow site (fcc assumed here)

SAMPLE PREPARATION (1 sample)
Treatment : Cs evaporation at RT; coverage from AES and saturation
Crystallinity:
Anal. methods:
Contamination: monitored by AES and LEED

COMMENTS
Ag-Cs bond length coverage dependent (cf 0.15ML structure)

DATA COLLECTION
Technique: SEXAFS
Dataset : SEXAFS total electron yield spectra for Cs L3 edge

THEORY/DATA TREATMENT
Comparison of Fourier filtered data and multishell curved wave calc; ph shs fit to EXAFS ph shs from CsBr and AgCl

QUALITY OF EXPERIMENT-THEORY FIT
Visual

2D UNIT CELLS (1 domain observed)

Cell	Ax (Å)	Ay (Å)	Bx (Å)	By (Å)	α (°)	Matrix	Pattern	Cell type
Bulk	2.887	0.000	1.443	2.500	60.0	(1.000, 0.000) (0.000, 1.000)	(1x1)	b: bulk lattice
Surface 1	4.330	2.500	-4.330	2.500	120.0	(1.000, 0.000) (0.000, 1.000)	disordered	nd1: non-recon. lattice-gas dis

3D COORDINATES

Cs1: disordered overlayer in 3-fold hollow sites (here fcc hollow assumed)

Dx/Dy in Å, or as a fraction of layer's unit cell vectors; atom 0 at the origin. Epir/subr are bulk repeat vectors.

No. of atoms: 3 Bulk z = 2.360 Å

Reg ion	Chem el.	At. no.	Cell type	Site occ.	Rel. to	Dx ± εx		Dy ± εy		Dz ± εz		Dz/Bz(%) ± εz/Bz
epir		-2					f		f		Å	
subr		-1				1.443	Å	0.833	Å	2.360	Å	
ovrl	Cs	1	nd1	.30	0	0.000	f	0.000	f	0.000	Å	0.0
intf	Ag	2	b	1.00	1	0.333	f	0.333	f	3.070 ± .030	Å	130.1 ± 1.3
subl	Ag	3	b	1.00	2	0.333	f	0.333	f	2.360	Å	100.0

BOND DISTANCES AND ANGLES

No. of distances/angles: 2

Interatomic dist. A-B (Å)	Atom A	Atom B	Atom C	Bond angle A-B-C (°)
3.500	Cs1	Ag2		
2.887	Ag2	Ag2(1,0)		

COMMON NAME : Ag(100)-(1x1)-Cu multilayer
CLASSIFICATION : 47.29.3
TECHNIQUE : SEXAFS
AUTHORS : D.T. Jiang, E.D. Crozier and B. Heinrich
REFERENCE : Phys. Rev., B44, 6401 (1991)

ILLUSTRATION: 83

SURFACE TYPE
Substrate : Ag
Crystal face: 100
Temperature : RT
Bulk lattice: fcc
2D bulk symm: p4m
2D surf symm: p4m

Adsorbate: Cu
Coverage : 8 Cu/Ag
Pattern : (1x1)
Matrix : (1.000, 0.000)
(0.000, 1.000)

STRUCTURE TYPE
About 8 epitaxial (1x1) monolayers, forming strained fcc Cu

SAMPLE PREPARATION (3 sample)
Treatment : Cu film grown in MBE chamber
Crystallinity:
Anal. methods: NEXAFS (XANES); coverage from AES
Contamination:

COMMENTS
Cu film is itself covered with 10ML Au film for transfer to SEXAFS chamber

DATA COLLECTION
Technique: SEXAFS; glancing angle EXAFS at SSRL
Dataset : XAFS fluorescence spectra near incident critical angle of 6.8mrad

THEORY/DATA TREATMENT
NEXAFS suggests bcc Cu structure; SEXAFS FT and curve-fitting determine 1st and 2nd neighbor shell distances

2D UNIT CELLS (1 domain observed)

Cell	Ax (Å)	Ay (Å)	Bx (Å)	By (Å)	α (°)	Matrix	Pattern	Cell type
Bulk	2.880	0.000	0.000	2.880	90.0	(1.000, 0.000) (0.000, 1.000)	(1x1)	b: bulk lattice
Surface 1	2.880	0.000	0.000	2.880	90.0	(1.000, 0.000) (0.000, 1.000)	(1x1)	s1: commens. superlattice

3D COORDINATES

coordinates are derived from bond distances

Dx/Dy in Å, or as a fraction of layer's unit cell vectors; atom 0 at the origin. Epir/subr are bulk repeat vectors.

No. of atoms: 4

Bulk z = 1.550 Å

Region	Chem el.	At. no.	Cell type	Site occ.	Rel. to	Dx ± εx		Dy ± εy		Dz ± εz		Dz/Bz(%) ± εz/Bz
epir		-2					f		f		Å	
subr		-1				-1.440	Å	-1.440	Å	1.550	Å	
intf	Cu	1	b	1.00	0	0.000	f	0.000	f	0.000	Å	0.0
intf	Cu	2	b	1.00	1	0.500	f	0.500	f	1.550 ± .020	Å	100.0 ± 1.3
intf	Cu	3	b	1.00	2	-0.500	f	-0.500	f	1.550 ± .020	Å	100.0 ± 1.3
subl	Cu	4	b	1.00	3	0.500	f	0.500	f	1.550 ± .020	Å	100.0 ± 1.3

BOND DISTANCES AND ANGLES

No. of distances/angles: 3

Interatomic dist. A-B (Å)	Atom A	Atom B	Atom C	Bond angle A-B-C (°)
2.880	Cu1	Cu1(1,0)	Cu2	55.8
2.559	Cu1	Cu2	Cu3	74.6
2.559	Cu2	Cu3	Cu4	74.6

ILLUSTRATION: 83

COMMON NAME : Ag(100)-(1x1)-2Cu
CLASSIFICATION : 47.29.2a
TECHNIQUE : LEED
AUTHORS : H. Li, D. Tian, J. Quinn, Y.S. Li, F. Jona and P.M. Marcus
REFERENCE : Phys. Rev., B43, 6342 (1991)

SURFACE TYPE
Substrate : Ag
Crystal face: 100
Temperature : RT
Bulk lattice: fcc
2D bulk symm: p4m
2D surf symm: p4m

Adsorbate: Cu
Coverage : 2.0 Cu/Ag
Pattern : (1x1)
Matrix : (1.000, 0.000)
(0.000, 1.000)

STRUCTURE TYPE
2 to 3 epitaxial (1x1) monolayers, forming metastable bcc Cu

SAMPLE PREPARATION (1 sample)
Treatment : Cu evaporated from Cu single crystal
Crystallinity: broad LEED beams, increased background
Anal. methods: ARPES; coverage from AES
Contamination: monitored by AES

COMMENTS
Limited amount of long-range order found: tabulated structure refers to ordered part only; coverages measured in layer-equivalents (i.e. giving equivalent AES signal as layer-by-layer growth); 2-layer model tabulated here; 3-layer model gives slightly worse R-factors

DATA COLLECTION
Technique: LEED; video LEED
Dataset : IV spectra for 3 beams at normal incidence: (10),(11),(20); 50<E<280 eV

THEORY/DATA TREATMENT
Dynamical LEED (CHANGE code): Moruzzi et al pots, 8 ph shs; Vor=-10 eV, Voi=-4eV; rms vibs 0.156Å

STRUCTURES EXAMINED
Fcc continuation; 1 to 4 monolayers and semi-infinite Cu with lateral Ag(100) lattice constant tried; spacings varied: 'bulk' Cu, Cu-Ag

QUALITY OF EXPERIMENT-THEORY FIT
RVHT=0.45, RPE=0.72, RZJ=0.25

2D UNIT CELLS (1 domain observed)

Cell	Ax (Å)	Ay (Å)	Bx (Å)	By (Å)	α (°)	Matrix	Pattern	Cell type
Bulk	2.889	0.000	0.000	2.889	90.0	(1.000, 0.000) (0.000, 1.000)	(1x1)	b: bulk lattice
Surface 1	2.889	0.000	0.000	2.889	90.0	(1.000, 0.000) (0.000, 1.000)	(1x1)	s1: commens. superlattice

3D COORDINATES

Cu1-Cu2: 2 (1x1) epitaxial monolayers, continuing flattened Ag fcc lattice with bcc-like strain

Dx/Dy in Å, or as a fraction of layer's unit cell vectors; atom 0 at the origin. Epir/subr are bulk repeat vectors.

No. of atoms: 4 Bulk z = 2.043 Å

Region	Chem el.	At. no.	Cell type	Site occ.	Rel. to	Dx ± εx		Dy ± εy		Dz ± εz		Dz/Bz(%) ± εz/Bz
epir		-2					f		f		Å	
subr		-1				-1.445	Å	-1.445	Å	2.043	Å	
ovrl	Cu	1	b	1.00	0	0.000	f	0.000	f	0.000	Å	0.0
ovrl	Cu	2	b	1.00	1	0.500	f	0.500	f	1.450 ± .060	Å	71.0 ± 2.9
intf	Ag	3	b	1.00	2	-0.500	f	-0.500	f	1.520 ± .060	Å	74.4 ± 2.9
subl	Ag	4	b	1.00	3	0.500	f	0.500	f	2.043	Å	100.0

BOND DISTANCES AND ANGLES

Bond distances and angles are derived from coordinates

No. of distances/angles: 3

Interatomic dist. A-B (Å)	Atom A	Atom B	Atom C	Bond angle A-B-C (°)
2.889	Cu1	Cu1(1,0)	Cu2	54.8
2.505	Cu1	Cu2	Ag3	72.0
2.546	Cu2	Ag3	Ag4	81.7

COMMON NAME : Ag(100)-(1x1)-5Fe (bcc)
CLASSIFICATION : 47.26.0a
TECHNIQUE : ARXPS
AUTHORS : Hong Li and B.P. Tonner
REFERENCE : Phys. Rev., B40, 10241 (1989)

ILLUSTRATION: 83

SURFACE TYPE
Substrate : Ag
Crystal face: 100
Temperature : RT
Bulk lattice: fcc
2D bulk symm: p4m
2D surf symm: p4m

Adsorbate: Fe
Coverage : 5 Fe/(1x1)
Pattern : (1x1)
Matrix : (1.000, 0.000)
(0.000, 1.000)

STRUCTURE TYPE
Fe grows in the bcc structure, with its [110] parallel to the [100] of the fcc Ag(100) lattice; 2 and 3.5 MLs show layer-by-layer growth and sharp (1x1) LEED patterns; at 5 ML, long-range order is lost, but the local bcc order is preserved

SAMPLE PREPARATION (1 sample)
Treatment : resistive heating of a 99.997-purity Fe wire
Crystallinity:
Anal. methods: LEED; XPS; AES; quartz microbalance for
Contamination: <2% of C and O

COMMENTS
ARXPS was used in the fingerprint mode to identify the overlayer crystallography: no attempt was made to optimize structural parameters

DATA COLLECTION
Technique: ARXPS
Dataset : Fe 2p3/2 polar intensity scans at two azimuths

THEORY/DATA TREATMENT

STRUCTURES EXAMINED
No structure optimization

2D UNIT CELLS (1 domain observed)

Cell	Ax (Å)	Ay (Å)	Bx (Å)	By (Å)	α (°)	Matrix	Pattern	Cell type
Bulk	2.892	0.000	0.000	2.892	90.0	(1.000, 0.000) (0.000, 1.000)	(1x1)	b: bulk lattice
Surface 1	2.892	0.000	0.000	2.892	90.0	(1.000, 0.000) (0.000, 1.000)	(1x1)	s1: commens. superlattice

3D COORDINATES

Fe1-Fe5: 5 bcc layers of Fe

Dx/Dy in Å, or as a fraction of layer's unit cell vectors; atom 0 at the origin. Epir/subr are bulk repeat vectors.

No. of atoms: 6

Bulk z = 2.892 Å

Region	Chem el.	At. no.	Cell type	Site occ.	Rel. to	Dx ± εx		Dy ± εy		Dz ± εz		Dz/Bz(%) ± εz/Bz
epir		-2					f		f		Å	
subr		-1				1.446	Å	1.446	Å	2.045	Å	
ovrl	Fe	1	s1	1.00	0	0.000	f	0.000	f	0.000	Å	0.0
ovrl	Fe	2	s1	1.00	1	0.500	f	0.500	f	1.446	Å	70.7
ovrl	Fe	3	s1	1.00	2	-0.500	f	-0.500	f	1.446	Å	70.7
ovrl	Fe	4	s1	1.00	3	0.500	f	0.500	f	1.446	Å	70.7
ovrl	Fe	5	s1	1.00	4	-0.500	f	-0.500	f	1.446	Å	70.7
subl	Ag	6	b	1.00	5	0.500	f	0.500	f	2.045	Å	100.0

BOND DISTANCES AND ANGLES

Bond distances and angles are derived from coordinates

No. of distances/angles: 3

Interatomic dist. A-B (Å)	Atom A	Atom B	Atom C	Bond angle A-B-C (°)
2.505	Fe1	Fe2	Fe3	90.0
2.892	Fe1	Fe1(1,0)	Fe1(1,1)	90.0
2.892	Fe5	Ag1	Ag1(1,1)	45.0

ILLUSTRATION: 83

COMMON NAME : Ag(100)-(1x1)-Fe multilayer
CLASSIFICATION : 47.26.1
TECHNIQUE : LEED
AUTHORS : H. Li, Y.S. Li, J. Quinn, D. Tian, J. Sokolov, F. Jona and P.M. Marcus
REFERENCE : Phys. Rev., B42, 9195 (1990)

SURFACE TYPE
Substrate : Ag
Crystal face: 100
Temperature : RT
Bulk lattice: fcc
2D bulk symm: p4m
2D surf symm: p4m

Adsorbate: Fe
Coverage : 25 Fe/Ag
Pattern : (1x1)
Matrix : (1.000, 0.000)
(0.000, 1.000)

STRUCTURE TYPE
About 25 epitaxial (1x1) monolayers, forming bct Fe (slightly distorted from bcc)

SAMPLE PREPARATION (1 sample)
Treatment :
Crystallinity:
Anal. methods: coverage from AES
Contamination: monitored by AES

COMMENTS
Bcc Fe structure cannot be excluded; it would imply a 0.8% different lateral lattice constant from Ag(100), a difference which LEED could not detect

DATA COLLECTION
Technique: LEED; video LEED
Dataset : IV spectra at normal incidence:
(10),(11),(20),(21); 60<E<360 eV

THEORY/DATA TREATMENT
Dynamical LEED

STRUCTURES EXAMINED
Semi-infinite Fe(100) with lateral Ag(100) lattice constant; spacings varied: 'bulk' Fe, top two Fe-Fe

QUALITY OF EXPERIMENT-THEORY FIT
RZJ=0.16

2D UNIT CELLS (1 domain observed)

Cell	Ax (Å)	Ay (Å)	Bx (Å)	By (Å)	α (°)	Matrix	Pattern	Cell type
Bulk	2.889	0.000	0.000	2.889	90.0	(1.000, 0.000) (0.000, 1.000)	(1x1)	b: bulk lattice
Surface 1	2.889	0.000	0.000	2.889	90.0	(1.000, 0.000) (0.000, 1.000)	(1x1)	s1: commens. superlattice

3D COORDINATES

1.42Å bulk spacing was fit, keeping lateral Ag distance

Dx/Dy in Å, or as a fraction of layer's unit cell vectors; atom 0 at the origin. Epir/subr are bulk repeat vectors.

No. of atoms: 4

Bulk z = 1.420 Å

Region	Chem el.	At. no.	Cell type	Site occ.	Rel. to	Dx ± єx		Dy ± єy		Dz ± єz		Dz/Bz(%) ± єz/Bz
epir		-2					f		f		Å	
subr		-1				-1.445	Å	-1.445	Å	1.420	Å	
intf	Fe	1	b	1.00	0	0.000	f	0.000	f	0.000	Å	0.0
intf	Fe	2	b	1.00	1	0.500	f	0.500	f	1.450 ± .030	Å	102.1 ± 2.1
intf	Fe	3	b	1.00	2	-0.500	f	-0.500	f	1.450 ± .030	Å	102.1 ± 2.1
subl	Fe	4	b	1.00	3	0.500	f	0.500	f	1.420 ± .030	Å	100.0 ± 2.1

BOND DISTANCES AND ANGLES

Bond distances and angles are derived from coordinates

No. of distances/angles: 3

Interatomic dist. A-B (Å)	Atom A	Atom B	Atom C	Bond angle A-B-C (°)
2.889	Fe1	Fe1(1,0)	Fe2	54.8
2.505	Fe1	Fe2	Fe3	70.7
2.505	Fe2	Fe3	Fe4	70.2

COMMON NAME : Ag(111)-(√3x√3)R30°-I
CLASSIFICATION : 47.53.1
TECHNIQUE : LEED
AUTHORS : F. Forstmann, W. Berndt and P. Buttner
REFERENCE : Phys. Rev. Lett., 30, 17 (1973)

ILLUSTRATION: 22,24

SURFACE TYPE
Substrate : Ag
Crystal face: 111
Temperature : RT*
Bulk lattice: fcc
2D bulk symm: p3m1
2D surf symm: p31m

Adsorbate: I
Coverage : 1/3 I/Ag
Pattern : (√3x√3)R30°
Matrix : (1.000, 1.000)
(-1.000, 2.000)

STRUCTURE TYPE
Atomic adsorption in 3-fold fcc hollow sites on unrelaxed substrate

SAMPLE PREPARATION (1 sample)
Treatment : 20E-6 L exposure from I2 gas at RT
Crystallinity:
Anal. methods:
Contamination:

COMMENTS
For iodine, atomic, ionic and averaged phase shifts were tried: atomic fit best

DATA COLLECTION
Technique: LEED; spot photometer and Faraday cup
Dataset : at least 3 independent beams at Θ=8° in [-211] azimuth; E range 10-140 eV

THEORY/DATA TREATMENT
Dynamical LEED (RFS): pots have HF Coulomb and SWW exchange; Vor=-11 eV (fit), Voi=-4eV; empirical Debye-Waller factor

STRUCTURES EXAMINED
Unrelaxed bulk; top, fcc-and hcp-hollow sites; variable I-Ag spacing

QUALITY OF EXPERIMENT-THEORY FIT
Visual: moderate

2D UNIT CELLS (1 domain observed)

Cell	Ax (Å)	Ay (Å)	Bx (Å)	By (Å)	α (°)	Matrix	Pattern	Cell type
Bulk	2.889	0.000	1.445	2.502	60.0	(1.000, 0.000) (0.000, 1.000)	(1x1)	b: bulk lattice
Surface 1	4.334	2.502	0.000	5.005	60.0	(1.000, 1.000) (-1.000, 2.000)	(√3x√3)R30°	s1: commens. superlattice

3D COORDINATES

I1: overlayer in fcc hollow sites

Dx/Dy in Å, or as a fraction of layer's unit cell vectors; atom 0 at the origin. Epir/subr are bulk repeat vectors.

No. of atoms: 3

Bulk z = 2.359 Å

Region	Chem el.	At. no.	Cell type	Site occ.	Rel. to	Dx ± εx		Dy ± εy		Dz ± εz		Dz/Bz(%) ± εz/Bz
epir		-2					f		f		Å	
subr		-1				1.445	Å	0.834	Å	2.359	Å	
ovrl	I	1	s1	.33	0	0.000	f	0.000	f	0.000	Å	0.0
intf	Ag	2	b	1.00	1	0.333	f	0.333	f	2.250 ± .174	Å	95.4 ± 7.7
subl	Ag	3	b	1.00	2	0.333	f	0.333	f	2.359	Å	100.0

BOND DISTANCES AND ANGLES

Bond distances and angles are derived from coordinates

No. of distances/angles: 3

Interatomic dist. A-B (Å)	Atom A	Atom B	Atom C	Bond angle A-B-C (°)
2.801	I1	Ag2	Ag2(1,0)	121.1
2.801	I1	Ag2	Ag3	180.0
2.889	Ag2	Ag2(1,0)		

COMMON NAME : Ag(111)-(√3x√3)R30°-I
CLASSIFICATION : 47.53.4
TECHNIQUE : LEED
AUTHORS : M. Maglietta, E. Zanazzi, U. Bardi, D. Sondericker and F. Jona
REFERENCE : Surf. Sci., 123, 141 (1982)

ILLUSTRATION: 22,24

SURFACE TYPE
Substrate : Ag
Crystal face: 111
Temperature : RT
Bulk lattice: fcc
2D bulk symm: p3m1
2D surf symm: p31m

Adsorbate: I
Coverage : 0.33 I/Ag
Pattern : (√3x√3)R30°
Matrix : (1.000, 1.000)
(-2.000, 1.000)

STRUCTURE TYPE
Atomic adsorption mostly (55-75%) in fcc hollows, but also in hcp hollows (only fcc structure is tabulated here)

SAMPLE PREPARATION (1 sample)
Treatment : exposure at 373-423 K to I vapor at 1E-2 torr, then annealed
Crystallinity:
Anal. methods:
Contamination: monitored by LEED and AES

COMMENTS
Mixed fcc and hcp hollows is in disagreement with earlier work of Forstmann et al, Phys. Rev. Lett. 30, 17 (1973), which favored fcc hollows

DATA COLLECTION
Technique: LEED
Dataset : I-V spectra: 3 beams at Θ=0,φ=141.5°, 6 beams at Θ=10,φ=141.5°, 5 beams at Θ=30,φ=-38.5 °, 40<E<160 eV

THEORY/DATA TREATMENT
Dynamical LEED (CHANGE): 8 phase shifts; 93 beams; Vor=-11 eV (best fit), Voi=-4eV, rms vibr ampl=0.16Å

STRUCTURES EXAMINED
1) 3-fold fcc-hollow sites: I-Ag spacing 2.15-2.35Å; 2) mixed Ag/I wurtzite type layer; substitutional model: 1/3 top layer Ag sites occupied by I; domains of I in fcc and hcp sites: concentration of fcc domains varied from 55 to 100%

QUALITY OF EXPERIMENT-THEORY FIT
RZJ=0.22

2D UNIT CELLS (1 domain observed)

Cell	Ax (Å)	Ay (Å)	Bx (Å)	By (Å)	α (°)	Matrix	Pattern	Cell type
Bulk	2.887	0.000	1.443	2.500	60.0	(1.000, 0.000) (0.000, 1.000)	(1x1)	b: bulk lattice
Surface 1	4.330	2.500	-4.330	2.500	120.0	(1.000, 1.000) (-2.000, 1.000)	(√3x√3)R30°	s1: commens. superlattice

3D COORDINATES

I1: overlayer in fcc hollows (coexisting with I in hcp hollows, not tabulated here)

Dx/Dy in Å, or as a fraction of layer's unit cell vectors; atom 0 at the origin. Epir/subr are bulk repeat vectors.

No. of atoms: 3 Bulk z = 2.360 Å

Region	Chem el.	At. no.	Cell type	Site occ.	Rel. to	Dx ± εx		Dy ± εy		Dz ± εz		Dz/Bz(%) ± εz/Bz
epir		-2					f		f		Å	
subr		-1				1.443	Å	0.833	Å	2.360	Å	
ovrl	I	1	s1	.33	0	0.000	f	0.000	f	0.000	Å	0.0
intf	Ag	2	b	1.00	1	0.333	f	0.333	f	2.290 ± .060	Å	97.0 ± 2.5
subl	Ag	3	b	1.00	2	0.333	f	0.333	f	2.360 ± .060	Å	100.0 ± 2.5

BOND DISTANCES AND ANGLES

Bond distances and angles are derived from coordinates

No. of distances/angles: 3

Interatomic dist. A-B (Å)	Atom A	Atom B	Atom C	Bond angle A-B-C (°)
2.832	I1	Ag2	Ag2(1,0)	120.6
2.832	I1	Ag2	Ag3	180.0
2.889	Ag2	Ag3		

COMMON NAME : Ag(110)-(2x1)-O
CLASSIFICATION : 47.8.4
TECHNIQUE : SEXAFS
AUTHORS : A. Puschmann and J. Haase
REFERENCE : Surf. Sci., 144, 559 (1984)

ILLUSTRATION: 35

SURFACE TYPE
Substrate : Ag
Crystal face: 110
Temperature : RT
Bulk lattice: fcc
2D bulk symm: pmm
2D surf symm: pmm

Adsorbate: O
Coverage : 0.5 O/Ag
Pattern : (2x1)
Matrix : (2.000, 0.000)
(0.000, 1.000)

STRUCTURE TYPE
Atomic adsorption in long-bridge sites (short-bridge sites w.r.t. 2nd Ag layer)

SAMPLE PREPARATION (1 sample)
Treatment : oxygen exposure at RT and pressures <1.0E-9 torr
Crystallinity:
Anal. methods:
Contamination: 2000L gave sharp (2x1) LEED pattern

COMMENTS
Use made of the effective coordination number of the central atom by the scatterers, determined from SEXAFS amplitude; bulk coordination for substrate was assumed

DATA COLLECTION
Technique: SEXAFS; oxygen K-edge SEXAFS
Dataset : SEXAFS spectra (500-800 eV photons) for E-vector parallel to [110], and for E parallel to [100]

THEORY/DATA TREATMENT
Fourier transform, using polar and azimuthal dependence of signal

STRUCTURES EXAMINED
Long-bridge site, 2-fold and 3-fold hollow sites, short-bridge site with O above or below top Ag plane

2D UNIT CELLS (1 domain observed)

Cell	Ax (Å)	Ay (Å)	Bx (Å)	By (Å)	α (°)	Matrix	Pattern	Cell type
Bulk	2.890	0.000	0.000	4.090	90.0	(1.000, 0.000) (0.000, 1.000)	(1x1)	b: bulk lattice
Surface 1	5.780	0.000	0.000	4.090	90.0	(2.000, 0.000) (0.000, 1.000)	(2x1)	s1: commens. superlattice

3D COORDINATES

O1: overlayer in long-bridge site, bonding to 2 Ag atoms in 1st Ag layer, and to 2 Ag atoms in 2nd Ag layer; 0.1Å error bar assumed for tabulation

Dx/Dy in Å, or as a fraction of layer's unit cell vectors; atom 0 at the origin. Epir/subr are bulk repeat vectors.

No. of atoms: 3

Bulk z = 1.445 Å

Reg ion	Chem el.	At. no.	Cell type	Site occ.	Rel. to	Dx ± εx		Dy ± εy		Dz ± εz		Dz/Bz(%) ± εz/Bz
epir		-2					f		f		Å	
subr		-1				1.445	Å	-2.045	Å	1.445	Å	
ovrl	O	1	s1	.50	0	0.000	f	0.000	f	0.000	Å	0.0
intf	Ag	2	b	1.00	1	0.000	f	0.500	f	0.200 ± .100	Å	13.8 ± 6.9
subl	Ag	3	b	1.00	2	0.500	f	-0.500	f	1.445	Å	100.0

BOND DISTANCES AND ANGLES

No. of distances/angles: 3

Interatomic dist. A-B (Å)	Atom A	Atom B	Atom C	Bond angle A-B-C (°)
2.055	O1	Ag2	O1(0,1)	168.8
2.190	O1	Ag3	Ag3(1,0)	131.3
2.891	Ag2	Ag3		

ILLUSTRATION: 28,29

COMMON NAME : Ag(100)-c(2x2)-Se
CLASSIFICATION : 47.34.1
TECHNIQUE : LEED
AUTHORS : A. Ignatiev, F. Jona, D.W. Jepsen and P.M. Marcus
REFERENCE : Surf. Sci., 40, 439 (1973)

SURFACE TYPE
Substrate : Ag
Crystal face: 100
Temperature : RT
Bulk lattice: fcc
2D bulk symm: p4m
2D surf symm: p4m

Adsorbate: Se
Coverage : 0.5 Se/Ag
Pattern : c(2x2)
Matrix : (1.000, 1.000)
(-1.000, 1.000)

STRUCTURE TYPE
Atomic adsorption in hollow sites

SAMPLE PREPARATION (1 sample)
Treatment : Se source was a pre-evacuated pyrex vial
Crystallinity:
Anal. methods:
Contamination: monitored by LEED

COMMENTS
Good agreement obtained between theory and experiment with the integral order spectra for 4-fold hollow site; agreement was not good for any of the models tested for the fractional order beams; work was reported to be in progress for testing a mixed Se-Ag layer model

DATA COLLECTION
Technique: LEED
Dataset : I-V curves: (0,0), (1,0), and (0.5,0.5) beams at Θ=5°, φ=0 and 90°

THEORY/DATA TREATMENT
Dynamical LEED (layer KKR): 8 phase shifts, 56 beams
ΘD=150 K(Se), 215K(Ag)

STRUCTURES EXAMINED
Hollow, top and bridge sites; Se-Ag spacing varied between 1.1 and 2.6Å

QUALITY OF EXPERIMENT-THEORY FIT
Visual

2D UNIT CELLS (1 domain observed)

Cell	Ax (Å)	Ay (Å)	Bx (Å)	By (Å)	α (°)	Matrix	Pattern	Cell type
Bulk	2.892	0.000	0.000	2.892	90.0	(1.000, 0.000) (0.000, 1.000)	(1x1)	b: bulk lattice
Surface 1	2.892	2.892	-2.892	2.892	90.0	(1.000, 1.000) (-1.000, 1.000)	c(2x2)	s1: commens. superlattice

3D COORDINATES

Se1: overlayer in hollow sites

Dx/Dy in Å, or as a fraction of layer's unit cell vectors; atom 0 at the origin. Epir/subr are bulk repeat vectors.

No. of atoms: 3

Bulk z = 2.045 Å

Region	Chem el.	At. no.	Cell type	Site occ.	Rel. to	Dx ± εx		Dy ± εy		Dz ± εz		Dz/Bz(%) ± εz/Bz
epir		-2					f		f		Å	
subr		-1				-1.446	Å	-1.446	Å	2.045	Å	
ovrl	Se	1	s1	.50	0	0.000	f	0.000	f	0.000	Å	0.0
intf	Ag	2	b	1.00	1	0.500	f	0.500	f	1.910 ± .040	Å	93.4 ± 2.0
subl	Ag	3	b	1.00	2	-0.500	f	-0.500	f	2.045	Å	100.0

BOND DISTANCES AND ANGLES

Bond distances and angles are derived from coordinates

No. of distances/angles: 3

Interatomic dist. A-B (Å)	Atom A	Atom B	Atom C	Bond angle A-B-C (°)
2.798	Se1	Ag2	Ag2(1,0)	121.1
2.798	Se1	Ag2	Ag3	88.1
2.892	Ag2	Ag3		

COMMON NAME : Ag(111)-Xe incommensurate
CLASSIFICATION : 47.54.1
TECHNIQUE : LEED
AUTHORS : P.I. Cohen, J. Unguris and M.B. Webb
REFERENCE : Surf. Sci., 58, 429 (1976)

ILLUSTRATION: 82

SURFACE TYPE
Substrate : Ag
Crystal face: 111
Temperature : 25 K
Bulk lattice: fcc
2D bulk symm: p3m1
2D surf symm: none

Adsorbate: Xe
Coverage : 0.42 (Xe/Ag)
Pattern : incommensurate
Matrix : (1.542, 0.000)
(0.000, 1.542)

STRUCTURE TYPE
Atomic physisorption in dense incommensurate hexagonal monolayer with variable overlayer lattice orientation; Xe 2D lattice constant 4.44±0.01Å; monolayer assumed planar

SAMPLE PREPARATION (1 sample)
Treatment : Xe admitted at 25 K until sharp diffraction ring appeared
Crystallinity:
Anal. methods:
Contamination: diffuse scattering: <0.002ML impurities

COMMENTS

DATA COLLECTION
Technique: LEED
Dataset : I-k curves: (00) beam at 10 polar angles, E range 40-380 eV

THEORY/DATA TREATMENT
Kinematic LEED (constant momentum transfer averaging and Fourier transform): ms vibr. ampl.=0.004575Å^2

STRUCTURES EXAMINED
Variation of Xe-Ag spacing, assuming no buckling

2D UNIT CELLS (0 domain observed)

Cell	Ax (Å)	Ay (Å)	Bx (Å)	By (Å)	α (°)	Matrix	Pattern	Cell type
Bulk	2.880	0.000	-1.440	2.494	120.0	(1.000, 0.000) (0.000, 1.000)	(1x1)	b: bulk lattice
Surface 1	4.440	0.000	-2.220	3.845	120.0	(1.542, 0.000) (0.000, 1.542)	incommensurate	i1: incomm. superlattice

3D COORDINATES

Xe1: incommensurate overlayer; 0.1Å error bar assumed for tabulation

Dx/Dy in Å, or as a fraction of layer's unit cell vectors; atom 0 at the origin. Epir/subr are bulk repeat vectors.

No. of atoms: 3

Bulk z = 2.350 Å

Region	Chem el.	At. no.	Cell type	Site occ.	Rel. to	Dx ± εx		Dy ± εy		Dz ± εz		Dz/Bz(%) ± εz/Bz
epir		-2					f		f		Å	
subr		-1				1.440	Å	0.831	Å	2.350	Å	
ovrl	Xe	1	i1	.42	0	0.000	f	0.000	f	0.000	Å	0.0
intf	Ag	2	b	1.00	1	0.000	f	0.000	f	3.500 ± .100	Å	148.9 ± 4.3
subl	Ag	3	b	1.00	2	0.667	f	0.333	f	2.350	Å	100.0

COMMON NAME : Ag(111)-Xe incommensurate
CLASSIFICATION : 47.54.2
TECHNIQUE : LEED
AUTHORS : N. Stoner, M.A. Van Hove, S.Y. Tong and M.B. Webb
REFERENCE : Phys. Rev. Lett., 40, 243 (1978)

ILLUSTRATION: 82

SURFACE TYPE
Substrate : Ag
Crystal face: 111
Temperature : 25 K
Bulk lattice: fcc
2D bulk symm: p3m1
2D surf symm: none

Adsorbate: Xe
Coverage : 0.42 (Xe/Ag)
Pattern : incommensurate
Matrix : (1.542, 0.000)
(0.000, 1.542)

STRUCTURE TYPE
Atomic physisorption in dense incommensurate hexagonal monolayer with variable overlayer lattice orientation; Xe 2D lattice constant 4.44±0.01Å; monolayer assumed planar

SAMPLE PREPARATION (1 sample)
Treatment : Xe admitted at 25 K until sharp diffraction ring appeared
Crystallinity:
Anal. methods:
Contamination: diffuse scattering: <0.002ML impurities

COMMENTS

DATA COLLECTION
Technique: LEED
Dataset : I-V curves

THEORY/DATA TREATMENT
Dynamical LEED: RFS for substrate, special perturbation method for substrate-adsorbate region

STRUCTURES EXAMINED
Xe-Ag spacing varied

QUALITY OF EXPERIMENT-THEORY FIT
Visual

2D UNIT CELLS (0 domain observed)

Cell	Ax (Å)	Ay (Å)	Bx (Å)	By (Å)	α (°)	Matrix	Pattern	Cell type
Bulk	2.880	0.000	-1.440	2.494	120.0	(1.000, 0.000) (0.000, 1.000)	(1x1)	b: bulk lattice
Surface 1	4.440	0.000	-2.220	3.845	120.0	(1.542, 0.000) (0.000, 1.542)	incommensurate	i1: incomm. superlattice

3D COORDINATES

Xe1: incommensurate overlayer

Dx/Dy in Å, or as a fraction of layer's unit cell vectors; atom 0 at the origin. Epir/subr are bulk repeat vectors.

No. of atoms: 3

Bulk z = 2.360 Å

Reg ion	Chem el.	At. no.	Cell type	Site occ.	Rel. to	Dx ± εx		Dy ± εy		Dz ± εz		Dz/Bz(%) ± εz/Bz
epir		-2					f		f		Å	
subr		-1				1.440	Å	0.831	Å	2.360	Å	
ovrl	Xe	1	i1	.42	0	0.000	f	0.000	f	0.000	Å	0.0
intf	Ag	2	b	1.00	1	0.000	f	0.000	f	3.550 ± .100	Å	150.4 ± 4.2
subl	Ag	3	b	1.00	2	0.667	f	0.333	f	2.360	Å	100.0

ILLUSTRATION: 149

COMMON NAME : AgBr(100)-(1x1)
CLASSIFICATION : 47.35.2a
TECHNIQUE : SEXAFS
AUTHORS : P. Tangyunyong, T.N. Rhodin, Y.T. Tan and K.J. Lushington
REFERENCE : Surf. Sci., 255, 259 (1991)

SURFACE TYPE
Substrate : AgBr
Crystal face: 100
Temperature : 30 K
Bulk lattice: NaCl
2D bulk symm: p4m
2D surf symm: p4m

Adsorbate:
Coverage :
Pattern : (1x1)
Matrix : (1.000, 0.000)
(0.000, 1.000)

STRUCTURE TYPE
Unreconstructed bulk-like termination without layer relaxation

SAMPLE PREPARATION (1 sample)
Treatment : sheet crystals grown with a gradient-growth technique
Crystallinity:
Anal. methods:
Contamination:

COMMENTS

DATA COLLECTION
Technique: SEXAFS
Dataset : SEXAFS data taken at Ag and Br adsorption edges at x-ray total external reflection condition

THEORY/DATA TREATMENT
Fourier filtering and multishell curve-fitting

STRUCTURES EXAMINED
No contractions in the first- and second nearest-neighbor distances were observed for both Br and Ag

2D UNIT CELLS (1 domain observed)

Cell	Ax (Å)	Ay (Å)	Bx (Å)	By (Å)	α (°)	Matrix	Pattern	Cell type
Bulk	4.083	0.000	0.000	4.083	90.0	(1.000, 0.000) (0.000, 1.000)	(1x1)	b: bulk lattice
Surface 1	4.083	0.000	0.000	4.083	90.0	(1.000, 0.000) (0.000, 1.000)	(1x1)	s1: commens. superlattice

3D COORDINATES

Ag1-Br2: non-buckled top layer; Ag3-Br4: periodically repeating bulk layers;

Dx/Dy in Å, or as a fraction of layer's unit cell vectors; atom 0 at the origin. Epir/subr are bulk repeat vectors.

No. of atoms: 4

Bulk z = 2.887 Å

Reg ion	Chem el.	At. no.	Cell type	Site occ.	Rel. to	Dx ± εx		Dy ± εy		Dz ± εz		Dz/Bz(%) ± εz/Bz
epir		-2					f		f		Å	
subr		-1				2.042	Å	2.042	Å	2.887	Å	
intf	Ag	1	b	1.00	0	0.000	f	0.000	f	0.000	Å	0.0
intf	Br	2	b	1.00	1	0.500	f	0.500	f	0.000	Å	0.0
subl	Ag	3	b	1.00	2	0.000	f	0.000	f	2.887 ± .010	Å	100.0 ± .3
subl	Br	4	b	1.00	3	0.500	f	0.500	f	0.000 ± .010	Å	0.0 ± .3

BOND DISTANCES AND ANGLES

No. of distances/angles: 4

Interatomic dist. A-B (Å)	Atom A	Atom B	Atom C	Bond angle A-B-C (°)
2.887	Ag1	Br2	Ag1(1,1)	180.0
2.887	Ag1	Br4	Ag3	90.0
2.887	Br2	Ag3	Br4(1,1)	90.0
2.887	Ag3	Br4(1,1)	Ag1(1,1)	90.0

COMMON NAME : AgBr(111)-(2x1)
CLASSIFICATION : 47.35.2b
TECHNIQUE : SEXAFS
AUTHORS : P. Tangyunyong, T.N. Rhodin, Y.T. Tan and K.J. Lushington
REFERENCE : Surf. Sci., 255, 259 (1991)

ILLUSTRATION: 151

SURFACE TYPE
Substrate : AgBr
Crystal face: 111
Temperature : 30 K
Bulk lattice: NaCl
2D bulk symm: p3m1
2D surf symm: pm

Adsorbate:
Coverage :
Pattern : (2x1)
Matrix : (2.000, 0.000)
(0.000, 1.000)

STRUCTURE TYPE
Ag-terminated surface with (2x1) reconstruction: every second row of Ag ions is missing in the top layer; first three interlayer spacings are contracted

SAMPLE PREPARATION (1 sample)
Treatment : sheet crystals grown with a gradient-growth technique
Crystallinity:
Anal. methods:
Contamination:

COMMENTS

DATA COLLECTION
Technique: SEXAFS
Dataset : SEXAFS data taken at Ag and Br adsorption edges at x-ray total external reflection condition

THEORY/DATA TREATMENT
Fourier filtering and multishell curve-fitting methods

STRUCTURES EXAMINED
Measured contractions in the nearest-neighbor Ag-Br and next-nearest-neighbor Ag-Ag and Br-Br distances were compared with those of two surface reconstructions predicted by theoretical calculation: alternate row model with Ag in the top layer agrees best with the data

2D UNIT CELLS (3 domains observed)

Cell	Ax (Å)	Ay (Å)	Bx (Å)	By (Å)	α (°)	Matrix	Pattern	Cell type
Bulk	4.083	0.000	2.042	3.536	60.0	(1.000, 0.000) (0.000, 1.000)	(1x1)	b: bulk lattice
Surface 1	8.166	0.000	2.042	3.536	60.0	(2.000, 0.000) (0.000, 1.000)	(2x1)	s1: commens. superlattice

3D COORDINATES

Ag1: rows in top layer with every other row missing; Br2: second hexagonal close packed layer; Ag3-Br4: periodically repeating bulk pair of layers

Dx/Dy in Å, or as a fraction of layer's unit cell vectors; atom 0 at the origin. Epir/subr are bulk repeat vectors.

No. of atoms: 6

Bulk z = 3.334 Å

Region	Chem el.	At. no.	Cell type	Site occ.	Rel. to	Dx ± εx		Dy ± εy		Dz ± εz		Dz/Bz(%) ± εz/Bz
epir		-2					f		f		Å	
subr		-1				2.042	Å	1.179	Å	3.334	Å	
intf	Ag	1	s1	.50	0	0.000	f	0.000	f	0.000	Å	0.0
intf	Br	2	b	1.00	1	0.667	f	0.667	f	1.632 ± .015	Å	49.0 ± .5
intf	Ag	3	b	1.00	2	-0.333	f	-0.333	f	1.650 ± .015	Å	49.5 ± .5
intf	Br	4	b	1.00	3	-0.333	f	-0.333	f	1.650 ± .015	Å	49.5 ± .5
subl	Ag	5	b	1.00	4	0.667	f	0.667	f	1.667	Å	50.0
subl	Br	6	b	1.00	5	-0.333	f	-0.333	f	1.667	Å	50.0

BOND DISTANCES AND ANGLES

No. of distances/angles: 7

Interatomic dist. A-B (Å)	Atom A	Atom B	Atom C	Bond angle A-B-C (°)
2.867	Ag1	Br2(0,-1)	Ag3	89.4
2.867	Br2	Ag1(1,1)	Br2(1,0)	90.8
2.877	Br2	Ag3	Br2(0,-1)	90.4
2.878	Ag3	Br2(0,-1)	Ag1	89.4

AgBr(111)-(2x1)
47.35.2b

Bond Distances and Angles - Continued

Interatomic dist. A-B (Å)	Atom A	Atom B	Atom C	Bond angle A-B-C (°)
2.877	Ag3	Br4	Ag3(0,-1)	90.4
2.887	Ag5	Br4(1,1)	Ag3(1,1)	180.0
2.887	Ag5	Br6	Ag1(1,0)	60.0

COMMON NAME : Al(100)-(1x1)
CLASSIFICATION : 13.15a
TECHNIQUE : LEED
AUTHORS : M.A. Van Hove, S.Y. Tong and N. Stoner
REFERENCE : Surf. Sci., 54, 259 (1976)

ILLUSTRATION: 2

SURFACE TYPE
Substrate : Al
Crystal face: 100
Temperature : 110 K
Bulk lattice: fcc
2D bulk symm: p4m
2D surf symm: p4m

Adsorbate:
Coverage :
Pattern : (1x1)
Matrix : (1.000, 0.000)
(0.000, 1.000)

STRUCTURE TYPE
Unrelaxed bulk termination

SAMPLE PREPARATION (1 sample)
Treatment : mech. polishing, then electropolishing and Xe sputtering
Crystallinity: sharp LEED pattern
Anal. methods: AES
Contamination: AES: <0.5% ML O

COMMENTS

DATA COLLECTION
Technique: LEED; spot photometer
Dataset :

THEORY/DATA TREATMENT
Dynamical LEED (RFS): Snow self consistent potential;
Vor=-7.5 eV, Voi=-4.1eV; ΘD=356 K

STRUCTURES EXAMINED
Top spacing varied from 1.725 to 2.225Å in 0.1Å steps

QUALITY OF EXPERIMENT-THEORY FIT
Visual

2D UNIT CELLS (1 domain observed)

Cell	Ax (Å)	Ay (Å)	Bx (Å)	By (Å)	α (°)	Matrix	Pattern	Cell type
Bulk	2.860	0.000	0.000	2.860	90.0	(1.000, 0.000) (0.000, 1.000)	(1x1)	b: bulk lattice
Surface 1	2.860	0.000	0.000	2.860	90.0	(1.000, 0.000) (0.000, 1.000)	(1x1)	s1: commens. superlattice

3D COORDINATES

Dx/Dy in Å, or as a fraction of layer's unit cell vectors; atom 0 at the origin. Epir/subr are bulk repeat vectors.

No. of atoms: 2

Bulk z = 2.025 Å

Region	Chem el.	At. no.	Cell type	Site occ.	Rel. to	Dx ± εx		Dy ± εy		Dz ± εz		Dz/Bz(%) ± εz/Bz
epir		-2					f		f		Å	
subr		-1				1.430	Å	1.430	Å	2.025	Å	
intf	Al	1	b	1.00	0	0.000	f	0.000	f	0.000	Å	0.0
subl	Al	2	b	1.00	1	0.500	f	0.500	f	2.025 ± .100	Å	100.0 ± 4.9

BOND DISTANCES AND ANGLES

Bond distances and angles are derived from coordinates

No. of distances/angles: 2

Interatomic dist. A-B (Å)	Atom A	Atom B	Atom C	Bond angle A-B-C (°)
2.860	Al1	Al1(1,0)		
2.862	Al1	Al2		

ILLUSTRATION: 2

COMMON NAME : Al(100)-(1x1)
CLASSIFICATION : 13.16a
TECHNIQUE : LEED
AUTHORS : Groupe d'Etude des Surfaces (Grenoble)
REFERENCE : Surf. Sci., 62, 567 (1977)

SURFACE TYPE

Substrate : Al
Crystal face: 100
Temperature : 293 K
Bulk lattice: fcc
2D bulk symm: p4m
2D surf symm: p4m

Adsorbate:
Coverage :
Pattern : (1x1)
Matrix : (1.000, 0.000)
(0.000, 1.000)

STRUCTURE TYPE

Unrelaxed bulk termination

SAMPLE PREPARATION (1 sample)

Treatment : crystal spark cut and given several hours Ar+ bombardment
Crystallinity: sharp LEED spots
Anal. methods:
Contamination: AES

COMMENTS

Voi was found to be energy dependent, decreasing from Voi=-3.4 eV at 10eV to Voi=-5eV at 120eV primary beam energy; best fit for Vor=-12±2 eV

DATA COLLECTION

Technique: LEED
Dataset : IV curves for (00) beam for a range of angles and primary beam energies of 10 to 190 eV

THEORY/DATA TREATMENT

Dyn. LEED (intralayer m/s treated exactly, interlayer m/s assumes 00 beam dominates forward scattering): ΘD = 380 K

STRUCTURES EXAMINED

Only truncated bulk

QUALITY OF EXPERIMENT-THEORY FIT

Visual

2D UNIT CELLS (1 domain observed)

Cell	Ax (Å)	Ay (Å)	Bx (Å)	By (Å)	α (°)	Matrix	Pattern	Cell type
Bulk	2.860	0.000	0.000	2.860	90.0	(1.000, 0.000) (0.000, 1.000)	(1x1)	b: bulk lattice
Surface 1	2.860	0.000	0.000	2.860	90.0	(1.000, 0.000) (0.000, 1.000)	(1x1)	s1: commens. superlattice

3D COORDINATES

Dx/Dy in Å, or as a fraction of layer's unit cell vectors; atom 0 at the origin. Epir/subr are bulk repeat vectors.

No. of atoms: 2

Bulk z = 2.022 Å

Reg ion	Chem el.	At. no.	Cell type	Site occ.	Rel. to	Dx ± εx		Dy ± εy		Dz ± εz		Dz/Bz(%) ± εz/Bz
epir		-2					f		f		Å	
subr		-1				1.430	Å	1.430	Å	2.022	Å	
intf	Al	1	b	1.00	0	0.000	f	0.000	f	0.000	Å	0.0
subl	Al	2	b	1.00	1	0.500	f	0.500	f	2.022	Å	100.0

BOND DISTANCES AND ANGLES

Bond distances and angles are derived from coordinates

No. of distances/angles: 2

Interatomic dist. A-B (Å)	Atom A	Atom B	Atom C	Bond angle A-B-C (°)
2.860	Al1	Al1(1,0)		
2.860	Al1	Al2		

COMMON NAME : Al(100)-(1x1)
CLASSIFICATION : 13.26
TECHNIQUE : MEED
AUTHORS : N. Masud, R. Baudoing, D. Aberdam and C. Gaubert
REFERENCE : Surf. Sci., 133, 580 (1983)

ILLUSTRATION: 2

SURFACE TYPE
Substrate : Al
Crystal face: 100
Temperature : 77 K
Bulk lattice: fcc
2D bulk symm: p4m
2D surf symm: p4m

Adsorbate:
Coverage :
Pattern : (1x1)
Matrix : (1.000, 0.000)
(0.000, 1.000)

STRUCTURE TYPE
Bulk termination with possible slight top contraction

SAMPLE PREPARATION (1 sample)
Treatment :
Crystallinity:
Anal. methods:
Contamination:

COMMENTS

Vor small because exchange potential is small at high energies
R-factor: mean square deviation of the peak energies in theory and experiment, averaged over 50 spectra

DATA COLLECTION
Technique: MEED
Dataset : rot. diagram at 980 eV; ϕ range 0-45°; Θ range 40-82°

THEORY/DATA TREATMENT
MEED (chain method with full mult. scattering in the layers, RFS between layers); 17 phase shifts; Vor=-5 eV, Voi=-3 eV

STRUCTURES EXAMINED
Top spacing relaxations of +10% +5% 0% -5% -10%, interpolated to best fit of -1.5%

QUALITY OF EXPERIMENT-THEORY FIT
See comments

2D UNIT CELLS (1 domain observed)

Cell	Ax (Å)	Ay (Å)	Bx (Å)	By (Å)	α (°)	Matrix	Pattern	Cell type
Bulk	2.860	0.000	0.000	2.860	90.0	(1.000, 0.000) (0.000, 1.000)	(1x1)	b: bulk lattice
Surface 1	2.860	0.000	0.000	2.860	90.0	(1.000, 0.000) (0.000, 1.000)	(1x1)	s1: commens. superlattice

3D COORDINATES

0.1Å error bar assumed for tabulation

Dx/Dy in Å, or as a fraction of layer's unit cell vectors; atom 0 at the origin. Epir/subr are bulk repeat vectors.

No. of atoms: 3

Bulk z = 2.022 Å

Reg ion	Chem el.	At. no.	Cell type	Site occ.	Rel. to	Dx ± εx		Dy ± εy		Dz ± εz		Dz/Bz(%) ± εz/Bz
epir		-2					f		f		Å	
subr		-1				-1.430	Å	-1.430	Å	2.022	Å	
intf	Al	1	b	1.00	0	0.000	f	0.000	f	0.000	Å	0.0
intf	Al	2	b	1.00	1	0.500	f	0.500	f	2.052 ± .100	Å	101.5 ± 5.0
subl	Al	3	b	1.00	2	-0.500	f	-0.500	f	2.022	Å	100.0

BOND DISTANCES AND ANGLES

Bond distances and angles are derived from coordinates

No. of distances/angles: 4

Interatomic dist. A-B (Å)	Atom A	Atom B	Atom C	Bond angle A-B-C (°)
2.860	Al1	Al1(1,0)	Al2	60.2
2.881	Al1	Al2	Al1(1,0)	59.5
2.881	Al1	Al2	Al2(1,0)	119.8
2.860	Al2	Al3	Al2(0,-1)	60.0

COMMON NAME : Al(110)-(1x1)
CLASSIFICATION : 13.16b
TECHNIQUE : LEED
AUTHORS : Groupe d'Etude des Surfaces (Grenoble)
REFERENCE : Surf. Sci., 62, 567 (1977)

ILLUSTRATION: 4

SURFACE TYPE
Substrate : Al
Crystal face: 110
Temperature : 293 K
Bulk lattice: fcc
2D bulk symm: pmm
2D surf symm: pmm

Adsorbate:
Coverage :
Pattern : (1x1)
Matrix : (1.000, 0.000)
(0.000, 1.000)

STRUCTURE TYPE
Relaxed bulk termination

SAMPLE PREPARATION (1 sample)
Treatment : crystal spark cut and given several hours Ar+ bombardment
Crystallinity: sharp LEED spots
Anal. methods:
Contamination: AES

COMMENTS
Standard LEED parameters,but Voi fit: Voi found E-dependent, decreasing from -3.4 eV at 10eV to Voi=-5eV at 120eV;
best fit Vor=-12±2 eV;
random surface steps in theory markedly improve fit

DATA COLLECTION
Technique: LEED
Dataset : I-V curves for (00) beam for a range of angles; E-range 10-190 eV

THEORY/DATA TREATMENT
Dyn. LEED (intralayer m/s treated exactly, interlayer m/s assumes 00 beam dominates forward scattering): ΘD=380 K

STRUCTURES EXAMINED
Contraction of outermost layer only by 0-15%

QUALITY OF EXPERIMENT-THEORY FIT
Visual

2D UNIT CELLS (1 domain observed)

Cell	Ax (Å)	Ay (Å)	Bx (Å)	By (Å)	α (°)	Matrix	Pattern	Cell type
Bulk	2.860	0.000	0.000	4.044	90.0	(1.000, 0.000) (0.000, 1.000)	(1x1)	b: bulk lattice
Surface 1	2.860	0.000	0.000	4.044	90.0	(1.000, 0.000) (0.000, 1.000)	(1x1)	s1: commens. superlattice

3D COORDINATES

0.1Å error bars assumed for tabulation

Dx/Dy in Å, or as a fraction of layer's unit cell vectors; atom 0 at the origin. Epir/subr are bulk repeat vectors.

No. of atoms: 3

Bulk z = 1.430 Å

Region	Chem el.	At. no.	Cell type	Site occ.	Rel. to	Dx ± εx		Dy ± εy		Dz ± εz		Dz/Bz(%) ± εz/Bz
epir		-2					f		f		Å	
subr		-1				1.430	Å	2.022	Å	1.430	Å	
intf	Al	1	b	1.00	0	0.000	f	0.000	f	0.000	Å	0.0
intf	Al	2	b	1.00	1	0.500	f	0.500	f	1.300 ± .100	Å	90.9 ± 7.0
subl	Al	3	b	1.00	2	-0.500	f	-0.500	f	1.430	Å	100.0

BOND DISTANCES AND ANGLES

Bond distances and angles are derived from coordinates

No. of distances/angles: 5

Interatomic dist. A-B (Å)	Atom A	Atom B	Atom C	Bond angle A-B-C (°)
2.860	Al1	Al1(1,0)	Al2	59.3
2.797	Al1	Al2	Al1(1,0)	61.5
2.797	Al1	Al2	Al3	57.7
2.730	Al1	Al3	Al2	60.0
2.860	Al2	Al3(1,1)	Al2(1,0)	60.0

COMMON NAME : Al(110)-(1x1)
CLASSIFICATION : 13.25
TECHNIQUE : LEED
AUTHORS : J.N. Andersen, H.B. Nielsen, L. Petersen and D.L. Adams
REFERENCE : J. Phys., C17, 173 (1984)

ILLUSTRATION: 4

SURFACE TYPE
Substrate : Al
Crystal face: 110
Temperature : 100 K
Bulk lattice: fcc
2D bulk symm: pmm
2D surf symm: pmm

Adsorbate:
Coverage :
Pattern : (1x1)
Matrix : (1.000, 0.000)
(0.000, 1.000)

STRUCTURE TYPE
Bulk termination with multilayer relaxation

SAMPLE PREPARATION (1 sample)
Treatment : cycles of long Ar+ bombardment and annealing
Crystallinity: sharp (1x1) LEED pattern
Anal. methods:
Contamination: AES: <0.02 monolayer Cu and O

COMMENTS

DATA COLLECTION
Technique: LEED
Dataset : I-V spectra for (01),(10),(11),(02),(12),(20),(03),(2,1) and (13) beams for 40-350 eV at normal inc

THEORY/DATA TREATMENT
Dynamical LEED: Vor=-9.3±0.8 eV, Voi=-3.9±0.6 eV, ΘD=625±125 K (all fit)

STRUCTURES EXAMINED
Various spacings between first 4 layers

QUALITY OF EXPERIMENT-THEORY FIT
Weighted R2=0.042

2D UNIT CELLS (1 domain observed)

Cell	Ax (Å)	Ay (Å)	Bx (Å)	By (Å)	α (°)	Matrix	Pattern	Cell type
Bulk	2.850	0.000	0.000	4.036	90.0	(1.000, 0.000) (0.000, 1.000)	(1x1)	b: bulk lattice
Surface 1	2.850	0.000	0.000	4.036	90.0	(1.000, 0.000) (0.000, 1.000)	(1x1)	s1: commens. superlattice

3D COORDINATES

Dx/Dy in Å, or as a fraction of layer's unit cell vectors; atom 0 at the origin. Epir/subr are bulk repeat vectors.

No. of atoms: 5

Bulk z = 1.425 Å

Region	Chem el.	At. no.	Cell type	Site occ.	Rel. to	Dx ± εx		Dy ± εy		Dz ± εz		Dz/Bz(%) ± εz/Bz
epir		-2					f		f		Å	
subr		-1				1.425	Å	2.018	Å	1.425	Å	
intf	Al	1	b	1.00	0	0.000	f	0.000	f	0.000	Å	0.0
intf	Al	2	b	1.00	1	0.500	f	0.500	f	1.304 ± .012	Å	91.5 ± .8
intf	Al	3	b	1.00	2	-0.500	f	-0.500	f	1.499 ± .015	Å	105.2 ± 1.1
intf	Al	4	b	1.00	3	0.500	f	0.500	f	1.404 ± .017	Å	98.5 ± 1.2
subl	Al	5	b	1.00	4	0.500	f	0.500	f	1.429 ± .018	Å	100.3 ± 1.3

BOND DISTANCES AND ANGLES

Bond distances and angles are derived from coordinates

No. of distances/angles: 18

Interatomic dist. A-B (Å)	Atom A	Atom B	Atom C	Bond angle A-B-C (°)
2.850	Al1	Al1(1,0)	Al2	59.3
2.890	Al2	Al3(1,1)	Al2(1,0)	59.1
2.890	Al2	Al3(1,1)	Al3(0,1)	60.5
2.890	Al2	Al3(1,1)	Al4(1,0)	90.4
2.890	Al2	Al3(1,0)	Al4	60.9
2.903	Al2	Al4	Al3	60.4

Al(110)-(1x1)
13.25

Bond Distances and Angles - Continued

Interatomic dist. A-B (Å)	Atom A	Atom B	Atom C	Bond angle A-B-C (°)
2.803	Al3	Al1	Al2	62.2
2.842	Al3	Al4	Al2	60.4
2.842	Al3	Al4	Al3(1,0)	60.2
2.842	Al3	Al4	Al4(1,0)	120.1
2.794	Al1	Al2	Al1(1,0)	61.3
2.794	Al1	Al2	Al2(1,0)	120.7
2.794	Al1	Al2	Al3(1,0)	89.4
2.794	Al1	Al2	Al3	59.1
2.794	Al1	Al2	Al4	117.8
2.803	Al1	Al3	Al2	58.8
2.803	Al1	Al3	Al4	119.6
2.890	Al2	Al3(1,1)	Al1(1,1)	58.8

COMMON NAME : Al(110)-(1x1)
CLASSIFICATION : 13.27
TECHNIQUE : LEED
AUTHORS : J.R. Noonan and H.L. Davis
REFERENCE : Phys. Rev., B29, 4349 (1984)

ILLUSTRATION: 4

SURFACE TYPE
Substrate : Al
Crystal face: 110
Temperature : 300 K
Bulk lattice: fcc
2D bulk symm: pmm
2D surf symm: pmm

Adsorbate:
Coverage :
Pattern : (1x1)
Matrix : (1.000, 0.000)
(0.000, 1.000)

STRUCTURE TYPE
Bulk termination with multilayer relaxation

SAMPLE PREPARATION (1 sample)
Treatment : electro-polishing, sputtering and annealing
Crystallinity: sharp LEED pattern
Anal. methods:
Contamination: AES: clean

COMMENTS
Two R-factors used, giving quantitatively similar results: RZJ and R2

DATA COLLECTION
Technique: LEED
Dataset : symmetrical LEED beams averaged; 8 beams for 50<E<100 eV at normal incidence

THEORY/DATA TREATMENT
Dynamical LEED (RFS): Moruzzi-Janak-Williams potential; Vor=-10.4ev, Voi=4.7ev; ΘD=470 K (fit)

STRUCTURES EXAMINED
First varied top 2 interlayer spacings to minimise RZJ, then optimized 3rd, then 4th spacings; non-structural parameters varied at the first stage only for RZJ: d12=-8.9%, d23=5.9%; for R2: d12=-8.1%, d23=5.2%

QUALITY OF EXPERIMENT-THEORY FIT
RZJ=0.032

2D UNIT CELLS (1 domain observed)

Cell	Ax (Å)	Ay (Å)	Bx (Å)	By (Å)	α (°)	Matrix	Pattern	Cell type
Bulk	2.860	0.000	0.000	4.040	90.0	(1.000, 0.000) (0.000, 1.000)	(1x1)	b: bulk lattice
Surface 1	2.860	0.000	0.000	4.040	90.0	(1.000, 0.000) (0.000, 1.000)	(1x1)	s1: commens. superlattice

3D COORDINATES

Dx/Dy in Å, or as a fraction of layer's unit cell vectors; atom 0 at the origin. Epir/subr are bulk repeat vectors.

No. of atoms: 5

Bulk z = 1.430 Å

Region	Chem el.	At. no.	Cell type	Site occ.	Rel. to	Dx ± εx		Dy ± εy		Dz ± εz		Dz/Bz(%) ± εz/Bz
epir		-2					f		f		Å	
subr		-1				1.430	Å	2.020	Å	1.430	Å	
intf	Al	1	b	1.00	0	0.000	f	0.000	f	0.000	Å	0.0
intf	Al	2	b	1.00	1	0.500	f	0.500	f	1.310 ± .014	Å	91.6 ± 1.0
intf	Al	3	b	1.00	2	-0.500	f	-0.500	f	1.510 ± .016	Å	105.6 ± 1.1
intf	Al	4	b	1.00	3	0.500	f	0.500	f	1.463 ± .019	Å	102.3 ± 1.3
subl	Al	5	b	1.00	4	-0.500	f	-0.500	f	1.455 ± .022	Å	101.8 ± 1.5

BOND DISTANCES AND ANGLES

Bond distances and angles are derived from coordinates

No. of distances/angles: 8

Interatomic dist. A-B (Å)	Atom A	Atom B	Atom C	Bond angle A-B-C (°)
2.860	Al1	Al1(1,0)	Al2	59.3
2.800	Al1	Al2	Al1(1,0)	61.4
2.800	Al1	Al2	Al3	59.3
2.800	Al1	Al2	Al4	117.9
2.899	Al2	Al3(1,1)	Al2(1,0)	59.1

Al(110)-(1x1)
13.27

Bond Distances and Angles - Continued

Interatomic dist. A-B (Å)	Atom A	Atom B	Atom C	Bond angle A-B-C (°)
2.899	Al2	Al3(1,1)	Al3(0,1)	60.5
2.899	Al2	Al3(1,1)	Al4	62.0
2.899	Al2	Al3(1,1)	Al5(1,1)	121.4

COMMON NAME : Al(111)-(1x1)
CLASSIFICATION : 13.19
TECHNIQUE : LEED
AUTHORS : F. Jona, D. Sondericker and P.M. Marcus
REFERENCE : J. Phys., C13, L155 (1980)

ILLUSTRATION: 1

SURFACE TYPE
Substrate : Al
Crystal face: 111
Temperature : RT*
Bulk lattice: fcc
2D bulk symm: p3m1
2D surf symm: p3m1

Adsorbate:
Coverage :
Pattern : (1x1)
Matrix : (1.000, 0.000)
(0.000, 1.000)

STRUCTURE TYPE
Bulk termination with expanded top spacing

SAMPLE PREPARATION (sample)
Treatment : see Jepsen et al, Phys Rev B6 3684 (1972) & B8 1786 (1973)
Crystallinity:
Anal. methods:
Contamination:

COMMENTS

DATA COLLECTION
Technique: LEED
Dataset : I-V curves for 12 beams at different angles of incidence; cumul. E range 2128 eV

THEORY/DATA TREATMENT
Dynamical LEED (program CHANGE): 8 ph sh, 31 beams; Vor=-8.7±0.6 eV (fit), Voi=-3eV; rms vib ampls 0.165Å

STRUCTURES EXAMINED
1st - 2nd layer spacing varied in range 2.038-2.638Å in steps of 0.1 Å using a bulk spacing of 2.338Å for subsequent layer spacings

QUALITY OF EXPERIMENT-THEORY FIT
RZJ=0.21

2D UNIT CELLS (1 domain observed)

Cell	Ax (Å)	Ay (Å)	Bx (Å)	By (Å)	α (°)	Matrix	Pattern	Cell type
Bulk	2.863	0.000	-1.432	2.479	120.0	(1.000, 0.000) (0.000, 1.000)	(1x1)	b: bulk lattice
Surface 1	2.863	0.000	-1.432	2.479	120.0	(1.000, 0.000) (0.000, 1.000)	(1x1)	s1: commens. superlattice

3D COORDINATES

Dx/Dy in Å, or as a fraction of layer's unit cell vectors; atom 0 at the origin. Epir/subr are bulk repeat vectors.

No. of atoms: 3

Bulk z = 2.338 Å

Region	Chem el.	At. no.	Cell type	Site occ.	Rel. to	Dx ± εx		Dy ± εy		Dz ± εz		Dz/Bz(%) ± εz/Bz
epir		-2					f		f		Å	
subr		-1				1.432	Å	-0.827	Å	2.338	Å	
intf	Al	1	b	1.00	0	0.000	f	0.000	f	0.000	Å	0.0
intf	Al	2	b	1.00	1	0.333	f	0.667	f	2.390 ± .030	Å	102.2 ± 1.3
subl	Al	3	b	1.00	2	0.333	f	-0.333	f	2.338	Å	100.0

BOND DISTANCES AND ANGLES

Bond distances and angles are derived from coordinates

No. of distances/angles: 2

Interatomic dist. A-B (Å)	Atom A	Atom B	Atom C	Bond angle A-B-C (°)
2.863	Al1	Al1(1,1)	Al2	60.5
2.906	Al1	Al2	Al1(1,1)	59.0

ILLUSTRATION: 1

COMMON NAME : Al(111)-(1x1)
CLASSIFICATION : 13.20a
TECHNIQUE : LEED
AUTHORS : V. Martinez, F. Soria, M.C. Munoz and J.L. Sacedon
REFERENCE : Surf. Sci., 128, 424 (1983)

SURFACE TYPE
Substrate : Al
Crystal face: 111
Temperature : 300 K
Bulk lattice: fcc
2D bulk symm: p3m1
2D surf symm: p3m1

Adsorbate:
Coverage :
Pattern : (1x1)
Matrix : (1.000, 0.000)
(0.000, 1.000)

STRUCTURE TYPE
Bulk termination with expanded top interlayer spacing

SAMPLE PREPARATION (1 sample)
Treatment : clean Al(111) grown in situ on mica substrate at 773 K
Crystallinity:
Anal. methods:
Contamination: AES: <0.01 monolayer of C and O

COMMENTS
Authors note that oxygen contaminated surface could lead to an anomalous interpretation as a first layer expansion of the clean surface

DATA COLLECTION
Technique: LEED
Dataset : I-V curves for (10), (01) beams at Θ=0°; (00) beam at Θ=5, ϕ=18°

THEORY/DATA TREATMENT
Dynamical LEED (CAVLEED package): Vor=-9 eV

STRUCTURES EXAMINED
-10% to +10% relaxation of top interlayer spacing

QUALITY OF EXPERIMENT-THEORY FIT
RZJ=0.23

2D UNIT CELLS (1 domain observed)

Cell	Ax (Å)	Ay (Å)	Bx (Å)	By (Å)	α (°)	Matrix	Pattern	Cell type
Bulk	2.864	0.000	-1.432	2.480	120.0	(1.000, 0.000) (0.000, 1.000)	(1x1)	b: bulk lattice
Surface 1	2.864	0.000	-1.432	2.480	120.0	(1.000, 0.000) (0.000, 1.000)	(1x1)	s1: commens. superlattice

3D COORDINATES

Dx/Dy in Å, or as a fraction of layer's unit cell vectors; atom 0 at the origin. Epir/subr are bulk repeat vectors.

No. of atoms: 2

Bulk z = 2.338 Å

Region	Chem el.	At. no.	Cell type	Site occ.	Rel. to	Dx ± εx		Dy ± εy		Dz ± εz		Dz/Bz(%) ± εz/Bz
epir		-2					f		f		Å	
subr		-1				1.432	Å	-0.827	Å	2.338	Å	
intf	Al	1	b	1.00	0	0.000	f	0.000	f	0.000	Å	0.0
subl	Al	2	b	1.00	1	0.333	f	0.667	f	2.410 ± .050	Å	103.1 ± 2.1

BOND DISTANCES AND ANGLES

Bond distances and angles are derived from coordinates

No. of distances/angles: 2

Interatomic dist. A-B (Å)	Atom A	Atom B	Atom C	Bond angle A-B-C (°)
2.864	Al1	Al1(1,1)	Al2	60.7
2.923	Al1	Al2	Al1(1,1)	58.7

COMMON NAME : Al(111)-(1x1)
CLASSIFICATION : 13.21
TECHNIQUE : LEED
AUTHORS : H.B. Nielsen and D.L. Adams
REFERENCE : J. Phys., C15, 615 (1982)

ILLUSTRATION: 1

SURFACE TYPE
Substrate : Al
Crystal face: 111
Temperature : 90 K
Bulk lattice: fcc
2D bulk symm: p3m1
2D surf symm: p3m1

Adsorbate:
Coverage :
Pattern : (1x1)
Matrix : (1.000, 0.000)
(0.000, 1.000)

STRUCTURE TYPE
Bulk termination with expanded top spacing

SAMPLE PREPARATION (1 sample)
Treatment : Ar+ bombardment at RT, with brief anneals to 700 K
Crystallinity: sharp (1x1) LEED pattern
Anal. methods:
Contamination: AES: clean

COMMENTS

DATA COLLECTION
Technique: LEED
Dataset : I-V curves for 5 non-equivalent normal incidence beams; 45<E<360 eV

THEORY/DATA TREATMENT
Dynamical LEED (RFS): 10 ph sh, 3 pots: Moruzzi et al, Snow, Herman-Skillman; Vor=-11.6 eV, Voi=-5.1eV; ΘD=490 K (all fit)

STRUCTURES EXAMINED
1st - 2nd layer spacing varied in range 2.1-2.6Å in steps of 0.05 Å using a bulk spacing of 2.3288Å for subsequent layer spacings

QUALITY OF EXPERIMENT-THEORY FIT
Weighted R2=0.063

2D UNIT CELLS (1 domain observed)

Cell	Ax (Å)	Ay (Å)	Bx (Å)	By (Å)	α (°)	Matrix	Pattern	Cell type
Bulk	2.852	0.000	-1.426	2.470	120.0	(1.000, 0.000) (0.000, 1.000)	(1x1)	b: bulk lattice
Surface 1	2.852	0.000	-1.426	2.470	120.0	(1.000, 0.000) (0.000, 1.000)	(1x1)	s1: commens. superlattice

3D COORDINATES

Dx/Dy in Å, or as a fraction of layer's unit cell vectors; atom 0 at the origin. Epir/subr are bulk repeat vectors.

No. of atoms: 2

Bulk z = 2.329 Å

Region	Chem el.	At. no.	Cell type	Site occ.	Rel. to	Dx ± εx		Dy ± εy		Dz ± εz		Dz/Bz(%) ± εz/Bz
epir		-2					f		f		Å	
subr		-1				1.426	Å	-0.823	Å	2.329	Å	
intf	Al	1	b	1.00	0	0.000	f	0.000	f	0.000	Å	0.0
subl	Al	2	b	1.00	1	0.333	f	0.667	f	2.350 ± .012	Å	100.9 ± .5

BOND DISTANCES AND ANGLES

Bond distances and angles are derived from coordinates

No. of distances/angles: 2

Interatomic dist. A-B (Å)	Atom A	Atom B	Atom C	Bond angle A-B-C (°)
2.852	Al1	Al1(1,1)	Al2	60.2
2.870	Al1	Al2	Al1(1,1)	59.6

ILLUSTRATION: 1

COMMON NAME : Al(111)-(1x1)
CLASSIFICATION : 13.21a
TECHNIQUE : LEED
AUTHORS : J. Neve, J. Rundgren and P. Westrin
REFERENCE : J. Phys., C15, 4391 (1982)

SURFACE TYPE
Substrate : Al
Crystal face: 111
Temperature : 293 K
Bulk lattice: fcc
2D bulk symm: p3m1
2D surf symm: p3m1

Adsorbate:
Coverage :
Pattern : (1x1)
Matrix : (1.000, 0.000)
(0.000, 1.000)

STRUCTURE TYPE
Bulk termination

SAMPLE PREPARATION (1 sample)
Treatment :
Crystallinity:
Anal. methods:
Contamination:

COMMENTS

DATA COLLECTION
Technique: LEED
Dataset : LEED data of Martinsson et al, Surf. Sci. 89, 102 (1979)

THEORY/DATA TREATMENT
Dynamical LEED (KKR): DCV method to generate pot with energy dependent excited state potential

STRUCTURES EXAMINED
Truncated bulk structure only

QUALITY OF EXPERIMENT-THEORY FIT
Visual

2D UNIT CELLS (1 domain observed)

Cell	Ax (Å)	Ay (Å)	Bx (Å)	By (Å)	α (°)	Matrix	Pattern	Cell type
Bulk	2.860	0.000	-1.430	2.477	120.0	(1.000, 0.000) (0.000, 1.000)	(1x1)	b: bulk lattice
Surface 1	2.860	0.000	-1.430	2.477	120.0	(1.000, 0.000) (0.000, 1.000)	(1x1)	s1: commens. superlattice

3D COORDINATES

Dx/Dy in Å, or as a fraction of layer's unit cell vectors; atom 0 at the origin. Epir/subr are bulk repeat vectors.

No. of atoms: 2

Bulk z = 2.340 Å

Region	Chem el.	At. no.	Cell type	Site occ.	Rel. to	Dx ± εx		Dy ± εy		Dz ± εz		Dz/Bz(%) ± εz/Bz
epir		-2					f		f		Å	
subr		-1				0.000	Å	1.651	Å	2.340	Å	
intf	Al	1	b	1.00	0	0.000	f	0.000	f	0.000	Å	0.0
subl	Al	2	b	1.00	1	0.333	f	0.667	f	2.340	Å	100.0

BOND DISTANCES AND ANGLES

Bond distances and angles are derived from coordinates

No. of distances/angles: 2

Interatomic dist. A-B (Å)	Atom A	Atom B	Atom C	Bond angle A-B-C (°)
2.860	Al1	Al1(1,1)	Al2(1,1)	120.0
2.864	Al1	Al2	Al1(1,1)	59.9

COMMON NAME : Al(111)-(1x1)
CLASSIFICATION : 13.41
TECHNIQUE : LEED
AUTHORS : J.R. Noonan and H.L. Davis
REFERENCE : J. Vac. Sci. Technol., A8, 2671 (1990)

ILLUSTRATION: 1

SURFACE TYPE
Substrate : Al
Crystal face: 111
Temperature : 160 K
Bulk lattice: fcc
2D bulk symm: p3m1
2D surf symm: p3m1

Adsorbate:
Coverage :
Pattern : (1x1)
Matrix : (1.000, 0.000)
(0.000, 1.000)

STRUCTURE TYPE
Multilayer relaxation: expansion of the 1st and 2nd interlayer spacing

SAMPLE PREPARATION (2 sample)
Treatment : sputtering and annealing at 525° C
Crystallinity: sharp LEED pattern
Anal. methods: AES
Contamination:

COMMENTS
Also done at RT with the same result

DATA COLLECTION
Technique: LEED; video LEED
Dataset : IV spectra for 7 non-equivalent beams, E range 50-380 eV

THEORY/DATA TREATMENT
Dynamical LEED: RFS

QUALITY OF EXPERIMENT-THEORY FIT
R2=0.0191

2D UNIT CELLS (1 domain observed)

Cell	Ax (Å)	Ay (Å)	Bx (Å)	By (Å)	α (°)	Matrix	Pattern	Cell type
Bulk	2.864	0.000	1.432	2.480	60.0	(1.000, 0.000) (0.000, 1.000)	(1x1)	b: bulk lattice
Surface 1	2.864	0.000	1.432	2.480	60.0	(1.000, 0.000) (0.000, 1.000)	(1x1)	s1: commens. superlattice

3D COORDINATES

Dx/Dy in Å, or as a fraction of layer's unit cell vectors; atom 0 at the origin. Epir/subr are bulk repeat vectors.

No. of atoms: 3 Bulk z = 2.338 Å

Region	Chem el.	At. no.	Cell type	Site occ.	Rel. to	Dx ± εx	Dy ± εy	Dz ± εz	Dz/Bz(%) ± εz/Bz
epir		-2				f	f	Å	
subr		-1				1.432 Å	0.827 Å	2.338 Å	
intf	Al	1	b	1.00	0	0.000 Å	0.000 Å	0.000 Å	0.0
intf	Al	2	b	1.00	0	1.432 Å	0.827 Å	2.378 ± .007 Å	101.7 ± .3
subl	Al	3	b	1.00	0	2.864 Å	1.673 Å	4.728 ± .016 Å	202.2 ± .7

BOND DISTANCES AND ANGLES

Bond distances and angles are derived from coordinates

No. of distances/angles: 1

Interatomic dist. A-B (Å)	Atom A	Atom B	Atom C	Bond angle A-B-C (°)
2.864	Al1	Al2		

COMMON NAME : Al(210)-(1x1) ILLUSTRATION: 9
CLASSIFICATION : 13.36
TECHNIQUE : LEED
AUTHORS : D.L. Adams, V. Jensen, X.F. Sun and J.H. Vollesen
REFERENCE : Phys. Rev., B38, 7913 (1988)

SURFACE TYPE
Substrate : Al
Crystal face: 210
Temperature : 135 K
Bulk lattice: fcc
2D bulk symm: cm
2D surf symm: cm

Adsorbate:
Coverage :
Pattern : (1x1)
Matrix : (1.000, 0.000)
(0.000, 1.000)

STRUCTURE TYPE
Bulk termination with multilayer relaxations perpendicular to surface (by -16, -1, +9, -4 and -1%) and parallel to surface (by 0, -3, +2, -2 and -1%) along symmetry plane

SAMPLE PREPARATION (1 sample)
Treatment : Ar sputturing at 3k eV, 8E-5torr, 15μA, annealing at 800 K
Crystallinity: surface <0.2° from (210) plane
Anal. methods:
Contamination: AES: small amounts of C and O

COMMENTS

DATA COLLECTION
Technique: LEED
Dataset : I-V curves for 14 symmetry-inequivalent beams; E range 40-340 eV

THEORY/DATA TREATMENT
Dynamical LEED (layer doubling): 12 phase shifts from Moruzzi et al potential; Vor=-7.9 eV, Voi=-4.0eV; ΘD=600 K

STRUCTURES EXAMINED
Variation of first 5 interlayer spacings and first 5 layer registries, giving 451 different structures

QUALITY OF EXPERIMENT-THEORY FIT
R2=0.1041

2D UNIT CELLS (1 domain observed)

Cell	Ax (Å)	Ay (Å)	Bx (Å)	By (Å)	α (°)	Matrix	Pattern	Cell type
Bulk	4.037	0.000	2.018	4.513	65.9	(1.000, 0.000) (0.000, 1.000)	(1x1)	b: bulk lattice
Surface 1	4.037	0.000	2.018	4.513	65.9	(1.000, 0.000) (0.000, 1.000)	(1x1)	s1: commens. superlattice

3D COORDINATES

Dx/Dy in Å, or as a fraction of layer's unit cell vectors; atom 0 at the origin. Epir/subr are bulk repeat vectors.

No. of atoms: 7 Bulk z = .903 Å

Region	Chem el.	At. no.	Cell type	Site occ.	Rel. to	Dx ± εx		Dy ± εy		Dz ± εz		Dz/Bz(%) ± εz/Bz
epir		-2					f		f		Å	
subr		-1				2.018	Å	1.805	Å	0.903	Å	
intf	Al	1	b	1.00	0	0.000	f	0.000	f	0.000	Å	0.0
intf	Al	2	b	1.00	1	0.300	f	0.400	f	0.758 ± .020	Å	84.0 ± 2.2
intf	Al	3	b	1.00	2	0.306	f	0.388	f	0.894 ± .030	Å	99.0 ± 3.3
intf	Al	4	b	1.00	3	0.296	f	-0.592	f	0.984 ± .030	Å	109.0 ± 3.3
intf	Al	5	b	1.00	4	-0.696	f	0.392	f	0.867 ± .040	Å	96.0 ± 4.4
intf	Al	6	b	1.00	5	0.302	f	0.396	f	0.894 ± .050	Å	99.0 ± 5.5
subl	Al	7	b	1.00	6	0.300	f	-0.600	f	0.903	Å	100.0

BOND DISTANCES AND ANGLES

Bond distances and angles are derived from coordinates

No. of distances/angles: 6

Interatomic dist. A-B (Å)	Atom A	Atom B	Atom C	Bond angle A-B-C (°)
2.812	Al1	Al2	Al3(0,-1)	58.1
2.812	Al1	Al2	Al4(-1,0)	58.2
2.812	Al1	Al2	Al5	116.6
2.818	Al2	Al3	Al4	61.7
2.818	Al2	Al3	Al5	60.5

Al(210)-(1x1)
13.36

Bond Distances and Angles - Continued

Interatomic dist. A-B (Å)	Atom A	Atom B	Atom C	Bond angle A-B-C (°)
2.847	Al3	Al4	Al5(1,0)	61.2

COMMON NAME : Al(311)-(1x1)
CLASSIFICATION : 13.30
TECHNIQUE : LEED
AUTHORS : J.R. Noonan, H.L. Davis, W. Erley
REFERENCE : Surf. Sci., 152/153, 142 (1985)

ILLUSTRATION: 8

SURFACE TYPE
Substrate : Al
Crystal face: 311
Temperature : 298 K
Bulk lattice: fcc
2D bulk symm: cm
2D surf symm: cm

Adsorbate:
Coverage :
Pattern : (1x1)
Matrix : (1.000, 0.000)
(0.000, 1.000)

STRUCTURE TYPE
Bulk termination with multilayer relaxation; no detectable lateral relaxation

SAMPLE PREPARATION (1 sample)
Treatment : electropolish in H2SO4/H3PO4; Ar+ sputter; 500C anneal
Crystallinity:
Anal. methods:
Contamination: AES: <0.05% ML Si

COMMENTS

DATA COLLECTION
Technique: LEED; Faraday cup
Dataset : 34 LEED beams (21 symmetry-inequivalent); energy range 50-300ev; normal incidence within 0.5°

THEORY/DATA TREATMENT
Dynamical LEED (Reverse Scattering Perturbation): Moruzzi-Janak-Williams potential; Voi=4.75 eV; ΘD=550 K

STRUCTURES EXAMINED
Relaxation of top two interlayer spacings and lateral displacement of top layer

QUALITY OF EXPERIMENT-THEORY FIT
RZJ=0.07, R2=0.083

2D UNIT CELLS (1 domain observed)

Cell	Ax (Å)	Ay (Å)	Bx (Å)	By (Å)	α (°)	Matrix	Pattern	Cell type
Bulk	2.860	0.000	-1.432	4.749	106.8	(1.000, 0.000) (0.000, 1.000)	(1x1)	b: bulk lattice
Surface 1	2.860	0.000	-1.432	4.749	106.8	(1.000, 0.000) (0.000, 1.000)	(1x1)	s1: commens. superlattice

3D COORDINATES

0.05Å error bars assumed for tabulation of lateral relaxation

Dx/Dy in Å, or as a fraction of layer's unit cell vectors; atom 0 at the origin. Epir/subr are bulk repeat vectors.

No. of atoms: 4 Bulk z = 1.227 Å

Region	Chem el.	At. no.	Cell type	Site occ.	Rel. to	Dx ± εx		Dy ± εy		Dz ± εz		Dz/Bz(%) ± εz/Bz
epir		-2					f		f		Å	
subr		-1				0.003	Å	-2.588	Å	1.227	Å	
intf	Al	1	b	1.00	0	0.000	f	0.000	f	0.000	Å	0.0
intf	Al	2	b	1.00	1	0.727 ± .023	f	0.454 ± .011	f	1.068 ± .010	Å	87.0 ± .8
intf	Al	3	b	1.00	2	-0.273	f	0.455	f	1.335 ± .020	Å	108.8 ± 1.6
subl	Al	4	b	1.00	3	-0.272	f	-0.545	f	1.227	Å	100.0 ± 8.2

BOND DISTANCES AND ANGLES

Bond distances and angles are derived from coordinates

No. of distances/angles: 14

Interatomic dist. A-B (Å)	Atom A	Atom B	Atom C	Bond angle A-B-C (°)
2.860	Al1	Al1(1,0)	Al2	59.3
2.916	Al2	Al3	Al1(0,1)	88.4
2.916	Al2	Al3	Al3(1,0)	60.6
2.830	Al3	Al1(1,1)	Al3(1,0)	60.7
2.912	Al3	Al2(0,1)	Al3(0,1)	116.7
2.916	Al3	Al2	Al3(1,0)	58.8

Al(311)-(1x1)
13.30

Bond Distances and Angles - Continued

Interatomic dist. A-B (Å)	Atom A	Atom B	Atom C	Bond angle A-B-C (°)
2.860	Al1	Al1(1,0)	Al3(0,-1)	59.6
2.798	Al1	Al2	Al1(1,0)	61.5
2.798	Al1	Al2	Al3	119.7
2.799	Al1	Al2(-1,0)	Al1(0,1)	124.5
2.799	Al1	Al2(-1,0)	Al3	119.7
2.804	Al2	Al1(1,1)	Al3	62.3
2.799	Al2	Al1(1,0)	Al3(0,-1)	62.3
2.914	Al2	Al3(1,0)	Al2(1,0)	58.8

COMMON NAME : Al(331)-(1x1) ILLUSTRATION: 10
CLASSIFICATION : 13.31
TECHNIQUE : LEED
AUTHORS : D.L. Adams and C.S. Sorensen
REFERENCE : Surf. Sci., 166, 495 (1986)

SURFACE TYPE
Substrate : Al
Crystal face: 331
Temperature : 115 K
Bulk lattice: fcc
2D bulk symm: cm
2D surf symm: cm
Adsorbate:
Coverage :
Pattern : (1x1)
Matrix : (1.000, 0.000)
(0.000, 1.000)

STRUCTURE TYPE
Bulk termination with multilayer relaxations perpendicular to surface (in 4 layers) and parallel to surface (in top layer) along symmetry plane

SAMPLE PREPARATION (1 sample)
Treatment : Ar sputturing at 3k eV, 8E-5torr, 15μA, annealing at 800 K
Crystallinity: surface <0.2° from (331) plane
Anal. methods:
Contamination: AES: trace amount of C

COMMENTS

DATA COLLECTION
Technique: LEED
Dataset : I-V curves for 15 symmetry-inequivalent beams; E range 50-300 eV

THEORY/DATA TREATMENT
Dynamical LEED (layer doubling): 10 phase shifts (Moruzzi et al pot); Vor=-10.3±1.6 eV, Voi=-4.4±1.4eV;ΘD=500±150 K

STRUCTURES EXAMINED
Variation of first 6 interlayer spacings and first 5 layer registries (and of non-structural parameters)

QUALITY OF EXPERIMENT-THEORY FIT
R2=0.065

2D UNIT CELLS (1 domain observed)

Cell	Ax (Å)	Ay (Å)	Bx (Å)	By (Å)	α (°)	Matrix	Pattern	Cell type
Bulk	2.853	0.000	1.427	6.219	77.1	(1.000, 0.000) (0.000, 1.000)	(1x1)	b: bulk lattice
Surface 1	2.853	0.000	1.427	6.219	77.1	(1.000, 0.000) (0.000, 1.000)	(1x1)	s1: commens. superlattice

3D COORDINATES

Dx/Dy in Å, or as a fraction of layer's unit cell vectors; atom 0 at the origin. Epir/subr are bulk repeat vectors.

No. of atoms: 7 Bulk z = .926 Å

Region	Chem el.	At. no.	Cell type	Site occ.	Rel. to	Dx ± εx		Dy ± εy		Dz ± εz		Dz/Bz(%) ± εz/Bz
epir		-2					f		f		Å	
subr		-1				1.427	Å	2.292	Å	0.926	Å	
intf	Al	1	b	1.00	0	0.000	f	0.000	f	0.000	Å	0.0
intf	Al	2	b	1.00	1	0.321	f	0.358	f	0.817 ± .021	Å	88.2 ± 2.3
intf	Al	3	b	1.00	2	0.316	f	0.368	f	0.888 ± .028	Å	95.9 ± 3.0
intf	Al	4	b	1.00	3	0.316	f	-0.632	f	1.022 ± .025	Å	110.4 ± 2.7
intf	Al	5	b	1.00	4	-0.684	f	0.368	f	0.881 ± .038	Å	95.1 ± 4.1
intf	Al	6	b	1.00	5	0.316	f	0.368	f	0.903 ± .049	Å	97.5 ± 5.3
subl	Al	7	b	1.00	6	0.316	f	-0.632	f	0.929 ± .051	Å	100.3 ± 5.5

BOND DISTANCES AND ANGLES

Bond distances and angles are derived from coordinates

No. of distances/angles: 8

Interatomic dist. A-B (Å)	Atom A	Atom B	Atom C	Bond angle A-B-C (°)
2.768	Al1	Al2	Al2(-1,0)	59.0
2.768	Al1	Al2	Al3(-1,0)	118.8
2.768	Al1	Al2	Al4(-1,0)	59.0
2.768	Al1	Al2(-1,0)	Al1(-1,0)	62.1
2.799	Al1	Al3(0,-1)	Al4(-1,0)	58.8

Al(331)-(1x1)
13.31

Bond Distances and Angles - Continued

Interatomic dist. A-B (Å)	Atom A	Atom B	Atom C	Bond angle A-B-C (°)
2.841	Al2	Al3	Al4(-1,1)	120.2
2.841	Al2	Al3	Al5	60.1
2.887	Al3	Al4(0,1)		

COMMON NAME : Al(100)-c(2x2)-Na
CLASSIFICATION : 13.11.1
TECHNIQUE : LEED
AUTHORS : B.A. Hutchins, T.N. Rhodin and J.E. Demuth
REFERENCE : Surf. Sci., 54, 419 (1976)

ILLUSTRATION: 28,29

SURFACE TYPE
Substrate : Al
Crystal face: 100
Temperature : <RT
Bulk lattice: fcc
2D bulk symm: p4m
2D surf symm: p4m

Adsorbate: Na
Coverage : 0.25 Na/Al
Pattern : c(2x2)
Matrix : (1.000, 1.000)
(-1.000, 1.000)

STRUCTURE TYPE
Atomic adsorption in hollow site

SAMPLE PREPARATION (sample)
Treatment : electropolished; Na vapor source, B.P. 1E-10 torr;
Crystallinity: Xe ion bombardment; anneal temp. 850K
Anal. methods:
Contamination: LEED/Auger: <5% O contamination

COMMENTS

DATA COLLECTION
Technique: LEED
Dataset : I-V curves for (00) beam at Θ=4.4,7,12° and for (-1,0) and (-1,-1) beams at Θ=0°

THEORY/DATA TREATMENT
Dynamical LEED: 8 ph sh (Snow pot for Al, metallic pot for Na), Voi=5.5 eV(Al), 3eV(Na),ΘD=339 K(blk)150K(Na)170K(Al)

STRUCTURES EXAMINED
Substrate undistorted variation of adsorbate site (top, bridge, hollow) and height (1.4-3.4Å)

QUALITY OF EXPERIMENT-THEORY FIT
Visual

2D UNIT CELLS (1 domain observed)

Cell	Ax (Å)	Ay (Å)	Bx (Å)	By (Å)	α (°)	Matrix	Pattern	Cell type
Bulk	2.864	0.000	0.000	2.864	90.0	(1.000, 0.000) (0.000, 1.000)	(1x1)	b: bulk lattice
Surface 1	2.864	2.864	-2.864	2.864	90.0	(1.000, 1.000) (-1.000, 1.000)	c(2x2)	s1: commens. superlattice

3D COORDINATES

Na1: overlayer in 4-fold hollow sites

Dx/Dy in Å, or as a fraction of layer's unit cell vectors; atom 0 at the origin. Epir/subr are bulk repeat vectors.

No. of atoms: 3

Bulk z = 2.020 Å

Reg ion	Chem el.	At. no.	Cell type	Site occ.	Rel. to	Dx ± εx	Dy ± εy	Dz ± εz	Dz/Bz(%) ± εz/Bz
epir		-2				f	f	Å	
subr		-1				1.432 Å	1.432 Å	2.020 Å	
ovrl	Na	1	s1	.50	0	0.000 f	0.000 f	0.000 Å	0.0
intf	Al	2	b	1.00	1	0.500 f	0.500 f	2.050 ± .100 Å	101.5 ± 5.0
subl	Al	3	b	1.00	2	-0.500 f	-0.500 f	2.020 ± .050 Å	100.0 ± 2.5

BOND DISTANCES AND ANGLES

Bond distances and angles are derived from coordinates

No. of distances/angles: 7

Interatomic dist. A-B (Å)	Atom A	Atom B	Atom C	Bond angle A-B-C (°)
2.882	Na1	Al2	Na1(1,0)	89.3
2.882	Na1	Al2	Al2(1,0)	119.8
2.882	Na1	Al2	Al3(1,0)	120.2
2.882	Na1	Al2	Al3	90.3
2.864	Al2	Al2(1,0)	Na1(1,0)	60.2

Al(100)-c(2x2)-Na
13.11.1

Bond Distances and Angles - Continued

Interatomic dist. A-B (Å)	Atom A	Atom B	Atom C	Bond angle A-B-C (°)
2.864	Al2	Al2(1,0)	Al3(2,0)	120.0
2.860	Al2	Al3(1,1)	Al2(1,0)	60.1

COMMON NAME : Al(100)-c(2x2)-Na
CLASSIFICATION : 13.11.2
TECHNIQUE : LEED
AUTHORS : M.A. Van Hove, S.Y. Tong and N. Stoner
REFERENCE : Surf. Sci., 54, 259 (1976)

ILLUSTRATION: 28,29

SURFACE TYPE
Substrate : Al
Crystal face: 100
Temperature : 110 K
Bulk lattice: fcc
2D bulk symm: p4m
2D surf symm: p4m

Adsorbate: Na
Coverage : 0.5 (Na/Al)
Pattern : c(2x2)
Matrix : (1.000, 1.000)
(-1.000, 1.000)

STRUCTURE TYPE
Atomic adsorption in 4-fold hollow sites

SAMPLE PREPARATION (sample)
Treatment : see B.M. Hutchins, T.N. Rhodin and J.E. Demuth, Surf. Sci.
Crystallinity: 45, 419 (1976)
Anal. methods:
Contamination:

COMMENTS

DATA COLLECTION
Technique: LEED
Dataset : I-V curves: (-10), (-1-1), (0.5,0.5) beams at Θ=0,φ=0; 00 beam at Θ=12,φ=0); cumulative E range: 530 eV

THEORY/DATA TREATMENT
Dynamical LEED (RFS); Snow pot for Al, superpos pot for Na; 8 phase shifts; Vor=-5 eV, Voi=-4.1eV, ΘD=356 K(Al),284K(Na)

STRUCTURES EXAMINED
Top site, bridge site, hollow site; Na-Al layer spacing varied from 1.7 to 2.9Å

QUALITY OF EXPERIMENT-THEORY FIT
Visual

2D UNIT CELLS (1 domain observed)

Cell	Ax (Å)	Ay (Å)	Bx (Å)	By (Å)	α (°)	Matrix	Pattern	Cell type
Bulk	2.860	0.000	0.000	2.860	90.0	(1.000, 0.000) (0.000, 1.000)	(1x1)	b: bulk lattice
Surface 1	2.860	2.860	-2.860	2.860	90.0	(1.000, 1.000) (-1.000, 1.000)	c(2x2)	s1: commens. superlattice

3D COORDINATES

Na1: overlayer in 4-fold hollow site

Dx/Dy in Å, or as a fraction of layer's unit cell vectors; atom 0 at the origin. Epir/subr are bulk repeat vectors.

No. of atoms: 3

Bulk z = 2.025 Å

Reg ion	Chem el.	At. no.	Cell type	Site occ.	Rel. to	Dx ± εx		Dy ± εy		Dz ± εz		Dz/Bz(%) ± εz/Bz
epir		-2					f		f		Å	
subr		-1				1.430	Å	1.430	Å	2.025	Å	
ovrl	Na	1	s1	.50	0	0.000	f	0.000	f	0.000	Å	0.0
intf	Al	2	b	1.00	1	0.500	f	0.500	f	2.080 ± .120	Å	102.7 ± 5.9
subl	Al	3	b	1.00	2	-0.500	f	-0.500	f	2.025 ± .100	Å	100.0 ± 4.9

BOND DISTANCES AND ANGLES

Bond distances and angles are derived from coordinates

No. of distances/angles: 7

Interatomic dist. A-B (Å)	Atom A	Atom B	Atom C	Bond angle A-B-C (°)
2.901	Na1	Al2	Na1(1,0)	88.4
2.901	Na1	Al2	Al2(1,0)	119.5
2.901	Na1	Al2	Al3(1,0)	120.5
2.901	Na1	Al2	Al3	90.8
2.860	Al2	Al2(1,0)	Na1(1,0)	60.5

Al(100)-c(2x2)-Na
13.11.2

Bond Distances and Angles - Continued

Interatomic dist. A-B (Å)	Atom A	Atom B	Atom C	Bond angle A-B-C (°)
2.860	Al2	Al2(1,0)	Al3(2,0)	120.0
2.862	Al2	Al3(1,1)	Al2(1,0)	60.0

COMMON NAME : Al(111)-(√3x√3)R30°-Na
CLASSIFICATION : 13.11.3
TECHNIQUE : SEXAFS
AUTHORS : A. Schmalz, S. Aminpirooz, L. Becker, J.Haase, J.Neugebauer, M. Scheffler, D.R. Batchelor, D.L. Adams and E
REFERENCE : Phys. Rev. Lett., 67, 2163 (1991)

ILLUSTRATION: 27

SURFACE TYPE
Substrate : Al
Crystal face: 111
Temperature : RT
Bulk lattice: fcc
2D bulk symm: p3m1
2D surf symm: p31m

Adsorbate: Na
Coverage : 0.16-0.33 ML
Pattern : (√3x√3)R30°
Matrix : (1.000, 1.000)
(-1.000, 2.000)

STRUCTURE TYPE
Na in a sixfold substitutional site 1.67A above Al surface

SAMPLE PREPARATION (1 sample)
Treatment : ion bombardment and annealing; Na from SAES getter sources
Crystallinity:
Anal. methods: LEED, AES, X-ray adsorption
Contamination: O < 0.005ML

COMMENTS
Measured at 0.16ML with the same structural result
total energy calculations also done

DATA COLLECTION
Technique: SEXAFS
Dataset : X-ray adsorption from 0-8Å-1
two polarizations

THEORY/DATA TREATMENT
Fourier-transform method (Excurve program)

STRUCTURES EXAMINED
Threefold hollow site, sixfold substitutional site

QUALITY OF EXPERIMENT-THEORY FIT
Visual

2D UNIT CELLS (1 domain observed)

Cell	Ax (Å)	Ay (Å)	Bx (Å)	By (Å)	α (°)	Matrix	Pattern	Cell type
Bulk	2.864	0.000	1.432	2.480	60.0	(1.000, 0.000) (0.000, 1.000)	(1x1)	b: bulk lattice
Surface 1	4.295	2.480	0.000	4.960	60.0	(1.000, 1.000) (-1.000, 2.000)	(√3x√3)R30°	s1: commens. superlattice

3D COORDINATES

Na1: in a 6-fold substitutional site Na1-Al2 bond determined within ±0.03A

Dx/Dy in Å, or as a fraction of layer's unit cell vectors; atom 0 at the origin. Epir/subr are bulk repeat vectors.

No. of atoms: 4

Bulk z = 2.338 Å

Region	Chem el.	At. no.	Cell type	Site occ.	Rel. to	Dx ± εx		Dy ± εy		Dz ± εz		Dz/Bz(%) ± εz/Bz
epir		-2					f		f		Å	
subr		-1				1.432	Å	0.827	Å	2.338	Å	
ovrl	Na	1	s1	.33	0	0.000	Å	0.000	Å	0.000	Å	0.0
intf	Al	2	s1	.33	0	1.432	Å	2.480	Å	1.670 ± .030	Å	71.4 ± 1.3
intf	Al	3	s1	.33	0	2.864	Å	4.960	Å	1.670 ± .030	Å	71.4 ± 1.3
subl	Al	4	b	1.00	0	1.432	Å	0.827	Å	4.008	Å	171.4

BOND DISTANCES AND ANGLES

No. of distances/angles: 1

Interatomic dist. A-B (Å)	Atom A	Atom B	Atom C	Bond angle A-B-C (°)
3.315	Na1	Al2		

COMMON NAME : Al(111)-O
CLASSIFICATION : 13.8.8
TECHNIQUE : SEXAFS
AUTHORS : R.Z. Bachrach, G.V. Hansson and R.S. Bauer
REFERENCE : Surf. Sci., 109, L560 (1981)

ILLUSTRATION: 22,23

SURFACE TYPE
Substrate : Al
Crystal face: 111
Temperature : RT*
Bulk lattice: fcc
2D bulk symm: p3m1
2D surf symm: none

Adsorbate: O
Coverage : 125L
Pattern : unspecified
Matrix : (1.000, 0.000)
(0.000, 1.000)

STRUCTURE TYPE
Atomic O in 3-fold hollow sites (which undetermined); ordering undetermined, here tabulated as (1x1); unrelaxed bulk

SAMPLE PREPARATION (1 sample)
Treatment : see Flodstrom et al, Phys. Rev. Lett. 40, 907 (1978)
Crystallinity:
Anal. methods:
Contamination:

COMMENTS
Authors find pressure dependent oxidation state: mol. oxygen chemisorbs for p<2E-7 torr, and atomic oxygen chemisorbs for p>1E-6 torr

DATA COLLECTION
Technique: SEXAFS
Dataset : oxygen K-edge EXAFS for 400-1000 eV above threshold; angle of incidence of p-polarised light was 85°

THEORY/DATA TREATMENT
Standard EXAFS Fourier transform

STRUCTURES EXAMINED
Those consistent with the derived O-Al bond length and the assumption of a 3-fold hollow site

QUALITY OF EXPERIMENT-THEORY FIT
Visual

2D UNIT CELLS (1 domain observed)

Cell	Ax (Å)	Ay (Å)	Bx (Å)	By (Å)	α (°)	Matrix	Pattern	Cell type
Bulk	2.860	0.000	-1.430	2.477	120.0	(1.000, 0.000) (0.000, 1.000)	(1x1)	b: bulk lattice
Surface 1	2.860	0.000	-1.430	2.477	120.0	(1.000, 0.000) (0.000, 1.000)	unspecified	s1: commens. superlattice

3D COORDINATES

O1: overlayer in one or the other 3-fold hollow site, here tabulated as fcc site; coverage assumed 1; coordinates are derived from O-Al bond length; error bar on O-Al bond length is 0.05Å

Dx/Dy in Å, or as a fraction of layer's unit cell vectors; atom 0 at the origin. Epir/subr are bulk repeat vectors.

No. of atoms: 3 Bulk z = 2.340 Å

Region	Chem el.	At. no.	Cell type	Site occ.	Rel. to	Dx ± εx		Dy ± εy		Dz ± εz		Dz/Bz(%) ± εz/Bz
epir		-2					f		f		Å	
subr		-1				0.000	Å	-1.651	Å	2.340	Å	
ovrl	O	1	b	1.00	0	0.000	f	0.000	f	0.000	Å	0.0
intf	Al	2	b	1.00	1	0.667	f	0.333	f	0.980	Å	41.9
subl	Al	3	b	1.00	2	-0.333	f	0.333	f	2.340	Å	100.0

BOND DISTANCES AND ANGLES

No. of distances/angles: 5

Interatomic dist. A-B (Å)	Atom A	Atom B	Atom C	Bond angle A-B-C (°)
1.920	O1	Al2	O1(1,0)	96.3
1.920	O1	Al2	Al2(1,0)	138.1
1.920	O1	Al2	Al3(1,0)	155.9
1.920	O1	Al2	Al3	99.7
2.860	Al2	Al2(1,1)	Al3(1,0)	60.1

COMMON NAME : Al(111)-(1x1)-O
CLASSIFICATION : 13.8.12
TECHNIQUE : LEED
AUTHORS : J. Neve, J. Rundgren and P. Westrin
REFERENCE : J. Phys., C15, 4391 (1982)

ILLUSTRATION: 22,23

SURFACE TYPE
Substrate : Al
Crystal face: 111
Temperature : 100 K
Bulk lattice: fcc
2D bulk symm: p3m1
2D surf symm: p3m1

Adsorbate: O
Coverage : 1.0 O/Al
Pattern : (1x1)
Matrix : (1.000, 0.000)
(0.000, 1.000)

STRUCTURE TYPE
Atomic oxygen in fcc hollow sites;
O-Al spacing temperature dependent:
0.7 and 1.3 Å at 100K and 293K

SAMPLE PREPARATION (1 sample)
Treatment :
Crystallinity:
Anal. methods:
Contamination:

COMMENTS
Conclusions based on visual comparison of (10) and (01) beams only;
different oxygen muffin-tin spheres used for calculations for 0.7 and 1.3 Å spacings

DATA COLLECTION
Technique: LEED
Dataset : LEED data of Martinsson et al, Surf. Sci. 89, 102 (1979)

THEORY/DATA TREATMENT
Dynamical LEED: DCV method to generate pots with E-dep. excited state pot, based on Moruzzi-Janak-Williams pot

STRUCTURES EXAMINED
O-Al spacings of 0.7 and 1.3 Å at 100K and 293K; O over hollow sites

2D UNIT CELLS (1 domain observed)

Cell	Ax (Å)	Ay (Å)	Bx (Å)	By (Å)	α (°)	Matrix	Pattern	Cell type
Bulk	2.860	0.000	-1.430	2.477	120.0	(1.000, 0.000) (0.000, 1.000)	(1x1)	b: bulk lattice
Surface 1	2.860	0.000	-1.430	2.477	120.0	(1.000, 0.000) (0.000, 1.000)	(1x1)	s1: commens. superlattice

3D COORDINATES

O1: overlayer in fcc hollows

Dx/Dy in Å, or as a fraction of layer's unit cell vectors; atom 0 at the origin. Epir/subr are bulk repeat vectors.

No. of atoms: 3

Bulk z = 2.340 Å

Reg ion	Chem el.	At. no.	Cell type	Site occ.	Rel. to	Dx ± εx		Dy ± εy		Dz ± εz		Dz/Bz(%) ± εz/Bz
epir		-2					f		f		Å	
subr		-1				1.430	Å	0.826	Å	2.340	Å	
ovrl	O	1	b	1.00	0	0.000	f	0.000	f	0.000	Å	0.0
intf	Al	2	b	1.00	1	0.667	f	0.333	f	0.700 ± .100	Å	29.9 ± 4.3
subl	Al	3	b	1.00	2	-0.333	f	0.333	f	2.340	Å	100.0

BOND DISTANCES AND ANGLES

Bond distances and angles are derived from coordinates

No. of distances/angles: 7

Interatomic dist. A-B (Å)	Atom A	Atom B	Atom C	Bond angle A-B-C (°)
1.794	O1	Al2	O1(1,1)	105.8
1.794	O1	Al2	Al2(1,1)	142.9
1.794	O1	Al2	Al2(-1,0)	37.1
1.794	O1	Al2	Al3(1,0)	148.2
1.794	O1	Al2	Al3	93.1
1.794	Al2	O1(1,1)	Al2(1,1)	105.8
2.860	Al2	Al2(1,1)	O1(1,1)	37.1

COMMON NAME : Al(111)-(1x1)-O ILLUSTRATION: 22,23
CLASSIFICATION : 13.8.15
TECHNIQUE : LEED
AUTHORS : V. Martinez, F. Soria, M.C. Munoz and J.L. Sacedon
REFERENCE : Surf. Sci., 128, 424 (1983)

SURFACE TYPE
Substrate : Al
Crystal face: 111
Temperature : 300 K
Bulk lattice: fcc
2D bulk symm: p3m1
2D surf symm: p3m1

Adsorbate: O
Coverage : 1.0 O/Al
Pattern : (1x1)
Matrix : (1.000, 0.000)
(0.000, 1.000)

STRUCTURE TYPE
Atomic overlayer in fcc 3-fold hollow sites; unrelaxed bulk termination

SAMPLE PREPARATION (1 sample)
Treatment : clean Al(111) grown on mica, then exposed to O2
Crystallinity:
Anal. methods:
Contamination: AES and LEED monitoring of coverage

COMMENTS
Measurements and calculations done for 3 O exposures; fit is weighted average, giving: 70% O in fcc sites, 10% O in tetrahedral underlayer sites, 20% clean Al(111)

DATA COLLECTION
Technique: LEED
Dataset : I-V data: (01) beam at Θ=0;
(10),(-10),(00) and (0-1) at Θ=5, φ=18°;
(00) and (-10) at Θ=12, φ=18°; E<=180 eV

THEORY/DATA TREATMENT
Dynamical LEED (CAVLEED package)

STRUCTURES EXAMINED
Substrate-adsorbate layer spacings between 0.6 and 0.9Å (see comments)

QUALITY OF EXPERIMENT-THEORY FIT
RZJ=0.18

2D UNIT CELLS (1 domain observed)

Cell	Ax (Å)	Ay (Å)	Bx (Å)	By (Å)	α (°)	Matrix	Pattern	Cell type
Bulk	2.864	0.000	-1.432	2.480	120.0	(1.000, 0.000) (0.000, 1.000)	(1x1)	b: bulk lattice
Surface 1	2.864	0.000	-1.432	2.480	120.0	(1.000, 0.000) (0.000, 1.000)	(1x1)	s1: commens. superlattice

3D COORDINATES

O1: overlayer in fcc hollow sites

Dx/Dy in Å, or as a fraction of layer's unit cell vectors; atom 0 at the origin. Epir/subr are bulk repeat vectors.

No. of atoms: 4 Bulk z = 2.338 Å

Reg ion	Chem el.	At. no.	Cell type	Site occ.	Rel. to	Dx ± εx		Dy ± εy		Dz ± εz		Dz/Bz(%) ± εz/Bz
epir		-2					f		f		Å	
subr		-1				-1.432	Å	-0.827	Å	2.338	Å	
ovrl	O	1	b	1.00	0	0.000	f	0.000	f	0.000	Å	0.0
intf	Al	2	b	1.00	1	0.333	f	0.667	f	0.700 ± .080	Å	29.9 ± 3.4
intf	Al	3	b	1.00	2	0.333	f	-0.333	f	2.338	Å	100.0
subl	Al	4	b	1.00	3	-0.667	f	-0.333	f	2.338	Å	100.0

BOND DISTANCES AND ANGLES

Bond distances and angles are derived from coordinates

No. of distances/angles: 4

Interatomic dist. A-B (Å)	Atom A	Atom B	Atom C	Bond angle A-B-C (°)
1.796	O1	Al2	O1(0,1)	105.8
1.796	O1	Al2	Al2(0,1)	142.9
1.796	O1	Al2	Al3(0,1)	148.2
1.796	O1	Al2	Al3	93.0

COMMON NAME : Al(111)-(1x1)-O
CLASSIFICATION : 13.8.6a
TECHNIQUE : SEXAFS
AUTHORS : D. Norman, S. Brennan, R. Jaeger and J. Stohr
REFERENCE : Surf. Sci., 105, L297 (1981)

ILLUSTRATION: 22,23

SURFACE TYPE
Substrate : Al
Crystal face: 111
Temperature : 300 K
Bulk lattice: fcc
2D bulk symm: p3m1
2D surf symm: p3m1

Adsorbate: O
Coverage : 50L O2
Pattern : (1x1)
Matrix : (1.000, 0.000)
(0.000, 1.000)

STRUCTURE TYPE
Atomic oxygen overlayer in 3-fold hollow sites (which undetermined)

SAMPLE PREPARATION (1 sample)
Treatment : exposure to 50L O2
Crystallinity:
Anal. methods:
Contamination:

COMMENTS
Bulk lattice constant of 2.86 Å was assumed; however, the measured O-O separation of 2.90±0.05 Å is consistent with this assumption

DATA COLLECTION
Technique: SEXAFS; SEXAFS at SSRL
Dataset : SEXAFS signal of O K-edge up to 900 eV above edge

THEORY/DATA TREATMENT
Fourier transform; phase shifts constructed from analysis of bulk Al2O3 where the bond length is known

STRUCTURES EXAMINED
Those consistent with the inferred SEXAFS O-Al bond length of 1.75±0.03Å, giving an interlayer spacing of 0.6±0.1Å

QUALITY OF EXPERIMENT-THEORY FIT
Visual

2D UNIT CELLS (1 domain observed)

Cell	Ax (Å)	Ay (Å)	Bx (Å)	By (Å)	α (°)	Matrix	Pattern	Cell type
Bulk	2.860	0.000	-1.430	2.477	120.0	(1.000, 0.000) (0.000, 1.000)	(1x1)	b: bulk lattice
Surface 1	2.860	0.000	-1.430	2.477	120.0	(1.000, 0.000) (0.000, 1.000)	(1x1)	s1: commens. superlattice

3D COORDINATES

coordinates are derived from O-Al bond length; error bar on O-Al bond length is 0.03Å
O coverage assumed 1

Dx/Dy in Å, or as a fraction of layer's unit cell vectors; atom 0 at the origin. Epir/subr are bulk repeat vectors.

No. of atoms: 4

Bulk z = 2.340 Å

Region	Chem el.	At. no.	Cell type	Site occ.	Rel. to	Dx ± εx		Dy ± εy		Dz ± εz		Dz/Bz(%) ± εz/Bz
epir		-2					f		f		Å	
subr		-1				-1.430	Å	-0.826	Å	2.340	Å	
ovrl	O	1	b	1.00	0	0.000	f	0.000	f	0.000	Å	0.0
intf	Al	2	b	1.00	1	0.333	f	0.667	f	0.600 ± .100	Å	25.6 ± 4.3
intf	Al	3	b	1.00	2	0.333	f	-0.333	f	2.340	Å	100.0
subl	Al	4	b	1.00	3	-0.667	f	-0.333	f	2.340	Å	100.0

BOND DISTANCES AND ANGLES

No. of distances/angles: 5

Interatomic dist. A-B (Å)	Atom A	Atom B	Atom C	Bond angle A-B-C (°)
1.757	O1	Al2	O1(0,1)	109.0
1.757	O1	Al2	Al2(1,1)	144.5
1.757	O1	Al2	Al3(0,1)	145.2
1.757	O1	Al2	Al3	90.5
2.860	Al2	Al2(1,1)	Al3(0,1)	60.1

COMMON NAME : Al(111)-(1x1)-O
CLASSIFICATION : 13.8.6b
TECHNIQUE : SEXAFS
AUTHORS : D. Norman, S. Brennan, R. Jaeger and J. Stohr
REFERENCE : Surf. Sci., 105, L297 (1981)

ILLUSTRATION: 26

SURFACE TYPE
Substrate : Al
Crystal face: 111
Temperature : RT*
Bulk lattice: fcc
2D bulk symm: p3m1
2D surf symm: p3m1

Adsorbate: O
Coverage : 50L O2
Pattern : (1x1)
Matrix : (1.000, 0.000)
(0.000, 1.000)

STRUCTURE TYPE
Atomic oxygen in tetrahedral sites below top Al layer, forming oxide-like surface;
Al possibly unrelaxed

SAMPLE PREPARATION (1 sample)
Treatment : exposure to 50L of O2, then heating to 473 K for 10 mins
Crystallinity:
Anal. methods:
Contamination:

COMMENTS

bulk lattice constant of 2.86Å was assumed; O placed below top Al on the evidence that: 1. sharp (1x1) LEED pattern observed up to 1000L of O2; 2. the site can be simultaneously occupied by oxide-like and chemisorbed O

DATA COLLECTION
Technique: SEXAFS; SEXAFS at SSRL
Dataset : SEXAFS signal for oxygen K-edge up to 900 eV above edge

THEORY/DATA TREATMENT
Fourier transform; phase shifts constructed from analysis of bulk Al2O3 where the bond length is known

STRUCTURES EXAMINED
Those consistent with the inferred SEXAFS O-Al bond length of 1.75±0.03Å, giving an interlayer spacing of 0.6±0.1Å

QUALITY OF EXPERIMENT-THEORY FIT
Visual

2D UNIT CELLS (1 domain observed)

Cell	Ax (Å)	Ay (Å)	Bx (Å)	By (Å)	α (°)	Matrix	Pattern	Cell type
Bulk	2.860	0.000	-1.430	2.477	120.0	(1.000, 0.000) (0.000, 1.000)	(1x1)	b: bulk lattice
Surface 1	2.860	0.000	-1.430	2.477	120.0	(1.000, 0.000) (0.000, 1.000)	(1x1)	s1: commens. superlattice

3D COORDINATES

coordinates are derived from O-Al bond length; error bar on O-Al bond length is 0.03Å;
O coverage assumed 1

Dx/Dy in Å, or as a fraction of layer's unit cell vectors; atom 0 at the origin. Epir/subr are bulk repeat vectors.

No. of atoms: 5 Bulk z = 2.340 Å

Reg ion	Chem el.	At. no.	Cell type	Site occ.	Rel. to	Dx ± εx		Dy ± εy		Dz ± εz		Dz/Bz(%) ± εz/Bz
epir		-2					f		f		Å	
subr		-1				0.000	Å	-1.651	Å	2.340	Å	
intf	Al	1	b	1.00	0	0.000	f	0.000	f	0.000	Å	0.0
intf	O	2	b	1.00	1	0.667	f	0.333	f	0.600 ± .100	Å	25.6 ± 4.3
intf	Al	3	b	1.00	2	0.000	f	0.000	f	1.740 ± .100	Å	74.4 ± 4.3
intf	Al	4	b	1.00	3	-0.333	f	0.333	f	2.340	Å	100.0
subl	Al	5	b	1.00	4	-0.333	f	-0.667	f	2.340	Å	100.0

BOND DISTANCES AND ANGLES

No. of distances/angles: 4

Interatomic dist. A-B (Å)	Atom A	Atom B	Atom C	Bond angle A-B-C (°)
2.860	Al1	Al1(1,1)	O2	35.5
2.860	Al1	Al1(1,1)	Al3	60.1
1.757	Al1	O2	Al1(1,0)	109.0
1.757	Al1	O2	Al3	110.0

COMMON NAME : AlP(110)-(1x1)
CLASSIFICATION : 13.15.2
TECHNIQUE : LEED
AUTHORS : C.B. Duke, A. Paton, A. Kahn and C.R. Bonapace
REFERENCE : Phys. Rev., B28, 852 (1983)

ILLUSTRATION: 116

SURFACE TYPE
Substrate : AlP
Crystal face: 110
Temperature : 300 K
Bulk lattice: zincblende
2D bulk symm: pm
2D surf symm: pm

Adsorbate:
Coverage :
Pattern : (1x1)
Matrix : (1.000, 0.000)
(0.000, 1.000)

STRUCTURE TYPE
Tilted topmost two AlP bilayers

SAMPLE PREPARATION (sample)
Treatment : GaP Ar+ bombarded to remove Ga, then Al deposition, anneal
Crystallinity:
Anal. methods:
Contamination: only high-E Ga AES peaks, so Ga deep

COMMENTS
LEED data taken below ΘD (=590 K), so lattice vibrations not important;
calculated spectra checked for sensitivity to inelastic mfp;

DATA COLLECTION
Technique: LEED
Dataset : I-V spectra for 14 beams up to 210 eV;
data taken at Al coverages of 16 and 75ML:
average compared to theory

THEORY/DATA TREATMENT
Dynamical LEED: 6 layer slab; 6 phase shifts (Hara exchange)
Vor=-15.5 to -13.2 eV (E=30eV-240 eV); mfp=10Å; ΘD=590 K

STRUCTURES EXAMINED
1. unreconstructed surface; 2. bond length conserving reconstructions with vertical relaxations of 1st 2 layers with relative displacements along (100) direction;
3. as in 2. without bond conservation

QUALITY OF EXPERIMENT-THEORY FIT
RZJ=0.19

2D UNIT CELLS (1 domain observed)

Cell	Ax (Å)	Ay (Å)	Bx (Å)	By (Å)	α (°)	Matrix	Pattern	Cell type
Bulk	3.850	0.000	0.000	5.450	90.0	(1.000, 0.000) (0.000, 1.000)	(1x1)	b: bulk lattice
Surface 1	3.850	0.000	0.000	5.450	90.0	(1.000, 0.000) (0.000, 1.000)	(1x1)	s1: commens. superlattice

3D COORDINATES

P1-Al2: top tilted layer; Al3-P4: next layer, slightly tilted in opposite sense;
Al5-P6: bulk planar bilayer; 0.1Å error bars assumed for tabulation

Dx/Dy in Å, or as a fraction of layer's unit cell vectors; atom 0 at the origin. Epir/subr are bulk repeat vectors.

No. of atoms: 6 Bulk z = 1.925 Å

Region	Chem el.	At. no.	Cell type	Site occ.	Rel. to	Dx ± εx	Dy ± εy	Dz ± εz	Dz/Bz(%) ± εz/Bz
epir		-2				f	f	Å	
subr		-1				1.925 Å	2.725 Å	1.925 Å	
intf	P	1	b	1.00	0	0.000 f	0.000 f	0.000 Å	0.0
intf	Al	2	b	1.00	1	0.500 ± .026 f	0.168 ± .018 f	0.630 ± .100 Å	32.7 ± 5.2
intf	Al	3	b	1.00	2	-0.500 ± .026 f	0.543 ± .018 f	1.327 ± .100 Å	68.9 ± 5.2
intf	P	4	b	1.00	3	0.500 ± .026 f	-0.250 ± .018 f	0.070 ± .100 Å	3.6 ± 5.2
subl	Al	5	b	1.00	4	0.000 ± .026 f	-0.250 ± .018 f	1.892 ± .100 Å	98.3 ± 5.2
subl	P	6	b	1.00	5	-0.500 f	0.750 f	0.000 Å	0.0

AlP(110)-(1x1)
13.15.2 BOND DISTANCES AND ANGLES

Bond distances and angles are derived from coordinates

No. of distances/angles: 9

Interatomic dist. A-B (Å)	Atom A	Atom B	Atom C	Bond angle A-B-C (°)
2.223	P1	Al2	P1(1,0)	120.0
2.223	P1	Al2	P4	119.8
2.512	P1	Al3(0,-1)	P4(0,-1)	112.7
2.223	Al2	P1	Al2(-1,0)	120.0
2.122	Al2	P4	Al3	114.5
2.359	Al3	P4	Al2	114.5
2.359	Al3	P4	Al3(1,0)	109.4
2.122	P4	Al2	P1	119.8
2.359	P4	Al3	P4(-1,0)	109.4

COMMON NAME : Au(100)-(1x1)
CLASSIFICATION : 79.8a
TECHNIQUE : LEED
AUTHORS : E. Lang, W. Grimm and K. Heinz
REFERENCE : Surf. Sci., 117, 169 (1982)

ILLUSTRATION: 2

SURFACE TYPE
Substrate : Au
Crystal face: 100
Temperature : 100 K
Bulk lattice: fcc
2D bulk symm: p4m
2D surf symm: p4m

Adsorbate:
Coverage :
Pattern : (1x1)
Matrix : (1.000, 0.000)
(0.000, 1.000)

STRUCTURE TYPE
Metastable unreconstructed surface

SAMPLE PREPARATION (1 sample)
Treatment : Ar+ sputtering followed by annealing above 250 K
Crystallinity:
Anal. methods:
Contamination: AES used to determine surface impurities

COMMENTS
Contraction of first layer uncertain due to shortcomings of theory

DATA COLLECTION
Technique: LEED
Dataset : IV curves at normal incidence for many beams; energy range 200-600 eV

THEORY/DATA TREATMENT
Quasidynamical LEED (no interlayer mult scatt, RFS, additional damping for convergence): ΘD=165 K

STRUCTURES EXAMINED
Variation of topmost interlayer spacing

QUALITY OF EXPERIMENT-THEORY FIT
RZJ=0.147

2D UNIT CELLS (1 domain observed)

Cell	Ax (Å)	Ay (Å)	Bx (Å)	By (Å)	α (°)	Matrix	Pattern	Cell type
Bulk	2.880	0.000	0.000	2.880	90.0	(1.000, 0.000) (0.000, 1.000)	(1x1)	b: bulk lattice
Surface 1	2.880	0.000	0.000	2.880	90.0	(1.000, 0.000) (0.000, 1.000)	(1x1)	s1: commens. superlattice

3D COORDINATES

0.1Å error bar assumed for tabulation

Dx/Dy in Å, or as a fraction of layer's unit cell vectors; atom 0 at the origin. Epir/subr are bulk repeat vectors.

No. of atoms: 3

Bulk z = 2.040 Å

Region	Chem el.	At. no.	Cell type	Site occ.	Rel. to	Dx ± εx		Dy ± εy		Dz ± εz		Dz/Bz(%) ± εz/Bz
epir		-2					f		f		Å	
subr		-1				1.440	Å	1.440	Å	2.040	Å	
intf	Au	1	b	1.00	0	0.000	f	0.000	f	0.000	Å	0.0
intf	Au	2	b	1.00	1	0.500	f	0.500	f	2.040 ± .100	Å	100.0 ± 4.9
subl	Au	3	b	1.00	2	-0.500	f	-0.500	f	2.040	Å	100.0

BOND DISTANCES AND ANGLES

Bond distances and angles are derived from coordinates

No. of distances/angles: 1

Interatomic dist. A-B (Å)	Atom A	Atom B	Atom C	Bond angle A-B-C (°)
2.880	Au1	Au1(1,0)		

COMMON NAME : Au(100)-hex incommensurate ILLUSTRATION: 3
CLASSIFICATION : 79.80
TECHNIQUE : XRD
AUTHORS : M.M. Ocko, D. Gibbs, K.G. Huang, D.M. Zehner and S.G.J. Mochrie
REFERENCE : Phys. Rev., B44, 6429 (1991)

SURFACE TYPE

Substrate : Au
Crystal face: 100
Temperature : 1100 K
Bulk lattice: fcc
2D bulk symm: p4m
2D surf symm: none

Adsorbate:
Coverage :
Pattern : incommensurate
Matrix : (.958, 0.000)
(.479, .829)

STRUCTURE TYPE

Incommensurate hexagonal top layer, with 20% expanded spacing to 2nd layer (due to variable registries); top 4 layers are found corrugated by 0.28, 0.14, 0.06, 0.02Å (= twice maximum excursion): this corrugation is not included in tabulation below, which assumes planar layers

SAMPLE PREPARATION (1 sample)

Treatment : see Gibbs et al, Phys. Rev. B42, 7330 (1990)
Crystallinity:
Anal. methods:
Contamination:

COMMENTS

At T<970 K, hexagonal top layer rotates 0.81° from alignment with substrate, without measurable change in lateral lattice constant, layer spacings or corrugations (but fit vibration amplitudes are reduced);
at T>1170 K, top layer disorders

DATA COLLECTION

Technique: XRD; synchrotron radiation (CHESS, NSLS)
Dataset : rocking curves in vertical scattering geometry with lambda= 1.59, 1.45 and 1.39Å

THEORY/DATA TREATMENT

Spacings, layer density and rms vibr ampl fit to specular data; corrugations fit to superlattice reflections

STRUCTURES EXAMINED

Variation of top 5 layer spacings and top layer density for non-corrugated layers; variation of corrugations in top 5 layers assuming commensurate model with 6 top-layer atoms per coincidence unit cell

QUALITY OF EXPERIMENT-THEORY FIT

Chi**2=0.005 (without corrugations)

2D UNIT CELLS (2 domains observed)

Cell	Ax (Å)	Ay (Å)	Bx (Å)	By (Å)	α (°)	Matrix	Pattern	Cell type
Bulk	2.885	0.000	0.000	2.885	90.0	(1.000, 0.000) (0.000, 1.000)	(1x1)	b: bulk lattice
Surface 1	2.762	0.000	1.381	2.392	60.0	(.958, 0.000) (.479, .829)	incommensurate	i1: incomm. superlattice

3D COORDINATES

Au1: incommensurate top layer (actually corrugated); Au2-Au3: (1x1) layers (actually slightly corrugated)

Dx/Dy in Å, or as a fraction of layer's unit cell vectors; atom 0 at the origin. Epir/subr are bulk repeat vectors.

No. of atoms: 4 Bulk z = 2.040 Å

Region	Chem el.	At. no.	Cell type	Site occ.	Rel. to	Dx ± εx		Dy ± εy		Dz ± εz		Dz/Bz(%) ± εz/Bz
epir		-2					f		f		Å	
subr		-1				1.443	Å	1.443	Å	2.040	Å	
intf	Au	1	i1	1.26	0	0.000	f	0.000	f	0.000	Å	0.0
intf	Au	2	b	1.00	1	0.000	f	0.000	f	2.448 ± .061	Å	120.0 ± 3.0
intf	Au	3	b	1.00	2	0.500	f	0.500	f	2.081 ± .061	Å	102.0 ± 3.0
subl	Au	4	b	1.00	3	0.500	f	0.500	f	2.040	Å	100.0

COMMON NAME : Au(110)-(1x2) ILLUSTRATION: 5
CLASSIFICATION : 79.25
TECHNIQUE : LEED
AUTHORS : W. Moritz and D. Wolf
REFERENCE : Surf. Sci., 163, L655 (1985)

SURFACE TYPE
Substrate : Au
Crystal face: 110
Temperature : RT
Bulk lattice: fcc
2D bulk symm: pmm
2D surf symm: pmm

Adsorbate:
Coverage :
Pattern : (1x2)
Matrix : (1.000, 0.000)
(0.000, 2.000)

STRUCTURE TYPE
Missing-row reconstruction with multilayer relaxation, 2nd row pairing and 3rd row buckling

SAMPLE PREPARATION (1 sample)
Treatment : see W. Moritz and D. Wolf, Surf. Sci. 88, l29 (1979)
Crystallinity:
Anal. methods:
Contamination:

COMMENTS

DATA COLLECTION
Technique: LEED
Dataset : I-V curves at normal incidence for 18 beams, E<=200 eV

THEORY/DATA TREATMENT
Dynamical LEED (matrix inversion, combined-space method): 9 phase shifts; ΘD=170 K

STRUCTURES EXAMINED
Missing-row model with variable relaxations of top three interlayer spacings, row-pairing in second layer and buckling in third layer

QUALITY OF EXPERIMENT-THEORY FIT
RPE=0.35, RZJ=0.25

2D UNIT CELLS (1 domain observed)

Cell	Ax (Å)	Ay (Å)	Bx (Å)	By (Å)	α (°)	Matrix	Pattern	Cell type
Bulk	2.880	0.000	0.000	4.080	90.0	(1.000, 0.000) (0.000, 1.000)	(1x1)	b: bulk lattice
Surface 1	2.880	0.000	0.000	8.160	90.0	(1.000, 0.000) (0.000, 2.000)	(1x2)	s1: commens. superlattice

3D COORDINATES

Au1: remaining row; Au2-Au3: paired 2nd layer;
Au4-Au5: buckled 3rd layer; 0.05Å error bars assumed for tabulation

Dx/Dy in Å, or as a fraction of layer's unit cell vectors; atom 0 at the origin. Epir/subr are bulk repeat vectors.

No. of atoms: 7 Bulk z = 1.440 Å

Region	Chem el.	At. no.	Cell type	Site occ.	Rel. to	Dx ± εx	Dy ± εy	Dz ± εz	Dz/Bz(%) ± εz/Bz
epir		-2				f	f	Å	
subr		-1				-1.440 Å	-2.040 Å	1.440 Å	
intf	Au	1	s1	1.00	0	0.000 f	0.000 f	0.000 Å	0.0
intf	Au	2	s1	1.00	1	0.500 f	0.259 ± .006 f	1.150 ± .050 Å	79.9 ± 3.5
intf	Au	3	s1	1.00	2	0.000 f	0.482 ± .006 f	0.000 Å	0.0
intf	Au	4	s1	1.00	3	-0.500 f	-0.241 ± .006 f	1.350 ± .050 Å	93.8 ± 3.5
intf	Au	5	s1	1.00	4	0.000 f	-0.500 f	0.240 ± .050 Å	16.7 ± 3.5
intf	Au	6	b	1.00	5	0.500 f	0.500 f	1.350 ± .050 Å	93.8 ± 3.5
subl	Au	7	b	1.00	6	-0.500 f	-0.500 f	1.440 Å	100.0

Au(110)-(1x2)
79.25

BOND DISTANCES AND ANGLES

Bond distances and angles are derived from coordinates

No. of distances/angles: 5

Interatomic dist. A-B (Å)	Atom A	Atom B	Atom C	Bond angle A-B-C (°)
2.880	Au1	Au1(1,0)		
2.804	Au1	Au2	Au4	117.7
2.740	Au1	Au5	Au6	118.4
2.786	Au2	Au4	Au6	61.5
3.011	Au2	Au5		

COMMON NAME : Au(110)-(1x2)
CLASSIFICATION : 79.32
TECHNIQUE : MEIS
AUTHORS : M. Copel and T. Gustafsson
REFERENCE : Phys. Rev. Lett., 57, 723 (1986)

ILLUSTRATION: 5

SURFACE TYPE
Substrate : Au
Crystal face: 110
Temperature : RT
Bulk lattice: fcc
2D bulk symm: pmm
2D surf symm: pmm

Adsorbate:
Coverage :
Pattern : (1x2)
Matrix : (1.000, 0.000)
(0.000, 2.000)

STRUCTURE TYPE
Missing-row reconstruction with multilayer relaxation, including 3rd-layer buckling

SAMPLE PREPARATION (1 sample)
Treatment :
Crystallinity: well-defined (1x2) LEED pattern
Anal. methods:
Contamination:

COMMENTS

DATA COLLECTION
Technique: MEIS
Dataset : medium-energy proton scattering in (-1,1,0), (-1,1,1), and (0,0,1) scattering planes: blocking and shadowing curves

THEORY/DATA TREATMENT
Qualitative analysis to support missing-row model; Monte Carlo simulations for coords; Å=170 K(bulk), 130K(surf)

STRUCTURES EXAMINED
Top two interlayer spacings varied; 3rd-layer buckling and 2nd-layer row-pairing considered: row-pairing ruled out

QUALITY OF EXPERIMENT-THEORY FIT
Visual

2D UNIT CELLS (1 domain observed)

Cell	Ax (Å)	Ay (Å)	Bx (Å)	By (Å)	α (°)	Matrix	Pattern	Cell type
Bulk	2.885	0.000	0.000	4.080	90.0	(1.000, 0.000) (0.000, 1.000)	(1x1)	b: bulk lattice
Surface 1	2.885	0.000	0.000	8.160	90.0	(1.000, 0.000) (0.000, 2.000)	(1x2)	s1: commens. superlattice

3D COORDINATES

Au1: remaining row; Pt2-Pt3: bulk-like 2nd layer;
Pt4-Pt5: buckled 3rd layer; 0.05Å/0.1Å perp/lateral error bars assumed for tabulation

Dx/Dy in Å, or as a fraction of layer's unit cell vectors; atom 0 at the origin. Epir/subr are bulk repeat vectors.

No. of atoms: 7 Bulk z = 1.440 Å

Reg ion	Chem el.	At. no.	Cell type	Site occ.	Rel. to	Dx ± εx		Dy ± εy		Dz ± εz		Dz/Bz(%) ± εz/Bz
epir		-2					f		f		Å	
subr		-1				-1.443	Å	-2.040	Å	1.440	Å	
intf	Au	1	s1	.50	0	0.000	f	0.000	f	0.000	Å	0.0
intf	Au	2	s1	.50	1	0.500	f	0.250 ± .012	f	1.180 ± .100	Å	81.9 ± 6.9
intf	Au	3	s1	.50	2	0.000	f	0.500 ± .012	f	0.000	Å	0.0
intf	Au	4	s1	.50	3	-0.500	f	-0.250 ± .012	f	1.399 ± .100	Å	97.2 ± 6.9
intf	Au	5	s1	.50	4	0.000	f	-0.500	f	0.202 ± .100	Å	14.0 ± 6.9
intf	Au	6	b	1.00	5	0.500	f	0.500	f	1.342 ± .100	Å	93.2 ± 6.9
subl	Au	7	b	1.00	6	-0.500	f	-0.500	f	1.440	Å	100.0

BOND DISTANCES AND ANGLES

Bond distances and angles are derived from coordinates

No. of distances/angles: 6

Interatomic dist. A-B (Å)	Atom A	Atom B	Atom C	Bond angle A-B-C (°)
2.885	Au1	Au1(1,0)		
2.763	Au1	Au2	Au3(1,-1)	121.5
2.763	Au1	Au2	Au4	118.1

Au(110)-(1x2)
79.32

Bond Distances and Angles - Continued

Interatomic dist. A-B (Å)	Atom A	Atom B	Atom C	Bond angle A-B-C (°)
2.781	Au1	Au5	Au6	118.2
2.864	Au2	Au4	Au5(0,1)	119.3
2.864	Au2	Au4	Au6	61.0

COMMON NAME : Au(110)-(1x2)
CLASSIFICATION : 79.34
TECHNIQUE : LEIS
AUTHORS : J. Moeller, K.J. Snowdon and W. Heiland
REFERENCE : Surf. Sci., 178, 475 (1986)

ILLUSTRATION: 5

SURFACE TYPE
Substrate : Au
Crystal face: 110
Temperature : RT
Bulk lattice: fcc
2D bulk symm: pmm
2D surf symm: pmm

Adsorbate:
Coverage :
Pattern : (1x2)
Matrix : (1.000, 0.000)
(0.000, 2.000)

STRUCTURE TYPE
Missing-row reconstruction with relaxation of top layer spacing

SAMPLE PREPARATION (1 sample)
Treatment : polishing, sputter etching, and annealing
Crystallinity:
Anal. methods: LEED pattern
Contamination: clean by ISS standards

COMMENTS

DATA COLLECTION
Technique: LEIS
Dataset : large angle (165°), low energy (2000 eV) Ne+/Ne0 backscattering along [1,-1,2] azimuth

THEORY/DATA TREATMENT
Calculation of low-energy ion scattering critical angles using Moliere screening function or universal ZBL potential

STRUCTURES EXAMINED
Top interlayer spacing varied

QUALITY OF EXPERIMENT-THEORY FIT
Visual

2D UNIT CELLS (1 domain observed)

Cell	Ax (Å)	Ay (Å)	Bx (Å)	By (Å)	α (°)	Matrix	Pattern	Cell type
Bulk	2.885	0.000	0.000	4.080	90.0	(1.000, 0.000) (0.000, 1.000)	(1x1)	b: bulk lattice
Surface 1	2.885	0.000	0.000	8.160	90.0	(1.000, 0.000) (0.000, 2.000)	(1x2)	s1: commens. superlattice

3D COORDINATES

Au1: remaining row

Dx/Dy in Å, or as a fraction of layer's unit cell vectors; atom 0 at the origin. Epir/subr are bulk repeat vectors.

No. of atoms: 4   Bulk z = 1.440 Å

Reg ion	Chem el.	At. no.	Cell type	Site occ.	Rel. to	Dx ± εx		Dy ± εy		Dz ± εz		Dz/Bz(%) ± εz/Bz
epir		-2					f		f		Å	
subr		-1				1.443	Å	2.040	Å	1.440	Å	
ovrl	Au	1	s1	.50	0	0.000	f	0.000	f	0.000	Å	0.0
intf	Au	2	b	1.00	1	0.500	f	0.500	f	1.240 ± .070	Å	86.1 ± 4.9
intf	Au	3	b	1.00	2	-0.500	f	-0.500	f	1.440	Å	100.0
subl	Au	4	b	1.00	3	0.500	f	0.500	f	1.440	Å	100.0

BOND DISTANCES AND ANGLES

Bond distances and angles are derived from coordinates

No. of distances/angles: 5

Interatomic dist. A-B (Å)	Atom A	Atom B	Atom C	Bond angle A-B-C (°)
2.885	Au1	Au1(1,0)		
2.789	Au1	Au2	Au3	56.4
2.680	Au1	Au3	Au4	120.0
2.884	Au2	Au3	Au4	59.9
2.884	Au3	Au4		

COMMON NAME : Au(110)-(1x2)
CLASSIFICATION : 79.66a
TECHNIQUE : XRD
AUTHORS : E. Vlieg, I.K. Robinson and K. Kern
REFERENCE : Surf. Sci., 233, 248 (1990)

ILLUSTRATION: 5

SURFACE TYPE
Substrate : Au
Crystal face: 110
Temperature : RT
Bulk lattice: fcc
2D bulk symm: pmm
2D surf symm: pmm

Adsorbate:
Coverage :
Pattern : (1x2)
Matrix : (1.000, 0.000)
(0.000, 2.000)

STRUCTURE TYPE
Missing-row reconstruction, with multilayer relaxations down to 4th layer

SAMPLE PREPARATION (1 sample)
Treatment : Ar sputtering, then 450C anneal
Crystallinity:
Anal. methods: AES
Contamination: AES: no contaminants

COMMENTS
Rms vibration amplitudes of topmost layers are found enhanced by a factor of nearly 2 over bulk values

DATA COLLECTION
Technique: XRD; synchrotron radiation (NSLS)
Dataset : 141 reflections from 23 (hk) rods; lambda=1.09Å

THEORY/DATA TREATMENT
Fitting of parameters by chi**2 minimization, including ΘD for layers 1, 2 and 4

STRUCTURES EXAMINED
Missing-row model with perpendicular relaxations of atoms in layer 1, pairing of layers 2 and 4

QUALITY OF EXPERIMENT-THEORY FIT
Chi**2=2.6

2D UNIT CELLS (1 domain observed)

Cell	Ax (Å)	Ay (Å)	Bx (Å)	By (Å)	α (°)	Matrix	Pattern	Cell type
Bulk	2.885	0.000	0.000	4.080	90.0	(1.000, 0.000) (0.000, 1.000)	(1x1)	b: bulk lattice
Surface 1	2.885	0.000	0.000	8.160	90.0	(1.000, 0.000) (0.000, 2.000)	(1x2)	s1: commens. superlattice

3D COORDINATES

Au1: ridge atoms; Au2-Au3: paired 2nd layer;
Au4-Au5: planar 3rd layer; Au6-Au7: paired 4th layer

Dx/Dy in Å, or as a fraction of layer's unit cell vectors; atom 0 at the origin. Epir/subr are bulk repeat vectors.

No. of atoms: 8

Bulk z = 1.443 Å

Region	Chem el.	At. no.	Cell type	Site occ.	Rel. to	Dx ± εx		Dy ± εy		Dz ± εz		Dz/Bz(%) ± εz/Bz
epir		-2					f		f		Å	
subr		-1				1.443	Å	2.040	Å	1.443	Å	
intf	Au	1	s1	.50	0	0.000	f	0.000	f	0.000	Å	0.0
intf	Au	2	s1	.50	1	0.500	f	-0.256 ± .001	f	1.123 ± .100	Å	77.8 ± 6.9
intf	Au	3	s1	.50	2	0.000	f	0.512 ± .001	f	0.000	Å	0.0
intf	Au	4	s1	.50	3	-0.500	f	-0.756 ± .001	f	1.443	Å	100.0
intf	Au	5	s1	.50	4	0.000	f	0.500	f	0.000	Å	0.0
intf	Au	6	s1	.50	5	0.500	f	-0.256 ± .001	f	1.443	Å	100.0
intf	Au	7	s1	.50	6	0.000	f	0.512 ± .001	f	0.000	Å	0.0
subl	Au	8	b	1.00	7	-0.500	f	-1.512 ± .002	f	1.443	Å	100.0

BOND DISTANCES AND ANGLES

Bond distances and angles are derived from coordinates

No. of distances/angles: 7

Interatomic dist. A-B (Å)	Atom A	Atom B	Atom C	Bond angle A-B-C (°)
2.885	Au1	Au1(1,0)	Au2(1,0)	121.3
2.777	Au1	Au2	Au2(1,0)	121.3
2.777	Au1	Au2	Au4	117.9

Au(110)-(1x2)
79.66a

Bond Distances and Angles - Continued

Interatomic dist. A-B (Å)	Atom A	Atom B	Atom C	Bond angle A-B-C (°)
2.850	Au2	Au4	Au4(1,0)	59.6
2.850	Au2	Au4	Au6(-1,0)	91.4
2.850	Au2	Au4	Au7(0,-1)	119.2
2.850	Au2	Au4	Au8	120.4

COMMON NAME : Au(110)-(1x3)
CLASSIFICATION : 79.55.2
TECHNIQUE : MEIS
AUTHORS : P. Haberle, P. Fenter and T. Gustafsson
REFERENCE : Phys. Rev., B39, 5810 (1989)

ILLUSTRATION: 7

SURFACE TYPE
Substrate : Au
Crystal face: 110
Temperature : RT*
Bulk lattice: fcc
2D bulk symm: pmm
2D surf symm: pmm

Adsorbate:
Coverage :
Pattern : (1x3)
Matrix : (1.000, 0.000)
(0.000, 3.000)

STRUCTURE TYPE
Missing-row reconstruction (facetting into 2nd layer), with relaxations down to 3rd layer; this rconstruction is stabilized by 0.05ML of Cs impurities

SAMPLE PREPARATION (1 sample)
Treatment : cycles of Ne+ sputt. and 800 K annealing, then Cs deposition
Crystallinity: clear (1x3) LEED pattern
Anal. methods: AES, LEED
Contamination: AES: 0.05(+0.06-0.02) ML Cs

COMMENTS
0.03ML of Cs produces a poorly ordered (1x5) structure

DATA COLLECTION
Technique: MEIS; 65 and 150keV protons
Dataset : polar angular yield distributions in several different azimuths

THEORY/DATA TREATMENT
Monte Carlo simulations with R-factor analysis

STRUCTURES EXAMINED
Models with varying numbers and depths of missing rows; in preferred (facetted) model, variations of 1st, 2nd and 3rd interlayer spacings and 2nd layer pairing

2D UNIT CELLS (1 domain observed)

Cell	Ax (Å)	Ay (Å)	Bx (Å)	By (Å)	α (°)	Matrix	Pattern	Cell type
Bulk	2.885	0.000	0.000	4.080	90.0	(1.000, 0.000) (0.000, 1.000)	(1x1)	b: bulk lattice
Surface 1	2.885	0.000	0.000	12.240	90.0	(1.000, 0.000) (0.000, 3.000)	(1x3)	s1: commens. superlattice

3D COORDINATES

Au1: ridge atom; Au2-Au3: coplanar 2nd layer atoms;
Au4-Au5-Au6: buckled 3rd layer

Dx/Dy in Å, or as a fraction of layer's unit cell vectors; atom 0 at the origin. Epir/subr are bulk repeat vectors.

No. of atoms: 8 Bulk z = 1.443 Å

Region	Chem el.	At. no.	Cell type	Site occ.	Rel. to	Dx ± εx		Dy ± εy		Dz ± εz		Dz/Bz(%) ± εz/Bz
epir		-2					f		f		Å	
subr		-1				1.443	Å	2.040	Å	1.443	Å	
intf	Au	1	s1	.33	0	0.000	f	0.000	f	0.000	Å	0.0
intf	Au	2	s1	.33	1	0.500	f	0.160 ± .007	f	1.125 ± .058	Å	78.0 ± 4.0
intf	Au	3	s1	.33	2	0.000	f	0.680 ± .007	f	0.000	Å	0.0
intf	Au	4	s1	.33	3	-0.500	f	-0.173 ± .007	f	1.243 ± .043	Å	86.2 ± 3.0
intf	Au	5	s1	.33	4	0.000	f	-0.333	f	0.000	Å	0.0
intf	Au	6	s1	.33	5	0.000	f	-0.333	f	0.140 ± .040	Å	9.7 ± 2.8
intf	Au	7	b	1.00	6	-0.500	f	0.500	f	1.329 ± .040	Å	92.2 ± 2.8
subl	Au	8	b	1.00	7	-0.500	f	-0.500	f	1.443	Å	100.0

BOND DISTANCES AND ANGLES

Bond distances and angles are derived from coordinates

No. of distances/angles: 12

Interatomic dist. A-B (Å)	Atom A	Atom B	Atom C	Bond angle A-B-C (°)
2.885	Au1	Au1(1,0)	Au2	57.4
2.850	Au2	Au5	Au7(-1,1)	175.3
2.798	Au2	Au6	Au7	57.6

Au(110)-(1x3)
79.55.2

Bond Distances and Angles - Continued

Interatomic dist. A-B (Å)	Atom A	Atom B	Atom C	Bond angle A-B-C (°)
2.713	Au2	Au7	Au6	60.6
2.680	Au1	Au2	Au1(1,0)	65.1
2.680	Au1	Au2	Au5	117.0
2.680	Au1	Au2	Au6	54.4
2.680	Au1	Au2	Au7	116.2
2.508	Au1	Au6	Au2	60.4
2.508	Au1	Au6	Au7	118.0
2.850	Au2	Au5	Au5(1,0)	59.6
2.850	Au2	Au5	Au7	56.3

COMMON NAME : Au(110)-c(2x2)-K
CLASSIFICATION : 79.19.3
TECHNIQUE : MEIS
AUTHORS : P. Haberle and T. Gustafsson
REFERENCE : Phys. Rev., B40, 8218 (1989)

ILLUSTRATION: 41

SURFACE TYPE
Substrate : Au
Crystal face: 110
Temperature : RT*
Bulk lattice: fcc
2D bulk symm: pmm
2D surf symm: cmm

Adsorbate: K
Coverage : 0.5 K/1x1
Pattern : c(2x2)
Matrix : (1.000,-1.000)
(1.000, 1.000)

STRUCTURE TYPE
Mixed Au/K top layer, inducing spacing relaxations and buckling in deeper Au layers

SAMPLE PREPARATION (1 sample)
Treatment : K deposited from resistively heated source
Crystallinity: good quality LEED pattern
Anal. methods: AES, TDS; ion scattering gave K
Contamination:

COMMENTS

DATA COLLECTION
Technique: MEIS; 50-65keV protons
Dataset : polar angular yield distributions in several different azimuths

THEORY/DATA TREATMENT
Monte Carlo simulations with R-factor analysis

STRUCTURES EXAMINED
Pure overlayer model and mixed top-layer model; in latter model, variation of K height, of first 2 Au-Au spacings and of 3rd Au-layer buckling

2D UNIT CELLS (1 domain observed)

Cell	Ax (Å)	Ay (Å)	Bx (Å)	By (Å)	α (°)	Matrix	Pattern	Cell type
Bulk	2.885	0.000	0.000	4.080	90.0	(1.000, 0.000) (0.000, 1.000)	(1x1)	b: bulk lattice
Surface 1	2.885	-4.080	2.885	4.080	109.5	(1.000,-1.000) (1.000, 1.000)	c(2x2)	s1: commens. superlattice

3D COORDINATES

K1-Au2: mixed non-planar top layer; Au3-Au4: planar 2nd Au layer;
Au5-Au6: buckled 3rd Au layer

Dx/Dy in Å, or as a fraction of layer's unit cell vectors; atom 0 at the origin. Epir/subr are bulk repeat vectors.

No. of atoms: 8

Bulk z = 1.443 Å

Region	Chem el.	At. no.	Cell type	Site occ.	Rel. to	Dx ± εx		Dy ± εy		Dz ± εz		Dz/Bz(%) ± εz/Bz
epir		-2					f		f		Å	
subr		-1				1.443	Å	2.040	Å	1.443	Å	
intf	K	1	s1	.50	0	0.000	f	0.000	f	0.000	Å	0.0
intf	Au	2	s1	.50	1	-0.500	f	0.500	f	1.050 ± .150	Å	72.8 ± 10.4
intf	Au	3	s1	.50	2	0.000	f	-0.500	f	1.255 ± .043	Å	87.0 ± 3.0
intf	Au	4	s1	.50	3	0.500	f	0.500	f	0.000	Å	0.0
intf	Au	5	s1	.50	4	0.000	f	-0.500	f	1.327 ± .072	Å	92.0 ± 5.0
intf	Au	6	s1	.50	5	-0.500	f	0.500	f	0.115 ± .043	Å	8.0 ± 3.0
intf	Au	7	b	1.00	6	-0.500	f	-0.500	f	1.385 ± .043	Å	96.0 ± 3.0
subl	Au	8	b	1.00	7	0.500	f	0.500	f	1.443	Å	100.0

BOND DISTANCES AND ANGLES

Bond distances and angles are derived from coordinates

No. of distances/angles: 11

Interatomic dist. A-B (Å)	Atom A	Atom B	Atom C	Bond angle A-B-C (°)
4.997	K1	K1(1,0)		
2.796	Au2	Au3	Au5	118.3
2.796	Au2	Au3	Au6	56.7

Au(110)-c(2x2)-K
79.19.3

Bond Distances and Angles - Continued

Interatomic dist. A-B (Å)	Atom A	Atom B	Atom C	Bond angle A-B-C (°)
3.070	K1	Au2(1,0)	K1(1,1)	140.0
3.070	K1	Au2(1,0)	K1(1,0)	85.1
3.070	K1	Au2(1,0)	Au3(1,0)	70.7
3.070	K1	Au2(1,0)	Au4	70.7
4.213	K1	Au2	Au3	53.5
3.399	K1	Au3	Au4	64.9
3.399	K1	Au3	Au5	70.7
2.796	Au2	Au3	Au4	58.9

COMMON NAME : AuCu3(100) disordered
CLASSIFICATION : 29.79.5
TECHNIQUE : XPD
AUTHORS : A. Stuck, J. Osterwalder, L. Schlapbach and H.C. Poon
REFERENCE : Surf. Sci., 251/252, 670 (1991)

ILLUSTRATION: 135

SURFACE TYPE
Substrate : AuCu3
Crystal face: 100
Temperature : 823 K
Bulk lattice: AuCu3
2D bulk symm: none
2D surf symm: none

Adsorbate:
Coverage :
Pattern : disordered
Matrix : (1.000, 0.000)
(0.000, 1.000)

STRUCTURE TYPE
Disordered alloy; 50% each of Au and Cu in top layer; 100% Cu in 2nd layer; 35% Au and 65% Cu in 3rd layer; deeper layers contain 25% Au and 75% Cu, reflecting the concentration of disordered bulk AuCu3

SAMPLE PREPARATION (1 sample)
Treatment : 800 eV Ar sputtering and annealing up to 920 K
Crystallinity: sharp LEED spots at RT
Anal. methods: XPS and LEED
Contamination: O and C were <15% of a monolayer

COMMENTS
For this tabulation, an ordered fcc lattice is assumed, with lattice constant of 3.7442Å

DATA COLLECTION
Technique: XPD; modified VG ESCALAB Mark II
Dataset : Au 4f and Cu 3p polar scans at 45° above surface

THEORY/DATA TREATMENT
Layer-type multiple-scattering calculations

STRUCTURES EXAMINED
A second structure was tested with 25% Au and 75% Cu in each layer to mimic complete disorder

QUALITY OF EXPERIMENT-THEORY FIT
Visual

2D UNIT CELLS (1 domain observed)

Cell	Ax (Å)	Ay (Å)	Bx (Å)	By (Å)	α (°)	Matrix	Pattern	Cell type
Bulk	2.648	0.000	0.000	2.648	90.0	(1.000, 0.000) (0.000, 1.000)	(1x1)	b: bulk lattice
Surface 1	5.295	0.000	0.000	5.295	90.0	(2.000, 0.000) (0.000, 2.000)	disordered	m1: randomly mixed layer

3D COORDINATES

Au1-Cu4: top random mixed layer, Au/Cu=50/50; Cu5-Cu8: 2nd layer, 100% Cu;
Au9-Cu12: 3rd random mixed layer, Au/Cu=35/65; Au13-Cu16: bulk random mixed layer, Au/Cu=25/75

Dx/Dy in Å, or as a fraction of layer's unit cell vectors; atom 0 at the origin. Epir/subr are bulk repeat vectors.

No. of atoms: 16

Bulk z = 1.872 Å

Region	Chem el.	At. no.	Cell type	Site occ.	Rel. to	Dx ± εx		Dy ± εy		Dz ± εz		Dz/Bz(%) ± εz/Bz
epir		-2					f		f		Å	
subr		-1				1.324	Å	1.324	Å	1.872	Å	
intf	Au	1	m1	.25	0	0.000	f	0.000	f	0.000	Å	0.0
intf	Au	2	m1	.25	1	0.500	f	0.500	f	0.000	Å	0.0
intf	Cu	3	m1	.25	1	0.500	f	0.000	f	0.000	Å	0.0
intf	Cu	4	m1	.25	1	0.000	f	0.500	f	0.000	Å	0.0
intf	Cu	5	m1	.25	1	0.250	f	0.250	f	1.872	Å	100.0
intf	Cu	6	m1	.25	5	0.500	f	0.500	f	0.000	Å	0.0
intf	Cu	7	m1	.25	5	0.500	f	0.000	f	0.000	Å	0.0
intf	Cu	8	m1	.25	5	0.000	f	0.500	f	0.000	Å	0.0
intf	Au	9	m1	.34	0	0.000	f	0.000	f	3.744	Å	200.0
intf	Cu	10	m1	.22	9	0.500	f	0.500	f	0.000	Å	0.0
intf	Cu	11	m1	.22	9	0.500	f	0.000	f	0.000	Å	0.0
intf	Cu	12	m1	.22	9	0.000	f	0.500	f	0.000	Å	0.0
subl	Au	13	m1	.25	9	0.250	f	0.250	f	1.872	Å	100.0
subl	Cu	14	m1	.25	13	0.500	f	0.000	f	0.000	Å	0.0
subl	Cu	15	m1	.25	13	0.000	f	0.500	f	0.000	Å	0.0
subl	Cu	16	m1	.25	13	0.500	f	0.500	f	0.000	Å	0.0

AuCu3(100) disordered
29.79.5

BOND DISTANCES AND ANGLES

Bond distances and angles are derived from coordinates

No. of distances/angles: 2

Interatomic dist. A-B (Å)	Atom A	Atom B	Atom C	Bond angle A-B-C (°)
2.648	Au1	Cu3	Au2	90.0
3.547	Au1	Au2	Cu5	45.0

COMMON NAME : C(0001)-(1x1) graphite
CLASSIFICATION : 6.4
TECHNIQUE : LEED
AUTHORS : N.J. Wu and A. Ignatiev
REFERENCE : Phys. Rev., B25, 2983 (1982)

ILLUSTRATION: 157

SURFACE TYPE
Substrate : C
Crystal face: 0001
Temperature : 300 K
Bulk lattice: graphite
2D bulk symm: p3m1
2D surf symm: p3m1

Adsorbate:
Coverage :
Pattern : (1x1)
Matrix : (1.000, 0.000)
(0.000, 1.000)

STRUCTURE TYPE
Bulk termination with top spacing contraction

SAMPLE PREPARATION (1 sample)
Treatment : natural crystal platelets cleaved in dry N2 gas stream
Crystallinity:
Anal. methods:
Contamination: AES: no impurity peaks in background

COMMENTS
3-fold symmetric LEED pattern indicates absence of steps

DATA COLLECTION
Technique: LEED
Dataset : normal incidence LEED I-V data for 13 beams; E-range 60-300 eV

THEORY/DATA TREATMENT
Dynamical LEED with RSP and RFS; 5 phase shifts;
22 symmetry reduced beams; Vor=-8.2 eV (fit); ΘD=973 K

STRUCTURES EXAMINED
Various 1st layer spacings for two stacking sequences: 1. ABABA... 2. AABAB...

QUALITY OF EXPERIMENT-THEORY FIT
RZJ=0.19

2D UNIT CELLS (1 domain observed)

Cell	Ax (Å)	Ay (Å)	Bx (Å)	By (Å)	α (°)	Matrix	Pattern	Cell type
Bulk	2.461	0.000	-1.231	2.131	120.0	(1.000, 0.000) (0.000, 1.000)	(1x1)	b: bulk lattice
Surface 1	2.461	0.000	-1.231	2.131	120.0	(1.000, 0.000) (0.000, 1.000)	(1x1)	s1: commens. superlattice

3D COORDINATES

C1,C2: topmost biatomic sheet; C3,C4: 1st biatomic sheet of bulk repeat unit;
C5,C6: 2nd biatomic sheet of bulk repeat unit; 0.1Å error bar assumed for tabulation

Dx/Dy in Å, or as a fraction of layer's unit cell vectors; atom 0 at the origin. Epir/subr are bulk repeat vectors.

No. of atoms: 6

Bulk z = 3.354 Å

Region	Chem el.	At. no.	Cell type	Site occ.	Rel. to	Dx ± εx		Dy ± εy		Dz ± εz		Dz/Bz(%) ± εz/Bz
epir		-2					f		f		Å	
subr		-1				0.000	Å	0.000	Å	6.708	Å	
intf	C	1	b	1.00	0	0.000	f	0.000	f	0.000	Å	0.0
intf	C	2	b	1.00	1	0.333	f	0.667	f	0.000	Å	0.0
subl	C	3	b	1.00	2	0.333	f	-0.333	f	3.304 ± .100	Å	98.5 ± 3.0
subl	C	4	b	1.00	3	-0.333	f	0.333	f	0.000	Å	0.0
subl	C	5	b	1.00	4	-0.333	f	-0.667	f	3.354	Å	100.0
subl	C	6	b	1.00	5	0.333	f	0.667	f	0.000	Å	0.0

BOND DISTANCES AND ANGLES

Bond distances and angles are derived from coordinates

No. of distances/angles: 1

Interatomic dist. A-B (Å)	Atom A	Atom B	Atom C	Bond angle A-B-C (°)
1.421	C1	C2	C1(1,1)	120.0

COMMON NAME : C(111)-(1x1) diamond
CLASSIFICATION : 6.5
TECHNIQUE : LEED
AUTHORS : W.S. Yang, J. Sokolov, F. Jona and P.M. Marcus
REFERENCE : Solid State Commun., 41, 191 (1982)

ILLUSTRATION: 166

SURFACE TYPE
Substrate : C
Crystal face: 111
Temperature : RT*
Bulk lattice: diamond
2D bulk symm: p3m1
2D surf symm: p3m1

Adsorbate:
Coverage :
Pattern : (1x1)
Matrix : (1.000, 0.000)
(0.000, 1.000)

STRUCTURE TYPE
Ideal bulk termination, within error bars;
(probably H-terminated)

SAMPLE PREPARATION (2 sample)
Treatment : sample 1 natural; sample 2 boron doped;
both acetone cleaned
Crystallinity: sharp (1x1) LEED pattern
Anal. methods:
Contamination:

COMMENTS
Sample 1 was insulating diamond (Lurie et al,
Surf. Sci. 65, 453 (1977));
sample 2 was semiconducting diamond (Himpsel et al,
J. Vac. Sci. Technol. 17 1085 (1980))

DATA COLLECTION
Technique: LEED
Dataset : 5 LEED I-V curves at off normal incidence;
E range 60-220 eV

THEORY/DATA TREATMENT
Dynamical LEED: 8 phase shifts; 61 beams; rms vibr. 0.05Å;
Vor=-10 eV, Voi=-3eV

STRUCTURES EXAMINED
Varied first interplanar spacing from bulk value of 0.515Å by contractions of 0.1 and 0.2Å, and relaxation of 0.02Å

QUALITY OF EXPERIMENT-THEORY FIT
Visual

2D UNIT CELLS (1 domain observed)

Cell	Ax (Å)	Ay (Å)	Bx (Å)	By (Å)	α (°)	Matrix	Pattern	Cell type
Bulk	2.524	0.000	-1.262	2.186	120.0	(1.000, 0.000) (0.000, 1.000)	(1x1)	b: bulk lattice
Surface 1	2.524	0.000	-1.262	2.186	120.0	(1.000, 0.000) (0.000, 1.000)	(1x1)	s1: commens. superlattice

3D COORDINATES

C1, C2: top bilayer; C3, C4: bulk bilayer;
0.1Å error bar assumed for tabulation

Dx/Dy in Å, or as a fraction of layer's unit cell vectors; atom 0 at the origin. Epir/subr are bulk repeat vectors.

No. of atoms: 4 Bulk z = 2.061 Å

Region	Chem el.	At. no.	Cell type	Site occ.	Rel. to	Dx ± εx		Dy ± εy		Dz ± εz		Dz/Bz(%) ± εz/Bz
epir		-2					f		f		Å	
subr		-1				1.262	Å	0.729	Å	2.061	Å	
intf	C	1	b	1.00	0	0.000	f	0.000	f	0.000	Å	0.0
intf	C	2	b	1.00	1	0.667	f	0.333	f	0.515 ± .100	Å	25.0 ± 4.9
subl	C	3	b	1.00	2	0.000	f	0.000	f	1.546	Å	75.0
subl	C	4	b	1.00	3	-0.333	f	0.333	f	0.515	Å	25.0

BOND DISTANCES AND ANGLES

Bond distances and angles are derived from coordinates

No. of distances/angles: 4

Interatomic dist. A-B (Å)	Atom A	Atom B	Atom C	Bond angle A-B-C (°)
1.546	C1	C2	C1(1,1)	109.5
1.546	C1	C2	C3	109.5
1.546	C2	C1(1,1)	C2(1,1)	109.5
1.546	C2	C3	C4	109.5

COMMON NAME : C(0001)-(2x2)-Cs
ILLUSTRATION: 162
CLASSIFICATION : 6.55.3
TECHNIQUE : LEED
AUTHORS : Z.P. Hu, Jia Li, N.J. Wu and A. Ignatiev
REFERENCE : Phys. Rev., B39, 13201 (1989)

SURFACE TYPE
Substrate : C
Crystal face: 0001
Temperature : RT*
Bulk lattice: graphite
2D bulk symm: p3m1
2D surf symm: p3m1

Adsorbate: Cs
Coverage : 0.25 Cs/1x1
Pattern : (2x2)
Matrix : (2.000, 0.000)
(0.000, 2.000)

STRUCTURE TYPE
Atomic Cs adsorbed in six-fold coord. hollow sites on unrelaxed bulk-like substrate

SAMPLE PREPARATION (1 sample)
Treatment : graphite cleaved
Crystallinity:
Anal. methods:
Contamination:

COMMENTS
Low step-density area chosen by monitoring 3-fold symmetry of LEED IV curves; R-factor used is average of ROS,R1,R2, RPE,RZJ

DATA COLLECTION
Technique: LEED
Dataset : I-V curves for 4 beams (55<E<200 eV) at normal incidence; cumulative E range 450 eV

THEORY/DATA TREATMENT
Dynamical LEED (RSP, RFS); 5 phase shifts, ΘD(Cs)=40 K

STRUCTURES EXAMINED
Cs in 1) hollow and 2) top sites with variation of Cs-C spacing and 1st C-C layer spacing; 3) (Cs hollow)A/BAB (/=disordered Cs); 4)(Cs)A/AB; 5) /A/AB; 6)/A/BA

QUALITY OF EXPERIMENT-THEORY FIT
R(5ave)=0.23 for 1)

2D UNIT CELLS (1 domain observed)

Cell	Ax (Å)	Ay (Å)	Bx (Å)	By (Å)	α (°)	Matrix	Pattern	Cell type
Bulk	2.460	0.000	-1.230	2.130	120.0	(1.000, 0.000) (0.000, 1.000)	(1x1)	b: bulk lattice
Surface 1	4.920	0.000	-2.460	4.261	120.0	(2.000, 0.000) (0.000, 2.000)	(2x2)	s1: commens. superlattice

3D COORDINATES

Cs1: overlayer in 6-fold coord. hollows; C2-C5: substrate repeat unit, with 2 biatomic sheets;

Dx/Dy in Å, or as a fraction of layer's unit cell vectors; atom 0 at the origin. Epir/subr are bulk repeat vectors.

No. of atoms: 5 Bulk z = 3.350 Å

Region	Chem el.	At. no.	Cell type	Site occ.	Rel. to	Dx ± εx		Dy ± εy		Dz ± εz		Dz/Bz(%) ± εz/Bz
epir		-2					f		f		Å	
subr		-1				0.000	Å	0.000	Å	6.700	Å	
intf	Cs	1	s1	.25	0	0.000	f	0.000	f	0.000 ± .100	Å	0.0 ± 3.0
subl	C	2	b	1.00	1	-0.333	f	-0.667	f	2.800 ± .100	Å	83.6 ± 3.0
subl	C	3	b	1.00	2	0.667	f	0.333	f	0.000	Å	0.0
subl	C	4	b	1.00	3	-0.333	f	0.333	f	3.350 ± .100	Å	100.0 ± 3.0
subl	C	5	b	1.00	4	0.333	f	-0.333	f	0.000	Å	0.0

BOND DISTANCES AND ANGLES

Bond distances and angles are derived from coordinates

No. of distances/angles: 2

Interatomic dist. A-B (Å)	Atom A	Atom B	Atom C	Bond angle A-B-C (°)
3.140	Cs1	C2		
1.420	C2	C3	C2(1,1)	120.0

COMMON NAME : C(0001)-(√3x√3)R30°-Cs ILLUSTRATION: 163
CLASSIFICATION : 6.55.2
TECHNIQUE : LEED
AUTHORS : Z.P. Hu, Jia Li, N.J. Wu and A. Ignatiev
REFERENCE : Surf. Sci., 218, 283 (1989)

SURFACE TYPE
Substrate : C
Crystal face: 0001
Temperature : RT*
Bulk lattice: graphite
2D bulk symm: p3m1
2D surf symm: p31m

Adsorbate: Cs
Coverage : 1/3 Cs/1x1
Pattern : (√3x√3)R30°
Matrix : (2.000, 1.000)
(-1.000, 1.000)

STRUCTURE TYPE
Atomic Cs adsorbed in six-fold coord. hollow sites; top graphite layer shifted to a (Cs)AABAB stacking, with expanded interlayer spacing between 2 top A layers

SAMPLE PREPARATION (1 sample)
Treatment : graphite cleaved
Crystallinity:
Anal. methods:
Contamination:

COMMENTS
Low step-density area chosen by monitoring 3-fold symmetry of LEED IV curves; R-factor used is average of ROS,R1,R2, RPE,RZJ; the only fractional beam is given double weight; d2(C-C first layer spacing) for (Cs)A/AB varied from 3.25-6.25Å; best-fit d2=3.85Å means no Cs intercalation

DATA COLLECTION
Technique: LEED
Dataset : I-V curves for 2 int. beams (55<E<200 eV) and 1 fract. beam (35<E<95 eV) at normal incidence

THEORY/DATA TREATMENT
Dynamical LEED (RSP, RFS): 5 phase shifts

STRUCTURES EXAMINED
Cs in 1) hollow and 2) top sites with variable height; 3) (Cs)A/BAB (/=disordered Cs); 4) (Cs)A/AB; 5) /A(Cs)AB; 6) (Cs)A(Cs)AB with relaxation of Cs-C spacing and 1st C-C layer spacing allowed; disordered Cs not modeled, but inferred from large C-C spacing, if present (see comment)

QUALITY OF EXPERIMENT-THEORY FIT
R(5ave)=0.18 for 4)

2D UNIT CELLS (1 domain observed)

Cell	Ax (Å)	Ay (Å)	Bx (Å)	By (Å)	α (°)	Matrix	Pattern	Cell type
Bulk	2.460	0.000	-1.230	2.130	120.0	(1.000, 0.000) (0.000, 1.000)	(1x1)	b: bulk lattice
Surface 1	3.690	2.130	-3.690	2.130	120.0	(2.000, 1.000) (-1.000, 1.000)	(√3x√3)R30°	s1: commens. superlattice

3D COORDINATES

Cs1: overlayer in 6-fold coord. hollows; C2-C3: shifted graphite sheet;
C4-C7: substrate repeat unit, with 2 biatomic sheets; 0.1Å error bar assumed for tabulation

Dx/Dy in Å, or as a fraction of layer's unit cell vectors; atom 0 at the origin. Epir/subr are bulk repeat vectors.

No. of atoms: 7 Bulk z = 3.350 Å

Region	Chem el.	At. no.	Cell type	Site occ.	Rel. to	Dx ± εx		Dy ± εy		Dz ± εz		Dz/Bz(%) ± εz/Bz
epir		-2					f		f		Å	
subr		-1				0.000	Å	0.000	Å	6.700	Å	
intf	Cs	1	s1	.33	0	0.000	f	0.000	f	0.000 ± .100	Å	0.0 ± 3.0
intf	C	2	b	1.00	1	-0.333	f	-0.667	f	2.650 ± .100	Å	78.1 ± 3.0
intf	C	3	b	1.00	2	0.667	f	0.333	f	0.000	Å	0.0
subl	C	4	b	1.00	1	-0.333	f	-0.667	f	6.500 ± .100	Å	194.0 ± 3.0
subl	C	5	b	1.00	4	0.667	f	0.333	f	0.000	Å	0.0
subl	C	6	b	1.00	5	-0.333	f	0.333	f	3.350 ± .100	Å	100.0 ± 3.0
subl	C	7	b	1.00	6	0.333	f	-0.333	f	0.000	Å	0.0

C(0001)-($\sqrt{3}$x$\sqrt{3}$)R30°-Cs
6.55.2

BOND DISTANCES AND ANGLES

Bond distances and angles are derived from coordinates

No. of distances/angles: 2

Interatomic dist. A-B (Å)	Atom A	Atom B	Atom C	Bond angle A-B-C (°)
3.010	Cs1	C2		
3.850	C2	C4	C5	90.0

COMMON NAME : C(111)-(1x1)-H (diamond)
CLASSIFICATION : 6.1.1a
TECHNIQUE : MEIS
AUTHORS : T.E. Derry, L. Smit and J.F. van der Veen
REFERENCE : Surf. Sci., 167, 502 (1986)

ILLUSTRATION: 166

SURFACE TYPE
Substrate : C
Crystal face: 111
Temperature : RT*
Bulk lattice: diamond
2D bulk symm: p3m1
2D surf symm: p3m1

Adsorbate: H
Coverage :
Pattern : (1x1)
Matrix : (1.000, 0.000)
(0.000, 1.000)

STRUCTURE TYPE
Unreconstructed bulk diamond termination between bilayers, probably stabilized by H, with minor C-C spacing contraction in top bilayer (H positions not determined, but probably terminate dangling bonds)

SAMPLE PREPARATION (1 sample)
Treatment : semicond. diamond: ultrasonic cleaning, annealing to 1073 K
Crystallinity: sharp (1x1) pattern; 4° misorientation
Anal. methods:
Contamination: O removed by annealing; H present

COMMENTS
Surface atoms have vibration amplitudes enhanced by a factor from 1.4 to 2 relative to bulk

DATA COLLECTION
Technique: MEIS; 98keV proton beam
Dataset : incidence along [-1-1-1] and [00-1]; detection in (1-10) plane; all angular channels monitored simultaneously

THEORY/DATA TREATMENT
Medium-energy ion scattering: blocking and shadowing interpretation by Monte Carlo simulation of ion trajectories

STRUCTURES EXAMINED
Various 1st layer spacings in conjunction with different enhancements of surface and subsurface vibrations (H ignored)

QUALITY OF EXPERIMENT-THEORY FIT
Determined by mean-square deviations

2D UNIT CELLS (1 domain observed)

Cell	Ax (Å)	Ay (Å)	Bx (Å)	By (Å)	α (°)	Matrix	Pattern	Cell type
Bulk	2.524	0.000	-1.262	2.186	120.0	(1.000, 0.000) (0.000, 1.000)	(1x1)	b: bulk lattice
Surface 1	2.524	0.000	-1.262	2.186	120.0	(1.000, 0.000) (0.000, 1.000)	(1x1)	s1: commens. superlattice

3D COORDINATES

C1-C2: top bilayer, with slightly contracted spacing; C3-C4: periodically repeating bulk bilayer

Dx/Dy in Å, or as a fraction of layer's unit cell vectors; atom 0 at the origin. Epir/subr are bulk repeat vectors.

No. of atoms: 4 Bulk z = 2.061 Å

Region	Chem el.	At. no.	Cell type	Site occ.	Rel. to	Dx ± εx		Dy ± εy		Dz ± εz		Dz/Bz(%) ± εz/Bz
epir		-2					f		f		Å	
subr		-1				-1.262	Å	0.729	Å	2.061	Å	
intf	C	1	b	1.00	0	0.000	f	0.000	f	0.000	Å	0.0
intf	C	2	b	1.00	1	0.667	f	0.333	f	0.510 ± .005	Å	24.8 ± .2
subl	C	3	b	1.00	2	0.000	f	0.000	f	1.546	Å	75.0
subl	C	4	b	1.00	3	-0.333	f	0.333	f	0.515	Å	25.0

BOND DISTANCES AND ANGLES

Bond distances and angles are derived from coordinates

No. of distances/angles: 3

Interatomic dist. A-B (Å)	Atom A	Atom B	Atom C	Bond angle A-B-C (°)
1.544	C1	C2	C1(1,0)	109.7
1.544	C1	C2	C3	109.3
1.546	C2	C3	C4	109.5

COMMON NAME : C(0001)-1K disordered underlayer ILLUSTRATION: 164
CLASSIFICATION : 6.19.2a
TECHNIQUE : LEED
AUTHORS : N.J. Wu and A. Ignatiev
REFERENCE : Phys. Rev., B28, 7288 (1983)

SURFACE TYPE
Substrate : C
Crystal face: 0001
Temperature : RT*
Bulk lattice: graphite
2D bulk symm: p3m1
2D surf symm: none

Adsorbate: K
Coverage :
Pattern : disordered
Matrix : (1.000, 0.000)
(0.000, 1.000)

STRUCTURE TYPE
Atomic K intercalated between first two graphite sheets;
K assumed in 6-fold hollows of 1st and 2nd C layers;
graphite has AABA.. stacking;
much expanded spacing between first two graphite sheets;
K assumed disordered with 0.25ML coverage

SAMPLE PREPARATION (1 sample)
Treatment : natural platelets cleaved in N2 stream; K evaporated
Crystallinity: monitored by LEED
Anal. methods: AES
Contamination: AES: 0.06 monolayer O

COMMENTS
With increasing K exposure, K induces shear shift of the graphite layers to A/A/A/A (/= disordered K) stacking, with AA spacing increased to 5.35 Å from bulk value of 3.35 Å;
K assumed random;
6-fold LEED pattern symmetry shows presence of steps

DATA COLLECTION
Technique: LEED
Dataset : normal incidence LEED I-V data 80<E<280 eV for 2 beams, taken at 3 different stages of intercalation

THEORY/DATA TREATMENT
Dynamical LEED (RSP, RFS): 5 phase shifts; K treated as incoherent scatterer

STRUCTURES EXAMINED
1. K adsorbed onto C(0001) in (1x1) pattern; 2. K assumed between first two carbon layers, with ABAB.. stacking sequence of C and dilated first layer spacing; 3. as in 2. but substrate in AABA.. stacking sequence; top C-C spacing varied 3.35-5.65Å; K ignored for 2., 3.

QUALITY OF EXPERIMENT-THEORY FIT
Visual

2D UNIT CELLS (1 domain observed)

Cell	Ax (Å)	Ay (Å)	Bx (Å)	By (Å)	α (°)	Matrix	Pattern	Cell type
Bulk	2.460	0.000	-1.230	2.130	120.0	(1.000, 0.000) (0.000, 1.000)	(1x1)	b: bulk lattice
Surface 1	2.460	0.000	-1.230	2.130	120.0	(1.000, 0.000) (0.000, 1.000)	disordered	nd1: non-recon. lattice-gas dis

3D COORDINATES

C1,C2: top graphite sheet; K3: disord. intercalate, coverage unknown (assumed 0.25);
C4-C7: substrate repeat unit, with 2 biatomic sheets; 0.1Å error bar assumed for tabulation

Dx/Dy in Å, or as a fraction of layer's unit cell vectors; atom 0 at the origin. Epir/subr are bulk repeat vectors.

No. of atoms: 7 Bulk z = 3.350 Å

Region	Chem el.	At. no.	Cell type	Site occ.	Rel. to	Dx ± εx		Dy ± εy		Dz ± εz		Dz/Bz(%) ± εz/Bz
epir		-2					f		f		Å	
subr		-1				0.000	Å	0.000	Å	6.700	Å	
intf	C	1	b	1.00	0	0.000	f	0.000	f	0.000	Å	0.0
intf	C	2	b	1.00	1	0.667	f	0.333	f	0.000	Å	0.0
intf	K	3	nd1	.25	2	-0.333	f	0.333	f	2.675 ± .100	Å	79.9 ± 3.0
subl	C	4	b	1.00	3	-0.333	f	-0.667	f	2.675 ± .100	Å	79.9 ± 3.0
subl	C	5	b	1.00	4	0.667	f	0.333	f	0.000	Å	0.0
subl	C	6	b	1.00	5	-0.333	f	0.333	f	3.350	Å	100.0
subl	C	7	b	1.00	6	0.333	f	-0.333	f	0.000	Å	0.0

C(0001)-1K disordered underlayer
6.19.2a

BOND DISTANCES AND ANGLES

Bond distances and angles are derived from coordinates

No. of distances/angles: 6

Interatomic dist. A-B (Å)	Atom A	Atom B	Atom C	Bond angle A-B-C (°)
1.420	C1	C2	K3	76.4
3.029	C1	K3	C2	27.1
3.029	C1	K3	K3(0,1)	114.0
3.029	C1	K3	C4(0,1)	152.9
3.029	C1	K3	C4	124.1
3.029	C1	K3	C5	132.1

COMMON NAME : C(0001)-2K disordered underlayers ILLUSTRATION: 165
CLASSIFICATION : 6.19.2b
TECHNIQUE : LEED
AUTHORS : N.J. Wu and A. Ignatiev
REFERENCE : Phys. Rev., B28, 7288 (1983)

SURFACE TYPE
Substrate : C
Crystal face: 0001
Temperature : RT*
Bulk lattice: graphite
2D bulk symm: p3m1
2D surf symm: none

Adsorbate: K
Coverage :
Pattern : disordered
Matrix : (1.000, 0.000)
(0.000, 1.000)

STRUCTURE TYPE
Atomic K intercalated between first 3 graphite sheets;
K assumed in 6-fold hollows of C layers;
graphite has AAABA.. stacking;
much expanded spacing between first 3 graphite sheets;
K assumed disordered

SAMPLE PREPARATION (1 sample)
Treatment : natural platelets cleaved in N2 stream;
K evaporated
Crystallinity: monitored by LEED
Anal. methods: AES
Contamination: AES: 0.06 monolayer O

COMMENTS
With increasing K exposure, K induces shear shift of the graphite layers to A/A/A/A (/= disordered K) stacking, with AA spacing increased to 5.35 Å from bulk value of 3.35 Å;
K assumed random;
6-fold LEED pattern symmetry shows presence of steps

DATA COLLECTION
Technique: LEED
Dataset : normal incidence LEED I-V data 80<E<280 eV for 2 beams, taken at 3 different stages of intercalation

THEORY/DATA TREATMENT
Dynamical LEED (RSP, RFS); 5 phase shifts; K treated as incoherent scatterer

STRUCTURES EXAMINED
1. K adsorbed onto C(0001) in (1x1) pattern; 2. K assumed between first three carbon layers, with ABAB.. stacking of C sheets and dilated first 2 layer spacings; 3. as in 2. but substrate in AAABA.. stacking sequence; top C-C spacings varied 3.35-5.65Å; K ignored for 2., 3.

QUALITY OF EXPERIMENT-THEORY FIT
Visual

2D UNIT CELLS (1 domain observed)

Cell	Ax (Å)	Ay (Å)	Bx (Å)	By (Å)	α (°)	Matrix	Pattern	Cell type
Bulk	2.460	0.000	-1.230	2.130	120.0	(1.000, 0.000) (0.000, 1.000)	(1x1)	b: bulk lattice
Surface 1	2.460	0.000	-1.230	2.130	120.0	(1.000, 0.000) (0.000, 1.000)	disordered	nd1: non-recon. lattice-gas dis

3D COORDINATES

C1,C2 and C3,C4: first and second graphite sheets; K3,K6: disord. intercalates (coverage assumed 2x0.25);
C7-C10: substrate repeat unit, with 2 biatomic sheets; 0.1Å error bar assumed for tabulation

Dx/Dy in Å, or as a fraction of layer's unit cell vectors; atom 0 at the origin. Epir/subr are bulk repeat vectors.

No. of atoms: 10 Bulk z = 3.350 Å

Region	Chem el.	At. no.	Cell type	Site occ.	Rel. to	Dx ± εx		Dy ± εy		Dz ± εz		Dz/Bz(%) ± εz/Bz
epir		-2					f		f		Å	
subr		-1				0.000	Å	0.000	Å	6.700	Å	
intf	C	1	b	1.00	0	0.000	f	0.000	f	0.000	Å	0.0
intf	C	2	b	1.00	1	0.667	f	0.333	f	0.000	Å	0.0
intf	K	3	nd1	.25	2	-0.333	f	0.333	f	2.675 ± .100	Å	79.9 ± 3.0
intf	C	4	b	1.00	3	-0.333	f	-0.667	f	2.675 ± .100	Å	79.9 ± 3.0
intf	C	5	b	1.00	4	0.667	f	0.333	f	0.000	Å	0.0
intf	K	6	nd2	.25	5	-0.333	f	0.333	f	2.675 ± .100	Å	79.9 ± 3.0
subl	C	7	b	1.00	6	-0.333	f	-0.667	f	2.675 ± .100	Å	79.9 ± 3.0
subl	C	8	b	1.00	7	0.667	f	0.333	f	0.000	Å	0.0
subl	C	9	b	1.00	8	-0.333	f	0.333	f	3.350	Å	100.0
subl	C	10	b	1.00	9	0.333	f	-0.333	f	0.000	Å	0.0

C(0001)-2K disordered underlayers
6.19.2b BOND DISTANCES AND ANGLES

Bond distances and angles are derived from coordinates

No. of distances/angles: 6

Interatomic dist. A-B (Å)	Atom A	Atom B	Atom C	Bond angle A-B-C (°)
1.420	C1	C2	K3	76.4
3.029	C1	K3	C2	27.1
3.029	C1	K3	K3(0,1)	114.0
3.029	C1	K3	C4(0,1)	152.9
3.029	C1	K3	C4	124.1
3.029	C1	K3	C5	132.1

COMMON NAME : CaO(100)-(1x1) ILLUSTRATION: 149
CLASSIFICATION : 20.8.1
TECHNIQUE : LEED
AUTHORS : M. Prutton, J.A. Ramsey, J.A. Walker and M.R. Welton Cook
REFERENCE : J. Phys., C12, 5271 (1979)

SURFACE TYPE
Substrate : CaO
Crystal face: 100
Temperature : 300 K
Bulk lattice: NaCl
2D bulk symm: p4m
2D surf symm: p4m

Adsorbate:
Coverage :
Pattern : (1x1)
Matrix : (1.000, 0.000)
(0.000, 1.000)

STRUCTURE TYPE
Unreconstructed CaO termination with contraction of top interlayer spacing and no buckling in mixed top layer

SAMPLE PREPARATION (2 sample)
Treatment : CaO crystal cleaved at room temperature at 2E-10 torr
Crystallinity:
Anal. methods:
Contamination: clean as-cleaved sample by EELS and AES

COMMENTS

DATA COLLECTION
Technique: LEED
Dataset : I-V spectra for 39 beams at incident angles Θ=0, 5, 10, and 20°, and φ=0 to 45°; 120<E<400 eV

THEORY/DATA TREATMENT
Dynamical LEED (CAVLEED package): 9 phase shifts from Clementi wavefunctions (equal radii for anion and cation)

STRUCTURES EXAMINED
±5% relaxations of first interlayer spacing with no cation/anion buckling; bucklings of 2% or 6% did not improve agreement with experiment

QUALITY OF EXPERIMENT-THEORY FIT
RZJ=0.156

2D UNIT CELLS (1 domain observed)

Cell	Ax (Å)	Ay (Å)	Bx (Å)	By (Å)	α (°)	Matrix	Pattern	Cell type
Bulk	3.403	0.000	0.000	3.403	90.0	(1.000, 0.000) (0.000, 1.000)	(1x1)	b: bulk lattice
Surface 1	3.403	0.000	0.000	3.403	90.0	(1.000, 0.000) (0.000, 1.000)	(1x1)	s1: commens. superlattice

3D COORDINATES

Ca1-O2: top mixed layer; Ca5-O6: repeating bulk mixed layer;

Dx/Dy in Å, or as a fraction of layer's unit cell vectors; atom 0 at the origin. Epir/subr are bulk repeat vectors.

No. of atoms: 6 Bulk z = 2.406 Å

Reg ion	Chem el.	At. no.	Cell type	Site occ.	Rel. to	Dx ± εx		Dy ± εy		Dz ± εz		Dz/Bz(%) ± εz/Bz
epir		-2					f		f		Å	
subr		-1				1.701	Å	1.701	Å	2.406	Å	
intf	Ca	1	b	1.00	0	0.000	f	0.000	f	0.000	Å	0.0
intf	O	2	b	1.00	1	0.500	f	0.500	f	0.000 ± .100	Å	0.0 ± 4.2
intf	O	3	b	1.00	2	-0.500	f	-0.500	f	2.376 ± .100	Å	98.8 ± 4.2
intf	Ca	4	b	1.00	3	0.500	f	0.500	f	0.000	Å	0.0
subl	Ca	5	b	1.00	4	-0.500	f	-0.500	f	2.406	Å	100.0
subl	O	6	b	1.00	5	0.500	f	0.500	f	0.000	Å	0.0

BOND DISTANCES AND ANGLES

Bond distances and angles are derived from coordinates

No. of distances/angles: 2

Interatomic dist. A-B (Å)	Atom A	Atom B	Atom C	Bond angle A-B-C (°)
3.403	Ca1	Ca1(1,0)	Ca4	59.8
2.376	Ca1	O3	Ca4	90.0

COMMON NAME : CdS(11-20)-(1x1) ILLUSTRATION: 125
CLASSIFICATION : 48.16.1
TECHNIQUE : LEED
AUTHORS : A. Kahn, C.B. Duke and Y.R. Wang
REFERENCE : Phys. Rev., B44, 5606 (1991)

SURFACE TYPE
Substrate : CdS
Crystal face: 11-20
Temperature : 50 K
Bulk lattice: wurtzite
2D bulk symm: p1
2D surf symm: p1

Adsorbate:
Coverage :
Pattern : (1x1)
Matrix : (1.000, 0.000)
(0.000, 1.000)

STRUCTURE TYPE
Top-layer atoms have relaxations both parallel and perpendicular to the surface, with a bond-length-conserving rotation of the surface Cd-S-Cd and S-Cd-S triplets by 30°, followed by a 0.1Å contraction of the first layer toward the bulk

SAMPLE PREPARATION (1 sample)
Treatment : cleavage in ultra high vacuum
Crystallinity:
Anal. methods:
Contamination:

COMMENTS

DATA COLLECTION
Technique: LEED; computer based aquisition system
Dataset : I-V curves for 15 beams, energy range 35-250 eV

THEORY/DATA TREATMENT
Dynamical LEED: 6 phase shifts, Vor=-10 eV; mfp=10Å

STRUCTURES EXAMINED
Varied were: a. tilt of plane defined by the S-Cd-S triplet; b. spacing between the Cd-subplanes in the first and second layers and buckling in second layer

QUALITY OF EXPERIMENT-THEORY FIT
RX=0.21, RI=0.083

2D UNIT CELLS (1 domain observed)

Cell	Ax (Å)	Ay (Å)	Bx (Å)	By (Å)	α (°)	Matrix	Pattern	Cell type
Bulk	6.749	0.000	0.000	7.162	90.0	(1.000, 0.000) (0.000, 1.000)	(1x1)	b: bulk lattice
Surface 1	6.749	0.000	0.000	7.162	90.0	(1.000, 0.000) (0.000, 1.000)	(1x1)	s1: commens. superlattice

3D COORDINATES

S1-S2-Cd3-Cd4: buckled top layer with S outermost; S5-S6-Cd7-Cd8: planar periodically repeating bulk layer; 0.1Å error bar assumed for tabulation

Dx/Dy in Å, or as a fraction of layer's unit cell vectors; atom 0 at the origin. Epir/subr are bulk repeat vectors.

No. of atoms: 8 Bulk z = 2.067 Å

Region	Chem el.	At. no.	Cell type	Site occ.	Rel. to	Dx ± εx		Dy ± εy		Dz ± εz		Dz/Bz(%) ± εz/Bz
epir		-2					f		f		Å	
subr		-1				0.000	Å	3.581	Å	2.067	Å	
intf	S	1	b	1.00	0	0.000	f	0.000	f	0.000	Å	0.0
intf	S	2	b	1.00	1	0.500	f	0.405	f	0.000	Å	0.0
intf	Cd	3	b	1.00	2	-0.286	f	-0.117	f	0.650 ± .100	Å	31.4 ± 4.8
intf	Cd	4	b	1.00	3	0.500	f	-0.216	f	0.000	Å	0.0
subl	S	5	b	1.00	1	0.000	f	0.500	f	2.150 ± .100	Å	104.0 ± 4.8
subl	S	6	b	1.00	5	0.500	f	0.333	f	0.000	Å	0.0
subl	Cd	7	b	1.00	6	-0.345	f	0.000	f	0.000	Å	0.0
subl	Cd	8	b	1.00	7	0.500	f	-0.333	f	0.000	Å	0.0

CdS(11-20)-(1x1)
48.16.1

BOND DISTANCES AND ANGLES

Bond distances and angles are derived from coordinates

No. of distances/angles: 10

Interatomic dist. A-B (Å)	Atom A	Atom B	Atom C	Bond angle A-B-C (°)
2.595	S1	Cd3	S2	135.6
2.595	Cd3	S1	Cd6(0,-1)	86.3
2.595	S1	Cd3	S5	107.7
2.102	S1	Cd4(-1,0)	S2(-1,0)	126.5
2.673	S1	Cd6(0,-1)	Cd3	46.0
2.205	S2	Cd3	S1	135.6
2.205	S2	Cd3	S5	115.9
2.863	S2	Cd4	S1(1,0)	126.5
2.863	S2	Cd4	S7(0,-1)	112.6
2.487	S2	Cd8	Cd3	35.9

COMMON NAME : CdSe(10-10)-(1x1) ILLUSTRATION: 124
CLASSIFICATION : 48.34.2
TECHNIQUE : LEED
AUTHORS : Y.R. Wang, C.B. Duke, A. Paton, K. Stiles and A. Kahn
REFERENCE : Phys. Rev., B36, 9406 (1987)

SURFACE TYPE
Substrate : CdSe
Crystal face: 10-10
Temperature : 125 K
Bulk lattice: wurtzite
2D bulk symm: pm
2D surf symm: pm

Adsorbate:
Coverage :
Pattern : (1x1)
Matrix : (1.000, 0.000)
(0.000, 1.000)

STRUCTURE TYPE
Relaxed bulk termination: top Cd-Se layer buckled (Se outward, Cd inward)

SAMPLE PREPARATION (1 sample)
Treatment : cleaved in situ
Crystallinity:
Anal. methods:
Contamination:

COMMENTS
Published coordinates give unusually long and short bond lengths (eds.)

DATA COLLECTION
Technique: LEED
Dataset : I-V spectra for 14 beams

THEORY/DATA TREATMENT
Dynamical LEED with R-factor minimisation

2D UNIT CELLS (1 domain observed)

Cell	Ax (Å)	Ay (Å)	Bx (Å)	By (Å)	α (°)	Matrix	Pattern	Cell type
Bulk	4.300	0.000	0.000	7.020	90.0	(1.000, 0.000) (0.000, 1.000)	(1x1)	b: bulk lattice
Surface 1	4.300	0.000	0.000	7.020	90.0	(1.000, 0.000) (0.000, 1.000)	(1x1)	s1: commens. superlattice

3D COORDINATES

Se1-Cd2: buckled top bilayer, Se outward; Cd5-Se6 and Cd7-Se8: 2 bulk bilayers, together forming repeating bulk set of layers

Dx/Dy in Å, or as a fraction of layer's unit cell vectors; atom 0 at the origin. Epir/subr are bulk repeat vectors.

No. of atoms: 8 Bulk z = 3.725 Å

Region	Chem el.	At. no.	Cell type	Site occ.	Rel. to	Dx ± εx		Dy ± εy		Dz ± εz		Dz/Bz(%) ± εz/Bz
epir		-2					f		f		Å	
subr		-1				-2.150	Å	0.000	Å	3.725	Å	
intf	Se	1	b	1.00	0	0.000	f	0.000	f	0.000	Å	0.0
intf	Cd	2	b	1.00	1	0.000	f	0.345 ± .028	f	1.030 ± .200	Å	27.7 ± 5.4
intf	Cd	3	b	1.00	2	0.500	f	0.584 ± .028	f	0.410 ± .200	Å	11.0 ± 5.4
intf	Se	4	b	1.00	3	0.000	f	-0.429 ± .028	f	0.000 ± .100	Å	0.0 ± 2.7
subl	Cd	5	b	1.00	4	0.000	f	-0.125	f	2.483	Å	66.7
subl	Se	6	b	1.00	5	0.000	f	-0.375	f	0.000	Å	0.0
subl	Cd	7	b	1.00	6	-0.500	f	0.875	f	1.242	Å	33.3
subl	Se	8	b	1.00	7	0.000	f	-0.375	f	0.000	Å	0.0

BOND DISTANCES AND ANGLES

Bond distances and angles are derived from coordinates

No. of distances/angles: 11

Interatomic dist. A-B (Å)	Atom A	Atom B	Atom C	Bond angle A-B-C (°)
2.632	Se1	Cd2	Se4	118.4
2.633	Cd5	Se6	Se8	90.0
2.633	Se6	Cd7(0,-1)	Se8	70.5
2.632	Se1	Cd2	Cd5	111.6
2.635	Se1	Cd3(0,-1)	Se4(0,-1)	100.9
2.635	Se1	Cd3(0,-1)	Se6	119.9
2.444	Cd2	Se4	Cd5	90.6
3.012	Cd3	Se4	Cd5	109.5

CdSe(10-10)-(1x1)
48.34.2

Bond Distances and Angles - Continued

Interatomic dist. A-B (Å)	Atom A	Atom B	Atom C	Bond angle A-B-C (°)
2.533	Cd3	Se6(0,1)	Cd7	113.4
2.634	Se4	Cd5	Se6	109.5
2.633	Cd5	Se6	Cd7(0,-1)	109.5

COMMON NAME : CdSe(10-10)-(1x1)
CLASSIFICATION : 48.34.4b
TECHNIQUE : LEPD
AUTHORS : C.B. Duke, D.E. Lessor, T.N. Horsky, G. Brandes, K.F. Canter, P.H. Lippel, A.P. Mills, A. Paton and Y.R. Wang
REFERENCE : J. Vac. Sci. Technol., A7, 2031 (1989)

ILLUSTRATION: 124

SURFACE TYPE
Substrate : CdSe
Crystal face: 10-10
Temperature : 300 K
Bulk lattice: wurtzite
2D bulk symm: pm
2D surf symm: pm

Adsorbate:
Coverage :
Pattern : (1x1)
Matrix : (1.000, 0.000)
(0.000, 1.000)

STRUCTURE TYPE
Top layer buckled with Se outermost, second layer shows small buckling with Cd outermost; also relaxations parallel to surface of the atoms in the top two layers

SAMPLE PREPARATION (1 sample)
Treatment : cleavage in ultra high vacuum
Crystallinity:
Anal. methods: LEED
Contamination:

COMMENTS

DATA COLLECTION
Technique: LEPD
Dataset : I-V curves for 14 beams, energy range 20-160 eV

THEORY/DATA TREATMENT
Dynamical LEPD: Vor=2 eV

STRUCTURES EXAMINED
Varied were: a. tilt of plane defined by the Se-Cd-Se triplet; b. spacing between the Cd-subplanes in the first and second layers and buckling in second layer

QUALITY OF EXPERIMENT-THEORY FIT
RX=0.08, RI=0.06

2D UNIT CELLS (1 domain observed)

Cell	Ax (Å)	Ay (Å)	Bx (Å)	By (Å)	α (°)	Matrix	Pattern	Cell type
Bulk	4.300	0.000	0.000	7.020	90.0	(1.000, 0.000) (0.000, 1.000)	(1x1)	b: bulk lattice
Surface 1	4.300	0.000	0.000	7.020	90.0	(1.000, 0.000) (0.000, 1.000)	(1x1)	s1: commens. superlattice

3D COORDINATES

Se1-Cd2: buckled top layer, Se outward; Cd3-Se4: buckled second layer, Cd outward;
Cd5-Se6-Cd7-Se8: 2 planar bulk layers repeated periodically; 0.1Å error bars assumed for tabulation

Dx/Dy in Å, or as a fraction of layer's unit cell vectors; atom 0 at the origin. Epir/subr are bulk repeat vectors.

No. of atoms: 8 Bulk z = 3.720 Å

Region	Chem el.	At. no.	Cell type	Site occ.	Rel. to	Dx ± єx		Dy ± єy		Dz ± єz		Dz/Bz(%) ± єz/Bz
epir		-2					f		f		Å	
subr		-1				-2.150	Å	0.000	Å	3.720	Å	
intf	Se	1	b	1.00	0	0.000	f	0.000	f	0.000	Å	0.0
intf	Cd	2	b	1.00	1	0.000	f	0.638 ± .014	f	0.680 ± .100	Å	18.3 ± 2.7
intf	Cd	3	b	1.00	2	0.500	f	-0.564 ± .014	f	0.650 ± .100	Å	17.5 ± 2.7
intf	Se	4	b	1.00	3	0.000	f	0.426 ± .014	f	0.050 ± .100	Å	1.3 ± 2.7
subl	Cd	5	b	1.00	4	0.000	f	0.155	f	2.480	Å	66.7
subl	Se	6	b	1.00	5	0.000	f	-0.655	f	0.000	Å	0.0
subl	Cd	7	b	1.00	6	-0.500	f	0.155	f	1.240	Å	33.3
subl	Se	8	b	1.00	7	0.000	f	0.345	f	0.000	Å	0.0

CdSe(10-10)-(1x1)
48.34.4b

BOND DISTANCES AND ANGLES

Bond distances and angles are derived from coordinates

No. of distances/angles: 8

Interatomic dist. A-B (Å)	Atom A	Atom B	Atom C	Bond angle A-B-C (°)
2.630	Se1	Cd2(0,-1)	Se4(0,-1)	117.0
2.581	Se1	Cd3	Se1(1,0)	112.8
2.581	Se1	Cd3	Se4	102.1
2.630	Cd2	Se1(0,1)	Cd3(0,1)	93.5
2.460	Cd2	Se4	Cd2(1,0)	121.8
2.581	Cd3	Se1(1,0)	Cd2(1,-1)	93.5
2.991	Cd3	Se4	Cd2(1,0)	112.9
2.460	Se4	Cd2(1,0)	Se1(1,1)	117.0

COMMON NAME : CdSe(11-20)-(1x1)
CLASSIFICATION : 48.34.4a
TECHNIQUE : LEPD
AUTHORS : C.B. Duke, D.E. Lessor, T.N. Horsky, G. Brandes, K.F. Canter, P.H. Lippel, A.P. Mills, A. Paton and Y.R. Wang
REFERENCE : J. Vac. Sci. Technol., A7, 2031 (1989)

ILLUSTRATION: 125

SURFACE TYPE
Substrate : CdSe
Crystal face: 11-20
Temperature : 105 K
Bulk lattice: wurtzite
2D bulk symm: p1
2D surf symm: p1

Adsorbate:
Coverage :
Pattern : (1x1)
Matrix : (1.000, 0.000)
(0.000, 1.000)

STRUCTURE TYPE
Top-layer atoms have relaxations with a bond-length-conserving rotation of the surface Cd-Se-Cd and Se-Cd-Se triplets by 27°, followed by a contraction of the first layer toward the bulk; 2nd layer also has a small buckling, with Cd outermost

SAMPLE PREPARATION (1 sample)
Treatment : cleavage in ultra high vacuum
Crystallinity:
Anal. methods: LEED
Contamination:

COMMENTS

DATA COLLECTION
Technique: LEPD
Dataset : I-V curves for 14 beams, energy range 20-160 eV

THEORY/DATA TREATMENT
Dynamical LEPD: Vor=2 eV

STRUCTURES EXAMINED
Varied were: a. tilt of plane defined by the Se-Cd-Se triplet; b. spacing between the Cd-subplanes in the first and second layers and buckling in second layer

QUALITY OF EXPERIMENT-THEORY FIT
RX=0.12, RI=0.04

2D UNIT CELLS (1 domain observed)

Cell	Ax (Å)	Ay (Å)	Bx (Å)	By (Å)	α (°)	Matrix	Pattern	Cell type
Bulk	7.020	0.000	0.000	7.448	90.0	(1.000, 0.000) (0.000, 1.000)	(1x1)	b: bulk lattice
Surface 1	7.020	0.000	0.000	7.448	90.0	(1.000, 0.000) (0.000, 1.000)	(1x1)	s1: commens. superlattice

3D COORDINATES

Se1-Se2-Cd3-Cd4: buckled top layer with Se outermost; Se5-Se6-Cd7-Cd8: buckled second layer; Se9-Se10-Cd11-Cd12: planar periodically repeated bulk layer; 0.1Å error bar assumed for tabulation

Dx/Dy in Å, or as a fraction of layer's unit cell vectors; atom 0 at the origin. Epir/subr are bulk repeat vectors.

No. of atoms: 12 Bulk z = 2.150 Å

Reg ion	Chem el.	At. no.	Cell type	Site occ.	Rel. to	Dx ± εx	Dy ± εy	Dz ± εz	Dz/Bz(%) ± εz/Bz
epir		-2				f	f	Å	
subr		-1				0.000 Å	3.724 Å	2.150 Å	
intf	Se	1	b	1.00	0	0.000 f	0.000 f	0.000 Å	0.0
intf	Se	2	b	1.00	1	0.500 ± .014 f	0.403 ± .013 f	0.000 Å	0.0
intf	Cd	3	b	1.00	2	-0.295 ± .014 f	-0.070 ± .013 f	0.610 ± .100 Å	28.4 ± 4.7
intf	Cd	4	b	1.00	3	0.438 ± .014 f	-0.264 ± .013 f	0.000 Å	0.0
intf	Se	5	b	1.00	1	0.000 f	0.500 f	2.230 ± .100 Å	103.7 ± 4.7
intf	Se	6	b	1.00	5	0.500 f	0.333 f	0.000 Å	0.0
intf	Cd	7	b	1.00	6	-0.345 f	0.000 f	0.120 Å	5.6
intf	Cd	8	b	1.00	7	0.500 f	-0.333 f	0.000 Å	0.0
subl	Se	9	b	1.00	5	0.000 f	0.500 f	2.170 ± .100 Å	100.9 ± 4.7
subl	Se	10	b	1.00	9	0.500 f	0.333 f	0.000 Å	0.0
subl	Cd	11	b	1.00	10	-0.345 f	0.000 f	0.000 Å	0.0
subl	Cd	12	b	1.00	11	0.500 f	-0.333 f	0.000 Å	0.0

CdSe(11-20)-(1x1)
48.34.4a

BOND DISTANCES AND ANGLES

Bond distances and angles are derived from coordinates

No. of distances/angles: 10

Interatomic dist. A-B (Å)	Atom A	Atom B	Atom C	Bond angle A-B-C (°)
2.933	Se1	Cd3	Se2	126.8
2.745	Cd4	Se2	Cd3	93.8
2.933	Se1	Cd3	Cd4	87.6
2.635	Se1	Cd4(-1,0)	Se2(-1,0)	118.3
2.872	Se1	Cd7(0,-1)	Se5(0,-1)	102.0
2.221	Se2	Cd3	Se5	126.8
2.745	Se2	Cd4	Cd3	37.5
2.688	Se2	Cd8	Cd4	48.5
2.221	Cd3	Se2	Cd4	93.8
2.497	Cd3	Se5	Cd7	104.6

COMMON NAME : CdTe(110)-(1x1)
CLASSIFICATION : 48.52.2
TECHNIQUE : LEED
AUTHORS : C.B. Duke, A. Paton, W.K. Ford, A. Kahn and G. Scott
REFERENCE : J. Vac. Sci. Technol., 20, 778 (1982)

ILLUSTRATION: 116

SURFACE TYPE
Substrate : CdTe
Crystal face: 110
Temperature : 110 K
Bulk lattice: zincblende
2D bulk symm: pm
2D surf symm: pm
Adsorbate:
Coverage :
Pattern : (1x1)
Matrix : (1.000, 0.000)
(0.000, 1.000)

STRUCTURE TYPE
Relaxed bulk termination with 30.5° tilt in top layer

SAMPLE PREPARATION (3 sample)
Treatment : 3 separate in situ cleaves: experimental data averaged
Crystallinity:
Anal. methods:
Contamination: monitored by LEED and AES

COMMENTS
Top layer is characterized by bond length conserving rotations of 30.5° and contraction of 0.05±0.05Å towards the substrate; observed reconstruction is almost identical with that observed in InSb(110); more complete account given in Duke et al, Phys. Rev. B24, 3310 (1981)

DATA COLLECTION
Technique: LEED
Dataset : I-V spectra: 12 beams at normal incidence

THEORY/DATA TREATMENT
Dynamical LEED: E-dependent exchange potentials from overlapping selfconsistent relativistic atomic charge densities

STRUCTURES EXAMINED
1. unreconstructed surface; 2. bond relaxations perpendicular to surface;
3. bond rotation top-layer reconstruction, 2nd layer shears

QUALITY OF EXPERIMENT-THEORY FIT
RX=0.2

2D UNIT CELLS (1 domain observed)

Cell	Ax (Å)	Ay (Å)	Bx (Å)	By (Å)	α (°)	Matrix	Pattern	Cell type
Bulk	4.582	0.000	0.000	6.480	90.0	(1.000, 0.000) (0.000, 1.000)	(1x1)	b: bulk lattice
Surface 1	4.582	0.000	0.000	6.480	90.0	(1.000, 0.000) (0.000, 1.000)	(1x1)	s1: commens. superlattice

3D COORDINATES

Te1-Cd2, Cd3-Te4: 2 bilayers with tilted Cd-Te chains; Cd7-Te8: periodically repeating bulk bilayer;
0.1Å error bars assumed for tabulation

Dx/Dy in Å, or as a fraction of layer's unit cell vectors; atom 0 at the origin. Epir/subr are bulk repeat vectors.

No. of atoms: 8 Bulk z = 2.291 Å

Region	Chem el.	At. no.	Cell type	Site occ.	Rel. to	Dx ± εx		Dy ± εy		Dz ± εz		Dz/Bz(%) ± εz/Bz
epir		-2					f		f		Å	
subr		-1				2.291	Å	3.240	Å	2.291	Å	
intf	Te	1	b	1.00	0	0.000	f	0.000	f	0.000	Å	0.0
intf	Cd	2	b	1.00	1	0.500	f	0.216 ± .015	f	0.820 ± .100	Å	35.8 ± 4.4
intf	Cd	3	b	1.00	2	-0.500	f	0.594 ± .015	f	1.560 ± .100	Å	68.1 ± 4.4
intf	Te	4	b	1.00	3	0.500	f	-0.250 ± .015	f	0.180 ± .100	Å	7.9 ± 4.4
intf	Cd	5	b	1.00	4	0.000	f	-0.250 ± .015	f	2.100 ± .100	Å	91.7 ± 4.4
intf	Te	6	b	1.00	5	-0.500	f	-0.250	f	0.000	Å	0.0
subl	Cd	7	b	1.00	6	0.000	f	0.750	f	2.291	Å	100.0
subl	Te	8	b	1.00	7	0.500	f	-0.250	f	0.000	Å	0.0

CdTe(110)-(1x1)
48.52.2 BOND DISTANCES AND ANGLES

Bond distances and angles are derived from coordinates

No. of distances/angles: 10

Interatomic dist. A-B (Å)	Atom A	Atom B	Atom C	Bond angle A-B-C (°)
2.807	Te1	Cd2	Te1(1,0)	109.4
2.797	Cd3	Te6(0,1)	Cd7	109.3
2.807	Te1	Cd2	Te4	124.9
2.680	Te1	Cd3(0,-1)	Te4(0,-1)	108.8
2.680	Te1	Cd3(0,-1)	Te6	117.3
2.828	Cd2	Te4	Cd3	114.5
2.828	Cd2	Te4	Cd5	90.3
2.812	Cd3	Te4	Cd3(1,0)	109.1
2.812	Cd3	Te4	Cd5	113.7
2.797	Cd3	Te6(0,1)	Cd5(0,1)	109.5

COMMON NAME : CdTe(110)-(1x1) ILLUSTRATION: 116
CLASSIFICATION : 48.52.6
TECHNIQUE : LEED
AUTHORS : P.G. Powell and V.E. de Carvalho
REFERENCE : J. Phys., C21, 2983 (1988)

SURFACE TYPE
Substrate : CdTe
Crystal face: 110
Temperature : RT
Bulk lattice: zincblende
2D bulk symm: pm
2D surf symm: pm

Adsorbate:
Coverage :
Pattern : (1x1)
Matrix : (1.000, 0.000)
(0.000, 1.000)

STRUCTURE TYPE
Relaxed bulk termination with 30° tilt in top layer

SAMPLE PREPARATION (1 sample)
Treatment : cleaved in situ
Crystallinity:
Anal. methods:
Contamination: monitored by AES and LEED

COMMENTS

DATA COLLECTION
Technique: LEED
Dataset : IV spectra: 12 inequivalent beams at Θ=0°, 6 at 40° and 5 at 25°; off normal data for incidence in the mirror plane

THEORY/DATA TREATMENT
Dynamical LEED: 10 phase shifts; 120 beams; Voi=-4 eV; Vor=-5.6 eV (fit); ΘD=250 K (surface), 600K (bulk), both fit

STRUCTURES EXAMINED
Top layer buckling, relaxation, anion and cation lateral shifts and second bilayer buckling and relaxation were optimized using an unconstrained optimisation algorithm

QUALITY OF EXPERIMENT-THEORY FIT
RPE=0.64

2D UNIT CELLS (1 domain observed)

Cell	Ax (Å)	Ay (Å)	Bx (Å)	By (Å)	α (°)	Matrix	Pattern	Cell type
Bulk	4.582	0.000	0.000	6.480	90.0	(1.000, 0.000) (0.000, 1.000)	(1x1)	b: bulk lattice
Surface 1	4.582	0.000	0.000	6.480	90.0	(1.000, 0.000) (0.000, 1.000)	(1x1)	s1: commens. superlattice

3D COORDINATES

Te1-Cd2, Cd3-Te4: 2 bilayers with tilted Cd-Te chains; Cd5-Te6: bulk bilayer;
Te7-Cd8: periodically repeating bulk bilayer; 0.1Å lateral error bars assumed for tabulation

Dx/Dy in Å, or as a fraction of layer's unit cell vectors; atom 0 at the origin. Epir/subr are bulk repeat vectors.

No. of atoms: 8 Bulk z = 2.291 Å

Region	Chem el.	At. no.	Cell type	Site occ.	Rel. to	Dx ± εx		Dy ± εy		Dz ± εz		Dz/Bz(%) ± εz/Bz
epir		-2					f		f		Å	
subr		-1				-2.291	Å	3.240	Å	2.291	Å	
intf	Te	1	b	1.00	0	0.000	f	0.000	f	0.000	Å	0.0
intf	Cd	2	b	1.00	1	0.500	f	0.216 ± .015	f	0.820 ± .100	Å	35.8 ± 4.4
intf	Cd	3	b	1.00	2	-0.500	f	0.560 ± .015	f	1.610 ± .150	Å	70.3 ± 6.5
intf	Te	4	b	1.00	3	0.500	f	-0.250	f	0.080 ± .060	Å	3.5 ± 2.6
intf	Cd	5	b	1.00	4	0.000	f	-0.250	f	2.240 ± .080	Å	97.8 ± 3.5
intf	Te	6	b	1.00	5	-0.500	f	-0.250	f	0.000	Å	0.0
subl	Te	7	b	1.00	6	0.500	f	0.500	f	2.291	Å	100.0
subl	Cd	8	b	1.00	7	-0.500	f	0.250	f	0.000	Å	0.0

CdTe(110)-(1x1)
48.52.6 BOND DISTANCES AND ANGLES

Bond distances and angles are derived from coordinates

No. of distances/angles: 11

Interatomic dist. A-B (Å)	Atom A	Atom B	Atom C	Bond angle A-B-C (°)
2.807	Te1	Cd2	Te1(1,0)	109.4
2.806	Cd5	Te6	Cd8(0,-1)	109.5
2.806	Cd5	Te7	Cd8	109.5
2.807	Te1	Cd2	Te4	124.7
2.831	Te1	Cd3(0,-1)	Te4(0,-1)	108.7
2.831	Te1	Cd3(0,-1)	Te6	114.2
2.625	Cd2	Te4	Cd5	94.2
2.807	Cd3	Te4	Cd5	111.2
2.830	Cd3	Te6(0,1)	Cd8	109.8
2.764	Te4	Cd5	Te6	109.8
2.764	Te4	Cd5	Te7	108.9

COMMON NAME : Co(0001)-(1x1) ILLUSTRATION: 19
CLASSIFICATION : 27.5a
TECHNIQUE : LEED
AUTHORS : B.W. Lee, R. Alsenz, A. Ignatiev and M.A. Van Hove
REFERENCE : Phys. Rev., B17, 1510 (1978)

SURFACE TYPE
Substrate : Co
Crystal face: 0001
Temperature : 300 K
Bulk lattice: hcp
2D bulk symm: p3m1
2D surf symm: p3m1

Adsorbate:
Coverage :
Pattern : (1x1)
Matrix : (1.000, 0.000)
(0.000, 1.000)

STRUCTURE TYPE
Bulk hcp termination

SAMPLE PREPARATION (1 sample)
Treatment : annealing below 623 K to avoid martensitic transformation
Crystallinity:
Anal. methods:
Contamination: C (main contaminant): in AES noise

COMMENTS
Bulk has (martensitic) phase transformation from hcp to fcc at 450C

DATA COLLECTION
Technique: LEED
Dataset : I-V spectra: (Θ=0,φ=0): 10 and 01 beams,
(Θ=6°,φ=0): 00, 10, 01, -10, -11, -21 beams

THEORY/DATA TREATMENT
Dynamical LEED (RFS): Moruzzi-Janak-Williams potential, 8 phase shifts; Vor=-16.0 eV, Voi=-5.0eV; ΘD=315 K

STRUCTURES EXAMINED
Hcp/fcc terminations on hcp/fcc bulk with top layer spacing varied from 1.85 to 2.15Å in steps of 0.05Å

QUALITY OF EXPERIMENT-THEORY FIT
Visual

2D UNIT CELLS (1 domain observed)

Cell	Ax (Å)	Ay (Å)	Bx (Å)	By (Å)	α (°)	Matrix	Pattern	Cell type
Bulk	2.520	0.000	1.260	2.182	60.0	(1.000, 0.000) (0.000, 1.000)	(1x1)	b: bulk lattice
Surface 1	2.520	0.000	1.260	2.182	60.0	(1.000, 0.000) (0.000, 1.000)	(1x1)	s1: commens. superlattice

3D COORDINATES

Dx/Dy in Å, or as a fraction of layer's unit cell vectors; atom 0 at the origin. Epir/subr are bulk repeat vectors.

No. of atoms: 4 Bulk z = 2.050 Å

Region	Chem el.	At. no.	Cell type	Site occ.	Rel. to	Dx ± εx		Dy ± εy		Dz ± εz		Dz/Bz(%) ± εz/Bz
epir		-2					f		f		Å	
subr		-1				0.000	Å	0.000	Å	4.100	Å	
intf	Co	1	b	1.00	0	0.000	f	0.000	f	0.000	Å	0.0
intf	Co	2	b	1.00	1	0.333	f	0.333	f	2.050 ± .050	Å	100.0 ± 2.4
subl	Co	3	b	1.00	2	-0.333	f	-0.333	f	2.050	Å	100.0
subl	Co	4	b	1.00	3	0.333	f	0.333	f	2.050	Å	100.0

BOND DISTANCES AND ANGLES

Bond distances and angles are derived from coordinates

No. of distances/angles: 5

Interatomic dist. A-B (Å)	Atom A	Atom B	Atom C	Bond angle A-B-C (°)
2.520	Co1	Co1(1,0)	Co2(1,0)	120.1
2.520	Co1	Co1(1,0)	Co2	59.9
2.514	Co1	Co2	Co2(1,0)	120.1
2.514	Co1	Co2	Co3(1,0)	146.4
2.514	Co1	Co2	Co3	109.3

COMMON NAME : Co(10-10)-(1x1)
CLASSIFICATION : 27.9a
TECHNIQUE : LEED
AUTHORS : M. Lindroos, C.J. Barnes, P. Hu and D.A. King
REFERENCE : Chem. Phys. Lett., 173, 92 (1990)

ILLUSTRATION: 20

SURFACE TYPE
Substrate : Co
Crystal face: 10-10
Temperature : RT*
Bulk lattice: hcp
2D bulk symm: pmm
2D surf symm: pmm

Adsorbate:
Coverage :
Pattern : (1x1)
Matrix : (1.000, 0.000)
(0.000, 1.000)

STRUCTURE TYPE
Relaxed bulk with lower-corrugation termination (of two possible terminations for a (10-10) hcp surface) and multilayer relaxations

SAMPLE PREPARATION (1 sample)
Treatment : cycles of sputtering/annealing, and oxidation/reduction
Crystallinity: sharp low-background LEED
Anal. methods:
Contamination:

COMMENTS

DATA COLLECTION
Technique: LEED; Auto-LEED system
Dataset : IV curves for 8 inequivalent beams, 50<E<250 eV, normal incidence

THEORY/DATA TREATMENT
Dynamical LEED (layer doubling, composite layers): E-indep. Vor, Voi (fit); ΘD=385 K

STRUCTURES EXAMINED
Two different bulk terminations, fit of top 3 interlayer distances for preferred termination

QUALITY OF EXPERIMENT-THEORY FIT
Several R-factors used

2D UNIT CELLS (1 domain observed)

Cell	Ax (Å)	Ay (Å)	Bx (Å)	By (Å)	α (°)	Matrix	Pattern	Cell type
Bulk	2.512	0.000	0.000	4.077	90.0	(1.000, 0.000) (0.000, 1.000)	(1x1)	b: bulk lattice
Surface 1	2.512	0.000	0.000	4.077	90.0	(1.000, 0.000) (0.000, 1.000)	(1x1)	s1: commens. superlattice

3D COORDINATES

Co1-Co4: have relaxed interlayer spacings; Co7-Co8: periodically repeating set of bulk layers

Dx/Dy in Å, or as a fraction of layer's unit cell vectors; atom 0 at the origin. Epir/subr are bulk repeat vectors.

No. of atoms: 8

Bulk z = 1.450 Å

Region	Chem el.	At. no.	Cell type	Site occ.	Rel. to	Dx ± εx		Dy ± εy		Dz ± εz		Dz/Bz(%) ± εz/Bz
epir		-2					f		f		Å	
subr		-1				0.000	Å	2.039	Å	2.175	Å	
intf	Co	1	b	1.00	0	0.000	f	0.500	f	0.000	Å	0.0
intf	Co	2	b	1.00	1	0.500	f	-0.500	f	0.678 ± .015	Å	46.8 ± 1.0
intf	Co	3	b	1.00	2	-0.500	f	0.000	f	1.465 ± .029	Å	101.0 ± 2.0
intf	Co	4	b	1.00	3	0.500	f	0.500	f	0.725 ± .015	Å	50.0 ± 1.0
intf	Co	5	b	1.00	4	-0.500	f	0.000	f	1.450	Å	100.0
intf	Co	6	b	1.00	5	0.500	f	-0.500	f	0.725	Å	50.0
subl	Co	7	b	1.00	6	-0.500	f	0.000	f	1.450	Å	100.0
subl	Co	8	b	1.00	7	0.500	f	0.500	f	0.725	Å	50.0

BOND DISTANCES AND ANGLES

Bond distances and angles are derived from coordinates

No. of distances/angles: 7

Interatomic dist. A-B (Å)	Atom A	Atom B	Atom C	Bond angle A-B-C (°)
2.512	Co1	Co1(1,0)	Co2(1,0)	120.3
2.489	Co1	Co2	Co3	83.0
2.957	Co1	Co3	Co4	69.4

Co(10-10)-(1x1)
27.9a

Bond Distances and Angles - Continued

Interatomic dist. A-B (Å)	Atom A	Atom B	Atom C	Bond angle A-B-C (°)
2.957	Co1	Co3	Co4(0,-1)	140.5
2.957	Co1	Co3	Co4(-1,0)	69.4
1.929	Co2	Co3	Co4	83.9
2.992	Co2	Co4	Co5	123.6

COMMON NAME : Co(10-10)-(1x1) ILLUSTRATION: 20
CLASSIFICATION : 27.10
TECHNIQUE : LEED
AUTHORS : H. Over, G. Kleinle, G. Ertl, W. Moritz, K.H. Ernst, H. Wohlgemuth, K. Christmann and E. Schwarz
REFERENCE : Surf. Sci., 254, L469 (1991)

SURFACE TYPE
Substrate : Co
Crystal face: 10-10
Temperature : RT*
Bulk lattice: hcp
2D bulk symm: pmm
2D surf symm: pmm

Adsorbate:
Coverage :
Pattern : (1x1)
Matrix : (1.000, 0.000)
(0.000, 1.000)

STRUCTURE TYPE
Relaxed bulk with lower-corrugation termination (of two possible terminations for a (10-10) hcp surface)

SAMPLE PREPARATION (1 sample)
Treatment : cycles of sputtering and annealing
Crystallinity:
Anal. methods: HREELS to control cleanliness
Contamination:

COMMENTS
Care was taken not to go beyond 680 K in the annealing so as not to cross hcp to fcc transition at 700K

DATA COLLECTION
Technique: LEED; Auto-Leed system
Dataset : IV curves for 8 inequivalent beams, 70<E<380 eV, normal incidence

THEORY/DATA TREATMENT
Dynamical LEED (layer doubling, composite layers): automatic search for fit; E-dep. Vor, Voi (fit); ΘD=450 K

STRUCTURES EXAMINED
Two different bulk terminations, fit of 7 interlayer distances

QUALITY OF EXPERIMENT-THEORY FIT
RPE=0.310, RDE=0.240

2D UNIT CELLS (1 domain observed)

Cell	Ax (Å)	Ay (Å)	Bx (Å)	By (Å)	α (°)	Matrix	Pattern	Cell type
Bulk	2.507	0.000	0.000	4.070	90.0	(1.000, 0.000) (0.000, 1.000)	(1x1)	b: bulk lattice
Surface 1	2.507	0.000	0.000	4.070	90.0	(1.000, 0.000) (0.000, 1.000)	(1x1)	s1: commens. superlattice

3D COORDINATES

Co1-Co7 have relaxed interlayer spacings; Co8-Co9 periodically repeating set of bulk layers:

Dx/Dy in Å, or as a fraction of layer's unit cell vectors; atom 0 at the origin. Epir/subr are bulk repeat vectors.

No. of atoms: 9 Bulk z = 1.436 Å

Region	Chem el.	At. no.	Cell type	Site occ.	Rel. to	Dx ± εx		Dy ± εy		Dz ± εz		Dz/Bz(%) ± εz/Bz
epir		-2					f		f		Å	
subr		-1				1.254	Å	0.000	Å	2.154	Å	
intf	Co	1	b	1.00	0	0.000	f	0.500	f	0.000	Å	0.0
intf	Co	2	b	1.00	1	0.500	f	-0.500	f	0.625 ± .003	Å	43.5 ± .2
intf	Co	3	b	1.00	2	0.000	f	0.500	f	1.458 ± .003	Å	101.5 ± .2
intf	Co	4	b	1.00	3	0.500	f	-0.500	f	0.722 ± .005	Å	50.3 ± .3
intf	Co	5	b	1.00	4	0.000	f	0.500	f	1.436 ± .010	Å	100.0 ± .7
intf	Co	6	b	1.00	5	0.500	f	-0.500	f	0.724 ± .016	Å	50.4 ± 1.1
intf	Co	7	b	1.00	6	0.000	f	0.500	f	1.403 ± .055	Å	97.7 ± 3.8
subl	Co	8	b	1.00	7	0.500	f	-0.500	f	0.730 ± .040	Å	50.8 ± 2.8
subl	Co	9	b	1.00	8	0.000	f	0.500	f	1.436	Å	100.0

Co(10-10)-(1x1)
27.10

BOND DISTANCES AND ANGLES

Bond distances and angles are derived from coordinates

No. of distances/angles: 6

Interatomic dist. A-B (Å)	Atom A	Atom B	Atom C	Bond angle A-B-C (°)
2.507	Co1	Co1(1,0)	Co2(1,1)	120.5
2.507	Co1	Co1(1,0)	Co2(0,1)	59.5
2.507	Co1	Co1(1,0)	Co3(1,1)	90.0
2.470	Co1	Co2(0,1)	Co1(1,1)	150.7
2.470	Co1	Co2(0,1)	Co1(1,0)	61.0
2.470	Co1	Co2(0,1)	Co1(0,1)	110.9

COMMON NAME : Co(11-20)-(1x1)
CLASSIFICATION : 27.8
TECHNIQUE : LEED
AUTHORS : M. Welz, W. Moritz and D. Wolf
REFERENCE : Surf. Sci., 125, 473 (1983)

ILLUSTRATION: 21

SURFACE TYPE
Substrate : Co
Crystal face: 11-20
Temperature : 600 K
Bulk lattice: hcp
2D bulk symm: pmg
2D surf symm: pmg

Adsorbate:
Coverage :
Pattern : (1x1)
Matrix : (1.000, 0.000)
(0.000, 1.000)

STRUCTURE TYPE
Bulk termination with top spacing contraction, but no registry shift

SAMPLE PREPARATION (1 sample)
Treatment : spark erosion, chemical polishing, >100 sputter/anneals
Crystallinity: sharp LEED pattern with low background
Anal. methods:
Contamination: C, S removed by sputtering and heating

COMMENTS
RZJ minimizes at top spacing of 1.16Å with Vor=-15.5 eV;
RPE minimizes at top spacing of 1.13Å with Vor=-16 eV;
sample temperature was maintained at all times below the hcp-fcc transition temperature;
hence large number of ion bombardment/anneal cycles

DATA COLLECTION
Technique: LEED
Dataset : I-V spectra for 6 beams at normal incidence up to 240 eV

THEORY/DATA TREATMENT
Dynamical LEED: Co potential from band structure calcs, 8 phase shifts, up to 102 symm. beams; Voi=-5 eV; ΘD=385 K

STRUCTURES EXAMINED
Top layer spacing varied from 1.25Å (bulk value) to 1.07Å in steps of 0.03Å; lateral shift of top layer atoms along (1-10) direction varied from -0.05 to 0.10Å in steps of 0.05Å

QUALITY OF EXPERIMENT-THEORY FIT
RPE=0.22, RZJ=0.09

2D UNIT CELLS (1 domain observed)

Cell	Ax (Å)	Ay (Å)	Bx (Å)	By (Å)	α (°)	Matrix	Pattern	Cell type
Bulk	4.070	0.000	0.000	4.340	90.0	(1.000, 0.000) (0.000, 1.000)	(1x1)	b: bulk lattice
Surface 1	4.070	0.000	0.000	4.340	90.0	(1.000, 0.000) (0.000, 1.000)	(1x1)	s1: commens. superlattice

3D COORDINATES

Co1-Co2, Co3-Co4, Co5-Co6: coplanar bilayers; 0.1Å lateral error bars assumed for tabulation

Dx/Dy in Å, or as a fraction of layer's unit cell vectors; atom 0 at the origin. Epir/subr are bulk repeat vectors.

No. of atoms: 6 Bulk z = 1.250 Å

Region	Chem el.	At. no.	Cell type	Site occ.	Rel. to	Dx ± εx		Dy ± εy		Dz ± εz		Dz/Bz(%) ± εz/Bz
epir		-2					f		f		Å	
subr		-1				0.000	Å	0.000	Å	2.500	Å	
intf	Co	1	b	1.00	0	0.000	f	0.000	f	0.000	Å	0.0
intf	Co	2	b	1.00	1	0.500	f	0.333	f	0.000	Å	0.0
subl	Co	3	b	1.00	2	-0.500	f	0.167 ± .023	f	1.140 ± .040	Å	91.2 ± 3.2
subl	Co	4	b	1.00	3	0.500	f	0.333	f	0.000	Å	0.0
subl	Co	5	b	1.00	4	-0.500	f	-0.833	f	1.250	Å	100.0
subl	Co	6	b	1.00	5	0.500	f	0.333	f	0.000	Å	0.0

BOND DISTANCES AND ANGLES

Bond distances and angles are derived from coordinates

No. of distances/angles: 7

Interatomic dist. A-B (Å)	Atom A	Atom B	Atom C	Bond angle A-B-C (°)
2.497	Co1	Co2	Co1(1,0)	109.2
2.497	Co1	Co2	Co3	59.5
2.497	Co1	Co2	Co4	120.9
2.451	Co1	Co3	Co1(0,1)	124.6

Co(11-20)-(1x1)
27.8

Bond Distances and Angles - Continued

Interatomic dist. A-B (Å)	Atom A	Atom B	Atom C	Bond angle A-B-C (°)
2.451	Co1	Co3	Co2	61.4
2.451	Co1	Co3	Co4	120.9
2.442	Co2	Co3	Co4	59.5

COMMON NAME : Co(100)-(1x1) ILLUSTRATION: 2
CLASSIFICATION : 27.4
TECHNIQUE : LEED
AUTHORS : M. Maglietta, E. Zanazzi, F. Jona, D.W. Jepsen and P.M. Marcus
REFERENCE : Appl. Phys., 15, 409 (1978)

SURFACE TYPE
Substrate : Co
Crystal face: 100
Temperature : 300 K
Bulk lattice: fcc
2D bulk symm: p4m
2D surf symm: p4m
Adsorbate:
Coverage :
Pattern : (1x1)
Matrix : (1.000, 0.000)
(0.000, 1.000)

STRUCTURE TYPE
Bulk termination with top spacing contraction

SAMPLE PREPARATION (1 sample)
Treatment : Ar+ bombardments and anneals, rapid cooling to RT
Crystallinity:
Anal. methods:
Contamination: AES: 0.1 ML C, O at noise level

COMMENTS
Large surface contraction may be due to incipient formation of C superstructure, though no direct experimental evidence for this; relatively poor agreement between theory and experiment for largest incident angles probably due to surface roughness

DATA COLLECTION
Technique: LEED
Dataset : I-V spectra for 12 beams at 3 angles of incidence, 30<E<150 eV

THEORY/DATA TREATMENT
Dynamical LEED: 49 beams, 8 phase shifts; Co pot from band struct calcs; Vor=-16.5 eV, Voi=-3.5eV; rms ampl 0.123Å

STRUCTURES EXAMINED
Truncated bulk structure with variations in the first interlayer spacing from 1.69 to 1.85Å.

QUALITY OF EXPERIMENT-THEORY FIT
RZJ=0.19

2D UNIT CELLS (1 domain observed)

Cell	Ax (Å)	Ay (Å)	Bx (Å)	By (Å)	α (°)	Matrix	Pattern	Cell type
Bulk	2.503	0.000	0.000	2.503	90.0	(1.000, 0.000) (0.000, 1.000)	(1x1)	b: bulk lattice
Surface 1	2.503	0.000	0.000	2.503	90.0	(1.000, 0.000) (0.000, 1.000)	(1x1)	s1: commens. superlattice

3D COORDINATES

0.05Å error bar assumed for tabulation

Dx/Dy in Å, or as a fraction of layer's unit cell vectors; atom 0 at the origin. Epir/subr are bulk repeat vectors.

No. of atoms: 3 Bulk z = 1.770 Å

Region	Chem el.	At. no.	Cell type	Site occ.	Rel. to	Dx ± εx		Dy ± εy		Dz ± εz		Dz/Bz(%) ± εz/Bz
epir		-2					f		f		Å	
subr		-1				-1.252	Å	-1.252	Å	1.770	Å	
intf	Co	1	b	1.00	0	0.000	f	0.000	f	0.000	Å	0.0
intf	Co	2	b	1.00	1	0.500	f	0.500	f	1.700 ± .050	Å	96.1 ± 2.3
subl	Co	3	b	1.00	2	-0.500	f	-0.500	f	1.770	Å	100.0

BOND DISTANCES AND ANGLES

Bond distances and angles are derived from coordinates

No. of distances/angles: 3

Interatomic dist. A-B (Å)	Atom A	Atom B	Atom C	Bond angle A-B-C (°)
2.503	Co1	Co1(1,0)	Co2	59.3
2.454	Co1	Co2	Co3	88.9
2.503	Co2	Co3		

COMMON NAME : Co(111)-(1x1)
CLASSIFICATION : 27.5b
TECHNIQUE : LEED
AUTHORS : B.W. Lee, R. Alsenz, A. Ignatiev and M.A. Van Hove
REFERENCE : Phys. Rev., B17, 1510 (1978)

ILLUSTRATION: 1

SURFACE TYPE
Substrate : Co
Crystal face: 111
Temperature : 730 K
Bulk lattice: fcc
2D bulk symm: p3m1
2D surf symm: p3m1

Adsorbate:
Coverage :
Pattern : (1x1)
Matrix : (1.000, 0.000)
(0.000, 1.000)

STRUCTURE TYPE
Bulk fcc termination

SAMPLE PREPARATION (1 sample)
Treatment : heating hcp(0001) through 450C to fcc(111)
Crystallinity:
Anal. methods:
Contamination: C (main contaminant): in AES noise

COMMENTS
Bulk has (martensitic) phase transformation from hcp to fcc at 450C

DATA COLLECTION
Technique: LEED
Dataset : I-V spectra: (Θ=6,φ=0°): 00, 10, 01, -10, -11, -21 beams

THEORY/DATA TREATMENT
Dynamical LEED (RFS): Moruzzi-Janak-Williams potential; 8 phase shifts; Vor=-16.0 eV, Voi=-5.0eV

STRUCTURES EXAMINED
Hcp/fcc terminations on hcp/fcc bulk with the top layer spacing varied from 1.95 to 2.15Å in steps of 0.1Å

QUALITY OF EXPERIMENT-THEORY FIT
Visual

2D UNIT CELLS (1 domain observed)

Cell	Ax (Å)	Ay (Å)	Bx (Å)	By (Å)	α (°)	Matrix	Pattern	Cell type
Bulk	2.510	0.000	1.255	2.174	60.0	(1.000, 0.000) (0.000, 1.000)	(1x1)	b: bulk lattice
Surface 1	2.510	0.000	1.255	2.174	60.0	(1.000, 0.000) (0.000, 1.000)	(1x1)	s1: commens. superlattice

3D COORDINATES

Dx/Dy in Å, or as a fraction of layer's unit cell vectors; atom 0 at the origin. Epir/subr are bulk repeat vectors.

No. of atoms: 3 Bulk z = 2.050 Å

Region	Chem el.	At. no.	Cell type	Site occ.	Rel. to	Dx ± εx		Dy ± εy		Dz ± εz		Dz/Bz(%) ± εz/Bz
epir		-2					f		f		Å	
subr		-1				1.255	Å	0.725	Å	2.050	Å	
intf	Co	1	b	1.00	0	0.000	f	0.000	f	0.000	Å	0.0
intf	Co	2	b	1.00	1	0.333	f	0.333	f	2.050 ± .050	Å	100.0 ± 2.4
subl	Co	3	b	1.00	2	0.333	f	0.333	f	2.050	Å	100.0

BOND DISTANCES AND ANGLES

Bond distances and angles are derived from coordinates

No. of distances/angles: 1

Interatomic dist. A-B (Å)	Atom A	Atom B	Atom C	Bond angle A-B-C (°)
2.510	Co1	Co1(1,0)		

COMMON NAME : Co(10-10)-c(2x2)-K
CLASSIFICATION : 27.19.1
TECHNIQUE : LEED
AUTHORS : C.J. Barnes, P. Hu, M. Lindroos and D.A. King
REFERENCE : Surf. Sci., 251, 561 (1991)

ILLUSTRATION: 60

SURFACE TYPE
Substrate : Co
Crystal face: 10-10
Temperature : RT
Bulk lattice: hcp
2D bulk symm: pmm
2D surf symm: cm

Adsorbate: K
Coverage : 0.5 K/1x1
Pattern : c(2x2)
Matrix : (1.000, 1.000)
(-1.000, 1.000)

STRUCTURE TYPE
Atomic adsorption at 4-fold hollow of bulk terminated structure of lower corrugation (see 27.10).Care taken not

SAMPLE PREPARATION (1 sample)
Treatment : crystal cleaned by cycles of sputtering and annealing
Crystallinity:
Anal. methods:
Contamination: checked by AES and XPS

COMMENTS
To go beyond 650 K in annealing not to cross hcp to fcc transition at 700 K; deep minima in R-factors for d(K-Co)=2.44 and 1.8Å, the first being slightly lower

DATA COLLECTION
Technique: LEED; Auto-LEED system
Dataset : IV curves for 8 inequivalent beams, 50<E<200 eV, normal incidence

THEORY/DATA TREATMENT
Dynamical LEED: Moruzzi phase shifts for Co and K; ΘD=385 K(Co),200K(K); Voi=-4 eV

STRUCTURES EXAMINED
For the two different bulk terminations: K in hollow, top, long bridge and short bridge sites; variation of K-Co and Co1-Co2 spacing

QUALITY OF EXPERIMENT-THEORY FIT
RPE=0.310, RDE=0.240

2D UNIT CELLS (1 domain observed)

Cell	Ax (Å)	Ay (Å)	Bx (Å)	By (Å)	α (°)	Matrix	Pattern	Cell type
Bulk	2.507	0.000	0.000	4.070	90.0	(1.000, 0.000) (0.000, 1.000)	(1x1)	b: bulk lattice
Surface 1	2.507	4.070	-2.507	4.070	63.3	(1.000, 1.000) (-1.000, 1.000)	c(2x2)	s1: commens. superlattice

3D COORDINATES

K1: atomic overlayer in 4-fold-coordinated hollow sites; Co3-Co4: periodically repeating set of bulk layers; error bar set to 0.1Å for tabulation

Dx/Dy in Å, or as a fraction of layer's unit cell vectors; atom 0 at the origin. Epir/subr are bulk repeat vectors.

No. of atoms: 6

Bulk z = 1.436 Å

Region	Chem el.	At. no.	Cell type	Site occ.	Rel. to	Dx ± εx		Dy ± εy		Dz ± εz		Dz/Bz(%) ± εz/Bz
epir		-2					f		f		Å	
subr		-1				0.000	Å	0.000	Å	4.308	Å	
ovrl	K	1	s1	.50	0	0.000	f	0.000	f	0.000	Å	0.0
intf	Co	2	b	1.00	1	0.500	f	0.500	f	2.440 ± .100	Å	169.9 ± 7.0
subl	Co	3	b	1.00	2	-0.500	f	-0.500	f	0.680 ± .100	Å	47.4 ± 7.0
subl	Co	4	b	1.00	3	0.000	f	0.500	f	1.436	Å	100.0
subl	Co	5	b	1.00	4	0.500	f	-0.500	f	0.718	Å	50.0
subl	Co	6	b	1.00	5	0.000	f	0.500	f	1.436	Å	100.0

BOND DISTANCES AND ANGLES

Bond distances and angles are derived from coordinates

No. of distances/angles: 10

Interatomic dist. A-B (Å)	Atom A	Atom B	Atom C	Bond angle A-B-C (°)
3.416	K1	Co2	K1(1,0)	88.8
2.459	Co2	Co4(1,0)	Co3(1,1)	60.3
3.416	K1	Co2	Co2(1,0)	111.5

Co(10-10)-c(2x2)-K
27.19.1

Bond Distances and Angles - Continued

Interatomic dist. A-B (Å)	Atom A	Atom B	Atom C	Bond angle A-B-C (°)
3.120	K1	Co3	Co2	74.1
3.120	K1	Co3	Co3(1,0)	90.0
2.507	Co2	Co2(1,0)	K1(1,0)	68.5
2.507	Co2	Co2(1,0)	Co2(2,0)	180.0
2.485	Co2	Co3(1,1)	K1(1,0)	74.1
2.485	Co2	Co3(1,1)	Co2(1,0)	60.6
2.459	Co2	Co4(1,0)	Co2(1,0)	61.3

COMMON NAME : Co(100)-c(2x2)-O
CLASSIFICATION : 27.8.1
TECHNIQUE : LEED
AUTHORS : M. Maglietta, E. Zanazzi, U. Bardi and F. Jona
REFERENCE : Surf. Sci., 77, 101 (1978)

ILLUSTRATION: 28,29

SURFACE TYPE
Substrate : Co
Crystal face: 100
Temperature : 300 K
Bulk lattice: fcc
2D bulk symm: p4m
2D surf symm: p4m

Adsorbate: O
Coverage : 0.5 O/Co
Pattern : c(2x2)
Matrix : (1.000, 1.000)
(-1.000, 1.000)

STRUCTURE TYPE
Atomic adsorption in 4-fold hollow sites

SAMPLE PREPARATION (1 sample)
Treatment : exposed to O2 gas at 1.0E-9 torr for 20 mins at 300 K
Crystallinity: diffuse LEED spots observed
Anal. methods:
Contamination: monitored by AES

COMMENTS
Insensitivity of the R-factor to the substrate interlayer spacings attributed to imperfect conditions on substrate; rms vibr ampls: 0.123Å

DATA COLLECTION
Technique: LEED
Dataset : 13 LEED I-V spectra: 3 at normal incidence, 10 at 2 off- normal angles; 40<E<140 eV

THEORY/DATA TREATMENT
Dynamical LEED (KKR): 8 ph shs; 58 beams; self consistent Co pot; O: overlap of atomic charges; Vor=-16 eV, Voi=-3eV

STRUCTURES EXAMINED
Truncated bulk varying O-Co spacing d: top sites 1.846<d<2.136 Å bridge sites 1.217<d<1.402 Å 4-fold hollows 0.770<d<1.068 Å

QUALITY OF EXPERIMENT-THEORY FIT
RZJ=0.21

2D UNIT CELLS (1 domain observed)

Cell	Ax (Å)	Ay (Å)	Bx (Å)	By (Å)	α (°)	Matrix	Pattern	Cell type
Bulk	2.503	0.000	0.000	2.503	90.0	(1.000, 0.000) (0.000, 1.000)	(1x1)	b: bulk lattice
Surface 1	2.503	2.503	-2.503	2.503	90.0	(1.000, 1.000) (-1.000, 1.000)	c(2x2)	s1: commens. superlattice

3D COORDINATES

O1: overlayer in 4-fold hollow sites 0.1Å error bar assumed for tabulation

Dx/Dy in Å, or as a fraction of layer's unit cell vectors; atom 0 at the origin. Epir/subr are bulk repeat vectors.

No. of atoms: 3 Bulk z = 1.770 Å

Reg ion	Chem el.	At. no.	Cell type	Site occ.	Rel. to	Dx ± εx		Dy ± εy		Dz ± εz		Dz/Bz(%) ± εz/Bz
epir		-2					f		f		Å	
subr		-1				-1.252	Å	-1.252	Å	1.770	Å	
ovrl	O	1	s1	.50	0	0.000	f	0.000	f	0.000	Å	0.0
intf	Co	2	b	1.00	1	0.500	f	0.500	f	0.800 ± .100	Å	45.2 ± 5.7
subl	Co	3	b	1.00	2	-0.500	f	-0.500	f	1.770	Å	100.0

BOND DISTANCES AND ANGLES

Bond distances and angles are derived from coordinates

No. of distances/angles: 5

Interatomic dist. A-B (Å)	Atom A	Atom B	Atom C	Bond angle A-B-C (°)
1.942	O1	Co2	O1(1,0)	131.4
1.942	O1	Co2	Co2(1,0)	130.1
1.942	O1	Co2	Co3	69.3
2.570	O1	Co3	Co2	45.0
2.503	Co2	Co2(1,0)	O1(1,0)	49.9

COMMON NAME : Co(100)-c(2x2)-S
CLASSIFICATION : 27.16.1
TECHNIQUE : LEED
AUTHORS : M. Maglietta
REFERENCE : Solid State Commun., 43, 395 (1982)

ILLUSTRATION: 28,29

SURFACE TYPE
Substrate : Co
Crystal face: 100
Temperature : 300 K
Bulk lattice: fcc
2D bulk symm: p4m
2D surf symm: p4m

Adsorbate: S
Coverage : 0.5 S/Co
Pattern : c(2x2)
Matrix : (1.000, 1.000)
(-1.000, 1.000)

STRUCTURE TYPE
Atomic adscrption in 4-fold hollow sites

SAMPLE PREPARATION (1 sample)
Treatment : see Maglietta et al, Surf. Sci. 71, 495 (1978)
Crystallinity:
Anal. methods:
Contamination:

COMMENTS

DATA COLLECTION
Technique: LEED
Dataset : I-V spectra for (0.5,0.5),(10) and (11) beams at normal incidence, E<=140 eV

THEORY/DATA TREATMENT
Visual comparison with calculated I-Vs for Ni(100)-c(2x2)-S, since spectra are very similar for clean Ni and Co surfaces

STRUCTURES EXAMINED
Only the 4-fold hollow sites were compatible with data; calc'd I-Vs taken from Demuth et al, Surf. Sci. 45, 249 (1974)

QUALITY OF EXPERIMENT-THEORY FIT
Visual

2D UNIT CELLS (1 domain observed)

Cell	Ax (Å)	Ay (Å)	Bx (Å)	By (Å)	α (°)	Matrix	Pattern	Cell type
Bulk	2.503	0.000	0.000	2.503	90.0	(1.000, 0.000) (0.000, 1.000)	(1x1)	b: bulk lattice
Surface 1	2.503	2.503	-2.503	2.503	90.0	(1.000, 1.000) (-1.000, 1.000)	c(2x2)	s1: commens. superlattice

3D COORDINATES

S1: overlayer in 4-fold hollow sites; 0.1Å error bar assumed for tabulation

Dx/Dy in Å, or as a fraction of layer's unit cell vectors; atom 0 at the origin. Epir/subr are bulk repeat vectors.

No. of atoms: 3 Bulk z = 1.770 Å

Region	Chem el.	At. no.	Cell type	Site occ.	Rel. to	Dx ± εx		Dy ± εy		Dz ± εz		Dz/Bz(%) ± εz/Bz
epir		-2					f		f		Å	
subr		-1				-1.252	Å	-1.252	Å	1.770	Å	
ovrl	S	1	s1	.50	0	0.000	f	0.000	f	0.000	Å	0.0
intf	Co	2	b	1.00	1	0.500	f	0.500	f	1.300 ± .100	Å	73.5 ± 5.7
subl	Co	3	b	1.00	2	-0.500	f	-0.500	f	1.770	Å	100.0

BOND DISTANCES AND ANGLES

Bond distances and angles are derived from coordinates

No. of distances/angles: 4

Interatomic dist. A-B (Å)	Atom A	Atom B	Atom C	Bond angle A-B-C (°)
2.196	S1	Co2	S1(1,0)	107.4
2.196	S1	Co2	Co2(1,0)	124.7
2.196	S1	Co2	Co3	81.3
2.503	Co2	Co2(1,0)	S1(1,0)	55.3

COMMON NAME : CoO(100)-(1x1)
CLASSIFICATION : 27.8.3
TECHNIQUE : LEED
AUTHORS : R.C. Felton, M. Prutton, S.P. Tear and M.R. Welton-Cook
REFERENCE : Surf. Sci., 88, 474 (1979)

ILLUSTRATION: 149

SURFACE TYPE
Substrate : CoO
Crystal face: 100
Temperature : 300 K
Bulk lattice: NaCl
2D bulk symm: p4m
2D surf symm: p4m

Adsorbate:
Coverage :
Pattern : (1x1)
Matrix : (1.000, 0.000)
(0.000, 1.000)

STRUCTURE TYPE
Non-buckled bulk termination in non-polar mixed CoO layer

SAMPLE PREPARATION (1 sample)
Treatment : CoO crystal cleaved in situ at RT and at 8.0E-10 torr
Crystallinity:
Anal. methods:
Contamination: AES and QMS to test for ESD

COMMENTS
Low R-factor value due to data smoothing and consequent reduction in noise level

DATA COLLECTION
Technique: LEED
Dataset : I-V spectra at normal incidence for the (11), (20) and (12) beams; 170<E<400 eV.

THEORY/DATA TREATMENT
Dynamical LEED: pots from Clementi charge density for Co++ and O--; Vor=-12 eV (optimized), Voi=-5eV (optimized)

STRUCTURES EXAMINED
First interlayer spacing relaxations of 0%, +3%, +6%

QUALITY OF EXPERIMENT-THEORY FIT
RZJ=0.101

2D UNIT CELLS (1 domain observed)

Cell	Ax (Å)	Ay (Å)	Bx (Å)	By (Å)	α (°)	Matrix	Pattern	Cell type
Bulk	3.016	0.000	0.000	3.016	90.0	(1.000, 0.000) (0.000, 1.000)	(1x1)	b: bulk lattice
Surface 1	3.016	0.000	0.000	3.016	90.0	(1.000, 0.000) (0.000, 1.000)	(1x1)	s1: commens. superlattice

3D COORDINATES

Co1-O2: mixed non-polar non-buckled top layer; Co3-O4: periodically repeating bulk layer

Dx/Dy in Å, or as a fraction of layer's unit cell vectors; atom 0 at the origin. Epir/subr are bulk repeat vectors.

No. of atoms: 4 Bulk z = 2.133 Å

Region	Chem el.	At. no.	Cell type	Site occ.	Rel. to	Dx ± εx		Dy ± εy		Dz ± εz		Dz/Bz(%) ± εz/Bz
epir		-2					f		f		Å	
subr		-1				1.508	Å	1.508	Å	2.133	Å	
intf	Co	1	b	1.00	0	0.000	f	0.000	f	0.000	Å	0.0
intf	O	2	b	1.00	1	0.500	f	0.500	f	0.000	Å	0.0
subl	Co	3	b	1.00	2	0.000	f	0.000	f	2.133 ± .080	Å	100.0 ± 3.8
subl	O	4	b	1.00	3	-0.500	f	-0.500	f	0.000	Å	0.0

BOND DISTANCES AND ANGLES

Bond distances are derived from coordinates

No. of distances/angles: 3

Interatomic dist. A-B (Å)	Atom A	Atom B	Atom C	Bond angle A-B-C (°)
3.016	Co1	Co1(1,0)		
2.133	Co1	O2		
3.016	Co1	Co3		

COMMON NAME : CoO(111)-(1x1)
CLASSIFICATION : 27.8.2
TECHNIQUE : LEED
AUTHORS : A. Ignatiev, B.W. Lee and M.A. Van Hove
REFERENCE : Proc 7th IVC and 3rd ICSS (Vienna), -, 2435 (1977)

ILLUSTRATION: 150

SURFACE TYPE
Substrate : CoO
Crystal face: 111
Temperature : 300 K
Bulk lattice: NaCl
2D bulk symm: p3m1
2D surf symm: p3m1

Adsorbate:
Coverage :
Pattern : (1x1)
Matrix : (1.000, 0.000)
(0.000, 1.000)

STRUCTURE TYPE
Bulk-like CoO termination with O atom at polar surface

SAMPLE PREPARATION (1 sample)
Treatment : 1000L of O2 at 773 K on Co(0001) and long anneal
Crystallinity:
Anal. methods:
Contamination: cleaned by ion bombardment and annealed

COMMENTS
Lattice is expanded in 3D by 3.65% from bulk CoO values; this fits pattern and IVs better than bulk lattice or 2D expanded lattice with contracted substrate layer spacings

DATA COLLECTION
Technique: LEED
Dataset : I-V spectra: 5 beams at Θ=0°, 5 beams at Θ=6°

THEORY/DATA TREATMENT
Dynamical LEED (RFS): Moruzzi-Janak-Williams pot for Co, O pot not given (Co+, O- pots also tested); Vor=-15 eV,Voi=-5eV

STRUCTURES EXAMINED
Fcc bulk with fcc/hcp termination in either O or Co; top layer spacing varied from 1.014 to 1.214Å in steps of 0.05Å

QUALITY OF EXPERIMENT-THEORY FIT
Visual

2D UNIT CELLS (1 domain observed)

Cell	Ax (Å)	Ay (Å)	Bx (Å)	By (Å)	α (°)	Matrix	Pattern	Cell type
Bulk	3.120	0.000	1.560	2.702	60.0	(1.000, 0.000) (0.000, 1.000)	(1x1)	b: bulk lattice
Surface 1	3.120	0.000	1.560	2.702	60.0	(1.000, 0.000) (0.000, 1.000)	(1x1)	s1: commens. superlattice

3D COORDINATES

O1-Co2: contracted top layer pair; O3-Co4: periodically repeating bulk layer pair;
0.1Å error bars assumed for tabulation

Dx/Dy in Å, or as a fraction of layer's unit cell vectors; atom 0 at the origin. Epir/subr are bulk repeat vectors.

No. of atoms: 4 Bulk z = 2.546 Å

Region	Chem el.	At. no.	Cell type	Site occ.	Rel. to	Dx ± εx		Dy ± εy		Dz ± εz		Dz/Bz(%) ± εz/Bz
epir		-2					f		f		Å	
subr		-1				-1.560	Å	-0.901	Å	2.546	Å	
intf	O	1	b	1.00	0	0.000	f	0.000	f	0.000	Å	0.0
intf	Co	2	b	1.00	1	0.333 ± .014	f	0.333 ± .037	f	1.060 ± .100	Å	41.6 ± 3.9
subl	O	3	b	1.00	2	0.333 ± .014	f	0.333 ± .037	f	1.273 ± .100	Å	50.0 ± 3.9
subl	Co	4	b	1.00	3	-0.667 ± .014	f	-0.667 ± .037	f	1.273 ± .100	Å	50.0 ± 3.9

BOND DISTANCES AND ANGLES

Bond distances and angles are derived from coordinates

No. of distances/angles: 4

Interatomic dist. A-B (Å)	Atom A	Atom B	Atom C	Bond angle A-B-C (°)
2.090	O1	Co2	O1(1,0)	96.6
2.090	O1	Co2	O3	175.2
2.090	O1	Co2	O3(-1,0)	86.6
2.206	Co2	O3	Co2(1,0)	90.0

COMMON NAME : CoSi2(111)-(1x1)
CLASSIFICATION : 14.27.1
TECHNIQUE : LEED
AUTHORS : S.C. Wu, Z.Q. Wang, Y.S. Li, F. Jona and P.M. Marcus
REFERENCE : Phys. Rev., B33, 2900 (1986)

ILLUSTRATION: 145

SURFACE TYPE
Substrate : CoSi2
Crystal face: 111
Temperature : RT
Bulk lattice: fluorite
2D bulk symm: p3m1
2D surf symm: p3m1

Adsorbate:
Coverage :
Pattern : (1x1)
Matrix : (1.000, 0.000)
(0.000, 1.000)

STRUCTURE TYPE
Unreconstructed termination of bulk fluorite structure between trilayers; top two interlayer spacings contracted

SAMPLE PREPARATION (1 sample)
Treatment : Co vaporized onto Si(111)-(7x7), then annealed at ≈973 K
Crystallinity:
Anal. methods:
Contamination:

COMMENTS
2-4 ML annealed at ≈873 K form (1x1) phase called pre-silicide (PRS); >8 ML annealed at 973 K form (2x2) or (2x1) phase, which converts to 1x1 after anneals at 1073 K and is called post-silicide (POS) phase: PRS and POS structures unknown; SiCoSi termination is inferred from paper

DATA COLLECTION
Technique: LEED
Dataset : I-V curves for 11 or 12 beams, corresponding to 5 or 6 non degenerate beams at each angle of incidence; 30<E<250

THEORY/DATA TREATMENT
Dynamical LEED analysis

STRUCTURES EXAMINED
Semi-infinite CoSi2(111) with relaxation of two topmost interlayer spacings

2D UNIT CELLS (1 domain observed)

Cell	Ax (Å)	Ay (Å)	Bx (Å)	By (Å)	α (°)	Matrix	Pattern	Cell type
Bulk	3.838	0.000	-1.919	3.324	120.0	(1.000, 0.000) (0.000, 1.000)	(1x1)	b: bulk lattice
Surface 1	3.838	0.000	-1.919	3.324	120.0	(1.000, 0.000) (0.000, 1.000)	(1x1)	s1: commens. superlattice

3D COORDINATES

Si1-Co2-Si3: relaxed topmost trilayer; Si4-Co5-Si6: periodically repeating bulk trilayer; 0.1Å error bars assumed for tabulation

Dx/Dy in Å, or as a fraction of layer's unit cell vectors; atom 0 at the origin. Epir/subr are bulk repeat vectors.

No. of atoms: 6 Bulk z = 3.092 Å

Reg ion	Chem el.	At. no.	Cell type	Site occ.	Rel. to	Dx ± εx		Dy ± εy		Dz ± εz		Dz/Bz(%) ± εz/Bz
epir		-2					f		f		Å	
subr		-1				-1.919	Å	-1.108	Å	3.092	Å	
intf	Si	1	b	1.00	0	0.000	f	0.000	f	0.000	Å	0.0
intf	Co	2	b	1.00	1	0.333	f	0.667	f	0.618 ± .100	Å	20.0 ± 3.2
intf	Si	3	b	1.00	2	0.333	f	-0.333	f	0.734 ± .100	Å	23.7 ± 3.2
subl	Si	4	b	1.00	3	-0.333	f	0.333	f	1.546	Å	50.0
subl	Co	5	b	1.00	4	0.333	f	-0.333	f	0.773	Å	25.0
subl	Si	6	b	1.00	5	-0.667	f	-0.333	f	0.773	Å	25.0

BOND DISTANCES AND ANGLES

Bond distances and angles are derived from coordinates

No. of distances/angles: 5

Interatomic dist. A-B (Å)	Atom A	Atom B	Atom C	Bond angle A-B-C (°)
2.300	Si1	Co2	Si3	68.1
2.300	Si1	Co2	Si4	105.6
2.334	Co2	Si3	Si4	53.2

CoSi2(111)-(1x1)
14.27.1

Bond Distances and Angles - Continued

Interatomic dist. A-B (Å)	Atom A	Atom B	Atom C	Bond angle A-B-C (°)
2.702	Si3	Si4	Co5	54.1
2.347	Si4	Co5	Si6	70.3

COMMON NAME : CoSi2(111)-(1x1) ILLUSTRATION: 145
CLASSIFICATION : 14.27.14
TECHNIQUE : XPD
AUTHORS : H.C. Poon, G. Grenet, S. Holmberg, Y. Jugnet and Tran Minh Duc
REFERENCE : Phys. Rev., B41, 12735 (1990)

SURFACE TYPE
Substrate : CoSi2
Crystal face: 111
Temperature : RT*
Bulk lattice: fluorite
2D bulk symm: p3m1
2D surf symm: p3m1

Adsorbate:
Coverage :
Pattern : (1x1)
Matrix : (1.000, 0.000)
(0.000, 1.000)

STRUCTURE TYPE
Unreconstructed termination of bulk fluorite structure

SAMPLE PREPARATION (1 sample)
Treatment : sample cleaned by cycles of Ar sputtering and 800C annealing
Crystallinity: checked by LEED
Anal. methods:
Contamination: no impurities by XPS

COMMENTS
No relaxations have been allowed; authors discriminate between 8 different models based on visual comparison of XPS exp. azimuthal distribution and theoretically computed distributions (at Θ=70°). Structure determined agrees with results for Co rich surface in 14.27.11

DATA COLLECTION
Technique: XPD; x-ray source Al Kα at 1486.6eV
Dataset : Co LVV and Si 2s azimuthal signal at various polar angles

THEORY/DATA TREATMENT
Multiple scattering analysis like LEED using RFS or CSM
1)bulk Si terminated, 2)bulk Co terminated,3) same as 1) but

STRUCTURES EXAMINED
With additional Si bilayer bonded to Co, 4)same as 3) but bilayer rotated 180°, 5) same as 1) but with additional Si bilayer bonded to Si, 6) same as 5) but rotated bilayer, 7) same as 2) but additional Si bilayer bonded to Co,8) same as 7) but rotated bilayer. 1) is the accepted structure

2D UNIT CELLS (1 domain observed)

Cell	Ax (Å)	Ay (Å)	Bx (Å)	By (Å)	α (°)	Matrix	Pattern	Cell type
Bulk	3.838	0.000	-1.919	3.324	120.0	(1.000, 0.000) (0.000, 1.000)	(1x1)	b: bulk lattice
Surface 1	3.838	0.000	-1.919	3.324	120.0	(1.000, 0.000) (0.000, 1.000)	(1x1)	s1: commens. superlattice

3D COORDINATES

Si1-Co2-Si3: topmost trilayer; Si4-Co5-Si6: periodically repeating bulk trilayer;
no error bar quoted because no relax. was allowed

Dx/Dy in Å, or as a fraction of layer's unit cell vectors; atom 0 at the origin. Epir/subr are bulk repeat vectors.

No. of atoms: 6 Bulk z = 3.092 Å

Region	Chem el.	At. no.	Cell type	Site occ.	Rel. to	Dx ± εx		Dy ± εy		Dz ± εz		Dz/Bz(%) ± εz/Bz
epir		-2					f		f		Å	
subr		-1				-1.919	Å	-1.108	Å	3.092	Å	
intf	Si	1	s1	1.00	0	0.000	f	0.000	f	0.000	Å	0.0
intf	Co	2	s1	1.00	1	0.333	f	0.667	f	0.773	Å	25.0
intf	Si	3	s1	1.00	2	0.333	f	-0.333	f	0.773	Å	25.0
subl	Si	4	b	1.00	3	-0.333	f	0.333	f	1.546	Å	50.0
subl	Co	5	b	1.00	4	0.333	f	-0.333	f	0.773	Å	25.0
subl	Si	6	b	1.00	5	-0.667	f	-0.333	f	0.773	Å	25.0

BOND DISTANCES AND ANGLES

Bond distances and angles are derived from coordinates

No. of distances/angles: 8

Interatomic dist. A-B (Å)	Atom A	Atom B	Atom C	Bond angle A-B-C (°)
2.347	Si1	Co2	Si1(1,1)	109.7
2.347	Si1	Co2	Si3	70.3
2.702	Si1	Si3	Si1(1,1)	90.5

CoSi2(111)-(1x1)
14.27.14

Bond Distances and Angles - Continued

Interatomic dist. A-B (Å)	Atom A	Atom B	Atom C	Bond angle A-B-C (°)
2.702	Si1	Si3	Co2(1,0)	125.9
2.702	Si1	Si3	Co2(0,-1)	54.9
2.347	Co2	Si1(1,1)	Co2(1,1)	109.7
2.347	Co2	Si1(1,1)	Si3(1,1)	125.9
2.347	Co2	Si1(1,1)	Si3(0,1)	54.9

COMMON NAME : CoSi2(111)-(1x1) Co-rich ILLUSTRATION: 145
CLASSIFICATION : 14.27.11b
TECHNIQUE : MEIS
AUTHORS : J. Vrijmoeth, A.G. Schins and J.F. van der Veen
REFERENCE : Phys. Rev., B40, 3121 (1989)

SURFACE TYPE
Substrate : CoSi2
Crystal face: 111
Temperature : RT*
Bulk lattice: fluorite
2D bulk symm: p3m1
2D surf symm: p3m1

Adsorbate:
Coverage :
Pattern : (1x1)
Matrix : (1.000, 0.000)
(0.000, 1.000)

STRUCTURE TYPE
Unreconstructed termination of bulk fluorite structure between trilayers; top two interlayer spacings contracted; this is model found by Wu et al, PR B33, 2900 (1986) (SSD 14.27.1)

SAMPLE PREPARATION (1 sample)
Treatment : sequential deposition of Co and Si at RT and annealing
Crystallinity:
Anal. methods:
Contamination: AES, ion scattering: substrate pure

COMMENTS
Thin epitaxial layer (16-30Å) grown on Si(111); see SSD 14.17.14 and compare with CoSi2 single crystal; both Co-rich and Si-rich surfaces prepared and analyzed: Si-rich see 14.27.11a

DATA COLLECTION
Technique: MEIS; 99.8keV protons collimated to 1°
Dataset : incidence along [22-1] direction in silicide;blocking patterns of Co backscatt. taken for exit angles 20-80°

THEORY/DATA TREATMENT
High resolution RBS compared with Monte Carlo simulations with Moliere potential; vib amps as for NiSi2

STRUCTURES EXAMINED
1) model found by Wu et al, PRB33,2900(1986), not optimized; 2) and 3) models proposed by Pirri et al, PR B33, 4108 (1986)

2D UNIT CELLS (1 domain observed)

Cell	Ax (Å)	Ay (Å)	Bx (Å)	By (Å)	α (°)	Matrix	Pattern	Cell type
Bulk	3.838	0.000	-1.919	3.324	120.0	(1.000, 0.000) (0.000, 1.000)	(1x1)	b: bulk lattice
Surface 1	3.838	0.000	-1.919	3.324	120.0	(1.000, 0.000) (0.000, 1.000)	(1x1)	s1: commens. superlattice

3D COORDINATES

Si1-Co2-Si3 form relaxed topmost trilayer; Si4-Co5-Si6 form periodically repeating bulk trilayer; 0.1Å error bars assumed for tabulation

Dx/Dy in Å, or as a fraction of layer's unit cell vectors; atom 0 at the origin. Epir/subr are bulk repeat vectors.

No. of atoms: 6 Bulk z = 3.092 Å

Region	Chem el.	At. no.	Cell type	Site occ.	Rel. to	Dx ± εx		Dy ± εy		Dz ± εz		Dz/Bz(%) ± εz/Bz
epir		-2					f		f		Å	
subr		-1				-1.919	Å	-1.108	Å	3.092	Å	
intf	Si	1	b	1.00	0	0.000	f	0.000	f	0.000	Å	0.0
intf	Co	2	b	1.00	1	0.333	f	0.667	f	0.618 ± .100	Å	20.0 ± 3.2
intf	Si	3	b	1.00	2	0.333	f	-0.333	f	0.734 ± .100	Å	23.7 ± 3.2
subl	Si	4	b	1.00	3	-0.333	f	0.333	f	1.546	Å	50.0
subl	Co	5	b	1.00	4	0.333	f	-0.333	f	0.773	Å	25.0
subl	Si	6	b	1.00	5	-0.667	f	-0.333	f	0.773	Å	25.0

BOND DISTANCES AND ANGLES

Bond distances and angles are derived from coordinates

No. of distances/angles: 5

Interatomic dist. A-B (Å)	Atom A	Atom B	Atom C	Bond angle A-B-C (°)
2.300	Si1	Co2	Si3	68.1
2.300	Si1	Co2	Si4	105.6
2.334	Co2	Si3	Si4	53.2
2.702	Si3	Si4	Co5	54.1

CoSi2(111)-(1x1) Co-rich
14.27.11b

Bond Distances and Angles - Continued

Interatomic dist. A-B (Å)	Atom A	Atom B	Atom C	Bond angle A-B-C (°)
2.347	Si4	Co5	Si6	70.3

COMMON NAME : CoSi2(111)-(1x1) Si-rich
CLASSIFICATION : 14.27.11a
TECHNIQUE : MEIS
AUTHORS : J. Vrijmoeth, A.G. Schins and J.F. van der Veen
REFERENCE : Phys. Rev., B40, 3121 (1989)

ILLUSTRATION: 146

SURFACE TYPE
Substrate : CoSi2
Crystal face: 111
Temperature : RT*
Bulk lattice: fluorite
2D bulk symm: p3m1
2D surf symm: p3m1

Adsorbate:
Coverage :
Pattern : (1x1)
Matrix : (1.000, 0.000)
(0.000, 1.000)

STRUCTURE TYPE
Si terminated bulk with additional Si bilayer on top

SAMPLE PREPARATION (1 sample)
Treatment : sequential deposition of Co and Si at RT and annealing
Crystallinity:
Anal. methods:
Contamination: AES, ion scattering: substrate pure

COMMENTS
Thin epitaxial layer (16-30Å) grown on Si(111); see SSD 14.17.14 and compare difference with CoSi2 single crystal; both Co-rich and Si-rich surfaces prepared and analyzed: Co-rich see 14.27.11b

DATA COLLECTION
Technique: MEIS; 99.8keV protons collimated to 1°
Dataset : incidence along [22-1] direction in silicide;blocking patterns of Co backscatt. taken for exit angles 20-80°

THEORY/DATA TREATMENT
High resolution RBS compared with Monte Carlo simulations with Moliere potential; vib amps as for NiSi2

STRUCTURES EXAMINED
1)bulk Si term. with additional Si bilayer bonded to Co, 2)bulk Si term. with additional Si bilayer bonded to Si, 3)bulk Co term. with additional Si bilayer bonded to Co, 4) same as 2) but bilayer rotated by 180°. 1) chosen, visual fit improved by relax. vertical position of top 3 Si atoms

2D UNIT CELLS (1 domain observed)

Cell	Ax (Å)	Ay (Å)	Bx (Å)	By (Å)	α (°)	Matrix	Pattern	Cell type
Bulk	3.838	0.000	-1.919	3.324	120.0	(1.000, 0.000) (0.000, 1.000)	(1x1)	b: bulk lattice
Surface 1	3.838	0.000	-1.919	3.324	120.0	(1.000, 0.000) (0.000, 1.000)	(1x1)	s1: commens. superlattice

3D COORDINATES

Si1-Si2: extra top bilayer; Si3-Co4-Si5: topmost trilayer, Si3-Co4 spacing contracted;
Si6-Co7-Si8: periodically repeating bulk trilayer

Dx/Dy in Å, or as a fraction of layer's unit cell vectors; atom 0 at the origin. Epir/subr are bulk repeat vectors.

No. of atoms: 8 Bulk z = 3.092 Å

Region	Chem el.	At. no.	Cell type	Site occ.	Rel. to	Dx ± εx		Dy ± εy		Dz ± εz		Dz/Bz(%) ± εz/Bz
epir		-2					f		f		Å	
subr		-1				-1.919	Å	-1.108	Å	3.092	Å	
intf	Si	1	s1	1.00	0	0.000	f	0.000	f	0.000	Å	0.0
intf	Si	2	s1	1.00	1	0.333	f	0.667	f	0.848 ± .056	Å	27.4 ± 1.8
intf	Si	3	s1	1.00	0	0.000	f	0.000	f	2.519 ± .070	Å	81.5 ± 2.3
intf	Co	4	s1	1.00	3	0.333	f	0.667	f	0.723 ± .050	Å	23.4 ± 1.6
intf	Si	5	s1	1.00	4	0.333	f	-0.333	f	0.773	Å	25.0
subl	Si	6	b	1.00	5	-0.333	f	0.333	f	1.546	Å	50.0
subl	Co	7	b	1.00	6	0.333	f	-0.333	f	0.773	Å	25.0
subl	Si	8	b	1.00	7	-0.667	f	-0.333	f	0.773	Å	25.0

CoSi2(111)-(1x1) Si-rich
14.27.11a

BOND DISTANCES AND ANGLES

Bond distances and angles are derived from coordinates

No. of distances/angles: 8

Interatomic dist. A-B (Å)	Atom A	Atom B	Atom C	Bond angle A-B-C (°)
2.373	Si1	Si2	Si1(1,1)	108.0
2.373	Si1	Si2	Si3(1,1)	126.0
2.373	Si1	Si2	Si3	58.0
2.519	Si1	Si3	Si2	53.0
2.775	Si2	Si3(1,1)	Si1(1,1)	53.0
2.394	Co4	Si2	Si3	53.0
2.331	Co4	Si3(1,1)	Si1(1,1)	108.1
2.347	Co4	Si5(0,1)	Si3(1,1)	54.9

COMMON NAME : Cr(100)-(1x1)-N
CLASSIFICATION : 24.7.1
TECHNIQUE : LEED
AUTHORS : Y. Joly, Y. Gauthier and R. Baudoing
REFERENCE : Phys. Rev., B40, 10119 (1989)

ILLUSTRATION: 48,49

SURFACE TYPE
Substrate : Cr
Crystal face: 100
Temperature : RT*
Bulk lattice: bcc
2D bulk symm: p4m
2D surf symm: p4m

Adsorbate: N
Coverage : 1.0 N/Cr
Pattern : (1x1)
Matrix : (1.000, 0.000)
(0.000, 1.000)

STRUCTURE TYPE
Atomic adsorption in 4-fold hollow sites, with large expansion of top Cr-Cr spacing by 25%

SAMPLE PREPARATION (1 sample)
Treatment : sputter/anneal cycles removed S,O, segregating N to surface
Crystallinity:
Anal. methods:
Contamination: AES: 'real nitride'

COMMENTS

DATA COLLECTION
Technique: LEED; display-type LEED
Dataset : IV spectra at 3 incident angles (Θ,φ)=(0,0),(10,0), (34,45)°: E range 30-225 eV; cumul. E range 3000eV

THEORY/DATA TREATMENT
Dynamical LEED (layer dblg; composite layers): <=9 ph shs (superpos pots); E-dep Vor; ΘD=485 K(Cr),600K(N)

STRUCTURES EXAMINED
N in hollow and bridge sites, also as underlayer; N in top and intermediate hollow-bridge sites; in hollow overlayer: N-Cr and top 2 Cr-Cr interlayer spacings varied

QUALITY OF EXPERIMENT-THEORY FIT
D1,2,2y,4,4y=11.65,4.13,2.46,7.58,4.39%

2D UNIT CELLS (1 domain observed)

Cell	Ax (Å)	Ay (Å)	Bx (Å)	By (Å)	α (°)	Matrix	Pattern	Cell type
Bulk	2.875	0.000	0.000	2.875	90.0	(1.000, 0.000) (0.000, 1.000)	(1x1)	b: bulk lattice
Surface 1	2.875	0.000	0.000	2.875	90.0	(1.000, 0.000) (0.000, 1.000)	(1x1)	s1: commens. superlattice

3D COORDINATES

O1 forms overlayer in 4-fold hollows with shorter O-Fe bond length to 2nd Fe layer than 1st

Dx/Dy in Å, or as a fraction of layer's unit cell vectors; atom 0 at the origin. Epir/subr are bulk repeat vectors.

No. of atoms: 5 Bulk z = 1.438 Å

Region	Chem el.	At. no.	Cell type	Site occ.	Rel. to	Dx ± εx		Dy ± εy		Dz ± εz		Dz/Bz(%) ± εz/Bz
epir		-2					f		f		Å	
subr		-1				1.438	Å	1.438	Å	1.438	Å	
ovrl	N	1	b	1.00	0	0.000	f	0.000	f	0.000	Å	0.0
intf	Cr	2	b	1.00	1	0.500	f	0.500	f	0.221 ± .016	Å	15.4 ± 1.1
intf	Cr	3	b	1.00	2	-0.500	f	-0.500	f	1.794 ± .008	Å	124.8 ± .6
intf	Cr	4	b	1.00	3	0.500	f	0.500	f	1.402 ± .011	Å	97.5 ± .8
subl	Cr	5	b	1.00	4	-0.500	f	-0.500	f	1.438	Å	100.0

BOND DISTANCES AND ANGLES

Bond distances and angles are derived from coordinates

No. of distances/angles: 6

Interatomic dist. A-B (Å)	Atom A	Atom B	Atom C	Bond angle A-B-C (°)
2.045	N1	Cr2	N1(1,0)	89.3
2.045	N1	Cr2	Cr2(1,0)	134.7
2.045	N1	Cr2	Cr3	47.6
2.015	N1	Cr3	Cr4	124.6

Cr(100)-(1x1)-N
24.7.1

Bond Distances and Angles - Continued

Interatomic dist. A-B (Å)	Atom A	Atom B	Atom C	Bond angle A-B-C (°)
2.875	Cr2	Cr2(1,0)	Cr3(1,0)	58.0
2.711	Cr2	Cr3	Cr4	76.0

COMMON NAME : Cr(100)-c(2x2)-S
CLASSIFICATION : 24.16.1
TECHNIQUE : ARPEFS
AUTHORS : L.J. Terminello, X.S. Zhang, Z.Q. Huang, S. Kim, A.E.Schach von Wittenau, K.T. Leung and D.A. Shirley
REFERENCE : Phys. Rev., B38, 3879 (1988)

ILLUSTRATION: 48,50

SURFACE TYPE
Substrate : Cr
Crystal face: 100
Temperature : RT*
Bulk lattice: bcc
2D bulk symm: p4m
2D surf symm: p4m

Adsorbate: S
Coverage : 0.5 S/Cr
Pattern : c(2x2)
Matrix : (1.000, 1.000)
(-1.000, 1.000)

STRUCTURE TYPE
Atomic adsorption in hollow sites with top Cr-Cr spacing relaxation (no detectable layer buckling)

SAMPLE PREPARATION (3 sample)
Treatment : few days of sputter-anneal cycles, then exposure to H2S
Crystallinity: sharp LEED pattern with no background
Anal. methods:
Contamination: AES: no contaminants

COMMENTS

DATA COLLECTION
Technique: ARPEFS; 2525-3025eV soft x-ray beam (1eV re
Dataset : ARPEFS spectra for two emission angles: [100], [110]; kinetic E range 50-550 eV

THEORY/DATA TREATMENT
Fourier transform; MSSW calcs: Moruzzi et al Cr pot; HF S pot; ΘD=470 K(bulk Fe), 332K(surf Fe), 423K(S)

STRUCTURES EXAMINED
Top, bridge and hollow site: FT and MSSW favor hollow; variation of S-Fe spacing, 1st and 3rd Fe layer buckling, top 4 Fe-Fe interlayer spacings, optimized by R-factor fitting (also fitting of emission directions and photon polarization angles)

2D UNIT CELLS (1 domain observed)

Cell	Ax (Å)	Ay (Å)	Bx (Å)	By (Å)	α (°)	Matrix	Pattern	Cell type
Bulk	2.880	0.000	0.000	2.880	90.0	(1.000, 0.000) (0.000, 1.000)	(1x1)	b: bulk lattice
Surface 1	2.880	2.880	-2.880	2.880	90.0	(1.000, 1.000) (-1.000, 1.000)	c(2x2)	s1: commens. superlattice

3D COORDINATES

S1: hollow-site overlayer

Dx/Dy in Å, or as a fraction of layer's unit cell vectors; atom 0 at the origin. Epir/subr are bulk repeat vectors.

No. of atoms: 7 Bulk z = 1.440 Å

Reg ion	Chem el.	At. no.	Cell type	Site occ.	Rel. to	Dx ± εx		Dy ± εy		Dz ± εz		Dz/Bz(%) ± εz/Bz
epir		-2					f		f		Å	
subr		-1				1.440	Å	1.440	Å	1.440	Å	
ovrl	S	1	s1	.50	0	0.000	f	0.000	f	0.000	Å	0.0
intf	Cr	2	b	1.00	1	0.500	f	0.500	f	1.170 ± .020	Å	81.3 ± 1.4
intf	Cr	3	b	1.00	2	-0.500	f	-0.500	f	1.310 ± .020	Å	91.0 ± 1.4
intf	Cr	4	b	1.00	3	0.500	f	0.500	f	1.470 ± .030	Å	102.1 ± 2.1
intf	Cr	5	b	1.00	4	-0.500	f	-0.500	f	1.420 ± .030	Å	98.6 ± 2.1
intf	Cr	6	b	1.00	5	0.500	f	0.500	f	1.410 ± .070	Å	97.9 ± 4.9
subl	Cr	7	b	1.00	6	-0.500	f	-0.500	f	1.440	Å	100.0

BOND DISTANCES AND ANGLES

Bond distances and angles are derived from coordinates

No. of distances/angles: 4

Interatomic dist. A-B (Å)	Atom A	Atom B	Atom C	Bond angle A-B-C (°)
2.349	S1	Cr2	S1(1,0)	120.2
2.349	S1	Cr2	Cr2(1,0)	127.8
2.349	S1	Cr2	Cr3	62.6

Cr(100)-c(2x2)-S
24.16.1

Bond Distances and Angles - Continued

Interatomic dist. A-B (Å)	Atom A	Atom B	Atom C	Bond angle A-B-C (°)
2.421	Cr2	Cr3	Cr4	68.6

COMMON NAME : Cu(100)-(1x1)
CLASSIFICATION : 29.25a
TECHNIQUE : LEED
AUTHORS : H.L. Davis and J.R. Noonan
REFERENCE : Surf. Sci., 126, 245 (1983)

ILLUSTRATION: 2

SURFACE TYPE
Substrate : Cu
Crystal face: 100
Temperature : RT
Bulk lattice: fcc
2D bulk symm: p4m
2D surf symm: p4m

Adsorbate:
Coverage :
Pattern : (1x1)
Matrix : (1.000, 0.000)
(0.000, 1.000)

STRUCTURE TYPE
Bulk termination with multilayer relaxation

SAMPLE PREPARATION (sample)
Treatment : see Davis and Noonan, J. Vac. Sci. Technol. 20. 842 (1982)
Crystallinity:
Anal. methods:
Contamination:

COMMENTS

DATA COLLECTION
Technique: LEED
Dataset : four inequivalent I-V spectra, equivalent beam averaging

THEORY/DATA TREATMENT
Dynamical LEED: ΘD=340 K

STRUCTURES EXAMINED
Variation of top three interlayer spacings

QUALITY OF EXPERIMENT-THEORY FIT
RZJ=0.039, R2=0.058, R5=0.145

2D UNIT CELLS (1 domain observed)

Cell	Ax (Å)	Ay (Å)	Bx (Å)	By (Å)	α (°)	Matrix	Pattern	Cell type
Bulk	2.553	0.000	0.000	2.553	90.0	(1.000, 0.000) (0.000, 1.000)	(1x1)	b: bulk lattice
Surface 1	2.553	0.000	0.000	2.553	90.0	(1.000, 0.000) (0.000, 1.000)	(1x1)	s1: commens. superlattice

3D COORDINATES

0.05Å error bars assumed for tabulation

Dx/Dy in Å, or as a fraction of layer's unit cell vectors; atom 0 at the origin. Epir/subr are bulk repeat vectors.

No. of atoms: 5 Bulk z = 1.805 Å

Region	Chem el.	At. no.	Cell type	Site occ.	Rel. to	Dx ± εx		Dy ± εy		Dz ± εz		Dz/Bz(%) ± εz/Bz
epir		-2					f		f		Å	
subr		-1				-1.277	Å	-1.277	Å	1.805	Å	
intf	Cu	1	b	1.00	0	0.000	f	0.000	f	0.000	Å	0.0
intf	Cu	2	b	1.00	1	0.500	f	0.500	f	1.785 ± .050	Å	98.9 ± 2.8
intf	Cu	3	b	1.00	2	-0.500	f	-0.500	f	1.836 ± .050	Å	101.7 ± 2.8
intf	Cu	4	b	1.00	3	0.500	f	0.500	f	1.832 ± .050	Å	101.5 ± 2.8
subl	Cu	5	b	1.00	4	-0.500	f	-0.500	f	1.805	Å	100.0

BOND DISTANCES AND ANGLES

Bond distances and angles are derived from coordinates

No. of distances/angles: 3

Interatomic dist. A-B (Å)	Atom A	Atom B	Atom C	Bond angle A-B-C (°)
2.553	Cu1	Cu1(1,0)		
2.539	Cu1	Cu2		
2.575	Cu2	Cu3		

COMMON NAME : Cu(100)-(1x1) ILLUSTRATION: 2
CLASSIFICATION : 29.36a
TECHNIQUE : LEED
AUTHORS : M.A. Abu-Joudeh, B.M. Davies and P.A. Montano
REFERENCE : Surf. Sci., 171, 331 (1986)

SURFACE TYPE
Substrate : Cu
Crystal face: 100
Temperature : RT*
Bulk lattice: fcc
2D bulk symm: p4m
2D surf symm: p4m

Adsorbate:
Coverage :
Pattern : (1x1)
Matrix : (1.000, 0.000)
(0.000, 1.000)

STRUCTURE TYPE
Bulk termination with multilayer relaxations

SAMPLE PREPARATION (1 sample)
Treatment : Ar+ bombardment and anneals at 673 K for 10min; O treatments
Crystallinity:
Anal. methods:
Contamination: monitored by AES and LEED

COMMENTS
RE=R-factor defined by Legg et al, J. Phys. C10, 937 (1977)

DATA COLLECTION
Technique: LEED
Dataset : IV curves for 3 non-equivalent beams: (00) (10) and (11); 30<E<450 eV

THEORY/DATA TREATMENT
Dynamical LEED (RFS): 57 beams, 8 phase shifts; Vor optimised to -9 eV; ΘD=344 K

STRUCTURES EXAMINED
Relaxations of top 2 interlayer spacings from -7.5 to +2.5%

QUALITY OF EXPERIMENT-THEORY FIT
RE=0.14 (see comments)

2D UNIT CELLS (1 domain observed)

Cell	Ax (Å)	Ay (Å)	Bx (Å)	By (Å)	α (°)	Matrix	Pattern	Cell type
Bulk	2.553	0.000	0.000	2.553	90.0	(1.000, 0.000) (0.000, 1.000)	(1x1)	b: bulk lattice
Surface 1	2.553	0.000	0.000	2.553	90.0	(1.000, 0.000) (0.000, 1.000)	(1x1)	s1: commens. superlattice

3D COORDINATES

Dx/Dy in Å, or as a fraction of layer's unit cell vectors; atom 0 at the origin. Epir/subr are bulk repeat vectors.

No. of atoms: 4 Bulk z = 1.810 Å

Region	Chem el.	At. no.	Cell type	Site occ.	Rel. to	Dx ± εx		Dy ± εy		Dz ± εz		Dz/Bz(%) ± εz/Bz
epir		-2					f		f		Å	
subr		-1				1.277	Å	1.277	Å	1.810	Å	
intf	Cu	1	b	1.00	0	0.000	f	0.000	f	0.000	Å	0.0
intf	Cu	2	b	1.00	1	0.500	f	0.500	f	1.780 ± .020	Å	98.3 ± 1.1
intf	Cu	3	b	1.00	2	-0.500	f	-0.500	f	1.830 ± .020	Å	101.1 ± 1.1
subl	Cu	4	b	1.00	3	0.500	f	0.500	f	1.810	Å	100.0

BOND DISTANCES AND ANGLES

Bond distances and angles are derived from coordinates

No. of distances/angles: 3

Interatomic dist. A-B (Å)	Atom A	Atom B	Atom C	Bond angle A-B-C (°)
2.553	Cu1	Cu1(1,0)	Cu2	59.8
2.535	Cu1	Cu2	Cu3	90.0
2.571	Cu2	Cu3	Cu4	90.5

COMMON NAME : Cu(100)-(1x1)
CLASSIFICATION : 29.43
TECHNIQUE : SPLEED
AUTHORS : D.M. Lind, F.B. Dunning, G.K. Walters and H.L. Davis
REFERENCE : Phys. Rev., B35, 9037 (1987)

ILLUSTRATION: 2

SURFACE TYPE
Substrate : Cu
Crystal face: 100
Temperature : RT*
Bulk lattice: fcc
2D bulk symm: p4m
2D surf symm: p4m

Adsorbate:
Coverage :
Pattern : (1x1)
Matrix : (1.000, 0.000)
(0.000, 1.000)

STRUCTURE TYPE
Bulk termination with multilayer relaxations

SAMPLE PREPARATION (1 sample)
Treatment : Ar+ sputtering for 30min followed by 30min anneals at 823 K
Crystallinity:
Anal. methods:
Contamination: monitored by AES and LEED

COMMENTS

DATA COLLECTION
Technique: SPLEED
Dataset : spin asymmetry parameter vs incident electron energy (A-V): 3 non-equivalent beams (10, 11, 20) at normal incidence

THEORY/DATA TREATMENT
Dynamical spin-polarized LEED: truncated free atom relativistic potential; Voi∝E**1/3; ΘD=330 K

STRUCTURES EXAMINED
Variation of top two interlayer spacings

QUALITY OF EXPERIMENT-THEORY FIT
RS=0.11 (single-beam R-factor)

2D UNIT CELLS (1 domain observed)

Cell	Ax (Å)	Ay (Å)	Bx (Å)	By (Å)	α (°)	Matrix	Pattern	Cell type
Bulk	2.553	0.000	0.000	2.553	90.0	(1.000, 0.000) (0.000, 1.000)	(1x1)	b: bulk lattice
Surface 1	2.553	0.000	0.000	2.553	90.0	(1.000, 0.000) (0.000, 1.000)	(1x1)	s1: commens. superlattice

3D COORDINATES

0.02Å error bars assumed for tabulation

Dx/Dy in Å, or as a fraction of layer's unit cell vectors; atom 0 at the origin. Epir/subr are bulk repeat vectors.

No. of atoms: 4 Bulk z = 1.805 Å

Region	Chem el.	At. no.	Cell type	Site occ.	Rel. to	Dx ± εx		Dy ± εy		Dz ± εz		Dz/Bz(%) ± εz/Bz
epir		-2					f		f		Å	
subr		-1				1.277	Å	1.277	Å	1.805	Å	
intf	Cu	1	b	1.00	0	0.000	f	0.000	f	0.000	Å	0.0
intf	Cu	2	b	1.00	1	0.500	f	0.500	f	1.783 ± .020	Å	98.8 ± 1.1
intf	Cu	3	b	1.00	2	-0.500	f	-0.500	f	1.821 ± .020	Å	100.9 ± 1.1
subl	Cu	4	b	1.00	3	0.500	f	0.500	f	1.805	Å	100.0

BOND DISTANCES AND ANGLES

Bond distances and angles are derived from coordinates

No. of distances/angles: 3

Interatomic dist. A-B (Å)	Atom A	Atom B	Atom C	Bond angle A-B-C (°)
2.553	Cu1	Cu1(0,1)	Cu2	59.8
2.537	Cu1	Cu2	Cu3	89.9
2.564	Cu2	Cu3	Cu4	90.3

COMMON NAME : Cu(100)-(1x1)
CLASSIFICATION : 29.59
TECHNIQUE : MEI scattering
AUTHORS : Q.T. Jiang, P. Fenter and T. Gustafsson
REFERENCE : Phys. Rev., B44, 5773 (1991)

ILLUSTRATION: 2

SURFACE TYPE
Substrate : Cu
Crystal face: 100
Temperature : RT*
Bulk lattice: fcc
2D bulk symm: p4m
2D surf symm: p4m

Adsorbate:
Coverage :
Pattern : (1x1)
Matrix : (1.000, 0.000)
(0.000, 1.000)

STRUCTURE TYPE
Multilayer ralaxation

SAMPLE PREPARATION (1 sample)
Treatment : sputtering and annealing cycles
Crystallinity: sharp (1x1) LEED pattern
Anal. methods: AES
Contamination:

COMMENTS
Anisotropy found in surface vibrations; checked for beam induced damage

DATA COLLECTION
Technique: MEI scattering
Dataset : 3 scattering geometries

THEORY/DATA TREATMENT
Monte Carlo simulations and R-factor analysis

STRUCTURES EXAMINED
Variation of 1st and 2nd interlayer spacings

2D UNIT CELLS (1 domain observed)

Cell	Ax (Å)	Ay (Å)	Bx (Å)	By (Å)	α (°)	Matrix	Pattern	Cell type
Bulk	2.556	0.000	0.000	2.556	90.0	(1.000, 0.000) (0.000, 1.000)	(1x1)	b: bulk lattice
Surface 1	2.556	0.000	0.000	2.556	90.0	(1.000, 0.000) (0.000, 1.000)	(1x1)	s1: commens. superlattice

3D COORDINATES

Dx/Dy in Å, or as a fraction of layer's unit cell vectors; atom 0 at the origin. Epir/subr are bulk repeat vectors.

No. of atoms: 3 Bulk z = 1.807 Å

Region	Chem el.	At. no.	Cell type	Site occ.	Rel. to	Dx ± εx	Dy ± εy	Dz ± εz	Dz/Bz(%) ± εz/Bz
epir		-2				f	f	Å	
subr		-1				1.278 Å	1.278 Å	1.807 Å	
intf	Cu	1	s1	1.00	0	0.000 Å	0.000 Å	0.000 Å	0.0
intf	Cu	2	s1	1.00	0	1.278 Å	1.278 Å	1.764 ± .015 Å	97.6 ± .8
subl	Cu	3	b	1.00	0	0.000 Å	0.000 Å	3.590 ± .018 Å	198.6 ± 1.0

BOND DISTANCES AND ANGLES

Bond distances and angles are derived from coordinates

No. of distances/angles: 2

Interatomic dist. A-B (Å)	Atom A	Atom B	Atom C	Bond angle A-B-C (°)
2.526	Cu1	Cu2		
2.569	Cu2	Cu3		

COMMON NAME : Cu(110)-(1x1)
CLASSIFICATION : 29.25b
TECHNIQUE : LEED
AUTHORS : H.L. Davis and J.R. Noonan
REFERENCE : Surf. Sci., 126, 245 (1983)

ILLUSTRATION: 4

SURFACE TYPE
Substrate : Cu
Crystal face: 110
Temperature : RT
Bulk lattice: fcc
2D bulk symm: pmm
2D surf symm: pmm

Adsorbate:
Coverage :
Pattern : (1x1)
Matrix : (1.000, 0.000)
(0.000, 1.000)

STRUCTURE TYPE
Bulk termination with multilayer relaxation

SAMPLE PREPARATION (sample)
Treatment : see Davis, Noonan and Jenkins, Surf. Sci. 83, 559 (1979)
Crystallinity:
Anal. methods:
Contamination:

COMMENTS

DATA COLLECTION
Technique: LEED
Dataset : four inequivalent I-V profiles, equivalent beam averaging

THEORY/DATA TREATMENT
Dynamical LEED; ΘD=340 K

STRUCTURES EXAMINED
Variation of top two interlayer spacings

QUALITY OF EXPERIMENT-THEORY FIT
RZJ=0.067, R2=0.039, R5=0.188

2D UNIT CELLS (1 domain observed)

Cell	Ax (Å)	Ay (Å)	Bx (Å)	By (Å)	α (°)	Matrix	Pattern	Cell type
Bulk	3.610	0.000	0.000	2.553	90.0	(1.000, 0.000) (0.000, 1.000)	(1x1)	b: bulk lattice
Surface 1	3.610	0.000	0.000	2.553	90.0	(1.000, 0.000) (0.000, 1.000)	(1x1)	s1: commens. superlattice

3D COORDINATES

0.05Å error bars assumed for tabulation

Dx/Dy in Å, or as a fraction of layer's unit cell vectors; atom 0 at the origin. Epir/subr are bulk repeat vectors.

No. of atoms: 4

Bulk z = 1.276 Å

Region	Chem el.	At. no.	Cell type	Site occ.	Rel. to	Dx ± εx		Dy ± εy		Dz ± εz		Dz/Bz(%) ± εz/Bz
epir		-2					f		f		Å	
subr		-1				1.805	Å	1.277	Å	1.276	Å	
intf	Cu	1	b	1.00	0	0.000	f	0.000	f	0.000	Å	0.0
intf	Cu	2	b	1.00	1	0.500	f	0.500	f	1.159 ± .050	Å	90.8 ± 3.9
intf	Cu	3	b	1.00	2	-0.500	f	-0.500	f	1.305 ± .050	Å	102.3 ± 3.9
subl	Cu	4	b	1.00	3	0.500	f	0.500	f	1.276	Å	100.0

BOND DISTANCES AND ANGLES

Bond distances and angles are derived from coordinates

No. of distances/angles: 4

Interatomic dist. A-B (Å)	Atom A	Atom B	Atom C	Bond angle A-B-C (°)
2.553	Cu1	Cu1(0,1)		
2.496	Cu1	Cu2		
2.567	Cu2	Cu3		
2.553	Cu3	Cu4		

COMMON NAME : Cu(110)-(1x1)
CLASSIFICATION : 29.28
TECHNIQUE : HEIS
AUTHORS : I. Stensgaard, R. Feidenhans'l and J.E. Sorensen
REFERENCE : Surf. Sci., 128, 281 (1983)

ILLUSTRATION: 4

SURFACE TYPE
Substrate : Cu
Crystal face: 110
Temperature : RT*
Bulk lattice: fcc
2D bulk symm: pmm
2D surf symm: pmm

Adsorbate:
Coverage :
Pattern : (1x1)
Matrix : (1.000, 0.000)
(0.000, 1.000)

STRUCTURE TYPE
Bulk termination with multilayer relaxation

SAMPLE PREPARATION (1 sample)
Treatment : numerous cycles of 1k eV Ar+ sputtering and annealing to 700 K
Crystallinity: sharp LEED pattern
Anal. methods:
Contamination: AES: clean

COMMENTS

DATA COLLECTION
Technique: HEIS
Dataset : energy spectra of 300k eV He+ beams

THEORY/DATA TREATMENT
Single alignment ion scattering (Rutherford backscattering); Moliere approximation to Thomas Fermi potential; ΘD=320 K

STRUCTURES EXAMINED
Variation of top two interlayer spacings

QUALITY OF EXPERIMENT-THEORY FIT
RIS=0.90

2D UNIT CELLS (1 domain observed)

Cell	Ax (Å)	Ay (Å)	Bx (Å)	By (Å)	α (°)	Matrix	Pattern	Cell type
Bulk	3.610	0.000	0.000	2.553	90.0	(1.000, 0.000) (0.000, 1.000)	(1x1)	b: bulk lattice
Surface 1	3.610	0.000	0.000	2.553	90.0	(1.000, 0.000) (0.000, 1.000)	(1x1)	s1: commens. superlattice

3D COORDINATES

Dx/Dy in Å, or as a fraction of layer's unit cell vectors; atom 0 at the origin. Epir/subr are bulk repeat vectors.

No. of atoms: 4 Bulk z = 1.278 Å

Region	Chem el.	At. no.	Cell type	Site occ.	Rel. to	Dx ± εx		Dy ± εy		Dz ± εz		Dz/Bz(%) ± εz/Bz
epir		-2					f		f		Å	
subr		-1				1.805	Å	1.277	Å	1.278	Å	
intf	Cu	1	b	1.00	0	0.000	f	0.000	f	0.000	Å	0.0
intf	Cu	2	b	1.00	1	0.500	f	0.500	f	1.210 ± .020	Å	94.7 ± 1.6
intf	Cu	3	b	1.00	2	-0.500	f	-0.500	f	1.320 ± .020	Å	103.3 ± 1.6
subl	Cu	4	b	1.00	3	0.500	f	0.500	f	1.278	Å	100.0

BOND DISTANCES AND ANGLES

Bond distances and angles are derived from coordinates

No. of distances/angles: 4

Interatomic dist. A-B (Å)	Atom A	Atom B	Atom C	Bond angle A-B-C (°)
2.553	Cu1	Cu1(0,1)		
2.520	Cu1	Cu2		
2.575	Cu2	Cu3		
2.554	Cu3	Cu4		

COMMON NAME : Cu(110)-(1x1)
CLASSIFICATION : 29.29
TECHNIQUE : LEED
AUTHORS : D.L. Adams, H.B. Nielsen and J.N. Andersen
REFERENCE : Surf. Sci., 128, 294 (1983)

ILLUSTRATION: 4

SURFACE TYPE
Substrate : Cu
Crystal face: 110
Temperature : RT*
Bulk lattice: fcc
2D bulk symm: pmm
2D surf symm: pmm

Adsorbate:
Coverage :
Pattern : (1x1)
Matrix : (1.000, 0.000)
(0.000, 1.000)

STRUCTURE TYPE
Bulk termination with multilayer relaxation

SAMPLE PREPARATION (1 sample)
Treatment : Ar+ ion bombardment (500 eV, 10μA beam current), anneal 900 K
Crystallinity:
Anal. methods:
Contamination: AES: no S, O, C contamination

COMMENTS

DATA COLLECTION
Technique: LEED; spot photometer
Dataset : I-V spectra: 9 symm.-inequivalent beams at normal incidence; E range 20-360 eV

THEORY/DATA TREATMENT
Dynamical LEED (layer doubling): 10 phase shifts (Moruzzi et al potential); ΘD=335 K

STRUCTURES EXAMINED
Variation of top two interlayer spacings

QUALITY OF EXPERIMENT-THEORY FIT
R2=0.023

2D UNIT CELLS (1 domain observed)

Cell	Ax (Å)	Ay (Å)	Bx (Å)	By (Å)	α (°)	Matrix	Pattern	Cell type
Bulk	3.610	0.000	0.000	2.553	90.0	(1.000, 0.000) (0.000, 1.000)	(1x1)	b: bulk lattice
Surface 1	3.610	0.000	0.000	2.553	90.0	(1.000, 0.000) (0.000, 1.000)	(1x1)	s1: commens. superlattice

3D COORDINATES

Dx/Dy in Å, or as a fraction of layer's unit cell vectors; atom 0 at the origin. Epir/subr are bulk repeat vectors.

No. of atoms: 4 Bulk z = 1.278 Å

Region	Chem el.	At. no.	Cell type	Site occ.	Rel. to	Dx ± εx		Dy ± εy		Dz ± εz		Dz/Bz(%) ± εz/Bz
epir		-2					f		f		Å	
subr		-1				1.805	Å	1.277	Å	1.278	Å	
intf	Cu	1	b	1.00	0	0.000	f	0.000	f	0.000	Å	0.0
intf	Cu	2	b	1.00	1	0.500	f	0.500	f	1.170 ± .008	Å	91.6 ± .6
intf	Cu	3	b	1.00	2	-0.500	f	-0.500	f	1.307 ± .010	Å	102.3 ± .8
subl	Cu	4	b	1.00	3	0.500	f	0.500	f	1.278	Å	100.0

BOND DISTANCES AND ANGLES

Bond distances and angles are derived from coordinates

No. of distances/angles: 6

Interatomic dist. A-B (Å)	Atom A	Atom B	Atom C	Bond angle A-B-C (°)
2.553	Cu1	Cu1(0,1)		
2.501	Cu1	Cu2		
2.477	Cu1	Cu3		
2.568	Cu2	Cu3		
2.585	Cu2	Cu4		
2.554	Cu3	Cu4		

COMMON NAME : Cu(110)-(1x1)
CLASSIFICATION : 29.37
TECHNIQUE : ICISS
AUTHORS : J.A. Yarmoff, D.M. Cyr, J.H. Huang, S. Kim and R.S. Williams
REFERENCE : Phys. Rev., B33, 3856 (1986)

ILLUSTRATION: 4

SURFACE TYPE
Substrate : Cu
Crystal face: 110
Temperature : RT
Bulk lattice: fcc
2D bulk symm: pmm
2D surf symm: pmm

Adsorbate:
Coverage :
Pattern : (1x1)
Matrix : (1.000, 0.000)
(0.000, 1.000)

STRUCTURE TYPE
Bulk termination with top layer contraction

SAMPLE PREPARATION (1 sample)
Treatment : cycles of 1 k eV Ar ion bombardment followed by annealing
Crystallinity:
Anal. methods:
Contamination: monitored by Auger and LEED

COMMENTS

DATA COLLECTION
Technique: ICISS
Dataset : polar scans along [1,-1,0], [1,-1,2], [0,0,1] azimuths

THEORY/DATA TREATMENT
Monte Carlo sim. at shadow and blocking cones: Moliere pot; fitted screening length; ΘD=343 K, x1.5 surface enhancem.

STRUCTURES EXAMINED
Top interlayer spacing varied

2D UNIT CELLS (1 domain observed)

Cell	Ax (Å)	Ay (Å)	Bx (Å)	By (Å)	α (°)	Matrix	Pattern	Cell type
Bulk	3.610	0.000	0.000	2.550	90.0	(1.000, 0.000) (0.000, 1.000)	(1x1)	b: bulk lattice
Surface 1	3.610	0.000	0.000	2.550	90.0	(1.000, 0.000) (0.000, 1.000)	(1x1)	s1: commens. superlattice

3D COORDINATES

Dx/Dy in Å, or as a fraction of layer's unit cell vectors; atom 0 at the origin. Epir/subr are bulk repeat vectors.

No. of atoms: 4 Bulk z = 1.280 Å

Region	Chem el.	At. no.	Cell type	Site occ.	Rel. to	Dx ± εx		Dy ± εy		Dz ± εz		Dz/Bz(%) ± εz/Bz
epir		-2					f		f		Å	
subr		-1				1.805	Å	1.275	Å	1.280	Å	
intf	Cu	1	b	1.00	0	0.000	f	0.000	f	0.000	Å	0.0
intf	Cu	2	b	1.00	1	0.500	f	0.500	f	1.150 ± .100	Å	89.8 ± 7.8
intf	Cu	3	b	1.00	2	-0.500	f	-0.500	f	1.280	Å	100.0
subl	Cu	4	b	1.00	3	0.500	f	0.500	f	1.280	Å	100.0

BOND DISTANCES AND ANGLES

Bond distances and angles are derived from coordinates

No. of distances/angles: 6

Interatomic dist. A-B (Å)	Atom A	Atom B	Atom C	Bond angle A-B-C (°)
2.550	Cu1	Cu1(0,1)	Cu2	59.2
2.491	Cu1	Cu2	Cu3	57.6
2.430	Cu1	Cu3	Cu4	120.1
2.554	Cu2	Cu3	Cu4	60.2
2.560	Cu2	Cu4		
2.554	Cu3	Cu4		

COMMON NAME : Cu(110)-(1x1)
CLASSIFICATION : 29.38
TECHNIQUE : MEIS
AUTHORS : M. Copel, T. Gustafsson, W.R. Graham and S.M. Yalisove
REFERENCE : Phys. Rev., B33, 8110 (1986)

ILLUSTRATION: 4

SURFACE TYPE
Substrate : Cu
Crystal face: 110
Temperature : 323 K
Bulk lattice: fcc
2D bulk symm: pmm
2D surf symm: pmm

Adsorbate:
Coverage :
Pattern : (1x1)
Matrix : (1.000, 0.000)
(0.000, 1.000)

STRUCTURE TYPE
Bulk termination with multilayer relaxations

SAMPLE PREPARATION (1 sample)
Treatment : Ne+ sputtering and annealing; final flash to 723 K
Crystallinity:
Anal. methods:
Contamination: monitored by AES and LEED

COMMENTS

DATA COLLECTION
Technique: MEIS
Dataset : angular scans for 2 geometries: beam in (-111) plane, [0-11] channeling [101] blocking; (100) pl. [-100] chann [010] bl.

THEORY/DATA TREATMENT
Comparison with Monte-Carlo calcs of chann. and blocking; fitted isotropic vibr. ampl. 0.072Å (+55% at surface)

STRUCTURES EXAMINED
Relaxations of top two interplanar spacings

QUALITY OF EXPERIMENT-THEORY FIT
RS=0.21 (scaled R-factor)

2D UNIT CELLS (1 domain observed)

Cell	Ax (Å)	Ay (Å)	Bx (Å)	By (Å)	α (°)	Matrix	Pattern	Cell type
Bulk	3.610	0.000	0.000	2.553	90.0	(1.000, 0.000) (0.000, 1.000)	(1x1)	b: bulk lattice
Surface 1	3.610	0.000	0.000	2.553	90.0	(1.000, 0.000) (0.000, 1.000)	(1x1)	s1: commens. superlattice

3D COORDINATES

Dx/Dy in Å, or as a fraction of layer's unit cell vectors; atom 0 at the origin. Epir/subr are bulk repeat vectors.

No. of atoms: 4

Bulk z = 1.278 Å

Region	Chem el.	At. no.	Cell type	Site occ.	Rel. to	Dx ± εx		Dy ± εy		Dz ± εz		Dz/Bz(%) ± εz/Bz
epir		-2					f		f		Å	
subr		-1				1.805	Å	1.277	Å	1.278	Å	
intf	Cu	1	b	1.00	0	0.000	f	0.000	f	0.000	Å	0.0
intf	Cu	2	b	1.00	1	0.500	f	0.500	f	1.182 ± .020	Å	92.5 ± 1.6
intf	Cu	3	b	1.00	2	-0.500	f	-0.500	f	1.310 ± .020	Å	102.5 ± 1.6
subl	Cu	4	b	1.00	3	0.500	f	0.500	f	1.278	Å	100.0

BOND DISTANCES AND ANGLES

Bond distances and angles are derived from coordinates

No. of distances/angles: 6

Interatomic dist. A-B (Å)	Atom A	Atom B	Atom C	Bond angle A-B-C (°)
2.553	Cu1	Cu1(0,1)	Cu2	59.4
2.507	Cu1	Cu2	Cu3	58.8
2.492	Cu1	Cu3	Cu4	120.0
2.570	Cu2	Cu3	Cu4	60.7
2.588	Cu2	Cu4		
2.554	Cu3	Cu4		

COMMON NAME : Cu(110)-(1x1)
CLASSIFICATION : 29.41a
TECHNIQUE : LEED
AUTHORS : A.P. Baddorf, I.-W. Lyo, E.W. Plummer and H.L. Davis
REFERENCE : J. Vac. Sci. Technol., A5, 782 (1987)

ILLUSTRATION: 4

SURFACE TYPE
Substrate : Cu
Crystal face: 110
Temperature : 90 K
Bulk lattice: fcc
2D bulk symm: pmm
2D surf symm: pmm

Adsorbate:
Coverage :
Pattern : (1x1)
Matrix : (1.000, 0.000)
(0.000, 1.000)

STRUCTURE TYPE
Bulk termination with multilayer relaxations

SAMPLE PREPARATION (sample)
Treatment :
Crystallinity:
Anal. methods:
Contamination: close attention to H coverage

COMMENTS

DATA COLLECTION
Technique: LEED
Dataset : IV curves for 6 non-equivalent beams: (10), (01), (11), (20), (02), (12); E range 50-430 eV

THEORY/DATA TREATMENT
Dynamical LEED (RSP): up to 11 phase shifts from truncated atomic potential, full Slater exchange; Voi=-4 eV; ΘD=340 K

STRUCTURES EXAMINED
Variation of top two interlayer spacings

QUALITY OF EXPERIMENT-THEORY FIT
R2=0.116

2D UNIT CELLS (1 domain observed)

Cell	Ax (Å)	Ay (Å)	Bx (Å)	By (Å)	α (°)	Matrix	Pattern	Cell type
Bulk	3.610	0.000	0.000	2.553	90.0	(1.000, 0.000) (0.000, 1.000)	(1x1)	b: bulk lattice
Surface 1	3.610	0.000	0.000	2.553	90.0	(1.000, 0.000) (0.000, 1.000)	(1x1)	s1: commens. superlattice

3D COORDINATES

0.02Å error bars assumed for tabulation

Dx/Dy in Å, or as a fraction of layer's unit cell vectors; atom 0 at the origin. Epir/subr are bulk repeat vectors.

No. of atoms: 4 Bulk z = 1.278 Å

Region	Chem el.	At. no.	Cell type	Site occ.	Rel. to	Dx ± εx	Dy ± εy	Dz ± εz	Dz/Bz(%) ± εz/Bz
epir		-2				f	f	Å	
subr		-1				1.805 Å	1.277 Å	1.278 Å	
intf	Cu	1	b	1.00	0	0.000 f	0.000 f	0.000 Å	0.0
intf	Cu	2	b	1.00	1	0.500 f	0.500 f	1.158 ± .020 Å	90.6 ± 1.6
intf	Cu	3	b	1.00	2	-0.500 f	-0.500 f	1.341 ± .020 Å	104.9 ± 1.6
subl	Cu	4	b	1.00	3	0.500 f	0.500 f	1.278 Å	100.0

BOND DISTANCES AND ANGLES

Bond distances and angles are derived from coordinates

No. of distances/angles: 6

Interatomic dist. A-B (Å)	Atom A	Atom B	Atom C	Bond angle A-B-C (°)
2.553	Cu1	Cu1(0,1)	Cu2	59.2
2.496	Cu1	Cu2	Cu3	58.9
2.499	Cu1	Cu3	Cu4	120.0
2.586	Cu2	Cu3	Cu4	61.3
2.619	Cu2	Cu4		
2.554	Cu3	Cu4		

COMMON NAME : Cu(110)-(1x1) ILLUSTRATION: 4
CLASSIFICATION : 29.48
TECHNIQUE : LEIS
AUTHORS : E. Van de Riet, J.B.J. Smeets, J.M. Fluit and A. Niehaus
REFERENCE : Surf. Sci., 214, 111 (1989)

SURFACE TYPE
Substrate : Cu
Crystal face: 110
Temperature : RT*
Bulk lattice: fcc
2D bulk symm: pmm
2D surf symm: pmm

Adsorbate:
Coverage :
Pattern : (1x1)
Matrix : (1.000, 0.000)
(0.000, 1.000)

STRUCTURE TYPE
Contraction of the 1st interlayer spacing

SAMPLE PREPARATION (1 sample)
Treatment : Ne+ bombardment followed by annealing
Crystallinity:
Anal. methods:
Contamination:

COMMENTS

DATA COLLECTION
Technique: LEIS
Dataset :

THEORY/DATA TREATMENT
Classical dynamics

STRUCTURES EXAMINED
Variation of the 1st interlayer spacing

2D UNIT CELLS (1 domain observed)

Cell	Ax (Å)	Ay (Å)	Bx (Å)	By (Å)	α (°)	Matrix	Pattern	Cell type
Bulk	2.556	0.000	0.000	3.615	90.0	(1.000, 0.000) (0.000, 1.000)	(1x1)	b: bulk lattice
Surface 1	2.556	0.000	0.000	3.615	90.0	(1.000, 0.000) (0.000, 1.000)	(1x1)	s1: commens. superlattice

3D COORDINATES

Dx/Dy in Å, or as a fraction of layer's unit cell vectors; atom 0 at the origin. Epir/subr are bulk repeat vectors.

No. of atoms: 3 Bulk z = 1.278 Å

Region	Chem el.	At. no.	Cell type	Site occ.	Rel. to	Dx ± εx	Dy ± εy	Dz ± εz	Dz/Bz(%) ± εz/Bz
epir		-2				f	f	Å	
subr		-1				1.278 Å	1.807 Å	1.278 Å	
ovrl	Cu	1	s1	1.00	0	0.000 Å	0.000 Å	0.000 Å	0.0
intf	Cu	2	b	1.00	0	1.278 Å	1.807 Å	1.238 ± .040 Å	97.0 ± 3.0
subl	Cu	3	b	1.00	0	0.000 Å	0.000 Å	2.516 Å	197.0

BOND DISTANCES AND ANGLES

Bond distances and angles are derived from coordinates

No. of distances/angles: 1

Interatomic dist. A-B (Å)	Atom A	Atom B	Atom C	Bond angle A-B-C (°)
2.556	Cu1	Cu2		

COMMON NAME : Cu(110)-(1x2)
CLASSIFICATION : 29.52
TECHNIQUE : LEED
AUTHORS : Z. P. Hu, B. C. Pan, W. C. Fan and A. Ignatiev
REFERENCE : Phys. Rev., B41, 9692 (1990)

ILLUSTRATION: 5

SURFACE TYPE
Substrate : Cu
Crystal face: 110
Temperature : RT*
Bulk lattice: fcc
2D bulk symm: pmm
2D surf symm: pmm

Adsorbate:
Coverage :
Pattern : (1x2)
Matrix : (1.000, 0.000)
(0.000, 2.000)

STRUCTURE TYPE
Alkali-impurity-induced missing-row reconstruction: top Cu interlayer spacing contracted, lateral pairing in the 2nd Cu layer

SAMPLE PREPARATION (2 sample)
Treatment : Ar ion bombardment and annealing
Crystallinity:
Anal. methods:
Contamination:

COMMENTS
K and Cs atoms are disordered

DATA COLLECTION
Technique: LEED; video LEED
Dataset : IV spectra for 3 integral- and 4 fractional order beams

THEORY/DATA TREATMENT
Dynamical LEED: layer-doubling; RFS

STRUCTURES EXAMINED
Missing-row, row pairing and sawtooth models

QUALITY OF EXPERIMENT-THEORY FIT
Average R=0.18

2D UNIT CELLS (1 domain observed)

Cell	Ax (Å)	Ay (Å)	Bx (Å)	By (Å)	α (°)	Matrix	Pattern	Cell type
Bulk	2.556	0.000	0.000	3.615	90.0	(1.000, 0.000) (0.000, 1.000)	(1x1)	b: bulk lattice
Surface 1	2.556	0.000	0.000	7.230	90.0	(1.000, 0.000) (0.000, 2.000)	(1x2)	s1: commens. superlattice

3D COORDINATES

top interlayer spacing contracted; lateral pairing in the 2nd layer

Dx/Dy in Å, or as a fraction of layer's unit cell vectors; atom 0 at the origin. Epir/subr are bulk repeat vectors.

No. of atoms: 5

Bulk z = 1.278 Å

Region	Chem el.	At. no.	Cell type	Site occ.	Rel. to	Dx ± εx	Dy ± εy	Dz ± εz	Dz/Bz(%) ± εz/Bz
epir		-2				f	f	Å	
subr		-1				1.278 Å	1.807 Å	1.278 Å	
intf	Cu	1	s1	1.00	0	0.000 Å	0.000 Å	0.000 Å	0.0
intf	Cu	2	s1	.50	0	1.278 Å	-1.857 Å	1.130 Å	88.4
intf	Cu	3	s1	.50	0	1.278 Å	1.857 Å	1.130 Å	88.4
intf	Cu	4	b	1.00	0	0.000 Å	0.000 Å	2.408 ± .050 Å	188.4 ± 6.4
subl	Cu	5	b	1.00	0	1.278 Å	1.807 Å	3.686 Å	288.4

BOND DISTANCES AND ANGLES

Bond distances and angles are derived from coordinates

No. of distances/angles: 1

Interatomic dist. A-B (Å)	Atom A	Atom B	Atom C	Bond angle A-B-C (°)
2.522	Cu1	Cu2		

COMMON NAME : Cu(110)-(1x2) (Li induced) ILLUSTRATION: 5
CLASSIFICATION : 29.3.1
TECHNIQUE : MEIS
AUTHORS : M. Copel, W.R. Graham, T. Gustafsson and S. Yalisove
REFERENCE : Solid State Commun., 54, 695 (1985)

SURFACE TYPE
Substrate : Cu
Crystal face: 110
Temperature : RT
Bulk lattice: fcc
2D bulk symm: pmm
2D surf symm: pmm

Adsorbate: Li
Coverage : 0.2 Li/Cu
Pattern : (1x2)
Matrix : (1.000, 0.000)
(0.000, 2.000)

STRUCTURE TYPE
Missing-row reconstruction induced by Li (Li position undetermined)

SAMPLE PREPARATION (1 sample)
Treatment : Ar+ sputtering and anneal; Li deposited by resist. heating
Crystallinity: sharp (1x2) LEED pattern
Anal. methods:
Contamination: AES: no contamination on clean surface

COMMENTS
A structure with a 5.3% contraction of 1st layer spacing and 3.3% expansion of 2nd layer spacing also agrees with the experimental data; a pairing model with small displacements (<0.12Å) is possible with stiffened vibrations

DATA COLLECTION
Technique: MEIS
Dataset : angular scans for [0-11] incidence, [101] blocking and [100] incidence, [010] blocking

THEORY/DATA TREATMENT
Monte Carlo simulation: E≈100k eV; ΘD=250 K

STRUCTURES EXAMINED
Missing row model; buckling/pairing model

QUALITY OF EXPERIMENT-THEORY FIT
Visual

2D UNIT CELLS (1 domain observed)

Cell	Ax (Å)	Ay (Å)	Bx (Å)	By (Å)	α (°)	Matrix	Pattern	Cell type
Bulk	2.550	0.000	0.000	3.610	90.0	(1.000, 0.000) (0.000, 1.000)	(1x1)	b: bulk lattice
Surface 1	2.550	0.000	0.000	7.220	90.0	(1.000, 0.000) (0.000, 2.000)	(1x2)	s1: commens. superlattice

3D COORDINATES

Cu1: remaining rows (ridges); 0.02Å error bars assumed for tabulation

Dx/Dy in Å, or as a fraction of layer's unit cell vectors; atom 0 at the origin. Epir/subr are bulk repeat vectors.

No. of atoms: 3 Bulk z = 1.278 Å

Region	Chem el.	At. no.	Cell type	Site occ.	Rel. to	Dx ± εx		Dy ± εy		Dz ± εz		Dz/Bz(%) ± εz/Bz
epir		-2					f		f		Å	
subr		-1				-1.275	Å	-1.805	Å	1.278	Å	
intf	Cu	1	s1	.50	0	0.000	f	0.000	f	0.000	Å	0.0
intf	Cu	2	b	1.00	1	0.500	f	0.500	f	1.230 ± .020	Å	96.2 ± 1.6
subl	Cu	3	b	1.00	2	-0.500	f	-0.500	f	1.278	Å	100.0

BOND DISTANCES AND ANGLES

Bond distances and angles are derived from coordinates

No. of distances/angles: 4

Interatomic dist. A-B (Å)	Atom A	Atom B	Atom C	Bond angle A-B-C (°)
2.550	Cu1	Cu1(1,0)	Cu2	59.7
2.529	Cu1	Cu2	Cu3	59.1
2.508	Cu1	Cu3	Cu2	60.0
2.553	Cu2	Cu3	Cu1	60.0

COMMON NAME : Cu(111)-(1x1) ILLUSTRATION: 1
CLASSIFICATION : 29.31
TECHNIQUE : LEED
AUTHORS : S.A. Lindgren, L. Wallden, J. Rundgren and P. Westrin
REFERENCE : Phys. Rev., B29, 576 (1984)

SURFACE TYPE
Substrate : Cu
Crystal face: 111
Temperature : RT*
Bulk lattice: fcc
2D bulk symm: p3m1
2D surf symm: p3m1

Adsorbate:
Coverage :
Pattern : (1x1)
Matrix : (1.000, 0.000)
(0.000, 1.000)

STRUCTURE TYPE
Relaxed bulk termination

SAMPLE PREPARATION (1 sample)
Treatment : Ar+ bombardment (1μA, 250 eV), heating to 750 K
Crystallinity:
Anal. methods:
Contamination:

COMMENTS

DATA COLLECTION
Technique: LEED
Dataset : I-V curves: 00 (Θ=5°), 10 and 01 (Θ=0°) beams, E range 16-190 eV

THEORY/DATA TREATMENT
Dynamical LEED (layer doubling): 8 phase shifts, energy-dep. Voi and Vor; ΘD=343 K(bulk),300K(top layer)

STRUCTURES EXAMINED
Variation of topmost interlayer spacing

QUALITY OF EXPERIMENT-THEORY FIT
Four metrics used as R-factors

2D UNIT CELLS (1 domain observed)

Cell	Ax (Å)	Ay (Å)	Bx (Å)	By (Å)	α (°)	Matrix	Pattern	Cell type
Bulk	2.560	0.000	1.280	2.217	60.0	(1.000, 0.000) (0.000, 1.000)	(1x1)	b: bulk lattice
Surface 1	2.560	0.000	1.280	2.217	60.0	(1.000, 0.000) (0.000, 1.000)	(1x1)	s1: commens. superlattice

3D COORDINATES

Dx/Dy in Å, or as a fraction of layer's unit cell vectors; atom 0 at the origin. Epir/subr are bulk repeat vectors.

No. of atoms: 3 Bulk z = 2.090 Å

Region	Chem el.	At. no.	Cell type	Site occ.	Rel. to	Dx ± εx		Dy ± εy		Dz ± εz		Dz/Bz(%) ± εz/Bz
epir		-2					f		f		Å	
subr		-1				1.280	Å	0.739	Å	2.090	Å	
intf	Cu	1	b	1.00	0	0.000	f	0.000	f	0.000	Å	0.0
intf	Cu	2	b	1.00	1	0.333	f	0.333	f	2.076 ± .020	Å	99.3 ± 1.0
subl	Cu	3	b	1.00	2	0.333	f	0.333	f	2.090	Å	100.0

BOND DISTANCES AND ANGLES

Bond distances and angles are derived from coordinates

No. of distances/angles: 3

Interatomic dist. A-B (Å)	Atom A	Atom B	Atom C	Bond angle A-B-C (°)
2.560	Cu1	Cu1(1,0)		
2.548	Cu1	Cu2		
2.560	Cu2	Cu2(1,0)		

COMMON NAME : Cu(311)-(1x1) ILLUSTRATION: 8
CLASSIFICATION : 29.11
TECHNIQUE : LEED
AUTHORS : R.W. Streater, W.T. Moore, P.R. Watson, D.C. Frost and K.A.R. Mitchell
REFERENCE : Surf. Sci., 72, 744 (1978)

SURFACE TYPE
Substrate : Cu
Crystal face: 311
Temperature : RT*
Bulk lattice: fcc
2D bulk symm: cm
2D surf symm: cm

Adsorbate:
Coverage :
Pattern : (1x1)
Matrix : (1.000, 0.000)
(0.000, 1.000)

STRUCTURE TYPE
Bulk termination with top layer relaxation

SAMPLE PREPARATION (sample)
Treatment : see Moore et al, Sol. St. Commun. 24, 139 (1977)
Crystallinity:
Anal. methods:
Contamination:

COMMENTS

DATA COLLECTION
Technique: LEED
Dataset : I-V spectra for 13 beams (incl. 4 equivalent pairs), energy up to 220 eV

THEORY/DATA TREATMENT
Dynamical LEED (layer doubling): 8 phase shifts from Burdick-Chodorow potential; Voi∝E**1/3

STRUCTURES EXAMINED
Variation of topmost interlayer spacing from 5% expansion to 15% contraction from bulk value

QUALITY OF EXPERIMENT-THEORY FIT
RZJ=0.088

2D UNIT CELLS (1 domain observed)

Cell	Ax (Å)	Ay (Å)	Bx (Å)	By (Å)	α (°)	Matrix	Pattern	Cell type
Bulk	2.553	0.000	-1.275	4.229	106.8	(1.000, 0.000) (0.000, 1.000)	(1x1)	b: bulk lattice
Surface 1	2.553	0.000	-1.275	4.229	106.8	(1.000, 0.000) (0.000, 1.000)	(1x1)	s1: commens. superlattice

3D COORDINATES

Dx/Dy in Å, or as a fraction of layer's unit cell vectors; atom 0 at the origin. Epir/subr are bulk repeat vectors.

No. of atoms: 3 Bulk z = 1.090 Å

Region	Chem el.	At. no.	Cell type	Site occ.	Rel. to	Dx ± εx		Dy ± εy		Dz ± εz		Dz/Bz(%) ± εz/Bz
epir		-2					f		f		Å	
subr		-1				-1.277	Å	1.924	Å	1.090	Å	
intf	Cu	1	b	1.00	0	0.000	f	0.000	f	0.000	Å	0.0
intf	Cu	2	b	1.00	1	0.727	f	0.454	f	1.035 ± .020	Å	95.0 ± 1.8
subl	Cu	3	b	1.00	2	-0.273	f	0.455	f	1.090	Å	100.0

BOND DISTANCES AND ANGLES

Bond distances and angles are derived from coordinates

No. of distances/angles: 5

Interatomic dist. A-B (Å)	Atom A	Atom B	Atom C	Bond angle A-B-C (°)
2.553	Cu1	Cu1(1,0)		
2.528	Cu1	Cu2		
2.508	Cu1	Cu3(0,-1)		
2.509	Cu1	Cu3(-1,-1)		
2.554	Cu2	Cu3		

COMMON NAME : Cu(311)-(1x1)
CLASSIFICATION : 29.46
TECHNIQUE : LEED
AUTHORS : P.R. Watson and K.A.R. Mitchell
REFERENCE : Surf. Sci., 203, 323 (1988)

ILLUSTRATION: 8

SURFACE TYPE
Substrate : Cu
Crystal face: 311
Temperature : RT*
Bulk lattice: fcc
2D bulk symm: cm
2D surf symm: cm

Adsorbate:
Coverage :
Pattern : (1x1)
Matrix : (1.000, 0.000)
(0.000, 1.000)

STRUCTURE TYPE
Bulk termination with multilayer relaxations

SAMPLE PREPARATION (1 sample)
Treatment : see Streater et al (class. no. 29.11)
Crystallinity:
Anal. methods:
Contamination:

COMMENTS

DATA COLLECTION
Technique: LEED
Dataset : I-V curves for 8 symmetrically inequivalent beams at normal incidence; E<=220 eV

THEORY/DATA TREATMENT
Dynamical LEED (layer doubling): 8 phase shifts from Burdick potential; VoiαE**1/3; ΘD=343 K

STRUCTURES EXAMINED
Variation of first 3 interlayer spacings (registries kept bulk-like)

QUALITY OF EXPERIMENT-THEORY FIT
R2=0.0399, RZJ=0.0420, RPE=0.1406

2D UNIT CELLS (1 domain observed)

Cell	Ax (Å)	Ay (Å)	Bx (Å)	By (Å)	α (°)	Matrix	Pattern	Cell type
Bulk	2.553	0.000	1.276	4.233	73.2	(1.000, 0.000) (0.000, 1.000)	(1x1)	b: bulk lattice
Surface 1	2.553	0.000	1.276	4.233	73.2	(1.000, 0.000) (0.000, 1.000)	(1x1)	s1: commens. superlattice

3D COORDINATES

Dx/Dy in Å, or as a fraction of layer's unit cell vectors; atom 0 at the origin. Epir/subr are bulk repeat vectors.

No. of atoms: 4

Bulk z = 1.089 Å

Region	Chem el.	At. no.	Cell type	Site occ.	Rel. to	Dx ± εx		Dy ± εy		Dz ± εz		Dz/Bz(%) ± εz/Bz
epir		-2					f		f		Å	
subr		-1				1.276	Å	1.924	Å	1.089	Å	
intf	Cu	1	b	1.00	0	0.000	f	0.000	f	0.000	Å	0.0
intf	Cu	2	b	1.00	1	0.273	f	0.455	f	1.010 ± .030	Å	92.8 ± 2.8
intf	Cu	3	b	1.00	2	0.273	f	0.455	f	1.130 ± .100	Å	103.8 ± 9.2
subl	Cu	4	b	1.00	3	0.273	f	-0.546	f	1.089 ± .100	Å	100.0 ± 9.2

BOND DISTANCES AND ANGLES

Bond distances and angles are derived from coordinates

No. of distances/angles: 3

Interatomic dist. A-B (Å)	Atom A	Atom B	Atom C	Bond angle A-B-C (°)
2.553	Cu1	Cu1(1,0)	Cu2	59.6
2.520	Cu1	Cu2	Cu3	180.0
2.570	Cu2	Cu3	Cu4	60.7

COMMON NAME : Cu(100)-c(2x2)-Au ILLUSTRATION: 33
CLASSIFICATION : 29.79.2
TECHNIQUE : LEED
AUTHORS : Z.Q. Wang, Y.S. Li, C.K.C. Lok, J. Quinn, F. Jona and P.M. Marcus
REFERENCE : Solid State Commun., 62, 181 (1987)

SURFACE TYPE
Substrate : Cu
Crystal face: 100
Temperature : RT*
Bulk lattice: fcc
2D bulk symm: p4m
2D surf symm: p4m

Adsorbate: Au
Coverage : 0.5 Au/Cu
Pattern : c(2x2)
Matrix : (1.000, 1.000)
(-1.000, 1.000)

STRUCTURE TYPE
Substitutional adsorption, forming buckled monolayer of mixed alloy

SAMPLE PREPARATION (1 sample)
Treatment : Au evaporated from Au ribbon
Crystallinity:
Anal. methods: PED normal emission spectra for energies of 40 to 280eV
Contamination: coverage and cleanliness checked by AES

COMMENTS

DATA COLLECTION
Technique: LEED
Dataset : IV spectra at normal incidence for 6 beams, at (Θ,φ)= (7,0)° for 7 beams

THEORY/DATA TREATMENT
Dynamical LEED and PED

STRUCTURES EXAMINED
Overlayer model with Au-Cu spacing from 1.23 to 2.03Å and relaxations of top Cu layer from 1.6 to 2.2Å; buckled mixed layer with variable first interlayer spacing; mixed layers in first and third atomic planes

2D UNIT CELLS (1 domain observed)

Cell	Ax (Å)	Ay (Å)	Bx (Å)	By (Å)	α (°)	Matrix	Pattern	Cell type
Bulk	2.560	0.000	0.000	2.560	90.0	(1.000, 0.000) (0.000, 1.000)	(1x1)	b: bulk lattice
Surface 1	2.560	2.560	-2.560	2.560	90.0	(1.000, 1.000) (-1.000, 1.000)	c(2x2)	s1: commens. superlattice

3D COORDINATES

Au1-Cu2: mixed buckled top layer; 0.05Å error bars assumed for tabulation

Dx/Dy in Å, or as a fraction of layer's unit cell vectors; atom 0 at the origin. Epir/subr are bulk repeat vectors.

No. of atoms: 4 Bulk z = 1.810 Å

Region	Chem el.	At. no.	Cell type	Site occ.	Rel. to	Dx ± εx		Dy ± εy		Dz ± εz		Dz/Bz(%) ± εz/Bz
epir		-2					f		f		Å	
subr		-1				-1.280	Å	-1.280	Å	1.810	Å	
intf	Au	1	s1	.50	0	0.000	f	0.000	f	0.000	Å	0.0
intf	Cu	2	s1	.50	1	0.500	f	0.500	f	0.100 ± .050	Å	5.5 ± 2.8
intf	Cu	3	b	1.00	2	0.500	f	-0.500	f	1.883 ± .050	Å	104.0 ± 2.8
subl	Cu	4	b	1.00	3	-0.500	f	-0.500	f	1.810	Å	100.0

BOND DISTANCES AND ANGLES

Bond distances and angles are derived from coordinates

No. of distances/angles: 3

Interatomic dist. A-B (Å)	Atom A	Atom B	Atom C	Bond angle A-B-C (°)
2.562	Au1	Cu2	Cu3	62.5
2.685	Au1	Cu3(-1,0)	Cu4	92.6
2.612	Cu2	Cu3	Cu4	120.7

COMMON NAME : Cu(100)-CH3O disordered
CLASSIFICATION : 29.6.1.8.10
TECHNIQUE : PED
AUTHORS : Th. Lindner, J. Somers, A.M. Bradshaw, A.L. Kilcoyne and D.P. Woodruff
REFERENCE : Surf. Sci., 203, 333 (1988)

ILLUSTRATION: 76

SURFACE TYPE
Substrate : Cu
Crystal face: 100
Temperature : 200 K
Bulk lattice: fcc
2D bulk symm: p4m
2D surf symm: none

Adsorbate: CH3O
Coverage : unknown
Pattern : disordered
Matrix : (1.000, 0.000)
(0.000, 1.000)

STRUCTURE TYPE
CH3O- (methoxy) adsorption with O end down (forming tilted bridge with 2 Cu atoms) and CH3 pointing away from and perp. to surface (H positions not determined)

SAMPLE PREPARATION (1 sample)
Treatment : dosing with 3E-5 mbar O2, then 3E-5 mbar CH3OH at 200 K
Crystallinity:
Anal. methods:
Contamination: monitored by LEED, AES and NEXAFS

COMMENTS
NEXAFS polarisation dependence was used to indicate that CO axis is tilted by less than 10° from the surface normal

DATA COLLECTION
Technique: PED
Dataset : NEXAFS: Auger electron yield at emission angles 0 to 90°; ARUPS: normal emission Ek 50 to 400 eV

THEORY/DATA TREATMENT
NEXAFS: polarization dependence of 6a(1) resonance;
ARUPS: curved wave calcs (500 atom cluster)

STRUCTURES EXAMINED
Adsorption sites between bridge and hollow; molecular tilts of 0 to 60°; adsorption heights of 0.8 to 1.6Å; substrate kept bulk-like; H positions not determined

2D UNIT CELLS (2 domains observed)

Cell	Ax (Å)	Ay (Å)	Bx (Å)	By (Å)	α (°)	Matrix	Pattern	Cell type
Bulk	2.553	0.000	0.000	2.553	90.0	(1.000, 0.000) (0.000, 1.000)	(1x1)	b: bulk lattice
Surface 1	2.553	0.000	0.000	2.553	90.0	(1.000, 0.000) (0.000, 1.000)	disordered	nd1: non-recon. lattice-gas dis

3D COORDINATES

C1-O2: perp. axis of OCH3, O pointing down to surface; 0.1Å error bars assumed for tabulation

Dx/Dy in Å, or as a fraction of layer's unit cell vectors; atom 0 at the origin. Epir/subr are bulk repeat vectors.

No. of atoms: 4

Bulk z = 1.810 Å

Region	Chem el.	At. no.	Cell type	Site occ.	Rel. to	Dx ± εx		Dy ± εy		Dz ± εz		Dz/Bz(%) ± εz/Bz
epir		-2					f		f		Å	
subr		-1				-1.276	Å	-1.276	Å	1.810	Å	
ovrl	C	1	nd1	.25	0	0.000	f	0.000	f	0.000	Å	0.0
ovrl	O	2	nd1	.25	1	0.000 ± .039	f	0.000 ± .039	f	1.430 ± .100	Å	79.0 ± 5.5
intf	Cu	3	b	1.00	2	0.656 ± .039	f	0.500 ± .039	f	1.800 ± .100	Å	99.5 ± 5.5
subl	Cu	4	b	1.00	3	-0.500	f	-0.500	f	1.810	Å	100.0

BOND DISTANCES AND ANGLES

Bond distances and angles are derived from coordinates

No. of distances/angles: 3

Interatomic dist. A-B (Å)	Atom A	Atom B	Atom C	Bond angle A-B-C (°)
1.430	C1	O2	Cu3(-1,0)	139.3
2.375	Cu3	O2(1,0)	Cu3(0,-1)	65.0
2.770	Cu3	O2		

COMMON NAME : Cu(100)-C2H2 disordered ILLUSTRATION: 73
CLASSIFICATION : 29.6.1.2a
TECHNIQUE : SEXAFS
AUTHORS : D. Arvanitis, L. Wenzel and K. Baberschke
REFERENCE : Phys. Rev. Lett., 59, 2435 (1987)

SURFACE TYPE
Substrate : Cu
Crystal face: 100
Temperature : 60 K
Bulk lattice: fcc
2D bulk symm: p4m
2D surf symm: none

Adsorbate: C2H2
Coverage : <1ML
Pattern : disordered
Matrix : (1.000, 0.000)
(0.000, 1.000)

STRUCTURE TYPE
Disordered molecular adsorption parallel to surface, with each C near a bridge site, and a stretched C-C bond of 1.42±0.05Å

SAMPLE PREPARATION (1 sample)
Treatment : see Arvanitis et al, Surf. Sci. 178, 696 (1986)
Crystallinity:
Anal. methods:
Contamination:

COMMENTS

DATA COLLECTION
Technique: SEXAFS; SEXAFS with total electron yield
Dataset : SEXAFS spectra at normal x-ray incidence; range of photoelectron wavevector 4-11Å-1

THEORY/DATA TREATMENT
Fourier transform using an experimental phase shift from a bulk Cu phthalocyanine standard

STRUCTURES EXAMINED
Various adsorption geometries on a bulk-like substrate

2D UNIT CELLS (4 domains observed)

Cell	Ax (Å)	Ay (Å)	Bx (Å)	By (Å)	α (°)	Matrix	Pattern	Cell type
Bulk	2.556	0.000	0.000	2.556	90.0	(1.000, 0.000) (0.000, 1.000)	(1x1)	b: bulk lattice
Surface 1	2.556	0.000	0.000	2.556	90.0	(1.000, 0.000) (0.000, 1.000)	disordered	nd1: non-recon. lattice-gas dis

3D COORDINATES

C1-C2: disordered molecular C2H2 layer

Dx/Dy in Å, or as a fraction of layer's unit cell vectors; atom 0 at the origin. Epir/subr are bulk repeat vectors.

No. of atoms: 4 Bulk z = 1.805 Å

Region	Chem el.	At. no.	Cell type	Site occ.	Rel. to	Dx ± εx	Dy ± εy	Dz ± εz	Dz/Bz(%) ± εz/Bz
epir		-2				f	f	Å	
subr		-1				1.278 Å	1.278 Å	1.805 Å	
ovrl	C	1	nd1	.25	0	0.054 f	0.446 f	0.000 Å	0.0
ovrl	C	2	nd1	.25	0	0.446 ± .014 f	0.054 ± .014 f	0.000 Å	0.0
intf	Cu	3	b	1.00	0	0.000 f	0.000 f	1.300 ± .100 Å	72.0 ± 5.5
subl	Cu	4	b	1.00	3	0.500 f	0.500 f	1.805 Å	100.0

BOND DISTANCES AND ANGLES

No. of distances/angles: 3

Interatomic dist. A-B (Å)	Atom A	Atom B	Atom C	Bond angle A-B-C (°)
1.420	C1	C2		
1.735	C1	Cu3		
2.554	Cu3	Cu4		

COMMON NAME : Cu(100)-C2H4 disordered ILLUSTRATION: 74
CLASSIFICATION : 29.6.1.2b
TECHNIQUE : SEXAFS
AUTHORS : D. Arvanitis, L. Wenzel and K. Baberschke
REFERENCE : Phys. Rev. Lett., 59, 2435 (1987)

SURFACE TYPE
Substrate : Cu
Crystal face: 100
Temperature : 60 K
Bulk lattice: fcc
2D bulk symm: p4m
2D surf symm: none

Adsorbate: C2H4
Coverage : <1ML
Pattern : disordered
Matrix : (1.000, 0.000)
(0.000, 1.000)

STRUCTURE TYPE
Disordered molecular adsorption parallel to surface, centered on hollow site, oriented along [110], with a stretched C-C bond of 1.47±0.05Å

SAMPLE PREPARATION (1 sample)
Treatment : see Arvanitis et al, Surf. Sci. 178, 696 (1986)
Crystallinity:
Anal. methods:
Contamination:

COMMENTS

DATA COLLECTION
Technique: SEXAFS; SEXAFS with total electron yield
Dataset : SEXAFS spectra at normal x-ray incidence; range of photoelectron wavevector 4-11Å-1

THEORY/DATA TREATMENT
Fourier transform using an experimental phase shift from a bulk Cu phthalocyanine standard

STRUCTURES EXAMINED
Various adsorption geometries on a bulk-like substrate

2D UNIT CELLS (2 domains observed)

Cell	Ax (Å)	Ay (Å)	Bx (Å)	By (Å)	α (°)	Matrix	Pattern	Cell type
Bulk	2.556	0.000	0.000	2.556	90.0	(1.000, 0.000) (0.000, 1.000)	(1x1)	b: bulk lattice
Surface 1	2.556	0.000	0.000	2.556	90.0	(1.000, 0.000) (0.000, 1.000)	disordered	nd1: non-recon. lattice-gas dis

3D COORDINATES

C1-C2: disordered molecular C2H4 layer

Dx/Dy in Å, or as a fraction of layer's unit cell vectors; atom 0 at the origin. Epir/subr are bulk repeat vectors.

No. of atoms: 4 Bulk z = 1.805 Å

Region	Chem el.	At. no.	Cell type	Site occ.	Rel. to	Dx ± εx		Dy ± εy		Dz ± εz		Dz/Bz(%) ± εz/Bz
epir		-2					f		f		Å	
subr		-1				1.278	Å	1.278	Å	1.805	Å	
ovrl	C	1	nd1	.25	0	0.212 ± .014	f	0.500	f	0.000	Å	0.0
ovrl	C	2	nd1	.25	0	0.788 ± .014	f	0.500	f	0.000	Å	0.0
intf	Cu	3	b	1.00	0	0.000	f	0.000	f	1.250 ± .100	Å	69.3 ± 5.5
subl	Cu	4	b	1.00	3	0.500	f	0.500	f	1.805	Å	100.0

BOND DISTANCES AND ANGLES

No. of distances/angles: 3

Interatomic dist. A-B (Å)	Atom A	Atom B	Atom C	Bond angle A-B-C (°)
1.470	C1	C2		
1.868	C1	Cu3		
2.554	Cu3	Cu4		

COMMON NAME : Cu(100)-C2H4 disordered
CLASSIFICATION : 29.6.1.3
TECHNIQUE : NEXAFS
AUTHORS : J. C. Tang, X. S. Feng, and J. F. Shen, T. Fujikawa and T. Okazawa
REFERENCE : Phys. Rev., B44, 13018 (1991)

ILLUSTRATION: -

SURFACE TYPE
Substrate : Cu
Crystal face: 100
Temperature : RT*
Bulk lattice: fcc
2D bulk symm: p4m
2D surf symm: none

Adsorbate: C2H4 (ethylene)
Coverage : 0.1ML
Pattern : disordered
Matrix : (1.000, 0.000)
(0.000, 1.000)

STRUCTURE TYPE
Intact molecular adsorption over a 4-fold hollow site with the C-C bond parallel to the [001] or [010] directions

SAMPLE PREPARATION (sample)
Treatment :
Crystallinity:
Anal. methods:
Contamination:

COMMENTS
Coverage assumed to be 0.1 ML for tabulation

DATA COLLECTION
Technique: NEXAFS
Dataset :

THEORY/DATA TREATMENT
Multiple-scattering cluster method of NEXAFS

2D UNIT CELLS (2 domains observed)

Cell	Ax (Å)	Ay (Å)	Bx (Å)	By (Å)	α (°)	Matrix	Pattern	Cell type
Bulk	2.556	0.000	0.000	2.556	90.0	(1.000, 0.000) (0.000, 1.000)	(1x1)	b: bulk lattice
Surface 1	2.556	0.000	0.000	2.556	90.0	(1.000, 0.000) (0.000, 1.000)	disordered	nd1: non-recon. lattice-gas dis

3D COORDINATES

C1-C2: ethylene over 4-fold hollow site; C-C bond parallel to the [001] or [010] directions

Dx/Dy in Å, or as a fraction of layer's unit cell vectors; atom 0 at the origin. Epir/subr are bulk repeat vectors.

No. of atoms: 4

Bulk z = 1.807 Å

Region	Chem el.	At. no.	Cell type	Site occ.	Rel. to	Dx ± εx		Dy ± εy		Dz ± εz		Dz/Bz(%) ± εz/Bz
epir		-2					f		f		Å	
subr		-1				1.278	Å	1.278	Å	1.807	Å	
ovrl	C	1	nd1	.10	0	-0.516	Å	-0.516	Å	0.000	Å	0.0
ovrl	C	2	nd1	.10	0	0.516	Å	0.516	Å	0.000	Å	0.0
intf	Cu	3	b	1.00	0	1.278	Å	1.278	Å	1.565 ± .030	Å	86.6 ± 1.7
subl	Cu	4	b	1.00	0	0.000	Å	0.000	Å	3.562	Å	186.6

BOND DISTANCES AND ANGLES

Bond distances and angles are derived from coordinates

No. of distances/angles: 3

Interatomic dist. A-B (Å)	Atom A	Atom B	Atom C	Bond angle A-B-C (°)
1.460	C1	C2		
1.900	C1	Cu3		
1.900	C2	Cu3		

COMMON NAME : Cu(100)-c(2x2)-CO
CLASSIFICATION : 29.6.8.0a
TECHNIQUE : LEED
AUTHORS : S. Andersson and J.B. Pendry
REFERENCE : J. Phys., C13, 3547 (1980)

ILLUSTRATION: 71

SURFACE TYPE
Substrate : Cu
Crystal face: 100
Temperature : 80 K
Bulk lattice: fcc
2D bulk symm: p4m
2D surf symm: p4m

Adsorbate: CO
Coverage : 1/2 (CO/Cu)
Pattern : c(2x2)
Matrix : (1.000, 1.000)
(-1.000, 1.000)

STRUCTURE TYPE
Molecular adsorption on top sites; C-O axis perpendicular to surface, bonding through C end

SAMPLE PREPARATION (1 sample)
Treatment : CO exposures of about 2.2L
Crystallinity:
Anal. methods:
Contamination:

COMMENTS

DATA COLLECTION
Technique: LEED
Dataset : I-V spectra for (1,0),(1,1),(1/2,1/2),(3/2,1/2) beams at normal incidence; E range 10-125 eV

THEORY/DATA TREATMENT
Dynamical LEED (layer doubling): 8 phase shifts (Chodorow pot for Cu, superposition pots for C and O); VoiαE**1/3

STRUCTURES EXAMINED
CO perpendicular to unrelaxed substrate in top site with variable Cu-C and C-O layer spacings

QUALITY OF EXPERIMENT-THEORY FIT
RPE=0.50

2D UNIT CELLS (1 domain observed)

Cell	Ax (Å)	Ay (Å)	Bx (Å)	By (Å)	α (°)	Matrix	Pattern	Cell type
Bulk	2.560	0.000	0.000	2.560	90.0	(1.000, 0.000) (0.000, 1.000)	(1x1)	b: bulk lattice
Surface 1	2.560	2.560	-2.560	2.560	90.0	(1.000, 1.000) (-1.000, 1.000)	c(2x2)	s1: commens. superlattice

3D COORDINATES

O1-C2: upright CO molecules in top sites

Dx/Dy in Å, or as a fraction of layer's unit cell vectors; atom 0 at the origin. Epir/subr are bulk repeat vectors.

No. of atoms: 4 Bulk z = 1.810 Å

Region	Chem el.	At. no.	Cell type	Site occ.	Rel. to	Dx ± εx		Dy ± εy		Dz ± εz		Dz/Bz(%) ± εz/Bz
epir		-2					f		f		Å	
subr		-1				1.280	Å	1.280	Å	1.810	Å	
ovrl	O	1	s1	.50	0	0.000	f	0.000	f	0.000	Å	0.0
ovrl	C	2	s1	.50	1	0.000	f	0.000	f	1.130 ± .100	Å	62.4 ± 5.5
intf	Cu	3	b	1.00	2	0.000	f	0.000	f	1.900 ± .100	Å	105.0 ± 5.5
subl	Cu	4	b	1.00	3	0.500	f	0.500	f	1.810	Å	100.0

BOND DISTANCES AND ANGLES

Bond distances and angles are derived from coordinates

No. of distances/angles: 4

Interatomic dist. A-B (Å)	Atom A	Atom B	Atom C	Bond angle A-B-C (°)
1.130	O1	C2		
1.900	C2	Cu3		
2.560	Cu3	Cu3(1,0)		
2.560	Cu3	Cu4		

COMMON NAME : Cu(100)-c(2x2)-CO
CLASSIFICATION : 29.6.8.1
TECHNIQUE : NEXAFS
AUTHORS : C.F. McConville, D.P. Woodruff, K.C. Prince, G. Paolucci, V. Chab, M. Surman and A.M. Bradshaw
REFERENCE : Surf. Sci., 166, 221 (1986)

ILLUSTRATION: 71

SURFACE TYPE
Substrate : Cu
Crystal face: 100
Temperature : 100 K
Bulk lattice: fcc
2D bulk symm: p4m
2D surf symm: p4m

Adsorbate: CO
Coverage : 1/2 CO/Cu
Pattern : c(2x2)
Matrix : (1.000, 1.000)
(-1.000, 1.000)

STRUCTURE TYPE
Molecular adsorption perp. to surface with C down on top site

SAMPLE PREPARATION (1 sample)
Treatment : ion bombardment and annealing; CO adsorption at 100 K
Crystallinity:
Anal. methods:
Contamination: monitored by LEED and AES

COMMENTS
PED (normal photoelectron diffraction) also used

DATA COLLECTION
Technique: NEXAFS
Dataset : NEXAFS: O(KVV) Auger electron detection (510-560 eV); PED: C1s and O1s peak region for 180-240eV

THEORY/DATA TREATMENT
NEXAFS of π^* feature with polarization dependence;
PED (single scatt): Vor=-11 eV, Voi=-7.5eV, therm. effects

STRUCTURES EXAMINED
PED: perpendicular CO in top, bridge and hollow sites on unrelaxed Cu(100) substrate;
CO bond length of 1.13Å assumed

2D UNIT CELLS (1 domain observed)

Cell	Ax (Å)	Ay (Å)	Bx (Å)	By (Å)	α (°)	Matrix	Pattern	Cell type
Bulk	2.560	0.000	0.000	2.560	90.0	(1.000, 0.000) (0.000, 1.000)	(1x1)	b: bulk lattice
Surface 1	2.560	2.560	-2.560	2.560	90.0	(1.000, 1.000) (-1.000, 1.000)	c(2x2)	s1: commens. superlattice

3D COORDINATES

01-C2: CO overlayer on top sites

Dx/Dy in Å, or as a fraction of layer's unit cell vectors; atom 0 at the origin. Epir/subr are bulk repeat vectors.

No. of atoms: 4 Bulk z = 1.810 Å

Region	Chem el.	At. no.	Cell type	Site occ.	Rel. to	Dx ± εx	Dy ± εy	Dz ± εz	Dz/Bz(%) ± εz/Bz
epir		-2				f	f	Å	
subr		-1				1.280 Å	1.280 Å	1.810 Å	
ovrl	O	1	s1	.50	0	0.000 f	0.000 f	0.000 Å	0.0
ovrl	C	2	s1	.50	1	0.000 f	0.000 f	1.130 Å	62.4
intf	Cu	3	b	1.00	2	0.000 f	0.000 f	1.920 ± .050 Å	106.1 ± 2.8
subl	Cu	4	b	1.00	3	0.500 f	0.500 f	1.810 Å	100.0

BOND DISTANCES AND ANGLES

Bond distances and angles are derived from coordinates

No. of distances/angles: 2

Interatomic dist. A-B (Å)	Atom A	Atom B	Atom C	Bond angle A-B-C (°)
1.130	O1	C2	Cu3	180.0
1.920	C2	Cu3	Cu4	135.0

COMMON NAME : Cu(100)-c(2x2)-Cl ILLUSTRATION: 28,29
CLASSIFICATION : 29.17.12
TECHNIQUE : XSW
AUTHORS : J.R. Patel, D.W. Berreman, F. Sette, P.H. Citrin, J.E. Rowe, P.L. Cowan, T. Jach and B. Karlin
REFERENCE : Phys. Rev., B40, 1330 (1989)

SURFACE TYPE
Substrate : Cu
Crystal face: 100
Temperature : RT*
Bulk lattice: fcc
2D bulk symm: p4m
2D surf symm: p4m

Adsorbate: Cl
Coverage : 0.5 (Cl/Cu)
Pattern : c(2x2)
Matrix : (1.000, 1.000)
(-1.000, 1.000)

STRUCTURE TYPE
Atomic adsorption in hollow site with expanded top Cu-Cu interlayer spacing

SAMPLE PREPARATION (1 sample)
Treatment : see PRL 49,1712(1982), PRL 61,1384(1988)
Crystallinity:
Anal. methods: SEXAFS for bond distances
Contamination:

COMMENTS

DATA COLLECTION
Technique: XSW; fluorescence yield at NSLS
Dataset : standing-wave data at fixed incidence angle as fct. of crystal rotation about [111]

THEORY/DATA TREATMENT
8x8 matrix dynamical x-ray theory, with 2 fit parameters: P = at. pos. above (111) plane, F = coherent fraction

STRUCTURES EXAMINED
XSW and SEXAFS data directly give 4-fold hollow site; difference between XSW and SEXAFS for Cl-Cu spacing indicates change in top Cu-Cu spacing from bulk, since XSW refers adsorbate position to deep bulk layers

2D UNIT CELLS (1 domain observed)

Cell	Ax (Å)	Ay (Å)	Bx (Å)	By (Å)	α (°)	Matrix	Pattern	Cell type
Bulk	2.556	0.000	0.000	2.556	90.0	(1.000, 0.000) (0.000, 1.000)	(1x1)	b: bulk lattice
Surface 1	2.556	2.556	-2.556	2.556	90.0	(1.000, 1.000) (-1.000, 1.000)	c(2x2)	s1: commens. superlattice

3D COORDINATES

Cl1 forms overlayer in hollow sites; coordinates are derived from bond distance and spacings

Dx/Dy in Å, or as a fraction of layer's unit cell vectors; atom 0 at the origin. Epir/subr are bulk repeat vectors.

No. of atoms: 4 Bulk z = 1.807 Å

Region	Chem el.	At. no.	Cell type	Site occ.	Rel. to	Dx ± εx		Dy ± εy		Dz ± εz		Dz/Bz(%) ± εz/Bz
epir		-2					f		f		Å	
subr		-1				-1.278	Å	-1.278	Å	1.807	Å	
ovrl	Cl	1	s1	.50	0	0.000	f	0.000	f	0.000	Å	0.0
intf	Cu	2	b	1.00	1	0.500	f	0.500	f	1.530 ± .020	Å	84.7 ± 1.1
intf	Cu	3	b	1.00	2	-0.500	f	-0.500	f	1.877 ± .040	Å	103.9 ± 2.2
subl	Cu	4	b	1.00	3	0.500	f	0.500	f	1.807	Å	100.0

BOND DISTANCES AND ANGLES

No. of distances/angles: 4

Interatomic dist. A-B (Å)	Atom A	Atom B	Atom C	Bond angle A-B-C (°)
2.368	Cl1	Cu2	Cl1(1,0)	99.5
2.368	Cl1	Cu2	Cu2(1,0)	122.7
2.556	Cu2	Cu2(1,0)	Cu3(1,0)	60.6
2.606	Cu2	Cu3	Cu4	91.1

COMMON NAME : Cu(100)-c(2x2)-Cl ILLUSTRATION: 28,29
CLASSIFICATION : 29.17.13
TECHNIQUE : ARPEFS
AUTHORS : L.-Q. Wang, A.E. Schach von Wittenau, Z.G. Ji, L.S. Wang, Z.Q. Huang and D.A. Shirley
REFERENCE : Phys. Rev., B44, 1292 (1991)

SURFACE TYPE
Substrate : Cu
Crystal face: 100
Temperature : 110 K
Bulk lattice: fcc
2D bulk symm: p4m
2D surf symm: p4m

Adsorbate: Cl
Coverage : 0.5 Cl/Cu
Pattern : c(2x2)
Matrix : (1.000, 1.000)
(-1.000, 1.000)

STRUCTURE TYPE
Atomic adsorption in hollow site, with multilayer relaxation relaxation and 2nd Cu layer buckling

SAMPLE PREPARATION (1 sample)
Treatment : sputter-anneal cycles, then exposure to Cl2 and 400 K anneal
Crystallinity: sharp LEED pattern with no background
Anal. methods:
Contamination: monitored by AES

COMMENTS
Same analysis performed both at 110 K and 300K (110K gives deeper information): average result reported

DATA COLLECTION
Technique: ARPEFS; 2870-3370eV soft x-ray beam (2eV re
Dataset : ARPEFS spectra for two emission angles: [100], [110]; kinetic E range 50-550 eV

THEORY/DATA TREATMENT
Fourier transform; MSSW calcs: 16 ph shs; HF Cl pot; ΘD=343 K(bulk Cu), 243K(surf Cu), 325K(Cl), then fit

STRUCTURES EXAMINED
Top, bridge and hollow site: FT and MSSW favor hollow; automated fitting of Cl-Cu spacing, 1st Cu layer buckling, top 3 Cu-Cu interlayer spacings, optimized by R-factor fitting (also fitting of emission directions, photon polarization angles, sample and Debye temperatures, Vor)

QUALITY OF EXPERIMENT-THEORY FIT
R2=0.06-0.15

2D UNIT CELLS (1 domain observed)

Cell	Ax (Å)	Ay (Å)	Bx (Å)	By (Å)	α (°)	Matrix	Pattern	Cell type
Bulk	2.547	0.000	0.000	2.547	90.0	(1.000, 0.000) (0.000, 1.000)	(1x1)	b: bulk lattice
Surface 1	2.547	2.547	-2.547	2.547	90.0	(1.000, 1.000) (-1.000, 1.000)	c(2x2)	s1: commens. superlattice

3D COORDINATES

Cl1: overlayer in hollow sites; Cu3-Cu4: buckled second Cu layer;
error bars are statistical precision, not accuracy

Dx/Dy in Å, or as a fraction of layer's unit cell vectors; atom 0 at the origin. Epir/subr are bulk repeat vectors.

No. of atoms: 7 Bulk z = 1.801 Å

Reg ion	Chem el.	At. no.	Cell type	Site occ.	Rel. to	Dx ± εx		Dy ± εy		Dz ± εz		Dz/Bz(%) ± εz/Bz
epir		-2					f		f		Å	
subr		-1				1.274	Å	1.274	Å	1.801	Å	
ovrl	Cl	1	s1	.50	0	0.000	f	0.000	f	0.000	Å	0.0
intf	Cu	2	b	1.00	1	0.500	f	0.500	f	1.604 ± .005	Å	89.1 ± .3
intf	Cu	3	s1	.50	2	0.500	f	0.000	f	1.808 ± .021	Å	100.4 ± 1.2
intf	Cu	4	s1	.50	3	-0.500	f	-0.500	f	0.041 ± .012	Å	2.3 ± .7
intf	Cu	5	b	1.00	4	0.500	f	0.500	f	1.769 ± .027	Å	98.2 ± 1.5
intf	Cu	6	b	1.00	5	-0.500	f	-0.500	f	1.810 ± .033	Å	100.5 ± 1.8
subl	Cu	7	b	1.00	6	0.500	f	0.500	f	1.801	Å	100.0

Cu(100)-c(2x2)-Cl
29.17.13

BOND DISTANCES AND ANGLES

Bond distances and angles are derived from coordinates

No. of distances/angles: 4

Interatomic dist. A-B (Å)	Atom A	Atom B	Atom C	Bond angle A-B-C (°)
2.412	Cl1	Cu2	Cl1(1,0)	96.6
2.412	Cl1	Cu2	Cu2(1,0)	121.9
2.412	Cl1	Cu2	Cu3	86.8
2.552	Cu2	Cu3	Cu4	60.8

COMMON NAME : Cu(100)-c(2x2)-Cl
CLASSIFICATION : 29.17.5
TECHNIQUE : LEED
AUTHORS : F. Jona, D. Westphal, A. Goldman and P.M. Marcus
REFERENCE : J. Phys., C16, 3001 (1983)

ILLUSTRATION: 28,29

SURFACE TYPE
Substrate : Cu
Crystal face: 100
Temperature : RT*
Bulk lattice: fcc
2D bulk symm: p4m
2D surf symm: p4m

Adsorbate: Cl
Coverage : 0.5 Cl/Cu
Pattern : c(2x2)
Matrix : (1.000, 1.000)
(-1.000, 1.000)

STRUCTURE TYPE
Atomic adsorption in hollow site of unreconstructed substrate

SAMPLE PREPARATION (1 sample)
Treatment : Cl exposure at 2.0E-9torr; c(2x2) pattern produced at 5-100L
Crystallinity:
Anal. methods:
Contamination: monitored by LEED, UPS and XPS

COMMENTS
R-factor is mean for combined normal and off-normal inc.; variation of Cl muffin-tin radius between 1.23 and 1.13Å did not significantly change the result of the structure determination

DATA COLLECTION
Technique: LEED
Dataset : I-V
spectra:Θ=0°:(10),(11),(20),(0.5,0.5),(1.5,
spectra:Θ=0°:(10),(11),(20),(0.5,0.5),(1.5,

THEORY/DATA TREATMENT
Dynamical LEED (CHANGE): 8 phase shifts; Vor=-9 eV,Voi=-3.5eV obtained by minimising R-factor; rms ampl 0.15Å

STRUCTURES EXAMINED
1. 4-fold hollow site; 2. various Cl-Cu interlayer spacings 1.55-1.65Å;
3. various top Cu-Cu layer spacings 1.81-1.86Å

QUALITY OF EXPERIMENT-THEORY FIT
RZJ=0.17 (see comments)

2D UNIT CELLS (1 domain observed)

Cell	Ax (Å)	Ay (Å)	Bx (Å)	By (Å)	α (°)	Matrix	Pattern	Cell type
Bulk	2.556	0.000	0.000	2.556	90.0	(1.000, 0.000) (0.000, 1.000)	(1x1)	b: bulk lattice
Surface 1	2.556	2.556	-2.556	2.556	90.0	(1.000, 1.000) (-1.000, 1.000)	c(2x2)	s1: commens. superlattice

3D COORDINATES

Cl1: overlayer in hollow sites

Dx/Dy in Å, or as a fraction of layer's unit cell vectors; atom 0 at the origin. Epir/subr are bulk repeat vectors.

No. of atoms: 4

Bulk z = 1.807 Å

Region	Chem el.	At. no.	Cell type	Site occ.	Rel. to	Dx ± εx		Dy ± εy		Dz ± εz		Dz/Bz(%) ± εz/Bz
epir		-2					f		f		Å	
subr		-1				1.278	Å	1.278	Å	1.807	Å	
ovrl	Cl	1	s1	.50	0	0.000	f	0.000	f	0.000	Å	0.0
intf	Cu	2	b	1.00	1	0.500	f	0.500	f	1.600 ± .030	Å	88.5 ± 1.7
intf	Cu	3	b	1.00	2	-0.500	f	-0.500	f	1.850 ± .030	Å	102.4 ± 1.7
subl	Cu	4	b	1.00	3	0.500	f	0.500	f	1.807	Å	100.0

BOND DISTANCES AND ANGLES

Bond distances and angles are derived from coordinates

No. of distances/angles: 5

Interatomic dist. A-B (Å)	Atom A	Atom B	Atom C	Bond angle A-B-C (°)
2.414	Cl1	Cu2	Cl1(1,0)	97.0
2.414	Cl1	Cu2	Cu2(1,0)	122.0
2.556	Cu2	Cu2(1,0)		

Cu(100)-c(2x2)-Cl
29.17.5

Bond Distances and Angles - Continued

Interatomic dist. A-B (Å)	Atom A	Atom B	Atom C	Bond angle A-B-C (°)
2.586	Cu2	Cu3		
2.556	Cu3	Cu4		

COMMON NAME : Cu(100)-c(2x2)-Cl
CLASSIFICATION : 29.17.7
TECHNIQUE : SEXAFS
AUTHORS : P.H. Citrin, D.R. Hamann, L.F. Mattheiss and J.E. Rowe
REFERENCE : Phys. Rev. Lett., 49, 1712 (1982)

ILLUSTRATION: 28,29

SURFACE TYPE
Substrate : Cu
Crystal face: 100
Temperature : RT*
Bulk lattice: fcc
2D bulk symm: p4m
2D surf symm: p4m

Adsorbate: Cl
Coverage : 1/2 (Cl/Cu)
Pattern : c(2x2)
Matrix : (1.000, 1.000)
(-1.000, 1.000)

STRUCTURE TYPE
Atomic adsorption in hollow site of unreconstructed substrate

SAMPLE PREPARATION (1 sample)
Treatment : exposure to 15L of Cl2 at RT, annealing at 373 K for 2 mins
Crystallinity:
Anal. methods:
Contamination: monitored by AES and LEED

COMMENTS

DATA COLLECTION
Technique: SEXAFS
Dataset : SEXAFS data for polarisation parallel (Θ=90°) and nearly perpendicular (Θ=5°) to surface

THEORY/DATA TREATMENT
Fourier transform

STRUCTURES EXAMINED
1. 1-fold top model; 2. 2-fold bridge model; 3. 4-fold hollow site model; 4. mixed layer model; models distinguished by comparing exp. and theor. effective coordination

2D UNIT CELLS (1 domain observed)

Cell	Ax (Å)	Ay (Å)	Bx (Å)	By (Å)	α (°)	Matrix	Pattern	Cell type
Bulk	2.556	0.000	0.000	2.556	90.0	(1.000, 0.000) (0.000, 1.000)	(1x1)	b: bulk lattice
Surface 1	2.556	2.556	-2.556	2.556	90.0	(1.000, 1.000) (-1.000, 1.000)	c(2x2)	s1: commens. superlattice

3D COORDINATES

Cl1: overlayer in hollow sites; coordinates are derived from bond distance

Dx/Dy in Å, or as a fraction of layer's unit cell vectors; atom 0 at the origin. Epir/subr are bulk repeat vectors.

No. of atoms: 3 Bulk z = 1.807 Å

Region	Chem el.	At. no.	Cell type	Site occ.	Rel. to	Dx ± εx		Dy ± εy		Dz ± εz		Dz/Bz(%) ± εz/Bz
epir		-2					f		f		Å	
subr		-1				-1.278	Å	-1.278	Å	1.807	Å	
ovrl	Cl	1	s1	.50	0	0.000	f	0.000	f	0.000	Å	0.0
intf	Cu	2	b	1.00	1	0.500	f	0.500	f	1.590 ± .020	Å	88.0 ± 1.1
subl	Cu	3	b	1.00	2	-0.500	f	-0.500	f	1.807	Å	100.0

BOND DISTANCES AND ANGLES

No. of distances/angles: 4

Interatomic dist. A-B (Å)	Atom A	Atom B	Atom C	Bond angle A-B-C (°)
2.407	Cl1	Cu2	Cl1(1,0)	97.3
2.407	Cl1	Cu2	Cu2(1,0)	122.1
2.556	Cu2	Cu2(1,0)		
2.556	Cu2	Cu3		

COMMON NAME : Cu(111)-(√3x√3)R30°-Cl ILLUSTRATION: 22,24
CLASSIFICATION : 29.17.8
TECHNIQUE : SEXAFS
AUTHORS : M.D. Crapper, C.E. Riley, P.J.J. Sweeney, C.F. McConville and D.P. Woodruff
REFERENCE : Surf. Sci., 182, 213 (1987)

SURFACE TYPE
Substrate : Cu
Crystal face: 111
Temperature : RT*
Bulk lattice: fcc
2D bulk symm: p3m1
2D surf symm: p31m

Adsorbate: Cl
Coverage : 0.333 Cl/Cu
Pattern : (√3x√3)R30°
Matrix : (1.000, 1.000)
(-2.000, 1.000)

STRUCTURE TYPE
Atomic adsorption in fcc hollow sites

SAMPLE PREPARATION (1 sample)
Treatment : Cl adsorbed from electrochemical AgCl-based cell
Crystallinity:
Anal. methods: photoelectron diffraction 100-400 eV
Contamination: monitored by AES and LEED

COMMENTS
Complementary photoelectron diffraction studies provided distinction between fcc and hcp hollow sites

DATA COLLECTION
Technique: SEXAFS
Dataset : SEXAFS spectra at 2 incidence angles, E range 2800-3200 eV

THEORY/DATA TREATMENT
SEXAFS: single shell Fourier filtering and multishell sim.; photoelectron diffraction: single scattering; ΘD: 180 K

STRUCTURES EXAMINED
Bridge, top, fcc and hcp hollow sites; Cl-Cu spacing varied

QUALITY OF EXPERIMENT-THEORY FIT
Visual

2D UNIT CELLS (1 domain observed)

Cell	Ax (Å)	Ay (Å)	Bx (Å)	By (Å)	α (°)	Matrix	Pattern	Cell type
Bulk	2.552	0.000	1.276	2.210	60.0	(1.000, 0.000) (0.000, 1.000)	(1x1)	b: bulk lattice
Surface 1	3.828	2.210	-3.828	2.210	120.0	(1.000, 1.000) (-2.000, 1.000)	(√3x√3)R30°	s1: commens. superlattice

3D COORDINATES

Cl1: overlayer in fcc hollow sites; coordinates are derived from bond distances and angles

Dx/Dy in Å, or as a fraction of layer's unit cell vectors; atom 0 at the origin. Epir/subr are bulk repeat vectors.

No. of atoms: 3 Bulk z = 2.080 Å

Reg ion	Chem el.	At. no.	Cell type	Site occ.	Rel. to	Dx ± εx		Dy ± εy		Dz ± εz		Dz/Bz(%) ± εz/Bz
epir		-2					f		f		Å	
subr		-1				1.276	Å	0.737	Å	2.080	Å	
ovrl	Cl	1	s1	.33	0	0.000	f	0.000	f	0.000	Å	0.0
intf	Cu	2	b	1.00	1	0.333	f	0.333	f	1.880 ± .020	Å	90.4 ± 1.0
subl	Cu	3	b	1.00	2	0.333	f	0.333	f	2.080	Å	100.0

BOND DISTANCES AND ANGLES

No. of distances/angles: 3

Interatomic dist. A-B (Å)	Atom A	Atom B	Atom C	Bond angle A-B-C (°)
2.389	Cl1	Cu2	Cu2(1,0)	122.3
2.389	Cl1	Cu2	Cu3	177.2
2.552	Cu2	Cu2(1,0)		

COMMON NAME : Cu(111)-(√3x√3)R30°-Cl   ILLUSTRATION: 22,24
CLASSIFICATION : 29.17.9
TECHNIQUE : XSW
AUTHORS : D.P. Woodruff, D.L. Seymour, C.F. McConville, C.E. Riley, M.D. Crapper, N.P. Prince and Robert G. Jones
REFERENCE : Phys. Rev. Lett., 58, 1460 (1987)

SURFACE TYPE
Substrate : Cu
Crystal face: 111
Temperature : RT*
Bulk lattice: fcc
2D bulk symm: p3m1
2D surf symm: p31m

Adsorbate: Cl
Coverage : 0.33 ML
Pattern : (√3x√3)R30°
Matrix : (1.000, 1.000)
(-2.000, 1.000)

STRUCTURE TYPE
Atomic adsorption in hollow site (believed fcc); slight contraction in top subtrate spacing

SAMPLE PREPARATION (1 sample)
Treatment : Ar ion bombardment
Crystallinity:
Anal. methods: LEED, AES
Contamination:

COMMENTS
Compatible with SEXAFS structure; no high degree of crystalline perfection

DATA COLLECTION
Technique: XSW; cylindrical mirror analyzer
Dataset : Auger electron emissions from Cu 2p and Cl 1s photon energy scanned though the (111) Bragg reflection

THEORY/DATA TREATMENT
XSW analysis

QUALITY OF EXPERIMENT-THEORY FIT
Visual

2D UNIT CELLS (1 domain observed)

Cell	Ax (Å)	Ay (Å)	Bx (Å)	By (Å)	α (°)	Matrix	Pattern	Cell type
Bulk	2.552	0.000	1.276	2.210	60.0	(1.000, 0.000) (0.000, 1.000)	(1x1)	b: bulk lattice
Surface 1	3.828	2.210	-3.828	2.210	120.0	(1.000, 1.000) (-2.000, 1.000)	(√3x√3)R30°	s1: commens. superlattice

3D COORDINATES

Cl1: atomic overlayer in hollow site (believed fcc)

Dx/Dy in Å, or as a fraction of layer's unit cell vectors; atom 0 at the origin. Epir/subr are bulk repeat vectors.

No. of atoms: 3   Bulk z = 2.080 Å

Region	Chem el.	At. no.	Cell type	Site occ.	Rel. to	Dx ± εx	Dy ± εy	Dz ± εz	Dz/Bz(%) ± εz/Bz
epir		-2				f	f	Å	
subr		-1				1.276 Å	0.737 Å	2.080 Å	0.0
ovrl	Cl	1	s1	.33	0	0.000 Å	0.000 Å	0.000 Å	0.0
intf	Cu	2	b	1.00	0	1.276 Å	0.738 Å	1.810 ± .050 Å	87.0 ± 2.4
subl	Cu	3	b	1.00	0	2.552 Å	1.475 Å	3.890 Å	187.0

BOND DISTANCES AND ANGLES

Bond distances and angles derived from coordinates

No. of distances/angles: 1

Interatomic dist. A-B (Å)	Atom A	Atom B	Atom C	Bond angle A-B-C (°)
2.334	Cl1	Cu2		

COMMON NAME : Cu(100)-(1x1)-1Co ILLUSTRATION: 83
CLASSIFICATION : 29.27.2a
TECHNIQUE : LEED
AUTHORS : A. Clarke, G. Jennings, R.F. Willis, P.J. Rous and J.B. Pendry
REFERENCE : Surf. Sci., 187, 327 (1987)

SURFACE TYPE
Substrate : Cu
Crystal face: 100
Temperature : RT
Bulk lattice: fcc
2D bulk symm: p4m
2D surf symm: p4m

Adsorbate: Co
Coverage : 1.0 Co/Cu
Pattern : (1x1)
Matrix : (1.000, 0.000)
(0.000, 1.000)

STRUCTURE TYPE
Epitaxial (1x1) monolayer, continuing fcc lattice

SAMPLE PREPARATION (1 sample)
Treatment : Co evaporated from Co wire filament
Crystallinity:
Anal. methods: Co coverage and purity monitored by AES
Contamination: AES: C contamination < 'a few percent'

COMMENTS
Voi varied from -3 to -7 eV; Co phase shifts constructed with Co in hcp and fcc (Cu) lattices: no significant difference in the best fit structure was found

DATA COLLECTION
Technique: LEED
Dataset : IV spectra at normal incidence: 3 inequivalent beams; E range 30-180 eV

THEORY/DATA TREATMENT
Dynamical LEED (RFS): 8 phase shifts, 57 beams;
Vor=-8 eV (fit); VoiαE**1/3; ΘD=343 K

STRUCTURES EXAMINED
Top, bridge and hollow sites; top two spacings relaxed by -15 to +3% of Cu bulk value

QUALITY OF EXPERIMENT-THEORY FIT
RPE=0.22

2D UNIT CELLS (1 domain observed)

Cell	Ax (Å)	Ay (Å)	Bx (Å)	By (Å)	α (°)	Matrix	Pattern	Cell type
Bulk	2.550	0.000	0.000	2.550	90.0	(1.000, 0.000) (0.000, 1.000)	(1x1)	b: bulk lattice
Surface 1	2.550	0.000	0.000	2.550	90.0	(1.000, 0.000) (0.000, 1.000)	(1x1)	s1: commens. superlattice

3D COORDINATES

Co1: (1x1) epitaxial monolayer, continuing fcc lattice

Dx/Dy in Å, or as a fraction of layer's unit cell vectors; atom 0 at the origin. Epir/subr are bulk repeat vectors.

No. of atoms: 4 Bulk z = 1.810 Å

Reg ion	Chem el.	At. no.	Cell type	Site occ.	Rel. to	Dx ± εx		Dy ± εy		Dz ± εz		Dz/Bz(%) ± εz/Bz
epir		-2					f		f		Å	
subr		-1				1.275	Å	1.275	Å	1.810	Å	
ovrl	Co	1	b	1.00	0	0.000	f	0.000	f	0.000	Å	0.0
intf	Cu	2	b	1.00	1	0.500	f	0.500	f	1.700 ± .020	Å	93.9 ± 1.1
intf	Cu	3	b	1.00	2	-0.500	f	-0.500	f	1.700 ± .040	Å	93.9 ± 2.2
subl	Cu	4	b	1.00	3	0.500	f	0.500	f	1.810	Å	100.0

BOND DISTANCES AND ANGLES

Bond distances and angles are derived from coordinates

No. of distances/angles: 3

Interatomic dist. A-B (Å)	Atom A	Atom B	Atom C	Bond angle A-B-C (°)
2.550	Co1	Co1(1,0)	Cu2	59.0
2.478	Co1	Cu2	Cu3	86.6
2.478	Cu2	Cu3		

COMMON NAME : Cu(100)-(1x1)-8Co ILLUSTRATION: 83
CLASSIFICATION : 29.27.2b
TECHNIQUE : LEED
AUTHORS : A. Clarke, G. Jennings, R.F. Willis, P.J. Rous and J.B. Pendry
REFERENCE : Surf. Sci., 187, 327 (1987)

SURFACE TYPE
Substrate : Cu
Crystal face: 100
Temperature : RT
Bulk lattice: fcc
2D bulk symm: p4m
2D surf symm: p4m

Adsorbate: Co
Coverage : 8.0 Co/Cu
Pattern : (1x1)
Matrix : (1.000, 0.000)
(0.000, 1.000)

STRUCTURE TYPE
8 epitaxial (1x1) monolayers, continuing fcc lattice

SAMPLE PREPARATION (1 sample)
Treatment : Co evaporated from Co wire filament
Crystallinity:
Anal. methods: Co coverage and purity monitored by AES
Contamination: AES: C contamination < 'a few percent'

COMMENTS
Voi varied from -3 to -7 eV; Co phase shifts constructed with Co in hcp and fcc (Cu) lattices: no significant difference in the best fit structure was found; determination insensitive to spacing beyond top two Co layers: contracted Co-Co spacing of 1.76Å slightly better

DATA COLLECTION
Technique: LEED
Dataset : IV spectra at normal incidence: 3 inequivalent beams; E range 30-180 eV

THEORY/DATA TREATMENT
Dynamical LEED (RFS): 8 phase shifts, 57 beams;
Vor=-8 eV (fit); Voi∝E**1/3; ΘD=343 K

STRUCTURES EXAMINED
Hollow (continuation) sites only; top 2 spacings varied from -11 to 0% of bulk Cu spacing; spacing of remaining Co layers varied from -11 to 0% of Cu spacing

QUALITY OF EXPERIMENT-THEORY FIT
RPE=0.25

2D UNIT CELLS (1 domain observed)

Cell	Ax (Å)	Ay (Å)	Bx (Å)	By (Å)	α (°)	Matrix	Pattern	Cell type
Bulk	2.550	0.000	0.000	2.550	90.0	(1.000, 0.000) (0.000, 1.000)	(1x1)	b: bulk lattice
Surface 1	2.550	0.000	0.000	2.550	90.0	(1.000, 0.000) (0.000, 1.000)	(1x1)	s1: commens. superlattice

3D COORDINATES

Co1-Co8: 8 (1x1) epitaxial monolayers, continuing fcc lattice

Dx/Dy in Å, or as a fraction of layer's unit cell vectors; atom 0 at the origin. Epir/subr are bulk repeat vectors.

No. of atoms: 10 Bulk z = 1.810 Å

Region	Chem el.	At. no.	Cell type	Site occ.	Rel. to	Dx ± εx		Dy ± εy		Dz ± εz		Dz/Bz(%) ± εz/Bz
epir		-2					f		f		Å	
subr		-1				1.275	Å	1.275	Å	1.810	Å	
ovrl	Co	1	b	1.00	0	0.000	f	0.000	f	0.000	Å	0.0
ovrl	Co	2	b	1.00	1	0.500	f	0.500	f	1.700 ± .020	Å	93.9 ± 1.1
ovrl	Co	3	b	1.00	2	-0.500	f	-0.500	f	1.750 ± .040	Å	96.7 ± 2.2
ovrl	Co	4	b	1.00	3	0.500	f	0.500	f	1.760 ± .040	Å	97.2 ± 2.2
ovrl	Co	5	b	1.00	4	-0.500	f	-0.500	f	1.760 ± .040	Å	97.2 ± 2.2
ovrl	Co	6	b	1.00	5	0.500	f	0.500	f	1.760 ± .040	Å	97.2 ± 2.2
ovrl	Co	7	b	1.00	6	-0.500	f	-0.500	f	1.760 ± .040	Å	97.2 ± 2.2
ovrl	Co	8	b	1.00	7	0.500	f	0.500	f	1.760 ± .040	Å	97.2 ± 2.2
intf	Cu	9	b	1.00	8	-0.500	f	-0.500	f	1.760 ± .040	Å	97.2 ± 2.2
subl	Cu	10	b	1.00	9	0.500	f	0.500	f	1.810	Å	100.0

Cu(100)-(1x1)-8Co
29.27.2b BOND DISTANCES AND ANGLES

Bond distances and angles are derived from coordinates

No. of distances/angles: 3

Interatomic dist. A-B (Å)	Atom A	Atom B	Atom C	Bond angle A-B-C (°)
2.550	Co1	Co1(1,0)	Co2	59.0
2.478	Co1	Co2	Co3	87.5
2.513	Co2	Co3	Co4	88.5

COMMON NAME : Cu(100)-(1x1)-20Co
CLASSIFICATION : 29.27.3
TECHNIQUE : ARXPS
AUTHORS : Hong Li and B.P. Tonner
REFERENCE : Surf. Sci., 237, 141 (1990)

ILLUSTRATION: 83

SURFACE TYPE
Substrate : Cu
Crystal face: 100
Temperature : RT
Bulk lattice: fcc
2D bulk symm: p4m
2D surf symm: p4m

Adsorbate: Co
Coverage : 20.0 Co/Cu
Pattern : (1x1)
Matrix : (1.000, 0.000)
(0.000, 1.000)

STRUCTURE TYPE
20 epitaxial (1x1) monolayers, continuing fcc lattice

SAMPLE PREPARATION (sample)
Treatment : Co evaporated from Co wire filament
Crystallinity: p(1x1) LEED pattern
Anal. methods: Co coverage monitored by quartz microbalance and AES/XPS
Contamination:

COMMENTS
No structure optimization; only the lattice type of the overlayer determined

DATA COLLECTION
Technique: ARXPS; see Rev. Sci. Instrum. 58 (1987) 116
Dataset : Cu and Co 2p3/2 ARXPS patterns for Co coverages ranging from 0 to 20 ML

THEORY/DATA TREATMENT

STRUCTURES EXAMINED
No theoretical analysis: ARXPS is used in the finger-print mode to show that Co overlayers are fcc.

2D UNIT CELLS (1 domain observed)

Cell	Ax (Å)	Ay (Å)	Bx (Å)	By (Å)	α (°)	Matrix	Pattern	Cell type
Bulk	2.550	0.000	0.000	2.550	90.0	(1.000, 0.000) (0.000, 1.000)	(1x1)	b: bulk lattice
Surface 1	2.550	0.000	0.000	2.550	90.0	(1.000, 0.000) (0.000, 1.000)	(1x1)	s1: commens. superlattice

3D COORDINATES

Co1-Co20: 20 (1x1) epitaxial monolayers, continuing fcc lattice;
Cu21 represents bulk

Dx/Dy in Å, or as a fraction of layer's unit cell vectors; atom 0 at the origin. Epir/subr are bulk repeat vectors.

No. of atoms: 21 Bulk z = 1.810 Å

Reg ion	Chem el.	At. no.	Cell type	Site occ.	Rel. to	Dx ± εx		Dy ± εy		Dz ± εz		Dz/Bz(%) ± εz/Bz
epir		-2					f		f		Å	
subr		-1				1.275	Å	1.275	Å	1.810	Å	
ovrl	Co	1	b	1.00	0	0.000	f	0.000	f	0.000	Å	0.0
ovrl	Co	2	b	1.00	1	0.500	f	0.500	f	1.810	Å	100.0
ovrl	Co	3	b	1.00	2	-0.500	f	-0.500	f	1.810	Å	100.0
ovrl	Co	4	b	1.00	3	0.500	f	0.500	f	1.810	Å	100.0
ovrl	Co	5	b	1.00	4	-0.500	f	-0.500	f	1.810	Å	100.0
ovrl	Co	6	b	1.00	5	0.500	f	0.500	f	1.810	Å	100.0
ovrl	Co	7	b	1.00	6	-0.500	f	-0.500	f	1.810	Å	100.0
ovrl	Co	8	b	1.00	7	0.500	f	0.500	f	1.810	Å	100.0
ovrl	Co	9	b	1.00	8	-0.500	f	-0.500	f	1.810	Å	100.0
ovrl	Co	10	b	1.00	9	0.500	f	0.500	f	1.810	Å	100.0
ovrl	Co	11	b	1.00	10	-0.500	f	-0.500	f	1.810	Å	100.0
ovrl	Co	12	b	1.00	11	0.500	f	0.500	f	1.810	Å	100.0
ovrl	Co	13	b	1.00	12	-0.500	f	-0.500	f	1.810	Å	100.0
ovrl	Co	14	b	1.00	13	0.500	f	0.500	f	1.810	Å	100.0
ovrl	Co	15	b	1.00	14	-0.500	f	-0.500	f	1.810	Å	100.0
ovrl	Co	16	b	1.00	15	0.500	f	0.500	f	1.810	Å	100.0
ovrl	Co	17	b	1.00	16	-0.500	f	-0.500	f	1.810	Å	100.0
ovrl	Co	18	b	1.00	17	0.500	f	0.500	f	1.810	Å	100.0
ovrl	Co	19	b	1.00	18	-0.500	f	-0.500	f	1.810	Å	100.0
ovrl	Co	20	b	1.00	19	0.500	f	0.500	f	1.810	Å	100.0
subl	Cu	21	b	1.00	20	-0.500	f	-0.500	f	1.810	Å	100.0

Cu(100)-(1x1)-20Co
29.27.3

BOND DISTANCES AND ANGLES

Bond distances and angles are derived from coordinates

No. of distances/angles: 3

Interatomic dist. A-B (Å)	Atom A	Atom B	Atom C	Bond angle A-B-C (°)
2.550	Co1	Co1(1,0)	Co2	60.0
2.550	Co1	Co2	Co3	90.0
2.550	Co2	Co3	Co4	90.0

COMMON NAME : Cu(111)-(1x1)-1Co
CLASSIFICATION : 29.27.1
TECHNIQUE : SEXAFS
AUTHORS : D. Chandesris, P. Roubin, G. Rossi and J. Lecante
REFERENCE : Surf. Sci., 169, 57 (1986)

ILLUSTRATION: 81

SURFACE TYPE
Substrate : Cu
Crystal face: 111
Temperature : RT
Bulk lattice: fcc
2D bulk symm: p3m1
2D surf symm: p3m1

Adsorbate: Co
Coverage : 1.0 Co/Cu
Pattern : (1x1)
Matrix : (1.000, 0.000)
(0.000, 1.000)

STRUCTURE TYPE
Epitaxial (1x1) monolayer, continuing fcc lattice

SAMPLE PREPARATION (1 sample)
Treatment :
Crystallinity:
Anal. methods:
Contamination:

COMMENTS
The Co-Cu phase shift was calculated by subtracting calculated Cu and Co shifts from the value for bulk Co; the analysis included a correction to the phase shifts which was fitted to the experimental data

DATA COLLECTION
Technique: SEXAFS
Dataset : SEXAFS spectra for polarisation normal and parallel to surface; photon energy from 50 to 450 eV above Co K edge

THEORY/DATA TREATMENT
Shell by shell fitting of total SEXAFS yield; backscattering amplitudes from bulk Cu and Co; phase shifts from bulk

2D UNIT CELLS (1 domain observed)

Cell	Ax (Å)	Ay (Å)	Bx (Å)	By (Å)	α (°)	Matrix	Pattern	Cell type
Bulk	2.550	0.000	1.275	2.208	60.0	(1.000, 0.000) (0.000, 1.000)	(1x1)	b: bulk lattice
Surface 1	2.550	0.000	1.275	2.208	60.0	(1.000, 0.000) (0.000, 1.000)	(1x1)	s1: commens. superlattice

3D COORDINATES

Co1: (1x1) epitaxial monolayer; coordinates are derived from bond distances

Dx/Dy in Å, or as a fraction of layer's unit cell vectors; atom 0 at the origin. Epir/subr are bulk repeat vectors.

No. of atoms: 3

Bulk z = 2.080 Å

Region	Chem el.	At. no.	Cell type	Site occ.	Rel. to	Dx ± εx	Dy ± εy	Dz ± εz	Dz/Bz(%) ± εz/Bz
epir		-2				f	f	Å	
subr		-1				-1.275 Å	-0.736 Å	2.080 Å	
ovrl	Co	1	b	1.00	0	0.000 f	0.000 f	0.000 Å	0.0
intf	Cu	2	b	1.00	1	0.667 f	0.667 f	1.980 ± .030 Å	95.2 ± 1.4
subl	Cu	3	b	1.00	2	-0.333 f	-0.333 f	2.080 ± .020 Å	100.0 ± 1.0

BOND DISTANCES AND ANGLES

No. of distances/angles: 3

Interatomic dist. A-B (Å)	Atom A	Atom B	Atom C	Bond angle A-B-C (°)
2.550	Co1	Co1(1,0)	Cu2(0,-1)	58.9
2.467	Co1	Cu2(-1,0)	Cu3	118.9
2.548	Cu2	Cu3		

COMMON NAME : Cu(111)-(2x2)-Cs
CLASSIFICATION : 29.55.1
TECHNIQUE : LEED
AUTHORS : S.A. Lindgren, L. Wallden, J. Rundgren, P. Westrin and J. Neve
REFERENCE : Phys. Rev., B28, 6707 (1983)

ILLUSTRATION: 22

SURFACE TYPE
Substrate : Cu
Crystal face: 111
Temperature : 300 K
Bulk lattice: fcc
2D bulk symm: p3m1
2D surf symm: p3m1

Adsorbate: Cs
Coverage : 0.25 Cs/Cu
Pattern : (2x2)
Matrix : (2.000, 0.000)
(0.000, 2.000)

STRUCTURE TYPE
Atomic adsorption in top sites

SAMPLE PREPARATION (1 sample)
Treatment : Cs evaporated to saturation coverage, giving (2x2) pattern
Crystallinity:
Anal. methods:
Contamination: workfunction and EELS measurements

COMMENTS

DATA COLLECTION
Technique: LEED
Dataset : I-V spectra for (1,0),(0,1),(1/2,0),(0,1/2) beams at Θ=0; 10<E<150 eV; better data obtained by Fourie

THEORY/DATA TREATMENT
Dynamical LEED (layer doubling): 10 phase shifts from Cu7Cs cluster and from LMTO band structure; ΘD=60 K(//),180K(perp)

STRUCTURES EXAMINED
Cu-Cs spacing varied from 2.5 to 4.5Å; bulk Cu assumed

QUALITY OF EXPERIMENT-THEORY FIT
Metric distances used

2D UNIT CELLS (1 domain observed)

Cell	Ax (Å)	Ay (Å)	Bx (Å)	By (Å)	α (°)	Matrix	Pattern	Cell type
Bulk	2.560	0.000	1.280	2.217	60.0	(1.000, 0.000) (0.000, 1.000)	(1x1)	b: bulk lattice
Surface 1	5.120	0.000	2.560	4.434	60.0	(2.000, 0.000) (0.000, 2.000)	(2x2)	s1: commens. superlattice

3D COORDINATES

Cs: overlayer in top sites

Dx/Dy in Å, or as a fraction of layer's unit cell vectors; atom 0 at the origin. Epir/subr are bulk repeat vectors.

No. of atoms: 3 Bulk z = 2.090 Å

Region	Chem el.	At. no.	Cell type	Site occ.	Rel. to	Dx ± εx	Dy ± εy	Dz ± εz	Dz/Bz(%) ± εz/Bz
epir		-2				f	f	Å	
subr		-1				2.560 Å	1.478 Å	2.090 Å	
ovrl	Cs	1	s1	.25	0	0.000 f	0.000 f	0.000 Å	0.0
intf	Cu	2	b	1.00	1	0.000 f	0.000 f	3.010 ± .050 Å	144.0 ± 2.4
subl	Cu	3	b	1.00	2	0.667 f	0.667 f	2.090 Å	100.0

BOND DISTANCES AND ANGLES

Bond distances are derived from coordinates

No. of distances/angles: 4

Interatomic dist. A-B (Å)	Atom A	Atom B	Atom C	Bond angle A-B-C (°)
5.120	Cs1	Cs1(1,0)		
3.010	Cs1	Cu2		
3.951	Cu2	Cs1(1,0)		
2.560	Cu2	Cu2(1,0)		

COMMON NAME : Cu(100)-(1x1)-1Fe
CLASSIFICATION : 29.26.2a
TECHNIQUE : LEED
AUTHORS : Y. Darici, J. Marcano, H. Min and P.A. Montano
REFERENCE : Surf. Sci., 182, 477 (1987)

ILLUSTRATION: 83

SURFACE TYPE
Substrate : Cu
Crystal face: 100
Temperature : RT
Bulk lattice: fcc
2D bulk symm: p4m
2D surf symm: p4m

Adsorbate: Fe
Coverage : 1.0 Fe/Cu
Pattern : (1x1)
Matrix : (1.000, 0.000)
(0.000, 1.000)

STRUCTURE TYPE
Epitaxial (1x1) monolayer, continuing fcc lattice

SAMPLE PREPARATION (1 sample)
Treatment : Fe evaporated from Knudsen cell onto substrate held a RT
Crystallinity: Fe grows 'approximately' layer by layer
Anal. methods:
Contamination: Fe coverage and purity monitored by AES

COMMENTS
Fe deposited at 463 K caused Cu segregation to the surface over 1hr;
note: initial growth now thought to be more complex (eds.)

DATA COLLECTION
Technique: LEED
Dataset : IV spectra at normal incidence: 3 inequivalent beams; E range 25 and 470 eV

THEORY/DATA TREATMENT
Dynamical LEED (RFS): 8 phase shifts, 57 beams; Voi const, Vor=-11 eV (fit); ΘD=344 K(Cu, assumed), 233K(Fe, fit)

STRUCTURES EXAMINED
Hollow site with adsorption height varied 1.68-1.83Å; relaxation of top Cu spacings varied 1.68-1.83Å

QUALITY OF EXPERIMENT-THEORY FIT
Visual

2D UNIT CELLS (1 domain observed)

Cell	Ax (Å)	Ay (Å)	Bx (Å)	By (Å)	α (°)	Matrix	Pattern	Cell type
Bulk	2.550	0.000	0.000	2.550	90.0	(1.000, 0.000) (0.000, 1.000)	(1x1)	b: bulk lattice
Surface 1	2.550	0.000	0.000	2.550	90.0	(1.000, 0.000) (0.000, 1.000)	(1x1)	s1: commens. superlattice

3D COORDINATES

Fe1: (1x1) epitaxial monolayer

Dx/Dy in Å, or as a fraction of layer's unit cell vectors; atom 0 at the origin. Epir/subr are bulk repeat vectors.

No. of atoms: 4 Bulk z = 1.810 Å

Reg ion	Chem el.	At. no.	Cell type	Site occ.	Rel. to	Dx ± εx		Dy ± εy		Dz ± εz		Dz/Bz(%) ± εz/Bz
epir		-2					f		f		Å	
subr		-1				1.275	Å	1.275	Å	1.810	Å	
ovrl	Fe	1	b	1.00	0	0.000	f	0.000	f	0.000	Å	0.0
intf	Cu	2	b	1.00	1	0.500	f	0.500	f	1.780 ± .020	Å	98.3 ± 1.1
intf	Cu	3	b	1.00	2	-0.500	f	-0.500	f	1.810 ± .020	Å	100.0 ± 1.1
subl	Cu	4	b	1.00	3	0.500	f	0.500	f	1.810	Å	100.0

BOND DISTANCES AND ANGLES

Bond distances and angles are derived from coordinates

No. of distances/angles: 3

Interatomic dist. A-B (Å)	Atom A	Atom B	Atom C	Bond angle A-B-C (°)
2.550	Fe1	Fe1(1,0)	Cu2	59.8
2.534	Fe1	Cu2	Cu3	89.7
2.555	Cu2	Cu3	Cu4	90.2

COMMON NAME : Cu(100)-(1x1)-1Fe ILLUSTRATION: 83
CLASSIFICATION : 29.26.3a
TECHNIQUE : LEED
AUTHORS : A. Clarke, P.J. Rous, M. Arnott, G. Jennings and R.F. Willis
REFERENCE : Surf. Sci., 192, L843 (1987)

SURFACE TYPE
Substrate : Cu
Crystal face: 100
Temperature : RT
Bulk lattice: fcc
2D bulk symm: p4m
2D surf symm: p4m

Adsorbate: Fe
Coverage : 1.0 Fe/Cu
Pattern : (1x1)
Matrix : (1.000, 0.000)
(0.000, 1.000)

STRUCTURE TYPE
Epitaxial (1x1) monolayer, continuing fcc lattice

SAMPLE PREPARATION (1 sample)
Treatment : Fe evaporated at RT and not annealed after deposition
Crystallinity: Fe grows layer by layer
Anal. methods:
Contamination: Fe coverage and purity monitored by AES

COMMENTS
Annealing the film after deposition promotes interdiffusion; note: initial growth now thought to be more complex (eds.)

DATA COLLECTION
Technique: LEED
Dataset : IV spectra at normal incidence: 3 inequivalent beams, E range 30-180 eV

THEORY/DATA TREATMENT
Dynamical LEED (RFS): 8 phase shifts, 57 beams; Voi=-4 eV, Vor=-11 eV (fit); ΘD=343 K

STRUCTURES EXAMINED
Fcc continuation: first two interlayer spacings relaxed by -15 to +5% relative to Cu bulk

QUALITY OF EXPERIMENT-THEORY FIT
RPE=0.26

2D UNIT CELLS (1 domain observed)

Cell	Ax (Å)	Ay (Å)	Bx (Å)	By (Å)	α (°)	Matrix	Pattern	Cell type
Bulk	2.550	0.000	0.000	2.550	90.0	(1.000, 0.000) (0.000, 1.000)	(1x1)	b: bulk lattice
Surface 1	2.550	0.000	0.000	2.550	90.0	(1.000, 0.000) (0.000, 1.000)	(1x1)	s1: commens. superlattice

3D COORDINATES

Fe1: (1x1) epitaxial monolayer

Dx/Dy in Å, or as a fraction of layer's unit cell vectors; atom 0 at the origin. Epir/subr are bulk repeat vectors.

No. of atoms: 3 Bulk z = 1.810 Å

Region	Chem el.	At. no.	Cell type	Site occ.	Rel. to	Dx ± εx		Dy ± εy		Dz ± εz		Dz/Bz(%) ± εz/Bz
epir		-2					f		f		Å	
subr		-1				1.275	Å	1.275	Å	1.810	Å	
ovrl	Fe	1	b	1.00	0	0.000	f	0.000	f	0.000	Å	0.0
intf	Cu	2	b	1.00	1	0.500	f	0.500	f	1.770 ± .020	Å	97.8 ± 1.1
subl	Cu	3	b	1.00	2	-0.500	f	-0.500	f	1.810 ± .040	Å	100.0 ± 2.2

BOND DISTANCES AND ANGLES

Bond distances and angles are derived from coordinates

No. of distances/angles: 3

Interatomic dist. A-B (Å)	Atom A	Atom B	Atom C	Bond angle A-B-C (°)
2.550	Fe1	Fe1(1,0)	Cu2	59.7
2.527	Fe1	Cu2	Cu3	89.6
2.555	Cu2	Cu3		

COMMON NAME : Cu(100)-(1x1)-2Fe ILLUSTRATION: 83
CLASSIFICATION : 29.26.3b
TECHNIQUE : LEED
AUTHORS : A. Clarke, P.J. Rous, M. Arnott, G. Jennings and R.F. Willis
REFERENCE : Surf. Sci., 192, L843 (1987)

SURFACE TYPE
Substrate : Cu
Crystal face: 100
Temperature : RT
Bulk lattice: fcc
2D bulk symm: p4m
2D surf symm: p4m
Adsorbate: Fe
Coverage : 2.0 Fe/Cu
Pattern : (1x1)
Matrix : (1.000, 0.000)
(0.000, 1.000)

STRUCTURE TYPE
2 epitaxial (1x1) monolayers, continuing fcc lattice

SAMPLE PREPARATION (1 sample)
Treatment : Fe evaporated at RT and not annealed after deposition
Crystallinity: Fe grows layer by layer
Anal. methods:
Contamination: Fe coverage and purity monitored by AES

COMMENTS
Annealing the film after deposition promotes interdiffusion; note: initial growth now thought to be more complex (eds.)

DATA COLLECTION
Technique: LEED
Dataset : IV spectra at normal incidence: 3 inequivalent beams, energy range 30-180 eV

THEORY/DATA TREATMENT
Dynamical LEED (RFS): 8 phase shifts, 57 beams; Voi=-4 eV, Vor=-11 eV (fit); ΘD=343 K

STRUCTURES EXAMINED
Fcc continuation: first two interlayer spacings relaxed by -15 to +5% relative to Cu bulk

QUALITY OF EXPERIMENT-THEORY FIT
RPE=0.30

2D UNIT CELLS (1 domain observed)

Cell	Ax (Å)	Ay (Å)	Bx (Å)	By (Å)	α (°)	Matrix	Pattern	Cell type
Bulk	2.550	0.000	0.000	2.550	90.0	(1.000, 0.000) (0.000, 1.000)	(1x1)	b: bulk lattice
Surface 1	2.550	0.000	0.000	2.550	90.0	(1.000, 0.000) (0.000, 1.000)	(1x1)	s1: commens. superlattice

3D COORDINATES

Fe1-Fe2: 2 (1x1) epitaxial monolayers, continuing fcc lattice

Dx/Dy in Å, or as a fraction of layer's unit cell vectors; atom 0 at the origin. Epir/subr are bulk repeat vectors.

No. of atoms: 4 Bulk z = 1.810 Å

Reg ion	Chem el.	At. no.	Cell type	Site occ.	Rel. to	Dx ± εx		Dy ± εy		Dz ± εz		Dz/Bz(%) ± εz/Bz
epir		-2					f		f		Å	
subr		-1				1.275	Å	1.275	Å	1.810	Å	
ovrl	Fe	1	b	1.00	0	0.000	f	0.000	f	0.000	Å	0.0
ovrl	Fe	2	b	1.00	1	0.500	f	0.500	f	1.830 ± .020	Å	101.1 ± 1.1
intf	Cu	3	b	1.00	2	-0.500	f	-0.500	f	1.860 ± .040	Å	102.8 ± 2.2
subl	Cu	4	b	1.00	3	0.500	f	0.500	f	1.810	Å	100.0

BOND DISTANCES AND ANGLES

Bond distances and angles are derived from coordinates

No. of distances/angles: 3

Interatomic dist. A-B (Å)	Atom A	Atom B	Atom C	Bond angle A-B-C (°)
2.550	Fe1	Fe1(1,0)	Fe2	60.3
2.569	Fe1	Fe2	Cu3	91.3
2.591	Fe2	Cu3	Cu4	91.0

COMMON NAME : Cu(100)-(1x1)-3Fe
CLASSIFICATION : 29.26.3c
TECHNIQUE : LEED
AUTHORS : A. Clarke, P.J. Rous, M. Arnott, G. Jennings and R.F. Willis
REFERENCE : Surf. Sci., 192, L843 (1987)

ILLUSTRATION: 83

SURFACE TYPE
Substrate : Cu
Crystal face: 100
Temperature : RT
Bulk lattice: fcc
2D bulk symm: p4m
2D surf symm: p4m

Adsorbate: Fe
Coverage : 3.0 Fe/Cu
Pattern : (1x1)
Matrix : (1.000, 0.000)
(0.000, 1.000)

STRUCTURE TYPE
3 epitaxial (1x1) monolayers, continuing fcc lattice

SAMPLE PREPARATION (1 sample)
Treatment : Fe evaporated at RT and not annealed after deposition
Crystallinity: Fe grows layer by layer
Anal. methods:
Contamination: Fe coverage and purity monitored by AES

COMMENTS
Annealing the film after deposition promotes interdiffusion; note: initial growth now thought to be more complex (eds.)

DATA COLLECTION
Technique: LEED
Dataset : IV spectra at normal incidence: 3 inequivalent beams, energy range 30-180 eV

THEORY/DATA TREATMENT
Dynamical LEED (RFS): 8 phase shifts, 57 beams; Voi=-4 eV, Vor=-11 eV (fit); ΘD=343 K

STRUCTURES EXAMINED
Fcc continuation: first 3 interlayer spacings relaxed by -15 to +5% relative to Cu bulk

QUALITY OF EXPERIMENT-THEORY FIT
RPE=0.29

2D UNIT CELLS (1 domain observed)

Cell	Ax (Å)	Ay (Å)	Bx (Å)	By (Å)	α (°)	Matrix	Pattern	Cell type
Bulk	2.550	0.000	0.000	2.550	90.0	(1.000, 0.000) (0.000, 1.000)	(1x1)	b: bulk lattice
Surface 1	2.550	0.000	0.000	2.550	90.0	(1.000, 0.000) (0.000, 1.000)	(1x1)	s1: commens. superlattice

3D COORDINATES

Fe1-Fe3: 3 (1x1) epitaxial monolayers, continuing fcc lattice

Dx/Dy in Å, or as a fraction of layer's unit cell vectors; atom 0 at the origin. Epir/subr are bulk repeat vectors.

No. of atoms: 5

Bulk z = 1.810 Å

Region	Chem el.	At. no.	Cell type	Site occ.	Rel. to	Dx ± εx		Dy ± εy		Dz ± εz		Dz/Bz(%) ± εz/Bz
epir		-2					f		f		Å	
subr		-1				-1.275	Å	-1.275	Å	1.810	Å	
ovrl	Fe	1	b	1.00	0	0.000	f	0.000	f	0.000	Å	0.0
ovrl	Fe	2	b	1.00	1	0.500	f	0.500	f	1.850 ± .020	Å	102.2 ± 1.1
ovrl	Fe	3	b	1.00	2	-0.500	f	-0.500	f	1.880 ± .040	Å	103.9 ± 2.2
intf	Cu	4	b	1.00	3	0.500	f	0.500	f	1.880 ± .040	Å	103.9 ± 2.2
subl	Cu	5	b	1.00	4	-0.500	f	-0.500	f	1.810	Å	100.0

BOND DISTANCES AND ANGLES

Bond distances and angles are derived from coordinates

No. of distances/angles: 3

Interatomic dist. A-B (Å)	Atom A	Atom B	Atom C	Bond angle A-B-C (°)
2.550	Fe1	Fe1(1,0)	Fe2	60.4
2.583	Fe1	Fe2	Fe3	91.9
2.605	Fe2	Fe3	Cu4	92.4

COMMON NAME : Cu(100)-(1x1)-4Fe
CLASSIFICATION : 29.26.3d
TECHNIQUE : LEED
AUTHORS : A. Clarke, P.J. Rous, M. Arnott, G. Jennings and R.F. Willis
REFERENCE : Surf. Sci., 192, L843 (1987)

ILLUSTRATION: 83

SURFACE TYPE
Substrate : Cu
Crystal face: 100
Temperature : RT
Bulk lattice: fcc
2D bulk symm: p4m
2D surf symm: p4m

Adsorbate: Fe
Coverage : 4.0 Fe/Cu
Pattern : (1x1)
Matrix : (1.000, 0.000)
(0.000, 1.000)

STRUCTURE TYPE
4 epitaxial (1x1) monolayers, continuing fcc lattice

SAMPLE PREPARATION (1 sample)
Treatment : Fe evaporated at RT and not annealed after deposition
Crystallinity: Fe grows layer by layer
Anal. methods:
Contamination: Fe coverage and purity monitored by AES

COMMENTS
Annealing the film after deposition promotes interdiffusion; note: initial growth now thought to be more complex (eds.)

DATA COLLECTION
Technique: LEED
Dataset : IV spectra at normal incidence: 3 inequivalent beams, energy range 30-180 eV

THEORY/DATA TREATMENT
Dynamical LEED (RFS): 8 phase shifts, 57 beams; Voi=-4 eV, Vor=-11 eV (fit); ΘD=343 K

STRUCTURES EXAMINED
Fcc continuation: first 4 interlayer spacings relaxed by -15 to +5% relative to Cu bulk

QUALITY OF EXPERIMENT-THEORY FIT
RPE=0.25

2D UNIT CELLS (1 domain observed)

Cell	Ax (Å)	Ay (Å)	Bx (Å)	By (Å)	α (°)	Matrix	Pattern	Cell type
Bulk	2.550	0.000	0.000	2.550	90.0	(1.000, 0.000) (0.000, 1.000)	(1x1)	b: bulk lattice
Surface 1	2.550	0.000	0.000	2.550	90.0	(1.000, 0.000) (0.000, 1.000)	(1x1)	s1: commens. superlattice

3D COORDINATES

Fe1-Fe4: 4 (1x1) epitaxial monolayers, continuing fcc lattice

Dx/Dy in Å, or as a fraction of layer's unit cell vectors; atom 0 at the origin. Epir/subr are bulk repeat vectors.

No. of atoms: 6 Bulk z = 1.810 Å

Region	Chem el.	At. no.	Cell type	Site occ.	Rel. to	Dx ± εx		Dy ± εy		Dz ± εz		Dz/Bz(%) ± εz/Bz
epir		-2					f		f		Å	
subr		-1				1.275	Å	1.275	Å	1.810	Å	
ovrl	Fe	1	b	1.00	0	0.000	f	0.000	f	0.000	Å	0.0
ovrl	Fe	2	b	1.00	1	0.500	f	0.500	f	1.870 ± .020	Å	103.3 ± 1.1
ovrl	Fe	3	b	1.00	2	-0.500	f	-0.500	f	1.900 ± .040	Å	105.0 ± 2.2
ovrl	Fe	4	b	1.00	3	0.500	f	0.500	f	1.900 ± .040	Å	105.0 ± 2.2
intf	Cu	5	b	1.00	4	-0.500	f	-0.500	f	1.900 ± .040	Å	105.0 ± 2.2
subl	Cu	6	b	1.00	5	0.500	f	0.500	f	1.810	Å	100.0

BOND DISTANCES AND ANGLES

Bond distances and angles are derived from coordinates

No. of distances/angles: 3

Interatomic dist. A-B (Å)	Atom A	Atom B	Atom C	Bond angle A-B-C (°)
2.550	Fe1	Fe1(1,0)	Fe2	60.6
2.598	Fe1	Fe2	Fe3	92.5
2.619	Fe2	Fe3	Fe4	93.0

COMMON NAME : Cu(100)-(1x1)-5Fe ILLUSTRATION: 83
CLASSIFICATION : 29.26.3e
TECHNIQUE : LEED
AUTHORS : A. Clarke, P.J. Rous, M. Arnott, G. Jennings and R.F. Willis
REFERENCE : Surf. Sci., 192, L843 (1987)

SURFACE TYPE
Substrate : Cu
Crystal face: 100
Temperature : RT
Bulk lattice: fcc
2D bulk symm: p4m
2D surf symm: p4m

Adsorbate: Fe
Coverage : 5.0 Fe/Cu
Pattern : (1x1)
Matrix : (1.000, 0.000)
(0.000, 1.000)

STRUCTURE TYPE
5 epitaxial (1x1) monolayers, continuing fcc lattice

SAMPLE PREPARATION (1 sample)
Treatment : Fe evaporated at RT and not annealed after deposition
Crystallinity: Fe grows layer by layer
Anal. methods:
Contamination: Fe coverage and purity monitored by AES

COMMENTS
Annealing the film after deposition promotes interdiffusion;
note: initial growth now thought to be more complex (eds.)

DATA COLLECTION
Technique: LEED
Dataset : IV spectra at normal incidence: 3 inequivalent beams, energy range 30-180 eV

THEORY/DATA TREATMENT
Dynamical LEED (RFS): 8 phase shifts, 57 beams; Voi=-4 eV,
Vor=-11 eV (fit); ΘD=343 K

STRUCTURES EXAMINED
Fcc continuation: first 4 interlayer spacings relaxed by -15 to +5% relative to Cu bulk

QUALITY OF EXPERIMENT-THEORY FIT
RPE=0.30

2D UNIT CELLS (1 domain observed)

Cell	Ax (Å)	Ay (Å)	Bx (Å)	By (Å)	α (°)	Matrix	Pattern	Cell type
Bulk	2.550	0.000	0.000	2.550	90.0	(1.000, 0.000) (0.000, 1.000)	(1x1)	b: bulk lattice
Surface 1	2.550	0.000	0.000	2.550	90.0	(1.000, 0.000) (0.000, 1.000)	(1x1)	s1: commens. superlattice

3D COORDINATES

Fe1-Fe5: 5 (1x1) epitaxial monolayers, continuing fcc lattice

Dx/Dy in Å, or as a fraction of layer's unit cell vectors; atom 0 at the origin. Epir/subr are bulk repeat vectors.

No. of atoms: 7 Bulk z = 1.810 Å

Region	Chem el.	At. no.	Cell type	Site occ.	Rel. to	Dx ± εx		Dy ± εy		Dz ± εz		Dz/Bz(%) ± εz/Bz
epir		-2					f		f		Å	
subr		-1				-1.275	Å	-1.275	Å	1.810	Å	
ovrl	Fe	1	b	1.00	0	0.000	f	0.000	f	0.000	Å	0.0
ovrl	Fe	2	b	1.00	1	0.500	f	0.500	f	1.880 ± .020	Å	103.9 ± 1.1
ovrl	Fe	3	b	1.00	2	-0.500	f	-0.500	f	1.920 ± .040	Å	106.1 ± 2.2
ovrl	Fe	4	b	1.00	3	0.500	f	0.500	f	1.920 ± .040	Å	106.1 ± 2.2
ovrl	Fe	5	b	1.00	4	-0.500	f	-0.500	f	1.920 ± .040	Å	106.1 ± 2.2
intf	Cu	6	b	1.00	5	0.500	f	0.500	f	1.920 ± .040	Å	106.1 ± 2.2
subl	Cu	7	b	1.00	6	-0.500	f	-0.500	f	1.810	Å	100.0

BOND DISTANCES AND ANGLES

Bond distances and angles are derived from coordinates

No. of distances/angles: 3

Interatomic dist. A-B (Å)	Atom A	Atom B	Atom C	Bond angle A-B-C (°)
2.550	Fe1	Fe1(1,0)	Fe2	60.7
2.605	Fe1	Fe2	Fe3	93.0
2.634	Fe2	Fe3	Fe4	93.6

COMMON NAME : Cu(100)-(1x1)-10Fe
CLASSIFICATION : 29.26.2b
TECHNIQUE : LEED
AUTHORS : Y. Darici, J. Marcano, H. Min and P.A. Montano
REFERENCE : Surf. Sci., 182, 477 (1987)

ILLUSTRATION: 83

SURFACE TYPE
Substrate : Cu
Crystal face: 100
Temperature : RT
Bulk lattice: fcc
2D bulk symm: p4m
2D surf symm: p4m

Adsorbate: Fe
Coverage : 10.0 Fe/Cu
Pattern : (1x1)
Matrix : (1.000, 0.000)
(0.000, 1.000)

STRUCTURE TYPE
10 epitaxial (1x1) monolayers, continuing fcc lattice

SAMPLE PREPARATION (1 sample)
Treatment : Fe evaporated from Knudsen cell onto substrate held a RT
Crystallinity: Fe grows 'approximately' layer by layer
Anal. methods:
Contamination: Fe coverage and purity monitored by AES

COMMENTS
Fe deposited at 463 K caused Cu segregation to the surface over 1hr

DATA COLLECTION
Technique: LEED
Dataset : IV spectra at normal incidence: 3 inequivalent beams; E range 25 and 470 eV

THEORY/DATA TREATMENT
Dynamical LEED (RFS): 8 phase shifts, 57 beams; Voi const, Vor=-11 eV (fit); ΘD=380 K(Cu, fit), 550K(Fe, fit)

STRUCTURES EXAMINED
Fe overlayers in Cu continuation (hollow) sites; top two interlayer spacing and common interlayer spacing of remaining Fe layers varied

QUALITY OF EXPERIMENT-THEORY FIT
Visual

2D UNIT CELLS (1 domain observed)

Cell	Ax (Å)	Ay (Å)	Bx (Å)	By (Å)	α (°)	Matrix	Pattern	Cell type
Bulk	2.550	0.000	0.000	2.550	90.0	(1.000, 0.000) (0.000, 1.000)	(1x1)	b: bulk lattice
Surface 1	2.550	0.000	0.000	2.550	90.0	(1.000, 0.000) (0.000, 1.000)	(1x1)	s1: commens. superlattice

3D COORDINATES

Fe1-Fe10: 10 (1x1) epitaxial monolayers

Dx/Dy in Å, or as a fraction of layer's unit cell vectors; atom 0 at the origin. Epir/subr are bulk repeat vectors.

No. of atoms: 12 Bulk z = 1.810 Å

Region	Chem el.	At. no.	Cell type	Site occ.	Rel. to	Dx ± εx	Dy ± εy	Dz ± εz	Dz/Bz(%) ± εz/Bz
epir		-2				f	f	Å	
subr		-1				1.275 Å	1.275 Å	1.810 Å	
ovrl	Fe	1	b	1.00	0	0.000 f	0.000 f	0.000 Å	0.0
ovrl	Fe	2	b	1.00	1	0.500 f	0.500 f	1.810 ± .010 Å	100.0 ± .6
ovrl	Fe	3	b	1.00	2	-0.500 f	-0.500 f	1.780 ± .020 Å	98.3 ± 1.1
ovrl	Fe	4	b	1.00	3	0.500 f	0.500 f	1.780 ± .020 Å	98.3 ± 1.1
ovrl	Fe	5	b	1.00	4	-0.500 f	-0.500 f	1.780 ± .020 Å	98.3 ± 1.1
ovrl	Fe	6	b	1.00	5	0.500 f	0.500 f	1.780 ± .020 Å	98.3 ± 1.1
ovrl	Fe	7	b	1.00	6	-0.500 f	-0.500 f	1.780 ± .020 Å	98.3 ± 1.1
ovrl	Fe	8	b	1.00	7	0.500 f	0.500 f	1.780 ± .020 Å	98.3 ± 1.1
ovrl	Fe	9	b	1.00	8	-0.500 f	-0.500 f	1.780 ± .020 Å	98.3 ± 1.1
ovrl	Fe	10	b	1.00	9	0.500 f	0.500 f	1.780 ± .020 Å	98.3 ± 1.1
intf	Cu	11	b	1.00	10	-0.500 f	-0.500 f	1.780 Å	98.3
subl	Cu	12	b	1.00	11	0.500 f	0.500 f	1.810 Å	100.0

Cu(100)-(1x1)-10Fe
29.26.2b

BOND DISTANCES AND ANGLES

Bond distances and angles are derived from coordinates

No. of distances/angles: 3

Interatomic dist. A-B (Å)	Atom A	Atom B	Atom C	Bond angle A-B-C (°)
2.550	Fe1	Fe1(1,0)	Fe2	60.1
2.555	Fe1	Fe2	Fe3	89.7
2.534	Fe2	Fe3	Fe4	89.3

COMMON NAME : Cu(110)-(1x1)-Fe
CLASSIFICATION : 29.26.11
TECHNIQUE : LEED
AUTHORS : J. Marcano, Y. Darici, H. Min, Y. Yin, and P.A. Montano
REFERENCE : Surf. Sci., 217, 1 (1989)

ILLUSTRATION: 85

SURFACE TYPE
Substrate : Cu
Crystal face: 110
Temperature : RT
Bulk lattice: fcc
2D bulk symm: pmm
2D surf symm: pmm

Adsorbate: Fe
Coverage : 1.0 Fe/Cu
Pattern : (1x1)
Matrix : (1.000, 0.000)
(0.000, 1.000)

STRUCTURE TYPE
Pseudomorphic Fe monolayer; Fe-Cu and Cu(1)-Cu(2) interlayer spacings are (within error bar) equal to Cu-Cu bulk interlayer spacing

SAMPLE PREPARATION (1 sample)
Treatment : Cu cleaned by cycles of Ar sputtering and annealing at 823 K
Crystallinity:
Anal. methods: AES for Fe thickness
Contamination:

COMMENTS
Fit fails for T=383 K and T=423K, probably because of surface segregation of Cu

DATA COLLECTION
Technique: LEED
Dataset : IV curves for 5 inequivalent beams: cumul. E range 1200 eV; off normal incidence

THEORY/DATA TREATMENT
Dynamical LEED: Burdick pot. for Cu; Vor, Voi, ΘD fit: Vor=-6 eV, Voi=-4eV, ΘD=344 K

STRUCTURES EXAMINED
Fe in Cu continuation sites with variation of top 3 interlayer spacings

QUALITY OF EXPERIMENT-THEORY FIT
RZJ=0.038

2D UNIT CELLS (1 domain observed)

Cell	Ax (Å)	Ay (Å)	Bx (Å)	By (Å)	α (°)	Matrix	Pattern	Cell type
Bulk	2.556	0.000	0.000	3.615	90.0	(1.000, 0.000) (0.000, 1.000)	(1x1)	b: bulk lattice
Surface 1	2.556	0.000	0.000	3.615	90.0	(1.000, 0.000) (0.000, 1.000)	(1x1)	s1: commens. superlattice

3D COORDINATES

Fe1: epitaxial monolayer in hollow sites

Dx/Dy in Å, or as a fraction of layer's unit cell vectors; atom 0 at the origin. Epir/subr are bulk repeat vectors.

No. of atoms: 4 Bulk z = 1.278 Å

Region	Chem el.	At. no.	Cell type	Site occ.	Rel. to	Dx ± εx		Dy ± εy		Dz ± εz		Dz/Bz(%) ± εz/Bz
epir		-2					f		f		Å	
subr		-1				1.278	Å	1.807	Å	1.278	Å	
ovrl	Fe	1	s1	1.00	0	0.000	f	0.000	f	-1.250 ± .025	Å	97.8 ± 2.0
ovrl	Cu	2	s1	1.00	0	0.500	f	0.500	f	0.000 ± .025	Å	0.0 ± 2.0
ovrl	Cu	3	s1	1.00	2	-0.500	f	-0.500	f	1.270 ± .025	Å	99.4 ± 2.0
subl	Cu	4	b	1.00	3	0.500	f	0.500	f	1.278	Å	100.0

BOND DISTANCES AND ANGLES

Bond distances and angles are derived from coordinates

No. of distances/angles: 7

Interatomic dist. A-B (Å)	Atom A	Atom B	Atom C	Bond angle A-B-C (°)
2.556	Fe1	Fe1(1,0)	Cu2	59.8
2.542	Fe1	Cu2	Fe1(1,1)	121.1
2.520	Fe1	Cu3	Cu2	60.2
2.520	Fe1	Cu3	Cu3(1,0)	90.0
2.556	Cu2	Cu2(1,0)	Fe1(1,1)	59.8

Cu(110)-(1x1)-Fe
29.26.11

Bond Distances and Angles - Continued

Interatomic dist. A-B (Å)	Atom A	Atom B	Atom C	Bond angle A-B-C (°)
2.520	Cu3	Fe1	Fe1(1,0)	90.0
2.520	Cu3	Fe1	Cu2(-1,-1)	60.6

ILLUSTRATION: 81

COMMON NAME : Cu(111)-(1x1)-1Fe
CLASSIFICATION : 29.26.4
TECHNIQUE : LEED
AUTHORS : Y. Darici, J. Marcano, H. Min and P.A. Montano
REFERENCE : Surf. Sci., 195, 566 (1988)

SURFACE TYPE
Substrate : Cu
Crystal face: 111
Temperature : RT
Bulk lattice: fcc
2D bulk symm: p3m1
2D surf symm: p3m1

Adsorbate: Fe
Coverage : 1.0 Fe/Cu
Pattern : (1x1)
Matrix : (1.000, 0.000)
(0.000, 1.000)

STRUCTURE TYPE
1 monolayer epitaxial growth, continuing fcc lattice

SAMPLE PREPARATION (1 sample)
Treatment : Fe evaporated from Knudsen cell onto substrate held a RT
Crystallinity:
Anal. methods: Fe coverage and purity monitored by AES
Contamination: AES: no evidence for C or S impurities

COMMENTS
LEED measurements performed at 373 K and 433K; segregation of Cu to the surface was seen at elevated temperatures; Fe grows 'approximately' layer by layer; five different muffin-tin potentials were considered

DATA COLLECTION
Technique: LEED
Dataset : IV spectra: 3 inequivalent beams at normal incidence averaged from 5 beams for energies between 40 and 400 eV

THEORY/DATA TREATMENT
Dynamical LEED (RFS): 8 phase shifts, 61 beams; Voi and and Vor optimised; ΘD=344 K (Cu)

STRUCTURES EXAMINED
Cu continuation sites with variation of top 3 interlayer spacings from 1.96 to 2.16Å

QUALITY OF EXPERIMENT-THEORY FIT
RZJ=0.035

2D UNIT CELLS (1 domain observed)

Cell	Ax (Å)	Ay (Å)	Bx (Å)	By (Å)	α (°)	Matrix	Pattern	Cell type
Bulk	2.550	0.000	1.275	2.208	60.0	(1.000, 0.000) (0.000, 1.000)	(1x1)	b: bulk lattice
Surface 1	2.550	0.000	1.275	2.208	60.0	(1.000, 0.000) (0.000, 1.000)	(1x1)	s1: commens. superlattice

3D COORDINATES

Fe1: (1x1) epitaxial monolayer

Dx/Dy in Å, or as a fraction of layer's unit cell vectors; atom 0 at the origin. Epir/subr are bulk repeat vectors.

No. of atoms: 5 Bulk z = 2.080 Å

Reg ion	Chem el.	At. no.	Cell type	Site occ.	Rel. to	Dx ± εx		Dy ± εy		Dz ± εz		Dz/Bz(%) ± εz/Bz
epir		-2					f		f		Å	
subr		-1				-1.275	Å	-0.736	Å	2.080	Å	
ovrl	Fe	1	b	1.00	0	0.000	f	0.000	f	0.000	Å	0.0
intf	Cu	2	b	1.00	1	0.667	f	0.667	f	2.020 ± .020	Å	97.1 ± 1.0
intf	Cu	3	b	1.00	2	-0.333	f	-0.333	f	2.060 ± .020	Å	99.0 ± 1.0
intf	Cu	4	b	1.00	3	-0.333	f	-0.333	f	2.080 ± .020	Å	100.0 ± 1.0
subl	Cu	5	b	1.00	4	0.667	f	0.667	f	2.080	Å	100.0

BOND DISTANCES AND ANGLES

Bond distances and angles are derived from coordinates

No. of distances/angles: 3

Interatomic dist. A-B (Å)	Atom A	Atom B	Atom C	Bond angle A-B-C (°)
2.550	Fe1	Fe1(1,0)	Cu2(1,-1)	120.7
2.500	Fe1	Cu2(-1,0)	Cu3	119.1
2.532	Cu2	Cu3	Cu4	180.0

COMMON NAME : Cu(100)-p(2x1)-Fe multilayer ILLUSTRATION: 83
CLASSIFICATION : 29.26.13
TECHNIQUE : LEED
AUTHORS : H.Landskron, G. Schmidt, K. Heinz, K. Muller, C. Stuhlmann, U. Beckers, M. Wuttig and H. Ibach
REFERENCE : Surf. Sci., 256, 115 (1991)

SURFACE TYPE
Substrate : Cu
Crystal face: 100
Temperature : 90 K
Bulk lattice: fcc
2D bulk symm: p4m
2D surf symm: pmg

Adsorbate: Fe
Coverage : 5.0 Fe/Cu
Pattern : (1x2)
Matrix : (1.000, 0.000) (0.000, 2.000)

STRUCTURE TYPE
4-6 epitaxial Fe layers: except for top Fe layer, atoms reside in nearly ideal fcc positions given by the Cu lattice; in top Fe layer adjacent close-packed Fe rows are shifted antiparallel with respect to each other by 0.14Å, forming zigzag rows with pmg symmetry

SAMPLE PREPARATION (1 sample)
Treatment : Cu cleaned by cycles of Ne sputtering and annealing at 900 K
Crystallinity:
Anal. methods: thickness calibrated with AES and MEED
Contamination: CO below 3 at.%

COMMENTS
Coverage measured with AES gives wrong thickness by a factor of 3

DATA COLLECTION
Technique: LEED; Auto-LEED optics
Dataset : IV curves for 15 beams: E=65-500 eV (fitted 65-350 range), off-normal incidence angle as determined for clean Cu

THEORY/DATA TREATMENT
Dynamical LEED: 11 phase shifts; angle of inc. Θ=1.25°; Voi=-5 eV, ΘD=343 K(Cu), 467K(Fe)

STRUCTURES EXAMINED
Variation of interlayer spacings Fe-Fe, Fe-Cu and Cu-Cu; lateral shift in top Fe layer; number of Fe monolayers (between 1 and 7): best fit between 4 and 6 layers

QUALITY OF EXPERIMENT-THEORY FIT
RPE=0.25

2D UNIT CELLS (2 domains observed)

Cell	Ax (Å)	Ay (Å)	Bx (Å)	By (Å)	α (°)	Matrix	Pattern	Cell type
Bulk	2.556	0.000	0.000	2.556	90.0	(1.000, 0.000) (0.000, 1.000)	(1x1)	b: bulk lattice
Surface 1	2.556	0.000	0.000	5.112	90.0	(1.000, 0.000) (0.000, 2.000)	(1x2)	s1: commens. superlattice

3D COORDINATES

Fe1-Fe2: planar top Fe monolayer with zigzag structure; Fe3-Fe5: next 3 epitaxial Fe monolayers

Dx/Dy in Å, or as a fraction of layer's unit cell vectors; atom 0 at the origin. Epir/subr are bulk repeat vectors.

No. of atoms: 8 Bulk z = 1.807 Å

Region	Chem el.	At. no.	Cell type	Site occ.	Rel. to	Dx ± εx		Dy ± εy		Dz ± εz		Dz/Bz(%) ± εz/Bz
epir		-2					f		f		Å	
subr		-1				1.278	Å	1.278	Å	1.807	Å	
ovrl	Fe	1	s1	.50	0	0.140 ± .090	Å	0.000	Å	-1.880 ± .020	Å	104.0 ± 1.1
ovrl	Fe	2	s1	.50	0	2.416 ± .090	Å	2.556	Å	-1.880 ± .020	Å	-104.0 ± 1.1
ovrl	Fe	3	b	1.00	0	0.500	f	0.500	f	0.000	Å	0.0
ovrl	Fe	4	b	1.00	3	-0.500	f	-0.500	f	1.808	Å	100.0
ovrl	Fe	5	b	1.00	4	0.500	f	0.500	f	1.808	Å	100.0
ovrl	Cu	6	b	1.00	5	-0.500	f	-0.500	f	1.810 ± .030	Å	100.1 ± 1.7
ovrl	Cu	7	b	1.00	6	0.500	f	0.500	f	1.770 ± .040	Å	97.9 ± 2.2
subl	Cu	8	b	1.00	7	-0.500	f	-0.500	f	1.808	Å	100.0

Cu(100)-p(2x1)-Fe multilayer
29.26.13 BOND DISTANCES AND ANGLES

Bond distances and angles are derived from coordinates

No. of distances/angles: 8

Interatomic dist. A-B (Å)	Atom A	Atom B	Atom C	Bond angle A-B-C (°)
2.556	Fe1	Fe1(1,0)	Fe2	83.8
2.556	Fe1	Fe1(1,0)	Fe3(1,0)	116.6
2.571	Fe1	Fe2(-1,0)	Fe1(0,1)	167.5
2.571	Fe1	Fe2(-1,0)	Fe2	83.8
2.542	Fe1	Fe3	Fe1(1,0)	58.6
2.542	Fe1	Fe3	Fe2	84.6
2.679	Fe1	Fe3(-1,0)	Fe1(-1,0)	58.6
2.679	Fe1	Fe3(-1,0)	Fe3(-1,1)	118.5

COMMON NAME : Cu(110)-(1x1)-H 2L
CLASSIFICATION : 29.1.2a
TECHNIQUE : LEED
AUTHORS : A.P. Baddorf, I.-W. Lyo, E.W. Plummer and H.L. Davis
REFERENCE : J. Vac. Sci. Technol., A5, 782 (1987)

ILLUSTRATION: 4

SURFACE TYPE
Substrate : Cu
Crystal face: 110
Temperature : 90 K
Bulk lattice: fcc
2D bulk symm: pmm
2D surf symm: pmm

Adsorbate: H
Coverage : 2L
Pattern : (1x1)
Matrix : (1.000, 0.000)
(0.000, 1.000)

STRUCTURE TYPE
Atomic adsorption inducing changes in top two Cu-Cu interlayer spacings (H position not determined)

SAMPLE PREPARATION (1 sample)
Treatment : exposure to atomic H dissociated over heated W filament
Crystallinity:
Anal. methods:
Contamination: close attention to H coverage

COMMENTS

DATA COLLECTION
Technique: LEED
Dataset : IV curves for 6 non-equivalent beams: (10), (01), (11), (20), (02), (12); E range 50-430 eV

THEORY/DATA TREATMENT
Dynamical LEED (RSP): up to 11 ph shs from truncated atomic pot (full Slater exchange); Voi=-4 eV; ΘD=340 K; H ignored;

STRUCTURES EXAMINED
Variation of top two Cu-Cu interlayer spacings

QUALITY OF EXPERIMENT-THEORY FIT
R2=0.110

2D UNIT CELLS (1 domain observed)

Cell	Ax (Å)	Ay (Å)	Bx (Å)	By (Å)	α (°)	Matrix	Pattern	Cell type
Bulk	3.610	0.000	0.000	2.553	90.0	(1.000, 0.000) (0.000, 1.000)	(1x1)	b: bulk lattice
Surface 1	3.610	0.000	0.000	2.553	90.0	(1.000, 0.000) (0.000, 1.000)	(1x1)	s1: commens. superlattice

3D COORDINATES

0.03Å error bars assumed for tabulation

Dx/Dy in Å, or as a fraction of layer's unit cell vectors; atom 0 at the origin. Epir/subr are bulk repeat vectors.

No. of atoms: 4

Bulk z = 1.278 Å

Region	Chem el.	At. no.	Cell type	Site occ.	Rel. to	Dx ± εx		Dy ± εy		Dz ± εz		Dz/Bz(%) ± εz/Bz
epir		-2					f		f		Å	
subr		-1				1.805	Å	1.277	Å	1.278	Å	
intf	Cu	1	b	1.00	0	0.000	f	0.000	f	0.000	Å	0.0
intf	Cu	2	b	1.00	1	0.500	f	0.500	f	1.176 ± .030	Å	92.0 ± 2.4
intf	Cu	3	b	1.00	2	-0.500	f	-0.500	f	1.334 ± .030	Å	104.4 ± 2.4
subl	Cu	4	b	1.00	3	0.500	f	0.500	f	1.278	Å	100.0

BOND DISTANCES AND ANGLES

Bond distances and angles are derived from coordinates

No. of distances/angles: 4

Interatomic dist. A-B (Å)	Atom A	Atom B	Atom C	Bond angle A-B-C (°)
2.553	Cu1	Cu1(0,1)	Cu2	59.4
2.504	Cu1	Cu2	Cu3	59.1
2.504	Cu1	Cu2	Cu4	118.0
2.582	Cu2	Cu3	Cu4	61.1

COMMON NAME : Cu(110)-(1x1)-H 10L
CLASSIFICATION : 29.1.2b
TECHNIQUE : LEED
AUTHORS : A.P. Baddorf, I.-W. Lyo, E.W. Plummer and H.L. Davis
REFERENCE : J. Vac. Sci. Technol., A5, 782 (1987)

ILLUSTRATION: 4

SURFACE TYPE
Substrate : Cu
Crystal face: 110
Temperature : 90 K
Bulk lattice: fcc
2D bulk symm: pmm
2D surf symm: pmm

Adsorbate: H
Coverage : 10L
Pattern : (1x1)
Matrix : (1.000, 0.000)
(0.000, 1.000)

STRUCTURE TYPE
Atomic adsorption inducing changes in top two Cu-Cu interlayer spacings (H position not determined)

SAMPLE PREPARATION (1 sample)
Treatment : exposure to atomic H dissociated over heated W filament
Crystallinity:
Anal. methods:
Contamination: close attention to H coverage

COMMENTS

DATA COLLECTION
Technique: LEED
Dataset : IV curves for 6 non-equivalent beams: (10), (01), (11), (20), (02), (12); energy range 50-430 eV

THEORY/DATA TREATMENT
Dynamical LEED (RSP): up to 11 ph shs from truncated atomic pot (full Slater exchange); Voi=-4 eV; ΘD=340 K; H ignored;

STRUCTURES EXAMINED
Variation of top two Cu-Cu interlayer spacings

QUALITY OF EXPERIMENT-THEORY FIT
R2=0.122

2D UNIT CELLS (1 domain observed)

Cell	Ax (Å)	Ay (Å)	Bx (Å)	By (Å)	α (°)	Matrix	Pattern	Cell type
Bulk	3.610	0.000	0.000	2.553	90.0	(1.000, 0.000) (0.000, 1.000)	(1x1)	b: bulk lattice
Surface 1	3.610	0.000	0.000	2.553	90.0	(1.000, 0.000) (0.000, 1.000)	(1x1)	s1: commens. superlattice

3D COORDINATES

0.03Å error bars assumed for tabulation

Dx/Dy in Å, or as a fraction of layer's unit cell vectors; atom 0 at the origin. Epir/subr are bulk repeat vectors.

No. of atoms: 4

Bulk z = 1.278 Å

Region	Chem el.	At. no.	Cell type	Site occ.	Rel. to	Dx ± εx		Dy ± εy		Dz ± εz		Dz/Bz(%) ± εz/Bz
epir		-2					f		f		Å	
subr		-1				1.805	Å	1.277	Å	1.278	Å	
intf	Cu	1	b	1.00	0	0.000	f	0.000	f	0.000	Å	0.0
intf	Cu	2	b	1.00	1	0.500	f	0.500	f	1.223 ± .030	Å	95.7 ± 2.4
intf	Cu	3	b	1.00	2	-0.500	f	-0.500	f	1.323 ± .030	Å	103.5 ± 2.4
subl	Cu	4	b	1.00	3	0.500	f	0.500	f	1.278	Å	100.0

BOND DISTANCES AND ANGLES

Bond distances and angles are derived from coordinates

No. of distances/angles: 4

Interatomic dist. A-B (Å)	Atom A	Atom B	Atom C	Bond angle A-B-C (°)
2.553	Cu1	Cu1(0,1)	Cu2	59.7
2.527	Cu1	Cu2	Cu3	59.9
2.527	Cu1	Cu2	Cu4	119.0
2.576	Cu2	Cu3	Cu4	60.9

COMMON NAME : Cu(110)-(1x1)-H 50L
CLASSIFICATION : 29.1.2c
TECHNIQUE : LEED
AUTHORS : A.P. Baddorf, I.-W. Lyo, E.W. Plummer and H.L. Davis
REFERENCE : J. Vac. Sci. Technol., A5, 782 (1987)

ILLUSTRATION: 4

SURFACE TYPE
Substrate : Cu
Crystal face: 110
Temperature : 90 K
Bulk lattice: fcc
2D bulk symm: pmm
2D surf symm: pmm

Adsorbate: H
Coverage : 50L
Pattern : (1x1)
Matrix : (1.000, 0.000)
(0.000, 1.000)

STRUCTURE TYPE
Atomic adsorption inducing changes in top two Cu-Cu interlayer spacings (H position not determined)

SAMPLE PREPARATION (1 sample)
Treatment : exposure to atomic H dissociated over heated W filament
Crystallinity:
Anal. methods:
Contamination: close attention to H coverage

COMMENTS

DATA COLLECTION
Technique: LEED
Dataset : IV curves for 6 non-equivalent beams: (10), (01), (11), (20), (02), (12); energy range 50-430 eV

THEORY/DATA TREATMENT
Dynamical LEED (RSP): up to 11 ph shs from truncated atomic pot (full Slater exchange); Voi=-4 eV; ΘD=340 K; H ignored;

STRUCTURES EXAMINED
Variation of top two Cu-Cu interlayer spacings

QUALITY OF EXPERIMENT-THEORY FIT
R2=0.128

2D UNIT CELLS (1 domain observed)

Cell	Ax (Å)	Ay (Å)	Bx (Å)	By (Å)	α (°)	Matrix	Pattern	Cell type
Bulk	3.610	0.000	0.000	2.553	90.0	(1.000, 0.000) (0.000, 1.000)	(1x1)	b: bulk lattice
Surface 1	3.610	0.000	0.000	2.553	90.0	(1.000, 0.000) (0.000, 1.000)	(1x1)	s1: commens. superlattice

3D COORDINATES

0.03Å error bars assumed for tabulation

Dx/Dy in Å, or as a fraction of layer's unit cell vectors; atom 0 at the origin. Epir/subr are bulk repeat vectors.

No. of atoms: 4

Bulk z = 1.278 Å

Region	Chem el.	At. no.	Cell type	Site occ.	Rel. to	Dx ± εx		Dy ± εy		Dz ± εz		Dz/Bz(%) ± εz/Bz
epir		-2					f		f		Å	
subr		-1				1.805	Å	1.277	Å	1.278	Å	
intf	Cu	1	b	1.00	0	0.000	f	0.000	f	0.000	Å	0.0
intf	Cu	2	b	1.00	1	0.500	f	0.500	f	1.246 ± .030	Å	97.5 ± 2.4
intf	Cu	3	b	1.00	2	-0.500	f	-0.500	f	1.313 ± .030	Å	102.7 ± 2.4
subl	Cu	4	b	1.00	3	0.500	f	0.500	f	1.278	Å	100.0

BOND DISTANCES AND ANGLES

Bond distances and angles are derived from coordinates

No. of distances/angles: 4

Interatomic dist. A-B (Å)	Atom A	Atom B	Atom C	Bond angle A-B-C (°)
2.553	Cu1	Cu1(0,1)	Cu2	59.8
2.538	Cu1	Cu2	Cu3	60.1
2.538	Cu1	Cu2	Cu4	119.4
2.571	Cu2	Cu3	Cu4	60.7

COMMON NAME : Cu(110)-(1x1)-H 200L
CLASSIFICATION : 29.1.2d
TECHNIQUE : LEED
AUTHORS : A.P. Baddorf, I.-W. Lyo, E.W. Plummer and H.L. Davis
REFERENCE : J. Vac. Sci. Technol., A5, 782 (1987)

ILLUSTRATION: 4

SURFACE TYPE
Substrate : Cu
Crystal face: 110
Temperature : 90 K
Bulk lattice: fcc
2D bulk symm: pmm
2D surf symm: pmm

Adsorbate: H
Coverage : 200L
Pattern : (1x1)
Matrix : (1.000, 0.000)
(0.000, 1.000)

STRUCTURE TYPE
Atomic adsorption inducing changes in top two Cu-Cu interlayer spacings (H position not determined)

SAMPLE PREPARATION (1 sample)
Treatment : exposure to atomic H dissociated over heated W filament
Crystallinity:
Anal. methods:
Contamination: close attention to H coverage

COMMENTS

DATA COLLECTION
Technique: LEED
Dataset : IV curves for 6 non-equivalent beams: (10), (01), (11), (20), (02), (12); energy range 50-430 eV

THEORY/DATA TREATMENT
Dynamical LEED (RSP): up to 11 ph shs from truncated atomic pot (full Slater exchange); Voi=-4 eV; ΘD=340 K; H ignored;

STRUCTURES EXAMINED
Variation of top two Cu-Cu interlayer spacings

QUALITY OF EXPERIMENT-THEORY FIT
R2=0.152

2D UNIT CELLS (1 domain observed)

Cell	Ax (Å)	Ay (Å)	Bx (Å)	By (Å)	α (°)	Matrix	Pattern	Cell type
Bulk	3.610	0.000	0.000	2.553	90.0	(1.000, 0.000) (0.000, 1.000)	(1x1)	b: bulk lattice
Surface 1	3.610	0.000	0.000	2.553	90.0	(1.000, 0.000) (0.000, 1.000)	(1x1)	s1: commens. superlattice

3D COORDINATES

0.03Å error bars assumed for tabulation

Dx/Dy in Å, or as a fraction of layer's unit cell vectors; atom 0 at the origin. Epir/subr are bulk repeat vectors.

No. of atoms: 4 Bulk z = 1.278 Å

Region	Chem el.	At. no.	Cell type	Site occ.	Rel. to	Dx ± εx		Dy ± εy		Dz ± εz		Dz/Bz(%) ± εz/Bz
epir		-2					f		f		Å	
subr		-1				1.805	Å	1.277	Å	1.278	Å	
intf	Cu	1	b	1.00	0	0.000	f	0.000	f	0.000	Å	0.0
intf	Cu	2	b	1.00	1	0.500	f	0.500	f	1.268 ± .030	Å	99.2 ± 2.4
intf	Cu	3	b	1.00	2	-0.500	f	-0.500	f	1.314 ± .030	Å	102.8 ± 2.4
subl	Cu	4	b	1.00	3	0.500	f	0.500	f	1.278	Å	100.0

BOND DISTANCES AND ANGLES

Bond distances and angles are derived from coordinates

No. of distances/angles: 4

Interatomic dist. A-B (Å)	Atom A	Atom B	Atom C	Bond angle A-B-C (°)
2.553	Cu1	Cu1(0,1)	Cu2	59.9
2.549	Cu1	Cu2	Cu3	60.6
2.549	Cu1	Cu2	Cu4	119.8
2.572	Cu2	Cu3	Cu4	60.8

COMMON NAME : Cu(110)-HCO2 disordered
CLASSIFICATION : 29.6.1.8.3
TECHNIQUE : SEXAFS
AUTHORS : A. Puschmann, J. Haase, M.D. Crapper, C.E. Riley and D.P. Woodruff
REFERENCE : Phys. Rev. Lett., 54, 2250 (1985)

ILLUSTRATION: 79

SURFACE TYPE
Substrate : Cu
Crystal face: 110
Temperature : 300 K
Bulk lattice: fcc
2D bulk symm: pmm
2D surf symm: none

Adsorbate: HCO2
Coverage :
Pattern : disordered
Matrix : (1.000, 0.000)
(0.000, 1.000)

STRUCTURE TYPE
Disordered HCO2 (formate) species with two oxygens over two adjacent short bridge sites (off-center toward each other), C bridging the two oxygens in a plane perpendicular to the surface (H assumed perpendicular above C)

SAMPLE PREPARATION (1 sample)
Treatment : formate species formed by exposure to formic acid at RT
Crystallinity:
Anal. methods: NEXAFS to determine C-O bond length and
Contamination:

COMMENTS
Cu-O bond length determined by SEXAFS is an average of two unequal lengths (1.88 and 2.08Å);
0.25ML coverage assumed here

DATA COLLECTION
Technique: SEXAFS
Dataset : NEXAFS of O KLL Auger line with 2 polarization orientations; SEXAFS of O K edge with similar polarization orientations

THEORY/DATA TREATMENT
NEXAFS: empirical correlations to get bond length and angle;
SEXAFS: calc. vs exp. amplitudes to get adsorption site

STRUCTURES EXAMINED
Formate plane perpendicular to surface along [1-10] azimuth with oxygens pointing towards the surface and C atom over a top or bridge site; H ignored

2D UNIT CELLS (1 domain observed)

Cell	Ax (Å)	Ay (Å)	Bx (Å)	By (Å)	α (°)	Matrix	Pattern	Cell type
Bulk	2.560	0.000	0.000	3.620	90.0	(1.000, 0.000) (0.000, 1.000)	(1x1)	b: bulk lattice
Surface 1	2.560	0.000	0.000	3.620	90.0	(1.000, 0.000) (0.000, 1.000)	disordered	nd1: non-recon. lattice-gas dis

3D COORDINATES

C1-O2-O3: disordered HCO2 molecule; 0.1Å error bars assumed for tabulation

Dx/Dy in Å, or as a fraction of layer's unit cell vectors; atom 0 at the origin. Epir/subr are bulk repeat vectors.

No. of atoms: 5

Bulk z = 1.280 Å

Region	Chem el.	At. no.	Cell type	Site occ.	Rel. to	Dx ± εx		Dy ± εy		Dz ± εz		Dz/Bz(%) ± εz/Bz
epir		-2					f		f		Å	
subr		-1				1.280	Å	1.810	Å	1.280	Å	
ovrl	C	1	nd1	.50	0	0.000	f	0.000	f	0.000	Å	0.0
ovrl	O	2	nd1	.50	1	0.442 ± .039	f	0.000 ± .028	f	0.530 ± .100	Å	41.4 ± 7.8
ovrl	O	3	nd1	.50	2	-0.884 ± .039	f	0.000 ± .028	f	0.000 ± .100	Å	0.0 ± 7.8
intf	Cu	4	b	1.00	3	-0.558 ± .039	f	0.000 ± .028	f	1.510 ± .100	Å	118.0 ± 7.8
subl	Cu	5	b	1.00	4	0.500	f	0.500	f	1.280	Å	100.0

BOND DISTANCES AND ANGLES

No. of distances/angles: 5

Interatomic dist. A-B (Å)	Atom A	Atom B	Atom C	Bond angle A-B-C (°)
1.250	C1	O2		
2.040	C1	Cu4		
1.887	O2	Cu4		
2.560	Cu4	Cu4(1,0)		
2.560	Cu4	Cu5		

COMMON NAME : Cu(100)-HCO2 disordered
CLASSIFICATION : 29.6.1.8.8a
TECHNIQUE : PED
AUTHORS : D.P. Woodruff, C.F. McConville, A.L.D. Kilcoyne, Th.Lindner, J. Somers, M. Surman, G. Paolucci and A.M. Brads
REFERENCE : Surf. Sci., 201, 228 (1988)

ILLUSTRATION: 75

SURFACE TYPE
Substrate : Cu
Crystal face: 100
Temperature : RT*
Bulk lattice: fcc
2D bulk symm: p4m
2D surf symm: none

Adsorbate: HCO2
Coverage : 0.25 HCO2/Cu
Pattern : disordered
Matrix : (1.000, 0.000)
(0.000, 1.000)

STRUCTURE TYPE
HCO2 (formate) adsorption with both O down forming OCO bridge between 2 Cu atoms (OCO plane perp. to surface); H position undetermined, probably perp. above C

SAMPLE PREPARATION (1 sample)
Treatment : exposure to formic acid to about 0.25ML
Crystallinity:
Anal. methods:
Contamination: monitored by XPS

COMMENTS

DATA COLLECTION
Technique: PED
Dataset : PED curves at normal emission with photon energy range of 80-380 eV above threshold

THEORY/DATA TREATMENT
Analysis with curved-wave double scattering calculations (H ignored)

STRUCTURES EXAMINED
Several adsorption sites, with two equal O-Cu bonds and O-C-O plane normal to surface (H ignored); Cu-O bond length and O-C-O bond angle varied; substrate kept bulk-like

2D UNIT CELLS (2 domains observed)

Cell	Ax (Å)	Ay (Å)	Bx (Å)	By (Å)	α (°)	Matrix	Pattern	Cell type
Bulk	2.560	0.000	0.000	2.560	90.0	(1.000, 0.000) (0.000, 1.000)	(1x1)	b: bulk lattice
Surface 1	2.560	0.000	0.000	2.560	90.0	(1.000, 0.000) (0.000, 1.000)	disordered	nd1: non-recon. lattice-gas dis

3D COORDINATES

C1-O2-O3: HCO2, C up, Os down near top sites over 2 Cu; 0.1Å lateral error bars assumed for tabulation

Dx/Dy in Å, or as a fraction of layer's unit cell vectors; atom 0 at the origin. Epir/subr are bulk repeat vectors.

No. of atoms: 5 Bulk z = 1.810 Å

Region	Chem el.	At. no.	Cell type	Site occ.	Rel. to	Dx ± εx		Dy ± εy		Dz ± εz		Dz/Bz(%) ± εz/Bz
epir		-2					f		f		Å	
subr		-1				1.280	Å	1.280	Å	1.810	Å	
ovrl	C	1	nd1	.25	0	0.000	f	0.000	f	0.000	Å	0.0
ovrl	O	2	nd1	.25	1	0.450 ± .039	f	0.000	f	0.488 ± .040	Å	27.0 ± 2.2
ovrl	O	3	nd1	.25	1	-0.450 ± .039	f	0.000	f	0.488 ± .040	Å	27.0 ± 2.2
intf	Cu	4	b	1.00	1	0.500	f	0.000	f	2.464 ± .100	Å	136.1 ± 5.5
subl	Cu	5	b	1.00	4	0.500	f	0.500	f	1.810	Å	100.0

BOND DISTANCES AND ANGLES

Bond distances and angles are derived from coordinates

No. of distances/angles: 3

Interatomic dist. A-B (Å)	Atom A	Atom B	Atom C	Bond angle A-B-C (°)
1.251	C1	O2	Cu4	116.7
1.980	O2	Cu4	Cu4(1,0)	93.7
1.980	O2	Cu4	Cu5	137.6

COMMON NAME : Cu(110)-HCO2 disordered
CLASSIFICATION : 29.6.1.8.8b
TECHNIQUE : PED
AUTHORS : D.P. Woodruff, C.F. McConville, A.L.D. Kilcoyne, Th.Lindner, J. Somers, M. Surman, G. Paolucci and A.M. Brads
REFERENCE : Surf. Sci., 201, 228 (1988)

ILLUSTRATION: 79

SURFACE TYPE
Substrate : Cu
Crystal face: 110
Temperature : RT*
Bulk lattice: fcc
2D bulk symm: pmm
2D surf symm: none

Adsorbate: HCO2
Coverage : 0.25 HCO2/Cu
Pattern : disordered
Matrix : (1.000, 0.000)
(0.000, 1.000)

STRUCTURE TYPE
HCO2 (formate) adsorption with both O down forming OCO bridge between 2 Cu atoms along ridge (OCO plane perp. to surface); H position undetermined, probably perp. above C

SAMPLE PREPARATION (1 sample)
Treatment : exposure to formic acid to about 0.25ML
Crystallinity:
Anal. methods:
Contamination: monitored by XPS

COMMENTS

DATA COLLECTION
Technique: PED
Dataset : PED curves at normal emission with photon energy range of 80-380 eV above threshold

THEORY/DATA TREATMENT
Analysis with curved-wave double scattering calculations (H ignored)

STRUCTURES EXAMINED
Several adsorption sites, with two equal O-Cu bonds and O-C-O plane normal to surface (H ignored); Cu-O bond length and O-C-O bond angle varied; substrate kept bulk-like

2D UNIT CELLS (1 domain observed)

Cell	Ax (Å)	Ay (Å)	Bx (Å)	By (Å)	α (°)	Matrix	Pattern	Cell type
Bulk	2.560	0.000	0.000	3.620	90.0	(1.000, 0.000) (0.000, 1.000)	(1x1)	b: bulk lattice
Surface 1	2.560	0.000	0.000	3.620	90.0	(1.000, 0.000) (0.000, 1.000)	disordered	nd1: non-recon. lattice-gas dis

3D COORDINATES

C1-O2-O3: HCO2, C up, Os down near top sites over 2 Cu; 0.1Å lateral error bars assumed for tabulation

Dx/Dy in Å, or as a fraction of layer's unit cell vectors; atom 0 at the origin. Epir/subr are bulk repeat vectors.

No. of atoms: 5

Bulk z = 1.280 Å

Region	Chem el.	At. no.	Cell type	Site occ.	Rel. to	Dx ± εx	Dy ± εy	Dz ± εz	Dz/Bz(%) ± εz/Bz
epir		-2				f	f	Å	
subr		-1				1.280 Å	1.810 Å	1.280 Å	
ovrl	C	1	nd1	.25	0	0.000 f	0.000 f	0.000 Å	0.0
ovrl	O	2	nd1	.25	1	0.450 ± .039 f	0.000 f	0.488 ± .040 Å	38.1 ± 3.1
ovrl	O	3	nd1	.25	1	-0.450 ± .039 f	0.000 f	0.488 ± .040 Å	38.1 ± 3.1
intf	Cu	4	b	1.00	1	0.500 f	0.000 f	2.464 ± .100 Å	192.5 ± 7.8
subl	Cu	5	b	1.00	4	0.500 f	0.500 f	1.280 Å	100.0

BOND DISTANCES AND ANGLES

Bond distances and angles are derived from coordinates

No. of distances/angles: 3

Interatomic dist. A-B (Å)	Atom A	Atom B	Atom C	Bond angle A-B-C (°)
1.251	C1	O2	Cu4	116.7
1.980	O2	Cu4	Cu4(1,0)	93.7
1.980	O2	Cu4	Cu5	122.1

COMMON NAME : Cu(100)-(2x2)-I
CLASSIFICATION : 29.53.2b
TECHNIQUE : SEXAFS
AUTHORS : P.H. Citrin, P. Eisenberger and R.C. Hewitt
REFERENCE : Phys. Rev. Lett., 45, 1948 (1980)

ILLUSTRATION: 28,30

SURFACE TYPE
Substrate : Cu
Crystal face: 100
Temperature : 300 K
Bulk lattice: fcc
2D bulk symm: p4m
2D surf symm: p4m

Adsorbate: I
Coverage : 0.25 (I/Cu)
Pattern : (2x2)
Matrix : (2.000, 0.000)
(0.000, 2.000)

STRUCTURE TYPE
Atomic adsorption in 4-fold hollow sites

SAMPLE PREPARATION (1 sample)
Treatment : differentially pumped iodine doser used as iodine source
Crystallinity:
Anal. methods:
Contamination: checked by LEED and AES

COMMENTS

DATA COLLECTION
Technique: SEXAFS; SEXAFS with synchrotron radiation
Dataset : light polarization angle parallel and 70° to surface

THEORY/DATA TREATMENT
Fourier transform and polarization dependence

STRUCTURES EXAMINED
Top, bridge, and hollow site

2D UNIT CELLS (1 domain observed)

Cell	Ax (Å)	Ay (Å)	Bx (Å)	By (Å)	α (°)	Matrix	Pattern	Cell type
Bulk	2.555	0.000	0.000	2.555	90.0	(1.000, 0.000) (0.000, 1.000)	(1x1)	b: bulk lattice
Surface 1	5.110	0.000	0.000	5.110	90.0	(2.000, 0.000) (0.000, 2.000)	(2x2)	s1: commens. superlattice

3D COORDINATES

I1: overlayer in 4-fold hollow sites coordinates are derived from bond distance

Dx/Dy in Å, or as a fraction of layer's unit cell vectors; atom 0 at the origin. Epir/subr are bulk repeat vectors.

No. of atoms: 3

Bulk z = 1.810 Å

Region	Chem el.	At. no.	Cell type	Site occ.	Rel. to	Dx ± εx	Dy ± εy	Dz ± εz	Dz/Bz(%) ± εz/Bz
epir		-2				f	f	Å	
subr		-1				-1.278 Å	-1.278 Å	1.810 Å	
ovrl	I	1	s1	.25	0	0.000 f	0.000 f	0.000 Å	0.0
intf	Cu	2	b	1.00	1	0.500 f	0.500 f	1.980 ± .020 Å	109.4 ± 1.1
subl	Cu	3	b	1.00	2	-0.500 f	-0.500 f	1.810 Å	100.0

BOND DISTANCES AND ANGLES

No. of distances/angles: 3

Interatomic dist. A-B (Å)	Atom A	Atom B	Atom C	Bond angle A-B-C (°)
2.680	I1	Cu2	Cu2(1,0)	118.5
2.555	Cu2	Cu2(1,0)		
2.557	Cu2	Cu3		

COMMON NAME : Cu(111)-(√3x√3)R30°-I
CLASSIFICATION : 29.53.2a
TECHNIQUE : SEXAFS
AUTHORS : P.H. Citrin, P. Eisenberger and R.C. Hewitt
REFERENCE : Phys. Rev. Lett., 45, 1948 (1980)

ILLUSTRATION: 22,24

SURFACE TYPE
Substrate : Cu
Crystal face: 111
Temperature : 300 K
Bulk lattice: fcc
2D bulk symm: p3m1
2D surf symm: p31m

Adsorbate: I
Coverage : 0.3333 (I/Cu)
Pattern : (√3x√3)R30°
Matrix : (1.000, 1.000)
(-1.000, 2.000)

STRUCTURE TYPE
Atomic adsorption in 3-fold hollow site (hcp assumed here)

SAMPLE PREPARATION (1 sample)
Treatment : differentially pumped iodine doser used as iodine source
Crystallinity:
Anal. methods:
Contamination: checked by LEED/AES

COMMENTS
Analysis could not discriminate between fcc and hcp hollow sites

DATA COLLECTION
Technique: SEXAFS; synchrotron radiation
Dataset : light polarization parallel and 70° to surface

THEORY/DATA TREATMENT
SEXAFS with Fourier transform and polarization dependence

STRUCTURES EXAMINED
Top, bridge, and hollow sites

QUALITY OF EXPERIMENT-THEORY FIT
Visual

2D UNIT CELLS (1 domain observed)

Cell	Ax (Å)	Ay (Å)	Bx (Å)	By (Å)	α (°)	Matrix	Pattern	Cell type
Bulk	2.558	0.000	1.279	2.215	60.0	(1.000, 0.000) (0.000, 1.000)	(1x1)	b: bulk lattice
Surface 1	3.837	2.215	.000	4.430	60.0	(1.000, 1.000) (-1.000, 2.000)	(√3x√3)R30°	s1: commens. superlattice

3D COORDINATES

I1: overlayer in 3-fold hollow sites (hcp site assumed here, but not determined); coordinates are derived from bond distance

Dx/Dy in Å, or as a fraction of layer's unit cell vectors; atom 0 at the origin. Epir/subr are bulk repeat vectors.

No. of atoms: 3 Bulk z = 2.080 Å

Region	Chem el.	At. no.	Cell type	Site occ.	Rel. to	Dx ± εx		Dy ± εy		Dz ± εz		Dz/Bz(%) ± εz/Bz
epir		-2					f		f		Å	
subr		-1				0.000	Å	-1.477	Å	2.080	Å	
ovrl	I	1	s1	.33	0	0.000	f	0.000	f	0.000	Å	0.0
intf	Cu	2	b	1.00	1	-0.333	f	0.667	f	2.210 ± .020	Å	106.3 ± 1.0
subl	Cu	3	b	1.00	2	0.333	f	-0.667	f	2.080	Å	100.0

BOND DISTANCES AND ANGLES

No. of distances/angles: 4

Interatomic dist. A-B (Å)	Atom A	Atom B	Atom C	Bond angle A-B-C (°)
2.658	I1	Cu2	Cu2(1,-1)	61.2
2.658	I1	Cu2	Cu2(0,1)	118.8
2.558	Cu2	Cu2(1,0)		
2.551	Cu2	Cu3		

COMMON NAME : Cu(100)-c(2x2)-N ILLUSTRATION: 28,29
CLASSIFICATION : 29.7.3
TECHNIQUE : LEED
AUTHORS : H.C. Zheng, R.N.S. Sodhi and K.A.R. Mitchell
REFERENCE : Surf. Sci., 188, 599 (1987)

SURFACE TYPE
Substrate : Cu
Crystal face: 100
Temperature : RT*
Bulk lattice: fcc
2D bulk symm: p4m
2D surf symm: p4m

Adsorbate: N
Coverage : 0.5 (N/Cu)
Pattern : c(2x2)
Matrix : (1.000, 1.000)
(-1.000, 1.000)

STRUCTURE TYPE
Atomic overlayer coplanar with top Cu layer in 4-fold hollow sites

SAMPLE PREPARATION (1 sample)
Treatment : exposure at 300 K for 50min to N2 at 5E-5 torr, then anneal
Crystallinity:
Anal. methods:
Contamination:

COMMENTS

DATA COLLECTION
Technique: LEED
Dataset : 3 IV curves at normal incidence and 7 at Θ=15°, φ=0° (symm. independent); energy range 50-220 eV

THEORY/DATA TREATMENT
Dynamical LEED (layer doubling): 8 ph shs (Cu Moruzzi et al, N at. superpos pot); Vor=-10 eV,Voi=-5eV;ΘD=343 K(Cu),731K(N)

STRUCTURES EXAMINED
Bulk substrate; N at top, bridge and hollow sites as overlayer;
variable top substrate interlayer spacing and N at hollow sites from 0.6Å above to 0.15Å below top Cu layer;
without N, buckling in second Cu layer.

QUALITY OF EXPERIMENT-THEORY FIT
RPE=0.46

2D UNIT CELLS (1 domain observed)

Cell	Ax (Å)	Ay (Å)	Bx (Å)	By (Å)	α (°)	Matrix	Pattern	Cell type
Bulk	2.556	0.000	0.000	2.556	90.0	(1.000, 0.000) (0.000, 1.000)	(1x1)	b: bulk lattice
Surface 1	2.556	2.556	-2.556	2.556	90.0	(1.000, 1.000) (-1.000, 1.000)	c(2x2)	s1: commens. superlattice

3D COORDINATES

N1-Cu2: coplanar layer (N intercalated in bulk structure);
0.05Å error bars assumed for tabulation

Dx/Dy in Å, or as a fraction of layer's unit cell vectors; atom 0 at the origin. Epir/subr are bulk repeat vectors.

No. of atoms: 4 Bulk z = 1.807 Å

Region	Chem el.	At. no.	Cell type	Site occ.	Rel. to	Dx ± εx		Dy ± εy		Dz ± εz		Dz/Bz(%) ± εz/Bz
epir		-2					f		f		Å	
subr		-1				1.278	Å	1.278	Å	1.807	Å	
ovrl	N	1	s1	.50	0	0.000	f	0.000	f	0.000	Å	0.0
intf	Cu	2	b	1.00	1	0.500	f	0.500	f	0.000 ± .050	Å	0.0 ± 2.8
intf	Cu	3	b	1.00	2	-0.500	f	-0.500	f	1.950 ± .050	Å	107.9 ± 2.8
subl	Cu	4	b	1.00	3	0.500	f	0.500	f	1.807	Å	100.0

BOND DISTANCES AND ANGLES

Bond distances and angles are derived from coordinates

No. of distances/angles: 4

Interatomic dist. A-B (Å)	Atom A	Atom B	Atom C	Bond angle A-B-C (°)
1.807	N1	Cu2	Cu3	47.2
1.950	N1	Cu3	Cu2	42.8
2.556	Cu2	Cu2(1,0)		

Cu(100)-c(2x2)-N
29.7.3

Bond Distances and Angles - Continued

Interatomic dist. A-B (Å)	Atom A	Atom B	Atom C	Bond angle A-B-C (°)
2.659	Cu2	Cu3	Cu4	92.2

COMMON NAME : Cu(100)-c(2x2)-N
CLASSIFICATION : 29.7.4
TECHNIQUE : LEED
AUTHORS : H.C. Zeng and K.A.R. Mitchell
REFERENCE : Langmuir, 5, 829 (1989)

ILLUSTRATION: 28,29

SURFACE TYPE
Substrate : Cu
Crystal face: 100
Temperature : RT*
Bulk lattice: fcc
2D bulk symm: p4m
2D surf symm: p4m

Adsorbate: N
Coverage : 0.5 (N/Cu)
Pattern : c(2x2)
Matrix : (1.000,-1.000)
(1.000, 1.000)

STRUCTURE TYPE
N almost coplanar with 1st Cu layer (N 0.06Å above); buckling in 2nd Cu layer (Cu below N pushed down 0.09Å)

SAMPLE PREPARATION (1 sample)
Treatment : same as Surf. Sci. 188, 599 (1987)
Crystallinity:
Anal. methods:
Contamination:

COMMENTS

DATA COLLECTION
Technique: LEED
Dataset : 3 beams at normal incidence (2 int. + 1 fract. order), 7 beams at 15° off-normal incidence (4+3)

THEORY/DATA TREATMENT
Dynamical LEED

STRUCTURES EXAMINED
Incidence angle fit (13-17°) clock rotation, registry shift (1st Cu + N layer), subsurface N (+ clock rotation in 2nd Cu layer) tested

QUALITY OF EXPERIMENT-THEORY FIT
RRZJ=0.27, RPE=0.37

2D UNIT CELLS (1 domain observed)

Cell	Ax (Å)	Ay (Å)	Bx (Å)	By (Å)	α (°)	Matrix	Pattern	Cell type
Bulk	2.556	0.000	0.000	2.556	90.0	(1.000, 0.000) (0.000, 1.000)	(1x1)	b: bulk lattice
Surface 1	2.556	-2.556	2.556	2.556	90.0	(1.000,-1.000) (1.000, 1.000)	c(2x2)	s1: commens. superlattice

3D COORDINATES

N1: adsorbate in 4-fold hollow sites; Cu2,3: 1st Cu layer, Cu4,5: 2nd Cu layer

Dx/Dy in Å, or as a fraction of layer's unit cell vectors; atom 0 at the origin. Epir/subr are bulk repeat vectors.

No. of atoms: 6

Bulk z = 1.807 Å

Region	Chem el.	At. no.	Cell type	Site occ.	Rel. to	Dx ± εx		Dy ± εy		Dz ± εz		Dz/Bz(%) ± εz/Bz
epir		-2					f		f		Å	
subr		-1				1.278	Å	1.278	Å	1.807	Å	
ovrl	N	1	s1	.50	0	0.000	Å	0.000	Å	0.000	Å	0.0
intf	Cu	2	s1	.50	1	1.278	Å	-1.278	Å	0.060	Å	3.3
intf	Cu	3	s1	.50	1	-1.278	Å	-1.278	Å	0.060	Å	3.3
intf	Cu	4	s1	.50	1	0.000	Å	-2.556	Å	1.910	Å	105.7
intf	Cu	5	s1	.50	1	0.000	Å	0.000	Å	2.000	Å	110.7
subl	Cu	6	b	1.00	1	-1.278	Å	-1.278	Å	3.810	Å	210.8

BOND DISTANCES AND ANGLES

Bond distances and angles are derived from coordinates

No. of distances/angles: 2

Interatomic dist. A-B (Å)	Atom A	Atom B	Atom C	Bond angle A-B-C (°)
1.808	N1	Cu2	N1(1,0)	176.2
2.000	N1	Cu5	Cu2(0,1)	43.0

COMMON NAME : Cu(100)-(1x1)-1Ni ILLUSTRATION: 83
CLASSIFICATION : 29.28.2a
TECHNIQUE : LEED
AUTHORS : M.A. Abu-Joudeh, B.M. Davies and P.A. Montano
REFERENCE : Surf. Sci., 171, 331 (1986)

SURFACE TYPE
Substrate : Cu
Crystal face: 100
Temperature : RT*
Bulk lattice: fcc
2D bulk symm: p4m
2D surf symm: p4m

Adsorbate: Ni
Coverage : 1.0 Ni/Cu
Pattern : (1x1)
Matrix : (1.000, 0.000)
(0.000, 1.000)

STRUCTURE TYPE
Epitaxial (1x1) monolayer, continuing fcc lattice

SAMPLE PREPARATION (1 sample)
Treatment : Ni evaporated from Knudsen cell at 473 K
Crystallinity:
Anal. methods: coverage calibrated by AES
Contamination: monitored by AES and LEED

COMMENTS
Authors state that interdiffusion was only a few percent during typical deposition cycles

DATA COLLECTION
Technique: LEED
Dataset : IV curves for 3 inequivalent beams for 50<E<450 eV at normal incidence and 7° off-normal

THEORY/DATA TREATMENT
Dynamical LEED (RFS): 57 beams, 8 phase shifts; Vor=-10 eV (fit); ΘD=268 K(Ni, fit), 344K(Cu)

STRUCTURES EXAMINED
Hollow (continuation) site; Ni-Cu and top 2 Cu-Cu spacings varied

QUALITY OF EXPERIMENT-THEORY FIT
RE=0.24

2D UNIT CELLS (1 domain observed)

Cell	Ax (Å)	Ay (Å)	Bx (Å)	By (Å)	α (°)	Matrix	Pattern	Cell type
Bulk	2.553	0.000	0.000	2.553	90.0	(1.000, 0.000) (0.000, 1.000)	(1x1)	b: bulk lattice
Surface 1	2.553	0.000	0.000	2.553	90.0	(1.000, 0.000) (0.000, 1.000)	(1x1)	s1: commens. superlattice

3D COORDINATES

Ni1: (1x1) epitaxial monolayer, continuing fcc lattice

Dx/Dy in Å, or as a fraction of layer's unit cell vectors; atom 0 at the origin. Epir/subr are bulk repeat vectors.

No. of atoms: 4 Bulk z = 1.810 Å

Reg ion	Chem el.	At. no.	Cell type	Site occ.	Rel. to	Dx ± εx		Dy ± εy		Dz ± εz		Dz/Bz(%) ± εz/Bz
epir		-2					f		f		Å	
subr		-1				1.277	Å	1.277	Å	1.810	Å	
ovrl	Ni	1	b	1.00	0	0.000	f	0.000	f	0.000	Å	0.0
intf	Cu	2	b	1.00	1	0.500	f	0.500	f	1.800 ± .020	Å	99.5 ± 1.1
intf	Cu	3	b	1.00	2	-0.500	f	-0.500	f	1.780 ± .020	Å	98.3 ± 1.1
subl	Cu	4	b	1.00	3	0.500	f	0.500	f	1.810 ± .020	Å	100.0 ± 1.1

BOND DISTANCES AND ANGLES

Bond distances and angles are derived from coordinates

No. of distances/angles: 3

Interatomic dist. A-B (Å)	Atom A	Atom B	Atom C	Bond angle A-B-C (°)
2.553	Ni1	Ni1(1,0)	Cu2	60.0
2.549	Ni1	Cu2	Cu3	89.5
2.535	Cu2	Cu3	Cu4	89.7

COMMON NAME : Cu(100)-(1x1)-2Ni ILLUSTRATION: 83
CLASSIFICATION : 29.28.4a
TECHNIQUE : HREELS
AUTHORS : Y. Chen, S.Y. Tong, J.-S. Kim, M.H. Mohamed and L.L. Kesmodel
REFERENCE : Phys. Rev., B43, 6788 (1991)

SURFACE TYPE
Substrate : Cu
Crystal face: 100
Temperature : RT
Bulk lattice: fcc
2D bulk symm: p4m
2D surf symm: p4m
Adsorbate: Ni
Coverage : 2.0 Ni/Cu
Pattern : (1x1)
Matrix : (1.000, 0.000)
(0.000, 1.000)

STRUCTURE TYPE
2 epitaxial (1x1) monolayers, continuing fcc lattice

SAMPLE PREPARATION (1 sample)
Treatment : Ni evaporated onto RT substrate, then annealed up to 500 K
Crystallinity: sharp (1x1) LEED pattern
Anal. methods: coverage from AES and quartz oscillator
Contamination: monitored by AES

COMMENTS
No interdiffusion of Ni and Cu observed by AES; see also 4ML structure 29.28.4b

DATA COLLECTION
Technique: HREELS
Dataset : HREELS spectra at various primary electron energies (incl. 160 eV)

THEORY/DATA TREATMENT
First fit of vibr. frequencies by phonon calc.; then dynamical LEED-like calc. with R-factor fit

STRUCTURES EXAMINED
Fcc continuation; Ni-Ni and Ni-Cu spacings varied

2D UNIT CELLS (1 domain observed)

Cell	Ax (Å)	Ay (Å)	Bx (Å)	By (Å)	α (°)	Matrix	Pattern	Cell type
Bulk	2.553	0.000	0.000	2.553	90.0	(1.000, 0.000) (0.000, 1.000)	(1x1)	b: bulk lattice
Surface 1	2.553	0.000	0.000	2.553	90.0	(1.000, 0.000) (0.000, 1.000)	(1x1)	s1: commens. superlattice

3D COORDINATES

Ni1-Ni2: 2 (1x1) epitaxial monolayers, continuing fcc lattice

Dx/Dy in Å, or as a fraction of layer's unit cell vectors; atom 0 at the origin. Epir/subr are bulk repeat vectors.

No. of atoms: 4 Bulk z = 1.800 Å

Reg ion	Chem el.	At. no.	Cell type	Site occ.	Rel. to	Dx ± εx	Dy ± εy	Dz ± εz	Dz/Bz(%) ± εz/Bz
epir		-2				f	f	Å	
subr		-1				-1.277 Å	-1.277 Å	1.800 Å	
ovrl	Ni	1	b	1.00	0	0.000 f	0.000 f	0.000 Å	0.0
ovrl	Ni	2	b	1.00	1	0.500 f	0.500 f	1.750 ± .100 Å	97.2 ± 5.6
intf	Cu	3	b	1.00	2	-0.500 f	-0.500 f	1.750 ± .100 Å	97.2 ± 5.6
subl	Cu	4	b	1.00	3	0.500 f	0.500 f	1.800 Å	100.0

BOND DISTANCES AND ANGLES

Bond distances and angles are derived from coordinates

No. of distances/angles: 3

Interatomic dist. A-B (Å)	Atom A	Atom B	Atom C	Bond angle A-B-C (°)
2.553	Ni1	Ni1(1,0)	Ni2	59.5
2.514	Ni1	Ni2	Cu3	88.2
2.514	Ni2	Cu3	Cu4	89.0

COMMON NAME : Cu(100)-(1x1)-3Ni
CLASSIFICATION : 29.28.2b
TECHNIQUE : LEED
AUTHORS : M.A. Abu-Joudeh, B.M. Davies and P.A. Montano
REFERENCE : Surf. Sci., 171, 331 (1986)

ILLUSTRATION: 83

SURFACE TYPE
Substrate : Cu
Crystal face: 100
Temperature : RT*
Bulk lattice: fcc
2D bulk symm: p4m
2D surf symm: p4m

Adsorbate: Ni
Coverage : 3.0 Ni/Cu
Pattern : (1x1)
Matrix : (1.000, 0.000)
(0.000, 1.000)

STRUCTURE TYPE
3 epitaxial (1x1) monolayers, continuing fcc lattice

SAMPLE PREPARATION (1 sample)
Treatment : Ni evaporated from Knudsen cell at 473 K
Crystallinity:
Anal. methods: coverage calibrated by AES
Contamination: monitored by AES and LEED

COMMENTS
Authors state that interdiffusion was only a few percent during typical deposition cycles

DATA COLLECTION
Technique: LEED
Dataset : IV curves for 3 inequivalent beams for 50<E<450 eV at normal incidence and 7° off-normal

THEORY/DATA TREATMENT
Dynamical LEED (RFS): 57 beams, 8 phase shifts; Vor=-10 eV (fit); ΘD=268 K(Ni, fit), 344K(Cu)

STRUCTURES EXAMINED
Hollow (continuation) site; top 2 Ni-Ni spacings varied

QUALITY OF EXPERIMENT-THEORY FIT
RE=0.14

2D UNIT CELLS (1 domain observed)

Cell	Ax (Å)	Ay (Å)	Bx (Å)	By (Å)	α (°)	Matrix	Pattern	Cell type
Bulk	2.553	0.000	0.000	2.553	90.0	(1.000, 0.000) (0.000, 1.000)	(1x1)	b: bulk lattice
Surface 1	2.553	0.000	0.000	2.553	90.0	(1.000, 0.000) (0.000, 1.000)	(1x1)	s1: commens. superlattice

3D COORDINATES

Ni1-Ni3: 3 (1x1) epitaxial monolayers, continuing fcc lattice

Dx/Dy in Å, or as a fraction of layer's unit cell vectors; atom 0 at the origin. Epir/subr are bulk repeat vectors.

No. of atoms: 5 Bulk z = 1.810 Å

Region	Chem el.	At. no.	Cell type	Site occ.	Rel. to	Dx ± εx		Dy ± εy		Dz ± εz		Dz/Bz(%) ± εz/Bz
epir		-2					f		f		Å	
subr		-1				-1.277	Å	-1.277	Å	1.810	Å	
ovrl	Ni	1	b	1.00	0	0.000	f	0.000	f	0.000	Å	0.0
ovrl	Ni	2	b	1.00	1	0.500	f	0.500	f	1.740 ± .020	Å	96.1 ± 1.1
ovrl	Ni	3	b	1.00	2	-0.500	f	-0.500	f	1.760 ± .020	Å	97.2 ± 1.1
intf	Cu	4	b	1.00	3	0.500	f	0.500	f	1.760 ± .020	Å	97.2 ± 1.1
subl	Cu	5	b	1.00	4	-0.500	f	-0.500	f	1.810	Å	100.0

BOND DISTANCES AND ANGLES

Bond distances and angles are derived from coordinates

No. of distances/angles: 3

Interatomic dist. A-B (Å)	Atom A	Atom B	Atom C	Bond angle A-B-C (°)
2.553	Ni1	Ni1(1,0)	Ni2	59.4
2.507	Ni1	Ni2	Ni3	88.2
2.521	Ni2	Ni3	Cu4	88.6

COMMON NAME : Cu(100)-(1x1)-4Ni
CLASSIFICATION : 29.28.4b
TECHNIQUE : HREELS
AUTHORS : Y. Chen, S.Y. Tong, J.-S. Kim, M.H. Mohamed and L.L. Kesmodel
REFERENCE : Phys. Rev., B43, 6788 (1991)

ILLUSTRATION: 83

SURFACE TYPE
Substrate : Cu
Crystal face: 100
Temperature : RT
Bulk lattice: fcc
2D bulk symm: p4m
2D surf symm: p4m

Adsorbate: Ni
Coverage : 4.0 Ni/Cu
Pattern : (1x1)
Matrix : (1.000, 0.000)
(0.000, 1.000)

STRUCTURE TYPE
4 epitaxial (1x1) monolayers, continuing fcc lattice

SAMPLE PREPARATION (1 sample)
Treatment : Ni evaporated onto RT substrate, then annealed up to 500 K
Crystallinity: sharp (1x1) LEED pattern
Anal. methods: coverage from AES and quartz oscillator
Contamination: monitored by AES

COMMENTS
No interdiffusion of Ni and Cu observed by AES; see also 2ML structure 29.28.4a

DATA COLLECTION
Technique: HREELS
Dataset : HREELS spectra at various primary electron energies (incl. 160 eV)

THEORY/DATA TREATMENT
First fit of vibr. frequencies by phonon calc.; then dynamical LEED-like calc. with R-factor fit

STRUCTURES EXAMINED
Fcc continuation; top Ni-Ni and common deeper spacings varied

2D UNIT CELLS (1 domain observed)

Cell	Ax (Å)	Ay (Å)	Bx (Å)	By (Å)	α (°)	Matrix	Pattern	Cell type
Bulk	2.553	0.000	0.000	2.553	90.0	(1.000, 0.000) (0.000, 1.000)	(1x1)	b: bulk lattice
Surface 1	2.553	0.000	0.000	2.553	90.0	(1.000, 0.000) (0.000, 1.000)	(1x1)	s1: commens. superlattice

3D COORDINATES

Ni1-Ni4: 4 (1x1) epitaxial monolayers, continuing fcc lattice

Dx/Dy in Å, or as a fraction of layer's unit cell vectors; atom 0 at the origin. Epir/subr are bulk repeat vectors.

No. of atoms: 6 Bulk z = 1.800 Å

Region	Chem el.	At. no.	Cell type	Site occ.	Rel. to	Dx ± εx		Dy ± εy		Dz ± εz		Dz/Bz(%) ± εz/Bz
epir		-2					f		f		Å	
subr		-1				-1.277	Å	-1.277	Å	1.800	Å	
ovrl	Ni	1	b	1.00	0	0.000	f	0.000	f	0.000	Å	0.0
ovrl	Ni	2	b	1.00	1	0.500	f	0.500	f	1.700 ± .100	Å	94.4 ± 5.6
ovrl	Ni	3	b	1.00	2	-0.500	f	-0.500	f	1.800 ± .100	Å	100.0 ± 5.6
ovrl	Ni	4	b	1.00	3	0.500	f	0.500	f	1.800 ± .100	Å	100.0 ± 5.6
intf	Cu	5	b	1.00	4	-0.500	f	-0.500	f	1.800 ± .100	Å	100.0 ± 5.6
subl	Cu	6	b	1.00	5	0.500	f	0.500	f	1.800	Å	100.0

BOND DISTANCES AND ANGLES

Bond distances and angles are derived from coordinates

No. of distances/angles: 3

Interatomic dist. A-B (Å)	Atom A	Atom B	Atom C	Bond angle A-B-C (°)
2.553	Ni1	Ni1(1,0)	Ni2	59.0
2.480	Ni1	Ni2	Ni3	88.2
2.549	Ni2	Ni3	Ni4	89.8

COMMON NAME : Cu(111)-(1x1)-Ni
CLASSIFICATION : 29.28.1
TECHNIQUE : LEED
AUTHORS : S.P. Tear and K. Roell
REFERENCE : J. Phys., C15, 5521 (1982)

ILLUSTRATION: 81

SURFACE TYPE
Substrate : Cu
Crystal face: 111
Temperature : 298 K
Bulk lattice: fcc
2D bulk symm: p3m1
2D surf symm: p3m1

Adsorbate: Ni
Coverage : 1.0 (Ni/Cu)
Pattern : (1x1)
Matrix : (1.000, 0.000)
(0.000, 1.000)

STRUCTURE TYPE
(1x1) Ni monolayer on Cu, continuing fcc lattice

SAMPLE PREPARATION (1 sample)
Treatment : Ni from heating Ni-wrapped W filament while Cu at 180C
Crystallinity:
Anal. methods:
Contamination: AES: up to 6%ML C present

COMMENTS

DATA COLLECTION
Technique: LEED
Dataset : I-V spectra: 27 beams at 3 diffraction geometries (Θ=0; Θ=10°,φ=0°;Θ=10°,φ=30°)

THEORY/DATA TREATMENT
Dynamical LEED (CAVLEED): ΘD=343 K(Cu), 242K(Ni)

STRUCTURES EXAMINED
Relaxation of topmost interlayer spacing; fcc lattice assumed to continue into Ni layer

QUALITY OF EXPERIMENT-THEORY FIT
RZJ=0.0545

2D UNIT CELLS (1 domain observed)

Cell	Ax (Å)	Ay (Å)	Bx (Å)	By (Å)	α (°)	Matrix	Pattern	Cell type
Bulk	2.553	0.000	1.277	2.211	60.0	(1.000, 0.000) (0.000, 1.000)	(1x1)	b: bulk lattice
Surface 1	2.553	0.000	1.277	2.211	60.0	(1.000, 0.000) (0.000, 1.000)	(1x1)	s1: commens. superlattice

3D COORDINATES

Ni1: monolayer continuing fcc substrate

Dx/Dy in Å, or as a fraction of layer's unit cell vectors; atom 0 at the origin. Epir/subr are bulk repeat vectors.

No. of atoms: 3 Bulk z = 2.080 Å

Reg ion	Chem el.	At. no.	Cell type	Site occ.	Rel. to	Dx ± εx		Dy ± εy		Dz ± εz		Dz/Bz(%) ± εz/Bz
epir		-2					f		f		Å	
subr		-1				-1.277	Å	-0.737	Å	2.080	Å	
intf	Ni	1	b	1.00	0	0.000	f	0.000	f	0.000	Å	0.0
intf	Cu	2	b	1.00	1	0.667	f	0.667	f	2.040 ± .020	Å	98.1 ± 1.0
subl	Cu	3	b	1.00	2	-0.333	f	-0.333	f	2.080	Å	100.0

BOND DISTANCES AND ANGLES

Bond distances are derived from coordinates

No. of distances/angles: 4

Interatomic dist. A-B (Å)	Atom A	Atom B	Atom C	Bond angle A-B-C (°)
2.553	Ni1	Ni1(1,0)		
2.517	Ni1	Cu2(-1,0)		
2.553	Cu2	Cu2(1,0)		
2.549	Cu2	Cu3		

COMMON NAME : Cu(100)-c(2x2)-O
CLASSIFICATION : 29.8.15
TECHNIQUE : SEXAFS
AUTHORS : U. Doebler, K. Baberschke, J. Stoehr and D.A. Outka
REFERENCE : Phys. Rev., B31, 2532 (1985)

ILLUSTRATION: 28,29

SURFACE TYPE
Substrate : Cu
Crystal face: 100
Temperature : 300 K
Bulk lattice: fcc
2D bulk symm: p4m
2D surf symm: p4m

Adsorbate: O
Coverage : 0.5 O/Ni
Pattern : c(2x2)
Matrix : (1.000, 1.000)
(-1.000, 1.000)

STRUCTURE TYPE
Atomic adsorption in hollow site of unreconstructed substrate

SAMPLE PREPARATION (1 sample)
Treatment : 300L O2 dosage at 300 K
Crystallinity:
Anal. methods:
Contamination: AES: no O, S, or C

COMMENTS
This structure is now believed to have a missing-row substrate reconstruction (eds.)
authors' estimate for upper limit for additional adsorption site such as (√2x2√2)R45° is 30%

DATA COLLECTION
Technique: SEXAFS; partial electron yield SEXAFS
Dataset : yield vs energy (500-900 eV) detected for polar angles 90 and 45°; SEXAFS amplitude ratio A(90°)/A(45°)=1.4(2)

THEORY/DATA TREATMENT
Fourier transform method

STRUCTURES EXAMINED
1. reconstructed surface layer with alternating Cu and O; 2. bridge site 1.4Å above first Cu layer;
3. hollow site with various O/Cu spacings 0.0-1.5Å

2D UNIT CELLS (1 domain observed)

Cell	Ax (Å)	Ay (Å)	Bx (Å)	By (Å)	α (°)	Matrix	Pattern	Cell type
Bulk	2.556	0.000	0.000	2.556	90.0	(1.000, 0.000) (0.000, 1.000)	(1x1)	b: bulk lattice
Surface 1	2.556	2.556	-2.556	2.556	90.0	(1.000, 1.000) (-1.000, 1.000)	c(2x2)	s1: commens. superlattice

3D COORDINATES

O1: overlayer in hollow sites coordinates are derived from bond distance

Dx/Dy in Å, or as a fraction of layer's unit cell vectors; atom 0 at the origin. Epir/subr are bulk repeat vectors.

No. of atoms: 3

Bulk z = 1.810 Å

Reg ion	Chem el.	At. no.	Cell type	Site occ.	Rel. to	Dx ± εx		Dy ± εy		Dz ± εz		Dz/Bz(%) ± εz/Bz
epir		-2					f		f		Å	
subr		-1				-1.278	Å	-1.278	Å	1.810	Å	
ovrl	O	1	s1	.50	0	0.000	f	0.000	f	0.000	Å	0.0
intf	Cu	2	b	1.00	1	0.500	f	0.500	f	0.700 ± .010	Å	38.7 ± .6
subl	Cu	3	b	1.00	2	-0.500	f	-0.500	f	1.810	Å	100.0

BOND DISTANCES AND ANGLES

No. of distances/angles: 4

Interatomic dist. A-B (Å)	Atom A	Atom B	Atom C	Bond angle A-B-C (°)
1.938	O1	Cu2		
1.938	O1	Cu2	Cu2(0,1)	131.3
2.556	Cu2	Cu2(1,0)		
2.558	Cu2	Cu3		

COMMON NAME : Cu(100)-c(2x2)-O
CLASSIFICATION : 29.8.2
TECHNIQUE : LEED
AUTHORS : J.H. Onuferko and D.P. Woodruff
REFERENCE : Surf. Sci., 95, 555 (1980)

ILLUSTRATION: 28,29

SURFACE TYPE
Substrate : Cu
Crystal face: 100
Temperature : RT
Bulk lattice: fcc
2D bulk symm: p4m
2D surf symm: p4m

Adsorbate: O
Coverage : 1/2 (O/Cu)
Pattern : c(2x2)
Matrix : (1.000, 1.000)
(-1.000, 1.000)

STRUCTURE TYPE
Atomic adsorption in bridge site of unreconstructed substrate

SAMPLE PREPARATION (2 sample)
Treatment : cycles of Ar+ bombardment and annealing
Crystallinity:
Anal. methods:
Contamination: AES: clean

COMMENTS
This structure is now believed to have a missing-row substrate reconstruction (eds.)

DATA COLLECTION
Technique: LEED
Dataset : I-V curves for (0,0),(1,0),(1/2,1/2),(1,1) beams at several incidence angles for energies 30-200 eV

THEORY/DATA TREATMENT
Dynamical LEED (reverse scattering perturbation of Zimmer and Holland): 6 phase shifts

STRUCTURES EXAMINED
Unrelaxed substrate, O in hollow and bridge sites with variable O-Cu layer spacing

QUALITY OF EXPERIMENT-THEORY FIT
Visual

2D UNIT CELLS (1 domain observed)

Cell	Ax (Å)	Ay (Å)	Bx (Å)	By (Å)	α (°)	Matrix	Pattern	Cell type
Bulk	2.560	0.000	0.000	2.560	90.0	(1.000, 0.000) (0.000, 1.000)	(1x1)	b: bulk lattice
Surface 1	2.560	2.560	-2.560	2.560	90.0	(1.000, 1.000) (-1.000, 1.000)	c(2x2)	s1: commens. superlattice

3D COORDINATES

O1: overlayer in bridge sites; 0.1Å error bar assumed for tabulation

Dx/Dy in Å, or as a fraction of layer's unit cell vectors; atom 0 at the origin. Epir/subr are bulk repeat vectors.

No. of atoms: 3 Bulk z = 1.810 Å

Region	Chem el.	At. no.	Cell type	Site occ.	Rel. to	Dx ± εx		Dy ± εy		Dz ± εz		Dz/Bz(%) ± εz/Bz
epir		-2					f		f		Å	
subr		-1				1.280	Å	1.280	Å	1.810	Å	
ovrl	O	1	s1	.50	0	0.000	f	0.000	f	0.000	Å	0.0
intf	Cu	2	b	1.00	1	0.000	f	0.500	f	1.400 ± .100	Å	77.4 ± 5.5
subl	Cu	3	b	1.00	2	0.500	f	0.500	f	1.810	Å	100.0

BOND DISTANCES AND ANGLES

Bond distances and angles are derived from coordinates

No. of distances/angles: 3

Interatomic dist. A-B (Å)	Atom A	Atom B	Atom C	Bond angle A-B-C (°)
1.897	O1	Cu2	Cu2(0,1)	132.4
2.560	Cu2	Cu2(1,0)		
2.560	Cu2	Cu3		

COMMON NAME : Cu(100)-c(2x2)-O
CLASSIFICATION : 29.8.7
TECHNIQUE : PED
AUTHORS : J.G. Tobin, L.E. Klebanoff, D.H. Rosenblatt, R.F. Davis, E.Umbach, A. Baca, D. Shirley, Y. Huang, W. Kang and S.Y. To
REFERENCE : Phys. Rev., B26, 7076 (1982)

ILLUSTRATION: 28,29

SURFACE TYPE
Substrate : Cu
Crystal face: 100
Temperature : RT*
Bulk lattice: fcc
2D bulk symm: p4m
2D surf symm: p4m

Adsorbate: O
Coverage : 0.5 O/Cu
Pattern : c(2x2)
Matrix : (1.000, 1.000)
(-1.000, 1.000)

STRUCTURE TYPE
Atomic adsorption in hollow site of unreconstructed substrate

SAMPLE PREPARATION (1 sample)
Treatment : exposures of 400L O2, then annealing to 375 K for 4min
Crystallinity:
Anal. methods:
Contamination: AES, PED, LEED: C/Cu AES ratio <0.005

COMMENTS
This structure is now believed to have a missing-row substrate reconstruction (eds.);
R-factor was normalized to Zanazzi-Jona R-factor;
R-factors show relative minimum at 0.1Å (hollow) and absolute minimum at 0.8Å (hollow) for O-Cu spacing

DATA COLLECTION
Technique: PED; normal photoelectron diffraction
Dataset : O(1s) cross section vs. kinetic energy (30-180 eV)

THEORY/DATA TREATMENT
Convergent multiple scattering and R-factor comparisons; also Fourier transform used independently

STRUCTURES EXAMINED
Various O/Cu spacings between 0 and 1.2Å

QUALITY OF EXPERIMENT-THEORY FIT
RZJ=0.16 (see comments)

2D UNIT CELLS (1 domain observed)

Cell	Ax (Å)	Ay (Å)	Bx (Å)	By (Å)	α (°)	Matrix	Pattern	Cell type
Bulk	2.556	0.000	0.000	2.556	90.0	(1.000, 0.000) (0.000, 1.000)	(1x1)	b: bulk lattice
Surface 1	2.556	2.556	-2.556	2.556	90.0	(1.000, 1.000) (-1.000, 1.000)	c(2x2)	s1: commens. superlattice

3D COORDINATES

O1: overlayer in hollow sites; 0.1Å error bar assumed for tabulation

Dx/Dy in Å, or as a fraction of layer's unit cell vectors; atom 0 at the origin. Epir/subr are bulk repeat vectors.

No. of atoms: 3

Bulk z = 1.810 Å

Region	Chem el.	At. no.	Cell type	Site occ.	Rel. to	Dx ± εx		Dy ± εy		Dz ± εz		Dz/Bz(%) ± εz/Bz
epir		-2					f		f		Å	
subr		-1				-1.278	Å	-1.278	Å	1.810	Å	
ovrl	O	1	s1	.50	0	0.000	f	0.000	f	0.000	Å	0.0
intf	Cu	2	b	1.00	1	0.500	f	0.500	f	0.800 ± .100	Å	44.2 ± 5.5
subl	Cu	3	b	1.00	2	-0.500	f	-0.500	f	1.810	Å	100.0

BOND DISTANCES AND ANGLES

Bond distances and angles are derived from coordinates

No. of distances/angles: 3

Interatomic dist. A-B (Å)	Atom A	Atom B	Atom C	Bond angle A-B-C (°)
1.977	O1	Cu2	Cu2(0,1)	130.3
2.556	Cu2	Cu2(1,0)		
2.558	Cu2	Cu3		

COMMON NAME : Cu(100)-(2√2x√2)R45°-2O
CLASSIFICATION : 29.8.39
TECHNIQUE : LEED
AUTHORS : H.C. Zeng and K.A.R. Mitchell
REFERENCE : Surf. Sci., 239, L571 (1990)

ILLUSTRATION: 34

SURFACE TYPE
Substrate : Cu
Crystal face: 100
Temperature : RT
Bulk lattice: fcc
2D bulk symm: p4m
2D surf symm: pmm

Adsorbate: O
Coverage : 0.5 (O/Cu)
Pattern : (2√2x√2)R45°
Matrix : (1.000,-1.000)
(2.000, 2.000)

STRUCTURE TYPE
O in 4-fold coordinated site with 1 Cu neighbor missing; missing Cu row in (1,-1) direction, lateral shift of top Cu atoms adjacent to rows (pairing, 0.3Å) and lifting up (0.01Å); buckling in 2nd Cu layer (0.1Å), Cu below O is lifted

SAMPLE PREPARATION (1 sample)
Treatment : oxygen exposure 10E-6 Torr, 3-5 min at 300C & flash 300C
Crystallinity:
Anal. methods:
Contamination:

COMMENTS
Same fit in joint STM paper (fewer details):
Ch.Woell, R.J.Wilson, S.Chiang, H.C.Zeng and K.A.R.Mitchell
Phys. Rev. B42, 11926 (1990)
Previous publication with experimental details: H.C.Zeng, R.A.McFarlane and K.A.R.Mitchell, Surf. Sci. 208, L7 (1989)

DATA COLLECTION
Technique: LEED; Video LEED
Dataset : 2 integer, 4 fractional order beams at normal incidence E range 50-220 eV

THEORY/DATA TREATMENT
dynamical calculation (RFS, composite layers, comb. space)

STRUCTURES EXAMINED
2 more complicated missing row models: missing row two layers deep, missing atoms crosswise in 1st and 2nd Cu layer

QUALITY OF EXPERIMENT-THEORY FIT
RRZJ=0.148, RPE=0.326

2D UNIT CELLS (2 domains observed)

Cell	Ax (Å)	Ay (Å)	Bx (Å)	By (Å)	α (°)	Matrix	Pattern	Cell type
Bulk	2.556	0.000	0.000	2.556	90.0	(1.000, 0.000) (0.000, 1.000)	(1x1)	b: bulk lattice
Surface 1	2.556	-2.556	5.112	5.112	90.0	(1.000,-1.000) (2.000, 2.000)	(2√2x√2)R45°	s1: commens. superlattice

3D COORDINATES

O1,O2: adsorbate in 4-fold coordinated sites next to missing Cu rows;
Cu3-5: 1st Cu layer, Cu6-9: 2nd Cu layer

Dx/Dy in Å, or as a fraction of layer's unit cell vectors; atom 0 at the origin. Epir/subr are bulk repeat vectors.

No. of atoms: 10 Bulk z = 1.807 Å

Region	Chem el.	At. no.	Cell type	Site occ.	Rel. to	Dx ± εx		Dy ± εy		Dz ± εz		Dz/Bz(%) ± εz/Bz
epir		-2					f		f		Å	
subr		-1				1.278	Å	1.278	Å	1.807	Å	
ovrl	O	1	s1	.25	0	0.000	Å	0.000	Å	0.000	Å	0.0
ovrl	O	2	s1	.25	1	2.556	Å	2.556	Å	0.000	Å	0.0
intf	Cu	3	s1	.25	1	1.490 ± .050	Å	4.046 ± .050	Å	0.100	Å	5.5
intf	Cu	4	s1	.25	1	3.622 ± .050	Å	6.178 ± .050	Å	0.100	Å	5.5
intf	Cu	5	s1	.25	1	1.278	Å	1.278	Å	0.200	Å	11.1
intf	Cu	6	s1	.25	1	0.000	Å	0.000	Å	2.140	Å	118.4
intf	Cu	7	s1	.25	1	2.556	Å	2.556	Å	2.140	Å	118.4
intf	Cu	8	s1	.25	1	0.000	Å	2.556	Å	2.240	Å	123.9
intf	Cu	9	s1	.25	1	2.556	Å	5.112	Å	2.240	Å	123.9
subl	Cu	10	b	1.00	1	-1.278	Å	1.278	Å	3.980	Å	220.2

Cu(100)-(2√2x√2)R45°-2O
29.8.39

BOND DISTANCES AND ANGLES

Bond distances and angles are derived from coordinates

No. of distances/angles: 6

Interatomic dist. A-B (Å)	Atom A	Atom B	Atom C	Bond angle A-B-C (°)
1.835	O1	Cu4(1,-1)	O1(1,0)	160.2
1.818	O1	Cu5	O2	167.4
2.140	O1	Cu6	Cu4(1,-1)	41.9
1.835	O2	Cu3(1,0)	O2(1,0)	160.2
1.818	O2	Cu5	O1	167.4
2.140	O2	Cu7	Cu3(1,0)	41.9

COMMON NAME : Cu(110)-(2x1)-O
CLASSIFICATION : 29.8.17
TECHNIQUE : SEXAFS
AUTHORS : M. Bader, A. Puschmann, C. Ocal and J. Haase
REFERENCE : Phys. Rev. Lett., 57, 3273 (1986)

ILLUSTRATION: 39

SURFACE TYPE
Substrate : Cu
Crystal face: 110
Temperature : 100 K
Bulk lattice: fcc
2D bulk symm: pmm
2D surf symm: pmm

Adsorbate: O
Coverage : 0.5 O/Cu
Pattern : (2x1)
Matrix : (1.000, 0.000)
(0.000, 2.000)

STRUCTURE TYPE
Atomic adsorption in long-bridge sites (O above top Cu layer), with missing Cu [001] rows

SAMPLE PREPARATION (1 sample)
Treatment : exposure to O2 at RT
Crystallinity:
Anal. methods:
Contamination: monitored by AES and LEED

COMMENTS
Strong anisotropy of both surface mean free path and surface Debye-Waller factor was found;
O-Cu interlayer spacing taken from U. Dobler et al, Phys. Rev. Lett. 52, 1437 (1984)

DATA COLLECTION
Technique: SEXAFS; oxygen K-edge SEXAFS
Dataset : data for normal incidence (Θ=90) with E-vector in [0,0,1] and [1,-1,0] azimuths, and Θ=45 with E in [1,-1,0] azimuth

THEORY/DATA TREATMENT
Fourier transform and polarization dependence

STRUCTURES EXAMINED
Buckled-row, missing-row and sawtooth reconstruction models examined by bond-length determination and SEXAFS amplitude comparisons

2D UNIT CELLS (1 domain observed)

Cell	Ax (Å)	Ay (Å)	Bx (Å)	By (Å)	α (°)	Matrix	Pattern	Cell type
Bulk	3.610	0.000	0.000	2.553	90.0	(1.000, 0.000) (0.000, 1.000)	(1x1)	b: bulk lattice
Surface 1	3.610	0.000	0.000	5.106	90.0	(1.000, 0.000) (0.000, 2.000)	(1x2)	s1: commens. superlattice

3D COORDINATES

O1: occupies long-bridge sites between remaining Cu2 rows; 0.03Å error bars assumed for tabulation

Dx/Dy in Å, or as a fraction of layer's unit cell vectors; atom 0 at the origin. Epir/subr are bulk repeat vectors.

No. of atoms: 4 Bulk z = 1.276 Å

Region	Chem el.	At. no.	Cell type	Site occ.	Rel. to	Dx ± εx		Dy ± εy		Dz ± εz		Dz/Bz(%) ± εz/Bz
epir		-2					f		f		Å	
subr		-1				1.805	Å	1.277	Å	1.276	Å	
intf	O	1	s1	.50	0	0.000	f	0.000	f	0.000	Å	0.0
intf	Cu	2	s1	.50	1	0.500	f	0.000	f	0.350 ± .030	Å	27.4 ± 2.4
intf	Cu	3	b	1.00	2	-0.500	f	0.500	f	1.276	Å	100.0
subl	Cu	4	b	1.00	3	0.500	f	0.500	f	1.276	Å	100.0

BOND DISTANCES AND ANGLES

No. of distances/angles: 4

Interatomic dist. A-B (Å)	Atom A	Atom B	Atom C	Bond angle A-B-C (°)
1.839	O1	Cu2	O1(1,0)	158.1
1.839	O1	Cu2	Cu3	53.2
1.839	O1	Cu2	Cu4(0,-1)	101.0
2.553	Cu2	Cu3	Cu4	90.0

COMMON NAME : Cu(110)-(2x1)-O
CLASSIFICATION : 29.8.18a
TECHNIQUE : ICISS
AUTHORS : J.A. Yarmoff, D.M. Cyr, J.H. Huang, S. Kim and R.S. Williams
REFERENCE : Phys. Rev., B33, 3856 (1986)

ILLUSTRATION: 39

SURFACE TYPE
Substrate : Cu
Crystal face: 110
Temperature : RT*
Bulk lattice: fcc
2D bulk symm: pmm
2D surf symm: pmm

Adsorbate: O
Coverage : 1/2 O/Cu
Pattern : (2x1)
Matrix : (1.000, 0.000)
(0.000, 2.000)

STRUCTURE TYPE
Atomic adsorption in long-bridge sites (O below top Cu layer), with missing Cu [001] rows

SAMPLE PREPARATION (1 sample)
Treatment : O2 leaked in at 10E-6 torr
Crystallinity:
Anal. methods:
Contamination: checked with AES

COMMENTS
O position taken from LEIS result of Dewit et al, Surf. Sci. 82, 177 (1979) and Hupkens and Fluit, Surf. Sci. 143, 267 (1984), giving unusual O-Cu distances

DATA COLLECTION
Technique: ICISS; ICISS with 5keV Li+ ions
Dataset : polar scans along three azimuths [1-10], [1-12] and [001]

THEORY/DATA TREATMENT
Comparison with Monte Carlo simulation: thermal vibrations 1.5 x rms bulk ampl in top Cu layer

STRUCTURES EXAMINED
Variation of top two Cu-Cu interlayer spacings, assuming missing-row model; oxygen shown to have negligible effect and ignored

2D UNIT CELLS (1 domain observed)

Cell	Ax (Å)	Ay (Å)	Bx (Å)	By (Å)	α (°)	Matrix	Pattern	Cell type
Bulk	3.610	0.000	0.000	2.553	90.0	(1.000, 0.000) (0.000, 1.000)	(1x1)	b: bulk lattice
Surface 1	3.610	0.000	0.000	5.106	90.0	(1.000, 0.000) (0.000, 2.000)	(1x2)	s1: commens. superlattice

3D COORDINATES

Cu1: remaining row; O2: occupies long-bridge sites between remaining Cu rows

Dx/Dy in Å, or as a fraction of layer's unit cell vectors; atom 0 at the origin. Epir/subr are bulk repeat vectors.

No. of atoms: 5

Bulk z = 1.278 Å

Reg ion	Chem el.	At. no.	Cell type	Site occ.	Rel. to	Dx ± εx		Dy ± εy		Dz ± εz		Dz/Bz(%) ± εz/Bz
epir		-2					f		f		Å	
subr		-1				1.805	Å	1.277	Å	1.278	Å	
intf	Cu	1	s1	.50	0	0.000	f	0.000	f	0.000	Å	0.0
intf	O	2	s1	.50	1	0.500	f	0.000	f	0.600	Å	47.0
intf	Cu	3	b	1.00	1	0.500	f	0.500	f	1.600 ± .160	Å	125.2 ± 12.5
intf	Cu	4	b	1.00	3	-0.500	f	-0.500	f	1.150 ± .060	Å	90.0 ± 4.7
subl	Cu	5	b	1.00	4	0.500	f	0.500	f	1.278	Å	100.0

BOND DISTANCES AND ANGLES

Bond distances and angles are derived from coordinates

No. of distances/angles: 6

Interatomic dist. A-B (Å)	Atom A	Atom B	Atom C	Bond angle A-B-C (°)
1.902	Cu1	O2	Cu1(1,0)	143.2
1.902	Cu1	O2	Cu3	101.2
2.729	Cu1	Cu3	Cu4	63.4
1.622	O2	Cu3	Cu4	83.2
1.622	O2	Cu3	Cu5	128.1
2.492	Cu3	Cu4	Cu5	57.5

COMMON NAME : Cu(110)-(2x1)-O
CLASSIFICATION : 29.8.30
TECHNIQUE : XRD
AUTHORS : R. Feidenhans'l, F. Grey, R.L. Johnson, S.G.J. Mochrie, J. Bohr and M. Nielsen
REFERENCE : Phys. Rev., B41, 5420 (1990)

ILLUSTRATION: 39

SURFACE TYPE
Substrate : Cu
Crystal face: 110
Temperature : 300 K
Bulk lattice: fcc
2D bulk symm: pmm
2D surf symm: pmm

Adsorbate: O
Coverage : 0.5 O/Cu
Pattern : (2x1)
Matrix : (2.000, 0.000)
(0.000, 1.000)

STRUCTURE TYPE
Atomic adsorption in long bridge sites with missing-row reconstruction and slight second-row pairing away from O sites

SAMPLE PREPARATION (1 sample)
Treatment : Ar+ sputtering, 730 K anneal; adsorption of 10L O2 at 300K
Crystallinity: crystal 1.4° from (110)
Anal. methods:
Contamination:

COMMENTS

DATA COLLECTION
Technique: XRD
Dataset : 15 fractional and 7 integer order symm.-inequivalent reflections, corrected for Lorentz factor

THEORY/DATA TREATMENT
Structure factor obtained by summing layers from a semi-infinite crystal

STRUCTURES EXAMINED
Performed a least squares analysis of intensities

QUALITY OF EXPERIMENT-THEORY FIT
Chi^2=2.6

2D UNIT CELLS (1 domain observed)

Cell	Ax (Å)	Ay (Å)	Bx (Å)	By (Å)	α (°)	Matrix	Pattern	Cell type
Bulk	2.556	0.000	0.000	3.615	90.0	(1.000, 0.000) (0.000, 1.000)	(1x1)	b: bulk lattice
Surface 1	5.112	0.000	0.000	3.615	90.0	(2.000, 0.000) (0.000, 1.000)	(2x1)	s1: commens. superlattice

3D COORDINATES

Cu1: remaining atoms in top Cu layer; O2: adatom in long bridge site, 0.34Å below top Cu;
Cu3-Cu4: slightly paired 2nd Cu layer rows;

Dx/Dy in Å, or as a fraction of layer's unit cell vectors; atom 0 at the origin. Epir/subr are bulk repeat vectors.

No. of atoms: 5 Bulk z = 1.278 Å

Region	Chem el.	At. no.	Cell type	Site occ.	Rel. to	Dx ± εx	Dy ± εy	Dz ± εz	Dz/Bz(%) ± εz/Bz
epir		-2				f	f	Å	
subr		-1				1.278 Å	1.807 Å	1.278 Å	
intf	Cu	1	s1	.50	0	0.000 f	0.000 f	0.000 Å	0.0
intf	O	2	s1	.50	1	0.000 f	0.500 f	0.340 ± .170 Å	26.6 ± 13.3
intf	Cu	3	s1	.50	2	0.256 ± .001 f	0.000 f	1.308 ± .050 Å	102.4 ± 3.9
intf	Cu	4	s1	.50	3	-0.512 ± .001 f	0.000 f	0.000 Å	0.0
subl	Cu	5	b	1.00	4	-0.488 f	-0.500 f	1.278 Å	100.0

BOND DISTANCES AND ANGLES

Bond distances and angles are derived from coordinates

No. of distances/angles: 7

Interatomic dist. A-B (Å)	Atom A	Atom B	Atom C	Bond angle A-B-C (°)
1.839	Cu1	O2	Cu1(0,1)	158.7
1.839	Cu1	O2	Cu3	97.5
2.774	Cu1	Cu3	Cu4	118.2

Cu(110)-(2x1)-O
29.8.30

Bond Distances and Angles - Continued

Interatomic dist. A-B (Å)	Atom A	Atom B	Atom C	Bond angle A-B-C (°)
2.926	Cu1	Cu5	Cu3	60.2
1.851	O2	Cu3	Cu1	41.1
1.851	O2	Cu3	Cu4	135.0
2.494	Cu3	Cu4	Cu1(1,0)	118.2

COMMON NAME : Cu(110)-(2x1)-O
CLASSIFICATION : 29.8.31
TECHNIQUE : LEED
AUTHORS : S.R. Parkin, H.C. Zeng, M.Y. Zhou and K.A.R. Mitchell
REFERENCE : Phys. Rev., B41, 5432 (1990)

ILLUSTRATION: 39

SURFACE TYPE
Substrate : Cu
Crystal face: 110
Temperature : RT*
Bulk lattice: fcc
2D bulk symm: pmm
2D surf symm: pmm

Adsorbate: O
Coverage : 0.5 O/Cu
Pattern : (2x1)
Matrix : (2.000, 0.000)
(0.000, 1.000)

STRUCTURE TYPE
Atomic adsorption in long bridge sites with missing-row reconstruction and slight second-row pairing by 0.03Å away from O sites

SAMPLE PREPARATION (1 sample)
Treatment : Ar+ sputtering/annealing; O2 leaked in at 4E-8 torr
Crystallinity:
Anal. methods:
Contamination: AES: clean

COMMENTS

DATA COLLECTION
Technique: LEED
Dataset : I-V curves for 9 diffracted beams; E range 50-250 eV

THEORY/DATA TREATMENT
Dynamical LEED (layer-doubling, composite layer for top O/Cu layers)

STRUCTURES EXAMINED
Tested missing-row and buckled-row models; varied interlayer spacings and lateral displacements for Cu and O-Cu bond length

QUALITY OF EXPERIMENT-THEORY FIT
RMZJ=0.228

2D UNIT CELLS (1 domain observed)

Cell	Ax (Å)	Ay (Å)	Bx (Å)	By (Å)	α (°)	Matrix	Pattern	Cell type
Bulk	2.556	0.000	0.000	3.615	90.0	(1.000, 0.000) (0.000, 1.000)	(1x1)	b: bulk lattice
Surface 1	5.112	0.000	0.000	3.615	90.0	(2.000, 0.000) (0.000, 1.000)	(2x1)	s1: commens. superlattice

3D COORDINATES

O1: adatom in long bridge site, slightly above top Cu; Cu2: remaining atoms in top Cu layer; Cu3-Cu4: paired 2nd Cu layer rows; Cu5-Cu6: buckled 3rd Cu layer; 0.05Å error bars assumed for tabulation

Dx/Dy in Å, or as a fraction of layer's unit cell vectors; atom 0 at the origin. Epir/subr are bulk repeat vectors.

No. of atoms: 7

Bulk z = 1.278 Å

Region	Chem el.	At. no.	Cell type	Site occ.	Rel. to	Dx ± εx	Dy ± εy	Dz ± εz	Dz/Bz(%) ± εz/Bz
epir		-2				f	f	Å	
subr		-1				1.278 Å	1.807 Å	1.278 Å	
ovrl	O	1	s1	.50	0	0.000 f	0.000 f	0.000 Å	0.0
intf	Cu	2	s1	.50	1	0.000 f	0.500 f	0.040 ± .050 Å	3.1 ± 3.9
intf	Cu	3	s1	.50	2	0.256 ± .010 f	-0.500 f	1.490 ± .050 Å	116.6 ± 3.9
intf	Cu	4	s1	.50	3	-0.512 ± .010 f	0.000 f	0.000 Å	0.0
intf	Cu	5	s1	.50	4	0.256 ± .010 f	0.500 f	1.210 ± .050 Å	94.7 ± 3.9
intf	Cu	6	s1	.50	5	0.500 ± .010 f	0.000 f	0.070 ± .050 Å	5.5 ± 3.9
subl	Cu	7	b	1.00	6	-0.500 f	-0.500 f	1.278 Å	100.0

Cu(110)-(2x1)-O
29.8.31

BOND DISTANCES AND ANGLES

Bond distances and angles are derived from coordinates

No. of distances/angles: 5

Interatomic dist. A-B (Å)	Atom A	Atom B	Atom C	Bond angle A-B-C (°)
1.808	O1	Cu2	O1(0,1)	180.0
1.808	O1	Cu2	Cu3	48.6
2.013	O1	Cu3	Cu4	130.5
2.683	Cu2	Cu3	Cu4	119.2
2.700	Cu2	Cu5	Cu6	91.6

COMMON NAME : Cu(110)-(2x1)-O
CLASSIFICATION : 29.8.51
TECHNIQUE : ICISS
AUTHORS : H. Duerr, Th. Fauster and R. Schneider
REFERENCE : Surf. Sci., 244, 237 (1991)

ILLUSTRATION: 39

SURFACE TYPE
Substrate : Cu
Crystal face: 110
Temperature : RT*
Bulk lattice: fcc
2D bulk symm: pmm
2D surf symm: pmm

Adsorbate: O
Coverage : 0.5 O/Cu
Pattern : (2x1)
Matrix : (2.000, 0.000)
(0.000, 1.000)

STRUCTURE TYPE
O in long bridge sites with missing row reconstruction; row pairing (away from O) of second Cu layer

SAMPLE PREPARATION (1 sample)
Treatment : Ar+ sputtering with annealing at 800 K
Crystallinity:
Anal. methods:
Contamination: ISS: no contamination

COMMENTS

DATA COLLECTION
Technique: ICISS
Dataset : data for scatt. angles 0-160°, ion beam of 5k eV Ne along the [1-12] direction at 55°, separated in TOF drift tube

THEORY/DATA TREATMENT
Also examined [001],[1-10],[3-32], and [1-11] directions calculated critical angles are in good agreement

STRUCTURES EXAMINED
Unreconstructed, buckling-row and missing-row substrates; for latter: top 2 Cu interlayer spacings, row-pairing in 2nd Cu layer, O-Cu spacing

2D UNIT CELLS (1 domain observed)

Cell	Ax (Å)	Ay (Å)	Bx (Å)	By (Å)	α (°)	Matrix	Pattern	Cell type
Bulk	2.556	0.000	0.000	3.615	90.0	(1.000, 0.000) (0.000, 1.000)	(1x1)	b: bulk lattice
Surface 1	5.112	0.000	0.000	3.615	90.0	(2.000, 0.000) (0.000, 1.000)	(2x1)	s1: commens. superlattice

3D COORDINATES

oxygen atom in long bridge

Dx/Dy in Å, or as a fraction of layer's unit cell vectors; atom 0 at the origin. Epir/subr are bulk repeat vectors.

No. of atoms: 6

Bulk z = 1.278 Å

Region	Chem el.	At. no.	Cell type	Site occ.	Rel. to	Dx ± εx		Dy ± εy		Dz ± εz		Dz/Bz(%) ± εz/Bz
epir		-2					f		f		Å	
subr		-1				1.278	Å	1.807	Å	1.278	Å	
intf	Cu	1	s1	.50	0	0.000	f	0.000	f	0.000	Å	0.0
intf	O	2	s1	.50	1	0.000	f	0.500	f	0.080 ± .150	Å	6.3 ± 11.7
intf	Cu	3	s1	.50	2	0.274 ± .014	f	0.000	f	1.430 ± .150	Å	111.9 ± 11.7
intf	Cu	4	s1	.50	3	0.453 ± .014	f	0.000	f	0.000	Å	0.0
intf	Cu	5	b	1.00	4	0.547 ± .027	f	0.500	f	1.280 ± .200	Å	100.2 ± 15.7
subl	Cu	6	b	1.00	5	-0.500	f	-0.500	f	1.278	Å	100.0

BOND DISTANCES AND ANGLES

Bond distances and angles are derived from coordinates

No. of distances/angles: 11

Interatomic dist. A-B (Å)	Atom A	Atom B	Atom C	Bond angle A-B-C (°)
1.809	Cu1	O2	Cu1(0,1)	174.9
2.499	Cu3	Cu5		
2.499	Cu4	Cu5		
1.809	Cu1	O2	Cu3	91.8
2.739	Cu1	Cu3	Cu1(0,1)	82.6
2.739	Cu1	Cu3	O2	41.3

Cu(110)-(2x1)-O
29.8.51

Bond Distances and Angles - Continued

Interatomic dist. A-B (Å)	Atom A	Atom B	Atom C	Bond angle A-B-C (°)
2.739	Cu1	Cu3	Cu4	120.7
1.809	O2	Cu1(0,1)	O2(0,1)	174.9
2.000	O2	Cu3	Cu1	41.3
2.000	O2	Cu3	Cu4	134.4
2.316	Cu3	Cu4		

COMMON NAME : Cu(410)-(1x1)-O
CLASSIFICATION : 29.8.12a
TECHNIQUE : PED
AUTHORS : K.A. Thompson and C.S. Fadley
REFERENCE : Surf. Sci., 146, 281 (1984)

ILLUSTRATION: 42

SURFACE TYPE
Substrate : Cu
Crystal face: 410
Temperature : RT*
Bulk lattice: fcc
2D bulk symm: cm
2D surf symm: cm

Adsorbate: O
Coverage : 1 (O/1x1)
Pattern : (1x1)
Matrix : (1.000, 0.000)
(0.000, 1.000)

STRUCTURE TYPE
Atomic O adsorption in 4-fold coord. step sites, 0.4Å above (100) terrace plane (this site is like 4-fold site on (100) terrace, but lacks one of the 4 metal surface atoms)

SAMPLE PREPARATION (1 sample)
Treatment : exposure to 1E3 L oxygen
Crystallinity:
Anal. methods:
Contamination:

COMMENTS

DATA COLLECTION
Technique: PED; Al Kα (1486.6eV) radiation source
Dataset : azimuthal XPD data obtained for O 1s emission at 9 polar angles from 7 to 45°

THEORY/DATA TREATMENT
Single scattering in clusters of ≈200 atoms: free-atom phase shifts; vibr ampl 0.0108 (surf), 0.0065Å**2 (subs)

STRUCTURES EXAMINED
Unrelaxed (410) substrate; one O in hollow site at terrace edge with variable height (-0.4 to 1.4Å) over terrace

QUALITY OF EXPERIMENT-THEORY FIT
Visual

2D UNIT CELLS (1 domain observed)

Cell	Ax (Å)	Ay (Å)	Bx (Å)	By (Å)	α (°)	Matrix	Pattern	Cell type
Bulk	3.620	0.000	1.810	7.464	76.4	(1.000, 0.000) (0.000, 1.000)	(1x1)	b: bulk lattice
Surface 1	3.620	0.000	1.810	7.464	76.4	(1.000, 0.000) (0.000, 1.000)	(1x1)	s1: commens. superlattice

3D COORDINATES

O1 in hollow site at step edge

Dx/Dy in Å, or as a fraction of layer's unit cell vectors; atom 0 at the origin. Epir/subr are bulk repeat vectors.

No. of atoms: 8

Bulk z = .439 Å

Region	Chem el.	At. no.	Cell type	Site occ.	Rel. to	Dx ± εx		Dy ± εy		Dz ± εz		Dz/Bz(%) ± εz/Bz
epir		-2					f		f		Å	
subr		-1				0.000	Å	-5.708	Å	0.439	Å	
ovrl	O	1	b	1.00	0	0.000	f	0.000	f	0.000	Å	0.0
intf	Cu	2	b	1.00	1	0.507 ± .042	f	0.987 ± .027	f	0.388 ± .200	Å	88.4 ± 45.6
intf	Cu	3	b	1.00	2	0.382	f	-0.765	f	0.439	Å	100.0
intf	Cu	4	b	1.00	3	-0.618	f	0.235	f	0.439	Å	100.0
intf	Cu	5	b	1.00	4	0.382	f	0.235	f	0.439	Å	100.0
intf	Cu	6	b	1.00	5	-0.618	f	0.235	f	0.439	Å	100.0
intf	Cu	7	b	1.00	6	0.382	f	-0.765	f	0.439	Å	100.0
intf	Cu	8	b	1.00	7	0.382	f	0.235	f	0.439	Å	100.0

BOND DISTANCES AND ANGLES

Bond distances and angles are derived from coordinates

No. of distances/angles: 5

Interatomic dist. A-B (Å)	Atom A	Atom B	Atom C	Bond angle A-B-C (°)
1.854	O1	Cu2(0,-1)	O1(1,0)	155.1
1.854	O1	Cu2(0,-1)	Cu3	133.7
1.854	O1	Cu3(-1,0)	Cu2(0,-1)	46.3
2.210	O1	Cu6(0,-1)	Cu2(0,-1)	45.0

Cu(410)-(1x1)-O
29.8.12a

Bond Distances and Angles - Continued

Interatomic dist. A-B (Å)	Atom A	Atom B	Atom C	Bond angle A-B-C (°)
2.560	Cu2	Cu3(0,1)		

COMMON NAME : Cu(410)-(1x1)-2O
CLASSIFICATION : 29.8.12b
TECHNIQUE : PED
AUTHORS : K.A. Thompson and C.S. Fadley
REFERENCE : Surf. Sci., 146, 281 (1984)

ILLUSTRATION: 43

SURFACE TYPE
Substrate : Cu
Crystal face: 410
Temperature : RT*
Bulk lattice: fcc
2D bulk symm: cm
2D surf symm: cm

Adsorbate: O
Coverage : 2 (O/1x1)
Pattern : (1x1)
Matrix : (1.000, 0.000)
(0.000, 1.000)

STRUCTURE TYPE
Atomic O adsorption in 4-fold coord. step sites, 0.4Å above (100) terrace plane (this site is like 4-fold site on (100) terrace, but lacks one of the 4 metal surface atoms); 2nd O per (1x1) cell forms c(2x2)-like structure on (100) terrace (also 0.4Å above terrace plane)

SAMPLE PREPARATION (1 sample)
Treatment : exposure to 1E3 L oxygen
Crystallinity:
Anal. methods:
Contamination:

COMMENTS

DATA COLLECTION
Technique: PED; Al Kα (1486.6eV) radiation source
Dataset : azimuthal XPD data obtained for O 1s emission at 9 polar angles from 7 to 45°

THEORY/DATA TREATMENT
Single scattering in clusters of ≈200 atoms: free-atom phase shifts; vibr ampl 0.0108 (surf), 0.0065Å**2 (subs)

STRUCTURES EXAMINED
Unrelaxed (410) substrate; one O in hollow site at terrace edge with height 0.4Å over terrace as determined for Cu(410)-(1x1)-O; other O over hollow in terrace with variable height (-0.4 to 1.4Å) over terrace; oxygens assumed to form c(2x2) lattice on (100) terrace

QUALITY OF EXPERIMENT-THEORY FIT
Visual

2D UNIT CELLS (1 domain observed)

Cell	Ax (Å)	Ay (Å)	Bx (Å)	By (Å)	α (°)	Matrix	Pattern	Cell type
Bulk	3.620	0.000	1.810	7.464	76.4	(1.000, 0.000) (0.000, 1.000)	(1x1)	b: bulk lattice
Surface 1	3.620	0.000	1.810	7.464	76.4	(1.000, 0.000) (0.000, 1.000)	(1x1)	s1: commens. superlattice

3D COORDINATES

O1 in hollow site at step edge; O4 in hollow site on (100) terrace
(O4 forms c(2x2) with O1)

Dx/Dy in Å, or as a fraction of layer's unit cell vectors; atom 0 at the origin. Epir/subr are bulk repeat vectors.

No. of atoms: 9

Bulk z = .439 Å

Region	Chem el.	At. no.	Cell type	Site occ.	Rel. to	Dx ± εx		Dy ± εy		Dz ± εz		Dz/Bz(%) ± εz/Bz
epir		-2					f		f		Å	
subr		-1				1.810	Å	1.756	Å	0.439	Å	
ovrl	O	1	b	1.00	0	0.000	f	0.000	f	0.000	Å	0.0
intf	Cu	2	b	1.00	1	0.507 ± .042	f	0.987 ± .027	f	0.388 ± .200	Å	88.4 ± 45.6
intf	Cu	3	b	1.00	2	0.382	f	-0.765	f	0.439	Å	100.0
ovrl	O	4	b	1.00	3	-0.124 ± .042	f	0.248 ± .027	f	0.051 ± .200	Å	11.6 ± 45.6
intf	Cu	5	b	1.00	4	-0.494 ± .042	f	-0.013 ± .027	f	0.388 ± .200	Å	88.4 ± 45.6
intf	Cu	6	b	1.00	5	0.382	f	0.235	f	0.439	Å	100.0
intf	Cu	7	b	1.00	6	-0.618	f	0.235	f	0.439	Å	100.0
intf	Cu	8	b	1.00	7	0.382	f	-0.765	f	0.439	Å	100.0
subl	Cu	9	b	1.00	8	0.382	f	0.235	f	0.439	Å	100.0

Cu(410)-(1x1)-2O
29.8.12b

BOND DISTANCES AND ANGLES

Bond distances and angles are derived from coordinates

No. of distances/angles: 7

Interatomic dist. A-B (Å)	Atom A	Atom B	Atom C	Bond angle A-B-C (°)
1.854	O1	Cu2(0,-1)	O1(1,0)	155.1
1.854	O1	Cu2(0,-1)	Cu3	133.7
1.854	O1	Cu2(0,-1)	Cu3(-1,0)	46.3
2.210	O1	Cu7(0,-1)	Cu2(0,-1)	45.0
2.560	Cu2	Cu3(0,1)		
1.854	O4	Cu3	O1(1,0)	155.1
2.210	O4	Cu9		

COMMON NAME : Cu(100)-c(2x2)-Pb
CLASSIFICATION : 29.82.1a
TECHNIQUE : LEED
AUTHORS : W. Hoesler and W. Moritz
REFERENCE : Surf. Sci., 117, 196 (1982)

ILLUSTRATION: 28,29

SURFACE TYPE
Substrate : Cu
Crystal face: 100
Temperature : 160 K
Bulk lattice: fcc
2D bulk symm: p4m
2D surf symm: p4m

Adsorbate: Pb
Coverage : 0.5 (Pb/Cu)
Pattern : c(2x2)
Matrix : (1.000, 1.000)
(-1.000, 1.000)

STRUCTURE TYPE
Atomic adsorption in 4-fold hollow sites

SAMPLE PREPARATION (1 sample)
Treatment : Pb evaporated from an rf heated source
Crystallinity:
Anal. methods:
Contamination: AES: no impurities

COMMENTS
ZJ R-factors do not resolve well between hollow, top and bridge sites: comparison of I-V peak heights provides additional means of discrimination

DATA COLLECTION
Technique: LEED
Dataset : I-V spectra for 7 beams at normal incidence

THEORY/DATA TREATMENT
Dynamical LEED: 8 phase shifts (modified Slater exchange); Voi∝E**1/3; ΘD=330 K(Cu),70K(Pb)

STRUCTURES EXAMINED
Variable Pb/Cu spacing in hollow, top and bridge sites

QUALITY OF EXPERIMENT-THEORY FIT
RZJ=0.17 (see comments)

2D UNIT CELLS (1 domain observed)

Cell	Ax (Å)	Ay (Å)	Bx (Å)	By (Å)	α (°)	Matrix	Pattern	Cell type
Bulk	2.560	0.000	0.000	2.560	90.0	(1.000, 0.000) (0.000, 1.000)	(1x1)	b: bulk lattice
Surface 1	2.560	2.560	-2.560	2.560	90.0	(1.000, 1.000) (-1.000, 1.000)	c(2x2)	s1: commens. superlattice

3D COORDINATES

Pb1: overlayer in 4-fold hollow sites

Dx/Dy in Å, or as a fraction of layer's unit cell vectors; atom 0 at the origin. Epir/subr are bulk repeat vectors.

No. of atoms: 3

Bulk z = 1.810 Å

Reg ion	Chem el.	At. no.	Cell type	Site occ.	Rel. to	Dx ± εx		Dy ± εy		Dz ± εz		Dz/Bz(%) ± εz/Bz
epir		-2					f		f		Å	
subr		-1				-1.280	Å	-1.280	Å	1.810	Å	
ovrl	Pb	1	s1	.50	0	0.000	f	0.000	f	0.000	Å	0.0
intf	Cu	2	b	1.00	1	0.500	f	0.500	f	2.400 ± .050	Å	132.6 ± 2.8
subl	Cu	3	b	1.00	2	-0.500	f	-0.500	f	1.810	Å	100.0

BOND DISTANCES AND ANGLES

Bond distances and angles are derived from coordinates

No. of distances/angles: 5

Interatomic dist. A-B (Å)	Atom A	Atom B	Atom C	Bond angle A-B-C (°)
3.620	Pb1	Pb1(1,0)		
3.006	Pb1	Cu2	Pb1(1,0)	74.0
3.006	Pb1	Cu2	Cu2(1,0)	115.2
2.560	Cu2	Cu2(1,0)		
2.560	Cu2	Cu3		

ILLUSTRATION: 28,29

COMMON NAME : Cu(100)-c(2x2)-Pb
CLASSIFICATION : 29.82.2
TECHNIQUE : LEED
AUTHORS : W. Hoesler, W. Moritz, E. Tamura and R. Feder
REFERENCE : Surf. Sci., 171, 55 (1986)

SURFACE TYPE
Substrate : Cu
Crystal face: 100
Temperature : 160 K
Bulk lattice: fcc
2D bulk symm: p4m
2D surf symm: p4m

Adsorbate: Pb
Coverage : 0.5 Pb/Cu
Pattern : c(2x2)
Matrix : (1.000, 1.000)
(-1.000, 1.000)

STRUCTURE TYPE
Atomic adsorption in hollow sites

SAMPLE PREPARATION (1 sample)
Treatment : Pb evaporated, then desorption/adsorption cycles
Crystallinity:
Anal. methods:
Contamination: coverage and cleanliness checked by AES

COMMENTS

DATA COLLECTION
Technique: LEED
Dataset : LEED and SPLEED IV and AV spectra: 50<E<280 eV; total E range 980eV; 7 non degenerate beams

THEORY/DATA TREATMENT
Dynamical LEED and SPLEED: spin averaged phase shifts (SWWLA scheme); Voi∝E**1/3. E-dep. Vor (fit); ΘD=100 K

STRUCTURES EXAMINED
Hollow site with variation of Pb-Cu and topmost Cu-Cu interlayer spacings

QUALITY OF EXPERIMENT-THEORY FIT
R=0.10

2D UNIT CELLS (1 domain observed)

Cell	Ax (Å)	Ay (Å)	Bx (Å)	By (Å)	α (°)	Matrix	Pattern	Cell type
Bulk	2.560	0.000	0.000	2.560	90.0	(1.000, 0.000) (0.000, 1.000)	(1x1)	b: bulk lattice
Surface 1	2.560	2.560	-2.560	2.560	90.0	(1.000, 1.000) (-1.000, 1.000)	c(2x2)	s1: commens. superlattice

3D COORDINATES

Pb1: overlayer in 4-fold hollow sites

Dx/Dy in Å, or as a fraction of layer's unit cell vectors; atom 0 at the origin. Epir/subr are bulk repeat vectors.

No. of atoms: 3

Bulk z = 1.810 Å

Reg ion	Chem el.	At. no.	Cell type	Site occ.	Rel. to	Dx ± εx		Dy ± εy		Dz ± εz		Dz/Bz(%) ± εz/Bz
epir		-2					f		f		Å	
subr		-1				-1.280	Å	-1.280	Å	1.810	Å	
ovrl	Pb	1	s1	.50	0	0.000	f	0.000	f	0.000	Å	0.0
intf	Cu	2	b	1.00	1	0.500	f	0.500	f	2.290 ± .040	Å	126.5 ± 2.2
subl	Cu	3	b	1.00	2	-0.500	f	-0.500	f	1.810 ± .030	Å	100.0 ± 1.7

BOND DISTANCES AND ANGLES

Bond distances and angles are derived from coordinates

No. of distances/angles: 3

Interatomic dist. A-B (Å)	Atom A	Atom B	Atom C	Bond angle A-B-C (°)
3.620	Pb1	Pb1(1,0)		
2.919	Pb1	Cu2	Cu3	96.7
2.560	Cu2	Cu3		

COMMON NAME : Cu(100)-c(2x2)-Pd
CLASSIFICATION : 29.46.2
TECHNIQUE : LEED
AUTHORS : Z.Q. Wu, S.H. Lu, Z.Q. Wang, C.K.C. Lok, J. Quinn, Y.S. Li, D. Tian, F. Jona and P.M. Marcus
REFERENCE : Phys. Rev., B38, 5363 (1988)

ILLUSTRATION: 33

SURFACE TYPE
Substrate : Cu
Crystal face: 100
Temperature : RT
Bulk lattice: fcc
2D bulk symm: p4m
2D surf symm: p4m

Adsorbate: Pd
Coverage : 0.5 Pd/Cu
Pattern : c(2x2)
Matrix : ((s0p16.67h8.5v0s3b0T 1.000, 1.000)
(-1.000, 1.000)

STRUCTURE TYPE
Substitutional adsorption, forming buckled monolayer of mixed alloy

SAMPLE PREPARATION (1 sample)
Treatment : Ar+ sputtering followed by 873 K annealing for 10 mins
Crystallinity:
Anal. methods:
Contamination: coverage and cleanliness checked by AES

COMMENTS
Authors state that additional Pd atoms may lie below the topmost atomic plane

DATA COLLECTION
Technique: LEED; PED
Dataset : IV spectra at normal incidence for 4 beams, at (Θ,ϕ)= (10,0)° for 5 beams

THEORY/DATA TREATMENT
Dynamical LEED and PED: 8 phase shifts, 45 beams; Vor=-10 eV (fit), VoiαE**1/3; vib ampl=0.15Å

STRUCTURES EXAMINED
Overlayer model at hollow sites for Pd-Cu spacing of 1.74 to 2.14Å; mixed top layer with buckling of 0.0, 0.05 or 0.1Å and first interlayer spacing of 1.607 to 2.000Å

2D UNIT CELLS (1 domain observed)

Cell	Ax (Å)	Ay (Å)	Bx (Å)	By (Å)	α (°)	Matrix	Pattern	Cell type
Bulk	2.560	0.000	0.000	2.560	90.0	(1.000, 0.000) (0.000, 1.000)	(1x1)	b: bulk lattice
Surface 1	2.560	2.560	-2.560	2.560	90.0	(1.000, 1.000) (-1.000, 1.000)	c(2x2)	s1: commens. superlattice

3D COORDINATES

Pd1-Cu2: mixed buckled top layer

Dx/Dy in Å, or as a fraction of layer's unit cell vectors; atom 0 at the origin. Epir/subr are bulk repeat vectors.

No. of atoms: 4

Bulk z = 1.810 Å

Region	Chem el.	At. no.	Cell type	Site occ.	Rel. to	Dx ± εx	Dy ± εy	Dz ± εz	Dz/Bz(%) ± εz/Bz
epir		-2				f	f	Å	
subr		-1				-1.280 Å	-1.280 Å	1.810 Å	
intf	Pd	1	s1	.50	0	0.000 f	0.000 f	0.000 Å	0.0
intf	Cu	2	s1	.50	1	0.500 f	0.500 f	0.020 ± .030 Å	1.1 ± 1.7
intf	Cu	3	b	1.00	2	0.500 f	-0.500 f	1.810 ± .030 Å	100.0 ± 1.7
subl	Cu	4	b	1.00	3	-0.500 f	-0.500 f	1.810 Å	100.0

BOND DISTANCES AND ANGLES

Bond distances and angles are derived from coordinates

No. of distances/angles: 3

Interatomic dist. A-B (Å)	Atom A	Atom B	Atom C	Bond angle A-B-C (°)
2.560	Pd1	Cu2	Cu3	60.4
2.574	Pd1	Cu3	Cu4	90.3
2.560	Cu2	Cu3	Cu4	120.0

ILLUSTRATION: 31

COMMON NAME : Cu(100)-c(5√2x√2)R45°-3Pb
CLASSIFICATION : 29.82.1b
TECHNIQUE : LEED
AUTHORS : W. Hoesler and W. Moritz
REFERENCE : Surf. Sci., 117, 196 (1982)

SURFACE TYPE
Substrate : Cu
Crystal face: 100
Temperature : 160 K
Bulk lattice: fcc
2D bulk symm: p4m
2D surf symm: cmm

Adsorbate: Pb
Coverage : 0.6 (Pb/Cu)
Pattern : c(5√2x√2)R45°
Matrix : (2.000, 3.000)
(-1.000, 1.000)

STRUCTURE TYPE
Atomic adsorption in 3-atom wide strip domains of c(2x2) structure, with lateral relaxation at domain boundaries (0.3Å displacements away from boundaries); central atoms in strips are on hollow sites, other 0.3Å from hollow sites

SAMPLE PREPARATION (1 sample)
Treatment : Pb evaporated from an rf heated source
Crystallinity:
Anal. methods:
Contamination: AES: no impurities

COMMENTS
Overlayer buckling is likely, but not explored in this study

DATA COLLECTION
Technique: LEED
Dataset : I-V spectra for 14 beams

THEORY/DATA TREATMENT
Dynamical LEED: 8 phase shifts (modified Slater exchange); Voi∝E**1/3; ΘD=330 K(Cu),70K(Pb)

STRUCTURES EXAMINED
Planar Pb layer as: 1) pseudo hexagonal layer
2) c(2x2) domains separated by denser boundaries, with near-boundary Pb shifted 0.0, 0.2, 0.4Å parallel to surface away from boundary

QUALITY OF EXPERIMENT-THEORY FIT
RZJ=0.41

2D UNIT CELLS (2 domains observed)

Cell	Ax (Å)	Ay (Å)	Bx (Å)	By (Å)	α (°)	Matrix	Pattern	Cell type
Bulk	2.560	0.000	0.000	2.560	90.0	(1.000, 0.000) (0.000, 1.000)	(1x1)	b: bulk lattice
Surface 1	5.120	7.680	-2.560	2.560	78.7	(2.000, 3.000) (-1.000, 1.000)	c(5√2x√2)R45°	s1: commens. superlattice

3D COORDINATES

Pb1: center row of 3-Pb-wide domains, in hollow site; Pb2-Pb3: 2 boundary rows, shifted 0.3Å from hollow sites; 0.1Å lateral error bars assumed for tabulation

Dx/Dy in Å, or as a fraction of layer's unit cell vectors; atom 0 at the origin. Epir/subr are bulk repeat vectors.

No. of atoms: 5 Bulk z = 1.810 Å

Reg ion	Chem el.	At. no.	Cell type	Site occ.	Rel. to	Dx ± εx		Dy ± εy		Dz ± εz		Dz/Bz(%) ± εz/Bz
epir		-2					f		f		Å	
subr		-1				1.280	Å	-1.280	Å	1.810	Å	
ovrl	Pb	1	s1	.20	0	0.000	f	0.000	f	0.000	Å	0.0
ovrl	Pb	2	s1	.20	1	0.367 ± .009	f	0.816 ± .028	f	0.000	Å	0.0
ovrl	Pb	3	s1	.20	1	0.633 ± .009	f	0.184 ± .028	f	0.000	Å	0.0
intf	Cu	4	b	1.00	1	-0.500	f	0.500	f	2.400 ± .050	Å	132.6 ± 2.8
subl	Cu	5	b	1.00	4	0.500	f	-0.500	f	1.810	Å	100.0

BOND DISTANCES AND ANGLES

Bond distances and angles are derived from coordinates

No. of distances/angles: 5

Interatomic dist. A-B (Å)	Atom A	Atom B	Atom C	Bond angle A-B-C (°)
3.620	Pb1	Pb1(0,1)		
3.323	Pb1	Pb2(0,-1)		
3.006	Pb1	Cu4	Pb1(0,1)	74.1
3.006	Pb1	Cu4	Cu4(0,1)	115.2

Cu(100)-c(5√2x√2)R45°-3Pb
29.82.1b

Bond Distances and Angles - Continued

Interatomic dist. A-B (Å)	Atom A	Atom B	Atom C	Bond angle A-B-C (°)
3.010	Pb2	Pb3	Pb2(0,-1)	74.0

ILLUSTRATION: 31

COMMON NAME : Cu(100)-c(5√2x√2)R45°-3Pb
CLASSIFICATION : 29.82.3
TECHNIQUE : LEED
AUTHORS : W. Hoesler and W. Moritz
REFERENCE : Surf. Sci., 175, 63 (1986)

SURFACE TYPE
Substrate : Cu
Crystal face: 100
Temperature : 160 K
Bulk lattice: fcc
2D bulk symm: p4m
2D surf symm: cmm

Adsorbate: Pb
Coverage : 0.6 Pb/Cu
Pattern : c(5√2x√2)R45°
Matrix : (3.000, 2.000)
(-1.000, 1.000)

STRUCTURE TYPE
Atomic adsorption (3Pb per supercell) at and near hollow sites, forming 3-Pb-wide strips (domains) of c(2x2) structure with compressed domain boundaries; no buckling in Pb layer or relaxations in substrate detected

SAMPLE PREPARATION (1 sample)
Treatment : Pb evaporated from RF heated crucible
Crystallinity:
Anal. methods:
Contamination: AES and LEED: S and C < 2% ML

COMMENTS

DATA COLLECTION
Technique: LEED
Dataset : IV spectra: 21 inequivalent beams;
cumulative E range 2150 eV

THEORY/DATA TREATMENT
Dynamical LEED (layer doubling): 10 phase shifts, Vor=-3.9 eV (fit); Voi∝E**1/2; ΘD=85 K(Pb), 330K(Cu)

STRUCTURES EXAMINED
Pseudo-hexagonal and domain boundary models with 3Pb per supercell in various lateral sites; buckling of Pb overlayer; variations of Pb-Cu spacing and relaxations of top Cu layer.

QUALITY OF EXPERIMENT-THEORY FIT
RPE=0.46

2D UNIT CELLS (2 domains observed)

Cell	Ax (Å)	Ay (Å)	Bx (Å)	By (Å)	α (°)	Matrix	Pattern	Cell type
Bulk	2.560	0.000	0.000	2.560	90.0	(1.000, 0.000) (0.000, 1.000)	(1x1)	b: bulk lattice
Surface 1	7.679	5.120	-2.560	2.560	101.3	(3.000, 2.000) (-1.000, 1.000)	c(5√2x√2)R45°	s1: commens. superlattice

3D COORDINATES

Pb1: positioned over 4-fold hollow sites; Pb2-Pb3: positioned 0.42Å laterally from 4-fold hollow sites, coplanar with Pb1; 0.1Å error bars assumed for tabulation

Dx/Dy in Å, or as a fraction of layer's unit cell vectors; atom 0 at the origin. Epir/subr are bulk repeat vectors.

No. of atoms: 5

Bulk z = 1.810 Å

Region	Chem el.	At. no.	Cell type	Site occ.	Rel. to	Dx ± εx		Dy ± εy		Dz ± εz		Dz/Bz(%) ± εz/Bz
epir		-2					f		f		Å	
subr		-1				-1.280	Å	-1.280	Å	1.810	Å	
ovrl	Pb	1	s1	.20	0	0.000	f	0.000	f	0.000	Å	0.0
ovrl	Pb	2	s1	.20	1	0.354	f	0.177	f	0.000 ± .100	Å	0.0 ± 5.5
ovrl	Pb	3	s1	.20	2	0.293	f	0.646	f	0.000	Å	0.0
intf	Cu	4	b	1.00	3	-0.616	f	-1.616	f	2.310 ± .100	Å	127.6 ± 5.5
subl	Cu	5	b	1.00	4	-0.500	f	-0.500	f	1.810 ± .100	Å	100.0 ± 5.5

BOND DISTANCES AND ANGLES

Bond distances and angles are derived from coordinates

No. of distances/angles: 4

Interatomic dist. A-B (Å)	Atom A	Atom B	Atom C	Bond angle A-B-C (°)
3.620	Pb1	Pb1(0,1)		
3.200	Pb1	Pb2	Pb3	145.7
2.935	Pb1	Cu4	Cu5	96.9
2.696	Pb2	Cu4	Cu5	166.0

COMMON NAME : Cu(100)-p(2x2)-S
CLASSIFICATION : 29.16.0a
TECHNIQUE : PED
AUTHORS : E.L. Bullock, C.S. Fadley and P.J. Orders
REFERENCE : Phys. Rev., B28, 4867 (1983)

ILLUSTRATION: 28,30

SURFACE TYPE
Substrate : Cu
Crystal face: 100
Temperature : RT*
Bulk lattice: fcc
2D bulk symm: p4m
2D surf symm: p4m

Adsorbate: S
Coverage : 0.25 (S/Cu)
Pattern : p(2x2)
Matrix : (2.000, 0.000)
(0.000, 2.000)

STRUCTURE TYPE
Atomic adsorption in hollow site of unreconstructed substrate

SAMPLE PREPARATION (1 sample)
Treatment : 75L H2S and heating to 500 K gave p(2x2) LEED pattern
Crystallinity:
Anal. methods:
Contamination: see Barton et al, PRL 51, 272 (1983)

COMMENTS

DATA COLLECTION
Technique: PED; E-dependent photo-yield
Dataset : yield at Θe-=45° so emission along [110]; Θhv= 30 so polarization has large component parallel to surface

THEORY/DATA TREATMENT
Single scattering theory, using LEED scattering factors; cluster sum over 60 S atoms and 1800 Cu atoms in 8 layers

STRUCTURES EXAMINED
No structural variation

QUALITY OF EXPERIMENT-THEORY FIT
Visual

2D UNIT CELLS (1 domain observed)

Cell	Ax (Å)	Ay (Å)	Bx (Å)	By (Å)	α (°)	Matrix	Pattern	Cell type
Bulk	2.553	0.000	0.000	2.553	90.0	(1.000, 0.000) (0.000, 1.000)	(1x1)	b: bulk lattice
Surface 1	5.105	0.000	0.000	5.105	90.0	(2.000, 0.000) (0.000, 2.000)	p(2x2)	s1: commens. superlattice

3D COORDINATES

S1: overlayer in hollow sites; 0.1Å error bar assumed for tabulation

Dx/Dy in Å, or as a fraction of layer's unit cell vectors; atom 0 at the origin. Epir/subr are bulk repeat vectors.

No. of atoms: 3

Bulk z = 1.810 Å

Region	Chem el.	At. no.	Cell type	Site occ.	Rel. to	Dx ± εx		Dy ± εy		Dz ± εz		Dz/Bz(%) ± εz/Bz
epir		-2					f		f		Å	
subr		-1				-1.276	Å	-1.276	Å	1.810	Å	
ovrl	S	1	s1	.25	0	0.000	f	0.000	f	0.000	Å	0.0
intf	Cu	2	b	1.00	1	0.500	f	0.500	f	1.390 ± .100	Å	76.8 ± 5.5
subl	Cu	3	b	1.00	2	-0.500	f	-0.500	f	1.810	Å	100.0

BOND DISTANCES AND ANGLES

Bond distances and angles are derived from coordinates

No. of distances/angles: 3

Interatomic dist. A-B (Å)	Atom A	Atom B	Atom C	Bond angle A-B-C (°)
2.278	S1	Cu2	Cu2(1,0)	124.1
2.553	Cu2	Cu2(1,0)		
2.556	Cu2	Cu3		

COMMON NAME : Cu(100)-p(2x2)-S
CLASSIFICATION : 29.16.1
TECHNIQUE : LEED
AUTHORS : H.C. Zeng, R.N.S. Sodhi and K.A.R. Mitchell
REFERENCE : Surf. Sci., 177, 329 (1986)

ILLUSTRATION: 28,30

SURFACE TYPE
Substrate : Cu
Crystal face: 100
Temperature : 553 K
Bulk lattice: fcc
2D bulk symm: p4m
2D surf symm: p4m

Adsorbate: S
Coverage : 1/4 (S/Cu)
Pattern : p(2x2)
Matrix : (2.000, 0.000)
(0.000, 2.000)

STRUCTURE TYPE
Atomic adsorption in 4-fold hollow sites on unrelaxed substrate

SAMPLE PREPARATION (1 sample)
Treatment : exposure at RT to H2S at 2E-8 torr
Crystallinity:
Anal. methods:
Contamination: checked by AES and LEED

COMMENTS
Only moderate theory-experiment agreement was found

DATA COLLECTION
Technique: LEED
Dataset : total of 16 independent I-V curves for beams at normal incidence and at Θ=14° off-normal incidence

THEORY/DATA TREATMENT
Dynamical LEED (RFS): 8 ph shs (Cu Moruzzi et al; S atomic superpos pot); Voi∝E**1/3; ΘD=343 K(Cu), 335K(S)

STRUCTURES EXAMINED
Cu fixed at bulk structure; S-Cu interlayer spacing varied: 1.20-1.60Å over hollow; 1.45-2.05Å over bridge; 1.70-2.30Å over top site

QUALITY OF EXPERIMENT-THEORY FIT
RPE=0.306

2D UNIT CELLS (1 domain observed)

Cell	Ax (Å)	Ay (Å)	Bx (Å)	By (Å)	α (°)	Matrix	Pattern	Cell type
Bulk	2.553	0.000	0.000	2.553	90.0	(1.000, 0.000) (0.000, 1.000)	(1x1)	b: bulk lattice
Surface 1	5.105	0.000	0.000	5.105	90.0	(2.000, 0.000) (0.000, 2.000)	p(2x2)	s1: commens. superlattice

3D COORDINATES

S1: overlayer in 4-fold hollow sites

Dx/Dy in Å, or as a fraction of layer's unit cell vectors; atom 0 at the origin. Epir/subr are bulk repeat vectors.

No. of atoms: 3

Bulk z = 1.810 Å

Region	Chem el.	At. no.	Cell type	Site occ.	Rel. to	Dx ± εx		Dy ± εy		Dz ± εz		Dz/Bz(%) ± εz/Bz
epir		-2					f		f		Å	
subr		-1				1.276	Å	1.276	Å	1.810	Å	
ovrl	S	1	s1	.25	0	0.000	f	0.000	f	0.000	Å	0.0
intf	Cu	2	b	1.00	1	0.500	f	0.500	f	1.320 ± .040	Å	72.9 ± 2.2
subl	Cu	3	b	1.00	2	0.500	f	0.500	f	1.810	Å	100.0

BOND DISTANCES AND ANGLES

Bond distances and angles are derived from coordinates

No. of distances/angles: 3

Interatomic dist. A-B (Å)	Atom A	Atom B	Atom C	Bond angle A-B-C (°)
2.236	S1	Cu2	Cu2(1,0)	124.8
2.236	S1	Cu2	Cu3	171.1
2.556	Cu2	Cu3		

COMMON NAME : Cu(100)-p(2x2)-S ILLUSTRATION: 28,30
CLASSIFICATION : 29.16.10
TECHNIQUE : XSW
AUTHORS : J.R. Patel, D.W. Berreman, F. Sette, P.H. Citrin, J.E. Rowe, P.L. Cowan, T. Jach and B. Karlin
REFERENCE : Phys. Rev., B40, 1330 (1989)

SURFACE TYPE
Substrate : Cu
Crystal face: 100
Temperature : RT*
Bulk lattice: fcc
2D bulk symm: p4m
2D surf symm: p4m
Adsorbate: S
Coverage : 0.25 (S/Cu)
Pattern : p(2x2)
Matrix : (2.000, 0.000)
(0.000, 2.000)

STRUCTURE TYPE
Atomic adsorption in hollow site of possibly relaxed substrate

SAMPLE PREPARATION (1 sample)
Treatment : see PRL 49,1712(1982), PRL 61,1384(1988)
Crystallinity:
Anal. methods: SEXAFS for bond distances
Contamination:

COMMENTS
No lateral relaxation allowed in topmost Cu layer; eds.: incompatibility found between XSW and SEXAFS results for relaxation of top Cu-Cu spacing may be caused by that assumption

DATA COLLECTION
Technique: XSW; fluorescence yield at NSLS
Dataset : standing-wave data at fixed incidence angle as fct. of crystal rotation about [111]

THEORY/DATA TREATMENT
8x8 matrix dynamical x-ray theory, with 2 fit parameters: P = at. pos. above (111) plane, F = coherent fraction

STRUCTURES EXAMINED
XSW and SEXAFS data directly give 4-fold hollow site; difference between XSW and SEXAFS for S-Cu spacing indicates change in top Cu-Cu spacing from bulk, since XSW refers adsorbate position to deep bulk layers; no lateral relaxation allowed in topmost Cu layer

2D UNIT CELLS (1 domain observed)

Cell	Ax (Å)	Ay (Å)	Bx (Å)	By (Å)	α (°)	Matrix	Pattern	Cell type
Bulk	2.553	0.000	0.000	2.553	90.0	(1.000, 0.000) (0.000, 1.000)	(1x1)	b: bulk lattice
Surface 1	5.105	0.000	0.000	5.105	90.0	(2.000, 0.000) (0.000, 2.000)	p(2x2)	s1: commens. superlattice

3D COORDINATES

S1 forms overlayer in hollow sites; coordinates are derived from bond distance and spacings

Dx/Dy in Å, or as a fraction of layer's unit cell vectors; atom 0 at the origin. Epir/subr are bulk repeat vectors.

No. of atoms: 4 Bulk z = 1.810 Å

Region	Chem el.	At. no.	Cell type	Site occ.	Rel. to	Dx ± εx		Dy ± εy		Dz ± εz		Dz/Bz(%) ± εz/Bz
epir		-2					f		f		Å	
subr		-1				-1.276	Å	-1.276	Å	1.810	Å	
ovrl	S	1	s1	.25	0	0.000	f	0.000	f	0.000	Å	0.0
intf	Cu	2	b	1.00	1	0.500	f	0.500	f	1.440 ± .020	Å	79.6 ± 1.1
intf	Cu	3	b	1.00	2	-0.500	f	-0.500	f	1.810 ± .040	Å	100.0 ± 2.2
subl	Cu	4	b	1.00	3	0.500	f	0.500	f	1.810	Å	100.0

BOND DISTANCES AND ANGLES

No. of distances/angles: 4

Interatomic dist. A-B (Å)	Atom A	Atom B	Atom C	Bond angle A-B-C (°)
2.309	S1	Cu2	Cu2(1,0)	123.6
2.309	S1	Cu2	Cu3	83.7
2.556	Cu2	Cu3	Cu4	90.2
2.556	Cu3	Cu4		

COMMON NAME : Cu(100)-p(2x2)-S
CLASSIFICATION : 29.16.2
TECHNIQUE : ARPEFS
AUTHORS : C.C. Bahr, J.J. Barton, Z. Hussain, S.W. Robey, J.G. Tobinand and D.A. Shirley
REFERENCE : Phys. Rev., B35, 3773 (1987)

ILLUSTRATION: 28,30

SURFACE TYPE
Substrate : Cu
Crystal face: 100
Temperature : RT*
Bulk lattice: fcc
2D bulk symm: p4m
2D surf symm: p4m

Adsorbate: S
Coverage : 0.25 S/Cu
Pattern : p(2x2)
Matrix : (2.000, 0.000)
(0.000, 2.000)

STRUCTURE TYPE
Atomic adsorption in 4-fold hollow sites on unrelaxed substrate

SAMPLE PREPARATION (1 sample)
Treatment : exposure to 40L of H2S and flashing to 500 K
Crystallinity:
Anal. methods:
Contamination: monitored by LEED

COMMENTS
Evidence found for second-layer buckling, with the Cu atoms under fourfold symmetric open sites lying 0.13Å deeper than the Cu atoms under S sites

DATA COLLECTION
Technique: ARPEFS
Dataset : 3 independent ARPEFS curves corresponding to emission angles aligned with [100], [110] and [111] directions

THEORY/DATA TREATMENT
Direct Fourier analysis and multiple-scattering spherical-wave calculation; ΘD=337 K(S),239K(surf Cu),343K(bulk Cu)

STRUCTURES EXAMINED
Top, bridge and hollow adsorption sites

QUALITY OF EXPERIMENT-THEORY FIT
Visual

2D UNIT CELLS (1 domain observed)

Cell	Ax (Å)	Ay (Å)	Bx (Å)	By (Å)	α (°)	Matrix	Pattern	Cell type
Bulk	2.553	0.000	0.000	2.553	90.0	(1.000, 0.000) (0.000, 1.000)	(1x1)	b: bulk lattice
Surface 1	5.105	0.000	0.000	5.105	90.0	(2.000, 0.000) (0.000, 2.000)	p(2x2)	s1: commens. superlattice

3D COORDINATES

S1: overlayer in 4-fold hollow sites

Dx/Dy in Å, or as a fraction of layer's unit cell vectors; atom 0 at the origin. Epir/subr are bulk repeat vectors.

No. of atoms: 3

Bulk z = 1.810 Å

Reg ion	Chem el.	At. no.	Cell type	Site occ.	Rel. to	Dx ± εx		Dy ± εy		Dz ± εz		Dz/Bz(%) ± εz/Bz
epir		-2					f		f		Å	
subr		-1				1.276	Å	1.276	Å	1.810	Å	
ovrl	S	1	s1	.25	0	0.000	f	0.000	f	0.000	Å	0.0
intf	Cu	2	b	1.00	1	0.500	f	0.500	f	1.360 ± .010	Å	75.1 ± .6
subl	Cu	3	b	1.00	2	0.500	f	0.500	f	1.810	Å	100.0

BOND DISTANCES AND ANGLES

Bond distances and angles are derived from coordinates

No. of distances/angles: 3

Interatomic dist. A-B (Å)	Atom A	Atom B	Atom C	Bond angle A-B-C (°)
2.260	S1	Cu2	Cu2(1,0)	124.4
2.260	S1	Cu2	Cu3	171.9
2.556	Cu2	Cu3		

COMMON NAME : Cu(100)-p(2x2)-S
CLASSIFICATION : 29.16.3
TECHNIQUE : HREELS
AUTHORS : Z.Q. Wu, M.L. Xu, Y. Chen, S.Y. Tong, M.H. Mohamed and L.L. Kesmodel
REFERENCE : Phys. Rev., B36, 9329 (1987)

ILLUSTRATION: 28,30

SURFACE TYPE
Substrate : Cu
Crystal face: 100
Temperature : RT
Bulk lattice: fcc
2D bulk symm: p4m
2D surf symm: p4m

Adsorbate: S
Coverage : 0.25 S/Cu
Pattern : p(2x2)
Matrix : (2.000, 0.000)
(0.000, 2.000)

STRUCTURE TYPE
Atomic adsorption in 4-fold hollow sites on unrelaxed substrate

SAMPLE PREPARATION (1 sample)
Treatment : exposure to 30L H2S at RT and anneal to 473 K for 10-15mins
Crystallinity:
Anal. methods:
Contamination:

COMMENTS

DATA COLLECTION
Technique: HREELS
Dataset : phonon loss spectra at 5 k// points.

THEORY/DATA TREATMENT
Ms HREELS cross-section analysis of bulk and surface phonons: nn lattice dynamical model for 31-layer slab

STRUCTURES EXAMINED
Hollow site with adsorption heights of 1.1 to 1.5Å in steps of 0.05Å

QUALITY OF EXPERIMENT-THEORY FIT
Visual

2D UNIT CELLS (1 domain observed)

Cell	Ax (Å)	Ay (Å)	Bx (Å)	By (Å)	α (°)	Matrix	Pattern	Cell type
Bulk	2.550	0.000	0.000	2.550	90.0	(1.000, 0.000) (0.000, 1.000)	(1x1)	b: bulk lattice
Surface 1	5.100	0.000	0.000	5.100	90.0	(2.000, 0.000) (0.000, 2.000)	p(2x2)	s1: commens. superlattice

3D COORDINATES

S1: overlayer in 4-fold hollow sites

Dx/Dy in Å, or as a fraction of layer's unit cell vectors; atom 0 at the origin. Epir/subr are bulk repeat vectors.

No. of atoms: 3 Bulk z = 1.807 Å

Region	Chem el.	At. no.	Cell type	Site occ.	Rel. to	Dx ± εx		Dy ± εy		Dz ± εz		Dz/Bz(%) ± εz/Bz
epir		-2					f		f		Å	
subr		-1				1.275	Å	1.275	Å	1.807	Å	
ovrl	S	1	s1	.25	0	0.000	f	0.000	f	0.000	Å	0.0
intf	Cu	2	b	1.00	1	0.500	f	0.500	f	1.300 ± .050	Å	71.9 ± 2.8
subl	Cu	3	b	1.00	2	0.500	f	0.500	f	1.807	Å	100.0

BOND DISTANCES AND ANGLES

Bond distances and angles are derived from coordinates

No. of distances/angles: 3

Interatomic dist. A-B (Å)	Atom A	Atom B	Atom C	Bond angle A-B-C (°)
2.223	S1	Cu2	Cu2(1,0)	125.0
2.223	S1	Cu2	Cu3	170.7
2.553	Cu2	Cu3		

COMMON NAME : Cu(100)-(2x2)-S
CLASSIFICATION : 29.16.12
TECHNIQUE : XRD
AUTHORS : E. Vlieg, I.K. Robinson and R. McGrath
REFERENCE : Phys. Rev., B41, 7896 (1990)

ILLUSTRATION: 28,30

SURFACE TYPE
Substrate : Cu
Crystal face: 100
Temperature : RT*
Bulk lattice: fcc
2D bulk symm: p4m
2D surf symm: p4m

Adsorbate: S
Coverage : 0.25 (S/Cu)
Pattern : (2x2)
Matrix : (2.000, 0.000)
(0.000, 2.000)

STRUCTURE TYPE
4-fold site (1.19Å spacing), lateral expansion of top Cu layer by 0.03Å

SAMPLE PREPARATION (1 sample)
Treatment : sputter/anneal, H2S exposure 1E-7 Torr at RT, flash 250C
Crystallinity: (0.5,0,0.1) reflex and background int.
Anal. methods:
Contamination: substrate AES clean

COMMENTS
Phase transition at 200-250C:
possibly order/disorder

DATA COLLECTION
Technique: XRD; x-ray diffr., Si(111) double monochr.
Dataset : 400 reflexes (84 nonequiv.) on 16 fract. 4 int order rods

THEORY/DATA TREATMENT
Fit: Lorentzian peak shape and linear background
structure factor calc. w. isotropic Debye-Waller-Factor

STRUCTURES EXAMINED
Lateral displacements of 1st Cu layer 'inwards/outwards'

QUALITY OF EXPERIMENT-THEORY FIT
Chi=0.75

2D UNIT CELLS (1 domain observed)

Cell	Ax (Å)	Ay (Å)	Bx (Å)	By (Å)	α (°)	Matrix	Pattern	Cell type
Bulk	2.556	0.000	0.000	2.556	90.0	(1.000, 0.000) (0.000, 1.000)	(1x1)	b: bulk lattice
Surface 1	5.112	0.000	0.000	5.112	90.0	(2.000, 0.000) (0.000, 2.000)		s1: commens. superlattice

3D COORDINATES

S1: adsorbate in 4-fold hollow sites; Cu2-5: 1st Cu layer

Dx/Dy in Å, or as a fraction of layer's unit cell vectors; atom 0 at the origin. Epir/subr are bulk repeat vectors.

No. of atoms: 6

Bulk z = 1.807 Å

Region	Chem el.	At. no.	Cell type	Site occ.	Rel. to	Dx ± εx	Dy ± εy	Dz ± εz	Dz/Bz(%) ± εz/Bz
epir		-2				f	f	Å	
subr		-1				1.278 Å	1.278 Å	1.807 Å	
ovrl	S	1	s1	.25	0	0.000 Å	0.000 Å	0.000 ± .140 Å	0.0
intf	Cu	2	s1	.25	1	1.299 ± .010 Å	1.299 ± .010 Å	1.190 Å	65.8
intf	Cu	3	s1	.25	1	3.813 ± .010 Å	3.813 ± .010 Å	1.190 Å	65.8
intf	Cu	4	s1	.25	1	1.299 ± .010 Å	3.813 ± .010 Å	1.190 Å	65.8
intf	Cu	5	s1	.25	1	3.813 ± .010 Å	1.299 ± .010 Å	1.190 Å	65.8
subl	Cu	6	b	1.00	1	0.000 Å	0.000 Å	2.997 Å	165.8

BOND DISTANCES AND ANGLES

Bond distances and angles are derived from coordinates

No. of distances/angles: 5

Interatomic dist. A-B (Å)	Atom A	Atom B	Atom C	Bond angle A-B-C (°)
2.189	S1	Cu2		
2.997	S1	Cu6		
2.189	Cu2	S1	Cu3(-1,-1)	114.1

Cu(100)-(2x2)-S
29.16.12

Bond Distances and Angles - Continued

Interatomic dist. A-B (Å)	Atom A	Atom B	Atom C	Bond angle A-B-C (°)
2.189	Cu2	S1	Cu4(0,-1)	72.8
2.189	Cu2	S1	Cu6	57.1

COMMON NAME : Cu(100)-(2x2)-S
CLASSIFICATION : 29.16.13
TECHNIQUE : MEIS
AUTHORS : Q.T. Jiang, P. Fenter and T. Gustafsson
REFERENCE : Phys. Rev., B42, 9291 (1990)

ILLUSTRATION: 28,30

SURFACE TYPE
Substrate : Cu
Crystal face: 100
Temperature : RT
Bulk lattice: fcc
2D bulk symm: p4m
2D surf symm: p4m

Adsorbate: S
Coverage : 0.25 (S/Cu)
Pattern : (2x2)
Matrix : (2.000, 0.000)
(0.000, 2.000)

STRUCTURE TYPE
S in 4-fold hollow (spacing 1.30Å);
site laterally expanded (shift of 1st layer Cu atoms by 0.03Å away from S); no vertical buckling in 2nd Cu layer

SAMPLE PREPARATION (1 sample)
Treatment : 22L H2S at RT, 300C anneal, gives 0.23+-0.05ML
Crystallinity: substrate AES clean, (1x1) sharp LEED
Anal. methods: AES
Contamination: AES cleanliness checked periodically

COMMENTS

DATA COLLECTION
Technique: MEIS; MEIS, channeling incidence, 100 keV P
Dataset : 3 incident scatt. angles, detection angles scanned

THEORY/DATA TREATMENT
Blocking angles indicate interatomic directions;
Monto Carlo simulations of scattering process

STRUCTURES EXAMINED
Blocking angles used for Cu positions, fit by R-factor (PR B38, 10197 (1988))

2D UNIT CELLS (1 domain observed)

Cell	Ax (Å)	Ay (Å)	Bx (Å)	By (Å)	α (°)	Matrix	Pattern	Cell type
Bulk	2.556	0.000	0.000	2.556	90.0	(1.000, 0.000) (0.000, 1.000)	(1x1)	b: bulk lattice
Surface 1	5.112	0.000	0.000	5.112	90.0	(2.000, 0.000) (0.000, 2.000)	(2x2)	s1: commens. superlattice

3D COORDINATES

S1: adsorbate in 4-fold hollow site; Cu2-5: laterally relaxed 1st substrate layer

Dx/Dy in Å, or as a fraction of layer's unit cell vectors; atom 0 at the origin. Epir/subr are bulk repeat vectors.

No. of atoms: 10

Bulk z = 1.807 Å

Region	Chem el.	At. no.	Cell type	Site occ.	Rel. to	Dx ± εx		Dy ± εy		Dz ± εz		Dz/Bz(%) ± εz/Bz
epir		-2					f		f		Å	
subr		-1				-1.278	Å	-1.278	Å	1.807	Å	
ovrl	S	1	s1	.25	0	0.000	Å	0.000	Å	0.000	Å	0.0
intf	Cu	3	s1	.25	1	1.299 ± .020	Å	1.299 ± .020	Å	1.300	Å	71.9
intf	Cu	4	s1	.25	1	3.813 ± .020	Å	1.299 ± .020	Å	1.300	Å	71.9
intf	Cu	5	s1	.25	1	3.813 ± .020	Å	3.813 ± .020	Å	1.300	Å	71.9
intf	Cu	6	s1	.25	1	0.000	Å	2.556	Å	3.110	Å	172.1
intf	Cu	7	s1	.25	1	2.556	Å	0.000	Å	3.110	Å	172.1
intf	Cu	8	s1	.25	1	2.556	Å	2.556	Å	3.110	Å	172.1
intf	Cu	9	s1	.25	1	0.000	Å	0.000	Å	3.110	Å	172.1
subl	Cu	10	b	1.00	1	1.278	Å	1.278	Å	4.930	Å	272.8

BOND DISTANCES AND ANGLES

Bond distances and angles are derived from coordinates

No. of distances/angles: 1

Interatomic dist. A-B (Å)	Atom A	Atom B	Atom C	Bond angle A-B-C (°)
2.251	S1	Cu2(0,-1)	Cu3	54.7

COMMON NAME : Cu(100)-(2x2)-S
CLASSIFICATION : 29.16.14
TECHNIQUE : LEED
AUTHORS : H.C. Zeng, R.A. McFarlane and K.A.R. Mitchell
REFERENCE : Can. J. Phys., 68, 353 (1990)

ILLUSTRATION: 28,30

SURFACE TYPE
Substrate : Cu
Crystal face: 100
Temperature : RT
Bulk lattice: fcc
2D bulk symm: p4m
2D surf symm: p4m

Adsorbate: S
Coverage : 0.25 (S/Cu)
Pattern : (2x2)
Matrix : (2.000, 0.000)
(0.000, 2.000)

STRUCTURE TYPE
4-fold hollow site for S, spacing to 1st Cu layer 1.28Å; lateral expansion of site (shift of Cu atoms 0.04Å away); buckling in 2nd layer (atom below S pushed down) by up to 0.03Å

SAMPLE PREPARATION (1 sample)
Treatment : H2S exposure at 1E-7 Torr
Crystallinity:
Anal. methods:
Contamination: sharpest LEED at 20L H2S

COMMENTS
Previous publication / normal incidence data only:
H.C. Zeng, R.A. McFarlane and K.A.R. Mitchell
Phys. Rev. B39, 8000 (1989)

DATA COLLECTION
Technique: LEED; Video LEED
Dataset : 5 beams (2 int. + 3 fract. order) at normal incidence, 7 beams (3+4) at 8° off-normal incidence

THEORY/DATA TREATMENT
Dynamical LEED (combined space, matrix inv, layer doubling): new phase shifts tried

STRUCTURES EXAMINED
Site contraction by 0.05Å: can be ruled out; off-normal angle fitted 6-10°

2D UNIT CELLS (1 domain observed)

Cell	Ax (Å)	Ay (Å)	Bx (Å)	By (Å)	α (°)	Matrix	Pattern	Cell type
Bulk	2.556	0.000	0.000	2.556	90.0	(1.000, 0.000) (0.000, 1.000)	(1x1)	b: bulk lattice
Surface 1	5.112	0.000	0.000	5.112	90.0	(2.000, 0.000) (0.000, 2.000)	(2x2)	s1: commens. superlattice

3D COORDINATES

S1: adsorbate in 4-fold hollow sites; Cu2-5: 1st layer Cu, Cu6-9: 2nd layer Cu

Dx/Dy in Å, or as a fraction of layer's unit cell vectors; atom 0 at the origin. Epir/subr are bulk repeat vectors.

No. of atoms: 10

Bulk z = 1.807 Å

Reg ion	Chem el.	At. no.	Cell type	Site occ.	Rel. to	Dx ± εx		Dy ± εy		Dz ± εz		Dz/Bz(%) ± εz/Bz
epir		-2					f		f		Å	
subr		-1				1.278	Å	1.278	Å	1.807	Å	
ovrl	S	1	s1	.25	0	0.000	Å	0.000	Å	0.000 ± .030	Å	0.0 ± 1.7
intf	Cu	2	s1	.25	1	1.306	Å	1.306	Å	1.280	Å	70.7
intf	Cu	3	s1	.25	1	3.806	Å	1.306	Å	1.280	Å	70.7
intf	Cu	4	s1	.25	1	1.306	Å	3.806	Å	1.280	Å	70.7
intf	Cu	5	s1	.25	1	3.806	Å	3.806	Å	1.280	Å	70.7
intf	Cu	6	s1	.25	1	2.556	Å	2.556	Å	3.090	Å	170.7
intf	Cu	7	s1	.25	1	0.000	Å	2.556	Å	3.110	Å	171.8
intf	Cu	8	s1	.25	1	2.556	Å	0.000	Å	3.110	Å	171.8
intf	Cu	9	s1	.25	1	0.000	Å	0.000	Å	3.120	Å	172.4
subl	Cu	10	b	1.00	1	1.278	Å	1.278	Å	4.930	Å	272.4

BOND DISTANCES AND ANGLES

Bond distances and angles are derived from coordinates

No. of distances/angles: 3

Interatomic dist. A-B (Å)	Atom A	Atom B	Atom C	Bond angle A-B-C (°)
2.248	S1	Cu2	Cu3	125.5

Cu(100)-(2x2)-S
29.16.14

Bond Distances and Angles - Continued

Interatomic dist. A-B (Å)	Atom A	Atom B	Atom C	Bond angle A-B-C (°)
2.248	Cu2	S1	Cu3(-1,0)	71.1
2.248	Cu2	S1	Cu5(-1,-1)	110.6

ILLUSTRATION: 28,30

COMMON NAME : Cu(100)-(2x2)-Te
CLASSIFICATION : 29.52.1
TECHNIQUE : LEED
AUTHORS : A. Salwen and J. Rundgren
REFERENCE : Surf. Sci., 53, 523 (1975)

SURFACE TYPE
Substrate : Cu
Crystal face: 100
Temperature : RT*
Bulk lattice: fcc
2D bulk symm: p4m
2D surf symm: p4m

Adsorbate: Te
Coverage : 0.25 (Te/Cu)
Pattern : (2x2)
Matrix : (2.000, 0.000)
(0.000, 2.000)

STRUCTURE TYPE
Atomic adsorption in 4-fold hollow sites

SAMPLE PREPARATION (sample)
Treatment : see D.E. Andersson, S. Andersson, Surf. Sci. 23. 311 (1970)
Crystallinity:
Anal. methods:
Contamination:

COMMENTS

DATA COLLECTION
Technique: LEED
Dataset : I-V spectra: Θ=3°: (0,0) beam (0-140 eV); Θ=0°: (0,1) beam (0-140 eV), (1,1) beam (0-70eV)

THEORY/DATA TREATMENT
Dynamical LEED: 5 different potentials for Te; TFA-HFP pot for Cu (Clementi w.f.); 6 phase shifts; ΘD=343 K

STRUCTURES EXAMINED
Top, bridge, hollow sites with various Te/Cu layer spacings

QUALITY OF EXPERIMENT-THEORY FIT
Visual

2D UNIT CELLS (1 domain observed)

Cell	Ax (Å)	Ay (Å)	Bx (Å)	By (Å)	α (°)	Matrix	Pattern	Cell type
Bulk	2.560	0.000	0.000	2.560	90.0	(1.000, 0.000) (0.000, 1.000)	(1x1)	b: bulk lattice
Surface 1	5.120	0.000	0.000	5.120	90.0	(2.000, 0.000) (0.000, 2.000)	(2x2)	s1: commens. superlattice

3D COORDINATES

Te1: overlayer in 4-fold hollow sites

Dx/Dy in Å, or as a fraction of layer's unit cell vectors; atom 0 at the origin. Epir/subr are bulk repeat vectors.

No. of atoms: 3

Bulk z = 1.810 Å

Region	Chem el.	At. no.	Cell type	Site occ.	Rel. to	Dx ± εx		Dy ± εy		Dz ± εz		Dz/Bz(%) ± εz/Bz
epir		-2					f		f		Å	
subr		-1				-1.280	Å	-1.280	Å	1.810	Å	
ovrl	Te	1	s1	.25	0	0.000	f	0.000	f	0.000	Å	0.0
intf	Cu	2	b	1.00	1	0.500	f	0.500	f	1.700 ± .150	Å	93.9 ± 8.3
subl	Cu	3	b	1.00	2	-0.500	f	-0.500	f	1.810	Å	100.0

BOND DISTANCES AND ANGLES

Bond distances and angles are derived from coordinates

No. of distances/angles: 3

Interatomic dist. A-B (Å)	Atom A	Atom B	Atom C	Bond angle A-B-C (°)
2.483	Te1	Cu2	Cu2(1,0)	121.0
2.560	Cu2	Cu2(1,0)		
2.560	Cu2	Cu3		

COMMON NAME : Cu(100)-(2x2)-Te
CLASSIFICATION : 29.52.2a
TECHNIQUE : SEXAFS
AUTHORS : F. Comin, P.H. Citrin, P. Eisenberger and J.E. Rowe
REFERENCE : Phys. Rev., B26, 7060 (1982)

ILLUSTRATION: 28,30

SURFACE TYPE
Substrate : Cu
Crystal face: 100
Temperature : 623 K
Bulk lattice: fcc
2D bulk symm: p4m
2D surf symm: p4m

Adsorbate: Te
Coverage : 0.25 (Te/Cu)
Pattern : (2x2)
Matrix : (2.000, 0.000)
(0.000, 2.000)

STRUCTURE TYPE
Atomic adsorption in 4-fold hollow sites

SAMPLE PREPARATION (1 sample)
Treatment : evaporation of Te onto previously cleaned surface
Crystallinity: checked by LEED
Anal. methods:
Contamination: checked by LEED and AES

COMMENTS

DATA COLLECTION
Technique: SEXAFS; total yield SEXAFS
Dataset : total yield from Te L(III) edge at different polarization directions

THEORY/DATA TREATMENT
Fourier transform and polarization dependence

STRUCTURES EXAMINED
Top, bridge, hollow and substitutional sites; theoretical surface atom coordination numbers tested against experimentally determined values

2D UNIT CELLS (1 domain observed)

Cell	Ax (Å)	Ay (Å)	Bx (Å)	By (Å)	α (°)	Matrix	Pattern	Cell type
Bulk	2.560	0.000	0.000	2.560	90.0	(1.000, 0.000) (0.000, 1.000)	(1x1)	b: bulk lattice
Surface 1	5.120	0.000	0.000	5.120	90.0	(2.000, 0.000) (0.000, 2.000)	(2x2)	s1: commens. superlattice

3D COORDINATES

Te1: overlayer in 4-fold hollow sites; coordinates are derived from bond distance

Dx/Dy in Å, or as a fraction of layer's unit cell vectors; atom 0 at the origin. Epir/subr are bulk repeat vectors.

No. of atoms: 3

Bulk z = 1.810 Å

Region	Chem el.	At. no.	Cell type	Site occ.	Rel. to	Dx ± εx		Dy ± εy		Dz ± εz		Dz/Bz(%) ± εz/Bz
epir		-2					f		f		Å	
subr		-1				-1.280	Å	-1.280	Å	1.810	Å	
ovrl	Te	1	s1	.25	0	0.000	f	0.000	f	0.000	Å	0.0
intf	Cu	2	b	1.00	1	0.500	f	0.500	f	1.900 ± .040	Å	105.0 ± 2.2
subl	Cu	3	b	1.00	2	-0.500	f	-0.500	f	1.810	Å	100.0

BOND DISTANCES AND ANGLES

No. of distances/angles: 3

Interatomic dist. A-B (Å)	Atom A	Atom B	Atom C	Bond angle A-B-C (°)
2.624	Te1	Cu2	Cu2(1,0)	119.2
2.560	Cu2	Cu2(1,0)		
2.560	Cu2	Cu3		

COMMON NAME : α-Cu(111)-16%Al-(√3x√3)R30°
CLASSIFICATION : 29.13.2
TECHNIQUE : LEED
AUTHORS : R.J. Baird, D.F. Ogletree, M.A. Van Hove and G.A. Somorjai
REFERENCE : Surf. Sci., 165, 345 (1986)

ILLUSTRATION: 133

SURFACE TYPE
Substrate : Cu(111)-16%Al
Crystal face: 111
Temperature : 150 K
Bulk lattice: fcc alloy
2D bulk symm: p3m1
2D surf symm: p31m

Adsorbate:
Coverage :
Pattern : (√3x√3)R30°
Matrix : (2.000,-1.000)
(1.000, 1.000)

STRUCTURE TYPE
(√3x√3)R30° ordered top layer in otherwise random Cu(1-x)Alx fcc alloy: top layer has Cu2Al1 composition in undistorted fcc lattice (Al substitution for Cu); second and deeper layers may have random distributions of Al and Cu atoms

SAMPLE PREPARATION (1 sample)
Treatment : Ar+ sputtering, heating to 670 K, final quenching to 150K
Crystallinity:
Anal. methods:
Contamination: AES: clean

COMMENTS

DATA COLLECTION
Technique: LEED
Dataset : I-V spectra for 7 indep. beams at normal incidence, 15 at Θ=15°; E range 20-200 eV

THEORY/DATA TREATMENT
Dynamical LEED (matrix inversion for mixed layers, RFS): ΘD(Al)=514 K, ΘD(Cu)=335K

STRUCTURES EXAMINED
1) 1/3ML Al in fcc and hcp hollows on Cu(111), Cu-Al spacing 1.707-2.337Å; 2) 1/3ML Al located substitutionally within top layer of Cu(111), with buckling and relaxation; 3) as 2) with hcp termination of Cu; 4) as 2) but with Al substituted in every other layer

QUALITY OF EXPERIMENT-THEORY FIT
RVH=0.218, RZJ=0.300, RPE=0.514

2D UNIT CELLS (1 domain observed)

Cell	Ax (Å)	Ay (Å)	Bx (Å)	By (Å)	α (°)	Matrix	Pattern	Cell type
Bulk	2.540	0.000	1.270	2.200	60.0	(1.000, 0.000) (0.000, 1.000)	(1x1)	b: bulk lattice
Surface 1	3.811	-2.200	3.811	2.200	60.0	(2.000,-1.000) (1.000, 1.000)	(√3x√3)R30°	s1: commens. superlattice

3D COORDINATES

Al1-Cu2-Cu3: planar mixed top layer of fcc(111) lattice

Dx/Dy in Å, or as a fraction of layer's unit cell vectors; atom 0 at the origin. Epir/subr are bulk repeat vectors.

No. of atoms: 5

Bulk z = 2.090 Å

Region	Chem el.	At. no.	Cell type	Site occ.	Rel. to	Dx ± εx	Dy ± εy	Dz ± εz	Dz/Bz(%) ± εz/Bz
epir		-2				f	f	Å	
subr		-1				1.270 Å	-0.733 Å	2.090 Å	
ovrl	Al	1	s1	.33	0	0.000 f	0.000 f	0.000 Å	0.0
ovrl	Cu	2	s1	.33	1	0.333 f	0.333 f	0.000 ± .050 Å	0.0 ± 2.4
ovrl	Cu	3	s1	.33	2	0.333 f	0.333 f	0.000 Å	0.0
intf	Cu	4	b	1.00	3	-1.333 f	-0.333 f	2.090 ± .050 Å	100.0 ± 2.4
subl	Cu	5	b	1.00	4	0.667 f	-0.333 f	2.090 ± .050 Å	100.0 ± 2.4

BOND DISTANCES AND ANGLES

Bond distances and angles are derived from coordinates

No. of distances/angles: 3

Interatomic dist. A-B (Å)	Atom A	Atom B	Atom C	Bond angle A-B-C (°)
2.540	Al1	Cu2		
2.553	Al1	Cu4		
2.540	Cu2	Cu3		

COMMON NAME : Cu0.85Pd0.15(110)-(2x1)
CLASSIFICATION : 29.46.4
TECHNIQUE : LEED
AUTHORS : M. Lindroos, C.J. Barnes, M. Bowker and D.A. King
REFERENCE : Springer Series in Surface Sciences, 24, 287 (1991)

ILLUSTRATION: 141

SURFACE TYPE
Substrate : Cu0.85Pd0.15
Crystal face: 110
Temperature : RT*
Bulk lattice: fcc
2D bulk symm: none
2D surf symm: none

Adsorbate:
Coverage :
Pattern : (2x1) and disorder
Matrix : (0.000, 0.000)
(0.000, 0.000)

STRUCTURE TYPE
Cu-rich top layer (70% Cu); chemically ordered (2x1) 2nd layer (50% Cu), buckled by 0.07Å (Cu outward); pure Cu 3rd layer; all deeper layers disordered with bulk compos. 85%Cu/15%Pd; 1st, 2nd spacings contracted by 4.7%, 0.8%

SAMPLE PREPARATION (1 sample)
Treatment : argon-ion sputtering and annealing to 700 K
Crystallinity:
Anal. methods:
Contamination:

COMMENTS
Here the disordered top and bulk layers are simulated as a (2x3) ordering

DATA COLLECTION
Technique: LEED; video camera
Dataset :

THEORY/DATA TREATMENT
Dynamical LEED (Van Hove/Tong package): substitutional disorder modelled by ATA approximation

STRUCTURES EXAMINED
Variation of number and position of ordered CuPd layers, first two interlayer spacings, buckling within composite CuPd layers and composition within the substitutionally disordered surface layers

QUALITY OF EXPERIMENT-THEORY FIT
RPE=0.3

2D UNIT CELLS (1 domain observed)

Cell	Ax (Å)	Ay (Å)	Bx (Å)	By (Å)	α (°)	Matrix	Pattern	Cell type
Bulk	2.602	0.000	0.000	3.680	90.0	(1.000, 0.000) (0.000, 1.000)	(1x1)	b: bulk lattice
Surface 1	5.204	0.000	0.000	11.040	90.0	(2.000, 0.000) (0.000, 3.000)	disordered	m1: randomly mixed layer

3D COORDINATES

Pd1-Pd2, Cu3-Cu6: disordered planar top layer (70% Cu); Cu7-Pd8: (2x1) ordered buckled second layer (50% Cu); Cu9: third layer (100% Cu); Pd10-Cu15: period. repeating disordered planar bulk layer (85% Cu)

Dx/Dy in Å, or as a fraction of layer's unit cell vectors; atom 0 at the origin. Epir/subr are bulk repeat vectors.

No. of atoms: 15

Bulk z = 1.301 Å

Region	Chem el.	At. no.	Cell type	Site occ.	Rel. to	Dx ± εx		Dy ± εy		Dz ± εz		Dz/Bz(%) ± εz/Bz
epir		-2					f		f		Å	
subr		-1				1.301	Å	1.840	Å	1.301	Å	
intf	Pd	1	m1	.12	0	0.000	f	0.000	f	0.000	Å	0.0
intf	Pd	2	m1	.12	1	0.500	f	0.667	f	0.000	Å	0.0
intf	Cu	3	m1	.12	2	0.000	f	-0.333	f	0.000	Å	0.0
intf	Cu	4	m1	.12	3	0.000	f	-0.333	f	0.000	Å	0.0
intf	Cu	5	m1	.12	4	-0.500	f	0.333	f	0.000	Å	0.0
intf	Cu	6	m1	.12	5	0.000	f	0.333	f	0.000	Å	0.0
intf	Cu	7	s1	.50	1	0.250	f	0.500	f	1.240 ± .040	Å	95.3 ± 3.0
intf	Pd	8	s1	.50	7	0.500	f	0.000	f	0.070 ± .040	Å	5.4 ± 3.1
intf	Cu	9	b	1.00	8	0.500	f	0.500	f	1.290 ± .040	Å	99.1 ± 3.1
subl	Pd	10	m1	.15	9	0.250	f	0.167	f	1.301	Å	100.0
subl	Cu	11	m1	.17	10	0.000	f	0.333	f	0.000	Å	0.0
subl	Cu	12	m1	.17	11	0.000	f	0.333	f	0.000	Å	0.0
subl	Cu	13	m1	.17	12	0.500	f	0.000	f	0.000	Å	0.0
subl	Cu	14	m1	.17	13	0.000	f	-0.333	f	0.000	Å	0.0
subl	Cu	15	m1	.17	14	0.000	f	-0.333	f	0.000	Å	0.0

Cu0.85Pd0.15(110)-(2x1)
29.46.4 BOND DISTANCES AND ANGLES

Bond distances and angles are derived from coordinates

No. of distances/angles: 6

Interatomic dist. A-B (Å)	Atom A	Atom B	Atom C	Bond angle A-B-C (°)
2.602	Pd1	Cu4	Pd1(1,0)	180.0
2.602	Pd1	Cu4	Cu7	59.6
2.572	Pd1	Cu7	Cu3	122.4
2.573	Pd1	Cu9(0,-1)	Pd2(0,-1)	122.4
2.602	Pd2	Cu6(1,0)	Pd2(1,0)	180.0
2.572	Pd2	Cu8	Cu3	91.3

COMMON NAME : Fe(100)-(1x1)
CLASSIFICATION : 26.4
TECHNIQUE : LEED
AUTHORS : K.O. Legg, F. Jona, D.W. Jepsen and P.M. Marcus
REFERENCE : J. Phys., C10, 937 (1977)

ILLUSTRATION: 12

SURFACE TYPE
Substrate : Fe
Crystal face: 100
Temperature : RT*
Bulk lattice: bcc
2D bulk symm: p4m
2D surf symm: p4m

Adsorbate:
Coverage :
Pattern : (1x1)
Matrix : (1.000, 0.000)
(0.000, 1.000)

STRUCTURE TYPE
Bulk termination with contraction of top layer spacing

SAMPLE PREPARATION (1 sample)
Treatment : polish, long Ar+ bombardment at 625-675 K, anneal at 725-775K
Crystallinity: sharp LEED pattern
Anal. methods:
Contamination: AES: O, C, S dominant <2-3% ML

COMMENTS
Non-structural parameters: Voi varied from -2 to -4 eV for energy range 50-150 eV; Vor=-11.5 eV;
not ZJ R-factor, but: weighted R-factor from sum on beams, taking into account both peak positions and intensities

DATA COLLECTION
Technique: LEED
Dataset : I-V spectra for 16 beams at 3 angles of incidence: Θ=0, 10, 20° at φ=0°, E range 50-150 eV

THEORY/DATA TREATMENT
LEED KKR: 8 phase shifts and 38 beams; self consistent Hartree-Fock-Slater Fe potential; rms = 0.115Å

STRUCTURES EXAMINED
Varied first interlayer spacing from 1.35 to 1.48Å

QUALITY OF EXPERIMENT-THEORY FIT
R=0.156 (see comments)

2D UNIT CELLS (1 domain observed)

Cell	Ax (Å)	Ay (Å)	Bx (Å)	By (Å)	α (°)	Matrix	Pattern	Cell type
Bulk	2.866	0.000	0.000	2.866	90.0	(1.000, 0.000) (0.000, 1.000)	(1x1)	b: bulk lattice
Surface 1	2.866	0.000	0.000	2.866	90.0	(1.000, 0.000) (0.000, 1.000)	(1x1)	s1: commens. superlattice

3D COORDINATES

Dx/Dy in Å, or as a fraction of layer's unit cell vectors; atom 0 at the origin. Epir/subr are bulk repeat vectors.

No. of atoms: 3

Bulk z = 1.433 Å

Reg ion	Chem el.	At. no.	Cell type	Site occ.	Rel. to	Dx ± εx	Dy ± εy	Dz ± εz	Dz/Bz(%) ± εz/Bz
epir		-2				f	f	Å	
subr		-1				-1.433 Å	-1.433 Å	1.433 Å	
intf	Fe	1	b	1.00	0	0.000 f	0.000 f	0.000 Å	0.0
intf	Fe	2	b	1.00	1	0.500 f	0.500 f	1.410 ± .040 Å	98.4 ± 2.8
subl	Fe	3	b	1.00	2	-0.500 f	-0.500 f	1.433 Å	100.0

BOND DISTANCES AND ANGLES

Bond distances and angles are derived from coordinates

No. of distances/angles: 6

Interatomic dist. A-B (Å)	Atom A	Atom B	Atom C	Bond angle A-B-C (°)
2.469	Fe1	Fe2	Fe1(1,0)	71.0
2.469	Fe1	Fe2	Fe2(1,0)	125.5
2.469	Fe1	Fe2	Fe3	70.1
2.482	Fe2	Fe3	Fe1	54.7
2.482	Fe2	Fe3	Fe2(-1,0)	70.5
2.482	Fe2	Fe3	Fe3(1,0)	54.7

ILLUSTRATION: 12

COMMON NAME : Fe(100)-(1x1)
CLASSIFICATION : 26.18
TECHNIQUE : LEED
AUTHORS : Z.Q. Wang, Y.S. Li, F. Jona and P.M. Marcus
REFERENCE : Solid State Commun., 61, 623 (1987)

SURFACE TYPE
Substrate : Fe
Crystal face: 100
Temperature : RT*
Bulk lattice: bcc
2D bulk symm: p4m
2D surf symm: p4m

Adsorbate:
Coverage :
Pattern : (1x1)
Matrix : (1.000, 0.000)
(0.000, 1.000)

STRUCTURE TYPE
Bulk termination with multilayer relaxation perpendicular to surface

SAMPLE PREPARATION (1 sample)
Treatment : Ar sputtering at 500C, followed by annealing at 500C
Crystallinity:
Anal. methods:
Contamination: monitored with AES

COMMENTS

DATA COLLECTION
Technique: LEED
Dataset : I-V curves at normal incidence (4beams) and at off-normal incidence (4 beams) in the range 50 to 300 eV

THEORY/DATA TREATMENT
Dynamical LEED (CHANGE program): 6 phase shifts, 49 beams

STRUCTURES EXAMINED
Variation of top two interlayer spacings

QUALITY OF EXPERIMENT-THEORY FIT
RZJ=0.07

2D UNIT CELLS (1 domain observed)

Cell	Ax (Å)	Ay (Å)	Bx (Å)	By (Å)	α (°)	Matrix	Pattern	Cell type
Bulk	2.866	0.000	0.000	2.866	90.0	(1.000, 0.000) (0.000, 1.000)	(1x1)	b: bulk lattice
Surface 1	2.866	0.000	0.000	2.866	90.0	(1.000, 0.000) (0.000, 1.000)	(1x1)	s1: commens. superlattice

3D COORDINATES

Dx/Dy in Å, or as a fraction of layer's unit cell vectors; atom 0 at the origin. Epir/subr are bulk repeat vectors.

No. of atoms: 4

Bulk z = 1.433 Å

Reg ion	Chem el.	At. no.	Cell type	Site occ.	Rel. to	Dx ± εx		Dy ± εy		Dz ± εz		Dz/Bz(%) ± εz/Bz
							f		f		Å	
epir		-2				1.433	Å	1.433	Å	1.433	Å	
subr		-1				0.000	f	0.000	f	0.000	Å	0.0
intf	Fe	1	b	1.00	0	0.500	f	0.500	f	1.363 ± .030	Å	95.1 ± 2.1
intf	Fe	2	b	1.00	1	-0.500	f	-0.500	f	1.503 ± .030	Å	104.9 ± 2.1
intf	Fe	3	b	1.00	2	0.500	f	0.500	f	1.433	Å	100.0
subl	Fe	4	b	1.00	3							

BOND DISTANCES AND ANGLES

Bond distances and angles are derived from coordinates

No. of distances/angles: 3

Interatomic dist. A-B (Å)	Atom A	Atom B	Atom C	Bond angle A-B-C (°)
2.866	Fe1	Fe1(1,0)		
2.442	Fe1	Fe2	Fe3	70.5
2.523	Fe2	Fe3	Fe4	71.8

ILLUSTRATION: 12

COMMON NAME : Fe(100)-(1x1)
CLASSIFICATION : 26.18a
TECHNIQUE : MEIS
AUTHORS : R.L. Headrick, P. Konarski, S.M. Yalisove and W.R. Graham
REFERENCE : Phys. Rev., B39, 5713 (1989)

SURFACE TYPE
Substrate : Fe
Crystal face: 100
Temperature : RT*
Bulk lattice: bcc
2D bulk symm: p4m
2D surf symm: p4m

Adsorbate:
Coverage :
Pattern : (1x1)
Matrix : (1.000, 0.000)
(0.000, 1.000)

STRUCTURE TYPE
Bulk termination with multilayer relaxation perpendicular to surface

SAMPLE PREPARATION (1 sample)
Treatment : Ar sputtering, repolish, recleaning with sputter-anneals
Crystallinity:
Anal. methods:
Contamination: AES: trace of O

COMMENTS

DATA COLLECTION
Technique: MEIS
Dataset : blocking curves in two geometries

THEORY/DATA TREATMENT
Monte Carlo simulations, incl. fit of ΘD (exponentially increasing from 255 K at surface to 420K in bulk)

STRUCTURES EXAMINED
Variation of top two interlayer spacings

QUALITY OF EXPERIMENT-THEORY FIT
R=0.30

2D UNIT CELLS (1 domain observed)

Cell	Ax (Å)	Ay (Å)	Bx (Å)	By (Å)	α (°)	Matrix	Pattern	Cell type
Bulk	2.860	0.000	0.000	2.860	90.0	(1.000, 0.000) (0.000, 1.000)	(1x1)	b: bulk lattice
Surface 1	2.860	0.000	0.000	2.860	90.0	(1.000, 0.000) (0.000, 1.000)	(1x1)	s1: commens. superlattice

3D COORDINATES

Dx/Dy in Å, or as a fraction of layer's unit cell vectors; atom 0 at the origin. Epir/subr are bulk repeat vectors.

No. of atoms: 4

Bulk z = 1.430 Å

Reg ion	Chem el.	At. no.	Cell type	Site occ.	Rel. to	Dx ± εx		Dy ± εy		Dz ± εz		Dz/Bz(%) ± εz/Bz
epir		-2					f		f		Å	
subr		-1				1.430	Å	1.430	Å	1.430	Å	
intf	Fe	1	b	1.00	0	0.000	f	0.000	f	0.000	Å	0.0
intf	Fe	2	b	1.00	1	0.500	f	0.500	f	1.380 ± .040	Å	96.5 ± 2.8
intf	Fe	3	b	1.00	2	-0.500	f	-0.500	f	1.480 ± .040	Å	103.5 ± 2.8
subl	Fe	4	b	1.00	3	0.500	f	0.500	f	1.430	Å	100.0

BOND DISTANCES AND ANGLES

Bond distances and angles are derived from coordinates

No. of distances/angles: 3

Interatomic dist. A-B (Å)	Atom A	Atom B	Atom C	Bond angle A-B-C (°)
2.860	Fe1	Fe1(1,0)	Fe2	54.3
2.448	Fe1	Fe2	Fe3	70.5
2.506	Fe2	Fe3	Fe4	71.5

COMMON NAME : Fe(100)-(1x1) epitaxial on Cu(100)
CLASSIFICATION : 29.26.7
TECHNIQUE : LEED
AUTHORS : S.H. Lu, J. Quinn, D. Tian, F. Jona and P.M. Marcus
REFERENCE : Surf. Sci., 209, 364 (1989)

ILLUSTRATION: 83

SURFACE TYPE
Substrate : Fe
Crystal face: 100
Temperature : RT
Bulk lattice: bct
2D bulk symm: p4m
2D surf symm: p4m

Adsorbate:
Coverage :
Pattern : (1x1)
Matrix : (1.000, 0.000)
(0.000, 1.000)

STRUCTURE TYPE
Epitaxial (1x1) multilayer (6 to 18 layers) grown on Cu(100) and Cu alloys, with lateral lattice constant of Cu(100); forms body-centered tetragonal lattice, related to fcc by 2% uniaxial contraction in growth direction

SAMPLE PREPARATION (3 sample)
Treatment : cycles of Ar+ bomb. and 600C anneals; Fe vapor-deposited
Crystallinity: LEED: some disorder and defects
Anal. methods: AES
Contamination: AES: no contaminants

COMMENTS
1-, 2- and 3-layer films gave poor experiment-theory fits, suggesting initial non-layer-by-layer growth

DATA COLLECTION
Technique: LEED; TV camera-microcomputer system
Dataset : IV spectra for 4 non-equiv. beams (10,11,20,21) at normal incidence, 5 (00,10,01,11,0-2) at Θ=10°,φ=0°

THEORY/DATA TREATMENT
Dynamical LEED (CHANGE program): 57 beams, 8 phase shifts; Voi=-4 eV; Cu or alloy substrate below 12 Fe layers

STRUCTURES EXAMINED
Fcc with variable 'bulk' spacing of 1.74-1.81Å perp. to surface (giving bct lattice), and variable top 2 interlayer spacings of 1.81-1.89Å and 1.73-1.89Å

QUALITY OF EXPERIMENT-THEORY FIT
RZJ=0.05, 0.13 for Θ=0°, Θ=10°

2D UNIT CELLS (1 domain observed)

Cell	Ax (Å)	Ay (Å)	Bx (Å)	By (Å)	α (°)	Matrix	Pattern	Cell type
Bulk	2.553	0.000	0.000	2.553	90.0	(1.000, 0.000) (0.000, 1.000)	(1x1)	b: bulk lattice
Surface 1	2.553	0.000	0.000	2.553	90.0	(1.000, 0.000) (0.000, 1.000)	(1x1)	s1: commens. superlattice

3D COORDINATES

Dx/Dy in Å, or as a fraction of layer's unit cell vectors; atom 0 at the origin. Epir/subr are bulk repeat vectors.

No. of atoms: 4

Bulk z = 1.770 Å

Region	Chem el.	At. no.	Cell type	Site occ.	Rel. to	Dx ± εx		Dy ± εy		Dz ± εz		Dz/Bz(%) ± εz/Bz
epir		-2					f		f		Å	
subr		-1				1.277	Å	1.277	Å	1.770	Å	
intf	Fe	1	b	1.00	0	0.000	f	0.000	f	0.000	Å	0.0
intf	Fe	2	b	1.00	1	0.500	f	0.500	f	1.840 ± .030	Å	104.0 ± 1.7
intf	Fe	3	b	1.00	2	-0.500	f	-0.500	f	1.760 ± .030	Å	99.4 ± 1.7
subl	Fe	4	b	1.00	3	0.500	f	0.500	f	1.770 ± .030	Å	100.0 ± 1.7

BOND DISTANCES AND ANGLES

Bond distances and angles are derived from coordinates

No. of distances/angles: 3

Interatomic dist. A-B (Å)	Atom A	Atom B	Atom C	Bond angle A-B-C (°)
2.553	Fe1	Fe1(1,0)	Fe2	60.3
2.578	Fe1	Fe2	Fe3	89.8
2.521	Fe2	Fe3	Fe4	88.7

ILLUSTRATION: 83

COMMON NAME : Fe(100)-(1x1) epitaxial on Cu(100)
CLASSIFICATION : 29.26.8
TECHNIQUE : LEED
AUTHORS : Y. Darici, J. Marcano, H. Min and P. Montano
REFERENCE : Surf. Sci., 217, 521 (1989)

SURFACE TYPE
Substrate : Fe
Crystal face: 100
Temperature : 190C
Bulk lattice: bct
2D bulk symm: p4m
2D surf symm: p4m

Adsorbate:
Coverage :
Pattern : (1x1)
Matrix : (1.000, 0.000)
(0.000, 1.000)

STRUCTURE TYPE
Epitaxial (1x1) multilayer (5 layers) grown on Cu(100), with lateral lattice constant of Cu(100); forms body-centered tetragonal lattice, related to fcc by 2% uniaxial contraction in growth direction

SAMPLE PREPARATION (1 sample)
Treatment : cycles of Ar+ bomb. and 550C anneals; Fe vapor-deposited
Crystallinity: clear and sharp LEED spots
Anal. methods: AES
Contamination: AES: no contaminants

COMMENTS
10-layer film at RT gave same result, except top spacing reduced from 1.83 to 1.81Å;
1- to 4-layer films gave poor experiment-theory fits, suggesting initial non-layer-by-layer growth and/or interdiffusion

DATA COLLECTION
Technique: LEED
Dataset : IV spectra for 3 non-equiv. beams (10,11,20) at normal incidence, 1 (00) at off-normal incidence

THEORY/DATA TREATMENT
Dynamical LEED (RFS): 57 beams, 8 phase shifts; Vor, Voi fit; ΘD=380 K(surf), 550K(bulk) fit

STRUCTURES EXAMINED
Fcc with variable 'bulk' spacing perp. to surface (giving bct lattice), and variable top 2 interlayer spacings

QUALITY OF EXPERIMENT-THEORY FIT
RZJ<0.1

2D UNIT CELLS (1 domain observed)

Cell	Ax (Å)	Ay (Å)	Bx (Å)	By (Å)	α (°)	Matrix	Pattern	Cell type
Bulk	2.553	0.000	0.000	2.553	90.0	(1.000, 0.000) (0.000, 1.000)	(1x1)	b: bulk lattice
Surface 1	2.553	0.000	0.000	2.553	90.0	(1.000, 0.000) (0.000, 1.000)	(1x1)	s1: commens. superlattice

3D COORDINATES

Dx/Dy in Å, or as a fraction of layer's unit cell vectors; atom 0 at the origin. Epir/subr are bulk repeat vectors.

No. of atoms: 4

Bulk z = 1.780 Å

Region	Chem el.	At. no.	Cell type	Site occ.	Rel. to	Dx ± εx		Dy ± εy		Dz ± εz		Dz/Bz(%) ± εz/Bz
epir		-2					f		f		Å	
subr		-1				1.277	Å	1.277	Å	1.780	Å	
intf	Fe	1	b	1.00	0	0.000	f	0.000	f	0.000	Å	0.0
intf	Fe	2	b	1.00	1	0.500	f	0.500	f	1.830 ± .020	Å	102.8 ± 1.1
intf	Fe	3	b	1.00	2	-0.500	f	-0.500	f	1.780 ± .020	Å	100.0 ± 1.1
subl	Fe	4	b	1.00	3	0.500	f	0.500	f	1.780 ± .020	Å	100.0 ± 1.1

BOND DISTANCES AND ANGLES

Bond distances and angles are derived from coordinates

No. of distances/angles: 3

Interatomic dist. A-B (Å)	Atom A	Atom B	Atom C	Bond angle A-B-C (°)
2.553	Fe1	Fe1(1,0)	Fe2	60.2
2.571	Fe1	Fe2	Fe3	90.0
2.535	Fe2	Fe3	Fe4	89.2

COMMON NAME : Fe(100)-(1x1) epitaxial on Ni(100)
CLASSIFICATION : 28.26.2c
TECHNIQUE : LEED
AUTHORS : S.H. Lu, Z.Q. Wang, D. Tian, Y.S. Li, F. Jona and P.M. Marcus
REFERENCE : Surf. Sci., 221, 35 (1989)

ILLUSTRATION: 83

SURFACE TYPE
Substrate : Fe
Crystal face: 100
Temperature : RT
Bulk lattice: bct
2D bulk symm: p4m
2D surf symm: p4m

Adsorbate:
Coverage :
Pattern : (1x1)
Matrix : (1.000, 0.000)
(0.000, 1.000)

STRUCTURE TYPE
Epitaxial (1x1) multilayer (6 to 25 layers) grown on Ni(100), with lateral lattice constant of Ni(100); forms body-centered tetragonal lattice, related to fcc by 10% uniaxial expansion in growth direction

SAMPLE PREPARATION (3 sample)
Treatment : cycles of Ar+ bomb. and 800-1000C anneals; Fe vapor-depos.
Crystallinity: LEED: weak extra spots
Anal. methods: AES
Contamination: AES: no contaminants

COMMENTS
LEED pattern shows weak extra spots, perhaps due to small bcc Fe(110) domains

DATA COLLECTION
Technique: LEED; TV camera-microcomputer system
Dataset : IV spectra for 4 beams (10,11,20,21) at normal incidence; cumul. E range 820 eV

THEORY/DATA TREATMENT
Dynamical LEED (CHANGE program): 49 beams, 6 phase shifts; E-dep. Vor and Voi

STRUCTURES EXAMINED
Fcc with variable 'bulk' spacing of 1.66-1.96Å perp. to surface (giving bct lattice), varying top 2 interlayer spacings by -0.05 to +0.10Å and by -0.12 to +0.12Å from bulk Ni spacing of 1.76Å

QUALITY OF EXPERIMENT-THEORY FIT
RZJ=0.09

2D UNIT CELLS (1 domain observed)

Cell	Ax (Å)	Ay (Å)	Bx (Å)	By (Å)	α (°)	Matrix	Pattern	Cell type
Bulk	2.489	0.000	0.000	2.489	90.0	(1.000, 0.000) (0.000, 1.000)	(1x1)	b: bulk lattice
Surface 1	2.489	0.000	0.000	2.489	90.0	(1.000, 0.000) (0.000, 1.000)	(1x1)	s1: commens. superlattice

3D COORDINATES

Dx/Dy in Å, or as a fraction of layer's unit cell vectors; atom 0 at the origin. Epir/subr are bulk repeat vectors.

No. of atoms: 4

Bulk z = 1.940 Å

Reg ion	Chem el.	At. no.	Cell type	Site occ.	Rel. to	Dx ± εx		Dy ± εy		Dz ± εz		Dz/Bz(%) ± εz/Bz
epir		-2					f		f		Å	
subr		-1				1.245	Å	1.245	Å	1.940	Å	
intf	Fe	1	b	1.00	0	0.000	f	0.000	f	0.000	Å	0.0
intf	Fe	2	b	1.00	1	0.500	f	0.500	f	2.040 ± .050	Å	105.0 ± 2.6
intf	Fe	3	b	1.00	2	-0.500	f	-0.500	f	1.955 ± .050	Å	100.8 ± 2.6
subl	Fe	4	b	1.00	3	0.500	f	0.500	f	1.940 ± .050	Å	100.0 ± 2.6

BOND DISTANCES AND ANGLES

Bond distances and angles are derived from coordinates

No. of distances/angles: 3

Interatomic dist. A-B (Å)	Atom A	Atom B	Atom C	Bond angle A-B-C (°)
2.489	Fe1	Fe1(1,0)	Fe2	62.5
2.694	Fe1	Fe2	Fe3	97.2
2.631	Fe2	Fe3	Fe4	95.8

ILLUSTRATION: 11

COMMON NAME : Fe(110)-(1x1)
CLASSIFICATION : 26.8
TECHNIQUE : LEED
AUTHORS : H.D. Shih, F. Jona, U. Bardi and P.M. Marcus
REFERENCE : J. Phys., C13, 3801 (1980)

SURFACE TYPE
Substrate : Fe
Crystal face: 110
Temperature : 300 K
Bulk lattice: bcc
2D bulk symm: cmm
2D surf symm: cmm

Adsorbate:
Coverage :
Pattern : (1x1)
Matrix : (1.000, 0.000)
(0.000, 1.000)

STRUCTURE TYPE
Bulk termination with possible slight expansion of topmost interlayer spacing

SAMPLE PREPARATION (1 sample)
Treatment : polished, Ar+ sputtered and 1123 K annealed
Crystallinity:
Anal. methods:
Contamination:

COMMENTS
Mean vibrational amplitude chosen as $\sqrt{<u*u>}$=0.115 Å; inner potential assumed independent of energy, with optimization yielding Vor=-11.5±0.4 eV

DATA COLLECTION
Technique: LEED
Dataset : I-V data: 2 sets of 7 and 9 beams at Θ=0°, 6 at (Θ,π)= (9,35.26)°, 5 at (26,-54.74)°, 7 at (26,-54.74)°

THEORY/DATA TREATMENT
Dynamical LEED: 8 phase shifts and 31 beams

STRUCTURES EXAMINED
Initially truncated bulk, and structure with top-layer atoms rolled into quasi 3-fold positions (which gave visibly worse agreement, therefore eliminated); then 1st interlayer spacing varied from 1.85 to 2.15Å in steps of 0.05 Å

QUALITY OF EXPERIMENT-THEORY FIT
RZJ=0.07

2D UNIT CELLS (1 domain observed)

Cell	Ax (Å)	Ay (Å)	Bx (Å)	By (Å)	α (°)	Matrix	Pattern	Cell type
Bulk	2.486	0.000	.829	2.343	70.5	(1.000, 0.000) (0.000, 1.000)	(1x1)	b: bulk lattice
Surface 1	2.486	0.000	.829	2.343	70.5	(1.000, 0.000) (0.000, 1.000)	(1x1)	s1: commens. superlattice

3D COORDINATES

Dx/Dy in Å, or as a fraction of layer's unit cell vectors; atom 0 at the origin. Epir/subr are bulk repeat vectors.

Bulk z = 2.029 Å

No. of atoms: 3

Reg ion	Chem el.	At. no.	Cell type	Site occ.	Rel. to	Dx ± εx		Dy ± εy		Dz ± εz		Dz/Bz(%) ± εz/Bz
epir		-2				1.657	f	1.172	f	2.029	Å	
subr		-1				0.000	Å	0.000	Å	0.000	Å	
intf	Fe	1	b	1.00	0	0.500	f	0.500	f	2.040 ± .040	Å	0.0
intf	Fe	2	b	1.00	1	-0.500	f	-0.500	f	2.029	Å	100.5 ± 2.0
subl	Fe	3	b	1.00	2		f		f		Å	100.0

BOND DISTANCES AND ANGLES

Bond distances and angles are derived from coordinates

No. of distances/angles: 4

Interatomic dist. A-B (Å)	Atom A	Atom B	Atom C	Bond angle A-B-C (°)
2.486	Fe1	Fe1(1,0)	Fe1(1,-1)	70.5
2.486	Fe1	Fe1(1,0)	Fe2	70.6
2.494	Fe1	Fe2(0,-1)	Fe3	109.6
2.486	Fe2	Fe2(0,-1)	Fe3(1,0)	54.7

COMMON NAME : Fe(111)-(1x1)
CLASSIFICATION : 26.17
TECHNIQUE : LEED
AUTHORS : J. Sokolov, F. Jona and P.M. Marcus
REFERENCE : Phys. Rev., B33, 1397 (1986)

ILLUSTRATION: 15

SURFACE TYPE
Substrate : Fe
Crystal face: 111
Temperature : RT*
Bulk lattice: bcc
2D bulk symm: p3m1
2D surf symm: p3m1

Adsorbate:
Coverage :
Pattern : (1x1)
Matrix : (1.000, 0.000)
(0.000, 1.000)

STRUCTURE TYPE
Bulk termination with multilayer relaxation perpendicular to surface

SAMPLE PREPARATION (1 sample)
Treatment : Ar+ ion bombardment followed by annealing at 850C
Crystallinity:
Anal. methods:
Contamination: monitored by AES

COMMENTS

DATA COLLECTION
Technique: LEED
Dataset : 14 non degenerate LEED spectra at two angles of incidence, 50<E<180 eV

THEORY/DATA TREATMENT
Dynamical LEED (layer KKR): 8 phase shifts from Hartree-Fock-Slater self-consistent potential; Voi=4 eV; rms 0.115Å

STRUCTURES EXAMINED
Varied first four interlayer spacings

QUALITY OF EXPERIMENT-THEORY FIT
RZJ=0.131

2D UNIT CELLS (1 domain observed)

Cell	Ax (Å)	Ay (Å)	Bx (Å)	By (Å)	α (°)	Matrix	Pattern	Cell type
Bulk	4.054	0.000	2.027	3.511	60.0	(1.000, 0.000) (0.000, 1.000)	(1x1)	b: bulk lattice
Surface 1	4.054	0.000	2.027	3.511	60.0	(1.000, 0.000) (0.000, 1.000)	(1x1)	s1: commens. superlattice

3D COORDINATES

Dx/Dy in Å, or as a fraction of layer's unit cell vectors; atom 0 at the origin. Epir/subr are bulk repeat vectors.

No. of atoms: 6

Bulk z = .827 Å

Region	Chem el.	At. no.	Cell type	Site occ.	Rel. to	Dx ± εx		Dy ± εy		Dz ± εz		Dz/Bz(%) ± εz/Bz
epir		-2					f		f		Å	
subr		-1				-2.027	Å	-1.170	Å	0.827	Å	
intf	Fe	1	b	1.00	0	0.000	f	0.000	f	0.000	Å	0.0
intf	Fe	2	b	1.00	1	0.667	f	0.667	f	0.690 ± .025	Å	83.4 ± 3.0
intf	Fe	3	b	1.00	2	-0.333	f	-0.333	f	0.750 ± .025	Å	90.7 ± 3.0
intf	Fe	4	b	1.00	3	-0.333	f	-0.333	f	0.860 ± .030	Å	104.0 ± 3.6
intf	Fe	5	b	1.00	4	0.667	f	0.667	f	0.810 ± .030	Å	97.9 ± 3.6
subl	Fe	6	b	1.00	5	-0.333	f	-0.333	f	0.827	Å	100.0

BOND DISTANCES AND ANGLES

Bond distances and angles are derived from coordinates

No. of distances/angles: 11

Interatomic dist. A-B (Å)	Atom A	Atom B	Atom C	Bond angle A-B-C (°)
2.440	Fe1	Fe2(-1,0)	Fe1(0,1)	112.3
2.458	Fe2	Fe3	Fe2(-1,0)	111.1
2.458	Fe3	Fe2	Fe4(1,0)	55.6
2.440	Fe1	Fe2(-1,0)	Fe3	68.3
2.440	Fe1	Fe2(-1,0)	Fe4	51.0
2.300	Fe1	Fe4	Fe2(-1,0)	55.5

Fe(111)-(1x1)
26.17

Bond Distances and Angles - Continued

Interatomic dist. A-B (Å)	Atom A	Atom B	Atom C	Bond angle A-B-C (°)
2.300	Fe1	Fe4	Fe3	69.8
2.440	Fe2	Fe1(1,0)	Fe2(0,-1)	112.3
2.440	Fe2	Fe1(1,0)	Fe3	56.2
2.440	Fe2	Fe1(1,0)	Fe4(1,0)	73.6
2.458	Fe2	Fe3	Fe1	130.6

COMMON NAME : Fe(111)-(1x1)
CLASSIFICATION : 26.9
TECHNIQUE : LEED
AUTHORS : H.D. Shih, F. Jona, D.W. Jepsen and P.M. Marcus
REFERENCE : Surf. Sci., 104, 39 (1981)

ILLUSTRATION: 15

SURFACE TYPE
Substrate : Fe
Crystal face: 111
Temperature : 300 K
Bulk lattice: bcc
2D bulk symm: p3m1
2D surf symm: p3m1

Adsorbate:
Coverage :
Pattern : (1x1)
Matrix : (1.000, 0.000)
(0.000, 1.000)

STRUCTURE TYPE
Bulk termination with top spacing contraction

SAMPLE PREPARATION (1 sample)
Treatment : polished, cleaned by Ar+ sputtering and annealed to 1125 K
Crystallinity:
Anal. methods:
Contamination: AES: <2-3% ML of O, C, S present

COMMENTS

DATA COLLECTION
Technique: LEED
Dataset : 2 sets of LEED IV curves: 5 beams at normal incidence and 9 beams at Θ=9°, π=0°; E<=180 eV

THEORY/DATA TREATMENT
Dynamical LEED (program THIN): Hartree-Fock-Slater potential for Fe; 8 phase shifts; 55 beams; Vor=-11.1±0.9 eV, Voi=-4eV

STRUCTURES EXAMINED
Varied spacing between 1st 2 layers from 0.577 to 0.827 in steps of 0.05Å. (preliminary tests revealed a contraction of the 1st layer spacing)

QUALITY OF EXPERIMENT-THEORY FIT
RZJ=0.15

2D UNIT CELLS (1 domain observed)

Cell	Ax (Å)	Ay (Å)	Bx (Å)	By (Å)	α (°)	Matrix	Pattern	Cell type
Bulk	4.054	0.000	-2.027	3.511	120.0	(1.000, 0.000) (0.000, 1.000)	(1x1)	b: bulk lattice
Surface 1	4.054	0.000	-2.027	3.511	120.0	(1.000, 0.000) (0.000, 1.000)	(1x1)	s1: commens. superlattice

3D COORDINATES

Dx/Dy in Å, or as a fraction of layer's unit cell vectors; atom 0 at the origin. Epir/subr are bulk repeat vectors.

No. of atoms: 3

Bulk z = .827 Å

Reg ion	Chem el.	At. no.	Cell type	Site occ.	Rel. to	Dx ± εx		Dy ± εy		Dz ± εz		Dz/Bz(%) ± εz/Bz
epir		-2					f		f		Å	
subr		-1				-2.027	Å	1.170	Å	0.827	Å	
intf	Fe	1	b	1.00	0	0.000	f	0.000	f	0.000	Å	0.0
intf	Fe	2	b	1.00	1	0.667	f	0.333	f	0.700 ± .030	Å	84.6 ± 3.6
subl	Fe	3	b	1.00	2	-0.333	f	0.333	f	0.827	Å	100.0

BOND DISTANCES AND ANGLES

Bond distances and angles are derived from coordinates

No. of distances/angles: 8

Interatomic dist. A-B (Å)	Atom A	Atom B	Atom C	Bond angle A-B-C (°)
2.443	Fe1	Fe2	Fe1(1,0)	112.1
2.443	Fe1	Fe2	Fe3	69.1
2.443	Fe2	Fe1	Fe2(-1,0)	112.1
2.443	Fe2	Fe1	Fe3	56.1
2.482	Fe2	Fe3	Fe1	54.8
2.482	Fe2	Fe3	Fe2(0,1)	109.5
2.482	Fe3	Fe2	Fe1	69.1

Fe(111)-(1x1)
26.9

Bond Distances and Angles - Continued

Interatomic dist. A-B (Å)	Atom A	Atom B	Atom C	Bond angle A-B-C (°)
2.482	Fe3	Fe2	Fe3(1,0)	109.5

COMMON NAME : Fe(210)-(1x1)
CLASSIFICATION : 26.16
TECHNIQUE : LEED
AUTHORS : J. Sokolov, F. Jona and P.M. Marcus
REFERENCE : Phys. Rev., B31, 1929 (1985)

ILLUSTRATION: 16

SURFACE TYPE
Substrate : Fe
Crystal face: 210
Temperature : 298 K
Bulk lattice: bcc
2D bulk symm: pm
2D surf symm: pm

Adsorbate:
Coverage :
Pattern : (1x1)
Matrix : (1.000, 0.000)
(0.000, 1.000)

STRUCTURE TYPE
Bulk termination with multilayer relaxation perpendicular and parallel to surface

SAMPLE PREPARATION (1 sample)
Treatment : polish, Ar+ sputt., 850C anneal, cycles of sputt./anneal
Crystallinity:
Anal. methods:
Contamination: AES: S,C,O are main impurities

COMMENTS

DATA COLLECTION
Technique: LEED; spot photometer
Dataset : 35 I-V curves (31 non-degenerate) at normal incidence and at 13.2° off normal

THEORY/DATA TREATMENT
Dynamical LEED: layer KKR with composite layers; 8 phase shifts; Voi=4ev; rms vibr ampl 0.115Å

STRUCTURES EXAMINED
Relaxations and registries in top four interlayer spacings, maintaining mirror plane

QUALITY OF EXPERIMENT-THEORY FIT
RZJ=0.103

2D UNIT CELLS (1 domain observed)

Cell	Ax (Å)	Ay (Å)	Bx (Å)	By (Å)	α (°)	Matrix	Pattern	Cell type
Bulk	2.870	0.000	0.000	6.418	90.0	(1.000, 0.000) (0.000, 1.000)	(1x1)	b: bulk lattice
Surface 1	2.870	0.000	0.000	6.418	90.0	(1.000, 0.000) (0.000, 1.000)	(1x1)	s1: commens. superlattice

3D COORDINATES

Fe1-Fe4 are relaxed

Dx/Dy in Å, or as a fraction of layer's unit cell vectors; atom 0 at the origin. Epir/subr are bulk repeat vectors.

No. of atoms: 6

Bulk z = .642 Å

Region	Chem el.	At. no.	Cell type	Site occ.	Rel. to	Dx ± εx		Dy ± εy		Dz ± εz		Dz/Bz(%) ± εz/Bz
epir		-2					f		f		Å	
subr		-1				1.435	Å	1.925	Å	0.642	Å	
intf	Fe	1	b	1.00	0	0.000	f	0.000	f	0.000	Å	0.0
intf	Fe	2	b	1.00	1	0.500	f	0.320 ± .008	f	0.500 ± .030	Å	77.9 ± 4.7
intf	Fe	3	b	1.00	2	-0.500	f	0.310 ± .008	f	0.570 ± .030	Å	88.8 ± 4.7
intf	Fe	4	b	1.00	3	0.500	f	0.290 ± .008	f	0.610 ± .030	Å	95.1 ± 4.7
intf	Fe	5	b	1.00	4	-0.500	f	-0.680 ± .008	f	0.642 ± .030	Å	100.0 ± 4.7
subl	Fe	6	b	1.00	5	0.500	f	0.300	f	0.642	Å	100.0

BOND DISTANCES AND ANGLES

Bond distances and angles are derived from coordinates

No. of distances/angles: 12

Interatomic dist. A-B (Å)	Atom A	Atom B	Atom C	Bond angle A-B-C (°)
2.870	Fe1	Fe1(1,0)	Fe2	55.8
2.518	Fe2	Fe3	Fe3(1,0)	55.3
2.518	Fe2	Fe3	Fe4	109.0
2.428	Fe3	Fe4	Fe4(1,0)	126.2
2.870	Fe1	Fe1(1,0)	Fe4(0,-1)	50.8

Fe(210)-(1x1)
26.16

Bond Distances and Angles - Continued

Interatomic dist. A-B (Å)	Atom A	Atom B	Atom C	Bond angle A-B-C (°)
2.555	Fe1	Fe2	Fe3	111.1
2.555	Fe1	Fe2	Fe4(0,-1)	49.6
2.604	Fe1	Fe3(0,-1)	Fe2(0,-1)	128.9
2.604	Fe1	Fe3(0,-1)	Fe4(0,-1)	53.4
2.268	Fe1	Fe4(0,-1)	Fe4(1,-1)	129.2
2.518	Fe2	Fe3	Fe1(0,1)	128.9
2.518	Fe2	Fe3	Fe2(-1,0)	69.5

COMMON NAME : Fe(211)-(1x1)
CLASSIFICATION : 26.14
TECHNIQUE : LEED
AUTHORS : J. Sokolov, H.D. Shih, U. Bardi, F. Jona and P.M. Marcus
REFERENCE : J. Phys., C17, 371 (1984)

ILLUSTRATION: 17

SURFACE TYPE
Substrate : Fe
Crystal face: 211
Temperature : 300 K
Bulk lattice: bcc
2D bulk symm: pm
2D surf symm: pm

Adsorbate:
Coverage :
Pattern : (1x1)
Matrix : (1.000, 0.000)
(0.000, 1.000)

STRUCTURE TYPE
Bulk termination with multilayer relaxation perpendicular and parallel to surface

SAMPLE PREPARATION (1 sample)
Treatment : polished, cycles of sputt./anneal
Crystallinity:
Anal. methods:
Contamination: AES: S,C,O present before cleaning

COMMENTS
LEED calculations based on scheme particularly suited to small interlayer spacings;
Vor was assumed energy dependent and was optimized along with the structural parameters: Vor=-11.3±0.5 eV

DATA COLLECTION
Technique: LEED
Dataset : LEED I-V data for 30 beams up to 210 eV: 12 beams at normal incidence, 11 at Θ=7.1,φ=90°; and 7 at Θ=10.7,φ=90°

THEORY/DATA TREATMENT
Dynamical LEED (see comments): 8 phase shifts; up to 55 beams; rms vibr ampl 0.115Å

STRUCTURES EXAMINED
Varied first 4 interlayer spacings and first 3 registries, maintaining mirror plane

QUALITY OF EXPERIMENT-THEORY FIT
RZJ=0.111

2D UNIT CELLS (1 domain observed)

Cell	Ax (Å)	Ay (Å)	Bx (Å)	By (Å)	α (°)	Matrix	Pattern	Cell type
Bulk	2.480	0.000	0.000	4.050	90.0	(1.000, 0.000) (0.000, 1.000)	(1x1)	b: bulk lattice
Surface 1	2.480	0.000	0.000	4.050	90.0	(1.000, 0.000) (0.000, 1.000)	(1x1)	s1: commens. superlattice

3D COORDINATES

Fe1-Fe3 are relaxed

Dx/Dy in Å, or as a fraction of layer's unit cell vectors; atom 0 at the origin. Epir/subr are bulk repeat vectors.

No. of atoms: 5

Bulk z = 1.172 Å

Region	Chem el.	At. no.	Cell type	Site occ.	Rel. to	Dx ± εx		Dy ± εy		Dz ± εz		Dz/Bz(%) ± εz/Bz
epir		-2				-0.827	f	2.025	f	1.172	Å	
subr		-1				0.000	Å	0.000	Å	0.000	Å	
intf	Fe	1	b	1.00	0	0.569 ± .012	f	0.500	f	1.050 ± .030	Å	0.0
intf	Fe	2	b	1.00	1	-0.219 ± .012	f	-0.500	f	1.230 ± .030	Å	89.6 ± 2.6
intf	Fe	3	b	1.00	2	-0.350 ± .012	f	0.500	f	1.150 ± .040	Å	105.0 ± 2.6
intf	Fe	4	b	1.00	3	-0.333	f	0.500	f	1.172	Å	98.2 ± 3.4
subl	Fe	5	b	1.00	4		f		f		Å	100.0

BOND DISTANCES AND ANGLES

Bond distances and angles are derived from coordinates

No. of distances/angles: 7

Interatomic dist. A-B (Å)	Atom A	Atom B	Atom C	Bond angle A-B-C (°)
2.480	Fe1	Fe1(1,0)	Fe2	64.9
2.480	Fe1	Fe1(1,0)	Fe3	54.7
2.682	Fe1	Fe2	Fe1(1,0)	56.9
2.682	Fe1	Fe2	Fe2(1,0)	121.7
2.682	Fe1	Fe2	Fe3	56.7
2.440	Fe1	Fe3	Fe3(1,0)	110.8

Fe(211)-(1x1)
26.14

Bond Distances and Angles - Continued

Interatomic dist. A-B (Å)	Atom A	Atom B	Atom C	Bond angle A-B-C (°)
2.431	Fe2	Fe3	Fe3(1,0)	77.1

COMMON NAME : Fe(310)-(1x1)
CLASSIFICATION : 26.13
TECHNIQUE : LEED
AUTHORS : J. Sokolov, F. Jona and P.M. Marcus
REFERENCE : Phys. Rev., B29, 5402 (1984)

ILLUSTRATION: 18

SURFACE TYPE
Substrate : Fe
Crystal face: 310
Temperature : 300 K
Bulk lattice: bcc
2D bulk symm: cm
2D surf symm: cm

Adsorbate:
Coverage :
Pattern : (1x1)
Matrix : (1.000, 0.000)
(0.000, 1.000)

STRUCTURE TYPE
Bulk termination with multilayer relaxation perpendicular and parallel to surface

SAMPLE PREPARATION (1 sample)
Treatment : polished, cycles of sputt./anneal
Crystallinity:
Anal. methods:
Contamination: AES: S,O,C primary contaminants

COMMENTS

DATA COLLECTION
Technique: LEED
Dataset : I-V spectra for 33 beams (21 non-degenerate): 14 atΘ=0; 10 at Θ=7,φ=-107.5°; 9 at Θ=14,φ=-107.5°; 40 eV<E

THEORY/DATA TREATMENT
Dynamical LEED: 8 phase shifts; up to 65 beams; rms vibr ampl 0.115Å

STRUCTURES EXAMINED
Varied first 4 interlayer spacings and first 3 registries, maintaining mirror plane

QUALITY OF EXPERIMENT-THEORY FIT
RZJ=0.115 (mean for 3 data sets)

2D UNIT CELLS (1 domain observed)

Cell	Ax (Å)	Ay (Å)	Bx (Å)	By (Å)	α (°)	Matrix	Pattern	Cell type
Bulk	2.870	0.000	1.432	4.529	72.5	(1.000, 0.000) (0.000, 1.000)	(1x1)	b: bulk lattice
Surface 1	2.870	0.000	1.432	4.529	72.5	(1.000, 0.000) (0.000, 1.000)	(1x1)	s1: commens. superlattice

3D COORDINATES

Fe1-Fe3 are relaxed

Dx/Dy in Å, or as a fraction of layer's unit cell vectors; atom 0 at the origin. Epir/subr are bulk repeat vectors.

No. of atoms: 5

Bulk z = .908 Å

Reg ion	Chem el.	At. no.	Cell type	Site occ.	Rel. to	Dx ± εx	Dy ± εy	Dz ± εz	Dz/Bz(%) ± εz/Bz
epir		-2				f	f	Å	
subr		-1				-1.436 Å	1.812 Å	0.908 Å	
intf	Fe	1	b	1.00	0	0.000 f	0.000 f	0.000 Å	0.0
intf	Fe	2	b	1.00	1	0.286 ± .018 f	0.424 ± .003 f	0.760 ± .030 Å	83.7 ± 3.3
intf	Fe	3	b	1.00	2	0.311 ± .018 f	0.383 ± .003 f	1.020 ± .030 Å	112.4 ± 3.3
intf	Fe	4	b	1.00	3	0.303 ± .018 f	-0.606 ± .003 f	0.870 ± .040 Å	95.9 ± 4.4
subl	Fe	5	b	1.00	4	-0.700 f	0.400 f	0.908 Å	100.0

BOND DISTANCES AND ANGLES

Bond distances and angles are derived from coordinates

No. of distances/angles: 6

Interatomic dist. A-B (Å)	Atom A	Atom B	Atom C	Bond angle A-B-C (°)
2.511	Fe1	Fe2	Fe1(1,0)	69.6
2.511	Fe1	Fe2	Fe2(1,0)	124.7
2.511	Fe1	Fe2	Fe3	172.7
2.475	Fe2	Fe3	Fe1(0,1)	67.0

Fe(310)-(1x1)
26.13

Bond Distances and Angles - Continued

Interatomic dist. A-B (Å)	Atom A	Atom B	Atom C	Bond angle A-B-C (°)
2.475	Fe2	Fe3	Fe2(1,0)	71.0
2.475	Fe2	Fe3	Fe3(1,0)	125.6

COMMON NAME : Fe(100)-c(2x2)-C+O disordered
CLASSIFICATION : 26.6.8.1
TECHNIQUE : LEED
AUTHORS : F. Jona, K.O. Legg, H.D. Shih, D.W. Jepsen and P.M. Marcus
REFERENCE : Phys. Rev. Lett., 40, 1466 (1978)

ILLUSTRATION: 48,50

SURFACE TYPE
Substrate : Fe
Crystal face: 100
Temperature : RT*
Bulk lattice: bcc
2D bulk symm: p4m
2D surf symm: none

Adsorbate: C;O
Coverage : 0.25 O,C/Fe
Pattern : (2x2)
Matrix : (2.000, 0.000)
(0.000, 2.000)

STRUCTURE TYPE
Decomposed CO as atomic C and O randomly positioned in hollow sites of a c(2x2) lattice; LEED shows c(2x2); (this structure is here modeled as alternating C and O atoms in a (2x2) structure)

SAMPLE PREPARATION (1 sample)
Treatment : dissociative adsorption of CO on Fe(100)
Crystallinity:
Anal. methods:
Contamination:

COMMENTS
To model the random occupation the c(2x2) structure, an atom with averaged C and O scattering properties was constructed (AES suggests C and O coverages of 0.4 and 0.6 ML)

DATA COLLECTION
Technique: LEED
Dataset : 23 LEED spectra

THEORY/DATA TREATMENT
Dynamical LEED: 5 phase shifts, 58 beams; see comments

STRUCTURES EXAMINED
1. random occupation of c(2x2) lattice; 2. 0.5 monolayer of O in c(2x2) structure with C lying in a) 4-fold vacant hollow sites, b) 2-fold bridge sites, c) top sites, d) tetrahedral interstitial sites, e) octahedral interstitial sites

QUALITY OF EXPERIMENT-THEORY FIT
Visual

2D UNIT CELLS (1 domain observed)

Cell	Ax (Å)	Ay (Å)	Bx (Å)	By (Å)	α (°)	Matrix	Pattern	Cell type
Bulk	2.870	0.000	0.000	2.870	90.0	(1.000, 0.000) (0.000, 1.000)	(1x1)	b: bulk lattice
Surface 1	5.740	0.000	0.000	5.740	90.0	(2.000, 0.000) (0.000, 2.000)	(2x2)	m1: randomly mixed layer

3D COORDINATES

C1-O1: in (randomized) c(2x2) lattice on hollow sites; 0.1Å error bars assumed for tabulation

Dx/Dy in Å, or as a fraction of layer's unit cell vectors; atom 0 at the origin. Epir/subr are bulk repeat vectors.

No. of atoms: 5

Bulk z = 1.435 Å

Region	Chem el.	At. no.	Cell type	Site occ.	Rel. to	Dx ± εx		Dy ± εy		Dz ± εz		Dz/Bz(%) ± εz/Bz
epir		-2					f		f		Å	
subr		-1				1.435	Å	1.435	Å	1.435	Å	
ovrl	C	1	m1	.50	0	0.000	f	0.000	f	0.000	Å	0.0
ovrl	O	2	m1	.50	1	0.500	f	0.500	f	0.000 ± .100	Å	0.0 ± 7.0
intf	Fe	3	b	1.00	2	-0.500	f	-0.500	f	0.480 ± .100	Å	33.5 ± 7.0
intf	Fe	4	b	1.00	3	-0.500	f	-0.500	f	1.435 ± .100	Å	100.0 ± 7.0
subl	Fe	5	b	1.00	4	0.500	f	0.500	f	1.435	Å	100.0

BOND DISTANCES AND ANGLES

Bond distances and angles are derived from coordinates

No. of distances/angles: 7

Interatomic dist. A-B (Å)	Atom A	Atom B	Atom C	Bond angle A-B-C (°)
2.085	C1	Fe3	O2	153.4
2.085	C1	Fe3	Fe3(1,0)	133.5
2.085	C1	Fe3	Fe4	48.6
1.915	C1	Fe4	Fe3	54.7
2.486	Fe3	Fe4	C1	54.7

Fe(100)-c(2x2)-C+O disordered
26.6.8.1

Bond Distances and Angles - Continued

Interatomic dist. A-B (Å)	Atom A	Atom B	Atom C	Bond angle A-B-C (°)
2.486	Fe3	Fe4	Fe3(-1,0)	70.5
2.486	Fe3	Fe4	Fe4(1,0)	54.7

COMMON NAME : Fe(100)-(1x1)-3Co
CLASSIFICATION : 26.27.1
TECHNIQUE : ARXPS/ARAES
AUTHORS : Hong Li and B.P. Tonner
REFERENCE : Phys. Rev., B40, 10241 (1989)

ILLUSTRATION: 86

SURFACE TYPE
Substrate : Fe
Crystal face: 100
Temperature : RT
Bulk lattice: bcc
2D bulk symm: p4m
2D surf symm: p4m

Adsorbate:
Coverage : 3 (Co/Fe)
Pattern : (1x1)
Matrix : (1.000, 0.000)
(0.000, 1.000)

SAMPLE PREPARATION (1 sample)
Treatment : Co and Fe evaporation
Crystallinity:
Anal. methods: LEED, quartz microbalance for thickness
Contamination:

DATA COLLECTION
Technique: ARXPS/ARAES; see Rev. Sci. Instrum. 58 (198
Dataset : several Fe 2p and Co LVV polar diffraction patterns

STRUCTURES EXAMINED
No structure optimization

STRUCTURE TYPE
3 epitaxial (1x1) Co monolayers, continuing bcc lattice

COMMENTS
The substrate was an Fe film (3.5 ML) grown on Ag(001); here Ag substrate is ignored and the Fe layer is assumed to be semi-infinite; no attempt was made to optimize the structure: only the lattice type of the overlayer reported

THEORY/DATA TREATMENT

2D UNIT CELLS (1 domain observed)

Cell	Ax (Å)	Ay (Å)	Bx (Å)	By (Å)	α (°)	Matrix	Pattern	Cell type
Bulk	2.866	0.000	0.000	2.866	90.0	(1.000, 0.000) (0.000, 1.000)	(1x1)	b: bulk lattice
Surface 1	2.866	0.000	0.000	2.866	90.0	(1.000, 0.000) (0.000, 1.000)	(1x1)	s1: commens. superlattice

3D COORDINATES

Co1-Co3: Co overlayer

Dx/Dy in Å, or as a fraction of layer's unit cell vectors; atom 0 at the origin. Epir/subr are bulk repeat vectors.

No. of atoms: 4

Bulk z = 1.433 Å

Region	Chem el.	At. no.	Cell type	Site occ.	Rel. to	Dx ± εx		Dy ± εy		Dz ± εz		Dz/Bz(%) ± εz/Bz
epir		-2					f		f		Å	
subr		-1				-1.433	Å	-1.433	Å	1.433	Å	
ovrl	Co	1	b	1.00	0	0.000	f	0.000	f	0.000	Å	0.0
ovrl	Co	2	b	1.00	1	0.500	f	0.500	f	1.433	Å	100.0
ovrl	Co	3	b	1.00	2	-0.500	f	-0.500	f	1.433	Å	100.0
subl	Fe	4	b	1.00	3	0.500	f	0.500	f	1.433	Å	100.0

BOND DISTANCES AND ANGLES

Bond distances and angles are derived from coordinates

No. of distances/angles: 3

Interatomic dist. A-B (Å)	Atom A	Atom B	Atom C	Bond angle A-B-C (°)
2.866	Co1	Co1(1,0)	Co2	60.0
2.866	Co1	Co2	Co3	90.0
2.866	Co2	Co3	Fe4	90.0

Fe(110)-(2x1)-H
26.1.1a

Bond Distances and Angles - Continued

Interatomic dist. A-B (Å)	Atom A	Atom B	Atom C	Bond angle A-B-C (°)
1.752	Fe2	H1(1,0)	Fe2(0,-1)	110.1
1.752	Fe2	H1(1,0)	Fe2(-1,0)	88.1

COMMON NAME : Fe(110)-(3x1)-2H
CLASSIFICATION : 26.1.1b
TECHNIQUE : LEED
AUTHORS : W. Moritz, R. Imbihl, R.J. Behm, G. Ertl and T. Matsushima
REFERENCE : J. Chem. Phys., 83, 1959 (1985)

ILLUSTRATION: 46

SURFACE TYPE
Substrate : Fe
Crystal face: 110
Temperature : 110 K
Bulk lattice: bcc
2D bulk symm: cmm
2D surf symm: cmm

Adsorbate: H
Coverage : 0.667 (H/Fe)
Pattern : (3x1)
Matrix : (3.000, 0.000)
(0.000, 1.000)

STRUCTURE TYPE
Atomic adsorption in 3-fold coordinated hollow sites, two H per unit cell

SAMPLE PREPARATION (1 sample)
Treatment : H2 exposure at 110 K and annealing to 280K
Crystallinity:
Anal. methods:
Contamination: monitored by AES and LEED; C,O < 5% ML

COMMENTS

DATA COLLECTION
Technique: LEED
Dataset : IV curves at Θ=0, 11.3° with φ=90°: 7 fract.-order beams and 4 integer-order beams; E-range 40-180 eV

THEORY/DATA TREATMENT
Dynamical LEED (layer doubling and RFS): free atom H pot. (radius 1.0Å); Moruzzi Fe pot.; Voi α E**1/3; ΘD=467 K

STRUCTURES EXAMINED
H in top, long, short-bridge and quasi-threefold hollow sites; first Fe-Fe spacing varied 1.9-2.5Å in steps of 0.05Å; H-Fe spacing varied 1.5-2.2Å

QUALITY OF EXPERIMENT-THEORY FIT
RZJ=0.4, RPE=0.58

2D UNIT CELLS (2 domains observed)

Cell	Ax (Å)	Ay (Å)	Bx (Å)	By (Å)	α (°)	Matrix	Pattern	Cell type
Bulk	2.480	0.000	1.663	2.339	54.6	(1.000, 0.000) (0.000, 1.000)	(1x1)	b: bulk lattice
Surface 1	7.440	0.000	1.663	2.339	54.6	(3.000, 0.000) (0.000, 1.000)	(3x1)	s1: commens. superlattice

3D COORDINATES

H1-H2: overlayer in quasi-threefold hollow sites; 0.05Å error bar assumed for tabulation

Dx/Dy in Å, or as a fraction of layer's unit cell vectors; atom 0 at the origin. Epir/subr are bulk repeat vectors.

No. of atoms: 4

Bulk z = 2.020 Å

Region	Chem el.	At. no.	Cell type	Site occ.	Rel. to	Dx ± εx		Dy ± εy		Dz ± εz		Dz/Bz(%) ± εz/Bz
epir		-2					f		f		Å	
subr		-1				0.831	Å	1.170	Å	2.020	Å	
ovrl	H	1	s1	.33	0	0.000	f	0.000	f	0.000	Å	0.0
ovrl	H	2	s1	.33	1	0.480	f	0.780	f	0.000 ± .100	Å	0.0 ± 5.0
intf	Fe	3	b	1.00	2	0.780	f	-0.390	f	1.000 ± .050	Å	49.5 ± 2.5
subl	Fe	4	b	1.00	3	0.000	f	0.500	f	2.000	Å	99.0

BOND DISTANCES AND ANGLES

Bond distances and angles are derived from coordinates

No. of distances/angles: 6

Interatomic dist. A-B (Å)	Atom A	Atom B	Atom C	Bond angle A-B-C (°)
1.805	Fe3	H2(1,0)	Fe3(1,0)	85.0
1.805	Fe3	H2(1,0)	Fe3(0,1)	105.3
1.804	Fe3	H2(1,-1)	Fe3(1,-1)	84.9
1.804	Fe3	H2(1,-1)	Fe3(0,-1)	105.3
1.867	Fe3	H2	Fe3(-1,1)	84.9

Fe(110)-(3x1)-2H
26.1.1b

Bond Distances and Angles - Continued

Interatomic dist. A-B (Å)	Atom A	Atom B	Atom C	Bond angle A-B-C (°)
1.867	Fe3	H2	Fe3(-1,0)	85.0

ILLUSTRATION: 48,50

COMMON NAME : Fe(100)-c(2x2)-N
CLASSIFICATION : 26.7.1
TECHNIQUE : LEED
AUTHORS : R. Imbihl, R.J. Behm, G. Ertl and W. Moritz
REFERENCE : Surf. Sci., 123, 129 (1982)

SURFACE TYPE
Substrate : Fe
Crystal face: 100
Temperature : 300 K
Bulk lattice: bcc
2D bulk symm: p4m
2D surf symm: p4m

Adsorbate: N
Coverage : 0.5 N/Fe
Pattern : c(2x2)
Matrix : (1.000,-1.000)
(1.000, 1.000)

STRUCTURE TYPE
Atomic adsorption in 4-fold hollows

SAMPLE PREPARATION (1 sample)
Treatment : oxidation and sputter/anneal cycles; N from decomposing NH3
Crystallinity: sharp LEED pattern after N adsorption
Anal. methods:
Contamination: AES: C, O <few % of a monolayer

COMMENTS
Results for clean Fe(100) indicated a top layer contraction of 1.5%;
potentials: Morruzzi-Janak-Williams for Fe; superposition of atomic charge densities for N

DATA COLLECTION
Technique: LEED
Dataset : I-V curves for 17 beans in energy range 40-200 eV at incident angles Θ=0, 6, 11.6° and φ=-90°

THEORY/DATA TREATMENT
Dynamical LEED (layer doubling): 7 phase shifts; Vor=-9 eV, Voi=-0.85 E**1/3 eV; ΘD=400 K(surf), 467K(bulk)

STRUCTURES EXAMINED
A) hollow sites with N radius varied from 0.5 to 1.0Å, 1st Fe-Fe spacing from 1.35 to 1.60Å; b) top sites; c) (domain averaged) bridge sites; d) underlayer

QUALITY OF EXPERIMENT-THEORY FIT
RZJ=0.20

2D UNIT CELLS (1 domain observed)

Cell	Ax (Å)	Ay (Å)	Bx (Å)	By (Å)	α (°)	Matrix	Pattern	Cell type
Bulk	2.864	0.000	0.000	2.864	90.0	(1.000, 0.000) (0.000, 1.000)	(1x1)	b: bulk lattice
Surface 1	2.864	-2.864	2.864	2.864	90.0	(1.000,-1.000) (1.000, 1.000)	c(2x2)	s1: commens. superlattice

3D COORDINATES

N1: hollow-site overlayer

Dx/Dy in Å, or as a fraction of layer's unit cell vectors; atom 0 at the origin. Epir/subr are bulk repeat vectors.

No. of atoms: 4

Bulk z = 1.430 Å

Region	Chem el.	At. no.	Cell type	Site occ.	Rel. to	Dx ± εx		Dy ± εy		Dz ± εz		Dz/Bz(%) ± εz/Bz
epir		-2					f		f		Å	
subr		-1				1.432	Å	1.432	Å	1.430	Å	
ovrl	N	1	s1	.50	0	0.000	f	0.000	f	0.000	Å	0.0
intf	Fe	2	b	1.00	1	0.500	f	-0.500	f	0.270 ± .050	Å	18.9 ± 3.5
intf	Fe	3	b	1.00	2	-0.500	f	0.500	f	1.540 ± .050	Å	107.7 ± 3.5
subl	Fe	4	b	1.00	3	0.500	f	-0.500	f	1.430	Å	100.0

BOND DISTANCES AND ANGLES

Bond distances and angles are derived from coordinates

No. of distances/angles: 7

Interatomic dist. A-B (Å)	Atom A	Atom B	Atom C	Bond angle A-B-C (°)
2.043	N1	Fe2	N1(0,1)	164.8
2.043	N1	Fe2	Fe2(1,0)	134.5
2.043	N1	Fe2	Fe3(1,0)	94.6
2.043	N1	Fe2	Fe3	44.9
2.043	N1	Fe2	Fe4	97.6

Fe(100)-c(2x2)-N
26.7.1

Bond Distances and Angles - Continued

Interatomic dist. A-B (Å)	Atom A	Atom B	Atom C	Bond angle A-B-C (°)
1.810	N1	Fe3	Fe2	52.8
1.810	N1	Fe3	Fe4	125.2

ILLUSTRATION: 86

COMMON NAME : Fe(100)-(1x1)-3Ni
CLASSIFICATION : 26.28.0a
TECHNIQUE : LEED
AUTHORS : Z.Q. Wang, Y.S. Li, F. Jona and P.M. Marcus
REFERENCE : Solid State Commun., 61, 623 (1987)

SURFACE TYPE
Substrate : Fe
Crystal face: 100
Temperature : RT*
Bulk lattice: bcc
2D bulk symm: p4m
2D surf symm: p4m

Adsorbate:
Coverage : 3 (Ni/Fe)
Pattern : (1x1)
Matrix : (1.000, 0.000)
(0.000, 1.000)

STRUCTURE TYPE
Triple (1x1) Ni overlayer in distorted bcc lattice (compressed perp. to surface), continuing Fe bcc lattice

SAMPLE PREPARATION (1 sample)
Treatment : Ni deposited by heating a Ni strip to about 1200C
Crystallinity:
Anal. methods:
Contamination: monitored with AES

COMMENTS
The Ni film has the same multilayer relaxation as the clean Fe(100) surface

DATA COLLECTION
Technique: LEED
Dataset : I-V curves at normal incidence (4 beams) and at off-normal incidence (4 beams) in E range 50 to 300 eV

THEORY/DATA TREATMENT
Dynamical LEED (CHANGE program): 6 phase shifts, 49 beams

STRUCTURES EXAMINED
Variation of top three interlayer spacings

QUALITY OF EXPERIMENT-THEORY FIT
RZJ=0.06

2D UNIT CELLS (1 domain observed)

Cell	Ax (Å)	Ay (Å)	Bx (Å)	By (Å)	α (°)	Matrix	Pattern	Cell type
Bulk	2.866	0.000	0.000	2.866	90.0	(1.000, 0.000) (0.000, 1.000)	(1x1)	b: bulk lattice
Surface 1	2.866	0.000	0.000	2.866	90.0	(1.000, 0.000) (0.000, 1.000)	(1x1)	s1: commens. superlattice

3D COORDINATES

Ni1-Ni2-Ni3: trilayer with distorted bcc lattice

Dx/Dy in Å, or as a fraction of layer's unit cell vectors; atom 0 at the origin. Epir/subr are bulk repeat vectors.

No. of atoms: 5

Bulk z = 1.433 Å

Region	Chem el.	At. no.	Cell type	Site occ.	Rel. to	Dx ± εx		Dy ± εy		Dz ± εz		Dz/Bz(%) ± εz/Bz
epir		-2					f		f		Å	
subr		-1				-1.433	Å	-1.433	Å	1.433	Å	
ovrl	Ni	1	b	1.00	0	0.000	f	0.000	f	0.000	Å	0.0
ovrl	Ni	2	b	1.00	1	0.500	f	0.500	f	1.363 ± .030	Å	95.1 ± 2.1
ovrl	Ni	3	b	1.00	2	-0.500	f	-0.500	f	1.503 ± .030	Å	104.9 ± 2.1
intf	Fe	4	b	1.00	3	0.500	f	0.500	f	1.433	Å	100.0
subl	Fe	5	b	1.00	4	-0.500	f	-0.500	f	1.433	Å	100.0

BOND DISTANCES AND ANGLES

Bond distances and angles are derived from coordinates

No. of distances/angles: 3

Interatomic dist. A-B (Å)	Atom A	Atom B	Atom C	Bond angle A-B-C (°)
2.866	Ni1	Ni1(1,0)	Ni2	54.1
2.442	Ni1	Ni2	Ni3	70.5
2.523	Ni2	Ni3	Fe4	71.8

COMMON NAME : Fe(100)-(1x1)-O
CLASSIFICATION : 26.8.10
TECHNIQUE : MEIS
AUTHORS : R.L. Headrick, P. Konarski, S.M. Yalisove and W.R. Graham
REFERENCE : Phys. Rev., B39, 5713 (1989)

ILLUSTRATION: 48,49

SURFACE TYPE
Substrate : Fe
Crystal face: 100
Temperature : RT*
Bulk lattice: bcc
2D bulk symm: p4m
2D surf symm: p4m

Adsorbate: O
Coverage : 1.0 O/Fe
Pattern : (1x1)
Matrix : (1.000, 0.000)
(0.000, 1.000)

STRUCTURE TYPE
Atomic adsorption in 4-fold hollow sites (assumed), with expansion of top 2 Fe-Fe spacings by 11% and 0.7%

SAMPLE PREPARATION (2 sample)
Treatment : Fe exposed to O2 at 1.0E-8 torr
Crystallinity:
Anal. methods:
Contamination:

COMMENTS
Coverage was measured to be 1.00±0.05ML from deposition, 0.98±0.05ML by segregation from bulk; both preparations give same results

DATA COLLECTION
Technique: MEIS
Dataset :

THEORY/DATA TREATMENT
Monte Carlo simulations, incl. fit of ΘD (exponentially increasing from 255 K at surface to 420K in bulk)

STRUCTURES EXAMINED
Variation of top two Fe-Fe spacings, keeping spacing of 1.97Å between O and 2nd-layer Fe (= sum of covalent radii)

2D UNIT CELLS (1 domain observed)

Cell	Ax (Å)	Ay (Å)	Bx (Å)	By (Å)	α (°)	Matrix	Pattern	Cell type
Bulk	2.860	0.000	0.000	2.860	90.0	(1.000, 0.000) (0.000, 1.000)	(1x1)	b: bulk lattice
Surface 1	2.860	0.000	0.000	2.860	90.0	(1.000, 0.000) (0.000, 1.000)	(1x1)	s1: commens. superlattice

3D COORDINATES

O1 forms overlayer in 4-fold hollows

Dx/Dy in Å, or as a fraction of layer's unit cell vectors; atom 0 at the origin. Epir/subr are bulk repeat vectors.

No. of atoms: 5

Bulk z = 1.430 Å

Region	Chem el.	At. no.	Cell type	Site occ.	Rel. to	Dx ± εx		Dy ± εy		Dz ± εz		Dz/Bz(%) ± εz/Bz
epir		-2					f		f		Å	
subr		-1				1.430	Å	1.430	Å	1.430	Å	
ovrl	O	1	b	1.00	0	0.000	f	0.000	f	0.000	Å	0.0
intf	Fe	2	b	1.00	1	0.500	f	0.500	f	0.380 ± .040	Å	26.6 ± 2.8
intf	Fe	3	b	1.00	2	-0.500	f	-0.500	f	1.590 ± .040	Å	111.2 ± 2.8
intf	Fe	4	b	1.00	3	0.500	f	0.500	f	1.440 ± .040	Å	100.7 ± 2.8
subl	Fe	5	b	1.00	4	-0.500	f	-0.500	f	1.430	Å	100.0

BOND DISTANCES AND ANGLES

Bond distances and angles are derived from coordinates

No. of distances/angles: 6

Interatomic dist. A-B (Å)	Atom A	Atom B	Atom C	Bond angle A-B-C (°)
2.058	O1	Fe2	O1(1,0)	88.1
2.058	O1	Fe2	Fe2(1,0)	134.0
2.058	O1	Fe2	Fe3	48.8
1.970	O1	Fe3	Fe4	125.5
2.860	Fe2	Fe2(1,0)	Fe3(1,0)	56.2
2.573	Fe2	Fe3	Fe4	73.6

ILLUSTRATION: 48,49

COMMON NAME : Fe(100)-(1x1)-O
CLASSIFICATION : 26.8.3
TECHNIQUE : LEED
AUTHORS : K.O. Legg, F. Jona, D.W. Jepsen and P.M. Marcus
REFERENCE : Phys. Rev., B16, 5271 (1977)

SURFACE TYPE
Substrate : Fe
Crystal face: 100
Temperature : RT*
Bulk lattice: bcc
2D bulk symm: p4m
2D surf symm: p4m

Adsorbate: O
Coverage : 1.0 O/Fe
Pattern : (1x1)
Matrix : (1.000, 0.000)
(0.000, 1.000)

STRUCTURE TYPE
Atomic adsorption deep in 4-fold hollow sites

SAMPLE PREPARATION (1 sample)
Treatment : clean sample exposed to 99.999% pure O2 at 1.0E-8 torr
Crystallinity:
Anal. methods:
Contamination: weak c(2x2)-CO pattern below 1ML O

COMMENTS
Non-structural parameters: Vor=-12 eV, Voi energy dependent ranging from -2 eV at 50eV to -4eV at 150eV;
R-factor (not RZJ) takes into account both peak positions and intensities: see Legg et al, J. Phys. C10, 937 (1977)

DATA COLLECTION
Technique: LEED
Dataset : 16 LEED I-V spectra: E range 40-160 eV at incident angles (Θ,ϕ)=(0,0),(10,0),(20,0)°

THEORY/DATA TREATMENT
Dynamical LEED (layer KKR method): 8 phase shifts; O pot from superpos of at ch dens in bcc latt; rms ampl=0.115Å

STRUCTURES EXAMINED
Varied O-Fe spacings from 0.27 to 0.8Å for O in 4-fold hollows; first Fe-Fe spacing simultaneously varied by ±10%

QUALITY OF EXPERIMENT-THEORY FIT
R=0.3 (see comments)

2D UNIT CELLS (1 domain observed)

Cell	Ax (Å)	Ay (Å)	Bx (Å)	By (Å)	α (°)	Matrix	Pattern	Cell type
Bulk	2.864	0.000	0.000	2.864	90.0	(1.000, 0.000) (0.000, 1.000)	(1x1)	b: bulk lattice
Surface 1	2.864	0.000	0.000	2.864	90.0	(1.000, 0.000) (0.000, 1.000)	(1x1)	s1: commens. superlattice

3D COORDINATES

O1: overlayer deep in 4-fold hollow sites

Dx/Dy in Å, or as a fraction of layer's unit cell vectors; atom 0 at the origin. Epir/subr are bulk repeat vectors.

No. of atoms: 4

Bulk z = 1.430 Å

Region	Chem el.	At. no.	Cell type	Site occ.	Rel. to	Dx ± εx		Dy ± εy		Dz ± εz		Dz/Bz(%) ± εz/Bz
epir		-2					f		f		Å	
subr		-1				1.432	Å	1.432	Å	1.430	Å	
ovrl	O	1	b	1.00	0	0.000	f	0.000	f	0.000	Å	0.0
intf	Fe	2	b	1.00	1	0.500	f	0.500	f	0.480 ± .100	Å	33.6 ± 7.0
intf	Fe	3	b	1.00	2	-0.500	f	-0.500	f	1.540 ± .100	Å	107.7 ± 7.0
subl	Fe	4	b	1.00	3	0.500	f	0.500	f	1.430	Å	100.0

BOND DISTANCES AND ANGLES

Bond distances and angles are derived from coordinates

No. of distances/angles: 7

Interatomic dist. A-B (Å)	Atom A	Atom B	Atom C	Bond angle A-B-C (°)
2.081	O1	Fe2	O1(1,0)	87.0
2.081	O1	Fe2	Fe2(1,0)	133.5
2.081	O1	Fe2	Fe3	50.6
2.081	O1	Fe2	Fe4	103.3
2.020	O1	Fe3	Fe2	52.8

Fe(100)-(1x1)-O
26.8.3

Bond Distances and Angles - Continued

Interatomic dist. A-B (Å)	Atom A	Atom B	Atom C	Bond angle A-B-C (°)
2.020	O1	Fe3	Fe4	125.2
2.544	Fe2	Fe3	Fe4	72.5

ILLUSTRATION: 48,49

COMMON NAME : Fe(100)-(1x1)-O
CLASSIFICATION : 26.8.7
TECHNIQUE : LEIS
AUTHORS : J.M. van Zoest, J.M. Fluit, T.J. Vink and B.A. van Hassel
REFERENCE : Surf. Sci., 182, 179 (1987)

SURFACE TYPE
Substrate : Fe
Crystal face: 100
Temperature : RT
Bulk lattice: bcc
2D bulk symm: p4m
2D surf symm: p4m

Adsorbate: O
Coverage : 1.0 O/Fe
Pattern : (1x1)
Matrix : (1.000, 0.000)
(0.000, 1.000)

STRUCTURE TYPE
Atomic adsorption in 4-fold hollow sites, with unrelaxed substrate assumed

SAMPLE PREPARATION (1 sample)
Treatment : Fe cleaned by cycles of heating >450 K, 5 K eV Ne+ ion bomb.
Crystallinity:
Anal. methods:
Contamination: clean Fe monitored with scattered Ne+

COMMENTS
LEIS insensitive to Fe-Fe spacings

DATA COLLECTION
Technique: LEIS
Dataset : low energy ion scattering: azimuthal distributions of Ne+ ions and neutrals at specular refl. and 45° scatt. angle

THEORY/DATA TREATMENT
O- recoil intensity as function of elevation angle of incidence at spec. refl. was used to calculate O-Fe distance

STRUCTURES EXAMINED
Data consistent with 4-fold symmetrical hollow site; O-Fe spacing was varied

QUALITY OF EXPERIMENT-THEORY FIT
Visual

2D UNIT CELLS (1 domain observed)

Cell	Ax (Å)	Ay (Å)	Bx (Å)	By (Å)	α (°)	Matrix	Pattern	Cell type
Bulk	2.864	0.000	0.000	2.864	90.0	(1.000, 0.000) (0.000, 1.000)	(1x1)	b: bulk lattice
Surface 1	2.864	0.000	0.000	2.864	90.0	(1.000, 0.000) (0.000, 1.000)	(1x1)	s1: commens. superlattice

3D COORDINATES

O1: overlayer in 4-fold hollows with shorter O-Fe bond length to 2nd Fe layer than to 1st

Dx/Dy in Å, or as a fraction of layer's unit cell vectors; atom 0 at the origin. Epir/subr are bulk repeat vectors.

No. of atoms: 5

Bulk z = 1.430 Å

Reg ion	Chem el.	At. no.	Cell type	Site occ.	Rel. to	Dx ± εx		Dy ± εy		Dz ± εz		Dz/Bz(%) ± εz/Bz
epir		-2					f		f		Å	
subr		-1				-1.432	Å	-1.432	Å	1.430	Å	
ovrl	O	1	b	1.00	0	0.000	f	0.000	f	0.000	Å	0.0
intf	Fe	2	b	1.00	1	0.500	f	0.500	f	0.560 ± .050	Å	39.2 ± 3.5
intf	Fe	3	b	1.00	2	-0.500	f	-0.500	f	1.430	Å	100.0
intf	Fe	4	b	1.00	3	0.500	f	0.500	f	1.430	Å	100.0
subl	Fe	5	b	1.00	4	-0.500	f	-0.500	f	1.430	Å	100.0

BOND DISTANCES AND ANGLES

Bond distances and angles are derived from coordinates

No. of distances/angles: 3

Interatomic dist. A-B (Å)	Atom A	Atom B	Atom C	Bond angle A-B-C (°)
2.101	O1	Fe2	Fe2(1,0)	133.0
2.101	O1	Fe2	Fe3	50.7
2.479	Fe2	Fe3	Fe4	70.5

COMMON NAME : Fe(100)-(1x1)-O
CLASSIFICATION : 26.8.9
TECHNIQUE : LEED
AUTHORS : F. Jona and P.M. Marcus
REFERENCE : Solid State Commun., 64, 667 (1987)

ILLUSTRATION: 48,49

SURFACE TYPE
Substrate : Fe
Crystal face: 100
Temperature : RT*
Bulk lattice: bcc
2D bulk symm: p4m
2D surf symm: p4m

Adsorbate: O
Coverage : 1.0 O/Fe
Pattern : (1x1)
Matrix : (1.000, 0.000)
(0.000, 1.000)

STRUCTURE TYPE
Atomic adsorption in 4-fold hollow sites, with expansion of top 2 Fe-Fe spacings by 8% and 3%

SAMPLE PREPARATION (1 sample)
Treatment : Fe exposed to 99.999% pure O2 at 1.0E-8 torr
Crystallinity:
Anal. methods:
Contamination: LEED: CO disappeared as O approached ML

COMMENTS
R-factors RVHT and RPE were also calculated

DATA COLLECTION
Technique: LEED
Dataset : 16 I-V spectra at 3 inc angles Θ=0,10,20,φ=0°: E range 30-300 eV; respective cumulative E ranges 887, 1463 an

THEORY/DATA TREATMENT
Dynamical LEED (CHANGE program): 8 ph shs (self-cons. Fe pot superpos O pot); Vor, E-dep Voi; rms vibs 0.115Å

STRUCTURES EXAMINED
O in 4-fold hollow site; O-Fe interlayer spacing varied 0.28-0.58Å in steps of 0.1Å; first Fe-Fe spacing varied 1.43(bulk)-1.87Å in steps of 0.055Å (later 0.04Å); 2nd Fe-Fe spacing varied 1.35-1.51Å in steps of 0.02Å

QUALITY OF EXPERIMENT-THEORY FIT
RZJ=0.079, 0.118, 0.262 (Θ=0,10,20°)

2D UNIT CELLS (1 domain observed)

Cell	Ax (Å)	Ay (Å)	Bx (Å)	By (Å)	α (°)	Matrix	Pattern	Cell type
Bulk	2.864	0.000	0.000	2.864	90.0	(1.000, 0.000) (0.000, 1.000)	(1x1)	b: bulk lattice
Surface 1	2.864	0.000	0.000	2.864	90.0	(1.000, 0.000) (0.000, 1.000)	(1x1)	s1: commens. superlattice

3D COORDINATES

O1: overlayer in 4-fold hollows with shorter O-Fe bond length to 2nd Fe layer than to 1st

Dx/Dy in Å, or as a fraction of layer's unit cell vectors; atom 0 at the origin. Epir/subr are bulk repeat vectors.

No. of atoms: 5

Bulk z = 1.432 Å

Region	Chem el.	At. no.	Cell type	Site occ.	Rel. to	Dx ± εx		Dy ± εy		Dz ± εz		Dz/Bz(%) ± εz/Bz
epir		-2					f		f		Å	
subr		-1				-1.432	Å	-1.432	Å	1.432	Å	
ovrl	O	1	b	1.00	0	0.000	f	0.000	f	0.000	Å	0.0
intf	Fe	2	b	1.00	1	0.500	f	0.500	f	0.450 ± .040	Å	31.4 ± 2.8
intf	Fe	3	b	1.00	2	-0.500	f	-0.500	f	1.550 ± .040	Å	108.2 ± 2.8
intf	Fe	4	b	1.00	3	0.500	f	0.500	f	1.470 ± .040	Å	102.7 ± 2.8
subl	Fe	5	b	1.00	4	-0.500	f	-0.500	f	1.432	Å	100.0

BOND DISTANCES AND ANGLES

Bond distances and angles are derived from coordinates

No. of distances/angles: 4

Interatomic dist. A-B (Å)	Atom A	Atom B	Atom C	Bond angle A-B-C (°)
2.075	O1	Fe2	Fe2(1,0)	133.7
2.075	O1	Fe2	Fe3	50.0

Fe(100)-(1x1)-O
26.8.9

Bond Distances and Angles - Continued

Interatomic dist. A-B (Å)	Atom A	Atom B	Atom C	Bond angle A-B-C (°)
2.000	O1	Fe3	Fe4	126.0
2.550	Fe2	Fe3	Fe4	73.4

COMMON NAME : Fe(211)-(2x1)-O
CLASSIFICATION : 26.8.8
TECHNIQUE : LEED
AUTHORS : J. Sokolov, F. Jona and P.M. Marcus
REFERENCE : Europhys. Lett., 1, 401 (1986)

ILLUSTRATION: 54

SURFACE TYPE
Substrate : Fe
Crystal face: 211
Temperature : 300 K
Bulk lattice: bcc
2D bulk symm: pm
2D surf symm: pm

Adsorbate: O
Coverage : 0.5 (O/Fe)
Pattern : (2x1)
Matrix : (1.000, 0.000)
(0.000, 2.000)

STRUCTURE TYPE
Atomic adsorption in long-bridge sites, forming -Fe-O-Fe-O- strings perpendicular to clean-surface ridges, in which half the Fe atoms are missing ('missing-row' model);

SAMPLE PREPARATION (1 sample)
Treatment : 7L exposure of O2, followed by 30min annealing at 300C
Crystallinity:
Anal. methods:
Contamination: checked by LEED and AES

COMMENTS
This structure is qualitatively similar to the corresponding missing-row structure of Ni and Cu(110)-(2x1)-O

DATA COLLECTION
Technique: LEED
Dataset : I-V curves for 36 (22 non-degenerate) beams: 16 at Θ=0°, 12 at Θ=10°,φ=90°, 8 at Θ=20°,φ=90°; 30<E<195 eV.

THEORY/DATA TREATMENT
Dynamical LEED: 8 phase shifts, <=47 beams;
Vor=-10.5±1 eV, Voi=-4eV

STRUCTURES EXAMINED
Unreconstructed models: O in various sites on ridges, in troughs and in between; reconstructed models: missing-row and sawtooth models with O in various sites; variation of O-Fe spacing, top Fe-Fe spacing and registry

QUALITY OF EXPERIMENT-THEORY FIT
RZJ=0.136

2D UNIT CELLS (1 domain observed)

Cell	Ax (Å)	Ay (Å)	Bx (Å)	By (Å)	α (°)	Matrix	Pattern	Cell type
Bulk	4.054	0.000	0.000	2.482	90.0	(1.000, 0.000) (0.000, 1.000)	(1x1)	b: bulk lattice
Surface 1	4.054	0.000	0.000	4.964	90.0	(1.000, 0.000) (0.000, 2.000)	(2x1)	s1: commens. superlattice

3D COORDINATES

O1-Fe2: nearly straight bonded -Fe-O-Fe-O- strings

Dx/Dy in Å, or as a fraction of layer's unit cell vectors; atom 0 at the origin. Epir/subr are bulk repeat vectors.

No. of atoms: 4 Bulk z = 1.170 Å

Region	Chem el.	At. no.	Cell type	Site occ.	Rel. to	Dx ± εx		Dy ± εy		Dz ± εz		Dz/Bz(%) ± εz/Bz
epir		-2					f		f		Å	
subr		-1				2.027	Å	0.828	Å	1.170	Å	
ovrl	O	1	s1	.50	0	0.000	f	0.000	f	0.000	Å	0.0
intf	Fe	2	s1	.50	1	0.500	f	0.984	f	0.260 ± .050	Å	22.2 ± 4.3
intf	Fe	3	b	1.00	2	-0.500	f	-0.533	f	1.090 ± .040	Å	93.2 ± 3.4
subl	Fe	4	b	1.00	3	0.500	f	0.333	f	1.170	Å	100.0

BOND DISTANCES AND ANGLES

Bond distances and angles are derived from coordinates

No. of distances/angles: 5

Interatomic dist. A-B (Å)	Atom A	Atom B	Atom C	Bond angle A-B-C (°)
2.045	O1	Fe2(0,-1)	O1(1,0)	164.7
2.045	O1	Fe2(0,-1)	Fe3(0,-1)	42.0
1.946	O1	Fe3(0,-1)	Fe4(0,-2)	95.0

Fe(211)-(2x1)-O
26.8.8

Bond Distances and Angles - Continued

Interatomic dist. A-B (Å)	Atom A	Atom B	Atom C	Bond angle A-B-C (°)
2.654	Fe2	Fe3	Fe4	53.4
2.313	Fe2	Fe4	Fe4(0,1)	77.7

COMMON NAME : Fe(100)-c(2x2)-S
CLASSIFICATION : 26.16.1
TECHNIQUE : LEED
AUTHORS : K.O. Legg, F. Jona, D.W. Jepsen and P.M. Marcus
REFERENCE : Surf. Sci., 66, 25 (1977)

ILLUSTRATION: 48,50

SURFACE TYPE
Substrate : Fe
Crystal face: 100
Temperature : 300 K
Bulk lattice: bcc
2D bulk symm: p4m
2D surf symm: p4m

Adsorbate: S
Coverage : 0.5 S/Fe
Pattern : c(2x2)
Matrix : (1.000, 1.000)
(-1.000, 1.000)

STRUCTURE TYPE
Atomic adsorption in hollow sites

SAMPLE PREPARATION (1 sample)
Treatment : S deposited from 99.999% pure S at 1.0E-10 torr
Crystallinity: sharp LEED pattern seen for overlayer
Anal. methods:
Contamination: AES: 2-4%ML C

COMMENTS
R-factor is product of intensity R-factor and mean peak deviation;
the best S-Fe spacing depended somewhat on the S muffin-tin sphere radius chosen, at the 0.02 Å level

DATA COLLECTION
Technique: LEED
Dataset : 24 I-V curves: 6 at normal incidence, 18 at 2 off-normal geometries; 50<E<150 eV

THEORY/DATA TREATMENT
Dynamical LEED (layer KKR): 8 phase shifts and 58 beams
rms vibr ampl 0.115 Å

STRUCTURES EXAMINED
Varied S-Fe spacing 0.93-1.19Å assuming bulk truncation, then adjusted 1st-2nd Fe spacing using optimised S-Fe; best fit for the unrelaxed substrate

QUALITY OF EXPERIMENT-THEORY FIT
See comments

2D UNIT CELLS (1 domain observed)

Cell	Ax (Å)	Ay (Å)	Bx (Å)	By (Å)	α (°)	Matrix	Pattern	Cell type
Bulk	2.864	0.000	0.000	2.864	90.0	(1.000, 0.000) (0.000, 1.000)	(1x1)	b: bulk lattice
Surface 1	2.864	2.864	-2.864	2.864	90.0	(1.000, 1.000) (-1.000, 1.000)	c(2x2)	s1: commens. superlattice

3D COORDINATES

S1: hollow-site overlayer

Dx/Dy in Å, or as a fraction of layer's unit cell vectors; atom 0 at the origin. Epir/subr are bulk repeat vectors.

No. of atoms: 3

Bulk z = 1.430 Å

Region	Chem el.	At. no.	Cell type	Site occ.	Rel. to	Dx ± εx		Dy ± εy		Dz ± εz		Dz/Bz(%) ± εz/Bz
epir		-2					f		f		Å	
subr		-1				1.432	Å	1.432	Å	1.430	Å	
ovrl	S	1	s1	.50	0	0.000	f	0.000	f	0.000	Å	0.0
intf	Fe	2	b	1.00	1	0.500	f	0.500	f	1.050 ± .050	Å	73.4 ± 3.5
subl	Fe	3	b	1.00	2	-0.500	f	-0.500	f	1.430	Å	100.0

BOND DISTANCES AND ANGLES

Bond distances and angles are derived from coordinates

No. of distances/angles: 6

Interatomic dist. A-B (Å)	Atom A	Atom B	Atom C	Bond angle A-B-C (°)
2.281	S1	Fe2	S1(1,0)	125.2
2.281	S1	Fe2	Fe2(1,0)	128.9
2.281	S1	Fe2	Fe3	62.6
2.480	S1	Fe3	Fe2	54.8
2.864	Fe2	Fe2(1,0)	Fe3(1,0)	54.7
2.479	Fe2	Fe3	Fe2(-1,0)	70.6

COMMON NAME : Fe(100)-c(2x2)-S
CLASSIFICATION : 26.16.6
TECHNIQUE : ARPEFS
AUTHORS : X.S. Zhang, L.J. Terminello, S. Kim, Z.Q. Huang, A.E. Schach von Wittenau and D.A. Shirley
REFERENCE : J. Chem. Phys., 89, 6538 (1988)

ILLUSTRATION: 48,50

SURFACE TYPE
Substrate : Fe
Crystal face: 100
Temperature : RT*
Bulk lattice: bcc
2D bulk symm: p4m
2D surf symm: p4m

Adsorbate: S
Coverage : 0.5 S/Fe
Pattern : c(2x2)
Matrix : (1.000, 1.000)
(-1.000, 1.000)

STRUCTURE TYPE
Atomic adsorption in hollow sites with multilayer relaxation (no detectable layer buckling)

SAMPLE PREPARATION (1 sample)
Treatment : 5 weeks of sputter-anneal cycles, then exposure to H2S
Crystallinity:
Anal. methods:
Contamination: AES: no contaminants

COMMENTS

DATA COLLECTION
Technique: ARPEFS; 2500-3000eV soft x-rays (1.5eV reso
Dataset : ARPEFS spectra for two emission angles: [100], [110]; kinetic E range 50-600 eV

THEORY/DATA TREATMENT
Fourier transform; MSSW calcs: 17 ph shs; Moruzzi et al Fe pot; HF S pot; ΘD=420 K(bulk Fe), 297K(surf Fe), 395K(S)

STRUCTURES EXAMINED
Top, bridge and hollow site: FT and MSSW favor hollow; variation of S-Fe spacing, 1st Fe layer buckling, top 2 Fe-Fe interlayer spacings, optimized by R-factor fitting

2D UNIT CELLS (1 domain observed)

Cell	Ax (Å)	Ay (Å)	Bx (Å)	By (Å)	α (°)	Matrix	Pattern	Cell type
Bulk	2.860	0.000	0.000	2.860	90.0	(1.000, 0.000) (0.000, 1.000)	(1x1)	b: bulk lattice
Surface 1	2.860	2.860	-2.860	2.860	90.0	(1.000, 1.000) (-1.000, 1.000)	c(2x2)	s1: commens. superlattice

3D COORDINATES

S1: hollow-site overlayer

Dx/Dy in Å, or as a fraction of layer's unit cell vectors; atom 0 at the origin. Epir/subr are bulk repeat vectors.

No. of atoms: 5 Bulk z = 1.430 Å

Reg ion	Chem el.	At. no.	Cell type	Site occ.	Rel. to	Dx ± εx		Dy ± εy		Dz ± εz		Dz/Bz(%) ± εz/Bz
epir		-2					f		f		Å	
subr		-1				1.430	Å	1.430	Å	1.430	Å	
ovrl	S	1	s1	.50	0	0.000	f	0.000	f	0.000	Å	0.0
intf	Fe	2	b	1.00	1	0.500	f	0.500	f	1.100 ± .020	Å	76.9 ± 1.4
intf	Fe	3	b	1.00	2	-0.500	f	-0.500	f	1.400 ± .020	Å	97.9 ± 1.4
intf	Fe	4	b	1.00	3	0.500	f	0.500	f	1.460 ± .030	Å	102.1 ± 2.1
subl	Fe	5	b	1.00	4	-0.500	f	-0.500	f	1.430	Å	100.0

BOND DISTANCES AND ANGLES

Bond distances and angles are derived from coordinates

No. of distances/angles: 4

Interatomic dist. A-B (Å)	Atom A	Atom B	Atom C	Bond angle A-B-C (°)
2.302	S1	Fe2	S1(1,0)	122.9
2.302	S1	Fe2	Fe2(1,0)	128.4
2.302	S1	Fe2	Fe3	63.2
2.460	Fe2	Fe3	Fe4	70.5

COMMON NAME : Fe(110)-p(2x2)-S
CLASSIFICATION : 26.16.4
TECHNIQUE : LEED
AUTHORS : H.D. Shih, F. Jona, D.W. Jepsen and P.M. Marcus
REFERENCE : Phys. Rev. Lett., 46, 731 (1981)

ILLUSTRATION: 47

SURFACE TYPE
Substrate : Fe
Crystal face: 110
Temperature : RT*
Bulk lattice: bcc
2D bulk symm: cmm
2D surf symm: cmm

Adsorbate: S
Coverage : 0.25
Pattern : (2x2)
Matrix : (2.000, 0.000)
(0.000, 2.000)

STRUCTURE TYPE
Atomic S at center site ('4-fold' hollow); top substrate layer laterally relaxed

SAMPLE PREPARATION (1 sample)
Treatment : see Shih et al, J. Phys. C13, 3801 (1980); S from capsule
Crystallinity:
Anal. methods:
Contamination: monitored by AES

COMMENTS
Other models also tested, including 3 fold hollow site

DATA COLLECTION
Technique: LEED
Dataset : LEED I-V spectra for 8 beams at normal incidence and 7 at Θ=9°, ϕ=35.26°; E<=170 eV

THEORY/DATA TREATMENT
Dynamical LEED: 8 phase shifts; 124 beams;
Vor=-11.5 eV, Voi=-4eV; vibr amps=0.115Å

STRUCTURES EXAMINED
1. S on 4 fold hollow sites and Fe-S interlayer spacing varied with Fe atoms in truncated bulk positions;
2. various reconstructions of the substrate with hard spheres with bulk radius in (2x2) cells, preserving symmetry and commensurability with the bulk

QUALITY OF EXPERIMENT-THEORY FIT
Visual

2D UNIT CELLS (1 domain observed)

Cell	Ax (Å)	Ay (Å)	Bx (Å)	By (Å)	α (°)	Matrix	Pattern	Cell type
Bulk	2.485	0.000	-.833	2.341	109.6	(1.000, 0.000) (0.000, 1.000)	(1x1)	b: bulk lattice
Surface 1	4.969	0.000	-1.665	4.682	109.6	(2.000, 0.000) (0.000, 2.000)	(2x2)	s1: commens. superlattice

3D COORDINATES

S1: overlayer in distorted '4-fold' hollows; Fe2-Fe5: laterally relaxed top substrate layer;
Fe6: periodic bulk layer; 0.1Å error bars assumed for tabulation

Dx/Dy in Å, or as a fraction of layer's unit cell vectors; atom 0 at the origin. Epir/subr are bulk repeat vectors.

No. of atoms: 6 Bulk z = 2.030 Å

Region	Chem el.	At. no.	Cell type	Site occ.	Rel. to	Dx ± εx		Dy ± εy		Dz ± εz		Dz/Bz(%) ± εz/Bz
epir		-2					f		f		Å	
subr		-1				3.311	Å	1.170	Å	2.030	Å	
ovrl	S	1	s1	.25	0	0.000	f	0.000	f	0.000	Å	0.0
intf	Fe	2	s1	.25	1	0.716 ± .027	f	0.716 ± .021	f	1.430 ± .100	Å	70.4 ± 4.9
intf	Fe	3	s1	.25	1	0.284 ± .027	f	0.284 ± .021	f	1.430 ± .100	Å	70.4 ± 4.9
intf	Fe	4	s1	.25	1	0.232 ± .027	f	0.768 ± .021	f	1.430 ± .100	Å	70.4 ± 4.9
intf	Fe	5	s1	.25	1	0.768 ± .027	f	0.232 ± .021	f	1.430 ± .100	Å	70.4 ± 4.9
subl	Fe	6	b	1.00	2	-1.432	f	-1.432	f	2.030 ± .100	Å	100.0 ± 4.9

Fe(110)-p(2x2)-S
26.16.4

BOND DISTANCES AND ANGLES

Bond distances and angles are derived from coordinates

No. of distances/angles: 16

Interatomic dist. A-B (Å)	Atom A	Atom B	Atom C	Bond angle A-B-C (°)
2.166	S1	Fe2(-1,-1)	Fe3(-1,-1)	138.7
2.489	Fe2	Fe4(1,0)	Fe5	55.0
2.489	Fe2	Fe4(1,0)	Fe6(2,1)	70.6
2.504	Fe2	Fe4	Fe3	59.3
2.504	Fe2	Fe4	Fe5(0,1)	54.6
2.504	Fe2	Fe4	Fe6(0,1)	103.5
2.489	Fe2	Fe5(0,1)	Fe3(1,1)	81.6
2.489	Fe2	Fe5(0,1)	Fe4(1,1)	136.7
2.166	S1	Fe2(-1,-1)	Fe4(0,-1)	60.6
2.166	S1	Fe2(-1,-1)	Fe6	92.6
2.365	S1	Fe4(0,-1)	Fe2(0,-1)	133.8
2.365	S1	Fe4(0,-1)	Fe5	94.6
2.365	S1	Fe4(0,-1)	Fe6	84.3
2.475	Fe2	Fe3	Fe4	60.4
2.475	Fe2	Fe3	Fe6(1,0)	93.9
2.475	Fe2	Fe3	Fe6	128.7

ILLUSTRATION: 116

COMMON NAME : GaAs(110)-(1x1)
CLASSIFICATION : 31.33.26
TECHNIQUE : LEED
AUTHORS : C.B. Duke and A. Paton
REFERENCE : J. Vac. Sci. Technol., B2, 327 (1984)

SURFACE TYPE
Substrate : GaAs
Crystal face: 110
Temperature : RT*
Bulk lattice: zincblende
2D bulk symm: pm
2D surf symm: pm

Adsorbate:
Coverage :
Pattern : (1x1)
Matrix : (1.000, 0.000)
(0.000, 1.000)

STRUCTURE TYPE
Relaxed bulk termination with 31.3° tilt in top layer

SAMPLE PREPARATION (sample)
Treatment : see Meyer et al,PRB19,5194(1979);Duke et al,SS127,L135(1983)
Crystallinity:
Anal. methods:
Contamination:

COMMENTS
Authors conclude parallel displacements in top bilayer are =<0.1Å;
this is re-examination of previously published data

DATA COLLECTION
Technique: LEED
Dataset : I-V curves: 30<E<240 eV

THEORY/DATA TREATMENT
Quasi-dynamical LEED: Vor=-10 eV; mfp=8Å; no thermal vibs

STRUCTURES EXAMINED
1. best-fit structure of Duke et al with lateral displacements of top Ga and As towards their unreconstructed lateral positions; 2. complete reoptimisation of structure: lateral displacements in the top bilayer to be zero - no improvement

QUALITY OF EXPERIMENT-THEORY FIT
RX=0.18

2D UNIT CELLS (1 domain observed)

Cell	Ax (Å)	Ay (Å)	Bx (Å)	By (Å)	α (°)	Matrix	Pattern	Cell type
Bulk	3.998	0.000	0.000	5.654	90.0	(1.000, 0.000) (0.000, 1.000)	(1x1)	b: bulk lattice
Surface 1	3.998	0.000	0.000	5.654	90.0	(1.000, 0.000) (0.000, 1.000)	(1x1)	s1: commens. superlattice

3D COORDINATES

As1-Ga2, Ga3-As4: 2 bilayers with tilted Ga-As chains; Ga5-As6: bulk bilayer;
Ga7-As8: periodically repeating bulk bilayer; 0.1Å error bars assumed for tabulation

Dx/Dy in Å, or as a fraction of layer's unit cell vectors; atom 0 at the origin. Epir/subr are bulk repeat vectors.

No. of atoms: 8 Bulk z = 1.999 Å

Reg ion	Chem el.	At. no.	Cell type	Site occ.	Rel. to	Dx ± εx		Dy ± εy		Dz ± εz		Dz/Bz(%) ± εz/Bz
epir		-2					f		f		Å	
subr		-1				1.999	Å	2.827	Å	1.999	Å	
intf	As	1	b	1.00	0	0.000	f	0.000	f	0.000	Å	0.0
intf	Ga	2	b	1.00	1	0.500 ± .025	f	0.201 ± .018	f	0.690 ± .100	Å	34.5 ± 5.0
intf	Ga	3	b	1.00	2	-0.500 ± .025	f	0.591 ± .018	f	1.440 ± .100	Å	72.0 ± 5.0
intf	As	4	b	1.00	3	0.500 ± .025	f	-0.250 ± .018	f	0.060 ± .100	Å	3.0 ± 5.0
intf	Ga	5	b	1.00	4	0.000 ± .025	f	-0.250 ± .018	f	1.970 ± .100	Å	98.6 ± 5.0
intf	As	6	b	1.00	5	-0.500	f	-0.250	f	0.000	Å	0.0
subl	Ga	7	b	1.00	6	0.000	f	0.750	f	1.999	Å	100.0
subl	As	8	b	1.00	7	0.500	f	-0.250	f	0.000	Å	0.0

GaAs(110)-(1x1)
31.33.26

BOND DISTANCES AND ANGLES

Bond distances and angles are derived from coordinates

No. of distances/angles: 10

Interatomic dist. A-B (Å)	Atom A	Atom B	Atom C	Bond angle A-B-C (°)
2.401	As1	Ga2	As1(1,0)	112.7
2.474	Ga3	As6(0,1)	Ga7	109.9
2.401	As1	Ga2	As4	123.4
2.433	As1	Ga3(0,-1)	As4(0,-1)	107.5
2.433	As1	Ga3(0,-1)	As6	116.3
2.443	Ga2	As4	Ga3	116.1
2.443	Ga2	As4	Ga5	92.2
2.449	Ga3	As4	Ga3(1,0)	109.4
2.449	Ga3	As4	Ga5	110.9
2.474	Ga3	As6(0,1)	Ga5(0,1)	109.3

ILLUSTRATION: 116

COMMON NAME : GaAs(110)-(1x1)
CLASSIFICATION : 31.33.27
TECHNIQUE : HEIS
AUTHORS : H.J. Grossman and W.M. Gibson
REFERENCE : J. Vac. Sci. Technol., B2, 343 (1984)

SURFACE TYPE
Substrate : GaAs
Crystal face: 110
Temperature : RT*
Bulk lattice: zincblende
2D bulk symm: pm
2D surf symm: pm

Adsorbate:
Coverage :
Pattern : (1x1)
Matrix : (1.000, 0.000)
(0.000, 1.000)

STRUCTURE TYPE
Relaxed bulk termination with 15.8° tilt in top layer

SAMPLE PREPARATION (1 sample)
Treatment : in situ sputtering and annealing at 620 K of wafer
Crystallinity:
Anal. methods:
Contamination: monitored by AES and LEED

COMMENTS
This technique has large error bars on normal displacements; an upper limit of 0.1 Å on the lateral displacements in the first layer was found

DATA COLLECTION
Technique: HEIS; surface peak intensities
Dataset : intensities for normal incidence as function of tilt angle about channel direction, and for <111> incidence

THEORY/DATA TREATMENT
High energy ion channeling with computer simulation of intensities; rms ampl=0.0895Å(Ga), 0.0928Å(As)

STRUCTURES EXAMINED
1) bulk terminated surface; 2) 27° 1st layer rotation model of Tong et al (1978); 3) 7° reconstruction of Duke et al (1983); possibility of lateral displacements was also examined

QUALITY OF EXPERIMENT-THEORY FIT
Visual

2D UNIT CELLS (1 domain observed)

Cell	Ax (Å)	Ay (Å)	Bx (Å)	By (Å)	α (°)	Matrix	Pattern	Cell type
Bulk	3.998	0.000	0.000	5.654	90.0	(1.000, 0.000) (0.000, 1.000)	(1x1)	b: bulk lattice
Surface 1	3.998	0.000	0.000	5.654	90.0	(1.000, 0.000) (0.000, 1.000)	(1x1)	s1: commens. superlattice

3D COORDINATES

As1-Ga2: top bilayer with tilted Ga-As chains; Ga3-As4, Ga5-As6: 2 bulk bilayers;
Ga7-As8: periodically repeating bulk bilayer;

Dx/Dy in Å, or as a fraction of layer's unit cell vectors; atom 0 at the origin. Epir/subr are bulk repeat vectors.

No. of atoms: 8

Bulk z = 1.999 Å

Region	Chem el.	At. no.	Cell type	Site occ.	Rel. to	Dx ± εx		Dy ± εy		Dz ± εz		Dz/Bz(%) ± εz/Bz
epir		-2					f		f		Å	
subr		-1				1.999	Å	2.827	Å	1.999	Å	
intf	As	1	b	1.00	0	0.000	f	0.000	f	0.000	Å	0.0
intf	Ga	2	b	1.00	1	0.500 ± .025	f	0.250 ± .018	f	0.400 ± .300	Å	20.0 ± 15.0
intf	Ga	3	b	1.00	2	-0.500 ± .025	f	0.500 ± .018	f	1.790 ± .300	Å	89.5 ± 15.0
intf	As	4	b	1.00	3	0.500	f	-0.250	f	0.000	Å	0.0
intf	Ga	5	b	1.00	4	0.000	f	-0.250	f	1.999	Å	100.0
intf	As	6	b	1.00	5	-0.500	f	-0.250	f	0.000	Å	0.0
subl	Ga	7	b	1.00	6	0.000	f	0.750	f	1.999	Å	100.0
subl	As	8	b	1.00	7	0.500	f	-0.250	f	0.000	Å	0.0

GaAs(110)-(1x1)
31.33.27

BOND DISTANCES AND ANGLES

Bond distances and angles are derived from coordinates

No. of distances/angles: 10

Interatomic dist. A-B (Å)	Atom A	Atom B	Atom C	Bond angle A-B-C (°)
2.481	As1	Ga2	As1(1,0)	107.4
2.448	Ga3	As6(0,1)	Ga7	109.5
2.481	As1	Ga2	As4	118.7
2.607	As1	Ga3(0,-1)	As4(0,-1)	108.3
2.607	As1	Ga3(0,-1)	As6	111.9
2.281	Ga2	As4	Ga3	111.0
2.281	Ga2	As4	Ga5	106.4
2.448	Ga3	As4	Ga3(1,0)	109.5
2.448	Ga3	As4	Ga5	109.5
2.448	Ga3	As6(0,1)	Ga5(0,1)	109.5

COMMON NAME : GaAs(110)-(1x1)
CLASSIFICATION : 31.33.29a
TECHNIQUE : LEED
AUTHORS : S.Y. Tong, W.M. Mei and G. Xu
REFERENCE : J. Vac. Sci. Technol., B2, 393 (1984)

ILLUSTRATION: 116

SURFACE TYPE
Substrate : GaAs
Crystal face: 110
Temperature : RT*
Bulk lattice: zincblende
2D bulk symm: pm
2D surf symm: pm

Adsorbate:
Coverage :
Pattern : (1x1)
Matrix : (1.000, 0.000)
(0.000, 1.000)

STRUCTURE TYPE
Relaxed bulk termination with 28° tilt in top layer

SAMPLE PREPARATION (sample)
Treatment :
Crystallinity:
Anal. methods:
Contamination:

COMMENTS

DATA COLLECTION
Technique: LEED
Dataset : I-V spectra: 14 beams from previously published exp. results: Tong et al, Phys. Rev. B17, 3303 (1978)

THEORY/DATA TREATMENT
Dynamical LEED (combined space method)

STRUCTURES EXAMINED
1. bond conserving tilt of top layer Ga-As bond; 2. lateral relaxations of top Ga and As

QUALITY OF EXPERIMENT-THEORY FIT
RN and RZJ used

2D UNIT CELLS (1 domain observed)

Cell	Ax (Å)	Ay (Å)	Bx (Å)	By (Å)	α (°)	Matrix	Pattern	Cell type
Bulk	3.998	0.000	0.000	5.654	90.0	(1.000, 0.000) (0.000, 1.000)	(1x1)	b: bulk lattice
Surface 1	3.998	0.000	0.000	5.654	90.0	(1.000, 0.000) (0.000, 1.000)	(1x1)	s1: commens. superlattice

3D COORDINATES

As1-Ga2, Ga3-As4: 2 bilayers with tilted Ga-As chains; Ga5-As6: bulk bilayer;
Ga7-As8: periodically repeating bulk bilayer; 0.1Å error bars assumed for tabulation

Dx/Dy in Å, or as a fraction of layer's unit cell vectors; atom 0 at the origin. Epir/subr are bulk repeat vectors.

No. of atoms: 8

Bulk z = 1.999 Å

Region	Chem el.	At. no.	Cell type	Site occ.	Rel. to	Dx ± εx		Dy ± εy		Dz ± εz		Dz/Bz(%) ± εz/Bz
epir		-2					f		f		Å	
subr		-1				1.999	Å	2.827	Å	1.999	Å	
intf	As	1	b	1.00	0	0.000	f	0.000	f	0.000	Å	0.0
intf	Ga	2	b	1.00	1	0.500 ± .025	f	0.229 ± .018	f	0.690 ± .100	Å	34.5 ± 5.0
intf	Ga	3	b	1.00	2	-0.500 ± .025	f	0.560 ± .018	f	1.470 ± .100	Å	73.5 ± 5.0
intf	As	4	b	1.00	3	0.500 ± .025	f	-0.250 ± .018	f	0.030 ± .100	Å	1.5 ± 5.0
intf	Ga	5	b	1.00	4	0.000 ± .025	f	-0.250 ± .018	f	1.980 ± .100	Å	99.1 ± 5.0
intf	As	6	b	1.00	5	-0.500	f	-0.250	f	0.000	Å	0.0
subl	Ga	7	b	1.00	6	0.000	f	0.750	f	1.999	Å	100.0
subl	As	8	b	1.00	7	0.500	f	-0.250	f	0.000	Å	0.0

GaAs(110)-(1x1)
31.33.29a

BOND DISTANCES AND ANGLES

Bond distances and angles are derived from coordinates

No. of distances/angles: 10

Interatomic dist. A-B (Å)	Atom A	Atom B	Atom C	Bond angle A-B-C (°)
2.480	As1	Ga2	As1(1,0)	107.5
2.457	Ga3	As6(0,1)	Ga7	109.6
2.480	As1	Ga2	As4	125.3
2.468	As1	Ga3(0,-1)	As4(0,-1)	106.9
2.468	As1	Ga3(0,-1)	As6	116.0
2.307	Ga2	As4	Ga3	115.5
2.307	Ga2	As4	Ga5	95.0
2.448	Ga3	As4	Ga3(1,0)	109.5
2.448	Ga3	As4	Ga5	110.2
2.457	Ga3	As6(0,1)	Ga5(0,1)	109.4

COMMON NAME : GaAs(110)-(1x1)
CLASSIFICATION : 31.33.68
TECHNIQUE : LEED
AUTHORS : W.K. Ford, T. Guo, D.L. Lessor and C.B. Duke
REFERENCE : Phys. Rev., B42, 8952 (1990)

ILLUSTRATION: 116

SURFACE TYPE
Substrate : GaAs
Crystal face: 110
Temperature : 1100 K
Bulk lattice: zincblende
2D bulk symm: pm
2D surf symm: pm

Adsorbate:
Coverage :
Pattern : (1x1)
Matrix : (1.000, 0.000)
(0.000, 1.000)

STRUCTURE TYPE
Relaxed bulk termination with 28.4° tilt in top layer and other relaxations in top two layers

SAMPLE PREPARATION (sample)
Treatment : cleavage in vacuum
Crystallinity:
Anal. methods:
Contamination:

COMMENTS

DATA COLLECTION
Technique: LEED
Dataset : 18 ineq. symmetry-averaged I-V curves: 50<E<300 eV

THEORY/DATA TREATMENT
Dynamical LEED (quasi-dyn. below 4th bilayer): 6 ph shs from relat. Hartree-Fock-Slater pot.; E-dep. mfp; no thermal vibs

STRUCTURES EXAMINED
Relaxation of 5 struct. parameters: 1st and 2nd bilayer rotation angles, Ga-As bond lengths in 1st bilayer and between 1st and 2nd bilayers

QUALITY OF EXPERIMENT-THEORY FIT
RX=0.204

2D UNIT CELLS (1 domain observed)

Cell	Ax (Å)	Ay (Å)	Bx (Å)	By (Å)	α (°)	Matrix	Pattern	Cell type
Bulk	3.997	0.000	0.000	5.653	90.0	(1.000, 0.000) (0.000, 1.000)	(1x1)	b: bulk lattice
Surface 1	3.997	0.000	0.000	5.653	90.0	(1.000, 0.000) (0.000, 1.000)	(1x1)	s1: commens. superlattice

3D COORDINATES

As1-Ga2, Ga3-As4: 2 bilayers with tilted Ga-As chains; As5-Ga6: unrelaxed bulk bilayer;
As7-Ga8: periodically repeating bulk bilayer; 0.1Å error bars assumed for tabulation

Dx/Dy in Å, or as a fraction of layer's unit cell vectors; atom 0 at the origin. Epir/subr are bulk repeat vectors.

No. of atoms: 8 Bulk z = 1.999 Å

Region	Chem el.	At. no.	Cell type	Site occ.	Rel. to	Dx ± εx		Dy ± εy		Dz ± εz		Dz/Bz(%) ± εz/Bz
epir		-2					f		f		Å	
subr		-1				1.999	Å	2.827	Å	1.999	Å	
intf	As	1	b	1.00	0	0.000	f	0.000	f	0.000	Å	0.0
intf	Ga	2	b	1.00	1	0.500	f	0.229 ± .018	f	0.698 ± .100	Å	34.9 ± 5.0
intf	Ga	3	b	1.00	2	-0.500	f	0.572 ± .018	f	1.444 ± .100	Å	72.2 ± 5.0
intf	As	4	b	1.00	3	0.500	f	-0.250 ± .018	f	0.084 ± .100	Å	4.2 ± 5.0
intf	As	5	b	1.00	4	-0.500	f	-0.499 ± .018	f	1.955 ± .100	Å	97.8 ± 5.0
intf	Ga	6	b	1.00	5	0.500	f	0.238	f	0.000	Å	0.0
subl	As	7	b	1.00	6	0.000	f	0.250	f	1.999	Å	100.0
subl	Ga	8	b	1.00	7	-0.500	f	0.250	f	0.000	Å	0.0

GaAs(110)-(1x1)
31.33.68

BOND DISTANCES AND ANGLES

Bond distances and angles are derived from coordinates

No. of distances/angles: 10

Interatomic dist. A-B (Å)	Atom A	Atom B	Atom C	Bond angle A-B-C (°)
2.480	As1	Ga2	As1(1,0)	107.4
2.485	Ga3	As5(0,1)	Ga8	108.6
2.480	As1	Ga2	As4	125.5
2.420	As1	Ga3(0,-1)	As4(-1,-1)	107.4
2.420	As1	Ga3(0,-1)	As5	117.4
2.380	Ga2	As4	Ga3	114.8
2.380	Ga2	As4	Ga6	92.9
2.448	Ga3	As4	Ga3(1,0)	109.5
2.448	Ga3	As4	Ga6	112.0
2.485	Ga3	As5(0,1)	Ga6(0,1)	108.7

COMMON NAME : GaAs(111)-(2x2) ILLUSTRATION: 117
CLASSIFICATION : 31.33.24
TECHNIQUE : LEED
AUTHORS : S.Y. Tong. G. Xu and W.M. Mei
REFERENCE : Phys. Rev. Lett., 52, 1693 (1984)

SURFACE TYPE
Substrate : GaAs
Crystal face: 111
Temperature : RT*
Bulk lattice: zincblende
2D bulk symm: p3m1
2D surf symm: p3m1

Adsorbate:
Coverage :
Pattern : (2x2)
Matrix : (2.000, 0.000)
(0.000, 2.000)

STRUCTURE TYPE
1 missing Ga per (2x2) unit cell in heavily relaxed top bilayer; top bilayer almost planar

SAMPLE PREPARATION (1 sample)
Treatment : Ar+ bombardment, then annealing at 500C at 3E-10 torr
Crystallinity:
Anal. methods:
Contamination:

COMMENTS
Sp3 bonds rehybridized to sp2 for Ga surface atoms
same analysis also given in Tong et al, JVST B2, 393 (1984);
the R-factor used is the 'normalised R-factor' of Tong et al

DATA COLLECTION
Technique: LEED
Dataset : I-V spectra for 5 integral order and 5 fractional order beams

THEORY/DATA TREATMENT
Dynamical LEED (combined space method)

STRUCTURES EXAMINED
M1: 1/4 layer vacancies in top layer + vertical compression; M2: same as M1 + lateral shifts in 1st layer;
M3: as for M2 but with additional vertical shifts; M4: structure given in the tables;
M5: M4 with the vacancy filled with an As atom

QUALITY OF EXPERIMENT-THEORY FIT
RN=0.127 (see comments)

2D UNIT CELLS (1 domain observed)

Cell	Ax (Å)	Ay (Å)	Bx (Å)	By (Å)	α (°)	Matrix	Pattern	Cell type
Bulk	3.998	0.000	1.999	3.462	60.0	(1.000, 0.000) (0.000, 1.000)	(1x1)	b: bulk lattice
Surface 1	7.996	0.000	3.998	6.925	60.0	(2.000, 0.000) (0.000, 2.000)	(2x2)	s1: commens. superlattice

3D COORDINATES

Ga1 through As7: reconstructed top bilayer; Ga8 through As15: slightly distorted bulk bilayer;
Ga16-As17: periodically repeating bulk bilayer; 0.1Å error bars assumed for tabulation

Dx/Dy in Å, or as a fraction of layer's unit cell vectors; atom 0 at the origin. Epir/subr are bulk repeat vectors.

No. of atoms: 17 Bulk z = 3.264 Å

Reg ion	Chem el.	At. no.	Cell type	Site occ.	Rel. to	Dx ± εx		Dy ± εy		Dz ± εz		Dz/Bz(%) ± εz/Bz
epir		-2					f		f		Å	
subr		-1				1.999	Å	1.154	Å	3.264	Å	
intf	Ga	1	s1	.25	0	0.522	f	0.000	f	0.000	Å	0.0
intf	Ga	2	s1	.25	1	-0.522 ± .005	f	0.000 ± .014	f	0.000	Å	0.0
intf	Ga	3	s1	.25	2	0.000 ± .005	f	0.522 ± .014	f	0.000	Å	0.0
intf	As	4	s1	.25	3	0.214 ± .005	f	0.132 ± .014	f	0.070 ± .100	Å	2.1 ± 3.1
intf	As	5	s1	.25	4	0.439 ± .005	f	-0.439 ± .014	f	0.000	Å	0.0
intf	As	6	s1	.25	5	0.000 ± .005	f	0.439 ± .014	f	0.000	Å	0.0
intf	As	7	s1	.25	6	-0.480 ± .005	f	-0.480 ± .014	f	0.130 ± .100	Å	4.0 ± 3.1
intf	Ga	8	s1	.25	7	0.500	f	0.000	f	2.350 ± .100	Å	72.0 ± 3.1
intf	Ga	9	s1	.25	8	-0.500	f	0.500	f	0.000	Å	0.0
intf	Ga	10	s1	.25	9	0.500	f	0.000	f	0.000	Å	0.0
intf	Ga	11	s1	.25	10	-0.500	f	-0.500	f	0.090 ± .100	Å	2.8 ± 3.1
intf	As	12	s1	.25	11	0.167	f	0.667	f	0.740 ± .100	Å	22.7 ± 3.1
intf	As	13	s1	.25	12	0.000	f	-0.500	f	0.000	Å	0.0
intf	As	14	s1	.25	13	0.500	f	0.000	f	0.000	Å	0.0
intf	As	15	s1	.25	14	0.000	f	0.500	f	0.000	Å	0.0
subl	Ga	16	b	1.00	15	0.000	f	0.000	f	2.448	Å	75.0
subl	As	17	b	1.00	16	-0.667	f	-0.667	f	0.816	Å	25.0

GaAs(111)-(2x2)
31.33.24 BOND DISTANCES AND ANGLES

Bond distances and angles are derived from coordinates

No. of distances/angles: 12

Interatomic dist. A-B (Å)	Atom A	Atom B	Atom C	Bond angle A-B-C (°)
2.422	Ga1	As5	Ga2(1,0)	104.3
2.453	Ga8	As13	Ga9	109.2
2.453	Ga8	As13	Ga11	110.2
2.453	Ga8	As13	Ga16(-1,-1)	109.8
2.422	Ga1	As5	Ga8	87.7
2.416	Ga1	As7	Ga2	119.3
2.416	Ga1	As7	Ga11	94.8
2.422	As4	Ga2(0,1)	As5(-1,1)	135.5
2.422	As4	Ga2(0,1)	As7(0,1)	111.9
2.496	As4	Ga9	As12	106.5
2.496	As4	Ga9	As12(-1,0)	116.2
2.440	As7	Ga11	As13	107.8

COMMON NAME : GaAs(311)-(1x1) As termination
ILLUSTRATION: 121
CLASSIFICATION : 31.33.53b
TECHNIQUE : LEED
AUTHORS : C.B. Duke, C. Mailhiot, A. Paton, A. Kahn and K. Stiles
REFERENCE : J. Vac. Sci. Technol., A4, 947 (1986)

SURFACE TYPE
Substrate : GaAs
Crystal face: 311
Temperature : 300 K
Bulk lattice: zincblende
2D bulk symm: pm
2D surf symm: pm

Adsorbate:
Coverage :
Pattern : (1x1)
Matrix : (1.000, 0.000)
(0.000, 1.000)

STRUCTURE TYPE
As-terminated bulk lattice with spacing relaxations between top 3 mono-atomic layers

SAMPLE PREPARATION (1 sample)
Treatment : crystal cut, polished, ion bombarded and vacuum annealed
Crystallinity:
Anal. methods: etching assessed termination before
Contamination:

COMMENTS
RX=x-ray R-factor, RI=integrated intensity R-factor

DATA COLLECTION
Technique: LEED
Dataset : I-V spectra: normal incidence, 15 beams, 40<E<240 eV

THEORY/DATA TREATMENT
Dynamical LEED: 6 phase shifts

STRUCTURES EXAMINED
1. relaxed clean surface - As termination; 2. 3-fold Ga adatom on relaxed clean surface with As term.; 3. relaxed clean surface - Ga termination

QUALITY OF EXPERIMENT-THEORY FIT
RX=0.24, RI=0.06

2D UNIT CELLS (1 domain observed)

Cell	Ax (Å)	Ay (Å)	Bx (Å)	By (Å)	α (°)	Matrix	Pattern	Cell type
Bulk	3.998	0.000	-1.999	6.630	106.8	(1.000, 0.000) (0.000, 1.000)	(1x1)	b: bulk lattice
Surface 1	3.998	0.000	-1.999	6.630	106.8	(1.000, 0.000) (0.000, 1.000)	(1x1)	s1: commens. superlattice

3D COORDINATES

As3-Ga4: periodically repeating set of bulk layers; 0.1Å error bars assumed for tabulation

Dx/Dy in Å, or as a fraction of layer's unit cell vectors; atom 0 at the origin. Epir/subr are bulk repeat vectors.

No. of atoms: 4 Bulk z = 1.710 Å

Region	Chem el.	At. no.	Cell type	Site occ.	Rel. to	Dx ± εx		Dy ± εy		Dz ± εz		Dz/Bz(%) ± εz/Bz
epir		-2					f		f		Å	
subr		-1				-0.000	Å	3.617	Å	1.710	Å	
intf	As	1	b	1.00	0	0.000	f	0.000	f	0.000	Å	0.0
intf	Ga	2	b	1.00	1	0.818	f	0.636	f	0.630 ± .100	Å	36.8 ± 5.9
subl	As	3	b	1.00	2	-0.546	f	-0.091	f	1.180 ± .100	Å	69.0 ± 5.9
subl	Ga	4	b	1.00	3	-0.182	f	-0.364	f	0.430	Å	25.2

BOND DISTANCES AND ANGLES

Bond distances and angles are derived from coordinates

No. of distances/angles: 3

Interatomic dist. A-B (Å)	Atom A	Atom B	Atom C	Bond angle A-B-C (°)
2.492	As1	Ga2(-1,-1)	As3(0,-1)	111.6
2.398	Ga2	As3	Ga4	109.5
2.449	As3	Ga4		

COMMON NAME : GaAs(311)-(1x1) Ga termination
CLASSIFICATION : 31.33.53a
TECHNIQUE : LEED
AUTHORS : C.B. Duke, C. Mailhiot, A. Paton, A. Kahn and K. Stiles
REFERENCE : J. Vac. Sci. Technol., A4, 947 (1986)

ILLUSTRATION: 122

SURFACE TYPE
Substrate : GaAs
Crystal face: 311
Temperature : 300 K
Bulk lattice: zincblende
2D bulk symm: pm
2D surf symm: pm

Adsorbate:
Coverage :
Pattern : (1x1)
Matrix : (1.000, 0.000)
(0.000, 1.000)

STRUCTURE TYPE
Ga-terminated bulk lattice with spacing relaxations between top 3 mono-atomic layers

SAMPLE PREPARATION (1 sample)
Treatment : crystal cut, polished, ion bombarded and vacuum annealed
Crystallinity:
Anal. methods: etching assessed termination before
Contamination:

COMMENTS
RX=x-ray R-factor, RI=integrated intensity R-factor

DATA COLLECTION
Technique: LEED
Dataset : I-V spectra: normal incidence, 15 beams, 40<E<240 eV

THEORY/DATA TREATMENT
Dynamical LEED: 6 phase shifts

STRUCTURES EXAMINED
1. relaxed clean surface - As termination; 2. 3-fold Ga adatom on relaxed clean surface with As term.;
3. relaxed clean surface - Ga termination

QUALITY OF EXPERIMENT-THEORY FIT
RX=0.24, RI=0.08

2D UNIT CELLS (1 domain observed)

Cell	Ax (Å)	Ay (Å)	Bx (Å)	By (Å)	α (°)	Matrix	Pattern	Cell type
Bulk	3.998	0.000	-1.999	6.630	106.8	(1.000, 0.000) (0.000, 1.000)	(1x1)	b: bulk lattice
Surface 1	3.998	0.000	-1.999	6.630	106.8	(1.000, 0.000) (0.000, 1.000)	(1x1)	s1: commens. superlattice

3D COORDINATES

As4-Ga5: periodically repeating set of bulk layers; 0.1Å error bars assumed for tabulation

Dx/Dy in Å, or as a fraction of layer's unit cell vectors; atom 0 at the origin. Epir/subr are bulk repeat vectors.

No. of atoms: 5

Bulk z = 1.710 Å

Region	Chem el.	At. no.	Cell type	Site occ.	Rel. to	Dx ± εx		Dy ± εy		Dz ± εz		Dz/Bz(%) ± εz/Bz
epir		-2					f		f		Å	
subr		-1				-0.000	Å	3.617	Å	1.710	Å	
intf	Ga	1	b	1.00	0	0.000	f	0.000	f	0.000	Å	0.0
intf	As	2	b	1.00	1	0.455	f	0.909	f	1.080 ± .100	Å	63.2 ± 5.9
intf	Ga	3	b	1.00	2	-0.182	f	-0.364	f	0.580 ± .100	Å	33.9 ± 5.9
subl	As	4	b	1.00	3	0.455	f	-0.091	f	1.280	Å	74.9
subl	Ga	5	b	1.00	4	-0.182	f	-0.364	f	0.430	Å	25.2

BOND DISTANCES AND ANGLES

Bond distances and angles are derived from coordinates

No. of distances/angles: 6

Interatomic dist. A-B (Å)	Atom A	Atom B	Atom C	Bond angle A-B-C (°)
2.351	Ga1	As2(0,-1)	Ga3(0,-1)	110.9
2.351	Ga1	As2(0,-1)	Ga5	106.7
2.480	As2	Ga3	As4	111.2
2.480	As2	Ga3	As4(-1,0)	111.2
2.588	As2	Ga5(0,1)	As4(0,1)	107.7

GaAs(311)-(1x1) Ga termination
31.33.53a

Bond Distances and Angles - Continued

Interatomic dist. A-B (Å)	Atom A	Atom B	Atom C	Bond angle A-B-C (°)
2.449	Ga3	As4	Ga5	109.5

COMMON NAME : GaAs(110)-(1x1)-1Al (low coverage) ILLUSTRATION: 127
CLASSIFICATION : 31.33.13.4a
TECHNIQUE : LEED
AUTHORS : A. Kahn, J. Carelli, D. Kanani, C.B. Duke, A. Paton and L. Brillson
REFERENCE : J. Vac. Sci. Technol., 19, 331 (1981)

SURFACE TYPE
Substrate : GaAs Adsorbate: Al
Crystal face: 110 Coverage : 0.5-1.0ML (Al/1x1)
Temperature : RT Pattern : (1x1)
Bulk lattice: zincblende Matrix : (1.000, 0.000) (0.000, 1.000)
2D bulk symm: pm
2D surf symm: pm

STRUCTURE TYPE
Substitution of Al in 2nd-layer Ga positions; otherwise same structure as GaAs(110) (class. no. 31.33.26), except for 0.1Å reduction of 1st-2nd layer spacing

SAMPLE PREPARATION (>2 sample)
Treatment : sputter-annealed n-type wafers; also cleaved surfaces
Crystallinity:
Anal. methods: Ga-3d, As-3d, Al-2p core level SXPS spectra
Contamination: monitored by LEED

COMMENTS
RT deposition of Al on GaAs(110) produces disordered overlayer with possibly some random Al-Ga replacements; annealing leads to ordered structures in which Al first replaces Ga in 2nd layer and then in 3rd, 1st and deeper layers (see class. nos. 31.33.13.4b and c)

DATA COLLECTION
Technique: LEED
Dataset : I-V spectra for beams: 01,0-1,11=-11,1-1=-1-1,10=-10,02=0-2, 1-2=-1-2,2-1=-2-1 at normal incidence

THEORY/DATA TREATMENT
Dynamical LEED: overlapping atomic potentials with Slater exchange

STRUCTURES EXAMINED
Substitutional replacements in GaAs(110): 0.5ML with Al replacing Ga in 1st or 2nd layer; 1ML with Al replacing Ga in 1st and 2nd, or 2nd and 3rd layers; large coverages with Al in 1st 3 layers, or 6 layers (pure AlAs); from GaAs(110) struc., relaxations of 1st-2nd layer spacing

QUALITY OF EXPERIMENT-THEORY FIT
RZJ=0.24, RX=0.25

2D UNIT CELLS (1 domain observed)

Cell	Ax (Å)	Ay (Å)	Bx (Å)	By (Å)	α (°)	Matrix	Pattern	Cell type
Bulk	4.000	0.000	0.000	5.660	90.0	(1.000, 0.000) (0.000, 1.000)	(1x1)	b: bulk lattice
Surface 1	4.000	0.000	0.000	5.660	90.0	(1.000, 0.000) (0.000, 1.000)	(1x1)	s1: commens. superlattice

3D COORDINATES

As1-Ga2, Al3-As4: 2 bilayers with tilted chains; Ga5-As6: bulk bilayer;
Ga7-As8: periodically repeating bulk bilayer; 0.1Å error bars assumed for tabulation

Dx/Dy in Å, or as a fraction of layer's unit cell vectors; atom 0 at the origin. Epir/subr are bulk repeat vectors.

No. of atoms: 8 Bulk z = 2.000 Å

Region	Chem el.	At. no.	Cell type	Site occ.	Rel. to	Dx ± εx		Dy ± εy		Dz ± εz		Dz/Bz(%) ± εz/Bz
epir		-2					f		f		Å	
subr		-1				2.000	Å	2.830	Å	2.000	Å	
intf	As	1	b	1.00	0	0.000	f	0.000	f	0.000	Å	0.0
intf	Ga	2	b	1.00	1	0.500 ± .025	f	0.201 ± .018	f	0.690 ± .100	Å	34.5 ± 5.0
intf	Al	3	b	1.00	2	-0.500 ± .025	f	0.591 ± .018	f	1.340 ± .100	Å	67.0 ± 5.0
intf	As	4	b	1.00	3	0.500 ± .025	f	-0.250 ± .018	f	0.060 ± .100	Å	3.0 ± 5.0
intf	Ga	5	b	1.00	4	0.000 ± .025	f	-0.250 ± .018	f	1.970 ± .100	Å	98.5 ± 5.0
intf	As	6	b	1.00	5	-0.500	f	-0.250	f	0.000	Å	0.0
subl	Ga	7	b	1.00	6	0.000	f	0.750	f	2.000	Å	100.0
subl	As	8	b	1.00	7	0.500	f	-0.250	f	0.000	Å	0.0

GaAs(110)-(1x1)-1Al (low coverage)
31.33.13.4a BOND DISTANCES AND ANGLES

Bond distances and angles are derived from coordinates

No. of distances/angles: 10

Interatomic dist. A-B (Å)	Atom A	Atom B	Atom C	Bond angle A-B-C (°)
2.402	As1	Ga2	As1(1,0)	112.7
2.475	Al3	As6(0,1)	Ga7	109.8
2.402	As1	Ga2	As4	123.5
2.347	As1	Al3(0,-1)	As4(0,-1)	108.1
2.347	As1	Al3(0,-1)	As6	115.0
2.384	Ga2	As4	Al3	116.9
2.384	Ga2	As4	Ga5	90.3
2.451	Al3	As4	Al3(1,0)	109.4
2.451	Al3	As4	Ga5	110.9
2.475	Al3	As6(0,1)	Ga5(0,1)	109.3

COMMON NAME : GaAs(110)-(1x1)-2Al (medium coverage) ILLUSTRATION: 127
CLASSIFICATION : 31.33.13.4b
TECHNIQUE : LEED
AUTHORS : A. Kahn, J. Carelli, D. Kanani, C.B. Duke, A. Paton and L. Brillson
REFERENCE : J. Vac. Sci. Technol., 19, 331 (1981)

SURFACE TYPE
Substrate : GaAs
Crystal face: 110
Temperature : RT
Bulk lattice: zincblende
2D bulk symm: pm
2D surf symm: pm
Adsorbate: Al
Coverage : 1.5-1.0ML (Al/1x1)
Pattern : (1x1)
Matrix : (1.000, 0.000)
(0.000, 1.000)

STRUCTURE TYPE
Substitution of Al in 2nd- and 3rd-layer Ga positions; otherwise same structure as GaAs(110) (class. no. 31.33.26), except for 0.1Å reduction of 1st-2nd layer spacing

SAMPLE PREPARATION (>2 sample)
Treatment : sputter-annealed n-type wafers; also cleaved surfaces
Crystallinity:
Anal. methods: Ga-3d, As-3d, Al-2p core level SXPS spectra
Contamination: monitored by LEED

COMMENTS
RT deposition of Al on GaAs(110) produces disordered overlayer with possibly some random Al-Ga replacements; annealing leads to ordered structures in which Al first replaces Ga in 2nd layer and then in 3rd, 1st and deeper layers (see class. nos. 31.33.13.4a and c)

DATA COLLECTION
Technique: LEED
Dataset : I-V spectra for beams:
01,0-1,11=-11,1-1=-1-1,10=-10,02=0-2,
1-2=-1-2,2-1=-2-1 at normal incidence

THEORY/DATA TREATMENT
Dynamical LEED: overlapping atomic potentials with Slater exchange

STRUCTURES EXAMINED
Substitutional replacements in GaAs(110): 0.5ML with Al replacing Ga in 1st or 2nd layer; 1ML with Al replacing Ga in 1st and 2nd or 2nd and 3rd layers; large coverages with Al in 1st 3 layers, or 6 layers (pure AlAs); from GaAs(110) struc., relaxations of 1st-2nd layer spacing

QUALITY OF EXPERIMENT-THEORY FIT
RZJ=0.29, RX=0.29

2D UNIT CELLS (1 domain observed)

Cell	Ax (Å)	Ay (Å)	Bx (Å)	By (Å)	α (°)	Matrix	Pattern	Cell type
Bulk	4.000	0.000	0.000	5.660	90.0	(1.000, 0.000) (0.000, 1.000)	(1x1)	b: bulk lattice
Surface 1	4.000	0.000	0.000	5.660	90.0	(1.000, 0.000) (0.000, 1.000)	(1x1)	s1: commens. superlattice

3D COORDINATES

As1-Ga2, Al3-As4: 2 bilayers with tilted chains; Al5-As6: bulk-like bilayer;
Ga7-As8: periodically repeating bulk bilayer; 0.1Å error bars assumed for tabulation

Dx/Dy in Å, or as a fraction of layer's unit cell vectors; atom 0 at the origin. Epir/subr are bulk repeat vectors.

No. of atoms: 8 Bulk z = 2.000 Å

Region	Chem el.	At. no.	Cell type	Site occ.	Rel. to	Dx ± εx		Dy ± εy		Dz ± εz		Dz/Bz(%) ± εz/Bz
epir		-2					f		f		Å	
subr		-1				2.000	Å	2.830	Å	2.000	Å	
intf	As	1	b	1.00	0	0.000	f	0.000	f	0.000	Å	0.0
intf	Ga	2	b	1.00	1	0.500 ± .025	f	0.201 ± .018	f	0.690 ± .100	Å	34.5 ± 5.0
intf	Al	3	b	1.00	2	-0.500 ± .025	f	0.591 ± .018	f	1.340 ± .100	Å	67.0 ± 5.0
intf	As	4	b	1.00	3	0.500 ± .025	f	-0.250 ± .018	f	0.060 ± .100	Å	3.0 ± 5.0
intf	Al	5	b	1.00	4	0.000 ± .025	f	-0.250 ± .018	f	1.970 ± .100	Å	98.5 ± 5.0
intf	As	6	b	1.00	5	-0.500	f	-0.250	f	0.000	Å	0.0
subl	Ga	7	b	1.00	6	0.000	f	0.750	f	2.000	Å	100.0
subl	As	8	b	1.00	7	0.500	f	-0.250	f	0.000	Å	0.0

GaAs(110)-(1x1)-2Al (medium coverage)
31.33.13.4b BOND DISTANCES AND ANGLES

Bond distances and angles are derived from coordinates

No. of distances/angles: 10

Interatomic dist. A-B (Å)	Atom A	Atom B	Atom C	Bond angle A-B-C (°)
2.402	As1	Ga2	As1(1,0)	112.7
2.475	Al3	As6(0,1)	Ga7	109.8
2.402	As1	Ga2	As4	123.5
2.347	As1	Al3(0,-1)	As4(0,-1)	108.1
2.347	As1	Al3(0,-1)	As6	115.0
2.384	Ga2	As4	Al3	116.9
2.384	Ga2	As4	Al5	90.3
2.451	Al3	As4	Al3(1,0)	109.4
2.451	Al3	As4	Al5	110.9
2.475	Al3	As6(0,1)	Al5(0,1)	109.3

COMMON NAME : GaAs(110)-(1x1)-3Al (high coverage) ILLUSTRATION: 127
CLASSIFICATION : 31.33.13.4c
TECHNIQUE : LEED
AUTHORS : A. Kahn, J. Carelli, D. Kanani, C.B. Duke, A. Paton and L. Brillson
REFERENCE : J. Vac. Sci. Technol., 19, 331 (1981)

SURFACE TYPE
Substrate : GaAs Adsorbate: Al
Crystal face: 110 Coverage : >=3.5ML (Al/1x1)
Temperature : RT Pattern : (1x1)
Bulk lattice: zincblende Matrix : (1.000, 0.000)
2D bulk symm: pm (0.000, 1.000)
2D surf symm: pm

STRUCTURE TYPE
Substitution of Al in 1st, 2nd- and 3rd-layer Ga positions; otherwise same structure as GaAs(110) (class. no. 31.33.26), except for 0.1Å reduction of 1st-2nd layer spacing; higher coverages cause Al substitution in deeper layers, e.g. at 8.5ML substitution down to at least 6th layer

SAMPLE PREPARATION (>2 sample)
Treatment : sputter-annealed n-type wafers; also cleaved surfaces
Crystallinity:
Anal. methods: Ga-3d, As-3d, Al-2p core level SXPS spectra
Contamination: monitored by LEED

COMMENTS
RT deposition of Al on GaAs(110) produces disordered overlayer with possibly some random Al-Ga replacements; annealing leads to ordered structures in which Al first replaces Ga in 2nd layer and then in 3rd, 1st and deeper layers (see class. nos. 31.33.13.4a and b)

DATA COLLECTION
Technique: LEED
Dataset : I-V spectra for beams:
01,0-1,11=-11,1-1=-1-1,10=-10,02=0-2,
1-2=-1-2,2-1=-2-1 at normal incidence

THEORY/DATA TREATMENT
Dynamical LEED: overlapping atomic potentials with Slater exchange

STRUCTURES EXAMINED
Substitutional replacements in GaAs(110): 0.5ML with Al replacing Ga in 1st or 2nd layer; 1ML with Al replacing Ga in 1st and 2nd or 2nd and 3rd layers; large coverages with Al in 1st 3 layers, or 6 layers (pure AlAs); from GaAs(110) struc., relaxations of 1st-2nd layer spacing

QUALITY OF EXPERIMENT-THEORY FIT
RZJ=0.29, RX=0.22

2D UNIT CELLS (1 domain observed)

Cell	Ax (Å)	Ay (Å)	Bx (Å)	By (Å)	α (°)	Matrix	Pattern	Cell type
Bulk	4.000	0.000	0.000	5.660	90.0	(1.000, 0.000) (0.000, 1.000)	(1x1)	b: bulk lattice
Surface 1	4.000	0.000	0.000	5.660	90.0	(1.000, 0.000) (0.000, 1.000)	(1x1)	s1: commens. superlattice

3D COORDINATES

As1-Al2, Al3-As4: 2 bilayers with tilted chains; Al5-As6: bulk-like bilayer;
Ga7 (or Al7)-As8: periodically repeating bulk bilayer; 0.1Å error bars assumed for tabulation

Dx/Dy in Å, or as a fraction of layer's unit cell vectors; atom 0 at the origin. Epir/subr are bulk repeat vectors.

No. of atoms: 8 Bulk z = 2.000 Å

Reg ion	Chem el.	At. no.	Cell type	Site occ.	Rel. to	Dx ± εx		Dy ± εy		Dz ± εz		Dz/Bz(%) ± εz/Bz
epir		-2					f		f		Å	
subr		-1				2.000	Å	2.830	Å	2.000	Å	
intf	As	1	b	1.00	0	0.000	f	0.000	f	0.000	Å	0.0
intf	Al	2	b	1.00	1	0.500 ± .025	f	0.200 ± .018	f	0.690 ± .100	Å	34.5 ± 5.0
intf	Al	3	b	1.00	2	-0.500 ± .025	f	0.592 ± .018	f	1.340 ± .100	Å	67.0 ± 5.0
intf	As	4	b	1.00	3	0.500 ± .025	f	-0.250 ± .018	f	0.060 ± .100	Å	3.0 ± 5.0
intf	Al	5	b	1.00	4	0.000 ± .025	f	-0.250 ± .018	f	1.970 ± .100	Å	98.5 ± 5.0
intf	As	6	b	1.00	5	-0.500	f	-0.250	f	0.000	Å	0.0
subl	Ga	7	b	1.00	6	0.000	f	0.750	f	2.000	Å	100.0
subl	As	8	b	1.00	7	0.500	f	-0.250	f	0.000	Å	0.0

GaAs(110)-(1x1)-3Al (high coverage)
31.33.13.4c BOND DISTANCES AND ANGLES

Bond distances and angles are derived from coordinates

No. of distances/angles: 10

Interatomic dist. A-B (Å)	Atom A	Atom B	Atom C	Bond angle A-B-C (°)
2.400	As1	Al2	As1(1,0)	112.9
2.475	Al3	As6(0,1)	Ga7	109.8
2.400	As1	Al2	As4	123.4
2.347	As1	Al3(0,-1)	As4(0,-1)	108.1
2.347	As1	Al3(0,-1)	As6	115.0
2.389	Al2	As4	Al3	117.0
2.389	Al2	As4	Al5	90.2
2.451	Al3	As4	Al3(1,0)	109.4
2.451	Al3	As4	Al5	110.9
2.475	Al3	As6(0,1)	Al5(0,1)	109.3

COMMON NAME : GaAs(110)-(1x1)-2Bi
CLASSIFICATION : 31.33.83.2
TECHNIQUE : LEED
AUTHORS : W.K. Ford, T. Guo, D.L. Lessor and C.B. Duke
REFERENCE : Phys. Rev., B42, 8952 (1990)

ILLUSTRATION: 126

SURFACE TYPE
Substrate : GaAs
Crystal face: 110
Temperature : <150 K
Bulk lattice: zincblende
2D bulk symm: pm
2D surf symm: pm
Adsorbate: Bi
Coverage : 2.0 Bi/(1x1)
Pattern : (1x1)
Matrix : (1.000, 0.000)
(0.000, 1.000)

STRUCTURE TYPE
Bi forms slightly tilted zigzag chains continuing GaAs lattice outward but with expanded Bi-Bi and Bi-substrate distances and with slight tilting of topmost GaAs chains

SAMPLE PREPARATION (sample)
Treatment : cleavage in vacuum; Bi deposited by sublimation
Crystallinity: LEED: (6x1) pattern
Anal. methods:
Contamination:

COMMENTS
Preparation-dependent (6x1) LEED pattern observed: (1x1) structure assumed

DATA COLLECTION
Technique: LEED
Dataset : 18 ineq. symmetry-averaged I-V curves: 50<E<300 eV

THEORY/DATA TREATMENT
Dynamical LEED (quasi-dyn. below 4th bilayer): 6 ph shs from relat. Hartree-Fock-Slater pot.; E-dep. mfp; no thermal vibs

STRUCTURES EXAMINED
Relaxation of 5 struct. parameters: 1st and 2nd bilayer rotation angles, bond lengths in 1st bilayer and between 1st and 2nd bilayers; also tested: disordered model (Bi ignored), Skeath model (dangling Bi chain)

QUALITY OF EXPERIMENT-THEORY FIT
RX=0.238

2D UNIT CELLS (1 domain observed)

Cell	Ax (Å)	Ay (Å)	Bx (Å)	By (Å)	α (°)	Matrix	Pattern	Cell type
Bulk	3.997	0.000	0.000	5.653	90.0	(1.000, 0.000) (0.000, 1.000)	(1x1)	b: bulk lattice
Surface 1	3.997	0.000	0.000	5.653	90.0	(1.000, 0.000) (0.000, 1.000)	(1x1)	s1: commens. superlattice

3D COORDINATES

Bi1-Bi2, As3-Ga4: 2 bilayers with tilted chains; As5-Ga6: unrelaxed bulk bilayer;
As7-Ga8: periodically repeating bulk bilayer; 0.1Å error bars assumed for tabulation

Dx/Dy in Å, or as a fraction of layer's unit cell vectors; atom 0 at the origin. Epir/subr are bulk repeat vectors.

No. of atoms: 8 Bulk z = 1.999 Å

Region	Chem el.	At. no.	Cell type	Site occ.	Rel. to	Dx ± εx		Dy ± εy		Dz ± εz		Dz/Bz(%) ± εz/Bz
epir		-2					f		f		Å	
subr		-1				1.999	Å	2.827	Å	1.999	Å	
ovrl	Bi	1	b	1.00	0	0.000	f	0.000	f	0.000	Å	0.0
ovrl	Bi	2	b	1.00	1	0.500	f	0.364 ± .018	f	0.093 ± .100	Å	4.7 ± 5.0
intf	Ga	3	b	1.00	2	-0.500	f	0.449 ± .018	f	2.424 ± .100	Å	121.3 ± 5.0
intf	As	4	b	1.00	3	0.500	f	-0.249 ± .018	f	0.106 ± .100	Å	5.3 ± 5.0
intf	As	5	b	1.00	4	-0.500	f	-0.514 ± .018	f	1.942 ± .100	Å	97.2 ± 5.0
intf	Ga	6	b	1.00	5	0.500	f	0.250	f	0.000	Å	0.0
subl	As	7	b	1.00	6	0.000	f	0.250	f	1.999	Å	100.0
subl	Ga	8	b	1.00	7	-0.500	f	0.250	f	0.000	Å	0.0

GaAs(110)-(1x1)-2Bi
31.33.83.2

BOND DISTANCES AND ANGLES

Bond distances and angles are derived from coordinates

No. of distances/angles: 8

Interatomic dist. A-B (Å)	Atom A	Atom B	Atom C	Bond angle A-B-C (°)
2.870	Bi1	Bi2	Bi1(1,0)	88.3
2.870	Bi1	Bi2	As4	108.8
2.730	Bi1	Ga3(0,-1)	As4(0,-1)	105.2
2.730	Bi1	Ga3(0,-1)	As5	124.0
2.770	Bi2	As4	Ga3	101.3
2.770	Bi2	As4	Ga6	118.5
2.448	Ga3	As4	Ga6	112.6
2.448	As4	Ga6	As7	107.2

COMMON NAME : GaAs(110)-(1x1)-2Sb
CLASSIFICATION : 31.33.51.2
TECHNIQUE : LEED
AUTHORS : C.B. Duke, A. Paton, W.K. Ford, A. Kahn and J. Carelli
REFERENCE : Phys. Rev., B26, 803 (1982)

ILLUSTRATION: 126

SURFACE TYPE
Substrate : GaAs
Crystal face: 110
Temperature : 300 K
Bulk lattice: zincblende
2D bulk symm: pm
2D surf symm: pm
Adsorbate: Sb
Coverage : 2.0 Sb/(1x1)
Pattern : (1x1)
Matrix : (1.000, 0.000) (0.000, 1.000)

STRUCTURE TYPE
Sb forms zigzag chains continuing GaAs lattice outward, but with expanded Sb-Sb and Sb-substrate distances, and with slight tilting of topmost GaAs chains

SAMPLE PREPARATION (2 sample)
Treatment : sputtering and annealing, or in situ cleavage; Sb exposure
Crystallinity:
Anal. methods:
Contamination:

COMMENTS

quasi-dynamical LEED: scattered amplitudes from 2 slabs, each consisting of 3 layers, are superposed; scattering within the uppermost slab is calculated exactly, but within the lower slab multiple scattering is neglected

DATA COLLECTION
Technique: LEED
Dataset : I-V data: 14 beams averaged from 2 sets of experimental data 60=<E=<210 eV

THEORY/DATA TREATMENT
Quasi-dynamical LEED: 6 bilayer slabs, 6 ph shs from superpos. atomic charge densities; mfp=8 Å; Vor=-10 eV (fit)

STRUCTURES EXAMINED
1) sp3 chain: Sb in Ga and As sites of the next monolayer; 2) overlapping chain: π-bonding between Sb and Ga, As; 3) single Sb defect model: 0.5ML Sb bonding to Ga substr.; 4) Sb2 dimer model: Sb2 π-orbitals bond to Ga pz orbitals; 5) p3 bonding model: Skeath et al, JVST 19, 556 (1981)

QUALITY OF EXPERIMENT-THEORY FIT
RX=0.2

2D UNIT CELLS (1 domain observed)

Cell	Ax (Å)	Ay (Å)	Bx (Å)	By (Å)	α (°)	Matrix	Pattern	Cell type
Bulk	4.000	0.000	0.000	5.650	90.0	(1.000, 0.000) (0.000, 1.000)	(1x1)	b: bulk lattice
Surface 1	4.000	0.000	0.000	5.650	90.0	(1.000, 0.000) (0.000, 1.000)	(1x1)	s1: commens. superlattice

3D COORDINATES

Sb1-Sb2: zigzag chain continuing GaAs lattice outward; As7-Ga8: periodically repeating set of bulk layers; 0.1Å error bars assumed for tabulation

Dx/Dy in Å, or as a fraction of layer's unit cell vectors; atom 0 at the origin. Epir/subr are bulk repeat vectors.

No. of atoms: 8 Bulk z = 2.000 Å

Region	Chem el.	At. no.	Cell type	Site occ.	Rel. to	Dx ± εx		Dy ± εy		Dz ± εz		Dz/Bz(%) ± εz/Bz
epir		-2					f		f		Å	
subr		-1				2.000	Å	2.825	Å	2.000	Å	
ovrl	Sb	1	b	1.00	0	0.000	f	0.000	f	0.000	Å	0.0
ovrl	Sb	2	b	1.00	1	0.500	f	0.297 ± .018	f	0.100 ± .100	Å	5.0 ± 5.0
intf	Ga	3	b	1.00	2	-0.500	f	0.520 ± .018	f	2.290 ± .100	Å	114.5 ± 5.0
intf	As	4	b	1.00	3	0.500	f	-0.250	f	0.100 ± .100	Å	5.0 ± 5.0
intf	As	5	b	1.00	4	-0.500	f	-0.500	f	1.900 ± .100	Å	95.0 ± 5.0
intf	Ga	6	b	1.00	5	0.500	f	0.250	f	0.000	Å	0.0
subl	As	7	b	1.00	6	0.000	f	0.250	f	2.000	Å	100.0
subl	Ga	8	b	1.00	7	-0.500	f	0.250	f	0.000	Å	0.0

GaAs(110)-(1x1)-2Sb
31.33.51.2 BOND DISTANCES AND ANGLES

Bond distances and angles are derived from coordinates

No. of distances/angles: 8

Interatomic dist. A-B (Å)	Atom A	Atom B	Atom C	Bond angle A-B-C (°)
2.613	Sb1	Sb2	Sb1(1,0)	99.9
2.613	Sb1	Sb2	As4	112.2
2.604	Sb1	Ga3(0,-1)	As4(0,-1)	105.5
2.604	Sb1	Ga3(0,-1)	As5	121.4
2.835	Sb2	As4	Ga3	106.0
2.835	Sb2	As4	Ga6	110.8
2.451	Ga3	As4	Ga6	112.1
2.368	As4	Ga6	As7	108.1

COMMON NAME : GaAs(110)-(1x1)-2Sb
CLASSIFICATION : 31.33.51.5
TECHNIQUE : LEED
AUTHORS : W.K. Ford, T. Guo, D.L. Lessor and C.B. Duke
REFERENCE : Phys. Rev., B42, 8952 (1990)

ILLUSTRATION: 126

SURFACE TYPE
Substrate : GaAs
Crystal face: 110
Temperature : <150 K
Bulk lattice: zincblende
2D bulk symm: pm
2D surf symm: pm

Adsorbate: Sb
Coverage : 2.0 Sb/(1x1)
Pattern : (1x1)
Matrix : (1.000, 0.000)
(0.000, 1.000)

STRUCTURE TYPE
Sb forms slightly tilted zigzag chains continuing GaAs lattice outward but with expanded Sb-Sb and Sb-substrate distances and with slight tilting of topmost GaAs chains

SAMPLE PREPARATION (sample)
Treatment : cleavage in vacuum; Sb deposited by sublimation
Crystallinity:
Anal. methods:
Contamination:

COMMENTS

DATA COLLECTION
Technique: LEED
Dataset : 18 ineq. symmetry-averaged I-V curves: 50<E<300 eV

THEORY/DATA TREATMENT
Dynamical LEED (quasi-dyn. below 4th bilayer): 6 ph shs from relat. Hartree-Fock-Slater pot.; E-dep. mfp; no thermal vibs

STRUCTURES EXAMINED
Relaxation of 5 struct. parameters: 1st and 2nd bilayer rotation angles, bond lengths in 1st bilayer and between 1st and 2nd bilayers; also tested: disordered model (Sb ignored), Skeath model (dangling Sb chain)

QUALITY OF EXPERIMENT-THEORY FIT
RX=0.199

2D UNIT CELLS (1 domain observed)

Cell	Ax (Å)	Ay (Å)	Bx (Å)	By (Å)	α (°)	Matrix	Pattern	Cell type
Bulk	3.997	0.000	0.000	5.653	90.0	(1.000, 0.000) (0.000, 1.000)	(1x1)	b: bulk lattice
Surface 1	3.997	0.000	0.000	5.653	90.0	(1.000, 0.000) (0.000, 1.000)	(1x1)	s1: commens. superlattice

3D COORDINATES

Sb1-Sb2, Ga3-As4: 2 bilayers with tilted chains; As5-Ga6: unrelaxed bulk bilayer;
As7-Ga8: periodically repeating bulk bilayer; 0.1Å error bars assumed for tabulation

Dx/Dy in Å, or as a fraction of layer's unit cell vectors; atom 0 at the origin. Epir/subr are bulk repeat vectors.

No. of atoms: 8 Bulk z = 1.999 Å

Region	Chem el.	At. no.	Cell type	Site occ.	Rel. to	Dx ± εx		Dy ± εy		Dz ± εz		Dz/Bz(%) ± εz/Bz
epir		-2					f		f		Å	
subr		-1				1.999	Å	2.827	Å	1.999	Å	
ovrl	Sb	1	b	1.00	0	0.000	f	0.000	f	0.000	Å	0.0
ovrl	Sb	2	b	1.00	1	0.500	f	0.339 ± .018	f	0.077 ± .100	Å	3.9 ± 5.0
intf	Ga	3	b	1.00	2	-0.500	f	0.453 ± .018	f	2.287 ± .100	Å	114.4 ± 5.0
intf	As	4	b	1.00	3	0.500	f	-0.249 ± .018	f	0.111 ± .100	Å	5.6 ± 5.0
intf	As	5	b	1.00	4	-0.500	f	-0.514 ± .018	f	1.939 ± .100	Å	97.0 ± 5.0
intf	Ga	6	b	1.00	5	0.500	f	0.250	f	0.000	Å	0.0
subl	As	7	b	1.00	6	0.000	f	0.250	f	1.999	Å	100.0
subl	Ga	8	b	1.00	7	-0.500	f	0.250	f	0.000	Å	0.0

GaAs(110)-(1x1)-2Sb
31.33.51.5 BOND DISTANCES AND ANGLES

Bond distances and angles are derived from coordinates

No. of distances/angles: 8

Interatomic dist. A-B (Å)	Atom A	Atom B	Atom C	Bond angle A-B-C (°)
2.770	Sb1	Sb2	Sb1(1,0)	92.4
2.770	Sb1	Sb2	As4	109.0
2.640	Sb1	Ga3(0,-1)	As4(0,-1)	107.3
2.640	Sb1	Ga3(0,-1)	As5	120.4
2.660	Sb2	As4	Ga3	102.0
2.660	Sb2	As4	Ga6	116.7
2.448	Ga3	As4	Ga6	112.8
2.448	As4	Ga6	As7	107.1

COMMON NAME : GaP(110)-(1x1)
CLASSIFICATION : 31.15.4
TECHNIQUE : LEED
AUTHORS : C.B. Duke, A. Paton and A. Kahn
REFERENCE : J. Vac. Sci. Technol., A2, 515 (1984)

ILLUSTRATION: 116

SURFACE TYPE
Substrate : GaP
Crystal face: 110
Temperature : 125 K
Bulk lattice: zincblende
2D bulk symm: pm
2D surf symm: pm
Adsorbate:
Coverage :
Pattern : (1x1)
Matrix : (1.000, 0.000)
(0.000, 1.000)

STRUCTURE TYPE
Relaxed bulk termination with 27.5° tilt in top layer

SAMPLE PREPARATION (1 sample)
Treatment : Ar+ bombardment for 20 mins, then anneal for 4 hrs at 823 K
Crystallinity:
Anal. methods:
Contamination: monitored by LEED and AES

COMMENTS
X-ray R-factor showed 2 minima corresponding to top-layer Ga-P tilts of 2.5° and 27.5°; integrated beam R-factor RI clearly distinguishes in favor of 27.5°

DATA COLLECTION
Technique: LEED
Dataset : I-V curves: 14 beams
10=-10,01=0-1,11=-11,1-1=-1-1,02,
20=-20,12=-12,1-2=-1-2,21=-21,2-1=-2-1,13=-

THEORY/DATA TREATMENT
Dynamical LEED: multiple scattering model of Meyer et al, Phys. Rev. B19, 5194 (1979); mfp=6Å; charge overlap pots

STRUCTURES EXAMINED
Bond length conserving rotation of top layer was used to determine Ga-P spacing; spacing between top 2 layers was varied; Ga-P spacing in 2nd-layer was varied; Ga and P registries in top layer were varied

QUALITY OF EXPERIMENT-THEORY FIT
RX=0.22, RI=0.07

2D UNIT CELLS (1 domain observed)

Cell	Ax (Å)	Ay (Å)	Bx (Å)	By (Å)	α (°)	Matrix	Pattern	Cell type
Bulk	3.854	0.000	0.000	5.451	90.0	(1.000, 0.000) (0.000, 1.000)	(1x1)	b: bulk lattice
Surface 1	3.854	0.000	0.000	5.451	90.0	(1.000, 0.000) (0.000, 1.000)	(1x1)	s1: commens. superlattice

3D COORDINATES

P1-Ga2: top bilayer with tilted Ga-P chains; Ga3-P4, Ga5-P6: 2 bulk bilayers;
Ga7-P8: periodically repeating bulk bilayer; 0.1Å error bars assumed for tabulation

Dx/Dy in Å, or as a fraction of layer's unit cell vectors; atom 0 at the origin. Epir/subr are bulk repeat vectors.

No. of atoms: 8

Bulk z = 1.927 Å

Region	Chem el.	At. no.	Cell type	Site occ.	Rel. to	Dx ± εx		Dy ± εy		Dz ± εz		Dz/Bz(%) ± εz/Bz
epir		-2					f		f		Å	
subr		-1				1.927	Å	2.726	Å	1.927	Å	
intf	P	1	b	1.00	0	0.000	f	0.000	f	0.000	Å	0.0
intf	Ga	2	b	1.00	1	0.500 ± .026	f	0.222 ± .018	f	0.630 ± .100	Å	32.7 ± 5.2
intf	Ga	3	b	1.00	2	-0.500 ± .026	f	0.586 ± .018	f	1.386 ± .100	Å	71.9 ± 5.2
intf	P	4	b	1.00	3	0.500	f	-0.250	f	0.000 ± .100	Å	0.0 ± 5.2
intf	Ga	5	b	1.00	4	0.000	f	-0.250	f	1.927 ± .100	Å	100.0 ± 5.2
intf	P	6	b	1.00	5	-0.500	f	-0.250	f	0.000	Å	0.0
subl	Ga	7	b	1.00	6	0.000	f	0.750	f	1.927	Å	100.0
subl	P	8	b	1.00	7	0.500	f	-0.250	f	0.000	Å	0.0

GaP(110)-(1x1)
31.15.4 BOND DISTANCES AND ANGLES

Bond distances and angles are derived from coordinates

No. of distances/angles: 10

Interatomic dist. A-B (Å)	Atom A	Atom B	Atom C	Bond angle A-B-C (°)
2.361	P1	Ga2	P1(1,0)	109.4
2.360	Ga3	P6(0,1)	Ga7	109.5
2.361	P1	Ga2	P4	124.7
2.272	P1	Ga3(0,-1)	P4(0,-1)	105.4
2.272	P1	Ga3(0,-1)	P6	117.3
2.297	Ga2	P4	Ga3	117.4
2.297	Ga2	P4	Ga5	91.9
2.360	Ga3	P4	Ga3(1,0)	109.5
2.360	Ga3	P4	Ga5	109.5
2.360	Ga3	P6(0,1)	Ga5(0,1)	109.5

COMMON NAME : GaP(111)-(2x2) ILLUSTRATION: 117
CLASSIFICATION : 31.15.5
TECHNIQUE : LEED
AUTHORS : G. Xu, W.Y. Hu, M.W. Puga, S.Y. Tong, J.L. Yeh, S.R. Wang and B.W. Lee
REFERENCE : Phys. Rev., B32, 8473 (1985)

SURFACE TYPE
Substrate : GaP
Crystal face: 111
Temperature : RT*
Bulk lattice: zincblende
2D bulk symm: p3m1
2D surf symm: p3m1
Adsorbate:
Coverage :
Pattern : (2x2)
Matrix : (2.000, 0.000) (0.000, 2.000)

STRUCTURE TYPE
1 missing Ga per (2x2) unit cell in heavily relaxed top bilayer; top bilayer almost planar

SAMPLE PREPARATION (1 sample)
Treatment : etch, clean Ga rich face by Ar+, anneal at 550C
Crystallinity:
Anal. methods:
Contamination: AES: trace (<1%) carbon

COMMENTS

DATA COLLECTION
Technique: LEED
Dataset : IV curves: 5 independent substrate and 5 independent super- lattice beams; E range 20-220 eV

THEORY/DATA TREATMENT
Dynamical LEED (matrix inversion and RFS): 6 phase shifts; 229 beams; VoiαE**1/3; rms ampl 0.13Å

STRUCTURES EXAMINED
Vary top Ga-As spacing 0-1.4Å; then vary Ga and P positions in first 2 bilayers, as per GaAs study (Tong et al, PRL 52 1693 (1984)), and remove 1/4 of Ga atoms from first bilayer

QUALITY OF EXPERIMENT-THEORY FIT
RVHT=0.21

2D UNIT CELLS (1 domain observed)

Cell	Ax (Å)	Ay (Å)	Bx (Å)	By (Å)	α (°)	Matrix	Pattern	Cell type
Bulk	3.854	0.000	1.927	3.338	60.0	(1.000, 0.000) (0.000, 1.000)	(1x1)	b: bulk lattice
Surface 1	7.708	0.000	3.854	6.675	60.0	(2.000, 0.000) (0.000, 2.000)	(2x2)	s1: commens. superlattice

3D COORDINATES

Ga1 through P7: reconstructed top bilayer; Ga8 through P15: slightly distorted bulk bilayer; Ga16-P17: periodically repeating bulk bilayer; 0.1Å error bars assumed for tabulation

Dx/Dy in Å, or as a fraction of layer's unit cell vectors; atom 0 at the origin. Epir/subr are bulk repeat vectors.

No. of atoms: 17 Bulk z = 3.147 Å

Region	Chem el.	At. no.	Cell type	Site occ.	Rel. to	Dx ± εx	Dy ± εy	Dz ± εz	Dz/Bz(%) ± εz/Bz
epir		-2				f	f	Å	
subr		-1				0.000 Å	-2.226 Å	3.147 Å	
intf	Ga	1	s1	.25	0	0.000 f	0.000 f	0.000 Å	0.0
intf	Ga	2	s1	.25	1	0.560 ± .006 f	0.000 ± .015 f	0.000 Å	0.0
intf	Ga	3	s1	.25	2	-0.560 ± .006 f	0.560 ± .015 f	0.000 Å	0.0
intf	P	4	s1	.25	3	0.653 ± .006 f	0.093 ± .015 f	0.000 ± .100 Å	0.0 ± 3.2
intf	P	5	s1	.25	4	-0.398 ± .006 f	0.000 ± .015 f	0.000 Å	0.0
intf	P	6	s1	.25	5	0.398 ± .006 f	-0.398 ± .015 f	0.000 Å	0.0
intf	P	7	s1	.25	6	-0.466 ± .006 f	-0.068 ± .015 f	0.090 ± .100 Å	2.9 ± 3.2
intf	Ga	8	s1	.25	7	0.500 f	0.500 f	2.291 ± .100 Å	72.8 ± 3.2
intf	Ga	9	s1	.25	8	-0.500 f	0.000 f	0.000 Å	0.0
intf	Ga	10	s1	.25	9	0.500 f	-0.500 f	0.000 Å	0.0
intf	Ga	11	s1	.25	10	-0.500 f	0.000 f	0.110 ± .100 Å	3.5 ± 3.2
intf	P	12	s1	.25	11	0.667 f	0.667 f	0.697 ± .100 Å	22.2 ± 3.2
intf	P	13	s1	.25	12	0.000 f	-0.500 f	0.000 Å	0.0
intf	P	14	s1	.25	13	-0.500 f	0.500 f	0.000 Å	0.0
intf	P	15	s1	.25	14	0.000 f	-0.500 f	0.000 Å	0.0
subl	Ga	16	b	1.00	15	0.000 f	0.000 f	2.360 Å	75.0
subl	P	17	b	1.00	16	0.333 f	-0.667 f	0.787 Å	25.0

GaP(111)-(2x2)
31.15.5 BOND DISTANCES AND ANGLES

Bond distances and angles are derived from coordinates

No. of distances/angles: 14

Interatomic dist. A-B (Å)	Atom A	Atom B	Atom C	Bond angle A-B-C (°)
2.400	Ga1	P5(0,-1)	Ga3(0,-1)	89.8
2.492	Ga2	P7	Ga11	92.1
2.424	P4	Ga8	P12	120.7
2.424	P4	Ga8	P13	104.3
2.401	P7	Ga11	P15	107.4
2.367	Ga8	P12	Ga9(1,0)	109.0
2.400	Ga1	P5(0,-1)	Ga9(0,-1)	82.4
2.498	Ga1	P7	Ga2	119.8
2.498	Ga1	P7	Ga11	92.1
2.398	Ga2	P4(0,-1)	Ga3(1,-1)	90.0
2.398	Ga2	P4(0,-1)	Ga8(0,-1)	82.4
2.406	Ga2	P6	Ga1(1,0)	89.8
2.406	Ga2	P6	Ga10	82.4
2.492	Ga2	P7	Ga3	120.1

COMMON NAME : GaSb(110)-(1x1)
CLASSIFICATION : 31.51.2
TECHNIQUE : LEED
AUTHORS : C.B. Duke, A. Paton and A. Kahn
REFERENCE : Phys. Rev., B27, 3436 (1983)

ILLUSTRATION: 116

SURFACE TYPE
Substrate : GaSb
Crystal face: 110
Temperature : 125 K
Bulk lattice: zincblende
2D bulk symm: pm
2D surf symm: pm

Adsorbate:
Coverage :
Pattern : (1x1)
Matrix : (1.000, 0.000)
(0.000, 1.000)

STRUCTURE TYPE
Relaxed bulk termination with 30° tilt in top layer

SAMPLE PREPARATION (1 sample)
Treatment : Ar+ bombardment and annealing at 430 K for 1 hour
Crystallinity:
Anal. methods:
Contamination: monitored by AES and LEED

COMMENTS

DATA COLLECTION
Technique: LEED
Dataset : I-V spectra: 14 beams at normal incidence; 30<E<220 eV; 2 sets of data taken and averaged to improve signal to noise

THEORY/DATA TREATMENT
Quasi-dynamical LEED: 6 atomic bilayers; 6 phase shifts; mfp=8Å; Vor optimised

STRUCTURES EXAMINED
Bond length conserving rotations in top bilayer of up to 35° and various vertical and lateral displacements of top bilayer; smaller second layer displacements also considered

QUALITY OF EXPERIMENT-THEORY FIT
RX=0.24

2D UNIT CELLS (1 domain observed)

Cell	Ax (Å)	Ay (Å)	Bx (Å)	By (Å)	α (°)	Matrix	Pattern	Cell type
Bulk	4.326	0.000	0.000	6.118	90.0	(1.000, 0.000) (0.000, 1.000)	(1x1)	b: bulk lattice
Surface 1	4.326	0.000	0.000	6.118	90.0	(1.000, 0.000) (0.000, 1.000)	(1x1)	s1: commens. superlattice

3D COORDINATES

Sb1-Ga2: top bilayer with tilted Ga-Sb chains; Ga3-Sb4, Ga5-Sb6: 2 bulk bilayers;
Ga7-Sb8: periodically repeating bulk bilayer

Dx/Dy in Å, or as a fraction of layer's unit cell vectors; atom 0 at the origin. Epir/subr are bulk repeat vectors.

No. of atoms: 8

Bulk z = 2.163 Å

Region	Chem el.	At. no.	Cell type	Site occ.	Rel. to	Dx ± εx	Dy ± εy	Dz ± εz	Dz/Bz(%) ± εz/Bz
epir		-2				f	f	Å	
subr		-1				2.163 Å	3.059 Å	2.163 Å	
intf	Sb	1	b	1.00	0	0.000 f	0.000 f	0.000 Å	0.0
intf	Ga	2	b	1.00	1	0.500 ± .023 f	0.217 ± .016 f	0.770 ± .050 Å	35.6 ± 2.3
intf	Ga	3	b	1.00	2	-0.500 ± .023 f	0.593 ± .016 f	1.615 ± .050 Å	74.7 ± 2.3
intf	Sb	4	b	1.00	3	0.500 ± .023 f	-0.250 ± .016 f	0.000 ± .050 Å	0.0 ± 2.3
intf	Ga	5	b	1.00	4	0.000 ± .023 f	-0.250 ± .016 f	2.163 ± .050 Å	100.0 ± 2.3
intf	Sb	6	b	1.00	5	-0.500 f	-0.250 f	0.000 Å	0.0
subl	Ga	7	b	1.00	6	0.000 f	0.750 f	2.163 Å	100.0
subl	Sb	8	b	1.00	7	0.500 f	-0.250 f	0.000 Å	0.0

GaSb(110)-(1x1)
31.51.2 BOND DISTANCES AND ANGLES

Bond distances and angles are derived from coordinates

No. of distances/angles: 10

Interatomic dist. A-B (Å)	Atom A	Atom B	Atom C	Bond angle A-B-C (°)
2.652	Sb1	Ga2	Sb1(1,0)	109.3
2.649	Ga3	Sb6(0,1)	Ga7	109.5
2.652	Sb1	Ga2	Sb4	125.0
2.653	Sb1	Ga3(0,-1)	Sb4(0,-1)	104.7
2.653	Sb1	Ga3(0,-1)	Sb6	118.8
2.648	Ga2	Sb4	Ga3	117.2
2.648	Ga2	Sb4	Ga5	92.3
2.649	Ga3	Sb4	Ga3(1,0)	109.5
2.649	Ga3	Sb4	Ga5	109.5
2.649	Ga3	Sb6(0,1)	Ga5(0,1)	109.5

COMMON NAME : GaSb(110)-(1x1)
CLASSIFICATION : 31.51.3
TECHNIQUE : MEIS
AUTHORS : L. Smit, R.M. Tromp and J.F. van der Veen
REFERENCE : Phys. Rev., B29, 4814 (1984)

ILLUSTRATION: 116

SURFACE TYPE
Substrate : GaSb
Crystal face: 110
Temperature : RT*
Bulk lattice: zincblende
2D bulk symm: pm
2D surf symm: pm

Adsorbate:
Coverage :
Pattern : (1x1)
Matrix : (1.000, 0.000)
(0.000, 1.000)

STRUCTURE TYPE
Relaxed bulk termination with 29° tilt in top layer

SAMPLE PREPARATION (1 sample)
Treatment : in situ cleavage
Crystallinity:
Anal. methods:
Contamination: monitored by LEED

COMMENTS
The bond rotation of 29(+7-4)° is in good agreement with the LEED determination and rules out the 7° model; vibrational amplitudes for the bulk were assumed to be Ga: √<u*u>=0.122, Sb: √<u*u>=0.104Å; a surface enhancement factor of 1.5±0.2 was found

DATA COLLECTION
Technique: MEIS; Rutherford back scattering of He+
Dataset : [-1-1 2] channeling direction; detection in (110) plane for exit angles >10°, <30°; 174k eV

THEORY/DATA TREATMENT
Monte Carlo simulations

STRUCTURES EXAMINED
Bond length conserving rotations in top bilayer of up to 40°; enhancement of surface vibrations was varied; subsurface reconstruction was not considered.

QUALITY OF EXPERIMENT-THEORY FIT
Visual

2D UNIT CELLS (1 domain observed)

Cell	Ax (Å)	Ay (Å)	Bx (Å)	By (Å)	α (°)	Matrix	Pattern	Cell type
Bulk	4.326	0.000	0.000	6.118	90.0	(1.000, 0.000) (0.000, 1.000)	(1x1)	b: bulk lattice
Surface 1	4.326	0.000	0.000	6.118	90.0	(1.000, 0.000) (0.000, 1.000)	(1x1)	s1: commens. superlattice

3D COORDINATES

Sb1-Ga2: top bilayer with tilted Ga-Sb chains; Ga3-Sb4, Ga5-Sb6: 2 bulk bilayers; Ga7-Sb8: periodically repeating bulk bilayer; 0.1Å error bars assumed for tabulation

Dx/Dy in Å, or as a fraction of layer's unit cell vectors; atom 0 at the origin. Epir/subr are bulk repeat vectors.

No. of atoms: 8

Bulk z = 2.163 Å

Region	Chem el.	At. no.	Cell type	Site occ.	Rel. to	Dx ± εx		Dy ± εy		Dz ± εz		Dz/Bz(%) ± εz/Bz
epir		-2					f		f		Å	
subr		-1				2.163	Å	3.059	Å	2.163	Å	
intf	Sb	1	b	1.00	0	0.000	f	0.000	f	0.000	Å	0.0
intf	Ga	2	b	1.00	1	0.500 ± .023	f	0.217 ± .016	f	0.770 ± .100	Å	35.6 ± 4.6
intf	Ga	3	b	1.00	2	-0.500 ± .023	f	0.593 ± .016	f	1.615 ± .100	Å	74.7 ± 4.6
intf	Sb	4	b	1.00	3	0.500	f	-0.250	f	0.000	Å	0.0
intf	Ga	5	b	1.00	4	0.000	f	-0.250	f	2.163	Å	100.0
intf	Sb	6	b	1.00	5	-0.500	f	-0.250	f	0.000	Å	0.0
subl	Ga	7	b	1.00	6	0.000	f	0.750	f	2.163	Å	100.0
subl	Sb	8	b	1.00	7	0.500	f	-0.250	f	0.000	Å	0.0

GaSb(110)-(1x1)
31.51.3

BOND DISTANCES AND ANGLES

Bond distances and angles are derived from coordinates

No. of distances/angles: 10

Interatomic dist. A-B (Å)	Atom A	Atom B	Atom C	Bond angle A-B-C (°)
2.652	Sb1	Ga2	Sb1(1,0)	109.3
2.649	Ga3	Sb6(0,1)	Ga7	109.5
2.652	Sb1	Ga2	Sb4	125.0
2.653	Sb1	Ga3(0,-1)	Sb4(0,-1)	104.7
2.653	Sb1	Ga3(0,-1)	Sb6	118.8
2.648	Ga2	Sb4	Ga3	117.2
2.648	Ga2	Sb4	Ga5	92.3
2.649	Ga3	Sb4	Ga3(1,0)	109.5
2.649	Ga3	Sb4	Ga5	109.5
2.649	Ga3	Sb6(0,1)	Ga5(0,1)	109.5

COMMON NAME : GaSb(110)-(1x1)
CLASSIFICATION : 31.51.3a
TECHNIQUE : MEIS
AUTHORS : L. Smit and J.F. van der Veen
REFERENCE : Surf. Sci., 166, 183 (1986)

ILLUSTRATION: 116

SURFACE TYPE
Substrate : GaSb
Crystal face: 110
Temperature : RT
Bulk lattice: zincblende
2D bulk symm: pm
2D surf symm: pm

Adsorbate:
Coverage :
Pattern : (1x1)
Matrix : (1.000, 0.000)
(0.000, 1.000)

STRUCTURE TYPE
Relaxed bulk termination with 28.5±2.6° tilt in top layer

SAMPLE PREPARATION (1 sample)
Treatment : in-situ cleavage monitored by laser light back scattering
Crystallinity: sharp (1x1) LEED pattern
Anal. methods:
Contamination:

COMMENTS
Thermal vibrations: bulk Ga rms amplitude 0.112Å, bulk Sb amplitude 0.104Å; surface layer vibrations were optimized to 1.6±0.1 times the bulk

DATA COLLECTION
Technique: MEIS; 174keV He+ Rutherford back scattering
Dataset : scattering aligned with [-1-1 2] channeling axis and exit angles > 10°, < 30°, ions incident from both directions

THEORY/DATA TREATMENT
Monte Carlo simulation: see comment

STRUCTURES EXAMINED
Bond length conserving rotations in top bilayer up to 40° and enhancement of surface layer vibration; bond length relaxation was permitted but substrate relaxation was not included

2D UNIT CELLS (1 domain observed)

Cell	Ax (Å)	Ay (Å)	Bx (Å)	By (Å)	α (°)	Matrix	Pattern	Cell type
Bulk	4.326	0.000	0.000	6.118	90.0	(1.000, 0.000) (0.000, 1.000)	(1x1)	b: bulk lattice
Surface 1	4.326	0.000	0.000	6.118	90.0	(1.000, 0.000) (0.000, 1.000)	(1x1)	s1: commens. superlattice

3D COORDINATES

Sb1-Ga2: top bilayer with tilted Ga-Sb chains; Sb3-Ga4: bulk bilayer;
Sb5-Ga6: periodically repeating bulk bilayer; 0.1Å error bars assumed for tabulation

Dx/Dy in Å, or as a fraction of layer's unit cell vectors; atom 0 at the origin. Epir/subr are bulk repeat vectors.

No. of atoms: 6

Bulk z = 2.163 Å

Reg ion	Chem el.	At. no.	Cell type	Site occ.	Rel. to	Dx ± εx		Dy ± εy		Dz ± εz		Dz/Bz(%) ± εz/Bz
epir		-2					f		f		Å	
subr		-1				2.163	Å	-3.059	Å	2.163	Å	
intf	Sb	1	b	1.00	0	0.000	f	0.000	f	0.000	Å	0.0
intf	Ga	2	b	1.00	1	0.500	f	0.220 ± .016	f	0.730 ± .100	Å	33.8 ± 4.6
intf	Sb	3	b	1.00	2	0.000	f	0.339 ± .016	f	1.647 ± .100	Å	76.1 ± 4.6
intf	Ga	4	b	1.00	3	-0.500	f	0.250	f	0.000	Å	0.0
subl	Sb	5	b	1.00	4	0.000	f	-0.750	f	2.163	Å	100.0
subl	Ga	6	b	1.00	5	0.500	f	0.250	f	0.000	Å	0.0

BOND DISTANCES AND ANGLES

Bond distances and angles are derived from coordinates

No. of distances/angles: 9

Interatomic dist. A-B (Å)	Atom A	Atom B	Atom C	Bond angle A-B-C (°)
2.650	Sb1	Ga2	Sb1(1,0)	109.4
2.650	Sb1	Ga2	Sb3	124.7
2.649	Sb1	Ga4(0,-1)	Sb3(0,-1)	104.8
2.649	Sb1	Ga4(0,-1)	Sb5	118.6

GaSb(110)-(1x1)
31.51.3a

Bond Distances and Angles - Continued

Interatomic dist. A-B (Å)	Atom A	Atom B	Atom C	Bond angle A-B-C (°)
2.648	Ga2	Sb3	Ga4	116.9
2.648	Ga2	Sb3	Ga6	93.2
2.649	Sb3	Ga4	Sb3(-1,0)	109.5
2.649	Sb3	Ga4	Sb5(0,1)	109.5
2.649	Sb3	Ga6	Sb5	109.5

COMMON NAME : GaSb(111)-(2x2) ILLUSTRATION: 117
CLASSIFICATION : 31.51.4
TECHNIQUE : XRD
AUTHORS : R. Feidenhans'l, M. Nielsen, F. Grey, R.L. Johnson and I.K. Robinson
REFERENCE : Surf. Sci., 186, 499 (1987)

SURFACE TYPE
Substrate : GaSb
Crystal face: 111
Temperature : RT*
Bulk lattice: zincblende
2D bulk symm: p3m1
2D surf symm: p3m1

Adsorbate:
Coverage :
Pattern : (2x2)
Matrix : (2.000, 0.000)
(0.000, 2.000)

STRUCTURE TYPE
1 missing Ga per (2x2) unit cell in heavily relaxed top bilayer; top bilayer relatively planar

SAMPLE PREPARATION (1 sample)
Treatment : cycles of 500 eV Ar+ bombardment, then annealing at 773 K
Crystallinity: sharp (2x2) LEED pattern
Anal. methods:
Contamination: photoemission: no trace of impurities

COMMENTS
Of 6 possible surface registries, preferred structure has chi²=20, for the rest chi²>250;
lateral displacement errors: in surface layer hexagon the Sb atoms were displaced radially outwards by 0.38±0.03Å and the Ga atoms by 0.17±0.06Å

DATA COLLECTION
Technique: XRD; grazing incidence x-ray diffraction
Dataset : 15 fract-order and 4 integer-order integrated intensities after symmetry averaging at lamda=1.242Å

THEORY/DATA TREATMENT
Model structure factors compared with exp. by chi-squared testing; vibr. ampl. squared = 0.0±1.0Å²

STRUCTURES EXAMINED
Similarity of exp. results to InSb(111)-(2x2) (PRL 54 1275) implied similar 7 atom model; combinations of surface cell atoms, position relaxation, and surface cell registry tried, and Ga to Sb layer distances of 0, 0.3 and 0.8Å in surface bilayer

QUALITY OF EXPERIMENT-THEORY FIT
Chi²=20

2D UNIT CELLS (1 domain observed)

Cell	Ax (Å)	Ay (Å)	Bx (Å)	By (Å)	α (°)	Matrix	Pattern	Cell type
Bulk	4.310	0.000	2.155	3.733	60.0	(1.000, 0.000) (0.000, 1.000)	(1x1)	b: bulk lattice
Surface 1	8.620	0.000	4.310	7.465	60.0	(2.000, 0.000) (0.000, 2.000)	(2x2)	s1: commens. superlattice

3D COORDINATES

Ga1 through Sb7: reconstructed top bilayer; Ga8-Sb9: undistorted bulk bilayer;
Ga10-Sb11: periodically repeating bulk bilayer; 0.1Å error bars assumed for tabulation (see comment)

Dx/Dy in Å, or as a fraction of layer's unit cell vectors; atom 0 at the origin. Epir/subr are bulk repeat vectors.

No. of atoms: 11 Bulk z = 3.520 Å

Region	Chem el.	At. no.	Cell type	Site occ.	Rel. to	Dx ± εx		Dy ± εy		Dz ± εz		Dz/Bz(%) ± εz/Bz
epir		-2					f		f		Å	
subr		-1				2.155	Å	1.244	Å	3.520	Å	
intf	Ga	1	s1	.25	0	0.000	f	0.000	f	0.000	Å	0.0
intf	Ga	2	s1	.25	1	0.534 ± .005	f	0.466 ± .013	f	0.000	Å	0.0
intf	Ga	3	s1	.25	2	-0.534 ± .005	f	0.000 ± .013	f	0.000	Å	0.0
intf	Sb	4	s1	.25	3	0.652 ± .005	f	-0.348 ± .013	f	0.000 ± .300	Å	0.0 ± 8.5
intf	Sb	5	s1	.25	4	0.000 ± .005	f	0.577 ± .013	f	0.000	Å	0.0
intf	Sb	6	s1	.25	5	-0.423 ± .005	f	-0.577 ± .013	f	0.000	Å	0.0
intf	Sb	7	s1	.25	6	-0.051 ± .005	f	0.526 ± .013	f	0.000	Å	0.0
intf	Ga	8	b	1.00	7	0.000	f	0.000	f	2.640	Å	75.0
intf	Sb	9	b	1.00	8	0.333	f	-0.667	f	0.880	Å	25.0
subl	Ga	10	b	1.00	9	0.000	f	0.000	f	2.640	Å	75.0
subl	Sb	11	b	1.00	10	0.333	f	0.333	f	0.880	Å	25.0

GaSb(111)-(2x2)
31.51.4 BOND DISTANCES AND ANGLES

Bond distances and angles are derived from coordinates

No. of distances/angles: 8

Interatomic dist. A-B (Å)	Atom A	Atom B	Atom C	Bond angle A-B-C (°)
2.642	Ga1	Sb4(-1,0)	Ga2(-1,0)	99.0
2.642	Ga1	Sb4(-1,0)	Ga8(-1,-1)	84.6
2.634	Ga1	Sb6	Ga3	99.2
2.634	Ga1	Sb6	Ga8(0,-1)	84.7
2.660	Ga1	Sb7(0,-1)	Ga2(0,-1)	119.8
2.660	Ga1	Sb7(0,-1)	Ga3(0,-1)	120.1
2.640	Sb7	Ga8	Sb9	109.5
2.640	Ga8	Sb9	Ga10	109.5

COMMON NAME : Ge(100)-(2x1)
CLASSIFICATION : 32.1
TECHNIQUE : XRD
AUTHORS : P. Eisenberger and W.C. Marra
REFERENCE : Phys. Rev. Lett., 46, 1081 (1981)

ILLUSTRATION: 94

SURFACE TYPE
Substrate : Ge
Crystal face: 100
Temperature : 300 K
Bulk lattice: diamond
2D bulk symm: pmm
2D surf symm: pmm

Adsorbate:
Coverage :
Pattern : (2x1)
Matrix : (2.000, 0.000)
(0.000, 1.000)

STRUCTURE TYPE
Tilted dimer reconstruction with lateral relaxation in 2nd layer

SAMPLE PREPARATION (1 sample)
Treatment : Ar+ sputtering followed by repeated anneals at 973 K
Crystallinity:
Anal. methods:
Contamination: monitored by LEED

COMMENTS
Each basic model was used to fix the direction and symmetry of atomic displacements from the unreconstructed positions; the magnitude of displacement for each model was then determined by comparison with experimental data

DATA COLLECTION
Technique: XRD; x-ray total external reflection (TRBD)
Dataset : intensities of (1/2,0) (3/2,0) (5/2,0) (1/2,-1) and (3/2,-1) Bragg reflections

THEORY/DATA TREATMENT
Least squares fit of integrated intensity to kinematic intensities; rms vibs = 0.24Å

STRUCTURES EXAMINED
1) dimer model: Farnsworth et al, J Appl Phys 29 1150 1958; 2) 2-layer model: Appelbaum et al Surf Sci 74 21 1978; 3) Chadi's asymmetric dimer model: Phys Rev Lett 43 43 1979; 1) gave poor fit and was eliminated; 2) and 3) yielded identical structures for the displacements examined

QUALITY OF EXPERIMENT-THEORY FIT
Least squares fit

2D UNIT CELLS (1 domain observed)

Cell	Ax (Å)	Ay (Å)	Bx (Å)	By (Å)	α (°)	Matrix	Pattern	Cell type
Bulk	3.995	0.000	0.000	3.995	90.0	(1.000, 0.000) (0.000, 1.000)	(1x1)	b: bulk lattice
Surface 1	7.990	0.000	0.000	3.995	90.0	(2.000, 0.000) (0.000, 1.000)	(2x1)	s1: commens. superlattice

3D COORDINATES

Ge1-Ge2: asymmetrical dimer; Ge3-Ge4: laterally relaxed planar 2nd layer;
Ge5-Ge6: periodically repeating bulk layer pair

Dx/Dy in Å, or as a fraction of layer's unit cell vectors; atom 0 at the origin. Epir/subr are bulk repeat vectors.

No. of atoms: 6

Bulk z = 1.410 Å

Region	Chem el.	At. no.	Cell type	Site occ.	Rel. to	Dx ± εx		Dy ± εy		Dz ± εz		Dz/Bz(%) ± εz/Bz
epir		-2					f		f		Å	
subr		-1				1.998	Å	-1.998	Å	2.820	Å	
intf	Ge	1	s1	.50	0	0.000	f	0.000	f	0.000	Å	0.0
intf	Ge	2	s1	.50	1	0.270 ± .013	f	0.000 ± .025	f	1.280 ± .040	Å	90.8 ± 2.8
intf	Ge	3	s1	.50	2	0.629 ± .013	f	0.500 ± .025	f	0.750 ± .040	Å	53.2 ± 2.8
intf	Ge	4	s1	.50	3	-0.529 ± .013	f	0.000 ± .025	f	0.000	Å	0.0
subl	Ge	5	b	1.00	4	-0.470	f	0.000	f	1.410	Å	100.0
subl	Ge	6	b	1.00	5	0.000	f	-0.500	f	1.410	Å	100.0

Ge(100)-(2x1)
32.1 BOND DISTANCES AND ANGLES

Bond distances and angles are derived from coordinates

No. of distances/angles: 10

Interatomic dist. A-B (Å)	Atom A	Atom B	Atom C	Bond angle A-B-C (°)
2.509	Ge1	Ge2	Ge1(1,0)	114.5
2.278	Ge2	Ge4	Ge3(-1,0)	110.5
2.509	Ge1	Ge2	Ge3(-1,0)	122.4
2.509	Ge1	Ge2	Ge4	118.0
2.509	Ge1	Ge2	Ge5	93.2
2.960	Ge1	Ge3(-1,0)	Ge1(0,1)	84.9
2.960	Ge1	Ge3(-1,0)	Ge2(-1,1)	118.1
2.960	Ge1	Ge3(-1,0)	Ge4(-1,0)	105.8
2.960	Ge1	Ge3(-1,0)	Ge5	101.1
2.960	Ge1	Ge3(-1,0)	Ge5(-1,0)	127.4

COMMON NAME : Ge(100)-(2x1)
CLASSIFICATION : 32.5
TECHNIQUE : XRD
AUTHORS : F. Grey, R.L. Johnson, J.S. Pederson, M. Nielsen and R. Feidenhans'l
REFERENCE : Springer Series in Surface Sciences, 11, 292 (1988)

ILLUSTRATION: 94

SURFACE TYPE
Substrate : Ge
Crystal face: 100
Temperature : RT*
Bulk lattice: diamond
2D bulk symm: pmm
2D surf symm: pmm

Adsorbate:
Coverage :
Pattern : (2x1)
Matrix : (2.000, 0.000)
(0.000, 1.000)

STRUCTURE TYPE
Symmetric dimer reconstruction (with relaxations down to 7th layer taken from theory)

SAMPLE PREPARATION (1 sample)
Treatment : cycles of Ar+ sputtering and annealing for 2 hours at 873 K
Crystallinity: sharp (2x1) LEED pattern
Anal. methods:
Contamination: photoemission: no sign of contamination

COMMENTS
Small dimer buckling cannot be ruled out

DATA COLLECTION
Technique: XRD; x-ray wavelength 1.36Å
Dataset : angle of incidence 0.52°: 26 fractional order beams, 12 of which were studied for two different polarizations

THEORY/DATA TREATMENT
Least squares fit of integrated intensity to theoretical kinematic intensities

STRUCTURES EXAMINED
Large data set eliminated all but the dimer model; various amounts of buckling tested; subsurface configuration was assumed to be mainly bond bending, not bond stretching, with relaxations chosen from J.A. Appelbaum and D.R. Hamann, Surf. Sci. 74, 21 (1978)

2D UNIT CELLS (1 domain observed)

Cell	Ax (Å)	Ay (Å)	Bx (Å)	By (Å)	α (°)	Matrix	Pattern	Cell type
Bulk	3.995	0.000	0.000	3.995	90.0	(1.000, 0.000) (0.000, 1.000)	(1x1)	b: bulk lattice
Surface 1	7.990	0.000	0.000	3.995	90.0	(2.000, 0.000) (0.000, 1.000)	(2x1)	s1: commens. superlattice

3D COORDINATES

Ge1-Ge2: symmetric dimer; Ge3-Ge16: relaxed substrate layers (from theory);
Ge17-Ge18: periodically repeating set of bulk layers; 0.1Å error bars assumed for tabulation

Dx/Dy in Å, or as a fraction of layer's unit cell vectors; atom 0 at the origin. Epir/subr are bulk repeat vectors.

No. of atoms: 18

Bulk z = 2.825 Å

Region	Chem el.	At. no.	Cell type	Site occ.	Rel. to	Dx ± εx		Dy ± εy		Dz ± εz		Dz/Bz(%) ± εz/Bz
epir		-2					f		f		Å	
subr		-1				1.998	Å	1.998	Å	2.825	Å	
intf	Ge	1	s1	.50	0	0.000	f	0.000	f	0.000	Å	0.0
intf	Ge	2	s1	.50	1	0.292 ± .013	f	0.000	f	0.000 ± .100	Å	0.0 ± 3.5
intf	Ge	3	s1	.50	2	0.609 ± .013	f	0.500	f	1.170 ± .100	Å	41.4 ± 3.5
intf	Ge	4	s1	.50	3	-0.510	f	0.000	f	0.000 ± .100	Å	0.0 ± 3.5
intf	Ge	5	s1	.50	4	0.255	f	0.000	f	1.360	Å	48.2
intf	Ge	6	s1	.50	5	-0.500	f	0.000	f	0.300	Å	10.6
intf	Ge	7	s1	.50	6	0.500	f	-0.500	f	1.110	Å	39.3
intf	Ge	8	s1	.50	7	-0.500	f	0.000	f	0.300	Å	10.6
intf	Ge	9	s1	.50	8	0.747	f	0.000	f	1.380	Å	48.9
intf	Ge	10	s1	.50	9	-0.494	f	0.000	f	0.000	Å	0.0
intf	Ge	11	s1	.50	10	0.494	f	0.500	f	1.410	Å	49.9
intf	Ge	12	s1	.50	11	-0.494	f	0.000	f	0.000	Å	0.0
intf	Ge	13	s1	.50	12	-0.253	f	0.000	f	1.380	Å	48.9
intf	Ge	14	s1	.50	13	0.500	f	0.000	f	0.120	Å	4.3
intf	Ge	15	s1	.50	14	-0.500	f	-0.500	f	1.290	Å	45.7
intf	Ge	16	s1	.50	15	0.500	f	0.000	f	0.120	Å	4.3
subl	Ge	17	b	1.00	16	0.500	f	0.000	f	1.412	Å	50.0
subl	Ge	18	b	1.00	17	0.000	f	0.500	f	1.412	Å	50.0

Ge(100)-(2x1)
32.5 BOND DISTANCES AND ANGLES

Bond distances and angles are derived from coordinates

No. of distances/angles: 6

Interatomic dist. A-B (Å)	Atom A	Atom B	Atom C	Bond angle A-B-C (°)
2.333	Ge1	Ge2	Ge3(-1,0)	110.6
2.446	Ge1	Ge3(-1,0)	Ge5(-2,0)	122.3
2.446	Ge1	Ge3(-1,0)	Ge6(-1,0)	123.6
2.450	Ge3	Ge5	Ge7	108.7
2.450	Ge3	Ge5	Ge8(1,0)	111.2
2.567	Ge3	Ge6(1,0)	Ge7(1,0)	108.3

COMMON NAME : Ge(111)-c(2x8)
CLASSIFICATION : 32.21
TECHNIQUE : LEED
AUTHORS : S.Y. Tong, H. Huang and C.M. Wei
REFERENCE : Chem. and Phys. of Sol. Surf. VIII, 22, 395 (1990)

ILLUSTRATION: 92

SURFACE TYPE
Substrate : Ge
Crystal face: 111
Temperature : RT*
Bulk lattice: diamond
2D bulk symm: p3m1
2D surf symm: pm

Adsorbate:
Coverage :
Pattern : c(2x8)
Matrix : (2.000, 0.000)
(-1.000, 4.000)

STRUCTURE TYPE
2 adatoms A and B in the top layer in T4 sites with identical local environments; relaxations found in the top 5 Ge layers (adatoms + 2 bilayers)

SAMPLE PREPARATION (1 sample)
Treatment :
Crystallinity:
Anal. methods:
Contamination:

COMMENTS
The positions of the adatoms make 1/4 order spots vanish in kinematic approximation; their intensities are due to multiple scattering and are two orders of magnitude smaller than for 1/2 order spots, in agreement with experiment
R-factor used not mentioned.

DATA COLLECTION
Technique: LEED
Dataset : IV curves for 7 beams: E=30-320 eV, normal incidence

THEORY/DATA TREATMENT
Quasi-dynamical LEED (Ge bilayers treated as composite layers):

STRUCTURES EXAMINED
1) dimers-chain model (K. Takayanagi et al. Phys. Rev. B34 10324 (1986)), 2) relaxed H3 adatoms, 3) relaxed T4 adatoms, 3) is preferred

QUALITY OF EXPERIMENT-THEORY FIT
1) R=0.34, 2) R=0.31, 3) R=0.27

2D UNIT CELLS (3 domains observed)

Cell	Ax (Å)	Ay (Å)	Bx (Å)	By (Å)	α (°)	Matrix	Pattern	Cell type
Bulk	4.000	0.000	2.000	3.464	60.0	(1.000, 0.000) (0.000, 1.000)	(1x1)	b: bulk lattice
Surface 1	8.000	0.000	4.000	13.857	73.9	(2.000, 0.000) (-1.000, 4.000)	c(2x8)	s1: commens. superlattice

3D COORDINATES

Ge1-Ge2: adatoms; Ge3-Ge10: top half of 1st bilayer; Ge11-Ge18: bottom half of 1st bilayer; Ge19-26 and Ge27-34: 2nd bilayer; Ge35-Ge36: bulk bilayer; 0.1Å error bars set for fitted coordinates

Dx/Dy in Å, or as a fraction of layer's unit cell vectors; atom 0 at the origin. Epir/subr are bulk repeat vectors.

No. of atoms: 36 Bulk z = 2.450 Å

Region	Chem el.	At. no.	Cell type	Site occ.	Rel. to	Dx ± εx	Dy ± εy	Dz ± εz	Dz/Bz(%) ± εz/Bz
epir		-2				f	f	Å	
subr		-1				2.000 Å	1.155 Å	3.267 Å	
intf	Ge	1	s1	.13	0	0.000 Å	0.000 Å	-8.506 ± .100 Å	-347.2 ± 4.1
intf	Ge	2	s1	.13	0	8.000 Å	6.932 Å	-8.506 Å	-347.2
intf	Ge	3	s1	.13	0	4.000 Å	2.311 Å	-7.356 Å	-300.3
intf	Ge	4	s1	.13	0	4.000 Å	9.243 Å	-7.356 Å	-300.3
intf	Ge	5	s1	.13	0	8.000 Å	2.011 ± .100 Å	-7.136 ± .100 Å	-291.3 ± 4.1
intf	Ge	6	s1	.13	0	8.000 Å	8.943 Å	-7.136 Å	-291.3
intf	Ge	7	s1	.13	0	5.741 Å	12.852 Å	-7.136 Å	-291.3
intf	Ge	8	s1	.13	0	1.741 Å	5.927 Å	-7.136 Å	-291.3
intf	Ge	9	s1	.13	0	10.260 Å	12.852 Å	-7.136 Å	-291.3
intf	Ge	10	s1	.13	0	6.259 Å	5.927 Å	-7.136 Å	-291.3
intf	Ge	11	s1	.13	0	4.000 Å	0.120 ± .100 Å	-6.639 ± .100 Å	-271.0 ± 4.1
intf	Ge	12	s1	.13	0	4.000 Å	7.052 Å	-6.639 Å	-271.0
intf	Ge	13	s1	.13	0	2.105 Å	3.406 Å	-6.639 Å	-271.0
intf	Ge	14	s1	.13	0	10.105 Å	10.338 Å	-6.639 Å	-271.0
intf	Ge	15	s1	.13	0	5.895 Å	3.406 Å	-6.639 Å	-271.0
intf	Ge	16	s1	.13	0	5.895 Å	10.338 Å	-6.639 Å	-271.0
intf	Ge	17	s1	.13	0	0.000 Å	0.000 Å	-5.919 ± .100 Å	-241.6 ± 4.1
intf	Ge	18	s1	.13	0	8.000 Å	6.932 Å	-5.919 Å	-241.6
intf	Ge	19	s1	.13	0	4.000 Å	0.000 Å	-4.126 ± .100 Å	-168.4 ± 4.1
intf	Ge	20	s1	.13	0	4.000 Å	6.932 Å	-4.126 Å	-168.4

Ge(111)-c(2x8)
32.21

3D Coordinates - Continued

Region	Chem el.	At. no.	Cell type	Site occ.	Rel. to	Dx ± εx	Dy ± εy	Dz ± εz	Dz/Bz(%) ± εz/Bz
intf	Ge	21	s1	.13	0	2.000 Å	3.466 Å	-4.126 Å	-168.4
intf	Ge	22	s1	.13	0	10.001 Å	10.398 Å	-4.126 Å	-168.4
intf	Ge	23	s1	.13	0	6.000 Å	3.466 Å	-4.126 Å	-168.4
intf	Ge	24	s1	.13	0	6.000 Å	10.398 Å	-4.126 Å	-168.4
intf	Ge	25	s1	.13	0	0.000 Å	0.000 Å	-3.746 ± .100 Å	-152.9 ± 4.1
intf	Ge	26	s1	.13	0	8.000 Å	6.932 Å	-3.746 Å	-152.9
intf	Ge	27	s1	.13	0	8.000 Å	4.621 Å	-3.199 ± .100 Å	-130.6 ± 4.1
intf	Ge	28	s1	.13	0	8.000 Å	11.553 Å	-3.199 Å	-130.6
intf	Ge	29	s1	.13	0	4.000 Å	4.621 Å	-3.199 Å	-130.6
intf	Ge	30	s1	.13	0	4.000 Å	11.553 Å	-3.199 Å	-130.6
intf	Ge	31	s1	.13	0	2.000 Å	1.155 Å	-3.199 Å	-130.6
intf	Ge	32	s1	.13	0	10.001 Å	8.087 Å	-3.199 Å	-130.6
intf	Ge	33	s1	.13	0	6.000 Å	1.155 Å	-3.199 Å	-130.6
intf	Ge	34	s1	.13	0	6.000 Å	8.087 Å	-3.199 Å	-130.6
subl	Ge	35	b	1.00	0	2.000 Å	1.153 Å	-0.816 Å	-33.3
subl	Ge	36	b	1.00	0	2.000 Å	-1.153 Å	0.000 Å	0.0

BOND DISTANCES AND ANGLES

Bond distances and angles are derived from coordinates

No. of distances/angles: 16

Interatomic dist. A-B (Å)	Atom A	Atom B	Atom C	Bond angle A-B-C (°)
2.433	Ge1	Ge5(-1,0)	Ge13	123.8
2.433	Ge2	Ge8(1,0)	Ge13(1,0)	127.6
2.433	Ge2	Ge8(1,0)	Ge18	65.5
2.305	Ge3	Ge11	Ge7(0,-1)	110.8
2.305	Ge3	Ge11	Ge9(0,-1)	110.8
2.305	Ge3	Ge11	Ge19	110.9
2.303	Ge3	Ge13	Ge5(-1,0)	110.8
2.303	Ge3	Ge15	Ge23	110.9
2.433	Ge1	Ge5(-1,0)	Ge15(-1,0)	123.8
2.433	Ge1	Ge5(-1,0)	Ge17	65.5
2.433	Ge1	Ge7(0,-1)	Ge11	123.8
2.433	Ge1	Ge7(0,-1)	Ge16(0,-1)	123.8
2.587	Ge1	Ge17	Ge7(0,-1)	58.8
2.587	Ge1	Ge17	Ge9(-1,-1)	58.8
2.587	Ge1	Ge17	Ge25	180.0
2.433	Ge2	Ge8(1,0)	Ge12(1,0)	123.8

COMMON NAME : Ge(111)-(√3x√3)R30°-Bi
CLASSIFICATION : 32.83.1
TECHNIQUE : LEED and AES
AUTHORS : K.J.Wan, W.K. Ford, G.J. Lapeyre and J.C. Hermanson
REFERENCE : Phys. Rev., B44, 6500 (1991)

ILLUSTRATION: 96,97

SURFACE TYPE
Substrate : Ge
Crystal face: 111
Temperature : 320C
Bulk lattice: diamond
2D bulk symm: p3m1
2D surf symm: p31m

Adsorbate: Bi
Coverage : 1/3 Bi/1x1
Pattern : (√3x√3)R30°
Matrix : (1.000, 1.000)
(-1.000, 2.000)

STRUCTURE TYPE
Atomic adsorption of Bi in T4 site (3-fold hollow above 2nd Ge layer) on unreconstructed, relaxed substrate

SAMPLE PREPARATION (2 sample)
Treatment : Bi depos. on hot-sample surface(320C) showing (1x1) pattern
Crystallinity: same IVs for different preparations
Anal. methods:
Contamination: AES used to monitor coverage

COMMENTS

DATA COLLECTION
Technique: LEED and AES
Dataset : IV curves for 11 ineq. beams: E=50-300 eV, cumul. E range 2000 eV

THEORY/DATA TREATMENT
Dynamical LEED : extension of Duke-Laramore codes
6 relativistic phase shifts; Voi, Vor fit

STRUCTURES EXAMINED
T4 site (Bi in 3-fold hollow above 2nd layer Ge atom); H3 site (in 3-fold hollow above 4th layer Ge); full structural optimization for T4 site

QUALITY OF EXPERIMENT-THEORY FIT
RX=0.244

2D UNIT CELLS (1 domain observed)

Cell	Ax (Å)	Ay (Å)	Bx (Å)	By (Å)	α (°)	Matrix	Pattern	Cell type
Bulk	3.464	-2.000	3.464	2.000	60.0	(1.000, 0.000) (0.000, 1.000)	(1x1)	b: bulk lattice
Surface 1	6.928	0.000	3.464	6.000	60.0	(1.000, 1.000) (-1.000, 2.000)	(√3x√3)R30°	s1: commens. superlattice

3D COORDINATES

Bi1: atomic overlayer in T4 sites; Ge2-Ge4, Ge5-Ge7, Ge8-Ge10: 1st, 2nd and 3rd planar half Ge bilayers; Ge11-Ge12: periodically repeating set of bulk layers; error bars of 0.1Å set for fitted coord.

Dx/Dy in Å, or as a fraction of layer's unit cell vectors; atom 0 at the origin. Epir/subr are bulk repeat vectors.

No. of atoms: 12 Bulk z = 3.267 Å

Region	Chem el.	At. no.	Cell type	Site occ.	Rel. to	Dx ± εx		Dy ± εy		Dz ± εz		Dz/Bz(%) ± εz/Bz
epir		-2					f		f		Å	
subr		-1				1.155	Å	2.000	Å	3.267	Å	
ovrl	Bi	1	s1	.33	0	0.000	Å	0.000	Å	0.000	Å	0.0
intf	Ge	2	s1	.33	0	6.120	Å	1.400	Å	1.324	Å	40.5
intf	Ge	3	s1	.33	0	1.617 ± .100	Å	0.000	Å	1.324 ± .100	Å	40.5 ± 3.1
intf	Ge	4	s1	.33	0	2.656	Å	4.600	Å	1.324	Å	40.5
intf	Ge	5	s1	.33	0	3.463	Å	2.000	Å	2.186 ± .100	Å	66.9 ± 3.1
intf	Ge	6	s1	.33	0	6.927	Å	4.000	Å	2.186	Å	66.9
intf	Ge	7	s1	.33	0	0.000	Å	0.000	Å	2.585 ± .100	Å	79.1 ± 3.1
intf	Ge	8	s1	.33	0	3.463	Å	2.000	Å	4.726 ± .100	Å	144.7 ± 3.1
intf	Ge	9	s1	.33	0	6.927	Å	4.000	Å	4.726	Å	144.7
intf	Ge	10	s1	.33	0	0.000	Å	0.000	Å	5.014 ± .100	Å	153.5 ± 3.1
subl	Ge	11	b	1.00	0	1.154	Å	2.000	Å	5.831	Å	178.5
subl	Ge	12	b	1.00	11	3.464	Å	-2.000	Å	2.450	Å	75.0

Ge(111)-(√3x√3)R30°-Bi
32.83.1

BOND DISTANCES AND ANGLES

Bond distances and angles are derived from coordinates

No. of distances/angles: 12

Interatomic dist. A-B (Å)	Atom A	Atom B	Atom C	Bond angle A-B-C (°)
2.090	Bi1	Ge2(-1,0)	Ge3	47.9
2.449	Ge10	Ge11	Ge8	106.3
2.560	Ge11	Ge8	Ge5	115.6
2.449	Ge11	Ge10	Ge7	109.5
2.585	Bi1	Ge7	Ge2(-1,0)	52.0
2.585	Bi1	Ge7	Ge10	180.0
2.800	Ge2	Ge3(1,0)	Bi1(1,0)	47.9
2.800	Ge2	Ge3(1,0)	Ge4(1,-1)	60.0
2.051	Ge3	Ge7	Bi1	52.1
2.051	Ge3	Ge7	Ge2(-1,0)	86.1
2.429	Ge10	Ge7	Bi1	180.0
2.429	Ge10	Ge7	Ge2(-1,0)	128.0

COMMON NAME : Ge(111)-(1x1)-Cl
CLASSIFICATION : 32.17.1
TECHNIQUE : SEXAFS
AUTHORS : P.H. Citrin, J.E. Rowe and P. Eisenberger
REFERENCE : Phys. Rev., B28, 2299 (1983)

ILLUSTRATION: 96,99

SURFACE TYPE
Substrate : Ge
Crystal face: 111
Temperature : RT*
Bulk lattice: diamond
2D bulk symm: p3m1
2D surf symm: p3m1

Adsorbate: Cl
Coverage : 1ML Cl/Si
Pattern : (1x1)
Matrix : (1.000, 0.000)
(0.000, 1.000)

STRUCTURE TYPE
Atomic adsortion in top sites on unrelaxed unreconstructed substrate

SAMPLE PREPARATION (1 sample)
Treatment : Ar+, resistive annealing and exposure to Cl2
Crystallinity: (1x1) LEED pattern after annealing
Anal. methods:
Contamination:

COMMENTS

DATA COLLECTION
Technique: SEXAFS; polarization dependent SEXAFS
Dataset : SEXAFS (Θ=10°, 90°): filtered data in range 1.0-2.8Å

THEORY/DATA TREATMENT
Fourier transform and polarization dependence of first-neighbor peak

STRUCTURES EXAMINED
Only top site is consistent with polarization dependence of nearest-neighbor peak in Fourier transformed spectrum

2D UNIT CELLS (1 domain observed)

Cell	Ax (Å)	Ay (Å)	Bx (Å)	By (Å)	α (°)	Matrix	Pattern	Cell type
Bulk	3.995	0.000	-1.998	3.460	120.0	(1.000, 0.000) (0.000, 1.000)	(1x1)	b: bulk lattice
Surface 1	3.995	0.000	-1.998	3.460	120.0	(1.000, 0.000) (0.000, 1.000)	(1x1)	s1: commens. superlattice

3D COORDINATES

Cl1: top-site overlayer; Ge4-Ge5: periodically repeating bulk bilayer

Dx/Dy in Å, or as a fraction of layer's unit cell vectors; atom 0 at the origin. Epir/subr are bulk repeat vectors.

No. of atoms: 5 Bulk z = 3.264 Å

Reg ion	Chem el.	At. no.	Cell type	Site occ.	Rel. to	Dx ± εx		Dy ± εy		Dz ± εz		Dz/Bz(%) ± εz/Bz
epir		-2					f		f		Å	
subr		-1				0.000	Å	2.307	Å	3.264	Å	
ovrl	Cl	1	b	1.00	0	0.000	f	0.000	f	0.000	Å	0.0
intf	Ge	2	b	1.00	1	0.000	f	0.000	f	2.070 ± .030	Å	63.4 ± .9
intf	Ge	3	b	1.00	2	0.333	f	0.667	f	0.816	Å	25.0
subl	Ge	4	b	1.00	3	0.000	f	0.000	f	2.448	Å	75.0
subl	Ge	5	b	1.00	4	0.333	f	-0.333	f	0.816	Å	25.0

BOND DISTANCES AND ANGLES

No. of distances/angles: 2

Interatomic dist. A-B (Å)	Atom A	Atom B	Atom C	Bond angle A-B-C (°)
2.070	Cl1	Ge2	Ge3	109.5
2.447	Ge2	Ge3	Ge4	109.5

COMMON NAME : Ge(111)-(1x1)-H ILLUSTRATION: 88
CLASSIFICATION : 32.1.1
TECHNIQUE : LEED
AUTHORS : R. Imbihl, J.E. Demuth, F.J. Himpsel, P.M. Marcus, W.A. Thompson and F. Jona
REFERENCE : Phys. Rev., B36, 5037 (1987)

SURFACE TYPE
Substrate : Ge
Crystal face: 111
Temperature : RT
Bulk lattice: diamond
2D bulk symm: p3m1
2D surf symm: p3m1

Adsorbate: H
Coverage : 1.0 H/Ge
Pattern : (1x1)
Matrix : (1.000, 0.000)
(0.000, 1.000)

STRUCTURE TYPE
Atomic adsorption stabilizing unreconstructed substrate with relaxed top two interlayer spacings (H position undetermined)

SAMPLE PREPARATION (1 sample)
Treatment : cleavage at RT; single domain surfaces exposed to atomic H
Crystallinity: (2x1) completely converted to (1x1)
Anal. methods:
Contamination:

COMMENTS
IV spectra obtained at 20-30 K displayed only minor differences, corresponding to structural changes within the error bars

DATA COLLECTION
Technique: LEED
Dataset : I-V spectra: 5 inequivalent beams at normal incidence; 70<E<220 eV

THEORY/DATA TREATMENT
Dynamical LEED: 6 phase shifts; Vor=-4±3 eV (fit), Voi=-4 eV; ΘD=298 K

STRUCTURES EXAMINED
1) Ge(111)-(1x1) without H, 1st & 2nd layer spacings varied; 2) Ge(111)-(1x1) with H: Ge-H = 1.53Å as in GeH4, GeClH3; 3) Ge(111)-(1x1) with H: Ge-H = 1.59Å as in GeH; analysis is found insensitive to H position

QUALITY OF EXPERIMENT-THEORY FIT
RZJ=0.25

2D UNIT CELLS (1 domain observed)

Cell	Ax (Å)	Ay (Å)	Bx (Å)	By (Å)	α (°)	Matrix	Pattern	Cell type
Bulk	4.000	0.000	-2.000	3.464	120.0	(1.000, 0.000) (0.000, 1.000)	(1x1)	b: bulk lattice
Surface 1	4.000	0.000	-2.000	3.464	120.0	(1.000, 0.000) (0.000, 1.000)	(1x1)	s1: commens. superlattice

3D COORDINATES

Ge1-Ge2: relaxed top bilayer; Ge3-Ge4: unrelaxed 2nd bilayer;
Ge5-Ge6: periodically repeating set of bulk layers

Dx/Dy in Å, or as a fraction of layer's unit cell vectors; atom 0 at the origin. Epir/subr are bulk repeat vectors.

No. of atoms: 6 Bulk z = 3.267 Å

Region	Chem el.	At. no.	Cell type	Site occ.	Rel. to	Dx ± εx		Dy ± εy		Dz ± εz		Dz/Bz(%) ± εz/Bz
epir		-2					f		f		Å	
subr		-1				-2.000	Å	-1.155	Å	3.267	Å	
intf	Ge	1	b	1.00	0	0.000	f	0.000	f	0.000	Å	0.0
intf	Ge	2	b	1.00	1	0.333	f	0.667	f	0.716 ± .050	Å	21.9 ± 1.5
intf	Ge	3	b	1.00	2	0.000	f	0.000	f	2.500 ± .050	Å	76.5 ± 1.5
intf	Ge	4	b	1.00	3	0.333	f	-0.333	f	0.816	Å	25.0
subl	Ge	5	b	1.00	4	0.000	f	0.000	f	2.450	Å	75.0
subl	Ge	6	b	1.00	5	-0.667	f	-0.333	f	0.817	Å	25.0

Ge(111)-(1x1)-H
32.1.1 BOND DISTANCES AND ANGLES

Bond distances and angles are derived from coordinates

No. of distances/angles: 4

Interatomic dist. A-B (Å)	Atom A	Atom B	Atom C	Bond angle A-B-C (°)
2.418	Ge1	Ge2	Ge3	107.2
2.500	Ge2	Ge3	Ge4	109.5
2.449	Ge3	Ge4	Ge5	109.5
2.450	Ge4	Ge5	Ge6	109.5

COMMON NAME : Ge(111)-(1x1)-I
CLASSIFICATION : 32.53.2
TECHNIQUE : XSW, SEXAFS
AUTHORS : M.J. Bedzyk, Q. Shen, M.E. Keeffe and G. Navrotski
REFERENCE : Surf. Sci., 220, 419 (1989)

ILLUSTRATION: 96,99

SURFACE TYPE
Substrate : Ge
Crystal face: 111
Temperature : RT*
Bulk lattice: diamond
2D bulk symm: p3m1
2D surf symm: p3m1

Adsorbate: I
Coverage : 1.0 I/1x1
Pattern : (1x1)
Matrix : (1.000, 0.000)
(0.000, 1.000)

STRUCTURE TYPE
Atomic adsorption of I in top sites on unreconstructed, relaxed substrate: first substrate interlayer spacing contracted by 10%

SAMPLE PREPARATION (1 sample)
Treatment : Ar sputt, annealing, then I deposited from electrol. source
Crystallinity:
Anal. methods:
Contamination: AES, LEED used to assess clean Ge(111)

COMMENTS
XSW data measured in this work used together with SEXAFS data previously recorded to determine structure

DATA COLLECTION
Technique: XSW, SEXAFS; XSW measurement, E(gamma)=6.0k
Dataset : I L3 peak at 1400 eV scanned over 5.3eV range

THEORY/DATA TREATMENT

2D UNIT CELLS (3 domains observed)

Cell	Ax (Å)	Ay (Å)	Bx (Å)	By (Å)	α (°)	Matrix	Pattern	Cell type
Bulk	4.000	0.000	-2.000	3.464	120.0	(1.000, 0.000) (0.000, 1.000)	(1x1)	b: bulk lattice
Surface 1	4.000	0.000	-2.000	3.464	120.0	(1.000, 0.000) (0.000, 1.000)	(1x1)	s1: commens. superlattice

3D COORDINATES

I1: atomic overlayer in top sites; Ge2-Ge3: unreconstructed, contracted top substrate bilayer;
Ge4-Ge5: periodically repeating set of bulk layers

Dx/Dy in Å, or as a fraction of layer's unit cell vectors; atom 0 at the origin. Epir/subr are bulk repeat vectors.

No. of atoms: 5 Bulk z = 3.267 Å

Region	Chem el.	At. no.	Cell type	Site occ.	Rel. to	Dx ± εx		Dy ± εy		Dz ± εz		Dz/Bz(%) ± εz/Bz
epir		-2					f		f		Å	
subr		-1				-2.000	Å	-1.155	Å	3.267	Å	
ovrl	I	1	s1	1.00	0	0.000	f	0.000	f	-2.500 ± .040	Å	-76.5 ± 1.2
intf	Ge	2	s1	1.00	0	0.000	f	0.000	f	0.000	Å	0.0
intf	Ge	3	s1	1.00	0	0.333	f	0.667	f	0.710 ± .100	Å	21.7 ± 3.1
subl	Ge	4	b	1.00	0	0.333	f	0.667	f	3.213	Å	98.4
subl	Ge	5	b	1.00	3	0.333	f	-0.333	f	3.267	Å	100.0

BOND DISTANCES AND ANGLES

Bond distances and angles are derived from coordinates

No. of distances/angles: 6

Interatomic dist. A-B (Å)	Atom A	Atom B	Atom C	Bond angle A-B-C (°)
2.500	I1	Ge2	Ge3	107.1
2.416	Ge2	Ge3	Ge2(0,1)	111.8
2.416	Ge2	Ge3	Ge4	107.1
2.503	Ge3	Ge4	Ge5(0,1)	108.3
2.503	Ge4	Ge3	Ge2(1,1)	107.1
2.432	Ge4	Ge5(0,1)	Ge4(1,1)	110.6

COMMON NAME : Ge(111)-(1x1)-PHx ILLUSTRATION: 110
CLASSIFICATION : 32.15.1.1
TECHNIQUE : ARPEFS
AUTHORS : L.J. Terminello, K.T. Leung, Z. Hussain, Y. Hayashi, X.S. Zhang and D.A. Shirley
REFERENCE : Phys. Rev., B41, 12787 (1990)

SURFACE TYPE
Substrate : Ge
Crystal face: 111
Temperature : RT*
Bulk lattice: diamond
2D bulk symm: p3m1
2D surf symm: pm

Adsorbate: PHx (x<=3)
Coverage : 1.0 PHx/1x1
Pattern : (1x1)
Matrix : (1.000, 0.000)
(0.000, 1.000)

STRUCTURE TYPE
Adsorption of partially dissociated PH3 in tilted top sites on unreconstructed, relaxed substrate; first Ge-Ge interlayer spacing contracted by 16%

SAMPLE PREPARATION (1 sample)
Treatment : exposure to PH3: saturation achieved after 3-4 cycles
Crystallinity: clear (1x1) with minimum background
Anal. methods:
Contamination: AES, LEED used to assess clean Ge(111)

COMMENTS
Indirect evidence that PH2 is the specie adsorbed

DATA COLLECTION
Technique: ARPEFS
Dataset : normal,off-normal emiss. fine struct. chi(E) 50<E<550 eV

THEORY/DATA TREATMENT
Curved wave multiple scattering theory with R-factor; non-structural parameters described in SSD 32.16.1

STRUCTURES EXAMINED
P in top (a), 3-fold (eclipsed(b) and hollow(c)) and substitutional(d) sites; displacement of P used to choose a over c; optimization of P position and first two Ge layer spacings for final structure determination; H neglected

2D UNIT CELLS (3 domains observed)

Cell	Ax (Å)	Ay (Å)	Bx (Å)	By (Å)	α (°)	Matrix	Pattern	Cell type
Bulk	4.000	0.000	-2.000	3.464	120.0	(1.000, 0.000) (0.000, 1.000)	(1x1)	b: bulk lattice
Surface 1	4.000	0.000	-2.000	3.464	120.0	(1.000, 0.000) (0.000, 1.000)	(1x1)	s1: commens. superlattice

3D COORDINATES

P1: forms PHx overlayer in tilted top sites; Ge2-Ge3: unreconstructed, contracted top substrate bilayer; Ge4-Ge5: periodically repeating set of bulk layers

Dx/Dy in Å, or as a fraction of layer's unit cell vectors; atom 0 at the origin. Epir/subr are bulk repeat vectors.

No. of atoms: 5 Bulk z = 3.267 Å

Region	Chem el.	At. no.	Cell type	Site occ.	Rel. to	Dx ± εx		Dy ± εy		Dz ± εz		Dz/Bz(%) ± εz/Bz
epir		-2					f		f		Å	
subr		-1				-2.000	Å	-1.155	Å	3.267	Å	
ovrl	P	1	s1	1.00	0	0.000	Å	0.630 ± .050	Å	-2.260 ± .040	Å	-69.2 ± 1.2
intf	Ge	2	s1	1.00	0	0.000	f	0.000	f	0.000	Å	0.0
intf	Ge	3	s1	1.00	0	0.333	f	0.667	f	0.680	Å	20.8
subl	Ge	4	b	1.00	0	0.333	f	0.667	f	3.340	Å	102.2
subl	Ge	5	b	1.00	3	0.333	f	-0.333	f	3.267	Å	100.0

BOND DISTANCES AND ANGLES

Bond distances and angles are derived from coordinates

No. of distances/angles: 8

Interatomic dist. A-B (Å)	Atom A	Atom B	Atom C	Bond angle A-B-C (°)
2.346	P1	Ge2	Ge3	90.8
2.346	P1	Ge2	Ge3(0,-1)	113.6
2.408	Ge2	Ge3	Ge2(1,1)	112.4
2.408	Ge2	Ge3	Ge4	106.4

Ge(111)-(1x1)-PHx
32.15.1.1

Bond Distances and Angles - Continued

Interatomic dist. A-B (Å)	Atom A	Atom B	Atom C	Bond angle A-B-C (°)
2.660	Ge3	Ge4	Ge5	104.7
2.660	Ge4	Ge3	Ge2	106.4
2.388	Ge4	Ge5	Ge4(1,0)	113.8
2.388	Ge4	Ge5	Ge4(0,-1)	113.8

COMMON NAME : Ge(111)-(√3x√3)R30°-Pb (1/3ML) ILLUSTRATION: 96,97
CLASSIFICATION : 32.82.6a
TECHNIQUE : LEED
AUTHORS : H. Huang, C.M. Wei, H. Li, B.P. Tonner and S.Y. Tong
REFERENCE : Phys. Rev. Lett., 62, 559 (1989)

SURFACE TYPE
Substrate : Ge
Crystal face: 111
Temperature : RT*
Bulk lattice: diamond
2D bulk symm: p3m1
2D surf symm: p31m

Adsorbate: Pb
Coverage : 1/3 Pb/1x1
Pattern : (√3x√3)R30°
Matrix : (2.000, 1.000)
(-1.000, 1.000)

STRUCTURE TYPE
α structure: atomic adsorption of Pb in T4 sites on unreconstructed, relaxed substrate: buckling of second and third Ge monolayers (first is planar)

SAMPLE PREPARATION (sample)
Treatment :
Crystallinity:
Anal. methods:
Contamination:

COMMENTS
Compare 4/3ML β structure: SSD 32.82.6b

DATA COLLECTION
Technique: LEED
Dataset : I-V spectra: 5 inequivalent beams at normal incidence; 50<E<300 eV
full relaxation of Pb position and first Ge layer consistent with p31m symmetry; only vertical relaxation (again p3m1) for 2nd and 3rd Ge layers

THEORY/DATA TREATMENT
Dynamical LEED

QUALITY OF EXPERIMENT-THEORY FIT
R(VHT)=0.28

2D UNIT CELLS (1 domain observed)

Cell	Ax (Å)	Ay (Å)	Bx (Å)	By (Å)	α (°)	Matrix	Pattern	Cell type
Bulk	2.000	-3.464	2.000	3.464	120.0	(1.000, 0.000) (0.000, 1.000)	(1x1)	b: bulk lattice
Surface 1	6.000	-3.464	0.000	6.928	120.0	(2.000, 1.000) (-1.000, 1.000)	(√3x√3)R30°	s1: commens. superlattice

3D COORDINATES

Pb1: atomic overlayer in T4 sites; Ge2-Ge4, Ge5-Ge7, Ge8-Ge10: 1st, 2nd and 3rd half bilayers; Ge11-Ge12 form period. repeated bulk set error bars of 0.1Å set for fitted coord.

Dx/Dy in Å, or as a fraction of layer's unit cell vectors; atom 0 at the origin. Epir/subr are bulk repeat vectors.

No. of atoms: 12 Bulk z = 3.267 Å

Region	Chem el.	At. no.	Cell type	Site occ.	Rel. to	Dx ± εx		Dy ± εy		Dz ± εz		Dz/Bz(%) ± εz/Bz
epir		-2					f		f		Å	
subr		-1				2.000	Å	1.155	Å	3.267	Å	
ovrl	Pb	1	s1	.33	0	0.000	Å	2.310	Å	-5.785 ± .100	Å	-177.1 ± 3.1
intf	Ge	2	s1	.33	0	0.000	Å	4.201 ± .100	Å	-4.085 ± .100	Å	-125.1 ± 3.1
intf	Ge	3	s1	.33	0	1.637	Å	1.365	Å	-4.085	Å	-125.1
intf	Ge	4	s1	.33	0	4.363	Å	-2.099	Å	-4.085	Å	-125.1
intf	Ge	5	s1	.33	0	2.000	Å	-1.155	Å	-3.368 ± .100	Å	-103.1 ± 3.1
intf	Ge	6	s1	.33	0	4.000	Å	2.309	Å	-3.368	Å	-103.1
intf	Ge	7	s1	.33	0	0.000	Å	2.310	Å	-2.918 ± .100	Å	-89.3 ± 3.1
intf	Ge	8	s1	.33	0	2.000	Å	-1.155	Å	-0.867 ± .100	Å	-26.5 ± 3.1
intf	Ge	9	s1	.33	0	4.000	Å	2.309	Å	-0.867	Å	-26.5
intf	Ge	10	s1	.33	0	0.000	Å	2.310	Å	-0.617 ± .100	Å	-18.9 ± 3.1
subl	Ge	11	b	1.00	0	0.000	Å	0.000	Å	0.000	Å	0.0
subl	Ge	12	b	1.00	11	0.000	Å	0.000	Å	2.450	Å	75.0

Ge(111)-(√3x√3)R30°-Pb (1/3ML)
32.82.6a BOND DISTANCES AND ANGLES

Bond distances and angles are derived from coordinates

No. of distances/angles: 14

Interatomic dist. A-B (Å)	Atom A	Atom B	Atom C	Bond angle A-B-C (°)
2.543	Pb1	Ge2	Ge5(0,1)	128.6
2.467	Ge8	Ge11	Ge9(-1,-1)	108.3
2.467	Ge8	Ge11	Ge10	111.2
2.450	Ge11	Ge12		
2.450	Ge12	Ge11	Ge8	110.6
2.450	Ge4	Ge5	Ge6	109.5
2.543	Pb1	Ge2	Ge7	73.6
2.867	Pb1	Ge7	Ge2	58.3
2.643	Ge2	Ge5(0,1)	Ge3(0,1)	112.9
2.643	Ge2	Ge5(0,1)	Ge4(0,1)	113.0
2.222	Ge2	Ge7	Pb1	58.3
2.222	Ge2	Ge7	Ge3	94.9
2.501	Ge5	Ge8	Ge11(1,1)	110.6
2.301	Ge7	Ge10	Ge11(0,1)	105.0

COMMON NAME : Ge(111)-(√3x√3)R30°-4Pb (4/3ML) ILLUSTRATION: 115
CLASSIFICATION : 32.82.6b
TECHNIQUE : LEED
AUTHORS : H. Huang, C.M. Wei, H. Li, B.P. Tonner and S.Y. Tong
REFERENCE : Phys. Rev. Lett., 62, 559 (1989)

SURFACE TYPE
Substrate : Ge
Crystal face: 111
Temperature : RT*
Bulk lattice: diamond
2D bulk symm: p3m1
2D surf symm: p31m

Adsorbate: Pb
Coverage : 4/3 Pb/1x1
Pattern : (√3x√3)R30°
Matrix : (2.000, 1.000)
(-1.000, 1.000)

STRUCTURE TYPE
ß structure: two Pb layers separated by 0.44Å on unreconstructed, relaxed substrate: inner layer has Pb in H3 sites (1/3ML); outer layer has Pb in off-centered T1 sites (1ML)

SAMPLE PREPARATION (sample)
Treatment :
Crystallinity:
Anal. methods:
Contamination:

COMMENTS
B.N.Dev et al analyzed same structure with XSW, finding 1.3Å spacing between Pb layers (instead of 0.44Å): this model gives R(VHT)=0.339 by LEED; compare ß structure with 1/3ML α structure: SSD 32.82.6a

DATA COLLECTION
Technique: LEED
Dataset : I-V spectra: 10 inequivalent beams at normal incidence; 30<E<300 eV
full relaxation for Pb atoms and first Ge layer consistent with p31m symmetry; only vertical relaxation (again p3m1) for 2nd layer

THEORY/DATA TREATMENT
Dynamical LEED

QUALITY OF EXPERIMENT-THEORY FIT
R(VHT)=0.275

2D UNIT CELLS (1 domain observed)

Cell	Ax (Å)	Ay (Å)	Bx (Å)	By (Å)	α (°)	Matrix	Pattern	Cell type
Bulk	2.000	-3.464	2.000	3.464	120.0	(1.000, 0.000) (0.000, 1.000)	(1x1)	b: bulk lattice
Surface 1	6.000	-3.464	0.000	6.928	120.0	(2.000, 1.000) (-1.000, 1.000)	(√3x√3)R30°	s1: commens. superlattice

3D COORDINATES

Pb1-Pb3: in off-center T1 (top) sites; Pb4: in H3 (hollow) Ge5-Ge7, Ge8-Ge10: 1st and 2nd Ge layers; Ge14 and Ge15: periodically repeating set of bulk layers; error bar of 0.1Å set for fitted coord.

Dx/Dy in Å, or as a fraction of layer's unit cell vectors; atom 0 at the origin. Epir/subr are bulk repeat vectors.

No. of atoms: 15 Bulk z = 3.267 Å

Reg ion	Chem el.	At. no.	Cell type	Site occ.	Rel. to	Dx ± εx		Dy ± εy		Dz ± εz		Dz/Bz(%) ± εz/Bz
epir		-2					f		f		Å	
subr		-1				2.000	Å	1.155	Å	3.267	Å	
ovrl	Pb	1	s1	.33	0	0.000	Å	3.896 ± .100	Å	-6.746 ± .100	Å	-206.5 ± 3.1
ovrl	Pb	2	s1	.33	0	2.626	Å	1.516	Å	-6.746	Å	-206.5
ovrl	Pb	3	s1	.33	0	3.374	Å	-1.948	Å	-6.746	Å	-206.5
ovrl	Pb	4	s1	.33	0	0.000	Å	0.000	Å	-6.305 ± .100	Å	-193.0 ± 3.1
intf	Ge	5	s1	.33	0	1.949	Å	1.185	Å	-4.085 ± .100	Å	-125.1 ± 3.1
intf	Ge	6	s1	.33	0	4.051	Å	-2.279	Å	-4.085	Å	-125.1
intf	Ge	7	s1	.33	0	0.000	Å	4.561	Å	-4.085	Å	-125.1
intf	Ge	8	s1	.33	0	2.000	Å	-1.155	Å	-3.218 ± .100	Å	-98.5 ± 3.1
intf	Ge	9	s1	.33	0	4.000	Å	2.309	Å	-3.218	Å	-98.5
intf	Ge	10	s1	.33	0	0.000	Å	2.310	Å	-3.218	Å	-98.5
intf	Ge	11	s1	.33	0	2.000	Å	-1.155	Å	-0.817	Å	-25.0
intf	Ge	12	s1	.33	0	4.000	Å	2.309	Å	-0.817	Å	-25.0
intf	Ge	13	s1	.33	0	0.000	Å	2.310	Å	-0.817	Å	-25.0
subl	Ge	14	b	1.00	0	0.000	Å	0.000	Å	0.000	Å	0.0
subl	Ge	15	b	1.00	14	0.000	Å	0.000	Å	2.450	Å	75.0

Ge(111)-(√3x√3)R30°-4Pb (4/3ML)
32.82.6b BOND DISTANCES AND ANGLES

Bond distances and angles are derived from coordinates

No. of distances/angles: 13

Interatomic dist. A-B (Å)	Atom A	Atom B	Atom C	Bond angle A-B-C (°)
3.544	Pb1	Pb2	Pb1(1,0)	155.6
3.183	Pb4	Ge5	Pb2	61.5
3.183	Pb4	Ge5	Ge8	84.6
2.496	Ge5	Ge8	Ge6	108.6
2.496	Ge5	Ge8	Ge11	110.3
3.544	Pb1	Pb2	Pb3(0,1)	60.0
3.064	Pb1	Pb4(0,1)	Pb1(0,1)	165.3
3.064	Pb1	Pb4(0,1)	Pb2(0,1)	118.0
3.921	Pb1	Pb4	Pb1(0,-1)	165.3
3.921	Pb1	Pb4	Pb2	59.5
2.743	Pb1	Ge7	Pb4(0,1)	60.9
2.743	Pb1	Ge7	Ge8(0,1)	117.1
2.766	Pb2	Ge5	Pb4	61.5

COMMON NAME : Ge(100)-(2x1)-S ILLUSTRATION: 104,106
CLASSIFICATION : 32.16.2
TECHNIQUE : ARPEFS
AUTHORS : K.T. Leung, L.J. Terminello, Z. Hussain, X.S. Zhang, Y. Hayashi and D.A. Shirley
REFERENCE : Phys. Rev., B38, 8241 (1988)

SURFACE TYPE
Substrate : Ge
Crystal face: 100
Temperature : RT*
Bulk lattice: diamond
2D bulk symm: pmm
2D surf symm: pmm

Adsorbate: S
Coverage : 0.5 S/Ge
Pattern : (2x1)
Matrix : (2.000, 0.000)
(0.000, 1.000)

STRUCTURE TYPE
Atomic adsorption in bridge (i.e. bulk continuation) sites of unreconstructed but relaxed substrate

SAMPLE PREPARATION (1 sample)
Treatment : exposure to H2S: saturation achieved after 3-4 cycles
Crystallinity: strong (2x1) LEED pattern when clean
Anal. methods:
Contamination: AES: no contamination on clean surface

COMMENTS
All structures incorporating the here tabulated bridge site give similar agreement; the here tabulated structure gives best agreement for E>100 eV where CWMS theory performs best

DATA COLLECTION
Technique: ARPEFS
Dataset : normal emission fine structure chi(E) 77<E<420 eV

THEORY/DATA TREATMENT
Curved wave multiple scattering theory with R-factor minimization; ΘD=300-380 K(bulk), 350-450K(surface)

STRUCTURES EXAMINED
S in top, 4-fold, or bridge site (a) above the 2nd layer Ge atom and (b) above the 4th layer Ge atom; for site (a) structures with S-S bonds parallel or perpendicular to the dangling bonds were considered, including variable Ge-S bond lengths and 1st Ge-Ge interlayer spacing

2D UNIT CELLS (1 domain observed)

Cell	Ax (Å)	Ay (Å)	Bx (Å)	By (Å)	α (°)	Matrix	Pattern	Cell type
Bulk	3.995	0.000	0.000	3.995	90.0	(1.000, 0.000) (0.000, 1.000)	(1x1)	b: bulk lattice
Surface 1	7.990	0.000	0.000	3.995	90.0	(2.000, 0.000) (0.000, 1.000)	(2x1)	s1: commens. superlattice

3D COORDINATES

S1: overlayer bridging Ge2-Ge3 pairs; Ge2-Ge3: planar relaxed layer;
Ge6-Ge7: periodically repeating set of bulk layers; 0.1Å error bars assumed for tabulation

Dx/Dy in Å, or as a fraction of layer's unit cell vectors; atom 0 at the origin. Epir/subr are bulk repeat vectors.

No. of atoms: 7 Bulk z = 2.825 Å

Reg ion	Chem el.	At. no.	Cell type	Site occ.	Rel. to	Dx ± εx		Dy ± εy		Dz ± εz		Dz/Bz(%) ± εz/Bz
epir		-2					f		f		Å	
subr		-1				1.998	Å	-1.998	Å	2.825	Å	
ovrl	S	1	s1	.50	0	0.000	f	0.000	f	0.000	Å	0.0
intf	Ge	2	s1	.50	1	0.737 ± .013	f	0.000	f	1.080 ± .100	Å	38.2 ± 3.5
intf	Ge	3	s1	.50	2	-0.474 ± .013	f	0.000	f	0.000	Å	0.0
intf	Ge	4	b	1.00	3	-0.026 ± .025	f	0.500	f	1.410 ± .100	Å	49.9 ± 3.5
intf	Ge	5	b	1.00	4	-0.500	f	0.000	f	1.460 ± .100	Å	51.7 ± 3.5
subl	Ge	6	b	1.00	5	0.000	f	-0.500	f	1.412	Å	50.0
subl	Ge	7	b	1.00	6	0.500	f	0.000	f	1.412	Å	50.0

Ge(100)-(2x1)-S
32.16.2

BOND DISTANCES AND ANGLES

Bond distances and angles are derived from coordinates

No. of distances/angles: 2

Interatomic dist. A-B (Å)	Atom A	Atom B	Atom C	Bond angle A-B-C (°)
2.363	S1	Ge3	Ge4	103.0
2.447	Ge3	Ge4	Ge5	112.0

COMMON NAME : Ge(111)-(2x2)-S ILLUSTRATION: 96,98
CLASSIFICATION : 32.16.1
TECHNIQUE : ARPEFS
AUTHORS : S.W. Robey, C.C. Bahr, Z. Hussain, J.J. Barton, K.T. Leung, J. Lou, A.E. Schach von Wittenau and D.A. Shirley
REFERENCE : Phys. Rev., B35, 5657 (1987)

SURFACE TYPE
Substrate : Ge
Crystal face: 111
Temperature : RT*
Bulk lattice: diamond
2D bulk symm: p3m1
2D surf symm: cm

Adsorbate: S
Coverage : 0.25 S/Ge
Pattern : (2x2)
Matrix : (2.000, 0.000)
(0.000, 2.000)

STRUCTURE TYPE
Atomic adsorption in bridge sites with shorter bond to 3rd Ge atom, with unreconstructed substrate relaxed perpendicular to surface

SAMPLE PREPARATION (1 sample)
Treatment : exposure to H2S and anneal to 573-623 K to desorb excess H
Crystallinity: sharp (2x2) pattern
Anal. methods:
Contamination: AES: no sign of impurities

COMMENTS
Technique found insensitive to substrate layer buckling

DATA COLLECTION
Technique: ARPEFS; monochromatic x-rays (2500-3000eV)
Dataset : angles of incidence chosen for greatest sensitivity to the various structures

THEORY/DATA TREATMENT
Multiple scattering calculations: ΘD=350-450 K

STRUCTURES EXAMINED
1) top site; 2) 3-fold eclipsed; 3) 3-fold hollow; 4); 2-fold hollow; 5) several models of subsurface incorporation

QUALITY OF EXPERIMENT-THEORY FIT
R=0.5

2D UNIT CELLS (3 domains observed)

Cell	Ax (Å)	Ay (Å)	Bx (Å)	By (Å)	α (°)	Matrix	Pattern	Cell type
Bulk	4.000	0.000	-2.000	3.464	120.0	(1.000, 0.000) (0.000, 1.000)	(1x1)	b: bulk lattice
Surface 1	8.000	0.000	-4.000	6.928	120.0	(2.000, 0.000) (0.000, 2.000)	(2x2)	s1: commens. superlattice

3D COORDINATES

S1: overlayer in bridge sites, bonding to Ge2 and Ge3; Ge2-Ge3 and Ge4-Ge5: slightly relaxed bulk layers; Ge6-Ge7: periodically repeating set of bulk layers

Dx/Dy in Å, or as a fraction of layer's unit cell vectors; atom 0 at the origin. Epir/subr are bulk repeat vectors.

No. of atoms: 7 Bulk z = 3.263 Å

Reg ion	Chem el.	At. no.	Cell type	Site occ.	Rel. to	Dx ± εx		Dy ± εy		Dz ± εz		Dz/Bz(%) ± εz/Bz
epir		-2					f		f		Å	
subr		-1				0.000	Å	2.309	Å	3.263	Å	
ovrl	S	1	s1	.25	0	0.000	f	0.000	f	0.000	Å	0.0
intf	Ge	2	b	1.00	1	0.000	f	0.500	f	1.030 ± .050	Å	31.6 ± 1.5
intf	Ge	3	b	1.00	2	0.333	f	-0.333	f	0.730 ± .050	Å	22.4 ± 1.5
intf	Ge	4	b	1.00	3	0.000	f	0.000	f	2.650 ± .070	Å	81.2 ± 2.1
intf	Ge	5	b	1.00	4	0.333	f	0.667	f	0.816	Å	25.0
subl	Ge	6	b	1.00	5	0.000	f	0.000	f	2.447	Å	75.0
subl	Ge	7	b	1.00	6	0.333	f	0.667	f	0.816	Å	25.0

Ge(111)-(2x2)-S
32.16.1

BOND DISTANCES AND ANGLES

Bond distances and angles are derived from coordinates

No. of distances/angles: 5

Interatomic dist. A-B (Å)	Atom A	Atom B	Atom C	Bond angle A-B-C (°)
2.250	S1	Ge2	Ge3	53.4
2.105	S1	Ge3	Ge4	146.7
2.422	Ge2	Ge3	Ge4	107.5
2.650	Ge3	Ge4	Ge5	109.5
2.449	Ge4	Ge5	Ge6	109.5

COMMON NAME : HfC(100)-(1x1) ILLUSTRATION: 149
CLASSIFICATION : 72.6.2
TECHNIQUE : LEED
AUTHORS : G.R. Gruzalski, D.M. Zehner, J.R. Noonan, H.L. Davis, R.A. DiDio and K. Mueller
REFERENCE : J. Vac. Sci. Technol., A7, 2054 (1989)

SURFACE TYPE
Substrate : HfC
Crystal face: 100
Temperature : RT*
Bulk lattice: NaCl
2D bulk symm: p4m
2D surf symm: p4m

Adsorbate:
Coverage :
Pattern : (1x1)
Matrix : (1.000, 0.000)
(0.000, 1.000)

STRUCTURE TYPE
Bulk termination with buckled top two mixed layers (C moves outward, Hf inward) and contraction of first interlayer spacing

SAMPLE PREPARATION (1 sample)
Treatment : Ar-ion bombardment and annealing to 2900 K
Crystallinity: sharp (1x1) LEED pattern
Anal. methods: ARUPS, XPS and AES
Contamination:

COMMENTS

DATA COLLECTION
Technique: LEED
Dataset : I-V spectra for 7 non-degenerate beams at normal incidence; 20<E<350 eV

THEORY/DATA TREATMENT
Dynamical LEED (RFS): 10 phase shifts, different ΘD for Hf and C

STRUCTURES EXAMINED
Variations in spacing between C and Hf in 1st and 2nd mixed layers

QUALITY OF EXPERIMENT-THEORY FIT
RZJ=0.081

2D UNIT CELLS (1 domain observed)

Cell	Ax (Å)	Ay (Å)	Bx (Å)	By (Å)	α (°)	Matrix	Pattern	Cell type
Bulk	3.280	0.000	0.000	3.280	90.0	(1.000, 0.000) (0.000, 1.000)	(1x1)	b: bulk lattice
Surface 1	3.280	0.000	0.000	3.280	90.0	(1.000, 0.000) (0.000, 1.000)	(1x1)	s1: commens. superlattice

3D COORDINATES

C1-Hf2: buckled top mixed layer; C3-Hf4: buckled 2nd mixed layer;
C5-Hf6: periodically repeating mixed bulk layer

Dx/Dy in Å, or as a fraction of layer's unit cell vectors; atom 0 at the origin. Epir/subr are bulk repeat vectors.

No. of atoms: 6 Bulk z = 2.320 Å

Region	Chem el.	At. no.	Cell type	Site occ.	Rel. to	Dx ± εx		Dy ± εy		Dz ± εz		Dz/Bz(%) ± εz/Bz
epir		-2					f		f		Å	
subr		-1				1.640	Å	1.640	Å	2.320	Å	
intf	C	1	b	1.00	0	0.000	f	0.000	f	0.000	Å	0.0
intf	Hf	2	b	1.00	1	0.500	f	0.500	f	0.110 ± .100	Å	4.7 ± 4.3
intf	C	3	b	1.00	2	0.000	f	0.000	f	2.220 ± .100	Å	95.7 ± 4.3
intf	Hf	4	b	1.00	3	-0.500	f	-0.500	f	0.030 ± .100	Å	1.3 ± 4.3
subl	C	5	b	1.00	4	0.000	f	0.000	f	2.320	Å	100.0
subl	Hf	6	b	1.00	5	0.500	f	0.500	f	0.000	Å	0.0

BOND DISTANCES AND ANGLES

Bond distances and angles are derived from coordinates

No. of distances/angles: 5

Interatomic dist. A-B (Å)	Atom A	Atom B	Atom C	Bond angle A-B-C (°)
2.322	C1	Hf2	C1(1,0)	89.9
2.322	C1	Hf2(0,-1)	C1(1,0)	89.9
2.360	C1	Hf4	Hf2	45.9

HfC(100)-(1x1)
72.6.2

Bond Distances and Angles - Continued

Interatomic dist. A-B (Å)	Atom A	Atom B	Atom C	Bond angle A-B-C (°)
2.220	Hf2	C3	Hf4(1,1)	90.7
2.120	Hf2	C3	Hf4	90.0

COMMON NAME : InAs(110)-(1x1) ILLUSTRATION: 116
CLASSIFICATION : 49.33.1
TECHNIQUE : LEED
AUTHORS : C.B. Duke, A. Paton, A. Kahn and C.R. Bonapace
REFERENCE : Phys. Rev., B27, 6189 (1983)

SURFACE TYPE
Substrate : InAs
Crystal face: 110
Temperature : 110 K
Bulk lattice: zincblende
2D bulk symm: pm
2D surf symm: pm
Adsorbate:
Coverage :
Pattern : (1x1)
Matrix : (1.000, 0.000)
(0.000, 1.000)

STRUCTURE TYPE
Relaxed bulk termination with 36.5±3° tilt in top layer

SAMPLE PREPARATION (1 sample)
Treatment : repeated cycles of Ar+ bombardment and heating to 693 K
Crystallinity:
Anal. methods:
Contamination: monitored by AES and LEED

COMMENTS
Quasi-dynamical LEED = scattering amplitudes from the top 3 bilayers calculated exactly, those from next 3 layers assumed no interlayer scattering; the amplitudes from these 2 slabs are superposed

DATA COLLECTION
Technique: LEED
Dataset : I-V spectra: 3 sets of data each of 14 beams: 30=<E=<240 eV

THEORY/DATA TREATMENT
Quasi-dynamical LEED(see comm.): 6 bilayers, 6 phase shifts; rel. Hara exch. pot.; mfp=12Å; Vor=-12 eV (fit); ΘD=300 K

STRUCTURES EXAMINED
1. unreconstructed geometry; 2. bond length conserving top layer rotations: 0 to 34.8°; 3. top intra-bilayer spacing: -0.2 to +0.2Å for 31°; 4. 2nd layer shears: -0.1 to 0.3Å for 31°; 5. various lateral registries

QUALITY OF EXPERIMENT-THEORY FIT
RX=0.23

2D UNIT CELLS (1 domain observed)

Cell	Ax (Å)	Ay (Å)	Bx (Å)	By (Å)	α (°)	Matrix	Pattern	Cell type
Bulk	4.270	0.000	0.000	6.040	90.0	(1.000, 0.000) (0.000, 1.000)	(1x1)	b: bulk lattice
Surface 1	4.270	0.000	0.000	6.040	90.0	(1.000, 0.000) (0.000, 1.000)	(1x1)	s1: commens. superlattice

3D COORDINATES

As1-In2, In3-As4: 2 bilayers with tilted In-As chains; In7-As8: periodically repeating bulk bilayer;
0.1Å error bars parallel to surface assumed for tabulation

Dx/Dy in Å, or as a fraction of layer's unit cell vectors; atom 0 at the origin. Epir/subr are bulk repeat vectors.

No. of atoms: 8 Bulk z = 2.135 Å

Region	Chem el.	At. no.	Cell type	Site occ.	Rel. to	Dx ± εx		Dy ± εy		Dz ± εz		Dz/Bz(%) ± εz/Bz
epir		-2					f		f		Å	
subr		-1				2.135	Å	3.020	Å	2.135	Å	
intf	As	1	b	1.00	0	0.000	f	0.000	f	0.000	Å	0.0
intf	In	2	b	1.00	1	0.500	f	0.174 ± .017	f	0.780 ± .050	Å	36.5 ± 2.3
intf	In	3	b	1.00	2	-0.500	f	0.596 ± .017	f	1.500 ± .080	Å	70.3 ± 3.7
intf	As	4	b	1.00	3	0.500	f	-0.250 ± .017	f	0.150 ± .200	Å	7.0 ± 9.4
intf	As	5	b	1.00	4	-0.500	f	-0.500 ± .017	f	2.060 ± .100	Å	96.5 ± 4.7
intf	In	6	b	1.00	5	0.500	f	0.250	f	0.000	Å	0.0
subl	In	7	b	1.00	6	-0.500	f	0.500	f	2.135	Å	100.0
subl	As	8	b	1.00	7	0.500	f	-0.250	f	0.000	Å	0.0

InAs(110)-(1x1)
49.33.1

BOND DISTANCES AND ANGLES

Bond distances and angles are derived from coordinates

No. of distances/angles: 9

Interatomic dist. A-B (Å)	Atom A	Atom B	Atom C	Bond angle A-B-C (°)
2.504	As1	In2	As1(1,0)	117.0
2.504	As1	In2	As4	121.5
2.670	As1	In3(0,-1)	As4(0,-1)	110.4
2.670	As1	In3(0,-1)	As5	114.3
2.663	In2	As4	In6	92.1
2.619	In3	As4	In3(1,0)	109.2
2.619	In3	As4	In6	112.8
2.677	In3	As5(0,1)	In6(0,1)	109.0
2.677	In3	As5(0,1)	In7	110.4

COMMON NAME : InAs(110)-(1x1)
CLASSIFICATION : 49.33.2
TECHNIQUE : MEIS
AUTHORS : L. Smit and J.F. van der Veen
REFERENCE : Surf. Sci., 166, 183 (1986)

ILLUSTRATION: 116

SURFACE TYPE
Substrate : InAs
Crystal face: 110
Temperature : RT
Bulk lattice: zincblende
2D bulk symm: pm
2D surf symm: pm

Adsorbate:
Coverage :
Pattern : (1x1)
Matrix : (1.000, 0.000)
(0.000, 1.000)

STRUCTURE TYPE
Relaxed bulk termination with 30° tilt in top layer

SAMPLE PREPARATION (1 sample)
Treatment : in-situ cleavage monitored by laser light back scattering
Crystallinity: sharp (1x1) LEED pattern
Anal. methods:
Contamination:

COMMENTS
Thermal vibrations: bulk In rms amplitude = 0.108Å, bulk As amplitude = 0.101Å; the surface layer vibrations were optimized to 1.7±0.1 times the bulk

DATA COLLECTION
Technique: MEIS; 174keV He+ Rutherford back scattering
Dataset : scattering aligned with [-1-1 2] channeling axis and exit angles >10°, <30°, ions incident from both directions

THEORY/DATA TREATMENT
Monte Carlo simulation: see comment

STRUCTURES EXAMINED
Bond length conserving rotations in top bilayer up to 40° and enhancement of surface layer vibration; bond length relaxation allowed but no substrate relaxation

2D UNIT CELLS (1 domain observed)

Cell	Ax (Å)	Ay (Å)	Bx (Å)	By (Å)	α (°)	Matrix	Pattern	Cell type
Bulk	4.270	0.000	0.000	6.040	90.0	(1.000, 0.000) (0.000, 1.000)	(1x1)	b: bulk lattice
Surface 1	4.270	0.000	0.000	6.040	90.0	(1.000, 0.000) (0.000, 1.000)	(1x1)	s1: commens. superlattice

3D COORDINATES

As1-In2: top bilayer with tilted In-As chain; In3-As4: bulk bilayer;
In5-As6: periodically repeating bulk bilayer; 0.1Å error bars assumed for tabulation

Dx/Dy in Å, or as a fraction of layer's unit cell vectors; atom 0 at the origin. Epir/subr are bulk repeat vectors.

No. of atoms: 6

Bulk z = 2.135 Å

Region	Chem el.	At. no.	Cell type	Site occ.	Rel. to	Dx ± εx		Dy ± εy		Dz ± εz		Dz/Bz(%) ± εz/Bz
epir		-2					f		f		Å	
subr		-1				2.135	Å	-3.020	Å	2.135	Å	
intf	As	1	b	1.00	0	0.000	f	0.000	f	0.000	Å	0.0
intf	In	2	b	1.00	1	0.500	f	0.217 ± .017	f	0.755 ± .100	Å	35.4 ± 4.7
intf	As	3	b	1.00	2	0.000	f	0.343 ± .017	f	1.594 ± .100	Å	74.7 ± 4.7
intf	In	4	b	1.00	3	-0.500	f	0.250	f	0.000	Å	0.0
subl	As	5	b	1.00	4	0.000	f	-0.750	f	2.135	Å	100.0
subl	In	6	b	1.00	5	0.500	f	0.250	f	0.000	Å	0.0

BOND DISTANCES AND ANGLES

Bond distances and angles are derived from coordinates

No. of distances/angles: 8

Interatomic dist. A-B (Å)	Atom A	Atom B	Atom C	Bond angle A-B-C (°)
2.617	As1	In2	As1(1,0)	109.4
2.617	As1	In2	As3	125.0
2.614	As1	In4(0,-1)	As5	118.7
2.614	In2	As3	In4	117.2

InAs(110)-(1x1)
49.33.2

Bond Distances and Angles - Continued

Interatomic dist. A-B (Å)	Atom A	Atom B	Atom C	Bond angle A-B-C (°)
2.614	In2	As3	In6	92.3
2.615	As3	In4	As5(0,1)	109.5
2.615	In4	As5(0,1)	In6(0,1)	109.5
2.615	As5	In6	As5(1,0)	109.5

COMMON NAME : InP(110)-(1x1)
CLASSIFICATION : 49.15.2
TECHNIQUE : LEED
AUTHORS : R.J. Meyer, C.B. Duke, A. Paton, J.C. Tsang, J.L. Yeh, A. Kahn and P. Mark
REFERENCE : Phys. Rev., B22, 6171 (1980)

ILLUSTRATION: 116

SURFACE TYPE
Substrate : InP
Crystal face: 110
Temperature : 150 K
Bulk lattice: zincblende
2D bulk symm: pm
2D surf symm: pm
Adsorbate:
Coverage :
Pattern : (1x1)
Matrix : (1.000, 0.000)
(0.000, 1.000)

STRUCTURE TYPE
Relaxed bulk termination with 27.3° tilt in top layer

SAMPLE PREPARATION (1 sample)
Treatment : Ar+ bombardment for 5 mins, then 16 hour anneal at 660 K
Crystallinity:
Anal. methods:
Contamination: monitored by LEED and AES

COMMENTS
Both x-ray and Zanazzi-Jona R-factors used, giving slightly different results

DATA COLLECTION
Technique: LEED
Dataset : I-V spectra: 14 beams at normal incidence; 30<E<250 eV

THEORY/DATA TREATMENT
First kinematic analysis (10 layers, 10 phase shifts), then quasi-dynamical (6 layers, 6 phase shifts); Vor and Voi fit

STRUCTURES EXAMINED
Perp. displacements of P (<=0.35Å) and In (<=0.75Å) in 1st bilayer, smaller displacements in 2nd and 3rd bilayers; displacements parallel to surface allowed in 1st layer

QUALITY OF EXPERIMENT-THEORY FIT
RZJ=0.25 (also RX used)

2D UNIT CELLS (1 domain observed)

Cell	Ax (Å)	Ay (Å)	Bx (Å)	By (Å)	α (°)	Matrix	Pattern	Cell type
Bulk	4.152	0.000	0.000	5.869	90.0	(1.000, 0.000) (0.000, 1.000)	(1x1)	b: bulk lattice
Surface 1	4.152	0.000	0.000	5.869	90.0	(1.000, 0.000) (0.000, 1.000)	(1x1)	s1: commens. superlattice

3D COORDINATES

P1-In2, In3-P4: 2 bilayers with tilted In-P chains; In7-P8: periodically repeating bulk bilayer

Dx/Dy in Å, or as a fraction of layer's unit cell vectors; atom 0 at the origin. Epir/subr are bulk repeat vectors.

No. of atoms: 8

Bulk z = 2.076 Å

Reg ion	Chem el.	At. no.	Cell type	Site occ.	Rel. to	Dx ± εx		Dy ± εy		Dz ± εz		Dz/Bz(%) ± εz/Bz
epir		-2					f		f		Å	
subr		-1				2.076	Å	2.935	Å	2.076	Å	
intf	P	1	b	1.00	0	0.000	f	0.000	f	0.000	Å	0.0
intf	In	2	b	1.00	1	0.500	f	0.228 ± .017	f	0.690 ± .100	Å	33.2 ± 4.8
intf	In	3	b	1.00	2	-0.500	f	0.588 ± .017	f	1.410 ± .100	Å	67.9 ± 4.8
intf	P	4	b	1.00	3	0.500	f	-0.250	f	0.070 ± .100	Å	3.4 ± 4.8
intf	In	5	b	1.00	4	0.000	f	-0.250	f	2.076 ± .100	Å	100.0 ± 4.8
intf	P	6	b	1.00	5	-0.500	f	-0.250	f	0.000	Å	0.0
subl	In	7	b	1.00	6	0.000	f	0.750	f	2.076	Å	100.0
subl	P	8	b	1.00	7	0.500	f	-0.250	f	0.000	Å	0.0

InP(110)-(1x1)
49.15.2

BOND DISTANCES AND ANGLES

Bond distances and angles are derived from coordinates

No. of distances/angles: 10

Interatomic dist. A-B (Å)	Atom A	Atom B	Atom C	Bond angle A-B-C (°)
2.565	P1	In2	P1(1,0)	108.1
2.600	In3	P6(0,1)	In7	110.4
2.565	P1	In2	P4	125.4
2.361	P1	In3(0,-1)	P4(0,-1)	106.8
2.361	P1	In3(0,-1)	P6	118.4
2.475	In2	P4	In3	116.5
2.475	In2	P4	In5	91.5
2.543	In3	P4	In3(1,0)	109.4
2.543	In3	P4	In5	110.8
2.600	In3	P6(0,1)	In5(0,1)	109.0

COMMON NAME : InP(110)-(1x1) ILLUSTRATION: 116
CLASSIFICATION : 49.15.3
TECHNIQUE : LEED
AUTHORS : S.P. Tear, M.R. Welton-Cook, M. Prutton and J.A. Walker
REFERENCE : Surf. Sci., 99, 598 (1980)

SURFACE TYPE
Substrate : InP
Crystal face: 110
Temperature : RT*
Bulk lattice: zincblende
2D bulk symm: pm
2D surf symm: pm
Adsorbate:
Coverage :
Pattern : (1x1)
Matrix : (1.000, 0.000)
(0.000, 1.000)

STRUCTURE TYPE
Relaxed bulk termination with 26.4° tilt in top layer

SAMPLE PREPARATION (1 sample)
Treatment : in situ cleavage at RT and 2.0E-10 torr
Crystallinity:
Anal. methods:
Contamination: monitored by AES

COMMENTS
Displacements considered only in outermost bilayer; theory-exp. agreement not good enough to trust results of Zanazzi-Jona R-factor analysis: visual inspection was used

DATA COLLECTION
Technique: LEED
Dataset : I-V spectra for total of 9 non-degenerate beams for incident polar angles of Θ=0, 1.5, 3.0, ±4.5°

THEORY/DATA TREATMENT
Dynamical LEED (2 atom/unit cell CAVLEED program): ph shs from superposed neutral atom charge densities

STRUCTURES EXAMINED
First only perpendicular 1st-bilayer displacements; then lateral shifts also considered, as predicted by Chadi, Phys. Rev. B19, 2074 (1979)

QUALITY OF EXPERIMENT-THEORY FIT
Visual

2D UNIT CELLS (1 domain observed)

Cell	Ax (Å)	Ay (Å)	Bx (Å)	By (Å)	α (°)	Matrix	Pattern	Cell type
Bulk	4.149	0.000	0.000	5.869	90.0	(1.000, 0.000) (0.000, 1.000)	(1x1)	b: bulk lattice
Surface 1	4.149	0.000	0.000	5.869	90.0	(1.000, 0.000) (0.000, 1.000)	(1x1)	s1: commens. superlattice

3D COORDINATES

P1-In2: top bilayer with tilted In-P chains; P7-In8: periodically repeating bulk bilayer;
0.1Å error bars assumed for tabulation

Dx/Dy in Å, or as a fraction of layer's unit cell vectors; atom 0 at the origin. Epir/subr are bulk repeat vectors.

No. of atoms: 8 Bulk z = 2.075 Å

Region	Chem el.	At. no.	Cell type	Site occ.	Rel. to	Dx ± εx		Dy ± εy		Dz ± εz		Dz/Bz(%) ± εz/Bz
epir		-2					f		f		Å	
subr		-1				2.075	Å	2.935	Å	2.075	Å	
intf	P	1	b	1.00	0	0.000	f	0.000	f	0.000	Å	0.0
intf	In	2	b	1.00	1	0.500	f	0.220 ± .017	f	0.640 ± .100	Å	30.8 ± 4.8
intf	P	3	b	1.00	2	0.000	f	0.317 ± .017	f	1.610 ± .100	Å	77.6 ± 4.8
intf	In	4	b	1.00	3	-0.500	f	0.250	f	0.000	Å	0.0
intf	P	5	b	1.00	4	0.000	f	-0.750	f	2.075	Å	100.0
intf	In	6	b	1.00	5	0.500	f	0.250	f	0.000	Å	0.0
subl	P	7	b	1.00	6	0.000	f	0.250	f	2.075	Å	100.0
subl	In	8	b	1.00	7	-0.500	f	0.250	f	0.000	Å	0.0

InP(110)-(1x1)
49.15.3

BOND DISTANCES AND ANGLES

Bond distances and angles are derived from coordinates

No. of distances/angles: 9

Interatomic dist. A-B (Å)	Atom A	Atom B	Atom C	Bond angle A-B-C (°)
2.526	P1	In2	P1(1,0)	110.4
2.526	P1	In2	P3	123.5
2.574	P1	In4(0,-1)	P3(0,-1)	106.3
2.574	P1	In4(0,-1)	P5	115.7
2.460	In2	P3	In4	115.9
2.460	In2	P3	In6	95.6
2.541	P3	In4	P3(-1,0)	109.5
2.541	P3	In6	P5	109.5
2.541	P3	In6	P7	109.5

COMMON NAME : InSb(110)-(1x1) ILLUSTRATION: 116
CLASSIFICATION : 49.51.1
TECHNIQUE : LEED
AUTHORS : R.J. Meyer, C.B. Duke, A. Paton, J.L. Yeh, J.C. Tsang, A. Kahn and P. Mark
REFERENCE : Phys. Rev., B21, 4740 (1980)

SURFACE TYPE
Substrate : InSb
Crystal face: 110
Temperature : 150 K
Bulk lattice: zincblende
2D bulk symm: pm
2D surf symm: pm

Adsorbate:
Coverage :
Pattern : (1x1)
Matrix : (1.000, 0.000)
(0.000, 1.000)

STRUCTURE TYPE
Relaxed bulk termination with 28.8° tilt in top layer

SAMPLE PREPARATION (1 sample)
Treatment : Ar+ bombardment and annealing at 640 K for 75min
Crystallinity:
Anal. methods:
Contamination: LEED and AES: <0.1% contamination

COMMENTS

DATA COLLECTION
Technique: LEED
Dataset : I-V spectra: 14 beams at Θ=0,ϕ=0°, averaged over 3 separate experimental runs

THEORY/DATA TREATMENT
Quasi-dynamical LEED: 6 layers, 6 phase shifts (superpos. of atomic charge densities); Vor=-8 eV; mfp=8Å

STRUCTURES EXAMINED
1. various rotationally relaxed structures for top layer rotations between 25.7 and 30.9°;
2. various second layer shears between 0.0Å and 0.2Å

QUALITY OF EXPERIMENT-THEORY FIT
Visual

2D UNIT CELLS (1 domain observed)

Cell	Ax (Å)	Ay (Å)	Bx (Å)	By (Å)	α (°)	Matrix	Pattern	Cell type
Bulk	4.581	0.000	0.000	6.478	90.0	(1.000, 0.000) (0.000, 1.000)	(1x1)	b: bulk lattice
Surface 1	4.581	0.000	0.000	6.478	90.0	(1.000, 0.000) (0.000, 1.000)	(1x1)	s1: commens. superlattice

3D COORDINATES

Sb1-In2, In3-Sb4: 2 bilayers with tilted In-Sb chains; In7-Sb8: periodically repeating bulk bilayer; 0.1Å error bars parallel to surface assumed for tabulation

Dx/Dy in Å, or as a fraction of layer's unit cell vectors; atom 0 at the origin. Epir/subr are bulk repeat vectors.

No. of atoms: 8 Bulk z = 2.290 Å

Region	Chem el.	At. no.	Cell type	Site occ.	Rel. to	Dx ± ϵx	Dy ± ϵy	Dz ± ϵz	Dz/Bz(%) ± ϵz/Bz
epir		-2				f	f	Å	
subr		-1				2.291 Å	3.239 Å	2.290 Å	
intf	Sb	1	b	1.00	0	0.000 f	0.000 f	0.000 Å	0.0
intf	In	2	b	1.00	1	0.500 f	0.219 ± .015 f	0.781 ± .100 Å	34.1 ± 4.4
intf	In	3	b	1.00	2	-0.500 f	0.590 ± .015 f	1.597 ± .100 Å	69.7 ± 4.4
intf	Sb	4	b	1.00	3	0.500 f	-0.250 ± .015 f	0.180 ± .100 Å	7.9 ± 4.4
intf	Sb	5	b	1.00	4	-0.500 f	-0.500 ± .015 f	2.201 ± .100 Å	96.1 ± 4.4
intf	In	6	b	1.00	5	0.500 f	0.250 f	0.000 Å	0.0
subl	In	7	b	1.00	6	-0.500 f	0.500 f	2.290 Å	100.0
subl	Sb	8	b	1.00	7	0.500 f	-0.250 f	0.000 Å	0.0

InSb(110)-(1x1)
49.51.1 BOND DISTANCES AND ANGLES

Bond distances and angles are derived from coordinates

No. of distances/angles: 8

Interatomic dist. A-B (Å)	Atom A	Atom B	Atom C	Bond angle A-B-C (°)
2.805	Sb1	In2	Sb1(1,0)	109.5
2.805	Sb1	In2	Sb4	124.6
2.681	Sb1	In3(0,-1)	Sb4(0,-1)	108.8
2.681	Sb1	In3(0,-1)	Sb5	118.3
2.830	In2	Sb4	In3	114.1
2.830	In2	Sb4	In6	92.6
2.880	In3	Sb5(0,1)	In6(0,1)	109.0
2.880	In3	Sb5(0,1)	In7	110.5

COMMON NAME : InSb(110)-(1x1) ILLUSTRATION: 116
CLASSIFICATION : 49.51.6
TECHNIQUE : LEED
AUTHORS : V.E. de Carvalho, M. Prutton and S.P. Tear
REFERENCE : Surf. Sci., 184, 198 (1987)

SURFACE TYPE
Substrate : InSb
Crystal face: 110
Temperature : RT
Bulk lattice: zincblende
2D bulk symm: pm
2D surf symm: pm
Adsorbate:
Coverage :
Pattern : (1x1)
Matrix : (1.000, 0.000)
(0.000, 1.000)

STRUCTURE TYPE
Relaxed bulk termination with 41° tilt in top layer

SAMPLE PREPARATION (1 sample)
Treatment : sample cleaved in situ at RT and 8E-11mbar
Crystallinity: sharp diffraction pattern
Anal. methods: surface topography measured with SEM
Contamination: AES: no contamination

COMMENTS
Non-structural parameters considered: ΘD=113, 160 K (bulk), 250, 400 K, and rigid lattice; exchange parameter α=1/3, 2/3

DATA COLLECTION
Technique: LEED
Dataset : I-V spectra for 36 beams at Θ=0.5°, 7.5°; 40<E<160 eV

THEORY/DATA TREATMENT
Dynamical LEED: 9 phase shifts; 104 beams; Voi=-4 eV; Pendry and x-ray R-factors; ΘD=160 K: see comment

STRUCTURES EXAMINED
Several bond length conserving rotations including 1st interlayer spacing relaxation; lateral displacements in optimum structures; finally surface bond length change and 2nd layer buckling and relaxation

QUALITY OF EXPERIMENT-THEORY FIT
RPE=0.49

2D UNIT CELLS (1 domain observed)

Cell	Ax (Å)	Ay (Å)	Bx (Å)	By (Å)	α (°)	Matrix	Pattern	Cell type
Bulk	4.581	0.000	0.000	6.478	90.0	(1.000, 0.000) (0.000, 1.000)	(1x1)	b: bulk lattice
Surface 1	4.581	0.000	0.000	6.478	90.0	(1.000, 0.000) (0.000, 1.000)	(1x1)	s1: commens. superlattice

3D COORDINATES

Sb1-In2, In3-Sb4: 2 bilayers with tilted In-Sb chains; In5-Sb6: bulk bilayer; In7-Sb8: periodically repeating bulk bilayer; 0.1Å lateral error bars assumed for tabulation

Dx/Dy in Å, or as a fraction of layer's unit cell vectors; atom 0 at the origin. Epir/subr are bulk repeat vectors.

No. of atoms: 8 Bulk z = 2.290 Å

Region	Chem el.	At. no.	Cell type	Site occ.	Rel. to	Dx ± εx		Dy ± εy		Dz ± εz		Dz/Bz(%) ± εz/Bz
epir		-2					f		f		Å	
subr		-1				2.291	Å	3.239	Å	2.290	Å	
intf	Sb	1	b	1.00	0	0.000	f	0.000	f	0.000	Å	0.0
intf	In	2	b	1.00	1	0.500	f	0.202 ± .015	f	0.810 ± .060	Å	35.4 ± 2.6
intf	In	3	b	1.00	2	-0.500	f	0.597 ± .015	f	1.590 ± .060	Å	69.4 ± 2.6
intf	Sb	4	b	1.00	3	0.500	f	-0.250	f	0.180 ± .030	Å	7.9 ± 1.3
intf	In	5	b	1.00	4	0.000	f	-0.250 ± .015	f	2.200 ± .030	Å	96.1 ± 1.3
intf	Sb	6	b	1.00	5	-0.500	f	-0.250	f	0.000	Å	0.0
subl	In	7	b	1.00	6	0.000	f	0.750	f	2.290	Å	100.0
subl	Sb	8	b	1.00	7	0.500	f	-0.250	f	0.000	Å	0.0

InSb(110)-(1x1)
49.51.6 BOND DISTANCES AND ANGLES

Bond distances and angles are derived from coordinates

No. of distances/angles: 10

Interatomic dist. A-B (Å)	Atom A	Atom B	Atom C	Bond angle A-B-C (°)
2.760	Sb1	In2	Sb1(1,0)	112.2
2.732	Sb4	In5	Sb8	108.4
2.760	Sb1	In2	Sb4	123.7
2.760	Sb1	In2	In5	111.4
2.731	Sb1	In3(0,-1)	Sb4(0,-1)	109.3
2.731	Sb1	In3(0,-1)	Sb6	117.3
2.861	In2	Sb4	In3	114.4
2.861	In2	Sb4	In5	91.9
2.811	In3	Sb4	In5	113.2
2.732	Sb4	In5	Sb6	110.0

COMMON NAME : Ir(100)-(1x1) ILLUSTRATION: 2
CLASSIFICATION : 77.10
TECHNIQUE : LEED
AUTHORS : K. Heinz and G. Besold
REFERENCE : Surf. Sci., 125, 515 (1983)

SURFACE TYPE
Substrate : Ir
Crystal face: 100
Temperature : 100 K
Bulk lattice: fcc
2D bulk symm: p4m
2D surf symm: p4m

Adsorbate:
Coverage :
Pattern : (1x1)
Matrix : (1.000, 0.000)
(0.000, 1.000)

STRUCTURE TYPE
Unreconstructed metastable surface with 3.6% top spacing contraction

SAMPLE PREPARATION (1 sample)
Treatment : see Besold et al, J. Vac. Sci. Technol. A1, 1473 (1983)
Crystallinity:
Anal. methods:
Contamination:

COMMENTS

DATA COLLECTION
Technique: LEED
Dataset : I-V curves for 6 beams at normal incidence; energy range 100-500 eV

THEORY/DATA TREATMENT
Quasidynamical LEED (RFS, no intralayer mult. scatt., with extra damping for convergence): ΘD 420 K

STRUCTURES EXAMINED
Variation of topmost interlayer spacing

QUALITY OF EXPERIMENT-THEORY FIT
RPE=<0.65

2D UNIT CELLS (1 domain observed)

Cell	Ax (Å)	Ay (Å)	Bx (Å)	By (Å)	α (°)	Matrix	Pattern	Cell type
Bulk	2.720	0.000	0.000	2.720	90.0	(1.000, 0.000) (0.000, 1.000)	(1x1)	b: bulk lattice
Surface 1	2.720	0.000	0.000	2.720	90.0	(1.000, 0.000) (0.000, 1.000)	(1x1)	s1: commens. superlattice

3D COORDINATES

Dx/Dy in Å, or as a fraction of layer's unit cell vectors; atom 0 at the origin. Epir/subr are bulk repeat vectors.

No. of atoms: 3 Bulk z = 1.920 Å

Region	Chem el.	At. no.	Cell type	Site occ.	Rel. to	Dx ± εx	Dy ± εy	Dz ± εz	Dz/Bz(%) ± εz/Bz
epir		-2				f	f	Å	
subr		-1				-1.360 Å	-1.360 Å	1.920 Å	
intf	Ir	1	b	1.00	0	0.000 f	0.000 f	0.000 Å	0.0
intf	Ir	2	b	1.00	1	0.500 f	0.500 f	1.850 ± .010 Å	96.4 ± .5
subl	Ir	3	b	1.00	2	-0.500 f	-0.500 f	1.920 Å	100.0

BOND DISTANCES AND ANGLES

Bond distances and angles are derived from coordinates

No. of distances/angles: 3

Interatomic dist. A-B (Å)	Atom A	Atom B	Atom C	Bond angle A-B-C (°)
2.720	Ir1	Ir1(1,0)	Ir2	59.4
2.669	Ir1	Ir2	Ir3	88.8
2.718	Ir2	Ir3		

COMMON NAME : Ir(100)-(1x5)
CLASSIFICATION : 77.11
TECHNIQUE : LEED
AUTHORS : E. Lang, K. Mueller, K. Heinz, M.A. Van Hove, R.J. Koestner and G.A. Somorjai
REFERENCE : Surf. Sci., 127, 347 (1983)

ILLUSTRATION: 3

SURFACE TYPE
Substrate : Ir
Crystal face: 100
Temperature : 100 K
Bulk lattice: fcc
2D bulk symm: p4m
2D surf symm: pmm

Adsorbate:
Coverage :
Pattern : (1x5)
Matrix : (1.000, 0.000)
(0.000, 5.000)

STRUCTURE TYPE
Quasi-hexagonal commensurate buckled top-layer reconstruction with 'two-bridges' registry

SAMPLE PREPARATION (1 sample)
Treatment : Ar+ bombardment and O2 treatments
Crystallinity:
Anal. methods:
Contamination: AES: clean

COMMENTS
Full buckling is preferred, defined by positions of touching balls with bulk radii on a (100) substrate

DATA COLLECTION
Technique: LEED
Dataset : I-V curves for 10 beams at normal incidence

THEORY/DATA TREATMENT
Dynamical LEED (RFS, RSP): 6 ph shs (Arbman/Hoernfelt and relat. pots); Vor=fitted, Voi=-5.0 eV; ΘD=236 K

STRUCTURES EXAMINED
56 structures of different types: hexagonal top-layer model with 2-bridges or top-hollow registry (with different bucklings), shifted-rows models (with 5-, 4- and 3-atom clusters) and charge-density-wave model

QUALITY OF EXPERIMENT-THEORY FIT
RPE=0.51, RZJ=0.34

2D UNIT CELLS (2 domains observed)

Cell	Ax (Å)	Ay (Å)	Bx (Å)	By (Å)	α (°)	Matrix	Pattern	Cell type
Bulk	2.715	0.000	0.000	2.715	90.0	(1.000, 0.000) (0.000, 1.000)	(1x1)	b: bulk lattice
Surface 1	2.715	0.000	0.000	13.575	90.0	(1.000, 0.000) (0.000, 5.000)	(1x5)	s1: commens. superlattice

3D COORDINATES

Ir1-Ir6: quasi-hexagonal buckled top layer in 'two-bridge sites' over substrate layer Ir7;
0.1Å lateral error bars assumed for tabulation

Dx/Dy in Å, or as a fraction of layer's unit cell vectors; atom 0 at the origin. Epir/subr are bulk repeat vectors.

No. of atoms: 8

Bulk z = 1.920 Å

Region	Chem el.	At. no.	Cell type	Site occ.	Rel. to	Dx ± εx		Dy ± εy		Dz ± εz		Dz/Bz(%) ± εz/Bz
epir		-2					f		f		Å	
subr		-1				1.358	Å	1.358	Å	1.920	Å	
intf	Ir	1	s1	.20	0	0.000	f	0.167 ± .007	f	0.000	Å	0.0
intf	Ir	2	s1	.20	0	0.000	f	0.833 ± .007	f	0.000	Å	0.0
intf	Ir	3	s1	.20	0	0.500	f	0.000 ± .007	f	0.340 ± .020	Å	17.7 ± 1.0
intf	Ir	4	s1	.20	0	0.000	f	0.500 ± .007	f	0.340 ± .020	Å	17.7 ± 1.0
intf	Ir	5	s1	.20	0	0.500	f	0.333 ± .007	f	0.480 ± .020	Å	25.0 ± 1.0
intf	Ir	6	s1	.20	0	0.500	f	0.667 ± .007	f	0.480 ± .020	Å	25.0 ± 1.0
intf	Ir	7	b	1.00	0	0.000	f	0.000 ± .037	f	2.500 ± .050	Å	130.2 ± 2.6
subl	Ir	8	b	1.00	7	0.500	f	0.500	f	1.920	Å	100.0

Ir(100)-(1x5)
77.11

BOND DISTANCES AND ANGLES

Bond distances and angles are derived from coordinates

No. of distances/angles: 4

Interatomic dist. A-B (Å)	Atom A	Atom B	Atom C	Bond angle A-B-C (°)
2.715	Ir1	Ir1(1,0)	Ir3	59.3
2.715	Ir1	Ir1(1,0)	Ir5	59.6
2.660	Ir1	Ir3	Ir1(1,0)	61.4
3.372	Ir1	Ir7	Ir1(0,-1)	52.4

COMMON NAME : Ir(100)-(1x5)
CLASSIFICATION : 77.6
TECHNIQUE : LEED
AUTHORS : M.A. Van Hove, R.J. Koestner, P.C. Stair, J.P. Biberian, L.L. Kesmodel, I. Bartos and G.A. Somorjai
REFERENCE : Surf. Sci., 103, 218 (1981)

ILLUSTRATION: 3

SURFACE TYPE
Substrate : Ir
Crystal face: 100
Temperature : 310 K
Bulk lattice: fcc
2D bulk symm: p4m
2D surf symm: pmm

Adsorbate:
Coverage :
Pattern : (1x5)
Matrix : (1.000, 0.000)
(0.000, 5.000)

STRUCTURE TYPE
Quasi-hexagonal commensurate buckled top-layer reconstruction with 'two-bridges' registry

SAMPLE PREPARATION (1 sample)
Treatment : Ar+ bombardment and O2 treatments
Crystallinity:
Anal. methods:
Contamination: AES: clean

COMMENTS
1/2 or 1/3 buckling preferred (full buckling defined by positions of touching hard balls with bulk radii on a (100) substrate), giving spacing of 0.477Å between most separated nuclear planes, with smallest distance from nuclear planes of buckled hexagonal layer to 1st substrate layer of 2.2Å

DATA COLLECTION
Technique: LEED
Dataset : I-V curves at Θ=0, 10, 20°, with φ=90°

THEORY/DATA TREATMENT
Dynamical LEED (RFS, RSP): 6 ph shs (Arbman/Hoernfelt and relat. pots); Vor=fitted, Voi=-5.0 eV; ΘD=236 K

STRUCTURES EXAMINED
Various spacings between top and 2nd layers for hexagonal top layer (planar, 1/2 or 1/3 or fully buckled, etc); hex. top layer with 1 missing row; shifted rows models; charge density wave model

QUALITY OF EXPERIMENT-THEORY FIT
Visual

2D UNIT CELLS (2 domains observed)

Cell	Ax (Å)	Ay (Å)	Bx (Å)	By (Å)	α (°)	Matrix	Pattern	Cell type
Bulk	2.715	0.000	0.000	2.715	90.0	(1.000, 0.000) (0.000, 1.000)	(1x1)	b: bulk lattice
Surface 1	2.715	0.000	0.000	13.575	90.0	(1.000, 0.000) (0.000, 5.000)	(1x5)	s1: commens. superlattice

3D COORDINATES

Ir1-Ir6: quasi-hexagonal buckled top layer in 'two-bridge sites' over substrate layer Ir7;
0.1Å lateral error bars assumed for tabulation

Dx/Dy in Å, or as a fraction of layer's unit cell vectors; atom 0 at the origin. Epir/subr are bulk repeat vectors.

No. of atoms: 8 Bulk z = 1.920 Å

Region	Chem el.	At. no.	Cell type	Site occ.	Rel. to	Dx ± εx		Dy ± εy		Dz ± εz		Dz/Bz(%) ± εz/Bz
epir		-2					f		f		Å	
subr		-1				1.358	Å	1.358	Å	1.920	Å	
intf	Ir	1	s1	.20	0	0.000	f	0.167 ± .007	f	0.000	Å	0.0
intf	Ir	2	s1	.20	0	0.000	f	0.833 ± .007	f	0.000	Å	0.0
intf	Ir	3	s1	.20	0	0.500	f	0.000 ± .007	f	0.140 ± .020	Å	7.3 ± 1.0
intf	Ir	4	s1	.20	0	0.000	f	0.500 ± .007	f	0.140 ± .020	Å	7.3 ± 1.0
intf	Ir	5	s1	.20	0	0.500	f	0.333 ± .007	f	0.200 ± .020	Å	10.4 ± 1.0
intf	Ir	6	s1	.20	0	0.500	f	0.667 ± .007	f	0.200 ± .020	Å	10.4 ± 1.0
intf	Ir	7	b	1.00	0	0.000	f	0.000 ± .037	f	2.200 ± .100	Å	114.6 ± 5.2
subl	Ir	8	b	1.00	7	0.500	f	0.500	f	1.920	Å	100.0

Ir(100)-(1x5)
77.6

BOND DISTANCES AND ANGLES

Bond distances and angles are derived from coordinates

No. of distances/angles: 4

Interatomic dist. A-B (Å)	Atom A	Atom B	Atom C	Bond angle A-B-C (°)
2.715	Ir1	Ir1(1,0)	Ir3	59.1
2.715	Ir1	Ir1(1,0)	Ir5	59.1
2.642	Ir1	Ir3	Ir1(1,0)	61.8
3.298	Ir1	Ir7	Ir1(0,-1)	54.0

COMMON NAME : Ir(110)-(1x1)
CLASSIFICATION : 77.3
TECHNIQUE : LEED
AUTHORS : C.M. Chan, S.L. Cunningham, K.L. Luke, W.H. Weinberg and S.P. Withrow
REFERENCE : Surf. Sci., 78, 15 (1978)

ILLUSTRATION: 4

SURFACE TYPE
Substrate : Ir
Crystal face: 110
Temperature : 573 K
Bulk lattice: fcc
2D bulk symm: pmm
2D surf symm: pmm

Adsorbate:
Coverage :
Pattern : (1x1)
Matrix : (1.000, 0.000)
(0.000, 1.000)

STRUCTURE TYPE
Impurity-stabilized unreconstructed surface with top layer spacing contraction by 7.4%

SAMPLE PREPARATION (1 sample)
Treatment : (1x2) surface treated for 2 mins at 1123 K in oxygen
Crystallinity:
Anal. methods:
Contamination: 0.25ML of disordered O

COMMENTS
(1x1) structure stabilized by 0.25ML of disordered oxygen to avoid (1x2) reconstruction

DATA COLLECTION
Technique: LEED
Dataset : I-V spectra at normal incidence for 01,10,11,12,21,02,20,22 beams; cumulative energy range 1140 eV

THEORY/DATA TREATMENT
Dynamical LEED (layer doubling): 8 phase shifts, Arbman/Hoernfelt atomic Ir potential (O ignored); Vor=-8 eV,Voi=-5eV

STRUCTURES EXAMINED
Variation of top interlayer spacing from 1.156 to 1.564Å

QUALITY OF EXPERIMENT-THEORY FIT
Visual

2D UNIT CELLS (1 domain observed)

Cell	Ax (Å)	Ay (Å)	Bx (Å)	By (Å)	α (°)	Matrix	Pattern	Cell type
Bulk	2.715	0.000	0.000	3.840	90.0	(1.000, 0.000) (0.000, 1.000)	(1x1)	b: bulk lattice
Surface 1	2.715	0.000	0.000	3.840	90.0	(1.000, 0.000) (0.000, 1.000)	(1x1)	s1: commens. superlattice

3D COORDINATES

Dx/Dy in Å, or as a fraction of layer's unit cell vectors; atom 0 at the origin. Epir/subr are bulk repeat vectors.

No. of atoms: 3 Bulk z = 1.360 Å

Region	Chem el.	At. no.	Cell type	Site occ.	Rel. to	Dx ± εx		Dy ± εy		Dz ± εz		Dz/Bz(%) ± εz/Bz
epir		-2					f		f		Å	
subr		-1				-1.358	Å	-1.920	Å	1.360	Å	
intf	Ir	1	b	1.00	0	0.000	f	0.000	f	0.000	Å	0.0
intf	Ir	2	b	1.00	1	0.500	f	0.500	f	1.260 ± .100	Å	92.7 ± 7.4
subl	Ir	3	b	1.00	2	-0.500	f	-0.500	f	1.360	Å	100.0 ± 7.4

BOND DISTANCES AND ANGLES

Bond distances and angles are derived from coordinates

No. of distances/angles: 3

Interatomic dist. A-B (Å)	Atom A	Atom B	Atom C	Bond angle A-B-C (°)
2.715	Ir1	Ir1(1,0)	Ir2	59.4
2.668	Ir1	Ir2	Ir3	58.2
2.716	Ir2	Ir3		

COMMON NAME : Ir(110)-(1x2) ILLUSTRATION: 5
CLASSIFICATION : 77.16
TECHNIQUE : LEED
AUTHORS : C.-M. Chan and M.A. Van Hove
REFERENCE : Surf. Sci., 171, 226 (1986)

SURFACE TYPE
Substrate : Ir
Crystal face: 110
Temperature : RT
Bulk lattice: fcc
2D bulk symm: pmm
2D surf symm: pmm

Adsorbate:
Coverage :
Pattern : (1x2)
Matrix : (1.000, 0.000)
(0.000, 2.000)

STRUCTURE TYPE
Missing-row reconstruction with multilayer relaxations, row-pairing in second layer and buckling in third layer

SAMPLE PREPARATION (1 sample)
Treatment : Ar+ bombardment, O treatments at 800 K, annealing at 1600K
Crystallinity:
Anal. methods:
Contamination:

COMMENTS

DATA COLLECTION
Technique: LEED
Dataset : I-V curves for 20 beams at normal incidence

THEORY/DATA TREATMENT
Dynamical LEED (layer doubling, RSP): 6 ph shs (Arbman/Hoernfelt pot); Voi=-5 eV; ΘD=280 K(bulk), 237K(surf)

STRUCTURES EXAMINED
Bonzel-Ferrer (sawtooth) and two similar models with relaxations in top two interlayer spacings;
missing-row model with relaxations in top three interlayer spacings, variable row-pairing in second layer and buckling in third layer

QUALITY OF EXPERIMENT-THEORY FIT
RVH=0.25

2D UNIT CELLS (1 domain observed)

Cell	Ax (Å)	Ay (Å)	Bx (Å)	By (Å)	α (°)	Matrix	Pattern	Cell type
Bulk	2.717	0.000	0.000	3.843	90.0	(1.000, 0.000) (0.000, 1.000)	(1x1)	b: bulk lattice
Surface 1	2.717	0.000	0.000	7.685	90.0	(1.000, 0.000) (0.000, 2.000)	(1x2)	s1: commens. superlattice

3D COORDINATES

Ir1: ridge (remaining row); Ir2-Ir3: paired rows in second layer;
Ir4-Ir5: buckled third layer; 0.1Å lateral error bars assumed for tabulation

Dx/Dy in Å, or as a fraction of layer's unit cell vectors; atom 0 at the origin. Epir/subr are bulk repeat vectors.

No. of atoms: 7 Bulk z = 1.359 Å

Region	Chem el.	At. no.	Cell type	Site occ.	Rel. to	Dx ± εx	Dy ± εy	Dz ± εz	Dz/Bz(%) ± εz/Bz
epir		-2				f	f	Å	
subr		-1				-1.359 Å	-1.921 Å	1.359 Å	
intf	Ir	1	s1	.50	0	0.000 f	0.000 f	0.000 Å	0.0
intf	Ir	2	s1	.50	1	0.500 f	0.255 ± .013 f	1.190 ± .070 Å	87.6 ± 5.2
intf	Ir	3	s1	.50	2	0.000 f	0.490 ± .013 f	0.000 Å	0.0
intf	Ir	4	s1	.50	3	-0.500 f	-0.245 ± .013 f	1.200 ± .070 Å	88.3 ± 5.2
intf	Ir	5	s1	.50	4	0.000 f	-0.500 f	0.230 ± .070 Å	16.9 ± 5.2
intf	Ir	6	b	1.00	5	0.500 f	0.500 f	1.280 ± .070 Å	94.2 ± 5.2
subl	Ir	7	b	1.00	6	-0.500 f	-0.500 f	1.359 Å	100.0

Ir(110)-(1x2)
77.16 BOND DISTANCES AND ANGLES

Bond distances and angles are derived from coordinates

No. of distances/angles: 4

Interatomic dist. A-B (Å)	Atom A	Atom B	Atom C	Bond angle A-B-C (°)
2.717	Ir1	Ir1(1,0)	Ir2	59.4
2.665	Ir1	Ir2	Ir4	118.0
2.620	Ir1	Ir5	Ir6	118.6
2.614	Ir2	Ir4	Ir6	60.0

COMMON NAME : Ir(110)-(1x3)
CLASSIFICATION : 77.26
TECHNIQUE : TOF-SARS
AUTHORS : M. Shi, H. Bu and J.W. Rabalais
REFERENCE : Phys. Rev., B42, 2852 (1990)

ILLUSTRATION: 7

SURFACE TYPE
Substrate : Ir
Crystal face: 110
Temperature : RT
Bulk lattice: fcc
2D bulk symm: pmm
2D surf symm: pmm

Adsorbate:
Coverage :
Pattern : (1x3)
Matrix : (1.000, 0.000)
(0.000, 3.000)

STRUCTURE TYPE
Missing-row reconstruction exposing (111) facets, with relaxations in first 2 layers

SAMPLE PREPARATION (1 sample)
Treatment : Ar+ sputtering and O2 treatment cycles, then 1400C annealing
Crystallinity: streaky fractional-order spots
Anal. methods: AES, LEED
Contamination: AES: no C, O; TOF-SARS: no H, C, O

COMMENTS
This structure coexists with the (1x1) termination in a mixed, partly disordered manner, forming streaked LEED spots; the authors claim that the (1x2) reconstruction does not exist

DATA COLLECTION
Technique: TOF-SARS; 4keV Ar+ ion beam
Dataset : time-of-flight scattering and recoiling spectra as fct. of incident and scattering angle in back and forward scattering

THEORY/DATA TREATMENT
Comparison with trajectory calculations with Biersack-Ziegler pot., using experimentally determined shadow cone

STRUCTURES EXAMINED
1) missing-row facetted (1x3) (preferred); 2) missing-row facetted (1x2); 3) (1x3) with 2 missing rows only in top layer; 4) sawtooth model; in model 1): variation of top two interlayer spacings and 2nd-layer registry

QUALITY OF EXPERIMENT-THEORY FIT
Visual

2D UNIT CELLS (1 domain observed)

Cell	Ax (Å)	Ay (Å)	Bx (Å)	By (Å)	α (°)	Matrix	Pattern	Cell type
Bulk	2.715	0.000	0.000	3.840	90.0	(1.000, 0.000) (0.000, 1.000)	(1x1)	b: bulk lattice
Surface 1	2.715	0.000	0.000	11.519	90.0	(1.000, 0.000) (0.000, 3.000)	(1x3)	s1: commens. superlattice

3D COORDINATES

Ir1: ridge atom; Ir2-Ir3: 2nd coplanar layer, laterally expanded towards troughs

Dx/Dy in Å, or as a fraction of layer's unit cell vectors; atom 0 at the origin. Epir/subr are bulk repeat vectors.

No. of atoms: 5 Bulk z = 1.358 Å

Region	Chem el.	At. no.	Cell type	Site occ.	Rel. to	Dx ± εx		Dy ± εy		Dz ± εz		Dz/Bz(%) ± εz/Bz
epir		-2					f		f		Å	
subr		-1				1.358	Å	1.920	Å	1.358	Å	
intf	Ir	1	s1	.33	0	0.000	f	0.000	f	0.000	Å	0.0
intf	Ir	2	s1	.33	1	0.500	f	0.823 ± .009	f	1.250 ± .100	Å	92.1 ± 7.4
intf	Ir	3	s1	.33	2	0.000	f	-0.646 ± .009	f	0.000	Å	0.0
intf	Ir	4	b	1.00	3	-0.500	f	0.469	f	1.358 ± .100	Å	100.0 ± 7.4
subl	Ir	5	b	1.00	4	0.500	f	0.500	f	1.358 ± .100	Å	100.0 ± 7.4

Ir(110)-(1x3)
77.26 BOND DISTANCES AND ANGLES

Bond distances and angles are derived from coordinates

No. of distances/angles: 7

Interatomic dist. A-B (Å)	Atom A	Atom B	Atom C	Bond angle A-B-C (°)
2.715	Ir1	Ir1(1,0)	Ir3	60.4
2.751	Ir1	Ir3	Ir1(1,0)	59.1
2.751	Ir1	Ir3	Ir3(1,0)	119.6
2.751	Ir1	Ir3	Ir5(0,-1)	114.9
2.632	Ir3	Ir4	Ir3(-1,0)	62.1
2.632	Ir3	Ir4	Ir4(1,0)	58.9
2.632	Ir3	Ir4	Ir5	118.9

COMMON NAME : Ir(111)-(1x1) ILLUSTRATION: 1
CLASSIFICATION : 77.2
TECHNIQUE : LEED
AUTHORS : C.M. Chan, S.L. Cunningham, M.A. Van Hove, W.H. Weinberg and S.P. Withrow
REFERENCE : Surf. Sci., 66, 394 (1977)

SURFACE TYPE
Substrate : Ir
Crystal face: 111
Temperature : 573 K
Bulk lattice: fcc
2D bulk symm: p3m1
2D surf symm: p3m1
Adsorbate:
Coverage :
Pattern : (1x1)
Matrix : (1.000, 0.000)
(0.000, 1.000)

STRUCTURE TYPE
Bulk termination with 2.6% contracted top spacing

SAMPLE PREPARATION (1 sample)
Treatment : annealed, cycles of Ar+ bombardment and heating, annealed
Crystallinity:
Anal. methods:
Contamination: AES: no impurities

COMMENTS

DATA COLLECTION
Technique: LEED
Dataset : I-V curves: 00 beam at Θ=7,13,25°, φ=0°; cumulative energy range 510 eV

THEORY/DATA TREATMENT
Dynamical LEED (RFS): Arbman/Hoernfelt atomic potential; 8 phase shifts; Vor=-11.0 eV, Voi=-5.0eV; ΘD=196 K

STRUCTURES EXAMINED
Variations of top layer spacing

QUALITY OF EXPERIMENT-THEORY FIT
Visual

2D UNIT CELLS (1 domain observed)

Cell	Ax (Å)	Ay (Å)	Bx (Å)	By (Å)	α (°)	Matrix	Pattern	Cell type
Bulk	2.715	0.000	1.358	2.351	60.0	(1.000, 0.000) (0.000, 1.000)	(1x1)	b: bulk lattice
Surface 1	2.715	0.000	1.358	2.351	60.0	(1.000, 0.000) (0.000, 1.000)	(1x1)	s1: commens. superlattice

3D COORDINATES

Dx/Dy in Å, or as a fraction of layer's unit cell vectors; atom 0 at the origin. Epir/subr are bulk repeat vectors.

No. of atoms: 3 Bulk z = 2.217 Å

Region	Chem el.	At. no.	Cell type	Site occ.	Rel. to	Dx ± εx		Dy ± εy		Dz ± εz		Dz/Bz(%) ± εz/Bz
epir		-2					f		f		Å	
subr		-1				1.358	Å	0.784	Å	2.217	Å	
intf	Ir	1	b	1.00	0	0.000	f	0.000	f	0.000	Å	0.0
intf	Ir	2	b	1.00	1	0.333	f	0.333	f	2.160 ± .100	Å	97.4 ± 4.5
subl	Ir	3	b	1.00	2	0.333	f	0.333	f	2.217	Å	100.0

BOND DISTANCES AND ANGLES

Bond distances and angles are derived from coordinates

No. of distances/angles: 3

Interatomic dist. A-B (Å)	Atom A	Atom B	Atom C	Bond angle A-B-C (°)
2.715	Ir1	Ir1(1,0)	Ir2	59.4
2.669	Ir1	Ir2	Ir3(-1,0)	119.4
2.715	Ir2	Ir3		

COMMON NAME : Ir(110)-c(2x2)-O ILLUSTRATION: 35
CLASSIFICATION : 77.8.1
TECHNIQUE : LEED
AUTHORS : C.-M. Chan, K.L. Luke, M.A. Van Hove, W.H. Weinberg and S.P. Withrow
REFERENCE : Surf. Sci., 78, 386 (1978)

SURFACE TYPE
Substrate : Ir
Crystal face: 110
Temperature : 573 K
Bulk lattice: fcc
2D bulk symm: pmm
2D surf symm: cmm

Adsorbate: O
Coverage : 0.5 (O/Ir)
Pattern : c(2x2)
Matrix : (1.000,-1.000)
(1.000, 1.000)

STRUCTURE TYPE
Atomic adsorption in short-bridge site on unreconstructed substrate with top Ir-Ir layer spacing contraction by 2%

SAMPLE PREPARATION (1 sample)
Treatment : 0.25ML O forms (1x1) from (1x2), then 2L of O added at RT
Crystallinity:
Anal. methods:
Contamination: Ir(110)(1x2) was AES-clean

COMMENTS
Additional 1/4 monolayer of disordered O present on surface to remove (1x2) reconstruction

DATA COLLECTION
Technique: LEED
Dataset : I-V curves at normal incidence: 01,11,02,20,1/2 1/2,1/2 3/2 3/2 1/2, 3/2 3/2 beams; cumulative energy range: 940 eV

THEORY/DATA TREATMENT
Dynamical LEED (RFS): 8 ph shs (Ir Arbman/Hoernfelt at pot, overlapping superpos pot for O; Vor=-10 eV, VoiαE**1/3

STRUCTURES EXAMINED
Top site (registry 0,0); short bridge site (0.25,0.25); long bridge site (0.75,0.25); all with variable O/Ir spacing; for the short bridge site the 1st Ir-Ir layer spacing varied from 1.22 to 1.43Å

QUALITY OF EXPERIMENT-THEORY FIT
Visual

2D UNIT CELLS (1 domain observed)

Cell	Ax (Å)	Ay (Å)	Bx (Å)	By (Å)	α (°)	Matrix	Pattern	Cell type
Bulk	2.715	0.000	0.000	3.840	90.0	(1.000, 0.000) (0.000, 1.000)	(1x1)	b: bulk lattice
Surface 1	2.715	-3.840	2.715	3.840	109.5	(1.000,-1.000) (1.000, 1.000)	c(2x2)	s1: commens. superlattice

3D COORDINATES

O1: overlayer in short-bridge site on unreconstructed relaxed substrate

Dx/Dy in Å, or as a fraction of layer's unit cell vectors; atom 0 at the origin. Epir/subr are bulk repeat vectors.

No. of atoms: 4 Bulk z = 1.360 Å

Region	Chem el.	At. no.	Cell type	Site occ.	Rel. to	Dx ± εx		Dy ± εy		Dz ± εz		Dz/Bz(%) ± εz/Bz
epir		-2					f		f		Å	
subr		-1				-1.358	Å	-1.920	Å	1.360	Å	
ovrl	O	1	s1	.50	0	0.000	f	0.000	f	0.000	Å	0.0
intf	Ir	2	b	1.00	1	0.500	f	0.000	f	1.370 ± .050	Å	100.7 ± 3.7
intf	Ir	3	b	1.00	2	0.500	f	0.500	f	1.330 ± .070	Å	97.8 ± 5.2
subl	Ir	4	b	1.00	3	-0.500	f	-0.500	f	1.360	Å	100.0

BOND DISTANCES AND ANGLES

Bond distances and angles are derived from coordinates

No. of distances/angles: 4

Interatomic dist. A-B (Å)	Atom A	Atom B	Atom C	Bond angle A-B-C (°)
1.929	O1	Ir2	Ir2(1,0)	134.7
1.929	O1	Ir2	Ir3	134.7

Ir(110)-c(2x2)-O
77.8.1

Bond Distances and Angles - Continued

Interatomic dist. A-B (Å)	Atom A	Atom B	Atom C	Bond angle A-B-C (°)
2.702	Ir2	Ir3	Ir4	59.5
2.717	Ir3	Ir4		

COMMON NAME : Ir(111)-(2x2)-O
CLASSIFICATION : 77.8.2
TECHNIQUE : LEED
AUTHORS : C.-M. Chan and W.H. Weinberg
REFERENCE : J. Chem. Phys., 71, 2788 (1979)

ILLUSTRATION: 22,25

SURFACE TYPE
Substrate : Ir
Crystal face: 111
Temperature : RT
Bulk lattice: fcc
2D bulk symm: p3m1
2D surf symm: p3m1

Adsorbate: O
Coverage : 1/4 or 1/2? (O/Ir)
Pattern : (2x2)
Matrix : (2.000, 0.000)
(0.000, 2.000)

STRUCTURE TYPE
Atomic adsorption in fcc-hollow sites

SAMPLE PREPARATION (1 sample)
Treatment : exposure to 20L of O2 at RT, yielding sharp LEED pattern
Crystallinity:
Anal. methods:
Contamination:

COMMENTS
It is not known whether surface has (2x2) or (2x1) periodicity with 1/4 or 1/2ML coverage (LEED analysis gave identical adsorption structure and R-factor either way): here (2x2) is assumed

DATA COLLECTION
Technique: LEED
Dataset : I-V curves for 3 integral-order and 7 half-order beams at normal incidence, E range 40-160 eV

THEORY/DATA TREATMENT
Dynamical LEED (RFS): 7 phase shifts (Arbman-Hoernfelt Ir pot, superpos pot for O); Voi=-5 eV; ΘD=280 K(Ir), 840K(O)

STRUCTURES EXAMINED
Adsorbate on top, hcp and fcc hollow sites at variable spacings from 1.1 to 2.2Å in both (2x2) and (2x1) periodicities

QUALITY OF EXPERIMENT-THEORY FIT
RZJ=0.22

2D UNIT CELLS (1 domain observed)

Cell	Ax (Å)	Ay (Å)	Bx (Å)	By (Å)	α (°)	Matrix	Pattern	Cell type
Bulk	2.717	0.000	-1.359	2.353	120.0	(1.000, 0.000) (0.000, 1.000)	(1x1)	b: bulk lattice
Surface 1	5.434	0.000	-2.717	4.706	120.0	(2.000, 0.000) (0.000, 2.000)	(2x2)	s1: commens. superlattice

3D COORDINATES

O1: overlayer in fcc-hollow sites

Dx/Dy in Å, or as a fraction of layer's unit cell vectors; atom 0 at the origin. Epir/subr are bulk repeat vectors.

No. of atoms: 3 Bulk z = 2.218 Å

Region	Chem el.	At. no.	Cell type	Site occ.	Rel. to	Dx ± εx		Dy ± εy		Dz ± εz		Dz/Bz(%) ± εz/Bz
epir		-2					f		f		Å	
subr		-1				0.000	Å	1.569	Å	2.218	Å	
ovrl	O	1	s1	.25	0	0.000	f	0.000	f	0.000	Å	0.0
intf	Ir	2	b	1.00	1	0.333	f	0.667	f	1.300 ± .050	Å	58.6 ± 2.3
subl	Ir	3	b	1.00	2	0.333	f	0.667	f	2.218	Å	100.0

BOND DISTANCES AND ANGLES

Bond distances and angles are derived from coordinates

No. of distances/angles: 3

Interatomic dist. A-B (Å)	Atom A	Atom B	Atom C	Bond angle A-B-C (°)
2.037	O1	Ir2	Ir3	164.9
2.717	Ir2	Ir2(1,1)	Ir3	60.0
2.717	Ir2	Ir3		

COMMON NAME : Ir(110)-(2x2)-2S
CLASSIFICATION : 77.16.2
TECHNIQUE : LEED
AUTHORS : C.-M. Chan and M.A. Van Hove
REFERENCE : Surf. Sci., 183, 303 (1987)

ILLUSTRATION: 40

SURFACE TYPE
Substrate : Ir
Crystal face: 110
Temperature : RT
Bulk lattice: fcc
2D bulk symm: pmm
2D surf symm: p2mg

Adsorbate: S
Coverage : 0.5 (S/Ir)
Pattern : (2x2)
Matrix : (2.000, 0.000)
(0.000, 2.000)

STRUCTURE TYPE
Missing-row structure of substrate; atomic S over outermost 3-fold fcc hollow sites in zig-zag arrangement (2 per cell), bonding to two top-layer Ir atoms and one second-layer Ir atom

SAMPLE PREPARATION (1 sample)
Treatment : H2S exposure on clean reconstructed Ir(110)-(1x2)
Crystallinity:
Anal. methods:
Contamination:

COMMENTS
P2mg symmetry implies zig-zag arrangement of S atoms along ridge direction

DATA COLLECTION
Technique: LEED
Dataset : IV curves for 3 integral-order and 7 half-order beams at normal incidence

THEORY/DATA TREATMENT
Dyn. LEED (comb. space with matrix inv., RFS): 6 phase shs
Ir band struct. pot, S superpos. pot; ΘD=200 K(Ir),343K(S)

STRUCTURES EXAMINED
Missing-row model of substrate, with top two layer spacings, second-layer row-pairing and third-layer buckling variable; S in 3 different 3-fold hollow sites of (111) facets, with variable distance from facet and, for best hollow, deflection towards metal ridges

QUALITY OF EXPERIMENT-THEORY FIT
RPE=0.628, RZJ=0.399, RVHT=0.290

2D UNIT CELLS (1 domain observed)

Cell	Ax (Å)	Ay (Å)	Bx (Å)	By (Å)	α (°)	Matrix	Pattern	Cell type
Bulk	2.717	0.000	0.000	3.842	90.0	(1.000, 0.000) (0.000, 1.000)	(1x1)	b: bulk lattice
Surface 1	5.434	0.000	0.000	7.685	90.0	(2.000, 0.000) (0.000, 2.000)	(2x2)	s1: commens. superlattice

3D COORDINATES

S1 and S2: zigzag chain overlayer straddling metal ridges in fcc hollow sites of (111) facets;
Ir3 and Ir4: ridge atoms; error bars of 0.1Å assumed for tabulation

Dx/Dy in Å, or as a fraction of layer's unit cell vectors; atom 0 at the origin. Epir/subr are bulk repeat vectors.

No. of atoms: 6 Bulk z = 1.359 Å

Region	Chem el.	At. no.	Cell type	Site occ.	Rel. to	Dx ± εx		Dy ± εy		Dz ± εz		Dz/Bz(%) ± εz/Bz
epir		-2					f		f		Å	
subr		-1				1.359	Å	1.921	Å	1.359	Å	
ovrl	S	1	s1	.25	0	0.000	f	0.000	f	0.000	Å	0.0
ovrl	S	2	s1	.25	1	0.500	f	0.552	f	0.000	Å	0.0
intf	Ir	3	s1	.25	2	0.250 ± .018	f	0.224 ± .013	f	0.936 ± .100	Å	68.9 ± 7.4
intf	Ir	4	s1	.25	3	0.500	f	0.000	f	0.000	Å	0.0
intf	Ir	5	b	1.00	4	0.500	f	0.500	f	1.314 ± .100	Å	96.7 ± 7.4
subl	Ir	6	b	1.00	5	0.500	f	0.500	f	1.377	Å	101.3

Ir(110)-(2x2)-2S
77.16.2 BOND DISTANCES AND ANGLES

Bond distances and angles are derived from coordinates

No. of distances/angles: 6

Interatomic dist. A-B (Å)	Atom A	Atom B	Atom C	Bond angle A-B-C (°)
2.384	S1	Ir3(-1,-1)	S2(0,-1)	133.8
2.384	S1	Ir3(-1,-1)	Ir3(0,-1)	124.7
2.384	S1	Ir3(-1,-1)	Ir5	52.4
2.384	S1	Ir3(-1,-1)	Ir6(-1,-2)	113.1
2.259	S1	Ir5	Ir3(-1,-1)	56.7
2.259	S1	Ir5	Ir6(-1,-1)	124.4

COMMON NAME : Ir(111)-(√3x√3)R30°-S
CLASSIFICATION : 77.16.1
TECHNIQUE : LEED
AUTHORS : C.-M. Chan and W.H. Weinberg
REFERENCE : J. Chem. Phys., 71, 3988 (1979)

ILLUSTRATION: 22,24

SURFACE TYPE
Substrate : Ir
Crystal face: 111
Temperature : RT
Bulk lattice: fcc
2D bulk symm: p3m1
2D surf symm: p31m

Adsorbate: S
Coverage : 1/3 (S/Ir)
Pattern : (√3x√3)R30°
Matrix : (2.000, 1.000)
(1.000, 2.000)

STRUCTURE TYPE
Atomic adsorption in fcc-hollow sites

SAMPLE PREPARATION (1 sample)
Treatment : decomposition of H2S at RT, annealing at 800 K
Crystallinity:
Anal. methods:
Contamination:

COMMENTS

DATA COLLECTION
Technique: LEED
Dataset : I-V curves for 11 beams at normal incidence; energy range 0-200 eV

THEORY/DATA TREATMENT
Dynamical LEED (RFS): 8 ph shs (Ir Arbman/Hoernfelt pot, overlapping at pot for S); Voi=-5 eV; ΘD=310 K(Ir), 649K(S)

STRUCTURES EXAMINED
Adsorbate on top, hcp and fcc hollow sites at variable spacings

QUALITY OF EXPERIMENT-THEORY FIT
RZJ=0.19

2D UNIT CELLS (1 domain observed)

Cell	Ax (Å)	Ay (Å)	Bx (Å)	By (Å)	α (°)	Matrix	Pattern	Cell type
Bulk	2.717	0.000	-1.359	2.353	120.0	(1.000, 0.000) (0.000, 1.000)	(1x1)	b: bulk lattice
Surface 1	4.076	2.353	0.000	4.706	60.0	(2.000, 1.000) (1.000, 2.000)	(√3x√3)R30°	s1: commens. superlattice

3D COORDINATES

S1: overlayer in fcc-hollow sites

Dx/Dy in Å, or as a fraction of layer's unit cell vectors; atom 0 at the origin. Epir/subr are bulk repeat vectors.

No. of atoms: 3 Bulk z = 2.218 Å

Region	Chem el.	At. no.	Cell type	Site occ.	Rel. to	Dx ± εx		Dy ± εy		Dz ± εz		Dz/Bz(%) ± εz/Bz
epir		-2					f		f		Å	
subr		-1				1.359	Å	0.784	Å	2.218	Å	
ovrl	S	1	s1	.33	0	0.000	f	0.000	f	0.000	Å	0.0
intf	Ir	2	b	1.00	1	0.667	f	0.333	f	1.650 ± .070	Å	74.4 ± 3.2
subl	Ir	3	b	1.00	2	0.667	f	0.333	f	2.218	Å	100.0

BOND DISTANCES AND ANGLES

Bond distances and angles are derived from coordinates

No. of distances/angles: 3

Interatomic dist. A-B (Å)	Atom A	Atom B	Atom C	Bond angle A-B-C (°)
2.277	S1	Ir2	Ir3	171.7
2.717	Ir2	Ir2(1,0)	Ir3	60.0
2.717	Ir2	Ir3		

COMMON NAME : MgO(100)-(1x1)
CLASSIFICATION : 12.8.4
TECHNIQUE : LEED
AUTHORS : M.R. Welton-Cook and W. Berndt
REFERENCE : J. Phys., C15, 569 (1982)

ILLUSTRATION: 149

SURFACE TYPE
Substrate : MgO
Crystal face: 100
Temperature : RT*
Bulk lattice: NaCl
2D bulk symm: p4m
2D surf symm: p4m

Adsorbate:
Coverage :
Pattern : (1x1)
Matrix : (1.000, 0.000)
(0.000, 1.000)

STRUCTURE TYPE
Bulk termination with 2% buckled top MgO mixed layer (O out) and no average first interlayer spacing relaxation

SAMPLE PREPARATION (1 sample)
Treatment : air cleaved sample of MgO; repeated oxidations at 800 K
Crystallinity:
Anal. methods:
Contamination: AES: submonolayer traces of S and P

COMMENTS
Lattice constant from Wyckoff (1963) crystal structure; spacing reliability based on double reliability factor analysis;

DATA COLLECTION
Technique: LEED
Dataset : LEED spectra for 99 beams for 23 diffraction geometries; 0<Θ<70°, 20<E<150 eV, 27 beams for 7 diffr. geometries

THEORY/DATA TREATMENT
Dynamical LEED (CAVLEED package); 8 ph shs from Clementi/Roetti charge dens, α=1/3; 39 beams; Vor=-13 eV, Voi=-3eV

STRUCTURES EXAMINED
-4% to +4% buckling and -4% to +4% relaxation of top interlayer spacing

QUALITY OF EXPERIMENT-THEORY FIT
RPE=0.2518

2D UNIT CELLS (1 domain observed)

Cell	Ax (Å)	Ay (Å)	Bx (Å)	By (Å)	α (°)	Matrix	Pattern	Cell type
Bulk	2.978	0.000	0.000	2.978	90.0	(1.000, 0.000) (0.000, 1.000)	(1x1)	b: bulk lattice
Surface 1	2.978	0.000	0.000	2.978	90.0	(1.000, 0.000) (0.000, 1.000)	(1x1)	s1: commens. superlattice

3D COORDINATES

O1-Mg2: buckled top mixed MgO layer, O outermost; Mg3-O4: repeating bulk mixed copplanar MgO layer

Dx/Dy in Å, or as a fraction of layer's unit cell vectors; atom 0 at the origin. Epir/subr are bulk repeat vectors.

No. of atoms: 4

Bulk z = 2.106 Å

Region	Chem el.	At. no.	Cell type	Site occ.	Rel. to	Dx ± εx		Dy ± εy		Dz ± εz		Dz/Bz(%) ± εz/Bz
epir		-2					f		f		Å	
subr		-1				1.489	Å	1.489	Å	2.106	Å	
intf	O	1	b	1.00	0	0.000	f	0.000	f	0.000	Å	0.0
intf	Mg	2	b	1.00	1	0.500	f	0.500	f	0.042 ± .050	Å	2.0 ± 2.4
subl	Mg	3	b	1.00	2	-0.500	f	-0.500	f	2.085 ± .070	Å	99.0 ± 3.3
subl	O	4	b	1.00	3	0.500	f	0.500	f	0.000	Å	0.0

BOND DISTANCES AND ANGLES

Bond distances and angles are derived from coordinates

No. of distances/angles: 8

Interatomic dist. A-B (Å)	Atom A	Atom B	Atom C	Bond angle A-B-C (°)
2.106	O1	Mg2	O1(1,0)	90.0
2.106	O1	Mg2	Mg2(1,0)	135.0
2.106	O1	Mg2	Mg3	45.9
2.106	O1	Mg2	O4	91.1
2.106	Mg2	O1(1,1)	Mg3(1,1)	88.9
2.978	Mg2	Mg2(1,0)	Mg3(2,0)	120.2

MgO(100)-(1x1)
12.8.4

Bond Distances and Angles - Continued

Interatomic dist. A-B (Å)	Atom A	Atom B	Atom C	Bond angle A-B-C (°)
2.963	Mg2	Mg3(1,1)	Mg3(1,0)	59.8
2.085	Mg2	O4	Mg3	90.0

ILLUSTRATION: 149

COMMON NAME : MgO(100)-(1x1)
CLASSIFICATION : 12.8.5
TECHNIQUE : LEED
AUTHORS : T. Urano, T. Kanaji and M. Kaburagi
REFERENCE : Surf. Sci., 134, 109 (1983)

SURFACE TYPE
Substrate : MgO
Crystal face: 100
Temperature : 300 K
Bulk lattice: NaCl
2D bulk symm: p4m
2D surf symm: p4m

Adsorbate:
Coverage :
Pattern : (1x1)
Matrix : (1.000, 0.000)
(0.000, 1.000)

STRUCTURE TYPE
Ideal truncation of bulk lattice with mixed MgO layers; no spacing relaxation within error bars

SAMPLE PREPARATION (3 sample)
Treatment : 3 sample preps: cleavage in air and various anneals
Crystallinity:
Anal. methods:
Contamination: in 1 prep: AES detected C before heating

COMMENTS
3 preparation procedures gave very similar LEED spectra; buckling cannot be ruled out

DATA COLLECTION
Technique: LEED
Dataset : LEED I-V curves for (10),(11) and (20) beams for 70<E<300 eV

THEORY/DATA TREATMENT
Dynamical LEED: 5 phase shifts, 29 beams

STRUCTURES EXAMINED
Varied first interplanar spacing in the range -5% to +5% in 2.5% steps

QUALITY OF EXPERIMENT-THEORY FIT
Visual

2D UNIT CELLS (1 domain observed)

Cell	Ax (Å)	Ay (Å)	Bx (Å)	By (Å)	α (°)	Matrix	Pattern	Cell type
Bulk	2.977	0.000	0.000	2.977	90.0	(1.000, 0.000) (0.000, 1.000)	(1x1)	b: bulk lattice
Surface 1	2.977	0.000	0.000	2.977	90.0	(1.000, 0.000) (0.000, 1.000)	(1x1)	s1: commens. superlattice

3D COORDINATES

Mg1-O2: topmost mixed coplanar layer; O3-Mg4; repeating bulk mixed coplanar layer; 0.1Å error bars assumed for tabulation

Dx/Dy in Å, or as a fraction of layer's unit cell vectors; atom 0 at the origin. Epir/subr are bulk repeat vectors.

No. of atoms: 4

Bulk z = 2.105 Å

Region	Chem el.	At. no.	Cell type	Site occ.	Rel. to	Dx ± εx		Dy ± εy		Dz ± εz		Dz/Bz(%) ± εz/Bz
epir		-2					f		f		Å	
subr		-1				1.489	Å	1.489	Å	2.105	Å	
intf	Mg	1	b	1.00	0	0.000	f	0.000	f	0.000	Å	0.0
intf	O	2	b	1.00	1	0.500	f	0.500	f	0.000	Å	0.0
subl	O	3	b	1.00	2	0.500	f	0.500	f	2.105 ± .100	Å	100.0 ± 4.8
subl	Mg	4	b	1.00	3	0.500	f	0.500	f	0.000	Å	0.0

BOND DISTANCES AND ANGLES

Bond distances and angles are derived from coordinates

No. of distances/angles: 5

Interatomic dist. A-B (Å)	Atom A	Atom B	Atom C	Bond angle A-B-C (°)
2.977	Mg1	Mg1(1,0)		
2.105	Mg1	O2		
2.105	Mg1	O3		
2.977	Mg1	Mg4		
2.105	O2	Mg4		

COMMON NAME : $MgO(100)-(1x1)$ ILLUSTRATION: 149
CLASSIFICATION : 12.8.8
TECHNIQUE : LEED
AUTHORS : D.L. Blanchard, D.L. Lessor, J.P. LaFemina, D.R. Baer, W.K. Ford and T. Guo
REFERENCE : J. Vac. Sci. Technol., A9, 1814 (1990)

SURFACE TYPE
Substrate : MgO
Crystal face: 100
Temperature : 165 K
Bulk lattice: NaCl
2D bulk symm: p4m
2D surf symm: p4m

Adsorbate:
Coverage :
Pattern : (1x1)
Matrix : (1.000, 0.000)
(0.000, 1.000)

STRUCTURE TYPE
Bulk termination with 5% buckled top MgO compound layer, where O moves towards the vacuum

SAMPLE PREPARATION (1 sample)
Treatment : air cleaved sample of MgO; repeated oxidations at 750 K
Crystallinity:
Anal. methods:
Contamination:

COMMENTS
At electron energies below 140 eV electrostatic charging of sample was prevented by pulsing the beam to 450 eV for 1/3 second, which discharged the sample for 2/3 seconds

DATA COLLECTION
Technique: LEED; back-view LEED, CCD-based video camer
Dataset : I-V curves of 6 symm. inequivalent beams at normal incidence; E range 60-350 eV

THEORY/DATA TREATMENT
Dynamical LEED: 7 phase shifts, 12 atomic layers diffraction from top 6 layers calculated exactly; Voi=-5 eV

STRUCTURES EXAMINED
Varied were: outward relaxation of top layer and buckling of O anions outward (relative to the Mg ions)

QUALITY OF EXPERIMENT-THEORY FIT
RX=0.10

2D UNIT CELLS (1 domain observed)

Cell	Ax (Å)	Ay (Å)	Bx (Å)	By (Å)	α (°)	Matrix	Pattern	Cell type
Bulk	2.978	0.000	0.000	2.978	90.0	(1.000, 0.000) (0.000, 1.000)	(1x1)	b: bulk lattice
Surface 1	2.978	0.000	0.000	2.978	90.0	(1.000, 0.000) (0.000, 1.000)	(1x1)	s1: commens. superlattice

3D COORDINATES

O1-Mg2: buckled top compound MgO layer, O outermost; Mg3-O4: bulk repeat compound MgO layer

Dx/Dy in Å, or as a fraction of layer's unit cell vectors; atom 0 at the origin. Epir/subr are bulk repeat vectors.

No. of atoms: 4 Bulk z = 2.106 Å

Region	Chem el.	At. no.	Cell type	Site occ.	Rel. to	Dx ± εx		Dy ± εy		Dz ± εz		Dz/Bz(%) ± εz/Bz
epir		-2					f		f		Å	
subr		-1				1.489	Å	1.489	Å	2.106	Å	
intf	O	1	b	1.00	0	0.000	f	0.000	f	0.000	Å	0.0
intf	Mg	2	b	1.00	1	0.500	f	0.500	f	0.105 ± .053	Å	5.0 ± 2.5
subl	Mg	3	b	1.00	2	-0.500	f	-0.500	f	2.064 ± .084	Å	98.0 ± 4.0
subl	O	4	b	1.00	3	0.500	f	0.500	f	0.000	Å	0.0

BOND DISTANCES AND ANGLES

Bond distances and angles are derived from coordinates

No. of distances/angles: 4

Interatomic dist. A-B (Å)	Atom A	Atom B	Atom C	Bond angle A-B-C (°)
2.106	O1	Mg2	O1(1,0)	90.0
2.106	O1	Mg2	Mg2(1,0)	135.0
2.106	O1	Mg2	Mg3	45.9
2.106	Mg2	O1(1,1)	Mg3(1,1)	88.9

ILLUSTRATION: 149

COMMON NAME : MgO(100)-(1x1)
CLASSIFICATION : 12.8.9
TECHNIQUE : EELFS
AUTHORS : M. De Crescenzi, N. Motta, F. Patella, A. Sgarlata, F. Arciprete, A. Balzarotti, M. Benfatto and C.R. Natoli
REFERENCE : Springer Series in Surface Sciences, 24, 665 (1991)

SURFACE TYPE
Substrate : MgO
Crystal face: 100
Temperature : RT
Bulk lattice: NaCl
2D bulk symm: p4m
2D surf symm: p4m

Adsorbate:
Coverage :
Pattern : (1x1)
Matrix : (1.000, 0.000)
(0.000, 1.000)

STRUCTURE TYPE
Bulk termination with no surface relaxations

SAMPLE PREPARATION (1 sample)
Treatment : repeated heating to 700 K in 1E-6 torr O2
Crystallinity:
Anal. methods:
Contamination: AES: no contaminants or decomposition

COMMENTS
EXAFS also used for surface structure determination

DATA COLLECTION
Technique: EELFS
Dataset : EELFS spectra at Mg K edge and O K edge; numerically integrated and FS extracted by background subtraction

THEORY/DATA TREATMENT
Fourier transform

STRUCTURES EXAMINED
Varied were: first, second and third nearest neighbor distances

2D UNIT CELLS (1 domain observed)

Cell	Ax (Å)	Ay (Å)	Bx (Å)	By (Å)	α (°)	Matrix	Pattern	Cell type
Bulk	2.978	0.000	0.000	2.978	90.0	(1.000, 0.000) (0.000, 1.000)	(1x1)	b: bulk lattice
Surface 1	2.978	0.000	0.000	2.978	90.0	(1.000, 0.000) (0.000, 1.000)	(1x1)	s1: commens. superlattice

3D COORDINATES

O1-Mg2: planar top compound MgO layer; Mg3-O4: bulk repeat compound MgO layer

Dx/Dy in Å, or as a fraction of layer's unit cell vectors; atom 0 at the origin. Epir/subr are bulk repeat vectors.

No. of atoms: 4 Bulk z = 2.106 Å

Region	Chem el.	At. no.	Cell type	Site occ.	Rel. to	Dx ± εx		Dy ± εy		Dz ± εz		Dz/Bz(%) ± εz/Bz
epir		-2					f		f		Å	
subr		-1				1.489	Å	1.489	Å	2.106	Å	
intf	O	1	b	1.00	0	0.000	f	0.000	f	0.000	Å	0.0
intf	Mg	2	b	1.00	1	0.500	f	0.500	f	0.000	Å	0.0
subl	Mg	3	b	1.00	2	-0.500	f	-0.500	f	2.106	Å	100.0
subl	O	4	b	1.00	3	0.500	f	0.500	f	0.000	Å	0.0

BOND DISTANCES AND ANGLES

Bond distances and angles are derived from coordinates

No. of distances/angles: 2

Interatomic dist. A-B (Å)	Atom A	Atom B	Atom C	Bond angle A-B-C (°)
2.106	O1	Mg2	O1(1,0)	90.0
2.106	Mg2	O1(1,1)	Mg3(1,1)	90.0

COMMON NAME : Mn(100)-(1x1) epitaxial on Pd(100) ILLUSTRATION: 83
CLASSIFICATION : 46.25.3
TECHNIQUE : LEED
AUTHORS : D. Tian, S.C. Wu, F. Jona and P.M. Marcus
REFERENCE : Solid State Commun., 70, 199 (1989)

SURFACE TYPE
Substrate : Mn
Crystal face: 100
Temperature : RT
Bulk lattice: bct
2D bulk symm: p4m
2D surf symm: p4m
Adsorbate:
Coverage :
Pattern : (1x1)
Matrix : (1.000, 0.000)
(0.000, 1.000)

STRUCTURE TYPE
Epitaxial (1x1) multilayer (10 layers) grown on Pd(100), with lateral lattice constant of Pd(100); forms body-centered tetragonal lattice, related to fcc by 12% uniaxial contraction in growth direction

SAMPLE PREPARATION (2 sample)
Treatment : cycles of Ar+ bomb. and long anneals; Mn vapor-deposited
Crystallinity: LEED: some disorder and defects
Anal. methods: AES
Contamination: AES: no S, C, O

COMMENTS
RZJ,RVHT,RPE = 0.12,0.3,0.26 and 0.21,0.35,0.27 for Θ=0° and Θ=8°, resp.;
films with 1 to 6 layers gave weak c(2x2) pattern, suggesting interdiffusion

DATA COLLECTION
Technique: LEED; TV camera-microcomputer system
Dataset : IV spectra for 3 non-equiv. beams (10,11,20) at normal incidence, 5 (00,10,01,11,02) at Θ=8°,ϕ=0°

THEORY/DATA TREATMENT
Dynamical LEED (CHANGE program): 69 beams, 8 ph shs (Moruzzi et al); Voi=-4 eV; Pd substrate below 10 Fe layers

STRUCTURES EXAMINED
Fcc with variable 'bulk' spacing of 1.625-1.775Å perp. to surface (giving bct lattice), and top 2 interlayer spacings varied by up to ±0.4Å from bulk spacing

QUALITY OF EXPERIMENT-THEORY FIT
See comments

2D UNIT CELLS (1 domain observed)

Cell	Ax (Å)	Ay (Å)	Bx (Å)	By (Å)	α (°)	Matrix	Pattern	Cell type
Bulk	2.750	0.000	0.000	2.750	90.0	(1.000, 0.000) (0.000, 1.000)	(1x1)	b: bulk lattice
Surface 1	2.750	0.000	0.000	2.750	90.0	(1.000, 0.000) (0.000, 1.000)	(1x1)	s1: commens. superlattice

3D COORDINATES

Dx/Dy in Å, or as a fraction of layer's unit cell vectors; atom 0 at the origin. Epir/subr are bulk repeat vectors.

No. of atoms: 4 Bulk z = 1.715 Å

Region	Chem el.	At. no.	Cell type	Site occ.	Rel. to	Dx ± εx		Dy ± εy		Dz ± εz		Dz/Bz(%) ± εz/Bz
epir		-2					f		f		Å	
subr		-1				1.375	Å	1.375	Å	1.715	Å	
intf	Mn	1	b	1.00	0	0.000	f	0.000	f	0.000	Å	0.0
intf	Mn	2	b	1.00	1	0.500	f	0.500	f	1.805 ± .030	Å	104.0 ± 1.8
intf	Mn	3	b	1.00	2	-0.500	f	-0.500	f	1.715 ± .040	Å	99.4 ± 2.3
subl	Mn	4	b	1.00	3	0.500	f	0.500	f	1.715 ± .030	Å	100.0 ± 1.8

BOND DISTANCES AND ANGLES

Bond distances and angles are derived from coordinates

No. of distances/angles: 3

Interatomic dist. A-B (Å)	Atom A	Atom B	Atom C	Bond angle A-B-C (°)
2.750	Mn1	Mn1(1,0)	Mn2	58.8
2.653	Mn1	Mn2	Mn3	84.3
2.593	Mn2	Mn3	Mn4	82.8

ILLUSTRATION: 12

COMMON NAME : Mo(100)-(1x1)
CLASSIFICATION : 42.4
TECHNIQUE : LEED
AUTHORS : L.J. Clarke
REFERENCE : Surf. Sci., 91, 131 (1980)

SURFACE TYPE
Substrate : Mo
Crystal face: 100
Temperature : RT*
Bulk lattice: bcc
2D bulk symm: p4m
2D surf symm: p4m

Adsorbate:
Coverage :
Pattern : (1x1)
Matrix : (1.000, 0.000)
(0.000, 1.000)

SAMPLE PREPARATION (1 sample)
Treatment : C removed above 2000 K in 10E-4Pa of O2 for over 100h
Crystallinity:
Anal. methods:
Contamination: AES: no impurities

DATA COLLECTION
Technique: LEED
Dataset : I-V curves for 16 non-degenerate beams at (Θ,ϕ)= (11.5,8.5)° and (7,0)°

STRUCTURES EXAMINED
Variation of top two layer spacings

QUALITY OF EXPERIMENT-THEORY FIT
RZJ=0.146

STRUCTURE TYPE
Bulk termination with multilayer relaxation

COMMENTS
This structure now viewed as disordered high-temperature version of low-temperature reconstruction, including lateral displacements (eds.)

THEORY/DATA TREATMENT
Dynamical LEED (layer doubling): 8 phase shifts with several Mattheiss-Loucks-type superposition pots; Voi=4 eV; ΘD=230 K

2D UNIT CELLS (1 domain observed)

Cell	Ax (Å)	Ay (Å)	Bx (Å)	By (Å)	α (°)	Matrix	Pattern	Cell type
Bulk	3.147	0.000	0.000	3.147	90.0	(1.000, 0.000) (0.000, 1.000)	(1x1)	b: bulk lattice
Surface 1	3.147	0.000	0.000	3.147	90.0	(1.000, 0.000) (0.000, 1.000)	(1x1)	s1: commens. superlattice

3D COORDINATES

Dx/Dy in Å, or as a fraction of layer's unit cell vectors; atom 0 at the origin. Epir/subr are bulk repeat vectors.

No. of atoms: 4

Bulk z = 1.574 Å

Region	Chem el.	At. no.	Cell type	Site occ.	Rel. to	Dx ± εx		Dy ± εy		Dz ± εz		Dz/Bz(%) ± εz/Bz
epir		-2					f		f		Å	
subr		-1				1.574	Å	1.574	Å	1.574	Å	
intf	Mo	1	b	1.00	0	0.000	f	0.000	f	0.000	Å	0.0
intf	Mo	2	b	1.00	1	0.500	f	0.500	f	1.424 ± .030	Å	90.5 ± 1.9
intf	Mo	3	b	1.00	2	-0.500	f	-0.500	f	1.558 ± .030	Å	99.0 ± 1.9
subl	Mo	4	b	1.00	3	0.500	f	0.500	f	1.574	Å	100.0

BOND DISTANCES AND ANGLES

Bond distances and angles are derived from coordinates

No. of distances/angles: 3

Interatomic dist. A-B (Å)	Atom A	Atom B	Atom C	Bond angle A-B-C (°)
3.147	Mo1	Mo1(1,0)	Mo2	53.4
2.642	Mo1	Mo2	Mo3	67.6
2.717	Mo2	Mo3	Mo4	70.3

COMMON NAME : Mo(100)-(1x1) disordered
CLASSIFICATION : 42.10
TECHNIQUE : LEED
AUTHORS : D.G. Kelly, R.F. Lin, M.A. Van Hove and G.A. Somorjai
REFERENCE : Surf. Sci., 224, 97 (1989)

ILLUSTRATION: 14

SURFACE TYPE
Substrate : Mo
Crystal face: 100
Temperature : 300 K
Bulk lattice: bcc
2D bulk symm: p4m
2D surf symm: none

Adsorbate:
Coverage :
Pattern : (1x1) disordered
Matrix : (1.000, 0.000)
(0.000, 1.000)

STRUCTURE TYPE
'high-temperature' disordered version of 'low-temperature' reconstructed Mo(100): topmost atoms laterally relaxed by 0.13Å randomly in diagonal unit cell directions; also multilayer relaxation perp. to surface

SAMPLE PREPARATION (1 sample)
Treatment : heat in 5E-9 torr O2 to 900 K for 30secs, then flash to 2000K
Crystallinity:
Anal. methods:
Contamination: AES: <1%ML C

COMMENTS
Disorder here modeled as (1x1) structure with uniformly displaced top layer atoms

DATA COLLECTION
Technique: LEED
Dataset : I-V curves for 6 independent beams: (10),(11),(20),(21), (22),(30); E range 70-300 eV

THEORY/DATA TREATMENT
Dynamical LEED (RFS); 8 phase shifts from cluster calculation; ΘD=380 K

STRUCTURES EXAMINED
Variation of top 3 interlayer spacings and of top layer registry in 2 dimensions

QUALITY OF EXPERIMENT-THEORY FIT
RVHT=0.18, RZJ=0.10, RPE=0.30

2D UNIT CELLS (4 domains observed)

Cell	Ax (Å)	Ay (Å)	Bx (Å)	By (Å)	α (°)	Matrix	Pattern	Cell type
Bulk	3.147	0.000	0.000	3.147	90.0	(1.000, 0.000) (0.000, 1.000)	(1x1)	b: bulk lattice
Surface 1	3.147	0.000	0.000	3.147	90.0	(1.000, 0.000) (0.000, 1.000)	(1x1) disordered	s1: commens. superlattice

3D COORDINATES

Mo1: laterally displaced top layer, representing disordered displacement direction

Dx/Dy in Å, or as a fraction of layer's unit cell vectors; atom 0 at the origin. Epir/subr are bulk repeat vectors.

No. of atoms: 5

Bulk z = 1.575 Å

Reg ion	Chem el.	At. no.	Cell type	Site occ.	Rel. to	Dx ± εx	Dy ± εy	Dz ± εz	Dz/Bz(%) ± εz/Bz
epir		-2				f	f	Å	
subr		-1				1.573 Å	1.573 Å	1.575 Å	
intf	Mo	1	b	1.00	0	0.000 f	0.000 f	0.000 Å	0.0
intf	Mo	2	b	1.00	1	0.470 ± .016 f	0.470 ± .016 f	1.490 ± .020 Å	94.6 ± 1.3
intf	Mo	3	b	1.00	2	0.500 f	0.500 f	1.600 ± .020 Å	101.6 ± 1.3
intf	Mo	4	b	1.00	3	-0.500 f	-0.500 f	1.580 ± .020 Å	100.3 ± 1.3
subl	Mo	5	b	1.00	4	0.500 f	0.500 f	1.575 Å	100.0

BOND DISTANCES AND ANGLES

Bond distances and angles are derived from coordinates

No. of distances/angles: 2

Interatomic dist. A-B (Å)	Atom A	Atom B	Atom C	Bond angle A-B-C (°)
2.568	Mo1	Mo2	Mo3	180.0
2.741	Mo2	Mo3	Mo4	71.1

ILLUSTRATION: 11

COMMON NAME : Mo(110)-(1x1)
CLASSIFICATION : 42.7
TECHNIQUE : LEED
AUTHORS : L. Morales de la Garza and L.J. Clarke
REFERENCE : J. Phys., C14, 5391 (1981)

SURFACE TYPE
Substrate : Mo
Crystal face: 110
Temperature : RT*
Bulk lattice: bcc
2D bulk symm: cmm
2D surf symm: cmm

Adsorbate:
Coverage :
Pattern : (1x1)
Matrix : (1.000, 0.000)
(0.000, 1.000)

STRUCTURE TYPE
Bulk termination with 1.7% contracted top layer spacing

SAMPLE PREPARATION (1 sample)
Treatment : 1800 K desorbs S, 1300K in O2 removes C, 1100K removes CO
Crystallinity:
Anal. methods:
Contamination: AES: no traces of contaminants

COMMENTS

DATA COLLECTION
Technique: LEED
Dataset : I-V curves for 43 beams, forming 14 non-equivalent sets

THEORY/DATA TREATMENT
Dynamical LEED (RFS): 9 phase shifts; 43 beams; Voi=-4 eV; Vor E-dependent; ΘD=295 K

STRUCTURES EXAMINED
Top layer spacing varied, and many non-structural parameters (Vor E-dependent, Voi, exchange parameter α, ΘD)

QUALITY OF EXPERIMENT-THEORY FIT
RZJ=0.085

2D UNIT CELLS (1 domain observed)

Cell	Ax (Å)	Ay (Å)	Bx (Å)	By (Å)	α (°)	Matrix	Pattern	Cell type
Bulk	2.728	0.000	.909	2.572	70.5	(1.000, 0.000) (0.000, 1.000)	(1x1)	b: bulk lattice
Surface 1	2.728	0.000	.909	2.572	70.5	(1.000, 0.000) (0.000, 1.000)	(1x1)	s1: commens. superlattice

3D COORDINATES

Dx/Dy in Å, or as a fraction of layer's unit cell vectors; atom 0 at the origin. Epir/subr are bulk repeat vectors.

No. of atoms: 3

Bulk z = 2.227 Å

Region	Chem el.	At. no.	Cell type	Site occ.	Rel. to	Dx ± εx		Dy ± εy		Dz ± εz		Dz/Bz(%) ± εz/Bz
epir		-2					f		f		Å	
subr		-1				1.819	Å	1.286	Å	2.227	Å	
intf	Mo	1	b	1.00	0	0.000	f	0.000	f	0.000	Å	0.0
intf	Mo	2	b	1.00	1	0.500	f	0.500	f	2.190 ± .040	Å	98.3 ± 1.8
subl	Mo	3	b	1.00	2	-0.500	f	-0.500	f	2.227	Å	100.0

BOND DISTANCES AND ANGLES

Bond distances and angles are derived from coordinates

No. of distances/angles: 3

Interatomic dist. A-B (Å)	Atom A	Atom B	Atom C	Bond angle A-B-C (°)
2.728	Mo1	Mo1(1,0)	Mo2	70.3
3.124	Mo1	Mo2	Mo1(1,1)	91.0
3.150	Mo2	Mo3	Mo3(1,0)	54.7

COMMON NAME : Mo(111)-(1x1)
CLASSIFICATION : 42.8
TECHNIQUE : LEIS
AUTHORS : S.H. Overbury
REFERENCE : Surf. Sci., 175, 123 (1986)

ILLUSTRATION: 15

SURFACE TYPE
Substrate : Mo
Crystal face: 111
Temperature : RT*
Bulk lattice: bcc
2D bulk symm: p3m1
2D surf symm: p3m1

Adsorbate:
Coverage :
Pattern : (1x1)
Matrix : (1.000, 0.000)
(0.000, 1.000)

STRUCTURE TYPE
Bulk termination with multilayer relaxation perp. to surface

SAMPLE PREPARATION (1 sample)
Treatment : Ar+ sputtering and repeated O treatment up to 1500 K
Crystallinity:
Anal. methods:
Contamination: AES: residual C

COMMENTS

DATA COLLECTION
Technique: LEIS; low energy alkali (Li+) ion scatterin
Dataset : scattering intensity as function of incident polar angle in [1,-2,1] and [2,-1,-1] azimuths for various scatt. angle

THEORY/DATA TREATMENT
Simulation of single scattering intensity

STRUCTURES EXAMINED
Variation of top two interlayer spacings

QUALITY OF EXPERIMENT-THEORY FIT
Visual

2D UNIT CELLS (1 domain observed)

Cell	Ax (Å)	Ay (Å)	Bx (Å)	By (Å)	α (°)	Matrix	Pattern	Cell type
Bulk	4.455	0.000	-2.228	3.858	120.0	(1.000, 0.000) (0.000, 1.000)	(1x1)	b: bulk lattice
Surface 1	4.455	0.000	-2.228	3.858	120.0	(1.000, 0.000) (0.000, 1.000)	(1x1)	s1: commens. superlattice

3D COORDINATES

Dx/Dy in Å, or as a fraction of layer's unit cell vectors; atom 0 at the origin. Epir/subr are bulk repeat vectors.

No. of atoms: 4

Bulk z = .909 Å

Region	Chem el.	At. no.	Cell type	Site occ.	Rel. to	Dx ± εx		Dy ± εy		Dz ± εz		Dz/Bz(%) ± εz/Bz
epir		-2					f		f		Å	
subr		-1				0.000	Å	-2.572	Å	0.909	Å	
intf	Mo	1	b	1.00	0	0.000	f	0.000	f	0.000	Å	0.0
intf	Mo	2	b	1.00	1	0.667	f	0.333	f	1.073 ± .020	Å	118.0 ± 2.2
intf	Mo	3	b	1.00	2	-0.333	f	0.333	f	0.946 ± .040	Å	104.1 ± 4.4
subl	Mo	4	b	1.00	3	-0.333	f	-0.667	f	0.909	Å	100.0

BOND DISTANCES AND ANGLES

Bond distances and angles are derived from coordinates

No. of distances/angles: 4

Interatomic dist. A-B (Å)	Atom A	Atom B	Atom C	Bond angle A-B-C (°)
2.787	Mo1	Mo2	Mo3	72.5
2.928	Mo1	Mo4		
2.741	Mo2	Mo3	Mo4	70.9
2.728	Mo3	Mo4		

ILLUSTRATION: 48,50

COMMON NAME : Mo(100)-c(2x2)-C
CLASSIFICATION : 42.6.2
TECHNIQUE : LEED
AUTHORS : P.J. Rous, D. Jentz, D.G. Kelly, R.Q. Hwang, M.A. Van Hove and G.A. Somorjai
REFERENCE : Springer Series in Surface Sciences, 24, 432 (1991)

SURFACE TYPE
Substrate : Mo
Crystal face: 100
Temperature : RT
Bulk lattice: bcc
2D bulk symm: p4m
2D surf symm: p4m

Adsorbate: C
Coverage : 0.5 C/Mo
Pattern : c(2x2)
Matrix : (1.000,-1.000)
(1.000, 1.000)

STRUCTURE TYPE
Atomic adsorption in hollow site, without buckling in 2nd Mo layer, with shortest C-Mo bond to 2nd Mo layer

SAMPLE PREPARATION (1 sample)
Treatment : heat-cleaned in O2, flash to 2000 K; C from hydrocarbons
Crystallinity:
Anal. methods: AES
Contamination:

COMMENTS

DATA COLLECTION
Technique: LEED; video camera
Dataset : IV spectra for 10 beams at normal inc.; E range 70-310 eV

THEORY/DATA TREATMENT
Dynamical LEED: tensor LEED with automated search, based on RFS

STRUCTURES EXAMINED
Hollow site with independent relaxation of 3 coordinates for all atoms in C and top 2 Mo layers (15 parameters + muffin-tin zero): result symmetrized a posteriori

QUALITY OF EXPERIMENT-THEORY FIT
RPE=0.464

2D UNIT CELLS (1 domain observed)

Cell	Ax (Å)	Ay (Å)	Bx (Å)	By (Å)	α (°)	Matrix	Pattern	Cell type
Bulk	3.150	0.000	0.000	3.150	90.0	(1.000, 0.000) (0.000, 1.000)	(1x1)	b: bulk lattice
Surface 1	3.150	-3.150	3.150	3.150	90.0	(1.000,-1.000) (1.000, 1.000)	c(2x2)	s1: commens. superlattice

3D COORDINATES

C1: overlayer in hollow sites; Mo2-Mo3: planar, laterally unrelaxed 1st Mo layer; Mo4-Mo5: planar, laterally unrelaxed 2nd Mo layer; 0.05Å error bars assumed for tabulation

Dx/Dy in Å, or as a fraction of layer's unit cell vectors; atom 0 at the origin. Epir/subr are bulk repeat vectors.

No. of atoms: 4

Bulk z = 1.575 Å

Region	Chem el.	At. no.	Cell type	Site occ.	Rel. to	Dx ± εx		Dy ± εy		Dz ± εz		Dz/Bz(%) ± εz/Bz
epir		-2					f		f		Å	
subr		-1				1.575	Å	1.575	Å	1.575	Å	
ovrl	C	1	s1	.50	0	0.000	f	0.000	f	0.000	Å	0.0
intf	Mo	2	b	1.00	1	0.500 ± .016	f	0.500 ± .016	f	0.430 ± .050	Å	27.3 ± 3.2
intf	Mo	3	b	1.00	2	-0.500 ± .016	f	-0.500 ± .016	f	1.560 ± .050	Å	99.1 ± 3.2
subl	Mo	4	b	1.00	3	0.500 ± .016	f	0.500 ± .016	f	1.575 ± .050	Å	100.0 ± 3.2

BOND DISTANCES AND ANGLES

Bond distances and angles are derived from coordinates

No. of distances/angles: 5

Interatomic dist. A-B (Å)	Atom A	Atom B	Atom C	Bond angle A-B-C (°)
2.269	C1	Mo2	C1(0,1)	158.2
2.269	C1	Mo2	Mo2(1,0)	134.0
2.269	C1	Mo2	Mo3	45.9
1.990	C1	Mo3	Mo4	125.3

Mo(100)-c(2x2)-C
42.6.2

Bond Distances and Angles - Continued

Interatomic dist. A-B (Å)	Atom A	Atom B	Atom C	Bond angle A-B-C (°)
2.719	Mo2	Mo3	Mo4	70.3

COMMON NAME : Mo(100)-(1x1)-2H (D)
CLASSIFICATION : 42.1.1
TECHNIQUE : LEED
AUTHORS : M.L. Hildner, R.S. Daley, T.E. Felter and P.J. Estrup
REFERENCE : J. Vac. Sci. Technol., A9, 1604 (1991)

ILLUSTRATION: 12

SURFACE TYPE
Substrate : Mo
Crystal face: 100
Temperature : 10 K
Bulk lattice: bcc
2D bulk symm: p4m
2D surf symm: p4m

Adsorbate: H (D)
Coverage : 2.0 H/Mo
Pattern : (1x1)
Matrix : (1.000, 0.000)
(0.000, 1.000)

STRUCTURE TYPE
Unrelaxed bulk termination, both at 10K and at 80K

SAMPLE PREPARATION (1 sample)
Treatment : cycles of annealing in O2 and flashing in UHV, then D2 dep.
Crystallinity:
Anal. methods:
Contamination: AES: no impurities

COMMENTS
H (D) positions not determined

DATA COLLECTION
Technique: LEED; 650keV He+ ions
Dataset : backscattering spectra in [100] and [111] channeling directions

THEORY/DATA TREATMENT
Monte Carlo simulations (VEGAS code): Moliere pot.; isotropic vibrations; ΘD=380 K(Mo bulk),240K(Mo surf)

STRUCTURES EXAMINED
Variation of top layer spacing and parallel displacement

QUALITY OF EXPERIMENT-THEORY FIT
Visual

2D UNIT CELLS (1 domain observed)

Cell	Ax (Å)	Ay (Å)	Bx (Å)	By (Å)	α (°)	Matrix	Pattern	Cell type
Bulk	3.147	0.000	0.000	3.147	90.0	(1.000, 0.000) (0.000, 1.000)	(1x1)	b: bulk lattice
Surface 1	3.147	0.000	0.000	3.147	90.0	(1.000, 0.000) (0.000, 1.000)	(1x1)	s1: commens. superlattice

3D COORDINATES

0.1Å error bars used for tabulation

Dx/Dy in Å, or as a fraction of layer's unit cell vectors; atom 0 at the origin. Epir/subr are bulk repeat vectors.

No. of atoms: 4

Bulk z = 1.574 Å

Region	Chem el.	At. no.	Cell type	Site occ.	Rel. to	Dx ± εx	Dy ± εy	Dz ± εz	Dz/Bz(%) ± εz/Bz
epir		-2				f	f	Å	
subr		-1				1.574 Å	1.574 Å	1.574 Å	
intf	Mo	1	b	1.00	0	0.000 f	0.000 f	0.000 Å	0.0
intf	Mo	2	b	1.00	1	0.500 ± .064 f	0.500 ± .064 f	1.574 ± .100 Å	100.0 ± 6.4
intf	Mo	3	b	1.00	2	-0.500 f	-0.500 f	1.574 Å	100.0
subl	Mo	4	b	1.00	3	0.500 f	0.500 f	1.574 Å	100.0

BOND DISTANCES AND ANGLES

Bond distances and angles are derived from coordinates

No. of distances/angles: 2

Interatomic dist. A-B (Å)	Atom A	Atom B	Atom C	Bond angle A-B-C (°)
3.147	Mo1	Mo1(1,0)		
2.725	Mo1	Mo2		

COMMON NAME : Mo(100)-c(2x2)-N
CLASSIFICATION : 42.7.1
TECHNIQUE : LEED
AUTHORS : A. Ignatiev, F. Jona, D.W. Jepsen and P.M. Marcus
REFERENCE : Surf. Sci., 49, 189 (1975)

ILLUSTRATION: 48,50

SURFACE TYPE
Substrate : Mo
Crystal face: 100
Temperature : RT*
Bulk lattice: bcc
2D bulk symm: p4m
2D surf symm: p4m

Adsorbate: N
Coverage : 0.5 N/Mo
Pattern : c(2x2)
Matrix : (1.000, 1.000)
(-1.000, 1.000)

STRUCTURE TYPE
Atomic adsorption in 4-fold hollow sites on unrelaxed substrate

SAMPLE PREPARATION (1 sample)
Treatment : exposure to N2 gas at low pressure (eg 1.0E-8 torr)
Crystallinity:
Anal. methods:
Contamination: monitored by LEED and AES

COMMENTS
Agreement between theory and experiment was not good for all beams: conclusions of this study must be viewed as tentative

DATA COLLECTION
Technique: LEED
Dataset : I-V curves: 6 integral order and 5 half-order beams at Θ=9°

THEORY/DATA TREATMENT
Dynamical LEED: Vor=-15 eV, Voi=-4eV; N potential from superpos. of at. charge dens. in fictitious bcc lattice; ΘD=360 K

STRUCTURES EXAMINED
Hollow, top, and 2-fold bridge sites; N-Mo layer spacing varied from about 0.8 to 1.4Å in steps of 0.05Å; top Mo-Mo spacing contracted 0, 5 and 10%

QUALITY OF EXPERIMENT-THEORY FIT
Visual

2D UNIT CELLS (1 domain observed)

Cell	Ax (Å)	Ay (Å)	Bx (Å)	By (Å)	α (°)	Matrix	Pattern	Cell type
Bulk	3.154	0.000	0.000	3.154	90.0	(1.000, 0.000) (0.000, 1.000)	(1x1)	b: bulk lattice
Surface 1	3.154	3.154	-3.154	3.154	90.0	(1.000, 1.000) (-1.000, 1.000)	c(2x2)	s1: commens. superlattice

3D COORDINATES

N1: overlayer in 4-fold hollow sites; 0.1Å error bars assumed for tabulation

Dx/Dy in Å, or as a fraction of layer's unit cell vectors; atom 0 at the origin. Epir/subr are bulk repeat vectors.

No. of atoms: 3

Bulk z = 1.580 Å

Region	Chem el.	At. no.	Cell type	Site occ.	Rel. to	Dx ± εx		Dy ± εy		Dz ± εz		Dz/Bz(%) ± εz/Bz
epir		-2					f		f		Å	
subr		-1				-1.577	Å	-1.577	Å	1.580	Å	
ovrl	N	1	s1	.50	0	0.000	f	0.000	f	0.000	Å	0.0
intf	Mo	2	b	1.00	1	0.500	f	0.500	f	1.020 ± .100	Å	64.6 ± 6.3
subl	Mo	3	b	1.00	2	-0.500	f	-0.500	f	1.580 ± .100	Å	100.0 ± 6.3

BOND DISTANCES AND ANGLES

Bond distances and angles are derived from coordinates

No. of distances/angles: 4

Interatomic dist. A-B (Å)	Atom A	Atom B	Atom C	Bond angle A-B-C (°)
2.452	N1	Mo2	Mo2(1,0)	130.0
2.600	N1	Mo3	Mo2	54.7
3.154	Mo2	Mo2(1,0)	N1(1,0)	50.0
2.733	Mo2	Mo3		

COMMON NAME : Mo(100)-c(2x2)-S
CLASSIFICATION : 42.16.10
TECHNIQUE : LEED
AUTHORS : P.J. Rous, D. Jentz, D.G. Kelly, R.Q. Hwang, M.A. Van Hove and G.A. Somorjai
REFERENCE : Springer Series in Surface Sciences, 24, 432 (1991)

ILLUSTRATION: 48,50

SURFACE TYPE
Substrate : Mo
Crystal face: 100
Temperature : RT
Bulk lattice: bcc
2D bulk symm: p4m
2D surf symm: p4m

Adsorbate: S
Coverage : 0.5 S/Mo
Pattern : c(2x2)
Matrix : (1.000,-1.000)
(1.000, 1.000)

STRUCTURE TYPE
Atomic adsorption in hollow site, inducing buckling toward S in 2nd Mo layer, with shortest S-Mo bond to 2nd Mo layer

SAMPLE PREPARATION (1 sample)
Treatment : heat-cleaned in O2, flash to 2000 K; S from el.chem. cell
Crystallinity:
Anal. methods: AES
Contamination:

COMMENTS

DATA COLLECTION
Technique: LEED; video camera
Dataset : IV spectra for 8 beams at normal incidence; E range 60-250 eV

THEORY/DATA TREATMENT
Dynamical LEED: tensor LEED with automated search, based on RFS

STRUCTURES EXAMINED
Hollow site with independent relaxation of 3 coordinates for all atoms in S and top 2 Mo layers (15 parameters + muffin-tin zero): result symmetrized a posteriori

QUALITY OF EXPERIMENT-THEORY FIT
RPE=0.234

2D UNIT CELLS (1 domain observed)

Cell	Ax (Å)	Ay (Å)	Bx (Å)	By (Å)	α (°)	Matrix	Pattern	Cell type
Bulk	3.150	0.000	0.000	3.150	90.0	(1.000, 0.000) (0.000, 1.000)	(1x1)	b: bulk lattice
Surface 1	3.150	-3.150	3.150	3.150	90.0	(1.000,-1.000) (1.000, 1.000)	c(2x2)	s1: commens. superlattice

3D COORDINATES

S1: overlayer in hollow sites; Mo2-Mo3: planar, laterally unrelaxed 1st Mo layer;
Mo4-Mo5: buckled, laterally unrelaxed 2nd Mo layer; 0.05Å error bars assumed for tabulation

Dx/Dy in Å, or as a fraction of layer's unit cell vectors; atom 0 at the origin. Epir/subr are bulk repeat vectors.

No. of atoms: 7 Bulk z = 1.575 Å

Reg ion	Chem el.	At. no.	Cell type	Site occ.	Rel. to	Dx ± εx		Dy ± εy		Dz ± εz		Dz/Bz(%) ± εz/Bz
epir		-2					f		f		Å	
subr		-1				1.575	Å	1.575	Å	1.575	Å	
ovrl	S	1	s1	.50	0	0.000	f	0.000	f	0.000	Å	0.0
intf	Mo	2	s1	.50	1	0.000 ± .011	f	0.500 ± .011	f	1.000 ± .050	Å	63.5 ± 3.2
intf	Mo	3	s1	.50	2	-0.500 ± .011	f	-0.500 ± .011	f	0.000 ± .050	Å	0.0 ± 3.2
intf	Mo	4	s1	.50	3	0.500 ± .011	f	0.000 ± .011	f	1.380 ± .050	Å	87.6 ± 3.2
intf	Mo	5	s1	.50	4	-0.500 ± .011	f	0.500 ± .011	f	0.160 ± .050	Å	10.2 ± 3.2
subl	Mo	6	b	1.00	5	-0.500 ± .016	f	-0.500 ± .016	f	1.570 ± .050	Å	99.7 ± 3.2
subl	Mo	7	b	1.00	6	1.000	f	0.000	f	0.000	Å	0.0

Mo(100)-c(2x2)-S
42.16.10

BOND DISTANCES AND ANGLES

Bond distances and angles are derived from coordinates

No. of distances/angles: 8

Interatomic dist. A-B (Å)	Atom A	Atom B	Atom C	Bond angle A-B-C (°)
2.442	S1	Mo2	S1(0,1)	131.6
2.442	S1	Mo2	Mo2(1,0)	130.2
2.442	S1	Mo2	Mo3	49.8
2.380	S1	Mo4	Mo2	58.2
2.380	S1	Mo4	Mo3	58.2
2.380	S1	Mo4	Mo5	92.9
3.150	Mo2	Mo3	Mo4	53.1
3.150	Mo2	Mo3	Mo5	54.4

COMMON NAME : Mo(100)-(1x1)-Si
CLASSIFICATION : 42.14.1
TECHNIQUE : LEED
AUTHORS : A. Ignatiev, F. Jona, D.W. Jepsen and P.M. Marcus
REFERENCE : Phys. Rev., B11, 4780 (1975)

ILLUSTRATION: 48,49

SURFACE TYPE
Substrate : Mo
Crystal face: 100
Temperature : RT*
Bulk lattice: bcc
2D bulk symm: p4m
2D surf symm: p4m

Adsorbate: Si
Coverage : 1.0 Si/Mo
Pattern : (1x1)
Matrix : (1.000, 0.000)
(0.000, 1.000)

STRUCTURE TYPE
Atomic adsorption in 4-fold hollow sites on unrelaxed substrate

SAMPLE PREPARATION (1 sample)
Treatment : Si exposure at 0.2-0.5ML per minute at 1 to 5E-10 torr
Crystallinity:
Anal. methods:
Contamination: LEED and AES

COMMENTS
Variation of the overlayer inner potential and ΘD had little effect upon the structure determination

DATA COLLECTION
Technique: LEED
Dataset : I-V spectra: Θ=8°, ϕ=0°: 6 non-degenerate beams; Θ=21°, ϕ=0°: 4 non-degenerate beams; 20<E<150 eV

THEORY/DATA TREATMENT
Dynamical LEED (layer KKR): 8 phase shifts, 38 or 48 beams
Vor=-16 eV, Voi=-4eV; ΘD=114-360 K(Si), 360K(Mo)

STRUCTURES EXAMINED
Hollow, bridge and top sites; various Si/Mo spacings; top Mo-Mo spacing contracted 0%, 5%, and 10%

QUALITY OF EXPERIMENT-THEORY FIT
Visual

2D UNIT CELLS (1 domain observed)

Cell	Ax (Å)	Ay (Å)	Bx (Å)	By (Å)	α (°)	Matrix	Pattern	Cell type
Bulk	3.146	0.000	0.000	3.146	90.0	(1.000, 0.000) (0.000, 1.000)	(1x1)	b: bulk lattice
Surface 1	3.146	0.000	0.000	3.146	90.0	(1.000, 0.000) (0.000, 1.000)	(1x1)	s1: commens. superlattice

3D COORDINATES

Si1: overlayer in 4-fold hollow sites

Dx/Dy in Å, or as a fraction of layer's unit cell vectors; atom 0 at the origin. Epir/subr are bulk repeat vectors.

No. of atoms: 3

Bulk z = 1.570 Å

Reg ion	Chem el.	At. no.	Cell type	Site occ.	Rel. to	Dx ± εx		Dy ± εy		Dz ± εz		Dz/Bz(%) ± εz/Bz
epir		-2					f		f		Å	
subr		-1				-1.573	Å	-1.573	Å	1.570	Å	
ovrl	Si	1	b	1.00	0	0.000	f	0.000	f	0.000	Å	0.0
intf	Mo	2	b	1.00	1	0.500	f	0.500	f	1.160 ± .100	Å	73.9 ± 6.4
subl	Mo	3	b	1.00	2	-0.500	f	-0.500	f	1.570 ± .100	Å	100.0 ± 6.4

BOND DISTANCES AND ANGLES

Bond distances and angles are derived from coordinates

No. of distances/angles: 4

Interatomic dist. A-B (Å)	Atom A	Atom B	Atom C	Bond angle A-B-C (°)
2.509	Si1	Mo2	Mo2(1,0)	128.8
2.730	Si1	Mo3	Mo2	54.8
3.146	Mo2	Mo2(1,0)	Si1(1,0)	51.2
2.723	Mo2	Mo3		

COMMON NAME : MoS2(0001)-(1x1) ILLUSTRATION: 159
CLASSIFICATION : 42.16.4b
TECHNIQUE : LEED
AUTHORS : B.J. Mrstik, R. Kaplan, T.L. Reinecke, M.A. Van Hove and S.Y. Tong
REFERENCE : Phys. Rev., B15, 897 (1977)

SURFACE TYPE
Substrate : MoS2
Crystal face: 0001
Temperature : 95 K
Bulk lattice: 2H-MoS2
2D bulk symm: p3m1
2D surf symm: p3m1

Adsorbate:
Coverage :
Pattern : (1x1)
Matrix : (1.000, 0.000)
(0.000, 1.000)

STRUCTURE TYPE
Bulk termination with complete S-Mo-S sandwich and slight layer spacing relaxations; bulk layer stacking maintained as BAB ABA BAB ABA ...

SAMPLE PREPARATION (1 sample)
Treatment : cut from natural molybdenite in air; no further treatment
Crystallinity:
Anal. methods:
Contamination: AES: <3% C

COMMENTS
Later R-factor comparison (M.A. Van Hove et al, Surf. Sci. 64, 85 (1977)) confirms this structure, with top spacings of 1.518±0.005, 1.593, 2.87±0.01Å, resp., and R2=0.09

DATA COLLECTION
Technique: LEED
Dataset : I-V curves: 10,01,11,20,02 beams at Θ=0°; 10<E<200 eV; cumulative energy range: 740eV (non-degenerate)

THEORY/DATA TREATMENT
Dynamical LEED (RFS): up to 55 beams, 8 phase shifts, Mattheiss band structure pots;Vor=-5.0 eV,Voi=-5.0eV,ΘD=350 K

STRUCTURES EXAMINED
Various spacings between layers 1 and 2, and between 3 and 4 also the registry between these layers (permutations of stacking sequence)

QUALITY OF EXPERIMENT-THEORY FIT
Visual (see comments)

2D UNIT CELLS (1 domain observed)

Cell	Ax (Å)	Ay (Å)	Bx (Å)	By (Å)	α (°)	Matrix	Pattern	Cell type
Bulk	3.160	0.000	1.580	2.737	60.0	(1.000, 0.000) (0.000, 1.000)	(1x1)	b: bulk lattice
Surface 1	3.160	0.000	1.580	2.737	60.0	(1.000, 0.000) (0.000, 1.000)	(1x1)	s1: commens. superlattice

3D COORDINATES

S1-Mo2-S3: top sandwich; S4-Mo5-S6 + S7-Mo8-S9: repeating set of bulk layers;
0.1Å error bars assumed for tabulation (see comments)

Dx/Dy in Å, or as a fraction of layer's unit cell vectors; atom 0 at the origin. Epir/subr are bulk repeat vectors.

No. of atoms: 9 Bulk z = 1.593 Å

Region	Chem el.	At. no.	Cell type	Site occ.	Rel. to	Dx ± εx	Dy ± εy	Dz ± εz	Dz/Bz(%) ± εz/Bz
epir		-2				f	f	Å	
subr		-1				0.000 Å	0.000 Å	12.290 Å	
intf	S	1	b	1.00	0	0.000 f	0.000 f	0.000 Å	0.0
intf	Mo	2	b	1.00	1	0.333 f	0.333 f	1.513 ± .100 Å	95.0 ± 6.3
intf	S	3	b	1.00	2	-0.333 f	-0.333 f	1.593 Å	100.0
subl	S	4	b	1.00	3	0.333 f	0.333 f	2.959 ± .100 Å	185.8 ± 6.3
subl	Mo	5	b	1.00	4	-0.333 f	-0.333 f	1.593 Å	100.0
subl	S	6	b	1.00	5	0.333 f	0.333 f	1.593 Å	100.0
subl	S	7	b	1.00	6	-0.333 f	-0.333 f	2.959 Å	185.8
subl	Mo	8	b	1.00	7	0.333 f	0.333 f	1.593 Å	100.0
subl	S	9	b	1.00	8	-0.333 f	-0.333 f	1.593 Å	100.0

MoS2(0001)-(1x1)
42.16.4b BOND DISTANCES AND ANGLES

Bond distances and angles are derived from coordinates

No. of distances/angles: 6

Interatomic dist. A-B (Å)	Atom A	Atom B	Atom C	Bond angle A-B-C (°)
2.370	S1	Mo2	S1(1,0)	83.6
2.370	S1	Mo2	Mo2(1,0)	131.8
2.370	S1	Mo2	S3(1,0)	135.2
2.422	Mo2	S3(1,0)	Mo2(1,0)	81.4
2.422	Mo2	S3	Mo2(-1,0)	81.4
2.422	S4	Mo5	S4(-1,0)	81.4

COMMON NAME : Na(0001)-(1x1)
CLASSIFICATION : 11.2
TECHNIQUE : LEED
AUTHORS : S.A. Lindgren, J. Paul, L. Wallden and P. Westrin
REFERENCE : J. Phys., C15, 6285 (1982)

ILLUSTRATION: 19

SURFACE TYPE
Substrate : Na
Crystal face: 0001
Temperature : 100 K
Bulk lattice: hcp
2D bulk symm: p3m1
2D surf symm: p3m1

Adsorbate:
Coverage :
Pattern : (1x1)
Matrix : (1.000, 0.000)
(0.000, 1.000)

STRUCTURE TYPE
Ideal bulk termination

SAMPLE PREPARATION (1 sample)
Treatment : Na evaporated onto Cu(111) in UHV
Crystallinity: tested by UPS of Cu 3d emission vs time
Anal. methods:
Contamination:

COMMENTS

DATA COLLECTION
Technique: LEED
Dataset : I-V spectra for: 6 (10)-type beams at normal incidence (10<E<110 eV); (00) beam at Θ=4° (5<E<110eV)

THEORY/DATA TREATMENT
Dynamical LEED (symm. KKR program): excited state pot. with Hedin-Lundqvist model for exch. and corr.; ΘD=140 K

STRUCTURES EXAMINED
Fcc and hcp only, assuming packing densities as for bcc Na, and c/a=1.63 for hcp structure

QUALITY OF EXPERIMENT-THEORY FIT
Visual

2D UNIT CELLS (1 domain observed)

Cell	Ax (Å)	Ay (Å)	Bx (Å)	By (Å)	α (°)	Matrix	Pattern	Cell type
Bulk	3.520	0.000	-1.760	3.048	120.0	(1.000, 0.000) (0.000, 1.000)	(1x1)	b: bulk lattice
Surface 1	3.520	0.000	-1.760	3.048	120.0	(1.000, 0.000) (0.000, 1.000)	(1,0/0,1) (1x1)	s1: commens. superlattice

3D COORDINATES

bulk termination

Dx/Dy in Å, or as a fraction of layer's unit cell vectors; atom 0 at the origin. Epir/subr are bulk repeat vectors.

No. of atoms: 3

Bulk z = 2.870 Å

Region	Chem el.	At. no.	Cell type	Site occ.	Rel. to	Dx ± εx		Dy ± εy		Dz ± εz		Dz/Bz(%) ± εz/Bz
epir		-2					f		f		Å	
subr		-1				0.000	Å	0.000	Å	5.740	Å	
intf	Na	1	b	1.00	0	0.000	f	0.000	f	0.000	Å	0.0
subl	Na	2	b	1.00	1	0.333	f	0.667	f	2.870	Å	100.0
subl	Na	3	b	1.00	2	-0.333	f	-0.667	f	2.870	Å	100.0

BOND DISTANCES AND ANGLES

Bond distances and angles are derived from coordinates

No. of distances/angles: 5

Interatomic dist. A-B (Å)	Atom A	Atom B	Atom C	Bond angle A-B-C (°)
3.520	Na1	Na1(1,1)	Na2	60.0
3.517	Na1	Na2	Na1(1,1)	60.1
3.517	Na1	Na2	Na2(1,1)	120.0
3.517	Na1	Na2	Na3(1,1)	146.4
3.517	Na1	Na2	Na3	109.4

COMMON NAME : Na(110)-(1x1)
CLASSIFICATION : 11.1
TECHNIQUE : LEED
AUTHORS : P.M. Echenique
REFERENCE : J. Phys., C9, 3193 (1976)

ILLUSTRATION: 11

SURFACE TYPE
Substrate : Na
Crystal face: 110
Temperature : 173 K
Bulk lattice: bcc
2D bulk symm: cmm
2D surf symm: cmm

Adsorbate:
Coverage :
Pattern : (1x1)
Matrix : (1.000, 0.000)
(-1.000, 2.000)

STRUCTURE TYPE
Bulk-like, but lattice constant slightly contracted in this film deposited on Ni(100);

SAMPLE PREPARATION (1 sample)
Treatment : film grown on clean Ni(100) from heated break-seal ampule
Crystallinity: growth monitored by LEED, EELS, and WFC
Anal. methods:
Contamination:

COMMENTS
Test of different potentials yielded best fit with expt. for Hartree-Fock core treatment and Hedin-Lundqvist-type conduction-electron treatment

DATA COLLECTION
Technique: LEED
Dataset : normal incidence I-V data for (00), (10), and (11) beams; E<=80 eV

THEORY/DATA TREATMENT
Dynamical LEED (layer doubling for E<40 eV, RFS for E>40eV): 8 phase shifts; ΘD=114 K (surf.), 128K (bulk)

STRUCTURES EXAMINED
Truncated bulk only

QUALITY OF EXPERIMENT-THEORY FIT
Visual

2D UNIT CELLS (1 domain observed)

Cell	Ax (Å)	Ay (Å)	Bx (Å)	By (Å)	α (°)	Matrix	Pattern	Cell type
Bulk	4.290	0.000	2.145	3.035	54.7	(1.000, 0.000) (0.000, 1.000)	(1x1)	b: bulk lattice
Surface 1	4.290	0.000	2.145	3.035	54.7	(1.000, 0.000) (0.000, 1.000)	(1x1)	s1: commens. superlattice

3D COORDINATES

ideal bulk termination, but contracted lattice constant

Dx/Dy in Å, or as a fraction of layer's unit cell vectors; atom 0 at the origin. Epir/subr are bulk repeat vectors.

No. of atoms: 2

Bulk z = 3.000 Å

Region	Chem el.	At. no.	Cell type	Site occ.	Rel. to	Dx ± εx	Dy ± εy	Dz ± εz	Dz/Bz(%) ± εz/Bz
epir		-2				f	f	Å	
subr		-1				2.145 Å	0.000 Å	3.000 Å	
intf	Na	1	b	1.00	0	0.000 f	0.000 f	0.000 Å	0.0
subl	Na	2	b	1.00	1	0.500 f	0.000 f	3.000 Å	100.0

BOND DISTANCES AND ANGLES

Bond distances and angles are derived from coordinates

No. of distances/angles: 7

Interatomic dist. A-B (Å)	Atom A	Atom B	Atom C	Bond angle A-B-C (°)
3.717	Na1	Na1(1,-1)	Na1(1,0)	70.5
3.717	Na1	Na1(1,-1)	Na2	54.5
3.717	Na1	Na1(1,-1)	Na2(0,-1)	70.4
3.688	Na1	Na2	Na1(1,0)	71.1
3.688	Na1	Na2	Na2(1,0)	125.6
3.688	Na1	Na2	Na2(0,1)	109.6
3.688	Na1	Na2	Na2(-1,0)	54.4

COMMON NAME : Na(110)-(1x1) ILLUSTRATION: 11
CLASSIFICATION : 11.1a
TECHNIQUE : LEED
AUTHORS : S. Andersson, J.B. Pendry, P.M. Echenique
REFERENCE : Surf. Sci., 65, 539 (1977)

SURFACE TYPE
Substrate : Na
Crystal face: 110
Temperature : 173 K
Bulk lattice: bcc
2D bulk symm: cmm
2D surf symm: cmm

Adsorbate:
Coverage :
Pattern : (1x1)
Matrix : (1.000, 0.000)
(-1.000, 2.000)

STRUCTURE TYPE
Bulk-like, but lattice constant slightly contracted in this film deposited on Ni(100);

SAMPLE PREPARATION (1 sample)
Treatment : film grown on clean Ni(100) from heated break-seal ampule
Crystallinity: growth monitored by LEED, EELS, and WFC
Anal. methods:
Contamination:

COMMENTS
Hedin-Lundqvist potential provided better agreement with experiment than Slater construction with α=1/3, especially at lower energies;
average interlayer spacing determined by Bragg's law for 00 beam

DATA COLLECTION
Technique: LEED
Dataset : normal incidence I-V data for (00), (10), and (11) beams; E<=80 eV

THEORY/DATA TREATMENT
Dynamical LEED (layer doubling for E<20 eV, RFS for E>20eV): up to 35 beams and 5 phase shifts; ΘD=107 K

STRUCTURES EXAMINED
Truncated bulk only

QUALITY OF EXPERIMENT-THEORY FIT
Visual

2D UNIT CELLS (1 domain observed)

Cell	Ax (Å)	Ay (Å)	Bx (Å)	By (Å)	α (°)	Matrix	Pattern	Cell type
Bulk	4.290	0.000	2.145	3.035	54.7	(1.000, 0.000) (0.000, 1.000)	(1x1)	b: bulk lattice
Surface 1	4.290	0.000	2.145	3.035	54.7	(1.000, 0.000) (0.000, 1.000)	(1x1)	s1: commens. superlattice

3D COORDINATES

ideal bulk termination, but contracted lattice constant

Dx/Dy in Å, or as a fraction of layer's unit cell vectors; atom 0 at the origin. Epir/subr are bulk repeat vectors.

No. of atoms: 2 Bulk z = 3.000 Å

Reg ion	Chem el.	At. no.	Cell type	Site occ.	Rel. to	Dx ± εx		Dy ± εy		Dz ± εz		Dz/Bz(%) ± εz/Bz
epir		-2					f		f		Å	
subr		-1				2.145	Å	0.000	Å	3.000	Å	
intf	Na	1	b	1.00	0	0.000	f	0.000	f	0.000	Å	0.0
subl	Na	2	b	1.00	1	0.500	f	0.000	f	3.000	Å	100.0

BOND DISTANCES AND ANGLES

Bond distances and angles are derived from coordinates

No. of distances/angles: 7

Interatomic dist. A-B (Å)	Atom A	Atom B	Atom C	Bond angle A-B-C (°)
3.717	Na1	Na1(1,-1)	Na1(1,0)	70.5
3.717	Na1	Na1(1,-1)	Na2	54.5
3.717	Na1	Na1(1,-1)	Na2(0,-1)	70.4
3.688	Na1	Na2	Na1(1,0)	71.1
3.688	Na1	Na2	Na2(1,0)	125.6
3.688	Na1	Na2	Na2(0,1)	109.6
3.688	Na1	Na2	Na2(-1,0)	54.4

COMMON NAME : Na2O(111)-(1x1) ILLUSTRATION: 152
CLASSIFICATION : 11.8.1
TECHNIQUE : LEED
AUTHORS : S. Andersson, J.B. Pendry and P.M. Echenique
REFERENCE : Surf. Sci., 65, 539 (1977)

SURFACE TYPE
Substrate : Na2O
Crystal face: 111
Temperature : 293 K
Bulk lattice: fluorite
2D bulk symm: p3m1
2D surf symm: p3m1

Adsorbate:
Coverage :
Pattern : (1x1)
Matrix : (1.000, 0.000)
(0.000, 1.000)

STRUCTURE TYPE
Bulk structure truncated at wide Na-Na spacing between two Na-O-Na sandwiches

SAMPLE PREPARATION (1 sample)
Treatment : oxidation of Na(110) surface grown on Ni(100)
Crystallinity: monitored growth by LEED and WFC
Anal. methods:
Contamination:

COMMENTS
Empirically determined form for Voi;
Na,O,Na termination dictated by electrostatic forces within the neutral (Na,O,Na) sandwiches

DATA COLLECTION
Technique: LEED
Dataset : normal incidence I-V data for (00), (10), (11) beams; E<=90 eV

THEORY/DATA TREATMENT
Dynamical LEED (layer doubling for E<20 eV, RFS for E>20eV): up to 31 beams and 5 phase shifts

STRUCTURES EXAMINED
3 terminations of the bulk stacking sequence: 1. Na,O,Na,Na,O,Na,...
2. O,Na,Na,O,Na,Na,... 3. Na,Na,O,Na,Na,O,...

QUALITY OF EXPERIMENT-THEORY FIT
Visual

2D UNIT CELLS (1 domain observed)

Cell	Ax (Å)	Ay (Å)	Bx (Å)	By (Å)	α (°)	Matrix	Pattern	Cell type
Bulk	3.920	0.000	-1.960	3.395	120.0	(1.000, 0.000) (0.000, 1.000)	(1x1)	b: bulk lattice
Surface 1	3.920	0.000	-1.960	3.395	120.0	(1.000, 0.000) (0.000, 1.000)	(1x1)	s1: commens. superlattice

3D COORDINATES

Na1-O2-Na3: repeating bulk trilayer; trilayers are spaced 1.6Å apart;
0.1Å error bars assumed for tabulation

Dx/Dy in Å, or as a fraction of layer's unit cell vectors; atom 0 at the origin. Epir/subr are bulk repeat vectors.

No. of atoms: 3 Bulk z = 3.200 Å

Reg ion	Chem el.	At. no.	Cell type	Site occ.	Rel. to	Dx ± εx		Dy ± εy		Dz ± εz		Dz/Bz(%) ± εz/Bz
epir		-2					f		f		Å	
subr		-1				0.000	Å	0.000	Å	3.200	Å	
subl	Na	1	b	1.00	0	0.000	f	0.000	f	0.000	Å	0.0
subl	O	2	b	1.00	1	0.333	f	0.667	f	0.800	Å	25.0
subl	Na	3	b	1.00	2	0.333	f	-0.333	f	0.800	Å	25.0

BOND DISTANCES AND ANGLES

Bond distances and angles are derived from coordinates

No. of distances/angles: 7

Interatomic dist. A-B (Å)	Atom A	Atom B	Atom C	Bond angle A-B-C (°)
3.920	Na1	Na1(1,1)	O2(1,1)	144.7
3.920	Na1	Na1(1,1)	O2	35.3
2.401	Na1	O2	Na1(1,1)	109.5
2.401	Na1	O2	Na3	70.5
2.772	Na1	Na3	Na1(1,1)	90.0
2.772	Na1	Na3	O2	54.7

Na2O(111)-(1x1)
11.8.1

Bond Distances and Angles - Continued

Interatomic dist. A-B (Å)	Atom A	Atom B	Atom C	Bond angle A-B-C (°)
2.401	O2	Na1	Na3	54.7

COMMON NAME : Nb(110)-(1x1)
CLASSIFICATION : 41.2
TECHNIQUE : HEIS
AUTHORS : Xu Mingde, C.N. Whang and R.J. Smith
REFERENCE : J. Vac. Sci. Technol., A8, 2501 (1990)

ILLUSTRATION: 11

SURFACE TYPE
Substrate : Nb
Crystal face: 110
Temperature : RT
Bulk lattice: bcc
2D bulk symm: cmm
2D surf symm: cmm

Adsorbate:
Coverage :
Pattern : (1x1)
Matrix : (1.000, 0.000)
(0.000, 1.000)

STRUCTURE TYPE
Bulk terminated structure with no detectable relaxations

SAMPLE PREPARATION (1 sample)
Treatment : sample cleaned by heating up to 2300C
Crystallinity:
Anal. methods: LEED and ion channeling to check quality of surface
Contamination:

COMMENTS
Ion beam incident along [-1-10] and [-1-11] directions; scattered ions collected at grazing exit angle (10°); reduced bulk vibrational amplitudes possibly related to correlated nature of vibrations

DATA COLLECTION
Technique: HEIS
Dataset : ion yield in Nb surf.peak versus ion energy (0.5-2M eV) for [-1-10] and [-1-11] incident directions

THEORY/DATA TREATMENT
Computer simulation of expected ion yield

STRUCTURES EXAMINED
Adjustable parameters: 1) surface atom vibration amplitude, 2) bulk atom vibr. ampl., 3) relaxation of first layer; best fit with no enhancement in surface vibr., no relaxation, reduced bulk vibr. (wrt 275K ΘD)

2D UNIT CELLS (1 domain observed)

Cell	Ax (Å)	Ay (Å)	Bx (Å)	By (Å)	α (°)	Matrix	Pattern	Cell type
Bulk	2.864	0.000	.955	2.700	70.5	(1.000, 0.000) (0.000, 1.000)	(1x1)	b: bulk lattice
Surface 1	2.864	0.000	.955	2.700	70.5	(1.000, 0.000) (0.000, 1.000)	(1x1)	s1: commens. superlattice

3D COORDINATES

0.1Å error bars assumed for tabulation

Dx/Dy in Å, or as a fraction of layer's unit cell vectors; atom 0 at the origin. Epir/subr are bulk repeat vectors.

No. of atoms: 2

Bulk z = 2.338 Å

Region	Chem el.	At. no.	Cell type	Site occ.	Rel. to	Dx ± εx	Dy ± εy	Dz ± εz	Dz/Bz(%) ± εz/Bz
epir		-2				f	f	Å	
subr		-1				1.909 Å	1.350 Å	2.338 Å	
intf	Nb	1	b	1.00	0	0.000 ± .023 f	0.000 ± .037 f	0.000 ± .100 Å	0.0 ± 4.3
subl	Nb	2	b	1.00	1	0.500 f	0.500 f	2.338 Å	100.0

BOND DISTANCES AND ANGLES

Bond distances and angles are derived from coordinates

No. of distances/angles: 4

Interatomic dist. A-B (Å)	Atom A	Atom B	Atom C	Bond angle A-B-C (°)
2.864	Nb1	Nb1(1,0)	Nb1(1,1)	109.5
2.864	Nb1	Nb1(1,0)	Nb1(1,-1)	70.5
3.307	Nb1	Nb2	Nb1(1,1)	90.0
3.307	Nb1	Nb2	Nb1(1,0)	54.7

COMMON NAME : NbSe2(0001)-(1x1)
CLASSIFICATION : 42.16.4a
TECHNIQUE : LEED
AUTHORS : B.J. Mrstik, R. Kaplan, T.L. Reinecke, M.A. Van Hove and S.Y. Tong
REFERENCE : Phys. Rev., B15, 897 (1977)

ILLUSTRATION: 160

SURFACE TYPE
Substrate : NbSe2
Crystal face: 0001
Temperature : 95 K
Bulk lattice: 2H-NbSe2
2D bulk symm: p3m1
2D surf symm: p3m1

Adsorbate:
Coverage :
Pattern : (1x1)
Matrix : (1.000, 0.000)
(0.000, 1.000)

STRUCTURE TYPE
Bulk termination with complete Se-Nb-Se sandwich and no layer spacing relaxations; bulk layer stacking maintained as CBC ABA CBC ABA ...

SAMPLE PREPARATION (1 sample)
Treatment : cleaved single crystal grown by I-vapor transport method
Crystallinity:
Anal. methods:
Contamination: AES: <3% C

COMMENTS
Later R-factor comparison (M.A. Van Hove et al, Surf. Sci. 64, 85 (1977)) confirms this structure, with top spacings of 1.66±0.02, 1.68, 2.89±0.02Å, resp., and R2=0.15

DATA COLLECTION
Technique: LEED
Dataset : I-V curves: 10,01,11,20,02 beams at Θ=0°; 10<E<200 eV; cumulative energy range: 740eV (non-degenerate)

THEORY/DATA TREATMENT
Dynamical LEED (RFS): up to 55 beams, 8 phase shifts, Mattheiss band structure pots;Vor=-5.0 eV,Voi=-5.0eV,ΘD=280 K

STRUCTURES EXAMINED
Various spacings between layers 1 and 2, and between 3 and 4 also the registry between these layers (permutations of stacking sequence)

QUALITY OF EXPERIMENT-THEORY FIT
Visual (see comments)

2D UNIT CELLS (1 domain observed)

Cell	Ax (Å)	Ay (Å)	Bx (Å)	By (Å)	α (°)	Matrix	Pattern	Cell type
Bulk	3.449	0.000	1.725	2.987	60.0	(1.000, 0.000) (0.000, 1.000)	(1x1)	b: bulk lattice
Surface 1	3.449	0.000	1.725	2.987	60.0	(1.000, 0.000) (0.000, 1.000)	(1x1)	s1: commens. superlattice

3D COORDINATES

Se1-Nb2-Se3: top sandwich; Se4-Nb5-Se6 + Se7-Nb8-Se9: repeating set of bulk layers;
0.1Å error bars assumed for tabulation (see comments)

Dx/Dy in Å, or as a fraction of layer's unit cell vectors; atom 0 at the origin. Epir/subr are bulk repeat vectors.

No. of atoms: 9 Bulk z = 1.680 Å

Region	Chem el.	At. no.	Cell type	Site occ.	Rel. to	Dx ± εx		Dy ± εy		Dz ± εz		Dz/Bz(%) ± εz/Bz
epir		-2					f		f		Å	
subr		-1				0.000	Å	0.000	Å	12.540	Å	
intf	Se	1	b	1.00	0	0.000	f	0.000	f	0.000	Å	0.0
intf	Nb	2	b	1.00	1	0.333	f	0.333	f	1.680 ± .100	Å	100.0 ± 6.0
intf	Se	3	b	1.00	2	-0.333	f	-0.333	f	1.680	Å	100.0
subl	Se	4	b	1.00	3	0.667	f	0.667	f	2.910 ± .100	Å	173.2 ± 6.0
subl	Nb	5	b	1.00	4	-0.333	f	-0.333	f	1.680	Å	100.0
subl	Se	6	b	1.00	5	0.333	f	0.333	f	1.680	Å	100.0
subl	Se	7	b	1.00	6	-0.667	f	-0.667	f	2.910	Å	173.2
subl	Nb	8	b	1.00	7	0.333	f	0.333	f	1.680	Å	100.0
subl	Se	9	b	1.00	8	-0.333	f	-0.333	f	1.680	Å	100.0

NbSe2(0001)-(1x1)
42.16.4a

BOND DISTANCES AND ANGLES

Bond distances and angles are derived from coordinates

No. of distances/angles: 4

Interatomic dist. A-B (Å)	Atom A	Atom B	Atom C	Bond angle A-B-C (°)
2.605	Se1	Nb2	Se1(1,0)	82.9
2.605	Se1	Nb2	Nb2(1,0)	131.5
2.605	Se1	Nb2	Se3(1,0)	135.1
2.605	Nb2	Se3	Nb2(0,-1)	82.9

COMMON NAME : Ni(100)-(1x1)
CLASSIFICATION : 28.16a
TECHNIQUE : MEIS
AUTHORS : J.W.M. Frenken, J.F. van der Veen and G. Allan
REFERENCE : Phys. Rev. Lett., 51, 1876 (1983)

ILLUSTRATION: 2

SURFACE TYPE
Substrate : Ni
Crystal face: 100
Temperature : 370 K
Bulk lattice: fcc
2D bulk symm: p4m
2D surf symm: p4m

Adsorbate:
Coverage :
Pattern : (1x1)
Matrix : (1.000, 0.000)
(0.000, 1.000)

STRUCTURE TYPE
Bulk termination with top spacing contraction

SAMPLE PREPARATION (1 sample)
Treatment : 'standard procedures'
Crystallinity:
Anal. methods:
Contamination:

COMMENTS
Using tight-binding model, authors deduce strengthening of Ni interlayer force constants relative to bulk values, accompanying layer contraction

DATA COLLECTION
Technique: MEIS; Rutherford back scattering
Dataset :

THEORY/DATA TREATMENT
Geometric interpretation of blocking curves

STRUCTURES EXAMINED
Expansion and contraction of top interlayer spacing

QUALITY OF EXPERIMENT-THEORY FIT
Visual

2D UNIT CELLS (1 domain observed)

Cell	Ax (Å)	Ay (Å)	Bx (Å)	By (Å)	α (°)	Matrix	Pattern	Cell type
Bulk	2.490	0.000	0.000	2.490	90.0	(1.000, 0.000) (0.000, 1.000)	(1x1)	b: bulk lattice
Surface 1	2.490	0.000	0.000	2.490	90.0	(1.000, 0.000) (0.000, 1.000)	(1x1)	s1: commens. superlattice

3D COORDINATES

Dx/Dy in Å, or as a fraction of layer's unit cell vectors; atom 0 at the origin. Epir/subr are bulk repeat vectors.

No. of atoms: 3

Bulk z = 1.760 Å

Region	Chem el.	At. no.	Cell type	Site occ.	Rel. to	Dx ± εx		Dy ± εy		Dz ± εz		Dz/Bz(%) ± εz/Bz
epir		-2					f		f		Å	
subr		-1				-1.245	Å	-1.245	Å	1.760	Å	
intf	Ni	1	b	1.00	0	0.000	f	0.000	f	0.000	Å	0.0
intf	Ni	2	b	1.00	1	0.500	f	0.500	f	1.604 ± .008	Å	91.1 ± .5
subl	Ni	3	b	1.00	2	-0.500	f	-0.500	f	1.760	Å	100.0

BOND DISTANCES AND ANGLES

Bond distances and angles are derived from coordinates

No. of distances/angles: 5

Interatomic dist. A-B (Å)	Atom A	Atom B	Atom C	Bond angle A-B-C (°)
2.382	Ni1	Ni2	Ni1(1,0)	63.0
2.382	Ni1	Ni2	Ni2(1,0)	121.5
2.382	Ni1	Ni2	Ni3	87.3
2.490	Ni2	Ni2(1,0)		
2.490	Ni2	Ni3		

COMMON NAME : Ni(100)-(1x1)
CLASSIFICATION : 28.29a
TECHNIQUE : LEED
AUTHORS : W. Oed, H. Lindner, U. Starke, K. Heinz, K. Mueller and J.B. Pendry
REFERENCE : Surf. Sci., 224, 179 (1989)

ILLUSTRATION: 2

SURFACE TYPE
Substrate : Ni
Crystal face: 100
Temperature : 120 K
Bulk lattice: fcc
2D bulk symm: p4m
2D surf symm: p4m

Adsorbate:
Coverage :
Pattern : (1x1)
Matrix : (1.000, 0.000)
(0.000, 1.000)

STRUCTURE TYPE
1% contraction of top interlayer spacing

SAMPLE PREPARATION (1 sample)
Treatment : cycles of sputtering and annealing in oxygen
Crystallinity: sharp LEED pattern
Anal. methods: AES: impurities below detection level
Contamination:

COMMENTS

DATA COLLECTION
Technique: LEED; video LEED
Dataset : IV curves at normal incidence for 7 integer beams; E<=500 eV; cumul. E range 1700eV

THEORY/DATA TREATMENT
Dynamical LEED: 11 phase shifts;
Vor=-5 eV (E-dependence tested)

QUALITY OF EXPERIMENT-THEORY FIT
RPE=0.22

2D UNIT CELLS (1 domain observed)

Cell	Ax (Å)	Ay (Å)	Bx (Å)	By (Å)	α (°)	Matrix	Pattern	Cell type
Bulk	2.492	0.000	0.000	2.492	90.0	(1.000, 0.000) (0.000, 1.000)	(1x1)	b: bulk lattice
Surface 1	2.492	0.000	0.000	2.492	90.0	(1.000, 0.000) (0.000, 1.000)	(1x1)	s1: commens. superlattice

3D COORDINATES

Dx/Dy in Å, or as a fraction of layer's unit cell vectors; atom 0 at the origin. Epir/subr are bulk repeat vectors.

No. of atoms: 3

Bulk z = 1.762 Å

Reg ion	Chem el.	At. no.	Cell type	Site occ.	Rel. to	Dx ± εx		Dy ± εy		Dz ± εz		Dz/Bz(%) ± εz/Bz
epir		-2					f		f		Å	
subr		-1				1.246	Å	1.246	Å	1.762	Å	
intf	Ni	1	b	1.00	0	0.000	f	0.000	f	0.000 ± .017	Å	0.0 ± 2.0
intf	Ni	2	b	1.00	1	-0.500	f	-0.500	f	1.742 ± .017	Å	98.9 ± 2.0
subl	Ni	3	b	1.00	1	0.000	f	0.000	f	3.505	Å	198.9

COMMON NAME : Ni(100)-(1x1)
CLASSIFICATION : 28.4b
TECHNIQUE : LEED
AUTHORS : J.E. Demuth, P.M. Marcus and D.W. Jepsen
REFERENCE : Phys. Rev., B11, 1460 (1975)

ILLUSTRATION: 2

SURFACE TYPE
Substrate : Ni
Crystal face: 100
Temperature : RT
Bulk lattice: fcc
2D bulk symm: p4m
2D surf symm: p4m

Adsorbate:
Coverage :
Pattern : (1x1)
Matrix : (1.000, 0.000)
(0.000, 1.000)

STRUCTURE TYPE
Bulk termination with slight top spacing expansion

SAMPLE PREPARATION (1 sample)
Treatment : see J.E. Demuth and T.N. Rhodin, Surf. Sci. 42, 261 (1974)
Crystallinity:
Anal. methods:
Contamination:

COMMENTS

DATA COLLECTION
Technique: LEED
Dataset : I-V curves include 00 beam at Θ=4, 10, 20° with φ=0,45° and 01, 11 beams at Θ=0°; E range 10-220 eV

THEORY/DATA TREATMENT
Dynamical LEED (layer KKR): 8 phase shifts (Wakoh pot), Voi α E**1/3 (different functions tested); ΘD=335 K

STRUCTURES EXAMINED
Variation of top layer spacing

QUALITY OF EXPERIMENT-THEORY FIT
Visual

2D UNIT CELLS (1 domain observed)

Cell	Ax (Å)	Ay (Å)	Bx (Å)	By (Å)	α (°)	Matrix	Pattern	Cell type
Bulk	2.490	0.000	0.000	2.490	90.0	(1.000, 0.000) (0.000, 1.000)	(1x1)	b: bulk lattice
Surface 1	2.490	0.000	0.000	2.490	90.0	(1.000, 0.000) (0.000, 1.000)	(1x1)	s1: commens. superlattice

3D COORDINATES

Dx/Dy in Å, or as a fraction of layer's unit cell vectors; atom 0 at the origin. Epir/subr are bulk repeat vectors.

No. of atoms: 3

Bulk z = 1.760 Å

Region	Chem el.	At. no.	Cell type	Site occ.	Rel. to	Dx ± εx		Dy ± εy		Dz ± εz		Dz/Bz(%) ± εz/Bz
epir		-2					f		f		Å	
subr		-1				-1.245	Å	-1.245	Å	1.760	Å	
intf	Ni	1	b	1.00	0	0.000	f	0.000	f	0.000	Å	0.0
intf	Ni	2	b	1.00	1	0.500	f	0.500	f	1.780 ± .020	Å	101.1 ± 1.1
subl	Ni	3	b	1.00	2	-0.500	f	-0.500	f	1.760	Å	100.0

BOND DISTANCES AND ANGLES

Bond distances and angles are derived from coordinates

No. of distances/angles: 2

Interatomic dist. A-B (Å)	Atom A	Atom B	Atom C	Bond angle A-B-C (°)
2.504	Ni1	Ni2	Ni1(1,0)	59.6
2.490	Ni2	Ni3	Ni3(1,0)	60.0

COMMON NAME : Ni(110)-(1x1) ILLUSTRATION: 4
CLASSIFICATION : 28.11a
TECHNIQUE : MEIS
AUTHORS : J.F. van der Veen, R.M. Tromp, R.G. Smeenk and F.W. Saris
REFERENCE : Surf. Sci., 82, 468 (1979)

SURFACE TYPE
Substrate : Ni
Crystal face: 110
Temperature : RT
Bulk lattice: fcc
2D bulk symm: pmm
2D surf symm: pmm

Adsorbate:
Coverage :
Pattern : (1x1)
Matrix : (1.000, 0.000)
(0.000, 1.000)

STRUCTURE TYPE
Bulk termination with top spacing contraction

SAMPLE PREPARATION (1 sample)
Treatment : Ar+ sputtering at 570 K followed by 4 min annealing at 1120K
Crystallinity:
Anal. methods:
Contamination: back scattering: S<.02, O<.05, C<.1ML

COMMENTS

DATA COLLECTION
Technique: MEIS
Dataset : ion beam scattering as a function of scattering angle for 100 k eV protons

THEORY/DATA TREATMENT
Medium energy ion beam scattering (blocking cones); ΘD=390 K

STRUCTURES EXAMINED
Relaxation of top layer spacing

QUALITY OF EXPERIMENT-THEORY FIT
Visual

2D UNIT CELLS (1 domain observed)

Cell	Ax (Å)	Ay (Å)	Bx (Å)	By (Å)	α (°)	Matrix	Pattern	Cell type
Bulk	3.520	0.000	0.000	2.490	90.0	(1.000, 0.000) (0.000, 1.000)	(1x1)	b: bulk lattice
Surface 1	3.520	0.000	0.000	2.490	90.0	(1.000, 0.000) (0.000, 1.000)	(1x1)	s1: commens. superlattice

3D COORDINATES

Dx/Dy in Å, or as a fraction of layer's unit cell vectors; atom 0 at the origin. Epir/subr are bulk repeat vectors.

No. of atoms: 4 Bulk z = 1.245 Å

Reg ion	Chem el.	At. no.	Cell type	Site occ.	Rel. to	Dx ± εx		Dy ± εy		Dz ± εz		Dz/Bz(%) ± εz/Bz
epir		-2					f		f		Å	
subr		-1				1.760	Å	1.245	Å	1.245	Å	
intf	Ni	1	b	1.00	0	0.000	f	0.000	f	0.000	Å	0.0
intf	Ni	2	b	1.00	1	0.500	f	0.500	f	1.195 ± .010	Å	96.0 ± .8
intf	Ni	3	b	1.00	2	-0.500	f	-0.500	f	1.245	Å	100.0
subl	Ni	4	b	1.00	3	0.500	f	0.500	f	1.245	Å	100.0

BOND DISTANCES AND ANGLES

Bond distances and angles are derived from coordinates

No. of distances/angles: 6

Interatomic dist. A-B (Å)	Atom A	Atom B	Atom C	Bond angle A-B-C (°)
2.490	Ni1	Ni1(0,1)	Ni2	59.7
2.465	Ni1	Ni2	Ni3	59.0
2.465	Ni1	Ni2	Ni4	119.0
2.440	Ni1	Ni3	Ni4	120.0
2.490	Ni2	Ni3	Ni4	60.0
2.490	Ni3	Ni4	Ni2	60.0

COMMON NAME : Ni(110)-(1x1)
CLASSIFICATION : 28.15
TECHNIQUE : HEIS
AUTHORS : R. Feidenhans'l, J.E. Sorensen and I. Stensgaard
REFERENCE : Surf. Sci., 134, 329 (1983)

ILLUSTRATION: 4

SURFACE TYPE
Substrate : Ni
Crystal face: 110
Temperature : 300 K
Bulk lattice: fcc
2D bulk symm: pmm
2D surf symm: pmm

Adsorbate:
Coverage :
Pattern : (1x1)
Matrix : (1.000, 0.000)
(0.000, 1.000)

SAMPLE PREPARATION (1 sample)
Treatment : many Ar+ sputtering and annealing cycles
Crystallinity: LEED: well ordered surface
Anal. methods:
Contamination: monitored by AES

DATA COLLECTION
Technique: HEIS; surface peak yield of 300 keV He+
Dataset : surface peak yield along (101), (100) axes; angular scans for each axis recorded 9 times, then averaged;

STRUCTURE TYPE
Bulk termination with multilayer relaxations

COMMENTS
R-factor topographs plotted for 1st-2nd and 2nd-3rd layer spacings (R-factor defined: Phys. Rev. Lett. 49,669 (1982)); ΘD optimised in range 300-375 K;
theory/experiment fit sensitive to vibrational parameters

THEORY/DATA TREATMENT
Computer simulation of surface peak yield (see Stensgaard et al, Surf. Sci. 77, 513 (1978)); ΘD (surf) = 325 K

STRUCTURES EXAMINED
First and second Ni interlayer spacings varied independently in the range -0.02 to 0.08Å

QUALITY OF EXPERIMENT-THEORY FIT
See comments

2D UNIT CELLS (1 domain observed)

Cell	Ax (Å)	Ay (Å)	Bx (Å)	By (Å)	α (°)	Matrix	Pattern	Cell type
Bulk	3.520	0.000	0.000	2.490	90.0	(1.000, 0.000) (0.000, 1.000)	(1x1)	b: bulk lattice
Surface 1	3.520	0.000	0.000	2.490	90.0	(1.000, 0.000) (0.000, 1.000)	(1x1)	s1: commens. superlattice

3D COORDINATES

Dx/Dy in Å, or as a fraction of layer's unit cell vectors; atom 0 at the origin. Epir/subr are bulk repeat vectors.

No. of atoms: 4

Bulk z = 1.245 Å

Region	Chem el.	At. no.	Cell type	Site occ.	Rel. to	Dx ± εx		Dy ± εy		Dz ± εz		Dz/Bz(%) ± εz/Bz
epir		-2					f		f		Å	
subr		-1				1.760	Å	1.245	Å	1.245	Å	
intf	Ni	1	b	1.00	0	0.000	f	0.000	f	0.000	Å	0.0
intf	Ni	2	b	1.00	1	0.500	f	0.500	f	1.180 ± .020	Å	94.8 ± 1.6
intf	Ni	3	b	1.00	2	-0.500	f	-0.500	f	1.270 ± .020	Å	102.0 ± 1.6
subl	Ni	4	b	1.00	3	0.500	f	0.500	f	1.245	Å	100.0

BOND DISTANCES AND ANGLES

Bond distances and angles are derived from coordinates

No. of distances/angles: 6

Interatomic dist. A-B (Å)	Atom A	Atom B	Atom C	Bond angle A-B-C (°)
2.490	Ni1	Ni1(0,1)	Ni2	59.6
2.458	Ni1	Ni2	Ni3	59.2
2.458	Ni1	Ni2	Ni4	118.7
2.450	Ni1	Ni3	Ni4	120.0
2.502	Ni2	Ni3	Ni4	60.5
2.490	Ni3	Ni4	Ni4(0,1)	120.0

COMMON NAME : Ni(110)-(1x1)
CLASSIFICATION : 28.17
TECHNIQUE : HEIS
AUTHORS : E. Tornqvist, E.D. Adams, M. Copel, T. Gustafsson and W.R. Graham
REFERENCE : J. Vac. Sci. Technol., A2, 939 (1984)

ILLUSTRATION: 4

SURFACE TYPE
Substrate : Ni
Crystal face: 110
Temperature : 300 K
Bulk lattice: fcc
2D bulk symm: pmm
2D surf symm: pmm

Adsorbate:
Coverage :
Pattern : (1x1)
Matrix : (1.000, 0.000)
(0.000, 1.000)

STRUCTURE TYPE
Bulk termination with top spacing contraction

SAMPLE PREPARATION (1 sample)
Treatment : cycles of ion bombardment and annealing at 2E-10 torr
Crystallinity: sharp (1x1) LEED pattern
Anal. methods:
Contamination:

COMMENTS
Vibrational amplitudes assumed the same as for Ni(111): see Surf. Sci. 114, 331 (1982);
authors speculate that discrepancy with LEED analysis may be due to deeper layer relaxations; analysis of CO-covered Ni(110)-(2x1) yields unrelaxed substrate

DATA COLLECTION
Technique: HEIS
Dataset : 100 k eV proton beam fired along [10-1], data collected over range of scattered angles

THEORY/DATA TREATMENT
Monte Carlo analysis of proton scattering intensities as functions of angle, using a screened Coulomb potential

STRUCTURES EXAMINED
Varied 1st layer spacing only

QUALITY OF EXPERIMENT-THEORY FIT
Visual

2D UNIT CELLS (1 domain observed)

Cell	Ax (Å)	Ay (Å)	Bx (Å)	By (Å)	α (°)	Matrix	Pattern	Cell type
Bulk	3.520	0.000	0.000	2.490	90.0	(1.000, 0.000) (0.000, 1.000)	(1x1)	b: bulk lattice
Surface 1	3.520	0.000	0.000	2.490	90.0	(1.000, 0.000) (0.000, 1.000)	(1x1)	s1: commens. superlattice

3D COORDINATES

Dx/Dy in Å, or as a fraction of layer's unit cell vectors; atom 0 at the origin. Epir/subr are bulk repeat vectors.

No. of atoms: 4

Bulk z = 1.245 Å

Region	Chem el.	At. no.	Cell type	Site occ.	Rel. to	Dx ± εx		Dy ± εy		Dz ± εz		Dz/Bz(%) ± εz/Bz
epir		-2					f		f		Å	
subr		-1				1.760	Å	1.245	Å	1.245	Å	
intf	Ni	1	b	1.00	0	0.000	f	0.000	f	0.000	Å	0.0
intf	Ni	2	b	1.00	1	0.500	f	0.500	f	1.194 ± .010	Å	95.9 ± .8
intf	Ni	3	b	1.00	2	-0.500	f	-0.500	f	1.245	Å	100.0
subl	Ni	4	b	1.00	3	0.500	f	0.500	f	1.245	Å	100.0

BOND DISTANCES AND ANGLES

Bond distances and angles are derived from coordinates

No. of distances/angles: 6

Interatomic dist. A-B (Å)	Atom A	Atom B	Atom C	Bond angle A-B-C (°)
2.490	Ni1	Ni1(0,1)	Ni2	59.7
2.464	Ni1	Ni2	Ni3	59.0
2.464	Ni1	Ni2	Ni4	119.0
2.439	Ni1	Ni3	Ni4	120.0

Ni(110)-(1x1)
28.17

Bond Distances and Angles - Continued

Interatomic dist. A-B (Å)	Atom A	Atom B	Atom C	Bond angle A-B-C (°)
2.490	Ni2	Ni3	Ni4	60.0
2.490	Ni3	Ni4	Ni4(0,1)	120.0

COMMON NAME : Ni(110)-(1x1)
CLASSIFICATION : 28.18
TECHNIQUE : LEED
AUTHORS : Y. Gauthier, R. Baudoing, Y. Joly, C. Gaubert and J. Rundgren
REFERENCE : J. Phys., C17, 4547 (1984)

ILLUSTRATION: 4

SURFACE TYPE
Substrate : Ni
Crystal face: 110
Temperature : RT*
Bulk lattice: fcc
2D bulk symm: pmm
2D surf symm: pmm

Adsorbate:
Coverage :
Pattern : (1x1)
Matrix : (1.000, 0.000)
(0.000, 1.000)

STRUCTURE TYPE
Bulk termination with multilayer relaxation

SAMPLE PREPARATION (1 sample)
Treatment :
Crystallinity:
Anal. methods:
Contamination:

COMMENTS
Analysis was based on metric distances (see Lindgren et al, Phys. Rev. B29, 576 (1984)); RPE and RZJ also calculated: RPE favors 1st layer contraction of 7%, RZJ favors 4% and metric distances favor 8%

DATA COLLECTION
Technique: LEED
Dataset : 35 I-V spectra for 7 different diffraction geometries for energy range 30<E<200 eV

THEORY/DATA TREATMENT
Dynamical LEED (layer doubling): Voi energy dependent (Gauthier, J. Phys. C15, 3231 (1982)); ΘD=335 K

STRUCTURES EXAMINED
+3% to -12% relaxation of 1st layer spacing; -8% to +8% relaxation of 2nd layer spacing; -3eV to +2eV variation of Vor

QUALITY OF EXPERIMENT-THEORY FIT
See comments

2D UNIT CELLS (1 domain observed)

Cell	Ax (Å)	Ay (Å)	Bx (Å)	By (Å)	α (°)	Matrix	Pattern	Cell type
Bulk	3.520	0.000	0.000	2.490	90.0	(1.000, 0.000) (0.000, 1.000)	(1x1)	b: bulk lattice
Surface 1	3.520	0.000	0.000	2.490	90.0	(1.000, 0.000) (0.000, 1.000)	(1x1)	s1: commens. superlattice

3D COORDINATES

Dx/Dy in Å, or as a fraction of layer's unit cell vectors; atom 0 at the origin. Epir/subr are bulk repeat vectors.

No. of atoms: 4

Bulk z = 1.245 Å

Region	Chem el.	At. no.	Cell type	Site occ.	Rel. to	Dx ± εx		Dy ± εy		Dz ± εz		Dz/Bz(%) ± εz/Bz
epir		-2					f		f		Å	
subr		-1				1.760	Å	1.245	Å	1.245	Å	
intf	Ni	1	b	1.00	0	0.000	f	0.000	f	0.000	Å	0.0
intf	Ni	2	b	1.00	1	0.500	f	0.500	f	1.140 ± .010	Å	91.6 ± .8
intf	Ni	3	b	1.00	2	-0.500	f	-0.500	f	1.284 ± .013	Å	103.1 ± 1.0
subl	Ni	4	b	1.00	3	0.500	f	0.500	f	1.245	Å	100.0

BOND DISTANCES AND ANGLES

Bond distances and angles are derived from coordinates

No. of distances/angles: 6

Interatomic dist. A-B (Å)	Atom A	Atom B	Atom C	Bond angle A-B-C (°)
2.490	Ni1	Ni1(0,1)	Ni2	59.3
2.439	Ni1	Ni2	Ni3	58.7
2.439	Ni1	Ni2	Ni4	117.9
2.424	Ni1	Ni3	Ni4	120.0
2.509	Ni2	Ni3	Ni4	60.8
2.490	Ni3	Ni4	Ni4(0,1)	120.0

COMMON NAME : Ni(110)-(1x1)
CLASSIFICATION : 28.22
TECHNIQUE : LEED
AUTHORS : D.L. Adams, L.E. Petersen and C.S. Sorensen
REFERENCE : J. Phys., C18, 1753 (1985)

ILLUSTRATION: 4

SURFACE TYPE
Substrate : Ni
Crystal face: 110
Temperature : RT
Bulk lattice: fcc
2D bulk symm: pmm
2D surf symm: pmm

Adsorbate:
Coverage :
Pattern : (1x1)
Matrix : (1.000, 0.000)
(0.000, 1.000)

STRUCTURE TYPE
Bulk termination with multilayer relaxation

SAMPLE PREPARATION (2 sample)
Treatment : Ar+ sputtering, annealing at 800-1100 K, flashing to 900K
Crystallinity:
Anal. methods:
Contamination: AES: <0.1ML C, <0.02ML S

COMMENTS
Optimization of Vor, Voi and ΘD carried out

DATA COLLECTION
Technique: LEED
Dataset : I-V curves at normal incidence: 9 inequivalent beams; E range 60-360 eV

THEORY/DATA TREATMENT
Dynamical LEED (layer doubling): 10 phase shifts (Moruzzi potential); Voi=-3.3 eV; ΘD=514 K

STRUCTURES EXAMINED
Relaxation of first 3 interlayer spacings

QUALITY OF EXPERIMENT-THEORY FIT
R2=0.02

2D UNIT CELLS (1 domain observed)

Cell	Ax (Å)	Ay (Å)	Bx (Å)	By (Å)	α (°)	Matrix	Pattern	Cell type
Bulk	3.520	0.000	0.000	2.490	90.0	(1.000, 0.000) (0.000, 1.000)	(1x1)	b: bulk lattice
Surface 1	3.520	0.000	0.000	2.490	90.0	(1.000, 0.000) (0.000, 1.000)	(1x1)	s1: commens. superlattice

3D COORDINATES

Dx/Dy in Å, or as a fraction of layer's unit cell vectors; atom 0 at the origin. Epir/subr are bulk repeat vectors.

No. of atoms: 5

Bulk z = 1.245 Å

Region	Chem el.	At. no.	Cell type	Site occ.	Rel. to	Dx ± εx		Dy ± εy		Dz ± εz		Dz/Bz(%) ± εz/Bz
epir		-2					f		f		Å	
subr		-1				-1.760	Å	-1.245	Å	1.245	Å	
intf	Ni	1	b	1.00	0	0.000	f	0.000	f	0.000	Å	0.0
intf	Ni	2	b	1.00	1	0.500	f	0.500	f	1.138 ± .006	Å	91.4 ± .5
intf	Ni	3	b	1.00	2	-0.500	f	-0.500	f	1.284 ± .007	Å	103.1 ± .6
intf	Ni	4	b	1.00	3	0.500	f	0.500	f	1.240 ± .009	Å	99.6 ± .7
subl	Ni	5	b	1.00	4	-0.500	f	-0.500	f	1.245	Å	100.0

BOND DISTANCES AND ANGLES

Bond distances and angles are derived from coordinates

No. of distances/angles: 6

Interatomic dist. A-B (Å)	Atom A	Atom B	Atom C	Bond angle A-B-C (°)
2.490	Ni1	Ni1(0,1)	Ni2	59.3
2.438	Ni1	Ni2	Ni3	58.6
2.438	Ni1	Ni2	Ni4	117.8
2.509	Ni2	Ni3	Ni4	60.7
2.509	Ni2	Ni3	Ni5	120.8
2.524	Ni2	Ni4	Ni5	120.0

ILLUSTRATION: 4

COMMON NAME : Ni(110)-(1x1)
CLASSIFICATION : 28.23
TECHNIQUE : LEED
AUTHORS : M.L. Xu and S.Y. Tong
REFERENCE : Phys. Rev., B31, 6332 (1985)

SURFACE TYPE
Substrate : Ni
Crystal face: 110
Temperature : RT
Bulk lattice: fcc
2D bulk symm: pmm
2D surf symm: pmm

Adsorbate:
Coverage :
Pattern : (1x1)
Matrix : (1.000, 0.000)
(0.000, 1.000)

STRUCTURE TYPE
Bulk termination with multilayer relaxation

SAMPLE PREPARATION (1 sample)
Treatment : Ar+ sputtering for 10 min, annealing at 800C for 1 min
Crystallinity:
Anal. methods:
Contamination: AES: < 1%ML of S, C, O

COMMENTS

DATA COLLECTION
Technique: LEED
Dataset : data by J.E. Demuth and T.N. Rhodin (Surf. Sci. 42, 261 (1974)): 3, 6 beams at normal-, off-normal inc.; 25<E<230 eV

THEORY/DATA TREATMENT
Dynamical LEED: 8 phase shifts; Voi α E**1/3

STRUCTURES EXAMINED
Variation of top two interlayer spacings

QUALITY OF EXPERIMENT-THEORY FIT
RVHT=0.165

2D UNIT CELLS (1 domain observed)

Cell	Ax (Å)	Ay (Å)	Bx (Å)	By (Å)	α (°)	Matrix	Pattern	Cell type
Bulk	3.520	0.000	0.000	2.490	90.0	(1.000, 0.000) (0.000, 1.000)	(1x1)	b: bulk lattice
Surface 1	3.520	0.000	0.000	2.490	90.0	(1.000, 0.000) (0.000, 1.000)	(1x1)	s1: commens. superlattice

3D COORDINATES

Dx/Dy in Å, or as a fraction of layer's unit cell vectors; atom 0 at the origin. Epir/subr are bulk repeat vectors.

No. of atoms: 4

Bulk z = 1.245 Å

Region	Chem el.	At. no.	Cell type	Site occ.	Rel. to	Dx ± εx		Dy ± εy		Dz ± εz		Dz/Bz(%) ± εz/Bz
epir		-2					f		f		Å	
subr		-1				1.760	Å	1.245	Å	1.245	Å	
intf	Ni	1	b	1.00	0	0.000	f	0.000	f	0.000	Å	0.0
intf	Ni	2	b	1.00	1	0.500	f	0.500	f	1.123 ± .020	Å	90.2 ± 1.6
intf	Ni	3	b	1.00	2	-0.500	f	-0.500	f	1.292 ± .020	Å	103.8 ± 1.6
subl	Ni	4	b	1.00	3	0.500	f	0.500	f	1.245	Å	100.0

BOND DISTANCES AND ANGLES

Bond distances and angles are derived from coordinates

No. of distances/angles: 6

Interatomic dist. A-B (Å)	Atom A	Atom B	Atom C	Bond angle A-B-C (°)
2.490	Ni1	Ni1(0,1)	Ni2	59.2
2.431	Ni1	Ni2	Ni3	58.5
2.431	Ni1	Ni2	Ni4	117.5
2.415	Ni1	Ni3	Ni4	120.0
2.513	Ni2	Ni3	Ni4	60.9
2.490	Ni3	Ni4	Ni4(0,1)	120.0

COMMON NAME : Ni(110)-(1x1)
CLASSIFICATION : 28.25
TECHNIQUE : MEIS
AUTHORS : S.M. Yalisove, W.R. Graham, E.D. Adams, M. Copel and T. Gustafsson
REFERENCE : Surf. Sci., 171, 400 (1986)

ILLUSTRATION: 4

SURFACE TYPE
Substrate : Ni
Crystal face: 110
Temperature : RT
Bulk lattice: fcc
2D bulk symm: pmm
2D surf symm: pmm

Adsorbate:
Coverage :
Pattern : (1x1)
Matrix : (1.000, 0.000)
(0.000, 1.000)

STRUCTURE TYPE
Bulk termination with multilayer relaxation

SAMPLE PREPARATION (1 sample)
Treatment : extensive sputter (670 K) and anneal (1070K or 1170K)
Crystallinity: sharp (1x1) LEED pattern
Anal. methods:
Contamination: ISS: no C, S, or O

COMMENTS
Previous discrepancies between LEED and ion scattering for Ni(110) surface attributed to level of surface cleanliness

DATA COLLECTION
Technique: MEIS
Dataset : 110 k eV proton beam in (001) and (010) scattering planes.

THEORY/DATA TREATMENT
Monte Carlo analysis of proton scattering intensities as a function of scattering angle; surface ΘD=395 K

STRUCTURES EXAMINED
Top two interlayer spacings varied

QUALITY OF EXPERIMENT-THEORY FIT
Visual

2D UNIT CELLS (1 domain observed)

Cell	Ax (Å)	Ay (Å)	Bx (Å)	By (Å)	α (°)	Matrix	Pattern	Cell type
Bulk	3.524	0.000	0.000	2.492	90.0	(1.000, 0.000) (0.000, 1.000)	(1x1)	b: bulk lattice
Surface 1	3.524	0.000	0.000	2.492	90.0	(1.000, 0.000) (0.000, 1.000)	(1x1)	s1: commens. superlattice

3D COORDINATES

Dx/Dy in Å, or as a fraction of layer's unit cell vectors; atom 0 at the origin. Epir/subr are bulk repeat vectors.

No. of atoms: 4

Bulk z = 1.246 Å

Region	Chem el.	At. no.	Cell type	Site occ.	Rel. to	Dx ± εx		Dy ± εy		Dz ± εz		Dz/Bz(%) ± εz/Bz
epir		-2					f		f		Å	
subr		-1				1.762	Å	1.246	Å	1.246	Å	
intf	Ni	1	b	1.00	0	0.000	f	0.000	f	0.000	Å	0.0
intf	Ni	2	b	1.00	1	0.500	f	0.500	f	1.134 ± .012	Å	91.0 ± 1.0
intf	Ni	3	b	1.00	2	-0.500	f	-0.500	f	1.290 ± .018	Å	103.5 ± 1.5
subl	Ni	4	b	1.00	3	0.500	f	0.500	f	1.246	Å	100.0

BOND DISTANCES AND ANGLES

Bond distances and angles are derived from coordinates

No. of distances/angles: 3

Interatomic dist. A-B (Å)	Atom A	Atom B	Atom C	Bond angle A-B-C (°)
2.438	Ni1	Ni2	Ni3	58.6
2.424	Ni1	Ni3	Ni4	120.0
2.514	Ni2	Ni3	Ni4	60.9

ILLUSTRATION: 4

COMMON NAME : Ni(110)-(1x1)
CLASSIFICATION : 28.26a
TECHNIQUE : LEED
AUTHORS : W. Reimer, V. Penka, M. Skottke, R.J. Behm, G. Ertl and W. Moritz
REFERENCE : Surf. Sci., 186, 45 (1987)

STRUCTURE TYPE
Bulk termination with multilayer relaxation

SURFACE TYPE
Substrate : Ni
Crystal face: 110
Temperature : RT*
Bulk lattice: fcc
2D bulk symm: pmm
2D surf symm: pmm

Adsorbate:
Coverage :
Pattern : (1x1)
Matrix : (1.000, 0.000)
(0.000, 1.000)

SAMPLE PREPARATION (1 sample)
Treatment :
Crystallinity:
Anal. methods:
Contamination:

COMMENTS

DATA COLLECTION
Technique: LEED
Dataset : I-V curves over E range 40-390 eV

THEORY/DATA TREATMENT
Dynamical LEED (layer doubling): 10 noneq. beams; Moruzzi et al pot, 8 ph shs; Vr∝E**-1/2, Vi∝E**1/3; ΘD=450 K

STRUCTURES EXAMINED
First 3 interlayer spacings varied

QUALITY OF EXPERIMENT-THEORY FIT
RPE=0.20, RZJ=0.04

2D UNIT CELLS (1 domain observed)

Cell	Ax (Å)	Ay (Å)	Bx (Å)	By (Å)	α (°)	Matrix	Pattern	Cell type
Bulk	3.520	0.000	0.000	2.490	90.0	(1.000, 0.000) (0.000, 1.000)	(1x1)	b: bulk lattice
Surface 1	3.520	0.000	0.000	2.490	90.0	(1.000, 0.000) (0.000, 1.000)	(1x1)	s1: commens. superlattice

3D COORDINATES

Dx/Dy in Å, or as a fraction of layer's unit cell vectors; atom 0 at the origin. Epir/subr are bulk repeat vectors.

No. of atoms: 5

Bulk z = 1.245 Å

Region	Chem el.	At. no.	Cell type	Site occ.	Rel. to	Dx ± εx		Dy ± εy		Dz ± εz		Dz/Bz(%) ± εz/Bz
epir		-2					f		f		Å	
subr		-1				-1.760	Å	-1.245	Å	1.245	Å	
intf	Ni	1	b	1.00	0	0.000	f	0.000	f	0.000	Å	0.0
intf	Ni	2	b	1.00	1	0.500	f	0.500	f	1.140 ± .020	Å	91.6 ± 1.6
intf	Ni	3	b	1.00	2	-0.500	f	-0.500	f	1.290 ± .020	Å	103.6 ± 1.6
intf	Ni	4	b	1.00	3	0.500	f	0.500	f	1.260 ± .020	Å	101.2 ± 1.6
subl	Ni	5	b	1.00	4	-0.500	f	-0.500	f	1.245	Å	100.0

BOND DISTANCES AND ANGLES

Bond distances and angles are derived from coordinates

No. of distances/angles: 3

Interatomic dist. A-B (Å)	Atom A	Atom B	Atom C	Bond angle A-B-C (°)
2.439	Ni1	Ni2	Ni3	58.8
2.430	Ni1	Ni3	Ni4	120.3
2.512	Ni2	Ni3	Ni4	61.2

COMMON NAME : Ni(111)-(1x1)
CLASSIFICATION : 28.11b
TECHNIQUE : HEIS
AUTHORS : T. Narusawa, W.M. Gibson and E. Tornqvist
REFERENCE : Phys. Rev. Lett., 47, 417 (1981)

ILLUSTRATION: 1

SURFACE TYPE
Substrate : Ni
Crystal face: 111
Temperature : 300 K
Bulk lattice: fcc
2D bulk symm: p3m1
2D surf symm: p3m1

Adsorbate:
Coverage :
Pattern : (1x1)
Matrix : (1.000, 0.000)
(0.000, 1.000)

STRUCTURE TYPE
Unrelaxed bulk termination

SAMPLE PREPARATION (1 sample)
Treatment : oxidation/reduction cycles, Ar+ bombardment, annealing
Crystallinity:
Anal. methods:
Contamination: in situ AES, 5 keV electron diffraction

COMMENTS
Results suggest 20% enhancement in thermal vibration amplitude of the first atomic layers, vs bulk

DATA COLLECTION
Technique: HEIS; high energy He+ ion scattering
Dataset :

THEORY/DATA TREATMENT
Computer simulations (see Barrett, PRB3,1527(1971), Picraux et al, PRB6,1382(1972), Stensgaard et al, SS 77,513(1978))

STRUCTURES EXAMINED
Bulk like structures with enhanced and unenhanced surface thermal vibration

QUALITY OF EXPERIMENT-THEORY FIT
Visual

2D UNIT CELLS (1 domain observed)

Cell	Ax (Å)	Ay (Å)	Bx (Å)	By (Å)	α (°)	Matrix	Pattern	Cell type
Bulk	2.490	0.000	1.245	2.156	60.0	(1.000, 0.000) (0.000, 1.000)	(1x1)	b: bulk lattice
Surface 1	2.490	0.000	1.245	2.156	60.0	(1.000, 0.000) (0.000, 1.000)	(1x1)	s1: commens. superlattice

3D COORDINATES

Dx/Dy in Å, or as a fraction of layer's unit cell vectors; atom 0 at the origin. Epir/subr are bulk repeat vectors.

No. of atoms: 3

Bulk z = 2.033 Å

Reg ion	Chem el.	At. no.	Cell type	Site occ.	Rel. to	Dx ± εx		Dy ± εy		Dz ± εz		Dz/Bz(%) ± εz/Bz
epir		-2					f		f		Å	
subr		-1				1.245	Å	0.719	Å	2.033	Å	
intf	Ni	1	b	1.00	0	0.000	f	0.000	f	0.000	Å	0.0
intf	Ni	2	b	1.00	1	0.333	f	0.333	f	2.033 ± .020	Å	100.0 ± 1.0
subl	Ni	3	b	1.00	2	0.333	f	0.333	f	2.033	Å	100.0

BOND DISTANCES AND ANGLES

Bond distances and angles are derived from coordinates

No. of distances/angles: 2

Interatomic dist. A-B (Å)	Atom A	Atom B	Atom C	Bond angle A-B-C (°)
2.490	Ni1	Ni2		
2.490	Ni2	Ni3		

ILLUSTRATION: 1

COMMON NAME : Ni(111)-(1x1)
CLASSIFICATION : 28.4a
TECHNIQUE : LEED
AUTHORS : J.E. Demuth, P.M. Marcus and D.W. Jepsen
REFERENCE : Phys. Rev., B11, 1460 (1975)

SURFACE TYPE
Substrate : Ni
Crystal face: 111
Temperature : RT
Bulk lattice: fcc
2D bulk symm: p3m1
2D surf symm: p3m1

Adsorbate:
Coverage :
Pattern : (1x1)
Matrix : (1.000, 0.000)
(0.000, 1.000)

STRUCTURE TYPE
Bulk termination with top spacing contraction

SAMPLE PREPARATION (1 sample)
Treatment : see J.E. Demuth and T.N. Rhodin, Surf. Sci. 42, 261 (1974)
Crystallinity:
Anal. methods:
Contamination:

COMMENTS

DATA COLLECTION
Technique: LEED
Dataset : I-V curves include 00 beam at Θ=4, 10, 20° with φ=0° and 10, -10 beams at Θ=0°; E range 10-220 eV

THEORY/DATA TREATMENT
Dynamical LEED (layer KKR): 8 phase shifts (Wakoh pot), Voi α E**1/3 (different functions tested); ΘD=335 K

STRUCTURES EXAMINED
Variation of top layer spacing

QUALITY OF EXPERIMENT-THEORY FIT
Visual

2D UNIT CELLS (1 domain observed)

Cell	Ax (Å)	Ay (Å)	Bx (Å)	By (Å)	α (°)	Matrix	Pattern	Cell type
Bulk	2.490	0.000	1.245	2.156	60.0	(1.000, 0.000) (0.000, 1.000)	(1x1)	b: bulk lattice
Surface 1	2.490	0.000	1.245	2.156	60.0	(1.000, 0.000) (0.000, 1.000)	(1x1)	s1: commens. superlattice

3D COORDINATES

Dx/Dy in Å, or as a fraction of layer's unit cell vectors; atom 0 at the origin. Epir/subr are bulk repeat vectors.

No. of atoms: 3

Bulk z = 2.030 Å

Reg ion	Chem el.	At. no.	Cell type	Site occ.	Rel. to	Dx ± εx		Dy ± εy		Dz ± εz		Dz/Bz(%) ± εz/Bz
epir		-2					f		f		Å	
subr		-1				1.245	Å	0.719	Å	2.030	Å	
intf	Ni	1	b	1.00	0	0.000	f	0.000	f	0.000	Å	0.0
intf	Ni	2	b	1.00	1	0.333	f	0.333	f	2.005 ± .025	Å	98.8 ± 1.2
subl	Ni	3	b	1.00	2	0.333	f	0.333	f	2.030	Å	100.0

BOND DISTANCES AND ANGLES

Bond distances and angles are derived from coordinates

No. of distances/angles: 4

Interatomic dist. A-B (Å)	Atom A	Atom B	Atom C	Bond angle A-B-C (°)
2.467	Ni1	Ni2	Ni1(1,0)	60.6
2.467	Ni1	Ni2	Ni2(1,0)	120.3
2.490	Ni2	Ni2(1,0)	Ni1(1,0)	59.7
2.488	Ni2	Ni3	Ni2(1,0)	60.1

COMMON NAME : Ni(311)-(1x1)
CLASSIFICATION : 28.12
TECHNIQUE : LEED
AUTHORS : W.T. Moore, S.J. White, D.C. Frost and K.A.R. Mitchell
REFERENCE : Surf. Sci., 116, 253 (1982)

ILLUSTRATION: 8

SURFACE TYPE
Substrate : Ni
Crystal face: 311
Temperature : RT*
Bulk lattice: fcc
2D bulk symm: cm
2D surf symm: cm

Adsorbate:
Coverage :
Pattern : (1x1)
Matrix : (1.000, 0.000)
(0.000, 1.000)

STRUCTURE TYPE
Bulk termination with contraction of first layer spacing

SAMPLE PREPARATION (1 sample)
Treatment : cycles of Ar+ bombardment and annealing
Crystallinity:
Anal. methods:
Contamination: AES: <0.02ML S

COMMENTS
ΘD determined from weighted average of vibrational amplitudes for 1st 5 layers (Clark et al, Phys. Rev. 139, 1860 (1965));
RZJ gives Vor=-10.8 eV and 14% contraction;
see also Surf. Sci. 116, 261 (1982)

DATA COLLECTION
Technique: LEED
Dataset : I-V spectra for 14 beams at normal incidence over energy range 50-230 eV in 2eV intervals

THEORY/DATA TREATMENT
Dynamical LEED: 8 phase shifts (Wakoh Ni potential); up to 51 non-eq. beams; Vor optimised; Voi α E**1/3; ΘD=380 K

STRUCTURES EXAMINED
Relaxation of 1st interlayer spacing from 0% to 24% relative to bulk spacing of 1.0626Å

QUALITY OF EXPERIMENT-THEORY FIT
RPE=0.344; RZJ=0.127

2D UNIT CELLS (1 domain observed)

Cell	Ax (Å)	Ay (Å)	Bx (Å)	By (Å)	α (°)	Matrix	Pattern	Cell type
Bulk	2.492	0.000	-1.248	4.132	106.8	(1.000, 0.000) (0.000, 1.000)	(1x1)	b: bulk lattice
Surface 1	2.492	0.000	-1.248	4.132	106.8	(1.000, 0.000) (0.000, 1.000)	(1x1)	s1: commens. superlattice

3D COORDINATES

0.1Å error bars assumed for tabulation

Dx/Dy in Å, or as a fraction of layer's unit cell vectors; atom 0 at the origin. Epir/subr are bulk repeat vectors.

No. of atoms: 4

Bulk z = 1.063 Å

Region	Chem el.	At. no.	Cell type	Site occ.	Rel. to	Dx ± εx		Dy ± εy		Dz ± εz		Dz/Bz(%) ± εz/Bz
epir		-2					f		f		Å	
subr		-1				0.002	Å	-2.252	Å	1.063	Å	
intf	Ni	1	b	1.00	0	0.000	f	0.000	f	0.000	Å	0.0
intf	Ni	2	b	1.00	1	0.727	f	0.454	f	0.915 ± .100	Å	86.1 ± 9.4
intf	Ni	3	b	1.00	2	-0.273	f	0.455	f	1.063 ± .100	Å	100.0 ± 9.4
subl	Ni	4	b	1.00	3	-0.272	f	-0.545	f	1.063	Å	100.0

BOND DISTANCES AND ANGLES

Bond distances and angles are derived from coordinates

No. of distances/angles: 8

Interatomic dist. A-B (Å)	Atom A	Atom B	Atom C	Bond angle A-B-C (°)
2.492	Ni1	Ni1(1,0)	Ni2	59.2
2.430	Ni1	Ni2	Ni3	119.1
2.430	Ni1	Ni2	Ni4	87.1
2.367	Ni1	Ni3(0,-1)	Ni4	88.5
2.494	Ni2	Ni3	Ni4	60.0
2.490	Ni2	Ni3(0,-1)	Ni4	60.0

Ni(311)-(1x1)
28.12

Bond Distances and Angles - Continued

Interatomic dist. A-B (Å)	Atom A	Atom B	Atom C	Bond angle A-B-C (°)
2.491	Ni2	Ni4	Ni3(0,-1)	59.9
2.490	Ni3	Ni4	Ni4(1,0)	90.1

COMMON NAME : Ni(311)-(1x1)
CLASSIFICATION : 28.21
TECHNIQUE : LEED
AUTHORS : D.L. Adams, W.T. Moore and K.A.R. Mitchell
REFERENCE : Surf. Sci., 149, 407 (1985)

ILLUSTRATION: 8

SURFACE TYPE
Substrate : Ni
Crystal face: 311
Temperature : 298 K
Bulk lattice: fcc
2D bulk symm: cm
2D surf symm: cm

Adsorbate:
Coverage :
Pattern : (1x1)
Matrix : (1.000, 0.000)
(0.000, 1.000)

STRUCTURE TYPE
Bulk termination with multilayer relaxations, mainly perpendicular to surface

SAMPLE PREPARATION (1 sample)
Treatment : Ar+ bombardment 2000-500 eV and annealing 600-700C
Crystallinity:
Anal. methods:
Contamination: AES: <2%ML S

COMMENTS

DATA COLLECTION
Technique: LEED; vidicon/photography
Dataset : I-V curves: 12 beams, 9 symmetry-inequivalent at normal incidence; energy range 60-230 eV

THEORY/DATA TREATMENT
Dynamical LEED (layer doubling): 10 phase shifts (Moruzzi et al potential); Voi=-4 eV; ΘD=450 K

STRUCTURES EXAMINED
Relaxation of first three interlayer vectors within symmetry constraints

QUALITY OF EXPERIMENT-THEORY FIT
R2=0.048

2D UNIT CELLS (1 domain observed)

Cell	Ax (Å)	Ay (Å)	Bx (Å)	By (Å)	α (°)	Matrix	Pattern	Cell type
Bulk	2.492	0.000	-1.246	4.132	106.8	(1.000, 0.000) (0.000, 1.000)	(1x1)	b: bulk lattice
Surface 1	2.492	0.000	-1.246	4.132	106.8	(1.000, 0.000) (0.000, 1.000)	(1x1)	s1: commens. superlattice

3D COORDINATES

Dx/Dy in Å, or as a fraction of layer's unit cell vectors; atom 0 at the origin. Epir/subr are bulk repeat vectors.

No. of atoms: 5

Bulk z = 1.063 Å

Region	Chem el.	At. no.	Cell type	Site occ.	Rel. to	Dx ± εx		Dy ± εy		Dz ± εz		Dz/Bz(%) ± εz/Bz
epir		-2					f		f		Å	
subr		-1				1.246	Å	1.876	Å	1.063	Å	
intf	Ni	1	b	1.00	0	0.000	f	0.000	f	0.000	Å	0.0
intf	Ni	2	b	1.00	1	0.727 ± .009	f	0.454 ± .019	f	0.894 ± .010	Å	84.1 ± .9
intf	Ni	3	b	1.00	2	-0.273 ± .009	f	0.455 ± .019	f	1.106 ± .016	Å	104.1 ± 1.5
intf	Ni	4	b	1.00	3	-0.272 ± .016	f	-0.545 ± .032	f	1.045 ± .017	Å	98.3 ± 1.6
subl	Ni	5	b	1.00	4	0.727	f	0.454	f	1.063	Å	100.0

BOND DISTANCES AND ANGLES

Bond distances and angles are derived from coordinates

No. of distances/angles: 7

Interatomic dist. A-B (Å)	Atom A	Atom B	Atom C	Bond angle A-B-C (°)
2.492	Ni1	Ni1(1,0)	Ni2	59.1
2.423	Ni1	Ni2	Ni3	119.1
2.423	Ni1	Ni2	Ni4	86.9
2.386	Ni1	Ni3(0,-1)	Ni4	88.4
2.513	Ni2	Ni3	Ni4	60.4
2.483	Ni3	Ni4	Ni4(1,0)	90.0

Ni(311)-(1x1)
28.21

Bond Distances and Angles - Continued

Interatomic dist. A-B (Å)	Atom A	Atom B	Atom C	Bond angle A-B-C (°)
2.483	Ni3	Ni4	Ni4(-1,0)	90.0

COMMON NAME : Ni(100)-Ag(111) multilayers
CLASSIFICATION : 28.47.1
TECHNIQUE : ARXPD
AUTHORS : W.F. Egelhoff, Jr.
REFERENCE : J. Vac. Sci. Technol., A3, 730 (1988)

ILLUSTRATION: 84

SURFACE TYPE
Substrate : Ni
Crystal face: 100
Temperature : RT
Bulk lattice: fcc
2D bulk symm: p4m
2D surf symm: p3m1

Adsorbate: Ag
Coverage :
Pattern : (111)
Matrix : (1.000, 0.000)
(0.000, 1.000)

STRUCTURE TYPE
Beyond 2ML coverage, Ag seems to grow as a conventional (111) surface; at 1 ML coverage a c(8x2) LEED pattern is observed

SAMPLE PREPARATION (1 sample)
Treatment :
Crystallinity: sharp LEED patterns
Anal. methods: thickness monitors and LEED
Contamination:

COMMENTS

DATA COLLECTION
Technique: ARXPD; UHV x-ray photoelectron spectrometer
Dataset : LEED patterns and ARXPS polar scans along the <100> azimuth

THEORY/DATA TREATMENT

2D UNIT CELLS (1 domain observed)

Cell	Ax (Å)	Ay (Å)	Bx (Å)	By (Å)	α (°)	Matrix	Pattern	Cell type
Bulk	2.893	0.000	1.446	2.505	60.0	(1.000, 0.000) (0.000, 1.000)	(1x1)	b: bulk lattice
Surface 1	2.893	0.000	1.446	2.505	60.0	(1.000, 0.000) (0.000, 1.000)	(1x1)	s1: commens. superlattice

3D COORDINATES

Dx/Dy in Å, or as a fraction of layer's unit cell vectors; atom 0 at the origin. Epir/subr are bulk repeat vectors.

No. of atoms: 1

Bulk z = 2.361 Å

Region	Chem el.	At. no.	Cell type	Site occ.	Rel. to	Dx ± εx		Dy ± εy		Dz ± εz		Dz/Bz(%) ± εz/Bz
epir		-2					f		f		Å	
subr		-1				1.446	Å	0.835	Å	2.361	Å	
subl	Ag	1	b	1.00	0	0.000	f	0.000	f	0.000	Å	0.0

BOND DISTANCES AND ANGLES

Bond distances and angles are derived from coordinates

No. of distances/angles: 1

Interatomic dist. A-B (Å)	Atom A	Atom B	Atom C	Bond angle A-B-C (°)
2.893	Ag1	Ag1(1,0)	Ag1(1,1)	120.0

ILLUSTRATION: 28,29

COMMON NAME : Ni(100)-c(2x2)-Bi
CLASSIFICATION : 28.83.1
TECHNIQUE : Transm. Channeling
AUTHORS : C. Klink, M. Foss, I. Stensgaard and F. Besenbacher
REFERENCE : Surf. Sci., 251/252, 841 (1991)

SURFACE TYPE
Substrate : Ni
Crystal face: 100
Temperature : 110 K
Bulk lattice: fcc
2D bulk symm: p4m
2D surf symm: p4m

Adsorbate: Bi
Coverage : 0.50 ML
Pattern : c(2x2)
Matrix : (1.000, 1.000)
(-1.000, 1.000)

STRUCTURE TYPE
Atomic adsorption in 4-fold hollow site

SAMPLE PREPARATION (1 sample)
Treatment : Ni(100) grown on NaCl; cleaned; Bi from an evaporator
Crystallinity: sharp LEED pattern
Anal. methods: HEIS, RBS
Contamination:

COMMENTS

DATA COLLECTION
Technique: Transm. Channeling
Dataset : 0-40° angular scans in [100],[011] and [111] directions

THEORY/DATA TREATMENT
Transmission channeling: multi-row continuum model

STRUCTURES EXAMINED
Hcp, bridge, fcc and top sites

QUALITY OF EXPERIMENT-THEORY FIT
Visual

2D UNIT CELLS (1 domain observed)

Cell	Ax (Å)	Ay (Å)	Bx (Å)	By (Å)	α (°)	Matrix	Pattern	Cell type
Bulk	2.489	0.000	0.000	2.489	90.0	(1.000, 0.000) (0.000, 1.000)	(1x1)	b: bulk lattice
Surface 1	2.489	2.489	-2.489	2.489	90.0	(1.000, 1.000) (-1.000, 1.000)	c(2x2)	s1: commens. superlattice

3D COORDINATES

Bi1: atomic overlayer in 4-fold hollow site

Dx/Dy in Å, or as a fraction of layer's unit cell vectors; atom 0 at the origin. Epir/subr are bulk repeat vectors.

No. of atoms: 3

Bulk z = 1.760 Å

Reg ion	Chem el.	At. no.	Cell type	Site occ.	Rel. to	Dx ± εx		Dy ± εy		Dz ± εz		Dz/Bz(%) ± εz/Bz
epir		-2					f		f		Å	
subr		-1				1.245	Å	1.245	Å	1.760	Å	
ovrl	Bi	1	s1	.50	0	0.000	Å	0.000	Å	0.000	Å	0.0
intf	Ni	2	b	1.00	0	1.245	Å	1.245	Å	2.050 ± .100	Å	116.5 ± 5.7
subl	Ni	3	b	1.00	0	0.000	Å	0.000	Å	3.810	Å	216.5

BOND DISTANCES AND ANGLES

Bond distances and angles are derived from coordinates

No. of distances/angles: 1

Interatomic dist. A-B (Å)	Atom A	Atom B	Atom C	Bond angle A-B-C (°)
2.702	Bi1	Ni2		

COMMON NAME : Ni(100)-p4g(2x2)-2C
CLASSIFICATION : 28.6.16
TECHNIQUE : LEED
AUTHORS : Y. Gauthier, R. Baudoing-Savois, K. Heinz and H. Landskron
REFERENCE : Surf. Sci., 251/252, 493 (1991)

ILLUSTRATION: 32

SURFACE TYPE
Substrate : Ni
Crystal face: 100
Temperature : RT
Bulk lattice: fcc
2D bulk symm: p4m
2D surf symm: p4g

Adsorbate: C
Coverage : 0.5 C/Ni
Pattern : (2x2)
Matrix : (2.000, 0.000)
(0.000, 2.000)

STRUCTURE TYPE
Carbon adsorbed in fcc hollow site;
clockwise rotation of top Ni layer atoms around hollow (shift in <010> directions by 0.45Å);
buckling of 2nd Ni layer (Ni directly below C moves up by 0.16Å)

SAMPLE PREPARATION (1 sample)
Treatment : C segregation after bulk doping or CO cracking with elec.
Crystallinity: well defined 2x2 after annealing
Anal. methods: AES and LEED-IV to check stability in
Contamination: no S or O by AES

COMMENTS

DATA COLLECTION
Technique: LEED; halfsphere optics, spot photometer
Dataset : IV curves for 2 integer, 4 fractional order beams: cumul. E range 805 eV

THEORY/DATA TREATMENT
Dynamical LEED (composite layers, layer doubling)
matrix inversion, 8 phase shifts

STRUCTURES EXAMINED
Bridge site ruled out

QUALITY OF EXPERIMENT-THEORY FIT
RPE = 0.22

2D UNIT CELLS (1 domain observed)

Cell	Ax (Å)	Ay (Å)	Bx (Å)	By (Å)	α (°)	Matrix	Pattern	Cell type
Bulk	2.492	0.000	0.000	2.492	90.0	(1.000, 0.000) (0.000, 1.000)	(1x1)	b: bulk lattice
Surface 1	4.984	0.000	0.000	4.984	90.0	(2.000, 0.000) (0.000, 2.000)	(2x2)	s1: commens. superlattice

3D COORDINATES

C1-C2: adatoms in symm. equivalent hollow sites; Ni3-Ni6: top Ni layer, clockwise rotation around C;
Ni7-Ni10: buckled 2nd Ni layer

Dx/Dy in Å, or as a fraction of layer's unit cell vectors; atom 0 at the origin. Epir/subr are bulk repeat vectors.

No. of atoms: 11

Bulk z = 1.762 Å

Reg ion	Chem el.	At. no.	Cell type	Site occ.	Rel. to	Dx ± εx	Dy ± εy	Dz ± εz	Dz/Bz(%) ± εz/Bz
epir		-2				f	f	Å	
subr		-1				1.246 Å	1.246 Å	1.762 Å	
ovrl	C	1	s1	.25	0	0.000 Å	0.000 Å	0.000 ± .040 Å	0.0 ± 2.3
ovrl	C	2	s1	.25	1	2.492 Å	2.492 Å	0.000 ± .040 Å	0.0 ± 2.3
intf	Ni	3	s1	.25	1	0.928 ± .050 Å	3.420 ± .050 Å	0.120 ± .040 Å	6.8 ± 2.3
intf	Ni	4	s1	.25	1	1.564 ± .050 Å	0.928 ± .050 Å	0.120 ± .040 Å	6.8 ± 2.3
intf	Ni	5	s1	.25	1	-0.928 ± .050 Å	1.564 ± .050 Å	0.120 ± .040 Å	6.8 ± 2.3
intf	Ni	6	s1	.25	1	3.420 ± .050 Å	-0.928 ± .050 Å	0.120 ± .040 Å	6.8 ± 2.3
intf	Ni	7	s1	.25	1	2.492 Å	2.492 Å	1.950 ± .080 Å	110.7 ± 4.5
intf	Ni	8	s1	.25	1	0.000 Å	0.000 Å	1.950 ± .080 Å	110.7 ± 4.5
intf	Ni	9	s1	.25	1	2.492 Å	0.000 Å	2.100 ± .070 Å	119.2 ± 4.0
intf	Ni	10	s1	.25	1	0.000 Å	2.492 Å	2.100 ± .070 Å	119.2 ± 4.0
subl	Ni	11	b	1.00	1	-1.246 Å	-1.246 Å	3.790 Å	215.1 ± 0.0

Ni(100)-p4g(2x2)-2C
28.6.16

BOND DISTANCES AND ANGLES

Bond distances and angles are derived from coordinates

No. of distances/angles: 8

Interatomic dist. A-B (Å)	Atom A	Atom B	Atom C	Bond angle A-B-C (°)
1.823	C1	Ni4		
1.950	C1	Ni8		
1.823	C2	Ni3		
1.950	C2	Ni7		
1.823	Ni3	C1(0,1)	Ni4(0,1)	89.8
1.823	Ni3	C1(0,1)	Ni5(0,1)	172.5
1.823	Ni3	C2	Ni4	89.8
1.823	Ni3	C2	Ni5(1,0)	172.5

COMMON NAME : Ni(100)-p4g(2x2)-2C
CLASSIFICATION : 28.6.17
TECHNIQUE : PED
AUTHORS : A.L.D. Kilcoyne, D.P. Woodruff, A.W. Robinson, Th. Lindner, J.S. Somers and A.M. Bradshaw
REFERENCE : Surf. Sci., 253, 107 (1991)

ILLUSTRATION: 32

SURFACE TYPE
Substrate : Ni
Crystal face: 100
Temperature : RT*
Bulk lattice: fcc
2D bulk symm: p4m
2D surf symm: p4g

Adsorbate: C
Coverage : 0.5 C/Ni
Pattern : p(2x2)
Matrix : (2.000, 0.000)
(0.000, 2.000)

STRUCTURE TYPE
Carbon adsorbed in hollow site, 0.10Å above 1st Ni layer; clock rotation of 4 Ni neighbors by 0.55Å; expansion of top Ni-Ni interlayer spacing by 0.15Å

SAMPLE PREPARATION (1 sample)
Treatment : ethylene exposure at elevated temperatures
Crystallinity:
Anal. methods:
Contamination:

COMMENTS

DATA COLLECTION
Technique: PED; photoelectron diffraction: BESSY
Dataset : energy scans (30 eV wide integration) above C 1s edge 45° incidence in <110> az.; data range 80-400 eV

THEORY/DATA TREATMENT
Double scattering cluster calculation (500 atoms)

STRUCTURES EXAMINED
Ideal termination, layer relaxation, rotation amplitude

2D UNIT CELLS (1 domain observed)

Cell	Ax (Å)	Ay (Å)	Bx (Å)	By (Å)	α (°)	Matrix	Pattern	Cell type
Bulk	2.492	0.000	0.000	2.492	90.0	(1.000, 0.000) (0.000, 1.000)	(1x1)	b: bulk lattice
Surface 1	4.984	0.000	0.000	4.984	90.0	(2.000, 0.000) (0.000, 2.000)	p(2x2)	s1: commens. superlattice

3D COORDINATES

C1-2 = adsorbate in symm. equivalent hollows Ni3-6 is clock rotated 4 fold hollow in 1st layer
Ni7 is bulk

Dx/Dy in Å, or as a fraction of layer's unit cell vectors; atom 0 at the origin. Epir/subr are bulk repeat vectors.

No. of atoms: 7 Bulk z = 1.762 Å

Region	Chem el.	At. no.	Cell type	Site occ.	Rel. to	Dx ± εx		Dy ± εy		Dz ± εz		Dz/Bz(%) ± εz/Bz
epir		-2					f		f		Å	
subr		-1				1.246	Å	1.246	Å	1.762	Å	
ovrl	C	1	s1	.25	0	0.000	Å	0.000	Å	0.000 ± .120	Å	0.0 ± 6.8
ovrl	C	2	s1	.25	1	-2.492	Å	-2.492	Å	0.000 ± .120	Å	0.0 ± 6.8
intf	Ni	3	s1	.25	1	-1.636 ± .140	Å	-0.856 ± .140	Å	0.100 ± .100	Å	5.7 ± 5.7
intf	Ni	4	s1	.25	1	1.636 ± .140	Å	0.856 ± .140	Å	0.100 ± .100	Å	5.7 ± 5.7
intf	Ni	5	s1	.25	1	-0.856 ± .140	Å	1.636 ± .140	Å	0.100 ± .100	Å	5.7 ± 5.7
intf	Ni	6	s1	.25	1	0.856 ± .140	Å	-1.636 ± .140	Å	0.100 ± .100	Å	5.7 ± 5.7
subl	Ni	7	b	1.00	1	-2.492	Å	-2.492	Å	1.990	Å	112.9 ± 0.0

BOND DISTANCES AND ANGLES

Bond distances and angles are derived from coordinates

No. of distances/angles: 5

Interatomic dist. A-B (Å)	Atom A	Atom B	Atom C	Bond angle A-B-C (°)
1.849	C1	Ni3		
1.849	C2	Ni3		
1.990	C2	Ni7		

Ni(100)-p4g(2x2)-2C
28.6.17

Bond Distances and Angles - Continued

Interatomic dist. A-B (Å)	Atom A	Atom B	Atom C	Bond angle A-B-C (°)
1.849	Ni3	C1	Ni4	173.8
1.849	Ni3	C1	Ni5	89.8

COMMON NAME : Ni(100)-p4g(2x2)-2C ILLUSTRATION: 32
CLASSIFICATION : 28.6.2
TECHNIQUE : LEED
AUTHORS : J.H. Onuferko, D.P. Woodruff and B.W. Holland
REFERENCE : Surf. Sci., 87, 357 (1979)

SURFACE TYPE
Substrate : Ni
Crystal face: 100
Temperature : RT*
Bulk lattice: fcc
2D bulk symm: p4m
2D surf symm: p4g

Adsorbate: C
Coverage : 0.5 C/Ni
Pattern : (2x2)
Matrix : (2.000, 0.000)
(0.000, 2.000)

STRUCTURE TYPE
Atomic adsorption in rotated, expanded 4-fold hollow sites; 2 C per (2x2) unit cell in c(2x2) positions, but with opposite Ni rotations

SAMPLE PREPARATION (1 sample)
Treatment : Ar+ bombardment and 1273 K annealing; C formed from C2H4
Crystallinity:
Anal. methods:
Contamination: AES: <0.05ML C, <0.02ML O

COMMENTS

DATA COLLECTION
Technique: LEED
Dataset : I-V spectra for 9 beams at 2 angles of incidence for 30<E<140 eV

THEORY/DATA TREATMENT
Dynamical LEED (reverse scattering perturbation): 6 ph shs; 18 subplanes (tested for clean Ni(100)); Vor=-11 eV; mfp=8Å

STRUCTURES EXAMINED
Various combinations of perp. and parallel displacements of first-layer Ni atoms with C in either bridge or 4-fold hollow sites at several perpendicular positions; preliminary calculation eliminated possibility of sublayer C

QUALITY OF EXPERIMENT-THEORY FIT
Visual

2D UNIT CELLS (1 domain observed)

Cell	Ax (Å)	Ay (Å)	Bx (Å)	By (Å)	α (°)	Matrix	Pattern	Cell type
Bulk	2.490	0.000	0.000	2.490	90.0	(1.000, 0.000) (0.000, 1.000)	(1x1)	b: bulk lattice
Surface 1	4.980	0.000	0.000	4.980	90.0	(2.000, 0.000) (0.000, 2.000)	(2x2)	s1: commens. superlattice

3D COORDINATES

C1-C2: atomic overlayer in equivalent 4-f hollow sites; Ni3-Ni6: planar, laterally relaxed top Ni layer; accuracy of lateral displacements 0.35Å (assumed 4-fold symmetrical)

Dx/Dy in Å, or as a fraction of layer's unit cell vectors; atom 0 at the origin. Epir/subr are bulk repeat vectors.

No. of atoms: 8 Bulk z = 1.760 Å

Reg ion	Chem el.	At. no.	Cell type	Site occ.	Rel. to	Dx ± εx		Dy ± εy		Dz ± εz		Dz/Bz(%) ± εz/Bz
epir		-2					f		f		Å	
subr		-1				1.245	Å	1.245	Å	1.760	Å	
ovrl	C	1	s1	.25	0	0.000	f	0.000	f	0.000	Å	0.0
ovrl	C	2	s1	.25	1	0.500	f	0.500	f	0.000	Å	0.0
intf	Ni	3	s1	.25	2	-0.300 ± .050	f	-0.200 ± .050	f	0.100 ± .100	Å	5.7 ± 5.7
intf	Ni	4	s1	.25	3	0.100 ± .050	f	0.500 ± .050	f	0.000	Å	0.0
intf	Ni	5	s1	.25	4	0.400 ± .050	f	-0.600 ± .050	f	0.000	Å	0.0
intf	Ni	6	s1	.25	5	0.100 ± .050	f	0.500 ± .050	f	0.000	Å	0.0
intf	Ni	7	b	1.00	6	-1.600 ± .099	f	-1.400 ± .099	f	1.960 ± .050	Å	111.4 ± 2.8
subl	Ni	8	b	1.00	7	0.500	f	0.500	f	1.760	Å	100.0

Ni(100)-p4g(2x2)-2C
28.6.2

BOND DISTANCES AND ANGLES

Bond distances and angles are derived from coordinates

No. of distances/angles: 8

Interatomic dist. A-B (Å)	Atom A	Atom B	Atom C	Bond angle A-B-C (°)
1.798	C1	Ni3	C2	156.5
1.798	C1	Ni3	Ni3(1,0)	123.6
1.798	C1	Ni3	Ni4	157.4
1.798	C1	Ni3	Ni5	112.3
1.798	C1	Ni3	Ni6(-1,0)	160.1
1.798	C1	Ni3	Ni7	50.7
2.060	C1	Ni7	Ni3	42.5
2.060	C1	Ni7	Ni3(-1,0)	47.2

COMMON NAME : Ni(110)-(2x1)-C
CLASSIFICATION : 28.6.18
TECHNIQUE : EELFS
AUTHORS : L.S. Caputi, A. Amoddeo, R. Tucci and L. Papagno
REFERENCE : Phys. Rev., B44, 1357 (1991)

ILLUSTRATION: 35

SURFACE TYPE
Substrate : Ni
Crystal face: 110
Temperature : 520 K
Bulk lattice: fcc
2D bulk symm: pmm
2D surf symm: pm

Adsorbate: carbidic carbon
Coverage : 0.5 ML
Pattern : (2x1)
Matrix : (2.000, 0.000)
(0.000, 1.000)

STRUCTURE TYPE
Atomic adsorption in threefold site within troughs in the [110] direction

SAMPLE PREPARATION (1 sample)
Treatment : sputtering and annealing; exposure to ethylene at 520 K
Crystallinity: sharp (1x1) LEED pattern
Anal. methods: AES
Contamination:

COMMENTS
Sharp (1x1) LEED pattern of the clean surface; poor (2x1) LEED pattern after exposure to ethylene

DATA COLLECTION
Technique: EELFS
Dataset : carbon K-edge from k=2-6.8Å-1

THEORY/DATA TREATMENT
Fourier analysis with phase shift corrections

STRUCTURES EXAMINED
Top, bridge, threefold and fourfold sites

QUALITY OF EXPERIMENT-THEORY FIT
Visual

2D UNIT CELLS (2 domains observed)

Cell	Ax (Å)	Ay (Å)	Bx (Å)	By (Å)	α (°)	Matrix	Pattern	Cell type
Bulk	2.492	0.000	0.000	3.524	90.0	(1.000, 0.000) (0.000, 1.000)	(1x1)	b: bulk lattice
Surface 1	4.984	0.000	0.000	3.524	90.0	(2.000, 0.000) (0.000, 1.000)	(2x1)	s1: commens. superlattice

3D COORDINATES

C1: in 3-fold hollow site inside the Ni(110) troughs;

Dx/Dy in Å, or as a fraction of layer's unit cell vectors; atom 0 at the origin. Epir/subr are bulk repeat vectors.

No. of atoms: 3 Bulk z = 1.246 Å

Region	Chem el.	At. no.	Cell type	Site occ.	Rel. to	Dx ± εx	Dy ± εy	Dz ± εz	Dz/Bz(%) ± εz/Bz
epir		-2				f	f	Å	
subr		-1				1.246 Å	1.762 Å	1.246 Å	
ovrl	C	1	s1	.50	0	0.398 Å	1.762 Å	0.000 Å	0.0
intf	Ni	2	b	1.00	0	0.000 Å	0.000 Å	0.399 ± .050 Å	32.0 ± 4.0
subl	Ni	3	b	1.00	0	1.246 Å	1.762 Å	1.645 Å	132.0

BOND DISTANCES AND ANGLES

Bond distances and angles are derived from coordinates

No. of distances/angles: 1

Interatomic dist. A-B (Å)	Atom A	Atom B	Atom C	Bond angle A-B-C (°)
1.850	C1	Ni2		

COMMON NAME : Ni(111)-(2x2)-C2H2 ILLUSTRATION: 64
CLASSIFICATION : 28.6.1.2
TECHNIQUE : LEED
AUTHORS : G. Casalone, M.G. Cattania, F. Merati and M. Simonetta
REFERENCE : Surf. Sci., 120, 171 (1982)

SURFACE TYPE
Substrate : Ni
Crystal face: 111
Temperature : 140 K
Bulk lattice: fcc
2D bulk symm: p3m1
2D surf symm: cm

Adsorbate: C2H2
Coverage : 0.25 (C2H2/Ni)
Pattern : (2x2)
Matrix : (2.000, 0.000)
(0.000, 2.000)

STRUCTURE TYPE
Molecular adsorption parallel to surface across bridge site, i.e. two C sitting over adjacent hollow sites (off-center); H neglected

SAMPLE PREPARATION (1 sample)
Treatment : Ar+ bombardment, 720 K anneal, 0.5L C2H2 at 250K
Crystallinity:
Anal. methods:
Contamination: checked by LEED and AES

COMMENTS

DATA COLLECTION
Technique: LEED
Dataset : I-V curves for 5 beams at normal incidence (30<E<90 eV)

THEORY/DATA TREATMENT
Dynamical LEED (layer doubling): 7 phase shifts (Wakoh Ni potential, Kesmodel et al for C, H neglected)

STRUCTURES EXAMINED
128 model structures tested with C-C perpendicular or parallel bonded to Ni (including top, bridge, and two hollow sites); Ni-C bond distance varied from 1.8-2.1Å, C-C bond length of 1.20Å tested for molecular axis parallel to surface.

QUALITY OF EXPERIMENT-THEORY FIT
RZJ=0.19

2D UNIT CELLS (3 domains observed)

Cell	Ax (Å)	Ay (Å)	Bx (Å)	By (Å)	α (°)	Matrix	Pattern	Cell type
Bulk	2.490	0.000	1.245	2.156	60.0	(1.000, 0.000) (0.000, 1.000)	(1x1)	b: bulk lattice
Surface 1	4.980	0.000	2.490	4.313	60.0	(2.000, 0.000) (0.000, 2.000)	(2x2)	s1: commens. superlattice

3D COORDINATES

C1-C2: C2(H2) molecule parallel surface, centered over and oriented across bridge site

Dx/Dy in Å, or as a fraction of layer's unit cell vectors; atom 0 at the origin. Epir/subr are bulk repeat vectors.

No. of atoms: 4 Bulk z = 2.033 Å

Region	Chem el.	At. no.	Cell type	Site occ.	Rel. to	Dx ± εx		Dy ± εy		Dz ± εz		Dz/Bz(%) ± εz/Bz
epir		-2					f		f		Å	
subr		-1				1.245	Å	0.719	Å	2.033	Å	
ovrl	C	1	s1	.25	0	0.000	f	0.000	f	0.000	Å	0.0
ovrl	C	2	s1	.25	1	0.140	f	0.140	f	0.000 ± .100	Å	0.0 ± 4.9
intf	Ni	3	b	1.00	2	0.360	f	0.360	f	1.500 ± .100	Å	73.8 ± 4.9
subl	Ni	4	b	1.00	3	0.333	f	0.333	f	2.033	Å	100.0

BOND DISTANCES AND ANGLES

Bond distances and angles are derived from coordinates

No. of distances/angles: 11

Interatomic dist. A-B (Å)	Atom A	Atom B	Atom C	Bond angle A-B-C (°)
1.208	C1	C2	Ni3	136.0
2.490	Ni3	Ni3(1,0)	C2(1,0)	124.1
2.490	Ni3	Ni4		
2.159	C2	Ni3	Ni3(1,0)	128.5
2.159	C2	Ni3	Ni4	169.3

Ni(111)-(2x2)-C2H2
28.6.1.2

Bond Distances and Angles - Continued

Interatomic dist. A-B (Å)	Atom A	Atom B	Atom C	Bond angle A-B-C (°)
2.041	Ni3	C1(1,0)	C2	72.2
2.041	Ni3	C1(1,0)	Ni3(0,-1)	72.7
2.041	Ni3	C2(1,0)	C1(1,0)	131.7
2.159	Ni3	C2	C1	136.0
2.159	Ni3	C2	Ni3(-1,0)	72.7
2.490	Ni3	Ni3(1,0)	C1(1,0)	92.8

COMMON NAME : Ni(100)-C6SH5 disordered ILLUSTRATION: 77
CLASSIFICATION : 28.6.16.1
TECHNIQUE : SEXAFS, XANES
AUTHORS : Y. Takata, T. Yokoyama, S. Yagi, N. Happo, H. Sato, K. Seki, T. Ohta, Y. Kitajima and H. Kuroda
REFERENCE : Surf. Sci., 259, 266 (1991)

SURFACE TYPE
Substrate : Ni
Crystal face: 100
Temperature : 120 K
Bulk lattice: fcc
2D bulk symm: p4m
2D surf symm: none

Adsorbate: C6SH5 (thiophenol)
Coverage : 0.17ML
Pattern : disordered
Matrix : (1.000, 0.000)
(0.000, 1.000)

STRUCTURE TYPE
Thiophenol adsorbs, losing a hydrogen, in a disordered fashion with the S in a 4-fold hollow site; S-C bond is normal to the surface

SAMPLE PREPARATION (1 sample)
Treatment : sputtering and annealing cycles; exposed to thiophenol
Crystallinity:
Anal. methods: AES, LEED
Contamination:

COMMENTS
Orientation of the benzene ring is unknown: not included in tabulation

DATA COLLECTION
Technique: SEXAFS, XANES
Dataset : S K-edge photoelectrons, XANES analysis from 2460-2510 eV, SEXAFS from 2400-2900eV

THEORY/DATA TREATMENT
Fourier transform technique

STRUCTURES EXAMINED
Hollow, bridge and top sites

QUALITY OF EXPERIMENT-THEORY FIT
Visual

2D UNIT CELLS (1 domain observed)

Cell	Ax (Å)	Ay (Å)	Bx (Å)	By (Å)	α (°)	Matrix	Pattern	Cell type
Bulk	2.492	0.000	0.000	2.492	90.0	(1.000, 0.000) (0.000, 1.000)	(1x1)	b: bulk lattice
Surface 1	2.492	0.000	0.000	2.492	90.0	(1.000, 0.000) (0.000, 1.000)	disordered	nd1: non-recon. lattice-gas dis

3D COORDINATES

C1: benzene C bonded to S; S-C bond normal to surface; S2: bonded to C and 4-fold hollow site

Dx/Dy in Å, or as a fraction of layer's unit cell vectors; atom 0 at the origin. Epir/subr are bulk repeat vectors.

No. of atoms: 4 Bulk z = 1.762 Å

Region	Chem el.	At. no.	Cell type	Site occ.	Rel. to	Dx ± εx		Dy ± εy		Dz ± εz		Dz/Bz(%) ± εz/Bz
epir		-2					f		f		Å	
subr		-1				1.246	Å	1.246	Å	1.762	Å	
ovrl	C	1	nd1	.17	0	0.000	Å	0.000	Å	0.000	Å	0.0
ovrl	S	2	nd1	.17	0	0.000	Å	0.000	Å	1.840 ± .050	Å	104.4 ± 2.8
intf	Ni	3	b	1.00	0	1.246	Å	1.246	Å	3.239 ± .020	Å	183.8 ± 1.1
subl	Ni	4	b	1.00	0	0.000	Å	0.000	Å	5.001	Å	283.8

BOND DISTANCES AND ANGLES

No. of distances/angles: 2

Interatomic dist. A-B (Å)	Atom A	Atom B	Atom C	Bond angle A-B-C (°)
1.840	C1	S2		
2.250	S2	Ni3		

COMMON NAME : Ni(100)-c(2x2)-CO
CLASSIFICATION : 28.6.8.11
TECHNIQUE : LEED
AUTHORS : K. Heinz, E. Lang and K. Mueller
REFERENCE : Surf. Sci., 87, 595 (1979)

ILLUSTRATION: 71

SURFACE TYPE
Substrate : Ni
Crystal face: 100
Temperature : 100 K
Bulk lattice: fcc
2D bulk symm: p4m
2D surf symm: p4m

Adsorbate: CO
Coverage : 1/2 CO/Ni
Pattern : c(2x2)
Matrix : (1.000, 1.000)
(-1.000, 1.000)

STRUCTURE TYPE
Molecular on-top adsorption, C bonding to Ni

SAMPLE PREPARATION (1 sample)
Treatment : CO exposure 1E-5Pa x 50sec
Crystallinity: LEED intensities match published data
Anal. methods:
Contamination: AES: very little C on clean surface

COMMENTS
Spectra were taken within 16 sec after termination of adsorption process

DATA COLLECTION
Technique: LEED
Dataset : I-V spectra: (1,1) (1,0) (1/2,1/2) beams at normal incidence (0,0) beam at Θ=4°,φ=0°; E range 30-400 eV

THEORY/DATA TREATMENT
Comparison with earlier dynamical LEED I-V spectra by Pendry

STRUCTURES EXAMINED
Linearly bonded CO perpendicular to surface; variation of C-O bond length: 0.9-1.2Å in steps of 0.05Å; variation of Ni-C bond length: 1.7, 1.8, 1.9Å; bulk interlayer spacings assumed in metal

QUALITY OF EXPERIMENT-THEORY FIT
RZJ=0.19

2D UNIT CELLS (1 domain observed)

Cell	Ax (Å)	Ay (Å)	Bx (Å)	By (Å)	α (°)	Matrix	Pattern	Cell type
Bulk	2.489	0.000	0.000	2.489	90.0	(1.000, 0.000) (0.000, 1.000)	(1x1)	b: bulk lattice
Surface 1	2.489	2.489	-2.489	2.489	90.0	(1.000, 1.000) (-1.000, 1.000)	c(2x2)	s1: commens. superlattice

3D COORDINATES

O1-C2: upright on-top molecule (C bonded to Ni3)

Dx/Dy in Å, or as a fraction of layer's unit cell vectors; atom 0 at the origin. Epir/subr are bulk repeat vectors.

No. of atoms: 4 Bulk z = 1.760 Å

Region	Chem el.	At. no.	Cell type	Site occ.	Rel. to	Dx ± εx		Dy ± εy		Dz ± εz		Dz/Bz(%) ± εz/Bz
epir		-2					f		f		Å	
subr		-1				1.245	Å	1.245	Å	1.760	Å	
ovrl	O	1	s1	.50	0	0.000	f	0.000	f	0.000	Å	0.0
ovrl	C	2	s1	.50	1	0.000	f	0.000	f	1.150 ± .100	Å	65.3 ± 5.7
intf	Ni	3	b	1.00	2	0.000	f	0.000	f	1.800 ± .100	Å	102.3 ± 5.7
subl	Ni	4	b	1.00	3	0.500	f	0.500	f	1.760	Å	100.0

BOND DISTANCES AND ANGLES

Bond distances and angles are derived from coordinates

No. of distances/angles: 4

Interatomic dist. A-B (Å)	Atom A	Atom B	Atom C	Bond angle A-B-C (°)
1.150	O1	C2	Ni3	180.0
1.800	C2	Ni3		
2.489	Ni3	Ni3(1,0)		
2.489	Ni3	Ni4		

COMMON NAME : Ni(100)-c(2x2)-CO ILLUSTRATION: 71
CLASSIFICATION : 28.6.8.12b
TECHNIQUE : PED
AUTHORS : S.D. Kevan, R.F. Davis, D.H. Rosenblatt, J.G. Tobin, M.G. Mason, D.A. Shirley, C.H. Li and S.Y. Tong
REFERENCE : Phys. Rev. Lett., 46, 1629 (1981)

SURFACE TYPE
Substrate : Ni
Crystal face: 100
Temperature : RT*
Bulk lattice: fcc
2D bulk symm: p4m
2D surf symm: p4m

Adsorbate: CO
Coverage : 1/2 CO/Ni
Pattern : c(2x2)
Matrix : (1.000, 1.000)
(-1.000, 1.000)

STRUCTURE TYPE
Molecular on-top adsorption, C bonding to Ni

SAMPLE PREPARATION (1 sample)
Treatment : 'standard procedures' (see Davis et al, PRL 45, 1877(1980))
Crystallinity:
Anal. methods:
Contamination:

COMMENTS

DATA COLLECTION
Technique: PED; normal photoelectron diffraction
Dataset : photoelectron yield on both C(1s) and O(1s) core levels in photon energy range from 300 to 650 eV

THEORY/DATA TREATMENT
Convergent multiple-scattering calculation: Wakoh Ni pot, Xα scattered-wave potentials for C and O

STRUCTURES EXAMINED
CO bond axis normal to surface with C end down; top, bridge and hollow sites;
various C-Ni interplanar distances and C-O bond distances between 1.6 and 2.3Å in steps of 0.1Å;
bulk Ni assumed

QUALITY OF EXPERIMENT-THEORY FIT
Visual

2D UNIT CELLS (1 domain observed)

Cell	Ax (Å)	Ay (Å)	Bx (Å)	By (Å)	α (°)	Matrix	Pattern	Cell type
Bulk	2.489	0.000	0.000	2.489	90.0	(1.000, 0.000) (0.000, 1.000)	(1x1)	b: bulk lattice
Surface 1	2.489	2.489	-2.489	2.489	90.0	(1.000, 1.000) (-1.000, 1.000)	c(2x2)	s1: commens. superlattice

3D COORDINATES

O1-C2: upright on-top molecule (C bonded to Ni3)

Dx/Dy in Å, or as a fraction of layer's unit cell vectors; atom 0 at the origin. Epir/subr are bulk repeat vectors.

No. of atoms: 4 Bulk z = 1.760 Å

Reg ion	Chem el.	At. no.	Cell type	Site occ.	Rel. to	Dx ± εx		Dy ± εy		Dz ± εz		Dz/Bz(%) ± εz/Bz
epir		-2					f		f		Å	
subr		-1				1.245	Å	1.245	Å	1.760	Å	
ovrl	O	1	s1	.50	0	0.000	f	0.000	f	0.000	Å	0.0
ovrl	C	2	s1	.50	1	0.000	f	0.000	f	1.130 ± .100	Å	64.2 ± 5.7
intf	Ni	3	b	1.00	2	0.000	f	0.000	f	1.800 ± .040	Å	102.3 ± 2.3
subl	Ni	4	b	1.00	3	0.500	f	0.500	f	1.760	Å	100.0

BOND DISTANCES AND ANGLES

Bond distances and angles are derived from coordinates

No. of distances/angles: 4

Interatomic dist. A-B (Å)	Atom A	Atom B	Atom C	Bond angle A-B-C (°)
1.130	O1	C2	Ni3	180.0
1.800	C2	Ni3		
2.489	Ni3	Ni3(1,0)		

Ni(100)-c(2x2)-CO
28.6.8.12b

Bond Distances and Angles - Continued

Interatomic dist. A-B (Å)	Atom A	Atom B	Atom C	Bond angle A-B-C (°)
2.489	Ni3	Ni4		

COMMON NAME : Ni(100)-c(2x2)-CO
CLASSIFICATION : 28.6.8.4
TECHNIQUE : LEED
AUTHORS : M. Passler, A. Ignatiev, F. Jona, D.W. Jepsen and P.M. Marcus
REFERENCE : Phys. Rev. Lett., 43, 360 (1979)

ILLUSTRATION: 71

SURFACE TYPE
Substrate : Ni
Crystal face: 100
Temperature : 100 K
Bulk lattice: fcc
2D bulk symm: p4m
2D surf symm: p4m

Adsorbate: CO
Coverage : 1/2 CO/Ni
Pattern : c(2x2)
Matrix : (1.000, 1.000)
(-1.000, 1.000)

STRUCTURE TYPE
Molecular on-top adsorption, C bonding to Ni

SAMPLE PREPARATION (1 sample)
Treatment : CO adsorbed to 2L; care to avoid electron-beam damage
Crystallinity:
Anal. methods:
Contamination:

COMMENTS

DATA COLLECTION
Technique: LEED
Dataset : normal incidence and off normal incidence (Θ=10°, ϕ=0°) IV spectra

THEORY/DATA TREATMENT
Dynamical LEED: O, C, top Ni layers in ang. mom. repres., bulk in beam repres.; 5 phase shifts; 58 beams

STRUCTURES EXAMINED
CO molecule perpendicular to the surface, carbon end down; various adsorption sites (hollow, bridge, top); varied Ni-C and C-O distances; bulk spacings assumed for Ni

QUALITY OF EXPERIMENT-THEORY FIT
Visual

2D UNIT CELLS (1 domain observed)

Cell	Ax (Å)	Ay (Å)	Bx (Å)	By (Å)	α (°)	Matrix	Pattern	Cell type
Bulk	2.489	0.000	0.000	2.489	90.0	(1.000, 0.000) (0.000, 1.000)	(1x1)	b: bulk lattice
Surface 1	2.489	2.489	-2.489	2.489	90.0	(1.000, 1.000) (-1.000, 1.000)	c(2x2)	s1: commens. superlattice

3D COORDINATES

O1-C2: upright on-top molecule (C bonded to Ni3); 0.1Å error bars assumed for tabulation

Dx/Dy in Å, or as a fraction of layer's unit cell vectors; atom 0 at the origin. Epir/subr are bulk repeat vectors.

No. of atoms: 4   Bulk z = 1.760 Å

Region	Chem el.	At. no.	Cell type	Site occ.	Rel. to	Dx ± εx		Dy ± εy		Dz ± εz		Dz/Bz(%) ± εz/Bz
epir		-2					f		f		Å	
subr		-1				1.245	Å	1.245	Å	1.760	Å	
ovrl	O	1	s1	.50	0	0.000	f	0.000	f	0.000	Å	0.0
ovrl	C	2	s1	.50	1	0.000	f	0.000	f	1.150 ± .100	Å	65.3 ± 5.7
intf	Ni	3	b	1.00	2	0.000	f	0.000	f	1.720 ± .100	Å	97.7 ± 5.7
subl	Ni	4	b	1.00	3	0.500	f	0.500	f	1.760	Å	100.0

BOND DISTANCES AND ANGLES

Bond distances and angles are derived from coordinates

No. of distances/angles: 4

Interatomic dist. A-B (Å)	Atom A	Atom B	Atom C	Bond angle A-B-C (°)
1.150	O1	C2	Ni3	180.0
1.720	C2	Ni3		
2.489	Ni3	Ni3(1,0)		
2.489	Ni3	Ni4		

COMMON NAME : Ni(100)-c(2x2)-CO ILLUSTRATION: 71
CLASSIFICATION : 28.6.8.6
TECHNIQUE : PED
AUTHORS : L.G. Petersson, S. Kono, N.F.T. Hall, C.S. Fadley and J.B. Pendry
REFERENCE : Phys. Rev. Lett., 42, 1545 (1979)

SURFACE TYPE
Substrate : Ni
Crystal face: 100
Temperature : RT*
Bulk lattice: fcc
2D bulk symm: p4m
2D surf symm: p4m

Adsorbate: CO
Coverage : 1/2 CO/Ni
Pattern : c(2x2)
Matrix : (1.000, 1.000)
(-1.000, 1.000)

STRUCTURE TYPE
Molecular on-top adsorption, C bonding to Ni

SAMPLE PREPARATION (1 sample)
Treatment : CO adsorbed to 2.0L (see PRL 41, 117 & 41, 1831 (1978))
Crystallinity:
Anal. methods:
Contamination: checked by ARXPS: <3%ML C

COMMENTS
Average orientation of CO is determined to be within 12° of normal

DATA COLLECTION
Technique: PED; angle resolved x-ray photoemission
Dataset : polar-angle scans of C(1s) and O(1s) intensities

THEORY/DATA TREATMENT
Single-scattering calculations for both a single CO molecule and a finite cluster

STRUCTURES EXAMINED
C-O and Ni-C distance were assumed 1.15Å and 1.8Å, resp.; various C-O tilt angles tested; bulk Ni layer spacings assumed

QUALITY OF EXPERIMENT-THEORY FIT
Visual

2D UNIT CELLS (1 domain observed)

Cell	Ax (Å)	Ay (Å)	Bx (Å)	By (Å)	α (°)	Matrix	Pattern	Cell type
Bulk	2.489	0.000	0.000	2.489	90.0	(1.000, 0.000) (0.000, 1.000)	(1x1)	b: bulk lattice
Surface 1	2.489	2.489	-2.489	2.489	90.0	(1.000, 1.000) (-1.000, 1.000)	c(2x2)	s1: commens. superlattice

3D COORDINATES

O1-C2: upright on-top molecule (C bonded to Ni3)

Dx/Dy in Å, or as a fraction of layer's unit cell vectors; atom 0 at the origin. Epir/subr are bulk repeat vectors.

No. of atoms: 4 Bulk z = 1.760 Å

Region	Chem el.	At. no.	Cell type	Site occ.	Rel. to	Dx ± εx		Dy ± εy		Dz ± εz		Dz/Bz(%) ± εz/Bz
epir		-2					f		f		Å	
subr		-1				1.245	Å	1.245	Å	1.760	Å	
ovrl	O	1	s1	.50	0	0.000	f	0.000	f	0.000	Å	0.0
ovrl	C	2	s1	.50	1	0.000 ± .088	f	0.000 ± .088	f	1.150	Å	65.3
intf	Ni	3	b	1.00	2	0.000	f	0.000	f	1.800	Å	102.3
subl	Ni	4	b	1.00	3	0.500	f	0.500	f	1.760	Å	100.0

BOND DISTANCES AND ANGLES

Bond distances and angles are derived from coordinates

No. of distances/angles: 4

Interatomic dist. A-B (Å)	Atom A	Atom B	Atom C	Bond angle A-B-C (°)
1.150	O1	C2	Ni3	180.0
1.800	C2	Ni3		
2.489	Ni3	Ni3(1,0)		
2.489	Ni3	Ni4		

COMMON NAME : Ni(100)-c(2x2)-CO
CLASSIFICATION : 28.6.8.7
TECHNIQUE : LEED
AUTHORS : S. Andersson and J.B. Pendry
REFERENCE : J. Phys., C13, 3547 (1980)

ILLUSTRATION: 71

SURFACE TYPE
Substrate : Ni
Crystal face: 100
Temperature : 120 K
Bulk lattice: fcc
2D bulk symm: p4m
2D surf symm: p4m

Adsorbate: CO
Coverage : 1/2 (CO/Ni)
Pattern : c(2x2)
Matrix : (1.000, 1.000)
(-1.000, 1.000)

STRUCTURE TYPE
Molecular on-top adsorption, C bonding to Ni

SAMPLE PREPARATION (1 sample)
Treatment : 2.3L CO exposure, compensates for electron beam damage
Crystallinity:
Anal. methods:
Contamination:

COMMENTS

DATA COLLECTION
Technique: LEED
Dataset : I-V spectra for (10),(11),(1/2,1/2),(3/2,1/2) beams at normal incidence; E range 10-125 eV

THEORY/DATA TREATMENT
Dynamical LEED (layer doubling): 8 ph shifts, Wakoh Ni pot superposition potentials for C and O; VoiαE**1/3

STRUCTURES EXAMINED
CO perpendicular to unrelaxed substrate in top site with variable Ni-C and C-O spacings

QUALITY OF EXPERIMENT-THEORY FIT
RPE=0.39

2D UNIT CELLS (1 domain observed)

Cell	Ax (Å)	Ay (Å)	Bx (Å)	By (Å)	α (°)	Matrix	Pattern	Cell type
Bulk	2.489	0.000	0.000	2.489	90.0	(1.000, 0.000) (0.000, 1.000)	(1x1)	b: bulk lattice
Surface 1	2.489	2.489	-2.489	2.489	90.0	(1.000, 1.000) (-1.000, 1.000)	c(2x2)	s1: commens. superlattice

3D COORDINATES

O1-C2: upright on-top molecule (C bonded to Ni3)

Dx/Dy in Å, or as a fraction of layer's unit cell vectors; atom 0 at the origin. Epir/subr are bulk repeat vectors.

No. of atoms: 4 Bulk z = 1.760 Å

Region	Chem el.	At. no.	Cell type	Site occ.	Rel. to	Dx ± εx	Dy ± εy	Dz ± εz	Dz/Bz(%) ± εz/Bz
epir		-2				f	f	Å	
subr		-1				1.245 Å	1.245 Å	1.760 Å	
ovrl	O	1	s1	.50	0	0.000 f	0.000 f	0.000 Å	0.0
ovrl	C	2	s1	.50	1	0.000 f	0.000 f	1.150 ± .100 Å	65.3 ± 5.7
intf	Ni	3	b	1.00	2	0.000 f	0.000 f	1.710 ± .100 Å	97.2 ± 5.7
subl	Ni	4	b	1.00	3	0.500 f	0.500 f	1.760 Å	100.0

BOND DISTANCES AND ANGLES

Bond distances and angles are derived from coordinates

No. of distances/angles: 4

Interatomic dist. A-B (Å)	Atom A	Atom B	Atom C	Bond angle A-B-C (°)
1.150	O1	C2	Ni3	180.0
1.710	C2	Ni3		
2.489	Ni3	Ni3(1,0)		
2.489	Ni3	Ni4		

COMMON NAME : Ni(100)-c(2x2)-CO ILLUSTRATION: 71
CLASSIFICATION : 28.6.8.8
TECHNIQUE : LEED
AUTHORS : S.Y. Tong, A. Maldonado, C.H. Li and M.A. Van Hove
REFERENCE : Surf. Sci., 94, 73 (1980)

SURFACE TYPE
Substrate : Ni
Crystal face: 100
Temperature : 100 K
Bulk lattice: fcc
2D bulk symm: p4m
2D surf symm: p4m

Adsorbate: CO
Coverage : 1/2 (CO/Ni)
Pattern : c(2x2)
Matrix : (1.000, 1.000)
(-1.000, 1.000)

STRUCTURE TYPE
Molecular on-top adsorption, C bonding to Ni

SAMPLE PREPARATION (sample)
Treatment : see Passler et al, Phys. Rev. Lett. 43, 360 (1979)
Crystallinity:
Anal. methods:
Contamination:

COMMENTS

DATA COLLECTION
Technique: LEED
Dataset : I-V curves: Θ=0°: 3 beams; Θ=10°, φ=[110]: 10 beams

THEORY/DATA TREATMENT
Dynamical LEED (layer doubling): Wakoh Ni pot, Xα CO pot, 8 ph sh;Vor=-11.85 eV;VoiαE**1/3;ΘD=371 K(Ni),688K(C),596K(O)

STRUCTURES EXAMINED
Ni-C and C-O layer spacings varied in top and bridge sites; CO kept perpendicular to surface

QUALITY OF EXPERIMENT-THEORY FIT
Visual

2D UNIT CELLS (1 domain observed)

Cell	Ax (Å)	Ay (Å)	Bx (Å)	By (Å)	α (°)	Matrix	Pattern	Cell type
Bulk	2.489	0.000	0.000	2.489	90.0	(1.000, 0.000) (0.000, 1.000)	(1x1)	b: bulk lattice
Surface 1	2.489	2.489	-2.489	2.489	90.0	(1.000, 1.000) (-1.000, 1.000)	c(2x2)	s1: commens. superlattice

3D COORDINATES

O1-C2: upright on-top molecule (C bonded to Ni3)

Dx/Dy in Å, or as a fraction of layer's unit cell vectors; atom 0 at the origin. Epir/subr are bulk repeat vectors.

No. of atoms: 4 Bulk z = 1.760 Å

Region	Chem el.	At. no.	Cell type	Site occ.	Rel. to	Dx ± εx		Dy ± εy		Dz ± εz		Dz/Bz(%) ± εz/Bz
epir		-2					f		f		Å	
subr		-1				1.245	Å	1.245	Å	1.760	Å	
ovrl	O	1	s1	.50	0	0.000	f	0.000	f	0.000	Å	0.0
ovrl	C	2	s1	.50	1	0.000	f	0.000	f	1.130 ± .100	Å	64.2 ± 5.7
intf	Ni	3	b	1.00	2	0.000	f	0.000	f	1.700 ± .100	Å	96.6 ± 5.7
subl	Ni	4	b	1.00	3	0.500	f	0.500	f	1.760	Å	100.0

BOND DISTANCES AND ANGLES

Bond distances and angles are derived from coordinates

No. of distances/angles: 4

Interatomic dist. A-B (Å)	Atom A	Atom B	Atom C	Bond angle A-B-C (°)
1.130	O1	C2	Ni3	180.0
1.700	C2	Ni3		
2.489	Ni3	Ni3(1,0)		
2.489	Ni3	Ni4		

COMMON NAME : Ni(110)-p(2x1)-2CO
CLASSIFICATION : 28.6.8.20
TECHNIQUE : LEED
AUTHORS : D.J. Hannaman and M.A. Passler
REFERENCE : Surf. Sci., 203, 449 (1988)

ILLUSTRATION: 78

SURFACE TYPE
Substrate : Ni
Crystal face: 110
Temperature : 125 K
Bulk lattice: fcc
2D bulk symm: pmm
2D surf symm: p2mg

Adsorbate: CO
Coverage : 1.0 (CO/Ni)
Pattern : p(2x1)
Matrix : (2.000, 0.000)
(0.000, 1.000)

STRUCTURE TYPE
Tilted molecular adsorption through C ends over short bridge sites in zigzag arrangement (Ni-C tilt of 27°, C-O tilt of 17° from normal; tilt is toward troughs)

SAMPLE PREPARATION (1 sample)
Treatment : CO exposure at 5E-8 torr at 125 K for 2-3min; 5-10L
Crystallinity:
Anal. methods:
Contamination:

COMMENTS

DATA COLLECTION
Technique: LEED
Dataset : IV curves for 9 non-equivalent beams at normal incidence

THEORY/DATA TREATMENT
Dynamical LEED (RFS and composite layer method): 5 phase shifts

STRUCTURES EXAMINED
Long and short bridge, hollow and top sites; variations of C-O bond length from 0.95 to 1.35Å; Ni-C length from 1.7 to 2.07Å; Ni top interplanar spacing relaxed from -3 to +7%; variations of Ni-C and C-O tilt angles from 0 to 45° from normal

QUALITY OF EXPERIMENT-THEORY FIT
Visual

2D UNIT CELLS (1 domain observed)

Cell	Ax (Å)	Ay (Å)	Bx (Å)	By (Å)	α (°)	Matrix	Pattern	Cell type
Bulk	2.490	0.000	0.000	3.520	90.0	(1.000, 0.000) (0.000, 1.000)	(1x1)	b: bulk lattice
Surface 1	4.980	0.000	0.000	3.520	90.0	(2.000, 0.000) (0.000, 1.000)	p(2x1)	s1: commens. superlattice

3D COORDINATES

O1-C3, O2-C4: 2 oppositely tilted CO molecules, C down; 0.1Å error bars assumed for tabulation

Dx/Dy in Å, or as a fraction of layer's unit cell vectors; atom 0 at the origin. Epir/subr are bulk repeat vectors.

No. of atoms: 6

Bulk z = 1.245 Å

Region	Chem el.	At. no.	Cell type	Site occ.	Rel. to	Dx ± εx		Dy ± εy		Dz ± εz		Dz/Bz(%) ± εz/Bz
epir		-2					f		f		Å	
subr		-1				-1.245	Å	-1.760	Å	1.245	Å	
ovrl	O	1	s1	.50	0	0.000	f	0.000	f	0.000	Å	0.0
ovrl	O	2	s1	.50	1	0.500	f	0.432 ± .028	f	0.000	Å	0.0
ovrl	C	3	s1	.50	2	-0.500	f	0.477 ± .028	f	1.070 ± .100	Å	85.9 ± 8.0
ovrl	C	4	s1	.50	3	0.500	f	-0.387 ± .028	f	0.000	Å	0.0
intf	Ni	5	b	1.00	4	-0.500	f	0.194 ± .028	f	1.336 ± .100	Å	107.3 ± 8.0
subl	Ni	6	b	1.00	5	-0.500	f	-0.500	f	1.245	Å	100.0

BOND DISTANCES AND ANGLES

Bond distances and angles are derived from coordinates

No. of distances/angles: 4

Interatomic dist. A-B (Å)	Atom A	Atom B	Atom C	Bond angle A-B-C (°)
1.117	O1	C3(0,-1)	Ni5(0,-1)	139.2
1.948	C3	Ni5	Ni6	105.7
2.490	Ni5	Ni5(1,0)		

Ni(110)-p(2x1)-2CO
28.6.8.20

Bond Distances and Angles - Continued

Interatomic dist. A-B (Å)	Atom A	Atom B	Atom C	Bond angle A-B-C (°)
2.490	Ni5	Ni6		

COMMON NAME : Ni(111)-(√3x√3)R30°-CO
CLASSIFICATION : 28.6.8.12a
TECHNIQUE : PED
AUTHORS : S.D. Kevan, R.F. Davis, D.H. Rosenblatt, J.G. Tobin, M.G. Mason, D.A. Shirley, C.H. Li and S.Y. Tong
REFERENCE : Phys. Rev. Lett., 46, 1629 (1981)

ILLUSTRATION: 61

SURFACE TYPE
Substrate : Ni
Crystal face: 111
Temperature : RT*
Bulk lattice: fcc
2D bulk symm: p3m1
2D surf symm: cm

Adsorbate: CO
Coverage : 0.33 CO/Ni
Pattern : (√3x√3)R30°
Matrix : (1.000, 1.000)
(-2.000, 1.000)

STRUCTURE TYPE
Molecular adsorption perpendicular to surface in bridge sites

SAMPLE PREPARATION (1 sample)
Treatment : 'standard procedures' (see comments)
Crystallinity:
Anal. methods:
Contamination:

COMMENTS
Prep: R.F. Davis, S.D. Kevan, D.H. Rosenblatt, M.G. Mason, J.G. Tobin and D.A. Shirley, Phys. Rev. Lett. 45,1877 (1980)

DATA COLLECTION
Technique: PED; normal photoelectron diffraction
Dataset : PED on both the C(1s) and O(1s) adsorbate core levels in the photon energy range from 300 to 600 eV

THEORY/DATA TREATMENT
Convergent multiple-scattering calculation: Wakoh Ni potential, Xα scattered-wave potentials for CO

STRUCTURES EXAMINED
CO axis normal to surface with carbon end down; various adsorption sites (top, bridge, and hollow); various C-Ni interplanar spacings between 1.6 and 2.3Å in steps of 0.1Å; C-O distance of 1.13Å fixed; bulk spacings assumed in Ni

QUALITY OF EXPERIMENT-THEORY FIT
Visual

2D UNIT CELLS (3 domains observed)

Cell	Ax (Å)	Ay (Å)	Bx (Å)	By (Å)	α (°)	Matrix	Pattern	Cell type
Bulk	2.488	0.000	1.244	2.155	60.0	(1.000, 0.000) (0.000, 1.000)	(1x1)	b: bulk lattice
Surface 1	3.733	2.155	-3.733	2.155	120.0	(1.000, 1.000) (-2.000, 1.000)	(√3x√3)R30°	s1: commens. superlattice

3D COORDINATES

O1-C2: overlayer of upright molecules in bridge sites

Dx/Dy in Å, or as a fraction of layer's unit cell vectors; atom 0 at the origin. Epir/subr are bulk repeat vectors.

No. of atoms: 4 Bulk z = 2.030 Å

Region	Chem el.	At. no.	Cell type	Site occ.	Rel. to	Dx ± εx	Dy ± εy	Dz ± εz	Dz/Bz(%) ± εz/Bz
epir		-2				f	f	Å	
subr		-1				0.000 Å	-1.437 Å	2.030 Å	
ovrl	O	1	s1	.33	0	0.000 f	0.000 f	0.000 Å	0.0
ovrl	C	2	s1	.33	1	0.000 f	0.000 f	1.130 Å	55.7
intf	Ni	3	b	1.00	2	-0.500 f	1.000 f	1.270 ± .500 Å	62.6 ± 3.0
subl	Ni	4	b	1.00	3	0.333 f	-0.667 f	2.030 Å	100.0

BOND DISTANCES AND ANGLES

Bond distances and angles are derived from coordinates

No. of distances/angles: 7

Interatomic dist. A-B (Å)	Atom A	Atom B	Atom C	Bond angle A-B-C (°)
1.130	O1	C2	Ni3	120.5
2.501	C2	Ni3	Ni3(0,-1)	41.8
2.501	C2	Ni3	Ni4	85.2
1.778	Ni3	C2(0,1)	Ni3(1,0)	88.8

Ni(111)-(√3x√3)R30°-CO
28.6.8.12a

Bond Distances and Angles - Continued

Interatomic dist. A-B (Å)	Atom A	Atom B	Atom C	Bond angle A-B-C (°)
1.778	Ni3	C2(0,1)	Ni3(0,1)	68.7
2.488	Ni3	Ni3(1,0)	C2(1,0)	134.4
2.487	Ni3	Ni4		

ILLUSTRATION: 28,29

COMMON NAME : Ni(100)-c(2x2)-Cl
CLASSIFICATION : 28.17.1
TECHNIQUE : SEXAFS and X-ray SW
AUTHORS : T. Yokoyama, Y. Takata, T. Ohta, M. Funabashi, Y. Kitajima and H. Kuroda
REFERENCE : Phys. Rev., B42, 7000 (1990)

SURFACE TYPE
Substrate : Ni
Crystal face: 100
Temperature : RT*
Bulk lattice: fcc
2D bulk symm: p4m
2D surf symm: p4m

Adsorbate: Cl
Coverage : 0.50 ML
Pattern : c(2x2)
Matrix : (1.000,-1.000)
(1.000, 1.000)

STRUCTURE TYPE
Atomic adsorption at a 4-fold hollow site
top Ni layer relaxes outward

SAMPLE PREPARATION (1 sample)
Treatment : ion bombardment followed by annealing; Cl2 from AgCl cell
Crystallinity: sharp LEED pattern
Anal. methods: AES
Contamination:

COMMENTS

DATA COLLECTION
Technique: SEXAFS and X-ray SW
Dataset : 3 polarizations of Cl K-edge SEXAFS spectrum, 2800-3200 eV XSW measurments around the Ni(200) Bragg reflection

THEORY/DATA TREATMENT
Fourier transform

STRUCTURES EXAMINED
Top, bridge and 4-fold sites

QUALITY OF EXPERIMENT-THEORY FIT
Visual

2D UNIT CELLS (1 domain observed)

Cell	Ax (Å)	Ay (Å)	Bx (Å)	By (Å)	α (°)	Matrix	Pattern	Cell type
Bulk	2.492	0.000	0.000	2.492	90.0	(1.000, 0.000) (0.000, 1.000)	(1x1)	b: bulk lattice
Surface 1	2.492	-2.492	2.492	2.492	90.0	(1.000,-1.000) (1.000, 1.000)	c(2x2)	s1: commens. superlattice

3D COORDINATES

C1: in a 4-fold site; small expansion of the 1st Ni interlayer spacing; coordinates derived from bond lengths and Ni-Cl layer spacing

Dx/Dy in Å, or as a fraction of layer's unit cell vectors; atom 0 at the origin. Epir/subr are bulk repeat vectors.

No. of atoms: 3 Bulk z = 1.762 Å

Region	Chem el.	At. no.	Cell type	Site occ.	Rel. to	Dx ± εx		Dy ± εy		Dz ± εz		Dz/Bz(%) ± εz/Bz
epir		-2					f		f		Å	
subr		-1				1.246	Å	1.246	Å	1.762	Å	
ovrl	Cl	1	s1	.50	0	0.000	Å	0.000	Å	0.000	Å	0.0
intf	Ni	2	b	1.00	0	1.246	Å	1.246	Å	1.600 ± .020	Å	90.8 ± 1.1
subl	Ni	3	b	1.00	0	0.000	Å	0.000	Å	3.562 ± .050	Å	202.2 ± 2.8

BOND DISTANCES AND ANGLES

Ni-Ni bond distances are derived from coordinates

No. of distances/angles: 2

Interatomic dist. A-B (Å)	Atom A	Atom B	Atom C	Bond angle A-B-C (°)
2.380	Cl1	Ni2		
2.637	Ni2	Ni3		

COMMON NAME : Ni(111)-(√3x√3)R30°-Cl
CLASSIFICATION : 28.17.2
TECHNIQUE : SEXAFS and X-ray SW
AUTHORS : M. Funabashi, T. Yokoyama, Y. Takata, T. Ohta, Y. Kitajima and H. Kuroda
REFERENCE : Surf. Sci., 242, 59 (1991)

ILLUSTRATION: 22,24

SURFACE TYPE
Substrate : Ni
Crystal face: 111
Temperature : RT*
Bulk lattice: fcc
2D bulk symm: p3m1
2D surf symm: p31m

Adsorbate: Cl
Coverage : 0.33 ML
Pattern : (√3x√3)R30°
Matrix : (1.000, 1.000)
(-2.000, 1.000)

SAMPLE PREPARATION (1 sample)
Treatment : Cl from electrochemical source; monitored by AES, LEED
Crystallinity: clean LEED pattern
Anal. methods:
Contamination:

DATA COLLECTION
Technique: SEXAFS and X-ray SW
Dataset : Cl K-edge from 0-6Å-1 at two incident angles XSW around normal incidence

STRUCTURES EXAMINED
Top, hollow and bridge sites

QUALITY OF EXPERIMENT-THEORY FIT
Visual

STRUCTURE TYPE
Atomic adsorption in 3-fold hollow site no surface relaxation; unsure as to fcc or hcp (fcc assumed in the tabulation)

COMMENTS

THEORY/DATA TREATMENT
Fourier transform analysis

2D UNIT CELLS (1 domain observed)

Cell	Ax (Å)	Ay (Å)	Bx (Å)	By (Å)	α (°)	Matrix	Pattern	Cell type
Bulk	2.492	0.000	1.246	2.158	60.0	(1.000, 0.000) (0.000, 1.000)	(1x1)	b: bulk lattice
Surface 1	3.738	2.158	-3.738	2.158	120.0	(1.000, 1.000) (-2.000, 1.000)	(√3x√3)R30°	s1: commens. superlattice

3D COORDINATES

Cl1: overlayer in fcc or hcp hollow sites (uncertain); error bars: ±-0.03 on the Cl-Ni bond length; coordinates are derived from bond distances and angles

Dx/Dy in Å, or as a fraction of layer's unit cell vectors; atom 0 at the origin. Epir/subr are bulk repeat vectors.

No. of atoms: 3

Bulk z = 2.035 Å

Reg ion	Chem el.	At. no.	Cell type	Site occ.	Rel. to	Dx ± εx	Dy ± εy	Dz ± εz	Dz/Bz(%) ± εz/Bz
epir		-2				f	f	Å	
subr		-1				1.246 Å	0.719 Å	2.035 Å	
ovrl	Cl	1	s1	.33	0	0.000 Å	0.000 Å	0.000 Å	0.0
intf	Ni	2	b	1.00	0	1.246 Å	0.719 Å	1.833 ± .020 Å	90.1 ± 1.0
subl	Ni	3	b	1.00	0	0.000 Å	0.000 Å	3.867 Å	190.1

BOND DISTANCES AND ANGLES

No. of distances/angles: 1

Interatomic dist. A-B (Å)	Atom A	Atom B	Atom C	Bond angle A-B-C (°)
2.330	Cl1	Ni2		

COMMON NAME : Ni(111)-(√3x√3)R30°-Cl
CLASSIFICATION : 28.17.3
TECHNIQUE : ARPEFS
AUTHORS : Li-Qiong Wang, Z. Hussain, Z.Q. Huang, A.E. Schach von Wittenau, D.W. Lindle and D.A. Shirley
REFERENCE : Phys. Rev., B44, 13711 (1991)

ILLUSTRATION: 22,24

SURFACE TYPE
Substrate : Ni
Crystal face: 111
Temperature : 120 K
Bulk lattice: fcc
2D bulk symm: p3m1
2D surf symm: p31m

Adsorbate: Cl
Coverage : 0.33 ML
Pattern : (√3x√3)R30°
Matrix : (1.000, 1.000)
(-2.000, 1.000)

STRUCTURE TYPE
Atomic adsorption in fcc hollow site with contraction between the 1st and 2nd Ni layers

SAMPLE PREPARATION (2 sample)
Treatment : exposed to Cl2 at RT followed by heating to 350 K
Crystallinity: sharp LEED pattern
Anal. methods: AES, LEED
Contamination:

COMMENTS
Also conducted at 300 K with same result

DATA COLLECTION
Technique: ARPEFS
Dataset : Cl 1s photoemission from 50-550 eV

THEORY/DATA TREATMENT
Fourier and MSSW

STRUCTURES EXAMINED
Fcc and hcp hollow sites

QUALITY OF EXPERIMENT-THEORY FIT
Chi**2=0.05-0.13

2D UNIT CELLS (1 domain observed)

Cell	Ax (Å)	Ay (Å)	Bx (Å)	By (Å)	α (°)	Matrix	Pattern	Cell type
Bulk	2.492	0.000	1.246	2.158	60.0	(1.000, 0.000) (0.000, 1.000)	(1x1)	b: bulk lattice
Surface 1	3.738	2.158	-3.738	2.158	120.0	(1.000, 1.000) (-2.000, 1.000)	(√3x√3)R30°	s1: commens. superlattice

3D COORDINATES

Cl1: atomic overlayer in fcc hollow site

Dx/Dy in Å, or as a fraction of layer's unit cell vectors; atom 0 at the origin. Epir/subr are bulk repeat vectors.

No. of atoms: 3

Bulk z = 2.035 Å

Region	Chem el.	At. no.	Cell type	Site occ.	Rel. to	Dx ± εx		Dy ± εy		Dz ± εz		Dz/Bz(%) ± εz/Bz
epir		-2					f		f		Å	
subr		-1				1.246	Å	0.719	Å	2.035	Å	
ovrl	Cl	1	s1	.33	0	0.000	Å	0.000	Å	0.000	Å	0.0
intf	Ni	2	b	1.00	0	1.246	Å	0.719	Å	1.837 ± .001	Å	90.0 ± .0
subl	Ni	3	b	1.00	0	2.492	Å	1.439	Å	3.968 ± .001	Å	195.0 ± .0

BOND DISTANCES AND ANGLES

Bond angles and distances are derived from coordinates

No. of distances/angles: 3

Interatomic dist. A-B (Å)	Atom A	Atom B	Atom C	Bond angle A-B-C (°)
2.332	Cl1	Ni1		
3.763	Cl1	Ni2		
1.926	Ni1	Ni2		

COMMON NAME : Ni(100)-(1x1)-3Co
CLASSIFICATION : 28.27.1
TECHNIQUE : ARAES
AUTHORS : S.A. Chambers, S.B. Anderson, H.-W. Chen and J.H. Weaver
REFERENCE : Phys. Rev., B35, 2592 (1987)

ILLUSTRATION: 83

SURFACE TYPE
Substrate : Ni
Crystal face: 100
Temperature : RT
Bulk lattice: fcc
2D bulk symm: p4m
2D surf symm: p4m

Adsorbate: Co
Coverage : 3 Co/(1x1)
Pattern : (1x1)
Matrix : (1.000, 0.000)
(0.000, 1.000)

STRUCTURE TYPE
Unstrained layers of fcc Co on fcc Ni(001) up to 30 ML (the highest studied); 1 and 2 ML films show island formation; only 3 ML case is tabulated here

SAMPLE PREPARATION (1 sample)
Treatment : resistive heating of purified Co
Crystallinity:
Anal. methods: quartz crystal oscillator to monitor Co layer thickness
Contamination:

COMMENTS
No theoretical interpretation of layers thicker than 3 ML; absence of strain was determined by noticing the absence of movements in peaks corresponding to forward-scattering directions

DATA COLLECTION
Technique: ARAES; angle-resolved Auger diffraction
Dataset : polar angle intensity distributions for Ni(100) and Co/Ni(100) interface over a wide range of coverages.

THEORY/DATA TREATMENT
Single scattering calculations for layers of 1-3 ML thickness

STRUCTURES EXAMINED
Strained structures were excluded as there was no movement in the forward scattering peaks

2D UNIT CELLS (1 domain observed)

Cell	Ax (Å)	Ay (Å)	Bx (Å)	By (Å)	α (°)	Matrix	Pattern	Cell type
Bulk	2.489	0.000	0.000	2.489	90.0	(1.000, 0.000) (0.000, 1.000)	(1x1)	b: bulk lattice
Surface 1	2.489	0.000	0.000	2.489	90.0	(1.000, 0.000) (0.000, 1.000)	(1x1)	s1: commens. superlattice

3D COORDINATES

Co1-Co3: 3 overlayers of Co

Dx/Dy in Å, or as a fraction of layer's unit cell vectors; atom 0 at the origin. Epir/subr are bulk repeat vectors.

No. of atoms: 4

Bulk z = 1.760 Å

Region	Chem el.	At. no.	Cell type	Site occ.	Rel. to	Dx ± εx		Dy ± εy		Dz ± εz		Dz/Bz(%) ± εz/Bz
epir		-2					f		f		Å	
subr		-1				1.245	Å	1.245	Å	1.760	Å	
ovrl	Co	1	s1	1.00	0	0.000	f	0.000	f	0.000	Å	0.0
ovrl	Co	2	s1	1.00	1	0.500	f	0.500	f	1.760	Å	100.0
ovrl	Co	3	s1	1.00	2	-0.500	f	-0.500	f	1.760	Å	100.0
subl	Ni	4	b	1.00	3	0.500	f	0.500	f	1.760	Å	100.0

BOND DISTANCES AND ANGLES

Bond distances and angles are derived from coordinates

No. of distances/angles: 3

Interatomic dist. A-B (Å)	Atom A	Atom B	Atom C	Bond angle A-B-C (°)
2.489	Co1	Co2	Co3	90.0
2.489	Co1	Co(1,0)	Co(1,1)	90.0
2.489	Co3	Ni1	Ni1(1,1)	90.0

ILLUSTRATION: 83

COMMON NAME : Ni(100)-(1x1)-Cu
CLASSIFICATION : 28.29.3
TECHNIQUE : LEED
AUTHORS : M. Abu-Joudeh, P.P. Vaishnava and P.A. Montano
REFERENCE : J. Phys., C17, 6899 (1984)

SURFACE TYPE
Substrate : Ni
Crystal face: 100
Temperature : RT*
Bulk lattice: fcc
2D bulk symm: p4m
2D surf symm: p4m

Adsorbate: Cu
Coverage : 1.0 (Cu/Ni)
Pattern : (1x1)
Matrix : (1.000, 0.000)
(0.000, 1.000)

STRUCTURE TYPE
Atomic adsorption in 4-fold hollow sites

SAMPLE PREPARATION (1 sample)
Treatment : Cu deposited by heating Knudsen cell to 1250C (1Å/min)
Crystallinity:
Anal. methods:
Contamination: AES: no impurities detected

COMMENTS

DATA COLLECTION
Technique: LEED
Dataset : I-V curves for 00,10,11,20 and 22 beams: 20<E<240 eV

THEORY/DATA TREATMENT
Dynamical LEED (RFS): 8 phase shifts, ΘD=440 K(Ni), 344K(Cu)

STRUCTURES EXAMINED
Variation of Cu-Ni interlayer spacing

QUALITY OF EXPERIMENT-THEORY FIT
Visual

2D UNIT CELLS (1 domain observed)

Cell	Ax (Å)	Ay (Å)	Bx (Å)	By (Å)	α (°)	Matrix	Pattern	Cell type
Bulk	2.489	0.000	0.000	2.489	90.0	(1.000, 0.000) (0.000, 1.000)	(1x1)	b: bulk lattice
Surface 1	2.489	0.000	0.000	2.489	90.0	(1.000, 0.000) (0.000, 1.000)	(1x1)	s1: commens. superlattice

3D COORDINATES

Cu1: overlayer, continuing fcc lattice

Dx/Dy in Å, or as a fraction of layer's unit cell vectors; atom 0 at the origin. Epir/subr are bulk repeat vectors.

No. of atoms: 3

Bulk z = 1.770 Å

Region	Chem el.	At. no.	Cell type	Site occ.	Rel. to	Dx ± εx		Dy ± εy		Dz ± εz		Dz/Bz(%) ± εz/Bz
epir		-2					f		f		Å	
subr		-1				-1.245	Å	-1.245	Å	1.770	Å	
ovrl	Cu	1	b	1.00	0	0.000	f	0.000	f	0.000	Å	0.0
intf	Ni	2	b	1.00	1	0.500	f	0.500	f	1.800 ± .030	Å	101.7 ± 1.7
subl	Ni	3	b	1.00	2	-0.500	f	-0.500	f	1.770	Å	100.0

BOND DISTANCES AND ANGLES

Bond distances and angles are derived from coordinates

No. of distances/angles: 4

Interatomic dist. A-B (Å)	Atom A	Atom B	Atom C	Bond angle A-B-C (°)
2.489	Cu1	Cu1(1,0)	Ni2	60.4
2.518	Cu1	Ni2	Ni3(1,0)	120.5
2.518	Cu1	Ni2	Ni3	90.8
2.496	Ni2			

COMMON NAME : Ni(100)-(1x1)-3Cu
CLASSIFICATION : 28.29.5a
TECHNIQUE : MEED
AUTHORS : S.A. Chambers, I.M. Vitomirov, S.B. Anderson and J.H. Weaver
REFERENCE : Phys. Rev., B35, 2490 (1987)

ILLUSTRATION: 83

SURFACE TYPE
Substrate : Ni
Crystal face: 100
Temperature : RT
Bulk lattice: fcc
2D bulk symm: p4m
2D surf symm: p4m

Adsorbate: Cu
Coverage : 3 Cu/(1x1)
Pattern : (1x1)
Matrix : (1.000, 0.000)
(0.000, 1.000)

STRUCTURE TYPE
For 12Å of pseudomorphic Cu on Ni(100), a tetragonal distortion perpendicular to the surface is observed; the new lattice constant for Cu in that direction is 3.75± 0.02Å (compared to 3.61Å for the bulk)

SAMPLE PREPARATION (1 sample)
Treatment :
Crystallinity:
Anal. methods:
Contamination:

COMMENTS
In the structure optimization, the Cu lattice constant in the plane of the interface was held at the value for Ni (3.52 Å)

DATA COLLECTION
Technique: MEED; medium-energy electron diffraction
Dataset : 1-k eV elastic-peak polar-angle distributions for Ni(100), 12Å Cu/Ni(001), and Cu(001) in the (100) azimuth

THEORY/DATA TREATMENT
Single-scattering plane-wave calculations: incident beam channeling neglected

STRUCTURES EXAMINED
Perpendicular lattice parameter varied from 3.6 to 3.9Å; best fits to experiment found at 3.73, 3.75 and 3.77Å

QUALITY OF EXPERIMENT-THEORY FIT
Visual

2D UNIT CELLS (1 domain observed)

Cell	Ax (Å)	Ay (Å)	Bx (Å)	By (Å)	α (°)	Matrix	Pattern	Cell type
Bulk	2.489	0.000	0.000	2.489	90.0	(1.000, 0.000) (0.000, 1.000)	(1x1)	b: bulk lattice
Surface 1	2.489	0.000	0.000	2.489	90.0	(1.000, 0.000) (0.000, 1.000)	(1x1)	s1: commens. superlattice

3D COORDINATES

Cu1-Cu3: 3 layers (app. 12 Å) of Cu

Dx/Dy in Å, or as a fraction of layer's unit cell vectors; atom 0 at the origin. Epir/subr are bulk repeat vectors.

No. of atoms: 4

Bulk z = 1.760 Å

Reg ion	Chem el.	At. no.	Cell type	Site occ.	Rel. to	Dx ± ϵx		Dy ± ϵy		Dz ± ϵz		Dz/Bz(%) ± ϵz/Bz
epir		-2					f		f		Å	
subr		-1				1.245	Å	1.245	Å	1.760	Å	
ovrl	Cu	1	s1	1.00	0	0.000	f	0.000	f	0.000	Å	0.0
ovrl	Cu	2	s1	1.00	1	0.500	f	0.500	f	1.875 ± .020	Å	106.5 ± 1.1
ovrl	Cu	3	s1	1.00	2	-0.500	f	-0.500	f	1.875 ± .020	Å	106.5 ± 1.1
subl	Ni	4	b	1.00	3	0.500	f	0.500	f	1.760	Å	100.0

BOND DISTANCES AND ANGLES

Bond distances and angles are derived from coordinates

No. of distances/angles: 3

Interatomic dist. A-B (Å)	Atom A	Atom B	Atom C	Bond angle A-B-C (°)
2.489	Cu1	Cu1(1,0)	Cu1(1,1)	90.0
2.572	Cu1	Cu2	Cu3	93.6
2.572	Cu2	Cu3	Ni4	91.8

ILLUSTRATION: 83

COMMON NAME : Ni(100)-(1x1)-Cu multilayer
CLASSIFICATION : 28.29.5
TECHNIQUE : AED
AUTHORS : S.A. Chambers, H.W. Chen, I.M. Vitomirov, S.B. Anderson and J.H. Weaver
REFERENCE : Phys. Rev., B33, 8810 (1986)

SURFACE TYPE
Substrate : Ni
Crystal face: 100
Temperature : RT*
Bulk lattice: fcc
2D bulk symm: p4m
2D surf symm: p4m

Adsorbate: Cu
Coverage : multilayer (<14Å)
Pattern : (1x1)
Matrix : (1.000, 0.000)
(0.000, 1.000)

STRUCTURE TYPE
Epitaxial strained (1x1) multilayer, with Cu lattice constant perpendicular to interface (=interlayer spacing x2) of 3.71Å, vs 3.52Å parallel to interface (same as Ni)

SAMPLE PREPARATION (1 sample)
Treatment : resistive evaporation of high-purity Cu
Crystallinity:
Anal. methods:
Contamination: monitored with LEED and AES

COMMENTS

DATA COLLECTION
Technique: AED
Dataset : polar profiles of Ni and Cu L(3)M(4,5)M(4,5) Auger intensities

THEORY/DATA TREATMENT
Comparison of exp. profiles to single scattering theory

STRUCTURES EXAMINED
Cu lattice constant perpendicular to interface examined

QUALITY OF EXPERIMENT-THEORY FIT
Visual

2D UNIT CELLS (1 domain observed)

Cell	Ax (Å)	Ay (Å)	Bx (Å)	By (Å)	α (°)	Matrix	Pattern	Cell type
Bulk	2.489	0.000	0.000	2.489	90.0	(1.000, 0.000) (0.000, 1.000)	(1x1)	b: bulk lattice
Surface 1	2.489	0.000	0.000	2.489	90.0	(1.000, 0.000) (0.000, 1.000)	(1x1)	s1: commens. superlattice

3D COORDINATES

Cu1-Cu3: beginning of strained Cu multilayer (thickness < 14Å);

Dx/Dy in Å, or as a fraction of layer's unit cell vectors; atom 0 at the origin. Epir/subr are bulk repeat vectors.

No. of atoms: 3

Bulk z = 1.860 Å

Reg ion	Chem el.	At. no.	Cell type	Site occ.	Rel. to	Dx ± εx		Dy ± εy		Dz ± εz		Dz/Bz(%) ± εz/Bz
epir		-2					f		f		Å	
subr		-1				1.245	Å	1.245	Å	1.860	Å	
intf	Cu	1	b	1.00	0	0.000	f	0.000	f	0.000	Å	0.0
intf	Cu	2	b	1.00	1	0.500	f	0.500	f	1.860 ± .015	Å	100.0 ± .8
subl	Cu	3	b	1.00	2	-0.500	f	-0.500	f	1.860 ± .015	Å	100.0 ± .8

BOND DISTANCES AND ANGLES

Bond distances and angles are derived from coordinates

No. of distances/angles: 2

Interatomic dist. A-B (Å)	Atom A	Atom B	Atom C	Bond angle A-B-C (°)
2.489	Cu1	Cu1(1,0)		
2.561	Cu1	Cu2	Cu3	93.2

COMMON NAME : Ni(100)-(1x1)-1Fe
CLASSIFICATION : 28.26.2a
TECHNIQUE : LEED
AUTHORS : S.H. Lu, Z.Q. Wang, D. Tian, Y.S. Li, F. Jona and P.M. Marcus
REFERENCE : Surf. Sci., 221, 35 (1989)

ILLUSTRATION: 83

SURFACE TYPE
Substrate : Ni
Crystal face: 100
Temperature : RT
Bulk lattice: fcc
2D bulk symm: p4m
2D surf symm: p4m

Adsorbate: Fe
Coverage : 1.0 Fe/Ni
Pattern : (1x1)
Matrix : (1.000, 0.000)
(0.000, 1.000)

STRUCTURE TYPE
Epitaxial (1x1) monolayer

SAMPLE PREPARATION (3 sample)
Treatment : cycles of Ar+ bomb. and 800-1000C anneals; Fe vapor-depos.
Crystallinity: LEED: some defects and disorder
Anal. methods: AES
Contamination: AES: no contaminants

COMMENTS

DATA COLLECTION
Technique: LEED; TV camera-microcomputer system
Dataset : IV spectra for 4 beams (10,11,20,21) at normal incidence; cumul. E range 820 eV

THEORY/DATA TREATMENT
Dynamical LEED (CHANGE program): 49 beams, 6 phase shifts from Moruzzi et al pot (Ni); E-dep. Vor and Voi

STRUCTURES EXAMINED
Fcc continuation, varying top 2 interlayer spacings between -12% and +12% of bulk Ni spacing of 1.76Å

QUALITY OF EXPERIMENT-THEORY FIT
RZJ=0.08

2D UNIT CELLS (1 domain observed)

Cell	Ax (Å)	Ay (Å)	Bx (Å)	By (Å)	α (°)	Matrix	Pattern	Cell type
Bulk	2.489	0.000	0.000	2.489	90.0	(1.000, 0.000) (0.000, 1.000)	(1x1)	b: bulk lattice
Surface 1	2.489	0.000	0.000	2.489	90.0	(1.000, 0.000) (0.000, 1.000)	(1x1)	s1: commens. superlattice

3D COORDINATES

Fe1: epitaxial overlayer in fcc continuation; 0.05Å error bars assumed for tabulation

Dx/Dy in Å, or as a fraction of layer's unit cell vectors; atom 0 at the origin. Epir/subr are bulk repeat vectors.

No. of atoms: 4

Bulk z = 1.760 Å

Region	Chem el.	At. no.	Cell type	Site occ.	Rel. to	Dx ± εx		Dy ± εy		Dz ± εz		Dz/Bz(%) ± εz/Bz
epir		-2					f		f		Å	
subr		-1				1.245	Å	1.245	Å	1.760	Å	
ovrl	Fe	1	s1	1.00	0	0.000	f	0.000	f	0.000	Å	0.0
intf	Ni	2	b	1.00	1	0.500	f	0.500	f	1.850 ± .050	Å	105.1 ± 2.8
intf	Ni	3	b	1.00	2	-0.500	f	-0.500	f	1.752 ± .050	Å	99.6 ± 2.8
subl	Ni	4	b	1.00	3	0.500	f	0.500	f	1.760	Å	100.0

BOND DISTANCES AND ANGLES

Bond distances and angles are derived from coordinates

No. of distances/angles: 3

Interatomic dist. A-B (Å)	Atom A	Atom B	Atom C	Bond angle A-B-C (°)
2.489	Fe1	Fe1(1,0)	Ni2	60.8
2.553	Fe1	Ni2	Ni2(1,0)	119.2
2.483	Ni2	Ni3	Ni4	89.9

COMMON NAME : Ni(100)-(1x1)-2Fe
CLASSIFICATION : 28.26.2b
TECHNIQUE : LEED
AUTHORS : S.H. Lu, Z.Q. Wang, D. Tian, Y.S. Li, F. Jona and P.M. Marcus
REFERENCE : Surf. Sci., 221, 35 (1989)

ILLUSTRATION: 83

SURFACE TYPE
Substrate : Ni
Crystal face: 100
Temperature : RT
Bulk lattice: fcc
2D bulk symm: p4m
2D surf symm: p4m

Adsorbate: Fe
Coverage : 2.0 Fe/Ni
Pattern : (1x1)
Matrix : (1.000, 0.000)
(0.000, 1.000)

STRUCTURE TYPE
Epitaxial (1x1) bilayer

SAMPLE PREPARATION (3 sample)
Treatment : cycles of Ar+ bomb. and 800-1000C anneals; Fe vapor-depos.
Crystallinity: LEED: some defects and disorder
Anal. methods: AES
Contamination: AES: no contaminants

COMMENTS

DATA COLLECTION
Technique: LEED; TV camera-microcomputer system
Dataset : IV spectra for 4 beams (10,11,20,21) at normal incidence; cumul. E range 820 eV

THEORY/DATA TREATMENT
Dynamical LEED (CHANGE program): 49 beams, 6 phase shifts from Moruzzi et al pot (Ni); E-dep. Vor and Voi

STRUCTURES EXAMINED
Fcc continuation, varying top 2 interlayer spacings between -12% and +12% of bulk Ni spacing of 1.76Å

QUALITY OF EXPERIMENT-THEORY FIT
RZJ=0.09

2D UNIT CELLS (1 domain observed)

Cell	Ax (Å)	Ay (Å)	Bx (Å)	By (Å)	α (°)	Matrix	Pattern	Cell type
Bulk	2.489	0.000	0.000	2.489	90.0	(1.000, 0.000) (0.000, 1.000)	(1x1)	b: bulk lattice
Surface 1	2.489	0.000	0.000	2.489	90.0	(1.000, 0.000) (0.000, 1.000)	(1x1)	s1: commens. superlattice

3D COORDINATES

Fe1-Fe2: epitaxial overlayers in fcc continuation; 0.05Å error bars assumed for tabulation

Dx/Dy in Å, or as a fraction of layer's unit cell vectors; atom 0 at the origin. Epir/subr are bulk repeat vectors.

No. of atoms: 4 Bulk z = 1.760 Å

Reg ion	Chem el.	At. no.	Cell type	Site occ.	Rel. to	Dx ± εx	Dy ± εy	Dz ± εz	Dz/Bz(%) ± εz/Bz
epir		-2				f	f	Å	
subr		-1				1.245 Å	1.245 Å	1.760 Å	
ovrl	Fe	1	s1	1.00	0	0.000 f	0.000 f	0.000 Å	0.0
ovrl	Fe	2	s1	1.00	1	0.500 f	0.500 f	1.900 ± .050 Å	108.0 ± 2.8
intf	Ni	3	b	1.00	2	-0.500 f	-0.500 f	1.752 ± .050 Å	99.6 ± 2.8
subl	Ni	4	b	1.00	3	0.500 f	0.500 f	1.760 Å	100.0

BOND DISTANCES AND ANGLES

Bond distances and angles are derived from coordinates

No. of distances/angles: 3

Interatomic dist. A-B (Å)	Atom A	Atom B	Atom C	Bond angle A-B-C (°)
2.489	Fe1	Fe1(1,0)	Fe2	61.3
2.590	Fe1	Fe2	Ni3	92.1
2.483	Fe2	Ni3	Ni4	89.9

COMMON NAME : Ni(100)-H (D) disordered
CLASSIFICATION : 28.1.17
TECHNIQUE : HEIS
AUTHORS : I. Stensgaard and F. Jakobsen
REFERENCE : Phys. Rev. Lett., 54, 711 (1985)

ILLUSTRATION: 28

SURFACE TYPE
Substrate : Ni
Crystal face: 100
Temperature : 120 K
Bulk lattice: fcc
2D bulk symm: p4m
2D surf symm: none

Adsorbate: H (D)
Coverage : unknown
Pattern : disordered
Matrix : (1.000, 0.000)
(0.000, 1.000)

STRUCTURE TYPE
Disordered atomic adsorption in 4-fold hollow sites

SAMPLE PREPARATION (1 sample)
Treatment : D2 exposure at 1E-10 torr at 170 K
Crystallinity:
Anal. methods:
Contamination: monitored by AES and LEED.

COMMENTS
Analysis resulted in the use of a fitted rms vibrational amplitude of 0.24Å parallel to the surface

DATA COLLECTION
Technique: HEIS; He(3) transmission channeling: E=800k
Dataset : angular scans along [100], [110] and [111] for angles of ±8° off normal.

THEORY/DATA TREATMENT
Comparison with angular scans from a multirow continuum model; rms vib ampls=0.21Å

STRUCTURES EXAMINED
Hollow and bridge sites: spacings of 0.3, 0.5 and 0.7Å

2D UNIT CELLS (1 domain observed)

Cell	Ax (Å)	Ay (Å)	Bx (Å)	By (Å)	α (°)	Matrix	Pattern	Cell type
Bulk	2.490	0.000	0.000	2.490	90.0	(1.000, 0.000) (0.000, 1.000)	(1x1)	b: bulk lattice
Surface 1	2.490	0.000	0.000	2.490	90.0	(1.000, 0.000) (0.000, 1.000)	disordered	nd1: non-recon. lattice-gas dis

3D COORDINATES

H1: disordered D overlayer in 4-fold hollow sites (coverage unknown: assumed 0.25 here)

Dx/Dy in Å, or as a fraction of layer's unit cell vectors; atom 0 at the origin. Epir/subr are bulk repeat vectors.

No. of atoms: 3

Bulk z = 1.760 Å

Reg ion	Chem el.	At. no.	Cell type	Site occ.	Rel. to	Dx ± εx		Dy ± εy		Dz ± εz		Dz/Bz(%) ± εz/Bz
epir		-2					f		f		Å	
subr		-1				-1.245	Å	-1.245	Å	1.760	Å	
ovrl	H	1	nd1	.25	0	0.000	f	0.000	f	0.000	Å	0.0
intf	Ni	2	b	1.00	1	0.500	f	0.500	f	0.500 ± .100	Å	28.4 ± 5.7
subl	Ni	3	b	1.00	2	-0.500	f	-0.500	f	1.760	Å	100.0

BOND DISTANCES AND ANGLES

Bond distances and angles are derived from coordinates

No. of distances/angles: 2

Interatomic dist. A-B (Å)	Atom A	Atom B	Atom C	Bond angle A-B-C (°)
1.830	H1	Ni2	Ni3	60.8
2.490	Ni2	Ni2(1,0)		

COMMON NAME : Ni(110)-(2x1)-2H
CLASSIFICATION : 28.1.23
TECHNIQUE : LEED
AUTHORS : W. Reimer, V. Penka, M. Skottke, R.J. Behm, G. Ertl and W. Moritz
REFERENCE : Surf. Sci., 186, 45 (1987)

ILLUSTRATION: 35,37

SURFACE TYPE
Substrate : Ni
Crystal face: 110
Temperature : 180 K
Bulk lattice: fcc
2D bulk symm: pmm
2D surf symm: pmg

Adsorbate: H
Coverage : 1 H/Ni
Pattern : (2x1)
Matrix : (2.000, 0.000)
(0.000, 1.000)

STRUCTURE TYPE
Atomic adsorption in fcc 3-fold hollows on (111) facets of unreconstructed substrate with multilayer relaxations perp. to surface

SAMPLE PREPARATION (1 sample)
Treatment : exposure to 0.8L H2 at 120 K
Crystallinity:
Anal. methods:
Contamination:

COMMENTS

DATA COLLECTION
Technique: LEED
Dataset : IV curves at normal and off-normal incidence; 40<E<180 eV

THEORY/DATA TREATMENT
Dynamical LEED (layer doubling): Moruzzi et al pot (Ni), H xtal pot, 8 ph shs; Vor∝E**-1/2, Voi∝E**1/3; ΘD=450 K

STRUCTURES EXAMINED
Top 3 interlayer spacings varied; positions of H atoms varied parallel and perpendicular to surface for:
1) long bridge sites; 2) fcc hollow sites on (111) facets;
also test of lateral displacement of top-layer Ni atoms

QUALITY OF EXPERIMENT-THEORY FIT
RPE=0.27, RZJ=0.17

2D UNIT CELLS (1 domain observed)

Cell	Ax (Å)	Ay (Å)	Bx (Å)	By (Å)	α (°)	Matrix	Pattern	Cell type
Bulk	2.490	0.000	0.000	3.520	90.0	(1.000, 0.000) (0.000, 1.000)	(1x1)	b: bulk lattice
Surface 1	4.980	0.000	0.000	3.520	90.0	(2.000, 0.000) (0.000, 1.000)	(2x1)	s1: commens. superlattice

3D COORDINATES

H1-H2: zigzag chains in Ni troughs over fcc hollow sites of (111) facets (each H bonds to 2 Ni3 and 1 Ni4 atoms)

Dx/Dy in Å, or as a fraction of layer's unit cell vectors; atom 0 at the origin. Epir/subr are bulk repeat vectors.

No. of atoms: 6

Bulk z = 1.245 Å

Region	Chem el.	At. no.	Cell type	Site occ.	Rel. to	Dx ± εx		Dy ± εy		Dz ± εz		Dz/Bz(%) ± εz/Bz
epir		-2					f		f		Å	
subr		-1				-1.245	Å	-1.760	Å	1.245	Å	
ovrl	H	1	s1	.50	0	0.000	f	0.000	f	0.000	Å	0.0
ovrl	H	2	s1	.50	1	0.500	f	0.374	f	0.000	Å	0.0
intf	Ni	3	b	1.00	2	-0.500	f	0.313	f	0.410 ± .100	Å	32.9 ± 8.0
intf	Ni	4	b	1.00	3	-0.500	f	-0.500	f	1.190 ± .020	Å	95.6 ± 1.6
intf	Ni	5	b	1.00	4	0.500	f	0.500	f	1.310 ± .020	Å	105.2 ± 1.6
subl	Ni	6	b	1.00	5	-0.500	f	-0.500	f	1.245	Å	100.0

BOND DISTANCES AND ANGLES

Bond distances and angles are derived from coordinates

No. of distances/angles: 4

Interatomic dist. A-B (Å)	Atom A	Atom B	Atom C	Bond angle A-B-C (°)
1.712	H1	Ni3(0,-1)	Ni4	44.6
1.730	H1	Ni4	Ni5	138.2
2.490	Ni3	Ni3(1,0)		

Ni(110)-(2x1)-2H
28.1.23

Bond Distances and Angles - Continued

Interatomic dist. A-B (Å)	Atom A	Atom B	Atom C	Bond angle A-B-C (°)
2.463	Ni3	Ni4	Ni5	60.2

COMMON NAME : Ni(111)-(2x2)-2H
CLASSIFICATION : 28.1.6
TECHNIQUE : LEED
AUTHORS : K. Christmann, R.J. Behm, G. Ertl, M.A. Van Hove and W.H. Weinberg
REFERENCE : J. Chem. Phys., 70, 4168 (1979)

ILLUSTRATION: 22,25

SURFACE TYPE
Substrate : Ni
Crystal face: 111
Temperature : 110 K
Bulk lattice: fcc
2D bulk symm: p3m1
2D surf symm: p3m1

Adsorbate: H
Coverage : 1/2 (H/Ni)
Pattern : (2x2)
Matrix : (2.000, 0.000)
(0.000, 2.000)

STRUCTURE TYPE
Atomic adsorption in both kinds of hollow sites simultaneously, at same height, forming honeycomb lattice

SAMPLE PREPARATION (1 sample)
Treatment : hydrogen exposure yields about 0.5 ML coverage
Crystallinity:
Anal. methods:
Contamination: AES: C-free; ultrapure H2 gas (99.95%)

COMMENTS

DATA COLLECTION
Technique: LEED
Dataset : I-V curves: (Θ=0,φ=0): 1/2 0,0 1/2,1/2 1/2 beams; integral orders insensitive to H; total E-range 220 eV

THEORY/DATA TREATMENT
Dynamical LEED: Wakoh Ni potential, H potential with Slater exchange; 5 phase shifts;Vor=-11.0 eV, Voi α E1/3; ΘD=314 K

STRUCTURES EXAMINED
1) 1/4 ML (2x2): 11 positions (2 hollows, bridge); 2) 1/2 ML (2x1): 45 positions (top, hollow, bridge, underlayer); 3) 1/2 ML (2x2): 25 positions (honeycomb, planar or buckled with different hollows or hollow/top combinations; 4) 1/2 ML (2x2) quasi molecular: 16 positions, parallel or tilted

QUALITY OF EXPERIMENT-THEORY FIT
Visual

2D UNIT CELLS (1 domain observed)

Cell	Ax (Å)	Ay (Å)	Bx (Å)	By (Å)	α (°)	Matrix	Pattern	Cell type
Bulk	2.490	0.000	1.245	2.156	60.0	(1.000, 0.000) (0.000, 1.000)	(1x1)	b: bulk lattice
Surface 1	4.980	0.000	2.490	4.313	60.0	(2.000, 0.000) (0.000, 2.000)	(2x2)	s1: commens. superlattice

3D COORDINATES

H1-H2: honeycomb overlayer in both kinds of 3-fold hollow sites

Dx/Dy in Å, or as a fraction of layer's unit cell vectors; atom 0 at the origin. Epir/subr are bulk repeat vectors.

No. of atoms: 4

Bulk z = 2.033 Å

Region	Chem el.	At. no.	Cell type	Site occ.	Rel. to	Dx ± εx		Dy ± εy		Dz ± εz		Dz/Bz(%) ± εz/Bz
epir		-2					f		f		Å	
subr		-1				0.000	Å	1.438	Å	2.033	Å	
ovrl	H	1	s1	.25	0	0.000	f	0.000	f	0.000	Å	0.0
ovrl	H	2	s1	.25	1	0.667	f	0.667	f	0.000 ± .100	Å	0.0 ± 4.9
intf	Ni	3	b	1.00	1	0.667	f	0.667	f	1.150 ± .100	Å	56.6 ± 4.9
subl	Ni	4	b	1.00	3	-0.333	f	0.667	f	2.033	Å	100.0

BOND DISTANCES AND ANGLES

Bond distances and angles are derived from coordinates

No. of distances/angles: 3

Interatomic dist. A-B (Å)	Atom A	Atom B	Atom C	Bond angle A-B-C (°)
1.841	Ni3	H1(1,1)	H2	141.3

Ni(111)-(2x2)-2H
28.1.6

Bond Distances and Angles - Continued

Interatomic dist. A-B (Å)	Atom A	Atom B	Atom C	Bond angle A-B-C (°)
1.841	Ni3	H1(1,1)	Ni3(1,0)	85.1
2.490	Ni3	Ni3(1,0)	H2	90.0

COMMON NAME : Ni(111)-(2x2)-2H (D)
CLASSIFICATION : 28.1.31
TECHNIQUE : HEIS
AUTHORS : K. Mortensen, F. Besenbacher, I. Stensgaard and W.R. Wampler
REFERENCE : Surf. Sci., 205, 433 (1988)

ILLUSTRATION: 22,25

SURFACE TYPE
Substrate : Ni
Crystal face: 111
Temperature : 140 K
Bulk lattice: fcc
2D bulk symm: p3m1
2D surf symm: p3m1

Adsorbate: H (D)
Coverage : 0.5 D/Ni
Pattern : (2x2)
Matrix : (2.000, 0.000)
(0.000, 2.000)

STRUCTURE TYPE
Atomic adsorption with equal occupation of fcc and hcp hollow sites

SAMPLE PREPARATION (1 sample)
Treatment : sputtering/heating and O/H reduction cycles; D deposition
Crystallinity:
Anal. methods:
Contamination:

COMMENTS

DATA COLLECTION
Technique: HEIS; transmission channeling of high-E ion
Dataset : angular scans around the [111], [110], [112] and [100] axial and across the {100} and {111} planar directions

THEORY/DATA TREATMENT
Multirow continuum calculation, including electr. and nucl. scattering; D detected through D to He4 nuclear reaction

STRUCTURES EXAMINED
Fcc and hcp hollow sites (equally occupied); hcp hollow site; fcc hollow site; top site;
D-Ni spacing varied, substrate bulk-like

QUALITY OF EXPERIMENT-THEORY FIT
Visual

2D UNIT CELLS (1 domain observed)

Cell	Ax (Å)	Ay (Å)	Bx (Å)	By (Å)	α (°)	Matrix	Pattern	Cell type
Bulk	2.490	0.000	-1.245	2.156	120.0	(1.000, 0.000) (0.000, 1.000)	(1x1)	b: bulk lattice
Surface 1	4.980	0.000	-2.490	4.313	120.0	(2.000, 0.000) (0.000, 2.000)	(2x2)	s1: commens. superlattice

3D COORDINATES

H1-H2: overlayer in both fcc and hcp hollow sites

Dx/Dy in Å, or as a fraction of layer's unit cell vectors; atom 0 at the origin. Epir/subr are bulk repeat vectors.

No. of atoms: 4 Bulk z = 2.030 Å

Region	Chem el.	At. no.	Cell type	Site occ.	Rel. to	Dx ± εx		Dy ± εy		Dz ± εz		Dz/Bz(%) ± εz/Bz
epir		-2					f		f		Å	
subr		-1				0.000	Å	1.438	Å	2.030	Å	
ovrl	H	1	s1	.25	0	0.000	f	0.000	f	0.000	Å	0.0
ovrl	H	2	s1	.25	1	0.333	f	0.667	f	0.000	Å	0.0
intf	Ni	3	b	1.00	2	-0.333	f	-0.667	f	0.800 ± .100	Å	39.4 ± 4.9
subl	Ni	4	b	1.00	3	0.333	f	0.667	f	2.030	Å	100.0

BOND DISTANCES AND ANGLES

Bond distances and angles are derived from coordinates

No. of distances/angles: 3

Interatomic dist. A-B (Å)	Atom A	Atom B	Atom C	Bond angle A-B-C (°)
1.645	H1	Ni3	H2	121.8
1.645	H1	Ni3	Ni4	154.4
2.490	Ni3	Ni3(1,0)		

COMMON NAME : Ni(100)-c(2x2)-Hg ILLUSTRATION: 28
CLASSIFICATION : 28.80.1
TECHNIQUE : XSW
AUTHORS : N.P. Prince, N.K. Singh, W. Walter, D.P. Woodruff and Robert G. Jones
REFERENCE : J. Phys. CM, 1, SB21 (1989)

SURFACE TYPE
Substrate : Ni
Crystal face: 100
Temperature : RT*
Bulk lattice: fcc
2D bulk symm: p4m
2D surf symm: p4m

Adsorbate: Hg
Coverage : 0.5 ML
Pattern : c(2x2)
Matrix : (1.000, 1.000)
(-1.000, 1.000)

STRUCTURE TYPE
Atomic adsorption in bridge site with Ni-Hg bond length characteristic of metallic radii or the NiHg compound

SAMPLE PREPARATION (1 sample)
Treatment : ion bombardment and annealing
Crystallinity: sharp LEED pattern
Anal. methods: LEED, AES
Contamination:

COMMENTS

DATA COLLECTION
Technique: XSW
Dataset : Auger yield of Ni and Hg as fct of E around normal incidence

THEORY/DATA TREATMENT
XSW analysis

STRUCTURES EXAMINED
Variation of adsorbate-substrate layer spacing bridge, top and hollow sites

QUALITY OF EXPERIMENT-THEORY FIT
Visual

2D UNIT CELLS (2 domains observed)

Cell	Ax (Å)	Ay (Å)	Bx (Å)	By (Å)	α (°)	Matrix	Pattern	Cell type
Bulk	2.492	0.000	0.000	2.492	90.0	(1.000, 0.000) (0.000, 1.000)	(1x1)	b: bulk lattice
Surface 1	2.492	2.492	-2.492	2.492	90.0	(1.000, 1.000) (-1.000, 1.000)	c(2x2)	s1: commens. superlattice

3D COORDINATES

Hg1: atomic overlayer in bridge site

Dx/Dy in Å, or as a fraction of layer's unit cell vectors; atom 0 at the origin. Epir/subr are bulk repeat vectors.

No. of atoms: 3 Bulk z = 1.762 Å

Region	Chem el.	At. no.	Cell type	Site occ.	Rel. to	Dx ± εx		Dy ± εy		Dz ± εz		Dz/Bz(%) ± εz/Bz
epir		-2					f		f		Å	
subr		-1				1.246	Å	-1.246	Å	1.762	Å	
ovrl	Hg	1	s1	.50	0	0.000	Å	0.000	Å	0.000	Å	0.0
intf	Ni	2	b	1.00	1	1.246	Å	0.000	Å	2.360 ± .100	Å	133.9 ± 5.7
subl	Ni	3	b	1.00	2	1.246	Å	-1.246	Å	1.762	Å	100.0

BOND DISTANCES AND ANGLES

Bond distances and angles are derived from coordinates

No. of distances/angles: 1

Interatomic dist. A-B (Å)	Atom A	Atom B	Atom C	Bond angle A-B-C (°)
2.669	Hg1	Ni2	Ni2(1,0)	64.1

COMMON NAME : Ni(100)-c(2x2)-I
CLASSIFICATION : 28.53.2
TECHNIQUE : SEXAFS
AUTHORS : R.G. Jones, S. Ainsworth, M.D. Crapper, C. Somerton and D.P. Woodruff
REFERENCE : Surf. Sci., 179, 425 (1987)

ILLUSTRATION: 28,29

SURFACE TYPE
Substrate : Ni
Crystal face: 100
Temperature : RT
Bulk lattice: fcc
2D bulk symm: p4m
2D surf symm: p4m

Adsorbate: I
Coverage : 0.5 I/Ni
Pattern : c(2x2)
Matrix : (1.000, 1.000)
(-1.000, 1.000)

STRUCTURE TYPE
Atomic adsorption in hollow sites

SAMPLE PREPARATION (1 sample)
Treatment : exposure to iodine vapor
Crystallinity:
Anal. methods:
Contamination: monitored by LEED and AES

COMMENTS

DATA COLLECTION
Technique: SEXAFS; synchroton radiation, 2 GeV electro
Dataset : SEXAFS

THEORY/DATA TREATMENT
Multi-shell simulation of SEXAFS

STRUCTURES EXAMINED
Iodine in hollow, bridge, and top sites with variable Ni-I bond length

2D UNIT CELLS (1 domain observed)

Cell	Ax (Å)	Ay (Å)	Bx (Å)	By (Å)	α (°)	Matrix	Pattern	Cell type
Bulk	2.489	0.000	0.000	2.489	90.0	(1.000, 0.000) (0.000, 1.000)	(1x1)	b: bulk lattice
Surface 1	2.489	2.489	-2.489	2.489	90.0	(1.000, 1.000) (-1.000, 1.000)	c(2x2)	s1: commens. superlattice

3D COORDINATES

I1: overlayer in 4-fold hollow sites; coordinates are derived from bond distances and angles

Dx/Dy in Å, or as a fraction of layer's unit cell vectors; atom 0 at the origin. Epir/subr are bulk repeat vectors.

No. of atoms: 3

Bulk z = 1.760 Å

Region	Chem el.	At. no.	Cell type	Site occ.	Rel. to	Dx ± εx		Dy ± εy		Dz ± εz		Dz/Bz(%) ± εz/Bz
epir		-2					f		f		Å	
subr		-1				1.245	Å	1.245	Å	1.760	Å	
ovrl	I	1	s1	.50	0	0.000	f	0.000	f	0.000	Å	0.0
intf	Ni	2	b	1.00	1	0.500	f	0.500	f	2.150 ± .020	Å	122.2 ± 1.1
subl	Ni	3	b	1.00	2	-0.500	f	-0.500	f	1.760	Å	100.0

BOND DISTANCES AND ANGLES

No. of distances/angles: 2

Interatomic dist. A-B (Å)	Atom A	Atom B	Atom C	Bond angle A-B-C (°)
2.779	I1	Ni2	Ni3	95.7
2.489	Ni2	Ni2(1,0)		

COMMON NAME : Ni(100)-p4g(2x2)-2N ILLUSTRATION: 32
CLASSIFICATION : 28.7.2
TECHNIQUE : SEXAFS
AUTHORS : L. Wenzel, D. Arvanitis, W. Daum, H.H. Rotermund, J. Stoehr, K. Baberschke and H. Ibach
REFERENCE : Phys. Rev., B36, 7689 (1987)

SURFACE TYPE
Substrate : Ni
Crystal face: 100
Temperature : 90 K
Bulk lattice: fcc
2D bulk symm: p4m
2D surf symm: p4g

Adsorbate: N
Coverage : 1/2 (N/Ni)
Pattern : (2x2)
Matrix : (2.000, 0.000)
(0.000, 2.000)

STRUCTURE TYPE
Atomic adsorption in rotated, expanded 4-fold hollow sites; 2 N per (2x2) unit cell in c(2x2) positions, but with opposite rotations (linear Ni displacement parallel to surface by 0.68Å)
perp. distance between N and 2nd Ni layer well determined,

SAMPLE PREPARATION (1 sample)
Treatment : see Daum, Lehwald, Ibach, Surf. Sci. 178, 528 (1986)
Crystallinity:
Anal. methods:
Contamination:

COMMENTS
But assumption of bulk-like spacing between 1st and 2nd Ni layers yields very small Ni-Ni bond distances (eds.); same result found at 90 K and 295K

DATA COLLECTION
Technique: SEXAFS
Dataset : SEXAFS spectra at normal incidence and 20° grazing inc.; two temperatures, 90 K and 295K, give same structural results

THEORY/DATA TREATMENT
SEXAFS with experimentally determined phase shifts from bulk NiO standard (N and O phase shift assumed identical)

STRUCTURES EXAMINED
Hollow site with p4g-compatible rotation of 4 Ni atoms around each hollow site; all Ni-Ni interlayer spacings assumed bulk-like; variable spacing between N and 1st Ni layer

2D UNIT CELLS (1 domain observed)

Cell	Ax (Å)	Ay (Å)	Bx (Å)	By (Å)	α (°)	Matrix	Pattern	Cell type
Bulk	2.492	0.000	0.000	2.492	90.0	(1.000, 0.000) (0.000, 1.000)	(1x1)	b: bulk lattice
Surface 1	4.984	0.000	0.000	4.984	90.0	(2.000, 0.000) (0.000, 2.000)	(2x2)	s1: commens. superlattice

3D COORDINATES

N1-N2: atomic overlayer in equivalent 4-f hollow sites; Ni3-Ni6: planar, laterally relaxed top Ni layer; coordinates are derived from bond distances and angles

Dx/Dy in Å, or as a fraction of layer's unit cell vectors; atom 0 at the origin. Epir/subr are bulk repeat vectors.

No. of atoms: 7 Bulk z = 1.762 Å

Region	Chem el.	At. no.	Cell type	Site occ.	Rel. to	Dx ± εx		Dy ± εy		Dz ± εz		Dz/Bz(%) ± εz/Bz
epir		-2					f		f		Å	
subr		-1				2.492	Å	2.492	Å	1.762	Å	
ovrl	N	1	s1	.25	0	0.000	f	0.000	f	0.000	Å	0.0
ovrl	N	2	s1	.25	1	0.500	f	0.500	f	0.000	Å	0.0
intf	Ni	3	s1	.25	2	-0.154	f	-0.347	f	0.110 ± .060	Å	6.2 ± 3.4
intf	Ni	4	s1	.25	3	0.500	f	0.193	f	0.000	Å	0.0
intf	Ni	5	s1	.25	4	-0.193	f	0.500	f	0.000	Å	0.0
intf	Ni	6	s1	.25	5	-0.500	f	-0.193	f	0.000	Å	0.0
subl	Ni	7	b	1.00	6	-0.307	f	-1.307	f	1.762	Å	100.0

Ni(100)-p4g(2x2)-2N
28.7.2

BOND DISTANCES AND ANGLES

No. of distances/angles: 8

Interatomic dist. A-B (Å)	Atom A	Atom B	Atom C	Bond angle A-B-C (°)
1.892	N1	Ni3	N2	137.3
1.892	N1	Ni3	Ni3(1,0)	155.9
1.892	N1	Ni3	Ni4	175.7
1.892	N1	Ni3	Ni5(0,-2)	111.1
1.892	N1	Ni3	Ni6(1,-1)	155.9
1.892	N1	Ni3	Ni7	46.3
1.872	N1	Ni7	Ni3	47.0
1.872	N1	Ni7	Ni3(-1,0)	31.6

COMMON NAME : Ni(100)-p4g(2x2)-2N
ILLUSTRATION: 32
CLASSIFICATION : 28.7.4
TECHNIQUE : PED
AUTHORS : A.L.D. Kilcoyne, D.P. Woodruff, A.W. Robinson, Th. Lindner, J.S.Somers and A.M.Bradshaw
REFERENCE : Surf. Sci., 253, 107 (1991)

SURFACE TYPE
Substrate : Ni
Crystal face: 100
Temperature : RT*
Bulk lattice: fcc
2D bulk symm: p4m
2D surf symm: p4g

Adsorbate: N
Coverage : 0.5 (N/Ni)
Pattern : p(2x2)
Matrix : (2.000, 0.000)
(0.000, 2.000)

STRUCTURE TYPE
N adsorbed in hollow site, 0.10Å above 1st Ni layer; clock rotation of 4 Ni neighbors by 0.55Å; top Ni-Ni interlayer expansion by 0.15 Å

SAMPLE PREPARATION (1 sample)
Treatment : 500 eV ion bombardment with N, annealing
Crystallinity:
Anal. methods:
Contamination:

COMMENTS

DATA COLLECTION
Technique: PED; photoelectron diffraction, BESSY
Dataset : energy scans (30 eV wide integration) above N 1s edge 45° incidence in <110> az.; data range 80-400 eV

THEORY/DATA TREATMENT
Double scattering cluster calculation (500 atoms)

STRUCTURES EXAMINED
Ideal termination, layer relaxation, rotation amplitude

2D UNIT CELLS (1 domain observed)

Cell	Ax (Å)	Ay (Å)	Bx (Å)	By (Å)	α (°)	Matrix	Pattern	Cell type
Bulk	2.492	0.000	0.000	2.492	90.0	(1.000, 0.000) (0.000, 1.000)	(1x1)	b: bulk lattice
Surface 1	4.984	0.000	0.000	4.984	90.0	(2.000, 0.000) (0.000, 2.000)	p(2x2)	s1: commens. superlattice

3D COORDINATES

N1-2 = adsorbate in symm. equivalent hollows Ni3-6 is clock rotated 4 fold hollow in 1st layer
Ni7 is bulk

Dx/Dy in Å, or as a fraction of layer's unit cell vectors; atom 0 at the origin. Epir/subr are bulk repeat vectors.

No. of atoms: 7 Bulk z = 1.762 Å

Region	Chem el.	At. no.	Cell type	Site occ.	Rel. to	Dx ± εx	Dy ± εy	Dz ± εz	Dz/Bz(%) ± εz/Bz
epir		-2				f	f	Å	
subr		-1				1.246 Å	1.246 Å	1.762 Å	
ovrl	N	1	s1	.25	0	0.000 Å	0.000 Å	0.000 ± .120 Å	0.0 ± 6.8
ovrl	N	2	s1	.25	1	-2.492 Å	-2.492 Å	0.000 ± .120 Å	0.0 ± 6.8
intf	Ni	3	s1	.25	1	-1.636 ± .140 Å	-0.856 ± .140 Å	0.100 ± .100 Å	5.7 ± 5.7
intf	Ni	4	s1	.25	1	1.636 ± .140 Å	0.856 ± .140 Å	0.100 ± .100 Å	5.7 ± 5.7
intf	Ni	5	s1	.25	1	-0.856 ± .140 Å	1.636 ± .140 Å	0.100 ± .100 Å	5.7 ± 5.7
intf	Ni	6	s1	.25	1	0.856 ± .140 Å	-1.636 ± .140 Å	0.100 ± .100 Å	5.7 ± 5.7
subl	Ni	7	b	1.00	1	-2.492 Å	-2.492 Å	1.990 Å	112.9 ± 0.0

BOND DISTANCES AND ANGLES

Bond distances and angles are derived from coordinates

No. of distances/angles: 5

Interatomic dist. A-B (Å)	Atom A	Atom B	Atom C	Bond angle A-B-C (°)
1.849	N1	Ni3		
1.849	N2	Ni3		
1.990	N2	Ni7		

Ni(100)-p4g(2x2)-2N
28.7.4

Bond Distances and Angles - Continued

Interatomic dist. A-B (Å)	Atom A	Atom B	Atom C	Bond angle A-B-C (°)
1.849	Ni3	N1	Ni4	173.8
1.849	Ni3	N1	Ni5	89.8

COMMON NAME : Ni(100)-c(2x2)-N+O disordered ILLUSTRATION: 28,29
CLASSIFICATION : 28.7.8.1
TECHNIQUE : LEED
AUTHORS : M.A. Passler, T.H. Lin and A. Ignatiev
REFERENCE : J. Vac. Sci. Technol., 18, 481 (1981)

SURFACE TYPE
Substrate : Ni
Crystal face: 100
Temperature : 300 K
Bulk lattice: fcc
2D bulk symm: p4m
2D surf symm: none

Adsorbate: N;O
Coverage : 0.25 (N,O/Ni)
Pattern : (2x2)
Matrix : (2.000, 0.000)
(0.000, 2.000)

STRUCTURE TYPE
Decomposed NO as atomic N and O randomly positioned in 4-fold hollow sites of a c(2x2) lattice;
(this structure is here modeled as alternating N and O atoms in a p(2x2) structure)

SAMPLE PREPARATION (1 sample)
Treatment : NO dep. at 10E-8 torr with ion pump on to remove C
Crystallinity: nitric oxide admitted
Anal. methods:
Contamination: checked by LEED/AES

COMMENTS
Top site yields similar best R-factor as hollow site, but with improbable inner potential

DATA COLLECTION
Technique: LEED
Dataset : I-V spectra for 3 beams at normal incidence, 7 beams at 10° off normal; E range 50-275 eV

THEORY/DATA TREATMENT
Dynamical LEED: 8 phase shifts with average t-matrix approximation

STRUCTURES EXAMINED
Two models for dissociated NO: filled site model; random occupation model

QUALITY OF EXPERIMENT-THEORY FIT
RZJ=0.22

2D UNIT CELLS (1 domain observed)

Cell	Ax (Å)	Ay (Å)	Bx (Å)	By (Å)	α (°)	Matrix	Pattern	Cell type
Bulk	2.489	0.000	0.000	2.489	90.0	(1.000, 0.000) (0.000, 1.000)	(1x1)	b: bulk lattice
Surface 1	4.978	0.000	0.000	4.978	90.0	(2.000, 0.000) (0.000, 2.000)	(2x2)	m1: randomly mixed layer

3D COORDINATES

N1-O2: in (randomized) c(2x2) lattice on hollow sites

Dx/Dy in Å, or as a fraction of layer's unit cell vectors; atom 0 at the origin. Epir/subr are bulk repeat vectors.

No. of atoms: 4 Bulk z = 1.760 Å

Reg ion	Chem el.	At. no.	Cell type	Site occ.	Rel. to	Dx ± εx		Dy ± εy		Dz ± εz		Dz/Bz(%) ± εz/Bz
epir		-2					f		f		Å	
subr		-1				-1.245	Å	-1.245	Å	1.760	Å	
ovrl	N	1	m1	1.00	0	0.000	f	0.000	f	0.000	Å	0.0
ovrl	O	2	m1	1.00	1	0.500	f	0.500	f	0.000 ± .100	Å	0.0 ± 5.7
intf	Ni	3	b	1.00	2	-0.500	f	-0.500	f	0.930 ± .100	Å	52.8 ± 5.7
subl	Ni	4	b	1.00	3	-0.500	f	-0.500	f	1.760	Å	100.0

BOND DISTANCES AND ANGLES

Bond distances and angles are derived from coordinates

No. of distances/angles: 7

Interatomic dist. A-B (Å)	Atom A	Atom B	Atom C	Bond angle A-B-C (°)
1.991	N1	Ni3	O2	124.3
1.991	N1	Ni3	Ni3(1,0)	128.7
1.991	N1	Ni3	Ni4(1,1)	162.9
1.991	N1	Ni3	Ni4(1,0)	109.3
1.991	N1	Ni3	Ni4	72.9
2.489	Ni3	Ni3(1,0)	N1(1,0)	128.7

Ni(100)-c(2x2)-N+O disordered
28.7.8.1

Bond Distances and Angles - Continued

Interatomic dist. A-B (Å)	Atom A	Atom B	Atom C	Bond angle A-B-C (°)
2.489	Ni3	Ni4		

COMMON NAME : Ni(100)-c(2x2)-Na ILLUSTRATION: 28,29
CLASSIFICATION : 28.11.3
TECHNIQUE : LEED
AUTHORS : S. Andersson and J.B. Pendry
REFERENCE : Solid State Commun., 16, 563 (1975)

SURFACE TYPE
Substrate : Ni
Crystal face: 100
Temperature : RT*
Bulk lattice: fcc
2D bulk symm: p4m
2D surf symm: p4m

Adsorbate: Na
Coverage : 0.5 (Na/Ni)
Pattern : c(2x2)
Matrix : (1.000, 1.000)
(-1.000, 1.000)

STRUCTURE TYPE
Atomic adsorption in 4-fold hollow sites

SAMPLE PREPARATION (1 sample)
Treatment : Ar+ sputtering, followed by evaporation from high purity Na
Crystallinity:
Anal. methods:
Contamination:

COMMENTS

DATA COLLECTION
Technique: LEED
Dataset : I-V spectra: 3 beams at normal incidence from emergence to 90 eV

THEORY/DATA TREATMENT
Dynamical LEED: 5 phase shifts; 25 beams;
Vor=-7 eV, Voi varied with energy from -1eV to -4eV

STRUCTURES EXAMINED
Various Na-Ni interlayer spacings in hollow and top sites

QUALITY OF EXPERIMENT-THEORY FIT
Visual

2D UNIT CELLS (1 domain observed)

Cell	Ax (Å)	Ay (Å)	Bx (Å)	By (Å)	α (°)	Matrix	Pattern	Cell type
Bulk	2.489	0.000	0.000	2.489	90.0	(1.000, 0.000) (0.000, 1.000)	(1x1)	b: bulk lattice
Surface 1	2.489	2.489	-2.489	2.489	90.0	(1.000, 1.000) (-1.000, 1.000)	c(2x2)	s1: commens. superlattice

3D COORDINATES

Na1: atomic overlayer in 4-fold hollow sites

Dx/Dy in Å, or as a fraction of layer's unit cell vectors; atom 0 at the origin. Epir/subr are bulk repeat vectors.

No. of atoms: 3 Bulk z = 1.760 Å

Region	Chem el.	At. no.	Cell type	Site occ.	Rel. to	Dx ± εx		Dy ± εy		Dz ± εz		Dz/Bz(%) ± εz/Bz
epir		-2					f		f		Å	
subr		-1				-1.245	Å	-1.245	Å	1.760	Å	
ovrl	Na	1	s1	.50	0	0.000	f	0.000	f	0.000	Å	0.0
intf	Ni	2	b	1.00	1	0.500	f	0.500	f	2.200 ± .100	Å	125.0 ± 5.7
subl	Ni	3	b	1.00	2	-0.500	f	-0.500	f	1.760	Å	100.0

BOND DISTANCES AND ANGLES

Bond distances and angles are derived from coordinates

No. of distances/angles: 4

Interatomic dist. A-B (Å)	Atom A	Atom B	Atom C	Bond angle A-B-C (°)
3.520	Na1	Na1(1,0)	Ni2(1,1)	128.7
2.817	Na1	Ni2	Ni2(1,0)	116.2
2.817	Na1	Ni2	Ni3	96.3
2.489	Ni2	Ni3		

COMMON NAME : Ni(100)-c(2x2)-Na ILLUSTRATION: 28,29
CLASSIFICATION : 28.11.4
TECHNIQUE : LEED
AUTHORS : J.E. Demuth, D.W. Jepsen and P.M. Marcus
REFERENCE : J. Phys., C8, L25 (1975)

SURFACE TYPE
Substrate : Ni
Crystal face: 100
Temperature : 298 K
Bulk lattice: fcc
2D bulk symm: p4m
2D surf symm: p4m
Adsorbate: Na
Coverage : 0.5 Na/Ni
Pattern : c(2x2)
Matrix : (1.000, 1.000)
(-1.000, 1.000)

STRUCTURE TYPE
Atomic adsorption in 4-fold hollow sites

SAMPLE PREPARATION (1 sample)
Treatment : Na deposited from heated breakseal glass ampule
Crystallinity:
Anal. methods:
Contamination:

COMMENTS

DATA COLLECTION
Technique: LEED
Dataset : from Andersson and Pendry, Sol. St. Commun. 16, 563 (1975)

THEORY/DATA TREATMENT
Dynamical LEED: 8 (5) phase shifts for ~~rigid~~ Ni (vibr. Na); up to 50 beams; Vor=-11 eV, Voi=-2.5eV; ΘD=100 to 800 K

STRUCTURES EXAMINED
Na in 4-fold hollow site: Na/Ni layer spacing varied from 1.17 to 3.07Å; effects of Na potential, surface barriers, non-structural parameters such as adsorbate scattering, adsorbate ΘD and Vor were examined

QUALITY OF EXPERIMENT-THEORY FIT
Visual

2D UNIT CELLS (1 domain observed)

Cell	Ax (Å)	Ay (Å)	Bx (Å)	By (Å)	α (°)	Matrix	Pattern	Cell type
Bulk	2.489	0.000	0.000	2.489	90.0	(1.000, 0.000) (0.000, 1.000)	(1x1)	b: bulk lattice
Surface 1	2.489	2.489	-2.489	2.489	90.0	(1.000, 1.000) (-1.000, 1.000)	c(2x2)	s1: commens. superlattice

3D COORDINATES

Na1: atomic overlayer in 4-fold hollow sites

Dx/Dy in Å, or as a fraction of layer's unit cell vectors; atom 0 at the origin. Epir/subr are bulk repeat vectors.

No. of atoms: 3 Bulk z = 1.760 Å

Reg ion	Chem el.	At. no.	Cell type	Site occ.	Rel. to	Dx ± εx		Dy ± εy		Dz ± εz		Dz/Bz(%) ± εz/Bz
epir		-2					f		f		Å	
subr		-1				-1.245	Å	-1.245	Å	1.760	Å	
ovrl	Na	1	s1	.50	0	0.000	f	0.000	f	0.000	Å	0.0
intf	Ni	2	b	1.00	1	0.500	f	0.500	f	2.230 ± .100	Å	126.7 ± 5.7
subl	Ni	3	b	1.00	2	-0.500	f	-0.500	f	1.760	Å	100.0

BOND DISTANCES AND ANGLES

Bond distances and angles are derived from coordinates

No. of distances/angles: 4

Interatomic dist. A-B (Å)	Atom A	Atom B	Atom C	Bond angle A-B-C (°)
3.520	Na1	Na1(1,0)	Ni2(1,1)	128.3
2.841	Na1	Ni2	Ni2(1,0)	116.0
2.841	Na1	Ni2	Ni3	96.7
2.489	Ni2	Ni3		

COMMON NAME : Ni(100)-c(2x2)-Na+S
CLASSIFICATION : 28.11.16.1a
TECHNIQUE : LEED
AUTHORS : S. Andersson and J.B. Pendry
REFERENCE : J. Phys., C9, 2721 (1976)

ILLUSTRATION: 28,29

SURFACE TYPE
Substrate : Ni
Crystal face: 100
Temperature : RT*
Bulk lattice: fcc
2D bulk symm: p4m
2D surf symm: p4m
Adsorbate: Na;S
Coverage : 0.5 Na,S/Ni
Pattern : c(2x2)
Matrix : (1.000, 1.000)
(-1.000, 1.000)

STRUCTURE TYPE
Atomic adsorption over hollow sites in mixed 50/50 layer; Na and S individually form c(2x2) superlattices

SAMPLE PREPARATION (1 sample)
Treatment : 1E-4L H2S decomp. at 523 K; high purity Na evaporation
Crystallinity:
Anal. methods:
Contamination:

COMMENTS

DATA COLLECTION
Technique: LEED
Dataset : I-V spectra: 3 beams at normal incidence from emergence to 100 eV: (00), (1/2,1/2), (01)

THEORY/DATA TREATMENT
Dynamical LEED: 5 phase shifts, 25 beams

STRUCTURES EXAMINED
Various Ni-S and S-Na spacings

QUALITY OF EXPERIMENT-THEORY FIT
Visual

2D UNIT CELLS (1 domain observed)

Cell	Ax (Å)	Ay (Å)	Bx (Å)	By (Å)	α (°)	Matrix	Pattern	Cell type
Bulk	2.489	0.000	0.000	2.489	90.0	(1.000, 0.000) (0.000, 1.000)	(1x1)	b: bulk lattice
Surface 1	2.489	2.489	-2.489	2.489	90.0	(1.000, 1.000) (-1.000, 1.000)	c(2x2)	s1: commens. superlattice

3D COORDINATES

Na1-S2: mixed overlayer in alternating hollow sites

Dx/Dy in Å, or as a fraction of layer's unit cell vectors; atom 0 at the origin. Epir/subr are bulk repeat vectors.

No. of atoms: 4 Bulk z = 1.760 Å

Region	Chem el.	At. no.	Cell type	Site occ.	Rel. to	Dx ± εx		Dy ± εy		Dz ± εz		Dz/Bz(%) ± εz/Bz
epir		-2					f		f		Å	
subr		-1				-1.245	Å	-1.245	Å	1.760	Å	
ovrl	Na	1	s1	.50	0	0.000	f	0.000	f	0.000	Å	0.0
ovrl	S	2	s1	.50	1	0.500	f	0.500	f	1.200 ± .100	Å	68.2 ± 5.7
intf	Ni	3	b	1.00	2	0.500	f	-0.500	f	1.300 ± .100	Å	73.9 ± 5.7
subl	Ni	4	b	1.00	3	-0.500	f	-0.500	f	1.760	Å	100.0

BOND DISTANCES AND ANGLES

Bond distances and angles are derived from coordinates

No. of distances/angles: 9

Interatomic dist. A-B (Å)	Atom A	Atom B	Atom C	Bond angle A-B-C (°)
2.763	Na1	S2	Na1(1,0)	79.1
2.763	Na1	S2	Ni3	75.3
3.057	Na1	Ni3	S2	60.9
3.057	Na1	Ni3	Ni3(1,0)	114.0
3.057	Na1	Ni3	Ni4	99.9
2.188	S2	Ni3	Ni3(1,0)	124.7

Ni(100)-c(2x2)-Na+S
28.11.16.1a

Bond Distances and Angles - Continued

Interatomic dist. A-B (Å)	Atom A	Atom B	Atom C	Bond angle A-B-C (°)
2.188	S2	Ni3	Ni4	114.8
2.489	Ni3	Ni3(1,0)		
2.489	Ni3	Ni4		

COMMON NAME : Ni(100)-p(2x2)-Na+2S
CLASSIFICATION : 28.11.16.1b
TECHNIQUE : LEED
AUTHORS : S. Andersson and J.B. Pendry
REFERENCE : J. Phys., C9, 2721 (1976)

ILLUSTRATION: 28,30

SURFACE TYPE
Substrate : Ni
Crystal face: 100
Temperature : RT*
Bulk lattice: fcc
2D bulk symm: p4m
2D surf symm: p4m

Adsorbate: Na;S
Coverage : 0.25Na/Ni, 0.5S/Ni
Pattern : p(2x2)
Matrix : (2.000, 0.000)
(0.000, 2.000)

STRUCTURE TYPE
Atomic adsorption over hollow sites in mixed layer; Na forms p(2x2) lattice; S forms c(2x2) lattice; each Na bonds to 4 S atoms in adjacent hollows; each S bonds to 2 Na atoms in adjacent hollows

SAMPLE PREPARATION (1 sample)
Treatment : 1E-4L H2S decomp. at 523 K; high purity Na evaporation
Crystallinity:
Anal. methods:
Contamination:

COMMENTS

DATA COLLECTION
Technique: LEED
Dataset : I-V spectra: 4 beams at normal incidence from emergence to 100 eV: (00), (1/2,0), (1/2,1/2), (01)

THEORY/DATA TREATMENT
Dynamical LEED: 5 phase shifts, 49 beams

STRUCTURES EXAMINED
Various Ni-S and S-Na spacings

QUALITY OF EXPERIMENT-THEORY FIT
Visual

2D UNIT CELLS (1 domain observed)

Cell	Ax (Å)	Ay (Å)	Bx (Å)	By (Å)	α (°)	Matrix	Pattern	Cell type
Bulk	2.490	0.000	0.000	2.490	90.0	(1.000, 0.000) (0.000, 1.000)	(1x1)	b: bulk lattice
Surface 1	4.980	0.000	0.000	4.980	90.0	(2.000, 0.000) (0.000, 2.000)	p(2x2)	s1: commens. superlattice

3D COORDINATES

Na1: p(2x2) overlayer in hollow sites; S2-S3: c(2x2) overlayer in hollow sites

Dx/Dy in Å, or as a fraction of layer's unit cell vectors; atom 0 at the origin. Epir/subr are bulk repeat vectors.

No. of atoms: 5

Bulk z = 1.760 Å

Reg ion	Chem el.	At. no.	Cell type	Site occ.	Rel. to	Dx ± εx		Dy ± εy		Dz ± εz		Dz/Bz(%) ± εz/Bz
epir		-2					f		f		Å	
subr		-1				-1.245	Å	-1.245	Å	1.760	Å	
ovrl	Na	1	s1	.25	0	0.000	f	0.000	f	0.000	Å	0.0
ovrl	S	2	s1	.25	1	0.500	f	0.000	f	1.200 ± .100	Å	68.2 ± 5.7
ovrl	S	3	s1	.25	2	-0.500	f	0.500	f	0.000	Å	0.0
intf	Ni	4	b	1.00	3	0.500	f	-0.500	f	1.300 ± .100	Å	73.9 ± 5.7
subl	Ni	5	b	1.00	4	-0.500	f	-0.500	f	1.760	Å	100.0

BOND DISTANCES AND ANGLES

Bond distances and angles are derived from coordinates

No. of distances/angles: 8

Interatomic dist. A-B (Å)	Atom A	Atom B	Atom C	Bond angle A-B-C (°)
2.764	Na1	S2	Ni4	75.3
3.058	Na1	Ni4	S2	61.0
3.058	Na1	Ni4	Ni4(1,0)	114.0
3.058	Na1	Ni4	Ni5	99.8
2.189	S2	Ni4	Ni4(1,0)	55.3

Ni(100)-p(2x2)-Na+2S
28.11.16.1b

Bond Distances and Angles - Continued

Interatomic dist. A-B (Å)	Atom A	Atom B	Atom C	Bond angle A-B-C (°)
2.189	S2	Ni4	Ni5	114.8
2.490	Ni4	Ni4(1,0)		
2.490	Ni4	Ni5		

COMMON NAME : Ni(100)-p(2x2)-Na+S
CLASSIFICATION : 28.11.16.1c
TECHNIQUE : LEED
AUTHORS : S. Andersson and J.B. Pendry
REFERENCE : J. Phys., C9, 2721 (1976)

ILLUSTRATION: 28,30

SURFACE TYPE
Substrate : Ni
Crystal face: 100
Temperature : RT*
Bulk lattice: fcc
2D bulk symm: p4m
2D surf symm: p4m

Adsorbate: Na;S
Coverage : 0.25 Na,S/Ni
Pattern : p(2x2)
Matrix : (2.000, 0.000)
(0.000, 2.000)

STRUCTURE TYPE
Atomic adsorption over hollow sites in mixed 50/50 layer;
Na and S individually form p(2x2) superlattices;
Na and S form 1D strings of bonded atoms in adjacent hollows

SAMPLE PREPARATION (1 sample)
Treatment : 1E-4L H2S decomp. at 523 K; high purity Na evaporation
Crystallinity:
Anal. methods:
Contamination:

COMMENTS

DATA COLLECTION
Technique: LEED
Dataset : I-V spectra: 4 beams at normal incidence from emergence to 100 eV: (00), (1/2,0), (1/2,1/2), (01)

THEORY/DATA TREATMENT
Dynamical LEED: 5 phase shifts, 49 beams

STRUCTURES EXAMINED
Various Ni-S and S-Na spacings; Na sites 2-fold and 4-fold S coordinated

QUALITY OF EXPERIMENT-THEORY FIT
Visual

2D UNIT CELLS (1 domain observed)

Cell	Ax (Å)	Ay (Å)	Bx (Å)	By (Å)	α (°)	Matrix	Pattern	Cell type
Bulk	2.490	0.000	0.000	2.490	90.0	(1.000, 0.000) (0.000, 1.000)	(1x1)	b: bulk lattice
Surface 1	4.980	0.000	0.000	4.980	90.0	(2.000, 0.000) (0.000, 2.000)	p(2x2)	s1: commens. superlattice

3D COORDINATES

Na1-S2: mixed overlayer, alternating in adjacent hollow sites

Dx/Dy in Å, or as a fraction of layer's unit cell vectors; atom 0 at the origin. Epir/subr are bulk repeat vectors.

No. of atoms: 4 Bulk z = 1.760 Å

Region	Chem el.	At. no.	Cell type	Site occ.	Rel. to	Dx ± εx		Dy ± εy		Dz ± εz		Dz/Bz(%) ± εz/Bz
epir		-2					f		f		Å	
subr		-1				-1.245	Å	-1.245	Å	1.760	Å	
ovrl	Na	1	s1	.25	0	0.000	f	0.000	f	0.000	Å	0.0
ovrl	S	2	s1	.25	1	0.000	f	0.500	f	1.200 ± .100	Å	68.2 ± 5.7
intf	Ni	3	b	1.00	2	0.500	f	-0.500	f	1.300 ± .100	Å	73.9 ± 5.7
subl	Ni	4	b	1.00	3	-0.500	f	-0.500	f	1.760	Å	100.0

BOND DISTANCES AND ANGLES

Bond distances and angles are derived from coordinates

No. of distances/angles: 7

Interatomic dist. A-B (Å)	Atom A	Atom B	Atom C	Bond angle A-B-C (°)
2.764	Na1	S2	Ni3	75.3
3.058	Na1	Ni3	Ni3(1,0)	114.0
3.058	Na1	Ni3	Ni4	99.8
2.189	S2	Ni3	Ni3(1,0)	124.7
2.189	S2	Ni3	Ni4	114.8
2.490	Ni3	Ni3(1,0)		

Ni(100)-p(2x2)-Na+S
28.11.16.1c

Bond Distances and Angles - Continued

Interatomic dist. A-B (Å)	Atom A	Atom B	Atom C	Bond angle A-B-C (°)
2.490	Ni3	Ni4		

COMMON NAME : Ni(100)-c(2x2)-O
CLASSIFICATION : 28.8.21
TECHNIQUE : PED
AUTHORS : D.H. Rosenblatt, J.G. Tobin, M.G. Mason, R.F. Davis, S.D. Kevan, D.A. Shirley, C.H. Li and S.Y. Tong
REFERENCE : Phys. Rev., B23, 3828 (1981)

ILLUSTRATION: 28,29

SURFACE TYPE
Substrate : Ni
Crystal face: 100
Temperature : 300 K
Bulk lattice: fcc
2D bulk symm: p4m
2D surf symm: p4m

Adsorbate: O
Coverage : 0.5 O/Ni
Pattern : c(2x2)
Matrix : (1.000, 1.000)
(-1.000, 1.000)

STRUCTURE TYPE
Atomic adsorption in 4-fold hollow sites

SAMPLE PREPARATION (1 sample)
Treatment : Ar+ sputtering, annealing to 875 K
Crystallinity:
Anal. methods:
Contamination: determined by sharp (1x1) LEED pattern

COMMENTS

DATA COLLECTION
Technique: PED
Dataset : normal PED at SSRL with x-rays at near grazing incidence (10°) measured to 195 eV above O(1s) edge

THEORY/DATA TREATMENT
Full dynamical calcs: initial state calculated from MS-Xα method for local O environment

STRUCTURES EXAMINED
4-fold hollow (d=0.90Å) and top (d=1.76Å) sites

QUALITY OF EXPERIMENT-THEORY FIT
Visual

2D UNIT CELLS (1 domain observed)

Cell	Ax (Å)	Ay (Å)	Bx (Å)	By (Å)	α (°)	Matrix	Pattern	Cell type
Bulk	2.489	0.000	0.000	2.489	90.0	(1.000, 0.000) (0.000, 1.000)	(1x1)	b: bulk lattice
Surface 1	2.489	2.489	-2.489	2.489	90.0	(1.000, 1.000) (-1.000, 1.000)	c(2x2)	s1: commens. superlattice

3D COORDINATES

O1: overlayer in 4-fold hollow sites

Dx/Dy in Å, or as a fraction of layer's unit cell vectors; atom 0 at the origin. Epir/subr are bulk repeat vectors.

No. of atoms: 3

Bulk z = 1.760 Å

Reg ion	Chem el.	At. no.	Cell type	Site occ.	Rel. to	Dx ± εx		Dy ± εy		Dz ± εz		Dz/Bz(%) ± εz/Bz
epir		-2					f		f		Å	
subr		-1				-1.245	Å	-1.245	Å	1.760	Å	
ovrl	O	1	s1	.50	0	0.000	f	0.000	f	0.000	Å	0.0
intf	Ni	2	b	1.00	1	0.500	f	0.500	f	0.900 ± .040	Å	51.1 ± 2.3
subl	Ni	3	b	1.00	2	-0.500	f	-0.500	f	1.760	Å	100.0

BOND DISTANCES AND ANGLES

Bond distances and angles are derived from coordinates

No. of distances/angles: 7

Interatomic dist. A-B (Å)	Atom A	Atom B	Atom C	Bond angle A-B-C (°)
1.977	O1	Ni2	O1(1,0)	125.8
1.977	O1	Ni2	Ni2(1,0)	129.0
1.977	O1	Ni2	Ni3(1,1)	162.1
1.977	O1	Ni2	Ni3(1,0)	108.8
1.977	O1	Ni2	Ni3	72.1

Ni(100)-c(2x2)-O
28.8.21

Bond Distances and Angles - Continued

Interatomic dist. A-B (Å)	Atom A	Atom B	Atom C	Bond angle A-B-C (°)
2.489	Ni2	Ni2(1,0)	O1(1,0)	51.0
2.489	Ni2	Ni3		

COMMON NAME : Ni(100)-c(2x2)-O
CLASSIFICATION : 28.8.30a
TECHNIQUE : SEXAFS
AUTHORS : J. Stoehr, R. Jaeger and T. Kendelewicz
REFERENCE : Phys. Rev. Lett., 49, 142 (1982)

ILLUSTRATION: 28,29

SURFACE TYPE
Substrate : Ni
Crystal face: 100
Temperature : 300 K
Bulk lattice: fcc
2D bulk symm: p4m
2D surf symm: p4m

Adsorbate: O
Coverage : 20L
Pattern : c(2x2)
Matrix : (1.000, 1.000)
(-1.000, 1.000)

STRUCTURE TYPE
Atomic adsorption in 4-fold hollow sites

SAMPLE PREPARATION (sample)
Treatment : see Brennan et al, Phys. Rev. B24, 4871 (1981)
Crystallinity:
Anal. methods:
Contamination: AES: <1%ML C, O, and S

COMMENTS
For analysis of the corresponding XANES see Norman et al Phys. Rev. Lett. 51, 2052 (1983)

DATA COLLECTION
Technique: SEXAFS
Dataset : SEXAFS at SSRL with x-ray incidence of 45°

THEORY/DATA TREATMENT
Conventional EXAFS with experimentally determined phase shifts from bulk NiO standard

STRUCTURES EXAMINED
Different O-Ni spacings in 4-fold hollow site; spacing 0.26Å ruled out by absence of polarization dep. of Ni-O bond length

QUALITY OF EXPERIMENT-THEORY FIT
Visual

2D UNIT CELLS (1 domain observed)

Cell	Ax (Å)	Ay (Å)	Bx (Å)	By (Å)	α (°)	Matrix	Pattern	Cell type
Bulk	2.489	0.000	0.000	2.489	90.0	(1.000, 0.000) (0.000, 1.000)	(1x1)	b: bulk lattice
Surface 1	2.489	2.489	-2.489	2.489	90.0	(1.000, 1.000) (-1.000, 1.000)	c(2x2)	s1: commens. superlattice

3D COORDINATES

O1: overlayer in 4-fold hollow sites; coordinates are derived from bond length

Dx/Dy in Å, or as a fraction of layer's unit cell vectors; atom 0 at the origin. Epir/subr are bulk repeat vectors.

No. of atoms: 3

Bulk z = 1.760 Å

Region	Chem el.	At. no.	Cell type	Site occ.	Rel. to	Dx ± εx		Dy ± εy		Dz ± εz		Dz/Bz(%) ± εz/Bz
epir		-2					f		f		Å	
subr		-1				-1.245	Å	-1.245	Å	1.760	Å	
ovrl	O	1	s1	.50	0	0.000	f	0.000	f	0.000	Å	0.0
intf	Ni	2	b	1.00	1	0.500	f	0.500	f	0.860 ± .070	Å	48.9 ± 4.0
subl	Ni	3	b	1.00	2	-0.500	f	-0.500	f	1.760	Å	100.0

BOND DISTANCES AND ANGLES

No. of distances/angles: 7

Interatomic dist. A-B (Å)	Atom A	Atom B	Atom C	Bond angle A-B-C (°)
1.959	O1	Ni2	O1(1,0)	127.9
1.959	O1	Ni2	Ni2(1,0)	129.4
1.959	O1	Ni2	Ni3(1,1)	161.0
1.959	O1	Ni2	Ni3(1,0)	108.1
1.959	O1	Ni2	Ni3	71.0
2.489	Ni2	Ni2(1,0)	O1(1,0)	50.6
2.489	Ni2	Ni3		

COMMON NAME : Ni(100)-c(2x2)-O
CLASSIFICATION : 28.8.32a
TECHNIQUE : XANES (NEXAFS)
AUTHORS : D. Norman, J. Stoehr, R. Jaeger, P.J. Durham and J.B. Pendry
REFERENCE : Phys. Rev. Lett., 51, 2052 (1983)

ILLUSTRATION: 28,29

SURFACE TYPE
Substrate : Ni
Crystal face: 100
Temperature : 300 K
Bulk lattice: fcc
2D bulk symm: p4m
2D surf symm: p4m

Adsorbate: O
Coverage : 0.5 (O/Ni)
Pattern : c(2x2)
Matrix : (1.000, 1.000)
(-1.000, 1.000)

STRUCTURE TYPE
Atomic adsorption in 4-fold hollow sites

SAMPLE PREPARATION (sample)
Treatment : see J. Stoehr and R. Jaeger, Phys. Rev. B26, 4111 (1982)
Crystallinity:
Anal. methods:
Contamination:

COMMENTS
Authors report small differences between p(2x2) and c(2x2) measured spectra

DATA COLLECTION
Technique: XANES (NEXAFS)
Dataset : XANES for normal and grazing x-ray incidence from threshold (approx. 530 eV) to 564eV with resolution 2.5eV

THEORY/DATA TREATMENT
Dynamical XANES: 30 atom cluster, Mattheis muffin-tin pots; Voi=-0.7 eV

STRUCTURES EXAMINED
O-Ni layer spacings of 0.2 and 0.9Å in the 4 fold hollow site; bridge and top sites with fixed O-Ni bond length of 1.98Å

QUALITY OF EXPERIMENT-THEORY FIT
Visual

2D UNIT CELLS (1 domain observed)

Cell	Ax (Å)	Ay (Å)	Bx (Å)	By (Å)	α (°)	Matrix	Pattern	Cell type
Bulk	2.489	0.000	0.000	2.489	90.0	(1.000, 0.000) (0.000, 1.000)	(1x1)	b: bulk lattice
Surface 1	2.489	2.489	-2.489	2.489	90.0	(1.000, 1.000) (-1.000, 1.000)	c(2x2)	s1: commens. superlattice

3D COORDINATES

O1: overlayer in 4-fold hollow sites;

Dx/Dy in Å, or as a fraction of layer's unit cell vectors; atom 0 at the origin. Epir/subr are bulk repeat vectors.

No. of atoms: 3 Bulk z = 1.760 Å

Region	Chem el.	At. no.	Cell type	Site occ.	Rel. to	Dx ± εx		Dy ± εy		Dz ± εz		Dz/Bz(%) ± εz/Bz
epir		-2					f		f		Å	
subr		-1				-1.245	Å	-1.245	Å	1.760	Å	
ovrl	O	1	s1	.50	0	0.000	f	0.000	f	0.000	Å	0.0
intf	Ni	2	b	1.00	1	0.500	f	0.500	f	0.900 ± .100	Å	51.1 ± 5.7
subl	Ni	3	b	1.00	2	-0.500	f	-0.500	f	1.760	Å	100.0

BOND DISTANCES AND ANGLES

Bond distances and angles are derived from coordinates

No. of distances/angles: 7

Interatomic dist. A-B (Å)	Atom A	Atom B	Atom C	Bond angle A-B-C (°)
1.977	O1	Ni2	O1(1,0)	125.8
1.977	O1	Ni2	Ni2(1,0)	129.0
1.977	O1	Ni2	Ni3(1,1)	162.1
1.977	O1	Ni2	Ni3(1,0)	108.8
1.977	O1	Ni2	Ni3	72.1
2.489	Ni2	Ni2(1,0)	O1(1,0)	51.0

Ni(100)-c(2x2)-O
28.8.32a

Bond Distances and Angles - Continued

Interatomic dist. A-B (Å)	Atom A	Atom B	Atom C	Bond angle A-B-C (°)
2.489	Ni2	Ni3		

ILLUSTRATION: 28

COMMON NAME : Ni(100)-c(2x2)-O
CLASSIFICATION : 28.8.35
TECHNIQUE : LEED
AUTHORS : J.E. Demuth, N.J. Dinardo and G.S. Cargill III
REFERENCE : Phys. Rev. Lett., 50, 1373 (1983)

SURFACE TYPE
Substrate : Ni
Crystal face: 100
Temperature : 160 K
Bulk lattice: fcc
2D bulk symm: p4m
2D surf symm: pm

Adsorbate: O
Coverage : 0.5 O/Ni
Pattern : c(2x2)
Matrix : (1.000, 1.000)
(-1.000, 1.000)

SAMPLE PREPARATION (1 sample)
Treatment : 25L O exposure at 2E-8 torr at 423 K, and briefly to 520K
Crystallinity:
Anal. methods:
Contamination: monitored by AES

DATA COLLECTION
Technique: LEED
Dataset : LEED I-V spectra: 60<E<260 eV; 4 inequivalent beams at normal incidence, (00) and (1/2,1/2) beams for Θ=5° and Θ=10°

STRUCTURE TYPE
Atomic adsorption 0.1Å laterally away from 4-fold hollow site towards bridge site

COMMENTS
Various oxide potentials were investigated but with little variation in the resulting spectra;
the asymmetric site is preferred independent of ΘD, Vor, Voi or the potential;
asymmetrical site is no longer thought correct (eds.)

THEORY/DATA TREATMENT
Dynamical LEED: 8 phase shifts; 58, 90, 138 beams for E>60, 128, 198 eV resp.; rms amps=0.009Å(Ni) 0.012Å(O)

STRUCTURES EXAMINED
O in 4-fold hollow site from 0.05 to 0.85Å above Ni layer; nearly coplanar O/Ni structures with substrate distortions; reconstr. surfaces with O replacing alternate top Ni atoms; mixed layers with interstitial O between 1st and 2nd layers

QUALITY OF EXPERIMENT-THEORY FIT
RZJ=0.17

2D UNIT CELLS (4 domains observed)

Cell	Ax (Å)	Ay (Å)	Bx (Å)	By (Å)	α (°)	Matrix	Pattern	Cell type
Bulk	2.489	0.000	0.000	2.489	90.0	(1.000, 0.000) (0.000, 1.000)	(1x1)	b: bulk lattice
Surface 1	2.489	2.489	-2.489	2.489	90.0	(1.000, 1.000) (-1.000, 1.000)	c(2x2)	s1: commens. superlattice

3D COORDINATES

O1: overlayer in asymmetrical site near Ni hollows

Dx/Dy in Å, or as a fraction of layer's unit cell vectors; atom 0 at the origin. Epir/subr are bulk repeat vectors.

No. of atoms: 4 Bulk z = 1.760 Å

Region	Chem el.	At. no.	Cell type	Site occ.	Rel. to	Dx ± εx		Dy ± εy		Dz ± εz		Dz/Bz(%) ± εz/Bz
epir		-2					f		f		Å	
subr		-1				-1.245	Å	-1.245	Å	1.760	Å	
ovrl	O	1	s1	.50	0	0.000	f	0.000	f	0.000	Å	0.0
intf	Ni	2	b	1.00	1	-0.380 ± .040	f	0.500 ± .040	f	0.800 ± .025	Å	45.5 ± 1.4
intf	Ni	3	b	1.00	2	0.500	f	0.500	f	1.810 ± .020	Å	102.8 ± 1.4
subl	Ni	4	b	1.00	3	-0.500	f	-0.500	f	1.760	Å	100.0

BOND DISTANCES AND ANGLES

Bond distances and angles are derived from coordinates

No. of distances/angles: 3

Interatomic dist. A-B (Å)	Atom A	Atom B	Atom C	Bond angle A-B-C (°)
1.756	O1	Ni2	Ni3	114.2
2.138	O1	Ni2(0,-1)	Ni3(0,-1)	68.0
2.489	Ni2	Ni2(1,0)		

COMMON NAME : Ni(100)-c(2x2)-O
CLASSIFICATION : 28.8.36
TECHNIQUE : SEELFS
AUTHORS : M. de Crescenzi, F. Antonangeli, C. Bellini and R. Rosei
REFERENCE : Phys. Rev. Lett., 50, 1949 (1983)

ILLUSTRATION: 28,29

SURFACE TYPE
Substrate : Ni
Crystal face: 100
Temperature : RT*
Bulk lattice: fcc
2D bulk symm: p4m
2D surf symm: p4m

Adsorbate: O
Coverage : 0.3 O/Ni
Pattern : c(2x2)
Matrix : (1.000, 1.000)
(-1.000, 1.000)

STRUCTURE TYPE
Atomic adsorption in 4-fold hollow sites

SAMPLE PREPARATION (1 sample)
Treatment : sputtering, heating (1000C), oxidation/reduction cycles
Crystallinity:
Anal. methods:
Contamination: well below 0.01ML impurities

COMMENTS
Analysis assumed validity of dipole approximation for the SEELFS matrix element;
data indicated large oxygen vibr. ampl.

DATA COLLECTION
Technique: SEELFS
Dataset : SEELFS above O K edge and Ni M(23) edge using primary electron energies between 1500-2000 eV

THEORY/DATA TREATMENT
EXAFS theory (see comments) with experimentally determined phase shifts from thermally grown NiO film on Ni(100)

STRUCTURES EXAMINED
Structures consistent with O-Ni nn distance of 1.96±0.03Å (from O K edge data) and an unperturbed substrate (from Ni M(23) data)

QUALITY OF EXPERIMENT-THEORY FIT
Visual

2D UNIT CELLS (1 domain observed)

Cell	Ax (Å)	Ay (Å)	Bx (Å)	By (Å)	α (°)	Matrix	Pattern	Cell type
Bulk	2.489	0.000	0.000	2.489	90.0	(1.000, 0.000) (0.000, 1.000)	(1x1)	b: bulk lattice
Surface 1	2.489	2.489	-2.489	2.489	90.0	(1.000, 1.000) (-1.000, 1.000)	c(2x2)	s1: commens. superlattice

3D COORDINATES

O1: overlayer in 4-fold hollow sites; coordinates are derived from bond length

Dx/Dy in Å, or as a fraction of layer's unit cell vectors; atom 0 at the origin. Epir/subr are bulk repeat vectors.

No. of atoms: 3

Bulk z = 1.760 Å

Region	Chem el.	At. no.	Cell type	Site occ.	Rel. to	Dx ± εx		Dy ± εy		Dz ± εz		Dz/Bz(%) ± εz/Bz
epir		-2					f		f		Å	
subr		-1				-1.245	Å	-1.245	Å	1.760	Å	
ovrl	O	1	s1	.50	0	0.000	f	0.000	f	0.000	Å	0.0
intf	Ni	2	b	1.00	1	0.500	f	0.500	f	0.850 ± .050	Å	48.3 ± 2.8
subl	Ni	3	b	1.00	2	-0.500	f	-0.500	f	1.760	Å	100.0

BOND DISTANCES AND ANGLES

No. of distances/angles: 7

Interatomic dist. A-B (Å)	Atom A	Atom B	Atom C	Bond angle A-B-C (°)
1.955	O1	Ni2	O1(1,0)	128.4
1.955	O1	Ni2	Ni2(1,0)	129.6
1.955	O1	Ni2	Ni3(1,1)	160.8
1.955	O1	Ni2	Ni3(1,0)	107.9
1.955	O1	Ni2	Ni3	70.8

Ni(100)-c(2x2)-O
28.8.36

Bond Distances and Angles - Continued

Interatomic dist. A-B (Å)	Atom A	Atom B	Atom C	Bond angle A-B-C (°)
2.489	Ni2	Ni2(1,0)	O1(1,0)	50.5
2.489	Ni2	Ni3		

COMMON NAME : Ni(100)-c(2x2)-O
CLASSIFICATION : 28.8.37
TECHNIQUE : PED
AUTHORS : S.Y Tong, W.M. Kang, D.H. Rosenblatt, J.G. Tobin and D.A. Shirley
REFERENCE : Phys. Rev., B27, 4632 (1983)

ILLUSTRATION: 28,29

SURFACE TYPE
Substrate : Ni
Crystal face: 100
Temperature : 300 K
Bulk lattice: fcc
2D bulk symm: p4m
2D surf symm: p4m

Adsorbate: O
Coverage : 0.5 O/Ni
Pattern : c(2x2)
Matrix : (1.000, 1.000)
(-1.000, 1.000)

STRUCTURE TYPE
Atomic adsorption in 4-fold hollow sites

SAMPLE PREPARATION (sample)
Treatment : see Rosenblatt et al, Phys. Rev. B23, 3828 (1981)
Crystallinity:
Anal. methods:
Contamination:

COMMENTS

DATA COLLECTION
Technique: PED
Dataset : normal PED data of Rosenblatt et al, Phys. Rev. B23, 3828 (1981)

THEORY/DATA TREATMENT
Mult. scatt. theory (combined space method of Tong and Van Hove, Phys. Rev. B16, 1459 (1977)): 5 phase shifts

STRUCTURES EXAMINED
Various O-Ni layer spacings (0.0 to 1.2Å) with O in 4-fold hollow site

QUALITY OF EXPERIMENT-THEORY FIT
R=0.2 (average of 6 R-factors)

2D UNIT CELLS (1 domain observed)

Cell	Ax (Å)	Ay (Å)	Bx (Å)	By (Å)	α (°)	Matrix	Pattern	Cell type
Bulk	2.489	0.000	0.000	2.489	90.0	(1.000, 0.000) (0.000, 1.000)	(1x1)	b: bulk lattice
Surface 1	2.489	2.489	-2.489	2.489	90.0	(1.000, 1.000) (-1.000, 1.000)	c(2x2)	s1: commens. superlattice

3D COORDINATES

O1: overlayer in 4-fold hollow sites;

Dx/Dy in Å, or as a fraction of layer's unit cell vectors; atom O at the origin. Epir/subr are bulk repeat vectors.

No. of atoms: 3

Bulk z = 1.760 Å

Region	Chem el.	At. no.	Cell type	Site occ.	Rel. to	Dx ± εx		Dy ± εy		Dz ± εz		Dz/Bz(%) ± εz/Bz
epir		-2					f		f		Å	
subr		-1				-1.245	Å	-1.245	Å	1.760	Å	
ovrl	O	1	s1	.50	0	0.000	f	0.000	f	0.000	Å	0.0
intf	Ni	2	b	1.00	1	0.500	f	0.500	f	0.850 ± .040	Å	48.3 ± 2.3
subl	Ni	3	b	1.00	2	-0.500	f	-0.500	f	1.760	Å	100.0

BOND DISTANCES AND ANGLES

Bond distances and angles are derived from coordinates

No. of distances/angles: 7

Interatomic dist. A-B (Å)	Atom A	Atom B	Atom C	Bond angle A-B-C (°)
1.955	O1	Ni2	O1(1,0)	128.4
1.955	O1	Ni2	Ni2(1,0)	129.6
1.955	O1	Ni2	Ni3(1,1)	160.8
1.955	O1	Ni2	Ni3(1,0)	107.9
1.955	O1	Ni2	Ni3	70.8

Ni(100)-c(2x2)-O
28.8.37

Bond Distances and Angles - Continued

Interatomic dist. A-B (Å)	Atom A	Atom B	Atom C	Bond angle A-B-C (°)
2.489	Ni2	Ni2(1,0)	O1(1,0)	50.5
2.489	Ni2	Ni3		

COMMON NAME : Ni(100)-c(2x2)-O
CLASSIFICATION : 28.8.42
TECHNIQUE : MEIS
AUTHORS : J.W.M. Frenken, J.F. van der Veen and G. Allan
REFERENCE : Phys. Rev. Lett., 51, 1876 (1983)

ILLUSTRATION: 28,29

SURFACE TYPE
Substrate : Ni
Crystal face: 100
Temperature : 370 K
Bulk lattice: fcc
2D bulk symm: p4m
2D surf symm: p4m

Adsorbate: O
Coverage : 0.46±0.4 O/Ni
Pattern : c(2x2)
Matrix : (1.000, 1.000)
(-1.000, 1.000)

STRUCTURE TYPE
Atomic adsorption in 4-fold hollow sites

SAMPLE PREPARATION (sample)
Treatment : 'standard procedures'
Crystallinity:
Anal. methods:
Contamination:

COMMENTS
Using tight-binding model, authors deduce strengthening of Ni interlayer force constants relative to bulk values, accompanying oxygen-induced expansion

DATA COLLECTION
Technique: MEIS; Rutherford back scattering
Dataset : blocking curves

THEORY/DATA TREATMENT
Geometric interpretation of blocking curves

STRUCTURES EXAMINED
Various O-Ni layer spacings only

QUALITY OF EXPERIMENT-THEORY FIT
Visual

2D UNIT CELLS (1 domain observed)

Cell	Ax (Å)	Ay (Å)	Bx (Å)	By (Å)	α (°)	Matrix	Pattern	Cell type
Bulk	2.489	0.000	0.000	2.489	90.0	(1.000, 0.000) (0.000, 1.000)	(1x1)	b: bulk lattice
Surface 1	2.489	2.489	-2.489	2.489	90.0	(1.000, 1.000) (-1.000, 1.000)	c(2x2)	s1: commens. superlattice

3D COORDINATES

O1: overlayer in 4-fold hollow sites;

Dx/Dy in Å, or as a fraction of layer's unit cell vectors; atom 0 at the origin. Epir/subr are bulk repeat vectors.

No. of atoms: 4 Bulk z = 1.760 Å

Reg ion	Chem el.	At. no.	Cell type	Site occ.	Rel. to	Dx ± εx		Dy ± εy		Dz ± εz		Dz/Bz(%) ± εz/Bz
epir		-2					f		f		Å	
subr		-1				-1.245	Å	-1.245	Å	1.760	Å	
ovrl	O	1	s1	.50	0	0.000	f	0.000	f	0.000	Å	0.0
intf	Ni	2	b	1.00	1	0.500	f	0.500	f	0.860 ± .100	Å	48.9 ± 5.7
intf	Ni	3	b	1.00	2	-0.500	f	-0.500	f	1.850 ± .100	Å	105.1 ± 5.7
subl	Ni	4	b	1.00	3	0.500	f	0.500	f	1.760	Å	100.0

BOND DISTANCES AND ANGLES

Bond distances and angles are derived from coordinates

No. of distances/angles: 8

Interatomic dist. A-B (Å)	Atom A	Atom B	Atom C	Bond angle A-B-C (°)
1.959	O1	Ni2	O1(1,0)	127.9
1.959	O1	Ni2	Ni2(1,0)	129.4
1.959	O1	Ni2	Ni3(1,1)	159.6
1.959	O1	Ni2	Ni3(1,0)	108.6
1.959	O1	Ni2	Ni3	72.5
2.489	Ni2	Ni2(1,0)	O1(1,0)	50.6

Ni(100)-c(2x2)-O
28.8.42

Bond Distances and Angles - Continued

Interatomic dist. A-B (Å)	Atom A	Atom B	Atom C	Bond angle A-B-C (°)
2.553	Ni2	Ni3	Ni4	91.4
2.489	Ni3	Ni4		

COMMON NAME : Ni(100)-c(2x2)-O
CLASSIFICATION : 28.8.48
TECHNIQUE : HREELS
AUTHORS : T.S. Rahman, D.L. Mills, J.E. Black, J.M. Szeftel, S. Lehwald and H. Ibach
REFERENCE : Phys. Rev., B30, 589 (1984)

ILLUSTRATION: 28,29

SURFACE TYPE
Substrate : Ni
Crystal face: 100
Temperature : RT*
Bulk lattice: fcc
2D bulk symm: p4m
2D surf symm: p4m

Adsorbate: O
Coverage :
Pattern : c(2x2)
Matrix : (1.000, 1.000)
(-1.000, 1.000)

STRUCTURE TYPE
Atomic adsorption in 4-fold hollow sites

SAMPLE PREPARATION (1 sample)
Treatment : Ne+ bombardment, cycles of sputtering and annealing
Crystallinity:
Anal. methods:
Contamination: checked by AES and HREELS

COMMENTS
Authors conclude that c(2x2) and p(2x2) adsorbate-substrate relationship is the same, though O-Ni force constant is greater in the c(2x2) structure by a factor of 0.55

DATA COLLECTION
Technique: HREELS
Dataset : off-specular HREELS

THEORY/DATA TREATMENT
Calc of surface phonon spectra for O and 1st three Ni layers with short-range central force, dipole-dipole interactions

STRUCTURES EXAMINED
O-Ni layer spacings of 0.9 and 0.26Å in 4 fold hollow site; pseudo-bridge adsorption site (0.3Å displacement along (100) direction)

QUALITY OF EXPERIMENT-THEORY FIT
Visual

2D UNIT CELLS (1 domain observed)

Cell	Ax (Å)	Ay (Å)	Bx (Å)	By (Å)	α (°)	Matrix	Pattern	Cell type
Bulk	2.489	0.000	0.000	2.489	90.0	(1.000, 0.000) (0.000, 1.000)	(1x1)	b: bulk lattice
Surface 1	2.489	2.489	-2.489	2.489	90.0	(1.000, 1.000) (-1.000, 1.000)	c(2x2)	s1: commens. superlattice

3D COORDINATES

O1: overlayer in 4-fold hollow sites 0.1Å error bars assumed for tabulation

Dx/Dy in Å, or as a fraction of layer's unit cell vectors; atom 0 at the origin. Epir/subr are bulk repeat vectors.

No. of atoms: 3

Bulk z = 1.760 Å

Region	Chem el.	At. no.	Cell type	Site occ.	Rel. to	Dx ± εx	Dy ± εy	Dz ± εz	Dz/Bz(%) ± εz/Bz
epir		-2				f	f	Å	
subr		-1				-1.245 Å	-1.245 Å	1.760 Å	
ovrl	O	1	s1	.50	0	0.000 f	0.000 f	0.000 Å	0.0
intf	Ni	2	b	1.00	1	0.500 f	0.500 f	0.900 ± .100 Å	51.1 ± 5.7
subl	Ni	3	b	1.00	2	-0.500 f	-0.500 f	1.760 Å	100.0

BOND DISTANCES AND ANGLES

Bond distances and angles are derived from coordinates

No. of distances/angles: 7

Interatomic dist. A-B (Å)	Atom A	Atom B	Atom C	Bond angle A-B-C (°)
1.977	O1	Ni2	O1(1,0)	125.8
1.977	O1	Ni2	Ni2(1,0)	129.0
1.977	O1	Ni2	Ni3(1,1)	162.1
1.977	O1	Ni2	Ni3(1,0)	108.8
1.977	O1	Ni2	Ni3	72.1
2.489	Ni2	Ni2(1,0)	O1(1,0)	51.0

Ni(100)-c(2x2)-O
28.8.48

Bond Distances and Angles - Continued

Interatomic dist. A-B (Å)	Atom A	Atom B	Atom C	Bond angle A-B-C (°)
2.489	Ni2	Ni3		

COMMON NAME : Ni(100)-c(2x2)-O ILLUSTRATION: 28,29
CLASSIFICATION : 28.8.59
TECHNIQUE : PED
AUTHORS : R. Saiki, A. Kaduwela, J. Osterwalder, M. Sagurton, C.S. Fadley and C.R. Brundle
REFERENCE : J. Vac. Sci. Technol., A5, 932 (1987)

SURFACE TYPE
Substrate : Ni
Crystal face: 100
Temperature : RT*
Bulk lattice: fcc
2D bulk symm: p4m
2D surf symm: p4m

Adsorbate: O
Coverage : 1/2 (O/Ni)
Pattern : c(2x2)
Matrix : (1.000, 1.000)
(-1.000, 1.000)

STRUCTURE TYPE
Atomic adsorption in hollow sites

SAMPLE PREPARATION (1 sample)
Treatment : see C.S. Fadley, Prog. Surf. Sci. 16, 275 (1984)
Crystallinity:
Anal. methods:
Contamination:

COMMENTS

DATA COLLECTION
Technique: PED; x-ray photelectron diffraction (XPD)
Dataset : azimuthal scans of O1s photoelectron distributions at polar angles of 45° and 8° (wrt surface plane)

THEORY/DATA TREATMENT
Comprison with single-scattering cluster approach including spher.-wave scattering and correlated vibrations

STRUCTURES EXAMINED
O in hollows at various heights above unrelaxed substrate; 'pseudo-bridge' position of Demuth et al, PRL 50 1373 (1983)

QUALITY OF EXPERIMENT-THEORY FIT
Visual

2D UNIT CELLS (1 domain observed)

Cell	Ax (Å)	Ay (Å)	Bx (Å)	By (Å)	α (°)	Matrix	Pattern	Cell type
Bulk	2.489	0.000	0.000	2.489	90.0	(1.000, 0.000) (0.000, 1.000)	(1x1)	b: bulk lattice
Surface 1	2.489	2.489	-2.489	2.489	90.0	(1.000, 1.000) (-1.000, 1.000)	c(2x2)	s1: commens. superlattice

3D COORDINATES

O1: overlayer in 4-fold hollows

Dx/Dy in Å, or as a fraction of layer's unit cell vectors; atom O at the origin. Epir/subr are bulk repeat vectors.

No. of atoms: 3 Bulk z = 1.760 Å

Region	Chem el.	At. no.	Cell type	Site occ.	Rel. to	Dx ± εx	Dy ± εy	Dz ± εz	Dz/Bz(%) ± εz/Bz
epir		-2				f	f	Å	
subr		-1				-1.245 Å	-1.245 Å	1.760 Å	
ovrl	O	1	s1	.50	0	0.000 f	0.000 f	0.000 Å	0.0
intf	Ni	2	b	1.00	1	0.500 f	0.500 f	0.850 ± .100 Å	48.3 ± 5.7
subl	Ni	3	b	1.00	2	-0.500 f	-0.500 f	1.760 Å	100.0

BOND DISTANCES AND ANGLES

Bond distances and angles are derived from coordinates

No. of distances/angles: 2

Interatomic dist. A-B (Å)	Atom A	Atom B	Atom C	Bond angle A-B-C (°)
1.955	O1	Ni2	Ni3	70.8
2.489	Ni2	Ni2(1,0)		

COMMON NAME : Ni(100)-c(2x2)-O
CLASSIFICATION : 28.8.61
TECHNIQUE : SEXAFS
AUTHORS : L. Wenzel, D. Arvanitis, W. Daum, H.H. Rotermund, J. Stoehr, K. Baberschke and H. Ibach
REFERENCE : Phys. Rev., B36, 7689 (1987)

ILLUSTRATION: 28,29

SURFACE TYPE
Substrate : Ni
Crystal face: 100
Temperature : 50 K
Bulk lattice: fcc
2D bulk symm: p4m
2D surf symm: p4m

Adsorbate: O
Coverage : 1/2 (O/Ni)
Pattern : c(2x2)
Matrix : (1.000, 1.000)
(-1.000, 1.000)

STRUCTURE TYPE
Atomic adorption in hollow site

SAMPLE PREPARATION (1 sample)
Treatment : 15L O2 adsorption at 75C
Crystallinity: sharp and clear LEED spots visible
Anal. methods:
Contamination:

COMMENTS
This result rules out an off-center hollow site; same result found at 50K and 295K

DATA COLLECTION
Technique: SEXAFS
Dataset : SEXAFS spectra at normal incidence and 20° grazing inc.; two temperatures, 50 K and 295K, give same structural results

THEORY/DATA TREATMENT
SEXAFS with experimentally determined phase shifts from bulk NiO standard

STRUCTURES EXAMINED
Hollow site and small shifts away from hollow site; variable O-Ni spacing, bulk-like substrate assumed

2D UNIT CELLS (1 domain observed)

Cell	Ax (Å)	Ay (Å)	Bx (Å)	By (Å)	α (°)	Matrix	Pattern	Cell type
Bulk	2.489	0.000	0.000	2.489	90.0	(1.000, 0.000) (0.000, 1.000)	(1x1)	b: bulk lattice
Surface 1	2.489	2.489	-2.489	2.489	90.0	(1.000, 1.000) (-1.000, 1.000)	c(2x2)	s1: commens. superlattice

3D COORDINATES

O1: overlayer in 4-fold hollow sites; coordinates are derived from bond distances and angles

Dx/Dy in Å, or as a fraction of layer's unit cell vectors; atom O at the origin. Epir/subr are bulk repeat vectors.

No. of atoms: 3

Bulk z = 1.760 Å

Region	Chem el.	At. no.	Cell type	Site occ.	Rel. to	Dx ± εx		Dy ± εy		Dz ± εz		Dz/Bz(%) ± εz/Bz
epir		-2					f		f		Å	
subr		-1				-1.245	Å	-1.245	Å	1.760	Å	
ovrl	O	1	s1	.50	0	0.000	f	0.000	f	0.000	Å	0.0
intf	Ni	2	b	1.00	1	0.500	f	0.500	f	0.880 ± .040	Å	50.0 ± 2.3
subl	Ni	3	b	1.00	2	-0.500	f	-0.500	f	1.760	Å	100.0

BOND DISTANCES AND ANGLES

No. of distances/angles: 2

Interatomic dist. A-B (Å)	Atom A	Atom B	Atom C	Bond angle A-B-C (°)
1.968	O1	Ni2	Ni3	71.6
2.489	Ni2	Ni2(1,0)		

COMMON NAME : Ni(100)-c(2x2)-O ILLUSTRATION: 28,29
CLASSIFICATION : 28.8.7
TECHNIQUE : LEED
AUTHORS : P.M. Marcus, J.E. Demuth and D.W. Jepsen
REFERENCE : Surf. Sci., 53, 501 (1975)

SURFACE TYPE
Substrate : Ni
Crystal face: 100
Temperature : 300 K
Bulk lattice: fcc
2D bulk symm: p4m
2D surf symm: p4m
Adsorbate: O
Coverage : 0.5 O/Ni
Pattern : c(2x2)
Matrix : (1.000, 1.000)
(-1.000, 1.000)

STRUCTURE TYPE
Atomic adsorption in 4-fold hollow sites

SAMPLE PREPARATION (sample)
Treatment : see J.E. Demuth and T.N. Rhodin, Surf. Sci. 45, 249 (1974)
Crystallinity:
Anal. methods:
Contamination:

COMMENTS

DATA COLLECTION
Technique: LEED
Dataset : I-V spectra for (0,0) (0,1) (1,1) (1/2,1/2) beams at Θ=0°

THEORY/DATA TREATMENT
Dynamical LEED (layer-KKR method, Bloch): Ni: Wakoh pot., Voi∝E**1/3, ΘD=420 K; O: superpos. pot., Voi=-3 eV, ΘD=335K

STRUCTURES EXAMINED
Bulk spacings assumed for Ni; O-Ni layer spacing varied from 0.815 to 1.760Å in steps of 0.105Å

QUALITY OF EXPERIMENT-THEORY FIT
Visual

2D UNIT CELLS (1 domain observed)

Cell	Ax (Å)	Ay (Å)	Bx (Å)	By (Å)	α (°)	Matrix	Pattern	Cell type
Bulk	2.489	0.000	0.000	2.489	90.0	(1.000, 0.000) (0.000, 1.000)	(1x1)	b: bulk lattice
Surface 1	2.489	2.489	-2.489	2.489	90.0	(1.000, 1.000) (-1.000, 1.000)	c(2x2)	s1: commens. superlattice

3D COORDINATES

O1: overlayer in 4-fold hollow sites; 0.1Å error bars assumed for tabulation

Dx/Dy in Å, or as a fraction of layer's unit cell vectors; atom 0 at the origin. Epir/subr are bulk repeat vectors.

No. of atoms: 3 Bulk z = 1.760 Å

Reg ion	Chem el.	At. no.	Cell type	Site occ.	Rel. to	Dx ± εx		Dy ± εy		Dz ± εz		Dz/Bz(%) ± εz/Bz
epir		-2					f		f		Å	
subr		-1				-1.245	Å	-1.245	Å	1.760	Å	
ovrl	O	1	s1	.50	0	0.000	f	0.000	f	0.000	Å	0.0
intf	Ni	2	b	1.00	1	0.500	f	0.500	f	0.920 ± .100	Å	52.3 ± 5.7
subl	Ni	3	b	1.00	2	-0.500	f	-0.500	f	1.760	Å	100.0

BOND DISTANCES AND ANGLES

Bond distances and angles are derived from coordinates

No. of distances/angles: 7

Interatomic dist. A-B (Å)	Atom A	Atom B	Atom C	Bond angle A-B-C (°)
1.986	O1	Ni2	O1(1,0)	124.8
1.986	O1	Ni2	Ni2(1,0)	128.8
1.986	O1	Ni2	Ni3(1,1)	162.6
1.986	O1	Ni2	Ni3(1,0)	109.1
1.986	O1	Ni2	Ni3	72.6

Ni(100)-c(2x2)-O
28.8.7

Bond Distances and Angles - Continued

Interatomic dist. A-B (Å)	Atom A	Atom B	Atom C	Bond angle A-B-C (°)
2.489	Ni2	Ni2(1,0)	O1(1,0)	51.2
2.489	Ni2	Ni3		

COMMON NAME : Ni(100)-c(2x2)-O
CLASSIFICATION : 28.8.71
TECHNIQUE : LEED
AUTHORS : W. Oed, H. Lindner, U. Starke, K. Heinz, K. Mueller and J.B. Pendry
REFERENCE : Surf. Sci., 224, 179 (1989)

ILLUSTRATION: 28,29

SURFACE TYPE
Substrate : Ni
Crystal face: 100
Temperature : 120 K
Bulk lattice: fcc
2D bulk symm: p4m
2D surf symm: p4m
Adsorbate: O
Coverage : 0.5 O/Ni
Pattern : c(2x2)
Matrix : (1.000,-1.000)
(1.000, 1.000)

STRUCTURE TYPE
Oxygen adsorbed in hollow site;
0.35Å buckling in 2nd Ni layer;
top Ni-Ni interlayer spacing expanded by 6 %

SAMPLE PREPARATION (1 sample)
Treatment : adsorption at 400 K, 6L oxygen exposure, annealing to 750K
Crystallinity: sharp c(2x2) LEED spots
Anal. methods: HREELS to control purity of phase, no
Contamination:

COMMENTS
No evidence for Demuth's pseudobridge model;
but also no sensitivity to O side shift up to 0.2Å;
local minimum for side shift occurs with bulk-like substrate, i.e. if 2nd Ni layer assumed planar

DATA COLLECTION
Technique: LEED; video LEED
Dataset : IV curves at normal incidence for 4 integer, 2 fractional order beams; cumul. E range 1330 eV

THEORY/DATA TREATMENT
Dynamical LEED (comb. space method, layer doubling):
11 phase shifts (E=50-350 eV)

STRUCTURES EXAMINED
Hollow site (incl. off center site), substr. reconstr. varied: O height, O off-center shift, 1st Ni-Ni interlayer spacing, 2nd Ni-Ni interlayer spacing, 2nd Ni layer buckling

QUALITY OF EXPERIMENT-THEORY FIT
RPE=0.28

2D UNIT CELLS (1 domain observed)

Cell	Ax (Å)	Ay (Å)	Bx (Å)	By (Å)	α (°)	Matrix	Pattern	Cell type
Bulk	2.492	0.000	0.000	2.492	90.0	(1.000, 0.000) (0.000, 1.000)	(1x1)	b: bulk lattice
Surface 1	2.492	-2.492	2.492	2.492	90.0	(1.000,-1.000) (1.000, 1.000)	c(2x2)	s1: commens. superlattice

3D COORDINATES

O1: overlayer in hollow sites; Ni2-Ni3: planar top Ni layer;
Ni4-Ni5: buckled 2nd Ni layer

Dx/Dy in Å, or as a fraction of layer's unit cell vectors; atom 0 at the origin. Epir/subr are bulk repeat vectors.

No. of atoms: 6 Bulk z = 1.762 Å

Region	Chem el.	At. no.	Cell type	Site occ.	Rel. to	Dx ± εx		Dy ± εy		Dz ± εz		Dz/Bz(%) ± εz/Bz
epir		-2					f		f		Å	
subr		-1				1.246	Å	1.246	Å	1.762	Å	
ovrl	O	1	s1	.50	0	0.000 ± .200	Å	0.000 ± .200	Å	0.000 ± .040	Å	0.0 ± 2.3
intf	Ni	2	s1	.50	1	0.500	f	0.000	f	0.770 ± .020	Å	43.7 ± 1.1
intf	Ni	3	s1	.50	1	0.000	f	-0.500	f	0.770 ± .020	Å	43.7 ± 1.1
intf	Ni	4	s1	.50	1	0.500	f	-0.500	f	2.615 ± .020	Å	148.4 ± 1.1
intf	Ni	5	s1	.50	1	0.000	f	0.000	f	2.650 ± .020	Å	150.4 ± 1.1
subl	Ni	6	b	1.00	1	-0.500	f	-0.500	f	4.390	Å	249.1

Ni(100)-c(2x2)-O
28.8.71

BOND DISTANCES AND ANGLES

Bond distances and angles are derived from coordinates

No. of distances/angles: 2

Interatomic dist. A-B (Å)	Atom A	Atom B	Atom C	Bond angle A-B-C (°)
1.923	O1	Ni2		
2.650	O1	Ni5		

COMMON NAME : Ni(100)-p(2x2)-O
CLASSIFICATION : 28.8.30b
TECHNIQUE : SEXAFS
AUTHORS : J. Stoehr, R. Jaeger and T. Kendelewicz
REFERENCE : Phys. Rev. Lett., 49, 142 (1982)

ILLUSTRATION: 28,30

SURFACE TYPE
Substrate : Ni
Crystal face: 100
Temperature : 300 K
Bulk lattice: fcc
2D bulk symm: p4m
2D surf symm: p4m
Adsorbate: O
Coverage : 1.5 L
Pattern : (2x2)
Matrix : (2.000, 0.000)
(0.000, 2.000)

STRUCTURE TYPE
Atomic adsorption in 4-fold hollow sites

SAMPLE PREPARATION (sample)
Treatment : see Brennan et al, Phys. Rev. B24 4871 (1981)
Crystallinity:
Anal. methods: LEED
Contamination: AES: <1%ML C, O, and S

COMMENTS
For analysis of the corresponding XANES see Norman et al, Phys. Rev. Lett. 51, 2052 (1983)

DATA COLLECTION
Technique: SEXAFS; SEXAFS at SSRL
Dataset : x-ray incidence at 45°

THEORY/DATA TREATMENT
Conventional EXAFS with experimentally determined phase shifts from bulk NiO standard

STRUCTURES EXAMINED
Different O-Ni layer spacings in the 4 fold hollow sites

QUALITY OF EXPERIMENT-THEORY FIT
Visual

2D UNIT CELLS (1 domain observed)

Cell	Ax (Å)	Ay (Å)	Bx (Å)	By (Å)	α (°)	Matrix	Pattern	Cell type
Bulk	2.490	0.000	0.000	2.490	90.0	(1.000, 0.000) (0.000, 1.000)	(1x1)	b: bulk lattice
Surface 1	4.980	0.000	0.000	4.980	90.0	(2.000, 0.000) (0.000, 2.000)	(2x2)	s1: commens. superlattice

3D COORDINATES

O1: atomic overlayer in 4-fold hollow sites; coordinates derived from bond length

Dx/Dy in Å, or as a fraction of layer's unit cell vectors; atom 0 at the origin. Epir/subr are bulk repeat vectors.

No. of atoms: 3 Bulk z = 1.760 Å

Region	Chem el.	At. no.	Cell type	Site occ.	Rel. to	Dx ± εx		Dy ± εy		Dz ± εz		Dz/Bz(%) ± εz/Bz
epir		-2					f		f		Å	
subr		-1				-1.245	Å	-1.245	Å	1.760	Å	
ovrl	O	1	s1	.25	0	0.000	f	0.000	f	0.000	Å	0.0
intf	Ni	2	b	1.00	1	0.500	f	0.500	f	0.860 ± .070	Å	48.9 ± 4.0
subl	Ni	3	b	1.00	2	-0.500	f	-0.500	f	1.760	Å	100.0

BOND DISTANCES AND ANGLES

No. of distances/angles: 6

Interatomic dist. A-B (Å)	Atom A	Atom B	Atom C	Bond angle A-B-C (°)
1.960	O1	Ni2	Ni2(1,0)	129.5
1.960	O1	Ni2	Ni3(1,1)	161.0
1.960	O1	Ni2	Ni3(1,0)	108.1
1.960	O1	Ni2	Ni3	71.0
2.620	O1	Ni3		
2.490	Ni2	Ni3		

COMMON NAME : Ni(100)-p(2x2)-O
CLASSIFICATION : 28.8.32b
TECHNIQUE : XANES (NEXAFS)
AUTHORS : D. Norman, J. Stoehr, R. Jaeger, P.J. Durham and J.B. Pendry
REFERENCE : Phys. Rev. Lett., 51, 2052 (1983)

ILLUSTRATION: 28,30

SURFACE TYPE
Substrate : Ni
Crystal face: 100
Temperature : 300 K
Bulk lattice: fcc
2D bulk symm: p4m
2D surf symm: p4m

Adsorbate: O
Coverage : 0.25 (O/Ni)
Pattern : (2x2)
Matrix : (2.000, 0.000)
(0.000, 2.000)

STRUCTURE TYPE
Atomic adsorption in 4-fold hollow sites

SAMPLE PREPARATION (sample)
Treatment : see J. Stoehr and R Jaeger, Phys. Rev. B26, 4111 (1982)
Crystallinity:
Anal. methods:
Contamination:

COMMENTS
Authors report small differences between p(2x2) and c(2x2) measured spectra

DATA COLLECTION
Technique: XANES (NEXAFS)
Dataset : XANES for normal and grazing x-ray incidence from threshold (approx. 530 eV) to 564eV with resolution 2.5eV

THEORY/DATA TREATMENT
Dynamical XANES: 30 atom cluster, Mattheis muffin-tin pots; Voi=-0.7 eV

STRUCTURES EXAMINED
O-Ni layer distances of 0.2 and 0.9Å in the 4-fold hollow site; bridge and top sites with fixed O-Ni bond length of 1.98Å

QUALITY OF EXPERIMENT-THEORY FIT
Visual

2D UNIT CELLS (1 domain observed)

Cell	Ax (Å)	Ay (Å)	Bx (Å)	By (Å)	α (°)	Matrix	Pattern	Cell type
Bulk	2.490	0.000	0.000	2.490	90.0	(1.000, 0.000) (0.000, 1.000)	(1x1)	b: bulk lattice
Surface 1	4.980	0.000	0.000	4.980	90.0	(2.000, 0.000) (0.000, 2.000)	(2x2)	s1: commens. superlattice

3D COORDINATES

O1: atomic overlayer in 4-fold hollow sites

Dx/Dy in Å, or as a fraction of layer's unit cell vectors; atom 0 at the origin. Epir/subr are bulk repeat vectors.

No. of atoms: 3 Bulk z = 1.760 Å

Region	Chem el.	At. no.	Cell type	Site occ.	Rel. to	Dx ± εx		Dy ± εy		Dz ± εz		Dz/Bz(%) ± εz/Bz
epir		-2					f		f		Å	
subr		-1				-1.245	Å	-1.245	Å	1.760	Å	
ovrl	O	1	s1	.25	0	0.000	f	0.000	f	0.000	Å	0.0
intf	Ni	2	b	1.00	1	0.500	f	0.500	f	0.900 ± .100	Å	51.1 ± 5.7
subl	Ni	3	b	1.00	2	-0.500	f	-0.500	f	1.760	Å	100.0

BOND DISTANCES AND ANGLES

Bond distances and angles are derived from coordinates

No. of distances/angles: 6

Interatomic dist. A-B (Å)	Atom A	Atom B	Atom C	Bond angle A-B-C (°)
1.977	O1	Ni2	Ni2(1,0)	129.0
1.977	O1	Ni2	Ni3(1,1)	162.1
1.977	O1	Ni2	Ni3(1,0)	108.8
1.977	O1	Ni2	Ni3	72.1

Ni(100)-p(2x2)-O
28.8.32b

Bond Distances and Angles - Continued

Interatomic dist. A-B (Å)	Atom A	Atom B	Atom C	Bond angle A-B-C (°)
2.660	O1	Ni3		
2.490	Ni2	Ni3		

COMMON NAME : Ni(100)-p(2x2)-O ILLUSTRATION: 28,30
CLASSIFICATION : 28.8.5
TECHNIQUE : LEED
AUTHORS : M.A. Van Hove and S.Y. Tong
REFERENCE : J. Vac. Sci. Technol., 12, 230 (1975)

SURFACE TYPE
Substrate : Ni
Crystal face: 100
Temperature : 300 K
Bulk lattice: fcc
2D bulk symm: p4m
2D surf symm: p4m

Adsorbate: O
Coverage : 1/4 (O/Ni)
Pattern : (2x2)
Matrix : (2.000, 0.000)
(0.000, 2.000)

STRUCTURE TYPE
Atomic adsorption in 4-fold hollow sites

SAMPLE PREPARATION (sample)
Treatment : see Demuth and Rhodin, Surf. Sci. 42, 261 & 45, 249 (1974)
Crystallinity:
Anal. methods:
Contamination:

COMMENTS

DATA COLLECTION
Technique: LEED
Dataset :

THEORY/DATA TREATMENT
Dynamical LEED: layer doubling; Wakoh Ni potential, overlapping O at. pot., 8 phase shifts; Vor=-11.2 eV,Voi∝E**1/3

STRUCTURES EXAMINED
8 to 10 Ni-O spacings, 0.2Å apart, at hollow, bridge and top sites

QUALITY OF EXPERIMENT-THEORY FIT
Visual

2D UNIT CELLS (1 domain observed)

Cell	Ax (Å)	Ay (Å)	Bx (Å)	By (Å)	α (°)	Matrix	Pattern	Cell type
Bulk	2.490	0.000	0.000	2.490	90.0	(1.000, 0.000) (0.000, 1.000)	(1x1)	b: bulk lattice
Surface 1	4.980	0.000	0.000	4.980	90.0	(2.000, 0.000) (0.000, 2.000)	(2x2)	s1: commens. superlattice

3D COORDINATES

O1: atomic overlayer in 4-fold hollow sites

Dx/Dy in Å, or as a fraction of layer's unit cell vectors; atom 0 at the origin. Epir/subr are bulk repeat vectors.

No. of atoms: 3 Bulk z = 1.760 Å

Region	Chem el.	At. no.	Cell type	Site occ.	Rel. to	Dx ± εx		Dy ± εy		Dz ± εz		Dz/Bz(%) ± εz/Bz
epir		-2					f		f		Å	
subr		-1				-1.245	Å	-1.245	Å	1.760	Å	
ovrl	O	1	s1	.25	0	0.000	f	0.000	f	0.000	Å	0.0
intf	Ni	2	b	1.00	1	0.500	f	0.500	f	0.900 ± .100	Å	51.1 ± 5.7
subl	Ni	3	b	1.00	2	-0.500	f	-0.500	f	1.760	Å	100.0

BOND DISTANCES AND ANGLES

Bond distances and angles are derived from coordinates

No. of distances/angles: 6

Interatomic dist. A-B (Å)	Atom A	Atom B	Atom C	Bond angle A-B-C (°)
1.977	O1	Ni2	Ni2(1,0)	129.0
1.977	O1	Ni2	Ni3(1,1)	162.1
1.977	O1	Ni2	Ni3(1,0)	108.8
1.977	O1	Ni2	Ni3	72.1
2.660	O1	Ni3		
2.490	Ni2	Ni3		

ILLUSTRATION: 28,30

COMMON NAME : Ni(100)-p(2x2)-O
CLASSIFICATION : 28.8.72
TECHNIQUE : LEED
AUTHORS : W. Oed, H. Lindner, U. Starke, K. Heinz, K. Mueller, D.K. Saldin, P.L. de Andres and J.B.Pendry
REFERENCE : Surf. Sci., 225, 242 (1990)

SURFACE TYPE
Substrate : Ni
Crystal face: 100
Temperature : 120 K
Bulk lattice: fcc
2D bulk symm: p4m
2D surf symm: p4m

Adsorbate: O
Coverage : 0.25 O/Ni
Pattern : p(2x2)
Matrix : (2.000, 0.000)
(0.000, 2.000)

STRUCTURE TYPE
Oxygen in hollow site; expansion of top Ni-Ni interlayer spacing by 2.5%; second Ni layer buckling 0.10Å

SAMPLE PREPARATION (1 sample)
Treatment : adsorption at 400 K, 2L oxygen exposure, annealing to 750K
Crystallinity: sharp p(2x2) LEED spots
Anal. methods: HREELS to control purity of phase, no
Contamination:

COMMENTS
No evidence for Demuth's pseudobridge model; but also no sensitivity to O side shift up to 0.4Å; local minimum for side shift occurs with bulk-like substrate, i.e. if 2nd Ni layer assumed planar

DATA COLLECTION
Technique: LEED; video LEED
Dataset : IV curves at normal incidence for 4 integer, 3 fractional order beams; cumul. E range 1730 eV

THEORY/DATA TREATMENT
Dynamical LEED (comb. space method, layer doubling): 11 phase shifts (E=50-350 eV)

STRUCTURES EXAMINED
Hollow site (incl. off center site), substr. reconstr. varied: O height, O off-center shift, 1st Ni-Ni interlayer spacing, 2nd Ni-Ni interlayer spacing, 2nd Ni layer buckling; 1st Ni layer lateral shifts

QUALITY OF EXPERIMENT-THEORY FIT
RPE=0.24, Var(R)=0.04

2D UNIT CELLS (1 domain observed)

Cell	Ax (Å)	Ay (Å)	Bx (Å)	By (Å)	α (°)	Matrix	Pattern	Cell type
Bulk	2.492	0.000	0.000	2.492	90.0	(1.000, 0.000) (0.000, 1.000)	(1x1)	b: bulk lattice
Surface 1	4.984	0.000	0.000	4.984	90.0	(2.000, 0.000) (0.000, 2.000)	p(2x2)	s1: commens. superlattice

3D COORDINATES

O1: overlayer in hollow sites; Ni2-Ni5: planar top Ni layer;
Ni6-Ni9: buckled 2nd Ni layer

Dx/Dy in Å, or as a fraction of layer's unit cell vectors; atom 0 at the origin. Epir/subr are bulk repeat vectors.

No. of atoms: 10 Bulk z = 1.762 Å

Region	Chem el.	At. no.	Cell type	Site occ.	Rel. to	Dx ± εx		Dy ± εy		Dz ± εz		Dz/Bz(%) ± εz/Bz
epir		-2					f		f		Å	
subr		-1				1.246	Å	1.246	Å	1.762	Å	
ovrl	O	1	s1	.25	0	0.000 ± .400	Å	0.000 ± .400	Å	0.000 ± .050	Å	0.0 ± 2.8
intf	Ni	2	s1	.25	1	0.250	f	-0.250	f	0.800 ± .050	Å	45.4 ± 2.8
intf	Ni	3	s1	.25	1	0.250	f	0.250	f	0.800 ± .050	Å	45.4 ± 2.8
intf	Ni	4	s1	.25	1	-0.250	f	0.250	f	0.800 ± .050	Å	45.4 ± 2.8
intf	Ni	5	s1	.25	1	-0.250	f	-0.250	f	0.800 ± .050	Å	45.4 ± 2.8
intf	Ni	6	s1	.25	5	0.750	f	0.750	f	1.745 ± .020	Å	99.0 ± 1.1
intf	Ni	7	s1	.25	5	0.750	f	0.250	f	1.805 ± .020	Å	102.4 ± 1.1
intf	Ni	8	s1	.25	5	0.250	f	0.750	f	1.805 ± .020	Å	102.4 ± 1.1
intf	Ni	9	s1	.25	5	0.250	f	0.250	f	1.845 ± .020	Å	104.7 ± 1.1
subl	Ni	10	b	1.00	9	-0.500	f	-0.500	f	1.705	Å	96.8

Ni(100)-p(2x2)-O
28.8.72

BOND DISTANCES AND ANGLES

Bond distances and angles are derived from coordinates

No. of distances/angles: 2

Interatomic dist. A-B (Å)	Atom A	Atom B	Atom C	Bond angle A-B-C (°)
1.935	O1	Ni2		
2.645	O1	Ni9		

COMMON NAME : Ni(100)-O disordered ILLUSTRATION: 28
CLASSIFICATION : 28.8.82
TECHNIQUE : LEED
AUTHORS : U. Starke, W. Oed, P. Bayer, F. Bothe, G. Fuerst, P.L. de Andres, K. Heinz and J.B. Pendry
REFERENCE : Springer Series in Surface Sciences, 24, 427 (1991)

SURFACE TYPE
Substrate : Ni
Crystal face: 100
Temperature : 80 K
Bulk lattice: fcc
2D bulk symm: p4m
2D surf symm: none
Adsorbate: O
Coverage : 0.1 O/Ni
Pattern : disordered
Matrix : (1.000, 0.000) (0.000, 1.000)

STRUCTURE TYPE
Disordered hollow site adsorption in off center position (pseudobridge) shifted by 0.45Å; 2nd layer buckled, sideshift of Ni atoms close to O of up to 0.15Å possible; local minimum for 4-fold-site with 1st layer buckling; disordered substrate relaxation modeled here as (3x3)

SAMPLE PREPARATION (1 sample)
Treatment : adsorption at 80 K
Crystallinity:
Anal. methods:
Contamination:

COMMENTS
Experimental data similar for 0.015-0.25ML coverage range: same structure can be assumed
diagonal lateral displacement of Ni2, Ni4 (away from O) up to 0.15Å possible

DATA COLLECTION
Technique: LEED; video (diffuse) LEED, Y-function maps
Dataset : intensities at 8 energies (48 eV-76eV) for 7 Y-function maps; average over 1000 video frames, symm-eq data, 2D smoothing

THEORY/DATA TREATMENT
Dynamical LEED (3-step DLEED method, tensor LEED): 6 phase shifts, Voi=-4 eV

STRUCTURES EXAMINED
Hollow site, off center oxygen, substrate relaxations Buckling of 4 NN's in 1st layer, buckling of 9 atoms in 2nd layer: A below oxygen, 4*B=NN of A, 4*C=diag. NN of A only A buckles

QUALITY OF EXPERIMENT-THEORY FIT
RPE=0.198 (=MSD of Y-functions)

2D UNIT CELLS (4 domains observed)

Cell	Ax (Å)	Ay (Å)	Bx (Å)	By (Å)	α (°)	Matrix	Pattern	Cell type
Bulk	2.492	0.000	0.000	2.492	90.0	(1.000, 0.000) (0.000, 1.000)	(1x1)	b: bulk lattice
Surface 1	7.476	0.000	0.000	7.476	90.0	(3.000, 0.000) (0.000, 3.000)	disordered	rd1: reconstr. lattice-gas dis

3D COORDINATES

O1: disordered adatom offcenter from hollow sites; Ni2-Ni10: top Ni layer, buckled, with possible lateral relaxations; Ni11-Ni19: 2nd Ni layer

Dx/Dy in Å, or as a fraction of layer's unit cell vectors; atom 0 at the origin. Epir/subr are bulk repeat vectors.

No. of atoms: 20 Bulk z = 1.762 Å

Region	Chem el.	At. no.	Cell type	Site occ.	Rel. to	Dx ± εx		Dy ± εy		Dz ± εz		Dz/Bz(%) ± εz/Bz
epir		-2					f		f		Å	
subr		-1				1.246	Å	1.246	Å	1.762	Å	
ovrl	O	1	rd1	.11	0	0.450 ± .100	Å	0.000	Å	0.000 ± .025	Å	0.0 ± 1.4
intf	Ni	2	rd1	.11	0	0.167	f	0.167	f	0.860 ± .030	Å	48.8 ± 1.7
intf	Ni	3	rd1	.11	0	0.167	f	0.500	f	0.860	Å	48.8
intf	Ni	4	rd1	.11	0	0.167	f	-0.167	f	0.860 ± .030	Å	48.8 ± 1.7
intf	Ni	5	rd1	.11	0	0.500	f	0.167	f	0.860	Å	48.8
intf	Ni	6	rd1	.11	0	-0.167	f	0.167	f	0.860 ± .030	Å	48.8 ± 1.7
intf	Ni	7	rd1	.11	0	0.500	f	0.500	f	0.860	Å	48.8
intf	Ni	8	rd1	.11	0	0.500	f	-0.167	f	0.860	Å	48.8
intf	Ni	9	rd1	.11	0	-0.167	f	0.500	f	0.860	Å	48.8
intf	Ni	10	rd1	.11	0	-0.167	f	-0.167	f	0.860 ± .030	Å	48.8 ± 1.7
intf	Ni	11	rd1	.11	0	0.000	f	0.000	f	2.530 ± .050	Å	143.6 ± 2.8
intf	Ni	12	rd1	.11	0	-0.333	f	0.333	f	2.600	Å	147.6
intf	Ni	13	rd1	.11	0	-0.333	f	-0.333	f	2.600	Å	147.6
intf	Ni	14	rd1	.11	0	-0.333	f	0.000	f	2.600	Å	147.6
intf	Ni	15	rd1	.11	0	0.000	f	-0.333	f	2.600	Å	147.6
intf	Ni	16	rd1	.11	0	0.000	f	0.333	f	2.600	Å	147.6
intf	Ni	17	rd1	.11	0	0.333	f	0.000	f	2.600	Å	147.6
intf	Ni	18	rd1	.11	0	0.333	f	0.333	f	2.600	Å	147.6

Ni(100)-O disordered
28.8.82

BOND DISTANCES AND ANGLES

Bond distances for no lateral shift in Ni layer

No. of distances/angles: 2

Interatomic dist. A-B (Å)	Atom A	Atom B	Atom C	Bond angle A-B-C (°)
1.710	O1	Ni2		
2.570	O1	Ni11		

COMMON NAME : Ni(110)-(2x1)-O ILLUSTRATION: 39
CLASSIFICATION : 28.8.53
TECHNIQUE : ALICISS
AUTHORS : H. Niehus and G. Comsa
REFERENCE : Surf. Sci., 151, L171 (1985)

SURFACE TYPE
Substrate : Ni
Crystal face: 110
Temperature : RT*
Bulk lattice: fcc
2D bulk symm: pmm
2D surf symm: pmm
Adsorbate: O
Coverage : 0.5 O/Ni
Pattern : (2x1)
Matrix : (2.000, 0.000)
(0.000, 1.000)

STRUCTURE TYPE
Missing row reconstruction, removing every other ridge atom, with O occupying bridge sites between remaining Ni atoms in adjacent ridges; substrate relaxations not analyzed

SAMPLE PREPARATION (1 sample)
Treatment : Ar+ bomb. and anneal, then 1mbar-s O deposited at 450 K
Crystallinity: sharp (2x1) LEED superstructure
Anal. methods: AES, ISS, LEED
Contamination: monitored by LEED

COMMENTS

DATA COLLECTION
Technique: ALICISS; alkali impact collision ion scatt
Dataset : Na+ of E=2000 eV scattered through 145° in 3 azimuthal planes containing <1-10>, <1-12>, <001> directions

THEORY/DATA TREATMENT
Fit of shadow cone to data and used to predict ALICISS profiles for different structural models

STRUCTURES EXAMINED
Unreconstructed, sawtooth and missing row models tested; with miss. row model, O height was examined, a bridge site being assumed

QUALITY OF EXPERIMENT-THEORY FIT
Visual

2D UNIT CELLS (1 domain observed)

Cell	Ax (Å)	Ay (Å)	Bx (Å)	By (Å)	α (°)	Matrix	Pattern	Cell type
Bulk	2.490	0.000	0.000	3.520	90.0	(1.000, 0.000) (0.000, 1.000)	(1x1)	b: bulk lattice
Surface 1	4.980	0.000	0.000	3.520	90.0	(2.000, 0.000) (0.000, 1.000)	(2x1)	s1: commens. superlattice

3D COORDINATES

O1-Ni2: form -O-Ni-O-Ni- string, with O higher than Ni; Ni2: remaining row of atoms in Ni ridges; 0.2Å error bar assumed for tabulation

Dx/Dy in Å, or as a fraction of layer's unit cell vectors; atom 0 at the origin. Epir/subr are bulk repeat vectors.

No. of atoms: 4 Bulk z = 1.245 Å

Region	Chem el.	At. no.	Cell type	Site occ.	Rel. to	Dx ± εx		Dy ± εy		Dz ± εz		Dz/Bz(%) ± εz/Bz
epir		-2					f		f		Å	
subr		-1				-1.245	Å	1.760	Å	1.245	Å	
ovrl	O	1	s1	.50	0	0.000	f	0.000	f	0.000	Å	0.0
intf	Ni	2	s1	.50	1	0.000	f	0.500	f	0.250 ± .200	Å	20.1 ± 16.1
intf	Ni	3	b	1.00	2	0.500	f	-0.500	f	1.245	Å	100.0
subl	Ni	4	b	1.00	3	-0.500	f	0.500	f	1.245	Å	100.0

BOND DISTANCES AND ANGLES

Bond distances and angles are derived from coordinates

No. of distances/angles: 5

Interatomic dist. A-B (Å)	Atom A	Atom B	Atom C	Bond angle A-B-C (°)
1.778	O1	Ni2	O1(0,1)	163.8
1.778	O1	Ni2	Ni3(0,1)	140.4
1.778	O1	Ni2	Ni3	51.0
1.778	O1	Ni2	Ni4	98.1

Ni(110)-(2x1)-O
28.8.53

Bond Distances and Angles - Continued

Interatomic dist. A-B (Å)	Atom A	Atom B	Atom C	Bond angle A-B-C (°)
2.490	Ni2	Ni3		

COMMON NAME : Ni(110)-(2x1)-O
CLASSIFICATION : 28.8.58
TECHNIQUE : SEXAFS
AUTHORS : K. Baberschke, U. Dobler, L. Wenzel and D. Arvanitis
REFERENCE : Phys. Rev., B33, 5910 (1986)

ILLUSTRATION:

SURFACE TYPE
Substrate : Ni
Crystal face: 110
Temperature : RT*
Bulk lattice: fcc
2D bulk symm: pmm
2D surf symm: pm

Adsorbate: O
Coverage : 0.5 (O/Ni)
Pattern : (2x1)
Matrix : (2.000, 0.000)
(0.000, 1.000)

STRUCTURE TYPE
Atomic adsorption in long-bridge site of saw-tooth reconstructed substrate

SAMPLE PREPARATION (1 sample)
Treatment : exposure to 0.8L oxygen at 463 K
Crystallinity: sharp (2x1) LEED pattern
Anal. methods:
Contamination:

COMMENTS
An unspecified tilt of each oxygen towards (100) facets of the reconstructed substrate is favored (tilt is not included in tabulated coordinates)

DATA COLLECTION
Technique: SEXAFS; oxygen-K-edge SEXAFS
Dataset : spectra at normal inc. with E-vector // [0,0,1], [-1,1,0], and off-normal inc. (Θ=45, 25°) with E in (001) plane

THEORY/DATA TREATMENT
SEXAFS with polarized x-ray and Fourier transform analysis

STRUCTURES EXAMINED
Distance to nearest backscatterer and amplitude ratio for normal incidence for different models (missing-row and sawtooth) and orientations were calculated

2D UNIT CELLS (2 domains observed)

Cell	Ax (Å)	Ay (Å)	Bx (Å)	By (Å)	α (°)	Matrix	Pattern	Cell type
Bulk	2.490	0.000	0.000	3.520	90.0	(1.000, 0.000) (0.000, 1.000)	(1x1)	b: bulk lattice
Surface 1	4.980	0.000	0.000	3.520	90.0	(2.000, 0.000) (0.000, 1.000)	(2x1)	s1: commens. superlattice

3D COORDINATES

O1: O-Ni-O-Ni strings with Ni2 (Ni2 is edge of sawtooth); Ni3: forms sloping side of sawtooth;
Ni4: periodically repeating bulk layer; 0.03Å error bars assumed for tabulation

Dx/Dy in Å, or as a fraction of layer's unit cell vectors; atom 0 at the origin. Epir/subr are bulk repeat vectors.

No. of atoms: 4 Bulk z = 1.245 Å

Reg ion	Chem el.	At. no.	Cell type	Site occ.	Rel. to	Dx ± εx		Dy ± εy		Dz ± εz		Dz/Bz(%) ± εz/Bz
epir		-2					f		f		Å	
subr		-1				1.245	Å	-1.760	Å	1.245	Å	
ovrl	O	1	s1	.50	0	0.000	f	0.000	f	0.000	Å	0.0
intf	Ni	2	s1	.50	1	0.000	f	0.500	f	0.560 ± .030	Å	45.0 ± .3
intf	Ni	3	s1	.50	2	0.750	f	-0.500	f	1.245	Å	100.0
subl	Ni	4	b	1.00	3	-0.500	f	0.500	f	1.245	Å	100.0

BOND DISTANCES AND ANGLES

No. of distances/angles: 4

Interatomic dist. A-B (Å)	Atom A	Atom B	Atom C	Bond angle A-B-C (°)
1.847	O1	Ni2	Ni3(-1,0)	58.5
2.193	O1	Ni3(-1,0)	Ni4(-1,0)	97.3
2.490	Ni2	Ni3(-1,0)	Ni4(-1,0)	60.0
2.490	Ni3	Ni4		

COMMON NAME : Ni(110)-(2x1)-O
CLASSIFICATION : 28.8.70
TECHNIQUE : LEED
AUTHORS : G. Kleinle, J. Wintterlin, G. Ertl, R.J. Behm, F. Jona and W. Moritz
REFERENCE : Surf. Sci., 225, 171 (1990)

ILLUSTRATION: 39

SURFACE TYPE
Substrate : Ni
Crystal face: 110
Temperature : 90 K
Bulk lattice: fcc
2D bulk symm: pmm
2D surf symm: pm

Adsorbate: O
Coverage : 0.50 ML
Pattern : (2x1)
Matrix : (2.000, 0.000) (0.000, 1.000)

STRUCTURE TYPE
Missing-row structure in which the O atoms are above long bridge sites in [001] direction with slight asymmetry; top 2 Ni layers have an expanded separation while the 2nd and 3rd have a contracted separation; the 3rd layer is slightly buckled

SAMPLE PREPARATION (1 sample)
Treatment : exposed to 0.80L at 90 K; annealed to 500K
Crystallinity: perfect (2x1) LEED pattern
Anal. methods: work function, LEED
Contamination:

COMMENTS

DATA COLLECTION
Technique: LEED; video LEED
Dataset : IV spectra for 8 non-equivalent beams: 5 integral and 3 fractional

THEORY/DATA TREATMENT
Dynamical LEED (layer doubling, composite layers)

STRUCTURES EXAMINED
Missing-row, sawtooth and buckled models for the substrate with the O in the bridge site

QUALITY OF EXPERIMENT-THEORY FIT
RPE=0.179

2D UNIT CELLS (2 domains observed)

Cell	Ax (Å)	Ay (Å)	Bx (Å)	By (Å)	α (°)	Matrix	Pattern	Cell type
Bulk	2.492	0.000	0.000	3.524	90.0	(1.000, 0.000) (0.000, 1.000)	(1x1)	b: bulk lattice
Surface 1	4.984	0.000	0.000	3.524	90.0	(2.000, 0.000) (0.000, 1.000)	(2x1)	s1: commens. superlattice

3D COORDINATES

O1: in tilted bridge site wrt Ni2; Ni2: remaining row in missing-row model

Dx/Dy in Å, or as a fraction of layer's unit cell vectors; atom 0 at the origin. Epir/subr are bulk repeat vectors.

No. of atoms: 7 Bulk z = 1.246 Å

Reg ion	Chem el.	At. no.	Cell type	Site occ.	Rel. to	Dx ± εx		Dy ± εy		Dz ± εz		Dz/Bz(%) ± εz/Bz
epir		-2					f		f		Å	
subr		-1				1.246	Å	1.762	Å	1.246	Å	
ovrl	O	1	s1	.50	0	0.100 ± .100	Å	1.762	Å	0.000	Å	0.0
intf	Ni	2	s1	.50	0	0.000	Å	0.000	Å	0.200	Å	16.1
intf	Ni	3	s1	.50	0	1.246	Å	1.762	Å	1.500	Å	120.4
intf	Ni	4	s1	.50	0	3.738	Å	1.762	Å	1.500	Å	120.4
intf	Ni	5	s1	.50	0	0.000	Å	0.000	Å	2.780	Å	223.1
intf	Ni	6	s1	.50	0	2.492	Å	0.000	Å	2.680	Å	215.1
subl	Ni	7	b	1.00	0	1.246	Å	1.762	Å	3.980	Å	319.4

BOND DISTANCES AND ANGLES

Bond distances and angles are derived from coordinates

No. of distances/angles: 3

Interatomic dist. A-B (Å)	Atom A	Atom B	Atom C	Bond angle A-B-C (°)
1.773	O1	Ni2		

Ni(110)-(2x1)-O
28.8.70

Bond Distances and Angles - Continued

Interatomic dist. A-B (Å)	Atom A	Atom B	Atom C	Bond angle A-B-C (°)
1.860	O1	Ni3		
2.040	O1	Ni4		

COMMON NAME : Ni(111)-p(2x2)-O
CLASSIFICATION : 28.8.75a
TECHNIQUE : LEED
AUTHORS : D.T. Vu Grimsby, Y.K. Wu and K.A.R. Mitchell
REFERENCE : Surf. Sci., 232, 51 (1990)

ILLUSTRATION: 22,25

SURFACE TYPE
Substrate : Ni
Crystal face: 111
Temperature : RT
Bulk lattice: fcc
2D bulk symm: p3m1
2D surf symm: p3m1

Adsorbate: O
Coverage : 0.25 O/Ni
Pattern : p(2x2)
Matrix : (2.000, 0.000)
(0.000, 2.000)

STRUCTURE TYPE
Oxygen adsorbed in fcc hollow sites; buckling and lateral shifts in 1st Ni layer: 3 Ni next to O are lifted (by 0.12Å) and rotated/outwards shifted (120°, 0.08Å); 1st Ni-Ni interlayer spacing contracts (to 1.95Å); deeper layers are bulk like

SAMPLE PREPARATION (1 sample)
Treatment : O2 exposure 1E-8 torr at RT
Crystallinity: well ordered 1x1 substrate (AES/LEED)
Anal. methods:
Contamination:

COMMENTS

DATA COLLECTION
Technique: LEED; video LEED
Dataset : IV curves for 3 integer, 7 fractional beams at normal incidence; cumul. E range 1300 eV

THEORY/DATA TREATMENT
Dynamical LEED (RFS, combined space method): 8 phase shifts from band structure for Ni, from Demuth for O

STRUCTURES EXAMINED
Fcc site, hcp site, graphitic (0.5ML) overlayer: fcc site favored, then variation of top interlayer spacing, buckling (1st Ni layer), rotation/expansion of hollow

QUALITY OF EXPERIMENT-THEORY FIT
RPE=0.1087, RZJ=0.1326, average optim.

2D UNIT CELLS (1 domain observed)

Cell	Ax (Å)	Ay (Å)	Bx (Å)	By (Å)	α (°)	Matrix	Pattern	Cell type
Bulk	2.492	0.000	1.246	2.158	60.0	(1.000, 0.000) (0.000, 1.000)	(1x1)	b: bulk lattice
Surface 1	4.984	0.000	2.492	4.316	60.0	(2.000, 0.000) (0.000, 2.000)	p(2x2)	s1: commens. superlattice

3D COORDINATES

O1: overlayer in fcc hollow sites; Ni2-Ni4: 3 Ni nearest O in top Ni layer (lifted, rotated);
Ni5: 4th Ni in top Ni layer; 0.03Å error bars are estimated from 'visual observation'

Dx/Dy in Å, or as a fraction of layer's unit cell vectors; atom O at the origin. Epir/subr are bulk repeat vectors.

No. of atoms: 6

Bulk z = 2.035 Å

Reg ion	Chem el.	At. no.	Cell type	Site occ.	Rel. to	Dx ± εx		Dy ± εy		Dz ± εz		Dz/Bz(%) ± εz/Bz
epir		-2					f		f		Å	
subr		-1				1.246	Å	0.720	Å	2.035	Å	
ovrl	O	1	s1	.25	0	0.000 ± .030	Å	0.000 ± .030	Å	0.000 ± .030	Å	0.0 ± 1.5
intf	Ni	2	s1	.25	1	0.183 ± .012	f	0.159 ± .012	f	1.090 ± .030	Å	53.6 ± 1.5
intf	Ni	3	s1	.25	1	0.659 ± .012	f	0.183 ± .012	f	1.090 ± .030	Å	53.6 ± 1.5
intf	Ni	4	s1	.25	1	0.159 ± .012	f	0.659 ± .012	f	1.090 ± .030	Å	53.6 ± 1.5
intf	Ni	5	s1	.25	1	0.667 ± .012	f	0.667 ± .012	f	1.210 ± .030	Å	59.5 ± 1.5
subl	Ni	6	b	1.00	5	0.334	f	0.333	f	1.950	Å	95.8

BOND DISTANCES AND ANGLES

Bond distances and angles are derived from coordinates

No. of distances/angles: 2

Interatomic dist. A-B (Å)	Atom A	Atom B	Atom C	Bond angle A-B-C (°)
1.834	O1	Ni2		
1.834	Ni3	O1(1,0)	Ni2(1,0)	88.3

COMMON NAME : Ni(111)-(√3x√3)R30°-O
CLASSIFICATION : 28.8.27
TECHNIQUE : HEIS
AUTHORS : T. Narusawa, W.M. Gibson and E. Tornqvist
REFERENCE : Surf. Sci., 114, 331 (1981)

ILLUSTRATION: 22,24

SURFACE TYPE
Substrate : Ni
Crystal face: 111
Temperature : 300 K
Bulk lattice: fcc
2D bulk symm: p3m1
2D surf symm: p31m

Adsorbate: O
Coverage : 1.3 (O/Ni)
Pattern : (√3x√3)R30°
Matrix : (1.000, 1.000)
(-1.000, 2.000)

STRUCTURE TYPE
Atomic adsorption (in undetermined hollow sites, fcc assumed here); expanded top Ni-Ni interlayer spacing

SAMPLE PREPARATION (1 sample)
Treatment : oxidation/reduction, Ar+ bombardment, anneal, 4L O2
Crystallinity:
Anal. methods:
Contamination: AES: impurities<0.1%

COMMENTS
Result assumes 1.2Å O-Ni layer separation derived from LEED (Marcus et al, Phys. Rev. B15, 1460 (1975), where fcc vs hcp site was also undetermined and substrate relaxation not tested); substrate result is insensitive to O position

DATA COLLECTION
Technique: HEIS; high energy He+ ion scattering
Dataset :

THEORY/DATA TREATMENT
Comparison to theory assuming a binary collision model; ΘD=330 K

STRUCTURES EXAMINED
Various uniform and non-uniform relaxations of the top layer in steps of 0.03Å

QUALITY OF EXPERIMENT-THEORY FIT
Visual

2D UNIT CELLS (1 domain observed)

Cell	Ax (Å)	Ay (Å)	Bx (Å)	By (Å)	α (°)	Matrix	Pattern	Cell type
Bulk	2.490	0.000	1.245	2.156	60.0	(1.000, 0.000) (0.000, 1.000)	(1x1)	b: bulk lattice
Surface 1	3.735	2.156	0.000	4.313	60.0	(1.000, 1.000) (-1.000, 2.000)	(√3x√3)R30°	s1: commens. superlattice

3D COORDINATES

O1: overlayer in hollow sites (assumed fcc here) at height determined by LEED

Dx/Dy in Å, or as a fraction of layer's unit cell vectors; atom 0 at the origin. Epir/subr are bulk repeat vectors.

No. of atoms: 4

Bulk z = 2.033 Å

Reg ion	Chem el.	At. no.	Cell type	Site occ.	Rel. to	Dx ± εx		Dy ± εy		Dz ± εz		Dz/Bz(%) ± εz/Bz
epir		-2					f		f		Å	
subr		-1				1.245	Å	0.719	Å	2.033	Å	
ovrl	O	1	s1	.33	0	0.000	f	0.000	f	0.000	Å	0.0
intf	Ni	2	b	1.00	1	0.333	f	0.333	f	1.200	Å	59.0
intf	Ni	3	b	1.00	2	0.333	f	0.333	f	2.180 ± .020	Å	107.2 ± 1.0
subl	Ni	4	b	1.00	3	-0.667	f	0.333	f	2.033	Å	100.0

BOND DISTANCES AND ANGLES

Bond distances and angles are derived from coordinates

No. of distances/angles: 3

Interatomic dist. A-B (Å)	Atom A	Atom B	Atom C	Bond angle A-B-C (°)
1.873	O1	Ni2	Ni3	163.3
2.490	Ni2	Ni2(1,0)	Ni3	61.5
2.611	Ni2	Ni3	Ni4	121.5

ILLUSTRATION: 22,24

COMMON NAME : Ni(111)-(√3x√3)R30°-O
CLASSIFICATION : 28.8.85
TECHNIQUE : LEED
AUTHORS : M.A. Mendez, W. Oed, A. Fricke, L. Hammer, K. Heinz and K. Mueller
REFERENCE : Surf. Sci., 253, 99 (1991)

SURFACE TYPE
Substrate : Ni
Crystal face: 111
Temperature : 80 K
Bulk lattice: fcc
2D bulk symm: p3m1
2D surf symm: p31m

Adsorbate: O
Coverage : 1/3 O/Ni
Pattern : (√3x√3)R30°
Matrix : (1.000, 1.000)
(-1.000, 2.000)

STRUCTURE TYPE
O adsorbed in fcc hollow site; substrate has expanded 1st interlayer spacing, no lateral shifts or buckling

SAMPLE PREPARATION (1 sample)
Treatment : 0.03L O at 220 K
Crystallinity:
Anal. methods:
Contamination:

COMMENTS

DATA COLLECTION
Technique: LEED; video LEED
Dataset : IV-curves at normal incidence for 5 integer, 4 fractional order beams; E<=300 eV, cumul. E range 1550eV

THEORY/DATA TREATMENT
Dynamical LEED (comb. space, RFS):
11 phase shifts; VoiαE**1/3, Vor optimized

STRUCTURES EXAMINED
Fcc hollow, hcp hollow and domain mixture of both; variation of 2 Ni-Ni layer spacings and oxygen height; lateral 1st layer substrate displacements: radial and rotational shifts of 3 next neighbors

QUALITY OF EXPERIMENT-THEORY FIT
RPE=0.16

2D UNIT CELLS (1 domain observed)

Cell	Ax (Å)	Ay (Å)	Bx (Å)	By (Å)	α (°)	Matrix	Pattern	Cell type
Bulk	2.492	0.000	1.246	2.158	60.0	(1.000, 0.000) (0.000, 1.000)	(1x1)	b: bulk lattice
Surface 1	3.738	2.158	0.000	4.316	60.0	(1.000, 1.000) (-1.000, 2.000)	(√3x√3)R30°	s1: commens. superlattice

3D COORDINATES

O1: overlayer in fcc hollow site

Dx/Dy in Å, or as a fraction of layer's unit cell vectors; atom 0 at the origin. Epir/subr are bulk repeat vectors.

No. of atoms: 4 Bulk z = 2.035 Å

Region	Chem el.	At. no.	Cell type	Site occ.	Rel. to	Dx ± εx		Dy ± εy		Dz ± εz		Dz/Bz(%) ± εz/Bz
epir		-2					f		f		Å	
subr		-1				0.000	Å	2.878	Å	2.035	Å	
ovrl	O	1	s1	.33	0	0.000	Å	0.000	Å	0.000 ± .020	Å	0.0 ± 1.0
intf	Ni	2	b	1.00	1	0.000 ± .030	Å	2.878 ± .030	Å	1.080 ± .020	Å	53.1 ± 1.0
intf	Ni	3	b	1.00	1	0.000	Å	1.439	Å	3.130 ± .020	Å	153.8 ± 1.0
subl	Ni	4	b	1.00	2	0.000	Å	0.000	Å	5.150	Å	253.1

BOND DISTANCES AND ANGLES

Bond distances and angles are derived from coordinates

No. of distances/angles: 1

Interatomic dist. A-B (Å)	Atom A	Atom B	Atom C	Bond angle A-B-C (°)
1.799	O1	Ni2(0,-1)		

COMMON NAME : Ni(100)-c(2x2)-S
CLASSIFICATION : 28.16.10a
TECHNIQUE : PED
AUTHORS : D.H. Rosenblatt, J.G. Tobin, M.G. Mason, R.F. Davis, S.D. Kevan, D.A. Shirley, C.H. Li and S.Y. Tong
REFERENCE : Phys. Rev., B23, 3828 (1981)

ILLUSTRATION: 28,29

SURFACE TYPE
Substrate : Ni
Crystal face: 100
Temperature : 120 K
Bulk lattice: fcc
2D bulk symm: p4m
2D surf symm: p4m

Adsorbate: S
Coverage : 0.5 (S/Ni)
Pattern : c(2x2)
Matrix : (1.000, 1.000)
(-1.000, 1.000)

STRUCTURE TYPE
Atomic adsorption in 4-fold hollow sites

SAMPLE PREPARATION (1 sample)
Treatment : S from effusive beam of H2S, 20-30L, 300 K
Crystallinity:
Anal. methods:
Contamination: monitored by AES and LEED

COMMENTS
Bulk spacing (1.76Å) obtained by averaging over two spacings of 1.72 and 1.81Å due to two peaks in data

DATA COLLECTION
Technique: PED
Dataset : E-dependent photoyield up to 200 eV above threshold

THEORY/DATA TREATMENT
Dynamical theory of normal photoelectron diffraction

STRUCTURES EXAMINED
Hollow, bridge, and top sites

QUALITY OF EXPERIMENT-THEORY FIT
Visual

2D UNIT CELLS (1 domain observed)

Cell	Ax (Å)	Ay (Å)	Bx (Å)	By (Å)	α (°)	Matrix	Pattern	Cell type
Bulk	2.489	0.000	0.000	2.489	90.0	(1.000, 0.000) (0.000, 1.000)	(1x1)	b: bulk lattice
Surface 1	2.489	2.489	-2.489	2.489	90.0	(1.000, 1.000) (-1.000, 1.000)	c(2x2)	s1: commens. superlattice

3D COORDINATES

S1: atomic overlayer in 4-fold hollow sites

Dx/Dy in Å, or as a fraction of layer's unit cell vectors; atom 0 at the origin. Epir/subr are bulk repeat vectors.

No. of atoms: 3

Bulk z = 1.760 Å

Reg ion	Chem el.	At. no.	Cell type	Site occ.	Rel. to	Dx ± єx	Dy ± єy	Dz ± єz	Dz/Bz(%) ± єz/Bz
epir		-2				f	f	Å	
subr		-1				-1.245 Å	-1.245 Å	1.760 Å	
ovrl	S	1	s1	.50	0	0.000 f	0.000 f	0.000 Å	0.0
intf	Ni	2	b	1.00	1	0.500 f	0.500 f	1.300 ± .040 Å	73.9 ± 2.3
subl	Ni	3	b	1.00	2	-0.500 f	-0.500 f	1.760 Å	100.0

BOND DISTANCES AND ANGLES

Bond distances and angles are derived from coordinates

No. of distances/angles: 6

Interatomic dist. A-B (Å)	Atom A	Atom B	Atom C	Bond angle A-B-C (°)
2.188	S1	Ni2	Ni2(1,0)	124.7
2.188	S1	Ni2	Ni3(1,1)	171.5
2.188	S1	Ni2	Ni3(1,0)	114.8
2.188	S1	Ni2	Ni3	81.5
2.489	Ni2	Ni2(1,0)	S1(1,0)	55.3
2.489	Ni2	Ni3		

ILLUSTRATION: 28,29

COMMON NAME : Ni(100)-c(2x2)-S
CLASSIFICATION : 28.16.13
TECHNIQUE : SEXAFS
AUTHORS : J. Stoehr, R Jaeger and S Brennan
REFERENCE : Surf. Sci., 117, 503 (1982)

SURFACE TYPE
Substrate : Ni
Crystal face: 100
Temperature : RT
Bulk lattice: fcc
2D bulk symm: p4m
2D surf symm: p4m

Adsorbate: S
Coverage : 0.5 (S/Ni)
Pattern : c(2x2)
Matrix : (1.000, 1.000)
(-1.000, 1.000)

STRUCTURE TYPE
Atomic adsorption in 4-fold hollow sites

SAMPLE PREPARATION (1 sample)
Treatment : dosed with H2S, then characterized by AES and LEED
Crystallinity:
Anal. methods:
Contamination: monitored by AES and LEED

COMMENTS
S-Ni distance found to be 2.23±0.02Å; site determined from second-nearest neighbor distance of 4.15±0.10Å; site also determined from comparison of experimental and calculated polarization dependent SEXAFS amplitude ratios

DATA COLLECTION
Technique: SEXAFS; SEXAFS: electron or ion yield detec
Dataset : photon flux 1E11 photons/s; radiation in 200-400 eV spectral range; x-ray incidence angles 90, 45 and 10° from surfac

THEORY/DATA TREATMENT

STRUCTURES EXAMINED
4-fold hollow, 2-fold bridge, 1-fold top, and bulk NiS

QUALITY OF EXPERIMENT-THEORY FIT
Visual

2D UNIT CELLS (1 domain observed)

Cell	Ax (Å)	Ay (Å)	Bx (Å)	By (Å)	α (°)	Matrix	Pattern	Cell type
Bulk	2.489	0.000	0.000	2.489	90.0	(1.000, 0.000) (0.000, 1.000)	(1x1)	b: bulk lattice
Surface 1	2.489	2.489	-2.489	2.489	90.0	(1.000, 1.000) (-1.000, 1.000)	c(2x2)	s1: commens. superlattice

3D COORDINATES

S1: atomic overlayer in 4-fold hollow sites; coordinates are derived from bond distances

Dx/Dy in Å, or as a fraction of layer's unit cell vectors; atom 0 at the origin. Epir/subr are bulk repeat vectors.

No. of atoms: 3

Bulk z = 1.760 Å

Reg ion	Chem el.	At. no.	Cell type	Site occ.	Rel. to	Dx ± εx	Dy ± εy	Dz ± εz	Dz/Bz(%) ± εz/Bz
epir		-2				f	f	Å	
subr		-1				-1.245 Å	-1.245 Å	1.760 Å	
ovrl	S	1	s1	.50	0	0.000 f	0.000 f	0.000 Å	0.0
intf	Ni	2	b	1.00	1	0.500 f	0.500 f	1.370 ± .040 Å	77.8 ± 2.3
subl	Ni	3	b	1.00	2	-0.500 f	-0.500 f	1.760 Å	100.0

BOND DISTANCES AND ANGLES

No. of distances/angles: 6

Interatomic dist. A-B (Å)	Atom A	Atom B	Atom C	Bond angle A-B-C (°)
2.230	S1	Ni2	Ni2(1,0)	123.9
2.230	S1	Ni2	Ni3(1,1)	172.9
2.230	S1	Ni2	Ni3(1,0)	115.7
2.230	S1	Ni2	Ni3	82.9
2.489	Ni2	Ni2(1,0)	S1(1,0)	56.1
2.489	Ni2	Ni3		

COMMON NAME : Ni(100)-c(2x2)-S
CLASSIFICATION : 28.16.15
TECHNIQUE : ARPEFS
AUTHORS : J.J. Barton, C.C. Bahr, Z. Hussain, S.W. Robey, L.E. Klebanoff and D.A. Shirley
REFERENCE : J. Vac. Sci. Technol., A2, 847 (1984)

ILLUSTRATION: 28,29

SURFACE TYPE
Substrate : Ni
Crystal face: 100
Temperature : RT*
Bulk lattice: fcc
2D bulk symm: p4m
2D surf symm: p4m

Adsorbate: S
Coverage : 0.5 (S/Ni)
Pattern : c(2x2)
Matrix : (1.000, 1.000)
(-1.000, 1.000)

STRUCTURE TYPE
Atomic adsorption in 4-fold hollow sites

SAMPLE PREPARATION (sample)
Treatment : clean crystals exposed to H2S at room temperature
Crystallinity: c(2x2) pattern after heating to 200C
Anal. methods:
Contamination: checked by AES

COMMENTS
Primary information is S-Ni bond length of 2.23±0.03Å

DATA COLLECTION
Technique: ARPEFS; soft x-ray beam (2000-3000 eV)
Dataset : data taken in [110] and [010] directions; 100-500 eV range

THEORY/DATA TREATMENT
Single scattering theory with published atomic phase shifts

STRUCTURES EXAMINED
S-Ni bond length variation; hollow site assumed (since data consistent with that site)

QUALITY OF EXPERIMENT-THEORY FIT
Visual

2D UNIT CELLS (1 domain observed)

Cell	Ax (Å)	Ay (Å)	Bx (Å)	By (Å)	α (°)	Matrix	Pattern	Cell type
Bulk	2.489	0.000	0.000	2.489	90.0	(1.000, 0.000) (0.000, 1.000)	(1x1)	b: bulk lattice
Surface 1	2.489	2.489	-2.489	2.489	90.0	(1.000, 1.000) (-1.000, 1.000)	c(2x2)	s1: commens. superlattice

3D COORDINATES

S1: atomic overlayer in 4-fold hollow sites; coordinates derived from bond distance

Dx/Dy in Å, or as a fraction of layer's unit cell vectors; atom 0 at the origin. Epir/subr are bulk repeat vectors.

No. of atoms: 3

Bulk z = 1.760 Å

Reg ion	Chem el.	At. no.	Cell type	Site occ.	Rel. to	Dx ± εx		Dy ± εy		Dz ± εz		Dz/Bz(%) ± εz/Bz
epir		-2					f		f		Å	
subr		-1				-1.245	Å	-1.245	Å	1.760	Å	
ovrl	S	1	s1	.50	0	0.000	f	0.000	f	0.000	Å	0.0
intf	Ni	2	b	1.00	1	0.500	f	0.500	f	1.370 ± .050	Å	77.8 ± 2.8
subl	Ni	3	b	1.00	2	-0.500	f	-0.500	f	1.760	Å	100.0

BOND DISTANCES AND ANGLES

No. of distances/angles: 6

Interatomic dist. A-B (Å)	Atom A	Atom B	Atom C	Bond angle A-B-C (°)
2.230	S1	Ni2	Ni2(1,0)	123.9
2.230	S1	Ni2	Ni3(1,1)	172.9
2.230	S1	Ni2	Ni3(1,0)	115.7
2.230	S1	Ni2	Ni3	82.9
2.489	Ni2	Ni2(1,0)	S1(1,0)	56.1
2.489	Ni2	Ni3		

COMMON NAME : Ni(100)-c(2x2)-S
CLASSIFICATION : 28.16.16
TECHNIQUE : PED
AUTHORS : E.L. Bullock, C.S. Fadley and P.J. Orders
REFERENCE : Phys. Rev., B28, 4867 (1983)

ILLUSTRATION: 28,29

SURFACE TYPE
Substrate : Ni
Crystal face: 100
Temperature : RT*
Bulk lattice: fcc
2D bulk symm: p4m
2D surf symm: p4m

Adsorbate: S
Coverage : 0.5 (S/Ni)
Pattern : c(2x2)
Matrix : (1.000, 1.000)
(-1.000, 1.000)

STRUCTURE TYPE
Atomic adsorption in 4-fold hollow sites

SAMPLE PREPARATION (sample)
Treatment : c(2x2) obtained by decomposition of H2S
Crystallinity:
Anal. methods:
Contamination: see Barton et al, PRL 51, 272 (1983)

COMMENTS

DATA COLLECTION
Technique: PED; off-normal diffraction (OPD)
Dataset : E-dependent photo-yield due to Barton et al: Θe-=45°, so emission along [110]; Θhv=45°, so pol. along emission

THEORY/DATA TREATMENT
Single scattering theory referenced to LEED to determine scattering factors; cluster with 120/1800 S/Ni atoms

STRUCTURES EXAMINED
No structural variation

QUALITY OF EXPERIMENT-THEORY FIT
Visual

2D UNIT CELLS (1 domain observed)

Cell	Ax (Å)	Ay (Å)	Bx (Å)	By (Å)	α (°)	Matrix	Pattern	Cell type
Bulk	2.489	0.000	0.000	2.489	90.0	(1.000, 0.000) (0.000, 1.000)	(1x1)	b: bulk lattice
Surface 1	2.489	2.489	-2.489	2.489	90.0	(1.000, 1.000) (-1.000, 1.000)	c(2x2)	s1: commens. superlattice

3D COORDINATES

S1: atomic overlayer in 4-fold hollow sites; 0.1Å error bar assumed for tabulation

Dx/Dy in Å, or as a fraction of layer's unit cell vectors; atom 0 at the origin. Epir/subr are bulk repeat vectors.

No. of atoms: 3 Bulk z = 1.760 Å

Reg ion	Chem el.	At. no.	Cell type	Site occ.	Rel. to	Dx ± εx		Dy ± εy		Dz ± εz		Dz/Bz(%) ± εz/Bz
epir		-2					f		f		Å	
subr		-1				-1.245	Å	-1.245	Å	1.760	Å	
ovrl	S	1	s1	.50	0	0.000	f	0.000	f	0.000	Å	0.0
intf	Ni	2	b	1.00	1	0.500	f	0.500	f	1.350 ± .100	Å	76.7 ± 5.7
subl	Ni	3	b	1.00	2	-0.500	f	-0.500	f	1.760	Å	100.0

BOND DISTANCES AND ANGLES

Bond distances and angles are derived from coordinates

No. of distances/angles: 6

Interatomic dist. A-B (Å)	Atom A	Atom B	Atom C	Bond angle A-B-C (°)
2.218	S1	Ni2	S1(1,0)	105.0
2.218	S1	Ni2	Ni3(1,1)	172.5
2.218	S1	Ni2	Ni3(1,0)	115.5
2.218	S1	Ni2	Ni3	82.5
2.489	Ni2	Ni2(1,0)	S1(1,0)	55.9
2.489	Ni2	Ni3		

COMMON NAME : Ni(100)-c(2x2)-S
CLASSIFICATION : 28.16.18
TECHNIQUE : PED
AUTHORS : P.J. Orders, B. Sinkovic, C.S. Fadley, R. Trehan, Z. Hussain and J. Lecante
REFERENCE : Phys. Rev., B30, 1838 (1984)

ILLUSTRATION: 28,29

SURFACE TYPE
Substrate : Ni
Crystal face: 100
Temperature : RT
Bulk lattice: fcc
2D bulk symm: p4m
2D surf symm: p4m

Adsorbate: S
Coverage : 0.5 (S/Ni)
Pattern : c(2x2)
Matrix : (1.000, 1.000)
(-1.000, 1.000)

STRUCTURE TYPE
Atomic atomic adsorption in 4-fold hollow sites

SAMPLE PREPARATION (1 sample)
Treatment : 30L exposure to H2S at RT to get c(2x2)
Crystallinity:
Anal. methods:
Contamination: checked by LEED

COMMENTS

DATA COLLECTION
Technique: PED; x-ray photoelectron diffraction
Dataset : azimuthal-dependent photo-yields at photon energies of 2672 and 2744 eV; Θe-=38.5°, Θhv=51.5°

THEORY/DATA TREATMENT
Single-scattering theory for S 1s emission at kinetic E of 200-300 eV and geometry emphasizing backscattering

STRUCTURES EXAMINED
4-fold hollow, 2-fold bridge, and top site; six different Ni-S spacings between 1.20 and 1.60Å

QUALITY OF EXPERIMENT-THEORY FIT
Visual

2D UNIT CELLS (1 domain observed)

Cell	Ax (Å)	Ay (Å)	Bx (Å)	By (Å)	α (°)	Matrix	Pattern	Cell type
Bulk	2.489	0.000	0.000	2.489	90.0	(1.000, 0.000) (0.000, 1.000)	(1x1)	b: bulk lattice
Surface 1	2.489	2.489	-2.489	2.489	90.0	(1.000, 1.000) (-1.000, 1.000)	c(2x2)	s1: commens. superlattice

3D COORDINATES

S1: atomic overlayer in 4-fold hollow sites

Dx/Dy in Å, or as a fraction of layer's unit cell vectors; atom 0 at the origin. Epir/subr are bulk repeat vectors.

No. of atoms: 3

Bulk z = 1.760 Å

Region	Chem el.	At. no.	Cell type	Site occ.	Rel. to	Dx ± εx		Dy ± εy		Dz ± εz		Dz/Bz(%) ± εz/Bz
epir		-2					f		f		Å	
subr		-1				-1.245	Å	-1.245	Å	1.760	Å	
ovrl	S	1	s1	.50	0	0.000	f	0.000	f	0.000	Å	0.0
intf	Ni	2	b	1.00	1	0.500	f	0.500	f	1.350 ± .100	Å	76.7 ± 5.7
subl	Ni	3	b	1.00	2	-0.500	f	-0.500	f	1.760	Å	100.0

BOND DISTANCES AND ANGLES

Bond distances and angles are derived from coordinates

No. of distances/angles: 6

Interatomic dist. A-B (Å)	Atom A	Atom B	Atom C	Bond angle A-B-C (°)
2.218	S1	Ni2	S1(1,0)	105.0
2.218	S1	Ni2	Ni3(1,1)	172.5
2.218	S1	Ni2	Ni3(1,0)	115.5
2.218	S1	Ni2	Ni3	82.5
2.489	Ni2	Ni2(1,0)	S1(1,0)	55.9
2.489	Ni2	Ni3		

COMMON NAME : Ni(100)-c(2x2)-S
CLASSIFICATION : 28.16.19
TECHNIQUE : SEXAFS
AUTHORS : J. Stoehr, E.B. Kollin, D.A. Fischer, J.B. Hastings, F. Zaera and F. Sette
REFERENCE : Phys. Rev. Lett., 55, 1468 (1985)

ILLUSTRATION: 28,29

SURFACE TYPE
Substrate : Ni
Crystal face: 100
Temperature : 100 K
Bulk lattice: fcc
2D bulk symm: p4m
2D surf symm: p4m

Adsorbate: S
Coverage : 0.5 (S/Ni)
Pattern : c(2x2)
Matrix : (1.000, 1.000)
(-1.000, 1.000)

STRUCTURE TYPE
Atomic adsorption in 4-fold hollow sites

SAMPLE PREPARATION (1 sample)
Treatment : 5L H2S deposited at 100 K, heated to 420K
Crystallinity: sharp LEED pattern
Anal. methods:
Contamination:

COMMENTS

DATA COLLECTION
Technique: SEXAFS; synchrotron radiation
Dataset : SEXAFS spectra: storage ring current of 50mA, flux of 2E10 photons/sec/2x4mm

THEORY/DATA TREATMENT
Polarization-dependent x-ray fluorescence yield and electron yield SEXAFS

STRUCTURES EXAMINED
Site and bond length determination for S; bulk Ni assumed

QUALITY OF EXPERIMENT-THEORY FIT
Visual

2D UNIT CELLS (1 domain observed)

Cell	Ax (Å)	Ay (Å)	Bx (Å)	By (Å)	α (°)	Matrix	Pattern	Cell type
Bulk	2.489	0.000	0.000	2.489	90.0	(1.000, 0.000) (0.000, 1.000)	(1x1)	b: bulk lattice
Surface 1	2.489	2.489	-2.489	2.489	90.0	(1.000, 1.000) (-1.000, 1.000)	c(2x2)	s1: commens. superlattice

3D COORDINATES

S1: atomic overlayer in 4-fold hollow sites; coordinates are derived from bond distances

Dx/Dy in Å, or as a fraction of layer's unit cell vectors; atom 0 at the origin. Epir/subr are bulk repeat vectors.

No. of atoms: 3

Bulk z = 1.760 Å

Region	Chem el.	At. no.	Cell type	Site occ.	Rel. to	Dx ± εx		Dy ± εy		Dz ± εz		Dz/Bz(%) ± εz/Bz
epir		-2					f		f		Å	
subr		-1				-1.245	Å	-1.245	Å	1.760	Å	
ovrl	S	1	s1	.50	0	0.000	f	0.000	f	0.000	Å	0.0
intf	Ni	2	b	1.00	1	0.500	f	0.500	f	1.360 ± .030	Å	77.3 ± 1.7
subl	Ni	3	b	1.00	2	-0.500	f	-0.500	f	1.760	Å	100.0

BOND DISTANCES AND ANGLES

No. of distances/angles: 1

Interatomic dist. A-B (Å)	Atom A	Atom B	Atom C	Bond angle A-B-C (°)
2.224	S1	Ni2		

COMMON NAME : Ni(100)-c(2x2)-S
CLASSIFICATION : 28.16.22
TECHNIQUE : ARPEFS
AUTHORS : J.J. Barton, C.C. Bahr, S.W. Robey, Z. Hussain, E. Umbach and D.A. Shirley
REFERENCE : Phys. Rev., B34, 3807 (1986)

ILLUSTRATION: 28,29

SURFACE TYPE
Substrate : Ni
Crystal face: 100
Temperature : 473 K
Bulk lattice: fcc
2D bulk symm: p4m
2D surf symm: p4m

Adsorbate: S
Coverage : 0.5 (S/Ni)
Pattern : c(2x2)
Matrix : (1.000, 1.000)
(-1.000, 1.000)

STRUCTURE TYPE
Atomic adsorption in hollow sites

SAMPLE PREPARATION (1 sample)
Treatment : exposure to 0.5L S and annealing to 1023 K
Crystallinity: sharp c(2x2) overlayer LEED pattern
Anal. methods:
Contamination: no AES or LEED of the clean surface

COMMENTS
A similar analysis using a generalized Ramsauer-Townsend resonance in ARPEFS (Barton et al, Phys. Rev. B35, 933 (1987)) confirms this result, with a S-Ni spacing of 1.32±0.04Å;
some evidence for buckling in the second Ni layer is found

DATA COLLECTION
Technique: ARPEFS
Dataset : S 1s ARPEFS measured along [011] and [001] directions

THEORY/DATA TREATMENT
Fourier analysis and multiple-scattering calculations

STRUCTURES EXAMINED
Fourier transform indicates 4-fold hollow site; S-Ni and first Ni-Ni spacing varied

QUALITY OF EXPERIMENT-THEORY FIT
Visual

2D UNIT CELLS (1 domain observed)

Cell	Ax (Å)	Ay (Å)	Bx (Å)	By (Å)	α (°)	Matrix	Pattern	Cell type
Bulk	2.489	0.000	0.000	2.489	90.0	(1.000, 0.000) (0.000, 1.000)	(1x1)	b: bulk lattice
Surface 1	2.489	2.489	-2.489	2.489	90.0	(1.000, 1.000) (-1.000, 1.000)	c(2x2)	s1: commens. superlattice

3D COORDINATES

S1: overlayer in 4-fold hollow sites

Dx/Dy in Å, or as a fraction of layer's unit cell vectors; atom 0 at the origin. Epir/subr are bulk repeat vectors.

No. of atoms: 4

Bulk z = 1.760 Å

Region	Chem el.	At. no.	Cell type	Site occ.	Rel. to	Dx ± εx	Dy ± εy	Dz ± εz	Dz/Bz(%) ± εz/Bz
epir		-2				f	f	Å	
subr		-1				1.245 Å	1.245 Å	1.760 Å	
ovrl	S	1	s1	.50	0	0.000 f	0.000 f	0.000 Å	0.0
intf	Ni	2	b	1.00	1	0.500 f	0.500 f	1.310 ± .030 Å	74.4 ± 1.7
intf	Ni	3	b	1.00	2	-0.500 f	-0.500 f	1.830 ± .030 Å	104.0 ± 1.7
subl	Ni	4	b	1.00	3	0.500 f	0.500 f	1.760 Å	100.0

BOND DISTANCES AND ANGLES

Bond distances and angles are derived from coordinates

No. of distances/angles: 2

Interatomic dist. A-B (Å)	Atom A	Atom B	Atom C	Bond angle A-B-C (°)
2.194	S1	Ni2	Ni3	82.8
2.489	Ni2	Ni2(1,0)		

ILLUSTRATION: 28,29

COMMON NAME : Ni(100)-c(2x2)-S
CLASSIFICATION : 28.16.23a
TECHNIQUE : ICISS
AUTHORS : Th. Fauster, H. Durr and D. Hartwig
REFERENCE : Surf. Sci., 178, 657 (1986)

SURFACE TYPE
Substrate : Ni
Crystal face: 100
Temperature : RT*
Bulk lattice: fcc
2D bulk symm: p4m
2D surf symm: p4m

Adsorbate: S
Coverage : 0.5 (S/Ni)
Pattern : c(2x2)
Matrix : (1.000, 1.000)
(-1.000, 1.000)

STRUCTURE TYPE
Atomic adsorption in hollow sites

SAMPLE PREPARATION (1 sample)
Treatment : S adsorption at RT from H2S
Crystallinity:
Anal. methods:
Contamination: monitored by ISS, AES, LEED

COMMENTS

DATA COLLECTION
Technique: ICISS; 5 KeV Ne+ ions
Dataset : polar angle scan from clean and S covered surface along [100] and [110] azimuths

THEORY/DATA TREATMENT
Experimentally calibrated shadow cone was used to determine surface structure

STRUCTURES EXAMINED
Critical angles for several adsorption geometries (top, bridge, hollow sites) were calculated and used to compare with the ICISS data

2D UNIT CELLS (1 domain observed)

Cell	Ax (Å)	Ay (Å)	Bx (Å)	By (Å)	α (°)	Matrix	Pattern	Cell type
Bulk	2.489	0.000	0.000	2.489	90.0	(1.000, 0.000) (0.000, 1.000)	(1x1)	b: bulk lattice
Surface 1	2.489	2.489	-2.489	2.489	90.0	(1.000, 1.000) (-1.000, 1.000)	c(2x2)	s1: commens. superlattice

3D COORDINATES

S1: overlayer in 4-fold hollow sites

Dx/Dy in Å, or as a fraction of layer's unit cell vectors; atom 0 at the origin. Epir/subr are bulk repeat vectors.

No. of atoms: 3

Bulk z = 1.760 Å

Reg ion	Chem el.	At. no.	Cell type	Site occ.	Rel. to	Dx ± εx		Dy ± εy		Dz ± εz		Dz/Bz(%) ± εz/Bz
epir		-2					f		f		Å	
subr		-1				-1.245	Å	-1.245	Å	1.760	Å	
ovrl	S	1	s1	.50	0	0.000	f	0.000	f	0.000	Å	0.0
intf	Ni	2	b	1.00	1	0.500	f	0.500	f	1.400 ± .050	Å	79.6 ± 2.8
subl	Ni	3	b	1.00	2	-0.500	f	-0.500	f	1.760	Å	100.0

BOND DISTANCES AND ANGLES

Bond distances and angles are derived from coordinates

No. of distances/angles: 2

Interatomic dist. A-B (Å)	Atom A	Atom B	Atom C	Bond angle A-B-C (°)
2.249	S1	Ni2	Ni3	83.5
2.489	Ni2	Ni2(1,0)		

COMMON NAME : Ni(100)-c(2x2)-S
CLASSIFICATION : 28.16.28
TECHNIQUE : SEXAFS
AUTHORS : Z. Yu
REFERENCE : Phys. Rev. Lett., B37, 9083 (1988)

ILLUSTRATION: 28,29

SURFACE TYPE
Substrate : Ni
Crystal face: 100
Temperature : RT
Bulk lattice: fcc
2D bulk symm: p4m
2D surf symm: p4m

Adsorbate: S
Coverage : 0.5 S/Ni
Pattern : c(2x2)
Matrix : (1.000, 1.000)
(-1.000, 1.000)

STRUCTURE TYPE
Atomic adsorption in hollow sites

SAMPLE PREPARATION (1 sample)
Treatment :
Crystallinity:
Anal. methods:
Contamination:

COMMENTS
Analysis includes an additional k-independent constant to compensate for error in the theoretical phase shifts: this quantity was fitted to experimental data

DATA COLLECTION
Technique: SEXAFS
Dataset : SEXAFS spectrum for S K-edge for photon energies from 2400 to 2900 eV; Auger electron yield mode

THEORY/DATA TREATMENT
Fitting to Fourier transformed SEXAFS signal: amplitude and phase shifts calculated using method of Teo and Lee

STRUCTURES EXAMINED
Hollow site assumed: interatomic distances determined as 2.22Å, 4.03Å and 4.15Å, resp.

2D UNIT CELLS (1 domain observed)

Cell	Ax (Å)	Ay (Å)	Bx (Å)	By (Å)	α (°)	Matrix	Pattern	Cell type
Bulk	2.489	0.000	0.000	2.489	90.0	(1.000, 0.000) (0.000, 1.000)	(1x1)	b: bulk lattice
Surface 1	2.489	2.489	-2.489	2.489	90.0	(1.000, 1.000) (-1.000, 1.000)	c(2x2)	s1: commens. superlattice

3D COORDINATES

S1: overlayer in 4-fold hollow sites; coordinates are derived from bond distances and angles

Dx/Dy in Å, or as a fraction of layer's unit cell vectors; atom 0 at the origin. Epir/subr are bulk repeat vectors.

No. of atoms: 3 Bulk z = 1.760 Å

Reg ion	Chem el.	At. no.	Cell type	Site occ.	Rel. to	Dx ± εx		Dy ± εy		Dz ± εz		Dz/Bz(%) ± εz/Bz
epir		-2					f		f		Å	
subr		-1				-1.245	Å	-1.245	Å	1.760	Å	
ovrl	S	1	s1	.50	0	0.000	f	0.000	f	0.000	Å	0.0
intf	Ni	2	b	1.00	1	0.500	f	0.500	f	1.380	Å	78.4
subl	Ni	3	b	1.00	2	-0.500	f	-0.500	f	1.760	Å	100.0

BOND DISTANCES AND ANGLES

No. of distances/angles: 2

Interatomic dist. A-B (Å)	Atom A	Atom B	Atom C	Bond angle A-B-C (°)
2.237	S1	Ni2	Ni3	83.1
2.489	Ni2	Ni2(1,0)		

COMMON NAME : Ni(100)-c(2x2)-S
CLASSIFICATION : 28.16.35
TECHNIQUE :
AUTHORS : U. Starke, F. Bothe, W. Oed and K. Heinz
REFERENCE : Surf. Sci., 232, 56 (1990)

ILLUSTRATION: 28,29

SURFACE TYPE
Substrate : Ni
Crystal face: 100
Temperature : 90 K
Bulk lattice: fcc
2D bulk symm: p4m
2D surf symm: p4m

Adsorbate: S
Coverage : 0.5 S/Ni
Pattern : c(2x2)
Matrix : (1.000,-1.000)
(1.000, 1.000)

STRUCTURE TYPE
Sulfur in hollow site; expansion of top Ni-Ni interlayer spacing by 2±1%; contraction of 2nd Ni-Ni spacing by 1±2%; second Ni layer buckling 0.01±0.03Å

SAMPLE PREPARATION (1 sample)
Treatment : 10L H2S at 90 K, 250K anneal, 10L more at 90K, 400K anneal
Crystallinity: sharp c(2x2) pattern
Anal. methods: TPD to control H desorption
Contamination:

COMMENTS

DATA COLLECTION
Technique: ; video LEED
Dataset : IV curves at normal incidence: E<=350 eV, cumul. E range 1665 eV

THEORY/DATA TREATMENT
Dynamical LEED (also direct method)
hollow site assumed from literature:

STRUCTURES EXAMINED
Variation of layer distances, 2nd layer buckling

QUALITY OF EXPERIMENT-THEORY FIT
RPE=0.25

2D UNIT CELLS (1 domain observed)

Cell	Ax (Å)	Ay (Å)	Bx (Å)	By (Å)	α (°)	Matrix	Pattern	Cell type
Bulk	2.492	0.000	0.000	2.492	90.0	(1.000, 0.000) (0.000, 1.000)	(1x1)	b: bulk lattice
Surface 1	2.492	-2.492	2.492	2.492	90.0	(1.000,-1.000) (1.000, 1.000)	c(2x2)	s1: commens. superlattice

3D COORDINATES

S1: overlayer in hollow sites; Ni2-Ni3: planar top Ni layer;
Ni4-Ni5: buckled 2nd Ni layer, Ni6 = bulk

Dx/Dy in Å, or as a fraction of layer's unit cell vectors; atom 0 at the origin. Epir/subr are bulk repeat vectors.

No. of atoms: 6

Bulk z = 1.762 Å

Region	Chem el.	At. no.	Cell type	Site occ.	Rel. to	Dx ± εx		Dy ± εy		Dz ± εz		Dz/Bz(%) ± εz/Bz
epir		-2					f		f		Å	
subr		-1				1.246	Å	1.246	Å	1.762	Å	
ovrl	S	1	s1	.50	0	0.000	Å	0.000	Å	0.000 ± .020	Å	0.0 ± 1.1
intf	Ni	2	s1	.50	1	0.500	f	0.000	f	1.300 ± .020	Å	73.8 ± 1.1
intf	Ni	3	s1	.50	1	0.000	f	-0.500	f	1.300 ± .020	Å	73.8 ± 1.1
intf	Ni	4	s1	.50	1	0.500	f	-0.500	f	3.085 ± .030	Å	175.1 ± 1.7
intf	Ni	5	s1	.50	1	0.000	f	0.000	f	3.095 ± .030	Å	175.6 ± 1.7
subl	Ni	6	b	1.00	1	-0.500	f	-0.500	f	4.830	Å	274.1

BOND DISTANCES AND ANGLES

Bond distances and angles are derived from coordinates

No. of distances/angles: 2

Interatomic dist. A-B (Å)	Atom A	Atom B	Atom C	Bond angle A-B-C (°)
2.190	S1	Ni2		
3.095	S1	Ni5		

COMMON NAME : Ni(100)-c(2x2)-S
CLASSIFICATION : 28.16.4c
TECHNIQUE : LEED
AUTHORS : P.M. Marcus, J.E. Demuth and D.W. Jepsen
REFERENCE : Surf. Sci., 53, 501 (1975)

ILLUSTRATION: 28,29

SURFACE TYPE
Substrate : Ni
Crystal face: 100
Temperature : 300 K
Bulk lattice: fcc
2D bulk symm: p4m
2D surf symm: p4m

Adsorbate: S
Coverage : 0.5 S/Ni
Pattern : c(2x2)
Matrix : (1.000, 1.000)
(-1.000, 1.000)

STRUCTURE TYPE
Atomic adsorption in 4-fold hollow sites

SAMPLE PREPARATION (sample)
Treatment :
Crystallinity:
Anal. methods:
Contamination:

COMMENTS
Vor was chosen by aligning positions of peaks in the experimental and calculated I-V spectra for Θ=φ=0°

DATA COLLECTION
Technique: LEED
Dataset : I-V spectra: 20<E<200 eV

THEORY/DATA TREATMENT
Dynamical LEED (layer KKR): 8 ph sh; Vor=-11.4 eV,VoiαE**1/3 (bulk), Voi=-3 eV (surface); ΘD(bulk)=420 K,ΘD(surface)=335K

STRUCTURES EXAMINED
1. 4-fold hollow site with Ni-S spacing 1.175-2.12Å; 2. other low symmetry adsorption sites;
3. coplanar alloy type structures

QUALITY OF EXPERIMENT-THEORY FIT
Visual

2D UNIT CELLS (1 domain observed)

Cell	Ax (Å)	Ay (Å)	Bx (Å)	By (Å)	α (°)	Matrix	Pattern	Cell type
Bulk	2.489	0.000	0.000	2.489	90.0	(1.000, 0.000) (0.000, 1.000)	(1x1)	b: bulk lattice
Surface 1	2.489	2.489	-2.489	2.489	90.0	(1.000, 1.000) (-1.000, 1.000)	c(2x2)	s1: commens. superlattice

3D COORDINATES

S1: atomic overlayer in 4-fold hollow sites 0.1Å error bar assumed for tabulation

Dx/Dy in Å, or as a fraction of layer's unit cell vectors; atom 0 at the origin. Epir/subr are bulk repeat vectors.

No. of atoms: 3 Bulk z = 1.760 Å

Region	Chem el.	At. no.	Cell type	Site occ.	Rel. to	Dx ± εx		Dy ± εy		Dz ± εz		Dz/Bz(%) ± εz/Bz
epir		-2					f		f		Å	
subr		-1				-1.245	Å	-1.245	Å	1.760	Å	
ovrl	S	1	s1	.50	0	0.000	f	0.000	f	0.000	Å	0.0
intf	Ni	2	b	1.00	1	0.500	f	0.500	f	1.280 ± .100	Å	72.7 ± 5.7
subl	Ni	3	b	1.00	2	-0.500	f	-0.500	f	1.760	Å	100.0

BOND DISTANCES AND ANGLES

Bond distances and angles are derived from coordinates

No. of distances/angles: 6

Interatomic dist. A-B (Å)	Atom A	Atom B	Atom C	Bond angle A-B-C (°)
2.176	S1	Ni2	S1(1,0)	108.0
2.176	S1	Ni2	Ni3(1,1)	171.0
2.176	S1	Ni2	Ni3(1,0)	114.6
2.176	S1	Ni2	Ni3	81.0
2.489	Ni2	Ni2(1,0)	S1(1,0)	55.1
2.489	Ni2	Ni3		

COMMON NAME : Ni(100)-c(2x2)-S
CLASSIFICATION : 28.16.9
TECHNIQUE : LEED
AUTHORS : Y. Gauthier, D. Aberdam and R. Baudoing
REFERENCE : Surf. Sci., 78, 339 (1978)

ILLUSTRATION: 28,29

SURFACE TYPE
Substrate : Ni
Crystal face: 100
Temperature : RT
Bulk lattice: fcc
2D bulk symm: p4m
2D surf symm: p4m

Adsorbate: S
Coverage : 0.5 (S/Ni)
Pattern : c(2x2)
Matrix : (1.000, 1.000)
(-1.000, 1.000)

STRUCTURE TYPE
Atomic adsorption in 4-fold hollow sites

SAMPLE PREPARATION (1 sample)
Treatment : 0.5ML S from annealing at 830C for a few minutes
Crystallinity:
Anal. methods:
Contamination: monitored by AES

COMMENTS

DATA COLLECTION
Technique: LEED
Dataset : I-V spectra: E range 10-110 eV (off-normal 10-170 eV); iso- intensity maps used to select optimum I-V spectra

THEORY/DATA TREATMENT
Dynamical LEED: Wakoh Ni pot.; cluster S overlap pot.;
ΘD = 420C (Ni), 335C (S)

STRUCTURES EXAMINED
4-fold hollow, 2-fold bridge, top site: Ni-S spacing varied from 1.1 to 1.8Å in steps of 0.1Å

QUALITY OF EXPERIMENT-THEORY FIT
Visual

2D UNIT CELLS (1 domain observed)

Cell	Ax (Å)	Ay (Å)	Bx (Å)	By (Å)	α (°)	Matrix	Pattern	Cell type
Bulk	2.489	0.000	0.000	2.489	90.0	(1.000, 0.000) (0.000, 1.000)	(1x1)	b: bulk lattice
Surface 1	2.489	2.489	-2.489	2.489	90.0	(1.000, 1.000) (-1.000, 1.000)	c(2x2)	s1: commens. superlattice

3D COORDINATES

S1: atomic overlayer in 4-fold hollow sites

Dx/Dy in Å, or as a fraction of layer's unit cell vectors; atom 0 at the origin. Epir/subr are bulk repeat vectors.

No. of atoms: 3

Bulk z = 1.760 Å

Reg ion	Chem el.	At. no.	Cell type	Site occ.	Rel. to	Dx ± εx		Dy ± εy		Dz ± εz		Dz/Bz(%) ± εz/Bz
epir		-2					f		f		Å	
subr		-1				-1.245	Å	-1.245	Å	1.760	Å	
ovrl	S	1	s1	.50	0	0.000	f	0.000	f	0.000	Å	0.0
intf	Ni	2	b	1.00	1	0.500	f	0.500	f	1.300 ± .050	Å	73.9 ± 2.8
subl	Ni	3	b	1.00	2	-0.500	f	-0.500	f	1.760	Å	100.0

BOND DISTANCES AND ANGLES

Bond distances and angles are derived from coordinates

No. of distances/angles: 6

Interatomic dist. A-B (Å)	Atom A	Atom B	Atom C	Bond angle A-B-C (°)
2.188	S1	Ni2	Ni2(1,0)	124.7
2.188	S1	Ni2	Ni3(1,1)	171.5
2.188	S1	Ni2	Ni3(1,0)	114.8
2.188	S1	Ni2	Ni3	81.5
2.489	Ni2	Ni2(1,0)	S1(1,0)	55.3
2.489	Ni2	Ni3		

COMMON NAME : Ni(100)-p(2x2)-S ILLUSTRATION: 28,30
CLASSIFICATION : 28.16.36a
TECHNIQUE :
AUTHORS : W. Oed, U. Starke, F. Bothe and K. Heinz
REFERENCE : Surf. Sci., 234, 72 (1990)

SURFACE TYPE
Substrate : Ni
Crystal face: 100
Temperature : 90 K
Bulk lattice: fcc
2D bulk symm: p4m
2D surf symm: p4m

Adsorbate: S
Coverage : 0.25 S/Ni
Pattern : p(2x2)
Matrix : (2.000, 0.000)
(0.000, 2.000)

STRUCTURE TYPE
Sulfur in hollow site;
1st Ni-Ni interlayer spacing expanded 0.5±1%;
second Ni layer buckling 0.07±0.05Å

SAMPLE PREPARATION (1 sample)
Treatment : 8L H2S at 90 K, 10s annealing at 500K, or desorbing c(2x2)
Crystallinity: streaks from (1/2,0) through (1/2,1/2)
Anal. methods:
Contamination:

COMMENTS

DATA COLLECTION
Technique: ; video LEED
Dataset : IV curves at normal incidence, E<=350 eV
cumul. E range 1415 eV

THEORY/DATA TREATMENT
Dynamical LEED

STRUCTURES EXAMINED
Hollow site assumed from literature: variation of interlayer spacings, 2nd Ni layer buckling

QUALITY OF EXPERIMENT-THEORY FIT
RPE=0.21

2D UNIT CELLS (1 domain observed)

Cell	Ax (Å)	Ay (Å)	Bx (Å)	By (Å)	α (°)	Matrix	Pattern	Cell type
Bulk	2.492	0.000	0.000	2.492	90.0	(1.000, 0.000) (0.000, 1.000)	(1x1)	b: bulk lattice
Surface 1	4.984	0.000	0.000	4.984	90.0	(2.000, 0.000) (0.000, 2.000)	p(2x2)	s1: commens. superlattice

3D COORDINATES

S1: overlayer in hollow sites; Ni2-Ni5: planar top Ni layer;
Ni6-Ni9: buckled 2nd Ni layer

Dx/Dy in Å, or as a fraction of layer's unit cell vectors; atom 0 at the origin. Epir/subr are bulk repeat vectors.

No. of atoms: 10 Bulk z = 1.762 Å

Region	Chem el.	At. no.	Cell type	Site occ.	Rel. to	Dx ± εx		Dy ± εy		Dz ± εz		Dz/Bz(%) ± εz/Bz
epir		-2					f		f		Å	
subr		-1				1.246	Å	1.246	Å	1.762	Å	
ovrl	S	1	s1	.25	0	0.000	Å	0.000	Å	0.000 ± .030	Å	0.0 ± 1.7
intf	Ni	2	s1	.25	1	0.250	f	0.250	f	1.250 ± .020	Å	70.9 ± 1.1
intf	Ni	3	s1	.25	1	-0.250	f	-0.250	f	1.250 ± .020	Å	70.9 ± 1.1
intf	Ni	4	s1	.25	1	-0.250	f	0.250	f	1.250 ± .020	Å	70.9 ± 1.1
intf	Ni	5	s1	.25	1	0.250	f	-0.250	f	1.250 ± .020	Å	70.9 ± 1.1
intf	Ni	6	s1	.25	1	0.500	f	0.000	f	2.990 ± .050	Å	169.7 ± 2.8
intf	Ni	7	s1	.25	1	0.500	f	0.500	f	2.990 ± .050	Å	169.7 ± 2.8
intf	Ni	8	s1	.25	1	0.000	f	0.500	f	2.990 ± .050	Å	169.7 ± 2.8
intf	Ni	9	s1	.25	1	0.000	f	0.000	f	3.060 ± .050	Å	173.7 ± 2.8
subl	Ni	10	b	1.00	1	0.500	f	0.500	f	4.780	Å	271.3

Ni(100)-p(2x2)-S
28.16.36a

BOND DISTANCES AND ANGLES

Bond distances and angles are derived from coordinates

No. of distances/angles: 2

Interatomic dist. A-B (Å)	Atom A	Atom B	Atom C	Bond angle A-B-C (°)
2.160	S1	Ni2		
3.060	S1	Ni9		

COMMON NAME : Ni(100)-p(2x2)-S
CLASSIFICATION : 28.16.4a
TECHNIQUE : LEED
AUTHORS : M.A. Van Hove and S.Y. Tong
REFERENCE : J. Vac. Sci. Technol., 12, 230 (1975)

ILLUSTRATION: 28,30

SURFACE TYPE
Substrate : Ni
Crystal face: 100
Temperature : 300 K
Bulk lattice: fcc
2D bulk symm: p4m
2D surf symm: p4m

Adsorbate: S
Coverage : 1/4 (S/Ni)
Pattern : (2x2)
Matrix : (2.000, 0.000)
(0.000, 2.000)

STRUCTURE TYPE
Atomic adsorption in 4-fold hollow sites

SAMPLE PREPARATION (sample)
Treatment : see Demuth and Rhodin, Surf. Sci. 42, 261 & 45, 249 (1974)
Crystallinity:
Anal. methods:
Contamination:

COMMENTS

DATA COLLECTION
Technique: LEED
Dataset :

THEORY/DATA TREATMENT
Dynamical LEED (layer doubling): Wakoh Ni potential, overlap atomic S potential, 8 phase shifts; Vor=-11.2 eV, VoiαE**1/3

STRUCTURES EXAMINED
8 to 10 Ni-S spacings, 0.2Å apart, at hollow, bridge and top sites.

QUALITY OF EXPERIMENT-THEORY FIT
Visual

2D UNIT CELLS (1 domain observed)

Cell	Ax (Å)	Ay (Å)	Bx (Å)	By (Å)	α (°)	Matrix	Pattern	Cell type
Bulk	2.490	0.000	0.000	2.490	90.0	(1.000, 0.000) (0.000, 1.000)	(1x1)	b: bulk lattice
Surface 1	4.980	0.000	0.000	4.980	90.0	(2.000, 0.000) (0.000, 2.000)	(2x2)	s1: commens. superlattice

3D COORDINATES

S1: atomic overlayer in 4-fold hollow sites

Dx/Dy in Å, or as a fraction of layer's unit cell vectors; atom 0 at the origin. Epir/subr are bulk repeat vectors.

No. of atoms: 3

Bulk z = 1.760 Å

Region	Chem el.	At. no.	Cell type	Site occ.	Rel. to	Dx ± εx		Dy ± εy		Dz ± εz		Dz/Bz(%) ± εz/Bz
epir		-2					f		f		Å	
subr		-1				-1.245	Å	-1.245	Å	1.760	Å	
ovrl	S	1	s1	.25	0	0.000	f	0.000	f	0.000	Å	0.0
intf	Ni	2	b	1.00	1	0.500	f	0.500	f	1.300 ± .100	Å	73.9 ± 5.7
subl	Ni	3	b	1.00	2	-0.500	f	-0.500	f	1.760	Å	100.0

BOND DISTANCES AND ANGLES

Bond distances and angles are derived from coordinates

No. of distances/angles: 6

Interatomic dist. A-B (Å)	Atom A	Atom B	Atom C	Bond angle A-B-C (°)
2.189	S1	Ni2	Ni2(1,0)	124.7
2.189	S1	Ni2	Ni3(1,1)	171.5
2.189	S1	Ni2	Ni3(1,0)	114.8
2.189	S1	Ni2	Ni3	81.4
2.490	Ni2	Ni2(1,0)	S1(1,0)	124.7
2.490	Ni2	Ni3		

COMMON NAME : Ni(100)-S disordered
CLASSIFICATION : 28.16.36b
TECHNIQUE : LEED
AUTHORS : U. Starke, W. Oed, P. Bayer, F. Bothe, G. Fuerst, P.L. de Andres, K. Heinz and J.B. Pendry
REFERENCE : Springer Series in Surface Sciences, 24, 427 (1991)

ILLUSTRATION: 28

SURFACE TYPE
Substrate : Ni
Crystal face: 100
Temperature : 80 K
Bulk lattice: fcc
2D bulk symm: p4m
2D surf symm: none

Adsorbate: S
Coverage : 0.1 S/Ni
Pattern : (1x1) disordered
Matrix : (1.000, 0.000)
(0.000, 1.000)

STRUCTURE TYPE
Disordered hollow site adsorption in off center position (pseudobridge) shifted by 0.45Å; 2nd layer buckled, sideshift of Ni atoms close to O of up to 0.15Å possible; local minimum for 4-fold-site with 1st layer buckling; disordered substrate relaxation modeled here as (3x3)

SAMPLE PREPARATION (1 sample)
Treatment : adsorption of H2S at 80 K
Crystallinity:
Anal. methods:
Contamination:

COMMENTS

experimental data similar for 0.015-0.25ML coverage range: same structure can be assumed

DATA COLLECTION
Technique: LEED; video (diffuse) LEED, Y-function maps
Dataset : intensities at 8 energies (48 eV-76eV) for 7 Y-function maps; average over 1000 video frames, symm-eq data, 2D smoothing

THEORY/DATA TREATMENT
Dynamical LEED (3-step DLEED method, tensor LEED):
6 phase shifts, Voi=-4 eV

STRUCTURES EXAMINED
Hollow site, off center oxygen, substrate relaxations Buckling of 4 NN's in 1st layer, buckling of 9 atoms in 2nd layer: A below oxygen, 4*B=NN of A, 4*C=diag. NN of A only A buckles

QUALITY OF EXPERIMENT-THEORY FIT
RPE=0.225 (=MSD of Y-functions)

2D UNIT CELLS (4 domains observed)

Cell	Ax (Å)	Ay (Å)	Bx (Å)	By (Å)	α (°)	Matrix	Pattern	Cell type
Bulk	2.492	0.000	0.000	2.492	90.0	(1.000, 0.000) (0.000, 1.000)	(1x1)	b: bulk lattice
Surface 1	7.476	0.000	0.000	7.476	90.0	(3.000, 0.000) (0.000, 3.000)	disordered	rd1: reconstr. lattice-gas dis

3D COORDINATES

S1: disordered adatom offcenter from hollow sites; Ni2-Ni10: top Ni layer, buckled, with possible lateral relaxations; Ni11-Ni19: 2nd Ni layer

Dx/Dy in Å, or as a fraction of layer's unit cell vectors; atom 0 at the origin. Epir/subr are bulk repeat vectors.

No. of atoms: 20

Bulk z = 1.762 Å

Reg ion	Chem el.	At. no.	Cell type	Site occ.	Rel. to	Dx ± εx		Dy ± εy		Dz ± εz		Dz/Bz(%) ± εz/Bz
epir		-2					f		f		Å	
subr		-1				1.246	Å	1.246	Å	1.762	Å	
ovrl	S	1	rd1	.11	0	0.600	Å	0.000	Å	0.000	Å	0.0
intf	Ni	2	rd1	.11	1	0.167	f	0.500	f	1.200	Å	68.1
intf	Ni	3	rd1	.11	1	0.500	f	-0.167	f	1.200	Å	68.1
intf	Ni	4	rd1	.11	1	0.500	f	0.167	f	1.200	Å	68.1
intf	Ni	5	rd1	.11	1	0.500	f	0.500	f	1.200	Å	68.1
intf	Ni	6	rd1	.11	1	-0.167	f	-0.167	f	1.200	Å	68.1
intf	Ni	7	rd1	.11	1	-0.167	f	0.167	f	1.200	Å	68.1
intf	Ni	8	rd1	.11	1	-0.167	f	0.500	f	1.200	Å	68.1
intf	Ni	9	rd1	.11	1	0.167	f	-0.167	f	1.200	Å	68.1
intf	Ni	10	rd1	.11	1	0.167	f	0.167	f	1.200	Å	68.1
intf	Ni	11	rd1	.11	1	-0.333	f	0.333	f	2.940	Å	166.9
intf	Ni	12	rd1	.11	1	-0.333	f	-0.333	f	2.940	Å	166.9
intf	Ni	13	rd1	.11	1	-0.333	f	0.000	f	2.940	Å	166.9
intf	Ni	14	rd1	.11	1	0.000	f	-0.333	f	2.940	Å	166.9
intf	Ni	15	rd1	.11	1	0.000	f	0.000	f	2.940	Å	166.9
intf	Ni	16	rd1	.11	1	0.000	f	0.333	f	2.940	Å	166.9
intf	Ni	17	rd1	.11	1	0.333	f	0.000	f	2.940	Å	166.9
intf	Ni	18	rd1	.11	1	0.333	f	0.333	f	2.940	Å	166.9

Ni(100)-S disordered
28.16.36b

BOND DISTANCES AND ANGLES

Bond distances and angles are derived from coordinates

No. of distances/angles: 2

Interatomic dist. A-B (Å)	Atom A	Atom B	Atom C	Bond angle A-B-C (°)
1.850	S1	Ni6		
3.000	S1	Ni15		

COMMON NAME : Ni(110)-c(2x2)-S
CLASSIFICATION : 28.16.17
TECHNIQUE ; AED
AUTHORS : R. Baudoing, E. Blanc, C. Gaubert, Y. Gauthier and N. Gnuchev
REFERENCE : Surf. Sci., 128, 22 (1983)

ILLUSTRATION: 35,36

SURFACE TYPE
Substrate : Ni
Crystal face: 110
Temperature : RT
Bulk lattice: fcc
2D bulk symm: pmm
2D surf symm: cmm

Adsorbate: S
Coverage : 0.5 S/Ni
Pattern : c(2x2)
Matrix : (1.000, 1.000)
(-1.000, 1.000)

STRUCTURE TYPE
Atomic adsorption in hollow (center) site

SAMPLE PREPARATION (sample)
Treatment : see Y. Gauthier et al., J. Phys. C15, 3223 (1982)
Crystallinity:
Anal. methods:
Contamination: AES: clean

COMMENTS
Only a test of theory, not supposed to be a good structure determination

DATA COLLECTION
Technique: AED; Auger electron diffraction intensities
Dataset : M(23)VV emission of Ni and L(23)VV emission of S from normal incidence to glancing

THEORY/DATA TREATMENT
Theoretical comparison with experimental angle resolved AED curves; use of angular momenta l=0 to 3

STRUCTURES EXAMINED
S-Ni spacing relaxation and long-bridge vs. hollow site for S; S-Ni spacing varied from 0.85 to 1.3Å

QUALITY OF EXPERIMENT-THEORY FIT
Visual

2D UNIT CELLS (1 domain observed)

Cell	Ax (Å)	Ay (Å)	Bx (Å)	By (Å)	α (°)	Matrix	Pattern	Cell type
Bulk	2.488	0.000	0.000	3.519	90.0	(1.000, 0.000) (0.000, 1.000)	(1x1)	b: bulk lattice
Surface 1	2.488	3.519	-2.488	3.519	70.5	(1.000, 1.000) (-1.000, 1.000)	c(2x2)	s1: commens. superlattice

3D COORDINATES

S1: overlayer in hollow (center) sites; 0.1Å error bar assumed for tabulation

Dx/Dy in Å, or as a fraction of layer's unit cell vectors; atom 0 at the origin. Epir/subr are bulk repeat vectors.

No. of atoms: 3

Bulk z = 1.245 Å

Region	Chem el.	At. no.	Cell type	Site occ.	Rel. to	Dx ± εx	Dy ± εy	Dz ± εz	Dz/Bz(%) ± εz/Bz
epir		-2				f	f	Å	
subr		-1				-1.244 Å	-1.760 Å	1.245 Å	
ovrl	S	1	s1	.50	0	0.000 f	0.000 f	0.000 Å	0.0
intf	Ni	2	b	1.00	1	0.500 f	0.500 f	0.900 ± .100 Å	72.3 ± 8.0
subl	Ni	3	b	1.00	2	-0.500 f	-0.500 f	1.245 Å	100.0

BOND DISTANCES AND ANGLES

Bond distances and angles are derived from coordinates

No. of distances/angles: 6

Interatomic dist. A-B (Å)	Atom A	Atom B	Atom C	Bond angle A-B-C (°)
2.335	S1	Ni2	S1(1,0)	134.7
2.335	S1	Ni2	Ni2(1,0)	122.2
2.335	S1	Ni2	Ni3	52.7
2.145	S1	Ni3	Ni2	60.0

Ni(110)-c(2x2)-S
28.16.17

Bond Distances and Angles - Continued

Interatomic dist. A-B (Å)	Atom A	Atom B	Atom C	Bond angle A-B-C (°)
2.488	Ni2	Ni2(1,0)		
2.489	Ni2	Ni3		

COMMON NAME : Ni(110)-c(2x2)-S
CLASSIFICATION : 28.16.20
TECHNIQUE : LEED
AUTHORS : R. Baudoing, Y. Gauthier and Y. Joly
REFERENCE : J. Phys., C18, 4061 (1985)

ILLUSTRATION: 35,36

SURFACE TYPE
Substrate : Ni
Crystal face: 110
Temperature : RT
Bulk lattice: fcc
2D bulk symm: pmm
2D surf symm: cmm

Adsorbate: S
Coverage : 0.5 S/Ni
Pattern : c(2x2)
Matrix : (1.000, 1.000)
(-1.000, 1.000)

STRUCTURE TYPE
Atomic adsorption in hollow (center) site

SAMPLE PREPARATION (1 sample)
Treatment : S segregated from bulk to surface by annealing
Crystallinity:
Anal. methods:
Contamination: AES: 0.5ML S

COMMENTS
S muffin tin radius of 1 and 0.94Å used: 1Å best;

DATA COLLECTION
Technique: LEED
Dataset : I-V spectra for 11 non-equivalent diffraction beams at normal incidence; E range 30-180 eV

THEORY/DATA TREATMENT
Dynamical LEED: combined space method and layer doubling; superposition potentials; ΘD=420 K (S), 335K (Ni)

STRUCTURES EXAMINED
S in top, short-bridge, long-bridge, hollow sites; relaxations in top two substrate interlayer spacings

QUALITY OF EXPERIMENT-THEORY FIT
Metric distances used

2D UNIT CELLS (1 domain observed)

Cell	Ax (Å)	Ay (Å)	Bx (Å)	By (Å)	α (°)	Matrix	Pattern	Cell type
Bulk	2.488	0.000	0.000	3.519	90.0	(1.000, 0.000) (0.000, 1.000)	(1x1)	b: bulk lattice
Surface 1	2.488	3.519	-2.488	3.519	70.5	(1.000, 1.000) (-1.000, 1.000)	c(2x2)	s1: commens. superlattice

3D COORDINATES

S1: overlayer in hollow (center) sites

Dx/Dy in Å, or as a fraction of layer's unit cell vectors; atom 0 at the origin. Epir/subr are bulk repeat vectors.

No. of atoms: 5 Bulk z = 1.245 Å

Region	Chem el.	At. no.	Cell type	Site occ.	Rel. to	Dx ± εx		Dy ± εy		Dz ± εz		Dz/Bz(%) ± εz/Bz
epir		-2					f		f		Å	
subr		-1				-1.244	Å	-1.760	Å	1.245	Å	
ovrl	S	1	s1	.50	0	0.000	f	0.000	f	0.000	Å	0.0
intf	Ni	2	b	1.00	1	0.500	f	0.500	f	0.840 ± .030	Å	67.5 ± 2.4
intf	Ni	3	b	1.00	2	-0.500	f	-0.500	f	1.372 ± .020	Å	110.2 ± 1.6
intf	Ni	4	b	1.00	3	0.500	f	0.500	f	1.201 ± .008	Å	96.5 ± .6
subl	Ni	5	b	1.00	4	-0.500	f	-0.500	f	1.245	Å	100.0

BOND DISTANCES AND ANGLES

Bond distances and angles are derived from coordinates

No. of distances/angles: 7

Interatomic dist. A-B (Å)	Atom A	Atom B	Atom C	Bond angle A-B-C (°)
2.313	S1	Ni2	S1(1,0)	137.4
2.313	S1	Ni2	Ni2(1,0)	122.5
2.313	S1	Ni2	Ni3	53.8
2.212	S1	Ni3	Ni4	119.1
2.488	Ni2	Ni2(1,0)		

Ni(110)-c(2x2)-S
28.16.20

Bond Distances and Angles - Continued

Interatomic dist. A-B (Å)	Atom A	Atom B	Atom C	Bond angle A-B-C (°)
2.555	Ni2	Ni3	Ni4	61.6
2.573	Ni2	Ni4	Ni3(1,1)	60.9

COMMON NAME : Ni(110)-c(2x2)-S
CLASSIFICATION : 28.16.23b
TECHNIQUE : ICISS
AUTHORS : Th. Fauster, H. Durr and D. Hartwig
REFERENCE : Surf. Sci., 178, 657 (1986)

ILLUSTRATION: 35,36

SURFACE TYPE
Substrate : Ni
Crystal face: 110
Temperature : RT*
Bulk lattice: fcc
2D bulk symm: pmm
2D surf symm: cmm

Adsorbate: S
Coverage : 0.5 (S/Ni)
Pattern : c(2x2)
Matrix : (1.000, 1.000)
(-1.000, 1.000)

STRUCTURE TYPE
Atomic adsorption in bulk continuation site (center of rectangle)

SAMPLE PREPARATION (1 sample)
Treatment : S adsorption from H2S
Crystallinity:
Anal. methods:
Contamination: monitored by ISS, AES, LEED

COMMENTS

DATA COLLECTION
Technique: ICISS; 5 KeV Ne+ ions
Dataset : ICISS polar angle scan for scattering from clean and S covered surface along [1,-1,2], [0,0,1], [1,-1,0] azimuths

THEORY/DATA TREATMENT
Experimentally calibrated shadow cone was used to determine the surface structure

STRUCTURES EXAMINED
Critical angles for several adsorption geometries (top, hollow, short- and long-bridge sites) were calculated and used to fit the ICISS data

2D UNIT CELLS (1 domain observed)

Cell	Ax (Å)	Ay (Å)	Bx (Å)	By (Å)	α (°)	Matrix	Pattern	Cell type
Bulk	2.488	0.000	0.000	3.519	90.0	(1.000, 0.000) (0.000, 1.000)	(1x1)	b: bulk lattice
Surface 1	2.488	3.519	-2.488	3.519	70.5	(1.000, 1.000) (-1.000, 1.000)	c(2x2)	s1: commens. superlattice

3D COORDINATES

S1: overlayer in center sites

Dx/Dy in Å, or as a fraction of layer's unit cell vectors; atom 0 at the origin. Epir/subr are bulk repeat vectors.

No. of atoms: 4

Bulk z = 1.245 Å

Region	Chem el.	At. no.	Cell type	Site occ.	Rel. to	Dx ± εx		Dy ± εy		Dz ± εz		Dz/Bz(%) ± εz/Bz
epir		-2					f		f		Å	
subr		-1				1.244	Å	1.760	Å	1.245	Å	
ovrl	S	1	s1	.50	0	0.000	f	0.000	f	0.000	Å	0.0
intf	Ni	2	b	1.00	1	0.500	f	0.500	f	0.890 ± .050	Å	71.5 ± 4.0
intf	Ni	3	b	1.00	2	-0.500	f	-0.500	f	1.307 ± .040	Å	105.0 ± 3.2
subl	Ni	4	b	1.00	3	0.500	f	0.500	f	1.245	Å	100.0

BOND DISTANCES AND ANGLES

Bond distances and angles are derived from coordinates

No. of distances/angles: 5

Interatomic dist. A-B (Å)	Atom A	Atom B	Atom C	Bond angle A-B-C (°)
2.332	S1	Ni2	Ni3	53.7
2.197	S1	Ni3	Ni4	120.0
2.488	Ni2	Ni2(1,0)		
2.520	Ni2	Ni3	Ni4	61.3
2.489	Ni3	Ni4		

COMMON NAME : Ni(110)-c(2x2)-S
CLASSIFICATION : 28.16.25
TECHNIQUE : ARPEFS
AUTHORS : S.W. Robey, J.J. Barton, C.C. Bahr, G. Liu and D.A. Shirley
REFERENCE : Phys. Rev., B35, 1108 (1987)

ILLUSTRATION: 35,36

SURFACE TYPE
Substrate : Ni
Crystal face: 110
Temperature : RT*
Bulk lattice: fcc
2D bulk symm: pmm
2D surf symm: cmm

Adsorbate: S
Coverage : 0.5 (S/Ni)
Pattern : c(2x2)
Matrix : (1.000, 1.000)
(-1.000, 1.000)

STRUCTURE TYPE
Atomic adsorption in bulk continuation site (center of rectangle) with buckling in 2nd Ni layer

SAMPLE PREPARATION (1 sample)
Treatment : exposure to 1.5-2.0L of H2S
Crystallinity:
Anal. methods:
Contamination: AES and LEED: no C, O, S on clean surf.

COMMENTS

DATA COLLECTION
Technique: ARPEFS; 2500-3000eV soft x-ray beam (2eV re
Dataset : ARPEFS measurements taken for four different experimental geometries with energy range 60-500 eV

THEORY/DATA TREATMENT
ARPEFS Fourier transform and backtransform, then multiple-scattering calculation with R-factor fit

STRUCTURES EXAMINED
Fourier transform analysis leads to the hollow site; filtering and backtransform analysis determine structural parameters; final comparison to mult.-scatt. calculation evaluated by R-factor

2D UNIT CELLS (1 domain observed)

Cell	Ax (Å)	Ay (Å)	Bx (Å)	By (Å)	α (°)	Matrix	Pattern	Cell type
Bulk	2.488	0.000	0.000	3.519	90.0	(1.000, 0.000) (0.000, 1.000)	(1x1)	b: bulk lattice
Surface 1	2.488	3.519	-2.488	3.519	70.5	(1.000, 1.000) (-1.000, 1.000)	c(2x2)	s1: commens. superlattice

3D COORDINATES

S1: overlayer in center sites; Ni2: bulk-like top Ni layer;
Ni3-Ni4: buckled 2nd Ni layer

Dx/Dy in Å, or as a fraction of layer's unit cell vectors; atom 0 at the origin. Epir/subr are bulk repeat vectors.

No. of atoms: 5 Bulk z = 1.245 Å

Region	Chem el.	At. no.	Cell type	Site occ.	Rel. to	Dx ± εx		Dy ± εy		Dz ± εz		Dz/Bz(%) ± εz/Bz
epir		-2					f		f		Å	
subr		-1				1.244	Å	1.760	Å	1.245	Å	
ovrl	S	1	s1	.50	0	0.000	f	0.000	f	0.000	Å	0.0
intf	Ni	2	b	1.00	1	0.500	f	0.500	f	0.820 ± .020	Å	65.9 ± 1.6
intf	Ni	3	s1	.50	2	0.000	f	0.500	f	1.250 ± .020	Å	100.4 ± 1.6
intf	Ni	4	s1	.50	3	-0.500	f	-0.500	f	0.130 ± .020	Å	10.4 ± 1.6
subl	Ni	5	b	1.00	4	0.500	f	0.500	f	1.245	Å	100.0

BOND DISTANCES AND ANGLES

Bond distances and angles are derived from coordinates

No. of distances/angles: 5

Interatomic dist. A-B (Å)	Atom A	Atom B	Atom C	Bond angle A-B-C (°)
2.306	S1	Ni2	Ni3	116.6
2.306	S1	Ni2	Ni4	53.5
2.200	S1	Ni4	Ni5	120.0
2.491	Ni2	Ni3	Ni5	62.7
2.625	Ni2	Ni5		

COMMON NAME : Ni(110)-c(2x2)-S
CLASSIFICATION : 28.16.26
TECHNIQUE : SEXAFS
AUTHORS : D.R. Warburton, G. Thornton, D. Norman, C.H. Richardson, R. McGrath and F. Sette
REFERENCE : Surf. Sci., 189/190, 495 (1987)

ILLUSTRATION: 35,36

SURFACE TYPE
Substrate : Ni
Crystal face: 110
Temperature : 293 K
Bulk lattice: fcc
2D bulk symm: pmm
2D surf symm: cmm

Adsorbate: S
Coverage : 0.5 (S/Ni)
Pattern : c(2x2)
Matrix : (1.000, 1.000)
(-1.000, 1.000)

STRUCTURE TYPE
Atomic adsorption in bulk continuation site (center of rectangle)

SAMPLE PREPARATION (1 sample)
Treatment : exposure to 8L of H2S at 293 K, then flashing to 393K
Crystallinity:
Anal. methods:
Contamination: monitored by AES and LEED

COMMENTS

DATA COLLECTION
Technique: SEXAFS; S KLL Auger-yield SEXAFS
Dataset : SEXAFS data for Θ=10° inc. with E-vector in [1,-1,0], [0,0,1] azimuths, and Θ=65° with E-vector in (001) plane

THEORY/DATA TREATMENT
Polarization-dependent SEXAFS data with Fourier transform method

STRUCTURES EXAMINED
Single-shell and two-shell analyses used to fit the SEXAFS data for possible adsorption sites: top, hollow, short- and long-bridge sites

2D UNIT CELLS (1 domain observed)

Cell	Ax (Å)	Ay (Å)	Bx (Å)	By (Å)	α (°)	Matrix	Pattern	Cell type
Bulk	2.488	0.000	0.000	3.519	90.0	(1.000, 0.000) (0.000, 1.000)	(1x1)	b: bulk lattice
Surface 1	2.488	3.519	-2.488	3.519	70.5	(1.000, 1.000) (-1.000, 1.000)	c(2x2)	s1: commens. superlattice

3D COORDINATES

S1: overlayer in center sites; coordinates are derived from bond distances and angles

Dx/Dy in Å, or as a fraction of layer's unit cell vectors; atom 0 at the origin. Epir/subr are bulk repeat vectors.

No. of atoms: 4 Bulk z = 1.245 Å

Region	Chem el.	At. no.	Cell type	Site occ.	Rel. to	Dx ± εx		Dy ± εy		Dz ± εz		Dz/Bz(%) ± εz/Bz
epir		-2					f		f		Å	
subr		-1				1.244	Å	1.760	Å	1.245	Å	
ovrl	S	1	s1	.50	0	0.000	f	0.000	f	0.000	Å	0.0
intf	Ni	2	b	1.00	1	0.500	f	0.500	f	0.830 ± .040	Å	66.7 ± 3.2
intf	Ni	3	b	1.00	2	-0.500	f	-0.500	f	1.400 ± .040	Å	112.5 ± 3.2
subl	Ni	4	b	1.00	3	0.500	f	0.500	f	1.245	Å	100.0

BOND DISTANCES AND ANGLES

No. of distances/angles: 5

Interatomic dist. A-B (Å)	Atom A	Atom B	Atom C	Bond angle A-B-C (°)
2.309	S1	Ni2	Ni3	54.1
2.230	S1	Ni3	Ni4	120.0
2.488	Ni2	Ni2(1,0)		
2.570	Ni2	Ni3	Ni4	63.0
2.489	Ni3	Ni4		

COMMON NAME : Ni(110)-c(2x2)-S
CLASSIFICATION : 28.16.7
TECHNIQUE : MEIS
AUTHORS : J.F. van der Veen, R.M. Tromp, R.G. Smeenk and F.W. Saris
REFERENCE : Surf. Sci., 82, 468 (1979)

ILLUSTRATION: 35,36

SURFACE TYPE
Substrate : Ni
Crystal face: 110
Temperature : RT
Bulk lattice: fcc
2D bulk symm: pmm
2D surf symm: cmm

Adsorbate: S
Coverage : 0.5 (S/Ni)
Pattern : c(2x2)
Matrix : (1.000, 1.000)
(-1.000, 1.000)

STRUCTURE TYPE
Atomic adsorption in hollow (center) site

SAMPLE PREPARATION (1 sample)
Treatment : H2S deposited at 1E-7 torr for 20-100s at 470 K
Crystallinity:
Anal. methods:
Contamination: check by back-scattering before H2S dep.

COMMENTS

DATA COLLECTION
Technique: MEIS; medium energy ion scatt. blocking con
Dataset : ion beam scattering as a function of scattering angle for 100 k eV protons

THEORY/DATA TREATMENT

STRUCTURES EXAMINED
Top, bridge and hollow sites

QUALITY OF EXPERIMENT-THEORY FIT
Visual

2D UNIT CELLS (1 domain observed)

Cell	Ax (Å)	Ay (Å)	Bx (Å)	By (Å)	α (°)	Matrix	Pattern	Cell type
Bulk	2.488	0.000	0.000	3.519	90.0	(1.000, 0.000) (0.000, 1.000)	(1x1)	b: bulk lattice
Surface 1	2.488	3.519	-2.488	3.519	70.5	(1.000, 1.000) (-1.000, 1.000)	c(2x2)	s1: commens. superlattice

3D COORDINATES

S1: overlayer in hollow (center) sites

Dx/Dy in Å, or as a fraction of layer's unit cell vectors; atom 0 at the origin. Epir/subr are bulk repeat vectors.

No. of atoms: 4 Bulk z = 1.245 Å

Region	Chem el.	At. no.	Cell type	Site occ.	Rel. to	Dx ± εx		Dy ± εy		Dz ± εz		Dz/Bz(%) ± εz/Bz
epir		-2					f		f		Å	
subr		-1				1.244	Å	1.760	Å	1.245	Å	
ovrl	S	1	s1	.50	0	0.000	f	0.000	f	0.000	Å	0.0
intf	Ni	2	b	1.00	1	0.500	f	0.500	f	0.870 ± .030	Å	69.9 ± 2.4
intf	Ni	3	b	1.00	2	-0.500	f	-0.500	f	1.310 ± .040	Å	105.2 ± 3.2
subl	Ni	4	b	1.00	3	0.500	f	0.500	f	1.245	Å	100.0

BOND DISTANCES AND ANGLES

Bond distances and angles are derived from coordinates

No. of distances/angles: 5

Interatomic dist. A-B (Å)	Atom A	Atom B	Atom C	Bond angle A-B-C (°)
2.324	S1	Ni2	S1(1,0)	136.0
2.324	S1	Ni2	Ni2(1,0)	122.4
2.324	S1	Ni2	Ni3	53.3
2.488	Ni2	Ni2(1,0)		
2.522	Ni2	Ni3	Ni4	61.3

COMMON NAME : Ni(110)-p(2x2)-S
CLASSIFICATION : 28.16.4d
TECHNIQUE : LEED
AUTHORS : J.E. Demuth, D.W. Jepsen and P.M. Marcus
REFERENCE : Phys. Rev. Lett., 32, 1182 (1974)

ILLUSTRATION: 35

SURFACE TYPE
Substrate : Ni
Crystal face: 110
Temperature : 300 K
Bulk lattice: fcc
2D bulk symm: pmm
2D surf symm: pmm

Adsorbate: S
Coverage : 0.25 (S/Ni)
Pattern : p(2x2)
Matrix : (2.000, 0.000)
(0.000, 0.000)

STRUCTURE TYPE
Atomic adsorption in hollow (center) site

SAMPLE PREPARATION (1 sample)
Treatment : thermal diffusion of bulk S contaminants
Crystallinity:
Anal. methods:
Contamination:

COMMENTS

DATA COLLECTION
Technique: LEED
Dataset : I-V spectra: (0,0), (1/2,1/2), (0,1), (1,0), (1,1) beams at normal incidence; 20<E<160 eV

THEORY/DATA TREATMENT
Dynamical LEED: 116 beams, 7 phase shifts; Wakoh Ni pot., S superpos. pot.; Vor=-11 eV + WFC

QUALITY OF EXPERIMENT-THEORY FIT
Visual

2D UNIT CELLS (1 domain observed)

Cell	Ax (Å)	Ay (Å)	Bx (Å)	By (Å)	α (°)	Matrix	Pattern	Cell type
Bulk	2.490	0.000	0.000	3.521	90.0	(1.000, 0.000) (0.000, 1.000)	(1x1)	b: bulk lattice
Surface 1	4.980	0.000	0.000	7.043	90.0	(2.000, 0.000) (0.000, 2.000)	p(2x2)	s1: commens. superlattice

3D COORDINATES

S1: overlayer in hollow (center) sites

Dx/Dy in Å, or as a fraction of layer's unit cell vectors; atom 0 at the origin. Epir/subr are bulk repeat vectors.

No. of atoms: 3 Bulk z = 1.245 Å

Region	Chem el.	At. no.	Cell type	Site occ.	Rel. to	Dx ± εx		Dy ± εy		Dz ± εz		Dz/Bz(%) ± εz/Bz
epir		-2					f		f		Å	
subr		-1				1.245	Å	-1.761	Å	1.245	Å	
ovrl	S	1	s1	.25	0	0.000	f	0.000	f	0.000	Å	0.0
intf	Ni	2	b	1.00	1	-0.500	f	0.500	f	0.930 ± .100	Å	74.7 ± 8.0
subl	Ni	3	b	1.00	2	0.500	f	-0.500	f	1.245	Å	100.0

BOND DISTANCES AND ANGLES

Bond distances and angles are derived from coordinates

No. of distances/angles: 5

Interatomic dist. A-B (Å)	Atom A	Atom B	Atom C	Bond angle A-B-C (°)
2.348	S1	Ni2	Ni2(1,0)	58.0
2.348	S1	Ni2	Ni3	53.3
2.175	S1	Ni3	Ni2	60.0
2.490	Ni2	Ni2(1,0)		
2.490	Ni2	Ni3		

COMMON NAME : Ni(111)-(2x2)-S
CLASSIFICATION : 28.16.23c
TECHNIQUE : ICISS
AUTHORS : Th. Fauster, H. Durr, D. Hartwig
REFERENCE : Surf. Sci., 178, 657 (1986)

ILLUSTRATION: 22,25

SURFACE TYPE
Substrate : Ni
Crystal face: 111
Temperature : RT*
Bulk lattice: fcc
2D bulk symm: p3m1
2D surf symm: p3m1

Adsorbate: S
Coverage : 0.25 (S/Ni)
Pattern : (2x2)
Matrix : (2.000, 0.000)
(0.000, 2.000)

STRUCTURE TYPE
Atomic adsorption in fcc-hollow site

SAMPLE PREPARATION (1 sample)
Treatment : S adsorption from H2S at room temperature
Crystallinity:
Anal. methods:
Contamination: monitored by ISS, AES, LEED

COMMENTS

DATA COLLECTION
Technique: ICISS; 5 keV Ne+ ions scattering
Dataset : ICISS polar angle scan for from clean and S-covered surface along [-2,1,1],[-1,-1,2] and [-1,0,1] azimuths

THEORY/DATA TREATMENT
Experimentally calibrated shadow cone used to determine surface structure

STRUCTURES EXAMINED
Critical angle for several adsorption geometries (top, bridge, hcp-hollow and fcc-hollow sites) was calculated and used to fit the ICISS data

QUALITY OF EXPERIMENT-THEORY FIT
Visual

2D UNIT CELLS (1 domain observed)

Cell	Ax (Å)	Ay (Å)	Bx (Å)	By (Å)	α (°)	Matrix	Pattern	Cell type
Bulk	2.490	0.000	1.245	2.156	60.0	(1.000, 0.000) (0.000, 1.000)	(1x1)	b: bulk lattice
Surface 1	4.980	0.000	2.490	4.313	60.0	(2.000, 0.000) (0.000, 2.000)	(2x2)	s1: commens. superlattice

3D COORDINATES

S1: overlayer in fcc-hollow sites

Dx/Dy in Å, or as a fraction of layer's unit cell vectors; atom 0 at the origin. Epir/subr are bulk repeat vectors.

No. of atoms: 3 Bulk z = 2.033 Å

Region	Chem el.	At. no.	Cell type	Site occ.	Rel. to	Dx ± εx		Dy ± εy		Dz ± εz		Dz/Bz(%) ± εz/Bz
epir		-2					f		f		Å	
subr		-1				-1.245	Å	-0.719	Å	2.033	Å	
ovrl	S	1	s1	.25	0	0.000	f	0.000	f	0.000	Å	0.0
intf	Ni	2	b	1.00	1	0.667	f	0.667	f	1.610 ± .060	Å	79.2 ± 3.0
subl	Ni	3	b	1.00	2	-0.333	f	-0.333	f	2.033	Å	100.0

BOND DISTANCES AND ANGLES

Bond distances and angles are derived from coordinates

No. of distances/angles: 1

Interatomic dist. A-B (Å)	Atom A	Atom B	Atom C	Bond angle A-B-C (°)
2.158	Ni2	S1(1,1)		

COMMON NAME : Ni(111)-(2x2)-S
CLASSIFICATION : 28.16.4b
TECHNIQUE : LEED
AUTHORS : J.E. Demuth, D.W. Jepsen and P.M. Marcus
REFERENCE : Phys. Rev. Lett., 32, 1182 (1974)

ILLUSTRATION: 22,25

SURFACE TYPE
Substrate : Ni
Crystal face: 111
Temperature : 300 K
Bulk lattice: fcc
2D bulk symm: p3m1
2D surf symm: p3m1

Adsorbate: S
Coverage : 0.25 (S/Ni)
Pattern : (2x2)
Matrix : (2.000, 0.000)
(0.000, 2.000)

STRUCTURE TYPE
Atomic adsorption over fcc 3-fold hollow sites

SAMPLE PREPARATION (1 sample)
Treatment : decomposition of H2S
Crystallinity:
Anal. methods:
Contamination:

COMMENTS

DATA COLLECTION
Technique: LEED
Dataset : I-V spectra: (1/2,0) (-1/2,0) (0,0) at normal incidence; 20<E<160 eV

THEORY/DATA TREATMENT
Dynamical LEED: 116 beams, 7 phase shifts;
Vor=-11 eV + measured work function change upon S adsorption

STRUCTURES EXAMINED
S atom in 3-fold sites with hcp and fcc termination for various adsorbate heights

QUALITY OF EXPERIMENT-THEORY FIT
Visual

2D UNIT CELLS (1 domain observed)

Cell	Ax (Å)	Ay (Å)	Bx (Å)	By (Å)	α (°)	Matrix	Pattern	Cell type
Bulk	2.490	0.000	1.245	2.156	60.0	(1.000, 0.000) (0.000, 1.000)	(1x1)	b: bulk lattice
Surface 1	4.980	0.000	2.490	4.313	60.0	(2.000, 0.000) (0.000, 2.000)	(2x2)	s1: commens. superlattice

3D COORDINATES

S1: atomic overlayer in 3-fold fcc hollow sites

Dx/Dy in Å, or as a fraction of layer's unit cell vectors; atom 0 at the origin. Epir/subr are bulk repeat vectors.

No. of atoms: 3

Bulk z = 2.030 Å

Reg ion	Chem el.	At. no.	Cell type	Site occ.	Rel. to	Dx ± εx		Dy ± εy		Dz ± εz		Dz/Bz(%) ± εz/Bz
epir		-2					f		f		Å	
subr		-1				1.245	Å	0.719	Å	2.030	Å	
ovrl	S	1	s1	.25	0	0.000	f	0.000	f	0.000	Å	0.0
intf	Ni	2	b	1.00	1	0.333	f	0.333	f	1.400 ± .100	Å	69.0
subl	Ni	3	b	1.00	2	0.333	f	0.333	f	2.030	Å	100.0

BOND DISTANCES AND ANGLES

Bond distances and angles are derived from coordinates

No. of distances/angles: 6

Interatomic dist. A-B (Å)	Atom A	Atom B	Atom C	Bond angle A-B-C (°)
2.007	S1	Ni2	Ni2(1,0)	128.4
2.007	S1	Ni2	Ni2(0,-1)	51.7
2.007	S1	Ni2	Ni3	169.6
2.007	S1	Ni2	Ni3(-1,0)	111.2
2.490	Ni2	Ni2(1,0)		
2.488	Ni2	Ni3		

COMMON NAME : Ni(111)-p(2x2)-S
CLASSIFICATION : 28.16.33
TECHNIQUE : LEED
AUTHORS : Y.K. Wu and K.A.R. Mitchell
REFERENCE : Can. J. Chem., 67, 1975 (1989)

ILLUSTRATION: 22,25

SURFACE TYPE
Substrate : Ni
Crystal face: 111
Temperature : RT
Bulk lattice: fcc
2D bulk symm: p3m1
2D surf symm: p3m1

Adsorbate: S
Coverage : 0.25 S/Ni
Pattern : p(2x2)
Matrix : (2.000, 0.000)
(0.000, 2.000)

STRUCTURE TYPE
Overlayer in 3-fold hollow fcc sites;
3% and 2% expansions of top 2 Ni-Ni interlayer spacings;
lateral radial expansion of 3-fold site 0.03Å

SAMPLE PREPARATION (1 sample)
Treatment : 2E-8 torr x 2 min of H2S at RT
Crystallinity: LEED sharp
Anal. methods:
Contamination: AES: clean substrate

COMMENTS

DATA COLLECTION
Technique: LEED; video LEED
Dataset : IV curves at normal incidence for 3 integer, 7 fractional order beams; cumul. E range 1020 eV

THEORY/DATA TREATMENT
Dynamical LEED (comb. space, RFS): 8 phase shifts

STRUCTURES EXAMINED
Fcc and hcp sites: lateral expansion and rotation of 3-fold hollow

QUALITY OF EXPERIMENT-THEORY FIT
RZJ=0.105 (RPE determined site)

2D UNIT CELLS (1 domain observed)

Cell	Ax (Å)	Ay (Å)	Bx (Å)	By (Å)	α (°)	Matrix	Pattern	Cell type
Bulk	2.492	0.000	1.246	2.158	60.0	(1.000, 0.000) (0.000, 1.000)	(1x1)	b: bulk lattice
Surface 1	4.984	0.000	2.492	4.316	60.0	(2.000, 0.000) (0.000, 2.000)	p(2x2)	s1: commens. superlattice

3D COORDINATES

S1: overlayer in fcc hollow sites; Ni2-5: top substrate layer; Ni2,4,5: expanded hollow
0.03Å error bars given without details

Dx/Dy in Å, or as a fraction of layer's unit cell vectors; atom 0 at the origin. Epir/subr are bulk repeat vectors.

No. of atoms: 7

Bulk z = 2.035 Å

Region	Chem el.	At. no.	Cell type	Site occ.	Rel. to	Dx ± εx	Dy ± εy	Dz ± εz	Dz/Bz(%) ± εz/Bz
epir		-2				f	f	Å	
subr		-1				1.246 Å	0.719 Å	2.035 Å	
ovrl	S	1	s1	.25	0	0.000 ± .030 Å	0.000 ± .030 Å	0.000 ± .030 Å	0.0 ± 1.5
intf	Ni	2	s1	.25	1	3.712 ± .030 Å	0.734 ± .030 Å	1.500 ± .030 Å	73.7 ± 1.5
intf	Ni	3	s1	.25	1	4.984 ± .030 Å	2.878 ± .030 Å	1.500 ± .030 Å	73.7 ± 1.5
intf	Ni	4	s1	.25	1	2.492 ± .030 Å	2.848 ± .030 Å	1.500 ± .030 Å	73.7 ± 1.5
intf	Ni	5	s1	.25	1	1.272 ± .030 Å	0.734 ± .030 Å	1.500 ± .030 Å	73.7 ± 1.5
intf	Ni	6	b	1.00	1	3.738 ± .030 Å	3.597 ± .030 Å	3.590 ± .030 Å	176.4 ± 1.5
subl	Ni	7	b	1.00	1	0.000 Å	0.000 Å	5.670 Å	278.7

BOND DISTANCES AND ANGLES

Bond distances and angles are derived from coordinates

No. of distances/angles: 2

Interatomic dist. A-B (Å)	Atom A	Atom B	Atom C	Bond angle A-B-C (°)
2.099	S1	Ni2(-1,0)		
2.099	Ni2	S1(1,0)	Ni4(1,-1)	74.6

COMMON NAME : Ni(100)-c(2x2)-Se
CLASSIFICATION : 28.34.0a
TECHNIQUE : LEED
AUTHORS : J.E. Demuth, D.W. Jepsen and P.M. Marcus
REFERENCE : Phys. Rev. Lett., 31, 540 (1973)

ILLUSTRATION: 28,29

SURFACE TYPE
Substrate : Ni
Crystal face: 100
Temperature : RT*
Bulk lattice: fcc
2D bulk symm: p4m
2D surf symm: p4m

Adsorbate: Se
Coverage : 1/2 (Se/Ni)
Pattern : c(2x2)
Matrix : (1.000, 1.000)
(-1.000, 1.000)

STRUCTURE TYPE
Atomic adsorption in 4-fold hollow sites

SAMPLE PREPARATION (sample)
Treatment : see J.E. Demuth and T.N. Rhodin, Surf. Sci. 45, 249 (1974)
Crystallinity:
Anal. methods:
Contamination:

COMMENTS

DATA COLLECTION
Technique: LEED
Dataset : I-V curves: E range 25-155 eV

THEORY/DATA TREATMENT
Dynamical LEED (layer KKR): 8 phase shifts, Wakoh Ni pot, Se superposition pot; Voi∝E**1/3; ΘD=420 K(Ni), 335K(Se)

STRUCTURES EXAMINED
Different adsorption sites with variable Se-Ni spacings

QUALITY OF EXPERIMENT-THEORY FIT
Visual

2D UNIT CELLS (1 domain observed)

Cell	Ax (Å)	Ay (Å)	Bx (Å)	By (Å)	α (°)	Matrix	Pattern	Cell type
Bulk	2.489	0.000	0.000	2.489	90.0	(1.000, 0.000) (0.000, 1.000)	(1x1)	b: bulk lattice
Surface 1	2.489	2.489	-2.489	2.489	90.0	(1.000, 1.000) (-1.000, 1.000)	c(2x2)	s1: commens. superlattice

3D COORDINATES

Se1: overlayer in 4-fold hollow sites

Dx/Dy in Å, or as a fraction of layer's unit cell vectors; atom 0 at the origin. Epir/subr are bulk repeat vectors.

No. of atoms: 3

Bulk z = 1.760 Å

Region	Chem el.	At. no.	Cell type	Site occ.	Rel. to	Dx ± εx		Dy ± εy		Dz ± εz		Dz/Bz(%) ± εz/Bz
epir		-2					f		f		Å	
subr		-1				-1.245	Å	-1.245	Å	1.760	Å	
ovrl	Se	1	s1	.50	0	0.000	f	0.000	f	0.000	Å	0.0
intf	Ni	2	b	1.00	1	0.500	f	0.500	f	1.450 ± .100	Å	82.4 ± 5.7
subl	Ni	3	b	1.00	2	-0.500	f	-0.500	f	1.760	Å	100.0

BOND DISTANCES AND ANGLES

Bond distances and angles are derived from coordinates

No. of distances/angles: 6

Interatomic dist. A-B (Å)	Atom A	Atom B	Atom C	Bond angle A-B-C (°)
2.280	Se1	Ni2	Ni2(1,0)	123.1
2.280	Se1	Ni2	Ni3(1,1)	174.5
2.280	Se1	Ni2	Ni3(1,0)	116.7
2.280	Se1	Ni2	Ni3	84.5
2.489	Ni2	Ni2(1,0)		
2.489	Ni2	Ni3		

COMMON NAME : Ni(100)-c(2x2)-Se
CLASSIFICATION : 28.34.1
TECHNIQUE : PED
AUTHORS : D.H. Rosenblatt, S.D. Kevan, J.G. Tobin, R. Davis, M.G. Mason, D.A. Shirley, J.C. Tang and S.Y. Tong
REFERENCE : Phys. Rev., B26, 3181 (1982)

ILLUSTRATION: 28,29

SURFACE TYPE
Substrate : Ni
Crystal face: 100
Temperature : 120 K
Bulk lattice: fcc
2D bulk symm: p4m
2D surf symm: p4m

Adsorbate: Se
Coverage : 1/2 (Se/Ni)
Pattern : c(2x2)
Matrix : (1.000, 1.000)
(-1.000, 1.000)

STRUCTURE TYPE
Atomic adsorption in 4-fold hollow sites

SAMPLE PREPARATION (1 sample)
Treatment : 20-30L dosing of H2Se at 500 K
Crystallinity:
Anal. methods:
Contamination:

COMMENTS

DATA COLLECTION
Technique: PED; normal photoelectron diffraction
Dataset : Se 3d spectra with kinetic energies in range 20-200 eV at normal and off-normal takeoff (total of 6 directions)

THEORY/DATA TREATMENT
Dynamical theory: Xα initial state in Ni5Se cluster (also gives Se phase shifts), Wakoh Ni phase shifts

STRUCTURES EXAMINED
Hollow site with overlayer spacing 1.45-1.75Å; top site with spacing 2.34Å (hollow site had been established previously)

QUALITY OF EXPERIMENT-THEORY FIT
RZJ=0.05

2D UNIT CELLS (1 domain observed)

Cell	Ax (Å)	Ay (Å)	Bx (Å)	By (Å)	α (°)	Matrix	Pattern	Cell type
Bulk	2.489	0.000	0.000	2.489	90.0	(1.000, 0.000) (0.000, 1.000)	(1x1)	b: bulk lattice
Surface 1	2.489	2.489	-2.489	2.489	90.0	(1.000, 1.000) (-1.000, 1.000)	c(2x2)	s1: commens. superlattice

3D COORDINATES

Se1: overlayer in 4-fold hollow sites; 0.1Å error bar assumed for tabulation

Dx/Dy in Å, or as a fraction of layer's unit cell vectors; atom 0 at the origin. Epir/subr are bulk repeat vectors.

No. of atoms: 3

Bulk z = 1.760 Å

Reg ion	Chem el.	At. no.	Cell type	Site occ.	Rel. to	Dx ± εx		Dy ± εy		Dz ± εz		Dz/Bz(%) ± εz/Bz
epir		-2					f		f		Å	
subr		-1				-1.245	Å	-1.245	Å	1.760	Å	
ovrl	Se	1	s1	.50	0	0.000	f	0.000	f	0.000	Å	0.0
intf	Ni	2	b	1.00	1	0.500	f	0.500	f	1.550 ± .100	Å	88.1 ± 5.7
subl	Ni	3	b	1.00	2	-0.500	f	-0.500	f	1.760	Å	100.0

BOND DISTANCES AND ANGLES

Bond distances and angles are derived from coordinates

No. of distances/angles: 6

Interatomic dist. A-B (Å)	Atom A	Atom B	Atom C	Bond angle A-B-C (°)
2.345	Se1	Ni2	Ni2(1,0)	122.1
2.345	Se1	Ni2	Ni3(1,1)	176.4
2.345	Se1	Ni2	Ni3(1,0)	117.9
2.345	Se1	Ni2	Ni3	86.4

Ni(100)-c(2x2)-Se
28.34.1

Bond Distances and Angles - Continued

Interatomic dist. A-B (Å)	Atom A	Atom B	Atom C	Bond angle A-B-C (°)
2.489	Ni2	Ni2(1,0)		
2.489	Ni2	Ni3		

COMMON NAME : Ni(100)-p(2x2)-Se
CLASSIFICATION : 28.34.0b
TECHNIQUE : LEED
AUTHORS : M.A. Van Hove and S.Y. Tong
REFERENCE : J. Vac. Sci. Technol., 12, 230 (1975)

ILLUSTRATION: 28,30

SURFACE TYPE
Substrate : Ni
Crystal face: 100
Temperature : 300 K
Bulk lattice: fcc
2D bulk symm: p4m
2D surf symm: p4m

Adsorbate: Se
Coverage : 1/4 (Se/Ni)
Pattern : (2x2)
Matrix : (2.000, 0.000)
(0.000, 2.000)

STRUCTURE TYPE
Atomic adsorption in 4-fold hollow sites

SAMPLE PREPARATION (sample)
Treatment : see Demuth and Rhodin, Surf. Sci. 42, 261 & 45, 249 (1974)
Crystallinity:
Anal. methods:
Contamination:

COMMENTS

DATA COLLECTION
Technique: LEED
Dataset :

THEORY/DATA TREATMENT
Dynamical LEED (RFS): Wakoh Ni potential, overlapping atomic potential for Se, 8 phase shifts; Vor=-11.2 eV, Voi∝E**1/3

STRUCTURES EXAMINED
8 to 10 Ni-Se spacings, 0.2Å apart, at hollow, bridge and top sites

QUALITY OF EXPERIMENT-THEORY FIT
Visual

2D UNIT CELLS (1 domain observed)

Cell	Ax (Å)	Ay (Å)	Bx (Å)	By (Å)	α (°)	Matrix	Pattern	Cell type
Bulk	2.490	0.000	0.000	2.490	90.0	(1.000, 0.000) (0.000, 1.000)	(1x1)	b: bulk lattice
Surface 1	4.980	0.000	0.000	4.980	90.0	(2.000, 0.000) (0.000, 2.000)	(2x2)	s1: commens. superlattice

3D COORDINATES

Se1: overlayer in 4-fold hollow sites

Dx/Dy in Å, or as a fraction of layer's unit cell vectors; atom 0 at the origin. Epir/subr are bulk repeat vectors.

No. of atoms: 3

Bulk z = 1.760 Å

Region	Chem el.	At. no.	Cell type	Site occ.	Rel. to	Dx ± εx		Dy ± εy		Dz ± εz		Dz/Bz(%) ± εz/Bz
epir		-2					f		f		Å	
subr		-1				-1.245	Å	-1.245	Å	1.760	Å	
ovrl	Se	1	s1	.25	0	0.000	f	0.000	f	0.000	Å	0.0
intf	Ni	2	b	1.00	1	0.500	f	0.500	f	1.550 ± .100	Å	88.1 ± 5.7
subl	Ni	3	b	1.00	2	-0.500	f	-0.500	f	1.760	Å	100.0

BOND DISTANCES AND ANGLES

Bond distances and angles are derived from coordinates

No. of distances/angles: 6

Interatomic dist. A-B (Å)	Atom A	Atom B	Atom C	Bond angle A-B-C (°)
2.346	Se1	Ni2	Ni2(1,0)	122.1
2.346	Se1	Ni2	Ni3(1,1)	176.4
2.346	Se1	Ni2	Ni3(1,0)	117.9
2.346	Se1	Ni2	Ni3	86.4
2.490	Ni2	Ni2(1,0)		
2.490	Ni2	Ni3		

COMMON NAME : Ni(110)-c(2x2)-Se
CLASSIFICATION : 28.34.0c
TECHNIQUE : PED
AUTHORS : D.H. Rosenblatt, S.D. Kevan, J.G. Tobin, R. Davis, M.G.Mason, D.R. Denley, D.A. Shirley, Y. Huang and S.Y. Tong
REFERENCE : Phys. Rev., B26, 1812 (1982)

ILLUSTRATION: 35,36

SURFACE TYPE
Substrate : Ni
Crystal face: 110
Temperature : 120 K
Bulk lattice: fcc
2D bulk symm: pmm
2D surf symm: cmm

Adsorbate: Se
Coverage : 1/2 (Se/Ni)
Pattern : c(2x2)
Matrix : (1.000, 1.000)
(-1.000, 1.000)

STRUCTURE TYPE
Atomic adsorption in hollow (center) site

SAMPLE PREPARATION (1 sample)
Treatment : 10-15L dosing of H2Se at 300 K
Crystallinity:
Anal. methods:
Contamination:

COMMENTS

DATA COLLECTION
Technique: PED; normal photoelectron diffraction
Dataset : Se 3d spectra with kinetic energies in range 20-200 eV at normal takeoff

THEORY/DATA TREATMENT
Dynamical theory: Xα initial state in Ni5Se cluster (also gives Se phase shifts), Wakoh Ni pot.

STRUCTURES EXAMINED
Top, hollow and bridge sites with variable Se-Ni spacing

QUALITY OF EXPERIMENT-THEORY FIT
Visual

2D UNIT CELLS (1 domain observed)

Cell	Ax (Å)	Ay (Å)	Bx (Å)	By (Å)	α (°)	Matrix	Pattern	Cell type
Bulk	2.488	0.000	0.000	3.519	90.0	(1.000, 0.000) (0.000, 1.000)	(1x1)	b: bulk lattice
Surface 1	2.488	3.519	-2.488	3.519	70.5	(1.000, 1.000) (-1.000, 1.000)	c(2x2)	s1: commens. superlattice

3D COORDINATES

Se1: overlayer in hollow (center) sites

Dx/Dy in Å, or as a fraction of layer's unit cell vectors; atom 0 at the origin. Epir/subr are bulk repeat vectors.

No. of atoms: 3

Bulk z = 1.245 Å

Region	Chem el.	At. no.	Cell type	Site occ.	Rel. to	Dx ± εx		Dy ± εy		Dz ± εz		Dz/Bz(%) ± εz/Bz
epir		-2					f		f		Å	
subr		-1				-1.244	Å	-1.760	Å	1.245	Å	
ovrl	Se	1	s1	.50	0	0.000	f	0.000	f	0.000	Å	0.0
intf	Ni	2	b	1.00	1	0.500	f	0.500	f	1.100 ± .040	Å	88.4 ± 3.2
subl	Ni	3	b	1.00	2	-0.500	f	-0.500	f	1.245	Å	100.0

BOND DISTANCES AND ANGLES

Bond distances and angles are derived from coordinates

No. of distances/angles: 6

Interatomic dist. A-B (Å)	Atom A	Atom B	Atom C	Bond angle A-B-C (°)
2.420	Se1	Ni2	Se1(1,0)	125.9
2.420	Se1	Ni2	Ni2(1,0)	121.0
2.420	Se1	Ni2	Ni3	57.1
2.345	Se1	Ni3	Ni2	60.0
2.488	Ni2	Ni2(1,0)		
2.489	Ni2	Ni3		

COMMON NAME : Ni(111)-p(2x2)-Se
CLASSIFICATION : 28.34.0d
TECHNIQUE : PED
AUTHORS : D.H. Rosenblatt, S.D. Kevan, J.G. Tobin, R. Davis, M.G.Mason, D.R. Denley, D.A. Shirley, Y. Huang and S.Y. Tong
REFERENCE : Phys. Rev., B26, 1812 (1982)

ILLUSTRATION: 22,25

SURFACE TYPE
Substrate : Ni
Crystal face: 111
Temperature : 120 K
Bulk lattice: fcc
2D bulk symm: p3m1
2D surf symm: p3m1

Adsorbate: Se
Coverage : 1/4 (Se/Ni)
Pattern : (2x2)
Matrix : (2.000, 0.000)
(0.000, 2.000)

STRUCTURE TYPE
Atomic adsorption in 3-fold fcc hollow sites

SAMPLE PREPARATION (1 sample)
Treatment : 2L dosing of H2Se at 120 K and annealing at 500K
Crystallinity:
Anal. methods:
Contamination:

COMMENTS
Low-coverage (0.1ML) disordered Se/Ni(111) also analyzed: same bonding geometry found

DATA COLLECTION
Technique: PED; normal photoelectron diffraction
Dataset : Se 3d spectra with kinetic energies in range 20-200 eV at normal takeoff

THEORY/DATA TREATMENT
Dynamical theory: Xα initial state in Ni5Se cluster (also gives Se phase shifts), Wakoh Ni phase shifts

STRUCTURES EXAMINED
Top, hollow and bridge sites with variable Se-Ni spacing

QUALITY OF EXPERIMENT-THEORY FIT
Visual

2D UNIT CELLS (1 domain observed)

Cell	Ax (Å)	Ay (Å)	Bx (Å)	By (Å)	α (°)	Matrix	Pattern	Cell type
Bulk	2.490	0.000	-1.245	2.156	120.0	(1.000, 0.000) (0.000, 1.000)	(1x1)	b: bulk lattice
Surface 1	4.980	0.000	-2.490	4.313	120.0	(2.000, 0.000) (0.000, 2.000)	(2x2)	s1: commens. superlattice

3D COORDINATES

Se1: overlayer in 3-fold fcc hollow sites

Dx/Dy in Å, or as a fraction of layer's unit cell vectors; atom 0 at the origin. Epir/subr are bulk repeat vectors.

No. of atoms: 3

Bulk z = 2.033 Å

Region	Chem el.	At. no.	Cell type	Site occ.	Rel. to	Dx ± εx		Dy ± εy		Dz ± εz		Dz/Bz(%) ± εz/Bz
epir		-2					f		f		Å	
subr		-1				1.245	Å	-0.719	Å	2.033	Å	
ovrl	Se	1	s1	.25	0	0.000	f	0.000	f	0.000	Å	0.0
intf	Ni	2	b	1.00	1	0.333	f	0.667	f	1.800 ± .040	Å	88.5 ± 2.0
subl	Ni	3	b	1.00	2	0.333	f	-0.333	f	2.033	Å	100.0

BOND DISTANCES AND ANGLES

Bond distances and angles are derived from coordinates

No. of distances/angles: 5

Interatomic dist. A-B (Å)	Atom A	Atom B	Atom C	Bond angle A-B-C (°)
2.304	Se1	Ni2	Ni2(0,1)	122.7
2.304	Se1	Ni2	Ni3(0,1)	176.7
2.304	Se1	Ni2	Ni3	117.3
2.490	Ni2	Ni2(1,0)		
2.490	Ni2	Ni3		

ILLUSTRATION: 28,29

COMMON NAME : Ni(100)-c(2x2)-Te
CLASSIFICATION : 28.52.1
TECHNIQUE : LEED
AUTHORS : J.E. Demuth, D.W. Jepsen and P.M. Marcus
REFERENCE : J. Phys., C6, L307 (1973)

SURFACE TYPE
Substrate : Ni
Crystal face: 100
Temperature : RT*
Bulk lattice: fcc
2D bulk symm: p4m
2D surf symm: p4m

Adsorbate: Te
Coverage : 1/2 (Te/Ni)
Pattern : c(2x2)
Matrix : (1.000, 1.000)
(-1.000, 1.000)

STRUCTURE TYPE
Atomic adsorption in 4-fold hollow sites

SAMPLE PREPARATION (sample)
Treatment : see J.E. Demuth and T.N. Rhodin, Surf. Sci. 45, 249 (1974)
Crystallinity:
Anal. methods:
Contamination:

COMMENTS

DATA COLLECTION
Technique: LEED
Dataset : I-V curves: E range 25-155 eV

THEORY/DATA TREATMENT
Dynamical LEED (layer KKR): 8 phase shifts, Wakoh Ni pot, Te superposition pot; Voi∝E**1/3; ΘD=420 K(Ni), 295K(Te)

STRUCTURES EXAMINED
Different adsorption sites with variable Te-Ni spacings

QUALITY OF EXPERIMENT-THEORY FIT
Visual

2D UNIT CELLS (1 domain observed)

Cell	Ax (Å)	Ay (Å)	Bx (Å)	By (Å)	α (°)	Matrix	Pattern	Cell type
Bulk	2.489	0.000	0.000	2.489	90.0	(1.000, 0.000) (0.000, 1.000)	(1x1)	b: bulk lattice
Surface 1	2.489	2.489	-2.489	2.489	90.0	(1.000, 1.000) (-1.000, 1.000)	c(2x2)	s1: commens. superlattice

3D COORDINATES

Te1: overlayer in 4-fold hollow sites

Dx/Dy in Å, or as a fraction of layer's unit cell vectors; atom 0 at the origin. Epir/subr are bulk repeat vectors.

No. of atoms: 3

Bulk z = 1.760 Å

Region	Chem el.	At. no.	Cell type	Site occ.	Rel. to	Dx ± εx		Dy ± εy		Dz ± εz		Dz/Bz(%) ± εz/Bz
epir		-2					f		f		Å	
subr		-1				-1.245	Å	-1.245	Å	1.760	Å	
ovrl	Te	1	s1	.50	0	0.000	f	0.000	f	0.000	Å	0.0
intf	Ni	2	b	1.00	1	0.500	f	0.500	f	1.900 ± .100	Å	108.0 ± 5.7
subl	Ni	3	b	1.00	2	-0.500	f	-0.500	f	1.760	Å	100.0

BOND DISTANCES AND ANGLES

Bond distances and angles are derived from coordinates

No. of distances/angles: 6

Interatomic dist. A-B (Å)	Atom A	Atom B	Atom C	Bond angle A-B-C (°)
2.590	Te1	Ni2	Ni2(1,0)	118.7
2.590	Te1	Ni2	Ni3(1,1)	180.0
2.590	Te1	Ni2	Ni3(1,0)	121.3
2.590	Te1	Ni2	Ni3	92.2
2.489	Ni2	Ni2(1,0)		
2.489	Ni2	Ni3		

COMMON NAME : Ni(100)-p(2x2)-Te
CLASSIFICATION : 28.52.1a
TECHNIQUE : LEED
AUTHORS : M.A. Van Hove and S.Y. Tong
REFERENCE : J. Vac. Sci. Technol., 12, 230 (1975)

ILLUSTRATION: 28,30

SURFACE TYPE
Substrate : Ni
Crystal face: 100
Temperature : 300 K
Bulk lattice: fcc
2D bulk symm: p4m
2D surf symm: p4m

Adsorbate: Te
Coverage : 1/4 (Te/Ni)
Pattern : (2x2)
Matrix : (2.000, 0.000)
(0.000, 2.000)

STRUCTURE TYPE
Atomic adsorption in 4-fold hollow sites

SAMPLE PREPARATION (sample)
Treatment : see Demuth and Rhodin, Surf. Sci. 42, 261 & 45, 249 (1974)
Crystallinity:
Anal. methods:
Contamination:

COMMENTS

DATA COLLECTION
Technique: LEED
Dataset :

THEORY/DATA TREATMENT
Dynamical LEED (RFS): Wakoh Ni potential, overlapping atomic potential for Te, 8 phase shifts; Vor=-11.2 eV, VoiαE**1/3

STRUCTURES EXAMINED
8 to 10 Ni-Te spacings, 0.2Å apart, at hollow, bridge and top sites

QUALITY OF EXPERIMENT-THEORY FIT
Visual

2D UNIT CELLS (1 domain observed)

Cell	Ax (Å)	Ay (Å)	Bx (Å)	By (Å)	α (°)	Matrix	Pattern	Cell type
Bulk	2.490	0.000	0.000	2.490	90.0	(1.000, 0.000) (0.000, 1.000)	(1x1)	b: bulk lattice
Surface 1	4.980	0.000	0.000	4.980	90.0	(2.000, 0.000) (0.000, 2.000)	(2x2)	s1: commens. superlattice

3D COORDINATES

Te1: overlayer in 4-fold hollow sites

Dx/Dy in Å, or as a fraction of layer's unit cell vectors; atom 0 at the origin. Epir/subr are bulk repeat vectors.

No. of atoms: 3

Bulk z = 1.760 Å

Region	Chem el.	At. no.	Cell type	Site occ.	Rel. to	Dx ± εx		Dy ± εy		Dz ± εz		Dz/Bz(%) ± εz/Bz
epir		-2					f		f		Å	
subr		-1				-1.245	Å	-1.245	Å	1.760	Å	
ovrl	Te	1	s1	.25	0	0.000	f	0.000	f	0.000	Å	0.0
intf	Ni	2	b	1.00	1	0.500	f	0.500	f	1.800 ± .100	Å	102.3 ± 5.7
subl	Ni	3	b	1.00	2	-0.500	f	-0.500	f	1.760	Å	100.0

BOND DISTANCES AND ANGLES

Bond distances and angles are derived from coordinates

No. of distances/angles: 6

Interatomic dist. A-B (Å)	Atom A	Atom B	Atom C	Bond angle A-B-C (°)
2.518	Te1	Ni2	Ni2(1,0)	119.6
2.518	Te1	Ni2	Ni3(1,1)	180.0
2.518	Te1	Ni2	Ni3(1,0)	120.4
2.518	Te1	Ni2	Ni3	90.6
2.490	Ni2	Ni2(1,0)		
2.490	Ni2	Ni3		

COMMON NAME : Ni3Al(100)-(1x1) ILLUSTRATION: 134
CLASSIFICATION : 28.13.5
TECHNIQUE : LEED
AUTHORS : D. Sondericker, F. Jona and P.M. Marcus
REFERENCE : Phys. Rev., B33, 900 (1986)

SURFACE TYPE
Substrate : Ni3Al
Crystal face: 100
Temperature : RT*
Bulk lattice: AuCu3
2D bulk symm: p4m
2D surf symm: p4m

Adsorbate:
Coverage :
Pattern : (1x1)
Matrix : (1.000, 0.000)
(0.000, 1.000)

STRUCTURE TYPE
Slightly buckled 50-50 mixed top NiAl layer with 4% average top spacing contraction; even-numbered layers are pure Ni, odd-numbered layers 50-50 mixed NiAl, as in bulk

SAMPLE PREPARATION (1 sample)
Treatment : Ar+ bombardment for 1 hour, then annealing at T = 1170 K
Crystallinity:
Anal. methods:
Contamination: AES: trace amounts of O and C

COMMENTS

DATA COLLECTION
Technique: LEED
Dataset : I-V curves (some degenerate): 15 beams at Θ,φ = 0,0°, 14 at 15.5,45°, 16 at 13.5,0°; 42<E<165 eV; cum E range 4688eV

THEORY/DATA TREATMENT
Dynamical LEED: 6 phase shifts from band structure potentials for pure Ni and Al; Voi=-3.5 eV

STRUCTURES EXAMINED
Variations of two topmost interlayer spacings and of top layer buckling

QUALITY OF EXPERIMENT-THEORY FIT
RZJ=0.105,0.185,0.179 (Θ=0,15.5,13.5°)

2D UNIT CELLS (1 domain observed)

Cell	Ax (Å)	Ay (Å)	Bx (Å)	By (Å)	α (°)	Matrix	Pattern	Cell type
Bulk	3.560	0.000	0.000	3.560	90.0	(1.000, 0.000) (0.000, 1.000)	(1x1)	b: bulk lattice
Surface 1	3.560	0.000	0.000	3.560	90.0	(1.000, 0.000) (0.000, 1.000)	(1x1)	s1: commens. superlattice

3D COORDINATES

Al1-Ni2: planar top mixed layer; Al5-Ni8: periodically repeating set of bulk layers

Dx/Dy in Å, or as a fraction of layer's unit cell vectors; atom 0 at the origin. Epir/subr are bulk repeat vectors.

No. of atoms: 8 Bulk z = 1.780 Å

Region	Chem el.	At. no.	Cell type	Site occ.	Rel. to	Dx ± εx		Dy ± εy		Dz ± εz		Dz/Bz(%) ± εz/Bz
epir		-2					f		f		Å	
subr		-1				0.000	Å	0.000	Å	3.560	Å	
intf	Al	1	b	1.00	0	0.000	f	0.000	f	0.000	Å	0.0
intf	Ni	2	b	1.00	1	0.500	f	0.500	f	0.020 ± .030	Å	1.1 ± 1.7
intf	Ni	3	b	1.00	2	0.000	f	-0.500	f	1.730 ± .030	Å	97.2 ± 1.7
intf	Ni	4	b	1.00	3	-0.500	f	0.500	f	0.000	Å	0.0
subl	Al	5	b	1.00	4	0.000	f	-0.500	f	1.780 ± .030	Å	100.0 ± 1.7
subl	Ni	6	b	1.00	5	0.500	f	0.500	f	0.000	Å	0.0
subl	Ni	7	b	1.00	6	0.000	f	-0.500	f	1.780	Å	100.0
subl	Ni	8	b	1.00	7	-0.500	f	0.500	f	0.000	Å	0.0

Ni3Al(100)-(1x1)
28.13.5

BOND DISTANCES AND ANGLES

Bond distances and angles are derived from coordinates

No. of distances/angles: 11

Interatomic dist. A-B (Å)	Atom A	Atom B	Atom C	Bond angle A-B-C (°)
3.560	Al1	Al1(1,0)	Ni3	44.5
2.482	Ni2	Ni3	Al5	119.5
2.482	Ni2	Ni3	Ni6	89.2
2.517	Al1	Ni2	Ni3	59.9
2.517	Al1	Ni2	Ni4	59.9
2.496	Al1	Ni3	Ni2	60.8
2.496	Al1	Ni3	Ni4	59.7
2.496	Al1	Ni3	Al5	89.5
2.496	Al1	Ni3	Ni6	119.7
3.530	Al1	Al5	Ni6	90.0
2.482	Ni2	Ni3	Ni4	59.5

COMMON NAME : Ni3Al(110)-(1x1)
CLASSIFICATION : 28.13.13
TECHNIQUE : LEED
AUTHORS : D. Sondericker, F. Jona and P.M. Marcus
REFERENCE : Phys. Rev., B34, 6775 (1986)

ILLUSTRATION: 137

SURFACE TYPE
Substrate : Ni3Al
Crystal face: 110
Temperature : RT*
Bulk lattice: Cu3Au
2D bulk symm: pmm
2D surf symm: pmm

Adsorbate:
Coverage :
Pattern : (1x1)
Matrix : (1.000, 0.000)
(0.000, 1.000)

STRUCTURE TYPE
Unreconstructed bulk termination with mixed NiAl layer (rather than pure Ni layer), with buckling in top mixed layer and relaxations in top two interlayer spacings

SAMPLE PREPARATION (1 sample)
Treatment : 10 cycles of Ar+ sputtering (1h) and 1223 K anneals (1h)
Crystallinity:
Anal. methods:
Contamination: AES monitoring of Ni/Al ratio

COMMENTS

DATA COLLECTION
Technique: LEED
Dataset : I-V curves: 24 degenerate (10 non-deg.) beams at Θ=0°, 18 (12) beams at Θ=15°,φ=90°; cum. E ranges 2958 eV, 2189eV

THEORY/DATA TREATMENT
Dynamical LEED: 6 phase shifts from band structure potentials for pure Ni and Al; Voi=-3.5 eV, Vor=-10eV

STRUCTURES EXAMINED
Variation of two topmost interlayer spacings and of buckling in topmost mixed NiAl layer, for both pure-Ni termination and mixed NiAl termination

QUALITY OF EXPERIMENT-THEORY FIT
RZJ=0.14 (Θ=0°), 0.13 (Θ=15°)

2D UNIT CELLS (1 domain observed)

Cell	Ax (Å)	Ay (Å)	Bx (Å)	By (Å)	α (°)	Matrix	Pattern	Cell type
Bulk	3.560	0.000	0.000	5.035	90.0	(1.000, 0.000) (0.000, 1.000)	(1x1)	b: bulk lattice
Surface 1	3.560	0.000	0.000	5.035	90.0	(1.000, 0.000) (0.000, 1.000)	(1x1)	s1: commens. superlattice

3D COORDINATES

Al1-Ni2: topmost buckled mixed NiAl layer; Ni3-Ni4: next pure Ni layer;
Al9-Ni10-Ni11-Ni12: periodically repeating bulk pair of mixed/pure layers

Dx/Dy in Å, or as a fraction of layer's unit cell vectors; atom 0 at the origin. Epir/subr are bulk repeat vectors.

No. of atoms: 12 Bulk z = 2.517 Å

Region	Chem el.	At. no.	Cell type	Site occ.	Rel. to	Dx ± εx		Dy ± εy		Dz ± εz		Dz/Bz(%) ± εz/Bz
epir		-2					f		f		Å	
subr		-1				0.000	Å	-2.517	Å	2.517	Å	
intf	Al	1	b	1.00	0	0.000	f	0.000	f	0.000	Å	0.0
intf	Ni	2	b	1.00	1	0.000	f	0.500	f	0.020 ± .030	Å	.8 ± 1.2
intf	Ni	3	b	1.00	2	0.500	f	0.250	f	1.110 ± .030	Å	44.1 ± 1.2
intf	Ni	4	b	1.00	3	0.000	f	-0.500	f	0.000	Å	0.0
intf	Al	5	b	1.00	4	-0.500	f	0.250	f	1.300 ± .030	Å	51.6 ± 1.2
intf	Ni	6	b	1.00	5	0.000	f	-0.500	f	0.000	Å	0.0
intf	Ni	7	b	1.00	6	0.500	f	0.750	f	1.259	Å	50.0
intf	Ni	8	b	1.00	7	0.000	f	-0.500	f	0.000	Å	0.0
subl	Al	9	b	1.00	8	-0.500	f	-0.250	f	1.259	Å	50.0
subl	Ni	10	b	1.00	9	0.000	f	0.500	f	0.000	Å	0.0
subl	Ni	11	b	1.00	10	0.500	f	-0.250	f	1.259	Å	50.0
subl	Ni	12	b	1.00	11	0.000	f	0.500	f	0.000	Å	0.0

Ni3Al(110)-(1x1)
28.13.13

BOND DISTANCES AND ANGLES

Bond distances and angles are derived from coordinates

No. of distances/angles: 4

Interatomic dist. A-B (Å)	Atom A	Atom B	Atom C	Bond angle A-B-C (°)
2.517	Al1	Ni2	Ni3	121.2
2.456	Al1	Ni3(0,-1)	Ni4	59.2
2.446	Ni2	Ni3	Al5	57.8
2.410	Ni2	Al5	Ni7	120.0

COMMON NAME : Ni3Al(111)-(1x1)
CLASSIFICATION : 28.13.12
TECHNIQUE : LEED
AUTHORS : D. Sondericker, F. Jona and P.M. Marcus
REFERENCE : Phys. Rev., B34, 6770 (1986)

ILLUSTRATION: 128

SURFACE TYPE
Substrate : Ni3Al
Crystal face: 111
Temperature : RT*
Bulk lattice: Cu3Au
2D bulk symm: p3m1
2D surf symm: p3m1

Adsorbate:
Coverage :
Pattern : (1x1)
Matrix : (1.000, 0.000)
(0.000, 1.000)

STRUCTURE TYPE
Unreconstructed bulk termination with buckling in first mixed Ni3Al layer (Al outward) and slight contraction of topmost interlayer spacing

SAMPLE PREPARATION (1 sample)
Treatment : 2x2mm² grain Ni3Al ingot, Ar+ sputtering and annealing
Crystallinity: sharp LEED pattern
Anal. methods:
Contamination: AES monitored Al/Ni ratio during anneals

COMMENTS

DATA COLLECTION
Technique: LEED
Dataset : I-V curves: 29 degenerate (10 non-deg.) beams at Θ=0°, 19 (13) beams at Θ=15°,φ=-60°; cum. E ranges 3255 eV, 2225eV

THEORY/DATA TREATMENT
Dynamical LEED: 6 phase shifts from band structure potentials for pure Ni and Al. Voi=-3.5 eV; Vor=-12eV

STRUCTURES EXAMINED
Variation of topmost interlayer spacing and of buckling in topmost layer

QUALITY OF EXPERIMENT-THEORY FIT
RZJ=0.128 (Θ=0°), 0.165 (Θ=15°)

2D UNIT CELLS (1 domain observed)

Cell	Ax (Å)	Ay (Å)	Bx (Å)	By (Å)	α (°)	Matrix	Pattern	Cell type
Bulk	5.035	0.000	2.517	4.360	60.0	(1.000, 0.000) (0.000, 1.000)	(1x1)	b: bulk lattice
Surface 1	5.035	0.000	2.517	4.360	60.0	(1.000, 0.000) (0.000, 1.000)	(1x1)	s1: commens. superlattice

3D COORDINATES

Al1-Ni2-Ni3-Ni4: buckled topmost layer; Al9-Ni10-Ni11-Ni12: periodically repeating bulk layers

Dx/Dy in Å, or as a fraction of layer's unit cell vectors; atom 0 at the origin. Epir/subr are bulk repeat vectors.

No. of atoms: 12 Bulk z = 2.055 Å

Region	Chem el.	At. no.	Cell type	Site occ.	Rel. to	Dx ± εx		Dy ± εy		Dz ± εz		Dz/Bz(%) ± εz/Bz
epir		-2					f		f		Å	
subr		-1				2.517	Å	1.453	Å	2.055	Å	
intf	Al	1	b	1.00	0	0.000	f	0.000	f	0.000	Å	0.0
intf	Ni	2	b	1.00	1	0.500	f	0.000	f	0.060 ± .030	Å	2.9 ± 1.5
intf	Ni	3	b	1.00	2	-0.500	f	0.500	f	0.000	Å	0.0
intf	Ni	4	b	1.00	3	0.500	f	0.000	f	0.000	Å	0.0
intf	Al	5	b	1.00	4	-0.167	f	-0.167	f	2.045 ± .030	Å	99.5 ± 1.5
intf	Ni	6	b	1.00	5	0.500	f	0.000	f	0.000	Å	0.0
intf	Ni	7	b	1.00	6	-0.500	f	0.500	f	0.000	Å	0.0
intf	Ni	8	b	1.00	7	0.500	f	0.000	f	0.000	Å	0.0
subl	Al	9	b	1.00	8	-0.167	f	-0.167	f	2.055	Å	100.0
subl	Ni	10	b	1.00	9	-0.500	f	0.000	f	0.000	Å	0.0
subl	Ni	11	b	1.00	10	0.500	f	-0.500	f	0.000	Å	0.0
subl	Ni	12	b	1.00	11	-0.500	f	0.000	f	0.000	Å	0.0

Ni3Al(111)-(1x1)
28.13.12

BOND DISTANCES AND ANGLES

Bond distances and angles are derived from coordinates

No. of distances/angles: 5

Interatomic dist. A-B (Å)	Atom A	Atom B	Atom C	Bond angle A-B-C (°)
2.518	Al1	Ni2	Al1(1,0)	177.3
2.518	Al1	Ni2	Ni3	60.0
2.518	Al1	Ni2	Ni7(0,-1)	61.2
2.517	Ni2	Ni3	Ni8(-1,0)	120.1
2.509	Ni2	Al5	Ni7	120.1

COMMON NAME : NiAl(100)-(1x1) ILLUSTRATION: 143
CLASSIFICATION : 28.13.15c
TECHNIQUE : LEED
AUTHORS : H.L. Davis and J.R. Noonan
REFERENCE : Springer Series in Surface Sciences, 11, 152 (1988)

SURFACE TYPE
Substrate : NiAl
Crystal face: 100
Temperature : RT
Bulk lattice: CsCl
2D bulk symm: p4m
2D surf symm: p4m

Adsorbate:
Coverage :
Pattern : (1x1)
Matrix : (1.000, 0.000)
(0.000, 1.000)

STRUCTURE TYPE
Unreconstructed bulk termination with relaxation in top two interlayer spacings; Al termination favored

SAMPLE PREPARATION (1 sample)
Treatment : mech. polish, Ar+ sputtering and annealing to about 870C
Crystallinity: sharp LEED pattern
Anal. methods:
Contamination: AES (CMA): clean, Al:Ni ratio ≈1:1

COMMENTS

DATA COLLECTION
Technique: LEED
Dataset : I-V curves at normal incidence: averaged within 6 symmetry sets; E range 50-300 eV

THEORY/DATA TREATMENT
Dynamical LEED (RFS): 8 phase shifts from Moruzzi et al pot for bulk NiAl; Voi=-4.5 eV; ΘD=350 K(Ni), 500K(Al)

STRUCTURES EXAMINED
For both Ni and Al terminations, variation of top two interlayer spacings: 1st from -25% to +10%, 2nd from -15% to +20%, in steps of 5%

QUALITY OF EXPERIMENT-THEORY FIT
RZJ=0.084

2D UNIT CELLS (1 domain observed)

Cell	Ax (Å)	Ay (Å)	Bx (Å)	By (Å)	α (°)	Matrix	Pattern	Cell type
Bulk	2.874	0.000	0.000	2.874	90.0	(1.000, 0.000) (0.000, 1.000)	(1x1)	b: bulk lattice
Surface 1	2.874	0.000	0.000	2.874	90.0	(1.000, 0.000) (0.000, 1.000)	(1x1)	s1: commens. superlattice

3D COORDINATES

0.05Å error bars assumed for tabulation

Dx/Dy in Å, or as a fraction of layer's unit cell vectors; atom 0 at the origin. Epir/subr are bulk repeat vectors.

No. of atoms: 6 Bulk z = 1.437 Å

Region	Chem el.	At. no.	Cell type	Site occ.	Rel. to	Dx ± εx		Dy ± εy		Dz ± εz		Dz/Bz(%) ± εz/Bz
epir		-2					f		f		Å	
subr		-1				0.000	Å	0.000	Å	2.874	Å	
intf	Al	1	b	1.00	0	0.000	f	0.000	f	0.000	Å	0.0
intf	Ni	2	b	1.00	1	0.500	f	0.500	f	1.315 ± .050	Å	91.5 ± 3.5
intf	Al	3	b	1.00	2	-0.500	f	-0.500	f	1.495 ± .050	Å	104.0 ± 3.5
intf	Ni	4	b	1.00	3	0.500	f	0.500	f	1.437	Å	100.0
subl	Al	5	b	1.00	4	-0.500	f	-0.500	f	1.437	Å	100.0
subl	Ni	6	b	1.00	5	0.500	f	0.500	f	1.437	Å	100.0

BOND DISTANCES AND ANGLES

Bond distances and angles are derived from coordinates

No. of distances/angles: 2

Interatomic dist. A-B (Å)	Atom A	Atom B	Atom C	Bond angle A-B-C (°)
2.421	Al1	Ni2	Al3	69.2
2.523	Ni2	Al3	Ni4	71.6

COMMON NAME : NiAl(110)-(1x1) ILLUSTRATION: 142
CLASSIFICATION : 28.13.11
TECHNIQUE : MEIS
AUTHORS : S.M. Yalisove and W.R. Graham
REFERENCE : Surf. Sci., 183, 556 (1987)

SURFACE TYPE
Substrate : NiAl
Crystal face: 110
Temperature : RT
Bulk lattice: CsCl
2D bulk symm: pmm
2D surf symm: pmm

Adsorbate:
Coverage :
Pattern : (1x1)
Matrix : (1.000, 0.000)
(0.000, 1.000)

STRUCTURE TYPE
Unreconstructed bulk termination with buckling and interlayer spacing relaxations in top two NiAl layers: Al buckles out/inward in 1st/2nd NiAl layers

SAMPLE PREPARATION (1 sample)
Treatment : NiAl single crystal sputtered and annealed at about 1073 K
Crystallinity: sharp (1x1) LEED pattern
Anal. methods:
Contamination:

COMMENTS

DATA COLLECTION
Technique: MEIS
Dataset : Ni and Al surface blocking curves in (100) zone

THEORY/DATA TREATMENT
MEIS (channeling and blocking, 100k eV protons): Monte Carlo sim. with R-factor fit; rms ampl 0.128Å(surf)-0.08Å(bulk)

STRUCTURES EXAMINED
Variation of first two spacings between mixed NiAl layers and of buckling of first two mixed layers

2D UNIT CELLS (1 domain observed)

Cell	Ax (Å)	Ay (Å)	Bx (Å)	By (Å)	α (°)	Matrix	Pattern	Cell type
Bulk	2.874	0.000	0.000	4.064	90.0	(1.000, 0.000) (0.000, 1.000)	(1x1)	b: bulk lattice
Surface 1	2.874	0.000	0.000	4.064	90.0	(1.000, 0.000) (0.000, 1.000)	(1x1)	s1: commens. superlattice

3D COORDINATES

Al1-Ni2 and Ni3-Al4: buckled top two layers; Al7-Ni8: periodically repeating bulk layers; layer spacings between like elements in different layers are accurate to 0.01Å

Dx/Dy in Å, or as a fraction of layer's unit cell vectors; atom 0 at the origin. Epir/subr are bulk repeat vectors.

No. of atoms: 8 Bulk z = 2.032 Å

Region	Chem el.	At. no.	Cell type	Site occ.	Rel. to	Dx ± εx		Dy ± εy		Dz ± εz		Dz/Bz(%) ± εz/Bz
epir		-2					f		f		Å	
subr		-1				0.000	Å	2.032	Å	2.032	Å	
intf	Al	1	b	1.00	0	0.000	f	0.000	f	0.000	Å	0.0
intf	Ni	2	b	1.00	1	0.500	f	0.500	f	0.244 ± .040	Å	12.0 ± 2.0
intf	Ni	3	b	1.00	2	0.000	f	-0.500	f	1.829 ± .010	Å	90.0 ± .5
intf	Al	4	b	1.00	3	-0.500	f	0.500	f	0.041 ± .040	Å	2.0 ± .5
intf	Al	5	b	1.00	4	0.000	f	-0.500	f	2.012 ± .010	Å	99.0 ± .5
intf	Ni	6	b	1.00	5	0.500	f	0.500	f	0.000	Å	0.0
subl	Al	7	b	1.00	6	-0.500	f	0.000	f	2.032	Å	100.0
subl	Ni	8	b	1.00	7	0.500	f	-0.500	f	0.000	Å	0.0

BOND DISTANCES AND ANGLES

Bond distances and angles are derived from coordinates

No. of distances/angles: 2

Interatomic dist. A-B (Å)	Atom A	Atom B	Atom C	Bond angle A-B-C (°)
2.501	Al1	Ni2	Ni3	57.4
2.734	Ni2	Ni3	Al4	53.4

COMMON NAME : NiAl(110)-(1x1) ILLUSTRATION: 142
CLASSIFICATION : 28.13.15b
TECHNIQUE : LEED
AUTHORS : H.L. Davis and J.R. Noonan
REFERENCE : Springer Series in Surface Sciences, 11, 152 (1988)

SURFACE TYPE
Substrate : NiAl
Crystal face: 110
Temperature : RT
Bulk lattice: CsCl
2D bulk symm: pmm
2D surf symm: pmm

Adsorbate:
Coverage :
Pattern : (1x1)
Matrix : (1.000, 0.000)
(0.000, 1.000)

STRUCTURE TYPE
Unreconstructed bulk termination with buckling and interlayer spacing relaxations in top two NiAl layers: Al buckles outward in both topmost NiAl layers

SAMPLE PREPARATION (1 sample)
Treatment : mech. polish, Ar ion sputtering and annealing to about 870C
Crystallinity: sharp LEED pattern
Anal. methods:
Contamination: AES (CMA): clean; Al:Ni ratio ≈1:1

COMMENTS

DATA COLLECTION
Technique: LEED
Dataset : I-V curves at normal incidence: averaged within 14 symmetry sets; E range 50-330 eV

THEORY/DATA TREATMENT
Dynamical LEED (RFS): 8 phase shifts from Moruzzi et al pot for bulk NiAl; Voi=-4.5 eV; ΘD=350 K(Ni), 500K(Al)

STRUCTURES EXAMINED
Variation of top two spacings between mixed NiAl layers and of buckling of top two mixed layers

QUALITY OF EXPERIMENT-THEORY FIT
RZJ=0.063

2D UNIT CELLS (1 domain observed)

Cell	Ax (Å)	Ay (Å)	Bx (Å)	By (Å)	α (°)	Matrix	Pattern	Cell type
Bulk	2.874	0.000	0.000	4.064	90.0	(1.000, 0.000) (0.000, 1.000)	(1x1)	b: bulk lattice
Surface 1	2.874	0.000	0.000	4.064	90.0	(1.000, 0.000) (0.000, 1.000)	(1x1)	s1: commens. superlattice

3D COORDINATES

Al1-Ni2 and Al3-Ni4: buckled top two layers; Al7-Ni8: periodically repeating bulk layers;
0.02Å error bars assumed for tabulation

Dx/Dy in Å, or as a fraction of layer's unit cell vectors; atom 0 at the origin. Epir/subr are bulk repeat vectors.

No. of atoms: 8 Bulk z = 2.032 Å

Region	Chem el.	At. no.	Cell type	Site occ.	Rel. to	Dx ± εx		Dy ± εy		Dz ± εz		Dz/Bz(%) ± εz/Bz
epir		-2					f		f		Å	
subr		-1				0.000	Å	2.032	Å	2.032	Å	
intf	Al	1	b	1.00	0	0.000	f	0.000	f	0.000	Å	0.0
intf	Ni	2	b	1.00	1	0.500	f	0.500	f	0.199 ± .020	Å	9.8 ± 1.0
intf	Al	3	b	1.00	2	-0.500	f	0.000	f	1.939 ± .020	Å	95.4 ± 1.0
intf	Ni	4	b	1.00	3	0.500	f	-0.500	f	0.020 ± .020	Å	1.0 ± 1.0
intf	Al	5	b	1.00	4	-0.500	f	0.000	f	2.053 ± .020	Å	101.0 ± 1.0
intf	Ni	6	b	1.00	5	0.500	f	0.500	f	0.000	Å	0.0
subl	Al	7	b	1.00	6	-0.500	f	0.000	f	2.032	Å	100.0
subl	Ni	8	b	1.00	7	0.500	f	-0.500	f	0.000	Å	0.0

NiAl(110)-(1x1)
28.13.15b BOND DISTANCES AND ANGLES

Bond distances and angles are derived from coordinates

No. of distances/angles: 2

Interatomic dist. A-B (Å)	Atom A	Atom B	Atom C	Bond angle A-B-C (°)
2.497	Al1	Ni2	Al3	73.8
2.413	Ni2	Al3	Ni4	70.3

COMMON NAME : NiAl(110)-(1x1)
CLASSIFICATION : 28.13.19
TECHNIQUE : LEIS
AUTHORS : D.R. Mullins and S.H. Overbury
REFERENCE : Surf. Sci., 199, 141 (1988)

ILLUSTRATION: 142

SURFACE TYPE
Substrate : NiAl
Crystal face: 110
Temperature : RT*
Bulk lattice: CsCl
2D bulk symm: pmm
2D surf symm: pmm

Adsorbate:
Coverage :
Pattern : (1x1)
Matrix : (1.000, 0.000)
(0.000, 1.000)

STRUCTURE TYPE
Unreconstructed bulk termination with buckling in top NiAl layer: Al buckles outward

SAMPLE PREPARATION (1 sample)
Treatment : Ar+ bombardment and annealing
Crystallinity:
Anal. methods:
Contamination: ISS: monitoring of Al/Ni ratio

COMMENTS
Second and deeper spacings are assumed bulk-like in this tabulation

DATA COLLECTION
Technique: LEIS
Dataset : polar-angle dependent scattering intensities in several azimuths

THEORY/DATA TREATMENT
LEIS using 500 eV Li+ ions, with shadow-cone analysis: scattering potentials obtained by self-calibration

STRUCTURES EXAMINED
Buckling of mixed Al/Ni top layer; technique not sensitive to deeper structure

2D UNIT CELLS (1 domain observed)

Cell	Ax (Å)	Ay (Å)	Bx (Å)	By (Å)	α (°)	Matrix	Pattern	Cell type
Bulk	2.890	0.000	0.000	4.087	90.0	(1.000, 0.000) (0.000, 1.000)	(1x1)	b: bulk lattice
Surface 1	2.890	0.000	0.000	4.087	90.0	(1.000, 0.000) (0.000, 1.000)	(1x1)	s1: commens. superlattice

3D COORDINATES

Al1-Ni2: buckled top layer; Al3-Ni4: periodically repeating bulk layers

Dx/Dy in Å, or as a fraction of layer's unit cell vectors; atom 0 at the origin. Epir/subr are bulk repeat vectors.

No. of atoms: 4

Bulk z = 2.043 Å

Reg ion	Chem el.	At. no.	Cell type	Site occ.	Rel. to	Dx ± εx		Dy ± εy		Dz ± εz		Dz/Bz(%) ± εz/Bz
epir		-2					f		f		Å	
subr		-1				0.000	Å	-2.044	Å	2.043	Å	
intf	Al	1	b	1.00	0	0.000	f	0.000	f	0.000	Å	0.0
intf	Ni	2	b	1.00	1	0.500	f	0.500	f	0.210 ± .050	Å	10.3 ± 2.5
subl	Al	3	b	1.00	2	-0.500	f	0.000	f	2.043	Å	100.0
subl	Ni	4	b	1.00	3	0.500	f	-0.500	f	0.000	Å	0.0

BOND DISTANCES AND ANGLES

Bond distances and angles are derived from coordinates

No. of distances/angles: 2

Interatomic dist. A-B (Å)	Atom A	Atom B	Atom C	Bond angle A-B-C (°)
2.512	Al1	Ni2	Al3	74.7
2.502	Ni2	Al3	Ni4	70.5

COMMON NAME : NiAl(110)-(1x1)
CLASSIFICATION : 28.13.4
TECHNIQUE : LEED
AUTHORS : H.L. Davis and J.R. Noonan
REFERENCE : Phys. Rev. Lett., 54, 566 (1985)

ILLUSTRATION: 142

SURFACE TYPE
Substrate : NiAl
Crystal face: 110
Temperature : RT*
Bulk lattice: CsCl
2D bulk symm: pmm
2D surf symm: pmm

Adsorbate:
Coverage :
Pattern : (1x1)
Matrix : (1.000, 0.000)
(0.000, 1.000)

STRUCTURE TYPE
Bulk termination with 10% buckled top 50-50 mixed layer; average top spacing contracted 0.5%

SAMPLE PREPARATION (sample)
Treatment :
Crystallinity: LEED: (1x1) sharper than Al(110)
Anal. methods:
Contamination: AES: ≈50%Al, 50%Ni

COMMENTS

DATA COLLECTION
Technique: LEED
Dataset : I-V curves for 5 symm. inequivalent beams from 55-300 eV; cumulative E range 1118eV

THEORY/DATA TREATMENT
Dynamical LEED (matrix inv., RFS): 8 phase shifts, Wakoh Ni pot, Snow Al pot; Voi=4.5 eV; Θ=550 K(Al), 335K(Ni)

STRUCTURES EXAMINED
Independent displacements perpendicular to surface of both atoms (Ni and Al) in top layer

QUALITY OF EXPERIMENT-THEORY FIT
RZJ=0.053

2D UNIT CELLS (1 domain observed)

Cell	Ax (Å)	Ay (Å)	Bx (Å)	By (Å)	α (°)	Matrix	Pattern	Cell type
Bulk	2.886	0.000	0.000	4.082	90.0	(1.000, 0.000) (0.000, 1.000)	(1x1)	b: bulk lattice
Surface 1	2.886	0.000	0.000	4.082	90.0	(1.000, 0.000) (0.000, 1.000)	(1x1)	s1: commens. superlattice

3D COORDINATES

Al1-Ni2: buckled top 50-50 mixed AlNi layer; Al5-Ni6: periodically repeating bulk layer; 0.1Å error bars assumed for tabulation

Dx/Dy in Å, or as a fraction of layer's unit cell vectors; atom 0 at the origin. Epir/subr are bulk repeat vectors.

No. of atoms: 6 Bulk z = 2.041 Å

Region	Chem el.	At. no.	Cell type	Site occ.	Rel. to	Dx ± εx		Dy ± εy		Dz ± εz		Dz/Bz(%) ± εz/Bz
epir		-2					f		f		Å	
subr		-1				0.000	Å	0.000	Å	4.082	Å	
intf	Al	1	b	1.00	0	0.000	f	0.000	f	0.000	Å	0.0
intf	Ni	2	b	1.00	1	0.500	f	0.500	f	0.220 ± .100	Å	10.8 ± 4.9
intf	Al	3	b	1.00	2	-0.500	f	0.000	f	1.920 ± .100	Å	94.1 ± 4.9
intf	Ni	4	b	1.00	3	0.500	f	-0.500	f	0.000	Å	0.0
subl	Al	5	b	1.00	4	-0.500	f	0.000	f	2.041	Å	100.0
subl	Ni	6	b	1.00	5	0.500	f	0.500	f	0.000	Å	0.0

BOND DISTANCES AND ANGLES

Bond distances and angles are derived from coordinates

No. of distances/angles: 3

Interatomic dist. A-B (Å)	Atom A	Atom B	Atom C	Bond angle A-B-C (°)
2.509	Al1	Ni2	Al1(1,0)	70.2
2.509	Al1	Ni2	Al3	74.0
2.581	Al1	Ni4	Al5	110.7

COMMON NAME : NiAl(111)-(1x1) Al-terminated ILLUSTRATION: 144
CLASSIFICATION : 28.13.14b
TECHNIQUE : LEED
AUTHORS : J.R. Noonan and H.L. Davis
REFERENCE : Phys. Rev. Lett., 59, 1714 (1987)

SURFACE TYPE
Substrate : NiAl
Crystal face: 111
Temperature : RT
Bulk lattice: CsCl
2D bulk symm: p3m1
2D surf symm: p3m1

Adsorbate:
Coverage :
Pattern : (1x1)
Matrix : (1.000, 0.000)
(0.000, 1.000)

STRUCTURE TYPE
Al-terminated, unreconstructed, slightly relaxed structure; this Al-termination occurs in ≈50-50 mixture with Ni-term.: see its separate structure (class. no. 28.13.14a)

SAMPLE PREPARATION (1 sample)
Treatment : mech. polish, Ar+ sputtering and annealing to about 870C
Crystallinity: sharp LEED pattern
Anal. methods:
Contamination: AES (CMA): clean, Al:Ni ratio ≈1:1

COMMENTS

DATA COLLECTION
Technique: LEED
Dataset : normal-incidence I-V curves averaged within 7 symmetry sets, energy range 50-300 eV

THEORY/DATA TREATMENT
Dynamical LEED (16-layer giant-matrix inv.): 8 ph shs from Moruzzi NiAl bulk pot; Voi=-4.5 eV; ΘD=350 K(Ni), 500K(Al)

STRUCTURES EXAMINED
For both Ni and Al terminations: variation of two topmost layer spacings: 1st from -80% to +80%, 2nd from -30% to +50% (-60% to +60%) for Ni- (Al-) termination; mixture of both terminations was tested by incoherent intensity addition

QUALITY OF EXPERIMENT-THEORY FIT
RZJ=.063

2D UNIT CELLS (1 domain observed)

Cell	Ax (Å)	Ay (Å)	Bx (Å)	By (Å)	α (°)	Matrix	Pattern	Cell type
Bulk	4.064	0.000	2.032	3.520	60.0	(1.000, 0.000) (0.000, 1.000)	(1x1)	b: bulk lattice
Surface 1	4.064	0.000	2.032	3.520	60.0	(1.000, 0.000) (0.000, 1.000)	(1x1)	s1: commens. superlattice

3D COORDINATES

Ni4-Ni5: periodically repeating bulk layers; 0.05Å error bars assumed for tabulation

Dx/Dy in Å, or as a fraction of layer's unit cell vectors; atom 0 at the origin. Epir/subr are bulk repeat vectors.

No. of atoms: 5 Bulk z = .830 Å

Reg ion	Chem el.	At. no.	Cell type	Site occ.	Rel. to	Dx ± εx		Dy ± εy		Dz ± εz		Dz/Bz(%) ± εz/Bz
epir		-2					f		f		Å	
subr		-1				2.032	Å	1.173	Å	1.659	Å	
intf	Al	1	b	1.00	0	0.000	f	0.000	f	0.000	Å	0.0
intf	Ni	2	b	1.00	1	0.667	f	0.667	f	0.789 ± .050	Å	95.1 ± 6.0
intf	Al	3	b	1.00	2	-0.333	f	-0.333	f	0.872 ± .050	Å	105.1 ± 6.0
subl	Ni	4	b	1.00	3	-0.333	f	-0.333	f	0.830	Å	100.0
subl	Al	5	b	1.00	4	0.667	f	0.667	f	0.830	Å	100.0

BOND DISTANCES AND ANGLES

Bond distances and angles are derived from coordinates

No. of distances/angles: 2

Interatomic dist. A-B (Å)	Atom A	Atom B	Atom C	Bond angle A-B-C (°)
2.475	Al1	Ni2(0,-1)	Al3	70.5
2.503	Ni2	Al3	Ni4	180.0

COMMON NAME : NiAl(111)-(1x1) Ni-terminated ILLUSTRATION: 144
CLASSIFICATION : 28.13.14a
TECHNIQUE : LEED
AUTHORS : J.R. Noonan and H.L. Davis
REFERENCE : Phys. Rev. Lett., 59, 1714 (1987)

SURFACE TYPE
Substrate : NiAl
Crystal face: 111
Temperature : RT
Bulk lattice: CsCl
2D bulk symm: p3m1
2D surf symm: p3m1

Adsorbate:
Coverage :
Pattern : (1x1)
Matrix : (1.000, 0.000)
(0.000, 1.000)

STRUCTURE TYPE
Ni-terminated, unreconstructed, strongly relaxed structure; this Ni-termination occurs in ≈50-50 mixture with Al-term.: see its separate structure (class. no. 28.13.14b)

SAMPLE PREPARATION (1 sample)
Treatment : mech. polish, Ar+ sputtering and annealing to about 870C
Crystallinity: sharp LEED pattern
Anal. methods:
Contamination: AES (CMA): clean, Al:Ni ratio ≈1:1

COMMENTS

DATA COLLECTION
Technique: LEED
Dataset : normal-incidence I-V curves averaged within 7 symmetry sets, E range 50-300 eV

THEORY/DATA TREATMENT
Dynamical LEED (16-layer giant-matrix inv.): 8 ph shs from Moruzzi NiAl bulk pot; Voi=-4.5 eV; ΘD=350 K(Ni), 500K(Al)

STRUCTURES EXAMINED
For both Ni and Al terminations: variation of two topmost layer spacings: 1st from -80% to +80%, 2nd from -30% to +50% (-60% to +60%) for Ni- (Al-) termination; mixture of both terminations was tested by incoherent intensity addition

QUALITY OF EXPERIMENT-THEORY FIT
RZJ=0.063

2D UNIT CELLS (1 domain observed)

Cell	Ax (Å)	Ay (Å)	Bx (Å)	By (Å)	α (°)	Matrix	Pattern	Cell type
Bulk	4.064	0.000	2.032	3.520	60.0	(1.000, 0.000) (0.000, 1.000)	(1x1)	b: bulk lattice
Surface 1	4.064	0.000	2.032	3.520	60.0	(1.000, 0.000) (0.000, 1.000)	(1x1)	s1: commens. superlattice

3D COORDINATES

Ni5-Ni6: periodically repeating bulk layers; 0.05Å error bars assumed for tabulation

Dx/Dy in Å, or as a fraction of layer's unit cell vectors; atom 0 at the origin. Epir/subr are bulk repeat vectors.

No. of atoms: 6 Bulk z = .830 Å

Region	Chem el.	At. no.	Cell type	Site occ.	Rel. to	Dx ± εx		Dy ± εy		Dz ± εz		Dz/Bz(%) ± εz/Bz
epir		-2					f		f		Å	
subr		-1				-4.064	Å	-2.347	Å	1.659	Å	
intf	Ni	1	b	1.00	0	0.000	f	0.000	f	0.000	Å	0.0
intf	Al	2	b	1.00	1	0.667	f	0.667	f	0.415 ± .050	Å	50.0 ± 6.0
intf	Ni	3	b	1.00	2	-0.333	f	-0.333	f	0.955 ± .050	Å	115.1 ± 6.0
intf	Al	4	b	1.00	3	-0.333	f	-0.333	f	0.830	Å	100.0
subl	Ni	5	b	1.00	4	0.667	f	0.667	f	0.830	Å	100.0
subl	Al	6	b	1.00	5	-0.333	f	-0.333	f	0.830	Å	100.0

BOND DISTANCES AND ANGLES

Bond distances and angles are derived from coordinates

No. of distances/angles: 2

Interatomic dist. A-B (Å)	Atom A	Atom B	Atom C	Bond angle A-B-C (°)
2.383	Ni1	Al2(0,-1)	Ni3	67.0
2.533	Al2	Ni3	Al4	177.3

COMMON NAME : NiO(100)-(1x1) ILLUSTRATION: 149
CLASSIFICATION : 28.8.20
TECHNIQUE : LEED
AUTHORS : M.R. Welton Cook and M. Prutton
REFERENCE : J. Phys., C13, 3993 (1980)

SURFACE TYPE
Substrate : NiO
Crystal face: 100
Temperature : 390 K
Bulk lattice: NaCl
2D bulk symm: p4m
2D surf symm: p4m
Adsorbate:
Coverage :
Pattern : (1x1)
Matrix : (1.000, 0.000)
(0.000, 1.000)

STRUCTURE TYPE
Bulk-like non-buckled non-polar termination with 2% top spacing contraction

SAMPLE PREPARATION (1 sample)
Treatment : cleaned in situ at room temperature
Crystallinity:
Anal. methods:
Contamination: AES: ≈1% Cl and possibility of <1% C

COMMENTS
R-factor for high energy (110<E<250 eV) also includes normal incidence data of Netzer and Prutton, J. Phys. C8, 2401 (1975)

DATA COLLECTION
Technique: LEED
Dataset : I-V curves for: (0,-1),(-1,1) at normal incidence; (0,0), (0,-1) at Θ,φ= 10.5,45°; (1-2) at 6.5,45°; see comment

THEORY/DATA TREATMENT
Dynamical LEED: superposition pots, Xα exchange, 8 ph shs; Vor varied, Voi=-5 eV; no vibs; α=2/3, but α=0 for E>110eV

STRUCTURES EXAMINED
0-4% relaxation of first spacing, -4 to +4% 1st-layer buckling

QUALITY OF EXPERIMENT-THEORY FIT
RZJ=0.29

2D UNIT CELLS (1 domain observed)

Cell	Ax (Å)	Ay (Å)	Bx (Å)	By (Å)	α (°)	Matrix	Pattern	Cell type
Bulk	2.947	0.000	0.000	2.947	90.0	(1.000, 0.000) (0.000, 1.000)	(1x1)	b: bulk lattice
Surface 1	2.947	0.000	0.000	2.947	90.0	(1.000, 0.000) (0.000, 1.000)	(1x1)	s1: commens. superlattice

3D COORDINATES

Ni1-O2: planar mixed top layer; Ni5-O6: periodically repeating bulk mixed layer;
0.05Å error bars assumed for tabulation

Dx/Dy in Å, or as a fraction of layer's unit cell vectors; atom 0 at the origin. Epir/subr are bulk repeat vectors.

No. of atoms: 6 Bulk z = 2.084 Å

Region	Chem el.	At. no.	Cell type	Site occ.	Rel. to	Dx ± εx		Dy ± εy		Dz ± εz		Dz/Bz(%) ± εz/Bz
epir		-2					f		f		Å	
subr		-1				1.474	Å	1.474	Å	2.084	Å	
intf	Ni	1	b	1.00	0	0.000	f	0.000	f	0.000	Å	0.0
intf	O	2	b	1.00	1	0.500	f	0.500	f	0.000 ± .050	Å	0.0 ± 2.4
intf	Ni	3	b	1.00	2	0.000	f	0.000	f	2.043 ± .050	Å	98.0 ± 2.4
intf	O	4	b	1.00	3	-0.500	f	-0.500	f	0.000	Å	0.0
subl	Ni	5	b	1.00	4	0.000	f	0.000	f	2.084	Å	100.0
subl	O	6	b	1.00	5	0.500	f	0.500	f	0.000	Å	0.0

BOND DISTANCES AND ANGLES

Bond distances and angles are derived from coordinates

No. of distances/angles: 3

Interatomic dist. A-B (Å)	Atom A	Atom B	Atom C	Bond angle A-B-C (°)
2.947	Ni1	Ni1(1,0)		
2.084	Ni1	O2		
2.043	O2	Ni3	Ni1	45.6

COMMON NAME : NiSi2(100)-(1x1)
CLASSIFICATION : 14.28.6
TECHNIQUE : LEED
AUTHORS : S.C. Wu, Z.Q. Wang, Y.S. Li, F. Jona and P.M. Marcus
REFERENCE : Solid State Commun., 57, 687 (1986)

ILLUSTRATION: 148

SURFACE TYPE
Substrate : NiSi2
Crystal face: 100
Temperature : RT*
Bulk lattice: fluorite
2D bulk symm: p4m
2D surf symm: p4m

Adsorbate:
Coverage :
Pattern : (1x1)
Matrix : (1.000, 0.000)
(0.000, 1.000)

STRUCTURE TYPE
Unreconstructed bulk termination at pure Si layer, with additional Si layer, 4-fold coordinated to Si below, with buckling in topmost pure Si layer and contraction of interlayer spacing between top pure Si layer and next Ni layer (buckling breaks bulk 4-fold rot. symmetry)

SAMPLE PREPARATION (1 sample)
Treatment : 16-20ML Ni deposited on Si(100)-(2x1), annealed at 1073 K
Crystallinity:
Anal. methods: AES to confirm the silicide phase
Contamination:

COMMENTS
Weak 1/2, 1/4, and 1/5 order spots visible in LEED pattern

DATA COLLECTION
Technique: LEED
Dataset : IV curves for 10, 11, 20, 21, 22, beams at normal incidence, 00, 10, 01, 11, 20, 02, 1-2 beams at Θ=9°, ϕ=0°; E<200 eV

THEORY/DATA TREATMENT
Dynamical LEED intensity analysis

STRUCTURES EXAMINED
Both Ni and Si bulk termination with top layer relaxation; Si termination with missing row (like (100) of Si lattice); Ni in 4-fold site above 2nd layer Ni; Si in 4-fold site above void in 2nd layer;
Si in 4-fold site above 2nd layer Ni + Si buckling

2D UNIT CELLS (1 domain observed)

Cell	Ax (Å)	Ay (Å)	Bx (Å)	By (Å)	α (°)	Matrix	Pattern	Cell type
Bulk	3.840	0.000	0.000	3.840	90.0	(1.000, 0.000) (0.000, 1.000)	(1x1)	b: bulk lattice
Surface 1	3.840	0.000	0.000	3.840	90.0	(1.000, 0.000) (0.000, 1.000)	(1x1)	s1: commens. superlattice

3D COORDINATES

Si1: overlayer in non-bulk positions; Si2-Si3 (buckled) and Ni4: 2 layers with contracted spacing;
Si5-Si6 and Ni7: periodically repeating pair of bulk layers; 0.1Å error bars assumed for tabulation

Dx/Dy in Å, or as a fraction of layer's unit cell vectors; atom 0 at the origin. Epir/subr are bulk repeat vectors.

No. of atoms: 7 Bulk z = 1.350 Å

Region	Chem el.	At. no.	Cell type	Site occ.	Rel. to	Dx ± εx		Dy ± εy		Dz ± εz		Dz/Bz(%) ± εz/Bz
epir		-2					f		f		Å	
subr		-1				1.920	Å	1.920	Å	2.700	Å	
intf	Si	1	b	1.00	0	0.000	f	0.000	f	0.000	Å	0.0
intf	Si	2	b	1.00	1	0.500	f	0.000	f	0.950 ± .100	Å	70.4 ± 7.4
intf	Si	3	b	1.00	2	-0.500	f	0.500	f	0.200 ± .100	Å	14.8 ± 7.4
intf	Ni	4	b	1.00	3	0.000	f	-0.500	f	1.250 ± .100	Å	92.6 ± 7.4
subl	Si	5	b	1.00	4	0.500	f	0.000	f	1.350	Å	100.0
subl	Si	6	b	1.00	5	-0.500	f	0.500	f	0.000	Å	0.0
subl	Ni	7	b	1.00	6	0.500	f	0.000	f	1.350	Å	100.0

NiSi2(100)-(1x1)
14.28.6 BOND DISTANCES AND ANGLES

Bond distances and angles are derived from coordinates

No. of distances/angles: 6

Interatomic dist. A-B (Å)	Atom A	Atom B	Atom C	Bond angle A-B-C (°)
2.142	Si1	Si2	Si1(1,0)	127.4
2.142	Si1	Si2	Si3	53.2
2.142	Si1	Si2	Ni4	63.4
2.723	Si2	Si3	Ni4	56.6
2.406	Si2	Ni4	Si5	72.2
2.291	Si3	Ni4	Si6	68.2

COMMON NAME : NiSi2(100)-(1x1) ILLUSTRATION: 147
CLASSIFICATION : 14.28.9
TECHNIQUE : LEIS
AUTHORS : J.H. Huang and R.S. Williams
REFERENCE : Surf. Sci., 186, 115 (1987)

SURFACE TYPE
Substrate : NiSi2
Crystal face: 100
Temperature : RT*
Bulk lattice: fluorite
2D bulk symm: p4m
2D surf symm: p4m
Adsorbate:
Coverage :
Pattern : (1x1)
Matrix : (1.000, 0.000)
(0.000, 1.000)

STRUCTURE TYPE
Unreconstructed bulk termination at pure Si layer, with contraction of top interlayer spacing

SAMPLE PREPARATION (1 sample)
Treatment : 18ML Ni deposited on Si(100)-(2x1), and annealed at ≈1073 K
Crystallinity: strong 1x1 LEED pattern weak extra spots
Anal. methods: LEED I-V curves match those of Wu et al
Contamination:

COMMENTS
In addition to 30% vacancies in top layer, also evidence for some buckling or disorder in topmost Ni layer

DATA COLLECTION
Technique: LEIS; Li+ yield from Ni elastic peak area
Dataset : Li+ yield detected along [100] and [110] azimuths

THEORY/DATA TREATMENT
Shadowing and blocking compared with computer simulations: Moliere pot; vibr amplitudes: Ni 0.1Å, Si 0.11Å

STRUCTURES EXAMINED
Si terminated bulk NiSi2 with relaxed top layer with 30% vacancies; also model of Wu et al (class. no. 14.28.6)

2D UNIT CELLS (1 domain observed)

Cell	Ax (Å)	Ay (Å)	Bx (Å)	By (Å)	α (°)	Matrix	Pattern	Cell type
Bulk	3.840	0.000	0.000	3.840	90.0	(1.000, 0.000) (0.000, 1.000)	(1x1)	b: bulk lattice
Surface 1	3.840	0.000	0.000	3.840	90.0	(1.000, 0.000) (0.000, 1.000)	(1x1)	s1: commens. superlattice

3D COORDINATES

Si1-Si2 and Ni3: 2 layers with contracted spacing; Si4-Si5 and Ni6: periodically repeating pair of bulk layers

Dx/Dy in Å, or as a fraction of layer's unit cell vectors; atom 0 at the origin. Epir/subr are bulk repeat vectors.

No. of atoms: 6 Bulk z = 1.350 Å

Region	Chem el.	At. no.	Cell type	Site occ.	Rel. to	Dx ± εx		Dy ± εy		Dz ± εz		Dz/Bz(%) ± εz/Bz
epir		-2					f		f		Å	
subr		-1				-1.920	Å	1.920	Å	2.700	Å	
intf	Si	1	b	1.00	0	0.000	f	0.000	f	0.000	Å	0.0
intf	Si	2	b	1.00	1	0.500	f	0.500	f	0.000	Å	0.0
intf	Ni	3	b	1.00	2	0.000	f	-0.500	f	1.050 ± .050	Å	77.8 ± 3.7
subl	Si	4	b	1.00	3	-0.500	f	0.000	f	1.350	Å	100.0
subl	Si	5	b	1.00	4	0.500	f	0.500	f	0.000	Å	0.0
subl	Ni	6	b	1.00	5	-0.500	f	0.000	f	1.350	Å	100.0

BOND DISTANCES AND ANGLES

Bond distances and angles are derived from coordinates

No. of distances/angles: 4

Interatomic dist. A-B (Å)	Atom A	Atom B	Atom C	Bond angle A-B-C (°)
2.715	Si1	Si2	Ni3	51.7
2.188	Si1	Ni3	Si1(1,0)	122.7
2.188	Si1	Ni3	Si4	63.8
2.347	Ni3	Si4	Ni6	109.3

COMMON NAME : NiSi2(111)-(1x1) ILLUSTRATION: 145
CLASSIFICATION : 14.28.1
TECHNIQUE : LEED
AUTHORS : W.S. Yang, F. Jona and P.M. Marcus
REFERENCE : Phys. Rev., B28, 7377 (1983)

SURFACE TYPE
Substrate : NiSi2
Crystal face: 111
Temperature : 300 K
Bulk lattice: fluorite
2D bulk symm: p3m1
2D surf symm: p3m1

Adsorbate:
Coverage :
Pattern : (1x1)
Matrix : (1.000, 0.000)
(0.000, 1.000)

STRUCTURE TYPE
Bulk termination in wide gap between SiNiSi sandwiches, exposing Si, with contracted top Si-Ni spacing

SAMPLE PREPARATION (1 sample)
Treatment : 1-2ML Ni evaporated on 10Ω n-type (7x7), then annealed
Crystallinity:
Anal. methods:
Contamination:

COMMENTS
Two phases observed 1) high coverage (hc) > 6 monolayers, 2) low coverage (lc) < 6 monolayers; LEED analysis on hc phase only: for this phase angular dependence of AES suggests Si termination

DATA COLLECTION
Technique: LEED
Dataset : 4 I-V spectra for 2 beams at normal incidence, 20<E<180 eV

THEORY/DATA TREATMENT
Dynamical LEED for semi-infinite NiSi2 substrate

STRUCTURES EXAMINED
Ni and Si terminations; first interlayer spacing varied

QUALITY OF EXPERIMENT-THEORY FIT
Visual

2D UNIT CELLS (1 domain observed)

Cell	Ax (Å)	Ay (Å)	Bx (Å)	By (Å)	α (°)	Matrix	Pattern	Cell type
Bulk	3.840	0.000	-1.920	3.326	120.0	(1.000, 0.000) (0.000, 1.000)	(1x1)	b: bulk lattice
Surface 1	3.840	0.000	-1.920	3.326	120.0	(1.000, 0.000) (0.000, 1.000)	(1x1)	s1: commens. superlattice

3D COORDINATES

Si1-Ni2-Si3: topmost sandwich; Si4-Ni5-Si6: periodically repeating bulk sandwich; 1.57Å is bulk Si-Si intersandwich spacing; 0.1Å error bars assumed for tabulation

Dx/Dy in Å, or as a fraction of layer's unit cell vectors; atom 0 at the origin. Epir/subr are bulk repeat vectors.

No. of atoms: 6 Bulk z = 1.570 Å

Region	Chem el.	At. no.	Cell type	Site occ.	Rel. to	Dx ± εx		Dy ± εy		Dz ± εz		Dz/Bz(%) ± εz/Bz
epir		-2					f		f		Å	
subr		-1				-1.920	Å	-1.109	Å	3.130	Å	
intf	Si	1	b	1.00	0	0.000	f	0.000	f	0.000	Å	0.0
intf	Ni	2	b	1.00	1	0.333	f	0.667	f	0.585 ± .100	Å	37.4 ± 6.4
intf	Si	3	b	1.00	2	0.333	f	-0.333	f	0.780	Å	49.8
subl	Si	4	b	1.00	3	-0.333	f	0.333	f	1.570	Å	100.3
subl	Ni	5	b	1.00	4	0.333	f	-0.333	f	0.780	Å	49.8
subl	Si	6	b	1.00	5	-0.667	f	-0.333	f	0.780	Å	49.8

BOND DISTANCES AND ANGLES

Bond distances and angles are derived from coordinates

No. of distances/angles: 8

Interatomic dist. A-B (Å)	Atom A	Atom B	Atom C	Bond angle A-B-C (°)
2.293	Si1	Ni2	Si1(0,1)	113.7
2.293	Si1	Ni2	Si3	68.2
2.293	Si1	Ni2	Si4	104.8
2.350	Ni2	Si3	Si4	54.7

NiSi2(111)-(1x1)
14.28.1

Bond Distances and Angles - Continued

Interatomic dist. A-B (Å)	Atom A	Atom B	Atom C	Bond angle A-B-C (°)
2.350	Ni2	Si3	Ni5	109.4
2.350	Ni2	Si4	Si6	125.1
2.350	Si3	Ni5		
2.350	Si4	Ni5	Si6	70.4

COMMON NAME : NiSi2(111)-(1x1)
CLASSIFICATION : 14.28.13
TECHNIQUE : MEIS
AUTHORS : J. Vrijmoeth, P.M. Zagwijn, J.W.M. Frenken and J.F. van der Veen
REFERENCE : Phys. Rev. Lett., 67, 1134 (1991)

ILLUSTRATION: 145

SURFACE TYPE
Substrate : NiSi2
Crystal face: 111
Temperature : RT*
Bulk lattice: fluorite
2D bulk symm: p3m1
2D surf symm: p3m1

Adsorbate:
Coverage :
Pattern : (1x1)
Matrix : (1.000, 0.000)
(0.000, 1.000)

STRUCTURE TYPE
Si terminated bulk

SAMPLE PREPARATION (2 sample)
Treatment :
Crystallinity:
Anal. methods:
Contamination:

COMMENTS
Epitaxial layer grown on Si(111);
A- and B-type growth give same results;
high energy resolution allows separation of the contributions from successive layers

DATA COLLECTION
Technique: MEIS; 100keV protons stable to 10eV
Dataset : incidence in (1-10) scattering plane, 22.0° wrt (111) plane blocking patterns for exit angles between 20° and 80°

THEORY/DATA TREATMENT
High resolution RBS compared with Monte Carlo simulations

STRUCTURES EXAMINED
1)bulk Si-terminated surface with vertical relaxation of two topmost layers; data exclude Si bilayers on top, as in CoSi2 (see 14.27.11), as well as Ni termination

2D UNIT CELLS (1 domain observed)

Cell	Ax (Å)	Ay (Å)	Bx (Å)	By (Å)	α (°)	Matrix	Pattern	Cell type
Bulk	3.840	0.000	-1.920	3.326	120.0	(1.000, 0.000) (0.000, 1.000)	(1x1)	b: bulk lattice
Surface 1	3.840	0.000	-1.920	3.326	120.0	(1.000, 0.000) (0.000, 1.000)	(1x1)	s1: commens. superlattice

3D COORDINATES

Si1-Ni2-Si3: topmost sandwich; Si1 is relaxed inward; Si4-Ni5-Si6: periodically repeating bulk sandwich; 1.57Å is bulk Si-Si intersandwich spacing;

Dx/Dy in Å, or as a fraction of layer's unit cell vectors; atom 0 at the origin. Epir/subr are bulk repeat vectors.

No. of atoms: 6 Bulk z = 1.570 Å

Region	Chem el.	At. no.	Cell type	Site occ.	Rel. to	Dx ± εx		Dy ± εy		Dz ± εz		Dz/Bz(%) ± εz/Bz
epir		-2					f		f		Å	
subr		-1				-1.920	Å	-1.109	Å	3.130	Å	
intf	Si	1	b	1.00	0	0.000	f	0.000	f	0.000 ± .020	Å	0.0 ± 7.6
intf	Ni	2	b	1.00	1	0.333	f	0.667	f	0.780	Å	49.8
intf	Si	3	b	1.00	2	0.333	f	-0.333	f	0.780	Å	49.8
subl	Si	4	b	1.00	3	-0.333	f	0.333	f	1.520 ± .020	Å	96.8 ± 1.3
subl	Ni	5	b	1.00	4	0.333	f	-0.333	f	0.780	Å	49.8
subl	Si	6	b	1.00	5	-0.667	f	-0.333	f	0.780	Å	49.8

BOND DISTANCES AND ANGLES

Bond distances and angles are derived from coordinates

No. of distances/angles: 8

Interatomic dist. A-B (Å)	Atom A	Atom B	Atom C	Bond angle A-B-C (°)
2.350	Si1	Ni2	Si1(1,1)	109.6
2.350	Si1	Ni2	Si3(0,1)	180.0
2.350	Si1	Ni2	Si3	70.4
2.350	Ni2	Si3(0,1)	Si1(1,1)	54.8

NiSi2(111)-(1x1)
14.28.13

Bond Distances and Angles - Continued

Interatomic dist. A-B (Å)	Atom A	Atom B	Atom C	Bond angle A-B-C (°)
2.350	Ni2	Si3(0,1)	Ni2(1,1)	109.6
2.350	Ni2	Si3(0,1)	Si4(1,1)	125.2
2.688	Si3	Si4(1,0)	Ni2(1,0)	55.6
2.688	Si3	Si4(1,0)	Si3(1,1)	91.2

ILLUSTRATION: 2

COMMON NAME : Pb(100)-(1x1)
CLASSIFICATION : 82.13
TECHNIQUE : LEED
AUTHORS : R.F. Lin, Y.S.Li, F. Jona and P.M. Marcus
REFERENCE : Phys. Rev., B42, 1150 (1990)

SURFACE TYPE
Substrate : Pb
Crystal face: 100
Temperature : 133 K
Bulk lattice: fcc
2D bulk symm: p4m
2D surf symm: p4m

Adsorbate:
Coverage :
Pattern : (1x1)
Matrix : (1.000, 0.000)
(0.000, 1.000)

STRUCTURE TYPE
Multilayer relaxation down to 4th interlayer spacing

SAMPLE PREPARATION (1 sample)
Treatment : cycles of 1h Ar bombardments and 0.5h 513 K anneals
Crystallinity: sharp, low-background (1x1) pattern
Anal. methods: AES
Contamination: AES: no S, O, C

COMMENTS

DATA COLLECTION
Technique: LEED; video camera
Dataset : IV curves at 2 inc. angles: 5 nondegenerate beams at θ=0° and 12 nondegenerate beams at θ=10°, φ=0°

THEORY/DATA TREATMENT
Dynamical LEED (layer doubling): 10 relat. ph shs by Moritz; Vor=-6±2 eV (fit), Voi=-4eV; ΘD=90 K(bulk), 64K(surf)

STRUCTURES EXAMINED
Relaxation of first 4 interlayer spacings

QUALITY OF EXPERIMENT-THEORY FIT
RZJ=0.12, RPE=0.13, RVHT=0.225

2D UNIT CELLS (1 domain observed)

Cell	Ax (Å)	Ay (Å)	Bx (Å)	By (Å)	α (°)	Matrix	Pattern	Cell type
Bulk	3.484	0.000	0.000	3.484	90.0	(1.000, 0.000) (0.000, 1.000)	(1x1)	b: bulk lattice
Surface 1	3.484	0.000	0.000	3.484	90.0	(1.000, 0.000) (0.000, 1.000)	(1x1)	s1: commens. superlattice

3D COORDINATES

Dx/Dy in Å, or as a fraction of layer's unit cell vectors; atom 0 at the origin. Epir/subr are bulk repeat vectors.

No. of atoms: 6

Bulk z = 2.463 Å

Region	Chem el.	At. no.	Cell type	Site occ.	Rel. to	Dx ± εx		Dy ± εy		Dz ± εz		Dz/Bz(%) ± εz/Bz
epir		-2					f		f		Å	
subr		-1				1.742	Å	1.742	Å	2.463	Å	
intf	Pb	1	b	1.00	0	0.000	f	0.000	f	0.000	Å	0.0
intf	Pb	2	b	1.00	1	0.500	f	0.500	f	2.266 ± .030	Å	92.0 ± 1.2
intf	Pb	3	b	1.00	2	-0.500	f	-0.500	f	2.540 ± .030	Å	103.1 ± 1.2
intf	Pb	4	b	1.00	3	0.500	f	0.500	f	2.389 ± .030	Å	97.0 ± 1.2
intf	Pb	5	b	1.00	4	-0.500	f	-0.500	f	2.414 ± .099	Å	98.0 ± 4.0
subl	Pb	6	b	1.00	5	0.500	f	0.500	f	2.463	Å	100.0

BOND DISTANCES AND ANGLES

Bond distances and angles are derived from coordinates

No. of distances/angles: 3

Interatomic dist. A-B (Å)	Atom A	Atom B	Atom C	Bond angle A-B-C (°)
3.484	Pb1	Pb1(1,0)	Pb2	58.6
3.347	Pb1	Pb2	Pb3	88.5
3.538	Pb2	Pb3	Pb4	90.0

COMMON NAME : Pb(110)-(1x1)
CLASSIFICATION : 82.1
TECHNIQUE : MEIS
AUTHORS : J.W.M. Frenken, J.F. van der Veen, R.N. Barnett, U. Landman and C.L. Cleveland
REFERENCE : Surf. Sci., 172, 319 (1986)

ILLUSTRATION: 4

SURFACE TYPE
Substrate : Pb
Crystal face: 110
Temperature : RT
Bulk lattice: fcc
2D bulk symm: pmm
2D surf symm: pmm

Adsorbate:
Coverage :
Pattern : (1x1)
Matrix : (1.000, 0.000)
(0.000, 1.000)

STRUCTURE TYPE
Bulk termination with top spacing contraction

SAMPLE PREPARATION (1 sample)
Treatment : Ar ion bombardment, annealed 1 hour at 590 K
Crystallinity: sharp (1x1) LEED pattern
Anal. methods:
Contamination: no impurities detected by AES and RBS

COMMENTS
In PRL 58, 401 (1987), the authors present further analyses at 29 K and 485K; the relaxation of the first interlayer spacing varies: -15.4±2.5% at 29 K, -15.8±2.5% at RT and -3±5% at 485 K, indicating a decreasing contraction with increasing temperature

DATA COLLECTION
Technique: MEIS; 50.6 and 97.6keV proton beams used
Dataset : Rutherford back scattering: blocking and channeling curves measured in (1,-1,1) scattering plane

THEORY/DATA TREATMENT
Monte Carlo simulation of channeling and blocking data: vib. amps. 0.27, 0.31Å in layers 1, 2

STRUCTURES EXAMINED
Top 3 interlayer spacings varied;
d23 and d34 satisfy: d23 + 0.75*d34 = (0.5±2.5)%

2D UNIT CELLS (1 domain observed)

Cell	Ax (Å)	Ay (Å)	Bx (Å)	By (Å)	α (°)	Matrix	Pattern	Cell type
Bulk	4.950	0.000	0.000	3.500	90.0	(1.000, 0.000) (0.000, 1.000)	(1x1)	b: bulk lattice
Surface 1	4.950	0.000	0.000	3.500	90.0	(1.000, 0.000) (0.000, 1.000)	(1x1)	s1: commens. superlattice

3D COORDINATES

Dx/Dy in Å, or as a fraction of layer's unit cell vectors; atom 0 at the origin. Epir/subr are bulk repeat vectors.

No. of atoms: 3

Bulk z = 1.750 Å

Region	Chem el.	At. no.	Cell type	Site occ.	Rel. to	Dx ± εx		Dy ± εy		Dz ± εz		Dz/Bz(%) ± εz/Bz
epir		-2					f		f		Å	
subr		-1				2.475	Å	1.750	Å	1.750	Å	
intf	Pb	1	b	1.00	0	0.000	f	0.000	f	0.000	Å	0.0
intf	Pb	2	b	1.00	1	0.500	f	0.500	f	1.474 ± .040	Å	84.2 ± 2.3
subl	Pb	3	b	1.00	2	-0.500	f	-0.500	f	1.750	Å	100.0

BOND DISTANCES AND ANGLES

Bond distances and angles are derived from coordinates

No. of distances/angles: 3

Interatomic dist. A-B (Å)	Atom A	Atom B	Atom C	Bond angle A-B-C (°)
3.500	Pb1	Pb1(0,1)		
3.371	Pb1	Pb2	Pb3	55.9
3.500	Pb2	Pb3		

ILLUSTRATION: 4

COMMON NAME : Pb(110)-(1x1)
CLASSIFICATION : 82.12
TECHNIQUE : LEED
AUTHORS : U. Breuer, K.C. Prince, H.P. Bonzel, W.Oed, K. Heinz, G. Schmidt and K. Mueller
REFERENCE : Surf. Sci., 239, L493 (1990)

SURFACE TYPE
Substrate : Pb
Crystal face: 110
Temperature : 300 K
Bulk lattice: fcc
2D bulk symm: pmm
2D surf symm: pmm

Adsorbate:
Coverage :
Pattern : (1x1)
Matrix : (1.000, 0.000)
(0.000, 1.000)

STRUCTURE TYPE
Multilayer relaxation down to 3rd interlayer spacing

SAMPLE PREPARATION (1 sample)
Treatment : cycles of Ar sputtering and annealing
Crystallinity: modestly sharp diffraction spots
Anal. methods: AES
Contamination: AES: clean

COMMENTS
Same analysis performed at 130 K gave same results within error bars, e.g. top layer spacing changes of -17.1±2.9%, +3.4±2.9%, -6.8±4.6% (RPE=0.38, cumul. E range 1700 eV for 20 beams);
R-factors are 'not satisfactory'

DATA COLLECTION
Technique: LEED; video camera
Dataset : IV curves for 10 beams up to 150 eV; incidence angle fit to Θ=0.5°, φ=10° in struct. analysis; cumul. E range 750 eV

THEORY/DATA TREATMENT
Dynamical LEED: 11 relativistic phase shifts; Voi∝E**1/3; ΘD=105 K in bulk (fit in surface layers)

STRUCTURES EXAMINED
Relaxation of first 3 interlayer spacings

QUALITY OF EXPERIMENT-THEORY FIT
RPE=0.26

2D UNIT CELLS (1 domain observed)

Cell	Ax (Å)	Ay (Å)	Bx (Å)	By (Å)	α (°)	Matrix	Pattern	Cell type
Bulk	3.500	0.000	0.000	4.950	90.0	(1.000, 0.000) (0.000, 1.000)	(1x1)	b: bulk lattice
Surface 1	3.500	0.000	0.000	4.950	90.0	(1.000, 0.000) (0.000, 1.000)	(1x1)	s1: commens. superlattice

3D COORDINATES

Dx/Dy in Å, or as a fraction of layer's unit cell vectors; atom 0 at the origin. Epir/subr are bulk repeat vectors.

No. of atoms: 5 Bulk z = 1.750 Å

Region	Chem el.	At. no.	Cell type	Site occ.	Rel. to	Dx ± εx		Dy ± εy		Dz ± εz		Dz/Bz(%) ± εz/Bz
epir		-2					f		f		Å	
subr		-1				1.750	Å	2.475	Å	1.750	Å	
intf	Pb	1	b	1.00	0	0.000	f	0.000	f	0.000	Å	0.0
intf	Pb	2	b	1.00	1	0.500	f	0.500	f	1.410 ± .050	Å	80.6 ± 2.9
intf	Pb	3	b	1.00	2	-0.500	f	-0.500	f	1.830 ± .080	Å	104.6 ± 4.6
intf	Pb	4	b	1.00	3	0.500	f	0.500	f	1.630 ± .050	Å	93.1 ± 2.9
subl	Pb	5	b	1.00	4	-0.500	f	-0.500	f	1.750	Å	100.0

BOND DISTANCES AND ANGLES

Bond distances and angles are derived from coordinates

No. of distances/angles: 5

Interatomic dist. A-B (Å)	Atom A	Atom B	Atom C	Bond angle A-B-C (°)
3.500	Pb1	Pb1(1,0)	Pb2	58.4
3.343	Pb1	Pb2	Pb3	56.1
3.343	Pb1	Pb2	Pb4	115.0

Pb(110)-(1x1)
82.12

Bond Distances and Angles - Continued

Interatomic dist. A-B (Å)	Atom A	Atom B	Atom C	Bond angle A-B-C (°)
3.240	Pb1	Pb3	Pb4	118.3
3.541	Pb2	Pb3	Pb4	59.4

ILLUSTRATION: 4

COMMON NAME : Pb(110)-(1x1)
CLASSIFICATION : 82.7
TECHNIQUE : LEED
AUTHORS : Y.S. Li, J. Quinn, F. Jona and P.M. Marcus
REFERENCE : Phys. Rev., B40, 8239 (1989)

SURFACE TYPE
Substrate : Pb
Crystal face: 110
Temperature : 133 K
Bulk lattice: fcc
2D bulk symm: pmm
2D surf symm: pmm

Adsorbate:
Coverage :
Pattern : (1x1)
Matrix : (1.000, 0.000)
(0.000, 1.000)

STRUCTURE TYPE
Multilayer relaxation down to 3rd interlayer spacing

SAMPLE PREPARATION (1 sample)
Treatment : cycles of 1h Ar bombardments and 0.5h 513 K anneals
Crystallinity: sharp, low-background (1x1) pattern
Anal. methods: AES
Contamination: AES: no S, O, C

COMMENTS
R-factor values indicate 'at best fair' agreement between theory and experiment

DATA COLLECTION
Technique: LEED; video camera
Dataset : IV curves at 2 inc. angles: 10 nondegenerate beams at Θ=0° and 15 nondegenerate beams at Θ=15°, φ=0°

THEORY/DATA TREATMENT
Dynamical LEED (CHANGE program): 10 ph shs from several pots
Vor=-6±2 eV (fit), Voi=-4eV, rms vibr ampl=0.14Å (fit)

STRUCTURES EXAMINED
Relaxation of first 3 interlayer spacings

QUALITY OF EXPERIMENT-THEORY FIT
RZJ=0.27(Θ=0°), 0.34(Θ=15°)

2D UNIT CELLS (1 domain observed)

Cell	Ax (Å)	Ay (Å)	Bx (Å)	By (Å)	α (°)	Matrix	Pattern	Cell type
Bulk	3.480	0.000	0.000	4.922	90.0	(1.000, 0.000) (0.000, 1.000)	(1x1)	b: bulk lattice
Surface 1	3.480	0.000	0.000	4.922	90.0	(1.000, 0.000) (0.000, 1.000)	(1x1)	s1: commens. superlattice

3D COORDINATES

Dx/Dy in Å, or as a fraction of layer's unit cell vectors; atom 0 at the origin. Epir/subr are bulk repeat vectors.

No. of atoms: 5

Bulk z = 1.740 Å

Region	Chem el.	At. no.	Cell type	Site occ.	Rel. to	Dx ± εx	Dy ± εy	Dz ± εz	Dz/Bz(%) ± εz/Bz
epir		-2				f	f	Å	
subr		-1				1.740 Å	2.461 Å	1.740 Å	
intf	Pb	1	b	1.00	0	0.000 f	0.000 f	0.000 Å	0.0
intf	Pb	2	b	1.00	1	0.500 f	0.500 f	1.455 ± .030 Å	83.6 ± 1.7
intf	Pb	3	b	1.00	2	-0.500 f	-0.500 f	1.800 ± .100 Å	103.5 ± 5.8
intf	Pb	4	b	1.00	3	0.500 f	0.500 f	1.664 ± .030 Å	95.6 ± 1.7
subl	Pb	5	b	1.00	4	-0.500 f	-0.500 f	1.740 Å	100.0

BOND DISTANCES AND ANGLES

Bond distances and angles are derived from coordinates

No. of distances/angles: 5

Interatomic dist. A-B (Å)	Atom A	Atom B	Atom C	Bond angle A-B-C (°)
3.480	Pb1	Pb1(1,0)	Pb2	58.7
3.347	Pb1	Pb2	Pb3	56.6
3.347	Pb1	Pb2	Pb4	115.8
3.255	Pb1	Pb3	Pb4	118.9
3.510	Pb2	Pb3	Pb4	59.8

COMMON NAME : Pb(111)-(1x1)
CLASSIFICATION : 82.16
TECHNIQUE : LEED
AUTHORS : Y.S. Li, F. Jona and P.M. Marcus
REFERENCE : Phys. Rev., B43, 6337 (1991)

ILLUSTRATION: 1

SURFACE TYPE
Substrate : Pb
Crystal face: 111
Temperature : 133 K
Bulk lattice: fcc
2D bulk symm: p3m1
2D surf symm: p3m1

Adsorbate:
Coverage :
Pattern : (1x1)
Matrix : (1.000, 0.000)
(0.000, 1.000)

STRUCTURE TYPE
Multilayer relaxation down to 3rd interlayer spacing

SAMPLE PREPARATION (1 sample)
Treatment : cycles of 1h Ar bombardments and 0.5h 513 K anneals
Crystallinity: sharp, low-background (1x1) pattern
Anal. methods: AES
Contamination: AES: no S, O, C

COMMENTS

DATA COLLECTION
Technique: LEED; video camera
Dataset : IV curves at 2 inc. angles: 5 nondeg. beams at Θ=0° and 10 nondeg. beams at Θ=7.5°, φ=0°; cumul. E range 1110 eV

THEORY/DATA TREATMENT
Dynamical LEED (layer doubling): 10 relat. ph shs by Moritz; Vor=-4±2 eV (fit), Voi=-4.5eV; ΘD=90 K(bulk), 64K(surf)

STRUCTURES EXAMINED
Relaxation of first 3 interlayer spacings

QUALITY OF EXPERIMENT-THEORY FIT
RZJ=0.19, RVHT=0.23

2D UNIT CELLS (1 domain observed)

Cell	Ax (Å)	Ay (Å)	Bx (Å)	By (Å)	α (°)	Matrix	Pattern	Cell type
Bulk	3.484	0.000	1.742	3.017	60.0	(1.000, 0.000) (0.000, 1.000)	(1x1)	b: bulk lattice
Surface 1	3.484	0.000	1.742	3.017	60.0	(1.000, 0.000) (0.000, 1.000)	(1x1)	s1: commens. superlattice

3D COORDINATES

Dx/Dy in Å, or as a fraction of layer's unit cell vectors; atom 0 at the origin. Epir/subr are bulk repeat vectors.

No. of atoms: 5

Bulk z = 2.844 Å

Reg ion	Chem el.	At. no.	Cell type	Site occ.	Rel. to	Dx ± εx		Dy ± εy		Dz ± εz		Dz/Bz(%) ± εz/Bz
epir		-2					f		f		Å	
subr		-1				1.742	Å	1.006	Å	2.844	Å	
intf	Pb	1	b	1.00	0	0.000	f	0.000	f	0.000	Å	0.0
intf	Pb	2	b	1.00	1	0.333	f	0.333	f	2.744 ± .040	Å	96.5 ± 1.1
intf	Pb	3	b	1.00	2	0.333	f	0.333	f	2.859 ± .040	Å	100.5 ± 1.4
intf	Pb	4	b	1.00	3	-0.667	f	-0.667	f	2.889 ± .050	Å	101.6 ± 1.8
subl	Pb	5	b	1.00	4	0.333	f	0.333	f	2.844	Å	100.0

BOND DISTANCES AND ANGLES

Bond distances and angles are derived from coordinates

No. of distances/angles: 3

Interatomic dist. A-B (Å)	Atom A	Atom B	Atom C	Bond angle A-B-C (°)
3.484	Pb1	Pb1(1,0)	Pb2	59.2
3.402	Pb1	Pb2	Pb3	180.0
3.496	Pb2	Pb3	Pb4(1,0)	120.5

COMMON NAME : Pb(311)-(1x1)
CLASSIFICATION : 82.18
TECHNIQUE : LEED
AUTHORS : Y.S. Li, F. Jona and P.M.Marcus
REFERENCE : Phys. Rev., B44, 8267 (1991)

ILLUSTRATION: 8

SURFACE TYPE
Substrate : Pb
Crystal face: 311
Temperature : 130 K
Bulk lattice: fcc
2D bulk symm: cm
2D surf symm: cm

Adsorbate:
Coverage :
Pattern : (1x1)
Matrix : (1.000, 0.000)
(0.000, 1.000)

STRUCTURE TYPE
Multilayer relxations perpendicular to the surface by -14.1, -2.0, -0.07, 5.4%, and parallel to surface by 3.0 and -1.9%

SAMPLE PREPARATION (1 sample)
Treatment : cleaned by Ar+ sputtering in 5E-5 torr Ar 375V/513 K
Crystallinity: surface < 0.5° from the (311) plane
Anal. methods:
Contamination: AES: no S or O

COMMENTS

DATA COLLECTION
Technique: LEED
Dataset : I-V curves for 14 non-symmetrical beams; E range 20-180 eV

THEORY/DATA TREATMENT
Dynamical LEED (layer doubling): 13 relat. phase shifts, Vor=-2 eV, Voi=-4.5eV

STRUCTURES EXAMINED
Varied the top 4 interlayer spacings and 2 registries

QUALITY OF EXPERIMENT-THEORY FIT
RPE=0.41, RVHT=0.21, RZJ=0.21

2D UNIT CELLS (1 domain observed)

Cell	Ax (Å)	Ay (Å)	Bx (Å)	By (Å)	α (°)	Matrix	Pattern	Cell type
Bulk	3.484	0.000	-1.742	5.804	106.7	(1.000, 0.000) (0.000, 1.000)	(1x1)	b: bulk lattice
Surface 1	3.484	0.000	-1.742	5.804	106.7	(1.000, 0.000) (0.000, 1.000)	(1x1)	s1: commens. superlattice

3D COORDINATES

Dx/Dy in Å, or as a fraction of layer's unit cell vectors; atom 0 at the origin. Epir/subr are bulk repeat vectors.

No. of atoms: 6 Bulk z = 1.485 Å

Region	Chem el.	At. no.	Cell type	Site occ.	Rel. to	Dx ± εx		Dy ± εy		Dz ± εz		Dz/Bz(%) ± εz/Bz
epir		-2					f		f		Å	
subr		-1				1.742	Å	2.626	Å	1.485	Å	
intf	Pb	1	b	1.00	0	0.000	Å	0.000	Å	0.000	Å	0.0
intf	Pb	2	b	1.00	1	1.742	Å	2.706 ± .050	Å	1.275 ± .030	Å	86.0 ± 2.0
intf	Pb	3	b	1.00	2	1.742	Å	2.576 ± .050	Å	1.455 ± .030	Å	98.0 ± 2.0
intf	Pb	4	b	1.00	3	1.742	Å	2.626	Å	1.475 ± .050	Å	99.3 ± 3.4
intf	Pb	5	b	1.00	4	1.742	Å	2.626	Å	1.565 ± .060	Å	105.4 ± 4.0
subl	Pb	6	b	1.00	5	1.742	Å	2.626	Å	1.485	Å	100.0

BOND DISTANCES AND ANGLES

Bond distances and angles are derived from coordinates

No. of distances/angles: 6

Interatomic dist. A-B (Å)	Atom A	Atom B	Atom C	Bond angle A-B-C (°)
3.484	Pb1	Pb1(1,0)		
3.462	Pb1	Pb2		
3.281	Pb1	Pb3		
3.433	Pb2	Pb3		

Pb(311)-(1x1)
82.18

Bond Distances and Angles - Continued

Interatomic dist. A-B (Å)	Atom A	Atom B	Atom C	Bond angle A-B-C (°)
3.462	Pb2	Pb4		
3.479	Pb3	Pb4		

ILLUSTRATION: 2

COMMON NAME : Pd(100)-(1x1)
CLASSIFICATION : 46.2
TECHNIQUE : LEED
AUTHORS : R.J. Behm, K. Christmann, G. Ertl and M.A. Van Hove
REFERENCE : J. Chem. Phys., 73, 2984 (1980)

SURFACE TYPE
Substrate : Pd
Crystal face: 100
Temperature : 350 K
Bulk lattice: fcc
2D bulk symm: p4m
2D surf symm: p4m

Adsorbate:
Coverage :
Pattern : (1x1)
Matrix : (1.000, 0.000)
(0.000, 1.000)

STRUCTURE TYPE
Essentially unrelaxed bulk termination

SAMPLE PREPARATION (1 sample)
Treatment : repeatedly annealed at 1000 K and Ar+ sputtered
Crystallinity:
Anal. methods:
Contamination: AES, TDS, WFC: no impurities

COMMENTS

DATA COLLECTION
Technique: LEED
Dataset : I-V curves: 00 beam at Θ=0,φ=0°; 10, -1-1, 0-2, 1-2, 2-2 beams at Θ=8.5,φ=12°; cumulative energy range: 980 eV

THEORY/DATA TREATMENT
Dynamical LEED (RFS): Moruzzi et al potential, 8 phase shifts; Vor=-10.0 eV, Voi∝E**1/3; ΘD= 193 K

STRUCTURES EXAMINED
Various top layer spacings

QUALITY OF EXPERIMENT-THEORY FIT
Visual

2D UNIT CELLS (1 domain observed)

Cell	Ax (Å)	Ay (Å)	Bx (Å)	By (Å)	α (°)	Matrix	Pattern	Cell type
Bulk	2.740	0.000	0.000	2.740	90.0	(1.000, 0.000) (0.000, 1.000)	(1x1)	b: bulk lattice
Surface 1	2.740	0.000	0.000	2.740	90.0	(1.000, 0.000) (0.000, 1.000)	(1x1)	s1: commens. superlattice

3D COORDINATES

Dx/Dy in Å, or as a fraction of layer's unit cell vectors; atom 0 at the origin. Epir/subr are bulk repeat vectors.

No. of atoms: 3

Bulk z = 1.945 Å

Region	Chem el.	At. no.	Cell type	Site occ.	Rel. to	Dx ± εx		Dy ± εy		Dz ± εz		Dz/Bz(%) ± εz/Bz
epir		-2					f		f		Å	
subr		-1				-1.370	Å	-1.370	Å	1.945	Å	
intf	Pd	1	b	1.00	0	0.000	f	0.000	f	0.000	Å	0.0
intf	Pd	2	b	1.00	1	0.500	f	0.500	f	1.950 ± .050	Å	100.3 ± 2.6
subl	Pd	3	b	1.00	2	-0.500	f	-0.500	f	1.945	Å	100.0

BOND DISTANCES AND ANGLES

Bond distances are derived from coordinates

No. of distances/angles: 2

Interatomic dist. A-B (Å)	Atom A	Atom B	Atom C	Bond angle A-B-C (°)
2.740	Pd1	Pd1(1,0)		
2.749	Pd1	Pd2		

COMMON NAME : Pd(100)-(1x1)
CLASSIFICATION : 46.8
TECHNIQUE : LEED
AUTHORS : J. Quinn, Y.S. Li, D. Tian, H. Li, F. Jona and P.M. Marcus
REFERENCE : Phys. Rev., B42, 11348 (1990)

ILLUSTRATION: 2

SURFACE TYPE
Substrate : Pd
Crystal face: 100
Temperature : RT*
Bulk lattice: fcc
2D bulk symm: p4m
2D surf symm: p4m

Adsorbate:
Coverage :
Pattern : (1x1)
Matrix : (1.000, 0.000)
(0.000, 1.000)

STRUCTURE TYPE
Multilayer relaxation

SAMPLE PREPARATION (1 sample)
Treatment : ion bombardment followed by annealing
Crystallinity: sharp LEED pattern
Anal. methods: AES
Contamination:

COMMENTS

DATA COLLECTION
Technique: LEED
Dataset : IV spectra for 5 non-equivalent beams at normal incidence and for 6 non-equivalent beams at 11° off-normal

THEORY/DATA TREATMENT
Dynamical LEED (CHANGE program)

STRUCTURES EXAMINED
Variation of 1st and 2nd interlayer spacings

QUALITY OF EXPERIMENT-THEORY FIT
PRE=0.35

2D UNIT CELLS (1 domain observed)

Cell	Ax (Å)	Ay (Å)	Bx (Å)	By (Å)	α (°)	Matrix	Pattern	Cell type
Bulk	2.751	0.000	0.000	2.751	90.0	(1.000, 0.000) (0.000, 1.000)	(1x1)	b: bulk lattice
Surface 1	2.751	0.000	0.000	2.751	90.0	(1.000, 0.000) (0.000, 1.000)	(1x1)	s1: commens. superlattice

3D COORDINATES

Dx/Dy in Å, or as a fraction of layer's unit cell vectors; atom 0 at the origin. Epir/subr are bulk repeat vectors.

No. of atoms: 3

Bulk z = 1.945 Å

Region	Chem el.	At. no.	Cell type	Site occ.	Rel. to	Dx ± εx		Dy ± εy		Dz ± εz		Dz/Bz(%) ± εz/Bz
epir		-2					f		f		Å	
subr		-1				1.376	Å	1.376	Å	1.945	Å	
ovrl	Pd	1	s1	1.00	0	0.000	Å	0.000	Å	0.000	Å	0.0
ovrl	Pd	2	s1	1.00	0	1.376	Å	1.376	Å	2.005 ± .030	Å	103.1 ± 1.5
subl	Pd	3	b	1.00	0	0.000	Å	0.000	Å	3.931 ± .030	Å	202.1 ± 1.5

BOND DISTANCES AND ANGLES

Bond distances and angles are derived from coordinates

No. of distances/angles: 2

Interatomic dist. A-B (Å)	Atom A	Atom B	Atom C	Bond angle A-B-C (°)
2.794	Pd1	Pd2		
2.737	Pd2	Pd3		

COMMON NAME : Pd(110)-(1x1)
CLASSIFICATION : 46.4a
TECHNIQUE : LEED
AUTHORS : C.J. Barnes, M.Q. Ding, M. Lindroos, R.D. Diehl and D.A. King
REFERENCE : Surf. Sci., 162, 59 (1985)

ILLUSTRATION: 4

SURFACE TYPE
Substrate : Pd
Crystal face: 110
Temperature : 300 K
Bulk lattice: fcc
2D bulk symm: pmm
2D surf symm: pmm

Adsorbate:
Coverage :
Pattern : (1x1)
Matrix : (1.000, 0.000)
(0.000, 1.000)

STRUCTURE TYPE
Bulk termination with multilayer relaxations

SAMPLE PREPARATION (1 sample)
Treatment : see Diehl et al, J. Phys. C18, 4069 (1985)
Crystallinity:
Anal. methods:
Contamination:

COMMENTS

DATA COLLECTION
Technique: LEED
Dataset : I-V spectra for 7 symmetrically inequivalent beams at normal incidence, energy range 20-250 eV

THEORY/DATA TREATMENT
Dynamical LEED (CAVLEED package)

STRUCTURES EXAMINED
Relaxations of top two interlayer spacings

QUALITY OF EXPERIMENT-THEORY FIT
RPE=0.155

2D UNIT CELLS (1 domain observed)

Cell	Ax (Å)	Ay (Å)	Bx (Å)	By (Å)	α (°)	Matrix	Pattern	Cell type
Bulk	2.750	0.000	0.000	3.890	90.0	(1.000, 0.000) (0.000, 1.000)	(1x1)	b: bulk lattice
Surface 1	2.750	0.000	0.000	3.890	90.0	(1.000, 0.000) (0.000, 1.000)	(1x1)	s1: commens. superlattice

3D COORDINATES

Dx/Dy in Å, or as a fraction of layer's unit cell vectors; atom 0 at the origin. Epir/subr are bulk repeat vectors.

No. of atoms: 4

Bulk z = 1.370 Å

Region	Chem el.	At. no.	Cell type	Site occ.	Rel. to	Dx ± εx		Dy ± εy		Dz ± εz		Dz/Bz(%) ± εz/Bz
epir		-2					f		f		Å	
subr		-1				1.375	Å	1.945	Å	1.370	Å	
intf	Pd	1	b	1.00	0	0.000	f	0.000	f	0.000	Å	0.0
intf	Pd	2	b	1.00	1	0.500	f	0.500	f	1.290 ± .030	Å	94.2 ± 2.2
intf	Pd	3	b	1.00	2	-0.500	f	-0.500	f	1.380 ± .030	Å	100.7 ± 2.2
subl	Pd	4	b	1.00	3	0.500	f	0.500	f	1.370	Å	100.0

BOND DISTANCES AND ANGLES

Bond distances and angles are derived from coordinates

No. of distances/angles: 3

Interatomic dist. A-B (Å)	Atom A	Atom B	Atom C	Bond angle A-B-C (°)
2.750	Pd1	Pd1(1,0)	Pd2	59.5
2.709	Pd1	Pd2	Pd3	58.5
2.753	Pd2	Pd3	Pd4	60.0

COMMON NAME : Pd(110)-(1x1)
CLASSIFICATION : 46.5b
TECHNIQUE : LEED
AUTHORS : M. Skottke, R.J. Behm, G. Ertl, V. Penka and W. Moritz
REFERENCE : J. Chem. Phys., 87, 6191 (1987)

ILLUSTRATION: 4

SURFACE TYPE
Substrate : Pd
Crystal face: 110
Temperature : RT*
Bulk lattice: fcc
2D bulk symm: pmm
2D surf symm: pmm

Adsorbate:
Coverage :
Pattern : (1x1)
Matrix : (1.000, 0.000)
(0.000, 1.000)

STRUCTURE TYPE
Relaxations in top two interlayer spacings

SAMPLE PREPARATION (1 sample)
Treatment : cycles of Ar+ bombardment and annealing; C removed by O
Crystallinity:
Anal. methods: cleanliness checked by H2 desorption
Contamination: clean by AES

COMMENTS

DATA COLLECTION
Technique: LEED
Dataset : I-V curves for 7(9) inequiv. beams at normal (off-normal) incidence; E range 40-240 eV

THEORY/DATA TREATMENT
Dynamical LEED (layer doubling): 8 spin-averaged ph shs from relativistic atomic potential; VoiαE**1/3; ΘD=274 K

STRUCTURES EXAMINED
Variation of top 3 interlayer spacings

QUALITY OF EXPERIMENT-THEORY FIT
RPE=0.22, RZJ=0.14

2D UNIT CELLS (1 domain observed)

Cell	Ax (Å)	Ay (Å)	Bx (Å)	By (Å)	α (°)	Matrix	Pattern	Cell type
Bulk	2.750	0.000	0.000	3.890	90.0	(1.000, 0.000) (0.000, 1.000)	(1x1)	b: bulk lattice
Surface 1	2.750	0.000	0.000	3.890	90.0	(1.000, 0.000) (0.000, 1.000)	(1x1)	s1: commens. superlattice

3D COORDINATES

Dx/Dy in Å, or as a fraction of layer's unit cell vectors; atom 0 at the origin. Epir/subr are bulk repeat vectors.

No. of atoms: 4

Bulk z = 1.370 Å

Region	Chem el.	At. no.	Cell type	Site occ.	Rel. to	Dx ± εx		Dy ± εy		Dz ± εz		Dz/Bz(%) ± εz/Bz
epir		-2					f		f		Å	
subr		-1				1.375	Å	1.945	Å	1.370	Å	
intf	Pd	1	b	1.00	0	0.000	f	0.000	f	0.000	Å	0.0
intf	Pd	2	b	1.00	1	0.500	f	0.500	f	1.300 ± .020	Å	94.9 ± 1.5
intf	Pd	3	b	1.00	2	-0.500	f	-0.500	f	1.410 ± .020	Å	102.9 ± 1.5
subl	Pd	4	b	1.00	3	0.500	f	0.500	f	1.370 ± .020	Å	100.0 ± 1.5

BOND DISTANCES AND ANGLES

Bond distances and angles are derived from coordinates

No. of distances/angles: 5

Interatomic dist. A-B (Å)	Atom A	Atom B	Atom C	Bond angle A-B-C (°)
2.750	Pd1	Pd1(1,0)	Pd2	59.6
2.714	Pd1	Pd2	Pd3	59.3
2.710	Pd1	Pd3	Pd4	119.9
2.768	Pd2	Pd3	Pd4	60.5
2.748	Pd3	Pd4		

ILLUSTRATION: 5

COMMON NAME : Pd(110)-(1x2)
CLASSIFICATION : 46.4b
TECHNIQUE : LEED
AUTHORS : C.J. Barnes, M.Q. Ding, M. Lindroos, R.D. Diehl and D.A. King
REFERENCE : Surf. Sci., 162, 59 (1985)

SURFACE TYPE
Substrate : Pd
Crystal face: 110
Temperature : 300 K
Bulk lattice: fcc
2D bulk symm: pmm
2D surf symm: pmm

Adsorbate:
Coverage :
Pattern : (1x2)
Matrix : (1.000, 0.000)
(0.000, 2.000)

STRUCTURE TYPE
Alkali-impurity stabilized missing-row reconstruction

SAMPLE PREPARATION (1 sample)
Treatment : 0.1ML Cs or 0.2ML Na (disordered) induces (1x2) reconstruct.
Crystallinity:
Anal. methods:
Contamination:

COMMENTS
Disordered alkali overlayer ignored in structural analysis; sawtooth model nearly as good as missing-row model, in which second-layer relaxations gave no improvement

DATA COLLECTION
Technique: LEED
Dataset : I-V spectra for 13 symmetrically inequivalent beams at normal incidence; energy range 20-250 eV

THEORY/DATA TREATMENT
Dynamical LEED (CAVLEED package): Voi=-4 eV

STRUCTURES EXAMINED
Missing-row model with row-pairing or buckling in second layer; sawtooth model; paired-row top layer; buckled top layer; latter three were kept bulk-like below first layer

QUALITY OF EXPERIMENT-THEORY FIT
RPE=0.37

2D UNIT CELLS (1 domain observed)

Cell	Ax (Å)	Ay (Å)	Bx (Å)	By (Å)	α (°)	Matrix	Pattern	Cell type
Bulk	2.750	0.000	0.000	3.890	90.0	(1.000, 0.000) (0.000, 1.000)	(1x1)	b: bulk lattice
Surface 1	2.750	0.000	0.000	7.780	90.0	(1.000, 0.000) (0.000, 2.000)	(1x2)	s1: commens. superlattice

3D COORDINATES

Pd1: remaining row of missing-row model

Dx/Dy in Å, or as a fraction of layer's unit cell vectors; atom 0 at the origin. Epir/subr are bulk repeat vectors.

No. of atoms: 3

Bulk z = 1.370 Å

Region	Chem el.	At. no.	Cell type	Site occ.	Rel. to	Dx ± εx		Dy ± εy		Dz ± εz		Dz/Bz(%) ± εz/Bz
epir		-2					f		f		Å	
subr		-1				-1.375	Å	-1.945	Å	1.370	Å	
intf	Pd	1	s1	.50	0	0.000	f	0.000	f	0.000	Å	0.0
intf	Pd	2	b	1.00	1	0.500	f	0.500	f	1.300 ± .030	Å	94.9 ± 2.2
subl	Pd	3	b	1.00	2	-0.500	f	-0.500	f	1.370	Å	100.0

BOND DISTANCES AND ANGLES

Bond distances and angles are derived from coordinates

No. of distances/angles: 4

Interatomic dist. A-B (Å)	Atom A	Atom B	Atom C	Bond angle A-B-C (°)
2.750	Pd1	Pd1(1,0)	Pd2	59.6
2.714	Pd1	Pd2	Pd3	58.5
2.670	Pd1	Pd3		
2.748	Pd2	Pd3		

COMMON NAME : Pd(110)-(1x2) ILLUSTRATION: 5
CLASSIFICATION : 46.6
TECHNIQUE : LEED
AUTHORS : C.J. Barnes, M. Lindroos and D.A. King
REFERENCE : Surf. Sci., 201, 108 (1988)

SURFACE TYPE
Substrate : Pd
Crystal face: 110
Temperature : RT*
Bulk lattice: fcc
2D bulk symm: pmm
2D surf symm: pmm

Adsorbate: Cs
Coverage : <0.09ML
Pattern : (1x2)
Matrix : (1.000, 0.000)
(0.000, 2.000)

STRUCTURE TYPE
Cs-stabilized missing-row reconstruction with multilayer relaxation (row pairing in 2nd Pd layer, buckling in 3rd)

SAMPLE PREPARATION (1 sample)
Treatment : Cs evaporated to saturation, then annealed to 800 K
Crystallinity: sharp (1x2) pattern
Anal. methods:
Contamination: clean by AES and LEED

COMMENTS

DATA COLLECTION
Technique: LEED
Dataset : I-V curves for 6 integral and 4 fractional beams at normal incidence; E range 20-220 eV

THEORY/DATA TREATMENT
Dynamical LEED (CAVLEED program: layer doubling): 9 phase shifts from Moruzzi et al pot (Cs ignored); Voi=-4 eV

STRUCTURES EXAMINED
1) missing-row model; 2) sawtooth model; 3) paired-row model; 4) buckled model;
initially only topmost interlayer spacing varied; then, in missing-row model, also 2nd and 3rd spacings, as well as 2nd layer pairing and 3rd layer buckling varied

QUALITY OF EXPERIMENT-THEORY FIT
RPE=0.28

2D UNIT CELLS (1 domain observed)

Cell	Ax (Å)	Ay (Å)	Bx (Å)	By (Å)	α (°)	Matrix	Pattern	Cell type
Bulk	2.750	0.000	0.000	3.890	90.0	(1.000, 0.000) (0.000, 1.000)	(1x1)	b: bulk lattice
Surface 1	2.750	0.000	0.000	7.780	90.0	(1.000, 0.000) (0.000, 2.000)	(1x2)	s1: commens. superlattice

3D COORDINATES

Pd1: remaining row; Pd2-Pd3: paired 2nd layer; Pd4-Pd5: buckled 3rd layer;
Pd6: periodically repeating bulk layer; 0.1/0.05Å lateral/perp. error bars assumed for tabulation

Dx/Dy in Å, or as a fraction of layer's unit cell vectors; atom 0 at the origin. Epir/subr are bulk repeat vectors.

No. of atoms: 6 Bulk z = 1.370 Å

Region	Chem el.	At. no.	Cell type	Site occ.	Rel. to	Dx ± εx		Dy ± εy		Dz ± εz		Dz/Bz(%) ± εz/Bz
epir		-2					f		f		Å	
subr		-1				-1.375	Å	-1.945	Å	1.370	Å	
intf	Pd	1	s1	.50	0	0.000	f	0.000	f	0.000	Å	0.0
intf	Pd	2	s1	.50	1	0.500	f	0.263 ± .013	f	1.247 ± .050	Å	91.0 ± 3.7
intf	Pd	3	s1	.50	2	0.000	f	0.474 ± .013	f	0.000	Å	0.0
intf	Pd	4	b	1.00	3	-0.500	f	-1.474	f	1.456 ± .050	Å	106.3 ± 3.7
intf	Pd	5	b	1.00	4	0.500	f	0.500	f	1.377 ± .050	Å	100.5 ± 3.7
subl	Pd	6	b	1.00	5	-0.500	f	-0.500	f	1.370	Å	100.0

Pd(110)-(1x2)
46.6

BOND DISTANCES AND ANGLES

Bond distances and angles are derived from coordinates

No. of distances/angles: 5

Interatomic dist. A-B (Å)	Atom A	Atom B	Atom C	Bond angle A-B-C (°)
2.750	Pd1	Pd1(1,0)		
2.763	Pd1	Pd2	Pd3(1,-1)	116.2
2.763	Pd1	Pd3(0,-1)	Pd4	57.4
2.863	Pd2	Pd4	Pd5	60.6
2.863	Pd2	Pd4	Pd6	120.6

COMMON NAME : Pd(111)-(1x1) ILLUSTRATION: 1
CLASSIFICATION : 46.3
TECHNIQUE : HEIS
AUTHORS : Y. Kuk, L.C. Feldman and P.J. Silverman
REFERENCE : Phys. Rev. Lett., 50, 511 (1983)

SURFACE TYPE
Substrate : Pd
Crystal face: 111
Temperature : 298 K
Bulk lattice: fcc
2D bulk symm: p3m1
2D surf symm: p3m1

Adsorbate:
Coverage :
Pattern : (1x1)
Matrix : (1.000, 0.000)
(0.000, 1.000)

STRUCTURE TYPE
Unrelaxed bulk termination

SAMPLE PREPARATION (1 sample)
Treatment : sample cleaned in situ by sputtering and annealing
Crystallinity:
Anal. methods:
Contamination: no impurities by AES or ion scattering

COMMENTS

DATA COLLECTION
Technique: HEIS; He ion scattering/channeling at 1.8Me
Dataset : ion scattering/channeling along [00-1], [-1-1-1], and random directions to obtain Pd surface peak

THEORY/DATA TREATMENT
Simulation of ion scattering/channeling; ΘD=274 K

STRUCTURES EXAMINED
Relaxations of top interlayer spacing

QUALITY OF EXPERIMENT-THEORY FIT
Visual

2D UNIT CELLS (1 domain observed)

Cell	Ax (Å)	Ay (Å)	Bx (Å)	By (Å)	α (°)	Matrix	Pattern	Cell type
Bulk	2.750	0.000	1.375	2.382	60.0	(1.000, 0.000) (0.000, 1.000)	(1x1)	b: bulk lattice
Surface 1	2.750	0.000	1.375	2.382	60.0	(1.000, 0.000) (0.000, 1.000)	(1x1)	s1: commens. superlattice

3D COORDINATES

0.1Å error bar assumed for tabulation

Dx/Dy in Å, or as a fraction of layer's unit cell vectors; atom 0 at the origin. Epir/subr are bulk repeat vectors.

No. of atoms: 3 Bulk z = 2.250 Å

Region	Chem el.	At. no.	Cell type	Site occ.	Rel. to	Dx ± εx	Dy ± εy	Dz ± εz	Dz/Bz(%) ± εz/Bz
epir		-2				f	f	Å	
subr		-1				-1.375 Å	-0.794 Å	2.250 Å	
intf	Pd	1	b	1.00	0	0.000 f	0.000 f	0.000 Å	0.0
intf	Pd	2	b	1.00	1	0.667 f	0.667 f	2.250 ± .100 Å	100.0 ± 4.4
subl	Pd	3	b	1.00	2	-0.333 f	-0.333 f	2.250 Å	100.0

BOND DISTANCES AND ANGLES

Bond distances and angles are derived from coordinates

No. of distances/angles: 2

Interatomic dist. A-B (Å)	Atom A	Atom B	Atom C	Bond angle A-B-C (°)
2.750	Pd1	Pd1(1,0)		
2.754	Pd1	Pd2(0,-1)		

COMMON NAME : Pd(111)-(1x1)
CLASSIFICATION : 46.5a
TECHNIQUE : LEED
AUTHORS : H. Ohtani, M.A. Van Hove and G.A. Somorjai
REFERENCE : Surf. Sci., 187, 372 (1987)

ILLUSTRATION: 1

SURFACE TYPE
Substrate : Pd
Crystal face: 111
Temperature : 300 K
Bulk lattice: fcc
2D bulk symm: p3m1
2D surf symm: p3m1

Adsorbate:
Coverage :
Pattern : (1x1)
Matrix : (1.000, 0.000)
(0.000, 1.000)

STRUCTURE TYPE
Slight relaxation of top two interlayer spacings

SAMPLE PREPARATION (1 sample)
Treatment : 773 K O treatment, Ar+ sputtering at RT and 873K, 773K anneal
Crystallinity:
Anal. methods:
Contamination: monitored by AES and LEED

COMMENTS

DATA COLLECTION
Technique: LEED
Dataset : I-V curves at normal incidence (5 beams) and off-normal (21 beams) in the range 20 to 300 eV

THEORY/DATA TREATMENT
Dynamical LEED (RFS): band-structure potential; Voi=-5 eV; ΘD=225 K

STRUCTURES EXAMINED
Relaxation of top three interlayer spacings; various stacking sequences (ideal fcc, ideal hcp, fcc monolayer on hcp, hcp monolayer on fcc)

QUALITY OF EXPERIMENT-THEORY FIT
RVHT=0.12, RZJ=0.09, RPE=0.22

2D UNIT CELLS (1 domain observed)

Cell	Ax (Å)	Ay (Å)	Bx (Å)	By (Å)	α (°)	Matrix	Pattern	Cell type
Bulk	2.750	0.000	1.375	2.382	60.0	(1.000, 0.000) (0.000, 1.000)	(1x1)	b: bulk lattice
Surface 1	2.750	0.000	1.375	2.382	60.0	(1.000, 0.000) (0.000, 1.000)	(1x1)	s1: commens. superlattice

3D COORDINATES

Dx/Dy in Å, or as a fraction of layer's unit cell vectors; atom 0 at the origin. Epir/subr are bulk repeat vectors.

No. of atoms: 4

Bulk z = 2.300 Å

Region	Chem el.	At. no.	Cell type	Site occ.	Rel. to	Dx ± εx		Dy ± εy		Dz ± εz		Dz/Bz(%) ± εz/Bz
epir		-2					f		f		Å	
subr		-1				-2.750	Å	-1.588	Å	2.300	Å	
intf	Pd	1	b	1.00	0	0.000	f	0.000	f	0.000	Å	0.0
intf	Pd	2	b	1.00	1	0.333	f	0.333	f	2.280 ± .030	Å	99.1 ± 1.3
intf	Pd	3	b	1.00	2	0.333	f	0.333	f	2.220 ± .030	Å	96.5 ± 1.3
subl	Pd	4	b	1.00	3	-0.667	f	-0.667	f	2.300 ± .030	Å	100.0 ± 1.3

BOND DISTANCES AND ANGLES

Bond distances and angles are derived from coordinates

No. of distances/angles: 3

Interatomic dist. A-B (Å)	Atom A	Atom B	Atom C	Bond angle A-B-C (°)
2.750	Pd1	Pd1(1,0)	Pd2	60.3
2.778	Pd1	Pd2	Pd3	180.0
2.729	Pd2	Pd3	Pd4(1,0)	120.3

COMMON NAME : Pd(111)-(1x1)-Au
CLASSIFICATION : 46.79.1
TECHNIQUE : HEIS
AUTHORS : Y. Kuk, L.C. Feldman and P.J. Silverman
REFERENCE : Phys. Rev. Lett., 50, 511 (1983)

ILLUSTRATION: 81

SURFACE TYPE
Substrate : Pd
Crystal face: 111
Temperature : 298 K
Bulk lattice: fcc
2D bulk symm: p3m1
2D surf symm: p3m1

Adsorbate: Au
Coverage : 1.0 (Au/Pd)
Pattern : (1x1)
Matrix : (1.000, 0.000)
(0.000, 1.000)

STRUCTURE TYPE
Overlayer continuing substrate fcc lattice

SAMPLE PREPARATION (1 sample)
Treatment : sample cleaned in situ by sputtering and annealing
Crystallinity: Au deposition at 0.2μm/min
Anal. methods:
Contamination: no impurities by AES or ion scattering

COMMENTS
Beyond 1ML Au overlayer loses registry with substrate; with 1ML Au several layer spacings expand slightly, but this analysis did not resolve individual layer spacing changes

DATA COLLECTION
Technique: HEIS; He ion scattering/channeling at 1.8Me
Dataset : ion scattering along [00-1], [-1-1-1], and random directions to obtain Au coverage, Pd surface peak, and Au-Au shadowing

THEORY/DATA TREATMENT
Simulation of ion scattering/channeling

STRUCTURES EXAMINED
Various Au thicknesses in range 0-10ML; average layer spacing was monitored, as well as loss of registry (incommensuration)

QUALITY OF EXPERIMENT-THEORY FIT
Visual

2D UNIT CELLS (1 domain observed)

Cell	Ax (Å)	Ay (Å)	Bx (Å)	By (Å)	α (°)	Matrix	Pattern	Cell type
Bulk	2.750	0.000	1.375	2.382	60.0	(1.000, 0.000) (0.000, 1.000)	(1x1)	b: bulk lattice
Surface 1	2.750	0.000	1.375	2.382	60.0	(1.000, 0.000) (0.000, 1.000)	(1x1)	s1: commens. superlattice

3D COORDINATES

Au1: overlayer in fcc hollow sites

Dx/Dy in Å, or as a fraction of layer's unit cell vectors; atom 0 at the origin. Epir/subr are bulk repeat vectors.

No. of atoms: 3 Bulk z = 2.250 Å

Region	Chem el.	At. no.	Cell type	Site occ.	Rel. to	Dx ± εx	Dy ± εy	Dz ± εz	Dz/Bz(%) ± εz/Bz
epir		-2				f	f	Å	
subr		-1				-1.375 Å	-0.794 Å	2.250 Å	
ovrl	Au	1	b	1.00	0	0.000 f	0.000 f	0.000 Å	0.0
intf	Pd	2	b	1.00	1	0.667 f	0.667 f	2.250 ± .190 Å	100.0 ± 8.4
subl	Pd	3	b	1.00	2	-0.333 f	-0.333 f	2.250 Å	100.0

BOND DISTANCES AND ANGLES

Bond distances and angles are derived from coordinates

No. of distances/angles: 3

Interatomic dist. A-B (Å)	Atom A	Atom B	Atom C	Bond angle A-B-C (°)
2.750	Au1	Au1(1,0)		
2.754	Au1	Pd2(-1,0)		
2.754	Pd2	Pd3		

COMMON NAME : Pd(111)-(3x3)-C6H6+2CO
CLASSIFICATION : 46.6.1.8.2
TECHNIQUE : LEED
AUTHORS : H. Ohtani, M.A. Van Hove and G.A. Somorjai
REFERENCE : J. Phys. Chem., 92, 3974 (1988)

ILLUSTRATION: 70

SURFACE TYPE
Substrate : Pd
Crystal face: 111
Temperature : 150 K
Bulk lattice: fcc
2D bulk symm: p3m1
2D surf symm: p3m1

Adsorbate: C6H6;CO
Coverage : 1/9 Bz/Pd, 2/9 CO/Pd
Pattern : (3x3)
Matrix : (3.000, 0.000)
(0.000, 3.000)

STRUCTURE TYPE
Molecular coadsorption of one flat-lying C6H6 (benzene) and 2 upright CO per cell, all centered over fcc hollow sites, both with relaxed bonds (H ignored), on unrelaxed substrate; top two Pd-Pd spacings found expanded by 0.05Å from bulk value of 2.25Å

SAMPLE PREPARATION (1 sample)
Treatment : exposure to CO and benzene at room temperature
Crystallinity:
Anal. methods:
Contamination: monitored by AES and LEED

COMMENTS

DATA COLLECTION
Technique: LEED
Dataset : I-V curves at normal incidence (16 beams) and off-normal incidence (29 beams); 20<E<200 eV

THEORY/DATA TREATMENT
Dynamical LEED (BSN, RFS, KSLA): 5 phase shifts; ΘD=225 K

STRUCTURES EXAMINED
Benzene in top, bridge, fcc, hcp sites with various high symmetry orientations and distortions including out of plane buckling, ring expansion and Kekule distortion; CO intact, stretched and normal to surface in same sites consistent with benzene orientations

QUALITY OF EXPERIMENT-THEORY FIT
RVHT=0.25, RZJ=0.49, RPE=0.48

2D UNIT CELLS (1 domain observed)

Cell	Ax (Å)	Ay (Å)	Bx (Å)	By (Å)	α (°)	Matrix	Pattern	Cell type
Bulk	2.750	0.000	1.375	2.382	60.0	(1.000, 0.000) (0.000, 1.000)	(1x1)	b: bulk lattice
Surface 1	8.250	0.000	4.125	7.145	60.0	(3.000, 0.000) (0.000, 3.000)	(3x3)	s1: commens. superlattice

3D COORDINATES

O1-C15, O2-C16: 2 upright stretched COs in fcc hollows; H3-8, C9-14: flat benzene over fcc site; substrate spacing found expanded from 2.25 to 2.30Å; 0.1Å lateral error bars assumed for tabulation

Dx/Dy in Å, or as a fraction of layer's unit cell vectors; atom 0 at the origin. Epir/subr are bulk repeat vectors.

No. of atoms: 18 Bulk z = 2.300 Å

Reg ion	Chem el.	At. no.	Cell type	Site occ.	Rel. to	Dx ± εx		Dy ± εy		Dz ± εz		Dz/Bz(%) ± εz/Bz
epir		-2					f		f		Å	
subr		-1				1.375	Å	0.794	Å	2.300	Å	
ovrl	O	1	s1	.11	0	0.000	f	0.000	f	0.000	Å	0.0
ovrl	O	2	s1	.11	1	0.667	f	0.667	f	0.000	Å	0.0
ovrl	H	3	s1	.11	1	0.636	f	0.026	f	0.029	Å	1.3
ovrl	H	4	s1	.11	3	0.000	f	0.312	f	0.000	Å	0.0
ovrl	H	5	s1	.11	3	-0.298	f	0.609	f	0.000	Å	0.0
ovrl	H	6	s1	.11	3	-0.609	f	0.609	f	0.000	Å	0.0
ovrl	H	7	s1	.11	3	-0.609	f	0.312	f	0.000	Å	0.0
ovrl	H	8	s1	.11	3	-0.298	f	0.000	f	0.000	Å	0.0
ovrl	C	9	s1	.11	1	0.336 ± .005	f	0.159 ± .014	f	0.220 ± .050	Å	9.6 ± 2.2
ovrl	C	10	s1	.11	9	0.169 ± .005	f	0.000 ± .014	f	0.000 ± .050	Å	0.0 ± 2.2
ovrl	C	11	s1	.11	9	0.000 ± .005	f	0.347 ± .014	f	0.000 ± .050	Å	0.0 ± 2.2
ovrl	C	12	s1	.11	9	0.169 ± .005	f	0.177 ± .014	f	0.000 ± .050	Å	0.0 ± 2.2
ovrl	C	13	s1	.11	9	-0.177 ± .005	f	0.177 ± .014	f	0.000 ± .050	Å	0.0 ± 2.2
ovrl	C	14	s1	.11	9	-0.177 ± .005	f	0.347 ± .014	f	0.000 ± .050	Å	0.0 ± 2.2
ovrl	C	15	s1	.11	1	0.000	f	0.000	f	1.170 ± .050	Å	50.9 ± 2.2
ovrl	C	16	s1	.11	2	0.000	f	0.000	f	1.170 ± .050	Å	50.9 ± 2.2
intf	Pd	17	b	1.00	15	0.333	f	0.333	f	1.300 ± .050	Å	56.5 ± 2.2
subl	Pd	18	b	1.00	17	0.333	f	0.333	f	2.300 ± .050	Å	100.0 ± 2.2

Pd(111)-(3x3)-C6H6+2CO
46.6.1.8.2 BOND DISTANCES AND ANGLES

Bond distances and angles are derived from coordinates

No. of distances/angles: 8

Interatomic dist. A-B (Å)	Atom A	Atom B	Atom C	Bond angle A-B-C (°)
1.170	O1	C15	Pd17	129.3
1.398	C9	C10	C12	120.0
1.398	C9	C10	Pd17(1,0)	72.9
1.463	C9	C13	C14	120.0
1.463	C9	C13	Pd17(0,1)	105.7
2.052	C15	Pd17	Pd18	163.9
2.750	Pd17	Pd17(1,0)		
2.795	Pd17	Pd18		

COMMON NAME : Pd(111)-(√3x√3)R30°-CO ILLUSTRATION: 61
CLASSIFICATION : 46.6.8.2
TECHNIQUE : LEED
AUTHORS : H. Ohtani, M.A. Van Hove and G.A. Somorjai
REFERENCE : Surf. Sci., 187, 372 (1987)

SURFACE TYPE
Substrate : Pd
Crystal face: 111
Temperature : 300 K
Bulk lattice: fcc
2D bulk symm: p3m1
2D surf symm: p31m

Adsorbate: CO
Coverage : 1/3 CO/Pd
Pattern : (√3x√3)R30°
Matrix : (1.000, 1.000)
(-2.000, 1.000)

STRUCTURE TYPE
Molecular upright adsorption (C down) in fcc hollow sites

SAMPLE PREPARATION (1 sample)
Treatment : Ar+ sputtering and oxygen treatment; exposure to CO at RT
Crystallinity:
Anal. methods:
Contamination: monitored by AES and LEED

COMMENTS

DATA COLLECTION
Technique: LEED
Dataset : I-V curves at normal incidence (8 beams) and off-normal incidence (29 beams): 20<E<200 eV

THEORY/DATA TREATMENT
Dynamical LEED (RFS); band-structure potential; Voi=-5 eV; ΘD=225 K

STRUCTURES EXAMINED
CO perpendicular to surface in top, bridge and 2 hollow sites, with various Pd-Pd, Pd-C and C-O spacings; for fcc-hollow site, lateral radial relaxations of Pd atoms about site

QUALITY OF EXPERIMENT-THEORY FIT
RVHT=0.30, RZJ=0.56, RPE=0.55

2D UNIT CELLS (1 domain observed)

Cell	Ax (Å)	Ay (Å)	Bx (Å)	By (Å)	α (°)	Matrix	Pattern	Cell type
Bulk	2.748	0.000	1.374	2.380	60.0	(1.000, 0.000) (0.000, 1.000)	(1x1)	b: bulk lattice
Surface 1	4.122	2.380	-4.122	2.380	120.0	(1.000, 1.000) (-2.000, 1.000)	(√3x√3)R30°	s1: commens. superlattice

3D COORDINATES

O1-C2: upright CO overlayer in fcc hollow sites

Dx/Dy in Å, or as a fraction of layer's unit cell vectors; atom 0 at the origin. Epir/subr are bulk repeat vectors.

No. of atoms: 5 Bulk z = 2.250 Å

Reg ion	Chem el.	At. no.	Cell type	Site occ.	Rel. to	Dx ± εx		Dy ± εy		Dz ± εz		Dz/Bz(%) ± εz/Bz
epir		-2					f		f		Å	
subr		-1				-1.374	Å	0.793	Å	2.250	Å	
ovrl	O	1	s1	.33	0	0.000	f	0.000	f	0.000	Å	0.0
ovrl	C	2	s1	.33	1	0.000	f	0.000	f	1.150 ± .050	Å	51.1 ± 2.2
intf	Pd	3	b	1.00	2	0.333	f	0.333	f	1.290 ± .050	Å	57.3 ± 2.2
intf	Pd	4	b	1.00	3	0.333	f	0.333	f	2.390 ± .050	Å	106.2 ± 2.2
subl	Pd	5	b	1.00	4	-0.667	f	0.333	f	2.250	Å	100.0

BOND DISTANCES AND ANGLES

Bond distances and angles are derived from coordinates

No. of distances/angles: 4

Interatomic dist. A-B (Å)	Atom A	Atom B	Atom C	Bond angle A-B-C (°)
1.150	O1	C2	Pd3	129.1
2.045	C2	Pd3	Pd4	162.7

Pd(111)-(√3x√3)R30°-CO
46.6.8.2

Bond Distances and Angles - Continued

Interatomic dist. A-B (Å)	Atom A	Atom B	Atom C	Bond angle A-B-C (°)
2.748	Pd3	Pd3(1,0)	Pd4	61.4
2.869	Pd3	Pd4	Pd5	121.4

COMMON NAME : Pd(100)-(2√2x√2)R45°-2CO
CLASSIFICATION : 46.6.8.0a
TECHNIQUE : LEED
AUTHORS : R.J. Behm, K. Christmann, G. Ertl and M.A. Van Hove
REFERENCE : J. Chem. Phys., 73, 2984 (1980)

ILLUSTRATION: 71,72

SURFACE TYPE
Substrate : Pd
Crystal face: 100
Temperature : 350 K
Bulk lattice: fcc
2D bulk symm: p4m
2D surf symm: pgg

Adsorbate: CO
Coverage : 1/2 (CO/Pd)
Pattern : (2√2x√2)R45°
Matrix : (2.000, 2.000)
(-1.000, 1.000)

STRUCTURE TYPE
Molecular adsorption in bridge sites: CO upright with C down with two differently oriented bridge sites occupied per cell

SAMPLE PREPARATION (1 sample)
Treatment :
Crystallinity:
Anal. methods:
Contamination: no impurities by AES, TDS and WFC

COMMENTS

DATA COLLECTION
Technique: LEED
Dataset : I-V curves: 00 beam at Θ=0°; 1 0,1/2 1/2,3/2 1/2, 1/4 -3/4,1/4 5/4 at Θ=8.5,ϕ=12°; cum. E range: 630 eV

THEORY/DATA TREATMENT
Dynamical LEED (RFS): Pd Moruzzi et al pot, CO Xα and Jona pots, 8 phase shifts; Vor=-10.0 eV,Voi∝E**1/3; ΘD= 193 K(Pd)

STRUCTURES EXAMINED
Bridge sites only; CO perpendicular to surface; variable top Pd-Pd, Pd-C and C-O spacings

QUALITY OF EXPERIMENT-THEORY FIT
Visual

2D UNIT CELLS (2 domains observed)

Cell	Ax (Å)	Ay (Å)	Bx (Å)	By (Å)	α (°)	Matrix	Pattern	Cell type
Bulk	2.740	0.000	0.000	2.740	90.0	(1.000, 0.000) (0.000, 1.000)	(1x1)	b: bulk lattice
Surface 1	5.480	5.480	-2.740	2.740	90.0	(2.000, 2.000) (-1.000, 1.000)	(2√2x√2)R45°	s1: commens. superlattice

3D COORDINATES

O1-C3 and O2-C4: 2 upright CO molecules in differently oriented bridge sites

Dx/Dy in Å, or as a fraction of layer's unit cell vectors; atom 0 at the origin. Epir/subr are bulk repeat vectors.

No. of atoms: 6

Bulk z = 1.945 Å

Region	Chem el.	At. no.	Cell type	Site occ.	Rel. to	Dx ± εx		Dy ± εy		Dz ± εz		Dz/Bz(%) ± εz/Bz
epir		-2					f		f		Å	
subr		-1				-1.370	Å	-1.370	Å	1.945	Å	
ovrl	O	1	s1	.25	0	0.000	f	0.000	f	0.000	Å	0.0
ovrl	O	2	s1	.25	1	0.500	f	0.500	f	0.000	Å	0.0
ovrl	C	3	s1	.25	1	0.000	f	0.000	f	1.150 ± .100	Å	59.1 ± 5.1
ovrl	C	4	s1	.25	2	0.000	f	0.000	f	1.150 ± .100	Å	59.1 ± 5.1
intf	Pd	5	b	1.00	4	0.000	f	-0.500	f	1.360 ± .100	Å	69.9 ± 5.1
subl	Pd	6	b	1.00	5	-0.500	f	-0.500	f	1.945	Å	100.0

BOND DISTANCES AND ANGLES

Bond distances and angles are derived from coordinates

No. of distances/angles: 3

Interatomic dist. A-B (Å)	Atom A	Atom B	Atom C	Bond angle A-B-C (°)
1.150	O1	C3	Pd5(0,-1)	134.8
1.930	C4	Pd5	Pd5(0,-1)	135.2
2.745	Pd5	Pd6		

COMMON NAME : Pd(100)-(1x1)-Cu multilayer
CLASSIFICATION : 46.29.1
TECHNIQUE : LEED
AUTHORS : H. Li, S.C. Wu, D. Tian, J. Quinn, Y.S. Li, F. Jona and P.M. Marcus
REFERENCE : Phys. Rev., B40, 5841 (1989)

ILLUSTRATION: 83

SURFACE TYPE
Substrate : Pd
Crystal face: 100
Temperature : RT
Bulk lattice: fcc
2D bulk symm: p4m
2D surf symm: p4m

Adsorbate: Cu
Coverage : 6 Cu/Pd
Pattern : (1x1)
Matrix : (1.000, 0.000)
(0.000, 1.000)

STRUCTURE TYPE
At least 6 epitaxial (1x1) monolayers, forming strained fcc Cu

SAMPLE PREPARATION (3 sample)
Treatment : Cu deposited from Cu wire onto RT substrate
Crystallinity: LEED: increased background
Anal. methods: coverage from AES
Contamination:

COMMENTS
RVHT, RPE and RZJ gave results different by up to 0.16Å, indicating imperfect structural model; average of the 3 optimized structures is used in tabulation, as proposed by authors

DATA COLLECTION
Technique: LEED; video LEED
Dataset : IV spectra for 5 non-deg. beams at normal incidence, 9 at Θ=10°,φ=0°; 30<E<360 eV

THEORY/DATA TREATMENT
Dynamical LEED (CHANGE code): Moruzzi et al pot, 8 ph shs; Vor=-6 eV (fit), Voi=-4eV; rms vibs 0.125Å

STRUCTURES EXAMINED
Semi-infinite Cu(100) with lateral Pt(100) lattice constant; spacings varied: 'bulk' Cu-Cu, top two Cu-Cu

QUALITY OF EXPERIMENT-THEORY FIT
RVHT=0.24, RPE=0.38, RZJ=0.06

2D UNIT CELLS (1 domain observed)

Cell	Ax (Å)	Ay (Å)	Bx (Å)	By (Å)	α (°)	Matrix	Pattern	Cell type
Bulk	2.779	0.000	0.000	2.779	90.0	(1.000, 0.000) (0.000, 1.000)	(1x1)	b: bulk lattice
Surface 1	2.779	0.000	0.000	2.779	90.0	(1.000, 0.000) (0.000, 1.000)	(1x1)	s1: commens. superlattice

3D COORDINATES

1.62Å bulk spacing was fit, keeping lateral Pt distance

Dx/Dy in Å, or as a fraction of layer's unit cell vectors; atom 0 at the origin. Epir/subr are bulk repeat vectors.

No. of atoms: 4 Bulk z = 1.620 Å

Reg ion	Chem el.	At. no.	Cell type	Site occ.	Rel. to	Dx ± εx		Dy ± εy		Dz ± εz		Dz/Bz(%) ± εz/Bz
epir		-2					f		f		Å	
subr		-1				-1.390	Å	-1.390	Å	1.620	Å	
intf	Cu	1	b	1.00	0	0.000	f	0.000	f	0.000	Å	0.0
intf	Cu	2	b	1.00	1	0.500	f	0.500	f	1.540 ± .050	Å	95.1 ± 3.1
intf	Cu	3	b	1.00	2	-0.500	f	-0.500	f	1.570 ± .050	Å	96.9 ± 3.1
subl	Cu	4	b	1.00	3	0.500	f	0.500	f	1.620 ± .040	Å	100.0 ± 2.5

BOND DISTANCES AND ANGLES

Bond distances and angles are derived from coordinates

No. of distances/angles: 3

Interatomic dist. A-B (Å)	Atom A	Atom B	Atom C	Bond angle A-B-C (°)
2.779	Cu1	Cu1(1,0)	Cu2	56.2
2.497	Cu1	Cu2	Cu3	76.7
2.515	Cu2	Cu3	Cu4	78.1

COMMON NAME : Pd(100)-(1x1)-H (D)
CLASSIFICATION : 46.1.13a
TECHNIQUE : Transm. Channeling
AUTHORS : F. Besenbacher, I. Stensgaard, and K. Mortensen
REFERENCE : Springer Series in Surface Sciences, 11, 195 (1988)

ILLUSTRATION: 28

SURFACE TYPE
Substrate : Pd
Crystal face: 100
Temperature : 140 K
Bulk lattice: fcc
2D bulk symm: p4m
2D surf symm: p4m

Adsorbate: H
Coverage : 1ML
Pattern : (1x1)
Matrix : (1.000, 0.000)
(0.000, 1.000)

STRUCTURE TYPE
Atomic adsorbed in a 4-fold hollow site

SAMPLE PREPARATION (1 sample)
Treatment : Pd crystal grown on NaCl; exposure to D
Crystallinity:
Anal. methods: nuclear reaction
Contamination:

COMMENTS
Saturation coverage (see 46.1.13b for lower exposure structure)

DATA COLLECTION
Technique: Transm. Channeling
Dataset : angular scans along the [001], [011], [111] directions

THEORY/DATA TREATMENT
Transmission channeling simulations

QUALITY OF EXPERIMENT-THEORY FIT
Visual

2D UNIT CELLS (1 domain observed)

Cell	Ax (Å)	Ay (Å)	Bx (Å)	By (Å)	α (°)	Matrix	Pattern	Cell type
Bulk	2.751	0.000	0.000	2.751	90.0	(1.000, 0.000) (0.000, 1.000)	(1x1)	b: bulk lattice
Surface 1	2.751	0.000	0.000	2.751	90.0	(1.000, 0.000) (0.000, 1.000)	(1x1)	s1: commens. superlattice

3D COORDINATES

H1: atomic overlayer in a 4-fold hollow site

Dx/Dy in Å, or as a fraction of layer's unit cell vectors; atom 0 at the origin. Epir/subr are bulk repeat vectors.

No. of atoms: 3 Bulk z = 1.945 Å

Region	Chem el.	At. no.	Cell type	Site occ.	Rel. to	Dx ± εx		Dy ± εy		Dz ± εz		Dz/Bz(%) ± εz/Bz
epir		-2					f		f		Å	
subr		-1				1.376	Å	1.376	Å	1.945	Å	
ovrl	H	1	s1	1.00	0	0.000	Å	0.000	Å	0.000	Å	0.0
intf	Pd	2	b	1.00	0	1.376	Å	1.376	Å	0.300 ± .050	Å	7.7 ± 1.3
subl	Pd	3	b 1	0.00	0	0.000	Å	0.000	Å	2.245	Å	115.4

BOND DISTANCES AND ANGLES

Bond distances and angles are derived from coordinates

No. of distances/angles: 1

Interatomic dist. A-B (Å)	Atom A	Atom B	Atom C	Bond angle A-B-C (°)
1.968	H1	Pd2		

COMMON NAME : Pd(100)-c(2x2)-H (D)
CLASSIFICATION : 46.1.13b
TECHNIQUE : Transm. Channeling
AUTHORS : F. Besenbacher, I. Stensgaard, and K. Mortensen
REFERENCE : Springer Series in Surface Sciences, 11, 195 (1988)

ILLUSTRATION: 28,29

SURFACE TYPE
Substrate : Pd
Crystal face: 100
Temperature : 140 K
Bulk lattice: fcc
2D bulk symm: p4m
2D surf symm: p4m

Adsorbate: H
Coverage : 0.5 ML
Pattern : c(2x2)
Matrix : (1.000, 1.000)
(-1.000, 1.000)

STRUCTURE TYPE
Atomic adsorbed in a 4-fold hollow site

SAMPLE PREPARATION (1 sample)
Treatment : Pd crystal grown on NaCl; exposure to D
Crystallinity:
Anal. methods: nuclear reaction
Contamination:

COMMENTS
Lower exposure structure (see 46.1.13a for higher-exposure structure)

DATA COLLECTION
Technique: Transm. Channeling
Dataset : angular scans along the [001], [011], [111] directions

THEORY/DATA TREATMENT
Transmission channeling simulations

QUALITY OF EXPERIMENT-THEORY FIT
Visual

2D UNIT CELLS (1 domain observed)

Cell	Ax (Å)	Ay (Å)	Bx (Å)	By (Å)	α (°)	Matrix	Pattern	Cell type
Bulk	2.751	0.000	0.000	2.751	90.0	(1.000, 0.000) (0.000, 1.000)	(1x1)	b: bulk lattice
Surface 1	2.751	2.751	-2.751	2.751	90.0	(1.000, 1.000) (-1.000, 1.000)	c(2x2)	s1: commens. superlattice

3D COORDINATES

H1: atomic adsorbed in 4-fold hollow site

Dx/Dy in Å, or as a fraction of layer's unit cell vectors; atom 0 at the origin. Epir/subr are bulk repeat vectors.

No. of atoms: 3 Bulk z = 1.945 Å

Region	Chem el.	At. no.	Cell type	Site occ.	Rel. to	Dx ± εx		Dy ± εy		Dz ± εz		Dz/Bz(%) ± εz/Bz
epir		-2					f		f		Å	
subr		-1				1.376	Å	1.376	Å	1.945	Å	
ovrl	H	1	s1	.50	0	0.000	Å	0.000	Å	0.000	Å	0.0
intf	Pd	2	b	1.00	1	1.376	Å	1.376	Å	0.450 ± .100	Å	23.1 ± 5.1
subl	Pd	3	b	1.00	2	1.376	Å	1.376	Å	1.945	Å	100.0

BOND DISTANCES AND ANGLES

Bond distances and angles are derived from coordinates

No. of distances/angles: 1

Interatomic dist. A-B (Å)	Atom A	Atom B	Atom C	Bond angle A-B-C (°)
1.997	H1	Pd2		

COMMON NAME : Pd(110)-(1x2)-H
CLASSIFICATION : 46.1.11a
TECHNIQUE : LEED
AUTHORS : G. Kleinle, M. Skottke, V. Penka, G. Ertl, R.J. Behm and W. Moritz
REFERENCE : Surf. Sci., 189/190, 177 (1987)

ILLUSTRATION: -

SURFACE TYPE
Substrate : Pd
Crystal face: 110
Temperature : RT*
Bulk lattice: fcc
2D bulk symm: pmm
2D surf symm: pmm

Adsorbate: H
Coverage : 1.5 H/Pd
Pattern : (1x2)
Matrix : (1.000, 0.000)
(0.000, 2.000)

STRUCTURE TYPE
H-induced row pairing reconstruction (H positions not determined)

SAMPLE PREPARATION (1 sample)
Treatment : clean (1x1) surface exposed to 0.8L of H2 at 120 K
Crystallinity:
Anal. methods:
Contamination: close attention to H coverage

COMMENTS

DATA COLLECTION
Technique: LEED
Dataset : IV curves for 8 non-equivalent beams

THEORY/DATA TREATMENT
Dynamical LEED with layer doubling; H ignored

STRUCTURES EXAMINED
Row pairing: variation of top 3 interlayer spacings, of row pairing in 1st layer and of buckling in 2nd layer; missing row: variation of top 3 interlayer spacings, of row pairing in 2nd layer and of buckling in 3rd layer; H ignored

QUALITY OF EXPERIMENT-THEORY FIT
RPE=0.50, RZJ=0.22

2D UNIT CELLS (1 domain observed)

Cell	Ax (Å)	Ay (Å)	Bx (Å)	By (Å)	α (°)	Matrix	Pattern	Cell type
Bulk	2.750	0.000	0.000	3.890	90.0	(1.000, 0.000) (0.000, 1.000)	(1x1)	b: bulk lattice
Surface 1	2.750	0.000	0.000	7.780	90.0	(1.000, 0.000) (0.000, 2.000)	(1x2)	s1: commens. superlattice

3D COORDINATES

Pd1-Pd2: paired top layer; Pd3-Pd4: buckled 2nd layer

Dx/Dy in Å, or as a fraction of layer's unit cell vectors; atom 0 at the origin. Epir/subr are bulk repeat vectors.

No. of atoms: 6 Bulk z = 1.370 Å

Reg ion	Chem el.	At. no.	Cell type	Site occ.	Rel. to	Dx ± εx		Dy ± εy		Dz ± εz		Dz/Bz(%) ± εz/Bz
epir		-2					f		f		Å	
subr		-1				1.375	Å	-1.945	Å	1.370	Å	
intf	Pd	1	s1	.50	0	0.000	f	0.000	f	0.000	Å	0.0
intf	Pd	2	s1	.50	1	0.000	f	0.448 ± .006	f	0.000	Å	0.0
intf	Pd	3	s1	.50	2	0.500	f	0.276 ± .006	f	1.220 ± .030	Å	89.1 ± 2.2
intf	Pd	4	s1	.50	3	0.000	f	-0.500	f	0.300 ± .050	Å	21.9 ± 3.7
intf	Pd	5	b	1.00	4	-0.500	f	0.500	f	1.220 ± .030	Å	89.1 ± 2.2
subl	Pd	6	b	1.00	5	0.500	f	-0.500	f	1.370 ± .030	Å	100.0 ± 2.2

Pd(110)-(1x2)-H
46.1.11a

BOND DISTANCES AND ANGLES

Bond distances and angles are derived from coordinates

No. of distances/angles: 4

Interatomic dist. A-B (Å)	Atom A	Atom B	Atom C	Bond angle A-B-C (°)
2.750	Pd1	Pd1(1,0)		
2.827	Pd1	Pd3(0,-1)	Pd4(0,-1)	115.6
2.690	Pd1	Pd4	Pd5	117.8
2.690	Pd1	Pd4	Pd6	124.4

COMMON NAME : Pd(110)-(2x1)-2H
CLASSIFICATION : 46.1.12
TECHNIQUE : LEED
AUTHORS : M. Skottke, R.J. Behm, G. Ertl, V. Penka and W. Moritz
REFERENCE : J. Chem. Phys., 87, 6191 (1987)

ILLUSTRATION: 37

SURFACE TYPE
Substrate : Pd
Crystal face: 110
Temperature : RT*
Bulk lattice: fcc
2D bulk symm: pmm
2D surf symm: pmg

Adsorbate: H
Coverage : 1 H/Pd
Pattern : (2x1)
Matrix : (2.000, 0.000)
(0.000, 1.000)

STRUCTURE TYPE
Atomic adsorption over outermost 3-fold coord. hollow sites over (111) facets of bulk-like substrate with interlayer spacing relaxations

SAMPLE PREPARATION (1 sample)
Treatment : 0.3L of H2 adsorbed at 130 K
Crystallinity:
Anal. methods: cleanliness checked by H2 desorption and WFC during H ads.
Contamination: clean by AES

COMMENTS

DATA COLLECTION
Technique: LEED
Dataset : I-V curves for total of 3(13) int-order and 1(5) fract-order inequivalent beams at Θ=0°(Θ>0°); 40<E<180 eV

THEORY/DATA TREATMENT
Dynamical LEED (layer doubling): 8 Pd ph shs from relat. pot for H from superpos; Voi∝E**1/3; ΘD=274 K (Pd and H)

STRUCTURES EXAMINED
Variation of top 3 interlayer spacings; H on mirror plane near 3-fold hollow site of (111) facets: position parallel and perpendicular to surface varied

QUALITY OF EXPERIMENT-THEORY FIT
RPE=0.37, RZJ=0.19

2D UNIT CELLS (1 domain observed)

Cell	Ax (Å)	Ay (Å)	Bx (Å)	By (Å)	α (°)	Matrix	Pattern	Cell type
Bulk	2.750	0.000	0.000	3.890	90.0	(1.000, 0.000) (0.000, 1.000)	(1x1)	b: bulk lattice
Surface 1	5.500	0.000	0.000	3.890	90.0	(2.000, 0.000) (0.000, 1.000)	(2x1)	s1: commens. superlattice

3D COORDINATES

H1-H2: overlayer over 3-fold coord. hollows of (111) facets of multilayer-relaxed unreconstructed substrate

Dx/Dy in Å, or as a fraction of layer's unit cell vectors; atom 0 at the origin. Epir/subr are bulk repeat vectors.

No. of atoms: 6 Bulk z = 1.370 Å

Region	Chem el.	At. no.	Cell type	Site occ.	Rel. to	Dx ± εx		Dy ± εy		Dz ± εz		Dz/Bz(%) ± εz/Bz
epir		-2					f		f		Å	
subr		-1				-1.375	Å	-1.945	Å	1.370	Å	
ovrl	H	1	s1	.50	0	0.000	f	0.000	f	0.000	Å	0.0
ovrl	H	2	s1	.50	1	0.500	f	0.234 ± .077	f	0.000	Å	0.0
intf	Pd	3	b	1.00	2	-0.500	f	0.383 ± .077	f	0.600 ± .100	Å	43.8 ± 7.3
intf	Pd	4	b	1.00	3	-0.500	f	-0.500	f	1.340 ± .020	Å	97.8 ± 1.5
intf	Pd	5	b	1.00	4	0.500	f	0.500	f	1.410 ± .020	Å	102.9 ± 1.5
subl	Pd	6	b	1.00	5	-0.500	f	-0.500	f	1.370	Å	100.0

BOND DISTANCES AND ANGLES

Bond distances and angles are derived from coordinates

No. of distances/angles: 7

Interatomic dist. A-B (Å)	Atom A	Atom B	Atom C	Bond angle A-B-C (°)
2.114	H1	Pd3(0,-1)	H2(0,-1)	147.0
2.114	H1	Pd3(0,-1)	Pd3(1,-1)	130.6
2.114	H1	Pd3(0,-1)	Pd4	46.4

Pd(110)-(2x1)-2H
46.1.12

Bond Distances and Angles - Continued

Interatomic dist. A-B (Å)	Atom A	Atom B	Atom C	Bond angle A-B-C (°)
1.993	H1	Pd4	Pd5	131.0
2.733	Pd3	Pd4	Pd5	60.0
2.733	Pd3	Pd4	Pd6	119.4
2.768	Pd4	Pd5	Pd6	60.5

COMMON NAME : Pd(100)-(1x1)-12Fe
CLASSIFICATION : 46.26.2a
TECHNIQUE : LEED
AUTHORS : J. Quinn, Y.S. Li, H. Li, D. Tian, F. Jona and P.M. Marcus
REFERENCE : Phys. Rev., B43, 3959 (1991)

ILLUSTRATION: 83

SURFACE TYPE
Substrate : Pd
Crystal face: 100
Temperature : RT
Bulk lattice: fcc
2D bulk symm: p4m
2D surf symm: p4m

Adsorbate: Fe
Coverage : 12 Fe/Pd
Pattern : (1x1)
Matrix : (1.000, 0.000)
(0.000, 1.000)

STRUCTURE TYPE
About 12 epitaxial (1x1) monolayers, forming bct Fe (distorted from bcc)

SAMPLE PREPARATION (2 sample)
Treatment : Fe deposited 'slowly' on RT or 120 K substrate
Crystallinity: broadening of LEED beams
Anal. methods: ARPES; coverage from AES
Contamination: AES: 2.5at% O, 11at% C

COMMENTS
C and O impurities present, but are not expected to affect structure; coverages measured in layer-equivalents (i.e. giving equivalent AES signal as layer-by-layer growth): actual film thickness uncertain by factor 3; see also 53 and 200ML structures 46.26.2b and c

DATA COLLECTION
Technique: LEED; video LEED
Dataset : IV spectra at normal incidence:
(10),(11),(20),(21),(22); 40<E<360 eV

THEORY/DATA TREATMENT
Dynamical LEED (CHANGE code): Moruzzi et al pot, 8 ph shs; Vor=-10 eV (then fit), Voi=-4eV; rms vibs 0.125Å

STRUCTURES EXAMINED
Semi-infinite Fe(100) with lateral Pd(100) lattice constant; spacings varied: 'bulk' Fe-Fe, top two Fe-Fe

QUALITY OF EXPERIMENT-THEORY FIT
RVHT=0.35, RPE=0.58, RZJ=0.13

2D UNIT CELLS (1 domain observed)

Cell	Ax (Å)	Ay (Å)	Bx (Å)	By (Å)	α (°)	Matrix	Pattern	Cell type
Bulk	2.750	0.000	0.000	2.750	90.0	(1.000, 0.000) (0.000, 1.000)	(1x1)	b: bulk lattice
Surface 1	2.750	0.000	0.000	2.750	90.0	(1.000, 0.000) (0.000, 1.000)	(1x1)	s1: commens. superlattice

3D COORDINATES

1.53Å bulk spacing was fit, keeping lateral Pd distance

Dx/Dy in Å, or as a fraction of layer's unit cell vectors; atom 0 at the origin. Epir/subr are bulk repeat vectors.

No. of atoms: 4 Bulk z = 1.530 Å

Region	Chem el.	At. no.	Cell type	Site occ.	Rel. to	Dx ± εx		Dy ± εy		Dz ± εz		Dz/Bz(%) ± εz/Bz
epir		-2					f		f		Å	
subr		-1				-1.375	Å	-1.375	Å	1.530	Å	
intf	Fe	1	b	1.00	0	0.000	f	0.000	f	0.000	Å	0.0
intf	Fe	2	b	1.00	1	0.500	f	0.500	f	1.585 ± .030	Å	103.6 ± 2.0
intf	Fe	3	b	1.00	2	-0.500	f	-0.500	f	1.550 ± .030	Å	101.3 ± 2.0
subl	Fe	4	b	1.00	3	0.500	f	0.500	f	1.530 ± .030	Å	100.0 ± 2.0

BOND DISTANCES AND ANGLES

Bond distances and angles are derived from coordinates

No. of distances/angles: 3

Interatomic dist. A-B (Å)	Atom A	Atom B	Atom C	Bond angle A-B-C (°)
2.750	Fe1	Fe1(1,0)	Fe2	56.8
2.509	Fe1	Fe2	Fe3	77.7
2.487	Fe2	Fe3	Fe4	76.8

COMMON NAME : Pd(100)-(1x1)-53Fe
CLASSIFICATION : 46.26.2b
TECHNIQUE : LEED
AUTHORS : J. Quinn, Y.S. Li, H. Li, D. Tian, F. Jona and P.M. Marcus
REFERENCE : Phys. Rev., B43, 3959 (1991)

ILLUSTRATION: 83

SURFACE TYPE
Substrate : Pd
Crystal face: 100
Temperature : RT
Bulk lattice: fcc
2D bulk symm: p4m
2D surf symm: p4m

Adsorbate: Fe
Coverage : 53 Fe/Pd
Pattern : (1x1)
Matrix : (1.000, 0.000)
(0.000, 1.000)

STRUCTURE TYPE
About 53 epitaxial (1x1) monolayers, forming bct Fe (distorted from bcc)

SAMPLE PREPARATION (2 sample)
Treatment : Fe deposited 'slowly' on RT or 120 K substrate
Crystallinity: broadening of LEED beams
Anal. methods: ARPES; coverage from AES
Contamination: AES: 2.5at% O, 11at% C

COMMENTS
C and O impurities present, but are not expected to affect structure; coverages measured in layer-equivalents (i.e. giving equivalent AES signal as layer-by-layer growth): actual film thickness uncertain by factor 3; see also 12 and 200ML structures 46.26.2a and c

DATA COLLECTION
Technique: LEED; video LEED
Dataset : IV spectra at normal incidence:
(10),(11),(20),(21),(22); 40<E<360 eV

THEORY/DATA TREATMENT
Dynamical LEED (CHANGE code): Moruzzi et al pot, 8 ph shs; Vor=-10 eV (then fit), Voi=-4eV; rms vibs 0.125Å

STRUCTURES EXAMINED
Semi-infinite Fe(100) with lateral Pd(100) lattice constant; spacings varied: 'bulk' Fe-Fe, top two Fe-Fe

QUALITY OF EXPERIMENT-THEORY FIT
RVHT=0.34, RPE=0.59, RZJ=0.14

2D UNIT CELLS (1 domain observed)

Cell	Ax (Å)	Ay (Å)	Bx (Å)	By (Å)	α (°)	Matrix	Pattern	Cell type
Bulk	2.750	0.000	0.000	2.750	90.0	(1.000, 0.000) (0.000, 1.000)	(1x1)	b: bulk lattice
Surface 1	2.750	0.000	0.000	2.750	90.0	(1.000, 0.000) (0.000, 1.000)	(1x1)	s1: commens. superlattice

3D COORDINATES

1.50Å bulk spacing was fit, keeping lateral Pd distance

Dx/Dy in Å, or as a fraction of layer's unit cell vectors; atom 0 at the origin. Epir/subr are bulk repeat vectors.

No. of atoms: 4

Bulk z = 1.500 Å

Region	Chem el.	At. no.	Cell type	Site occ.	Rel. to	Dx ± εx		Dy ± εy		Dz ± εz		Dz/Bz(%) ± εz/Bz
epir		-2					f		f		Å	
subr		-1				-1.375	Å	-1.375	Å	1.500	Å	
intf	Fe	1	b	1.00	0	0.000	f	0.000	f	0.000	Å	0.0
intf	Fe	2	b	1.00	1	0.500	f	0.500	f	1.405 ± .030	Å	93.7 ± 2.0
intf	Fe	3	b	1.00	2	-0.500	f	-0.500	f	1.445 ± .030	Å	96.3 ± 2.0
subl	Fe	4	b	1.00	3	0.500	f	0.500	f	1.500 ± .030	Å	100.0 ± 2.0

BOND DISTANCES AND ANGLES

Bond distances and angles are derived from coordinates

No. of distances/angles: 3

Interatomic dist. A-B (Å)	Atom A	Atom B	Atom C	Bond angle A-B-C (°)
2.750	Fe1	Fe1(1,0)	Fe2	55.0
2.399	Fe1	Fe2	Fe3	72.5
2.423	Fe2	Fe3	Fe4	74.3

COMMON NAME : Pd(100)-(1x1)-200Fe
CLASSIFICATION : 46.26.2c
TECHNIQUE : LEED
AUTHORS : J. Quinn, Y.S. Li, H. Li, D. Tian, F. Jona and P.M. Marcus
REFERENCE : Phys. Rev., B43, 3959 (1991)

ILLUSTRATION: 83

SURFACE TYPE
Substrate : Pd
Crystal face: 100
Temperature : RT
Bulk lattice: fcc
2D bulk symm: p4m
2D surf symm: p4m

Adsorbate: Fe
Coverage : 200 Fe/Pd
Pattern : (1x1)
Matrix : (1.000, 0.000)
(0.000, 1.000)

STRUCTURE TYPE
About 200 epitaxial monolayers, forming slightly distorted bcc Fe

SAMPLE PREPARATION (2 sample)
Treatment : Fe deposited 'fast' on RT or 120 K substrate
Crystallinity: broadening of LEED beams
Anal. methods: ARPES; coverage from AES
Contamination:

COMMENTS
C and O impurities present, and may affect structure; coverages measured in layer-equivalents (i.e. giving equivalent AES signal as layer-by-layer growth): actual film thickness uncertain by factor 3; see also 20 and 53ML structures 46.26.2a and b

DATA COLLECTION
Technique: LEED; video LEED
Dataset : IV spectra at normal incidence:
(10),(11),(20),(21),(22); 40<E<360 eV

THEORY/DATA TREATMENT
Dynamical LEED (CHANGE code): Moruzzi et al pot, 8 ph shs; Vor=-10 eV (then fit), Voi=-4eV; rms vibs 0.125Å

STRUCTURES EXAMINED
Semi-infinite Fe(100) with lateral lattice constant expanded from Pd(100) (from 2.75Å to 2.90±0.08Å; best-fit gave 2.87Å = bcc Fe value); spacings varied: 'bulk' Fe-Fe, top Fe-Fe

QUALITY OF EXPERIMENT-THEORY FIT
RVHT=0.25, RPE=0.39, RZJ=0.10

2D UNIT CELLS (1 domain observed)

Cell	Ax (Å)	Ay (Å)	Bx (Å)	By (Å)	α (°)	Matrix	Pattern	Cell type
Bulk	2.870	0.000	0.000	2.870	90.0	(1.000, 0.000) (0.000, 1.000)	(1x1)	b: bulk lattice
Surface 1	2.870	0.000	0.000	2.870	90.0	(1.000, 0.000) (0.000, 1.000)	(1x1)	s1: commens. superlattice

3D COORDINATES

1.48Å bulk spacing was fit

Dx/Dy in Å, or as a fraction of layer's unit cell vectors; atom 0 at the origin. Epir/subr are bulk repeat vectors.

No. of atoms: 4

Bulk z = 1.500 Å

Region	Chem el.	At. no.	Cell type	Site occ.	Rel. to	Dx ± εx		Dy ± εy		Dz ± εz		Dz/Bz(%) ± εz/Bz
epir		-2					f		f		Å	
subr		-1				-1.435	Å	-1.435	Å	1.480	Å	
intf	Fe	1	b	1.00	0	0.000	f	0.000	f	0.000	Å	0.0
intf	Fe	2	b	1.00	1	0.500	f	0.500	f	1.480 ± .030	Å	100.0 ± 2.0
intf	Fe	3	b	1.00	2	-0.500	f	-0.500	f	1.480 ± .030	Å	100.0 ± 2.0
subl	Fe	4	b	1.00	3	0.500	f	0.500	f	1.480 ± .030	Å	100.0 ± 2.0

BOND DISTANCES AND ANGLES

Bond distances and angles are derived from coordinates

No. of distances/angles: 3

Interatomic dist. A-B (Å)	Atom A	Atom B	Atom C	Bond angle A-B-C (°)
2.870	Fe1	Fe1(1,0)	Fe2	55.2
2.512	Fe1	Fe2	Fe3	72.2
2.512	Fe2	Fe3	Fe4	72.2

COMMON NAME : Pd(100)-c(2x2)-Mn / Pd3Mn(100)-(1x1)
CLASSIFICATION : 46.25.5b
TECHNIQUE : LEED
AUTHORS : D. Tian, R.F. Lin, F. Jona and P.M. Marcus
REFERENCE : Solid State Commun., 74, 1017 (1990)

ILLUSTRATION: 134

SURFACE TYPE
Substrate : Pd
Crystal face: 100
Temperature : RT
Bulk lattice: fcc
2D bulk symm: p4m
2D surf symm: p4m

Adsorbate: Mn
Coverage : 1.0 Mn/Pd
Pattern : (1x1)
Matrix : (1.000, 0.000)
(0.000, 1.000)

STRUCTURE TYPE
Mixed top layer, with Mn buckled inward; presumed alloying to at least 4th layer as Pd3Mn(100);
this tabulation uses Pd3Mn(100) as reference substrate (as used in LEED calculation): hence (1x1) surface lattice

SAMPLE PREPARATION (1 sample)
Treatment : Mn deposited both at Rt and -120C, giving same LEED int.
Crystallinity: high LEED background
Anal. methods:
Contamination:

COMMENTS
Compare with unheated version of this structure (class. no. 46.25.5a), which has opposite buckling in top mixed layer, but no deeper alloying;
no evidence of magnetism, despite use of spin-resolved Mn scattering potential

DATA COLLECTION
Technique: LEED
Dataset : IV spectra for 3 beams (10),(11),(.5,.5) (in c(2x2) notation) at normal incidence; E<=360 eV

THEORY/DATA TREATMENT
Dyn. LEED (CHANGE): 69 beams, 8 ph shs from Moruzzi et al (Pd) and relat. pot (Mn); Vor=-10 eV, Voi=-4eV; rms=0.12Å

STRUCTURES EXAMINED
Pd3Mn bulk terminating in mixed layer, with variable top layer buckling and spacing to next layer

QUALITY OF EXPERIMENT-THEORY FIT
Visual: 'moderately good fit'

2D UNIT CELLS (1 domain observed)

Cell	Ax (Å)	Ay (Å)	Bx (Å)	By (Å)	α (°)	Matrix	Pattern	Cell type
Bulk	3.890	0.000	0.000	3.890	90.0	(1.000, 0.000) (0.000, 1.000)	(1x1)	b: bulk lattice
Surface 1	3.890	0.000	0.000	3.890	90.0	(1.000, 0.000) (0.000, 1.000)	(1x1)	s1: commens. superlattice

3D COORDINATES

Mn1-Pd2: mixed buckled top layer; Pd3-Pd4: pure Pd 2nd layer;
Mn5-Pd6-Pd7-Pd8: period. repeating pair of bulk layers; 0.05Å error bars assumed for tabulation

Dx/Dy in Å, or as a fraction of layer's unit cell vectors; atom 0 at the origin. Epir/subr are bulk repeat vectors.

No. of atoms: 8

Bulk z = 1.945 Å

Region	Chem el.	At. no.	Cell type	Site occ.	Rel. to	Dx ± εx		Dy ± εy		Dz ± εz		Dz/Bz(%) ± εz/Bz
epir		-2					f		f		Å	
subr		-1				0.000	Å	0.000	Å	3.890	Å	
intf	Pd	1	s1	1.00	0	0.000	f	0.000	f	0.000	Å	0.0
intf	Mn	2	s1	1.00	1	-0.500	f	-0.500	f	0.200 ± .050	Å	10.3 ± 2.6
intf	Pd	3	s1	1.00	2	0.000	f	0.500	f	1.645 ± .050	Å	84.6 ± 2.6
intf	Pd	4	s1	1.00	3	0.500	f	-0.500	f	0.000	Å	0.0
subl	Mn	5	b	1.00	4	-0.500	f	0.000	f	1.945	Å	100.0
subl	Pd	6	b	1.00	5	0.500	f	0.500	f	0.000	Å	0.0
subl	Pd	7	b	1.00	6	-0.500	f	0.000	f	1.945	Å	100.0
subl	Pd	8	b	1.00	7	0.500	f	-0.500	f	0.000	Å	0.0

Pd(100)-c(2x2)-Mn / Pd3Mn(100)-(1x1)
46.25.5b BOND DISTANCES AND ANGLES

Bond distances and angles are derived from coordinates

No. of distances/angles: 6

Interatomic dist. A-B (Å)	Atom A	Atom B	Atom C	Bond angle A-B-C (°)
2.758	Pd1	Mn2	Pd1(-1,0)	89.7
2.758	Pd1	Mn2	Pd3	60.6
2.681	Pd1	Pd3	Pd1(-1,0)	93.0
2.681	Pd1	Pd3	Mn2	63.6
2.681	Pd1	Pd3	Pd4	59.1
2.547	Mn2	Pd3	Pd4	57.3

COMMON NAME : Pd(100)-c(2x2)-Mn mixed top layer
CLASSIFICATION : 46.25.5a
TECHNIQUE : LEED
AUTHORS : D. Tian, R.F. Lin, F. Jona and P.M. Marcus
REFERENCE : Solid State Commun., 74, 1017 (1990)

ILLUSTRATION: 33

SURFACE TYPE
Substrate : Pd
Crystal face: 100
Temperature : RT
Bulk lattice: fcc
2D bulk symm: p4m
2D surf symm: p4m

Adsorbate: Mn
Coverage : 1.0 Mn/Pd
Pattern : c(2x2)
Matrix : (1.000,-1.000)
(1.000, 1.000)

STRUCTURE TYPE
Mixed top layer, with Mn buckled outward

SAMPLE PREPARATION (1 sample)
Treatment : Mn deposited both at Rt and -120C, giving same LEED int.
Crystallinity: high LEED background
Anal. methods:
Contamination:

COMMENTS
Compare with annealed version of this structure (class. no. 46.25.5b), which has opposite buckling in top mixed layer, as well as deeper alloying;
no evidence of magnetism, despite use of spin-resolved Mn scattering potential

DATA COLLECTION
Technique: LEED
Dataset : IV spectra for 3 beams (10),(11),(.5,.5) at normal incidence; E<=360 eV

THEORY/DATA TREATMENT
Dyn. LEED (CHANGE): 69 beams, 8 ph shs from Moruzzi et al (Pd) and relat. pot (Mn); Vor=-10 eV, Voi=-4eV; rms=0.12Å

STRUCTURES EXAMINED
Magnetic full Mn monolayer on Pd(100); mixed 50/50 Mn/Pd top layer on Pd(100) with variable top layer buckling and spacing to next layer

QUALITY OF EXPERIMENT-THEORY FIT
Visual: 'moderately good fit'

2D UNIT CELLS (1 domain observed)

Cell	Ax (Å)	Ay (Å)	Bx (Å)	By (Å)	α (°)	Matrix	Pattern	Cell type
Bulk	2.751	0.000	0.000	2.751	90.0	(1.000, 0.000) (0.000, 1.000)	(1x1)	b: bulk lattice
Surface 1	2.751	-2.751	2.751	2.751	90.0	(1.000,-1.000) (1.000, 1.000)	c(2x2)	s1: commens. superlattice

3D COORDINATES

Mn1-Pd2: mixed buckled top layer; 0.05Å error bars assumed for tabulation

Dx/Dy in Å, or as a fraction of layer's unit cell vectors; atom 0 at the origin. Epir/subr are bulk repeat vectors.

No. of atoms: 4

Bulk z = 1.945 Å

Region	Chem el.	At. no.	Cell type	Site occ.	Rel. to	Dx ± εx		Dy ± εy		Dz ± εz		Dz/Bz(%) ± εz/Bz
epir		-2					f		f		Å	
subr		-1				1.375	Å	1.375	Å	1.945	Å	
intf	Mn	1	s1	.50	0	0.000	f	0.000	f	0.000	Å	0.0
intf	Pd	2	s1	.50	1	-0.500	f	0.500	f	0.200 ± .050	Å	10.3 ± 2.6
intf	Pd	3	b	1.00	2	-0.500	f	-0.500	f	1.645 ± .050	Å	84.6 ± 2.6
subl	Pd	4	b	1.00	3	0.500	f	0.500	f	1.945	Å	100.0

BOND DISTANCES AND ANGLES

Bond distances and angles are derived from coordinates

No. of distances/angles: 6

Interatomic dist. A-B (Å)	Atom A	Atom B	Atom C	Bond angle A-B-C (°)
2.758	Mn1	Pd2	Mn1(0,1)	89.7
2.758	Mn1	Pd2	Pd3	60.6
2.681	Mn1	Pd3	Mn1(-1,0)	93.0
2.681	Mn1	Pd3	Pd2	63.6

Pd(100)-c(2x2)-Mn mixed top layer
46.25.5a

Bond Distances and Angles - Continued

Interatomic dist. A-B (Å)	Atom A	Atom B	Atom C	Bond angle A-B-C (°)
2.681	Mn1	Pd3	Pd4	119.1
2.547	Pd2	Pd3	Pd4	85.2

COMMON NAME : Pd(100)-c(2x2)-S
CLASSIFICATION : 46.16.1
TECHNIQUE : LEED
AUTHORS : W. Berndt, R. Hora and M. Scheffler
REFERENCE : Surf. Sci., 117, 188 (1982)

ILLUSTRATION: 28,29

SURFACE TYPE
Substrate : Pd
Crystal face: 100
Temperature : RT*
Bulk lattice: fcc
2D bulk symm: p4m
2D surf symm: p4m

Adsorbate: S
Coverage : 0.5 (S/Pd)
Pattern : c(2x2)
Matrix : (1.000, 1.000)
(-1.000, 1.000)

STRUCTURE TYPE
Atomic adsorption in hollow sites

SAMPLE PREPARATION (1 sample)
Treatment : O2 at 500C for 30 min, Ar ion bombardment for 10 min
Crystallinity:
Anal. methods:
Contamination:

COMMENTS

DATA COLLECTION
Technique: LEED
Dataset : I-V curves: 4, 14 non-equivalent beams at Θ=0, 5°, resp. along [100] azimuth; energy range 20-200 eV

THEORY/DATA TREATMENT
Dynamical LEED (layer KKR with layer doubling): 5 ph shs, Pd Moruzzi et al pot, SCASW pot for S; Voi=-4 eV

STRUCTURES EXAMINED
Hollow, bridge and top sites with variable S/Pd spacing

QUALITY OF EXPERIMENT-THEORY FIT
RPE=0.3

2D UNIT CELLS (1 domain observed)

Cell	Ax (Å)	Ay (Å)	Bx (Å)	By (Å)	α (°)	Matrix	Pattern	Cell type
Bulk	2.751	0.000	0.000	2.751	90.0	(1.000, 0.000) (0.000, 1.000)	(1x1)	b: bulk lattice
Surface 1	2.751	2.751	-2.751	2.751	90.0	(1.000, 1.000) (-1.000, 1.000)	c(2x2)	s1: commens. superlattice

3D COORDINATES

S1: overlayer in hollow sites

Dx/Dy in Å, or as a fraction of layer's unit cell vectors; atom 0 at the origin. Epir/subr are bulk repeat vectors.

No. of atoms: 3

Bulk z = 1.940 Å

Region	Chem el.	At. no.	Cell type	Site occ.	Rel. to	Dx ± εx		Dy ± εy		Dz ± εz		Dz/Bz(%) ± εz/Bz
epir		-2					f		f		Å	
subr		-1				-1.375	Å	-1.375	Å	1.940	Å	
ovrl	S	1	s1	.50	0	0.000	f	0.000	f	0.000	Å	0.0
intf	Pd	2	b	1.00	1	0.500	f	0.500	f	1.300 ± .050	Å	67.0 ± 2.6
subl	Pd	3	b	1.00	2	-0.500	f	-0.500	f	1.940	Å	100.0

BOND DISTANCES AND ANGLES

Bond distances and angles are derived from coordinates

No. of distances/angles: 3

Interatomic dist. A-B (Å)	Atom A	Atom B	Atom C	Bond angle A-B-C (°)
2.339	S1	Pd2	Pd2(1,0)	126.0
2.339	S1	Pd2	Pd3	78.7
2.747	Pd2	Pd3		

ILLUSTRATION: 22,24

COMMON NAME : Pd(111)-(√3x√3)R30°-S
CLASSIFICATION : 46.16.2
TECHNIQUE : LEED
AUTHORS : F. Maca, M. Scheffler and W. Berndt
REFERENCE : Surf. Sci., 160, 467 (1985)

SURFACE TYPE
Substrate : Pd
Crystal face: 111
Temperature : RT
Bulk lattice: fcc
2D bulk symm: p3m1
2D surf symm: p31m

Adsorbate: S
Coverage : 0.33 S/Pd
Pattern : (√3x√3)R30°
Matrix : (1.000, 1.000)
(-2.000, 1.000)

STRUCTURE TYPE
Atomic adsorption in fcc hollow sites

SAMPLE PREPARATION (1 sample)
Treatment : exposure to H_2S
Crystallinity:
Anal. methods:
Contamination: AES: no C, O and S detected

COMMENTS
Temperature effects were neglected

DATA COLLECTION
Technique: LEED
Dataset : I-V curves: 25 non equivalent beams at Θ=0° and 5°; energy range 30-200 eV
fcc and hcp hollow and top sites; S-Pd spacing varied from 1.1 to 1.8Å

THEORY/DATA TREATMENT
Dynamical LEED (RFS): 5 phase shifts (Pd Moruzzi et al pot, S potential for SPd_3 crystal); Voi=-4 eV

QUALITY OF EXPERIMENT-THEORY FIT
RPE=0.245

2D UNIT CELLS (1 domain observed)

Cell	Ax (Å)	Ay (Å)	Bx (Å)	By (Å)	α (°)	Matrix	Pattern	Cell type
Bulk	2.748	0.000	1.374	2.380	60.0	(1.000, 0.000) (0.000, 1.000)	(1x1)	b: bulk lattice
Surface 1	4.122	2.380	-4.122	2.380	120.0	(1.000, 1.000) (-2.000, 1.000)	(√3x√3)R30°	s1: commens. superlattice

3D COORDINATES

S1: overlayer in fcc hollow sites

Dx/Dy in Å, or as a fraction of layer's unit cell vectors; atom 0 at the origin. Epir/subr are bulk repeat vectors.

No. of atoms: 3

Bulk z = 2.250 Å

Reg ion	Chem el.	At. no.	Cell type	Site occ.	Rel. to	Dx ± εx		Dy ± εy		Dz ± εz		Dz/Bz(%) ± εz/Bz
epir		-2					f		f		Å	
subr		-1				1.374	Å	0.793	Å	2.250	Å	
ovrl	S	1	s1	.33	0	0.000	f	0.000	f	0.000	Å	0.0
intf	Pd	2	b	1.00	1	0.333	f	0.333	f	1.530 ± .050	Å	68.0 ± 2.2
subl	Pd	3	b	1.00	2	0.333	f	0.333	f	2.250	Å	100.0

BOND DISTANCES AND ANGLES

Bond distances and angles are derived from coordinates

No. of distances/angles: 3

Interatomic dist. A-B (Å)	Atom A	Atom B	Atom C	Bond angle A-B-C (°)
2.204	S1	Pd2	Pd2(1,0)	128.6
2.204	S1	Pd2	Pd3	169.2
2.753	Pd2	Pd3		

COMMON NAME : Pt(100)-(1x1)
CLASSIFICATION : 78.16
TECHNIQUE : HEIS
AUTHORS : J.A. Davies, T.E. Jackman, D.P. Jackson and P.R. Norton
REFERENCE : Surf. Sci., 109, 20 (1981)

ILLUSTRATION: 2

SURFACE TYPE
Substrate : Pt
Crystal face: 100
Temperature : 175 K
Bulk lattice: fcc
2D bulk symm: p4m
2D surf symm: p4m

Adsorbate:
Coverage :
Pattern : (1x1)
Matrix : (1.000, 0.000)
(0.000, 1.000)

STRUCTURE TYPE
Unreconstructed metastable structure with slight top spacing expansion

SAMPLE PREPARATION (1 sample)
Treatment : Ar+ sputtering at high T, then O2 treatments and annealing
Crystallinity:
Anal. methods:
Contamination: nuclear microanalysis: <0.02ML D, C, O

COMMENTS
Unreconstructed Pt(100) stable only in presence of 0.05-0.1ML H2; complete removal of H resulted in reconstruction; top layer spacing varies from 1.960 to 1.965Å with different impurities

DATA COLLECTION
Technique: HEIS; Rutherford backscatt. of 2MeV 4He+ io
Dataset : surface peak area as function of rotation about [100] axis

THEORY/DATA TREATMENT
Simulation of channeling/blocking: ΘD=115 K

STRUCTURES EXAMINED
Top interlayer spacing contractions and expansions

QUALITY OF EXPERIMENT-THEORY FIT
Visual

2D UNIT CELLS (1 domain observed)

Cell	Ax (Å)	Ay (Å)	Bx (Å)	By (Å)	α (°)	Matrix	Pattern	Cell type
Bulk	2.770	0.000	0.000	2.770	90.0	(1.000, 0.000) (0.000, 1.000)	(1x1)	b: bulk lattice
Surface 1	2.770	0.000	0.000	2.770	90.0	(1.000, 0.000) (0.000, 1.000)	(1x1)	s1: commens. superlattice

3D COORDINATES

0.05Å error bar assumed for tabulation

Dx/Dy in Å, or as a fraction of layer's unit cell vectors; atom 0 at the origin. Epir/subr are bulk repeat vectors.

No. of atoms: 3

Bulk z = 1.960 Å

Region	Chem el.	At. no.	Cell type	Site occ.	Rel. to	Dx ± εx		Dy ± εy		Dz ± εz		Dz/Bz(%) ± εz/Bz
epir		-2					f		f		Å	
subr		-1				-1.385	Å	-1.385	Å	1.960	Å	
intf	Pt	1	b	1.00	0	0.000	f	0.000	f	0.000	Å	0.0
intf	Pt	2	b	1.00	1	0.500	f	0.500	f	1.963 ± .050	Å	100.2 ± 2.6
subl	Pt	3	b	1.00	2	-0.500	f	-0.500	f	1.960	Å	100.0

BOND DISTANCES AND ANGLES

Bond distances and angles are derived from coordinates

No. of distances/angles: 3

Interatomic dist. A-B (Å)	Atom A	Atom B	Atom C	Bond angle A-B-C (°)
2.770	Pt1	Pt1(1,0)	Pt2	60.0
2.773	Pt1	Pt2	Pt3	90.1
2.771	Pt2	Pt3		

ILLUSTRATION: 2

COMMON NAME : Pt(100)-(1x1)
CLASSIFICATION : 78.16a
TECHNIQUE : LEED
AUTHORS : E. Lang, W. Grimm and K. Heinz
REFERENCE : Surf. Sci., 117, 169 (1982)

SURFACE TYPE
Substrate : Pt
Crystal face: 100
Temperature : 100 K
Bulk lattice: fcc
2D bulk symm: p4m
2D surf symm: p4m

Adsorbate:
Coverage :
Pattern : (1x1)
Matrix : (1.000, 0.000)
(0.000, 1.000)

STRUCTURE TYPE
Unreconstructed unrelaxed metastable surface

SAMPLE PREPARATION (1 sample)
Treatment : ion bombardment and 1000 K heating in O2, then 1300K flashes
Crystallinity:
Anal. methods:
Contamination: AES: no impurities

COMMENTS

DATA COLLECTION
Technique: LEED
Dataset : IV curves obtained for many beams at normal incidence, energy range 200-600 eV

THEORY/DATA TREATMENT
Quasidynamical LEED (RFS, no intralayer mult. scattering, additional damping for convergence): ΘD=240 K

STRUCTURES EXAMINED
Contractions and expansions of top interlayer spacing

QUALITY OF EXPERIMENT-THEORY FIT
RZJ=0.127

2D UNIT CELLS (1 domain observed)

Cell	Ax (Å)	Ay (Å)	Bx (Å)	By (Å)	α (°)	Matrix	Pattern	Cell type
Bulk	2.770	0.000	0.000	2.770	90.0	(1.000, 0.000) (0.000, 1.000)	(1x1)	b: bulk lattice
Surface 1	2.770	0.000	0.000	2.770	90.0	(1.000, 0.000) (0.000, 1.000)	(1x1)	s1: commens. superlattice

3D COORDINATES

0.1Å error bar assumed for tabulation

Dx/Dy in Å, or as a fraction of layer's unit cell vectors; atom 0 at the origin. Epir/subr are bulk repeat vectors.

No. of atoms: 3

Bulk z = 1.960 Å

Region	Chem el.	At. no.	Cell type	Site occ.	Rel. to	Dx ± εx		Dy ± εy		Dz ± εz		Dz/Bz(%) ± εz/Bz
epir		-2					f		f		Å	
subr		-1				1.385	Å	1.385	Å	1.960	Å	
intf	Pt	1	b	1.00	0	0.000	f	0.000	f	0.000	Å	0.0
intf	Pt	2	b	1.00	1	0.500	f	0.500	f	1.960 ± .100	Å	100.0 ± 5.1
subl	Pt	3	b	1.00	2	-0.500	f	-0.500	f	1.960	Å	100.0

BOND DISTANCES AND ANGLES

Bond distances are derived from coordinates

No. of distances/angles: 2

Interatomic dist. A-B (Å)	Atom A	Atom B	Atom C	Bond angle A-B-C (°)
2.770	Pt1	Pt1(1,0)		
2.771	Pt1	Pt2		

COMMON NAME : Pt(100)-(1x1)
CLASSIFICATION : 78.6
TECHNIQUE : SPLEED
AUTHORS : R. Feder
REFERENCE : Surf. Sci., 68, 229 (1977)

ILLUSTRATION: 2

SURFACE TYPE
Substrate : Pt
Crystal face: 100
Temperature : 300 K
Bulk lattice: fcc
2D bulk symm: p4m
2D surf symm: p4m

Adsorbate:
Coverage :
Pattern : (1x1)
Matrix : (1.000, 0.000)
(0.000, 1.000)

STRUCTURE TYPE
Unreconstructed metastable surface (assumed unrelaxed)

SAMPLE PREPARATION (1 sample)
Treatment : impurity stabilized
Crystallinity:
Anal. methods:
Contamination:

COMMENTS

DATA COLLECTION
Technique: SPLEED
Dataset : SPLEED I-V and P-V curves; E range 0-200 eV

THEORY/DATA TREATMENT
Dynamical SPLEED: 8 relat. phase shifts from relat. atomic charge densities; Voi=-4 eV; ΘD=178 K

STRUCTURES EXAMINED
Bulk termination assumed

QUALITY OF EXPERIMENT-THEORY FIT
Visual

2D UNIT CELLS (1 domain observed)

Cell	Ax (Å)	Ay (Å)	Bx (Å)	By (Å)	α (°)	Matrix	Pattern	Cell type
Bulk	2.770	0.000	0.000	2.770	90.0	(1.000, 0.000) (0.000, 1.000)	(1x1)	b: bulk lattice
Surface 1	2.770	0.000	0.000	2.770	90.0	(1.000, 0.000) (0.000, 1.000)	(1x1)	s1: commens. superlattice

3D COORDINATES

bulk structure assumed

Dx/Dy in Å, or as a fraction of layer's unit cell vectors; atom 0 at the origin. Epir/subr are bulk repeat vectors.

No. of atoms: 3

Bulk z = 1.960 Å

Reg ion	Chem el.	At. no.	Cell type	Site occ.	Rel. to	Dx ± εx		Dy ± εy		Dz ± εz		Dz/Bz(%) ± εz/Bz
epir		-2					f		f		Å	
subr		-1				1.385	Å	1.385	Å	1.960	Å	
intf	Pt	1	b	1.00	0	0.000	f	0.000	f	0.000	Å	0.0
intf	Pt	2	b	1.00	1	0.500	f	0.500	f	1.960	Å	100.0
subl	Pt	3	b	1.00	2	-0.500	f	-0.500	f	1.960	Å	100.0

BOND DISTANCES AND ANGLES

Bond distances are derived from coordinates

No. of distances/angles: 1

Interatomic dist. A-B (Å)	Atom A	Atom B	Atom C	Bond angle A-B-C (°)
2.770	Pt1	Pt1(1,0)		

ILLUSTRATION: 5

COMMON NAME : Pt(110)-(1x2)
CLASSIFICATION : 78.19
TECHNIQUE : ALICISS
AUTHORS : H. Niehus
REFERENCE : Surf. Sci., 145, 407 (1984)

SURFACE TYPE
Substrate : Pt
Crystal face: 110
Temperature : 300 K
Bulk lattice: fcc
2D bulk symm: pmm
2D surf symm: pmm

Adsorbate:
Coverage :
Pattern : (1x2)
Matrix : (1.000, 0.000)
(0.000, 2.000)

STRUCTURE TYPE
Missing-row model (bulk atom positions assumed)

SAMPLE PREPARATION (1 sample)
Treatment : Ar+ bombardment, annealing in O2
Crystallinity: sharp LEED pattern
Anal. methods:
Contamination: AES: no contaminants

COMMENTS

DATA COLLECTION
Technique: ALICISS; scattering cross-section for 2keV
Dataset : 7 azimuthal incident directions at polar incidence angles in range 0-90°; scattering angle 145°

THEORY/DATA TREATMENT
Low-energy alkali impact collision ion scattering spectr.: exp. determined shadow cone used to locate surface atoms

STRUCTURES EXAMINED
Missing-row, buckled-row, paired-rows and sawtooth models; unrelaxed bulk atomic positions assumed

QUALITY OF EXPERIMENT-THEORY FIT
Visual

2D UNIT CELLS (1 domain observed)

Cell	Ax (Å)	Ay (Å)	Bx (Å)	By (Å)	α (°)	Matrix	Pattern	Cell type
Bulk	2.770	0.000	0.000	3.917	90.0	(1.000, 0.000) (0.000, 1.000)	(1x1)	b: bulk lattice
Surface 1	2.770	0.000	0.000	7.835	90.0	(1.000, 0.000) (0.000, 2.000)	(1x2)	s1: commens. superlattice

3D COORDINATES

Pt1: remaining row of missing-row model

Dx/Dy in Å, or as a fraction of layer's unit cell vectors; atom 0 at the origin. Epir/subr are bulk repeat vectors.

No. of atoms: 3 Bulk z = 1.385 Å

Region	Chem el.	At. no.	Cell type	Site occ.	Rel. to	Dx ± εx		Dy ± εy		Dz ± εz		Dz/Bz(%) ± εz/Bz
epir		-2					f		f		Å	
subr		-1				1.385	Å	1.959	Å	1.385	Å	
intf	Pt	1	s1	.50	0	0.000	f	0.000	f	0.000	Å	0.0
intf	Pt	2	b	1.00	1	0.500	f	0.500	f	1.385	Å	100.0
subl	Pt	3	b	1.00	2	-0.500	f	-0.500	f	1.385	Å	100.0

BOND DISTANCES AND ANGLES

Bond distance is derived from coordinates

No. of distances/angles: 1

Interatomic dist. A-B (Å)	Atom A	Atom B	Atom C	Bond angle A-B-C (°)
2.770	Pt1	Pt2		

COMMON NAME : Pt(110)-(1x2)
CLASSIFICATION : 78.32
TECHNIQUE : LEED
AUTHORS : E.C. Sowa, M.A. Van Hove and D.L. Adams
REFERENCE : Surf. Sci., 199, 174 (1988)

ILLUSTRATION: 5

SURFACE TYPE
Substrate : Pt
Crystal face: 110
Temperature : RT
Bulk lattice: fcc
2D bulk symm: pmm
2D surf symm: pmm

Adsorbate:
Coverage :
Pattern : (1x2)
Matrix : (1.000, 0.000)
(0.000, 2.000)

STRUCTURE TYPE
Missing-row reconstruction with multilayer relaxation, including 2nd- and 4th-layer pairing and 3rd-layer buckling

SAMPLE PREPARATION (1 sample)
Treatment : cycles of Ar+ bombardment and annealing at 1273 K
Crystallinity: no facetting in LEED pattern
Anal. methods:
Contamination: AES: no impurities

COMMENTS

DATA COLLECTION
Technique: LEED
Dataset : I-V curves for 6 half-order and 4 integral-order beams

THEORY/DATA TREATMENT
Dynamical LEED: bulk ΘD=302 K, surface enhancement x 1.4

STRUCTURES EXAMINED
Missing-row model: top 4 interlayer spacings varied; row- pairing in layers 2 and 4, vertical buckling in layer 3

QUALITY OF EXPERIMENT-THEORY FIT
Average of RRZJ,RPE,ROS,R1,R2=0.239

2D UNIT CELLS (1 domain observed)

Cell	Ax (Å)	Ay (Å)	Bx (Å)	By (Å)	α (°)	Matrix	Pattern	Cell type
Bulk	2.775	0.000	0.000	3.924	90.0	(1.000, 0.000) (0.000, 1.000)	(1x1)	b: bulk lattice
Surface 1	2.775	0.000	0.000	7.848	90.0	(1.000, 0.000) (0.000, 2.000)	(1x2)	s1: commens. superlattice

3D COORDINATES

Pt1: remaining row; Pt2-Pt3, Pt6-Pt7: row-paired 2nd, 4th layers;
Pt4-Pt5: buckled 3rd layer; 0.1Å lateral error bars assumed for tabulation

Dx/Dy in Å, or as a fraction of layer's unit cell vectors; atom 0 at the origin. Epir/subr are bulk repeat vectors.

No. of atoms: 9

Bulk z = 1.387 Å

Reg ion	Chem el.	At. no.	Cell type	Site occ.	Rel. to	Dx ± εx	Dy ± εy	Dz ± εz	Dz/Bz(%) ± εz/Bz
epir		-2				f	f	Å	
subr		-1				1.388 Å	1.962 Å	1.387 Å	
intf	Pt	1	s1	.50	0	0.000 f	0.000 f	0.000 Å	0.0
intf	Pt	2	s1	.50	1	0.500 f	0.742 ± .013 f	1.131 ± .030 Å	81.5 ± 2.2
intf	Pt	3	s1	.50	2	0.000 f	-0.484 ± .013 f	0.000 Å	0.0
intf	Pt	4	s1	.50	3	-0.500 f	0.242 ± .013 f	1.052 ± .060 Å	75.8 ± 4.3
intf	Pt	5	s1	.50	4	0.000 f	-0.500 f	0.320 ± .060 Å	23.1 ± 4.3
intf	Pt	6	s1	.50	5	0.500 f	0.735 ± .013 f	1.106 ± .060 Å	79.7 ± 4.3
intf	Pt	7	s1	.50	6	0.000 f	-0.470 ± .013 f	0.000 Å	0.0
intf	Pt	8	b	1.00	7	-0.500 f	-0.530 ± .026 f	1.375 ± .060 Å	99.1 ± 4.3
subl	Pt	9	b	1.00	8	0.500 f	0.500 f	1.387 Å	100.0

Pt(110)-(1x2)
78.32

BOND DISTANCES AND ANGLES

Bond distances and angles are derived from coordinates

No. of distances/angles: 10

Interatomic dist. A-B (Å)	Atom A	Atom B	Atom C	Bond angle A-B-C (°)
2.775	Pt1	Pt1(1,0)		
2.577	Pt2	Pt4	Pt6	55.8
2.703	Pt1	Pt2(0,-1)	Pt3(0,-1)	41.5
2.703	Pt1	Pt2(0,-1)	Pt4(0,-1)	47.9
2.703	Pt1	Pt2(0,-1)	Pt5	53.9
2.703	Pt1	Pt3	Pt4	116.5
2.703	Pt1	Pt3	Pt5	53.9
2.703	Pt1	Pt3	Pt6(0,-1)	111.3
2.703	Pt1	Pt3	Pt7	115.8
2.503	Pt1	Pt5	Pt6(0,-1)	115.6

COMMON NAME : Pt(110)-(1x2)
CLASSIFICATION : 78.33a
TECHNIQUE : LEED
AUTHORS : P. Fery, W. Moritz and D. Wolf
REFERENCE : Phys. Rev., B38, 7275 (1988)

ILLUSTRATION: 5

SURFACE TYPE
Substrate : Pt
Crystal face: 110
Temperature : RT*
Bulk lattice: fcc
2D bulk symm: pmm
2D surf symm: pmm

Adsorbate:
Coverage :
Pattern : (1x2)
Matrix : (1.000, 0.000)
(0.000, 2.000)

STRUCTURE TYPE
Missing-row reconstruction with multilayer relaxation, including 2nd- and 4th-layer pairing, and 3rd- and 5th-layer buckling

SAMPLE PREPARATION (1 sample)
Treatment : Ar+ sputtering (2h), then annealing (10mins) at 900 K in O2
Crystallinity:
Anal. methods:
Contamination: AES: impurities below 5% ML

COMMENTS
9 structural parameters were optimised by a block refinement procedure

DATA COLLECTION
Technique: LEED
Dataset : IV spectra: 12 beams at normal incidence; 30<E<200 eV

THEORY/DATA TREATMENT
Dynamical LEED (layer doubling, matrix inversion): VoiαE**1/3; VorαE**1/2 (fit); ΘD=270 K

STRUCTURES EXAMINED
Missing row: variations of first 5 interlayer spacings, 2nd- and 4th-layer row pairing, 3rd- and 5th-layer buckling

QUALITY OF EXPERIMENT-THEORY FIT
RPE=0.36, RZJ=0.26

2D UNIT CELLS (1 domain observed)

Cell	Ax (Å)	Ay (Å)	Bx (Å)	By (Å)	α (°)	Matrix	Pattern	Cell type
Bulk	2.770	0.000	0.000	3.920	90.0	(1.000, 0.000) (0.000, 1.000)	(1x1)	b: bulk lattice
Surface 1	2.770	0.000	0.000	7.840	90.0	(1.000, 0.000) (0.000, 2.000)	(1x2)	s1: commens. superlattice

3D COORDINATES

Pt1: remaining row; Pt2-Pt3, Pt6-Pt7: row-paired 2nd, 4th layers;
Pt4-Pt5, Pt8-Pt9: buckled 3rd, 5th layers; 0.05Å/0.1Å perp/lateral error bars assumed for tabulation

Dx/Dy in Å, or as a fraction of layer's unit cell vectors; atom 0 at the origin. Epir/subr are bulk repeat vectors.

No. of atoms: 11

Bulk z = 1.390 Å

Region	Chem el.	At. no.	Cell type	Site occ.	Rel. to	Dx ± εx	Dy ± εy	Dz ± εz	Dz/Bz(%) ± εz/Bz
epir		-2				f	f	Å	
subr		-1				-1.385 Å	-1.960 Å	1.390 Å	
intf	Pt	1	s1	.50	0	0.000 f	0.000 f	0.000 Å	0.0
intf	Pt	2	s1	.50	1	0.500 f	0.256 ± .013 f	1.100 ± .050 Å	79.1 ± 3.6
intf	Pt	3	s1	.50	2	0.000 f	0.488 ± .013 f	0.000 Å	0.0
intf	Pt	4	s1	.50	3	-0.500 f	-0.244 ± .013 f	1.290 ± .050 Å	92.8 ± 3.6
intf	Pt	5	s1	.50	4	0.000 f	-0.500 f	0.030 ± .050 Å	2.2 ± 3.6
intf	Pt	6	s1	.50	5	0.500 f	0.256 ± .013 f	1.290 ± .050 Å	92.8 ± 3.6
intf	Pt	7	s1	.50	6	0.000 f	0.488 ± .013 f	0.000 Å	0.0
intf	Pt	8	s1	.50	7	-0.500 f	-0.244 ± .013 f	1.380 ± .050 Å	99.3 ± 3.6
intf	Pt	9	s1	.50	8	0.000 f	-0.500 f	0.030 ± .050 Å	2.2 ± 3.6
intf	Pt	10	b	1.00	9	0.500 f	0.500 f	1.360 ± .050 Å	97.8 ± 3.6
subl	Pt	11	b	1.00	10	-0.500 f	-0.500 f	1.390 Å	100.0

Pt(110)-(1x2)
78.33a

BOND DISTANCES AND ANGLES

Bond distances and angles are derived from coordinates

No. of distances/angles: 8

Interatomic dist. A-B (Å)	Atom A	Atom B	Atom C	Bond angle A-B-C (°)
2.770	Pt1	Pt1(1,0)		
2.675	Pt1	Pt2	Pt3(0,-1)	41.4
2.675	Pt1	Pt2	Pt4	117.6
2.675	Pt1	Pt2	Pt5	55.2
2.560	Pt1	Pt5	Pt6	117.9
2.560	Pt1	Pt5	Pt7(0,-1)	118.6
2.691	Pt2	Pt4	Pt5(0,1)	118.6
2.691	Pt2	Pt4	Pt6	60.4

COMMON NAME : Pt(110)-(1x2)
CLASSIFICATION : 78.34
TECHNIQUE : MEIS
AUTHORS : P. Fenter and T. Gustafsson
REFERENCE : Phys. Rev., 38, 10197 (1988)

ILLUSTRATION: 5

SURFACE TYPE
Substrate : Pt
Crystal face: 110
Temperature : RT
Bulk lattice: fcc
2D bulk symm: pmm
2D surf symm: pmm

Adsorbate:
Coverage :
Pattern : (1x2)
Matrix : (1.000, 0.000)
(0.000, 2.000)

STRUCTURE TYPE
Missing-row reconstruction with multilayer relaxation, including 3rd-layer buckling

SAMPLE PREPARATION (sample)
Treatment :
Crystallinity:
Anal. methods:
Contamination:

COMMENTS

DATA COLLECTION
Technique: MEIS
Dataset : proton scattering in (-1,1,1) and (0,0,1) scattering planes: channeling and blocking data

THEORY/DATA TREATMENT
Monte-Carlo simulation of channeling and blocking data

STRUCTURES EXAMINED
Top two interlayer spacings; 2nd-layer row-pairing and 3rd-layer, all within missing-row model

2D UNIT CELLS (1 domain observed)

Cell	Ax (Å)	Ay (Å)	Bx (Å)	By (Å)	α (°)	Matrix	Pattern	Cell type
Bulk	2.780	0.000	0.000	3.925	90.0	(1.000, 0.000) (0.000, 1.000)	(1x1)	b: bulk lattice
Surface 1	2.780	0.000	0.000	7.850	90.0	(1.000, 0.000) (0.000, 2.000)	(1x2)	s1: commens. superlattice

3D COORDINATES

Pt1: remaining ridge row; Pt2-Pt3: bulk-like 2nd layer;
Pt4-Pt5: buckled 3rd layer; 0.1Å error bars assumed for tabulation

Dx/Dy in Å, or as a fraction of layer's unit cell vectors; atom 0 at the origin. Epir/subr are bulk repeat vectors.

No. of atoms: 7 Bulk z = 1.390 Å

Region	Chem el.	At. no.	Cell type	Site occ.	Rel. to	Dx ± εx		Dy ± εy		Dz ± εz		Dz/Bz(%) ± εz/Bz
epir		-2					f		f		Å	
subr		-1				-1.390	Å	-1.963	Å	1.390	Å	
intf	Pt	1	s1	.50	0	0.000	f	0.000	f	0.000	Å	0.0
intf	Pt	2	s1	.50	1	0.500	f	0.250 ± .013	f	1.168 ± .100	Å	84.0 ± 7.2
intf	Pt	3	s1	.50	2	0.000	f	0.500 ± .013	f	0.000	Å	0.0
intf	Pt	4	s1	.50	3	-0.500	f	-0.250 ± .013	f	1.396 ± .100	Å	100.4 ± 7.2
intf	Pt	5	s1	.50	4	0.000	f	-0.500	f	0.100 ± .100	Å	7.2 ± 7.2
intf	Pt	6	b	1.00	5	0.500	f	0.500	f	1.340 ± .100	Å	96.4 ± 7.2
subl	Pt	7	b	1.00	6	-0.500	f	-0.500	f	1.390	Å	100.0

BOND DISTANCES AND ANGLES

Bond distances and angles are derived from coordinates

No. of distances/angles: 6

Interatomic dist. A-B (Å)	Atom A	Atom B	Atom C	Bond angle A-B-C (°)
2.780	Pt1	Pt1(1,0)		
2.674	Pt1	Pt2	Pt3(1,-1)	121.3
2.674	Pt1	Pt2	Pt4	118.5
2.674	Pt1	Pt2	Pt5	57.7
2.660	Pt1	Pt5	Pt6	119.1

Pt(110)-(1x2)
78.34

Bond Distances and Angles - Continued

Interatomic dist. A-B (Å)	Atom A	Atom B	Atom C	Bond angle A-B-C (°)
2.778	Pt2	Pt4	Pt6	60.9

COMMON NAME : Pt(110)-(1x2) ILLUSTRATION: 5
CLASSIFICATION : 78.41
TECHNIQUE : XRD
AUTHORS : E. Vlieg, I.K. Robinson and K. Kern
REFERENCE : Surf. Sci., 233, 248 (1990)

SURFACE TYPE
Substrate : Pt
Crystal face: 110
Temperature : RT
Bulk lattice: fcc
2D bulk symm: pmm
2D surf symm: pmm

Adsorbate:
Coverage :
Pattern : (1x2)
Matrix : (1.000, 0.000)
(0.000, 2.000)

STRUCTURE TYPE
Missing-row reconstruction, with multilayer relaxations down to 4th layer

SAMPLE PREPARATION (1 sample)
Treatment : extensive anneals and sputterings, then 900C anneals
Crystallinity:
Anal. methods: AES
Contamination: AES: no contaminants

COMMENTS
Rms vibration amplitudes of topmost layers are found enhanced by a factor of nearly 2 over bulk values

DATA COLLECTION
Technique: XRD; synchrotron radiation (NSLS)
Dataset : 88 reflections from 23 (hk) rods; lambda=1.09Å

THEORY/DATA TREATMENT
Fitting of parameters by chi**2 minimization, including ΘD for layers 1, 2 and 4

STRUCTURES EXAMINED
Missing-row model with perpendicular relaxations of atoms in layers 1 and 2, pairing of layers 2 and 4

QUALITY OF EXPERIMENT-THEORY FIT
Chi**2=2.1

2D UNIT CELLS (1 domain observed)

Cell	Ax (Å)	Ay (Å)	Bx (Å)	By (Å)	α (°)	Matrix	Pattern	Cell type
Bulk	2.775	0.000	0.000	3.924	90.0	(1.000, 0.000) (0.000, 1.000)	(1x1)	b: bulk lattice
Surface 1	2.775	0.000	0.000	7.848	90.0	(1.000, 0.000) (0.000, 2.000)	(1x2)	s1: commens. superlattice

3D COORDINATES

Pt1: ridge atoms; Pt2-Pt3: paired 2nd layer;
Pt4-Pt5: planar 3rd layer; Pt6-Pt7: paired 4th layer

Dx/Dy in Å, or as a fraction of layer's unit cell vectors; atom 0 at the origin. Epir/subr are bulk repeat vectors.

No. of atoms: 8 Bulk z = 1.387 Å

Region	Chem el.	At. no.	Cell type	Site occ.	Rel. to	Dx ± εx		Dy ± εy		Dz ± εz		Dz/Bz(%) ± εz/Bz
epir		-2					f		f		Å	
subr		-1				1.387	Å	1.962	Å	1.387	Å	
intf	Pt	1	s1	.50	0	0.000	f	0.000	f	0.000	Å	0.0
intf	Pt	2	s1	.50	1	0.500	f	-0.256 ± .001	f	1.117 ± .100	Å	80.5 ± 7.2
intf	Pt	3	s1	.50	2	0.000	f	0.513 ± .001	f	0.000	Å	0.0
intf	Pt	4	s1	.50	3	-0.500	f	-0.756 ± .001	f	1.277 ± .080	Å	92.1 ± 5.8
intf	Pt	5	s1	.50	4	0.000	f	0.500	f	0.000	Å	0.0
intf	Pt	6	s1	.50	5	0.500	f	0.255 ± .001	f	1.387	Å	100.0
intf	Pt	7	s1	.50	6	0.000	f	-0.510 ± .001	f	0.000	Å	0.0
subl	Pt	8	b	1.00	7	-0.500	f	-0.490 ± .003	f	1.387	Å	100.0

Pt(110)-(1x2)
78.41

BOND DISTANCES AND ANGLES

Bond distances and angles are derived from coordinates

No. of distances/angles: 7

Interatomic dist. A-B (Å)	Atom A	Atom B	Atom C	Bond angle A-B-C (°)
2.775	Pt1	Pt1(1,0)	Pt2(1,0)	121.1
2.687	Pt1	Pt2	Pt2(1,0)	121.1
2.687	Pt1	Pt2	Pt4	117.7
2.685	Pt2	Pt4	Pt4(1,0)	58.9
2.685	Pt2	Pt4	Pt6(0,-1)	118.5
2.685	Pt2	Pt4	Pt7	58.7
2.685	Pt2	Pt4	Pt8	118.4

COMMON NAME : Pt(110)-(1x2) ILLUSTRATION: 5
CLASSIFICATION : 78.47
TECHNIQUE : TOF-SARS
AUTHORS : F. Masson and J.W. Rabalais
REFERENCE : Surf. Sci., 253, 245 (1991)

SURFACE TYPE
Substrate : Pt
Crystal face: 110
Temperature : RT
Bulk lattice: fcc
2D bulk symm: pmm
2D surf symm: pmm

Adsorbate:
Coverage :
Pattern : (1x2)
Matrix : (1.000, 0.000)
(0.000, 2.000)

STRUCTURE TYPE
Missing-row reconstruction, with multilayer relaxations down to 3rd layer

SAMPLE PREPARATION (1 sample)
Treatment : long 1300C anneal, cycles of O2, Ar sputt., 1000C anneals
Crystallinity:
Anal. methods: AES, XPS
Contamination: no H, C, O in recoil spectra

COMMENTS

DATA COLLECTION
Technique: TOF-SARS; 2keV Ne+ ion beam
Dataset : time-of-flight scattering and recoiling spectra as fct. of incident and scattering angle in back and forward scattering

THEORY/DATA TREATMENT
Comparison with trajectory calcs with Ziegler-Biersack-Littmark pot., using experimentally determined shadow cone

STRUCTURES EXAMINED
Missing-row model with perpendicular relaxation in top layer, pairing in 2nd layer and buckling in 3rd layer

QUALITY OF EXPERIMENT-THEORY FIT
Visual

2D UNIT CELLS (1 domain observed)

Cell	Ax (Å)	Ay (Å)	Bx (Å)	By (Å)	α (°)	Matrix	Pattern	Cell type
Bulk	2.775	0.000	0.000	3.924	90.0	(1.000, 0.000) (0.000, 1.000)	(1x1)	b: bulk lattice
Surface 1	2.775	0.000	0.000	7.848	90.0	(1.000, 0.000) (0.000, 2.000)	(1x2)	s1: commens. superlattice

3D COORDINATES

Pt1: ridge atoms; Pt2-Pt3: paired 2nd layer;
Pt4-Pt5: buckled 3rd layer

Dx/Dy in Å, or as a fraction of layer's unit cell vectors; atom 0 at the origin. Epir/subr are bulk repeat vectors.

No. of atoms: 6 Bulk z = 1.387 Å

Region	Chem el.	At. no.	Cell type	Site occ.	Rel. to	Dx ± εx		Dy ± εy		Dz ± εz		Dz/Bz(%) ± εz/Bz
epir		-2					f		f		Å	
subr		-1				-1.387	Å	-1.962	Å	1.387	Å	
intf	Pt	1	s1	.50	0	0.000	f	0.000	f	0.000	Å	0.0
intf	Pt	2	s1	.50	1	0.500	f	0.259 ± .009	f	0.980 ± .070	Å	70.6 ± 5.1
intf	Pt	3	s1	.50	2	0.000	f	0.482 ± .009	f	0.000	Å	0.0
intf	Pt	4	s1	.50	3	-0.500	f	-0.241 ± .009	f	1.292 ± .130	Å	93.2 ± 9.4
intf	Pt	5	s1	.50	4	0.000	f	-0.500	f	0.190 ± .130	Å	13.7 ± 9.4
subl	Pt	6	b	1.00	5	0.500	f	0.500	f	1.292 ± .130	Å	93.2 ± 9.4

BOND DISTANCES AND ANGLES

Bond distances and angles are derived from coordinates

No. of distances/angles: 5

Interatomic dist. A-B (Å)	Atom A	Atom B	Atom C	Bond angle A-B-C (°)
2.775	Pt1	Pt1(1,0)	Pt2(1,0)	121.6
2.648	Pt1	Pt2	Pt2(1,0)	121.6
2.648	Pt1	Pt2	Pt4	116.7

Pt(110)-(1x2)
78.47

Bond Distances and Angles - Continued

Interatomic dist. A-B (Å)	Atom A	Atom B	Atom C	Bond angle A-B-C (°)
2.678	Pt2	Pt4	Pt4(1,0)	58.8
2.678	Pt2	Pt4	Pt6(0,1)	119.3

COMMON NAME : Pt(110)-(1x2) ILLUSTRATION: 5
CLASSIFICATION : 78.49
TECHNIQUE : PED
AUTHORS : S. Holmberg, H.C. Poon, Y. Jugnet, G. Grenet and Tran Minh Duc
REFERENCE : Surf. Sci., 254, L475 (1991)

SURFACE TYPE
Substrate : Pt
Crystal face: 110
Temperature : RT*
Bulk lattice: fcc
2D bulk symm: pmm
2D surf symm: pmm

Adsorbate:
Coverage :
Pattern : (1x2)
Matrix : (1.000, 0.000)
(0.000, 2.000)

STRUCTURE TYPE
Missing-row reconstruction, with top spacing contraction

SAMPLE PREPARATION (1 sample)
Treatment : cycles of Ar sputtering and 1000C anneals
Crystallinity: sharp LEED pattern
Anal. methods: XPS, LEED
Contamination: XPS: no contamination

COMMENTS
Two other structures (with top layer buckling and with top layer pairing) gave only slightly worse R-factors (0.38 for both vs. 0.36 for missing-row model)

DATA COLLECTION
Technique: PED; forward focusing of photoelectrons
Dataset : azimuthal and polar distributions of photoyield from Pt 4p3/2 core level

THEORY/DATA TREATMENT
Quasi-dynamical theory: all multiple scatt. between layers, none within layers

STRUCTURES EXAMINED
Missing-row, top-layer buckling and top-layer pairing models with relaxation of topmost spacing (and buckling)

QUALITY OF EXPERIMENT-THEORY FIT
RPE=0.36

2D UNIT CELLS (1 domain observed)

Cell	Ax (Å)	Ay (Å)	Bx (Å)	By (Å)	α (°)	Matrix	Pattern	Cell type
Bulk	2.775	0.000	0.000	3.924	90.0	(1.000, 0.000) (0.000, 1.000)	(1x1)	b: bulk lattice
Surface 1	2.775	0.000	0.000	7.848	90.0	(1.000, 0.000) (0.000, 2.000)	(1x2)	s1: commens. superlattice

3D COORDINATES

Pt1: ridge atoms; Pt2: planar 2nd layer

Dx/Dy in Å, or as a fraction of layer's unit cell vectors; atom 0 at the origin. Epir/subr are bulk repeat vectors.

No. of atoms: 3 Bulk z = 1.387 Å

Region	Chem el.	At. no.	Cell type	Site occ.	Rel. to	Dx ± εx		Dy ± εy		Dz ± εz		Dz/Bz(%) ± εz/Bz
epir		-2					f		f		Å	
subr		-1				1.387	Å	1.962	Å	1.387	Å	
intf	Pt	1	s1	.50	0	0.000	f	0.000	f	0.000	Å	0.0
intf	Pt	2	b	1.00	1	-0.500	f	-1.500	f	0.971 ± .250	Å	70.0 ± 18.0
subl	Pt	3	b	1.00	2	0.500	f	0.500	f	1.387	Å	100.0

BOND DISTANCES AND ANGLES

Bond distances and angles are derived from coordinates

No. of distances/angles: 3

Interatomic dist. A-B (Å)	Atom A	Atom B	Atom C	Bond angle A-B-C (°)
2.775	Pt1	Pt1(1,0)	Pt2(1,1)	57.6
2.592	Pt1	Pt2(0,1)	Pt1(-1,0)	64.7
2.592	Pt1	Pt2(0,1)	Pt2(1,2)	57.6

COMMON NAME : Pt(110)-(1x3) ILLUSTRATION: 7
CLASSIFICATION : 78.33b
TECHNIQUE : LEED
AUTHORS : P. Fery, W. Moritz and D. Wolf
REFERENCE : Phys. Rev., B38, 7275 (1988)

SURFACE TYPE
Substrate : Pt
Crystal face: 110
Temperature : RT*
Bulk lattice: fcc
2D bulk symm: pmm
2D surf symm: pmm

Adsorbate:
Coverage :
Pattern : (1x3)
Matrix : (1.000, 0.000)
(0.000, 3.000)

STRUCTURE TYPE
(1x3) reconstruction, probably impurity-stabilized; (111)-facetted missing-row reconstruction with multilayer relaxation, including 2nd-layer pairing, and 3rd- and 4th-layer buckling

SAMPLE PREPARATION (1 sample)
Treatment : Ar+ sputtering (2h), then annealing (10mins) at 900 K in O2
Crystallinity:
Anal. methods:
Contamination: AES: impurities below 5% ML

COMMENTS

DATA COLLECTION
Technique: LEED
Dataset : IV spectra: 19 beams at normal incidence; 30<E<200 eV

THEORY/DATA TREATMENT
Dynamical LEED (layer doubling, matrix inversion): VoiαE**1/3; VorαE**1/2 (fit); ΘD=270 K

STRUCTURES EXAMINED
Facetted missing rows: variations of first 5 interlayer spacings, 2nd-row pairing, 3rd- and 4th-layer buckling

QUALITY OF EXPERIMENT-THEORY FIT
RPE=0.35, RZJ=0.28

2D UNIT CELLS (1 domain observed)

Cell	Ax (Å)	Ay (Å)	Bx (Å)	By (Å)	α (°)	Matrix	Pattern	Cell type
Bulk	2.770	0.000	0.000	3.920	90.0	(1.000, 0.000) (0.000, 1.000)	(1x1)	b: bulk lattice
Surface 1	2.770	0.000	0.000	11.760	90.0	(1.000, 0.000) (0.000, 3.000)	(1x3)	s1: commens. superlattice

3D COORDINATES

Pt1: remaining top row; Pt2-Pt3: row-paired 2nd layer;
Pt4-Pt5-Pt6, Pt7-Pt8-Pt9: buckled 3rd, 4th layers; 0.05Å/0.1Å perp/lateral error bars assumed for tabulation

Dx/Dy in Å, or as a fraction of layer's unit cell vectors; atom 0 at the origin. Epir/subr are bulk repeat vectors.

No. of atoms: 11 Bulk z = 1.390 Å

Region	Chem el.	At. no.	Cell type	Site occ.	Rel. to	Dx ± εx		Dy ± εy		Dz ± εz		Dz/Bz(%) ± εz/Bz
epir		-2					f		f		Å	
subr		-1				1.385	Å	1.960	Å	1.390	Å	
intf	Pt	1	s1	.33	0	0.000	f	0.000	f	0.000	Å	0.0
intf	Pt	2	s1	.33	1	0.500	f	0.170 ± .009	f	1.100 ± .050	Å	79.1 ± 3.6
intf	Pt	3	s1	.33	2	0.000	f	0.660 ± .009	f	0.000	Å	0.0
intf	Pt	4	s1	.33	3	-0.500	f	-0.497 ± .009	f	1.320 ± .050	Å	95.0 ± 3.6
intf	Pt	5	s1	.33	4	0.000	f	0.333	f	0.000	Å	0.0
intf	Pt	6	s1	.33	5	0.000	f	-0.667	f	0.180 ± .050	Å	13.0 ± 3.6
intf	Pt	7	s1	.33	6	0.500	f	0.500	f	1.360 ± .050	Å	97.8 ± 3.6
intf	Pt	8	s1	.33	7	0.000	f	-0.333	f	0.040 ± .050	Å	2.9 ± 3.6
intf	Pt	9	s1	.33	8	0.000	f	0.667	f	0.000	Å	0.0
intf	Pt	10	b	1.00	9	-0.500	f	-2.500	f	1.380 ± .050	Å	99.3 ± 3.6
subl	Pt	11	b	1.00	10	0.500	f	0.500	f	1.390	Å	100.0

Pt(110)-(1x3)
78.33b

BOND DISTANCES AND ANGLES

Bond distances and angles are derived from coordinates

No. of distances/angles: 4

Interatomic dist. A-B (Å)	Atom A	Atom B	Atom C	Bond angle A-B-C (°)
2.770	Pt1	Pt1(1,0)		
2.669	Pt1	Pt2	Pt4	117.9
2.669	Pt1	Pt2	Pt6	56.0
2.600	Pt1	Pt6	Pt7(0,-1)	119.5

COMMON NAME : Pt(110)-(1x3)
CLASSIFICATION : 78.48
TECHNIQUE : TOF-SARS
AUTHORS : F. Masson and J.W. Rabalais
REFERENCE : Surf. Sci., 253, 258 (1991)

ILLUSTRATION: 6

SURFACE TYPE
Substrate : Pt
Crystal face: 110
Temperature : RT
Bulk lattice: fcc
2D bulk symm: pmm
2D surf symm: pmm

Adsorbate:
Coverage :
Pattern : (1x3)
Matrix : (1.000, 0.000)
(0.000, 3.000)

STRUCTURE TYPE
2-missing-rows reconstruction, leaving partial low 2nd-layer ridge within 3-wide trough; this structure is thought to be impurity-stabilized

SAMPLE PREPARATION (1 sample)
Treatment : prolonged 1300C annealing
Crystallinity:
Anal. methods: AES, XPS
Contamination: several %ML of Ca, K and perhaps P

COMMENTS
Ridge within trough distinguishes this structure from the (1x3) 'facetted' reconstruction, in which the trough penetrates the 2nd layer as a missing row;
authors suggest that this ridge within trough is only 40± 20% complete, i.e. coexists with facetted reconstruction

DATA COLLECTION
Technique: TOF-SARS; 2keV Ne+ ion beam
Dataset : time-of-flight scattering and recoiling spectra as fct. of incident and scattering angle in back and forward scattering

THEORY/DATA TREATMENT
Comparison with trajectory calcs with Ziegler-Biersack-Littmark pot., using experimentally determined shadow cone

STRUCTURES EXAMINED
4 models differing in number and depth of missing rows; for preferred 2-missing-row model, variation of first two interlayer spacings and 2nd layer buckling

QUALITY OF EXPERIMENT-THEORY FIT
Visual

2D UNIT CELLS (1 domain observed)

Cell	Ax (Å)	Ay (Å)	Bx (Å)	By (Å)	α (°)	Matrix	Pattern	Cell type
Bulk	2.775	0.000	0.000	3.924	90.0	(1.000, 0.000) (0.000, 1.000)	(1x1)	b: bulk lattice
Surface 1	2.775	0.000	0.000	11.772	90.0	(1.000, 0.000) (0.000, 3.000)	(1x3)	s1: commens. superlattice

3D COORDINATES

Pt1: upper ridge atom; Pt2-Pt3: 2nd layer atoms below upper ridge Pt1;
Pt4: ridge in trough, buckled down wrt Pt2-Pt3

Dx/Dy in Å, or as a fraction of layer's unit cell vectors; atom 0 at the origin. Epir/subr are bulk repeat vectors.

No. of atoms: 5 Bulk z = 1.387 Å

Region	Chem el.	At. no.	Cell type	Site occ.	Rel. to	Dx ± εx		Dy ± εy		Dz ± εz		Dz/Bz(%) ± εz/Bz
epir		-2					f		f		Å	
subr		-1				1.387	Å	1.962	Å	1.387	Å	
intf	Pt	1	s1	.33	0	0.000	f	0.000	f	0.000	Å	0.0
intf	Pt	2	s1	.33	1	0.500	f	0.167	f	1.150 ± .080	Å	82.9 ± 5.8
intf	Pt	3	s1	.33	2	0.000	f	0.667	f	0.000	Å	0.0
intf	Pt	4	s1	.33	3	0.000	f	-0.333	f	0.460 ± .110	Å	33.2 ± 7.9
subl	Pt	5	b	1.00	4	-0.500	f	-1.500	f	0.890 ± .110	Å	64.2 ± 7.9

BOND DISTANCES AND ANGLES

Bond distances and angles are derived from coordinates

No. of distances/angles: 7

Interatomic dist. A-B (Å)	Atom A	Atom B	Atom C	Bond angle A-B-C (°)
2.775	Pt1	Pt1(1,0)	Pt2(1,0)	121.4
2.664	Pt1	Pt2	Pt1(1,0)	62.8
2.664	Pt1	Pt2	Pt5(0,1)	118.3

Pt(110)-(1x3)
78.48

Bond Distances and Angles - Continued

Interatomic dist. A-B (Å)	Atom A	Atom B	Atom C	Bond angle A-B-C (°)
2.664	Pt1	Pt2	Pt5	54.9
2.500	Pt1	Pt5	Pt2	60.7
2.756	Pt2	Pt5	Pt5(1,0)	59.8
2.562	Pt5	Pt4(0,-1)		

COMMON NAME : Pt(111)-(1x1)   ILLUSTRATION: 1
CLASSIFICATION : 78.12
TECHNIQUE : LEED
AUTHORS : D.L. Adams, H.B. Nielsen and M.A. Van Hove
REFERENCE : Phys. Rev., B20, 4789 (1979)

SURFACE TYPE
Substrate : Pt
Crystal face: 111
Temperature : 85 K
Bulk lattice: fcc
2D bulk symm: p3m1
2D surf symm: p3m1

Adsorbate:
Coverage :
Pattern : (1x1)
Matrix : (1.000, 0.000)
(0.000, 1.000)

STRUCTURE TYPE
Bulk termination with top spacing expansion by 1.1%

SAMPLE PREPARATION (1 sample)
Treatment : heated at 1100C in 10E-7 torr O2, then at 1250C in vacuum
Crystallinity:
Anal. methods:
Contamination: AES: <1% ML of impurities

COMMENTS

DATA COLLECTION
Technique: LEED
Dataset : I-V curves: total of 49 non-degenerate beams at Θ=0, ±4, ±10, ±16° (φ=0°); cumulative E range: 6822 eV

THEORY/DATA TREATMENT
Dynamical LEED (RFS): 8 ph shs, different non-rel. and rel. potentials used; Vor=-5.42 eV, Voi=-5.18eV; ΘD=302 K

STRUCTURES EXAMINED
Top layer spacing varied from 1.9 to 3.6Å in 0.05Å steps

QUALITY OF EXPERIMENT-THEORY FIT
R1=0.40

2D UNIT CELLS (1 domain observed)

Cell	Ax (Å)	Ay (Å)	Bx (Å)	By (Å)	α (°)	Matrix	Pattern	Cell type
Bulk	2.775	0.000	1.388	2.403	60.0	(1.000, 0.000) (0.000, 1.000)	(1x1)	b: bulk lattice
Surface 1	2.775	0.000	1.388	2.403	60.0	(1.000, 0.000) (0.000, 1.000)	(1x1)	s1: commens. superlattice

3D COORDINATES

Dx/Dy in Å, or as a fraction of layer's unit cell vectors; atom 0 at the origin. Epir/subr are bulk repeat vectors.

No. of atoms: 3   Bulk z = 2.265 Å

Reg ion	Chem el.	At. no.	Cell type	Site occ.	Rel. to	Dx ± εx		Dy ± εy		Dz ± εz		Dz/Bz(%) ± εz/Bz
epir		-2					f		f		Å	
subr		-1				1.388	Å	0.801	Å	2.265	Å	
intf	Pt	1	b	1.00	0	0.000	f	0.000	f	0.000	Å	0.0
intf	Pt	2	b	1.00	1	0.333	f	0.333	f	2.290 ± .100	Å	101.1 ± 4.4
subl	Pt	3	b	1.00	2	0.333	f	0.333	f	2.265	Å	100.0

BOND DISTANCES AND ANGLES

Bond distances and angles are derived from coordinates

No. of distances/angles: 3

Interatomic dist. A-B (Å)	Atom A	Atom B	Atom C	Bond angle A-B-C (°)
2.775	Pt1	Pt1(1,0)	Pt2	60.2
2.795	Pt1	Pt2	Pt3	180.0
2.774	Pt2	Pt3		

COMMON NAME : Pt(111)-(1x1)
CLASSIFICATION : 78.15
TECHNIQUE : SPLEED
AUTHORS : R. Feder, H. Pleyer, P. Bauer and N. Mueller
REFERENCE : Surf. Sci., 109, 419 (1981)

ILLUSTRATION: 1

SURFACE TYPE
Substrate : Pt
Crystal face: 111
Temperature : RT*
Bulk lattice: fcc
2D bulk symm: p3m1
2D surf symm: p3m1

Adsorbate:
Coverage :
Pattern : (1x1)
Matrix : (1.000, 0.000)
(0.000, 1.000)

STRUCTURE TYPE
Bulk termination with top spacing expansion by 0.5%

SAMPLE PREPARATION (1 sample)
Treatment : cycles of 1100 K heating in 10E-6 torr O2 and Ar+ sputtering
Crystallinity:
Anal. methods:
Contamination: checked with AES

COMMENTS

DATA COLLECTION
Technique: SPLEED
Dataset : transverse and longitudinal spin polarization spectra with incident energies of 60, 80 and 95 eV for 00, 10, -1

THEORY/DATA TREATMENT
Dynamical spin polarized LEED (layer doubling): 8 ph shs from 3 relativistic scattering potentials; ΘD=230 K

STRUCTURES EXAMINED
Variation of topmost interlayer spacing

QUALITY OF EXPERIMENT-THEORY FIT
Visual

2D UNIT CELLS (1 domain observed)

Cell	Ax (Å)	Ay (Å)	Bx (Å)	By (Å)	α (°)	Matrix	Pattern	Cell type
Bulk	2.775	0.000	1.388	2.403	60.0	(1.000, 0.000) (0.000, 1.000)	(1x1)	b: bulk lattice
Surface 1	2.775	0.000	1.388	2.403	60.0	(1.000, 0.000) (0.000, 1.000)	(1x1)	s1: commens. superlattice

3D COORDINATES

Dx/Dy in Å, or as a fraction of layer's unit cell vectors; atom 0 at the origin. Epir/subr are bulk repeat vectors.

No. of atoms: 3 Bulk z = 2.265 Å

Region	Chem el.	At. no.	Cell type	Site occ.	Rel. to	Dx ± εx		Dy ± εy		Dz ± εz		Dz/Bz(%) ± εz/Bz
epir		-2					f		f		Å	
subr		-1				1.388	Å	0.801	Å	2.265	Å	
intf	Pt	1	b	1.00	0	0.000	f	0.000	f	0.000	Å	0.0
intf	Pt	2	b	1.00	1	0.333	f	0.333	f	2.276 ± .020	Å	100.5 ± .9
subl	Pt	3	b	1.00	2	0.333	f	0.333	f	2.265	Å	100.0

BOND DISTANCES AND ANGLES

Bond distances and angles are derived from coordinates

No. of distances/angles: 3

Interatomic dist. A-B (Å)	Atom A	Atom B	Atom C	Bond angle A-B-C (°)
2.775	Pt1	Pt1(1,0)	Pt2	60.1
2.783	Pt1	Pt2	Pt3	180.0
2.774	Pt2	Pt3		

COMMON NAME : Pt(111)-(1x1) ILLUSTRATION: 1
CLASSIFICATION : 78.20a
TECHNIQUE : LEED
AUTHORS : K. Hayek, H. Glassl, A. Gutmann, H. Leonhard, M. Prutton, S.P. Tear and M.R. Welton-Cook
REFERENCE : Surf. Sci., 152, 419 (1985)

SURFACE TYPE
Substrate : Pt
Crystal face: 111
Temperature : 293 K
Bulk lattice: fcc
2D bulk symm: p3m1
2D surf symm: p3m1

Adsorbate:
Coverage :
Pattern : (1x1)
Matrix : (1.000, 0.000)
(0.000, 1.000)

STRUCTURE TYPE
Unrelaxed bulk termination

SAMPLE PREPARATION (1 sample)
Treatment :
Crystallinity:
Anal. methods:
Contamination: monitored by AES

COMMENTS

DATA COLLECTION
Technique: LEED
Dataset : 5 non-degenerate I-V curves at normal incidence

THEORY/DATA TREATMENT
Dynamical LEED (CAVLEED program): 8 phase shifts (Herman-Skillman superposition potential); Voi=-5 eV

STRUCTURES EXAMINED
Top layer relaxations

QUALITY OF EXPERIMENT-THEORY FIT
RPE=0.43

2D UNIT CELLS (1 domain observed)

Cell	Ax (Å)	Ay (Å)	Bx (Å)	By (Å)	α (°)	Matrix	Pattern	Cell type
Bulk	2.770	0.000	1.385	2.399	60.0	(1.000, 0.000) (0.000, 1.000)	(1x1)	b: bulk lattice
Surface 1	2.770	0.000	1.385	2.399	60.0	(1.000, 0.000) (0.000, 1.000)	(1x1)	s1: commens. superlattice

3D COORDINATES

Dx/Dy in Å, or as a fraction of layer's unit cell vectors; atom 0 at the origin. Epir/subr are bulk repeat vectors.

No. of atoms: 3 Bulk z = 2.265 Å

Region	Chem el.	At. no.	Cell type	Site occ.	Rel. to	Dx ± εx		Dy ± εy		Dz ± εz		Dz/Bz(%) ± εz/Bz
epir		-2					f		f		Å	
subr		-1				1.385	Å	0.800	Å	2.265	Å	
intf	Pt	1	b	1.00	0	0.000	f	0.000	f	0.000	Å	0.0
intf	Pt	2	b	1.00	1	0.333	f	0.333	f	2.265 ± .050	Å	100.0 ± 2.2
subl	Pt	3	b	1.00	2	0.333	f	0.333	f	2.265	Å	100.0

BOND DISTANCES AND ANGLES

Bond distances and angles are derived from coordinates

No. of distances/angles: 3

Interatomic dist. A-B (Å)	Atom A	Atom B	Atom C	Bond angle A-B-C (°)
2.770	Pt1	Pt1(1,0)	Pt2	60.0
2.773	Pt1	Pt2	Pt3	180.0
2.773	Pt2	Pt3		

COMMON NAME : Pt(111)-(1x1) ILLUSTRATION: 1
CLASSIFICATION : 78.8
TECHNIQUE : MEIS
AUTHORS : J.F. van Der Veen, R.G. Smeenk, R.M. Tromp and F.W. Saris
REFERENCE : Surf. Sci., 79, 219 (1979)

SURFACE TYPE
Substrate : Pt
Crystal face: 111
Temperature : 420 K
Bulk lattice: fcc
2D bulk symm: p3m1
2D surf symm: p3m1

Adsorbate:
Coverage :
Pattern : (1x1)
Matrix : (1.000, 0.000)
(0.000, 1.000)

STRUCTURE TYPE
Bulk termination with top spacing expansion by 1.5%

SAMPLE PREPARATION (1 sample)
Treatment : sputtering with 2.8k eV Ar+ followed by annealing at 500C
Crystallinity:
Anal. methods:
Contamination: MEIS: <10% ML C and <5% ML O

COMMENTS

DATA COLLECTION
Technique: MEIS; MEIS using 173 keV protons
Dataset : blocking spectra along [11-6] and [001] axes, with angular detection from 60 to 90°

THEORY/DATA TREATMENT
Theoretical simulation of channeling and blocking; rms ampls = 0.16Å

STRUCTURES EXAMINED
Variations in top layer spacing

QUALITY OF EXPERIMENT-THEORY FIT
Visual

2D UNIT CELLS (1 domain observed)

Cell	Ax (Å)	Ay (Å)	Bx (Å)	By (Å)	α (°)	Matrix	Pattern	Cell type
Bulk	2.778	0.000	1.389	2.406	60.0	(1.000, 0.000) (0.000, 1.000)	(1x1)	b: bulk lattice
Surface 1	2.778	0.000	1.389	2.406	60.0	(1.000, 0.000) (0.000, 1.000)	(1x1)	s1: commens. superlattice

3D COORDINATES

Dx/Dy in Å, or as a fraction of layer's unit cell vectors; atom 0 at the origin. Epir/subr are bulk repeat vectors.

No. of atoms: 3 Bulk z = 2.268 Å

Region	Chem el.	At. no.	Cell type	Site occ.	Rel. to	Dx ± εx		Dy ± εy		Dz ± εz		Dz/Bz(%) ± εz/Bz
epir		-2					f		f		Å	
subr		-1				1.389	Å	0.802	Å	2.268	Å	
intf	Pt	1	b	1.00	0	0.000	f	0.000	f	0.000	Å	0.0
intf	Pt	2	b	1.00	1	0.333	f	0.333	f	2.300 ± .020	Å	101.4 ± .9
subl	Pt	3	b	1.00	2	0.333	f	0.333	f	2.268	Å	100.0

BOND DISTANCES AND ANGLES

Bond distances and angles are derived from coordinates

No. of distances/angles: 3

Interatomic dist. A-B (Å)	Atom A	Atom B	Atom C	Bond angle A-B-C (°)
2.778	Pt1	Pt1(1,0)	Pt2	60.3
2.804	Pt1	Pt2	Pt3	180.0
2.778	Pt2	Pt3		

COMMON NAME : Pt(210)-(1x1)
CLASSIFICATION : 78.43
TECHNIQUE : LEED
AUTHORS : X.G. Zhang, M.A. Van Hove, G.A. Somorjai, P.J. Rous, D. Tobin, A. Gonis, J.M. MacLaren, K.Heinz, M.Michl, et al
REFERENCE : Phys. Rev. Lett., 67, 1298 (1991)

ILLUSTRATION: 9

SURFACE TYPE
Substrate : Pt
Crystal face: 210
Temperature : 100 K
Bulk lattice: fcc
2D bulk symm: cm
2D surf symm: cm

Adsorbate:
Coverage :
Pattern : (1x1)
Matrix : (1.000, 0.000)
(0.000, 1.000)

STRUCTURE TYPE
Bulk termination with multilayer relaxations by -23%, -12%, +4%, and -3% in 4 top interlayer spacings

SAMPLE PREPARATION (1 sample)
Treatment : Ar+ sputtering, 1200 K anneal, O2 treatment, 1000K anneal
Crystallinity: surface <0.5° from (210) plane
Anal. methods:
Contamination:

COMMENTS
Et al = H. Lindner, K. Mueller, M. Ehsasi and J.H. Block

DATA COLLECTION
Technique: LEED; no impurities by AES
Dataset : I-V curves for 11 symmetry inequivalent beams; E range 15-120 eV

THEORY/DATA TREATMENT
Dynamical LEED (real-space ms, automated tensor LEED): 7 phase shifts from relat pot, Vor=-10 eV, ΘD=230 K

STRUCTURES EXAMINED
Variation of top 4 interlayer spacings and registries, keeping mirror plane

QUALITY OF EXPERIMENT-THEORY FIT
RPE=0.218

2D UNIT CELLS (1 domain observed)

Cell	Ax (Å)	Ay (Å)	Bx (Å)	By (Å)	α (°)	Matrix	Pattern	Cell type
Bulk	3.920	0.000	1.960	4.383	65.9	(1.000, 0.000) (0.000, 1.000)	(1x1)	b: bulk lattice
Surface 1	3.920	0.000	1.960	4.383	65.9	(1.000, 0.000) (0.000, 1.000)	(1x1)	s1: commens. superlattice

3D COORDINATES

Dx/Dy in Å, or as a fraction of layer's unit cell vectors; atom 0 at the origin. Epir/subr are bulk repeat vectors.

No. of atoms: 6

Bulk z = .877 Å

Region	Chem el.	At. no.	Cell type	Site occ.	Rel. to	Dx ± εx		Dy ± εy		Dz ± εz		Dz/Bz(%) ± εz/Bz
epir		-2					f		f		Å	
subr		-1				0.000	Å	-2.630	Å	0.877	Å	
intf	Pt	1	b	1.00	0	0.000	Å	0.000	Å	0.000	Å	0.0
intf	Pt	2	b	1.00	1	0.000	Å	-2.650 ± .070	Å	0.675 ± .035	Å	77.0 ± 4.0
intf	Pt	3	b	1.00	2	0.000	Å	-2.600 ± .080	Å	0.771 ± .044	Å	88.0 ± 5.0
intf	Pt	4	b	1.00	3	0.000	Å	-2.550 ± .100	Å	0.912 ± .061	Å	104.0 ± 7.0
intf	Pt	5	b	1.00	4	0.000	Å	-2.620 ± .100	Å	0.850 ± .061	Å	97.0 ± 7.0
subl	Pt	6	b	1.00	5	0.000	Å	-2.630	Å	0.877	Å	100.0

BOND DISTANCES AND ANGLES

Bond distances and angles are derived from coordinates

No. of distances/angles: 12

Interatomic dist. A-B (Å)	Atom A	Atom B	Atom C	Bond angle A-B-C (°)
2.702	Pt1	Pt2		
2.708	Pt3	Pt4		
2.834	Pt3	Pt4(-1,1)		
2.750	Pt3	Pt5		
2.585	Pt1	Pt3		

Pt(210)-(1x1)
78.43

Bond Distances and Angles - Continued

Interatomic dist. A-B (Å)	Atom A	Atom B	Atom C	Bond angle A-B-C (°)
2.548	Pt1	Pt4		
2.760	Pt2	Pt3		
2.712	Pt2	Pt3(0,-1)		
2.760	Pt2	Pt3(-1,0)		
2.695	Pt2	Pt4		
2.722	Pt2	Pt5		
2.834	Pt3	Pt4(0,1)		

COMMON NAME : Pt(111)-(1x1)-H (D)
CLASSIFICATION : 78.1.7
TECHNIQUE : Transm. Channeling
AUTHORS : K. Mortensen, F. Besenbacher, I. Stensgaard and C. Klink
REFERENCE : Surf. Sci., 211/212, 813 (1989)

ILLUSTRATION: 22,23

SURFACE TYPE
Substrate : Pt
Crystal face: 111
Temperature : 110 K
Bulk lattice: fcc
2D bulk symm: p3m1
2D surf symm: p3m1

Adsorbate: Deuterium
Coverage : 1.0ML
Pattern : (1x1)
Matrix : (1.000, 0.000)
(0.000, 1.000)

STRUCTURE TYPE
Atomic overlayer in 3-fold hollow site (no occupation of sub-surface octahedral sites)

SAMPLE PREPARATION (1 sample)
Treatment : thin Pt crystals grown on NaCl; cleaned, exposed to D
Crystallinity: good quality LEED pattern
Anal. methods: forward recoil detection
Contamination:

COMMENTS
RMS surface displacement parallel to the surface determined to be 0.29A

DATA COLLECTION
Technique: Transm. Channeling
Dataset : 0-40° angular scans in [111],[011] and [001] directions

THEORY/DATA TREATMENT
Transmission channeling: multi-row continuum model

STRUCTURES EXAMINED
Top, hcp- and fcc-hollow and bridge sites

QUALITY OF EXPERIMENT-THEORY FIT
Visual

2D UNIT CELLS (1 domain observed)

Cell	Ax (Å)	Ay (Å)	Bx (Å)	By (Å)	α (°)	Matrix	Pattern	Cell type
Bulk	2.775	0.000	1.388	2.403	60.0	(1.000, 0.000) (0.000, 1.000)	(1x1)	b: bulk lattice
Surface 1	2.775	0.000	1.388	2.403	60.0	(1.000, 0.000) (0.000, 1.000)	(1x1)	s1: commens. superlattice

3D COORDINATES

H1: D overlayer in fcc hollows

Dx/Dy in Å, or as a fraction of layer's unit cell vectors; atom 0 at the origin. Epir/subr are bulk repeat vectors.

No. of atoms: 3

Bulk z = 2.260 Å

Region	Chem el.	At. no.	Cell type	Site occ.	Rel. to	Dx ± εx		Dy ± εy		Dz ± εz		Dz/Bz(%) ± εz/Bz
epir		-2					f		f		Å	
subr		-1				1.388	Å	0.801	Å	2.260	Å	
ovrl	H	1	s1	1.00	0	0.000	Å	0.000	Å	0.000	Å	0.0
intf	Pt	2	b	1.00	0	1.388	Å	0.801	Å	0.580 ± .040	Å	25.7 ± 1.8
subl	Pt	3	b	1.00	0	2.775	Å	1.602	Å	2.840	Å	125.7

BOND DISTANCES AND ANGLES

Bond distances and angles are derived from coordinates

No. of distances/angles: 1

Interatomic dist. A-B (Å)	Atom A	Atom B	Atom C	Bond angle A-B-C (°)
1.704	D1	Pt2		

COMMON NAME : Pt(111)-(2x2)-C2H3
CLASSIFICATION : 78.6.1.5
TECHNIQUE : LEED
AUTHORS : L.L. Kesmodel, L.H. Dubois and G.A. Somorjai
REFERENCE : J. Chem. Phys., 70, 2180 (1979)

ILLUSTRATION: 65

SURFACE TYPE
Substrate : Pt
Crystal face: 111
Temperature : 300 K
Bulk lattice: fcc
2D bulk symm: p3m1
2D surf symm: p3m1

Adsorbate: ethylidyne CCH3
Coverage : 0.25 (C2H3/Pt)
Pattern : (2x2)
Matrix : (2.000, 0.000)
(0.000, 2.000)

STRUCTURE TYPE
Ethylidyne species (CCH3 = C2H3) formed from ethylene (C2H4) with upright C-C axis: lower C in fcc hollow site, upper C forms methyl group (H positions not determined, but presence derived from HREELS data)

SAMPLE PREPARATION (1 sample)
Treatment : adsorption of ≈1L ethylene
Crystallinity:
Anal. methods:
Contamination: AES: no impurities

COMMENTS
Methyl group may rotate freely about C-C axis

DATA COLLECTION
Technique: LEED
Dataset : IV curves at Θ=0,4,8°, φ=0° and Θ=30°, φ=30°; energy range 10-100 eV

THEORY/DATA TREATMENT
Dynamical LEED (layer doubling, combined space method): Andersen Pt pot, Pt-C superpos pot for C

STRUCTURES EXAMINED
Top, bridge, hcp, and fcc hollow sites; C-C length varied 1.20-1.54Å, Pt-C length varied 1.94-2.30Å;
C-C axis tilted 0-55° from normal and parallel to surface (for tilt angles >0° C-C length fixed at 1.34 or 1.54Å);

QUALITY OF EXPERIMENT-THEORY FIT
Visual

2D UNIT CELLS (1 domain observed)

Cell	Ax (Å)	Ay (Å)	Bx (Å)	By (Å)	α (°)	Matrix	Pattern	Cell type
Bulk	2.770	0.000	-1.385	2.399	120.0	(1.000, 0.000) (0.000, 1.000)	(1x1)	b: bulk lattice
Surface 1	5.540	0.000	-2.770	4.798	120.0	(2.000, 0.000) (0.000, 2.000)	(2x2)	s1: commens. superlattice

3D COORDINATES

H1-H2-H3-C4: methyl group, C4 pointing down to C5; H1 through C5: ethylidyne species adsorbed in fcc hollow (H positions assumed to form ideal methyl group)

Dx/Dy in Å, or as a fraction of layer's unit cell vectors; atom 0 at the origin. Epir/subr are bulk repeat vectors.

No. of atoms: 7

Bulk z = 2.260 Å

Region	Chem el.	At. no.	Cell type	Site occ.	Rel. to	Dx ± εx		Dy ± εy		Dz ± εz		Dz/Bz(%) ± εz/Bz
epir		-2					f		f		Å	
subr		-1				1.385	Å	0.800	Å	2.260	Å	
ovrl	H	1	s1	.25	0	0.786	f	0.893	f	0.000	Å	0.0
ovrl	H	2	s1	.25	0	0.107	f	0.893	f	0.000	Å	0.0
ovrl	H	3	s1	.25	0	0.107	f	0.214	f	0.000	Å	0.0
ovrl	C	4	s1	.25	0	0.000	f	0.000	f	0.363	Å	16.1
ovrl	C	5	s1	.25	4	0.000	f	0.000	f	1.500 ± .050	Å	66.4 ± 2.2
intf	Pt	6	b	1.00	5	0.667	f	0.333	f	1.200 ± .050	Å	53.1 ± 2.2
subl	Pt	7	b	1.00	6	0.667	f	0.333	f	2.260	Å	100.0

BOND DISTANCES AND ANGLES

Bond distances and angles are derived from coordinates

No. of distances/angles: 3

Interatomic dist. A-B (Å)	Atom A	Atom B	Atom C	Bond angle A-B-C (°)
1.500	C4	C5	Pt6	126.9

Pt(111)-(2x2)-C2H3
78.6.1.5

Bond Distances and Angles - Continued

Interatomic dist. A-B (Å)	Atom A	Atom B	Atom C	Bond angle A-B-C (°)
1.999	C5	Pt6	Pt6(1,0)	133.8
1.999	C5	Pt6	Pt7	162.2

COMMON NAME : Pt(111)-C6H6 disordered ILLUSTRATION: 67
CLASSIFICATION : 78.6.1.18
TECHNIQUE : DLEED
AUTHORS : A. Wander, G. Held, R.Q. Hwang, G.S. Blackman, M.L. Xu, P. de Andres, M.A. Van Hove and G.A. Somorjai
REFERENCE : Surf. Sci., 249, 21 (1991)

SURFACE TYPE
Substrate : Pt
Crystal face: 111
Temperature : 170 K
Bulk lattice: fcc
2D bulk symm: p3m1
2D surf symm: none

Adsorbate: C6H6 (benzene)
Coverage : 0.15 ML
Pattern : disordered
Matrix : (1.000, 0.000)
(0.000, 1.000)

STRUCTURE TYPE
Intact benzene overlayer centered over bridge site and buckled with parallel distortions

SAMPLE PREPARATION (1 sample)
Treatment : benzene adsorption of 2L from 300 K-170K after flash to 1250K
Crystallinity:
Anal. methods: AES
Contamination:

COMMENTS
Benzene rotated 30° with respect to the site found for the ordered phase in the presence of coadsorbed CO coverage assuming a saturated overlayer

DATA COLLECTION
Technique: DLEED; digital LEED
Dataset : 2 2D angular scans (in the form of Y functions) at 83 eV and 134eV

THEORY/DATA TREATMENT
Dynamical LEED (extension of beam set neglect, H ignored): 11x11 grid of exit directions used to generate Y functions

STRUCTURES EXAMINED
8 different sites (2 C2v sites, 4 C3v and 2 C6 sites): distortion of benzene ring consistent with local symmetry

QUALITY OF EXPERIMENT-THEORY FIT
RPE=0.049

2D UNIT CELLS (3 domains observed)

Cell	Ax (Å)	Ay (Å)	Bx (Å)	By (Å)	α (°)	Matrix	Pattern	Cell type
Bulk	2.774	0.000	-1.387	2.402	120.0	(1.000, 0.000) (0.000, 1.000)	(1x1)	b: bulk lattice
Surface 1	2.774	0.000	-1.387	2.402	120.0	(2.000, 0.000) (0.000, 2.000)	disordered	nd1: non-recon. lattice-gas dis

3D COORDINATES

C1-C6: distorted, buckled benzene centered over bridge site

Dx/Dy in Å, or as a fraction of layer's unit cell vectors; atom 0 at the origin. Epir/subr are bulk repeat vectors.

No. of atoms: 8 Bulk z = 2.265 Å

Region	Chem el.	At. no.	Cell type	Site occ.	Rel. to	Dx ± εx		Dy ± εy		Dz ± εz		Dz/Bz(%) ± εz/Bz
epir		-2					f		f		Å	
subr		-1				1.387	Å	0.801	Å	2.265	Å	
ovrl	C	1	nd1	.15	0	-0.063	Å	0.000	Å	0.000	Å	0.0
ovrl	C	2	nd1	.15	0	2.900	Å	0.000	Å	0.000 ± .020	Å	0.0 ± 4.5
ovrl	C	3	nd1	.15	0	0.566	Å	1.315	Å	-0.160 ± .020	Å	7.1 ± 4.5
ovrl	C	4	nd1	.15	0	0.566	Å	-1.315	Å	-0.160 ± .020	Å	7.1 ± 4.5
ovrl	C	5	nd1	.15	0	2.208	Å	1.315	Å	-0.160 ± .020	Å	7.1 ± 4.5
ovrl	C	6	nd1	.15	0	2.208	Å	-1.315	Å	-0.160 ± .020	Å	7.1 ± 4.5
intf	Pt	7	b	1.00	0	0.000	Å	0.000	Å	2.020	Å	89.2
subl	Pt	8	b	1.00	0	1.387	Å	0.801	Å	4.285	Å	189.2

Pt(111)-C6H6 disordered
78.6.1.18

BOND DISTANCES AND ANGLES

Bond distances and angles are derived from coordinates

No. of distances/angles: 5

Interatomic dist. A-B (Å)	Atom A	Atom B	Atom C	Bond angle A-B-C (°)
1.466	C1	C3		
1.643	C3	C5		
2.021	C1	Pt7	C3	97.0
2.571	C3	Pt(1,1)		
2.608	C3	Pt7		

COMMON NAME : Pt(111)-(2√3x4)rect-2C6H6+4CO
CLASSIFICATION : 78.6.1.8.1
TECHNIQUE : LEED
AUTHORS : D.F. Ogletree, M.A. Van Hove and G.A. Somorjai
REFERENCE : Surf. Sci., 183, 1 (1987)

ILLUSTRATION: 68

SURFACE TYPE
Substrate : Pt
Crystal face: 111
Temperature : 150 K
Bulk lattice: fcc
2D bulk symm: p3m1
2D surf symm: pg

Adsorbate: C6H6;CO
Coverage : 1/8(Bz), 1/4(CO)
Pattern : (2√3x4)rect
Matrix : (4.000, 0.000)
(2.000, 4.000)

STRUCTURE TYPE
Molecular C6H6 (benzene) and CO coadsorption: C6H6 parallel to surface, strongly expanded, centered over bridge sites; CO over bridge sites (site somewhat uncertain); 2 C6H6 and 4 CO molecules per unit cell

SAMPLE PREPARATION (1 sample)
Treatment : sputter, react with oxygen and anneal
Crystallinity:
Anal. methods:
Contamination: AES: no impurities

COMMENTS

DATA COLLECTION
Technique: LEED
Dataset : IV curves at Θ=0° for 3 substrate and 8 overlayer beams; energy range 20-150 eV

THEORY/DATA TREATMENT
Dynamical LEED (BSN, matrix inversion, KSLA, CSM, RFS): 5 ph shs (pot: Wang Pt, Li CO, Kesmodel C); Voi∝E**1/3; ΘD=300 K

STRUCTURES EXAMINED
C6H6 in top, hollow, bridge sites; in bridge site: boat, Kekule, ring distortions (H ignored); 2 and 4 CO's per cell in top, bridge, and hollow sites

QUALITY OF EXPERIMENT-THEORY FIT
RVH=0.28, RPE=0.54, RZJ=0.42

2D UNIT CELLS (3 domains observed)

Cell	Ax (Å)	Ay (Å)	Bx (Å)	By (Å)	α (°)	Matrix	Pattern	Cell type
Bulk	2.773	0.000	-1.386	2.401	120.0	(1.000, 0.000) (0.000, 1.000)	(1x1)	b: bulk lattice
Surface 1	11.090	0.000	0.000	9.602	90.0	(4.000, 0.000) (2.000, 4.000)	(2√3x4)rect	s1: commens. superlattice

3D COORDINATES

O1-C17, O2-C18, O3-C19, O4-C20: 4 identical upright CO molecules in bridge sites;
C5--C10, C11--C16: 2 identical C6H6 molecules centered on bridge sites (each molecule has 2 normal mirror planes)

Dx/Dy in Å, or as a fraction of layer's unit cell vectors; atom 0 at the origin. Epir/subr are bulk repeat vectors.

No. of atoms: 22

Bulk z = 2.260 Å

Region	Chem el.	At. no.	Cell type	Site occ.	Rel. to	Dx ± εx	Dy ± εy	Dz ± εz	Dz/Bz(%) ± εz/Bz
epir		-2				f	f	Å	
subr		-1				0.000 Å	1.600 Å	2.260 Å	
ovrl	O	1	s1	.13	0	0.000 f	0.000 f	0.000 Å	0.0
ovrl	O	2	s1	.13	1	0.250 f	0.000 f	0.000 Å	0.0
ovrl	O	3	s1	.13	1	0.625 f	0.500 f	0.000 Å	0.0
ovrl	O	4	s1	.13	3	0.250 f	0.000 f	0.000 Å	0.0
ovrl	C	5	s1	.13	3	-0.005 ± .009 f	-0.321 ± .010 f	0.500 ± .100 Å	22.1 ± 4.4
ovrl	C	6	s1	.13	5	0.139 ± .009 f	-0.090 ± .010 f	0.000 ± .100 Å	0.0 ± 4.4
ovrl	C	7	s1	.13	6	-0.002 ± .009 f	0.817 ± .010 f	0.000 ± .100 Å	0.0 ± 4.4
ovrl	C	8	s1	.13	7	-0.128 ± .009 f	-0.085 ± .010 f	0.000 ± .100 Å	0.0 ± 4.4
ovrl	C	9	s1	.13	8	-0.139 ± .009 f	0.090 ± .010 f	0.000 ± .100 Å	0.0 ± 4.4
ovrl	C	10	s1	.13	9	0.002 ± .009 f	-0.817 ± .010 f	0.000 ± .100 Å	0.0 ± 4.4
ovrl	C	11	s1	.13	3	-0.371 ± .009 f	0.179 ± .010 f	0.500 ± .100 Å	22.1 ± 4.4
ovrl	C	12	s1	.13	11	0.128 ± .009 f	-0.090 ± .010 f	0.000 ± .100 Å	0.0 ± 4.4
ovrl	C	13	s1	.13	12	0.002 ± .009 f	-0.183 ± .010 f	0.000 ± .100 Å	0.0 ± 4.4
ovrl	C	14	s1	.13	13	-0.139 ± .009 f	-0.085 ± .010 f	0.000 ± .100 Å	0.0 ± 4.4
ovrl	C	15	s1	.13	14	-0.128 ± .009 f	0.090 ± .010 f	0.000 ± .100 Å	0.0 ± 4.4
ovrl	C	16	s1	.13	15	-0.002 ± .009 f	0.183 ± .010 f	0.000 ± .100 Å	0.0 ± 4.4
ovrl	C	17	s1	.13	1	0.000 f	0.000 f	1.150 ± .100 Å	50.9 ± 4.4
ovrl	C	18	s1	.13	2	0.000 f	0.000 f	1.150 ± .100 Å	50.9 ± 4.4
ovrl	C	19	s1	.13	3	0.000 f	0.000 f	1.150 ± .100 Å	50.9 ± 4.4
ovrl	C	20	s1	.13	4	0.000 f	0.000 f	1.150 ± .100 Å	50.9 ± 4.4

Pt(111)-(2√3x4)rect-2C6H6+4CO
78.6.1.8.1

BOND DISTANCES AND ANGLES

Bond distances and angles are derived from coordinates

No. of distances/angles: 9

Interatomic dist. A-B (Å)	Atom A	Atom B	Atom C	Bond angle A-B-C (°)
1.150	O1	C17	Pt21	136.3
1.764	C5	C6	C7(0,-1)	118.5
1.640	C5	C10	C9(0,-1)	120.6
1.760	C6	C7(0,-1)	C8(0,-1)	120.6
1.640	C7	C8	C9	120.9
1.764	C8	C9	C10(0,1)	118.5
1.760	C9	C10(0,1)	C5(0,1)	120.6
2.006	C17	Pt21	Pt21(1,1)	133.7
2.769	Pt21	Pt22		

COMMON NAME : Pt(111)+CO 1/3ML disordered
CLASSIFICATION : 78.6.8.7
TECHNIQUE : DLEED
AUTHORS : G.S. Blackman, M.-L. Xu, M.A. Van Hove and G.A. Somorjai
REFERENCE : Phys. Rev. Lett., 61, 2352 (1988)

ILLUSTRATION: 61

SURFACE TYPE
Substrate : Pt
Crystal face: 111
Temperature : 160 K
Bulk lattice: fcc
2D bulk symm: p3m1
2D surf symm: none

Adsorbate: CO
Coverage : 0.33 CO/Pt
Pattern : disordered
Matrix : (1.000, 0.000)
(0.000, 1.000)

STRUCTURE TYPE
Disordered molecular adsorption, ≈88% in top sites and ≈12% in bridge sites: upright CO molecules (C down); unrelaxed substrate

SAMPLE PREPARATION (1 sample)
Treatment : dosing with 5E-8 torr CO at 150 K
Crystallinity:
Anal. methods:
Contamination: monitored by AES and LEED

COMMENTS

DATA COLLECTION
Technique: DLEED
Dataset : diffuse patterns at 80 and 130 eV (256x256 grid) for all angles up to 50° off-normal; normal incidence

THEORY/DATA TREATMENT
Dynamical diffuse LEED (BSN)

STRUCTURES EXAMINED
CO molecules perp. to surface with bond length of 1.15Å; independent variations of C-Pt spacings at top and bridge sites as a function of relative occupancy; bulk Pt termination

QUALITY OF EXPERIMENT-THEORY FIT
RPE=0.55

2D UNIT CELLS (3 domains observed)

Cell	Ax (Å)	Ay (Å)	Bx (Å)	By (Å)	α (°)	Matrix	Pattern	Cell type
Bulk	2.775	0.000	1.388	2.403	60.0	(1.000, 0.000) (0.000, 1.000)	(1x1)	b: bulk lattice
Surface 1	2.775	0.000	1.388	2.403	60.0	(1.000, 0.000) (0.000, 1.000)	disordered	nd1: non-recon. lattice-gas dis

3D COORDINATES

O1-C3: disordered top-site upright CO (88% of 1/3ML); O2-C4: disordered bridge-site upright CO (12% of 1/3ML); 0.1Å error bars assumed for tabulation

Dx/Dy in Å, or as a fraction of layer's unit cell vectors; atom 0 at the origin. Epir/subr are bulk repeat vectors.

No. of atoms: 6

Bulk z = 2.260 Å

Region	Chem el.	At. no.	Cell type	Site occ.	Rel. to	Dx ± εx		Dy ± εy		Dz ± εz		Dz/Bz(%) ± εz/Bz
epir		-2					f		f		Å	
subr		-1				1.388	Å	0.801	Å	2.260	Å	
ovrl	O	1	nd1	.29	0	0.000	f	0.000	f	0.000	Å	0.0
ovrl	O	2	nd2	.04	1	0.500	f	0.500	f	0.300 ± .100	Å	13.3 ± 4.4
ovrl	C	3	nd1	.29	1	0.000	f	0.000	f	1.150 ± .100	Å	50.9 ± 4.4
ovrl	C	4	nd2	.04	2	0.000	f	0.000	f	1.150 ± .100	Å	50.9 ± 4.4
intf	Pt	5	b	1.00	3	0.000	f	0.000	f	1.850 ± .100	Å	81.9 ± 4.4
subl	Pt	6	b	1.00	5	0.333	f	0.333	f	2.260	Å	100.0

BOND DISTANCES AND ANGLES

Bond distances and angles are derived from coordinates

No. of distances/angles: 5

Interatomic dist. A-B (Å)	Atom A	Atom B	Atom C	Bond angle A-B-C (°)
1.150	O1	C3	Pt5	180.0
1.150	O2	C4	Pt5(1,0)	138.2
1.850	C3	Pt5	Pt6	144.7

Pt(111)+CO 1/3ML disordered
78.6.8.7

Bond Distances and Angles - Continued

Interatomic dist. A-B (Å)	Atom A	Atom B	Atom C	Bond angle A-B-C (°)
2.080	C4	Pt5(1,0)	Pt6	105.9
2.770	Pt5	Pt6		

COMMON NAME : Pt(111)-c(4x2)-2CO
CLASSIFICATION : 78.6.8.4
TECHNIQUE : LEED
AUTHORS : D.F. Ogletree, M.A. Van Hove and G.A. Somorjai
REFERENCE : Surf. Sci., 173, 351 (1986)

ILLUSTRATION: 61,62

SURFACE TYPE
Substrate : Pt
Crystal face: 111
Temperature : 150 K
Bulk lattice: fcc
2D bulk symm: p3m1
2D surf symm: pm

Adsorbate: CO
Coverage : 0.5 (CO/Pt)
Pattern : c(4x2)
Matrix : (2.000, 0.000)
(1.000, 2.000)

STRUCTURE TYPE
Molecular CO adsorption in both top and bridge sites (one each per unit cell), perpendicular to surface

SAMPLE PREPARATION (1 sample)
Treatment : sputter, react with oxygen, and anneal
Crystallinity:
Anal. methods:
Contamination: AES: no impurities

COMMENTS

DATA COLLECTION
Technique: LEED
Dataset : IV curves at Θ=0,5,15°, ϕ=0° from 20-200 eV: 9 beams at Θ=0° and 14 beams at the other angles

THEORY/DATA TREATMENT
Dynamical LEED (combined space method for CO, RFS): 6 ph shs (Wang relat. Pt pot, Li CO pot); Voi∝E**1/3; ΘD=302(Pt)

STRUCTURES EXAMINED
Inequivalent molecules in following sites: top/bridge, top/hollow (fcc and hcp), top/disordered; all for intact CO perpendicular to surface, bonded through carbon atom on ideally terminated bulk structure.

QUALITY OF EXPERIMENT-THEORY FIT
RVH=0.29, RZJ=0.50, RPE=0.61

2D UNIT CELLS (3 domains observed)

Cell	Ax (Å)	Ay (Å)	Bx (Å)	By (Å)	α (°)	Matrix	Pattern	Cell type
Bulk	2.770	0.000	-1.385	2.400	120.0	(1.000, 0.000) (0.000, 1.000)	(1x1)	b: bulk lattice
Surface 1	5.540	0.000	0.000	4.800	90.0	(2.000, 0.000) (1.000, 2.000)	c(4x2)	s1: commens. superlattice

3D COORDINATES

O1-C3: bridge-site CO molecules; O2-: top-site CO molecules;

Dx/Dy in Å, or as a fraction of layer's unit cell vectors; atom 0 at the origin. Epir/subr are bulk repeat vectors.

No. of atoms: 6

Bulk z = 2.260 Å

Region	Chem el.	At. no.	Cell type	Site occ.	Rel. to	Dx ± εx		Dy ± εy		Dz ± εz		Dz/Bz(%) ± εz/Bz
epir		-2					f		f		Å	
subr		-1				1.385	Å	0.800	Å	2.260	Å	
ovrl	O	1	s1	.25	0	0.000	f	0.000	f	0.000	Å	0.0
ovrl	O	2	s1	.25	1	0.500	f	0.500	f	0.300 ± .100	Å	13.3 ± 4.4
ovrl	C	3	s1	.25	1	0.000	f	0.000	f	1.150 ± .050	Å	50.9 ± 2.2
ovrl	C	4	s1	.25	2	0.000	f	0.000	f	1.150 ± .050	Å	50.9 ± 2.2
intf	Pt	5	b	1.00	3	0.000	f	0.000	f	1.850 ± .025	Å	81.9 ± 1.1
subl	Pt	6	b	1.00	5	0.667	f	0.333	f	2.260	Å	100.0

BOND DISTANCES AND ANGLES

Bond distances and angles are derived from coordinates

No. of distances/angles: 3

Interatomic dist. A-B (Å)	Atom A	Atom B	Atom C	Bond angle A-B-C (°)
1.150	O1	C3	Pt5	180.0
1.150	O2	C4	Pt5(1,1)	138.2
2.769	Pt5	Pt6		

COMMON NAME : Pt(100)-(1x1)-Cu multilayer
CLASSIFICATION : 78.29.2
TECHNIQUE : LEED
AUTHORS : Y.S. Li, J. Quinn, H. Li, D. Tian, F. Jona and P.M. Marcus
REFERENCE : Phys. Rev., B44, 8261 (1991)

ILLUSTRATION: 83

SURFACE TYPE
Substrate : Pt
Crystal face: 100
Temperature : RT
Bulk lattice: fcc
2D bulk symm: p4m
2D surf symm: p4m

Adsorbate: Cu
Coverage : 10 Cu/Pd
Pattern : (1x1)
Matrix : (1.000, 0.000)
(0.000, 1.000)

STRUCTURE TYPE
About 10 epitaxial (1x1) monolayers, forming strained fcc Cu

SAMPLE PREPARATION (3 sample)
Treatment : Cu deposited from Cu single crystal onto RT substrate
Crystallinity: LEED: increased background
Anal. methods: coverage from AES
Contamination: AES: 2.5at% O, 11at% C

COMMENTS

DATA COLLECTION
Technique: LEED; video LEED
Dataset : IV spectra at normal incidence: (10),(11),(20); 40<E<320 eV; 3 film thicknesses: 4, 5, 6 layers

THEORY/DATA TREATMENT
Dynamical LEED (CHANGE code): Moruzzi et al pot, 8 ph shs; Vor=-8±2 eV (fit), Voi=-4eV; rms vibs 0.12Å

STRUCTURES EXAMINED
Semi-infinite Cu(100) with lateral Pd(100) lattice constant; spacings varied: 'bulk' Cu-Cu, top two Cu-Cu

QUALITY OF EXPERIMENT-THEORY FIT
RVHT=0.35, RPE=0.58, RZJ=0.13

2D UNIT CELLS (1 domain observed)

Cell	Ax (Å)	Ay (Å)	Bx (Å)	By (Å)	α (°)	Matrix	Pattern	Cell type
Bulk	2.750	0.000	0.000	2.750	90.0	(1.000, 0.000) (0.000, 1.000)	(1x1)	b: bulk lattice
Surface 1	2.750	0.000	0.000	2.750	90.0	(1.000, 0.000) (0.000, 1.000)	(1x1)	s1: commens. superlattice

3D COORDINATES

1.62Å bulk spacing was fit, keeping lateral Pd distance

Dx/Dy in Å, or as a fraction of layer's unit cell vectors; atom 0 at the origin. Epir/subr are bulk repeat vectors.

No. of atoms: 4   Bulk z = 1.620 Å

Region	Chem el.	At. no.	Cell type	Site occ.	Rel. to	Dx ± εx		Dy ± εy		Dz ± εz		Dz/Bz(%) ± εz/Bz
epir		-2					f		f		Å	
subr		-1				-1.375	Å	-1.375	Å	1.620	Å	
intf	Cu	1	b	1.00	0	0.000	f	0.000	f	0.000	Å	0.0
intf	Cu	2	b	1.00	1	0.500	f	0.500	f	1.570 ± .030	Å	96.9 ± 1.9
intf	Cu	3	b	1.00	2	-0.500	f	-0.500	f	1.620 ± .030	Å	100.0 ± 1.9
subl	Cu	4	b	1.00	3	0.500	f	0.500	f	1.620 ± .030	Å	100.0 ± 1.9

BOND DISTANCES AND ANGLES

Bond distances and angles are derived from coordinates

No. of distances/angles: 3

Interatomic dist. A-B (Å)	Atom A	Atom B	Atom C	Bond angle A-B-C (°)
2.750	Cu1	Cu1(1,0)	Cu2	56.6
2.499	Cu1	Cu2	Cu3	78.7
2.531	Cu2	Cu3	Cu4	79.6

COMMON NAME : Pt(111)-(√3x√3)R30°-S
CLASSIFICATION : 78.16.1
TECHNIQUE : LEED
AUTHORS : K. Hayek, H. Glassl, A. Gutmann, H. Leonhard, M. Prutton, S.P. Tear and M.R. Welton-Cook
REFERENCE : Surf. Sci., 152, 419 (1985)

ILLUSTRATION: 22,24

SURFACE TYPE
Substrate : Pt
Crystal face: 111
Temperature : 293 K
Bulk lattice: fcc
2D bulk symm: p3m1
2D surf symm: p31m

Adsorbate: S
Coverage : 0.33 S/Pt
Pattern : (√3x√3)R30°
Matrix : (1.000, 1.000)
(-2.000, 1.000)

STRUCTURE TYPE
Atomic adorption in fcc hollow site

SAMPLE PREPARATION (1 sample)
Treatment : exposure at 293 K to S2 then brief heating to 653K
Crystallinity:
Anal. methods:
Contamination: AES and radiotracer to check S coverage

COMMENTS

DATA COLLECTION
Technique: LEED
Dataset : 9 non-degenerate I-V curves at normal incidence

THEORY/DATA TREATMENT
Dynamical LEED (CAVLEED): 8 phase shifts (Pt Herman-Skillman superpos pot, S Clementi-Roetti pot); Voi=-4 eV; ΘD 234 K

STRUCTURES EXAMINED
3-fold fcc and hcp hollow sites and top site; relaxations of top Pt-Pt layer spacing from +6 to -6%

QUALITY OF EXPERIMENT-THEORY FIT
RPE=0.48

2D UNIT CELLS (1 domain observed)

Cell	Ax (Å)	Ay (Å)	Bx (Å)	By (Å)	α (°)	Matrix	Pattern	Cell type
Bulk	2.771	0.000	1.386	2.400	60.0	(1.000, 0.000) (0.000, 1.000)	(1x1)	b: bulk lattice
Surface 1	4.157	2.400	-4.157	2.400	120.0	(1.000, 1.000) (-2.000, 1.000)	(√3x√3)R30°	s1: commens. superlattice

3D COORDINATES

S1: overlayer in fcc hollow sites

Dx/Dy in Å, or as a fraction of layer's unit cell vectors; atom 0 at the origin. Epir/subr are bulk repeat vectors.

No. of atoms: 3

Bulk z = 2.260 Å

Region	Chem el.	At. no.	Cell type	Site occ.	Rel. to	Dx ± εx		Dy ± εy		Dz ± εz		Dz/Bz(%) ± εz/Bz
epir		-2					f		f		Å	
subr		-1				1.386	Å	0.800	Å	2.260	Å	
ovrl	S	1	s1	.33	0	0.000	f	0.000	f	0.000	Å	0.0
intf	Pt	2	b	1.00	1	0.333	f	0.333	f	1.620 ± .050	Å	71.7 ± 2.2
subl	Pt	3	b	1.00	2	0.333	f	0.333	f	2.260	Å	100.0

BOND DISTANCES AND ANGLES

Bond distances and angles are derived from coordinates

No. of distances/angles: 3

Interatomic dist. A-B (Å)	Atom A	Atom B	Atom C	Bond angle A-B-C (°)
2.277	S1	Pt2	Pt2(1,0)	127.5
2.277	S1	Pt2	Pt3	170.7
2.769	Pt2	Pt3		

COMMON NAME : Pt(111)-(2x2)-Sn
CLASSIFICATION : 78.50.1a
TECHNIQUE : ALICISS
AUTHORS : S.H. Overbury, D.R. Mullins, M.F. Paffett and B.E. Koel
REFERENCE : Surf. Sci., 254, 45 (1991)

ILLUSTRATION: 27

SURFACE TYPE
Substrate : Pt
Crystal face: 111
Temperature : RT*
Bulk lattice: fcc
2D bulk symm: p3m1
2D surf symm: p3m1

Adsorbate: Sn
Coverage : 0.25 Sn/Pt
Pattern : (2x2)
Matrix : (2.000, 0.000)
(0.000, 2.000)

STRUCTURE TYPE
Atomic substitutional adsorption in fcc hollow site, approximating Pt3Sn(111) termination

SAMPLE PREPARATION (1 sample)
Treatment : Sn deposited at RT using 2 diff. evaporators, then annealed
Crystallinity:
Anal. methods: LEED, AES
Contamination:

COMMENTS

DATA COLLECTION
Technique: ALICISS; spherical sector electrostatic ana
Dataset : Li+ ion scattering yield as fct. of polar and azimuthal angles

THEORY/DATA TREATMENT
Shadowing and blocking simulation; distances from critical angles; Thomas-Fermi-Moliere potential

STRUCTURES EXAMINED
Overlayer model; substitutional (in top Pt layer) model: variation of Sn-Pt buckling in top layer

2D UNIT CELLS (1 domain observed)

Cell	Ax (Å)	Ay (Å)	Bx (Å)	By (Å)	α (°)	Matrix	Pattern	Cell type
Bulk	2.771	0.000	1.386	2.400	60.0	(1.000, 0.000) (0.000, 1.000)	(1x1)	b: bulk lattice
Surface 1	5.543	0.000	2.771	4.800	60.0	(2.000, 0.000) (0.000, 2.000)	(2x2)	s1: commens. superlattice

3D COORDINATES

Dx/Dy in Å, or as a fraction of layer's unit cell vectors; atom 0 at the origin. Epir/subr are bulk repeat vectors.

No. of atoms: 5

Bulk z = 2.260 Å

Reg ion	Chem el.	At. no.	Cell type	Site occ.	Rel. to	Dx ± εx		Dy ± εy		Dz ± εz		Dz/Bz(%) ± εz/Bz
epir		-2					f		f		Å	
subr		-1				1.386	Å	0.800	Å	2.260	Å	
intf	Sn	1	s1	.25	0	0.000	f	0.000	f	0.000	Å	0.0
intf	Pt	2	s1	.25	1	0.500	f	0.000	f	0.220 ± .050	Å	9.7 ± 2.2
intf	Pt	3	s1	.25	2	0.000	f	0.500	f	0.000	Å	0.0
intf	Pt	4	s1	.25	3	-0.500	f	0.000	f	0.000	Å	0.0
subl	Pt	5	b	1.00	4	0.333	f	0.333	f	2.260	Å	100.0

BOND DISTANCES AND ANGLES

Bond distances and angles are derived from coordinates

No. of distances/angles: 2

Interatomic dist. A-B (Å)	Atom A	Atom B	Atom C	Bond angle A-B-C (°)
2.780	Sn1	Pt2	Sn1(1,0)	170.9
2.780	Sn1	Pt2	Pt3	119.9

COMMON NAME : Pt(111)-(√3x√3)R30°-Sn
CLASSIFICATION : 78.50.1b
TECHNIQUE : ALICISS
AUTHORS : S.H. Overbury, D.R. Mullins, M.F. Paffett and B.E. Koel
REFERENCE : Surf. Sci., 254, 45 (1991)

ILLUSTRATION: 27

SURFACE TYPE
Substrate : Pt
Crystal face: 111
Temperature : RT*
Bulk lattice: fcc
2D bulk symm: p3m1
2D surf symm: p31m

Adsorbate: Sn
Coverage : 0.33 Sn/Pt
Pattern : (√3x√3)R30°
Matrix : (1.000, 1.000)
(-2.000, 1.000)

STRUCTURE TYPE
Atomic substitutional adsorption in fcc hollow site, approximating Pt3Sn(111) termination

SAMPLE PREPARATION (1 sample)
Treatment : Sn deposited at RT using 2 diff. evaporators, then annealed
Crystallinity:
Anal. methods: LEED, AES
Contamination:

COMMENTS

DATA COLLECTION
Technique: ALICISS; spherical sector electrostatic ana
Dataset : Li+ ion scattering yield as fct. of polar and azimuthal angles

THEORY/DATA TREATMENT
Shadowing and blocking simulation; distances from critical angles; Thomas-Fermi-Moliere potential

STRUCTURES EXAMINED
Overlayer model; substitutional (in top Pt layer) model: variation of Sn-Pt buckling in top layer

2D UNIT CELLS (1 domain observed)

Cell	Ax (Å)	Ay (Å)	Bx (Å)	By (Å)	α (°)	Matrix	Pattern	Cell type
Bulk	2.771	0.000	1.386	2.400	60.0	(1.000, 0.000) (0.000, 1.000)	(1x1)	b: bulk lattice
Surface 1	4.157	2.400	-4.157	2.400	120.0	(1.000, 1.000) (-2.000, 1.000)	(√3x√3)R30°	s1: commens. superlattice

3D COORDINATES

Sn1: substitutional in top Pt layer

Dx/Dy in Å, or as a fraction of layer's unit cell vectors; atom 0 at the origin. Epir/subr are bulk repeat vectors.

No. of atoms: 4

Bulk z = 2.260 Å

Region	Chem el.	At. no.	Cell type	Site occ.	Rel. to	Dx ± εx		Dy ± εy		Dz ± εz		Dz/Bz(%) ± εz/Bz
epir		-2					f		f		Å	
subr		-1				1.386	Å	0.800	Å	2.260	Å	
intf	Sn	1	s1	.33	0	0.000	f	0.000	f	0.000	Å	0.0
intf	Pt	2	s1	.33	1	0.667	f	0.333	f	0.220 ± .050	Å	9.7 ± 2.2
intf	Pt	3	s1	.33	2	-0.333	f	0.333	f	0.000	Å	0.0
subl	Pt	4	b	1.00	3	0.333	f	0.333	f	2.260	Å	100.0

BOND DISTANCES AND ANGLES

Bond distances and angles are derived from coordinates

No. of distances/angles: 2

Interatomic dist. A-B (Å)	Atom A	Atom B	Atom C	Bond angle A-B-C (°)
2.780	Sn1	Pt2	Sn1(1,0)	170.9
2.780	Sn1	Pt2	Pt3	119.9

COMMON NAME : Pt0.8Fe0.2(110)-(1x2)
CLASSIFICATION : 78.26.2
TECHNIQUE : LEED
AUTHORS : R. Baudoing-Savois, Y. Gauthier and W. Moritz
REFERENCE : Phys. Rev., B38, 12977 (1991)

ILLUSTRATION: 140

SURFACE TYPE
Substrate : Pt0.8Fe0.2
Crystal face: 110
Temperature : RT*
Bulk lattice: fcc
2D bulk symm: none
2D surf symm: none

Adsorbate:
Coverage :
Pattern : (1x2) and disorder
Matrix : (1.000, 0.000)
(0.000, 2.000)

STRUCTURE TYPE
Missing-row reconstruction with multilayer relaxation, including 2nd- and 4th-layer pairing, and 3rd- and 5th-layer buckling similar to Pt(110)-(1x2); Pt concentrations in layer 1 to 5 are: 82%, 84%, 68%, 81% and 44%, respectively; chemical ordering is described by (2x2) superstructure cell

SAMPLE PREPARATION (1 sample)
Treatment : repeated cycles of Ar+ sputtering and annealing above 1000 K
Crystallinity:
Anal. methods:
Contamination: AES: no impurities

COMMENTS
Bulk structure is chemically ordered and can be described as alternating (110) layers with 100% Pt and 60% Pt/40% Fe; the bulk mixed layers are also simulated here with a (2x2) superstructure cell; chemical ordering is absent in external layers but gradually recovers over five to six layers deep

DATA COLLECTION
Technique: LEED
Dataset : IV spectra for 26/21 beams at normal/off-normal incidence; 30<E<280 eV; cumul. E range 5000/4640eV

THEORY/DATA TREATMENT
Dynamical LEED (layer doubling, matrix inversion): up to 10 phase shifts, ATA approximation to model composition

STRUCTURES EXAMINED
Only the missing row model of Pt(110) has been considered. varied were: top 4 interlayer spacings, row pairing in layers 2 and 4, buckling in all odd layers; compositions of the first 5 layers were optimized

QUALITY OF EXPERIMENT-THEORY FIT
RDE=0.36, RPE=0.34, RZJ=0.14

2D UNIT CELLS (1 domain observed)

Cell	Ax (Å)	Ay (Å)	Bx (Å)	By (Å)	α (°)	Matrix	Pattern	Cell type
Bulk	2.760	0.000	0.000	3.900	90.0	(1.000, 0.000) (0.000, 1.000)	(1x1)	b: bulk lattice
Surface 1	5.520	0.000	0.000	7.800	90.0	(2.000, 0.000) (0.000, 2.000)	(2x2)	s1: commens. superlattice

3D COORDINATES

Fe1-Pt2: disordered remaining row in top layer, 82%Pt/18%Fe; Fe3-Pt6: row-paired 2nd layer, chem. disord., 84%Pt/16%Fe; Pt7-Pt10, Fe15-Pt18: buckled 3rd layer (68%Pt), 5th (44%Pt); Fe11-Pt14: row-paired 4th layer, chem. disord., 81%Pt/19%Fe

Dx/Dy in Å, or as a fraction of layer's unit cell vectors; atom 0 at the origin. Epir/subr are bulk repeat vectors.

No. of atoms: 23

Bulk z = 1.380 Å

Region	Chem el.	At. no.	Cell type	Site occ.	Rel. to	Dx ± εx		Dy ± εy		Dz ± εz		Dz/Bz(%) ± εz/Bz
epir		-2					f		f		Å	
subr		-1				0.000	Å	0.000	Å	2.760	Å	
intf	Fe	1	m1	.18	0	0.000	f	0.000	f	0.000	Å	0.0
intf	Pt	2	m1	.82	1	0.500	f	0.000	f	0.000	Å	0.0
intf	Fe	3	m1	.16	1	0.250	f	0.254 ± .013	f	1.200 ± .100	Å	87.0 ± 7.2
intf	Pt	4	m1	.28	3	0.500	f	0.000	f	0.000	Å	0.0
intf	Pt	5	m1	.28	3	0.000	f	0.492 ± .013	f	0.000	Å	0.0
intf	Pt	6	m1	.28	3	0.500	f	0.492 ± .013	f	0.000	Å	0.0
intf	Pt	7	m1	.23	1	0.000	f	0.000	f	2.680 ± .100	Å	194.2 ± 7.2
intf	Pt	8	m1	.23	7	0.000	f	0.500	f	0.000	Å	2.2
intf	Fe	9	m1	.36	1	0.500	f	0.000	f	2.520 ± .100	Å	182.6 ± 7.2
intf	Pt	10	m1	.23	9	0.000	f	0.500	f	0.000	Å	0.0
intf	Fe	11	m1	.19	1	0.250	f	0.251 ± .013	f	4.000 ± .100	Å	289.9 ± 7.2
intf	Pt	12	m1	.27	11	0.500	f	0.000	f	0.000	Å	0.0
intf	Pt	13	m1	.27	11	0.000	f	0.497 ± .013	f	0.000	Å	0.0
intf	Pt	14	m1	.27	11	0.500	f	0.497 ± .013	f	0.000	Å	0.0
intf	Fe	15	m1	.22	1	0.000	f	0.000	f	5.440 ± .100	Å	394.2 ± 7.2
intf	Fe	16	m1	.22	15	0.000	f	0.500	f	0.000	Å	0.0
intf	Pt	17	m1	.28	1	0.500	f	0.000	f	5.380 ± .100	Å	389.9 ± 7.2
intf	Pt	18	m1	.28	17	0.000	f	0.500	f	0.000	Å	0.0
subl	Pt	19	b	1.00	1	0.500	f	0.500	f	6.810 ± .100	Å	493.5 ± 7.2

Pt0.8Fe0.2(110)-(1x2)
78.26.2

3D Coordinates - Continued

Region	Chem el.	At. no.	Cell type	Site occ.	Rel. to	Dx ± εx		Dy ± εy		Dz ± εz		Dz/Bz(%) ± εz/Bz
subl	Fe	20	s1	.20	1	0.000	f	0.000	f	8.180	Å	592.8
subl	Pt	21	s1	.30	20	0.500	f	0.000	f	0.000	Å	0.0
subl	Fe	22	s1	.20	20	0.000	f	0.500	f	0.000	Å	0.0
subl	Pt	23	s1	.30	20	0.500	f	0.500	f	0.000	Å	0.0

BOND DISTANCES AND ANGLES

Bond distances and angles are derived from coordinates

No. of distances/angles: 9

Interatomic dist. A-B (Å)	Atom A	Atom B	Atom C	Bond angle A-B-C (°)
2.760	Fe1	Pt2	Fe1(1,0)	180.0
2.760	Fe1	Pt2	Fe3	59.2
2.760	Fe1	Pt2	Pt4	120.8
2.695	Fe1	Fe3	Pt2	61.6
2.696	Fe1	Pt5(0,-1)	Pt2	61.6
2.680	Fe1	Pt9	Fe3	58.5
2.760	Fe3	Pt4	Fe1(1,0)	120.8
2.760	Fe3	Pt4	Pt2	59.2
2.708	Fe3	Pt7	Pt4	61.3

COMMON NAME : Pt0.8Fe0.2(111)-(1x1) ILLUSTRATION: 132
CLASSIFICATION : 78.26.1
TECHNIQUE : LEED
AUTHORS : P. Beccat, Y. Gauthier, R. Baudoing-Savois and J.C. Bertolini
REFERENCE : Surf. Sci., 238, 105 (1990)

SURFACE TYPE
Substrate : Pt0.8Fe0.2
Crystal face: 111
Temperature : RT
Bulk lattice: fcc
2D bulk symm: none
2D surf symm: none

Adsorbate:
Coverage :
Pattern : (1x1) and disorder
Matrix : (1.000, 0.000)
(0.000, 1.000)

STRUCTURE TYPE
Unreconstructed, relaxed, segregated bulk termination: top layer buckled by 0.09Å, where every Pt atom with 3 underlying Pt atoms moves outward; Pt concentrations in top 3 layers: 96%, 88%, 84%; top 2 spacings expanded 0.3% and contracted 0.6%; the bulk is chemically ordered.

SAMPLE PREPARATION (1 sample)
Treatment : Ar-ion bombardment and annealing to 1200 K
Crystallinity:
Anal. methods: AES
Contamination: AES: no impurities

COMMENTS
The chemically ordered bulk structure is approximately described with a (2x2) supercell, corresponding to Pt75Fe25, and the chemically disordered surface region is also described approximately with a (2x2) supercell

DATA COLLECTION
Technique: LEED
Dataset : I-V spectra for 12 non-equivalent beams at normal incidence; cumulative energy range 1612 eV

THEORY/DATA TREATMENT
Dynamical LEED: averaged-t-matrix approximation (ATA), 10 phase shifts

STRUCTURES EXAMINED
Varied were: first two interlayer spacings, buckling in top layer; compositions of first three layers; only the (111)(1x1) structure was considered

2D UNIT CELLS (1 domain observed)

Cell	Ax (Å)	Ay (Å)	Bx (Å)	By (Å)	α (°)	Matrix	Pattern	Cell type
Bulk	2.772	0.000	1.386	2.401	60.0	(1.000, 0.000) (0.000, 1.000)	(1x1)	b: bulk lattice
Surface 1	5.544	0.000	2.772	4.801	60.0	(2.000, 0.000) (0.000, 2.000)	(2x2)	s1: commens. superlattice

3D COORDINATES

Pt1-Pt4: top layer, Pt1 buckled outward; Fe5-Pt8: planar 2nd layer;
Fe9-Pt12: set of periodically repeating bulk layers

Dx/Dy in Å, or as a fraction of layer's unit cell vectors; atom 0 at the origin. Epir/subr are bulk repeat vectors.

No. of atoms: 12 Bulk z = 2.260 Å

Region	Chem el.	At. no.	Cell type	Site occ.	Rel. to	Dx ± εx		Dy ± εy		Dz ± εz		Dz/Bz(%) ± εz/Bz
epir		-2					f		f		Å	
subr		-1				2.770	Å	1.600	Å	2.260	Å	
intf	Pt	1	s1	.25	0	0.000	f	0.000	f	0.000	Å	0.0
intf	Pt	2	s1	.25	1	0.500	f	0.000	f	0.090 ± .020	Å	4.0 ± .9
intf	Pt	3	s1	.25	2	0.000	f	0.500	f	0.000	Å	0.0
intf	Pt	4	s1	.25	3	-0.500	f	0.000	f	0.000	Å	0.0
intf	Fe	5	s1	.25	4	0.333	f	-0.167	f	2.267 ± .025	Å	100.3 ± 1.1
intf	Pt	6	s1	.25	5	0.500	f	0.000	f	0.000	Å	0.0
intf	Pt	7	s1	.25	6	0.000	f	0.500	f	0.000	Å	0.0
intf	Pt	8	s1	.25	7	-0.500	f	0.000	f	0.000	Å	0.0
subl	Fe	9	s1	.25	8	0.333	f	-0.167	f	2.246 ± .041	Å	99.4 ± 1.8
subl	Pt	10	s1	.25	9	-0.500	f	0.000	f	0.000	Å	0.0
subl	Pt	11	s1	.25	10	0.000	f	-0.500	f	0.000	Å	0.0
subl	Pt	12	s1	.25	11	0.500	f	0.000	f	0.000	Å	0.0

Pt0.8Fe0.2(111)-(1x1)
78.26.1

BOND DISTANCES AND ANGLES

Bond distances and angles are derived from coordinates

No. of distances/angles: 5

Interatomic dist. A-B (Å)	Atom A	Atom B	Atom C	Bond angle A-B-C (°)
2.773	Pt1	Pt2	Pt1(1,0)	176.3
2.773	Pt1	Pt2	Pt3	120.0
2.849	Pt1	Pt6(-1,0)	Pt3(-1,0)	59.1
2.772	Pt2	Pt3	Pt1(1,0)	60.0
2.775	Pt2	Fe5	Pt3	59.9

ILLUSTRATION: 136

COMMON NAME : Pt0.1Ni0.9(100)-(1x1)
CLASSIFICATION : 78.28.8
TECHNIQUE : LEED
AUTHORS : Y. Gauthier, W. Hoffmann and M. Wuttig
REFERENCE : Surf. Sci., 233, 239 (1990)

SURFACE TYPE
Substrate : Pt0.1Ni0.9
Crystal face: 100
Temperature : RT
Bulk lattice: fcc
2D bulk symm: none
2D surf symm: none

Adsorbate:
Coverage :
Pattern : (1x1) and disorder
Matrix : (1.000, 0.000)
(0.000, 1.000)

STRUCTURE TYPE
Relaxed bulk termination with 24.3at.% Pt in the first layer and 6.4at.% Pt in the second layer; first, second and third interlayer spacings are relaxed by +2.0%, -1.2% and +1.6%, respectively; layer compositions are here simulated approximately with an ordered (3x2) superstructure cell

SAMPLE PREPARATION (1 sample)
Treatment : repeated cycles of Ar-ion bombardment and annealing to 1100 K
Crystallinity:
Anal. methods: AES, EDX and Laue x-ray diffraction
Contamination: AES: no impurities

COMMENTS
Bulk and surface are substitutionally disordered

DATA COLLECTION
Technique: LEED
Dataset : I-V spectra for 4 non-equivalent beams at normal incidence, E range 80-330 eV.

THEORY/DATA TREATMENT
Dynamical LEED: averaged-t-matrix approximation (ATA)

STRUCTURES EXAMINED
Only the (100)-(1x1) structure was considered; varied were: first 3 interlayer spacings and the compositions of the first 2 layers

2D UNIT CELLS (1 domain observed)

Cell	Ax (Å)	Ay (Å)	Bx (Å)	By (Å)	α (°)	Matrix	Pattern	Cell type
Bulk	2.522	0.000	0.000	2.522	90.0	(1.000, 0.000) (0.000, 1.000)	(1x1)	b: bulk lattice
Surface 1	7.567	0.000	0.000	5.044	90.0	(3.000, 0.000) (0.000, 2.000)	disordered	m1: randomly mixed layer

3D COORDINATES

Pt1-Ni6: 1st layer containing 24.3at.% Pt; Pt7-Ni12: 2nd layer containing 6.4at.% Pt;
Pt13-Ni18: 3rd layer containing 10at.% Pt (bulk value); Pt19-Ni24: repeating set of bulk layers

Dx/Dy in Å, or as a fraction of layer's unit cell vectors; atom 0 at the origin. Epir/subr are bulk repeat vectors.

No. of atoms: 24 Bulk z = 1.784 Å

Region	Chem el.	At. no.	Cell type	Site occ.	Rel. to	Dx ± εx		Dy ± εy		Dz ± εz		Dz/Bz(%) ± εz/Bz
epir		-2					f		f		Å	
subr		-1				1.261	Å	1.261	Å	1.784	Å	
intf	Pt	1	m1	.12	0	0.000	f	0.000	f	0.000	Å	0.0
intf	Ni	2	m1	.19	1	0.333	f	0.000	f	0.000	Å	0.0
intf	Ni	3	m1	.19	1	0.667	f	0.000	f	0.000	Å	0.0
intf	Ni	4	m1	.19	1	0.000	f	0.500	f	0.000	Å	0.0
intf	Pt	5	m1	.12	1	0.333	f	0.500	f	0.000	Å	0.0
intf	Ni	6	m1	.19	1	0.667	f	0.500	f	0.000	Å	0.0
intf	Pt	7	m1	.06	1	0.167	f	0.250	f	1.819 ± .005	Å	102.0 ± .3
intf	Ni	8	m1	.19	7	0.333	f	0.000	f	0.000	Å	0.0
intf	Ni	9	m1	.19	7	0.667	f	0.000	f	0.000	Å	0.0
intf	Ni	10	m1	.19	7	0.000	f	0.500	f	0.000	Å	0.0
intf	Ni	11	m1	.19	7	0.333	f	0.500	f	0.000	Å	0.0
intf	Ni	12	m1	.19	7	0.667	f	0.500	f	0.000	Å	0.0
intf	Pt	13	m1	.10	7	0.167	f	0.250	f	1.762 ± .007	Å	98.8 ± .4
intf	Ni	14	m1	.18	13	0.333	f	0.000	f	0.000	Å	0.0
intf	Ni	15	m1	.18	13	0.667	f	0.000	f	0.000	Å	0.0
intf	Ni	16	m1	.18	13	0.000	f	0.500	f	0.000	Å	0.0
intf	Ni	17	m1	.18	13	0.333	f	0.500	f	0.000	Å	0.0
intf	Ni	18	m1	.18	13	0.667	f	0.500	f	0.000	Å	0.0
subl	Pt	19	m1	.10	13	0.167	f	0.250	f	1.812 ± .016	Å	101.6 ± .9
subl	Ni	20	m1	.18	19	0.333	f	0.000	f	0.000	Å	0.0

Pt0.1Ni0.9(100)-(1x1)
78.28.8

3D Coordinates - Continued

Region	Chem el.	At. no.	Cell type	Site occ.	Rel. to	Dx ± εx		Dy ± εy		Dz ± εz		Dz/Bz(%) ± εz/Bz
subl	Ni	21	m1	.18	19	0.667	f	0.000	f	0.000	Å	0.0
subl	Ni	22	m1	.18	19	0.000	f	0.500	f	0.000	Å	0.0
subl	Ni	23	m1	.18	19	0.333	f	0.500	f	0.000	Å	0.0
subl	Ni	24	m1	.18	19	0.667	f	0.500	f	0.000	Å	0.0

BOND DISTANCES AND ANGLES

Bond distances and angles are derived from coordinates

No. of distances/angles: 2

Interatomic dist. A-B (Å)	Atom A	Atom B	Atom C	Bond angle A-B-C (°)
2.522	Pt1	Ni2	Pt7	60.3
2.522	Ni3	Ni2	Pt1	180.0

COMMON NAME : Pt0.1Ni0.9(110)-(1x1)
CLASSIFICATION : 78.28.7
TECHNIQUE : LEED
AUTHORS : Y. Gauthier, R. Baudoing and J. Jupille
REFERENCE : Phys. Rev., B40, 1500 (1989)

ILLUSTRATION: 138

SURFACE TYPE
Substrate : Pt0.1Ni0.9
Crystal face: 110
Temperature : RT
Bulk lattice: fcc
2D bulk symm: none
2D surf symm: none

Adsorbate:
Coverage :
Pattern : (1x1) and disorder
Matrix : (1.000, 0.000)
(0.000, 1.000)

STRUCTURE TYPE
Bulk termination with relaxations of top 2 interlayer spacings of -4.5% and +3.6%, respectively; oscillatory composition in top 2 layers: 6.4at.% Pt and 52.3at.% Pt, resp.

SAMPLE PREPARATION (1 sample)
Treatment : repeated argon bombardment and annealing up to 1200 K
Crystallinity: x-ray diffraction pattern
Anal. methods:
Contamination: AES: no impurities

COMMENTS
Bulk is substitutionally disordered, which is simulated with an ordered (4x2) structure here

DATA COLLECTION
Technique: LEED
Dataset : I-V spectra for 6 and 12 nonequivalent beams at normal and 20° off-normal incidence, E range 30-280 eV

THEORY/DATA TREATMENT
Dynamical LEED (layer doubling, averaged t-matrix approx.)

STRUCTURES EXAMINED
Only the (111)-(1x1) structure was considered; varied were first 3 interlayer spacings and compositions of the first 3 layers

2D UNIT CELLS (1 domain observed)

Cell	Ax (Å)	Ay (Å)	Bx (Å)	By (Å)	α (°)	Matrix	Pattern	Cell type
Bulk	2.526	0.000	0.000	3.572	90.0	(1.000, 0.000) (0.000, 1.000)	(1x1)	b: bulk lattice
Surface 1	10.103	0.000	0.000	7.144	90.0	(4.000, 0.000) (0.000, 2.000)	disordered	m1: randomly mixed layer

3D COORDINATES

Pt1-Ni8: top layer with 6.4at.% Pt; Pt9-Pt16: second layer with 52.3at.% Pt;
Pt17-Ni24: third bulklike layer with 10at.% Pt

Dx/Dy in Å, or as a fraction of layer's unit cell vectors; atom 0 at the origin. Epir/subr are bulk repeat vectors.

No. of atoms: 24 Bulk z = 1.263 Å

Region	Chem el.	At. no.	Cell type	Site occ.	Rel. to	Dx ± εx		Dy ± εy		Dz ± εz		Dz/Bz(%) ± εz/Bz
epir		-2					f		f		Å	
subr		-1				1.263	Å	1.786	Å	1.263	Å	
intf	Pt	1	m1	.06	0	0.000	f	0.000	f	0.000	Å	0.0
intf	Ni	2	m1	.13	1	0.250	f	0.000	f	0.000	Å	0.0
intf	Ni	3	m1	.13	1	0.500	f	0.000	f	0.000	Å	0.0
intf	Ni	4	m1	.13	1	0.750	f	0.000	f	0.000	Å	0.0
intf	Ni	5	m1	.13	1	0.000	f	0.500	f	0.000	Å	0.0
intf	Ni	6	m1	.13	1	0.250	f	0.500	f	0.000	Å	0.0
intf	Ni	7	m1	.13	1	0.500	f	0.500	f	0.000	Å	0.0
intf	Ni	8	m1	.13	1	0.750	f	0.500	f	0.000	Å	0.0
intf	Pt	9	m1	.13	1	0.125	f	0.250	f	1.206 ± .090	Å	95.5 ± 7.1
intf	Ni	10	m1	.12	9	0.250	f	0.000	f	0.000	Å	0.0
intf	Pt	11	m1	.13	9	0.500	f	0.000	f	0.000	Å	0.0
intf	Ni	12	m1	.12	9	0.750	f	0.000	f	0.000	Å	0.0
intf	Ni	13	m1	.12	9	0.000	f	0.500	f	0.000	Å	0.0
intf	Pt	14	m1	.13	9	0.250	f	0.500	f	0.000	Å	0.0
intf	Ni	15	m1	.12	9	0.500	f	0.500	f	0.000	Å	0.0
intf	Pt	16	m1	.13	9	0.750	f	0.500	f	0.000	Å	0.0
subl	Pt	17	m1	.10	9	0.125	f	0.250	f	1.308 ± .140	Å	103.6 ± 11.1
subl	Ni	18	m1	.13	17	0.250	f	0.000	f	0.000	Å	0.0
subl	Ni	19	m1	.13	17	0.500	f	0.000	f	0.000	Å	0.0
subl	Ni	20	m1	.13	17	0.750	f	0.000	f	0.000	Å	0.0
subl	Ni	21	m1	.13	17	0.000	f	0.500	f	0.000	Å	0.0

Pt0.1Ni0.9(110)-(1x1)
78.28.7

3D Coordinates - Continued

Reg ion	Chem el.	At. no.	Cell type	Site occ.	Rel. to	Dx ± εx		Dy ± εy		Dz ± εz		Dz/Bz(%) ± εz/Bz
subl	Ni	22	m1	.13	17	0.250	f	0.500	f	0.000	Å	0.0
subl	Ni	23	m1	.13	17	0.500	f	0.500	f	0.000	Å	0.0
subl	Ni	24	m1	.13	17	0.750	f	0.500	f	0.000	Å	0.0

BOND DISTANCES AND ANGLES

Bond distances and angles are derived from coordinates

No. of distances/angles: 6

Interatomic dist. A-B (Å)	Atom A	Atom B	Atom C	Bond angle A-B-C (°)
2.526	Pt1	Ni2	Pt9	59.6
2.526	Pt1	Ni2	Ni10	120.4
2.498	Pt1	Pt9	Ni2	60.7
2.498	Pt1	Pt9	Ni5	91.3
2.514	Pt1	Ni23	Pt9	59.1
2.514	Pt1	Ni23	Ni12(-1,0)	59.1

COMMON NAME : Pt0.1Ni0.9(111)-(1x1) ILLUSTRATION: 129
CLASSIFICATION : 78.28.2
TECHNIQUE : LEED
AUTHORS : R. Baudoing, Y. Gauthier, M. Lundberg and J. Rundgren
REFERENCE : J. Phys., C19, 2825 (1986)

SURFACE TYPE
Substrate : Pt0.1Ni0.9
Crystal face: 111
Temperature : RT
Bulk lattice: fcc
2D bulk symm: none
2D surf symm: none

Adsorbate:
Coverage :
Pattern : (1x1) and disorder
Matrix : (1.000, 0.000)
(0.000, 1.000)

STRUCTURE TYPE
Relaxed bulk termination with 30at.% Pt in first layer and 5at.% Pt in the second; second interlayer spacing contracted by 0.8%; layer compositions are here simulated approximately with an ordered (3x2) superstructure cell

SAMPLE PREPARATION (1 sample)
Treatment : cycles of argon-ion bombardment and annealing to 1200 K
Crystallinity:
Anal. methods: AES
Contamination: AES: no impurities

COMMENTS
Bulk and surface are substitutionally disordered

DATA COLLECTION
Technique: LEED
Dataset : I-V spectra for 27 non-equivalent beams at normal and non-normal incidence; E ranges 30-250 eV and 30-230eV, resp.

THEORY/DATA TREATMENT
Dynamical LEED: averaged-t-matrix approximation (ATA)

STRUCTURES EXAMINED
Only the (111)-(1x1) structure was considered; varied were: first 2 interlayer distances and the compositions of the first 2 layers

2D UNIT CELLS (1 domain observed)

Cell	Ax (Å)	Ay (Å)	Bx (Å)	By (Å)	α (°)	Matrix	Pattern	Cell type
Bulk	2.526	0.000	1.263	2.187	60.0	(1.000, 0.000) (0.000, 1.000)	(1x1)	b: bulk lattice
Surface 1	7.577	0.000	2.526	4.375	60.0	(3.000, 0.000) (0.000, 2.000)	disordered	m1: randomly mixed layer

3D COORDINATES

Pt1-Ni6: top layer with 30at.% Pt; Pt7-Ni12: second layer with 5at.% Pt;
Pt13-Ni18: repeating bulk layers with 10at.% Pt

Dx/Dy in Å, or as a fraction of layer's unit cell vectors; atom 0 at the origin. Epir/subr are bulk repeat vectors.

No. of atoms: 18 Bulk z = 2.062 Å

Region	Chem el.	At. no.	Cell type	Site occ.	Rel. to	Dx ± εx	Dy ± εy	Dz ± εz	Dz/Bz(%) ± εz/Bz
epir		-2				f	f	Å	
subr		-1				2.526 Å	1.458 Å	2.062 Å	
intf	Pt	1	m1	.15	0	0.000 f	0.000 f	0.000 Å	0.0
intf	Ni	2	m1	.18	1	0.333 f	0.000 f	0.000 Å	0.0
intf	Ni	3	m1	.18	1	0.667 f	0.000 f	0.000 Å	0.0
intf	Ni	4	m1	.18	1	0.000 f	0.500 f	0.000 Å	0.0
intf	Pt	5	m1	.15	1	0.333 f	0.500 f	0.000 Å	0.0
intf	Ni	6	m1	.18	1	0.667 f	0.500 f	0.000 Å	0.0
intf	Pt	7	m1	.05	1	0.222 f	0.333 f	2.062 ± .004 Å	100.0 ± .2
intf	Ni	8	m1	.19	7	0.333 f	0.000 f	0.000 Å	0.0
intf	Ni	9	m1	.19	7	0.667 f	0.000 f	0.000 Å	0.0
intf	Ni	10	m1	.19	7	0.000 f	0.500 f	0.000 Å	0.0
intf	Ni	11	m1	.19	7	0.333 f	0.500 f	0.000 Å	0.0
intf	Ni	12	m1	.19	7	0.667 f	0.500 f	0.000 Å	0.0
subl	Pt	13	m1	.10	7	0.222 f	0.333 f	2.046 ± .019 Å	99.2 ± .9
subl	Ni	14	m1	.18	13	0.333 f	0.000 f	0.000 Å	0.0
subl	Ni	15	m1	.18	13	0.667 f	0.000 f	0.000 Å	0.0
subl	Ni	16	m1	.18	13	0.000 f	0.500 f	0.000 Å	0.0
subl	Ni	17	m1	.18	13	0.333 f	0.500 f	0.000 Å	0.0
subl	Ni	18	m1	.18	13	0.667 f	0.500 f	0.000 Å	0.0

Pt0.1Ni0.9(111)-(1x1)
78.28.2 BOND DISTANCES AND ANGLES

Bond distance derived from coordinates

No. of distances/angles: 1

Interatomic dist. A-B (Å)	Atom A	Atom B	Atom C	Bond angle A-B-C (°)
2.525	Ni1	Ni2		

COMMON NAME : Pt0.5Ni0.5(110)-(1x1)
CLASSIFICATION : 78.28.3
TECHNIQUE : LEED
AUTHORS : Y. Gauthier, R. Baudoing, M. Lundberg and J. Rundgren
REFERENCE : Phys. Rev., B35, 7867 (1987)

ILLUSTRATION: 139

SURFACE TYPE
Substrate : Pt0.5Ni0.5
Crystal face: 110
Temperature : RT
Bulk lattice: fcc
2D bulk symm: none
2D surf symm: none

Adsorbate:
Coverage :
Pattern : (1x1) and disorder
Matrix : (1.000, -0.000)
(0.000, 1.000)

STRUCTURE TYPE
Bulk termination with relaxations of top 3 interlayer spacings of -19%, +11% and -1%, respectively; oscillatory composition in top 3 layers: 100at.% Ni, 95at.% Pt and 83at.% Ni, resp.

SAMPLE PREPARATION (1 sample)
Treatment : repeated argon bombardment and annealing up to 1200 K
Crystallinity: x-ray diffraction pattern
Anal. methods:
Contamination: AES: no impurities

COMMENTS
Bulk is substitutionally disordered, which is simulated with an ordered (2x2) structure here

DATA COLLECTION
Technique: LEED
Dataset : I-V spectra for (10), (13), (11) and (12) beams at four angles of incidence; E range 30-250 eV

THEORY/DATA TREATMENT
Dynamical LEED (layer doubling, averaged t-matrix approx.); Vor and Voi optimized

STRUCTURES EXAMINED
Only (110)-(1x1) structure was considered; first 3 interlayer spacings and the vacancy concentration in the top layer were varied, as well as the compositions of the first four layers

2D UNIT CELLS (1 domain observed)

Cell	Ax (Å)	Ay (Å)	Bx (Å)	By (Å)	α (°)	Matrix	Pattern	Cell type
Bulk	2.652	0.000	0.000	3.750	90.0	(1.000, 0.000) (0.000, 1.000)	(1x1)	b: bulk lattice
Surface 1	5.303	0.000	0.000	7.500	90.0	(2.000, 0.000) (0.000, 2.000)	disordered	m1: randomly mixed layer

3D COORDINATES

Ni1: topmost Ni-enriched layer with 100at.% Ni. Pt2: 95at.% Pt-enriched second layer
Pt3-Ni6: third layer with 83at.% Ni. Pt7-Pt10: bulklike substitutionally disordered layer

Dx/Dy in Å, or as a fraction of layer's unit cell vectors; atom 0 at the origin. Epir/subr are bulk repeat vectors.

No. of atoms: 10 Bulk z = 1.325 Å

Reg ion	Chem el.	At. no.	Cell type	Site occ.	Rel. to	Dx ± εx		Dy ± εy		Dz ± εz		Dz/Bz(%) ± εz/Bz
epir		-2					f		f		Å	
subr		-1				1.326	Å	1.875	Å	1.325	Å	
intf	Ni	1	b	1.00	0	0.000	f	0.000	f	0.000	Å	0.0
intf	Pt	2	b	.95	1	0.500	f	0.500	f	1.070 ± .010	Å	80.8 ± .8
intf	Pt	3	m1	.17	2	0.250	f	0.250	f	1.470 ± .020	Å	110.9 ± 1.5
intf	Ni	4	m1	.27	3	0.500	f	0.000	f	0.000	Å	0.0
intf	Ni	5	m1	.27	3	0.000	f	0.500	f	0.000	Å	0.0
intf	Ni	6	m1	.27	3	0.500	f	0.500	f	0.000	Å	0.0
subl	Pt	7	m1	.24	3	0.250	f	0.250	f	1.310 ± .020	Å	98.9 ± 1.5
subl	Ni	8	m1	.26	7	0.500	f	0.000	f	0.000	Å	0.0
subl	Ni	9	m1	.26	7	0.000	f	0.500	f	0.000	Å	0.0
subl	Pt	10	m1	.24	7	0.500	f	0.500	f	0.000	Å	0.0

Pt0.5Ni0.5(110)-(1x1)
78.28.3 BOND DISTANCES AND ANGLES

Bond distances and angles are derived from coordinates

No. of distances/angles: 5

Interatomic dist. A-B (Å)	Atom A	Atom B	Atom C	Bond angle A-B-C (°)
2.652	Ni1	Ni1(1,0)	Pt5(1,-1)	121.6
2.652	Ni1	Ni1(1,0)	Pt5(0,-2)	58.4
2.533	Ni1	Pt5(0,-1)	Ni1(1,1)	130.0
2.533	Ni1	Pt5(0,-1)	Ni1(1,0)	63.1
2.533	Ni1	Pt5(0,-1)	Ni1(0,1)	95.5

COMMON NAME : Pt0.5Ni0.5(111)-(1x1) ILLUSTRATION: 130
CLASSIFICATION : 78.28.1b
TECHNIQUE : LEED
AUTHORS : Y. Gauthier, R. Baudoing, Y. Joly, J. Rundgren, J.C. Bertolini and J. Massardier
REFERENCE : Surf. Sci., 162, 342 (1985)

SURFACE TYPE
Substrate : Pt0.5Ni0.5
Crystal face: 111
Temperature : RT
Bulk lattice: fcc
2D bulk symm: none
2D surf symm: none

Adsorbate:
Coverage :
Pattern : (1x1) and disorder
Matrix : (1.000, 0.000)
(0.000, 1.000)

STRUCTURE TYPE
Unrelaxed bulk termination with 88at.% Pt in first layer, 9at.% Pt in the second and 65at.% Pt in the third; layer compositions are here simulated approximately with an ordered (3x2) superstructure cell

SAMPLE PREPARATION (1 sample)
Treatment : cycles of Ar-ion bombardment and annealing to 1200 K
Crystallinity:
Anal. methods: AES
Contamination: AES: no impurities

COMMENTS
Bulk and surface are substitutionally disordered

DATA COLLECTION
Technique: LEED
Dataset : I-V spectra for 21 non-equivalent beams at normal incidence and at 10 and 20° off-normal incidence

THEORY/DATA TREATMENT
Dynamical LEED: averaged-t-matrix approximation (ATA)

STRUCTURES EXAMINED
Only the (111)-(1x1) structure was considered; varied were: first 3 interlayer distances and the compositions of the first 2 layers

2D UNIT CELLS (1 domain observed)

Cell	Ax (Å)	Ay (Å)	Bx (Å)	By (Å)	α (°)	Matrix	Pattern	Cell type
Bulk	2.650	0.000	1.325	2.295	60.0	(1.000, 0.000) (0.000, 1.000)	(1x1)	b: bulk lattice
Surface 1	7.950	0.000	2.650	4.590	60.0	(3.000, 0.000) (0.000, 2.000)	disordered	m1: randomly mixed layer

3D COORDINATES

Pt1-Pt6: top layer with 88at.% Pt; Pt7-Ni12: second layer with 9at.% Pt;
Pt13-Ni18: third layer with 65at.% Pt; Pt19-Ni24: repeating bulk layers with 50at.% Pt

Dx/Dy in Å, or as a fraction of layer's unit cell vectors; atom 0 at the origin. Epir/subr are bulk repeat vectors.

No. of atoms: 24 Bulk z = 2.164 Å

Region	Chem el.	At. no.	Cell type	Site occ.	Rel. to	Dx ± εx		Dy ± εy		Dz ± εz		Dz/Bz(%) ± εz/Bz
epir		-2					f		f		Å	
subr		-1				2.650	Å	1.530	Å	2.164	Å	
intf	Pt	1	m1	.18	0	0.000	f	0.000	f	0.000	Å	0.0
intf	Ni	2	m1	.12	1	0.333	f	0.000	f	0.000	Å	0.0
intf	Pt	3	m1	.18	1	0.667	f	0.000	f	0.000	Å	0.0
intf	Pt	4	m1	.18	1	0.000	f	0.500	f	0.000	Å	0.0
intf	Pt	5	m1	.18	1	0.333	f	0.500	f	0.000	Å	0.0
intf	Pt	6	m1	.18	1	0.667	f	0.500	f	0.000	Å	0.0
intf	Pt	7	m1	.09	1	0.222	f	0.333	f	2.164	Å	100.0
intf	Ni	8	m1	.18	7	0.333	f	0.000	f	0.000	Å	0.0
intf	Ni	9	m1	.18	7	0.667	f	0.000	f	0.000	Å	0.0
intf	Ni	10	m1	.18	7	0.000	f	0.500	f	0.000	Å	0.0
intf	Ni	11	m1	.18	7	0.333	f	0.500	f	0.000	Å	0.0
intf	Ni	12	m1	.18	7	0.667	f	0.500	f	0.000	Å	0.0
intf	Pt	13	m1	.16	7	0.222	f	0.333	f	2.164	Å	100.0
intf	Ni	14	m1	.18	13	0.333	f	0.000	f	0.000	Å	0.0
intf	Pt	15	m1	.16	13	0.667	f	0.000	f	0.000	Å	0.0
intf	Pt	16	m1	.16	13	0.000	f	0.500	f	0.000	Å	0.0
intf	Pt	17	m1	.16	13	0.333	f	0.500	f	0.000	Å	0.0
intf	Ni	18	m1	.18	13	0.667	f	0.500	f	0.000	Å	0.0
subl	Pt	19	m1	.17	13	0.222	f	0.333	f	2.164	Å	100.0
subl	Ni	20	m1	.17	19	0.333	f	0.000	f	0.000	Å	0.0

Pt0.5Ni0.5(111)-(1x1)
78.28.1b

3D Coordinates - Continued

Reg ion	Chem el.	At. no.	Cell type	Site occ.	Rel. to	Dx ± ϵx		Dy ± ϵy		Dz ± ϵz		Dz/Bz(%) ± ϵz/Bz
subl	Pt	21	m1	.17	19	0.667	f	0.000	f	0.000	Å	0.0
subl	Ni	22	m1	.17	19	0.000	f	0.500	f	0.000	Å	0.0
subl	Pt	23	m1	.17	19	0.333	f	0.500	f	0.000	Å	0.0
subl	Ni	24	m1	.17	19	0.667	f	0.500	f	0.000	Å	0.0

BOND DISTANCES AND ANGLES

Bond distances and angles are derived from coordinates

No. of distances/angles: 1

Interatomic dist. A-B (Å)	Atom A	Atom B	Atom C	Bond angle A-B-C (°)
2.650	Ni1	Pt2		

COMMON NAME : Pt0.5Ni0.5(111)-(1x1)
CLASSIFICATION : 78.28.9
TECHNIQUE : MEIS-SB
AUTHORS : S. Deckers, F.H.P.M. Habraken, W.F. van der Weg, A.W. Denier van der Gon, B. Pluis, J.F. van der Veen and R. Baudo
REFERENCE : Phys. Rev., B42, 3253 (1990)

ILLUSTRATION: 130

SURFACE TYPE
Substrate : Pt0.5Ni0.5
Crystal face: 111
Temperature : RT
Bulk lattice: fcc
2D bulk symm: none
2D surf symm: none

Adsorbate:
Coverage :
Pattern : (1x1) and disorder
Matrix : (1.000, 0.000) (0.000, 1.000)

STRUCTURE TYPE
Relaxed bulk termination with 75at.% Pt in first layer, 27at.% Pt in the second and 53at.% Pt in the third; contraction of first and second interlayer spacings by 2%; layer compositions are here simulated approximately with an ordered (3x2) superstructure cell

SAMPLE PREPARATION (1 sample)
Treatment : cycles of Ar-ion bombardment and annealing to 1270 K
Crystallinity:
Anal. methods: AES, LEED
Contamination: AES: no impurities

COMMENTS
Bulk and surface are substitutionally disordered

DATA COLLECTION
Technique: MEIS-SB; MEIS with shadowing and blocking
Dataset : 64 angular yield spectra covering 20° exit angle range

THEORY/DATA TREATMENT
Pt- and Ni-surface peak areas extracted from spectra and converted into number of visible Pt and Ni

STRUCTURES EXAMINED
Only the (111)-(1x1) structure was considered; varied were: first 2 interlayer distances and the compositions of the first 3 layers; thermal vibration amplitudes of the first 3 layer atoms were optimized.

2D UNIT CELLS (1 domain observed)

Cell	Ax (Å)	Ay (Å)	Bx (Å)	By (Å)	α (°)	Matrix	Pattern	Cell type
Bulk	2.650	0.000	1.325	2.295	60.0	(1.000, 0.000) (0.000, 1.000)	(1x1)	b: bulk lattice
Surface 1	7.950	0.000	2.650	4.590	60.0	(3.000, 0.000) (0.000, 2.000)	disordered	m1: randomly mixed layer

3D COORDINATES

Pt1-Pt6: top layer with 75at.% Pt; Ni7-Ni12: second layer with 27at.% Pt;
Pt13-Ni18: third layer with 53at.% Pt

Dx/Dy in Å, or as a fraction of layer's unit cell vectors; atom 0 at the origin. Epir/subr are bulk repeat vectors.

No. of atoms: 18 Bulk z = 2.164 Å

Region	Chem el.	At. no.	Cell type	Site occ.	Rel. to	Dx ± εx	Dy ± εy	Dz ± εz	Dz/Bz(%) ± εz/Bz
epir		-2				f	f	Å	
subr		-1				2.650 Å	1.530 Å	2.164 Å	
intf	Pt	1	m1	.19	0	0.000 f	0.000 f	0.000 Å	0.0
intf	Ni	2	m1	.13	1	0.333 f	0.000 f	0.000 Å	0.0
intf	Pt	3	m1	.19	1	0.667 f	0.000 f	0.000 Å	0.0
intf	Ni	4	m1	.13	1	0.000 f	0.500 f	0.000 Å	0.0
intf	Pt	5	m1	.19	1	0.333 f	0.500 f	0.000 Å	0.0
intf	Pt	6	m1	.19	1	0.667 f	0.500 f	0.000 Å	0.0
intf	Ni	7	m1	.18	1	0.222 f	0.333 f	2.120 ± .011 Å	98.0 ± .5
intf	Pt	8	m1	.14	7	0.333 f	0.000 f	0.000 Å	0.0
intf	Ni	9	m1	.18	7	0.667 f	0.000 f	0.000 Å	0.0
intf	Pt	10	m1	.14	7	0.000 f	0.500 f	0.000 Å	0.0
intf	Ni	11	m1	.18	7	0.333 f	0.500 f	0.000 Å	0.0
intf	Ni	12	m1	.18	7	0.667 f	0.500 f	0.000 Å	0.0
subl	Pt	13	m1	.18	7	0.222 f	0.333 f	2.120 ± .011 Å	98.0 ± .5
subl	Ni	14	m1	.16	13	0.333 f	0.000 f	0.000 Å	0.0
subl	Pt	15	m1	.18	13	0.667 f	0.000 f	0.000 Å	0.0
subl	Ni	16	m1	.16	13	0.000 f	0.500 f	0.000 Å	0.0
subl	Pt	17	m1	.18	13	0.333 f	0.500 f	0.000 Å	0.0
subl	Ni	18	m1	.16	13	0.667 f	0.500 f	0.000 Å	0.0

Pt0.5Ni0.5(111)-(1x1)
78.28.9

BOND DISTANCES AND ANGLES

Bond distance derived from coordinates

No. of distances/angles: 1

Interatomic dist. A-B (Å)	Atom A	Atom B	Atom C	Bond angle A-B-C (°)
2.650	Pt1	Ni2		

ILLUSTRATION: 131

COMMON NAME : Pt0.78Ni0.22(111)-(1x1)
CLASSIFICATION : 78.28.1a
TECHNIQUE : LEED
AUTHORS : Y. Gauthier, R. Baudoing, Y. Joly, J. Rundgren, J.C. Bertolini and J. Massardier
REFERENCE : Surf. Sci., 162, 342 (1985)

SURFACE TYPE
Substrate : Pt0.78Ni0.22
Crystal face: 111
Temperature : RT
Bulk lattice: fcc
2D bulk symm: none
2D surf symm: none

Adsorbate:
Coverage :
Pattern : (1x1) and disorder
Matrix : (1.000, 0.000)
(0.000, 1.000)

STRUCTURE TYPE
Unrelaxed bulk termination with 99at.% Pt in first layer, 30at.% Pt in the second and 87at.% Pt in the third; layer compositions are here simulated approximately with an ordered (3x2) superstructure cell

SAMPLE PREPARATION (1 sample)
Treatment : cycles of Ar-ion bombardment and annealing to 1200 K
Crystallinity:
Anal. methods: AES
Contamination: AES: no impurities

COMMENTS
Bulk and surface are substitutionally disordered

DATA COLLECTION
Technique: LEED
Dataset : I-V spectra for 25 non-equivalent beams at normal incidence and at 10 and 20° off-normal incidence

THEORY/DATA TREATMENT
Dynamical LEED: averaged-t-matrix approximation (ATA)

STRUCTURES EXAMINED
Only the (111)-(1x1) structure was considered; varied were: first 3 interlayer spacings and the compositions of the first 2 layers

2D UNIT CELLS (1 domain observed)

Cell	Ax (Å)	Ay (Å)	Bx (Å)	By (Å)	α (°)	Matrix	Pattern	Cell type
Bulk	2.720	0.000	1.360	2.356	60.0	(1.000, 0.000) (0.000, 1.000)	(1x1)	b: bulk lattice
Surface 1	8.160	0.000	2.720	4.711	60.0	(3.000, 0.000) (0.000, 2.000)	disordered	m1: randomly mixed layer

3D COORDINATES

Pt1: top layer with 99at.% Pt; Pt2-Ni7: second layer with 30at.% Pt;
Pt8-Pt13: third layer with 87at.% Pt; Pt14-Pt19: repeating bulk layers with 78at.% Pt

Dx/Dy in Å, or as a fraction of layer's unit cell vectors; atom 0 at the origin. Epir/subr are bulk repeat vectors.

No. of atoms: 19

Bulk z = 2.221 Å

Region	Chem el.	At. no.	Cell type	Site occ.	Rel. to	Dx ± εx		Dy ± εy		Dz ± εz		Dz/Bz(%) ± εz/Bz
epir		-2					f		f		Å	
subr		-1				2.720	Å	1.570	Å	2.221	Å	
intf	Pt	1	b	.99	0	0.000	f	0.000	f	0.000	Å	0.0
intf	Pt	2	m1	.15	1	0.222	f	0.333	f	2.221	Å	100.0
intf	Ni	3	m1	.18	2	0.333	f	0.000	f	0.000	Å	0.0
intf	Ni	4	m1	.18	2	0.667	f	0.000	f	0.000	Å	0.0
intf	Ni	5	m1	.18	2	0.000	f	0.500	f	0.000	Å	0.0
intf	Pt	6	m1	.15	2	0.333	f	0.500	f	0.000	Å	0.0
intf	Ni	7	m1	.18	2	0.667	f	0.500	f	0.000	Å	0.0
intf	Pt	8	m1	.17	2	0.222	f	0.333	f	2.221	Å	100.0
intf	Ni	9	m1	.13	8	0.333	f	0.000	f	0.000	Å	0.0
intf	Pt	10	m1	.17	8	0.667	f	0.000	f	0.000	Å	0.0
intf	Pt	11	m1	.17	8	0.000	f	0.500	f	0.000	Å	0.0
intf	Pt	12	m1	.17	8	0.333	f	0.500	f	0.000	Å	0.0
intf	Pt	13	m1	.17	8	0.667	f	0.500	f	0.000	Å	0.0
subl	Pt	14	m1	.16	8	0.222	f	0.333	f	2.221	Å	100.0
subl	Ni	15	m1	.22	14	0.333	f	0.000	f	0.000	Å	0.0
subl	Pt	16	m1	.16	14	0.667	f	0.000	f	0.000	Å	0.0
subl	Pt	17	m1	.16	14	0.000	f	0.500	f	0.000	Å	0.0
subl	Pt	18	m1	.16	14	0.333	f	0.500	f	0.000	Å	0.0
subl	Pt	19	m1	.16	14	0.667	f	0.500	f	0.000	Å	0.0

Pt0.78Ni0.22(111)-(1x1)
78.28.1a

BOND DISTANCES AND ANGLES

Bond distances and angles are derived from coordinates

No. of distances/angles: 1

Interatomic dist. A-B (Å)	Atom A	Atom B	Atom C	Bond angle A-B-C (°)
2.720	Pt1	Pt1(1,0)		

COMMON NAME : Re(10-10)-(1x1)
CLASSIFICATION : 75.2
TECHNIQUE : LEED
AUTHORS : H.L. Davis and D.M. Zehner
REFERENCE : J. Vac. Sci. Technol., 17, 190 (1980)

ILLUSTRATION: 20

SURFACE TYPE
Substrate : Re
Crystal face: 10-10
Temperature : RT
Bulk lattice: hcp
2D bulk symm: pmm
2D surf symm: pmm

Adsorbate:
Coverage :
Pattern : (1x1)
Matrix : (1.000, 0.000)
(0.000, 1.000)

STRUCTURE TYPE
Bulk termination with top interlayer spacing contraction; second interlayer spacing may be expanded 1 to 2%; termination between widely spaced layers

SAMPLE PREPARATION (1 sample)
Treatment : see Zehner and Farnsworth, Surf. Sci. 30, 335 (1972)
Crystallinity:
Anal. methods:
Contamination:

COMMENTS

DATA COLLECTION
Technique: LEED
Dataset : I-V spectra for (10) and (11) beams at normal incidence; energy range 50-200 eV

THEORY/DATA TREATMENT
Dynamical LEED (giant matrix inversion for 8-layer slab): 8 phase shifts, free atom potential; Voi=-5 eV; ΘD=320 K

STRUCTURES EXAMINED
Two possible lattice terminations; variations of two topmost layer spacings

QUALITY OF EXPERIMENT-THEORY FIT
RPE=0.117

2D UNIT CELLS (1 domain observed)

Cell	Ax (Å)	Ay (Å)	Bx (Å)	By (Å)	α (°)	Matrix	Pattern	Cell type
Bulk	2.760	0.000	0.000	4.460	90.0	(1.000, 0.000) (0.000, 1.000)	(1x1)	b: bulk lattice
Surface 1	2.760	0.000	0.000	4.460	90.0	(1.000, 0.000) (0.000, 1.000)	(1x1)	s1: commens. superlattice

3D COORDINATES

Re1-Re2: narrowly spaced (contracted) pair of layers; Re3-Re4 and Re5-Re6: 2 pairs of narrowly spaced layers, together forming periodically repeating set of bulk layers; 0.1Å error bars assumed for tabulation

Dx/Dy in Å, or as a fraction of layer's unit cell vectors; atom 0 at the origin. Epir/subr are bulk repeat vectors.

No. of atoms: 6

Bulk z = 2.400 Å

Reg ion	Chem el.	At. no.	Cell type	Site occ.	Rel. to	Dx ± εx		Dy ± εy		Dz ± εz		Dz/Bz(%) ± εz/Bz
epir		-2					f		f		Å	
subr		-1				0.000	Å	0.000	Å	4.800	Å	
intf	Re	1	b	1.00	0	0.000	f	0.000	f	0.000	Å	0.0
intf	Re	2	b	1.00	1	0.500	f	0.500	f	0.670 ± .100	Å	27.9 ± 4.2
subl	Re	3	b	1.00	2	0.000	f	-0.500	f	1.600 ± .100	Å	66.7 ± 4.2
subl	Re	4	b	1.00	3	-0.500	f	0.500	f	0.800	Å	33.3
subl	Re	5	b	1.00	4	0.000	f	-0.500	f	1.600	Å	66.7
subl	Re	6	b	1.00	5	0.500	f	0.500	f	0.800	Å	33.3

BOND DISTANCES AND ANGLES

Bond distances and angles are derived from coordinates

No. of distances/angles: 5

Interatomic dist. A-B (Å)	Atom A	Atom B	Atom C	Bond angle A-B-C (°)
2.760	Re1	Re1(1,0)	Re2	59.4
2.707	Re1	Re2	Re3	58.3
2.657	Re1	Re3	Re4	89.3
2.745	Re2	Re3	Re4	60.6

Re(10-10)-(1x1)
75.2

Bond Distances and Angles - Continued

Interatomic dist. A-B (Å)	Atom A	Atom B	Atom C	Bond angle A-B-C (°)
2.769	Re2	Re4	Re5	120.4

COMMON NAME : Rh(100)-(1x1)
CLASSIFICATION : 45.7b
TECHNIQUE : LEED
AUTHORS : S. Hengrasmee, K.A.R. Mitchell, P.R. Watson and S.J. White
REFERENCE : Can. J. of Phys., 58, 200 (1980)

ILLUSTRATION: 2

SURFACE TYPE
Substrate : Rh
Crystal face: 100
Temperature : RT*
Bulk lattice: fcc
2D bulk symm: p4m
2D surf symm: p4m

Adsorbate:
Coverage :
Pattern : (1x1)
Matrix : (1.000, 0.000)
(0.000, 1.000)

STRUCTURE TYPE
Unrelaxed bulk termination

SAMPLE PREPARATION (1 sample)
Treatment : Ar sputtering at 1000 K followed by annealing to 1300K
Crystallinity:
Anal. methods:
Contamination: AES: no detectable impurities

COMMENTS

DATA COLLECTION
Technique: LEED
Dataset : IV curves at normal incidence and at Θ=20°,φ=10° in range 0<E<300 eV

THEORY/DATA TREATMENT
Dynamical LEED (RFS): 2 different pots: Moruzzi et al, and superposition pot; 8 phase shifts; ΘD=402 K; VoiαE**1/3

STRUCTURES EXAMINED
Variation in top layer spacing from -10% to 10% from bulk spacing

QUALITY OF EXPERIMENT-THEORY FIT
RZJ=0.09

2D UNIT CELLS (1 domain observed)

Cell	Ax (Å)	Ay (Å)	Bx (Å)	By (Å)	α (°)	Matrix	Pattern	Cell type
Bulk	2.690	0.000	0.000	2.690	90.0	(1.000, 0.000) (0.000, 1.000)	(1x1)	b: bulk lattice
Surface 1	2.690	0.000	0.000	2.690	90.0	(1.000, 0.000) (0.000, 1.000)	(1x1)	s1: commens. superlattice

3D COORDINATES

first layer spacing shown is average of those obtained using two different potential models

Dx/Dy in Å, or as a fraction of layer's unit cell vectors; atom 0 at the origin. Epir/subr are bulk repeat vectors.

No. of atoms: 3 Bulk z = 1.900 Å

Reg ion	Chem el.	At. no.	Cell type	Site occ.	Rel. to	Dx ± εx		Dy ± εy		Dz ± εz		Dz/Bz(%) ± εz/Bz
epir		-2					f		f		Å	
subr		-1				-1.345	Å	-1.345	Å	1.900	Å	
intf	Rh	1	b	1.00	0	0.000	f	0.000	f	0.000	Å	0.0
intf	Rh	2	b	1.00	1	0.500	f	0.500	f	1.910 ± .020	Å	100.5 ± 1.1
subl	Rh	3	b	1.00	2	-0.500	f	-0.500	f	1.900	Å	100.0

BOND DISTANCES AND ANGLES

Bond distances are derived from coordinates

No. of distances/angles: 3

Interatomic dist. A-B (Å)	Atom A	Atom B	Atom C	Bond angle A-B-C (°)
2.690	Rh1	Rh1(1,0)		
2.696	Rh1	Rh2		
2.689	Rh2	Rh3		

COMMON NAME : Rh(110)-(1x1)
CLASSIFICATION : 45.7c
TECHNIQUE : LEED
AUTHORS : S. Hengrasmee, K.A.R. Mitchell, P.R. Watson and S.J. White
REFERENCE : Can. J. of Phys., 58, 200 (1980)

ILLUSTRATION: 4

SURFACE TYPE
Substrate : Rh
Crystal face: 110
Temperature : RT*
Bulk lattice: fcc
2D bulk symm: pmm
2D surf symm: pmm

Adsorbate:
Coverage :
Pattern : (1x1)
Matrix : (1.000, 0.000)
(0.000, 1.000)

STRUCTURE TYPE
Unrelaxed bulk termination

SAMPLE PREPARATION (1 sample)
Treatment : Ar sputtering at 1000 K followed by annealing to 1300K
Crystallinity:
Anal. methods:
Contamination: AES: no detectable impurities

COMMENTS

DATA COLLECTION
Technique: LEED
Dataset : IV curves at normal incidence (5 beams) and at Θ=10°,φ=135° (8 beams) in range 50<E<250 eV

THEORY/DATA TREATMENT
Dynamical LEED (RFS): 2 different pots: Moruzzi et al, and superposition pot; 8 phase shifts; ΘD=402 K; Voi∝E**1/3

STRUCTURES EXAMINED
-5% to -15% variation in top layer spacing from bulk

QUALITY OF EXPERIMENT-THEORY FIT
RZJ=0.09

2D UNIT CELLS (1 domain observed)

Cell	Ax (Å)	Ay (Å)	Bx (Å)	By (Å)	α (°)	Matrix	Pattern	Cell type
Bulk	2.690	0.000	0.000	3.800	90.0	(1.000, 0.000) (0.000, 1.000)	(1x1)	b: bulk lattice
Surface 1	2.690	0.000	0.000	3.800	90.0	(1.000, 0.000) (0.000, 1.000)	(1x1)	s1: commens. superlattice

3D COORDINATES

first layer spacing shown is average of those obtained using two different potential models

Dx/Dy in Å, or as a fraction of layer's unit cell vectors; atom 0 at the origin. Epir/subr are bulk repeat vectors.

No. of atoms: 3

Bulk z = 1.340 Å

Region	Chem el.	At. no.	Cell type	Site occ.	Rel. to	Dx ± εx	Dy ± εy	Dz ± εz	Dz/Bz(%) ± εz/Bz
epir		-2				f	f	Å	
subr		-1				-1.345 Å	-1.900 Å	1.340 Å	
intf	Rh	1	b	1.00	0	0.000 f	0.000 f	0.000 Å	0.0
intf	Rh	2	b	1.00	1	0.500 f	0.500 f	1.330 ± .020 Å	99.3 ± 1.5
subl	Rh	3	b	1.00	2	-0.500 f	-0.500 f	1.340 Å	100.0 ± 7.5

BOND DISTANCES AND ANGLES

Bond distances and angles are derived from coordinates

No. of distances/angles: 4

Interatomic dist. A-B (Å)	Atom A	Atom B	Atom C	Bond angle A-B-C (°)
2.690	Rh1	Rh1(1,0)	Rh2	59.9
2.681	Rh1	Rh2	Rh3	59.7
2.670	Rh1	Rh3	Rh2	60.1
2.686	Rh2	Rh3		

ILLUSTRATION: 4

COMMON NAME : Rh(110)-(1x1)
CLASSIFICATION : 45.9a
TECHNIQUE : LEED
AUTHORS : W. Nichtl, N. Bickel, L. Hammer, K. Heinz and K. Mueller
REFERENCE : Surf. Sci., 188, L729 (1987)

SURFACE TYPE
Substrate : Rh
Crystal face: 110
Temperature : RT*
Bulk lattice: fcc
2D bulk symm: pmm
2D surf symm: pmm

Adsorbate:
Coverage :
Pattern : (1x1)
Matrix : (1.000, 0.000)
(0.000, 1.000)

STRUCTURE TYPE
Unreconstructed surface with relaxations of top two interlayer spacings

SAMPLE PREPARATION (1 sample)
Treatment : Ar ion bombardment, then heating in O2 and H2
Crystallinity: sharp LEED pattern with poor background
Anal. methods:
Contamination: special attention to H

COMMENTS

DATA COLLECTION
Technique: LEED
Dataset : IV curves for 11 non-equivalent beams: E range 50-512 eV cumulative E range 4400eV

THEORY/DATA TREATMENT
Dynamical LEED (RFS): 11 phase shifts; Vor=-12 eV, Voi∝E**1/3; ΘD=480 K

STRUCTURES EXAMINED
Variation of two topmost interlayer spacings from 1.22 to 1.37Å

QUALITY OF EXPERIMENT-THEORY FIT
RPE=0.31

2D UNIT CELLS (1 domain observed)

Cell	Ax (Å)	Ay (Å)	Bx (Å)	By (Å)	α (°)	Matrix	Pattern	Cell type
Bulk	2.690	0.000	0.000	3.800	90.0	(1.000, 0.000) (0.000, 1.000)	(1x1)	b: bulk lattice
Surface 1	2.690	0.000	0.000	3.800	90.0	(1.000, 0.000) (0.000, 1.000)	(1x1)	s1: commens. superlattice

3D COORDINATES

Dx/Dy in Å, or as a fraction of layer's unit cell vectors; atom 0 at the origin. Epir/subr are bulk repeat vectors.

No. of atoms: 4

Bulk z = 1.345 Å

Region	Chem el.	At. no.	Cell type	Site occ.	Rel. to	Dx ± εx		Dy ± εy		Dz ± εz		Dz/Bz(%) ± εz/Bz
epir		-2					f		f		Å	
subr		-1				1.345	Å	1.900	Å	1.345	Å	
intf	Rh	1	b	1.00	0	0.000	f	0.000	f	0.000	Å	0.0
intf	Rh	2	b	1.00	1	0.500	f	0.500	f	1.252 ± .013	Å	93.1 ± 1.0
intf	Rh	3	b	1.00	2	-0.500	f	-0.500	f	1.371 ± .013	Å	101.9 ± 1.0
subl	Rh	4	b	1.00	3	0.500	f	0.500	f	1.345	Å	100.0

BOND DISTANCES AND ANGLES

Bond distances and angles are derived from coordinates

No. of distances/angles: 7

Interatomic dist. A-B (Å)	Atom A	Atom B	Atom C	Bond angle A-B-C (°)
2.690	Rh1	Rh1(1,0)		
2.643	Rh1	Rh2	Rh3	58.8
2.643	Rh1	Rh2	Rh4	118.3
2.623	Rh1	Rh3	Rh4	120.0
2.702	Rh2	Rh3	Rh4	60.5
2.716	Rh2	Rh4		
2.689	Rh3	Rh4		

COMMON NAME : Rh(100)-(1x1)
CLASSIFICATION : 45.9b
TECHNIQUE : LEED
AUTHORS : W. Oed, B. Doetsch, L. Hammer, K. Heinz and K. Mueller
REFERENCE : Surf. Sci., 207, 55 (1988)

ILLUSTRATION: 2

SURFACE TYPE
Substrate : Rh
Crystal face: 100
Temperature : 130 K
Bulk lattice: fcc
2D bulk symm: p4m
2D surf symm: p4m

Adsorbate:
Coverage :
Pattern : (1x1)
Matrix : (1.000, 0.000)
(0.000, 1.000)

STRUCTURE TYPE
Bulk termination with possible slight top interlayer expansion

SAMPLE PREPARATION (1 sample)
Treatment : Ar+ sputtering with ox/red cycles, then 1400 K anneal
Crystallinity:
Anal. methods:
Contamination:

COMMENTS

DATA COLLECTION
Technique: LEED
Dataset : I-V curves for 8 symmetry-inequivalent beams; E range 40-600 eV for theory and experiment

THEORY/DATA TREATMENT
Dynamical LEED (RFS): 11 relativistic phase shifts; Vor=-9 eV, ΘD=480 K

STRUCTURES EXAMINED
Variation of top two interlayer spacings

QUALITY OF EXPERIMENT-THEORY FIT
RPE=0.32

2D UNIT CELLS (1 domain observed)

Cell	Ax (Å)	Ay (Å)	Bx (Å)	By (Å)	α (°)	Matrix	Pattern	Cell type
Bulk	2.690	0.000	0.000	2.690	90.0	(1.000, 0.000) (0.000, 1.000)	(1x1)	b: bulk lattice
Surface 1	2.690	0.000	0.000	2.690	90.0	(1.000, 0.000) (0.000, 1.000)	(1x1)	s1: commens. superlattice

3D COORDINATES

Dx/Dy in Å, or as a fraction of layer's unit cell vectors; atom 0 at the origin. Epir/subr are bulk repeat vectors.

No. of atoms: 3

Bulk z = 1.902 Å

Region	Chem el.	At. no.	Cell type	Site occ.	Rel. to	Dx ± εx		Dy ± εy		Dz ± εz		Dz/Bz(%) ± εz/Bz
epir		-2					f		f		Å	
subr		-1				1.345	Å	1.345	Å	1.902	Å	
intf	Rh	1	b	1.00	0	0.000	f	0.000	f	0.000	Å	0.0
intf	Rh	2	b	1.00	1	0.500	f	0.500	f	1.912 ± .019	Å	100.5 ± 1.0
subl	Rh	3	b	1.00	2	-0.500	f	-0.500	f	1.902 ± .029	Å	100.0 ± 1.5

BOND DISTANCES AND ANGLES

Bond distances and angles are derived from coordinates

No. of distances/angles: 2

Interatomic dist. A-B (Å)	Atom A	Atom B	Atom C	Bond angle A-B-C (°)
2.690	Rh1	Rh1(1,0)	Rh2	60.1
2.697	Rh1	Rh2	Rh1(1,0)	59.8

ILLUSTRATION: 1

COMMON NAME : Rh(111)-(1x1)
CLASSIFICATION : 45.7a
TECHNIQUE : LEED
AUTHORS : S. Hengrasmee, K.A.R. Mitchell, P.R. Watson and S.J. White
REFERENCE : Can. J. of Phys., 58, 200 (1980)

SURFACE TYPE
Substrate : Rh
Crystal face: 111
Temperature : RT*
Bulk lattice: fcc
2D bulk symm: p3m1
2D surf symm: p3m1

Adsorbate:
Coverage :
Pattern : (1x1)
Matrix : (1.000, 0.000)
(0.000, 1.000)

STRUCTURE TYPE
Slightly relaxed bulk termination

SAMPLE PREPARATION (1 sample)
Treatment : Ar sputtering at 1000 K followed by annealing to 1300K
Crystallinity:
Anal. methods:
Contamination: AES: no detectable impurities

COMMENTS
This paper corrects an error in the first layer spacing previously reported in F.R. Shepherd, P.R. Watson, D.C. Frost and K.A.R. Mitchell, J. Phys. C11, 4591 (1978)

DATA COLLECTION
Technique: LEED
Dataset : IV curves at normal incidence (5 beams) and at Θ=10,φ=109° in range 0<E<250 eV

THEORY/DATA TREATMENT
Dynamical LEED (RFS): 2 different pots: Moruzzi et al, and superposition pot; 8 phase shifts; ΘD=402 K; Voi∝E**1/3

STRUCTURES EXAMINED
-10% to 10% variation in top layer spacing in increments of 2.5%

QUALITY OF EXPERIMENT-THEORY FIT
RZJ=0.12

2D UNIT CELLS (1 domain observed)

Cell	Ax (Å)	Ay (Å)	Bx (Å)	By (Å)	α (°)	Matrix	Pattern	Cell type
Bulk	2.690	0.000	1.345	2.330	60.0	(1.000, 0.000) (0.000, 1.000)	(1x1)	b: bulk lattice
Surface 1	2.690	0.000	1.345	2.330	60.0	(1.000, 0.000) (0.000, 1.000)	(1x1)	s1: commens. superlattice

3D COORDINATES

first layer spacing shown is average of those obtained using two different potential models

Dx/Dy in Å, or as a fraction of layer's unit cell vectors; atom 0 at the origin. Epir/subr are bulk repeat vectors.

No. of atoms: 3

Bulk z = 2.190 Å

Region	Chem el.	At. no.	Cell type	Site occ.	Rel. to	Dx ± εx		Dy ± εy		Dz ± εz		Dz/Bz(%) ± εz/Bz
epir		-2					f		f		Å	
subr		-1				1.345	Å	0.777	Å	2.190	Å	
intf	Rh	1	b	1.00	0	0.000	f	0.000	f	0.000	Å	0.0
intf	Rh	2	b	1.00	1	0.333	f	0.333	f	2.160 ± .020	Å	98.6 ± .9
subl	Rh	3	b	1.00	2	0.333	f	0.333	f	2.190	Å	100.0

BOND DISTANCES AND ANGLES

Bond distances and angles are derived from coordinates

No. of distances/angles: 3

Interatomic dist. A-B (Å)	Atom A	Atom B	Atom C	Bond angle A-B-C (°)
2.690	Rh1	Rh1(1,0)	Rh2	59.6
2.660	Rh1	Rh2		
2.685	Rh2	Rh3		

COMMON NAME : Rh(111)-(1x1)
CLASSIFICATION : 45.8
TECHNIQUE : LEED
AUTHORS : M.A. Van Hove and R.J. Koestner
REFERENCE : Detn. Surf. Struc. by LEED; Plenum, , 357 (1984)

ILLUSTRATION: 1

SURFACE TYPE
Substrate : Rh
Crystal face: 111
Temperature : 300 K
Bulk lattice: fcc
2D bulk symm: p3m1
2D surf symm: p3m1

Adsorbate:
Coverage :
Pattern : (1x1)
Matrix : (1.000, 0.000)
(0.000, 1.000)

STRUCTURE TYPE
Unrelaxed bulk termination

SAMPLE PREPARATION (1 sample)
Treatment : weeks of Ar+ bomb., anneals and O2 treatments
Crystallinity:
Anal. methods:
Contamination: AES: small amounts of S,B,C,Cl

COMMENTS

DATA COLLECTION
Technique: LEED; photographic method
Dataset : I-V curves: (Θ=0,φ=0): 10,01,11,20 beams; 45<E<200 eV cumulative energy range: 370eV

THEORY/DATA TREATMENT
Dynamical LEED (RFS): Moruzzi-Janak-Williams potential, 8 phase shifts; Vor=-10.0 eV, VoiαE**1/3; ΘD=406 K

STRUCTURES EXAMINED
Fcc and hcp terminations on fcc and hcp bulk; top layer spacing varied

QUALITY OF EXPERIMENT-THEORY FIT
RPE=0.20, RZJ=0.15

2D UNIT CELLS (1 domain observed)

Cell	Ax (Å)	Ay (Å)	Bx (Å)	By (Å)	α (°)	Matrix	Pattern	Cell type
Bulk	2.680	0.000	1.340	2.321	60.0	(1.000, 0.000) (0.000, 1.000)	(1x1)	b: bulk lattice
Surface 1	2.680	0.000	1.340	2.321	60.0	(1.000, 0.000) (0.000, 1.000)	(1x1)	s1: commens. superlattice

3D COORDINATES

Dx/Dy in Å, or as a fraction of layer's unit cell vectors; atom 0 at the origin. Epir/subr are bulk repeat vectors.

No. of atoms: 3

Bulk z = 2.192 Å

Region	Chem el.	At. no.	Cell type	Site occ.	Rel. to	Dx ± εx	Dy ± εy	Dz ± εz	Dz/Bz(%) ± εz/Bz
epir		-2				f	f	Å	
subr		-1				1.340 Å	0.774 Å	2.192 Å	
intf	Rh	1	b	1.00	0	0.000 f	0.000 f	0.000 Å	0.0
intf	Rh	2	b	1.00	1	0.333 f	0.333 f	2.192 ± .100 Å	100.0 ± 4.6
subl	Rh	3	b	1.00	2	0.333 f	0.333 f	2.192 Å	100.0

BOND DISTANCES AND ANGLES

Bond distances are derived from coordinates

No. of distances/angles: 3

Interatomic dist. A-B (Å)	Atom A	Atom B	Atom C	Bond angle A-B-C (°)
2.680	Rh1	Rh1(1,0)		
2.683	Rh1	Rh2		
2.683	Rh2	Rh3		

ILLUSTRATION: 8

COMMON NAME : Rh(311)-(1x1)
CLASSIFICATION : 45.11
TECHNIQUE : LEED
AUTHORS : S. Liepold, N. Elbel, M. Michl, W. Nichtl-Pecher, K. Heinz and K. Mueller
REFERENCE : Surf. Sci., 240, 81 (1990)

SURFACE TYPE
Substrate : Rh
Crystal face: 311
Temperature : 85 K
Bulk lattice: fcc
2D bulk symm: cm
2D surf symm: cm

Adsorbate:
Coverage :
Pattern : (1x1)
Matrix : (1.000, 0.000)
(0.000, 1.000)

STRUCTURE TYPE
Unreconstructed surface with multilayer relaxations by -14.5%, +4.9%, -1.0% in top 3 interlayer spacings

SAMPLE PREPARATION (1 sample)
Treatment : Ar+ sputtering, O2 and H2 treatment
Crystallinity:
Anal. methods:
Contamination: clean by AES

COMMENTS
Annealing to T>1100 K results in a partially disordered (1x2)

DATA COLLECTION
Technique: LEED
Dataset : I-V curves for 15 inequivalent beams; E range 50-230 eV

THEORY/DATA TREATMENT
Dynamical LEED (layer doubling): 10 relat. ph. shs.
Vor=-7 eV + 0.03(E-150eV), VoiαE**1/3; ΘD(Rh)=480 K

STRUCTURES EXAMINED
Variation of top 3 interlayer spacings and top 2 interlayer registries, keeping mirror plane

QUALITY OF EXPERIMENT-THEORY FIT
RPE=0.174

2D UNIT CELLS (1 domain observed)

Cell	Ax (Å)	Ay (Å)	Bx (Å)	By (Å)	α (°)	Matrix	Pattern	Cell type
Bulk	2.690	0.000	-1.345	4.460	106.8	(1.000, 0.000) (0.000, 1.000)	(1x1)	b: bulk lattice
Surface 1	2.690	0.000	1.345	4.460	106.8	(1.000, 0.000) (0.000, 1.000)	(1x1)	s1: commens. superlattice

3D COORDINATES

Dx/Dy in Å, or as a fraction of layer's unit cell vectors; atom 0 at the origin. Epir/subr are bulk repeat vectors.

No. of atoms: 5

Bulk z = 1.147 Å

Region	Chem el.	At. no.	Cell type	Site occ.	Rel. to	Dx ± εx		Dy ± εy		Dz ± εz		Dz/Bz(%) ± εz/Bz
epir		-2					f		f		Å	
subr		-1				1.345	Å	2.028	Å	1.147	Å	
intf	Rh	1	b	1.00	0	0.000	Å	0.000	Å	0.000	Å	0.0
intf	Rh	2	b	1.00	1	1.345	Å	2.028 ± .050	Å	0.981 ± .021	Å	85.5 ± 1.8
intf	Rh	3	b	1.00	2	1.345	Å	2.058 ± .070	Å	1.203 ± .023	Å	104.9 ± 2.0
intf	Rh	4	b	1.00	3	1.345	Å	2.028	Å	1.125 ± .023	Å	98.1 ± 2.0
subl	Rh	5	b	1.00	4	1.345	Å	2.028	Å	1.147	Å	100.0

BOND DISTANCES AND ANGLES

Bond distances and angles are derived from coordinates

No. of distances/angles: 6

Interatomic dist. A-B (Å)	Atom A	Atom B	Atom C	Bond angle A-B-C (°)
2.690	Rh1	Rh1(1,0)		
2.623	Rh1	Rh2		
2.592	Rh1	Rh3		
2.737	Rh2	Rh3		
2.715	Rh2	Rh4		
2.681	Rh3	Rh4		

COMMON NAME : Rh(111)-(2x2)-C2H3 ILLUSTRATION: 65
CLASSIFICATION : 45.6.1.11
TECHNIQUE : LEED
AUTHORS : A. Wander, M.A. Van Hove and G.A. Somorjai
REFERENCE : Phys. Rev. Lett., 67, 626 (1991)

SURFACE TYPE
Substrate : Rh
Crystal face: 111
Temperature : 240 K
Bulk lattice: fcc
2D bulk symm: p3m1
2D surf symm: p3m1

Adsorbate: ethylidyne CCH3
Coverage : 1/4 (C2H3/Rh)
Pattern : (2x2)
Matrix : (2.000, 0.000)
(0.000, 2.000)

STRUCTURE TYPE
Ethylidyne species (CCH3=C2H3) formed from ethylene (C2H4) with upright C-C axis: lower C in hcp hollow site, upper C forms methyl group (H positions not determined, but presence derived from HREELS data); buckling in top 2 Rh layers

SAMPLE PREPARATION (1 sample)
Treatment : exposure to C2H4
Crystallinity:
Anal. methods:
Contamination: monitored by AES

COMMENTS
Methyl group may rotate freely about C-C axis; reanalysis of data of SSD 45.6.1.3 with tensor LEED: all coords of 2 C and 8 Rh atoms in top two metal layers left free in automated search, later symmetrized according to p3m1 symmetry; H ignored in LEED calculation

DATA COLLECTION
Technique: LEED
Dataset : 8 and 11 independent I-V curves at normal and 31° incidence (total E ranges 732 and 1008 eV)

THEORY/DATA TREATMENT
Dynamical LEED (RFS): 8 phase shifts, Moruzzi et al Rh pot, superposition pot for C

STRUCTURES EXAMINED
Full relaxation of C-C and two layers of the substrate, starting from C-C axis perp. to surface in hcp hollow site (30 struct. param. fitted; H ignored); final structure checked with off-normal incidence data

QUALITY OF EXPERIMENT-THEORY FIT
RPE=0.32

2D UNIT CELLS (1 domain observed)

Cell	Ax (Å)	Ay (Å)	Bx (Å)	By (Å)	α (°)	Matrix	Pattern	Cell type
Bulk	2.680	0.000	1.340	2.321	60.0	(1.000, 0.000) (0.000, 1.000)	(1x1)	b: bulk lattice
Surface 1	5.360	0.000	2.680	4.642	60.0	(2.000, 0.000) (0.000, 2.000)	(2x2)	s1: commens. superlattice

3D COORDINATES

C1-C2: ethylidyne species adsorbed in hcp hollow; Rh3-Rh6: buckled 1st Rh layer;
Rh7-Rh10: buckled 2nd Rh layer Error bars set to 0.1 A fitted coord.

Dx/Dy in Å, or as a fraction of layer's unit cell vectors; atom 0 at the origin. Epir/subr are bulk repeat vectors.

No. of atoms: 11 Bulk z = 2.188 Å

Region	Chem el.	At. no.	Cell type	Site occ.	Rel. to	Dx ± εx	Dy ± εy	Dz ± εz	Dz/Bz(%) ± εz/Bz
epir		-2				f	f	Å	
subr		-1				0.000 Å	1.547 Å	2.188 Å	
ovrl	C	1	s1	.25	0	0.000 ± .100 Å	0.000 ± .100 Å	0.000 ± .100 Å	0.0 ± 4.6
ovrl	C	2	s1	.25	1	0.000 ± .100 Å	0.000 ± .100 Å	1.480 ± .100 Å	67.6 ± 4.6
intf	Rh	3	s1	.25	2	4.063 ± .100 Å	3.843 ± .100 Å	1.300 ± .100 Å	59.4 ± 4.6
intf	Rh	4	s1	.25	2	6.657 ± .100 Å	3.843 ± .100 Å	1.300 ± .100 Å	59.4 ± 4.6
intf	Rh	5	s1	.25	2	5.360 ± .100 Å	1.597 ± .100 Å	1.300 ± .100 Å	59.4 ± 4.6
intf	Rh	6	s1	.25	2	2.680 ± .100 Å	1.547 ± .100 Å	1.420 ± .100 Å	64.9 ± 4.6
intf	Rh	7	s1	.25	2	5.360 ± .100 Å	3.095 ± .100 Å	3.380 ± .100 Å	154.5 ± 4.6
intf	Rh	8	s1	.25	2	1.340 ± .100 Å	0.774 ± .100 Å	3.480 ± .100 Å	159.1 ± 4.6
intf	Rh	9	s1	.25	2	4.020 ± .100 Å	0.774 ± .100 Å	3.480 ± .100 Å	159.1 ± 4.6
intf	Rh	10	s1	.25	2	2.680 ± .100 Å	3.095 ± .100 Å	3.480 ± .100 Å	159.1 ± 4.6
subl	Rh	11	b	1.00	10	-2.680 Å	-3.095 Å	2.188 Å	100.0

Rh(111)-(2x2)-C2H3
45.6.1.11

BOND DISTANCES AND ANGLES

Bond distances and angles are derived from coordinates

No. of distances/angles: 8

Interatomic dist. A-B (Å)	Atom A	Atom B	Atom C	Bond angle A-B-C (°)
1.480	C1	C2	Rh3(0,-1)	129.1
1.480	C1	C2	Rh4(-1,-1)	129.1
1.480	C1	C2	Rh5(-1,0)	129.1
2.059	C2	Rh3(0,-1)	Rh4(0,-1)	132.2
2.593	Rh3	Rh4	C2(1,1)	132.2
2.767	Rh3	Rh4(-1,0)	C2(0,1)	47.8
2.767	Rh3	Rh4(-1,0)	Rh5(-1,1)	60.0
2.767	Rh3	Rh4(-1,0)	Rh6	59.0

COMMON NAME : Rh(111)-(2x2)-C2H3
CLASSIFICATION : 45.6.1.3
TECHNIQUE : LEED
AUTHORS : R.J. Koestner, M.A. Van Hove and G.A. Somorjai
REFERENCE : Surf. Sci., 121, 321 (1982)

ILLUSTRATION: 65

SURFACE TYPE
Substrate : Rh
Crystal face: 111
Temperature : 240 K
Bulk lattice: fcc
2D bulk symm: p3m1
2D surf symm: p3m1

Adsorbate: ethylidyne CCH3
Coverage : 1/4 (C2H3/Rh)
Pattern : (2x2)
Matrix : (2.000, 0.000)
(0.000, 2.000)

STRUCTURE TYPE
Ethylidyne species (CCH3 = C2H3) formed from ethylene (C2H4) with upright C-C axis: lower C in hcp hollow site, upper C forms methyl group (H positions not determined, but presence derived from HREELS data)

SAMPLE PREPARATION (1 sample)
Treatment : exposure to C2H4
Crystallinity:
Anal. methods:
Contamination: monitored by AES

COMMENTS
Methyl group may rotate freely about C-C axis

DATA COLLECTION
Technique: LEED
Dataset : total of 48 independent I-V curves at various angles of incidence

THEORY/DATA TREATMENT
Dynamical LEED (RFS): 8 phase shifts, Moruzzi et al Rh pot, superposition pot for C (H ignored): VoiαE**1/3

STRUCTURES EXAMINED
Unrelaxed substrate: top, bridge, fcc and hcp hollow sites with C-C axis perp. to surface (or also tilted over hcp-hollow site); also single C atoms at 0.25 monolayer in hcp- hollow sites; total of 220 structures

QUALITY OF EXPERIMENT-THEORY FIT
RPE=0.52, RZJ=0.49

2D UNIT CELLS (1 domain observed)

Cell	Ax (Å)	Ay (Å)	Bx (Å)	By (Å)	α (°)	Matrix	Pattern	Cell type
Bulk	2.680	0.000	1.340	2.321	60.0	(1.000, 0.000) (0.000, 1.000)	(1x1)	b: bulk lattice
Surface 1	5.360	0.000	2.680	4.642	60.0	(2.000, 0.000) (0.000, 2.000)	(2x2)	s1: commens. superlattice

3D COORDINATES

H1-H2-H3-C4: methyl group, C4 pointing down to C5; H1 through C5: ethylidyne species adsorbed in hcp hollow (H positions assumed to form ideal methyl group)

Dx/Dy in Å, or as a fraction of layer's unit cell vectors; atom 0 at the origin. Epir/subr are bulk repeat vectors.

No. of atoms: 7

Bulk z = 2.188 Å

Region	Chem el.	At. no.	Cell type	Site occ.	Rel. to	Dx ± εx		Dy ± εy		Dz ± εz		Dz/Bz(%) ± εz/Bz
epir		-2					f		f		Å	
subr		-1				0.000	Å	-1.547	Å	2.188	Å	
ovrl	H	1	s1	.25	0	0.332	f	0.000	f	0.000	Å	0.0
ovrl	H	2	s1	.25	1	-0.332	f	0.000	f	0.000	Å	0.0
ovrl	H	3	s1	.25	2	0.332	f	-0.332	f	0.000	Å	0.0
ovrl	C	4	s1	.25	3	-0.111	f	0.221	f	0.363	Å	16.6
ovrl	C	5	s1	.25	4	0.000 ± .016	f	0.000 ± .022	f	1.450 ± .100	Å	66.3 ± 4.6
intf	Rh	6	b	1.00	5	0.667	f	-0.333	f	1.310 ± .100	Å	59.9 ± 4.6
subl	Rh	7	b	1.00	6	0.333	f	-0.667	f	2.188	Å	100.0

BOND DISTANCES AND ANGLES

Bond distances and angles are derived from coordinates

No. of distances/angles: 3

Interatomic dist. A-B (Å)	Atom A	Atom B	Atom C	Bond angle A-B-C (°)
1.450	C4	C5	Rh6	130.3

Rh(111)-(2x2)-C2H3
45.6.1.3

Bond Distances and Angles - Continued

Interatomic dist. A-B (Å)	Atom A	Atom B	Atom C	Bond angle A-B-C (°)
2.027	C5	Rh6	Rh6(1,0)	131.4
2.027	C5	Rh6	Rh7	138.4

COMMON NAME : Rh(111)-c(4x2)-C2H3+CO ILLUSTRATION: 66
CLASSIFICATION : 45.6.1.8.4a
TECHNIQUE : LEED
AUTHORS : G.S. Blackman, C.T. Kao, B.E. Bent, C.M. Mate, M.A. Van Hove and G.A. Somorjai
REFERENCE : Surf. Sci., 207, 66 (1988)

SURFACE TYPE
Substrate : Rh
Crystal face: 111
Temperature : 300 K
Bulk lattice: fcc
2D bulk symm: p3m1
2D surf symm: pm

Adsorbate: C2H3;CO
Coverage : 1/4C2H3/Rh, 1/4CO/Rh
Pattern : c(4x2)
Matrix : (2.000, 0.000)
(-1.000, 2.000)

STRUCTURE TYPE
Molecular coadsorption: CO over hcp, C2H3 (ethylidyne) over fcc hollow site, both perpendicular to surface
(H positions guessed)

SAMPLE PREPARATION (1 sample)
Treatment : cycles of heating in O2, Ar+ sputtering and annealing
Crystallinity:
Anal. methods:
Contamination: checked by AES, HREELS and LEED

COMMENTS
Other R-factors: RZJ=0.304 (Θ=0°), 0.348 (Θ=21°);
RPE=0.512 (Θ=0°), 0.517 (Θ=21°)

DATA COLLECTION
Technique: LEED
Dataset : I-V curves for 12 inequiv. beams at Θ=0°, 25 at Θ=21°, φ=0°; cumul. E ranges 1134 eV and 1000eV, resp.

THEORY/DATA TREATMENT
Dynamical LEED (BSN, KSLA, RFS): 6 phase shifts (H ignored);
VoiαE**1/3; ΘD=284 K (Rh), double rms ampls for all adatoms

STRUCTURES EXAMINED
396 structures: variation of adsorption sites of both molecules in the unit cell, and independently of the layer spacings of the overlayer; some tilting of both molecules explored

QUALITY OF EXPERIMENT-THEORY FIT
RVH=0.242 (average); see comments

2D UNIT CELLS (3 domains observed)

Cell	Ax (Å)	Ay (Å)	Bx (Å)	By (Å)	α (°)	Matrix	Pattern	Cell type
Bulk	2.690	0.000	1.345	2.330	60.0	(1.000, 0.000) (0.000, 1.000)	(1x1)	b: bulk lattice
Surface 1	5.380	0.000	0.000	4.659	90.0	(2.000, 0.000) (-1.000, 2.000)	c(4x2)	s1: commens. superlattice

3D COORDINATES

H1,2,3-C4-C6: ethylidyne (C2H3) perp. surf. in fcc site; O5-C7: CO perp. surf. in hcp site (C down)

Dx/Dy in Å, or as a fraction of layer's unit cell vectors; atom 0 at the origin. Epir/subr are bulk repeat vectors.

No. of atoms: 9 Bulk z = 2.196 Å

Region	Chem el.	At. no.	Cell type	Site occ.	Rel. to	Dx ± εx		Dy ± εy		Dz ± εz		Dz/Bz(%) ± εz/Bz
epir		-2					f		f		Å	
subr		-1				1.345	Å	0.777	Å	2.196	Å	
ovrl	H	1	s1	.25	0	0.000	f	0.000	f	0.000	Å	0.0
ovrl	H	2	s1	.25	1	0.168	f	0.333	f	0.000	Å	0.0
ovrl	H	3	s1	.25	2	0.168	f	-0.330	f	0.000	Å	0.0
ovrl	C	4	s1	.25	3	-0.168	f	0.108	f	0.369	Å	16.8
ovrl	O	5	s1	.25	4	0.500	f	0.333	f	0.420 ± .050	Å	19.1 ± 2.3
ovrl	C	6	s1	.25	4	0.000	f	0.000	f	1.450 ± .050	Å	66.0 ± 2.3
ovrl	C	7	s1	.25	5	0.000	f	0.000	f	1.180 ± .050	Å	53.7 ± 2.3
intf	Rh	8	b	1.00	7	-0.333	f	0.667	f	1.300 ± .050	Å	59.2 ± 2.3
subl	Rh	9	b	1.00	8	0.333	f	0.333	f	2.196	Å	100.0

Rh(111)-c(4x2)-C2H3+CO
45.6.1.8.4a

BOND DISTANCES AND ANGLES

Bond distances and angles are derived from coordinates

No. of distances/angles: 6

Interatomic dist. A-B (Å)	Atom A	Atom B	Atom C	Bond angle A-B-C (°)
1.106	H1	C4	H2	109.3
1.450	C4	C6	Rh8(0,-1)	133.0
1.180	O5	C7	Rh8	129.9
2.025	C7	Rh8	Rh9	138.2
2.125	Rh8	C6(0,1)	Rh8(-1,0)	78.6
2.690	Rh8	Rh8(1,0)		

COMMON NAME : Rh(111)-c(4x2)-C2H3+NO
CLASSIFICATION : 45.6.7.8.1.1
TECHNIQUE : LEED
AUTHORS : G.S. Blackman, C.T. Kao, B.E. Bent, C.M. Mate, M.A. Van Hove and G.A. Somorjai
REFERENCE : Surf. Sci., 207, 66 (1988)

ILLUSTRATION: 66

SURFACE TYPE
Substrate : Rh
Crystal face: 111
Temperature : 300 K
Bulk lattice: fcc
2D bulk symm: p3m1
2D surf symm: pm

Adsorbate: C2H3;NO
Coverage : 1/4C2H3/Rh, 1/4NO/Rh
Pattern : c(4x2)
Matrix : (2.000, 0.000)
(-1.000, 2.000)

STRUCTURE TYPE
Molecular coadsorption: NO over fcc, C2H3 (ethylidyne) over hcp hollow site, both perpendicular to surface
(H positions guessed)

SAMPLE PREPARATION (1 sample)
Treatment : cycles of heating in O2, Ar+ sputtering and annealing
Crystallinity:
Anal. methods:
Contamination: checked by AES, HREELS and LEED

COMMENTS

DATA COLLECTION
Technique: LEED
Dataset : I-V curves for 13 inequiv. beams at Θ=0°; cumul. E range 1068 eV

THEORY/DATA TREATMENT
Dynamical LEED (BSN, KSLA, RFS): 6 phase shifts (H ignored); Voi∝E**1/3; ΘD=284 K (Rh), double rms ampls for all adatoms

STRUCTURES EXAMINED
576 structures: variation of adsorption sites of both molecules in the unit cell, and independently of the layer spacings of the overlayer; some tilting of NO explored

QUALITY OF EXPERIMENT-THEORY FIT
RZJ=0.348, RPE=0.688, RVH=0.294

2D UNIT CELLS (3 domains observed)

Cell	Ax (Å)	Ay (Å)	Bx (Å)	By (Å)	α (°)	Matrix	Pattern	Cell type
Bulk	2.690	0.000	1.345	2.330	60.0	(1.000, 0.000) (0.000, 1.000)	(1x1)	b: bulk lattice
Surface 1	5.380	0.000	0.000	4.659	90.0	(2.000, 0.000) (-1.000, 2.000)	c(4x2)	s1: commens. superlattice

3D COORDINATES

H1,2,3-C4-C6: ethylidyne (C2H3) perp. surf. in hcp site; O5-N7: NO perp. surf. in fcc site (N down)

Dx/Dy in Å, or as a fraction of layer's unit cell vectors; atom 0 at the origin. Epir/subr are bulk repeat vectors.

No. of atoms: 9

Bulk z = 2.196 Å

Reg ion	Chem el.	At. no.	Cell type	Site occ.	Rel. to	Dx ± εx		Dy ± εy		Dz ± εz		Dz/Bz(%) ± εz/Bz
epir		-2					f		f		Å	
subr		-1				1.345	Å	0.777	Å	2.196	Å	
ovrl	H	1	s1	.25	0	0.000	f	0.000	f	0.000	Å	0.0
ovrl	H	2	s1	.25	1	-0.168	f	-0.333	f	0.000	Å	0.0
ovrl	H	3	s1	.25	2	-0.168	f	0.330	f	0.000	Å	0.0
ovrl	C	4	s1	.25	3	0.168	f	-0.108	f	0.369	Å	16.8
ovrl	O	5	s1	.25	4	-0.500	f	-0.333	f	0.320 ± .050	Å	14.6 ± 2.3
ovrl	C	6	s1	.25	4	0.000	f	0.000	f	1.450 ± .050	Å	66.0 ± 2.3
ovrl	N	7	s1	.25	5	0.000	f	0.000	f	1.180 ± .050	Å	53.7 ± 2.3
intf	Rh	8	b	1.00	6	-0.333	f	0.667	f	1.350 ± .050	Å	61.5 ± 2.3
subl	Rh	9	b	1.00	8	0.333	f	0.333	f	2.196	Å	100.0

Rh(111)-c(4x2)-C2H3+NO
45.6.7.8.1.1 BOND DISTANCES AND ANGLES

Bond distances and angles are derived from coordinates

No. of distances/angles: 6

Interatomic dist. A-B (Å)	Atom A	Atom B	Atom C	Bond angle A-B-C (°)
1.106	H1	C4	H2	109.3
1.450	C4	C6	Rh8	131.0
1.180	O5	N7	Rh8(0,-1)	129.9
2.058	C6	Rh8	Rh9	138.9
2.025	Rh8	N7(0,1)	Rh8(0,-1)	83.2
2.690	Rh8	Rh8(1,0)		

COMMON NAME : Rh(111)-(3x3)-C6H6+2CO
CLASSIFICATION : 45.6.1.8.3
TECHNIQUE : LEED
AUTHORS : M.A. Van Hove, R.F. Lin, G.S. Blackman and G.A. Somorjai
REFERENCE : Acta. Crys., B43, 368 (1987)

ILLUSTRATION: 70

SURFACE TYPE
Substrate : Rh
Crystal face: 111
Temperature : 300 K
Bulk lattice: fcc
2D bulk symm: p3m1
2D surf symm: p3m1

Adsorbate: C6H6;CO
Coverage : 2/9 CO/Rh, 1/9 Bz/Rh
Pattern : (3x3)
Matrix : (3.000, 0.000)
(0.000, 3.000)

STRUCTURE TYPE
Molecular coadsorption of one flat-lying C6H6 (benzene) and 2 upright CO per cell, all centered over hcp hollow sites, both with relaxed bonds (H ignored), on unrelaxed substrate

SAMPLE PREPARATION (1 sample)
Treatment : sputter with 500 eV Ar+, react with O, anneal at 1300 K
Crystallinity:
Anal. methods:
Contamination: clean by AES before adsorption

COMMENTS

DATA COLLECTION
Technique: LEED
Dataset : 14 inequivalent IV curves at normal incidence with E range of 20-200 eV

THEORY/DATA TREATMENT
Dynamical LEED (BSN, RFS, KSLA): 6 phase shifts (Moruzzi Rh pot, Li CO pot, Kesmodel C pot); VoiαE**1/3

STRUCTURES EXAMINED
Benzene in top, bridge, fcc, hcp sites with various high symmetry orientations and distortions including out of plane buckling, ring expansion and Kekule distortion; CO intact, stretched and normal to surface in same sites consistent with benzene orientations

QUALITY OF EXPERIMENT-THEORY FIT
RVH=0.21, RZJ=0.34, RPE=0.41

2D UNIT CELLS (1 domain observed)

Cell	Ax (Å)	Ay (Å)	Bx (Å)	By (Å)	α (°)	Matrix	Pattern	Cell type
Bulk	2.690	0.000	1.345	2.330	60.0	(1.000, 0.000) (0.000, 1.000)	(1x1)	b: bulk lattice
Surface 1	8.070	0.000	4.035	6.989	60.0	(3.000, 0.000) (0.000, 3.000)	(3x3)	s1: commens. superlattice

3D COORDINATES

O7-C15, O8-C16: 2 upright stretched COs in hcp hollows; H1-6, C9-14: Kekule-distorted flat benzene over hcp site; 0.1Å lateral error bars assumed for tabulation

Dx/Dy in Å, or as a fraction of layer's unit cell vectors; atom 0 at the origin. Epir/subr are bulk repeat vectors.

No. of atoms: 18

Bulk z = 2.200 Å

Region	Chem el.	At. no.	Cell type	Site occ.	Rel. to	Dx ± εx		Dy ± εy		Dz ± εz		Dz/Bz(%) ± εz/Bz
epir		-2					f		f		Å	
subr		-1				-2.690	Å	-1.553	Å	2.200	Å	
ovrl	H	1	s1	.11	0	0.642	f	0.327	f	-0.280	Å	-12.7
ovrl	H	2	s1	.11	1	0.000	f	-0.294	f	0.000	Å	0.0
ovrl	H	3	s1	.11	1	-0.609	f	0.001	f	0.000	Å	0.0
ovrl	H	4	s1	.11	1	-0.609	f	0.315	f	0.000	Å	0.0
ovrl	H	5	s1	.11	1	-0.315	f	0.315	f	0.000	Å	0.0
ovrl	H	6	s1	.11	1	-0.314	f	-0.294	f	0.000	Å	0.0
ovrl	O	7	s1	.11	0	0.667	f	0.667	f	0.000	Å	0.0
ovrl	O	8	s1	.11	0	0.000	f	0.000	f	0.000	Å	0.0
ovrl	C	9	s1	.11	0	0.522 ± .005	f	0.329 ± .014	f	0.270 ± .050	Å	12.3 ± 2.3
ovrl	C	10	s1	.11	9	-0.193 ± .005	f	0.193 ± .014	f	0.000 ± .050	Å	0.0 ± 2.3
ovrl	C	11	s1	.11	9	-0.373 ± .005	f	0.000 ± .014	f	0.000 ± .050	Å	0.0 ± 2.3
ovrl	C	12	s1	.11	9	-0.373 ± .005	f	0.193 ± .014	f	0.000 ± .050	Å	0.0 ± 2.3
ovrl	C	13	s1	.11	9	0.000 ± .005	f	-0.180 ± .014	f	0.000 ± .050	Å	0.0 ± 2.3
ovrl	C	14	s1	.11	9	-0.193 ± .005	f	-0.180 ± .014	f	0.000 ± .050	Å	0.0 ± 2.3
ovrl	C	15	s1	.11	7	0.000	f	0.000	f	1.170 ± .050	Å	53.2 ± 2.3
ovrl	C	16	s1	.11	8	0.000	f	0.000	f	1.170 ± .050	Å	53.2 ± 2.3
intf	Rh	17	b	1.00	16	0.667	f	0.667	f	1.300 ± .050	Å	59.1 ± 2.3
subl	Rh	18	b	1.00	17	-0.667	f	-0.667	f	2.200	Å	100.0

Rh(111)-(3x3)-C6H6+2CO
45.6.1.8.3

BOND DISTANCES AND ANGLES

Bond distances and angles are derived from coordinates

No. of distances/angles: 7

Interatomic dist. A-B (Å)	Atom A	Atom B	Atom C	Bond angle A-B-C (°)
1.556	C9	C10	C12	120.1
1.556	C9	C10	Rh17(0,1)	104.1
1.453	C9	C13	C14	119.9
1.453	C9	C13	Rh17(1,0)	71.9
2.330	C11	Rh17	C14	36.4
2.025	Rh17	C15(-1,-1)		
2.690	Rh17	Rh17(1,0)		

COMMON NAME : Rh(111)-c(2√3x4)rect-C6H6+CO ILLUSTRATION: 69
CLASSIFICATION : 45.6.1.8.2
TECHNIQUE : LEED
AUTHORS : M.A. Van Hove, R.F. Lin and G.A. Somorjai
REFERENCE : J. Am. Chem. Soc., 108, 2532 (1986)

SURFACE TYPE
Substrate : Rh
Crystal face: 111
Temperature : 300 K
Bulk lattice: fcc
2D bulk symm: p3m1
2D surf symm: cm

Adsorbate: C6H6;CO
Coverage : 1/8 Bz/Rh, 1/8 CO/upright CO, both centered over hcp hollow sites, both with relaxed bonds (H ignored), on unrelaxed substrate
Pattern : c(2√3x4)rect
Matrix : (3.000,-2.000)
(1.000, 2.000)

STRUCTURE TYPE
Molecular coadsorption of flat-lying C6H6 (benzene) and

SAMPLE PREPARATION (1 sample)
Treatment :
Crystallinity: first benzene saturated, then CO added
Anal. methods:
Contamination: clean by AES

COMMENTS

DATA COLLECTION
Technique: LEED
Dataset : IV curves for 13 nonequivalent beams at normal incidence: E range 20-150 eV, cumulative E range 1224 eV

THEORY/DATA TREATMENT
Dynamical LEED (BSN, RFS, KSLA): 5 then 6 phase shifts; H neglected

STRUCTURES EXAMINED
1 upright CO (C end down) and one flat benzene in: top, fcc hcp, bridge sites; C-O varied 1.15-1.25A; height of molecule varied independently; Rh(111) bulk-like; benzene: Kekule distortion and buckling, then heights and Rh-Rh spacing varied; 440 coadsorption and 960 CO-free structures tested

QUALITY OF EXPERIMENT-THEORY FIT
RZJ=0.40, RPE=0.66, RVH=0.31

2D UNIT CELLS (3 domains observed)

Cell	Ax (Å)	Ay (Å)	Bx (Å)	By (Å)	α (°)	Matrix	Pattern	Cell type
Bulk	2.690	0.000	1.345	2.330	60.0	(1.000, 0.000) (0.000, 1.000)	(1x1)	b: bulk lattice
Surface 1	5.380	-4.659	5.380	4.659	81.8	(3.000,-2.000) (1.000, 2.000)	c(2√3x4)rect	s1: commens. superlattice

3D COORDINATES

O1-C14: upright stretched CO molecule in hcp hollows; H2-7, C8-13: Kekule-distorted flat benzene over hcp site; 0.1Å lateral error bars assumed for tabulation

Dx/Dy in Å, or as a fraction of layer's unit cell vectors; atom 0 at the origin. Epir/subr are bulk repeat vectors.

No. of atoms: 16 Bulk z = 2.196 Å

Region	Chem el.	At. no.	Cell type	Site occ.	Rel. to	Dx ± εx		Dy ± εy		Dz ± εz		Dz/Bz(%) ± εz/Bz
epir		-2					f		f		Å	
subr		-1				-1.345	Å	0.777	Å	2.196	Å	
ovrl	O	1	s1	.13	0	0.000	f	0.000	f	0.000	Å	0.0
ovrl	H	2	s1	.13	1	0.590	f	0.131	f	0.084	Å	3.8
ovrl	H	3	s1	.13	2	0.176	f	0.587	f	0.000	Å	0.0
ovrl	H	4	s1	.13	2	-0.447	f	0.521	f	0.000	Å	0.0
ovrl	H	5	s1	.13	2	0.279	f	0.279	f	0.000	Å	0.0
ovrl	H	6	s1	.13	2	-0.308	f	0.103	f	0.000	Å	0.0
ovrl	H	7	s1	.13	2	-0.242	f	0.726	f	0.000	Å	0.0
ovrl	C	8	s1	.13	1	0.284 ± .012	f	0.592 ± .014	f	0.460 ± .050	Å	20.9 ± 2.3
ovrl	C	9	s1	.13	8	0.270 ± .012	f	-0.315 ± .014	f	0.000 ± .050	Å	0.0 ± 2.3
ovrl	C	10	s1	.13	8	0.377 ± .012	f	0.040 ± .014	f	0.000 ± .050	Å	0.0 ± 2.3
ovrl	C	11	s1	.13	8	0.084 ± .012	f	-0.253 ± .014	f	0.000 ± .050	Å	0.0 ± 2.3
ovrl	C	12	s1	.13	8	0.439 ± .012	f	-0.146 ± .014	f	0.000 ± .050	Å	0.0 ± 2.3
ovrl	C	13	s1	.13	8	0.124 ± .012	f	0.124 ± .014	f	0.000 ± .050	Å	0.0 ± 2.3
ovrl	C	14	s1	.13	1	0.000	f	0.000	f	1.210 ± .050	Å	55.1 ± 2.3
intf	Rh	15	b	1.00	14	0.667	f	-0.333	f	1.500 ± .050	Å	68.3 ± 2.3
subl	Rh	16	b	1.00	15	-0.667	f	0.333	f	2.196 ± .050	Å	100.0 ± 2.3

Rh(111)-c(2√3x4)rect-C6H6+CO
45.6.1.8.2 BOND DISTANCES AND ANGLES

Bond distances and angles are derived from coordinates

No. of distances/angles: 9

Interatomic dist. A-B (Å)	Atom A	Atom B	Atom C	Bond angle A-B-C (°)
1.210	O1	C14	Rh15	134.0
1.812	C8	C11	C9	120.0
1.812	C8	C11	Rh15(1,0)	100.8
1.336	C8	C13	C10	120.0
1.336	C8	C13	Rh15(1,1)	73.5
2.159	C14	Rh15	Rh16	98.7
2.350	Rh15	C9(-1,0)		
2.159	Rh15	C14(1,0)	Rh15(1,0)	77.1
2.690	Rh15	Rh15(1,0)		

COMMON NAME : Rh(111)-(2x2)-3CO ILLUSTRATION: 63
CLASSIFICATION : 45.6.8.4
TECHNIQUE : LEED
AUTHORS : M.A. Van Hove, R.J. Koestner, J.C. Frost and G.A. Somorjai
REFERENCE : Surf. Sci., 129, 482 (1983)

SURFACE TYPE
Substrate : Rh
Crystal face: 111
Temperature : 240 K
Bulk lattice: fcc
2D bulk symm: p3m1
2D surf symm: cm

Adsorbate: CO
Coverage : 3/4 (CO/Rh)
Pattern : (2x2)
Matrix : (2.000, 0.000)
(0.000, 2.000)

STRUCTURE TYPE
Densely packed molecular adsorption: 1 upright CO on bridge and 2 essentially upright COs near top sites per (2x2) cell, forming nearly hexagonal buckled CO lattice (C down)

SAMPLE PREPARATION (1 sample)
Treatment : background CO P≈10**-6 to 10**-5 torr to maintain (2x2)
Crystallinity:
Anal. methods:
Contamination: AES: small S, C, B and Cl contamination

COMMENTS

DATA COLLECTION
Technique: LEED
Dataset : I-V curves for 5 independent beams (1 0,0 1,1/2 0,0 1/2, 1/2 1/2) at normal incidence, energy range 40-150 eV

THEORY/DATA TREATMENT
Dynamical LEED (combined space method): Moruzzi et al Rh pot, cluster CO pot, 5 phase shifts; Voi∝E**1/3

STRUCTURES EXAMINED
Unrelaxed substrate: 1/4 ML CO in bridge sites (perpendicu- lar to surface), 1/2 ML CO near top sites (perpendicular to surface or tilted); 5 struc. variables: 2 Rh-C spacings, 1 C-O spacing, variable near-top registry and variable CO tilt angle; total of about 250 structures

QUALITY OF EXPERIMENT-THEORY FIT
RZJ=0.25, RPE=0.47, RVH=0.19

2D UNIT CELLS (3 domains observed)

Cell	Ax (Å)	Ay (Å)	Bx (Å)	By (Å)	α (°)	Matrix	Pattern	Cell type
Bulk	2.680	0.000	1.340	2.321	60.0	(1.000, 0.000) (0.000, 1.000)	(1x1)	b: bulk lattice
Surface 1	5.360	0.000	2.680	4.642	60.0	(2.000, 0.000) (0.000, 2.000)	(2x2)	s1: commens. superlattice

3D COORDINATES

O1-C4, O2-C5: 2 near-top-site upright COs; O3-C6: 1 bridge-site upright CO

Dx/Dy in Å, or as a fraction of layer's unit cell vectors; atom 0 at the origin. Epir/subr are bulk repeat vectors.

No. of atoms: 8 Bulk z = 2.188 Å

Region	Chem el.	At. no.	Cell type	Site occ.	Rel. to	Dx ± εx		Dy ± εy		Dz ± εz		Dz/Bz(%) ± εz/Bz
epir		-2					f		f		Å	
subr		-1				1.340	Å	-0.774	Å	2.188	Å	
ovrl	O	1	s1	.25	0	0.000	f	0.000	f	0.000	Å	0.0
ovrl	O	2	s1	.25	1	0.386 ± .008	f	0.386 ± .022	f	0.000 ± .100	Å	0.0 ± 4.6
ovrl	O	3	s1	.25	2	0.307	f	0.307	f	0.350 ± .100	Å	16.0 ± 4.6
ovrl	C	4	s1	.25	1	0.000 ± .008	f	0.000 ± .022	f	1.150 ± .100	Å	52.6 ± 4.6
ovrl	C	5	s1	.25	2	0.000 ± .008	f	0.000 ± .022	f	1.150 ± .100	Å	52.6 ± 4.6
ovrl	C	6	s1	.25	3	0.000	f	0.000	f	1.150 ± .100	Å	52.6 ± 4.6
intf	Rh	7	b	1.00	6	-0.500	f	0.500	f	1.520 ± .100	Å	69.5 ± 4.6
subl	Rh	8	b	1.00	7	0.667	f	-0.333	f	2.188	Å	100.0

Rh(111)-(2x2)-3CO
45.6.8.4 BOND DISTANCES AND ANGLES

Bond distances and angles are derived from coordinates

No. of distances/angles: 4

Interatomic dist. A-B (Å)	Atom A	Atom B	Atom C	Bond angle A-B-C (°)
1.150	O1	C4	Rh7(-1,-2)	164.2
1.150	O2	C5	Rh7(0,-1)	164.2
1.150	O3	C6	Rh7	138.6
2.026	C6	Rh7	Rh8	106.4

COMMON NAME : Rh(111)-(√3x√3)R30°-CO
CLASSIFICATION : 45.6.8.1
TECHNIQUE : LEED
AUTHORS : R.J. Koestner, M.A. Van Hove and G.A. Somorjai
REFERENCE : Surf. Sci., 107, 439 (1981)

ILLUSTRATION: 61

SURFACE TYPE
Substrate : Rh
Crystal face: 111
Temperature : 240 K
Bulk lattice: fcc
2D bulk symm: p3m1
2D surf symm: p31m

Adsorbate: CO
Coverage : 1/3 (CO/Rh)
Pattern : (√3x√3)R30°
Matrix : (1.000, 1.000)
(-1.000, 2.000)

STRUCTURE TYPE
Molecular adsorption at top sites: CO upright, C down

SAMPLE PREPARATION (1 sample)
Treatment : adsorption of CO or CO2, resulting in same CO adsorbate
Crystallinity:
Anal. methods:
Contamination: AES: small amounts of S, B, C, Cl

COMMENTS

DATA COLLECTION
Technique: LEED
Dataset : I-V curves: 4 beams at Θ=0,φ=0°, 11 beams at Θ=10,φ=0° 7 beams at Θ=20,φ=0°; cum E range 1880 eV (non-degenerate)

THEORY/DATA TREATMENT
Dynamical LEED (RFS): Moruzzi Rh pot, Xα, Jona and Pendry pots for CO; 8 ph shs; Vor=-10.0 eV,Voi∝E**1/3; ΘD=284 K(Rh)

STRUCTURES EXAMINED
CO perpendicular to surface in top, bridge and 2 hollow sites, with various Rh-Rh, Rh-C, and C-O spacings

QUALITY OF EXPERIMENT-THEORY FIT
RPE=0.50, RZJ=0.40

2D UNIT CELLS (1 domain observed)

Cell	Ax (Å)	Ay (Å)	Bx (Å)	By (Å)	α (°)	Matrix	Pattern	Cell type
Bulk	2.679	0.000	1.340	2.320	60.0	(1.000, 0.000) (0.000, 1.000)	(1x1)	b: bulk lattice
Surface 1	4.018	2.320	0.000	4.640	60.0	(1.000, 1.000) (-1.000, 2.000)	(√3x√3)R30°	s1: commens. superlattice

3D COORDINATES

O1-C2: upright CO overlayer in top sites

Dx/Dy in Å, or as a fraction of layer's unit cell vectors; atom 0 at the origin. Epir/subr are bulk repeat vectors.

No. of atoms: 4

Bulk z = 2.192 Å

Reg ion	Chem el.	At. no.	Cell type	Site occ.	Rel. to	Dx ± εx		Dy ± εy		Dz ± εz		Dz/Bz(%) ± εz/Bz
epir		-2					f		f		Å	
subr		-1				1.339	Å	0.773	Å	2.192	Å	
ovrl	O	1	s1	.33	0	0.000	f	0.000	f	0.000	Å	0.0
ovrl	C	2	s1	.33	1	0.000	f	0.000	f	1.070 ± .100	Å	48.8 ± 4.6
intf	Rh	3	b	1.00	2	0.000	f	0.000	f	1.950 ± .100	Å	89.0 ± 4.6
subl	Rh	4	b	1.00	3	0.333	f	0.333	f	2.192	Å	100.0

BOND DISTANCES AND ANGLES

Bond distances and angles are derived from coordinates

No. of distances/angles: 2

Interatomic dist. A-B (Å)	Atom A	Atom B	Atom C	Bond angle A-B-C (°)
1.070	O1	C2	Rh3	180.0
1.950	C2	Rh3	Rh4	144.8

COMMON NAME : Rh(100)-c(4x2)-Cs
CLASSIFICATION : 45.55.1
TECHNIQUE : LEED
AUTHORS : C. von Eggeling, G. Schmidt, G. Besold, L. Hammer, K. Heinz and K. Mueller
REFERENCE : Surf. Sci., 221, 11 (1989)

ILLUSTRATION: 28,30

SURFACE TYPE
Substrate : Rh
Crystal face: 100
Temperature : 120 K
Bulk lattice: fcc
2D bulk symm: p4m
2D surf symm: cmm

Adsorbate: Cs
Coverage : 0.25 ML
Pattern : c(4x2)
Matrix : (2.000,-1.000)
(0.000, 2.000)

STRUCTURE TYPE
4-fold symmetric hollow site adsorption;

SAMPLE PREPARATION (1 sample)
Treatment : sputtering and annealing in oxygen; exposure to Cs
Crystallinity: sharp LEED pattern
Anal. methods: AES
Contamination:

COMMENTS

DATA COLLECTION
Technique: LEED; video LEED
Dataset : IV spectra for 4 integer and 3 fractional order beams

THEORY/DATA TREATMENT
Quasi-dynamical and full dynamical LEED: RFS

STRUCTURES EXAMINED
Bridge, top and hollow sites; variation of the 1st interlayer spacing; (use energy dependent inner potential

QUALITY OF EXPERIMENT-THEORY FIT
RPE=0.31

2D UNIT CELLS (2 domains observed)

Cell	Ax (Å)	Ay (Å)	Bx (Å)	By (Å)	α (°)	Matrix	Pattern	Cell type
Bulk	2.690	0.000	0.000	2.690	90.0	(1.000, 0.000) (0.000, 1.000)	(1x1)	b: bulk lattice
Surface 1	5.380	-2.690	0.000	5.380	116.6	(2.000,-1.000) (0.000, 2.000)	c(4x2)	s1: commens. superlattice

3D COORDINATES

Cs1: atomic overlayer in 4-fold hollow site

Dx/Dy in Å, or as a fraction of layer's unit cell vectors; atom 0 at the origin. Epir/subr are bulk repeat vectors.

No. of atoms: 3 Bulk z = 1.902 Å

Reg ion	Chem el.	At. no.	Cell type	Site occ.	Rel. to	Dx ± εx		Dy ± εy		Dz ± εz		Dz/Bz(%) ± εz/Bz
epir		-2					f		f		Å	
subr		-1				1.345	Å	1.345	Å	1.902	Å	
ovrl	Cs	1	s1	.25	0	0.000	Å	0.000	Å	0.000	Å	0.0
intf	Rh	2	b	1.00	0	1.345	Å	1.345	Å	2.870 ± .060	Å	147.2 ± 3.2
subl	Rh	3	b	1.00	0	0.000	Å	0.000	Å	4.760 ± .040	Å	250.3 ± 2.1

BOND DISTANCES AND ANGLES

Bond distances and angles are derived from coordinates

No. of distances/angles: 2

Interatomic dist. A-B (Å)	Atom A	Atom B	Atom C	Bond angle A-B-C (°)
3.443	Cs1	Rh2		
2.681	Rh2	Rh3		

COMMON NAME : Rh(110)-(1x1)-2H ILLUSTRATION: 37
CLASSIFICATION : 45.1.10
TECHNIQUE : LEED
AUTHORS : K. Heinz, W Nichtl-Pecher, W. Oed, H. Landskron, M. Michl and K. Mueller
REFERENCE : Springer Series in Surface Sciences, 24, 401 (1991)

SURFACE TYPE
Substrate : Rh
Crystal face: 110
Temperature : 90 K
Bulk lattice: fcc
2D bulk symm: pmm
2D surf symm: pmm

Adsorbate: H
Coverage : 2 H/Rh
Pattern : (1x1)
Matrix : (1.000, 0.000)
(0.000, 1.000)

STRUCTURE TYPE
Atomic adsorption of hydrogen in 3-fold hollows on either side of Rh ridges, reducing clean-surface relaxation

SAMPLE PREPARATION (1 sample)
Treatment : Ar+ sputt.; O2 and H2 treatment; 1300 K anneal
Crystallinity:
Anal. methods:
Contamination: AES: clean

COMMENTS

DATA COLLECTION
Technique: LEED
Dataset : 7 symmetry inequivalent beams; E range 50-206 eV

THEORY/DATA TREATMENT
Dynamical LEED (layer doubling):
Vor=-12 eV; Voi=-5eV; ΘD(Rh)=480 K; ΘD(H)=3400K

STRUCTURES EXAMINED
Variation of H positions near 3-fold hollows, keeping both mirror planes, and varying top 2 Rh-Rh interlayer spacing

QUALITY OF EXPERIMENT-THEORY FIT
RPE=0.28

2D UNIT CELLS (2 domains observed)

Cell	Ax (Å)	Ay (Å)	Bx (Å)	By (Å)	α (°)	Matrix	Pattern	Cell type
Bulk	2.690	0.000	0.000	3.804	90.0	(1.000, 0.000) (0.000, 1.000)	(1x1)	b: bulk lattice
Surface 1	2.690	0.000	0.000	3.804	90.0	(1.000, 0.000) (0.000, 1.000)	(1x1)	s1: commens. superlattice

3D COORDINATES

H1-H2: adatoms in 3-fold sites on both sides of ridges 0.05Å error bars assumed for tabulation

Dx/Dy in Å, or as a fraction of layer's unit cell vectors; atom 0 at the origin. Epir/subr are bulk repeat vectors.

No. of atoms: 5 Bulk z = 1.345 Å

Region	Chem el.	At. no.	Cell type	Site occ.	Rel. to	Dx ± εx		Dy ± εy		Dz ± εz		Dz/Bz(%) ± εz/Bz
epir		-2					f		f		Å	
subr		-1				1.345	Å	1.902	Å	1.345	Å	
ovrl	H	1	s1	1.00	0	0.000	f	0.000	f	0.000	Å	0.0
ovrl	H	2	s1	1.00	1	0.000	f	0.515 ± .013	f	0.000	Å	0.0
intf	Rh	3	b	1.00	2	0.500	f	-0.258 ± .010	f	0.780 ± .050	Å	58.0 ± 3.7
intf	Rh	4	b	1.00	3	-0.500	f	-0.500	f	1.328 ± .050	Å	98.7 ± 3.7
subl	Rh	5	b	1.00	4	0.500	f	0.500	f	1.345	Å	100.0

BOND DISTANCES AND ANGLES

Bond distances and angles are derived from coordinates

No. of distances/angles: 6

Interatomic dist. A-B (Å)	Atom A	Atom B	Atom C	Bond angle A-B-C (°)
1.838	H1	Rh3	H1(1,0)	94.1
1.838	H1	Rh3	H2(1,0)	129.8
1.838	H1	Rh3	H2	64.4
1.838	H1	Rh3	Rh3(1,0)	137.0

Rh(110)-(1x1)-2H
45.1.10

Bond Distances and Angles - Continued

Interatomic dist. A-B (Å)	Atom A	Atom B	Atom C	Bond angle A-B-C (°)
2.300	H1	Rh4(0,-1)	Rh5	100.1
2.681	Rh3	Rh4(1,0)	Rh5	59.7

COMMON NAME : Rh(110)-(1x1)-2H
CLASSIFICATION : 45.1.4
TECHNIQUE : LEED
AUTHORS : W. Oed, W. Puchta, N. Bickel, K. Heinz, W. Nichtl and K. Mueller
REFERENCE : J. Phys., C21, 237 (1988)

ILLUSTRATION: 37

SURFACE TYPE
Substrate : Rh
Crystal face: 110
Temperature : RT*
Bulk lattice: fcc
2D bulk symm: pmm
2D surf symm: pmm

Adsorbate: H
Coverage : 2 H/Rh
Pattern : (1x1)
Matrix : (1.000, 0.000)
(0.000, 1.000)

STRUCTURE TYPE
Atomic adsorption over each outer 3-fold coordinated site on (111) facets of slightly relaxed substrate

SAMPLE PREPARATION (1 sample)
Treatment : Ar ion bombardment, heating in O2 and H2, final 1330 K flash
Crystallinity: sharp LEED pattern with poor background
Anal. methods:
Contamination: special attention to H coverage

COMMENTS

DATA COLLECTION
Technique: LEED
Dataset : I-V curves for 7 non-equivalent beams; E range 50-206 eV

THEORY/DATA TREATMENT
Dynamical LEED (composite layer, layer doubling): 8 phase shifts; Voi=-7 eV; ΘD=480 K(Rh), 3400K(H)

STRUCTURES EXAMINED
Variation of H-Rh and 1st Rh-Rh interlayer spacings; 2nd Rh-Rh spacing taken from earlier higher-energy analysis (see Nichtl et al, Surf. Sci. 188 L279 (1987)); H kept in mirror plane near 3-fold hollow site of (111) facets, varying spacing and position parallel to surface

QUALITY OF EXPERIMENT-THEORY FIT
RPE=0.28

2D UNIT CELLS (1 domain observed)

Cell	Ax (Å)	Ay (Å)	Bx (Å)	By (Å)	α (°)	Matrix	Pattern	Cell type
Bulk	2.690	0.000	0.000	3.800	90.0	(1.000, 0.000) (0.000, 1.000)	(1x1)	b: bulk lattice
Surface 1	2.690	0.000	0.000	3.800	90.0	(1.000, 0.000) (0.000, 1.000)	(1x1)	s1: commens. superlattice

3D COORDINATES

H1-H2: overlayer in 3-fold coord. hollows of (111) facets

Dx/Dy in Å, or as a fraction of layer's unit cell vectors; atom 0 at the origin. Epir/subr are bulk repeat vectors.

No. of atoms: 6

Bulk z = 1.345 Å

Region	Chem el.	At. no.	Cell type	Site occ.	Rel. to	Dx ± εx		Dy ± εy		Dz ± εz		Dz/Bz(%) ± εz/Bz
epir		-2					f		f		Å	
subr		-1				-1.345	Å	-1.900	Å	1.345	Å	
ovrl	H	1	b	1.00	0	0.000	f	0.000	f	0.000	Å	0.0
ovrl	H	2	b	1.00	1	0.000	f	0.484 ± .079	f	0.000	Å	0.0
intf	Rh	3	b	1.00	2	0.500	f	0.258 ± .079	f	0.780 ± .300	Å	58.0 ± 22.3
intf	Rh	4	b	1.00	3	-0.500	f	-0.500	f	1.320 ± .050	Å	98.1 ± 3.7
intf	Rh	5	b	1.00	4	0.500	f	0.500	f	1.347	Å	100.2
subl	Rh	6	b	1.00	5	-0.500	f	-0.500	f	1.345	Å	100.0

Rh(110)-(1x1)-2H
45.1.4 BOND DISTANCES AND ANGLES

Bond distances and angles are derived from coordinates

No. of distances/angles: 6

Interatomic dist. A-B (Å)	Atom A	Atom B	Atom C	Bond angle A-B-C (°)
1.838	H1	Rh3(0,-1)	H1(1,0)	94.1
1.838	H1	Rh3(0,-1)	H2(0,-1)	64.5
1.838	H1	Rh3(0,-1)	Rh4	57.5
2.293	H1	Rh4	Rh5	137.9
2.676	Rh3	Rh4	Rh5	59.6
2.667	Rh3	Rh5	Rh6	120.0

COMMON NAME : Rh(110)-(1x2)-H
CLASSIFICATION : 45.1.8
TECHNIQUE : LEED
AUTHORS : W. Puchta, W. Nichtl, W. Oed, N. Bickel, K. Heinz and K. Mueller
REFERENCE : Phys. Rev., B39, 1020 (1989)

ILLUSTRATION: 38

SURFACE TYPE
Substrate : Rh
Crystal face: 110
Temperature : 90 K
Bulk lattice: fcc
2D bulk symm: pmm
2D surf symm: pm

Adsorbate: H
Coverage : 0.5 H/Rh
Pattern : (1x2)
Matrix : (1.000, 0.000)
(0.000, 2.000)

STRUCTURE TYPE
Atomic adsorption in nearly 3-fold sites on (111) facets on side of troughs (with long H-Rh distance to 2nd-Rh-layer atoms); top-Rh-rows to which H is bonded are buckled out and laterally shifted towards H ('shift-buckling')

SAMPLE PREPARATION (1 sample)
Treatment : Ar+ sputt, heating in O2 and H2, 1300 K anneal, H2 dosed 90K
Crystallinity:
Anal. methods:
Contamination: AES: clean

COMMENTS

DATA COLLECTION
Technique: LEED
Dataset : I-V curves for 8 independent beams: E range 50-206 eV

THEORY/DATA TREATMENT
Dynamical LEED (layer-doubling):ΘD(Rh)=480 K, (H)3400K

STRUCTURES EXAMINED
Examined symmetric row pairing and shift-buckling

QUALITY OF EXPERIMENT-THEORY FIT
RPE=0.44

2D UNIT CELLS (1 domain observed)

Cell	Ax (Å)	Ay (Å)	Bx (Å)	By (Å)	α (°)	Matrix	Pattern	Cell type
Bulk	2.690	0.000	0.000	3.804	90.0	(1.000, 0.000) (0.000, 1.000)	(1x1)	b: bulk lattice
Surface 1	2.690	0.000	0.000	7.608	90.0	(1.000, 0.000) (0.000, 2.000)	(1x2)	s1: commens. superlattice

3D COORDINATES

H in 3-fold site on side of trough; Rh2: ridge atoms displaced up and laterally toward H, relative to Rh3

Dx/Dy in Å, or as a fraction of layer's unit cell vectors; atom 0 at the origin. Epir/subr are bulk repeat vectors.

No. of atoms: 4 Bulk z = 1.345 Å

Region	Chem el.	At. no.	Cell type	Site occ.	Rel. to	Dx ± εx		Dy ± εy		Dz ± εz		Dz/Bz(%) ± εz/Bz
epir		-2					f		f		Å	
subr		-1				1.345	Å	1.902	Å	1.345	Å	
ovrl	H	1	s1	.50	0	0.000	f	0.000	f	0.000	Å	0.0
intf	Rh	2	s1	.50	1	0.500	f	-0.137 ± .013	f	0.800 ± .100	Å	59.5 ± 7.4
intf	Rh	3	s1	.50	2	0.000	f	0.500 ± .004	f	0.040 ± .010	Å	3.0 ± .7
subl	Rh	4	b	1.00	3	-0.500	f	-0.500	f	1.271 ± .027	Å	94.5 ± 2.0

BOND DISTANCES AND ANGLES

Bond distances and angles are derived from coordinates

No. of distances/angles: 7

Interatomic dist. A-B (Å)	Atom A	Atom B	Atom C	Bond angle A-B-C (°)
1.879	H1	Rh2(0,-1)	H1(1,0)	91.4
1.879	H1	Rh2(0,-1)	Rh2(1,-1)	135.7
2.279	H1	Rh4	Rh2(0,-1)	43.7
2.279	H1	Rh4	Rh3	80.0
2.690	Rh2	Rh2(1,0)	H1(1,1)	44.3

Rh(110)-(1x2)-H
45.1.8

Bond Distances and Angles - Continued

Interatomic dist. A-B (Å)	Atom A	Atom B	Atom C	Bond angle A-B-C (°)
2.670	Rh2	Rh4(1,1)	H1(1,1)	43.7
2.654	Rh3	Rh4		

COMMON NAME : Rh(110)-(1x2)-3H
CLASSIFICATION : 45.1.6
TECHNIQUE : LEED
AUTHORS : M. Michl, W, Nichtl-Pecher, W. Oed, H. Landskron, K. Heinz and K. Mueller
REFERENCE : Surf. Sci., 220, 59 (1989)

ILLUSTRATION: 38

SURFACE TYPE
Substrate : Rh
Crystal face: 110
Temperature : 90 K
Bulk lattice: fcc
2D bulk symm: pmm
2D surf symm: pm

Adsorbate: H
Coverage : 1.5 H/Rh
Pattern : (1x2)
Matrix : (1.000, 0.000)
(0.000, 2.000)

STRUCTURE TYPE
Atomic adsorption of hydrogen with top layer relaxation (slight buckling of doubly occupied Rh atoms), all H in 3 inequivalent quasi-3-fold sites in sides of troughs of unreconstructed substrate

SAMPLE PREPARATION (1 sample)
Treatment : Ar+ sputter, ox/red cycles of O and H, then anneal at 1300 K
Crystallinity:
Anal. methods:
Contamination: AES: clean

COMMENTS

DATA COLLECTION
Technique: LEED
Dataset : I-V curves for 8 symmetry inequivalent beams; E range 50-350 eV

THEORY/DATA TREATMENT
Dynamical LEED (exc. no H-H scatt.): 8 phase shifts; Vor=-12 eV, Voi=-5eV

STRUCTURES EXAMINED
Quasi-3-fold H sites on relaxed, buckled substrate, varying H heights and lateral positions, keeping one mirror plane

QUALITY OF EXPERIMENT-THEORY FIT
RPE=0.33

2D UNIT CELLS (2 domains observed)

Cell	Ax (Å)	Ay (Å)	Bx (Å)	By (Å)	α (°)	Matrix	Pattern	Cell type
Bulk	2.690	0.000	0.000	3.804	90.0	(1.000, 0.000) (0.000, 1.000)	(1x1)	b: bulk lattice
Surface 1	2.690	0.000	0.000	7.608	90.0	(1.000, 0.000) (0.000, 2.000)	(1x2)	s1: commens. superlattice

3D COORDINATES

all H in 3-fold sites in sides of troughs with long bonds to 2nd Rh layer;
Rh4-Rh5: buckled top Rh layer

Dx/Dy in Å, or as a fraction of layer's unit cell vectors; atom 0 at the origin. Epir/subr are bulk repeat vectors.

No. of atoms: 7

Bulk z = 1.345 Å

Reg ion	Chem el.	At. no.	Cell type	Site occ.	Rel. to	Dx ± εx		Dy ± εy		Dz ± εz		Dz/Bz(%) ± εz/Bz
epir		-2					f		f		Å	
subr		-1				1.345	Å	1.902	Å	1.345	Å	
ovrl	H	1	s1	.50	0	0.000	Å	0.000	Å	0.000	Å	0.0
ovrl	H	2	s1	.50	1	0.000	Å	-2.204 ± .200	Å	0.150 ± .100	Å	11.2 ± 7.4
ovrl	H	3	s1	.50	2	0.000	Å	-2.040 ± .200	Å	0.290 ± .100	Å	21.6 ± 7.4
intf	Rh	4	s1	.50	3	0.500	f	1.090 ± .200	Å	0.710 ± .100	Å	52.8 ± 7.4
intf	Rh	5	s1	.50	4	0.000	f	0.500	f	0.030 ± .020	Å	2.2 ± 1.5
intf	Rh	6	b	1.00	5	0.500	f	0.500	f	1.294 ± .014	Å	96.2 ± 1.0
subl	Rh	7	b	1.00	6	0.500	f	0.500	f	1.345 ± .014	Å	100.0 ± 1.0

Rh(110)-(1x2)-3H
45.1.6

BOND DISTANCES AND ANGLES

Bond distances and angles are derived from coordinates

No. of distances/angles: 11

Interatomic dist. A-B (Å)	Atom A	Atom B	Atom C	Bond angle A-B-C (°)
1.904	H1	Rh5	H1(1,0)	89.9
2.669	Rh4	Rh7	Rh7(1,0)	90.0
2.665	Rh5	Rh6	Rh6(1,0)	59.7
1.904	H1	Rh5	Rh5(1,0)	135.0
1.927	H2	Rh4	H2(1,0)	88.6
1.927	H2	Rh4	Rh4(1,0)	134.3
1.871	H3	Rh4	H3(1,0)	91.9
1.871	H3	Rh4	Rh4(1,0)	136.0
2.190	H3	Rh6	Rh6(1,0)	90.0
2.690	Rh4	Rh4(1,0)	Rh6(1,0)	59.9
2.680	Rh4	Rh6	Rh6(1,0)	59.9

COMMON NAME : Rh(110)-(1x3)-H
CLASSIFICATION : 45.1.5
TECHNIQUE : LEED
AUTHORS : K. Lehnberger, W. Nichtl-Pecher, W. Oed, K. Heinz and K. Mueller
REFERENCE : Surf. Sci., 217, 511 (1989)

ILLUSTRATION: 38

SURFACE TYPE
Substrate : Rh
Crystal face: 110
Temperature : 90 K
Bulk lattice: fcc
2D bulk symm: pmm
2D surf symm: pm

Adsorbate: H
Coverage : 0.33 H/Rh
Pattern : (1x3)
Matrix : (1.000, 0.000)
(0.000, 3.000)

STRUCTURE TYPE
Atomic adsorption of H in 3-fold sites on side of troughs, slight buckling of top layer

SAMPLE PREPARATION (1 sample)
Treatment : Ar+ sputtering, heating in O2 and H2, then flashing to 1300 K
Crystallinity:
Anal. methods:
Contamination: clean as determined by AES

COMMENTS

DATA COLLECTION
Technique: LEED
Dataset : I-V curves for 9 symmetry inequivalent beams; E range 50-200 eV

THEORY/DATA TREATMENT
Dynamical LEED (comp. layer, lay. dblg): 8 phase shifts

STRUCTURES EXAMINED
Top and bridge sites over ridge, top and bridge sites over trough, 3-fold hollow on side of trough; H height variation; for 3-fold site: Rh 2-row-pairing and 1-row-shift-buckling tried (latter is best); then, keeping one mirror plane: H, near Rh ridge varied in 2D, and var. of top 2 Rh-Rh spacings

QUALITY OF EXPERIMENT-THEORY FIT
RPE=0.25

2D UNIT CELLS (1 domain observed)

Cell	Ax (Å)	Ay (Å)	Bx (Å)	By (Å)	α (°)	Matrix	Pattern	Cell type
Bulk	2.690	0.000	0.000	3.804	90.0	(1.000, 0.000) (0.000, 1.000)	(1x1)	b: bulk lattice
Surface 1	2.690	0.000	0.000	11.413	90.0	(1.000, 0.000) (0.000, 3.000)	(1x3)	s1: commens. superlattice

3D COORDINATES

H: atomic adsorption in 3-fold site in side of trough with long bonds to 2nd Rh layer;
Rh2-4: buckled top metal layer (Rh2 out and shifted);

Dx/Dy in Å, or as a fraction of layer's unit cell vectors; atom 0 at the origin. Epir/subr are bulk repeat vectors.

No. of atoms: 7

Bulk z = 1.345 Å

Region	Chem el.	At. no.	Cell type	Site occ.	Rel. to	Dx ± εx		Dy ± εy		Dz ± εz		Dz/Bz(%) ± εz/Bz
epir		-2					f		f		Å	
subr		-1				1.345	Å	1.902	Å	1.345	Å	
ovrl	H	1	s1	.33	0	0.000	f	0.000	f	0.000	Å	0.0
intf	Rh	2	s1	.33	1	0.500	f	0.898 ± .018	f	0.510 ± .100	Å	37.9 ± 7.5
intf	Rh	3	s1	.33	2	0.000	f	-0.670 ± .003	f	0.030 ± .020	Å	2.2 ± 1.5
intf	Rh	4	s1	.33	3	0.000	f	0.333	f	0.000	Å	0.0
intf	Rh	5	b	1.00	4	-0.500	f	-0.500	f	1.260 ± .020	Å	93.5 ± 1.5
intf	Rh	6	b	1.00	5	0.500	f	0.500	f	1.360 ± .020	Å	101.0 ± 1.5
subl	Rh	7	b	1.00	6	-0.500	f	-0.500	f	1.358	Å	101.0

Rh(110)-(1x3)-H
45.1.5

BOND DISTANCES AND ANGLES

Bond distances and angles are derived from coordinates

No. of distances/angles: 5

Interatomic dist. A-B (Å)	Atom A	Atom B	Atom C	Bond angle A-B-C (°)
1.848	H1	Rh2(0,-1)	H1(1,0)	93.4
1.848	H1	Rh2(0,-1)	Rh2(1,-1)	136.7
2.690	Rh2	Rh2(1,0)	Rh5(1,2)	59.3
2.648	Rh3	Rh5	Rh5(1,0)	59.5
1.932	Rh5	H1(0,1)	Rh3	88.0

COMMON NAME : Rh(111)-(2x2)-3NO
CLASSIFICATION : 45.7.8.1
TECHNIQUE : LEED
AUTHORS : C.T. Kao, G.S. Blackman, M.A. Van Hove, G.A. Somorjai and C.-M. Chan
REFERENCE : Surf. Sci., 224, 77 (1989)

ILLUSTRATION: 63

SURFACE TYPE
Substrate : Rh
Crystal face: 111
Temperature : 40 K
Bulk lattice: fcc
2D bulk symm: p3m1
2D surf symm: pm

Adsorbate: NO
Coverage : 0.75 NO/Rh
Pattern : (2x2)
Matrix : (2.000, 0.000)
(0.000, 2.000)

STRUCTURE TYPE
Densely packed molecular adsorption: 1 upright NO on bridge and 2 upright NOs near top sites per (2x2) cell, forming roughly hexagonal buckled NO lattice (N down)

SAMPLE PREPARATION (1 sample)
Treatment : dosing of 2.2L of NO at 40 K, followed by annealing to 220K
Crystallinity:
Anal. methods:
Contamination: AES: no detectable C, S, O or B

COMMENTS

DATA COLLECTION
Technique: LEED
Dataset : I-V curves for 15 independent beams at normal incidence: E range 20-200 eV

THEORY/DATA TREATMENT
Dynamical LEED (RFS, BSN, KSLA): 5 ph shs (Moruzzi Rh pot cluster NO pot); VoiαE**1/3; ΘD=406 K(Rh),920K(N),861K(O)

STRUCTURES EXAMINED
1. 1 NO on bridge, 2 near top sites, varying Rh-N and N-O spacings and lateral NO-NO distance; some tilting allowed; 2. same with 1 NO on top and 2 near bridge sites; 3. 1 NO on top, bridge or hollow sites, using Rh-N-O spacings

QUALITY OF EXPERIMENT-THEORY FIT
RVHT=0.28, RZJ=0.26, RPE=0.67

2D UNIT CELLS (3 domains observed)

Cell	Ax (Å)	Ay (Å)	Bx (Å)	By (Å)	α (°)	Matrix	Pattern	Cell type
Bulk	2.680	0.000	1.340	2.321	60.0	(1.000, 0.000) (0.000, 1.000)	(1x1)	b: bulk lattice
Surface 1	5.360	0.000	2.680	4.642	60.0	(2.000, 0.000) (0.000, 2.000)	(2x2)	s1: commens. superlattice

3D COORDINATES

O1-N4, O2-N5: 2 near-top-site upright NOs; O3-N6: 1 bridge-site upright NO

Dx/Dy in Å, or as a fraction of layer's unit cell vectors; atom 0 at the origin. Epir/subr are bulk repeat vectors.

No. of atoms: 8 Bulk z = 2.188 Å

Region	Chem el.	At. no.	Cell type	Site occ.	Rel. to	Dx ± εx		Dy ± εy		Dz ± εz		Dz/Bz(%) ± εz/Bz
epir		-2					f		f		Å	
subr		-1				1.340	Å	-0.774	Å	2.188	Å	
ovrl	O	1	s1	.25	0	0.000	f	0.000	f	0.000	Å	0.0
ovrl	O	2	s1	.25	1	0.445 ± .008	f	0.445 ± .022	f	0.000 ± .100	Å	0.0 ± 4.6
ovrl	O	3	s1	.25	2	0.278	f	0.278	f	0.500 ± .100	Å	22.9 ± 4.6
ovrl	N	4	s1	.25	1	0.000 ± .008	f	0.000 ± .022	f	1.150 ± .100	Å	52.6 ± 4.6
ovrl	N	5	s1	.25	2	0.000 ± .008	f	0.000 ± .022	f	1.150 ± .100	Å	52.6 ± 4.6
ovrl	N	6	s1	.25	3	0.000	f	0.000	f	1.150 ± .100	Å	52.6 ± 4.6
intf	Rh	7	b	1.00	6	-0.500	f	0.500	f	1.550 ± .100	Å	70.8 ± 4.6
subl	Rh	8	b	1.00	7	0.667	f	-0.333	f	2.188	Å	100.0

Rh(111)-(2x2)-3NO
45.7.8.1

BOND DISTANCES AND ANGLES

Bond distances and angles are derived from coordinates

No. of distances/angles: 4

Interatomic dist. A-B (Å)	Atom A	Atom B	Atom C	Bond angle A-B-C (°)
1.150	O1	N4	Rh7(-1,-2)	172.9
1.150	O2	N5	Rh7(0,-1)	172.9
1.150	O3	N6	Rh7	139.2
2.049	N6	Rh7	Rh8	106.9

COMMON NAME : Rh(100)-(2x2)-O
CLASSIFICATION : 45.8.2
TECHNIQUE : LEED
AUTHORS : W. Oed, B. Doetsch, L. Hammer, K. Heinz and K. Mueller
REFERENCE : Surf. Sci., 207, 55 (1988)

ILLUSTRATION: 28,30

SURFACE TYPE
Substrate : Rh
Crystal face: 100
Temperature : 130 K
Bulk lattice: fcc
2D bulk symm: p4m
2D surf symm: p4m

Adsorbate: O
Coverage : 0.25 O/Rh
Pattern : (2x2)
Matrix : (2.000, 0.000)
(0.000, 2.000)

STRUCTURE TYPE
Hollow site adsorption with slight top Rh-Rh interlayer contraction

SAMPLE PREPARATION (1 sample)
Treatment : Ar+ sputtering with ox/red cycles, then 1400 K anneal
Crystallinity:
Anal. methods:
Contamination:

COMMENTS

DATA COLLECTION
Technique: LEED
Dataset : I-V curves for 9 symmetry-inequivalent beams; E range 25-350/202 eV for quasi-dyn./dyn. analysis

THEORY/DATA TREATMENT
Quasi-dyn. then dynamical LEED (lay. dblg): <=14 relat. phase shifts; Vor=-9 eV,Voi=-5eV; ΘD=480 K(Rh),843K(O)

STRUCTURES EXAMINED
Variation of top two interlayer spacings for hollow, bridge and top adsorption sites

QUALITY OF EXPERIMENT-THEORY FIT
RPE=0.27

2D UNIT CELLS (1 domain observed)

Cell	Ax (Å)	Ay (Å)	Bx (Å)	By (Å)	α (°)	Matrix	Pattern	Cell type
Bulk	2.690	0.000	0.000	2.690	90.0	(1.000, 0.000) (0.000, 1.000)	(1x1)	b: bulk lattice
Surface 1	5.380	0.000	0.000	5.380	90.0	(2.000, 0.000) (0.000, 2.000)	(2x2)	s1: commens. superlattice

3D COORDINATES

Dx/Dy in Å, or as a fraction of layer's unit cell vectors; atom 0 at the origin. Epir/subr are bulk repeat vectors.

No. of atoms: 4

Bulk z = 1.902 Å

Region	Chem el.	At. no.	Cell type	Site occ.	Rel. to	Dx ± εx		Dy ± εy		Dz ± εz		Dz/Bz(%) ± εz/Bz
epir		-2					f		f		Å	
subr		-1				1.345	Å	1.345	Å	1.902	Å	
ovrl	O	1	s1	.25	0	0.000	f	0.000	f	0.000	Å	0.0
intf	Rh	2	b	1.00	1	0.500	f	0.500	f	0.950 ± .040	Å	49.9 ± 2.1
intf	Rh	3	b	1.00	2	-0.500	f	-0.500	f	1.890 ± .040	Å	99.4 ± 2.1
subl	Rh	4	b	1.00	3	0.500	f	0.500	f	1.902	Å	100.0

BOND DISTANCES AND ANGLES

Bond distances and angles are derived from coordinates

No. of distances/angles: 4

Interatomic dist. A-B (Å)	Atom A	Atom B	Atom C	Bond angle A-B-C (°)
2.126	O1	Rh2	Rh2(1,0)	129.2
2.126	O1	Rh2	Rh3	71.4
2.690	Rh2	Rh2(1,0)	Rh3(1,0)	59.9
2.681	Rh2	Rh3	Rh3(1,0)	59.9

ILLUSTRATION: 22,25

COMMON NAME : Rh(111)-(2x2)-O
CLASSIFICATION : 45.8.1
TECHNIQUE : LEED
AUTHORS : P.C. Wong, K.C. Hui, M.Y. Zhou and K.A.R. Mitchell
REFERENCE : Surf. Sci., 165, L21 (1986)

SURFACE TYPE
Substrate : Rh
Crystal face: 111
Temperature : 215 K
Bulk lattice: fcc
2D bulk symm: p3m1
2D surf symm: p3m1

Adsorbate: O
Coverage : 0.25 O/Rh
Pattern : (2x2)
Matrix : (2.000, 0.000)
(0.000, 2.000)

STRUCTURE TYPE
Atomic adsorption in fcc hollow sites of unrelaxed substrate

SAMPLE PREPARATION (1 sample)
Treatment : exposure to 6L O at room temperature
Crystallinity:
Anal. methods:
Contamination:

COMMENTS

DATA COLLECTION
Technique: LEED
Dataset : I-V curves: E range 60-250 eV in increments of 2 eV

THEORY/DATA TREATMENT
Dynamical LEED (RFS): 7 phase shifts, Rh band structure pot, O pot from Demuth et al; ΘD=480 K(Rh), 843K(O)

STRUCTURES EXAMINED
1) fcc hollow site; 2) hcp hollow site;
3) fcc and hcp hollow sites ('graphitic' model); O-Rh spacing varied, substrate bulk-like

QUALITY OF EXPERIMENT-THEORY FIT
RPE=0.394

2D UNIT CELLS (1 domain observed)

Cell	Ax (Å)	Ay (Å)	Bx (Å)	By (Å)	α (°)	Matrix	Pattern	Cell type
Bulk	2.685	0.000	-1.343	2.325	120.0	(1.000, 0.000) (0.000, 1.000)	(1x1)	b: bulk lattice
Surface 1	5.370	0.000	-2.685	4.651	120.0	(2.000, 0.000) (0.000, 2.000)	(2x2)	s1: commens. superlattice

3D COORDINATES

O1: overlayer in fcc hollow sites

Dx/Dy in Å, or as a fraction of layer's unit cell vectors; atom 0 at the origin. Epir/subr are bulk repeat vectors.

No. of atoms: 3

Bulk z = 2.195 Å

Reg ion	Chem el.	At. no.	Cell type	Site occ.	Rel. to	Dx ± εx		Dy ± εy		Dz ± εz		Dz/Bz(%) ± εz/Bz
epir		-2					f		f		Å	
subr		-1				1.343	Å	0.775	Å	2.195	Å	
ovrl	O	1	s1	.25	0	0.000	f	0.000	f	0.000	Å	0.0
intf	Rh	2	b	1.00	1	0.667	f	0.333	f	1.230 ± .090	Å	56.0 ± 4.1
subl	Rh	3	b	1.00	2	0.667	f	0.333	f	2.195	Å	100.0

BOND DISTANCES AND ANGLES

Bond distances and angles are derived from coordinates

No. of distances/angles: 3

Interatomic dist. A-B (Å)	Atom A	Atom B	Atom C	Bond angle A-B-C (°)
1.979	O1	Rh2	Rh2(1,0)	132.7
1.979	O1	Rh2	Rh3	163.7
2.685	Rh2	Rh2(1,0)		

COMMON NAME : Rh(100)-(2x2)-S
CLASSIFICATION : 45.16.1
TECHNIQUE : LEED
AUTHORS : S. Hengrasmee, P.R. Watson, D.C. Frost and K.A.R. Mitchell
REFERENCE : Surf. Sci., 87, L249 (1979)

ILLUSTRATION: 28,30

SURFACE TYPE
Substrate : Rh
Crystal face: 100
Temperature : RT
Bulk lattice: fcc
2D bulk symm: p4m
2D surf symm: p4m

Adsorbate: S
Coverage : 1/4 (S/Rh)
Pattern : (2x2)
Matrix : (2.000, 0.000)
(0.000, 2.000)

STRUCTURE TYPE
Atomic adsorption in hollow sites

SAMPLE PREPARATION (2 sample)
Treatment : exposure to 1E-8 torr H2S and anneal (same from S segreg.)
Crystallinity:
Anal. methods:
Contamination: no impurities by AES

COMMENTS

DATA COLLECTION
Technique: LEED; photographic vidicon method
Dataset : I-V curves: 4 integral-order and 5 fractional-order beams at normal incidence; energy range 40-200 eV

THEORY/DATA TREATMENT
Dynamical LEED (layer doubling): 8 ph shs (Rh band struc. pot, S superpos pot); VoiαE**1/3; ΘD=406 K(Rh), 236K(S)

STRUCTURES EXAMINED
Unrelaxed substrate: S in top, bridge and hollow sites with variable S-Rh layer spacing

QUALITY OF EXPERIMENT-THEORY FIT
RZJ=0.33

2D UNIT CELLS (1 domain observed)

Cell	Ax (Å)	Ay (Å)	Bx (Å)	By (Å)	α (°)	Matrix	Pattern	Cell type
Bulk	2.680	0.000	0.000	2.680	90.0	(1.000, 0.000) (0.000, 1.000)	(1x1)	b: bulk lattice
Surface 1	5.360	0.000	0.000	5.360	90.0	(2.000, 0.000) (0.000, 2.000)	(2x2)	s1: commens. superlattice

3D COORDINATES

S1: overlayer in 4-fold hollow sites; 0.1Å error bar assumed for tabulation

Dx/Dy in Å, or as a fraction of layer's unit cell vectors; atom 0 at the origin. Epir/subr are bulk repeat vectors.

No. of atoms: 3

Bulk z = 1.902 Å

Region	Chem el.	At. no.	Cell type	Site occ.	Rel. to	Dx ± εx		Dy ± εy		Dz ± εz		Dz/Bz(%) ± εz/Bz
epir		-2					f		f		Å	
subr		-1				-1.340	Å	-1.340	Å	1.902	Å	
ovrl	S	1	s1	.25	0	0.000	f	0.000	f	0.000	Å	0.0
intf	Rh	2	b	1.00	1	0.500	f	0.500	f	1.290 ± .100	Å	67.8 ± 5.3
subl	Rh	3	b	1.00	2	-0.500	f	-0.500	f	1.902	Å	100.0

BOND DISTANCES AND ANGLES

Bond distances and angles are derived from coordinates

No. of distances/angles: 3

Interatomic dist. A-B (Å)	Atom A	Atom B	Atom C	Bond angle A-B-C (°)
2.292	S1	Rh2	Rh3	79.4
2.680	Rh2	Rh2(1,0)		
2.685	Rh2	Rh3		

ILLUSTRATION: 35,36

COMMON NAME : Rh(110)-c(2x2)-S
CLASSIFICATION : 45.16.2
TECHNIQUE : LEED
AUTHORS : S. Hengrasmee, P.R. Watson, D.C. Frost and K.A.R. Mitchell
REFERENCE : Surf. Sci., 92, 71 (1980)

SURFACE TYPE
Substrate : Rh
Crystal face: 110
Temperature : RT
Bulk lattice: fcc
2D bulk symm: pmm
2D surf symm: cmm

Adsorbate: S
Coverage : 1/2 (S/Rh)
Pattern : c(2x2)
Matrix : (1.000, 1.000)
(-1.000, 1.000)

STRUCTURE TYPE
Atomic adsorption in center site, on top of 2nd layer Rh

SAMPLE PREPARATION (1 sample)
Treatment : exposure to 5E-7 torr H2S for 3 min at 300C
Crystallinity:
Anal. methods:
Contamination: no impurities by AES

COMMENTS

DATA COLLECTION
Technique: LEED; photographic vidicon method
Dataset : I-V curves: 9 integral-order and 5 half-order beams at normal incidence; energy range 40-220 eV

THEORY/DATA TREATMENT
Dynamical LEED: 8 phase shifts (Rh band structure pot, S superpos pot); VoiαE**1/3; ΘD=406 K(Rh), 236K(S)

STRUCTURES EXAMINED
Unrelaxed substrate: S in top, hollow, long- and short- bridge sites with variable S-Rh layer spacing

QUALITY OF EXPERIMENT-THEORY FIT
RZJ=0.165

2D UNIT CELLS (1 domain observed)

Cell	Ax (Å)	Ay (Å)	Bx (Å)	By (Å)	α (°)	Matrix	Pattern	Cell type
Bulk	2.679	0.000	0.000	3.789	90.0	(1.000, 0.000) (0.000, 1.000)	(1x1)	b: bulk lattice
Surface 1	2.679	3.789	-2.679	3.789	70.5	(1.000, 1.000) (-1.000, 1.000)	c(2x2)	s1: commens. superlattice

3D COORDINATES

S1: overlayer in center site, on top above 2nd Rh layer; 0.1Å error bar assumed for tabulation

Dx/Dy in Å, or as a fraction of layer's unit cell vectors; atom 0 at the origin. Epir/subr are bulk repeat vectors.

No. of atoms: 3

Bulk z = 1.345 Å

Region	Chem el.	At. no.	Cell type	Site occ.	Rel. to	Dx ± εx		Dy ± εy		Dz ± εz		Dz/Bz(%) ± εz/Bz
epir		-2					f		f		Å	
subr		-1				-1.340	Å	-1.894	Å	1.345	Å	
ovrl	S	1	s1	.50	0	0.000	f	0.000	f	0.000	Å	0.0
intf	Rh	2	b	1.00	1	0.500	f	0.500	f	0.770 ± .100	Å	57.3 ± 7.4
subl	Rh	3	b	1.00	2	-0.500	f	-0.500	f	1.345	Å	100.0

BOND DISTANCES AND ANGLES

Bond distances and angles are derived from coordinates

No. of distances/angles: 3

Interatomic dist. A-B (Å)	Atom A	Atom B	Atom C	Bond angle A-B-C (°)
2.444	S1	Rh2	Rh2(1,0)	123.2
2.444	S1	Rh2	Rh3	48.5
2.115	S1	Rh3	Rh2	59.9

COMMON NAME : Rh(111)-(√3x√3)R30°-S
CLASSIFICATION : 45.16.3
TECHNIQUE : LEED
AUTHORS : P.C. Wong, M.Y. Zhou, K.C. Hui and K.A.R. Mitchell
REFERENCE : Surf. Sci., 163, 172 (1985)

ILLUSTRATION: 22,24

SURFACE TYPE
Substrate : Rh
Crystal face: 111
Temperature : RT
Bulk lattice: fcc
2D bulk symm: p3m1
2D surf symm: p31m

Adsorbate: S
Coverage : 0.33 S/Rh
Pattern : (√3x√3)R30°
Matrix : (1.000, 1.000)
(-2.000, 1.000)

STRUCTURE TYPE
Atomic adsorption in fcc hollow sites

SAMPLE PREPARATION (1 sample)
Treatment : RT exposure to about 25L of H2S and annealed to 200C
Crystallinity:
Anal. methods:
Contamination: AES: S(151eV)/Rh(304eV)≈0.75

COMMENTS

DATA COLLECTION
Technique: LEED
Dataset : normal incidence I-V curves for 9 beams, E range 50-250 eV; symm. equivalent beams averaged with equal weights

THEORY/DATA TREATMENT
Dynamical LEED (RFS): 8 phase shifts, Rh band structure pot, S Demuth superpos. pot.; Voi∝E**1/3; ΘD=480 K(Rh),335K(S)

STRUCTURES EXAMINED
3-fold fcc and hcp hollow sites, 2-fold bridge site and 1-fold top site with variable S-Rh spacing on bulk-like Rh

QUALITY OF EXPERIMENT-THEORY FIT
RPE=0.27

2D UNIT CELLS (1 domain observed)

Cell	Ax (Å)	Ay (Å)	Bx (Å)	By (Å)	α (°)	Matrix	Pattern	Cell type
Bulk	2.690	0.000	1.345	2.330	60.0	(1.000, 0.000) (0.000, 1.000)	(1x1)	b: bulk lattice
Surface 1	4.036	2.330	-4.036	2.330	120.0	(1.000, 1.000) (-2.000, 1.000)	(√3x√3)R30°	s1: commens. superlattice

3D COORDINATES

S1: overlayer in fcc hollows; 0.1Å error bar assumed for tabulation

Dx/Dy in Å, or as a fraction of layer's unit cell vectors; atom 0 at the origin. Epir/subr are bulk repeat vectors.

No. of atoms: 3

Bulk z = 2.200 Å

Reg ion	Chem el.	At. no.	Cell type	Site occ.	Rel. to	Dx ± εx		Dy ± εy		Dz ± εz		Dz/Bz(%) ± εz/Bz
epir		-2					f		f		Å	
subr		-1				1.345	Å	0.777	Å	2.200	Å	
ovrl	S	1	s1	.33	0	0.000	f	0.000	f	0.000	Å	0.0
intf	Rh	2	b	1.00	1	0.333	f	0.333	f	1.530 ± .100	Å	69.6 ± 4.6
subl	Rh	3	b	1.00	2	0.333	f	0.333	f	2.200	Å	100.0

BOND DISTANCES AND ANGLES

Bond distances and angles are derived from coordinates

No. of distances/angles: 4

Interatomic dist. A-B (Å)	Atom A	Atom B	Atom C	Bond angle A-B-C (°)
2.180	S1	Rh2	Rh2(1,0)	128.1
2.180	S1	Rh2	Rh3	169.8
2.690	Rh2	Rh2(1,0)	Rh3	60.0
2.693	Rh2	Rh3		

ILLUSTRATION: 19

COMMON NAME : Ru(0001)-(1x1)
CLASSIFICATION : 44.1
TECHNIQUE : LEED
AUTHORS : G. Michalk, W. Moritz, H. Pfnur and D. Menzel
REFERENCE : Surf. Sci., 129, 92 (1983)

SURFACE TYPE
Substrate : Ru
Crystal face: 0001
Temperature : RT*
Bulk lattice: hcp
2D bulk symm: p3m1
2D surf symm: p3m1

Adsorbate:
Coverage :
Pattern : (1x1)
Matrix : (1.000, 0.000)
(0.000, 1.000)

STRUCTURE TYPE
Bulk termination with 2% top spacing contraction

SAMPLE PREPARATION (1 sample)
Treatment : mechanical polishing with diamond pastes
Crystallinity:
Anal. methods:
Contamination: monitored by AES and vibrating capacitor

COMMENTS

DATA COLLECTION
Technique: LEED
Dataset : I-V curves: 5 non-degenerate beams at normal incidence: 10, 11, 20, 21 and 30 beams

THEORY/DATA TREATMENT
Dynamical LEED (RFS, layer doubling): 8 phase shifts; Vor=-12 eV (fit), Voi=-0.85*E**1/3 eV; ΘD=410 K

STRUCTURES EXAMINED
Variation of top spacing; intensities averaged over ABAB.. and BABA.. terminations

QUALITY OF EXPERIMENT-THEORY FIT
RZJ=0.041, RPE=0.16

2D UNIT CELLS (1 domain observed)

Cell	Ax (Å)	Ay (Å)	Bx (Å)	By (Å)	α (°)	Matrix	Pattern	Cell type
Bulk	2.700	0.000	-1.350	2.338	120.0	(1.000, 0.000) (0.000, 1.000)	(1x1)	b: bulk lattice
Surface 1	2.700	0.000	-1.350	2.338	120.0	(1.000, 0.000) (0.000, 1.000)	(1x1)	s1: commens. superlattice

3D COORDINATES

Dx/Dy in Å, or as a fraction of layer's unit cell vectors; atom 0 at the origin. Epir/subr are bulk repeat vectors.

No. of atoms: 4

Bulk z = 2.140 Å

Region	Chem el.	At. no.	Cell type	Site occ.	Rel. to	Dx ± εx		Dy ± εy		Dz ± εz		Dz/Bz(%) ± εz/Bz
epir		-2					f		f		Å	
subr		-1				0.000	Å	0.000	Å	4.280	Å	
intf	Ru	1	b	1.00	0	0.000	f	0.000	f	0.000	Å	0.0
intf	Ru	2	b	1.00	1	0.333	f	0.667	f	2.100 ± .020	Å	98.1 ± .9
subl	Ru	3	b	1.00	2	-0.333	f	-0.667	f	2.140	Å	100.0
subl	Ru	4	b	1.00	3	0.333	f	0.667	f	2.140	Å	100.0

BOND DISTANCES AND ANGLES

Bond distances and angles are derived from coordinates

No. of distances/angles: 3

Interatomic dist. A-B (Å)	Atom A	Atom B	Atom C	Bond angle A-B-C (°)
2.700	Ru1	Ru1(1,0)	Ru2(1,-1)	121.1
2.615	Ru1	Ru2	Ru3	107.3
2.648	Ru2	Ru3	Ru4	107.9

COMMON NAME : Ru(0001)-(1x1)-H
CLASSIFICATION : 44.1.1
TECHNIQUE : LEED
AUTHORS : M. Lindroos, H. Pfnur, P. Feulner and D. Menzel
REFERENCE : Surf. Sci., 180, 237 (1987)

ILLUSTRATION: 55,56

SURFACE TYPE
Substrate : Ru
Crystal face: 0001
Temperature : 150 K
Bulk lattice: hcp
2D bulk symm: p3m1
2D surf symm: p3m1

Adsorbate: H
Coverage : 1 H/Ru
Pattern : (1x1)
Matrix : (1.000, 0.000)
(0.000, 1.000)

STRUCTURE TYPE
Atomic adsorption in fcc hollow sites, with slight contraction of topmost Ru-Ru interlayer spacing

SAMPLE PREPARATION (1 sample)
Treatment : cleaned by cycles of O at 500-1550 K, heated in UHV to 1570K
Crystallinity:
Anal. methods:
Contamination:

COMMENTS

DATA COLLECTION
Technique: LEED; VLEED (very low energies)
Dataset : E range 7-32 eV: normal and off-normal incidence beams used (up to 300 beams used)

THEORY/DATA TREATMENT
Dynamical LEED: Moruzzi pot for Ru, Mattheiss pot for H; Vor=-14 eV, Voi=-0.6eV below plasmon energy, Voi=-2eV above

STRUCTURES EXAMINED
1) bridge sites; 2) 3-fold hcp sites; 3) 3-fold fcc sites; 4) top sites; 5) both hcp and fcc sites with independent H heights; 6) underlayer positions; H-Ru and first Ru-Ru spacings varied

QUALITY OF EXPERIMENT-THEORY FIT
Visual

2D UNIT CELLS (1 domain observed)

Cell	Ax (Å)	Ay (Å)	Bx (Å)	By (Å)	α (°)	Matrix	Pattern	Cell type
Bulk	2.710	0.000	-1.355	2.347	120.0	(1.000, 0.000) (0.000, 1.000)	(1x1)	b: bulk lattice
Surface 1	2.710	0.000	-1.355	2.347	120.0	(1.000, 0.000) (0.000, 1.000)	(1x1)	s1: commens. superlattice

3D COORDINATES

H1: overlayer in fcc hollow sites; Ru4-Ru5: periodically repeating set of bulk layers

Dx/Dy in Å, or as a fraction of layer's unit cell vectors; atom 0 at the origin. Epir/subr are bulk repeat vectors.

No. of atoms: 5

Bulk z = 2.180 Å

Region	Chem el.	At. no.	Cell type	Site occ.	Rel. to	Dx ± εx		Dy ± εy		Dz ± εz		Dz/Bz(%) ± εz/Bz
epir		-2					f		f		Å	
subr		-1				0.000	Å	0.000	Å	4.360	Å	
ovrl	H	1	b	1.00	0	0.000	f	0.000	f	0.000	Å	0.0
intf	Ru	2	b	1.00	1	0.667	f	0.333	f	1.100 ± .060	Å	50.5 ± 2.8
intf	Ru	3	b	1.00	2	-0.333	f	0.333	f	2.140 ± .030	Å	98.2 ± 1.4
subl	Ru	4	b	1.00	3	0.333	f	-0.333	f	2.180	Å	100.0
subl	Ru	5	b	1.00	4	-0.333	f	0.333	f	2.180	Å	100.0

BOND DISTANCES AND ANGLES

Bond distances and angles are derived from coordinates

No. of distances/angles: 3

Interatomic dist. A-B (Å)	Atom A	Atom B	Atom C	Bond angle A-B-C (°)
1.913	H1	Ru2	Ru3	102.9
2.651	Ru2	Ru3	Ru4	108.2
2.683	Ru3	Ru4		

COMMON NAME : Ru(0001)-(√3x√3)R30°-CO
ILLUSTRATION: 80
CLASSIFICATION : 44.6.8.1
TECHNIQUE : LEED
AUTHORS : G. Michalk, W. Moritz, H. Pfnur and D. Menzel
REFERENCE : Surf. Sci., 129, 92 (1983)

SURFACE TYPE
Substrate : Ru
Crystal face: 0001
Temperature : 150
Bulk lattice: hcp
2D bulk symm: p3m1
2D surf symm: p31m

Adsorbate: CO
Coverage : 0.3333 (CO/Ru)
Pattern : (√3x√3)R30°
Matrix : (1.000, 1.000)
(-2.000, 1.000)

STRUCTURE TYPE
Molecular adsorption perp. to surface over top sites, C end bonded to Ru

SAMPLE PREPARATION (1 sample)
Treatment : CO dosed at 2E-8 mbar to maximize 1/3,1/3 spot int. at 360 K
Crystallinity:
Anal. methods:
Contamination: monitored by AES and LEED

COMMENTS
The Ru-C and C-O distances were correlated for the on top site, their sum having a much higher accuracy (3.12±0.02Å)

DATA COLLECTION
Technique: LEED
Dataset : I-V curves: 7 orders of non-equivalent beams; E range 40-400 eV; normal incidence

THEORY/DATA TREATMENT
Dynamical LEED (RFS, layer doubling): 8 phase shifts;
Vor=-12 eV, Voi=-0.85*E**1/3 eV; ΘD=410 K

STRUCTURES EXAMINED
Top, bridge, hcp and fcc 3-fold hollow sites; CO assumed normal to surface, C atom down;
C-Ru spacing varied from 1.65 to 2.25Å (top), from 0.95 to 2.10Å (hollows), from 1.10 to 1.7Å (bridge site);
CO bond length varied from 0.9 to 1.2Å

QUALITY OF EXPERIMENT-THEORY FIT
RPE=0.51, RZJ=0.21

2D UNIT CELLS (1 domain observed)

Cell	Ax (Å)	Ay (Å)	Bx (Å)	By (Å)	α (°)	Matrix	Pattern	Cell type
Bulk	2.700	0.000	1.350	2.338	60.0	(1.000, 0.000) (0.000, 1.000)	(1x1)	b: bulk lattice
Surface 1	4.050	2.338	-4.050	2.338	120.0	(1.000, 1.000) (-2.000, 1.000)	(√3x√3)R30°	s1: commens. superlattice

3D COORDINATES

O1-C2: upright CO molecular overlayer; Ru4-Ru5: periodically repeating set of bulk layers

Dx/Dy in Å, or as a fraction of layer's unit cell vectors; atom 0 at the origin. Epir/subr are bulk repeat vectors.

No. of atoms: 5

Bulk z = 2.140 Å

Region	Chem el.	At. no.	Cell type	Site occ.	Rel. to	Dx ± εx		Dy ± εy		Dz ± εz		Dz/Bz(%) ± εz/Bz
epir		-2					f		f		Å	
subr		-1				0.000	Å	0.000	Å	4.280	Å	
ovrl	O	1	s1	.33	0	0.000	f	0.000	f	0.000	Å	0.0
ovrl	C	2	s1	.33	1	0.000	f	0.000	f	1.090 ± .100	Å	50.9 ± 4.7
intf	Ru	3	b	1.00	2	0.000	f	0.000	f	2.000 ± .100	Å	93.5 ± 4.7
subl	Ru	4	b	1.00	3	0.333	f	0.333	f	2.140	Å	100.0
subl	Ru	5	b	1.00	4	-0.333	f	-0.333	f	2.140	Å	100.0

BOND DISTANCES AND ANGLES

Bond distances and angles are derived from coordinates

No. of distances/angles: 4

Interatomic dist. A-B (Å)	Atom A	Atom B	Atom C	Bond angle A-B-C (°)
1.090	O1	C2		
2.000	C2	Ru3	Ru4	143.9

Ru(0001)-(√3x√3)R30°-CO
44.6.8.1

Bond Distances and Angles - Continued

Interatomic dist. A-B (Å)	Atom A	Atom B	Atom C	Bond angle A-B-C (°)
2.700	Ru3	Ru3(1,0)	Ru4	59.3
2.648	Ru3	Ru4	Ru5	107.9

COMMON NAME : Ru(0001)-CO disordered
CLASSIFICATION : 44.6.8.2
TECHNIQUE : DLEED
AUTHORS : P. Piercy, P.A. Heimann, G. Michalk, D. Menzel
REFERENCE : Surf. Sci., 219, 189 (1989)

ILLUSTRATION: 80

SURFACE TYPE
Substrate : Ru
Crystal face: 0001
Temperature : 120 K
Bulk lattice: hcp
2D bulk symm: p3m1
2D surf symm: none

Adsorbate: CO
Coverage : 0.10 ML
Pattern : disordered
Matrix : (1.000, 0.000)
(0.000, 1.000)

STRUCTURE TYPE
Disordered CO linearly bonded at top site

SAMPLE PREPARATION (2 sample)
Treatment : heating and cooling cycles in O2
Crystallinity: LEED pattern
Anal. methods: AES, LEED
Contamination:

COMMENTS
Same result obtained at 0.05, 0.10, 0.20ML
same result obtained at 120K and 330K
close agreement with the ordered (√3x√3)R30°
structure at 0.33ML

DATA COLLECTION
Technique: DLEED; multichannel electron analyzer
Dataset : 11 azimuthal angles between the (1,1) and
(1,0) directions at 4 energies

THEORY/DATA TREATMENT
DLEED and TAUMOL programs

STRUCTURES EXAMINED
Top, bridge, fcc, hcp sites

QUALITY OF EXPERIMENT-THEORY FIT
RPE=0.18

2D UNIT CELLS (1 domain observed)

Cell	Ax (Å)	Ay (Å)	Bx (Å)	By (Å)	α (°)	Matrix	Pattern	Cell type
Bulk	2.706	0.000	1.353	2.343	60.0	(1.000, 0.000) (0.000, 1.000)	(1x1)	b: bulk lattice
Surface 1	2.706	0.000	1.353	2.343	60.0	(1.000, 0.000) (0.000, 1.000)	disordered	nd1: non-recon. lattice-gas dis

3D COORDINATES

O1-C2: disordered CO linearly bonded at top site

Dx/Dy in Å, or as a fraction of layer's unit cell vectors; atom 0 at the origin. Epir/subr are bulk repeat vectors.

No. of atoms: 4 Bulk z = 2.141 Å

Reg ion	Chem el.	At. no.	Cell type	Site occ.	Rel. to	Dx ± εx		Dy ± εy		Dz ± εz		Dz/Bz(%) ± εz/Bz
epir		-2					f		f		Å	
subr		-1				0.000	Å	0.000	Å	4.282	Å	
ovrl	O	1	nd1	.10	0	0.000	Å	0.000	Å	0.000	Å	0.0
ovrl	C	2	nd1	.10	0	0.000	Å	0.000	Å	1.100 ± .100	Å	51.4 ± 5.0
subl	Ru	3	b	1.00	0	0.000	Å	0.000	Å	3.100 ± .150	Å	144.8 ± 7.0
subl	Ru	4	b	1.00	0	1.353	Å	0.781	Å	5.241	Å	244.8

BOND DISTANCES AND ANGLES

Bond distances and angles are derived from coordinates

No. of distances/angles: 2

Interatomic dist. A-B (Å)	Atom A	Atom B	Atom C	Bond angle A-B-C (°)
1.100	C2	O1		
2.000	C2	Ru3		

COMMON NAME : Ru(0001)-(1x1)-1Fe
CLASSIFICATION : 44.26.1a
TECHNIQUE : LEED
AUTHORS : D. Tian, H. Li, F. Jona and P.M. Marcus
REFERENCE : Solid State Commun., 80, 783 (1991)

ILLUSTRATION: 87

SURFACE TYPE
Substrate : Ru
Crystal face: 0001
Temperature : RT
Bulk lattice: hcp
2D bulk symm: p3m1
2D surf symm: p3m1

Adsorbate: Fe
Coverage : 1.0 Fe/Ru
Pattern : (1x1)
Matrix : (1.000, 0.000)
(0.000, 1.000)

STRUCTURE TYPE
Epitaxial monolayer continuing hcp lattice

SAMPLE PREPARATION (1 sample)
Treatment : cycles of Ar+ bomb., 1400C anneals, Fe from vapor
Crystallinity: Fe increases LEED background slightly
Anal. methods: AES
Contamination: AES: no C,N,O,S before Fe adsorption

COMMENTS
Exp. Fe coverage is ≈4ML: the excess of ≈3ML is supposed to form small islands that do not contribute to the LEED

DATA COLLECTION
Technique: LEED
Dataset : IV spectra for 10,11,20 beams at normal incidence;

THEORY/DATA TREATMENT
Dyn. LEED (program CHANGE)

STRUCTURES EXAMINED
1-, 2-, 3- and 4-layer films with ABAB... or ABCABC... stacking in thicker films, with interlayer spacing relaxations

QUALITY OF EXPERIMENT-THEORY FIT
RZJ=0.11, RPE=0.29, RVHT=0.18

2D UNIT CELLS (1 domain observed)

Cell	Ax (Å)	Ay (Å)	Bx (Å)	By (Å)	α (°)	Matrix	Pattern	Cell type
Bulk	2.706	0.000	1.353	2.344	60.0	(1.000, 0.000) (0.000, 1.000)	(1x1)	b: bulk lattice
Surface 1	2.706	0.000	1.353	2.344	60.0	(1.000, 0.000) (0.000, 1.000)	(1x1)	s1: commens. superlattice

3D COORDINATES

Fe1: epitaxial layer continuing hcp lattice of Ru 0.05Å error bar assumed for tabulation

Dx/Dy in Å, or as a fraction of layer's unit cell vectors; atom 0 at the origin. Epir/subr are bulk repeat vectors.

No. of atoms: 4

Bulk z = 2.140 Å

Region	Chem el.	At. no.	Cell type	Site occ.	Rel. to	Dx ± εx		Dy ± εy		Dz ± εz		Dz/Bz(%) ± εz/Bz
epir		-2					f		f		Å	
subr		-1				0.000	Å	0.000	Å	4.280	Å	
ovrl	Fe	1	s1	1.00	0	0.000	f	0.000	f	0.000	Å	0.0
intf	Ru	2	b	1.00	1	0.667	f	0.667	f	2.050 ± .050	Å	95.8 ± 2.3
subl	Ru	3	b	1.00	2	-0.667	f	-0.667	f	2.140	Å	100.0
subl	Ru	4	b	1.00	3	0.667	f	0.667	f	2.140	Å	100.0

BOND DISTANCES AND ANGLES

Bond distances and angles are derived from coordinates

No. of distances/angles: 3

Interatomic dist. A-B (Å)	Atom A	Atom B	Atom C	Bond angle A-B-C (°)
2.706	Fe1	Fe1(1,0)	Ru2(0,-1)	58.3
2.578	Fe1	Ru2(0,-1)	Ru3	106.6
2.650	Ru2	Ru3(1,0)	Ru4	107.7

COMMON NAME : Ru(0001)-p(2x1)-O
CLASSIFICATION : 44.8.1
TECHNIQUE : LEED
AUTHORS : H. Pfnur, G. Held, M. Lindroos and D. Menzel
REFERENCE : Surf. Sci., 220, 43 (1989)

ILLUSTRATION: 55,57

SURFACE TYPE
Substrate : Ru
Crystal face: 0001
Temperature : 200 K
Bulk lattice: hcp
2D bulk symm: p3m1
2D surf symm: pm

Adsorbate: O
Coverage : 0.5 O/Ru
Pattern : p(2x1)
Matrix : (1.000, 0.000)
(0.000, 2.000)

STRUCTURE TYPE
Atomic adsorption in hcp hollow sites; top 2 Ru layers buckled; bulk-like first and second Ru-Ru interlayer spacings (measured wrt center of mass of the layers)

SAMPLE PREPARATION (1 sample)
Treatment : cleaned by cycles at 400-1550 K in O, heated in UHV
Crystallinity:
Anal. methods:
Contamination:

COMMENTS

DATA COLLECTION
Technique: LEED; 4-grid LEED system with Faraday cup
Dataset : IV curves for 84 beams: E range 26-300 eV, normal incidence, average over symm. equiv. beams

THEORY/DATA TREATMENT
Dynamical LEED: Moruzzi pot for Ru, Tong pot for O(CO), Vor=-14 eV, Voi=-4eV,ΘD=410 K(Ru),843K(O)

STRUCTURES EXAMINED
1) p(2x1) hcp site; 2) p(2x1) fcc site; 3) p(2x1)top site; 4) p(2x1) bridge; 5)honeycomb top/fcc; 6)honeycomb top/hcp; 7)honeycomb fcc/hcp; O-Ru spacing varied to choose site; relaxation of top two Ru layers compatible with pm symm. allowed for p(2x1) hcp site

QUALITY OF EXPERIMENT-THEORY FIT
RPE=0.185 for 1)

2D UNIT CELLS (3 domains observed)

Cell	Ax (Å)	Ay (Å)	Bx (Å)	By (Å)	α (°)	Matrix	Pattern	Cell type
Bulk	2.706	0.000	-1.353	2.343	120.0	(1.000, 0.000) (0.000, 1.000)	(1x1)	b: bulk lattice
Surface 1	2.706	0.000	-2.706	4.687	120.0	(1.000, 0.000) (0.000, 2.000)	p(2x1)	s1: commens. superlattice

3D COORDINATES

O1: atomic overlayer in hcp hollow sites; Ru2-3, Ru4-5: two buckled substrate layers; Ru6-7: bulk hcp layers

Dx/Dy in Å, or as a fraction of layer's unit cell vectors; atom 0 at the origin. Epir/subr are bulk repeat vectors.

No. of atoms: 7 Bulk z = 2.141 Å

Region	Chem el.	At. no.	Cell type	Site occ.	Rel. to	Dx ± εx		Dy ± εy		Dz ± εz		Dz/Bz(%) ± εz/Bz
epir		-2					f		f		Å	
subr		-1				0.000	Å	0.000	Å	4.282	Å	
ovrl	O	1	s1	.50	0	0.000	Å	4.627 ± .050	Å	-1.240 ± .020	Å	-57.9 ± .9
intf	Ru	2	s1	.50	0	1.353	Å	0.831 ± .050	Å	-0.040 ± .020	Å	-1.9 ± .9
intf	Ru	3	s1	.50	0	0.000	Å	3.054 ± .050	Å	0.030 ± .020	Å	1.4 ± .9
intf	Ru	4	s1	.50	0	-1.353	Å	2.383 ± .050	Å	2.121 ± .020	Å	99.1 ± .9
intf	Ru	5	s1	.50	0	0.000	Å	0.050 ± .050	Å	2.151 ± .020	Å	100.5 ± .9
subl	Ru	6	b	1.00	0	1.353	Å	0.781	Å	4.282	Å	200.0
subl	Ru	7	b	1.00	0	0.000	Å	0.000	Å	6.422	Å	300.0

Ru(0001)-p(2x1)-O
44.8.1 BOND DISTANCES AND ANGLES

Bond distances and angles are derived from coordinates

No. of distances/angles: 8

Interatomic dist. A-B (Å)	Atom A	Atom B	Atom C	Bond angle A-B-C (°)
2.016	O1	Ru2(1,1)	O1(1,0)	84.3
2.016	O1	Ru2(1,1)	Ru2(0,1)	47.9
2.016	O1	Ru2(1,1)	Ru3(1,1)	92.6
2.021	O1	Ru3	Ru2(1,1)	45.8
2.706	Ru2	Ru2(1,0)	O1(1,-1)	132.2
2.706	Ru2	Ru2(1,0)	O1(0,-1)	47.9
2.604	Ru2	Ru3(1,0)	O1(1,0)	130.4
2.604	Ru2	Ru3(1,0)	Ru2(1,1)	119.8

COMMON NAME : Ru(0001)-p(2x2)-O
CLASSIFICATION : 44.8.2
TECHNIQUE : LEED
AUTHORS : M. Lindroos, H. Pfnur, G. Held, and D. Menzel
REFERENCE : Surf. Sci., 222, 451 (1989)

ILLUSTRATION: 55,57

SURFACE TYPE
Substrate : Ru
Crystal face: 0001
Temperature : 200 K
Bulk lattice: hcp
2D bulk symm: p3m1
2D surf symm: p3m1

Adsorbate: O
Coverage : 0.25 O/Ru
Pattern : p(2x2)
Matrix : (2.000, 0.000)
(0.000, 2.000)

STRUCTURE TYPE
Atomic adsorption in hcp hollow sites, slight contraction of first Ru-Ru interlayer spacing while second spacing essentially bulk-like; buckling of first and second Ru layers

SAMPLE PREPARATION (1 sample)
Treatment : cleaned by cycles at 400-1550 K in O, heated in UHV
Crystallinity:
Anal. methods:
Contamination:

COMMENTS

DATA COLLECTION
Technique: LEED; 4-grid LEED system with Faraday cup
Dataset : IV curves for 78 beams: E range 26-300 eV, normal incidence, average over symm. equiv. beams
O-Ru spacing varied to choose the site; relaxation of top two Ru layers compatible with p3m1 symmetry allowed for hcp site

THEORY/DATA TREATMENT
Dynamical LEED: Moruzzi pot for Ru, Tong pot for O(CO); Vor, Voi, ΘD: probably same as in SSD 44.8.1

QUALITY OF EXPERIMENT-THEORY FIT
R(Pe)=0.20

2D UNIT CELLS (1 domain observed)

Cell	Ax (Å)	Ay (Å)	Bx (Å)	By (Å)	α (°)	Matrix	Pattern	Cell type
Bulk	2.706	0.000	-1.353	2.343	120.0	(1.000, 0.000) (0.000, 1.000)	(1x1)	b: bulk lattice
Surface 1	5.412	0.000	-2.706	4.687	120.0	(2.000, 0.000) (0.000, 2.000)	p(2x2)	s1: commens. superlattice

3D COORDINATES

O1: overlayer in hcp hollow sites; Ru2-5, Ru6-9: two buckled substate layers; Ru10-11: bulk hcp layers

Dx/Dy in Å, or as a fraction of layer's unit cell vectors; atom 0 at the origin. Epir/subr are bulk repeat vectors.

No. of atoms: 11

Bulk z = 2.141 Å

Region	Chem el.	At. no.	Cell type	Site occ.	Rel. to	Dx ± εx	Dy ± εy	Dz ± εz	Dz/Bz(%) ± εz/Bz
epir		-2				f	f	Å	
subr		-1				0.000 Å	0.000 Å	4.282 Å	
ovrl	O	1	s1	.25	0	0.000 f	0.000 f	0.000 ± .030 Å	0.0
ovrl	Ru	2	s1	.25	1	0.176 ± .007 f	0.351 ± .014 f	1.180 ± .030 Å	55.1
ovrl	Ru	3	s1	.25	1	-0.351 ± .007 f	-0.176 ± .014 f	1.180 ± .030 Å	55.1
ovrl	Ru	4	s1	.25	1	0.176 ± .007 f	-0.176 ± .014 f	1.180 ± .030 Å	55.1
ovrl	Ru	5	s1	.25	1	-0.333 f	0.333 f	1.250 ± .030 Å	58.4
ovrl	Ru	6	s1	.25	5	-0.167 ± .007 f	-0.335 ± .014 f	2.020 ± .030 Å	94.4
ovrl	Ru	7	s1	.25	5	0.335 ± .007 f	0.167 ± .014 f	2.020 ± .030 Å	94.4
ovrl	Ru	8	s1	.25	5	-0.167 ± .007 f	0.167 ± .014 f	2.020 ± .030 Å	94.4
ovrl	Ru	9	s1	.25	5	0.333 f	-0.333 f	2.100 ± .030 Å	98.1
subl	Ru	10	b	1.00	9	0.333 f	0.667 f	2.060 Å	96.2
subl	Ru	11	b	1.00	9	0.000 f	0.000 f	4.202 Å	196.3

Ru(0001)-p(2x2)-O
44.8.2 BOND DISTANCES AND ANGLES

Bond distances and angles are derived from coordinates

No. of distances/angles: 2

Interatomic dist. A-B (Å)	Atom A	Atom B	Atom C	Bond angle A-B-C (°)
2.025	O1	Ru2	Ru3	45.3
2.561	Ru2	Ru3	Ru4	60.0

COMMON NAME : Sc(0001)-(1x1) ILLUSTRATION: 19
CLASSIFICATION : 21.2
TECHNIQUE : LEED
AUTHORS : S. Tougaard, A. Ignatiev and D.L. Adams
REFERENCE : Surf. Sci., 115, 270 (1982)

SURFACE TYPE
Substrate : Sc
Crystal face: 0001
Temperature : RT*
Bulk lattice: hcp
2D bulk symm: p3m1
2D surf symm: p3m1
Adsorbate:
Coverage :
Pattern : (1x1)
Matrix : (1.000, 0.000)
(0.000, 1.000)

STRUCTURE TYPE
Bulk-like termination with contraction of top interlayer spacing

SAMPLE PREPARATION (1 sample)
Treatment : cleaned by Kr+ ion bombardments, annealed to 773-1173 K
Crystallinity: annealing until no changes in LEED IV's
Anal. methods:
Contamination: O impurity 5% monolayer

COMMENTS
Shallow R-factor minimum as a function of vertical displacement; possibility of some H contamination on the surface

DATA COLLECTION
Technique: LEED
Dataset : I-V data for 7 beams, 2 at normal incidence, 5 at Θ=10°; 50<E<225 eV

THEORY/DATA TREATMENT
Dynamical LEED: Moruzzi et al potential;
Vor=-9.5 eV, Voi=-3eV

STRUCTURES EXAMINED
Top layer spacing varied from 2.44 to 2.74Å in steps of 0.1Å;
hcp termination yields lowest R-factor

QUALITY OF EXPERIMENT-THEORY FIT
RZJ=0.23

2D UNIT CELLS (1 domain observed)

Cell	Ax (Å)	Ay (Å)	Bx (Å)	By (Å)	α (°)	Matrix	Pattern	Cell type
Bulk	3.310	0.000	-1.655	2.867	120.0	(1.000, 0.000) (0.000, 1.000)	(1x1)	b: bulk lattice
Surface 1	3.310	0.000	-1.655	2.867	120.0	(1.000, 0.000) (0.000, 1.000)	(1x1)	s1: commens. superlattice

3D COORDINATES

Sc2-Sc3: repeating bulk pair of layers

Dx/Dy in Å, or as a fraction of layer's unit cell vectors; atom 0 at the origin. Epir/subr are bulk repeat vectors.

No. of atoms: 3 Bulk z = 2.640 Å

Reg ion	Chem el.	At. no.	Cell type	Site occ.	Rel. to	Dx ± єx	Dy ± єy	Dz ± єz	Dz/Bz(%) ± єz/Bz
epir		-2				f	f	Å	
subr		-1				0.000 Å	0.000 Å	5.280 Å	
intf	Sc	1	b	1.00	0	0.000 f	0.000 f	0.000 Å	0.0
subl	Sc	2	b	1.00	1	0.333 f	0.667 f	2.590 ± .020 Å	98.1 ± .8
subl	Sc	3	b	1.00	2	-0.333 f	-0.667 f	2.640 Å	100.0

BOND DISTANCES AND ANGLES

Bond distances and angles are derived from coordinates

No. of distances/angles: 7

Interatomic dist. A-B (Å)	Atom A	Atom B	Atom C	Bond angle A-B-C (°)
3.310	Sc1	Sc1(1,1)	Sc2	59.1
3.219	Sc1	Sc2	Sc1(0,1)	61.9
3.219	Sc1	Sc2	Sc2(0,-1)	59.1
3.219	Sc1	Sc2	Sc3	107.7
3.310	Sc2	Sc2(1,1)	Sc3(1,1)	59.5

Sc(0001)-(1x1)
21.2

Bond Distances and Angles - Continued

Interatomic dist. A-B (Å)	Atom A	Atom B	Atom C	Bond angle A-B-C (°)
3.259	Sc2	Sc3	Sc2(0,-1)	61.0
3.259	Sc2	Sc3	Sc3(0,1)	59.5

COMMON NAME : Si(100)-(2x1)
CLASSIFICATION : 14.170
TECHNIQUE : GIXD
AUTHORS : N. Jedrecy, M. Sauvage-Simkin, R. Pinchaux, J. Massies, N. Greiser and V. H. Etgens
REFERENCE : Surf. Sci., 230, 197 (1990)

ILLUSTRATION: 94

SURFACE TYPE
Substrate : Si
Crystal face: 100
Temperature : RT
Bulk lattice: diamond
2D bulk symm: pmm
2D surf symm: pm

Adsorbate:
Coverage :
Pattern : (2x1)
Matrix : (1.000, 0.000)
(0.000, 2.000)

STRUCTURE TYPE
Clean surface with buckled dimers

SAMPLE PREPARATION (1 sample)
Treatment : Si cut within 0.2° of (001), Shiraki etched, annealed
Crystallinity: (2x1) LEED pattern
Anal. methods:
Contamination:

COMMENTS

DATA COLLECTION
Technique: GIXD
Dataset : x-ray diffraction peaks at grazing incidence (3.8E-3 rad); wavelength 1.488Å

THEORY/DATA TREATMENT
X-ray diffraction

STRUCTURES EXAMINED
Symmetric dimer model; buckled dimer model; relaxations in 1st 2 layers

QUALITY OF EXPERIMENT-THEORY FIT
Chi^2 (2 models) = 5.10, 1.79

2D UNIT CELLS (2 domains observed)

Cell	Ax (Å)	Ay (Å)	Bx (Å)	By (Å)	α (°)	Matrix	Pattern	Cell type
Bulk	3.840	0.000	0.000	3.840	90.0	(1.000, 0.000) (0.000, 1.000)	(1x1)	b: bulk lattice
Surface 1	3.840	0.000	0.000	7.680	90.0	(1.000, 0.000) (0.000, 2.000)	(2x1)	s1: commens. superlattice

3D COORDINATES

Si1-Si2: buckled dimer; Si3-Si4: relaxed 2nd layer;
0.1Å error bars assumed for tabulation

Dx/Dy in Å, or as a fraction of layer's unit cell vectors; atom 0 at the origin. Epir/subr are bulk repeat vectors.

No. of atoms: 6

Bulk z = 1.358 Å

Region	Chem el.	At. no.	Cell type	Site occ.	Rel. to	Dx ± εx		Dy ± εy		Dz ± εz		Dz/Bz(%) ± εz/Bz
epir		-2					f		f		Å	
subr		-1				-1.920	Å	1.920	Å	2.715	Å	
intf	Si	1	s1	.50	0	0.000	f	0.694 ± .013	f	0.000 ± .100	Å	0.0 ± 7.4
intf	Si	2	s1	.50	1	0.000	f	-0.300 ± .013	f	0.310 ± .100	Å	22.8 ± 7.4
intf	Si	3	s1	.50	2	0.500	f	0.354 ± .013	f	0.990 ± .100	Å	72.9 ± 7.4
intf	Si	4	s1	.50	3	0.000	f	-0.474 ± .013	f	0.000 ± .100	Å	0.0 ± 7.4
subl	Si	5	b	1.00	4	0.000	f	-0.550	f	1.360	Å	100.2
subl	Si	6	b	1.00	5	-0.500	f	0.000	f	1.358	Å	100.0

Si(100)-(2x1)
14.170 BOND DISTANCES AND ANGLES

Bond distances and angles are derived from coordinates

No. of distances/angles: 8

Interatomic dist. A-B (Å)	Atom A	Atom B	Atom C	Bond angle A-B-C (°)
2.321	Si1	Si2	Si4	116.4
2.356	Si1	Si3	Si5(0,2)	117.6
2.356	Si1	Si3	Si5(0,1)	100.1
2.321	Si2	Si1	Si3	95.9
2.347	Si2	Si4	Si5(0,1)	87.3
2.347	Si2	Si4	Si5	123.9
2.347	Si2	Si4(-1,0)	Si2(-1,0)	109.8
2.356	Si3	Si1(1,0)	Si3(1,0)	109.2

COMMON NAME : Si(100)-(2x1) ILLUSTRATION: 94
CLASSIFICATION : 14.182a
TECHNIQUE : KLEED
AUTHORS : R.G. Zhao, Jinfeng Jia, Yanfang Li and W. S. Yang
REFERENCE : Springer Series in Surface Sciences, 24, 517 (1991)

SURFACE TYPE
Substrate : Si
Crystal face: 100
Temperature : RT
Bulk lattice: diamond
2D bulk symm: pmm
2D surf symm: pm

Adsorbate:
Coverage :
Pattern : (2x1)
Matrix : (2.000, 0.000)
(0.000, 1.000)

STRUCTURE TYPE
2 coexisting phases: 75% (2x1) and 25% c(4x2);
(2x1) has weakly buckled dimers;
c(4x2) has strongly buckled dimers: see structure 14.182b

SAMPLE PREPARATION (1 sample)
Treatment :
Crystallinity: (2x1) LEED pattern
Anal. methods:
Contamination:

COMMENTS

DATA COLLECTION
Technique: KLEED
Dataset : constant-momentum-transfer averaging over E-range 30-250 eV

THEORY/DATA TREATMENT
Kinematic LEED: Vor=-12.5 eV (fit); mfp=4Å

STRUCTURES EXAMINED
(2x1) buckled dimers; c(4x2) buckled dimers;
75%/25% mixture of (2x1) and c(4x2); x and z relaxations down to 6th Si layer

QUALITY OF EXPERIMENT-THEORY FIT
RVHT=0.22 (incl. c(4x2))

2D UNIT CELLS (2 domains observed)

Cell	Ax (Å)	Ay (Å)	Bx (Å)	By (Å)	α (°)	Matrix	Pattern	Cell type
Bulk	3.840	0.000	0.000	3.840	90.0	(1.000, 0.000) (0.000, 1.000)	(1x1)	b: bulk lattice
Surface 1	7.679	0.000	0.000	3.840	90.0	(2.000, 0.000) (0.000, 1.000)	(2x1)	s1: commens. superlattice

3D COORDINATES

Si1-Si2: weakly buckled dimer; Si3-Si12: relaxed layers;
Si14-Si15: form periodically repeating pair of bulk layers; 0.1Å error bars assumed for tabulation

Dx/Dy in Å, or as a fraction of layer's unit cell vectors; atom 0 at the origin. Epir/subr are bulk repeat vectors.

No. of atoms: 15 Bulk z = 1.358 Å

Region	Chem el.	At. no.	Cell type	Site occ.	Rel. to	Dx ± εx		Dy ± εy		Dz ± εz		Dz/Bz(%) ± εz/Bz
epir		-2					f		f		Å	
subr		-1				1.920	Å	1.920	Å	2.715	Å	
intf	Si	1	s1	.50	0	0.415 ± .013	f	0.000	f	0.000 ± .100	Å	0.0 ± 7.4
intf	Si	2	s1	.50	1	-0.310 ± .013	f	0.000	f	0.140 ± .100	Å	10.3 ± 7.4
intf	Si	3	s1	.50	2	0.375 ± .013	f	0.500	f	1.188 ± .100	Å	87.5 ± 7.4
intf	Si	4	s1	.50	3	-0.455 ± .013	f	0.000	f	0.050 ± .100	Å	3.7 ± 7.4
intf	Si	5	s1	.50	4	0.725 ± .013	f	0.000	f	1.158 ± .100	Å	85.3 ± 7.4
intf	Si	6	s1	.50	5	-0.500 ± .013	f	0.000	f	0.380 ± .100	Å	28.0 ± 7.4
intf	Si	7	s1	.50	6	0.500 ± .013	f	-0.500	f	1.008 ± .100	Å	74.2 ± 7.4
intf	Si	8	s1	.50	7	-0.500 ± .013	f	0.000	f	0.400 ± .100	Å	29.5 ± 7.4
intf	Si	9	s1	.50	8	0.735 ± .013	f	0.000	f	1.118 ± .100	Å	82.3 ± 7.4
intf	Si	10	s1	.50	9	-0.470 ± .013	f	0.000	f	0.000 ± .100	Å	0.0 ± 7.4
intf	Si	11	s1	.50	10	-0.005 ± .013	f	0.500	f	1.348 ± .100	Å	99.3 ± 7.4
intf	Si	12	s1	.50	11	0.480 ± .013	f	0.000	f	0.000 ± .100	Å	0.0 ± 7.4
intf	Si	13	b	1.00	12	-1.480 ± .026	f	0.000	f	1.338 ± .100	Å	98.5 ± 7.4
subl	Si	14	b	1.00	13	0.000	f	-0.500	f	1.398 ± .100	Å	103.0 ± 7.4
subl	Si	15	b	1.00	14	0.500	f	0.000	f	1.358	Å	100.0 ± 7.4

Si(100)-(2x1)
14.182a

BOND DISTANCES AND ANGLES

Bond distances and angles are derived from coordinates

No. of distances/angles: 15

Interatomic dist. A-B (Å)	Atom A	Atom B	Atom C	Bond angle A-B-C (°)
2.385	Si1	Si2	Si4	106.9
2.399	Si3	Si5	Si7	107.1
2.375	Si3	Si6	Si4	96.4
2.375	Si3	Si6	Si8	113.3
2.408	Si4	Si5(-1,0)	Si3(-1,0)	121.1
2.408	Si4	Si5(-1,0)	Si7(-1,0)	106.4
2.313	Si4	Si6	Si8	113.1
2.387	Si1	Si3	Si1(0,1)	107.1
2.387	Si1	Si3	Si5	117.4
2.387	Si1	Si3	Si6	102.5
2.385	Si2	Si1	Si3	100.1
2.365	Si2	Si4	Si2(0,1)	108.5
2.365	Si2	Si4	Si5(-1,0)	118.6
2.365	Si2	Si4	Si6	98.9
2.399	Si3	Si5	Si4(1,0)	121.1

ILLUSTRATION: 94

COMMON NAME : Si(100)-(2x1)
CLASSIFICATION : 14.75
TECHNIQUE : MEIS
AUTHORS : R.M. Tromp, R.G. Smeenk, F.W. Saris and D.J. Chadi
REFERENCE : Surf. Sci., 133, 137 (1983)

SURFACE TYPE
Substrate : Si
Crystal face: 100
Temperature : RT
Bulk lattice: diamond
2D bulk symm: pmm
2D surf symm: pm

Adsorbate:
Coverage :
Pattern : (2x1)
Matrix : (2.000, 0.000)
(0.000, 1.000)

STRUCTURE TYPE
Buckled dimer with multilayer relaxations

SAMPLE PREPARATION (1 sample)
Treatment : cleaned by direct current heating to 1250C for 30s
Crystallinity: sharp (2x1) LEED patterns down to 10eV
Anal. methods:
Contamination: monitored by AES and ISS

COMMENTS
Authors conclude that although (2x1) buckled dimer yields best agreement with MEIS data, the (2x1), (2x2) and c(4x2) geometries probably coexist on the clean Si(100) surface; energy minimisation calculations used for all coordinates (Hellman-Feynman forces in tight binding calculations)

DATA COLLECTION
Technique: MEIS
Dataset : MEIS channeling and blocking, with 50, 100 and 150 k eV protons in 3 scattering geometries

THEORY/DATA TREATMENT
Monte Carlo simulations of surface blocking minima and total energy minimisation calculations; vibs=0.14Å top 2 layers

STRUCTURES EXAMINED
(2x1) buckled dimer models of Appelbaum/Hamann (Surf. Sci. 74 21(1978)), Chadi (Phys. Rev. Lett. 43 43(1979)) and Yin/Cohen (Phys. Rev. B24 2303 (1981)); dimer models with (2x1), p(2x2), c(2x2) and c(4x2) symmetries calculated by total E minimisation of TB model (Chadi, JVST 16 1290(1979)

QUALITY OF EXPERIMENT-THEORY FIT
Visual

2D UNIT CELLS (2 domains observed)

Cell	Ax (Å)	Ay (Å)	Bx (Å)	By (Å)	α (°)	Matrix	Pattern	Cell type
Bulk	3.840	0.000	0.000	3.840	90.0	(1.000, 0.000) (0.000, 1.000)	(1x1)	b: bulk lattice
Surface 1	7.680	0.000	0.000	3.840	90.0	(2.000, 0.000) (0.000, 1.000)	(2x1)	s1: commens. superlattice

3D COORDINATES

Si1-Si2: tilted dimerized top layer; Si3-12: relaxed subsurface layers;
Si13-14: bulk pair of repeating layers; 0.1Å error bars assumed for tabulation (not determined)

Dx/Dy in Å, or as a fraction of layer's unit cell vectors; atom 0 at the origin. Epir/subr are bulk repeat vectors.

No. of atoms: 14 Bulk z = 1.358 Å

Region	Chem el.	At. no.	Cell type	Site occ.	Rel. to	Dx ± εx		Dy ± εy		Dz ± εz		Dz/Bz(%) ± εz/Bz
epir		-2					f		f		Å	
subr		-1				1.920	Å	1.920	Å	2.715	Å	
intf	Si	1	s1	.50	0	0.000	f	0.000	f	0.000	Å	0.0
intf	Si	2	s1	.50	1	0.299 ± .013	f	0.000 ± .026	f	0.559 ± .100	Å	41.2 ± 7.4
intf	Si	3	s1	.50	2	0.651 ± .013	f	0.500 ± .026	f	0.824 ± .100	Å	60.7 ± 7.4
intf	Si	4	s1	.50	3	-0.524 ± .013	f	0.000 ± .026	f	0.086 ± .100	Å	6.3 ± 7.4
intf	Si	5	s1	.50	4	0.258 ± .013	f	0.000 ± .026	f	1.215 ± .100	Å	89.5 ± 7.4
intf	Si	6	s1	.50	5	-0.499 ± .013	f	0.000 ± .026	f	0.238 ± .100	Å	17.5 ± 7.4
intf	Si	7	s1	.50	6	0.502 ± .013	f	-0.500 ± .026	f	1.156 ± .100	Å	85.1 ± 7.4
intf	Si	8	s1	.50	7	-0.500 ± .013	f	0.000 ± .026	f	0.184 ± .100	Å	13.6 ± 7.4
intf	Si	9	s1	.50	8	0.256 ± .013	f	0.000 ± .026	f	1.261 ± .100	Å	92.9 ± 7.4
intf	Si	10	s1	.50	9	0.488 ± .013	f	0.000 ± .026	f	0.006 ± .100	Å	.4 ± 7.4
intf	Si	11	s1	.50	10	-0.491 ± .013	f	0.500 ± .026	f	1.362 ± .100	Å	100.3 ± 7.4
intf	Si	12	s1	.50	11	0.495 ± .013	f	0.000 ± .026	f	0.003 ± .100	Å	.2 ± 7.4
subl	Si	13	b	1.00	12	-0.500	f	0.000	f	1.358	Å	100.0
subl	Si	14	b	1.00	13	0.000	f	-0.500	f	1.358	Å	100.0

Si(100)-(2x1)
14.75

BOND DISTANCES AND ANGLES

Bond distances and angles are derived from coordinates

No. of distances/angles: 12

Interatomic dist. A-B (Å)	Atom A	Atom B	Atom C	Bond angle A-B-C (°)
2.363	Si1	Si2	Si1(1,0)	146.4
2.338	Si4	Si2	Si1	119.8
2.338	Si4	Si2	Si3(-1,-1)	108.9
2.338	Si4	Si2	Si4(0,-1)	110.4
2.363	Si1	Si2	Si3(-1,0)	123.6
2.363	Si1	Si2	Si4	119.8
2.397	Si1	Si3(-1,0)	Si1(0,1)	106.4
2.397	Si1	Si3(-1,0)	Si2(-1,1)	121.5
2.397	Si1	Si3(-1,0)	Si4(-1,0)	112.9
2.338	Si2	Si4	Si1(1,1)	120.0
2.338	Si2	Si4	Si2(0,1)	110.4
2.338	Si2	Si4	Si3(-1,0)	102.3

COMMON NAME : Si(100)-(2x1)
CLASSIFICATION : 14.85
TECHNIQUE : LEED
AUTHORS : B.W. Holland, C.B. Duke and A. Paton
REFERENCE : Surf. Sci., 140, L269 (1984)

ILLUSTRATION: 94

SURFACE TYPE
Substrate : Si
Crystal face: 100
Temperature : RT*
Bulk lattice: diamond
2D bulk symm: pmm
2D surf symm: pm

Adsorbate:
Coverage :
Pattern : (2x1)
Matrix : (2.000, 0.000)
(0.000, 1.000)

STRUCTURE TYPE
Buckled dimer with multilayer relaxations

SAMPLE PREPARATION (2 sample)
Treatment :
Crystallinity:
Anal. methods:
Contamination:

COMMENTS
The data of Yang, Jona and Marcus previously thought to favour a twisted dimer model were here shown to be more compatible with the models of Chadi, and Yin and Cohen

DATA COLLECTION
Technique: LEED
Dataset : LEED I-V curves measured by (1) Yang et al, Phys. Rev. B28, 2049(1983) and (2) Ignatiev et al, J. Phys. C10, 1109(1977)

THEORY/DATA TREATMENT
Dynamical LEED: 8-layer slab; mfp=8Å

STRUCTURES EXAMINED
Starting from average between Chadi and Yin-Cohen models and keeping 1 mirror plane: 2D displacements of dimer atoms; displacements normal to the surface of the atoms in the 2nd and 3rd layers

QUALITY OF EXPERIMENT-THEORY FIT
RX=0.13

2D UNIT CELLS (2 domains observed)

Cell	Ax (Å)	Ay (Å)	Bx (Å)	By (Å)	α (°)	Matrix	Pattern	Cell type
Bulk	3.840	0.000	0.000	3.840	90.0	(1.000, 0.000) (0.000, 1.000)	(1x1)	b: bulk lattice
Surface 1	7.680	0.000	0.000	3.840	90.0	(2.000, 0.000) (0.000, 1.000)	(2x1)	s1: commens. superlattice

3D COORDINATES

Si1-Si2: tilted dimerized top layer; Si3-8: relaxed subsurface layers;
Si9-10: periodically repeating pair of bulk layers

Dx/Dy in Å, or as a fraction of layer's unit cell vectors; atom 0 at the origin. Epir/subr are bulk repeat vectors.

No. of atoms: 10 Bulk z = 1.358 Å

Region	Chem el.	At. no.	Cell type	Site occ.	Rel. to	Dx ± εx	Dy ± εy	Dz ± εz	Dz/Bz(%) ± εz/Bz
epir		-2				f	f	Å	
subr		-1				1.920 Å	1.920 Å	2.715 Å	
intf	Si	1	s1	.50	0	0.000 f	0.000 f	0.000 Å	0.0
intf	Si	2	s1	.50	1	0.318 f	0.000 f	0.364 ± .200 Å	26.8 ± 14.7
intf	Si	3	s1	.50	2	0.103 f	0.500 f	0.691 Å	50.9
intf	Si	4	s1	.50	3	0.526 f	0.000 f	0.077 Å	5.7
intf	Si	5	s1	.50	4	-0.262 f	0.000 f	1.207 Å	88.9
intf	Si	6	s1	.50	5	-0.502 f	0.000 f	0.277 Å	20.4
intf	Si	7	s1	.50	6	0.498 f	-0.500 f	1.114 Å	82.1
intf	Si	8	s1	.50	7	-0.493 f	0.000 f	0.212 Å	15.6
subl	Si	9	b	1.00	8	-0.500 f	0.000 f	1.358 Å	100.0
subl	Si	10	b	1.00	9	0.000 f	-0.500 f	1.358 Å	100.0

Si(100)-(2x1)
14.85 BOND DISTANCES AND ANGLES

Bond distances and angles are derived from coordinates

No. of distances/angles: 12

Interatomic dist. A-B (Å)	Atom A	Atom B	Atom C	Bond angle A-B-C (°)
2.469	Si1	Si2	Si1(1,0)	156.9
2.189	Si2	Si3	Si4(-1,0)	116.8
2.189	Si3	Si2	Si3(0,-1)	122.6
2.189	Si3	Si2	Si4(-1,-1)	118.2
2.469	Si1	Si2	Si3	113.8
2.469	Si1	Si2	Si4(-1,0)	118.5
2.266	Si1	Si4(-1,0)	Si1(0,1)	115.9
2.266	Si1	Si4(-1,0)	Si2(-1,1)	128.3
2.469	Si2	Si1	Si2(-1,0)	156.9
2.469	Si2	Si1	Si3(-1,0)	101.3
2.189	Si2	Si3	Si1(1,1)	133.7
2.189	Si2	Si3	Si2(0,1)	122.6

ILLUSTRATION: 95

COMMON NAME : Si(100)-c(4x2)
CLASSIFICATION : 14.182b
TECHNIQUE : KLEED
AUTHORS : R.G. Zhao, Jinfeng Jia, Yanfang Li and W. S. Yang
REFERENCE : Springer Series in Surface Sciences, 24, 517 (1991)

SURFACE TYPE
Substrate : Si
Crystal face: 100
Temperature : RT
Bulk lattice: diamond
2D bulk symm: pmm
2D surf symm: cm

Adsorbate:
Coverage :
Pattern : c(4x2)
Matrix : (4.000, 0.000)
(2.000, 1.000)

STRUCTURE TYPE
2 coexisting phases, 75% (2x1) and 25% c(4x2);
c(4x2) has strongly buckled dimers;
(2x1) has weakly buckled dimers: see structure 14.182a

SAMPLE PREPARATION (1 sample)
Treatment :
Crystallinity: (2x1) LEED pattern
Anal. methods:
Contamination:

COMMENTS

DATA COLLECTION
Technique: KLEED
Dataset : constant-momentum-transfer averaging over E-range 30-250 eV

THEORY/DATA TREATMENT
Kinematic LEED: Vor=-12.5 eV (fit); mfp=4Å

STRUCTURES EXAMINED
(2x1) buckled dimers; c(4x2) buckled dimers, pairwise symmetrical;
75%/25% mixture of (2x1) and c(4x2); x and z relaxations down to 6th Si layer

QUALITY OF EXPERIMENT-THEORY FIT
RVHT=0.22 (incl. (2x1))

2D UNIT CELLS (1 domain observed)

Cell	Ax (Å)	Ay (Å)	Bx (Å)	By (Å)	α (°)	Matrix	Pattern	Cell type
Bulk	3.840	0.000	0.000	3.840	90.0	(1.000, 0.000) (0.000, 1.000)	(1x1)	b: bulk lattice
Surface 1	15.358	0.000	7.679	3.840	21.6	(4.000, 0.000) (2.000, 1.000)	c(4x2)	s1: commens. superlattice

3D COORDINATES

Si1-Si4: 2 strongly buckled dimers; Si5-Si24: relaxed layers;
Si26-Si27: form periodically repeating pair of bulk layers; 0.1Å error bars assumed for tabulation

Dx/Dy in Å, or as a fraction of layer's unit cell vectors; atom 0 at the origin. Epir/subr are bulk repeat vectors.

No. of atoms: 27

Bulk z = 1.358 Å

Region	Chem el.	At. no.	Cell type	Site occ.	Rel. to	Dx ± εx		Dy ± εy		Dz ± εz		Dz/Bz(%) ± εz/Bz
epir		-2					f		f		Å	
subr		-1				1.920	Å	1.920	Å	2.715	Å	
intf	Si	1	s1	.25	0	0.000 ± .007	f	0.000	f	0.000 ± .100	Å	0.0 ± 7.4
intf	Si	2	s1	.25	1	0.313 ± .007	f	0.000	f	0.000 ± .100	Å	0.0 ± 7.4
intf	Si	3	s1	.25	2	0.542 ± .007	f	0.000	f	0.560 ± .100	Å	41.3 ± 7.4
intf	Si	4	s1	.25	3	-0.396 ± .007	f	0.000	f	0.000 ± .100	Å	0.0 ± 7.4
intf	Si	5	s1	.25	4	0.311 ± .007	f	0.500	f	0.818 ± .100	Å	60.2 ± 7.4
intf	Si	6	s1	.25	5	-0.725 ± .007	f	0.000	f	0.000 ± .100	Å	0.0 ± 7.4
intf	Si	7	s1	.25	6	0.500 ± .007	f	0.000	f	0.000 ± .100	Å	0.0 ± 7.4
intf	Si	8	s1	.25	7	-0.275 ± .007	f	0.000	f	0.000 ± .100	Å	0.0 ± 7.4
intf	Si	9	s1	.25	8	0.637 ± .007	f	0.000	f	1.288 ± .100	Å	94.8 ± 7.4
intf	Si	10	s1	.25	9	-0.500 ± .007	f	0.000	f	0.000 ± .100	Å	0.0 ± 7.4
intf	Si	11	s1	.25	10	0.250 ± .007	f	0.000	f	0.460 ± .100	Å	33.9 ± 7.4
intf	Si	12	s1	.25	11	-0.500 ± .007	f	0.000	f	0.000 ± .100	Å	0.0 ± 7.4
intf	Si	13	s1	.25	12	0.000 ± .007	f	-0.500	f	1.168 ± .100	Å	86.0 ± 7.4
intf	Si	14	s1	.25	13	0.500 ± .007	f	0.000	f	0.000 ± .100	Å	0.0 ± 7.4
intf	Si	15	s1	.25	14	0.250 ± .007	f	0.000	f	0.360 ± .100	Å	26.5 ± 7.4
intf	Si	16	s1	.25	15	-0.500 ± .007	f	0.000	f	0.000 ± .100	Å	0.0 ± 7.4
intf	Si	17	s1	.25	16	-0.368 ± .007	f	0.000	f	1.148 ± .100	Å	84.5 ± 7.4
intf	Si	18	s1	.25	17	0.236 ± .007	f	0.000	f	0.000 ± .100	Å	0.0 ± 7.4
intf	Si	19	s1	.25	18	0.500 ± .007	f	0.000	f	0.000 ± .100	Å	0.0 ± 7.4
intf	Si	20	s1	.25	19	-0.236 ± .007	f	0.000	f	0.000 ± .100	Å	0.0 ± 7.4
intf	Si	21	s1	.25	20	0.246 ± .007	f	0.500	f	1.368 ± .100	Å	100.7 ± 7.4

Si(100)-c(4x2)
14.182b

3D Coordinates - Continued

Reg ion	Chem el.	At. no.	Cell type	Site occ.	Rel. to	Dx ± εx	Dy ± εy	Dz ± εz	Dz/Bz(%) ± εz/Bz
intf	Si	22	s1	.25	21	-0.757 ± .007 f	0.000 f	0.000 ± .100 Å	0.0 ± 7.4
intf	Si	23	s1	.25	22	0.500 ± .007 f	0.000 f	0.000 ± .100 Å	0.0 ± 7.4
intf	Si	24	s1	.25	23	-0.243 ± .007 f	0.000 f	0.000 ± .100 Å	0.0 ± 7.4
intf	Si	25	b	1.00	24	-0.513 ± .026 f	0.000 f	1.378 ± .100 Å	101.5 ± 7.4
subl	Si	26	b	1.00	25	0.000 ± .026 f	-0.500 f	1.398 ± .100 Å	103.0 ± 7.4
subl	Si	27	b	1.00	26	-0.500 ± .026 f	0.000 f	1.358 ± .100 Å	100.0 ± 7.4

BOND DISTANCES AND ANGLES

Bond distances and angles are derived from coordinates

No. of distances/angles: 6

Interatomic dist. A-B (Å)	Atom A	Atom B	Atom C	Bond angle A-B-C (°)
2.308	Si1	Si3(-1,0)	Si6(0,-1)	118.8
2.381	Si1	Si5(-1,0)	Si4(-1,1)	114.9
2.381	Si1	Si5(-1,0)	Si9(-1,0)	113.9
2.284	Si3	Si6(1,-1)	Si2(1,-1)	114.9
2.472	Si5	Si9	Si6(1,0)	117.2
2.459	Si5	Si11		

ILLUSTRATION: 89

COMMON NAME : Si(111)-(2x1)
CLASSIFICATION : 14.120
TECHNIQUE : LEED
AUTHORS : H. Sakama, A. Kawazu and K. Ueda
REFERENCE : Phys. Rev., B34, 1367 (1986)

SURFACE TYPE
Substrate : Si
Crystal face: 111
Temperature : RT*
Bulk lattice: diamond
2D bulk symm: p3m1
2D surf symm: pm

Adsorbate:
Coverage :
Pattern : (2x1)
Matrix : (1.000, 1.000)
(-1.000, 1.000)

STRUCTURE TYPE
Tilted π-bonded chain model with relaxations down to 4th bilayer

SAMPLE PREPARATION (1 sample)
Treatment :
Crystallinity:
Anal. methods:
Contamination:

COMMENTS

DATA COLLECTION
Technique: LEED
Dataset : I-V spectra at 3 incidence angles; 40<E<160 eV; 13 integral- and fractional-order beams

THEORY/DATA TREATMENT
Dynamical LEED

STRUCTURES EXAMINED
Variations in 7 geometrical parameters concerning atomic positions down to 4th atomic layer

QUALITY OF EXPERIMENT-THEORY FIT
RZJ=0.40

2D UNIT CELLS (3 domains observed)

Cell	Ax (Å)	Ay (Å)	Bx (Å)	By (Å)	α (°)	Matrix	Pattern	Cell type
Bulk	3.838	0.000	1.919	3.324	60.0	(1.000, 0.000) (0.000, 1.000)	(1x1)	b: bulk lattice
Surface 1	5.757	3.324	-1.919	3.324	90.0	(1.000, 1.000) (-1.000, 1.000)	(2x1)	s1: commens. superlattice

3D COORDINATES

Si1-Si2: tilted π-bonded chain; Si3-Si4, Si5-Si6, Si7-Si8: relaxed bilayers;
Si9-Si10: repeating bulk bilayer; 0.1Å error bars assumed for tabulation

Dx/Dy in Å, or as a fraction of layer's unit cell vectors; atom 0 at the origin. Epir/subr are bulk repeat vectors.

No. of atoms: 10

Bulk z = 3.060 Å

Region	Chem el.	At. no.	Cell type	Site occ.	Rel. to	Dx ± εx		Dy ± εy		Dz ± εz		Dz/Bz(%) ± εz/Bz
epir		-2					f		f		Å	
subr		-1				1.877	Å	1.084	Å	3.060	Å	
intf	Si	1	s1	.50	0	0.000	f	0.000	f	0.000	Å	0.0
intf	Si	2	s1	.50	1	0.168	f	0.500	f	0.350 ± .100	Å	11.4 ± 3.3
intf	Si	3	s1	.50	2	0.330	f	0.000	f	0.860 ± .100	Å	28.1 ± 3.3
intf	Si	4	s1	.50	3	0.204	f	-0.500	f	0.050 ± .100	Å	1.6 ± 3.3
intf	Si	5	s1	.50	4	-0.371	f	0.500	f	2.050 ± .100	Å	67.0 ± 3.3
intf	Si	6	s1	.50	5	0.479	f	-0.500	f	0.150 ± .100	Å	4.9 ± 3.3
intf	Si	7	s1	.50	6	-0.656	f	0.000	f	0.530 ± .100	Å	17.3 ± 3.3
intf	Si	8	s1	.50	7	0.496	f	0.500	f	0.320 ± .100	Å	10.5 ± 3.3
subl	Si	9	b	1.00	8	-0.007	f	-1.007	f	2.280 ± .100	Å	74.5 ± 3.3
subl	Si	10	b	1.00	9	0.333	f	0.333	f	0.780	Å	25.5 ± 3.3

Si(111)-(2x1)
14.120 BOND DISTANCES AND ANGLES

Bond distances and angles are derived from coordinates

No. of distances/angles: 4

Interatomic dist. A-B (Å)	Atom A	Atom B	Atom C	Bond angle A-B-C (°)
2.248	Si1	Si2	Si3	121.3
2.356	Si2	Si3	Si4	123.0
2.350	Si3	Si4	Si6	101.5
2.315	Si4	Si6		

ILLUSTRATION: -

COMMON NAME : Si(111)-(2x1)
CLASSIFICATION : 14.25
TECHNIQUE : LEED
AUTHORS : R. Feder, W. Monch and P.P. Auer
REFERENCE : J. Phys., C12, L179 (1979)

SURFACE TYPE
Substrate : Si
Crystal face: 111
Temperature : RT*
Bulk lattice: diamond
2D bulk symm: p3m1
2D surf symm: pm

Adsorbate:
Coverage :
Pattern : (2x1)
Matrix : (1.000, 1.000)
(-1.000, 1.000)

STRUCTURE TYPE
Buckled top layer

SAMPLE PREPARATION (1 sample)
Treatment : cleaved p-type crystal
Crystallinity: single domain orientation chosen
Anal. methods:
Contamination:

COMMENTS
This model is no longer accepted as correct

DATA COLLECTION
Technique: LEED
Dataset : I-V curves for 11 beams at normal and near-normal incidence; energy range 25-150 eV

THEORY/DATA TREATMENT
Dynamical LEED (layer KKR); at. pot. from at. charge density superpos. (7 ph. sh.); Voi=3 eV; ΘD=625 K(bulk), 300K(surf)

STRUCTURES EXAMINED
1) pairing models; 2) buckling models with variable buckling of first two planes of atoms, variable first-to-second and second-to-third interplanar distances and variable lateral shift in second atomic plane

QUALITY OF EXPERIMENT-THEORY FIT
Visual

2D UNIT CELLS (3 domains observed)

Cell	Ax (Å)	Ay (Å)	Bx (Å)	By (Å)	α (°)	Matrix	Pattern	Cell type
Bulk	3.838	0.000	1.919	3.324	60.0	(1.000, 0.000) (0.000, 1.000)	(1x1)	b: bulk lattice
Surface 1	5.757	3.324	-1.919	3.324	90.0	(1.000, 1.000) (-1.000, 1.000)	(2x1)	s1: commens. superlattice

3D COORDINATES

Si1-Si2: buckled top half bilayer; buckling of 2x0.08Å; Si3-Si4: next half bilayer, with lateral shifts of 0.10Å; Si7-Si8: periodically repeating bulk bilayer

Dx/Dy in Å, or as a fraction of layer's unit cell vectors; atom 0 at the origin. Epir/subr are bulk repeat vectors.

No. of atoms: 8 Bulk z = 3.130 Å

Reg ion	Chem el.	At. no.	Cell type	Site occ.	Rel. to	Dx ± εx		Dy ± εy		Dz ± εz		Dz/Bz(%) ± εz/Bz
epir		-2					f		f		Å	
subr		-1				1.920	Å	1.108	Å	3.130	Å	
intf	Si	1	s1	.50	0	0.000	f	0.000	f	0.000	Å	0.0
intf	Si	2	s1	.50	1	0.500	f	0.500	f	0.300 ± .050	Å	9.6 ± 1.6
intf	Si	3	s1	.50	2	-0.352	f	0.000	f	0.700 ± .050	Å	22.4 ± 1.6
intf	Si	4	s1	.50	3	0.537	f	-0.500	f	0.000 ± .040	Å	0.0 ± 1.3
intf	Si	5	b	1.00	4	-1.018	f	-0.018	f	2.270 ± .020	Å	72.5 ± .6
intf	Si	6	b	1.00	5	0.667	f	-0.333	f	0.780	Å	24.9
subl	Si	7	b	1.00	6	0.000	f	0.000	f	2.348	Å	75.0
subl	Si	8	b	1.00	7	0.667	f	-0.333	f	0.783	Å	25.0

Si(111)-(2x1)
14.25 BOND DISTANCES AND ANGLES

Bond distances and angles are derived from coordinates

No. of distances/angles: 10

Interatomic dist. A-B (Å)	Atom A	Atom B	Atom C	Bond angle A-B-C (°)
2.377	Si1	Si3	Si1(0,1)	102.4
2.443	Si3	Si2	Si4	114.2
2.377	Si1	Si3	Si2	106.0
2.377	Si1	Si3	Si4(-1,0)	114.5
2.321	Si1	Si4(-1,0)	Si1(-1,0)	105.8
2.321	Si1	Si4(-1,0)	Si2(0,-2)	109.8
2.443	Si2	Si3	Si1	106.0
2.443	Si2	Si3	Si2(-1,0)	109.1
2.384	Si2	Si4	Si1(1,0)	104.3
2.384	Si2	Si4	Si2(1,0)	113.3

ILLUSTRATION: 89

COMMON NAME : Si(111)-(2x1)
CLASSIFICATION : 14.89
TECHNIQUE : LEED
AUTHORS : F.J. Himpsel, P.M. Marcus, R. Tromp, I.P. Batra, M.R. Cook, F. Jona and H. Liu
REFERENCE : Phys. Rev., B30, 2257 (1984)

SURFACE TYPE
Substrate : Si
Crystal face: 111
Temperature : RT
Bulk lattice: diamond
2D bulk symm: p3m1
2D surf symm: pm

Adsorbate:
Coverage :
Pattern : (2x1)
Matrix : (1.000, 1.000)
(-1.000, 1.000)

STRUCTURE TYPE
Tilted π-bonded chain model with overall compression

SAMPLE PREPARATION (1 sample)
Treatment :
Crystallinity:
Anal. methods:
Contamination:

COMMENTS

DATA COLLECTION
Technique: LEED
Dataset : I-V curves for 12 beams at normal incidence

THEORY/DATA TREATMENT
Dynamical LEED with Keating-type strain energy minimisation; Vor=-10 eV,Voi=-3.5 eV; rms 0.3Å outermost chain 0.1Å rest

STRUCTURES EXAMINED
Chain structures of Pandey and Northrup and Cohen; the 8 z-coords of all atoms down to the 4th layer were optimised; 5th and 6th layers and all coords parallel to mirror plane determined by Keating like strain energy minimisation; one mirror plane maintained

QUALITY OF EXPERIMENT-THEORY FIT
RZJ=0.42

2D UNIT CELLS (3 domains observed)

Cell	Ax (Å)	Ay (Å)	Bx (Å)	By (Å)	α (°)	Matrix	Pattern	Cell type
Bulk	3.840	0.000	1.920	3.326	60.0	(1.000, 0.000) (0.000, 1.000)	(1x1)	b: bulk lattice
Surface 1	5.760	3.326	-1.920	3.326	90.0	(1.000, 1.000) (-1.000, 1.000)	(2x1)	s1: commens. superlattice

3D COORDINATES

Si1-Si2: tilted π-bonded chain; Si3-Si4: tilted lower chain;
Si13-Si14: periodically repeating bulk bilayer

Dx/Dy in Å, or as a fraction of layer's unit cell vectors; atom 0 at the origin. Epir/subr are bulk repeat vectors.

No. of atoms: 14

Bulk z = 3.130 Å

Region	Chem el.	At. no.	Cell type	Site occ.	Rel. to	Dx ± εx		Dy ± εy		Dz ± εz		Dz/Bz(%) ± εz/Bz
epir		-2					f		f		Å	
subr		-1				1.920	Å	1.108	Å	3.130	Å	
intf	Si	1	s1	.50	0	0.000	f	0.000	f	0.000	Å	0.0
intf	Si	2	s1	.50	1	0.168	f	0.500	f	0.380 ± .080	Å	12.1 ± 2.6
intf	Si	3	s1	.50	2	0.322	f	0.000	f	0.810	Å	25.9
intf	Si	4	s1	.50	3	0.209	f	-0.500	f	0.070	Å	2.2
intf	Si	5	s1	.50	4	-0.338	f	0.500	f	2.130	Å	68.1
intf	Si	6	s1	.50	5	0.474	f	-0.500	f	0.070	Å	2.2
intf	Si	7	s1	.50	6	-0.654	f	0.000	f	0.600	Å	19.2
intf	Si	8	s1	.50	7	0.501	f	0.500	f	0.200	Å	6.4
intf	Si	9	s1	.50	8	-0.501	f	-0.500	f	2.190	Å	70.0
intf	Si	10	s1	.50	9	0.499	f	0.500	f	0.130	Å	4.2
intf	Si	11	s1	.50	10	-0.168	f	-0.500	f	0.690	Å	22.0
intf	Si	12	s1	.50	11	-0.495	f	0.500	f	0.030	Å	1.0
subl	Si	13	b	1.00	12	0.000	f	0.000	f	2.348	Å	75.0
subl	Si	14	b	1.00	13	-0.333	f	-0.333	f	0.783	Å	25.0

Si(111)-(2x1)
14.89

BOND DISTANCES AND ANGLES

Bond distances and angles are derived from coordinates

No. of distances/angles: 9

Interatomic dist. A-B (Å)	Atom A	Atom B	Atom C	Bond angle A-B-C (°)
2.255	Si1	Si2	Si1(0,1)	117.4
2.255	Si1	Si2	Si3	121.6
2.255	Si1	Si2	Si4(-1,0)	106.1
2.364	Si1	Si4(-1,0)	Si3(0,-2)	118.7
2.289	Si2	Si3	Si2(1,0)	110.0
2.289	Si2	Si3	Si4	124.0
2.289	Si3	Si2	Si3(-1,0)	110.0
2.371	Si3	Si4	Si1(1,0)	106.2
2.371	Si3	Si4	Si3(1,0)	125.8

ILLUSTRATION: 89

COMMON NAME : Si(111)-(2x1)
CLASSIFICATION : 14.96
TECHNIQUE : MEIS
AUTHORS : L. Smit, R.M. Tromp and J.F. van der Veen
REFERENCE : Surf. Sci., 163, 315 (1985)

SURFACE TYPE
Substrate : Si
Crystal face: 111
Temperature : RT*
Bulk lattice: diamond
2D bulk symm: p3m1
2D surf symm: pm

Adsorbate:
Coverage :
Pattern : (2x1)
Matrix : (1.000, 1.000)
(-1.000, 1.000)

STRUCTURE TYPE
Tilted π-bonded chain model, with tilt in lower chain in same direction as tilt in upper chain;

SAMPLE PREPARATION (4 sample)
Treatment : in-situ cleavage, ion-beam dose restricted
Crystallinity: single-domain pattern checked with LEED
Anal. methods:
Contamination:

COMMENTS

DATA COLLECTION
Technique: MEIS
Dataset : MEIS with shadowing and blocking; surface blocking patterns from 4 cleaves

THEORY/DATA TREATMENT
Monte Carlo simulations of Rutherford backscattering, Moliere approximation to Thomas-Fermi potential

STRUCTURES EXAMINED
Bulk-terminated surface; buckling model; π-bonded chain model with Keating relaxations and fit of tilt of first two chains and of height of first chain, keeping bond in upper chain to 2.25Å; π-bonded molecule model

QUALITY OF EXPERIMENT-THEORY FIT
Visual

2D UNIT CELLS (3 domains observed)

Cell	Ax (Å)	Ay (Å)	Bx (Å)	By (Å)	α (°)	Matrix	Pattern	Cell type
Bulk	3.841	0.000	1.920	3.326	60.0	(1.000, 0.000) (0.000, 1.000)	(1x1)	b: bulk lattice
Surface 1	5.761	3.326	-1.921	3.326	90.0	(1.000, 1.000) (-1.000, 1.000)	(2x1)	s1: commens. superlattice

3D COORDINATES

Si1-Si2: tilted π-bonded chain; Si3-Si4: tilted lower chain;
Si13-Si14: periodically repeating bulk bilayer; 0.1Å error bars assumed for tabulation (not determined)

Dx/Dy in Å, or as a fraction of layer's unit cell vectors; atom 0 at the origin. Epir/subr are bulk repeat vectors.

No. of atoms: 14

Bulk z = 3.130 Å

Region	Chem el.	At. no.	Cell type	Site occ.	Rel. to	Dx ± εx		Dy ± εy		Dz ± εz		Dz/Bz(%) ± εz/Bz
epir		-2					f		f		Å	
subr		-1				1.920	Å	1.109	Å	3.130	Å	
intf	Si	1	s1	.50	0	0.000	f	0.000	f	0.000	Å	0.0
intf	Si	2	s1	.50	1	0.169 ± .015	f	0.500 ± .026	f	0.300 ± .100	Å	9.6 ± 3.2
intf	Si	3	s1	.50	2	0.322 ± .015	f	0.000 ± .026	f	0.980 ± .100	Å	31.3 ± 3.2
intf	Si	4	s1	.50	3	0.215 ± .015	f	-0.500 ± .026	f	0.150 ± .100	Å	4.8 ± 3.2
intf	Si	5	s1	.50	4	-0.340 ± .015	f	0.500 ± .026	f	2.080 ± .100	Å	66.5 ± 3.2
intf	Si	6	s1	.50	5	0.475 ± .015	f	-0.500 ± .026	f	0.100 ± .100	Å	3.2 ± 3.2
intf	Si	7	s1	.50	6	-0.654 ± .015	f	0.000 ± .026	f	0.600 ± .100	Å	19.2 ± 3.2
intf	Si	8	s1	.50	7	0.499 ± .015	f	0.500 ± .026	f	0.270 ± .100	Å	8.6 ± 3.2
intf	Si	9	s1	.50	8	-0.499 ± .015	f	-0.500 ± .026	f	2.140 ± .100	Å	68.4 ± 3.2
intf	Si	10	s1	.50	9	0.499 ± .015	f	0.500 ± .026	f	0.140 ± .100	Å	4.5 ± 3.2
intf	Si	11	s1	.50	10	-0.665 ± .015	f	0.000 ± .026	f	0.710 ± .100	Å	22.7 ± 3.2
intf	Si	12	s1	.50	11	0.498 ± .015	f	-0.500 ± .026	f	0.000 ± .100	Å	0.0 ± 3.2
subl	Si	13	b	1.00	12	0.000	f	0.000	f	2.348	Å	75.0
subl	Si	14	b	1.00	13	-0.333	f	-0.333	f	0.783	Å	25.0

Si(111)-(2x1)
14.96

BOND DISTANCES AND ANGLES

Bond distances and angles are derived from coordinates

No. of distances/angles: 8

Interatomic dist. A-B (Å)	Atom A	Atom B	Atom C	Bond angle A-B-C (°)
2.246	Si1	Si2	Si1(0,1)	118.6
2.246	Si1	Si2	Si3	120.7
2.246	Si1	Si2	Si4(-1,0)	103.7
2.423	Si1	Si4(-1,0)	Si3(0,-2)	116.4
2.246	Si2	Si1	Si4(0,-1)	106.1
2.356	Si2	Si3	Si2(1,0)	105.8
2.356	Si2	Si3	Si4	124.6
2.399	Si3	Si4	Si1(1,0)	103.2

ILLUSTRATION: 88

COMMON NAME : Si(111) laser annealed
CLASSIFICATION : 14.108
TECHNIQUE : LEED
AUTHORS : F. Jona, P.M. Marcus, H.L. Davis and J.R. Noonan
REFERENCE : Phys. Rev., B33, 4005 (1986)

SURFACE TYPE
Substrate : Si
Crystal face: 111
Temperature : RT*
Bulk lattice: diamond
2D bulk symm: p3m1
2D surf symm: p3m1

Adsorbate:
Coverage :
Pattern : (1x1)
Matrix : (1.000, 0.000)
(0.000, 1.000)

STRUCTURE TYPE
Unreconstructed bulk termination with multilayer relaxations perpendicular to surface

SAMPLE PREPARATION (1 sample)
Treatment : see Zehner et al, J. Vac. Sci. Technol. 18, 852 (1981)
Crystallinity:
Anal. methods:
Contamination:

COMMENTS
Note: LEED spectra for models RB and GL are very similar; model GL gives: RZJ=0.1514, RPE=0.4959

DATA COLLECTION
Technique: LEED
Dataset : I-V curves for 6 normal incidence beams from 6-240 eV

THEORY/DATA TREATMENT
Dynamical LEED: 10 layer slab; 8 phase shifts; 60<E<240 eV; Voi=-4.25 eV (also mfp=8Å tested)

STRUCTURES EXAMINED
Relaxed bulk (RB): 25.5% contraction of 1st interlayer spacing, 3.2% expansion of 2nd, 5% expansion of 3rd; graphite-like reconstruction (GL) of Jones and Holland (Sol. St. Commun. 53, 45 (1985), class. no. 14.99)

QUALITY OF EXPERIMENT-THEORY FIT
RZJ=0.1134, RPE=0.4187

2D UNIT CELLS (1 domain observed)

Cell	Ax (Å)	Ay (Å)	Bx (Å)	By (Å)	α (°)	Matrix	Pattern	Cell type
Bulk	3.840	0.000	-1.920	3.326	120.0	(1.000, 0.000) (0.000, 1.000)	(1x1)	b: bulk lattice
Surface 1	3.840	0.000	-1.920	3.326	120.0	(1.000, 0.000) (0.000, 1.000)	(1x1)	s1: commens. superlattice

3D COORDINATES

Si1-Si2 and Si3-Si4: top 2 bilayers; Si5-Si6: repeating bulk bilayer

Dx/Dy in Å, or as a fraction of layer's unit cell vectors; atom 0 at the origin. Epir/subr are bulk repeat vectors.

No. of atoms: 6

Bulk z = 3.130 Å

Region	Chem el.	At. no.	Cell type	Site occ.	Rel. to	Dx ± εx		Dy ± εy		Dz ± εz		Dz/Bz(%) ± εz/Bz
epir		-2					f		f		Å	
subr		-1				0.000	Å	-2.217	Å	3.130	Å	
intf	Si	1	b	1.00	0	0.000	f	0.000	f	0.000	Å	0.0
intf	Si	2	b	1.00	1	0.667	f	0.333	f	0.581	Å	18.6
intf	Si	3	b	1.00	2	0.000	f	0.000	f	2.470	Å	78.9
intf	Si	4	b	1.00	3	-0.333	f	0.333	f	0.820	Å	26.2
subl	Si	5	b	1.00	4	0.000	f	0.000	f	2.350	Å	75.1
subl	Si	6	b	1.00	5	-0.333	f	-0.667	f	0.780	Å	24.9

BOND DISTANCES AND ANGLES

Bond distances and angles are derived from coordinates

No. of distances/angles: 4

Interatomic dist. A-B (Å)	Atom A	Atom B	Atom C	Bond angle A-B-C (°)
2.292	Si1	Si2	Si3	104.7
2.470	Si2	Si3	Si4	110.3

Si(111) laser annealed
14.108

Bond Distances and Angles - Continued

Interatomic dist. A-B (Å)	Atom A	Atom B	Atom C	Bond angle A-B-C (°)
2.364	Si3	Si4	Si5	110.3
2.350	Si4	Si5	Si6	109.4

COMMON NAME : Si(111)-(1x1) laser-annealed
CLASSIFICATION : 14.99
TECHNIQUE : LEED
AUTHORS : G.J.R. Jones and B.W. Holland
REFERENCE : Solid State Commun., 53, 45 (1985)

ILLUSTRATION: 88

SURFACE TYPE
Substrate : Si
Crystal face: 111
Temperature : RT*
Bulk lattice: diamond
2D bulk symm: p3m1
2D surf symm: p3m1

Adsorbate:
Coverage :
Pattern : (1x1)
Matrix : (1.000, 0.000)
(0.000, 1.000)

STRUCTURE TYPE
Unreconstructed bulk termination with relaxations perpendicular to surface

SAMPLE PREPARATION (1 sample)
Treatment : see Zehner et al, J. Vac. Sci. Technol. 18, 852 (1981)
Crystallinity:
Anal. methods:
Contamination:

COMMENTS
Best structure has nearly coplanar top bilayer, with 1st inter-bilayer bond lengths of 2.22Å and 2nd one of 2.95Å (bulk value 2.35Å); model of Zehner et al (J. Vac. Sci. Technol. 18, 852 (1981)) gives RPE=0.40

DATA COLLECTION
Technique: LEED
Dataset : I-V curves for 6 beams at normal incidence measured by Zehner et al

THEORY/DATA TREATMENT
Dynamical LEED (reverse scattering perturbation): 8-layer slab, 6 phase shifts, mfp=8Å

STRUCTURES EXAMINED
20 combinations of relaxations of first and second interlayer spacings

QUALITY OF EXPERIMENT-THEORY FIT
RPE=0.38

2D UNIT CELLS (1 domain observed)

Cell	Ax (Å)	Ay (Å)	Bx (Å)	By (Å)	α (°)	Matrix	Pattern	Cell type
Bulk	3.838	0.000	1.919	3.324	60.0	(1.000, 0.000) (0.000, 1.000)	(1x1)	b: bulk lattice
Surface 1	3.838	0.000	1.919	3.324	60.0	(1.000, 0.000) (0.000, 1.000)	(1x1)	s1: commens. superlattice

3D COORDINATES

Si1-Si2: nearly coplanar topmost bilayer; Si5-Si6: bulk bilayer

Dx/Dy in Å, or as a fraction of layer's unit cell vectors; atom 0 at the origin. Epir/subr are bulk repeat vectors.

No. of atoms: 6 Bulk z = 3.130 Å

Region	Chem el.	At. no.	Cell type	Site occ.	Rel. to	Dx ± εx		Dy ± εy		Dz ± εz		Dz/Bz(%) ± εz/Bz
epir		-2					f		f		Å	
subr		-1				1.919	Å	1.108	Å	3.130	Å	
intf	Si	1	b	1.00	0	0.000	f	0.000	f	0.000	Å	0.0
intf	Si	2	b	1.00	1	0.333	f	0.333	f	0.080 ± .020	Å	2.6 ± .6
intf	Si	3	b	1.00	2	0.000	f	0.000	f	2.950 ± .200	Å	94.3 ± 6.4
intf	Si	4	b	1.00	3	0.333	f	0.333	f	0.780	Å	24.9
subl	Si	5	b	1.00	4	0.000	f	0.000	f	2.348	Å	75.0
subl	Si	6	b	1.00	5	0.333	f	0.333	f	0.783	Å	25.0

BOND DISTANCES AND ANGLES

Bond distances and angles are derived from coordinates

No. of distances/angles: 4

Interatomic dist. A-B (Å)	Atom A	Atom B	Atom C	Bond angle A-B-C (°)
2.217	Si1	Si2	Si1(1,0)	119.9
2.217	Si1	Si2	Si3	92.1
2.950	Si2	Si3	Si4	109.4
2.349	Si3	Si4	Si3(1,0)	109.6

COMMON NAME : Si(111)-(7x7) ILLUSTRATION: 91
CLASSIFICATION : 14.132
TECHNIQUE : LEED
AUTHORS : S.Y. Tong, H. Huang, C.M. Wei, W.E. Packard, F.K. Men, G. Glander and M.B. Webb
REFERENCE : J. Vac. Sci. Technol., A6, 615 (1988)

SURFACE TYPE
Substrate : Si
Crystal face: 111
Temperature : RT*
Bulk lattice: diamond
2D bulk symm: p3m1
2D surf symm: p3m1

Adsorbate:
Coverage :
Pattern : (7x7)
Matrix : (7.000, 0.000)
(0.000, 7.000)

STRUCTURE TYPE
Optimized DAS (dimer-adatom-stacking fault) model

SAMPLE PREPARATION (1 sample)
Treatment :
Crystallinity:
Anal. methods:
Contamination:

COMMENTS

DATA COLLECTION
Technique: LEED
Dataset : 3 integer-order and 17 fractional-order non-equivalent beams at normal incidence; E range 30-250 eV

THEORY/DATA TREATMENT
Dynamical LEED (full symm. in both direct and recipr. space)
5 phase shifts; Vor=-10 eV, Voi=-4.24eV

STRUCTURES EXAMINED
Over 100 variations of following 7 models: DAS, DAS + (1x1) faulted layer, DPCS, DPCS+(1x1) faulted layer, symmetric trimer, Yang + Zhao, relaxed Yang + Zhao

QUALITY OF EXPERIMENT-THEORY FIT
RVHT=0.34

2D UNIT CELLS (1 domain observed)

Cell	Ax (Å)	Ay (Å)	Bx (Å)	By (Å)	α (°)	Matrix	Pattern	Cell type
Bulk	3.840	0.000	1.920	3.325	60.0	(1.000, 0.000) (0.000, 1.000)	(1x1)	b: bulk lattice
Surface 1	26.878	0.000	13.439	23.277	60.0	(7.000, 0.000) (0.000, 7.000)		s1: commens. superlattice

3D COORDINATES

Si1-12=adatoms;13-18=restatoms;19-54=top 1/2 top bilayer; 55-72&91-102=lower 1/2 top bilayer;73-90=dimer atoms; 103-151=top 1/2 2nd bilayer;140-200=lower 1/2 2nd bilay; 201-202=bulk; 0.1Å error bars assumed for tabulation

Dx/Dy in Å, or as a fraction of layer's unit cell vectors; atom 0 at the origin. Epir/subr are bulk repeat vectors.

No. of atoms: 202 Bulk z = 3.130 Å

Region	Chem el.	At. no.	Cell type	Site occ.	Rel. to	Dx ± εx		Dy ± εy		Dz ± εz		Dz/Bz(%) ± εz/Bz
epir		-2					f		f		Å	
subr		-1				1.920	Å	1.108	Å	3.130	Å	
intf	Si	1	s1	.02	0	0.000 ± .002	f	0.000 ± .004	f	0.000 ± .100	Å	0.0 ± 3.2
intf	Si	2	s1	.02	1	0.571 ± .002	f	0.000 ± .004	f	0.000 ± .100	Å	0.0 ± 3.2
intf	Si	3	s1	.02	2	-0.572 ± .002	f	0.572 ± .004	f	0.000 ± .100	Å	0.0 ± 3.2
intf	Si	4	s1	.02	3	0.000 ± .002	f	-0.286 ± .004	f	0.040 ± .100	Å	1.3 ± 3.2
intf	Si	5	s1	.02	4	0.286 ± .002	f	-0.286 ± .004	f	0.000 ± .100	Å	0.0 ± 3.2
intf	Si	6	s1	.02	5	0.000 ± .002	f	0.286 ± .004	f	0.000 ± .100	Å	0.0 ± 3.2
intf	Si	7	s1	.02	6	0.429 ± .002	f	-0.143 ± .004	f	0.040 ± .100	Å	1.3 ± 3.2
intf	Si	8	s1	.02	7	0.000 ± .002	f	0.572 ± .004	f	0.000 ± .100	Å	0.0 ± 3.2
intf	Si	9	s1	.02	8	-0.571 ± .002	f	0.000 ± .004	f	0.000 ± .100	Å	0.0 ± 3.2
intf	Si	10	s1	.02	9	0.572 ± .002	f	-0.286 ± .004	f	0.040 ± .100	Å	1.3 ± 3.2
intf	Si	11	s1	.02	10	-0.286 ± .002	f	0.286 ± .004	f	0.000 ± .100	Å	0.0 ± 3.2
intf	Si	12	s1	.02	11	0.000 ± .002	f	-0.286 ± .004	f	0.000 ± .100	Å	0.0 ± 3.2
intf	Si	13	s1	.02	12	-0.333 ± .002	f	-0.048 ± .004	f	1.010 ± .100	Å	32.3 ± 3.2
intf	Si	14	s1	.02	13	0.000 ± .002	f	-0.286 ± .004	f	0.000 ± .100	Å	0.0 ± 3.2
intf	Si	15	s1	.02	14	0.286 ± .002	f	0.000 ± .004	f	0.000 ± .100	Å	0.0 ± 3.2
intf	Si	16	s1	.02	15	-0.048 ± .002	f	0.524 ± .004	f	0.050 ± .100	Å	1.6 ± 3.2
intf	Si	17	s1	.02	16	0.286 ± .002	f	-0.286 ± .004	f	0.000 ± .100	Å	0.0 ± 3.2
intf	Si	18	s1	.02	17	0.000 ± .002	f	0.286 ± .004	f	0.000 ± .100	Å	0.0 ± 3.2
intf	Si	19	s1	.02	18	0.338 ± .002	f	-0.376 ± .004	f	0.080 ± .100	Å	2.6 ± 3.2

Si(111)-(7x7)
14.132

3D Coordinates - Continued

Region	Chem el.	At. no.	Cell type	Site occ.	Rel. to	Dx ± εx	Dy ± εy	Dz ± εz	Dz/Bz(%) ± εz/Bz
intf	Si	20	s1	.02	19	-0.000 ± .002 f	0.129 ± .004 f	0.000 ± .100 Å	0.0 ± 3.2
intf	Si	21	s1	.02	20	0.000 ± .002 f	-0.286 ± .004 f	0.000 ± .100 Å	0.0 ± 3.2
intf	Si	22	s1	.02	21	0.000 ± .002 f	0.443 ± .004 f	0.000 ± .100 Å	0.0 ± 3.2
intf	Si	23	s1	.02	22	-0.714 ± .002 f	0.429 ± .004 f	0.000 ± .100 Å	0.0 ± 3.2
intf	Si	24	s1	.02	23	0.000 ± .002 f	-0.586 ± .004 f	0.000 ± .100 Å	0.0 ± 3.2
intf	Si	25	s1	.02	24	0.129 ± .002 f	-0.129 ± .004 f	0.000 ± .100 Å	0.0 ± 3.2
intf	Si	26	s1	.02	25	0.157 ± .002 f	0.714 ± .004 f	0.000 ± .100 Å	0.0 ± 3.2
intf	Si	27	s1	.02	26	-0.157 ± .002 f	0.000 ± .004 f	0.000 ± .100 Å	0.0 ± 3.2
intf	Si	28	s1	.02	27	-0.129 ± .002 f	-0.714 ± .004 f	0.000 ± .100 Å	0.0 ± 3.2
intf	Si	29	s1	.02	28	-0.157 ± .002 f	0.000 ± .004 f	0.000 ± .100 Å	0.0 ± 3.2
intf	Si	30	s1	.02	29	0.000 ± .002 f	0.714 ± .004 f	0.000 ± .100 Å	0.0 ± 3.2
intf	Si	31	s1	.02	30	0.000 ± .002 f	-0.429 ± .004 f	0.000 ± .100 Å	0.0 ± 3.2
intf	Si	32	s1	.02	31	0.443 ± .002 f	-0.443 ± .004 f	0.000 ± .100 Å	0.0 ± 3.2
intf	Si	33	s1	.02	32	-0.286 ± .002 f	0.000 ± .004 f	0.000 ± .100 Å	0.0 ± 3.2
intf	Si	34	s1	.02	33	0.714 ± .002 f	0.871 ± .004 f	0.040 ± .100 Å	1.3 ± 3.2
intf	Si	35	s1	.02	34	-0.000 ± .002 f	-0.299 ± .004 f	0.000 ± .100 Å	0.0 ± 3.2
intf	Si	36	s1	.02	35	-0.299 ± .002 f	0.299 ± .004 f	0.000 ± .100 Å	0.0 ± 3.2
intf	Si	37	s1	.02	36	0.099 ± .002 f	-0.485 ± .004 f	0.010 ± .100 Å	.3 ± 3.2
intf	Si	38	s1	.02	37	-0.571 ± .002 f	0.157 ± .004 f	0.000 ± .100 Å	0.0 ± 3.2
intf	Si	39	s1	.02	38	0.443 ± .002 f	-0.443 ± .004 f	0.000 ± .100 Å	0.0 ± 3.2
intf	Si	40	s1	.02	39	0.129 ± .002 f	0.000 ± .004 f	0.000 ± .100 Å	0.0 ± 3.2
intf	Si	41	s1	.02	40	0.000 ± .002 f	0.443 ± .004 f	0.000 ± .100 Å	0.0 ± 3.2
intf	Si	42	s1	.02	41	-0.414 ± .002 f	0.129 ± .004 f	0.000 ± .100 Å	0.0 ± 3.2
intf	Si	43	s1	.02	42	0.000 ± .002 f	-0.286 ± .004 f	0.000 ± .100 Å	0.0 ± 3.2
intf	Si	44	s1	.02	43	0.129 ± .002 f	-0.129 ± .004 f	0.000 ± .100 Å	0.0 ± 3.2
intf	Si	45	s1	.02	44	0.286 ± .002 f	0.000 ± .004 f	0.000 ± .100 Å	0.0 ± 3.2
intf	Si	46	s1	.02	45	-0.129 ± .002 f	0.414 ± .004 f	0.000 ± .100 Å	0.0 ± 3.2
intf	Si	47	s1	.02	46	-0.157 ± .002 f	0.000 ± .004 f	0.000 ± .100 Å	0.0 ± 3.2
intf	Si	48	s1	.02	47	0.000 ± .002 f	-0.286 ± .004 f	0.000 ± .100 Å	0.0 ± 3.2
intf	Si	49	s1	.02	48	0.157 ± .002 f	0.000 ± .004 f	0.000 ± .100 Å	0.0 ± 3.2
intf	Si	50	s1	.02	49	-0.443 ± .002 f	0.286 ± .004 f	0.000 ± .100 Å	0.0 ± 3.2
intf	Si	51	s1	.02	50	0.286 ± .002 f	-0.129 ± .004 f	0.000 ± .100 Å	0.0 ± 3.2
intf	Si	52	s1	.02	51	-0.415 ± .002 f	0.129 ± .004 f	0.040 ± .100 Å	1.3 ± 3.2
intf	Si	53	s1	.02	52	0.701 ± .002 f	-0.701 ± .004 f	0.000 ± .100 Å	0.0 ± 3.2
intf	Si	54	s1	.02	53	0.000 ± .002 f	0.701 ± .004 f	0.000 ± .100 Å	0.0 ± 3.2
intf	Si	55	s1	.02	54	-0.759 ± .002 f	-0.614 ± .004 f	0.484 ± .100 Å	15.5 ± 3.2
intf	Si	56	s1	.02	55	0.000 ± .002 f	0.286 ± .004 f	0.000 ± .100 Å	0.0 ± 3.2
intf	Si	57	s1	.02	56	0.431 ± .002 f	-0.286 ± .004 f	0.000 ± .100 Å	0.0 ± 3.2
intf	Si	58	s1	.02	57	-0.145 ± .002 f	0.000 ± .004 f	0.000 ± .100 Å	0.0 ± 3.2
intf	Si	59	s1	.02	58	0.145 ± .002 f	-0.145 ± .004 f	0.000 ± .100 Å	0.0 ± 3.2
intf	Si	60	s1	.02	58	-0.140 ± .002 f	0.140 ± .004 f	0.000 ± .100 Å	0.0 ± 3.2
intf	Si	61	s1	.02	60	0.000 ± .002 f	-0.140 ± .004 f	0.000 ± .100 Å	0.0 ± 3.2
intf	Si	62	s1	.02	61	0.000 ± .002 f	-0.145 ± .004 f	0.000 ± .100 Å	0.0 ± 3.2
intf	Si	63	s1	.02	61	-0.000 ± .002 f	0.286 ± .004 f	0.000 ± .100 Å	0.0 ± 3.2
intf	Si	64	s1	.02	63	0.427 ± .002 f	0.287 ± .004 f	0.050 ± .100 Å	1.6 ± 3.2
intf	Si	65	s1	.02	64	-0.140 ± .002 f	-0.145 ± .004 f	0.000 ± .100 Å	0.0 ± 3.2
intf	Si	66	s1	.02	65	0.140 ± .002 f	-0.140 ± .004 f	0.000 ± .100 Å	0.0 ± 3.2
intf	Si	67	s1	.02	66	-0.286 ± .002 f	0.286 ± .004 f	0.000 ± .100 Å	0.0 ± 3.2
intf	Si	68	s1	.02	67	-0.002 ± .002 f	-0.141 ± .004 f	0.000 ± .100 Å	0.0 ± 3.2
intf	Si	69	s1	.02	68	0.288 ± .002 f	-0.290 ± .004 f	0.000 ± .100 Å	0.0 ± 3.2
intf	Si	70	s1	.02	69	0.145 ± .002 f	0.000 ± .004 f	0.000 ± .100 Å	0.0 ± 3.2
intf	Si	71	s1	.02	70	0.000 ± .002 f	0.286 ± .004 f	0.000 ± .100 Å	0.0 ± 3.2
intf	Si	72	s1	.02	71	-0.145 ± .002 f	0.000 ± .004 f	0.000 ± .100 Å	0.0 ± 3.2
intf	Si	73	s1	.02	72	0.118 ± .002 f	-0.545 ± .004 f	0.170 ± .100 Å	5.4 ± 3.2
intf	Si	74	s1	.02	73	0.169 ± .002 f	0.000 ± .004 f	0.000 ± .100 Å	0.0 ± 3.2
intf	Si	75	s1	.02	74	0.000 ± .002 f	0.663 ± .004 f	0.000 ± .100 Å	0.0 ± 3.2
intf	Si	76	s1	.02	75	0.000 ± .002 f	-0.286 ± .004 f	0.000 ± .100 Å	0.0 ± 3.2
intf	Si	77	s1	.02	76	-0.740 ± .002 f	0.454 ± .004 f	0.000 ± .100 Å	0.0 ± 3.2
intf	Si	78	s1	.02	77	0.000 ± .002 f	-0.260 ± .004 f	0.000 ± .100 Å	0.0 ± 3.2
intf	Si	79	s1	.02	78	0.480 ± .002 f	-0.480 ± .004 f	0.000 ± .100 Å	0.0 ± 3.2
intf	Si	80	s1	.02	79	0.260 ± .002 f	0.000 ± .004 f	0.000 ± .100 Å	0.0 ± 3.2
intf	Si	81	s1	.02	80	-0.454 ± .002 f	0.740 ± .004 f	0.000 ± .100 Å	0.0 ± 3.2
intf	Si	82	s1	.02	81	0.195 ± .002 f	0.000 ± .004 f	0.000 ± .100 Å	0.0 ± 3.2
intf	Si	83	s1	.02	82	0.091 ± .002 f	0.000 ± .004 f	0.000 ± .100 Å	0.0 ± 3.2
intf	Si	84	s1	.02	83	-0.377 ± .002 f	0.000 ± .004 f	0.000 ± .100 Å	0.0 ± 3.2
intf	Si	85	s1	.02	84	0.000 ± .002 f	-0.454 ± .004 f	0.000 ± .100 Å	0.0 ± 3.2
intf	Si	86	s1	.02	85	0.091 ± .002 f	-0.091 ± .004 f	0.000 ± .100 Å	0.0 ± 3.2
intf	Si	87	s1	.02	86	0.454 ± .002 f	0.000 ± .004 f	0.000 ± .100 Å	0.0 ± 3.2
intf	Si	88	s1	.02	87	-0.831 ± .002 f	0.546 ± .004 f	0.000 ± .100 Å	0.0 ± 3.2
intf	Si	89	s1	.02	88	0.000 ± .002 f	-0.169 ± .004 f	0.000 ± .100 Å	0.0 ± 3.2
intf	Si	90	s1	.02	89	0.831 ± .002 f	-0.091 ± .004 f	0.000 ± .100 Å	0.0 ± 3.2
intf	Si	91	s1	.02	90	-0.571 ± .002 f	-0.312 ± .004 f	0.510 ± .100 Å	16.3 ± 3.2

Si(111)-(7x7)
14.132

3D Coordinates - Continued

Region	Chem el.	At. no.	Cell type	Site occ.	Rel. to	Dx ± εx	Dy ± εy	Dz ± εz	Dz/Bz(%) ± εz/Bz
intf	Si	92	s1	.02	91	-0.286 ± .002 f	0.000 ± .004 f	0.000 ± .100 Å	0.0 ± 3.2
intf	Si	93	s1	.02	92	0.000 ± .002 f	-0.286 ± .004 f	0.000 ± .100 Å	0.0 ± 3.2
intf	Si	94	s1	.02	93	0.286 ± .002 f	0.000 ± .004 f	0.000 ± .100 Å	0.0 ± 3.2
intf	Si	95	s1	.02	94	-0.286 ± .002 f	0.572 ± .004 f	0.000 ± .100 Å	0.0 ± 3.2
intf	Si	96	s1	.02	95	0.572 ± .002 f	-0.572 ± .004 f	0.000 ± .100 Å	0.0 ± 3.2
intf	Si	97	s1	.02	96	-0.429 ± .002 f	0.714 ± .004 f	0.050 ± .100 Å	1.6 ± 3.2
intf	Si	98	s1	.02	97	0.286 ± .002 f	-0.286 ± .004 f	0.000 ± .100 Å	0.0 ± 3.2
intf	Si	99	s1	.02	98	0.286 ± .002 f	0.000 ± .004 f	0.000 ± .100 Å	0.0 ± 3.2
intf	Si	100	s1	.02	99	-0.000 ± .002 f	-0.286 ± .004 f	0.000 ± .100 Å	0.0 ± 3.2
intf	Si	101	s1	.02	100	-0.286 ± .002 f	0.572 ± .004 f	0.000 ± .100 Å	0.0 ± 3.2
intf	Si	102	s1	.02	101	0.286 ± .002 f	0.000 ± .004 f	0.000 ± .100 Å	0.0 ± 3.2
intf	Si	103	s1	.02	102	-0.429 ± .002 f	-0.572 ± .004 f	1.610 ± .100 Å	51.4 ± 3.2
intf	Si	104	s1	.02	103	0.143 ± .002 f	-0.143 ± .004 f	0.000 ± .100 Å	0.0 ± 3.2
intf	Si	105	s1	.02	104	-0.000 ± .002 f	0.572 ± .004 f	0.000 ± .100 Å	0.0 ± 3.2
intf	Si	106	s1	.02	105	0.143 ± .002 f	-0.143 ± .004 f	0.000 ± .100 Å	0.0 ± 3.2
intf	Si	107	s1	.02	106	0.000 ± .002 f	0.286 ± .004 f	0.000 ± .100 Å	0.0 ± 3.2
intf	Si	108	s1	.02	107	-0.429 ± .002 f	-0.714 ± .004 f	0.000 ± .100 Å	0.0 ± 3.2
intf	Si	109	s1	.02	108	0.429 ± .002 f	0.572 ± .004 f	0.000 ± .100 Å	0.0 ± 3.2
intf	Si	110	s1	.02	109	-0.286 ± .002 f	0.143 ± .004 f	0.000 ± .100 Å	0.0 ± 3.2
intf	Si	111	s1	.02	110	0.000 ± .002 f	-0.143 ± .004 f	0.000 ± .100 Å	0.0 ± 3.2
intf	Si	112	s1	.02	111	0.286 ± .002 f	-0.286 ± .004 f	0.000 ± .100 Å	0.0 ± 3.2
intf	Si	113	s1	.02	112	0.143 ± .002 f	0.000 ± .004 f	0.000 ± .100 Å	0.0 ± 3.2
intf	Si	114	s1	.02	113	0.000 ± .002 f	0.286 ± .004 f	0.000 ± .100 Å	0.0 ± 3.2
intf	Si	115	s1	.02	114	-0.572 ± .002 f	-0.286 ± .004 f	0.000 ± .100 Å	0.0 ± 3.2
intf	Si	116	s1	.02	115	0.000 ± .002 f	0.143 ± .004 f	0.000 ± .100 Å	0.0 ± 3.2
intf	Si	117	s1	.02	116	0.286 ± .002 f	-0.286 ± .004 f	0.000 ± .100 Å	0.0 ± 3.2
intf	Si	118	s1	.02	117	-0.429 ± .002 f	0.000 ± .004 f	0.000 ± .100 Å	0.0 ± 3.2
intf	Si	119	s1	.02	118	0.000 ± .002 f	0.286 ± .004 f	0.000 ± .100 Å	0.0 ± 3.2
intf	Si	120	s1	.02	119	0.143 ± .002 f	-0.286 ± .004 f	0.000 ± .100 Å	0.0 ± 3.2
intf	Si	121	s1	.02	120	0.714 ± .002 f	0.714 ± .004 f	0.020 ± .100 Å	.6 ± 3.2
intf	Si	122	s1	.02	121	-0.571 ± .002 f	-0.429 ± .004 f	0.050 ± .100 Å	1.6 ± 3.2
intf	Si	123	s1	.02	122	0.143 ± .002 f	-0.143 ± .004 f	0.000 ± .100 Å	0.0 ± 3.2
intf	Si	124	s1	.02	123	0.429 ± .002 f	0.000 ± .004 f	0.000 ± .100 Å	0.0 ± 3.2
intf	Si	125	s1	.02	124	0.000 ± .002 f	0.143 ± .004 f	0.000 ± .100 Å	0.0 ± 3.2
intf	Si	126	s1	.02	125	0.000 ± .002 f	0.286 ± .004 f	0.000 ± .100 Å	0.0 ± 3.2
intf	Si	127	s1	.02	126	-0.714 ± .002 f	0.143 ± .004 f	0.000 ± .100 Å	0.0 ± 3.2
intf	Si	128	s1	.02	127	0.000 ± .002 f	-0.286 ± .004 f	0.000 ± .100 Å	0.0 ± 3.2
intf	Si	129	s1	.02	128	0.429 ± .002 f	-0.429 ± .004 f	0.000 ± .100 Å	0.0 ± 3.2
intf	Si	130	s1	.02	129	0.286 ± .002 f	0.000 ± .004 f	0.000 ± .100 Å	0.0 ± 3.2
intf	Si	131	s1	.02	130	-0.143 ± .002 f	0.714 ± .004 f	0.000 ± .100 Å	0.0 ± 3.2
intf	Si	132	s1	.02	131	-0.143 ± .002 f	0.000 ± .004 f	0.000 ± .100 Å	0.0 ± 3.2
intf	Si	133	s1	.02	132	-0.143 ± .002 f	0.000 ± .004 f	0.000 ± .100 Å	0.0 ± 3.2
intf	Si	134	s1	.02	133	-0.429 ± .002 f	0.000 ± .004 f	0.000 ± .100 Å	0.0 ± 3.2
intf	Si	135	s1	.02	134	-0.000 ± .002 f	-0.143 ± .004 f	0.000 ± .100 Å	0.0 ± 3.2
intf	Si	136	s1	.02	135	0.714 ± .002 f	-0.714 ± .004 f	0.000 ± .100 Å	0.0 ± 3.2
intf	Si	137	s1	.02	136	0.143 ± .002 f	0.000 ± .004 f	0.000 ± .100 Å	0.0 ± 3.2
intf	Si	138	s1	.02	137	0.000 ± .002 f	0.572 ± .004 f	0.000 ± .100 Å	0.0 ± 3.2
intf	Si	139	s1	.02	138	-0.571 ± .002 f	0.286 ± .004 f	0.000 ± .100 Å	0.0 ± 3.2
intf	Si	140	s1	.02	139	-0.286 ± .002 f	-0.286 ± .004 f	0.450 ± .100 Å	14.4 ± 3.2
intf	Si	141	s1	.02	140	0.286 ± .002 f	-0.286 ± .004 f	0.000 ± .100 Å	0.0 ± 3.2
intf	Si	142	s1	.02	141	-0.286 ± .002 f	0.000 ± .004 f	0.000 ± .100 Å	0.0 ± 3.2
intf	Si	143	s1	.02	142	0.000 ± .002 f	-0.286 ± .004 f	0.000 ± .100 Å	0.0 ± 3.2
intf	Si	144	s1	.02	143	0.714 ± .002 f	0.429 ± .004 f	0.000 ± .100 Å	0.0 ± 3.2
intf	Si	145	s1	.02	144	-0.143 ± .002 f	-0.429 ± .004 f	0.000 ± .100 Å	0.0 ± 3.2
intf	Si	146	s1	.02	145	-0.286 ± .002 f	0.000 ± .004 f	0.000 ± .100 Å	0.0 ± 3.2
intf	Si	147	s1	.02	146	-0.143 ± .002 f	0.714 ± .004 f	0.000 ± .100 Å	0.0 ± 3.2
intf	Si	148	s1	.02	147	0.571 ± .002 f	-0.572 ± .004 f	0.000 ± .100 Å	0.0 ± 3.2
intf	Si	149	s1	.02	148	-0.286 ± .002 f	0.286 ± .004 f	0.000 ± .100 Å	0.0 ± 3.2
intf	Si	150	s1	.02	149	0.286 ± .002 f	0.286 ± .004 f	0.000 ± .100 Å	0.0 ± 3.2
intf	Si	151	s1	.02	150	-0.286 ± .002 f	0.000 ± .004 f	0.000 ± .100 Å	0.0 ± 3.2
intf	Si	152	s1	.02	151	-0.191 ± .002 f	0.095 ± .004 f	0.274 ± .100 Å	8.8 ± 3.2
intf	Si	153	s1	.02	152	0.000 ± .002 f	-0.286 ± .004 f	0.000 ± .100 Å	0.0 ± 3.2
intf	Si	154	s1	.02	153	0.286 ± .002 f	-0.286 ± .004 f	0.000 ± .100 Å	0.0 ± 3.2
intf	Si	155	s1	.02	154	0.286 ± .002 f	0.000 ± .004 f	0.000 ± .100 Å	0.0 ± 3.2
intf	Si	156	s1	.02	155	0.000 ± .002 f	0.286 ± .004 f	0.000 ± .100 Å	0.0 ± 3.2
intf	Si	157	s1	.02	156	-0.286 ± .002 f	0.286 ± .004 f	0.000 ± .100 Å	0.0 ± 3.2
intf	Si	158	s1	.02	157	-0.429 ± .002 f	-0.429 ± .004 f	0.010 ± .100 Å	.3 ± 3.2
intf	Si	159	s1	.02	158	0.000 ± .002 f	-0.286 ± .004 f	0.000 ± .100 Å	0.0 ± 3.2
intf	Si	160	s1	.02	159	0.429 ± .002 f	0.429 ± .004 f	0.000 ± .100 Å	0.0 ± 3.2
intf	Si	161	s1	.02	160	-0.143 ± .002 f	-0.429 ± .004 f	0.000 ± .100 Å	0.0 ± 3.2
intf	Si	162	s1	.02	161	0.429 ± .002 f	0.857 ± .004 f	0.010 ± .100 Å	.3 ± 3.2
intf	Si	163	s1	.02	162	0.143 ± .002 f	-0.143 ± .004 f	0.000 ± .100 Å	0.0 ± 3.2

Si(111)-(7x7)
14.132

3D Coordinates - Continued

Region	Chem el.	At. no.	Cell type	Site occ.	Rel. to	Dx ± εx	Dy ± εy	Dz ± εz	Dz/Bz(%) ± εz/Bz
intf	Si	164	s1	.02	163	-0.143 ± .002 f	0.000 ± .004 f	0.000 ± .100 Å	0.0 ± 3.2
intf	Si	165	s1	.02	164	-0.429 ± .002 f	-0.286 ± .004 f	0.060 ± .100 Å	1.9 ± 3.2
intf	Si	166	s1	.02	165	0.143 ± .002 f	-0.143 ± .004 f	0.000 ± .100 Å	0.0 ± 3.2
intf	Si	167	s1	.02	166	0.143 ± .002 f	0.000 ± .004 f	0.000 ± .100 Å	0.0 ± 3.2
intf	Si	168	s1	.02	167	0.000 ± .002 f	0.143 ± .004 f	0.000 ± .100 Å	0.0 ± 3.2
intf	Si	169	s1	.02	168	-0.571 ± .002 f	-0.286 ± .004 f	0.000 ± .100 Å	0.0 ± 3.2
intf	Si	170	s1	.02	169	0.286 ± .002 f	0.429 ± .004 f	0.000 ± .100 Å	0.0 ± 3.2
intf	Si	171	s1	.02	170	-0.143 ± .002 f	-0.571 ± .004 f	0.000 ± .100 Å	0.0 ± 3.2
intf	Si	172	s1	.02	171	0.286 ± .002 f	0.571 ± .004 f	0.000 ± .100 Å	0.0 ± 3.2
intf	Si	173	s1	.02	172	-0.286 ± .002 f	0.000 ± .004 f	0.000 ± .100 Å	0.0 ± 3.2
intf	Si	174	s1	.02	173	0.429 ± .002 f	-0.429 ± .004 f	0.000 ± .100 Å	0.0 ± 3.2
intf	Si	175	s1	.02	174	-0.429 ± .002 f	0.000 ± .004 f	0.000 ± .100 Å	0.0 ± 3.2
intf	Si	176	s1	.02	175	0.429 ± .002 f	0.429 ± .004 f	0.000 ± .100 Å	0.0 ± 3.2
intf	Si	177	s1	.02	176	0.286 ± .002 f	-0.571 ± .004 f	0.030 ± .100 Å	1.0 ± 3.2
intf	Si	178	s1	.02	177	0.000 ± .002 f	0.429 ± .004 f	0.000 ± .100 Å	0.0 ± 3.2
intf	Si	179	s1	.02	178	-0.714 ± .002 f	0.429 ± .004 f	0.000 ± .100 Å	0.0 ± 3.2
intf	Si	180	s1	.02	179	0.000 ± .002 f	-0.571 ± .004 f	0.000 ± .100 Å	0.0 ± 3.2
intf	Si	181	s1	.02	180	0.143 ± .002 f	-0.143 ± .004 f	0.000 ± .100 Å	0.0 ± 3.2
intf	Si	182	s1	.02	181	0.571 ± .002 f	0.000 ± .004 f	0.000 ± .100 Å	0.0 ± 3.2
intf	Si	183	s1	.02	182	0.000 ± .002 f	0.143 ± .004 f	0.000 ± .100 Å	0.0 ± 3.2
intf	Si	184	s1	.02	183	-0.857 ± .002 f	0.143 ± .004 f	0.000 ± .100 Å	0.0 ± 3.2
intf	Si	185	s1	.02	184	0.429 ± .002 f	0.429 ± .004 f	0.000 ± .100 Å	0.0 ± 3.2
intf	Si	186	s1	.02	185	-0.143 ± .002 f	0.000 ± .004 f	0.000 ± .100 Å	0.0 ± 3.2
intf	Si	187	s1	.02	186	-0.286 ± .002 f	0.000 ± .004 f	0.000 ± .100 Å	0.0 ± 3.2
intf	Si	188	s1	.02	187	0.429 ± .002 f	-0.857 ± .004 f	0.000 ± .100 Å	0.0 ± 3.2
intf	Si	189	s1	.02	188	0.429 ± .002 f	0.857 ± .004 f	0.020 ± .100 Å	.6 ± 3.2
intf	Si	190	s1	.02	189	-0.286 ± .002 f	0.000 ± .004 f	0.000 ± .100 Å	0.0 ± 3.2
intf	Si	191	s1	.02	190	0.286 ± .002 f	-0.286 ± .004 f	0.000 ± .100 Å	0.0 ± 3.2
intf	Si	192	s1	.02	191	-0.143 ± .002 f	0.000 ± .004 f	0.020 ± .100 Å	.6 ± 3.2
intf	Si	193	s1	.02	192	0.000 ± .002 f	-0.571 ± .004 f	0.000 ± .100 Å	0.0 ± 3.2
intf	Si	194	s1	.02	193	-0.714 ± .002 f	0.571 ± .004 f	0.000 ± .100 Å	0.0 ± 3.2
intf	Si	195	s1	.02	194	0.571 ± .002 f	-0.571 ± .004 f	0.000 ± .100 Å	0.0 ± 3.2
intf	Si	196	s1	.02	195	-0.286 ± .002 f	0.714 ± .004 f	0.000 ± .100 Å	0.0 ± 3.2
intf	Si	197	s1	.02	196	0.286 ± .002 f	0.000 ± .004 f	0.000 ± .100 Å	0.0 ± 3.2
intf	Si	198	s1	.02	197	-0.286 ± .002 f	-0.429 ± .004 f	0.000 ± .100 Å	0.0 ± 3.2
intf	Si	199	s1	.02	198	0.429 ± .002 f	0.000 ± .004 f	0.000 ± .100 Å	0.0 ± 3.2
intf	Si	200	s1	.02	199	-0.714 ± .002 f	0.429 ± .004 f	0.000 ± .100 Å	0.0 ± 3.2
subl	Si	201	b	1.00	200	-0.000 f	0.000 f	2.350 Å	75.1
subl	Si	202	b	1.00	201	0.333 f	0.333 f	0.780 Å	24.9

COMMON NAME : Si(111)-(√3x√3)R30° ILLUSTRATION: 90
CLASSIFICATION : 14.152
TECHNIQUE : LEED
AUTHORS : W.C. Fan, A. Ignatiev, H. Huang and S.Y. Tong
REFERENCE : Phys. Rev. Lett., 62, 1516 (1989)

SURFACE TYPE
Substrate : Si
Crystal face: 111
Temperature : RT
Bulk lattice: diamond
2D bulk symm: p3m1
2D surf symm: p31m
Adsorbate:
Coverage :
Pattern : (√3x√3)R30°
Matrix : (1.000, 1.000)
(-2.000, 1.000)

STRUCTURE TYPE
A first layer Si atom in each (√3x√3) unit cell missing; interlayer spacings: 1st-2nd layer 0.28Å; 2nd-3rd layer 2.28Å; 3rd-4th layer 0.64Å; 2nd layer Si atoms relax towards the vacancy by 0.65Å; similar relaxation in 3rd layer by 0.24Å

SAMPLE PREPARATION (1 sample)
Treatment : 1k eV Ar bombardment followed by annealing at 1000C
Crystallinity: (√3x√3)R30° after annealing
Anal. methods:
Contamination: <1% of Ar

COMMENTS
See C.T. Chang and K.M. Ho, Phys. Rev. Lett. 64, 491 (1990) for the result of a first-principles total-energy calculation on this system: they find a larger compression of the first double layer and a smaller lateral displacement for the 2nd layer atoms

DATA COLLECTION
Technique: LEED; four-grid LEED optics, video camera
Dataset : IV spectra: 30<E<215 eV;
(01),(10),(1/3,1/3),(2/3,2/3)

THEORY/DATA TREATMENT
Dynamical LEED (incl. symmetries in real and reciprocal spaces): 7 phase shifts; Vor=-12 eV

STRUCTURES EXAMINED
Tested previously suggested models and new models: vacancy model provided the best fit; all previously suggested models are ruled out

QUALITY OF EXPERIMENT-THEORY FIT
RVHT=0.29

2D UNIT CELLS (1 domain observed)

Cell	Ax (Å)	Ay (Å)	Bx (Å)	By (Å)	α (°)	Matrix	Pattern	Cell type
Bulk	3.839	0.000	1.920	3.325	60.0	(1.000, 0.000) (0.000, 1.000)	(1x1)	b: bulk lattice
Surface 1	5.759	3.325	-5.759	3.325	120.0	(1.000, 1.000) (-2.000, 1.000)	(√3x√3)R30°	s1: commens. superlattice

3D COORDINATES

Si1-Si2: remainder of top layer (top half of top bilayer); Si3-Si8: bulk-like substrate layers;
Si9-Si10: periodic bulk layers; 0.1Å error bars assumed for tabulation

Dx/Dy in Å, or as a fraction of layer's unit cell vectors; atom 0 at the origin. Epir/subr are bulk repeat vectors.

No. of atoms: 9 Bulk z = 3.130 Å

Region	Chem el.	At. no.	Cell type	Site occ.	Rel. to	Dx ± εx	Dy ± εy	Dz ± εz	Dz/Bz(%) ± εz/Bz
epir		-2				f	f	Å	
subr		-1				1.920 Å	1.108 Å	3.130 Å	
intf	Si	1	s1	.33	0	-3.839 ± .100 Å	0.000 ± .100 Å	0.000 ± .100 Å	0.0 ± 3.2
intf	Si	2	s1	.33	1	-1.920 ± .100 Å	1.758 ± .100 Å	0.280 ± .100 Å	9.0 ± 3.2
intf	Si	3	s1	.33	1	-0.563 ± .100 Å	-2.542 ± .100 Å	0.280 ± .100 Å	9.0 ± 3.2
intf	Si	4	s1	.33	1	0.563 ± .100 Å	-2.542 ± .100 Å	0.280 ± .100 Å	9.0 ± 3.2
intf	Si	5	s1	.33	1	-1.920 ± .100 Å	1.348 ± .100 Å	2.560 ± .100 Å	81.8 ± 3.2
intf	Si	6	s1	.33	1	-0.208 ± .100 Å	-2.337 ± .100 Å	2.560 ± .100 Å	81.8 ± 3.2
intf	Si	7	s1	.33	1	0.208 ± .100 Å	-2.337 ± .100 Å	2.560 ± .100 Å	81.8 ± 3.2
subl	Si	8	b	1.00	1	-1.920 Å	-1.108 Å	3.200 Å	102.2
subl	Si	9	b	1.00	1	-1.920 Å	-1.108 Å	5.550 Å	177.3

Si(111)-(√3x√3)R30°
14.152

BOND DISTANCES AND ANGLES

Bond distances and angles are derived from coordinates

No. of distances/angles: 3

Interatomic dist. A-B (Å)	Atom A	Atom B	Atom C	Bond angle A-B-C (°)
2.618	Si1	Si3	Si2	94.3
2.618	Si1	Si3	Si6	89.2
2.317	Si3	Si6	Si10(-1,0)	114.8

COMMON NAME : Si(111)-(√3x√3)R30°-Al ILLUSTRATION: 96,97
CLASSIFICATION : 14.13.12
TECHNIQUE : LEED
AUTHORS : H. Huang, S.Y. Tong, W.S. Yang, H.D. Shih and F. Jona
REFERENCE : Phys. Rev., B42, 7483 (1990)

SURFACE TYPE
Substrate : Si Adsorbate: Al
Crystal face: 111 Coverage : 0.333 Al/1x1
Temperature : RT Pattern : (√3x√3)R30°
Bulk lattice: diamond Matrix : (1.000, 1.000)
2D bulk symm: p3m1 (-2.000, 1.000)
2D surf symm: p31m

STRUCTURE TYPE
Al adsorbed at T4 site; the three first-layer Si atoms are moved radially inwards and up; the Si below the T4 site is moved down pushing the Si right below it downwards; other 2nd-and 3rd-layer Si atoms below them are moved upwards

SAMPLE PREPARATION (1 sample)
Treatment : Ag evaporation on Si(111); heat to 1000C for 30 min
Crystallinity: sharp LEED pattern
Anal. methods:
Contamination:

COMMENTS
Photoelectron diffraction confirms the T4 site (Daimon et al., Surf. Sci. 221, 244 (1989)); PEXAFS results agree with some bond lengths (Mangat et al., Phys. Rev. B44, 6284 (1991))

DATA COLLECTION
Technique: LEED; TV camera
Dataset : IV spectra for 8 beams (5 integer, 3 fractional) at normal incidence: 25<E<225 eV

THEORY/DATA TREATMENT
Dynamical LEED (incl. symmetries in real and reciprocal spaces): 7 phase shifts; Vor=-12 eV

STRUCTURES EXAMINED
Two overlayer models with Al at the T4 and H3 sites; substitutional model with 1/3 of surface Si atoms replaced by Al; R-factors optimized for each model; model with Al at T4 site gave best fit

QUALITY OF EXPERIMENT-THEORY FIT
RVHT=0.177

2D UNIT CELLS (1 domain observed)

Cell	Ax (Å)	Ay (Å)	Bx (Å)	By (Å)	α (°)	Matrix	Pattern	Cell type
Bulk	3.839	0.000	1.920	3.325	60.0	(1.000, 0.000) (0.000, 1.000)	(1x1)	b: bulk lattice
Surface 1	5.759	3.325	-5.759	3.325	120.0	(1.000, 1.000) (-2.000, 1.000)	(√3x√3)R30°	s1: commens. superlattice

3D COORDINATES

Al1: adsorbate layer at T4 site; Si2-Si4: first Si layer (half bilayer); Si5-Si7: second Si layer; Si8-10: third Si layer; Si11-Si12: repeating bulk layers; 0.1Å error bars assumed for tabulation

Dx/Dy in Å, or as a fraction of layer's unit cell vectors; atom 0 at the origin. Epir/subr are bulk repeat vectors.

No. of atoms: 12 Bulk z = 3.130 Å

Region	Chem el.	At. no.	Cell type	Site occ.	Rel. to	Dx ± εx	Dy ± εy	Dz ± εz	Dz/Bz(%) ± εz/Bz
epir		-2				f	f	Å	
subr		-1				1.920 Å	1.108 Å	3.130 Å	
intf	Al	1	s1	.33	0	0.000 ± .100 Å	0.000 ± .100 Å	0.000 ± .100 Å	0.0 ± 3.2
intf	Si	2	s1	.33	1	1.790 ± .100 Å	-1.033 ± .100 Å	1.390 ± .100 Å	42.1 ± 3.2
intf	Si	3	s1	.33	2	3.969 ± .100 Å	-0.225 ± .100 Å	0.000 ± .100 Å	0.0 ± 3.2
intf	Si	4	s1	.33	3	3.969 ± .100 Å	0.225 ± .100 Å	0.000 ± .100 Å	0.0 ± 3.2
intf	Si	5	s1	.33	1	0.000 ± .100 Å	0.000 ± .100 Å	2.630 ± .100 Å	84.0 ± 3.2
intf	Si	6	s1	.33	5	3.839 ± .100 Å	0.000 ± .100 Å	-0.600 ± .100 Å	-19.2 ± 3.2
intf	Si	7	s1	.33	6	3.839 ± .100 Å	0.000 ± .100 Å	0.000 ± .100 Å	0.0 ± 3.2
intf	Si	8	s1	.33	5	0.000 ± .100 Å	0.000 ± .100 Å	2.230 ± .100 Å	71.3 ± 3.2
intf	Si	9	s1	.33	6	0.000 ± .100 Å	0.000 ± .100 Å	2.430 ± .100 Å	77.6 ± 3.2
intf	Si	10	s1	.33	7	0.000 ± .100 Å	0.000 ± .100 Å	2.430 ± .100 Å	77.6 ± 3.2
subl	Si	11	b	1.00	1	1.920 Å	1.108 Å	5.320 Å	170.0
subl	Si	12	b	1.00	11	0.000 Å	0.000 Å	2.350 Å	75.1

Si(111)-(√3x√3)R30°-Al
14.13.12

BOND DISTANCES AND ANGLES

Bond distances and angles are derived from coordinates

No. of distances/angles: 3

Interatomic dist. A-B (Å)	Atom A	Atom B	Atom C	Bond angle A-B-C (°)
2.490	Si2	Al1	Si3(0,1)	91.9
2.630	Al1	Si5	Si2	59.0
2.380	Si2	Si6	Si9	105.6

COMMON NAME : Si(111)-(1x1)-As
CLASSIFICATION : 14.33.10
TECHNIQUE : MEIS
AUTHORS : M. Copel and R.M. Tromp
REFERENCE : Phys. Rev., B37, 2766 (1987)

ILLUSTRATION: 102

SURFACE TYPE
Substrate : Si
Crystal face: 111
Temperature : RT*
Bulk lattice: diamond
2D bulk symm: p3m1
2D surf symm: p3m1

Adsorbate: As
Coverage : 0.92±0.05 As/Si
Pattern : (1x1)
Matrix : (1.000, 0.000)
(0.000, 1.000)

STRUCTURE TYPE
Atomic substitutional replacement of top half of top Si bilayer: otherwise unreconstructed, unrelaxed substrate

SAMPLE PREPARATION (1 sample)
Treatment : exposure to high As flux, flash to 1323 K, cool to 623K
Crystallinity: sharp LEED pattern
Anal. methods: STM
Contamination: AES: C/Si < 1E-3

COMMENTS
Alternative phase was also found with different preparation procedure: authors conclude that the ordered (1x1) surface includes 20% of second phase; this allows a smaller As vibrational amplitude in their models; STM suggests that Headrick et al As deficiency model is incorrect

DATA COLLECTION
Technique: MEIS
Dataset : 94-101k eV protons in [00-1] direction: comparison with theory at scattering angle of 45°-65° ≈ [11-1] direction

THEORY/DATA TREATMENT
Comparison to Monte Carlo channeling and blocking calculations; As rms vibs = 0.109Å

STRUCTURES EXAMINED
As in perfect bulk termination with variable As-Si layer spacing; best match for 80% of perfect termination mixed with 20% of the exp. results for the alternative phase (see comment)

2D UNIT CELLS (1 domain observed)

Cell	Ax (Å)	Ay (Å)	Bx (Å)	By (Å)	α (°)	Matrix	Pattern	Cell type
Bulk	3.840	0.000	-1.920	3.326	120.0	(1.000, 0.000) (0.000, 1.000)	(1x1)	b: bulk lattice
Surface 1	3.840	0.000	-1.920	3.326	120.0	(1.000, 0.000) (0.000, 1.000)	(1x1)	s1: commens. superlattice

3D COORDINATES

As1 substitutionally replaces top half of 1st bilayer of Si lattice

Dx/Dy in Å, or as a fraction of layer's unit cell vectors; atom 0 at the origin. Epir/subr are bulk repeat vectors.

No. of atoms: 4 Bulk z = 3.130 Å

Reg ion	Chem el.	At. no.	Cell type	Site occ.	Rel. to	Dx ± εx		Dy ± εy		Dz ± εz		Dz/Bz(%) ± εz/Bz
epir		-2					f		f		Å	
subr		-1				-1.920	Å	1.109	Å	3.130	Å	
ovrl	As	1	b	1.00	0	0.000	f	0.000	f	0.000	Å	0.0
intf	Si	2	b	1.00	1	0.667	f	0.333	f	0.990 ± .100	Å	31.6 ± 3.2
subl	Si	3	b	1.00	2	0.000	f	0.000	f	2.350	Å	75.1
subl	Si	4	b	1.00	3	-0.333	f	0.333	f	0.780	Å	24.9

BOND DISTANCES AND ANGLES

Bond distances and angles are derived from coordinates

No. of distances/angles: 2

Interatomic dist. A-B (Å)	Atom A	Atom B	Atom C	Bond angle A-B-C (°)
2.428	As1	Si2	Si3	114.1
2.350	Si2	Si3	Si4	109.4

COMMON NAME : Si(111)-(1x1)-As
CLASSIFICATION : 14.33.7
TECHNIQUE : XSW
AUTHORS : J.R. Patel, J.A. Golovchenko, P.E. Freeland and H.J. Grossman
REFERENCE : Phys. Rev., B36, 7715 (1987)

ILLUSTRATION: 102

SURFACE TYPE
Substrate : Si
Crystal face: 111
Temperature : RT
Bulk lattice: diamond
2D bulk symm: p3m1
2D surf symm: p3m1

Adsorbate: As
Coverage : 0.93 As/Si
Pattern : (1x1)
Matrix : (1.000, 0.000)
(0.000, 1.000)

STRUCTURE TYPE
Atomic substitutional replacement of top half of top Si bilayer: otherwise unreconstructed, unrelaxed substrate

SAMPLE PREPARATION (1 sample)
Treatment : exposure to precleaned As effusion cell at 533 K
Crystallinity: sharp LEED pattern
Anal. methods:
Contamination: AES: trace of C, no O

COMMENTS

DATA COLLECTION
Technique: XSW; x-ray standing wave method
Dataset : fluorescence yield from As Kα line: angle of incidence varied from -3° to +3°

THEORY/DATA TREATMENT

STRUCTURES EXAMINED
As position found directly from peak position in As Kα fluorescence

2D UNIT CELLS (1 domain observed)

Cell	Ax (Å)	Ay (Å)	Bx (Å)	By (Å)	α (°)	Matrix	Pattern	Cell type
Bulk	3.840	0.000	-1.920	3.326	120.0	(1.000, 0.000) (0.000, 1.000)	(1x1)	b: bulk lattice
Surface 1	3.840	0.000	-1.920	3.326	120.0	(1.000, 0.000) (0.000, 1.000)	(1x1)	s1: commens. superlattice

3D COORDINATES

As1 substitutionally replaces top half of 1st bilayer of Si lattice

Dx/Dy in Å, or as a fraction of layer's unit cell vectors; atom 0 at the origin. Epir/subr are bulk repeat vectors.

No. of atoms: 4 Bulk z = 3.130 Å

Region	Chem el.	At. no.	Cell type	Site occ.	Rel. to	Dx ± εx		Dy ± εy		Dz ± εz		Dz/Bz(%) ± εz/Bz
epir		-2					f		f		Å	
subr		-1				-1.920	Å	1.109	Å	3.130	Å	
ovrl	As	1	b	1.00	0	0.000	f	0.000	f	0.000	Å	0.0
intf	Si	2	b	1.00	1	0.667	f	0.333	f	0.960 ± .030	Å	30.7 ± 1.0
subl	Si	3	b	1.00	2	0.000	f	0.000	f	2.350	Å	75.1
subl	Si	4	b	1.00	3	-0.333	f	0.333	f	0.780	Å	24.9

BOND DISTANCES AND ANGLES

Bond distances and angles are derived from coordinates

No. of distances/angles: 2

Interatomic dist. A-B (Å)	Atom A	Atom B	Atom C	Bond angle A-B-C (°)
2.416	As1	Si2	Si3	113.4
2.350	Si2	Si3	Si4	109.4

COMMON NAME : Si(111)-(1x1)-As
CLASSIFICATION : 14.33.8
TECHNIQUE : MEIS
AUTHORS : R.L. Headrik and W.R. Graham
REFERENCE : J. Vac. Sci. Technol., A6, 637 (1987)

ILLUSTRATION: 102

SURFACE TYPE
Substrate : Si
Crystal face: 111
Temperature : RT*
Bulk lattice: diamond
2D bulk symm: p3m1
2D surf symm: p3m1

Adsorbate: As
Coverage : 0.93±0.04 As/Si
Pattern : (1x1)
Matrix : (1.000, 0.000)
(0.000, 1.000)

STRUCTURE TYPE
Atomic substitutional replacement of top half of top Si bilayer: otherwise unreconstructed, unrelaxed substrate

SAMPLE PREPARATION (1 sample)
Treatment : (7x7) exposed for 1min to As4 beam at 1123 K and 10E-6 torr
Crystallinity: sharp LEED pattern
Anal. methods:
Contamination: AES: C/Si ratio < 5E-3; no O found

COMMENTS
A better R-factor was found (R=0.51) if 7-14% of the surface As is deficient;
the sample properties, especially resistance to contamination, were found to be highly dependent on surface preparation

DATA COLLECTION
Technique: MEIS
Dataset : 60, 100, 140, 180k eV protons at normal incidence

THEORY/DATA TREATMENT
Comparison to Monte Carlo channeling and blocking calculations; As rms vibs = 0.14±0.02Å

STRUCTURES EXAMINED
As in perfect bulk termination with variable As-Si layer spacing; various defect structures to account for the <1ML coverage

QUALITY OF EXPERIMENT-THEORY FIT
R=0.59

2D UNIT CELLS (1 domain observed)

Cell	Ax (Å)	Ay (Å)	Bx (Å)	By (Å)	α (°)	Matrix	Pattern	Cell type
Bulk	3.840	0.000	-1.920	3.326	120.0	(1.000, 0.000) (0.000, 1.000)	(1x1)	b: bulk lattice
Surface 1	3.840	0.000	-1.920	3.326	120.0	(1.000, 0.000) (0.000, 1.000)	(1x1)	s1: commens. superlattice

3D COORDINATES

As1 substitutionally replaces top half of 1st bilayer of Si lattice

Dx/Dy in Å, or as a fraction of layer's unit cell vectors; atom 0 at the origin. Epir/subr are bulk repeat vectors.

No. of atoms: 4

Bulk z = 3.130 Å

Region	Chem el.	At. no.	Cell type	Site occ.	Rel. to	Dx ± εx		Dy ± εy		Dz ± εz		Dz/Bz(%) ± εz/Bz
epir		-2					f		f		Å	
subr		-1				-1.920	Å	1.109	Å	3.130	Å	
ovrl	As	1	b	1.00	0	0.000	f	0.000	f	0.000	Å	0.0
intf	Si	2	b	1.00	1	0.667	f	0.333	f	1.020 ± .060	Å	32.6 ± 1.9
subl	Si	3	b	1.00	2	0.000	f	0.000	f	2.350	Å	75.1
subl	Si	4	b	1.00	3	-0.333	f	0.333	f	0.780	Å	24.9

BOND DISTANCES AND ANGLES

Bond distances and angles are derived from coordinates

No. of distances/angles: 2

Interatomic dist. A-B (Å)	Atom A	Atom B	Atom C	Bond angle A-B-C (°)
2.440	As1	Si2	Si3	114.7
2.350	Si2	Si3	Si4	109.4

COMMON NAME : Si(111)-(√3x√3)R30°-B
CLASSIFICATION : 14.5.6
TECHNIQUE : LEED
AUTHORS : H. Huang, S.Y. Tong, J. Quinn and F. Jona
REFERENCE : Phys. Rev., B41, 3276 (1990)

ILLUSTRATION: 101

SURFACE TYPE
Substrate : Si
Crystal face: 111
Temperature : RT
Bulk lattice: diamond
2D bulk symm: p3m1
2D surf symm: p31m

Adsorbate: B
Coverage : 0.333 B/1x1
Pattern : (√3x√3)R30°
Matrix : (1.000, 1.000)
(-2.000, 1.000)

STRUCTURE TYPE
B atom replaces a second layer Si atom, which becomes an adatom at the T4 site over the B atom; this B position is the B5 site; first-layer Si atoms moved towards the B atom and downwards; second Si layer moved up, B moved down, and third Si layer also moved down

SAMPLE PREPARATION (1 sample)
Treatment : Ar bombardment of B-doped Si wafers followed by annealing
Crystallinity: excellent (√3x√3)R30° LEED pattern
Anal. methods: AES for contamination
Contamination: AES: C, O near noise level

COMMENTS
The same structure with different parameters was proposed previously by R.L. Headrick, I.K. Robinson, E. Vlieg and L.C. Feldman, Phys. Rev. Lett. 63, 1253 (1989), but the structure reported here has more parameters experimentally determined

DATA COLLECTION
Technique: LEED; video camera
Dataset : IV spectra for 18 (9 integral, 9 fractional) beams at normal incidence: 30<E<260 eV

THEORY/DATA TREATMENT
Dynamical LEED (incl. symmetries in real and reciprocal spaces): 6 phase shifts; Vor=-6 eV

STRUCTURES EXAMINED
Surface models in which B atom was placed as an adatom at H3, T1 and T4 sites, plus the B5 site; all others gave modified Pendry R factors at least 25% larger compared to that for B5 site

QUALITY OF EXPERIMENT-THEORY FIT
Modified RPE=0.244

2D UNIT CELLS (1 domain observed)

Cell	Ax (Å)	Ay (Å)	Bx (Å)	By (Å)	α (°)	Matrix	Pattern	Cell type
Bulk	3.839	0.000	1.920	3.325	60.0	(1.000, 0.000) (0.000, 1.000)	(1x1)	b: bulk lattice
Surface 1	5.759	3.325	-5.759	3.325	120.0	(1.000, 1.000) (-2.000, 1.000)	(√3x√3)R30°	s1: commens. superlattice

3D COORDINATES

Si1: adatom layer; Si2-Si4: 1st Si layer (half bilayer); Si6-Si7: 2nd Si layer; B5: B layer in B5 sites; Si8-10: 3rd Si layer; Si11-Si12: repeating bulk layers

Dx/Dy in Å, or as a fraction of layer's unit cell vectors; atom 0 at the origin. Epir/subr are bulk repeat vectors.

No. of atoms: 12 Bulk z = 3.130 Å

Reg ion	Chem el.	At. no.	Cell type	Site occ.	Rel. to	Dx ± εx	Dy ± εy	Dz ± εz	Dz/Bz(%) ± εz/Bz
epir		-2				f	f	Å	
subr		-1				1.920 Å	1.108 Å	3.130 Å	
intf	Si	1	s1	.33	0	0.000 ± .200 Å	0.000 ± .200 Å	0.000 ± .100 Å	0.0 ± 3.2
intf	Si	2	s1	.33	1	1.660 ± .200 Å	-0.958 ± .200 Å	1.340 ± .100 Å	42.8 ± 3.2
intf	Si	3	s1	.33	2	4.099 ± .200 Å	-0.450 ± .200 Å	0.000 ± .100 Å	0.0 ± 3.2
intf	Si	4	s1	.33	3	4.099 ± .200 Å	0.450 ± .200 Å	0.000 ± .100 Å	0.0 ± 3.2
intf	B	5	s1	.33	1	0.000 ± .200 Å	0.000 ± .200 Å	2.320 ± .100 Å	74.1 ± 3.2
intf	Si	6	s1	.33	5	3.839 ± .200 Å	0.000 ± .200 Å	-0.550 ± .100 Å	-17.6 ± 3.2
intf	Si	7	s1	.33	6	3.839 ± .200 Å	0.000 ± .200 Å	0.000 ± .100 Å	0.0 ± 3.2
intf	Si	8	s1	.33	5	0.000 ± .200 Å	0.000 ± .200 Å	2.190 ± .100 Å	70.0 ± 3.2
intf	Si	9	s1	.33	6	0.000 ± .200 Å	0.000 ± .200 Å	2.400 ± .100 Å	76.8 ± 3.2
intf	Si	10	s1	.33	7	0.000 ± .200 Å	0.000 ± .200 Å	2.400 ± .100 Å	76.8 ± 3.2
subl	Si	11	b	1.00	8	1.920 Å	1.108 Å	0.440 Å	14.1
subl	Si	12	b	1.00	11	0.000 Å	0.000 Å	2.350 Å	75.1

Si(111)-(√3x√3)R30°-B
14.5.6 BOND DISTANCES AND ANGLES

Bond distances and angles are derived from coordinates

No. of distances/angles: 3

Interatomic dist. A-B (Å)	Atom A	Atom B	Atom C	Bond angle A-B-C (°)
2.320	Si1	B5	Si2	62.9
2.190	Si8	B5	Si1	180.0
2.154	Si2	B5	Si2(0,1)	100.9

COMMON NAME : Si(111)-(√3x√3)R30°-Bi
CLASSIFICATION : 14.83.2
TECHNIQUE : XRD
AUTHORS : T. Takahashi, S. Nakatani, T. Ishikawa and S. Kikuta
REFERENCE : Surf. Sci., 191, L825 (1987)

ILLUSTRATION: 103

SURFACE TYPE
Substrate : Si
Crystal face: 111
Temperature : RT*
Bulk lattice: diamond
2D bulk symm: p3m1
2D surf symm: p31m

Adsorbate: Bi
Coverage : 1 Bi/Si
Pattern : (√3x√3)R30°
Matrix : (1.000, 1.000)
(-2.000, 1.000)

STRUCTURE TYPE
Triangles of 3 Bi replace every third Si in top layer, which is lower half of a bilayer

SAMPLE PREPARATION (1 sample)
Treatment : Bi evaporated from Knudsen cell at RT, annealing at 613 K
Crystallinity: etched wafer had good (7x7) LEED pattern
Anal. methods:
Contamination:

COMMENTS
Good fit was also obtained with Si 0.8Å below Bi, but the bond angles suggest that the reported structure is preferred

DATA COLLECTION
Technique: XRD
Dataset : x-ray I(E) curves: 1.0Å<lambda<2.5Å; (00),(10),(-10),(02) rods

THEORY/DATA TREATMENT
Data corrected for collector efficiency and diffr. geometry; x-ray theory (interference between Bi, surface and bulk Si)

STRUCTURES EXAMINED
Bi to next layer spacing found from (00) beam, then 6 models studied: Tij denotes Bi atom above ith Si layer with lateral displacement to jth Si layer; Tij=T12, T14, T21, T24, T41, T42; substrate reconstruction examined with <=6 Si atoms; relaxation of 1st and 2nd substrate Si layers <0.15Å

QUALITY OF EXPERIMENT-THEORY FIT
RX=0.12

2D UNIT CELLS (1 domain observed)

Cell	Ax (Å)	Ay (Å)	Bx (Å)	By (Å)	α (°)	Matrix	Pattern	Cell type
Bulk	3.838	0.000	1.919	3.324	60.0	(1.000, 0.000) (0.000, 1.000)	(1x1)	b: bulk lattice
Surface 1	5.757	3.324	-5.757	3.324	120.0	(1.000, 1.000) (-2.000, 1.000)	(√3x√3)R30°	s1: commens. superlattice

3D COORDINATES

Si1-Si2: remainder of top 1/2 of a bilayer; Bi3-Bi4-Bi5: triangles substituting in top 1/2 of bilayer; Si6-Si7: bulk-like bilayer; Si8-Si9: periodically repeating bulk bilayer

Dx/Dy in Å, or as a fraction of layer's unit cell vectors; atom 0 at the origin. Epir/subr are bulk repeat vectors.

No. of atoms: 9 Bulk z = 3.130 Å

Region	Chem el.	At. no.	Cell type	Site occ.	Rel. to	Dx ± εx		Dy ± εy		Dz ± εz		Dz/Bz(%) ± εz/Bz
epir		-2					f		f		Å	
subr		-1				0.000	Å	2.216	Å	3.130	Å	
intf	Si	1	s1	.33	0	0.000	f	0.000	f	0.000	Å	0.0
intf	Si	2	s1	.33	1	0.667	f	0.333	f	0.000	Å	0.0
intf	Bi	3	s1	.33	2	-0.065	f	0.333	f	0.800 ± .200	Å	25.6 ± 6.4
intf	Bi	4	s1	.33	3	-0.268	f	0.268	f	0.000	Å	0.0
intf	Bi	5	s1	.33	4	-0.268	f	-0.536	f	0.000	Å	0.0
intf	Si	6	b	1.00	5	1.065	f	-0.131	f	2.680 ± .200	Å	85.6 ± 6.4
intf	Si	7	b	1.00	6	-0.333	f	-0.333	f	0.780	Å	24.9
subl	Si	8	b	1.00	7	0.000	f	0.000	f	2.350	Å	75.1
subl	Si	9	b	1.00	8	-0.333	f	0.667	f	0.780	Å	24.9

Si(111)-(√3x√3)R30°-Bi
14.83.2 BOND DISTANCES AND ANGLES

Bond distances and angles are derived from coordinates

No. of distances/angles: 4

Interatomic dist. A-B (Å)	Atom A	Atom B	Atom C	Bond angle A-B-C (°)
2.589	Si2	Bi3	Si1(1,1)	95.7
2.589	Si2	Bi3	Bi5	98.4
3.086	Bi3	Bi4	Si6(-2,1)	98.0
2.349	Si6	Si7	Si8	109.4

COMMON NAME : Si(111)-(√3x√3)R30°-Bi (1/3 ML) ILLUSTRATION: 96,97
CLASSIFICATION : 14.83.3a
TECHNIQUE : LEED
AUTHORS : K.J. Wan, T. Guo, W.K. Ford and J.C. Hermanson
REFERENCE : Phys. Rev., B44, 3471 (1991)

SURFACE TYPE
Substrate : Si
Crystal face: 111
Temperature : RT
Bulk lattice: diamond
2D bulk symm: p3m1
2D surf symm: p31m

Adsorbate: Bi
Coverage : 0.333 Bi/1x1
Pattern : (√3x√3)R30°
Matrix : (1.000, 1.000)
(-2.000, 1.000)

STRUCTURE TYPE
Bi atoms are at T4 site; nearest Si neighbors squeezed together; Si atom directly below Bi and third layer Si atom directly below are pushed downwards; second-layer Si atoms surrounding the hollow H3 site are moved upwards

SAMPLE PREPARATION (1 sample)
Treatment : deposit 3ML of Bi on Si(111)-(7x7), then anneal
Crystallinity: sharp LEED pattern
Anal. methods: AES
Contamination: AES: no inpurity signal

COMMENTS
AES used to monitor the Bi coverage.

DATA COLLECTION
Technique: LEED; video optics
Dataset : IV spectra for 9 (5-integer, 4-fractional) beams at normal incidence; 30<E<280 eV

THEORY/DATA TREATMENT
Dynamical LEED (matrix inversion): 6 phase shifts

STRUCTURES EXAMINED
T4, H3 and top-layer models; latter two models were excluded based on x-ray R-factors

QUALITY OF EXPERIMENT-THEORY FIT
Average RX=0.233

2D UNIT CELLS (1 domain observed)

Cell	Ax (Å)	Ay (Å)	Bx (Å)	By (Å)	α (°)	Matrix	Pattern	Cell type
Bulk	3.839	0.000	1.920	3.325	60.0	(1.000, 0.000) (0.000, 1.000)	(1x1)	b: bulk lattice
Surface 1	5.759	3.325	-5.759	3.325	120.0	(1.000, 1.000) (-2.000, 1.000)	(√3x√3)R30°	s1: commens. superlattice

3D COORDINATES

Bi1: at T4 site; Si2-Si4: top layer Si (half bilayer); Si5-Si7: second layer; Si8-Si10: third layer; Si11-Si12: repeating bulk layers; 0.1Å error bars assumed for tabulation

Dx/Dy in Å, or as a fraction of layer's unit cell vectors; atom 0 at the origin. Epir/subr are bulk repeat vectors.

No. of atoms: 12 Bulk z = 3.130 Å

Reg ion	Chem el.	At. no.	Cell type	Site occ.	Rel. to	Dx ± εx	Dy ± εy	Dz ± εz	Dz/Bz(%) ± εz/Bz
epir		-2				f	f	Å	
subr		-1				1.920 Å	1.108 Å	3.130 Å	
intf	Bi	1	s1	.33	0	0.000 ± .100 Å	0.000 ± .100 Å	0.000 ± .100 Å	0.0 ± 3.2
intf	Si	2	s1	.33	1	1.834 ± .100 Å	-1.059 ± .100 Å	1.107 ± .100 Å	35.4 ± 3.2
intf	Si	3	s1	.33	2	3.926 ± .100 Å	-0.100 ± .100 Å	0.000 ± .100 Å	0.0 ± 3.2
intf	Si	4	s1	.33	3	3.926 ± .100 Å	0.100 ± .100 Å	0.000 ± .100 Å	0.0 ± 3.2
intf	Si	5	s1	.33	1	0.000 ± .100 Å	0.000 ± .100 Å	2.445 ± .100 Å	78.1 ± 3.2
intf	Si	6	s1	.33	5	3.839 ± .100 Å	0.000 ± .100 Å	-0.684 ± .100 Å	-21.9 ± 3.2
intf	Si	7	s1	.33	6	3.839 ± .100 Å	0.000 ± .100 Å	0.000 ± .100 Å	0.0 ± 3.2
intf	Si	8	s1	.33	5	0.000 ± .100 Å	0.000 ± .100 Å	2.239 ± .100 Å	71.5 ± 3.2
intf	Si	9	s1	.33	6	0.000 ± .100 Å	0.000 ± .100 Å	2.476 ± .100 Å	79.1 ± 3.2
intf	Si	10	s1	.33	7	0.000 ± .100 Å	0.000 ± .100 Å	2.476 ± .100 Å	79.1 ± 3.2
subl	Si	11	b	1.00	10	1.920 Å	1.108 Å	0.780 Å	24.9
subl	Si	12	b	1.00	11	0.000 Å	0.000 Å	2.350 Å	75.1

Si(111)-(√3x√3)R30°-Bi (1/3 ML)
14.83.3a BOND DISTANCES AND ANGLES

Bond distances and angles are derived from coordinates

No. of distances/angles: 3

Interatomic dist. A-B (Å)	Atom A	Atom B	Atom C	Bond angle A-B-C (°)
2.388	Si2	Bi1	Si3(-1,0)	100.3
2.388	Bi1	Si2	Si3	69.8
2.239	Si5	Si8	Si11(-1,0)	98.5

COMMON NAME : Si(111)-(√3x√3)R30°-Bi (1 ML) ILLUSTRATION: 100
CLASSIFICATION : 14.83.3b
TECHNIQUE : LEED
AUTHORS : K.J. Wan, T. Guo, W.K. Ford and J.C. Hermanson
REFERENCE : Phys. Rev., B44, 3471 (1991)

SURFACE TYPE
Substrate : Si
Crystal face: 111
Temperature : RT
Bulk lattice: diamond
2D bulk symm: p3m1
2D surf symm: p31m

Adsorbate: Bi
Coverage : 1.0 Bi/1x1
Pattern : (√3x√3)R30°
Matrix : (1.000, 1.000)
(-2.000, 1.000)

STRUCTURE TYPE
Bi trimers centered at T4 site; nearest Si neighbors squeezed together; Si atom directly below Bi and third layer Si atom directly below pushed downwards; second-layer Si atoms surrounding the hollow H3 site are moved upwards

SAMPLE PREPARATION (1 sample)
Treatment : deposit 3ML of Bi on Si(111)-(7x7), then anneal
Crystallinity: sharp LEED pattern
Anal. methods: AES
Contamination: AES: no inpurity signal

COMMENTS
AES used to monitor the Bi coverage.

DATA COLLECTION
Technique: LEED; video optics
Dataset : IV spectra for 9 (5-integer, 4-fractional) beams at normal incidence: 30<E<280 eV

THEORY/DATA TREATMENT
Dynamical LEED (matrix inversion): 6 phase shifts

STRUCTURES EXAMINED
Takahashi model (Surf. Sci. 191, L825 (1987)): center of trimer was placed on both the T4 and the H3 sites; T4 site produced the best fit

QUALITY OF EXPERIMENT-THEORY FIT
Average RX=0.225

2D UNIT CELLS (1 domain observed)

Cell	Ax (Å)	Ay (Å)	Bx (Å)	By (Å)	α (°)	Matrix	Pattern	Cell type
Bulk	3.839	0.000	1.920	3.325	60.0	(1.000, 0.000) (0.000, 1.000)	(1x1)	b: bulk lattice
Surface 1	5.759	3.325	-5.759	3.325	120.0	(1.000, 1.000) (-2.000, 1.000)	(√3x√3)R30°	s1: commens. superlattice

3D COORDINATES

Bi1-Bi3: trimer centered at T4 site; Si4-Si6: top Si layer (half bilayer); Si7-Si9: second Si layer; Si10-Si12: third Si layer; Si13-Si14: periodic bulk layers; 0.1Å error bars assumed for tabulation

Dx/Dy in Å, or as a fraction of layer's unit cell vectors; atom 0 at the origin. Epir/subr are bulk repeat vectors.

No. of atoms: 14 Bulk z = 3.130 Å

Region	Chem el.	At. no.	Cell type	Site occ.	Rel. to	Dx ± εx	Dy ± εy	Dz ± εz	Dz/Bz(%) ± εz/Bz
epir		-2				f	f	Å	
subr		-1				1.920 Å	1.108 Å	3.130 Å	
intf	Bi	1	s1	.33	0	1.464 ± .100 Å	-0.845 ± .100 Å	0.000 ± .100 Å	0.0 ± 3.2
intf	Bi	2	s1	.33	0	-1.464 ± .100 Å	-0.845 ± .100 Å	0.000 ± .100 Å	0.0 ± 3.2
intf	Bi	3	s1	.33	0	0.000 ± .100 Å	1.691 ± .100 Å	0.000 ± .100 Å	0.0 ± 3.2
intf	Si	4	s1	.33	0	1.818 ± .100 Å	-1.049 ± .100 Å	2.210 ± .100 Å	70.6 ± 3.2
intf	Si	5	s1	.33	4	3.941 ± .100 Å	-0.118 ± .100 Å	0.000 ± .100 Å	0.0 ± 3.2
intf	Si	6	s1	.33	5	3.941 ± .100 Å	0.118 ± .100 Å	0.000 ± .100 Å	0.0 ± 3.2
intf	Si	7	s1	.33	0	0.000 ± .100 Å	0.000 ± .100 Å	3.326 ± .100 Å	106.3 ± 3.2
intf	Si	8	s1	.33	7	3.839 ± .100 Å	0.000 ± .100 Å	-0.343 ± .100 Å	-11.0 ± 3.2
intf	Si	9	s1	.33	8	3.839 ± .100 Å	0.000 ± .100 Å	0.000 ± .100 Å	0.0 ± 3.2
intf	Si	10	s1	.33	7	0.000 ± .100 Å	0.000 ± .100 Å	2.239 ± .100 Å	71.5 ± 3.2
intf	Si	11	s1	.33	8	0.000 ± .100 Å	0.000 ± .100 Å	2.357 ± .100 Å	75.3 ± 3.2
intf	Si	12	s1	.33	9	0.000 ± .100 Å	0.000 ± .100 Å	2.357 ± .100 Å	75.3 ± 3.2
subl	Si	13	b	1.00	11	1.920 Å	1.108 Å	0.780 Å	24.9
subl	Si	14	b	1.00	13	0.000 Å	0.000 Å	2.350 Å	75.1

Si(111)-($\sqrt{3}$x$\sqrt{3}$)R30°-Bi (1 ML)
14.83.3b

BOND DISTANCES AND ANGLES

Bond distances and angles are derived from coordinates

No. of distances/angles: 5

Interatomic dist. A-B (Å)	Atom A	Atom B	Atom C	Bond angle A-B-C (°)
2.928	Bi1	Bi2	Bi3	60.0
2.247	Bi1	Si4	Si7	107.5
2.377	Si4	Si7	Si10	92.0
2.322	Si4	Si8	Si11	109.5
2.239	Si7	Si10	Si13(-1,0)	101.6

COMMON NAME : Si(111)-Br 0.25ML
CLASSIFICATION : 14.35.2
TECHNIQUE : fluorescence XRD
AUTHORS : G. Materlik, A. Frohm and M.J. Bedzyk
REFERENCE : Phys. Rev. Lett., 52, 441 (1984)

ILLUSTRATION: 96,99

SURFACE TYPE
Substrate : Si
Crystal face: 111
Temperature : RT*
Bulk lattice: diamond
2D bulk symm: p3m1
2D surf symm: p3m1

Adsorbate: Br
Coverage : 0.25 ML
Pattern : (1x1)
Matrix : (1.000, 0.000)
(0.000, 1.000)

STRUCTURE TYPE
Atomic adsorption on top sites;
modeled here as (1x1) despite 0.25 ML

SAMPLE PREPARATION (1 sample)
Treatment : chemical cleaning
Crystallinity:
Anal. methods:
Contamination: not checked (in-air experiment)

COMMENTS
Data recorded in air: state of surface poorly determined; Laue case geometry (entrance and exit surfaces of incident and diffracted beams are different) allows direct determination of atomic positions in a plane parallel to the surface, although only the local geometry is determined here

DATA COLLECTION
Technique: fluorescence XRD; Br fluorescence
Dataset : fluorescent signals from x-ray standing waves from (220) reflections

THEORY/DATA TREATMENT
Dynamical x-ray diffraction

STRUCTURES EXAMINED
Top and 3-fold adsorption sites of Br

QUALITY OF EXPERIMENT-THEORY FIT
Visual

2D UNIT CELLS (1 domain observed)

Cell	Ax (Å)	Ay (Å)	Bx (Å)	By (Å)	α (°)	Matrix	Pattern	Cell type
Bulk	3.840	0.000	-1.920	3.326	120.0	(1.000, 0.000) (0.000, 1.000)	(1x1)	b: bulk lattice
Surface 1	3.840	0.000	-1.920	3.326	120.0	(1.000, 0.000) (0.000, 1.000)	(1x1)	s1: commens. superlattice

3D COORDINATES

Br1: top-site overlayer; Si2-Si3: repeating bulk bilayer

Dx/Dy in Å, or as a fraction of layer's unit cell vectors; atom 0 at the origin. Epir/subr are bulk repeat vectors.

No. of atoms: 3

Bulk z = 3.130 Å

Reg ion	Chem el.	At. no.	Cell type	Site occ.	Rel. to	Dx ± εx		Dy ± εy		Dz ± εz		Dz/Bz(%) ± εz/Bz
epir		-2					f		f		Å	
subr		-1				0.000	Å	2.217	Å	3.130	Å	
ovrl	Br	1	b	1.00	0	0.000	f	0.000	f	0.000	Å	0.0
subl	Si	2	b	1.00	1	0.000	f	0.000	f	2.180 ± .060	Å	69.7 ± 1.9
subl	Si	3	b	1.00	2	0.333	f	0.667	f	0.780	Å	24.9

BOND DISTANCES AND ANGLES

No. of distances/angles: 2

Interatomic dist. A-B (Å)	Atom A	Atom B	Atom C	Bond angle A-B-C (°)
2.180	Br1	Si2	Si3	109.4
2.350	Si2	Si3	Si2(0,1)	109.6

COMMON NAME : Si(111)-Br 0.67ML
CLASSIFICATION : 14.35.1
TECHNIQUE : fluorescence XRD
AUTHORS : J.A. Golovchenko, J.R. Patel, D.R. Kaplan, P.L. Cowan and M.J. Bedzyk
REFERENCE : Phys. Rev. Lett., 49, 1560 (1982)

ILLUSTRATION: 96,99

SURFACE TYPE
Substrate : Si
Crystal face: 111
Temperature : 300 K
Bulk lattice: diamond
2D bulk symm: p3m1
2D surf symm: p3m1

Adsorbate: Br
Coverage : 0.67 ML
Pattern : (1x1)
Matrix : (1.000, 0.000)
(0.000, 1.000)

STRUCTURE TYPE
Atomic adsorption on top sites;
modeled here as (1x1) despite 0.67 ML

SAMPLE PREPARATION (1 sample)
Treatment : chemical cleaning
Crystallinity:
Anal. methods:
Contamination: not checked (in-air experiment)

COMMENTS
Data recorded in air: state of surface poorly determined;
sample and results stable for days;
substrate relaxation found to be <0.03Å

DATA COLLECTION
Technique: fluorescence XRD; Br fluorescence
Dataset : reflectivity for (111) and (220) Bragg diffraction of Si(111) over rocking curve widths of 2°

THEORY/DATA TREATMENT
Minimum of fluorescence signal vs. theory gives interplanar spacing for that Bragg direction

STRUCTURES EXAMINED
The (111) and (220) reflections uniquely determine the local top site

QUALITY OF EXPERIMENT-THEORY FIT
Visual

2D UNIT CELLS (1 domain observed)

Cell	Ax (Å)	Ay (Å)	Bx (Å)	By (Å)	α (°)	Matrix	Pattern	Cell type
Bulk	3.840	0.000	-1.920	3.326	120.0	(1.000, 0.000) (0.000, 1.000)	(1x1)	b: bulk lattice
Surface 1	3.840	0.000	-1.920	3.326	120.0	(1.000, 0.000) (0.000, 1.000)	(1x1)	s1: commens. superlattice

3D COORDINATES

Br1: top-site overlayer; Si2-Si3: repeating bulk bilayer

Dx/Dy in Å, or as a fraction of layer's unit cell vectors; atom 0 at the origin. Epir/subr are bulk repeat vectors.

No. of atoms: 3 Bulk z = 3.130 Å

Region	Chem el.	At. no.	Cell type	Site occ.	Rel. to	Dx ± εx		Dy ± εy		Dz ± εz		Dz/Bz(%) ± εz/Bz
epir		-2					f		f		Å	
subr		-1				0.000	Å	2.217	Å	3.130	Å	
ovrl	Br	1	b	1.00	0	0.000	f	0.000	f	0.000	Å	0.0 ± 3.2
subl	Si	2	b	1.00	1	0.000	f	0.000	f	2.140 ± .100	Å	68.4 ± 1.3
subl	Si	3	b	1.00	2	0.333	f	0.667	f	0.780	Å	24.9

BOND DISTANCES AND ANGLES

No. of distances/angles: 2

Interatomic dist. A-B (Å)	Atom A	Atom B	Atom C	Bond angle A-B-C (°)
2.140	Br1	Si2	Si3	109.4
2.350	Si2	Si3	Si2(0,1)	109.6

COMMON NAME : Si(111)-(1x1)-Cl
CLASSIFICATION : 14.17.4b
TECHNIQUE : SEXAFS
AUTHORS : P.H. Citrin, J.E. Rowe and P. Eisenberger
REFERENCE : Phys. Rev., B28, 2299 (1983)

ILLUSTRATION: 96,99

SURFACE TYPE
Substrate : Si
Crystal face: 111
Temperature : RT*
Bulk lattice: diamond
2D bulk symm: p3m1
2D surf symm: p3m1

Adsorbate: Cl
Coverage : saturation
Pattern : (1x1)
Matrix : (1.000, 0.000)
(0.000, 1.000)

STRUCTURE TYPE
Atomic on-top adsorption;
(1x1) structure inferred from LEED before annealing Cl-covered (7x7); see structure 14.17.4a obtained with final annealing; coverage uncertain (assumed 1.0 ML here)

SAMPLE PREPARATION (1 sample)
Treatment : Si(111)-(7x7) exposed to Cl2, but not reannealed
Crystallinity: (1x1) pattern observed in LEED
Anal. methods:
Contamination:

COMMENTS

DATA COLLECTION
Technique: SEXAFS; polarisation dependent SEXAFS
Dataset : filtered SEXAFS data in the range 1.0 to 2.5Å; incidence at Θ=10° and 90°

THEORY/DATA TREATMENT
Standard SEXAFS for n and nn neighbor distances; top site deduced from polarisation dependence of 1st neighbor signal

STRUCTURES EXAMINED
Only the top site is consistent with the polarisation dependence of the nearest neighbour peak in the Fourier transformed spectrum

QUALITY OF EXPERIMENT-THEORY FIT
Visual

2D UNIT CELLS (1 domain observed)

Cell	Ax (Å)	Ay (Å)	Bx (Å)	By (Å)	α (°)	Matrix	Pattern	Cell type
Bulk	3.840	0.000	-1.920	3.326	120.0	(1.000, 0.000) (0.000, 1.000)	(1x1)	b: bulk lattice
Surface 1	3.840	0.000	-1.920	3.326	120.0	(1.000, 0.000) (0.000, 1.000)	(1x1)	s1: commens. superlattice

3D COORDINATES

Cl1: top-site overlayer; Si2-Si3: repeating bulk bilayer (assumed structure);
coordinates are derived from bond lengths

Dx/Dy in Å, or as a fraction of layer's unit cell vectors; atom 0 at the origin. Epir/subr are bulk repeat vectors.

No. of atoms: 3 Bulk z = 3.130 Å

Region	Chem el.	At. no.	Cell type	Site occ.	Rel. to	Dx ± εx		Dy ± εy		Dz ± εz		Dz/Bz(%) ± εz/Bz
epir		-2					f		f		Å	
subr		-1				0.000	Å	2.217	Å	3.130	Å	
ovrl	Cl	1	b	1.00	0	0.000	f	0.000	f	0.000	Å	0.0
subl	Si	2	b	1.00	1	0.000	f	0.000	f	1.980 ± .040	Å	63.3 ± 1.3
subl	Si	3	b	1.00	2	0.333	f	0.667	f	0.780	Å	24.9

BOND DISTANCES AND ANGLES

No. of distances/angles: 2

Interatomic dist. A-B (Å)	Atom A	Atom B	Atom C	Bond angle A-B-C (°)
1.980	Cl1	Si2	Si3	109.4
2.350	Si2	Si3	Si2(0,1)	109.6

COMMON NAME : Si(111)-(7x7)-Cl
CLASSIFICATION : 14.17.4a
TECHNIQUE : SEXAFS
AUTHORS : P.H. Citrin, J.E. Rowe and P. Eisenberger
REFERENCE : Phys. Rev., B28, 2299 (1983)

ILLUSTRATION: 96,99

SURFACE TYPE
Substrate : Si
Crystal face: 111
Temperature : RT*
Bulk lattice: diamond
2D bulk symm: p3m1
2D surf symm: p3m1

Adsorbate: Cl
Coverage : uncertain
Pattern : (1x1)
Matrix : (1.000, 0.000)
(0.000, 1.000)

STRUCTURE TYPE
Atomic on-top adsorption;
(7x7) structure inferred from LEED, results from annealing Cl-covered (7x7); see structure 14.17.4b obtained without final annealing; coverage uncertain (assumed 1.0 ML here); relationship with (7x7) structure unknown; only local

SAMPLE PREPARATION (1 sample)
Treatment : Si(111)-(7x7) exposed to Cl2, reannealed at 673 K for 5 mins
Crystallinity: (7x7) pattern observed in LEED
Anal. methods:
Contamination: (7x7) pattern unaffected by Cl & anneal

COMMENTS
Geometry near adsorbate determined
adsorption site geometry in excellent agreement with parameter-free total-energy calculations of Bachelet and Schluter, Phys. Rev. B20, 2302 (1983)

DATA COLLECTION
Technique: SEXAFS; polarisation dependent SEXAFS
Dataset : filtered SEXAFS data in the range 1.0 to 2.5Å; incidence at Θ=10° and 90°

THEORY/DATA TREATMENT
Standard SEXAFS for n and nn neighbor distances; top site deduced from polarisation dependence of 1st neighbor signal

STRUCTURES EXAMINED
Only the top site is consistent with the polarisation dependence of the nearest neighbour peak in the Fourier transformed spectrum

QUALITY OF EXPERIMENT-THEORY FIT
Visual

2D UNIT CELLS (1 domain observed)

Cell	Ax (Å)	Ay (Å)	Bx (Å)	By (Å)	α (°)	Matrix	Pattern	Cell type
Bulk	3.840	0.000	-1.920	3.326	120.0	(1.000, 0.000) (0.000, 1.000)	(1x1)	b: bulk lattice
Surface 1	3.840	0.000	-1.920	3.326	120.0	(1.000, 0.000) (0.000, 1.000)	(1x1)	s1: commens. superlattice

3D COORDINATES

Cl1: top-site overlayer; Si2-Si3: repeating bulk bilayer (assumed structure);
coordinates are derived from bond lengths

Dx/Dy in Å, or as a fraction of layer's unit cell vectors; atom 0 at the origin. Epir/subr are bulk repeat vectors.

No. of atoms: 3

Bulk z = 3.130 Å

Region	Chem el.	At. no.	Cell type	Site occ.	Rel. to	Dx ± εx		Dy ± εy		Dz ± εz		Dz/Bz(%) ± εz/Bz
epir		-2					f		f		Å	
subr		-1				0.000	Å	2.217	Å	3.130	Å	
ovrl	Cl	1	b	1.00	0	0.000	f	0.000	f	0.000	Å	0.0
subl	Si	2	b	1.00	1	0.000	f	0.000	f	2.030 ± .030	Å	64.9 ± 1.0
subl	Si	3	b	1.00	2	0.333	f	0.667	f	0.780	Å	24.9

BOND DISTANCES AND ANGLES

No. of distances/angles: 2

Interatomic dist. A-B (Å)	Atom A	Atom B	Atom C	Bond angle A-B-C (°)
2.030	Cl1	Si2	Si3	109.4
2.350	Si2	Si3	Si2(0,1)	109.6

COMMON NAME : Si(100)-Co 0.4ML
CLASSIFICATION : 14.27.16
TECHNIQUE : SEXAFS
AUTHORS : H.L. Meyerheim, U. Dobler and A. Puschmann
REFERENCE : Phys. Rev., B44, 5738 (1991)

ILLUSTRATION: 104,105

SURFACE TYPE
Substrate : Si
Crystal face: 100
Temperature : RT*
Bulk lattice: diamond
2D bulk symm: pmm
2D surf symm: none

Adsorbate: Co
Coverage : 0.4 Co/1x1
Pattern : (2x1)
Matrix : (2.000, 0.000)
(0.000, 1.000)

STRUCTURE TYPE
Atomic adsorption in-plane with top layer of unreconstructed Si(100), at center of square of top-layer Si atoms

SAMPLE PREPARATION (1 sample)
Treatment : Co dep. by resist. heating of Co wire onto Si(100)-(2x1)
Crystallinity: reduced brightness of (2x1) LEED spots
Anal. methods: coverage determined by RBS, AES and
Contamination: AES

COMMENTS
Slight relaxations <0.1Å of substrate expected, but not determined

DATA COLLECTION
Technique: SEXAFS; polariz.-dep. SEXAFS at BESSY
Dataset : SEXAFS spectra for k range 30-120 nm-1

THEORY/DATA TREATMENT
Fourier transform and simulations with single-scattering plane-wave formalism

STRUCTURES EXAMINED
Adsorption at various sites on unreconstructed Si(100), and interstitial and substitutional sites

QUALITY OF EXPERIMENT-THEORY FIT
Visual

2D UNIT CELLS (1 domain observed)

Cell	Ax (Å)	Ay (Å)	Bx (Å)	By (Å)	α (°)	Matrix	Pattern	Cell type
Bulk	3.838	0.000	0.000	3.838	90.0	(1.000, 0.000) (0.000, 1.000)	(1x1)	b: bulk lattice
Surface 1	7.675	0.000	0.000	3.838	90.0	(2.000, 0.000) (0.000, 1.000)	(2x1)	s1: commens. superlattice

3D COORDINATES

Co1: adatom in '4-fold' sites coplanar with Si2;

Dx/Dy in Å, or as a fraction of layer's unit cell vectors; atom 0 at the origin. Epir/subr are bulk repeat vectors.

No. of atoms: 3 Bulk z = 1.357 Å

Region	Chem el.	At. no.	Cell type	Site occ.	Rel. to	Dx ± εx		Dy ± εy		Dz ± εz		Dz/Bz(%) ± εz/Bz
epir		-2					f		f		Å	
subr		-1				-1.919	Å	-1.919	Å	2.714	Å	
intf	Co	1	s1	.50	0	0.000	f	0.000	f	0.000	Å	0.0
subl	Si	2	b	1.00	1	0.500	f	0.500	f	0.000 ± .015	Å	0.0 ± 1.1
subl	Si	3	b	1.00	2	-0.500	f	0.000	f	1.357	Å	100.0

BOND DISTANCES AND ANGLES

Measured Co-Si distances are 2.80±0.03 and 2.30±0.05Å

No. of distances/angles: 2

Interatomic dist. A-B (Å)	Atom A	Atom B	Atom C	Bond angle A-B-C (°)
2.714	Co1	Si2	Si3	54.7
2.350	Co1	Si3	Si2	70.5

COMMON NAME : Si(111)-(1x1)-CoSi2(111) interface ILLUSTRATION: 111
CLASSIFICATION : 14.27.2
TECHNIQUE : XSW
AUTHORS : A.E.M.J. Fischer, E. Vlieg, J.F. van der Veen, M. Clausnitzer and G. Materlik
REFERENCE : Phys. Rev., B36, 4769 (1987)

SURFACE TYPE
Substrate : Si Adsorbate: CoSi2
Crystal face: 111 Coverage : 9-28Å
Temperature : RT* Pattern : (1x1)
Bulk lattice: diamond Matrix : (1.000, 0.000) (0.000, 1.000)
2D bulk symm: p3m1
2D surf symm: p3m1

STRUCTURE TYPE
B-type epitaxial (1x1) growth, with 5-fold coordinated Co atoms at interface (same as obtained by MEIS: class. no. 14.27.3; see other comments there); bulk CoSi2=fluorite

SAMPLE PREPARATION (1 sample)
Treatment : Co deposited at RT, then same thickness of Si and annealing
Crystallinity: sharp (1x1) RHEED pattern
Anal. methods: composition of CoSi2 layers checked by Ion Back Scatt., TEM
Contamination:

COMMENTS
B-type epitaxy = (111) orientation with 180° rotation about substrate normal; silicide films are unstrained (lattice mismatch strain accommodated by defects: TEM image showed defects at interface); for thick films annealing may have been insufficient to order lattice completely

DATA COLLECTION
Technique: XSW; fluorescence detected by scint. counte
Dataset : fluorescence from (111) Bragg refl. with 14.2k eV x-rays: rocking curve scanned by varying beam E with fixed sample

THEORY/DATA TREATMENT
X-ray standing wave analysis with Rutherford back scattering to determine number of Co layers

STRUCTURES EXAMINED
Either 5-fold or 7-fold coordinated Co atoms at interface

QUALITY OF EXPERIMENT-THEORY FIT
Visual

2D UNIT CELLS (1 domain observed)

Cell	Ax (Å)	Ay (Å)	Bx (Å)	By (Å)	α (°)	Matrix	Pattern	Cell type
Bulk	3.838	0.000	-1.919	3.324	120.0	(1.000, 0.000) (0.000, 1.000)	(1x1)	b: bulk lattice
Surface 1	3.838	0.000	-1.919	3.324	120.0	(1.000, 0.000) (0.000, 1.000)	(1x1)	s1: commens. superlattice

3D COORDINATES

Si1-Co2-Si3: periodically repeating bulk CoSi2 trilayer; Si4-Co5: 2/3 of CoSi2 trilayer; Co5-Si6: interface plane; Si10-Si11: periodically repeating bulk substrate

Dx/Dy in Å, or as a fraction of layer's unit cell vectors; atom 0 at the origin. Epir/subr are bulk repeat vectors.

No. of atoms: 11 Bulk z = 3.130 Å

Region	Chem el.	At. no.	Cell type	Site occ.	Rel. to	Dx ± εx		Dy ± εy		Dz ± εz		Dz/Bz(%) ± εz/Bz
epir		-2				1.919	Å	-1.108	Å	3.092	Å	
subr		-1				1.919	Å	1.108	Å	3.130	Å	
epil	Si	1	b	1.00	0	0.000	f	0.000	f	0.000	Å	0.0
epil	Co	2	b	1.00	1	0.333	f	0.667	f	0.773	Å	24.7
epil	Si	3	b	1.00	2	0.333	f	-0.333	f	0.773	Å	24.7
intf	Si	4	b	1.00	3	-0.333	f	0.333	f	1.546	Å	49.4
intf	Co	5	b	1.00	4	0.333	f	-0.333	f	0.773	Å	24.7
intf	Si	6	b	1.00	5	0.000	f	0.000	f	2.370 ± .030	Å	75.7 ± 1.0
intf	Si	7	b	1.00	6	-0.333	f	0.333	f	0.820	Å	26.2
intf	Si	8	b	1.00	7	0.000	f	0.000	f	2.350	Å	75.1
intf	Si	9	b	1.00	8	-0.333	f	-0.667	f	0.780	Å	24.9
subl	Si	10	b	1.00	9	0.000	f	0.000	f	2.350	Å	75.1
subl	Si	11	b	1.00	10	0.667	f	0.333	f	0.780	Å	24.9

Si(111)-(1x1)-CoSi2(111) interface
14.27.2

BOND DISTANCES AND ANGLES

Bond distances and angles are derived from coordinates

No. of distances/angles: 2

Interatomic dist. A-B (Å)	Atom A	Atom B	Atom C	Bond angle A-B-C (°)
2.347	Si1	Co2	Si3	70.3
2.347	Si1	Co2	Si4	109.2

COMMON NAME : Si(111)-(1x1)-CoSi2(111) interface
CLASSIFICATION : 14.27.3
TECHNIQUE : MEIS
AUTHORS : A.E.M.J. Fischer, T. Gustafsson and J.F. van der Veen
REFERENCE : Phys. Rev., B37, 6305 (1988)

ILLUSTRATION: 111

SURFACE TYPE
Substrate : Si
Crystal face: 111
Temperature : RT*
Bulk lattice: diamond
2D bulk symm: p3m1
2D surf symm: p3m1

Adsorbate: CoSi2
Coverage :
Pattern : (1x1)
Matrix : (1.000, 0.000)
(0.000, 1.000)

STRUCTURE TYPE
B-type epitaxial (1x1) growth, with 5-fold coordinated Co atoms at interface (same as obtained by XSW: class. no. 14.27.2; see other comments there); bulk CoSi2=fluorite

SAMPLE PREPARATION (1 sample)
Treatment : Co deposited at RT, then same thickness of Si and annealing
Crystallinity:
Anal. methods: composition of CoSi2 layers checked by ion back scatt., TEM
Contamination: AES and ISS: substrate pure

COMMENTS
Co-Si distance taken from Phys. Rev. B36, 4769 (1987) (14.27.2); exp. data were mult. by 2 for comparison with Monte Carlo simulations: discrepancy may be due to defects; 8-fold coord. Co at interface possible: not distinguished from 5-fold; NiSi2 epilayer has 7-fold coord. Ni: 14.28.8

DATA COLLECTION
Technique: MEIS; RBS with 98keV protons collimated to
Dataset : incidence along [00-1] channels; detection along [110]; 7Å depth resolution; 3 incidence angles: 34.8°, 35.2°, 35.5°

THEORY/DATA TREATMENT
High resolution RBS compared with Monte Carlo simulations with Moliere potential; vib amps as for NiSi2

STRUCTURES EXAMINED
Interfacial Co atoms either 5-fold or 7-fold coordinated; lateral displacements of interface atoms along [11-2] wrt substrate considered for the 5-fold coordination but best fit was for no translation

QUALITY OF EXPERIMENT-THEORY FIT
Visual

2D UNIT CELLS (1 domain observed)

Cell	Ax (Å)	Ay (Å)	Bx (Å)	By (Å)	α (°)	Matrix	Pattern	Cell type
Bulk	3.838	0.000	-1.919	3.324	120.0	(1.000, 0.000) (0.000, 1.000)	(1x1)	b: bulk lattice
Surface 1	3.838	0.000	-1.919	3.324	120.0	(1.000, 0.000) (0.000, 1.000)	(1x1)	s1: commens. superlattice

3D COORDINATES

Si1-Co2-Si3: periodically repeating bulk CoSi2 trilayer; Si4-Co5: 2/3 of CoSi2 trilayer; Co5-Si6: interface plane; Si10-Si11: periodically repeating bulk substrate

Dx/Dy in Å, or as a fraction of layer's unit cell vectors; atom 0 at the origin. Epir/subr are bulk repeat vectors.

No. of atoms: 11

Bulk z = 3.130 Å

Region	Chem el.	At. no.	Cell type	Site occ.	Rel. to	Dx ± εx		Dy ± εy		Dz ± εz		Dz/Bz(%) ± εz/Bz
epir		-2				1.919	Å	-1.108	Å	3.092	Å	
subr		-1				1.919	Å	1.108	Å	3.130	Å	
epil	Si	1	b	1.00	0	0.000	f	0.000	f	0.000	Å	0.0
epil	Co	2	b	1.00	1	0.333	f	0.667	f	0.773	Å	24.7
epil	Si	3	b	1.00	2	0.333	f	-0.333	f	0.773	Å	24.7
intf	Si	4	b	1.00	3	-0.333	f	0.333	f	1.546	Å	49.4
intf	Co	5	b	1.00	4	0.333	f	-0.333	f	0.773	Å	24.7
intf	Si	6	b	1.00	5	0.000	f	0.000	f	2.370 ± .030	Å	75.7 ± 1.0
intf	Si	7	b	1.00	6	-0.333	f	0.333	f	0.820	Å	26.2
intf	Si	8	b	1.00	7	0.000	f	0.000	f	2.350	Å	75.1
intf	Si	9	b	1.00	8	-0.333	f	-0.667	f	0.780	Å	24.9
subl	Si	10	b	1.00	9	0.000	f	0.000	f	2.350	Å	75.1
subl	Si	11	b	1.00	10	0.667	f	0.333	f	0.780	Å	24.9

Si(111)-(1x1)-CoSi2(111) interface
14.27.3

BOND DISTANCES AND ANGLES

Bond distances and angles are derived from coordinates

No. of distances/angles: 2

Interatomic dist. A-B (Å)	Atom A	Atom B	Atom C	Bond angle A-B-C (°)
2.347	Si1	Co2	Si3	70.3
2.347	Si1	Co2	Si4	109.2

COMMON NAME : Si(111)-(1x1)-CoSi2(111) interface ILLUSTRATION: 112
CLASSIFICATION : 14.27.8
TECHNIQUE : SEXAFS
AUTHORS : G. Rossi, X. Jin, A. Santinello, P. DePadova and D. Chandesris
REFERENCE : Phys. Rev. Lett., 62, 191 (1989)

SURFACE TYPE
Substrate : Si Adsorbate: CoSi2
Crystal face: 111 Coverage :
Temperature : RT* Pattern : (1x1)
Bulk lattice: diamond Matrix : (1.000, 0.000)
2D bulk symm: p3m1 (0.000, 1.000)
2D surf symm: p3m1

STRUCTURE TYPE
B-type interface (assumed, referring to layer stacking across interface), with Co 8-fold coordinated; note discrepancy with 14.27.1-2; bulk CoSi2=fluorite

SAMPLE PREPARATION (1 sample)
Treatment : Co deposited on (7x7) and annealed at 630C
Crystallinity:
Anal. methods: characterization of CoSi2 layers by Auger, LEED, SEXAFS
Contamination:

COMMENTS
B or A type could not be discriminated in this experiment; B type is assumed; structure determination done on an ultra thin layer (2 to 3 full layers of silicide); 2.5% reduction of silicide layer spacing is possibly related to thinness of the CoSi2 layer

DATA COLLECTION
Technique: SEXAFS; pol.-dep. SEXAFS on Co K edge (7707
Dataset :

THEORY/DATA TREATMENT

1) fivefold Co, 2) sevenfold Co, 3) eightfold Co

STRUCTURES EXAMINED
3) is preferred structure, 1) and 2) ruled out

2D UNIT CELLS (1 domain observed)

Cell	Ax (Å)	Ay (Å)	Bx (Å)	By (Å)	α (°)	Matrix	Pattern	Cell type
Bulk	3.838	0.000	-1.919	3.324	120.0	(1.000, 0.000) (0.000, 1.000)	(1x1)	b: bulk lattice
Surface 1	3.838	0.000	-1.919	3.324	120.0	(1.000, 0.000) (0.000, 1.000)	(1x1)	s1: commens. superlattice

3D COORDINATES

Si1-Co2-Si3: periodically repeating bulk CoSi2 trilayer; Si4-Co5-Si6: last CoSi2 trilayer; Co5-Si6-Si7: interface plane; Si11-Si12: periodically repeating bulk substrate bilayer

Dx/Dy in Å, or as a fraction of layer's unit cell vectors; atom 0 at the origin. Epir/subr are bulk repeat vectors.

No. of atoms: 12 Bulk z = 3.130 Å

Region	Chem el.	At. no.	Cell type	Site occ.	Rel. to	Dx ± εx	Dy ± εy	Dz ± εz	Dz/Bz(%) ± εz/Bz
epir		-2				1.919 Å	-1.108 Å	3.092 Å	
subr		-1				1.919 Å	1.108 Å	3.130 Å	
epil	Si	1	b	1.00	0	0.000 f	0.000 f	0.000 Å	0.0
epil	Co	2	b	1.00	1	0.333 f	0.667 f	0.773 Å	24.7
epil	Si	3	b	1.00	2	0.333 f	-0.333 f	0.773 Å	24.7
intf	Si	4	b	1.00	3	-0.333 f	0.333 f	1.487 Å	47.5
intf	Co	5	b	1.00	4	0.333 ± .005 f	-0.333 ± .005 f	0.773 ± .030 Å	24.7 ± 1.0
intf	Si	6	b	1.00	5	-0.667 f	-0.333 f	0.773 Å	24.7
intf	Si	7	b	1.00	5	0.000 f	0.000 f	2.350 Å	75.1
intf	Si	8	b	1.00	7	-0.333 f	0.333 f	0.780 Å	24.9
intf	Si	9	b	1.00	8	0.000 f	0.000 f	2.350 Å	75.1
intf	Si	10	b	1.00	9	-0.333 f	-0.667 f	0.780 Å	24.9
subl	Si	11	b	1.00	10	0.000 f	0.000 f	2.350 Å	75.1
subl	Si	12	b	1.00	11	0.667 f	0.333 f	0.780 Å	24.9

Si(111)-(1x1)-CoSi2(111) interface
14.27.8

BOND DISTANCES AND ANGLES

No. of distances/angles: 7

Interatomic dist. A-B (Å)	Atom A	Atom B	Atom C	Bond angle A-B-C (°)
2.347	Si1	Co2	Si1(1,1)	109.7
2.668	Si3	Si4	Co2	56.1
2.668	Si3	Si4	Si3(0,1)	92.0
2.668	Si3	Si4	Co5(0,1)	125.1
2.347	Si4	Co5(0,1)	Si3(0,1)	70.8
2.347	Si4	Co5(0,1)	Si4(1,1)	109.7
2.347	Si4	Co5(0,1)	Si6(1,1)	70.3

COMMON NAME : Si(111)-(√3x√3)R30°-Ga
CLASSIFICATION : 14.31.4
TECHNIQUE : LEED
AUTHORS : A. Kawazu and H. Sakama
REFERENCE : Phys. Rev., B37, 2704 (1988)

ILLUSTRATION: 96,97

SURFACE TYPE
Substrate : Si
Crystal face: 111
Temperature : RT*
Bulk lattice: diamond
2D bulk symm: p3m1
2D surf symm: p31m

Adsorbate: Ga
Coverage : 0.3 Ga/Si
Pattern : (√3x√3)R30°
Matrix : (1.000, 1.000)
(-2.000, 1.000)

STRUCTURE TYPE
Atomic adsorption in 4-fold coordinated T4 'top' site over top bilayer, with relaxations down into 2nd bilayer

SAMPLE PREPARATION (1 sample)
Treatment : Ga molecular beam from Knudsen cell at 740 K
Crystallinity: sharp LEED pattern
Anal. methods:
Contamination:

COMMENTS
R-factors for the structures examined were, respectively: 0.34, 0.45, 0.45, 0.25, 0.15

DATA COLLECTION
Technique: LEED
Dataset : I-V spectra: 9 beams at normal incidence, 30<E<190 eV

THEORY/DATA TREATMENT
Dynamical LEED

STRUCTURES EXAMINED
Ga atoms in hollow site above the 4th layer Si; Ga atoms on 3 atom Si clusters centered above 2nd layer Si; clusters centered above 4th layer Si; Ga substituted for 1/3 of top layer Si; T4 'top' site; in each case structural parameters were varied (see comments)

QUALITY OF EXPERIMENT-THEORY FIT
RZJ=0.15

2D UNIT CELLS (1 domain observed)

Cell	Ax (Å)	Ay (Å)	Bx (Å)	By (Å)	α (°)	Matrix	Pattern	Cell type
Bulk	3.839	0.000	1.920	3.325	60.0	(1.000, 0.000) (0.000, 1.000)	(1x1)	b: bulk lattice
Surface 1	5.759	3.325	-5.759	3.325	120.0	(1.000, 1.000) (-2.000, 1.000)	(√3x√3)R30°	s1: commens. superlattice

3D COORDINATES

Ga1: in T4 'top' site over Si7; Si2-Si7 and Si8-Si13: relaxed top 2 bilayers;
Si14-Si15: repeating bulk substrate layers; 0.1Å error bars assumed for tabulation

Dx/Dy in Å, or as a fraction of layer's unit cell vectors; atom 0 at the origin. Epir/subr are bulk repeat vectors.

No. of atoms: 15

Bulk z = 3.130 Å

Region	Chem el.	At. no.	Cell type	Site occ.	Rel. to	Dx ± εx		Dy ± εy		Dz ± εz		Dz/Bz(%) ± εz/Bz
epir		-2					f		f		Å	
subr		-1				1.920	Å	1.108	Å	3.130	Å	
ovrl	Ga	1	s1	.33	0	0.000 ± .024	f	0.000 ± .017	f	0.000 ± .100	Å	0.0 ± 3.2
intf	Si	2	s1	.33	1	0.316 ± .024	f	0.316 ± .017	f	1.350 ± .100	Å	43.1 ± 3.2
intf	Si	3	s1	.33	2	0.368 ± .024	f	-0.316 ± .017	f	0.000 ± .100	Å	0.0 ± 3.2
intf	Si	4	s1	.33	3	-0.684 ± .024	f	0.684 ± .017	f	0.000 ± .100	Å	0.0 ± 3.2
intf	Si	5	s1	.33	4	0.667 ± .024	f	-0.351 ± .017	f	0.580 ± .100	Å	18.5 ± 3.2
intf	Si	6	s1	.33	5	-0.333 ± .024	f	0.333 ± .017	f	0.000 ± .100	Å	0.0 ± 3.2
intf	Si	7	s1	.33	6	-0.333 ± .024	f	-0.667 ± .017	f	0.640 ± .100	Å	20.5 ± 3.2
intf	Si	8	s1	.33	7	0.667 ± .024	f	0.333 ± .017	f	1.800 ± .100	Å	57.5 ± 3.2
intf	Si	9	s1	.33	8	-0.333 ± .024	f	0.333 ± .017	f	0.000 ± .100	Å	0.0 ± 3.2
intf	Si	10	s1	.33	9	-0.333 ± .024	f	-0.667 ± .017	f	0.340 ± .100	Å	10.9 ± 3.2
intf	Si	11	s1	.33	10	0.341 ± .024	f	0.000 ± .017	f	0.510 ± .100	Å	16.3 ± 3.2
intf	Si	12	s1	.33	11	-0.341 ± .024	f	0.341 ± .017	f	0.000 ± .100	Å	0.0 ± 3.2
intf	Si	13	s1	.33	12	0.659 ± .024	f	0.317 ± .017	f	0.000 ± .100	Å	0.0 ± 3.2
subl	Si	14	b	1.00	10	0.333 ± .011	f	0.333 ± .030	f	2.860 ± .100	Å	91.4 ± 3.2
subl	Si	15	b	1.00	14	0.333 ± .011	f	-0.667 ± .030	f	0.780 ± .100	Å	24.9 ± 3.2

Si(111)-(√3x√3)R30°-Ga
14.31.4

BOND DISTANCES AND ANGLES

Bond distances and angles are derived from coordinates

No. of distances/angles: 5

Interatomic dist. A-B (Å)	Atom A	Atom B	Atom C	Bond angle A-B-C (°)
2.497	Ga1	Si2	Si5	124.9
2.570	Ga1	Si7	Si2	59.9
2.350	Si2	Si5	Si3	114.1
2.350	Si2	Si6	Si9	104.3
2.429	Si2	Si7	Si10	120.2

COMMON NAME : Si(111)-(7x7)-I
CLASSIFICATION : 14.53.2
TECHNIQUE : SEXAFS
AUTHORS : P.H. Citrin, P. Eisenberger and J.E. Rowe
REFERENCE : Phys. Rev. Lett., 48, 802 (1982)

ILLUSTRATION: 96,99

SURFACE TYPE
Substrate : Si
Crystal face: 111
Temperature : RT*
Bulk lattice: diamond
2D bulk symm: p3m1
2D surf symm: p3m1

Adsorbate: I
Coverage : 1 ML
Pattern : (1x1)
Matrix : (1.000, 0.000)
(0.000, 1.000)

STRUCTURE TYPE
Atomic on-top adsorption;
(7x7) structure inferred from LEED; modeled as (1x1) here; relationship with (7x7) structure unknown; only local geometry near adsorbate determined

SAMPLE PREPARATION (1 sample)
Treatment : Ar+ bomb. and annealed, exposed to I2, annealed at 773 K
Crystallinity: LEED showed a (7x7) pattern
Anal. methods: AES determined 1 ML I coverage
Contamination:

COMMENTS

DATA COLLECTION
Technique: SEXAFS; total yield SEXAFS
Dataset : SEXAFS from I LIII edge vs polar angle and polarisation

THEORY/DATA TREATMENT
Fourier transform analysis of SEXAFS data with SiI(CH3)3 standard combined with analysis of amplitude ratios

STRUCTURES EXAMINED
Structure determined directly from first neighbor bond lengths and amplitude ratios

QUALITY OF EXPERIMENT-THEORY FIT
Visual

2D UNIT CELLS (1 domain observed)

Cell	Ax (Å)	Ay (Å)	Bx (Å)	By (Å)	α (°)	Matrix	Pattern	Cell type
Bulk	3.840	0.000	-1.920	3.326	120.0	(1.000, 0.000) (0.000, 1.000)	(1x1)	b: bulk lattice
Surface 1	3.840	0.000	-1.920	3.326	120.0	(1.000, 0.000) (0.000, 1.000)	(1x1)	s1: commens. superlattice

3D COORDINATES

I1: top-site overlayer; Si2-Si3: expanded bilayer;
Si4-Si5: repeating bulk bilayer (assumed structure); coordinates are derived from bond lengths

Dx/Dy in Å, or as a fraction of layer's unit cell vectors; atom 0 at the origin. Epir/subr are bulk repeat vectors.

No. of atoms: 5 Bulk z = 3.130 Å

Reg ion	Chem el.	At. no.	Cell type	Site occ.	Rel. to	Dx ± εx		Dy ± εy		Dz ± εz		Dz/Bz(%) ± εz/Bz
epir		-2					f		f		Å	
subr		-1				0.000	Å	2.217	Å	3.130	Å	
ovrl	I	1	b	1.00	0	0.000	f	0.000	f	0.000	Å	0.0
intf	Si	2	b	1.00	1	0.000	f	0.000	f	2.440 ± .030	Å	78.0 ± 1.0
intf	Si	3	b	1.00	2	0.333	f	0.667	f	0.900 ± .050	Å	28.8 ± 1.6
subl	Si	4	b	1.00	3	0.000	f	0.000	f	2.350	Å	75.1
subl	Si	5	b	1.00	4	0.333	f	-0.333	f	0.780	Å	24.9

BOND DISTANCES AND ANGLES

No. of distances/angles: 5

Interatomic dist. A-B (Å)	Atom A	Atom B	Atom C	Bond angle A-B-C (°)
2.440	I1	Si2	Si3	112.1
2.393	Si2	Si3	Si2(0,1)	106.7
2.393	Si2	Si3	Si4	112.1
2.350	Si3	Si4	Si5	109.4
2.350	Si4	Si5	Si4(1,0)	109.6

COMMON NAME : Si(100)-(2x1)-2K
CLASSIFICATION : 14.19.9
TECHNIQUE : LEED
AUTHORS : T. Urano, Y. Uchida, S. Hongo and T. Kanaji
REFERENCE : Surf. Sci., 242, 39 (1991)

ILLUSTRATION: 108

SURFACE TYPE
Substrate : Si
Crystal face: 100
Temperature : RT*
Bulk lattice: diamond
2D bulk symm: pmm
2D surf symm: pmm

Adsorbate: K
Coverage : 1 K/1x1
Pattern : (2x1)
Matrix : (2.000, 0.000)
(0.000, 1.000)

STRUCTURE TYPE
Atomic adsorption as 'double layer' model, with K at both the pedestal and cave sites on a dimerized Si(100)-(2x1) substrate

SAMPLE PREPARATION (1 sample)
Treatment : K dep. onto AES-clean Si(100-(2x1) from chromate dispenser
Crystallinity:
Anal. methods: AES; K coverage not calibrated
Contamination:

COMMENTS
The two other models tested produced similarly poor R-factors: RZJ=0.54, 0.56 for pedestal and cave sites, vs. 0.51 for double layer model; since only 4 phase shifts were used and no substrate relaxations below the Si dimers were allowed, this result must be considered uncertain

DATA COLLECTION
Technique: LEED; video camera
Dataset : IV curves for (10), (11), (20), (0.5,0), (0.5,1), (1.5,0) and (1.5,1) beams at normal incidence; E range 40-210 eV

THEORY/DATA TREATMENT
Dynamical LEED (CSM, RFS): 4 phase shifts from Herman-Skillman wave functions

STRUCTURES EXAMINED
Adsorption on symmetrically dimerized Si(100)-(2x1) in pedestal and cave sites, and as double layer (both pedestal and cave sites); relaxation of K heights, and dimer length and height

QUALITY OF EXPERIMENT-THEORY FIT
RZJ=0.51

2D UNIT CELLS (1 domain observed)

Cell	Ax (Å)	Ay (Å)	Bx (Å)	By (Å)	α (°)	Matrix	Pattern	Cell type
Bulk	3.838	0.000	0.000	3.838	90.0	(1.000, 0.000) (0.000, 1.000)	(1x1)	b: bulk lattice
Surface 1	7.675	0.000	0.000	3.838	90.0	(2.000, 0.000) (0.000, 1.000)	(2x1)	s1: commens. superlattice

3D COORDINATES

K1-K2: adatoms in pedestal and cave sites; Si3-Si4: symmetric Si dimer;
Si5-Si6: periodically repeating bulk pair of layers; 0.1Å error bars assumed for tabulation

Dx/Dy in Å, or as a fraction of layer's unit cell vectors; atom 0 at the origin. Epir/subr are bulk repeat vectors.

No. of atoms: 6 Bulk z = 1.357 Å

Region	Chem el.	At. no.	Cell type	Site occ.	Rel. to	Dx ± εx		Dy ± εy		Dz ± εz		Dz/Bz(%) ± εz/Bz
epir		-2					f		f		Å	
subr		-1				1.919	Å	1.919	Å	2.714	Å	
ovrl	K	1	s1	.50	0	0.000	f	0.000	f	0.000	Å	0.0
ovrl	K	2	s1	.50	1	-0.500	f	0.000	f	0.400 ± .100	Å	29.5 ± 7.4
intf	Si	3	s1	.50	2	0.652 ± .013	f	0.500	f	1.350 ± .100	Å	99.5 ± 7.4
intf	Si	4	s1	.50	3	-0.305 ± .013	f	0.000	f	0.000	Å	0.0
subl	Si	5	b	1.00	4	-0.195	f	-0.500	f	1.157 ± .100	Å	85.3 ± 7.4
subl	Si	6	b	1.00	5	-0.500	f	0.000	f	1.357	Å	100.0

Si(100)-(2x1)-2K
14.19.9 BOND DISTANCES AND ANGLES

Bond distances and angles are derived from coordinates

No. of distances/angles: 16

Interatomic dist. A-B (Å)	Atom A	Atom B	Atom C	Bond angle A-B-C (°)
3.838	K1	K1(0,1)	Si3(0,1)	132.4
3.553	K2	Si4	Si3	138.7
3.553	K2	Si4	Si5	60.6
3.157	K2	Si5	Si4	78.7
3.157	K2	Si5	Si6	87.8
2.338	Si3	Si4	K1	65.8
2.338	Si3	Si4	K2	138.7
2.338	Si3	Si4	Si5	108.5
3.838	K1	K1(0,1)	Si3	47.6
3.858	K1	K2	K1(-1,0)	168.1
3.858	K1	K2	Si3(-1,0)	141.9
2.848	K1	Si3	Si4	65.8
2.848	K1	Si3	Si5(1,0)	83.3
2.848	K1	Si4	K1(0,1)	84.7
2.848	K1	Si4	K2	73.2
3.553	K2	Si4	K1	73.2

ILLUSTRATION: 107

COMMON NAME : Si(100)-(2x1)-Na
CLASSIFICATION : 14.11.4
TECHNIQUE : LEED
AUTHORS : C.M. Wei, H. Huang, S.Y. Tong, G.S. Glander and M.B. Webb
REFERENCE : Phys. Rev., B42, 11284 (1990)

SURFACE TYPE
Substrate : Si
Crystal face: 100
Temperature : 165C
Bulk lattice: diamond
2D bulk symm: pmm
2D surf symm: pmm

Adsorbate: Na
Coverage : 1/2 Na/1x1
Pattern : (2x1)
Matrix : (2.000, 0.000)
(0.000, 1.000)

STRUCTURE TYPE
Atomic adsorption in 'pedestal site', bridging pairs of adjacent Si dimers, which are stretched wrt clean Si(100)-(2x1)

SAMPLE PREPARATION (1 sample)
Treatment : Na dosing at 60-170C onto clean (2x1); AES calibrated cov.
Crystallinity:
Anal. methods: AES
Contamination:

COMMENTS
2nd model could be ruled out (except its 1ML coverage is double the measured amount): it has, in addition to 0.5ML Na in the pedestal sites, also 0.5ML Na in the 'valley bridge sites'; both models have equal best R-factor

DATA COLLECTION
Technique: LEED; Faraday cup
Dataset : I-V curves at normal incidence for 2 integer-order and 3 fractional-order beams

THEORY/DATA TREATMENT
Dynamical LEED (fully symm.): 7 phase shifts from bulk Si superpos. pot.; Voi=-4.25 eV

STRUCTURES EXAMINED
6 models (all with distorted symm. Si dimers): Na on top, bridge, pedestal, cave or valley bridge sites (0.5ML), or both on pedestal and valley bridge sites (double layer model) (1ML); substrate relaxations down to 4th Si layer

QUALITY OF EXPERIMENT-THEORY FIT
RPE=0.253

2D UNIT CELLS (1 domain observed)

Cell	Ax (Å)	Ay (Å)	Bx (Å)	By (Å)	α (°)	Matrix	Pattern	Cell type
Bulk	3.840	0.000	0.000	3.840	90.0	(1.000, 0.000) (0.000, 1.000)	(1x1)	b: bulk lattice
Surface 1	7.680	0.000	0.000	3.840	90.0	(2.000, 0.000) (0.000, 1.000)	(2x1)	s1: commens. superlattice

3D COORDINATES

Na1: overlayer on pedestal site. bonding to 4 atoms Si2-Si3; Si2-Si3: stretched Si dimers; Si4-Si9: relaxed substrate; Si10-Si11: periodically repeating bulk pair of layers; 0.1Å error bars assumed for tabulation

Dx/Dy in Å, or as a fraction of layer's unit cell vectors; atom 0 at the origin. Epir/subr are bulk repeat vectors.

No. of atoms: 11

Bulk z = 1.358 Å

Region	Chem el.	At. no.	Cell type	Site occ.	Rel. to	Dx ± εx		Dy ± εy		Dz ± εz		Dz/Bz(%) ± εz/Bz
epir		-2					f		f		Å	
subr		-1				1.920	Å	1.920	Å	2.715	Å	
ovrl	Na	1	s1	.50	0	0.000	f	0.000	f	0.000	Å	0.0
intf	Si	2	s1	.50	1	-0.172 ± .013	f	-0.500 ± .026	f	1.850 ± .100	Å	136.3 ± 7.4
intf	Si	3	s1	.50	2	0.344 ± .013	f	0.000 ± .026	f	0.000 ± .100	Å	0.0 ± 7.4
intf	Si	4	s1	.50	3	-0.389 ± .013	f	0.500 ± .026	f	1.358 ± .100	Å	100.0 ± 7.4
intf	Si	5	s1	.50	4	0.435 ± .013	f	0.000 ± .026	f	0.000 ± .100	Å	0.0 ± 7.4
intf	Si	6	s1	.50	5	-0.717 ± .013	f	0.000 ± .026	f	1.233 ± .100	Å	90.8 ± 7.4
intf	Si	7	s1	.50	6	0.500 ± .013	f	0.000 ± .026	f	0.250 ± .100	Å	18.4 ± 7.4
intf	Si	8	s1	.50	7	-0.500 ± .013	f	-0.500 ± .026	f	1.068 ± .100	Å	78.7 ± 7.4
intf	Si	9	s1	.50	8	0.500 ± .013	f	0.000 ± .026	f	0.250 ± .100	Å	18.4 ± 7.4
subl	Si	10	b	1.00	9	-1.500 ± .026	f	0.000 ± .026	f	1.233 ± .100	Å	90.8 ± 7.4
subl	Si	11	b	1.00	10	0.000 ± .026	f	0.500 ± .026	f	1.358 ± .100	Å	100.0 ± 7.4

Si(100)-(2x1)-Na
14.11.4 BOND DISTANCES AND ANGLES

Bond distances and angles are derived from coordinates

No. of distances/angles: 7

Interatomic dist. A-B (Å)	Atom A	Atom B	Atom C	Bond angle A-B-C (°)
2.975	Na1	Si2	Na1(0,-1)	80.4
2.975	Na1	Si2	Si3	63.7
2.975	Na1	Si2	Si4	84.2
2.640	Si2	Si3	Si5	98.5
2.640	Si2	Si3	Si5(0,-1)	98.5
2.377	Si3	Si5	Si6(1,0)	114.2
2.377	Si3	Si5	Si7	105.6

COMMON NAME : Si(111)-(1x1)-NiSi2(111) interface
CLASSIFICATION : 14.28.12a
TECHNIQUE : XSW
AUTHORS : J. Zegenhagen, K.G. Huang, and W. M. Gibson
REFERENCE : Phys. Rev., B39, 10254 (1989)

ILLUSTRATION: 114

SURFACE TYPE
Substrate : Si
Crystal face: 111
Temperature : RT*
Bulk lattice: diamond
2D bulk symm: p3m1
2D surf symm: p3m1

Adsorbate: NiSi2
Coverage :
Pattern : (1x1)
Matrix : (1.000, 0.000)
(0.000, 1.000)

STRUCTURE TYPE
B-type, Ni at interface 7-fold coordinated;
note difference with 14.27.2; bulk NiSi2=fluorite;
A-type grows with substrate orientation, B-type is rotated 180° with respect to substrate

SAMPLE PREPARATION (3 sample)
Treatment : NiSi2 epitaxially grown on Si(111),thickness 60-220-970Å
Crystallinity:
Anal. methods: thickness (5% accuracy) by X-ray induced fluorescence
Contamination: see Appl. Phys. Lett. 51, 1176 (1987)

COMMENTS
A-type could also be grown: see 14.28.12b;
XSW measured Zd, the distance between Ni layer at interface and the average position of the first Si layer; reasonable geometrical models permit to exclude 5-, 8-fold coordination of the Ni atom at the interface

DATA COLLECTION
Technique: XSW
Dataset : Ni K fluorescence and reflectivity as a function of reflection angle Θ (32 angular intervals)

THEORY/DATA TREATMENT
1) fivefold Ni, 2) sevenfold Ni, 3) eightfold Ni

STRUCTURES EXAMINED
2) is preferred structure, 1) and 3) ruled out; Zd constant with thickness

2D UNIT CELLS (1 domain observed)

Cell	Ax (Å)	Ay (Å)	Bx (Å)	By (Å)	α (°)	Matrix	Pattern	Cell type
Bulk	3.838	0.000	-1.919	3.324	120.0	(1.000, 0.000) (0.000, 1.000)	(1x1)	b: bulk lattice
Surface 1	3.838	0.000	-1.919	3.324	120.0	(1.000, 0.000) (0.000, 1.000)	(1x1)	s1: commens. superlattice

3D COORDINATES

Si1-Ni2-Si3: periodically repeating bulk NiSi2 trilayer; Si4-Ni5-Si6: last NiSi2 trilayer;
Ni5-Si6-Si7: interface plane; 225Å sample used for data; Si11-Si12: periodically repeating bulk substrate bilayer

Dx/Dy in Å, or as a fraction of layer's unit cell vectors; atom 0 at the origin. Epir/subr are bulk repeat vectors.

No. of atoms: 12

Bulk z = 3.130 Å

Region	Chem el.	At. no.	Cell type	Site occ.	Rel. to	Dx ± εx		Dy ± εy		Dz ± εz		Dz/Bz(%) ± εz/Bz
epir		-2				1.919	Å	-1.108	Å	3.130	Å	
subr		-1				1.919	Å	1.108	Å	3.130	Å	
epil	Si	1	b	1.00	0	0.000	f	0.000	f	0.000	Å	0.0
epil	Ni	2	b	1.00	1	0.333	f	0.667	f	0.783	Å	25.0
epil	Si	3	b	1.00	2	0.333	f	-0.333	f	0.783	Å	25.0
intf	Si	4	b	1.00	3	-0.333	f	0.333	f	1.565	Å	50.0
intf	Ni	5	b	1.00	4	0.333	f	-0.333	f	0.783	Å	25.0
intf	Si	6	b	1.00	5	-0.667	f	-0.333	f	0.783	Å	25.0
intf	Si	7	b	1.00	6	0.000	f	0.000	f	2.300 ± .040	Å	73.5 ± 1.3
intf	Si	8	b	1.00	7	-0.333	f	0.333	f	0.783	Å	25.0
intf	Si	9	b	1.00	8	0.000	f	0.000	f	2.350	Å	75.1
intf	Si	10	b	1.00	9	-0.333	f	-0.667	f	0.783	Å	25.0
subl	Si	11	b	1.00	10	0.000	f	0.000	f	2.350	Å	75.1
subl	Si	12	b	1.00	11	0.667	f	0.333	f	0.783	Å	25.0

Si(111)-(1x1)-NiSi2(111) interface
14.28.12a BOND DISTANCES AND ANGLES

No. of distances/angles: 5

Interatomic dist. A-B (Å)	Atom A	Atom B	Atom C	Bond angle A-B-C (°)
2.350	Si1	Ni2	Si1(1,1)	109.5
2.350	Si1	Ni2	Si3(0,1)	180.0
2.350	Si1	Ni2	Si3	70.5
2.350	Si4	Ni5(0,1)	Si3(0,1)	70.6
2.350	Si4	Ni5(0,1)	Si4(1,1)	109.5

COMMON NAME : Si(111)-(1x1)-NiSi2(111) interface ILLUSTRATION: 113
CLASSIFICATION : 14.28.12b
TECHNIQUE : XSW
AUTHORS : J. Zegenhagen, K.G. Huang and W. M. Gibson
REFERENCE : Phys. Rev., B39, 10254 (1989)

SURFACE TYPE
Substrate : Si
Crystal face: 111
Temperature : RT*
Bulk lattice: diamond
2D bulk symm: p3m1
2D surf symm: p3m1

Adsorbate: NiSi2
Coverage :
Pattern : (1x1)
Matrix : (1.000, 0.000)
(0.000, 1.000)

STRUCTURE TYPE
A-type, Ni at interface 7-fold coordinated
note difference with 14.27.2; bulk NiSi2=fluorite;
A-type grows with substrate orientation, B-type is rotated 180° with respect to substrate

SAMPLE PREPARATION (3 sample)
Treatment : NiSi2 epitaxially grown on Si(111),thickness 230-220-530Å
Crystallinity:
Anal. methods: thickness (5% accuracy) by X-ray induced fluorescence
Contamination: see Appl. Phys. Lett. 51, 1176 (1987)

COMMENTS
B-type could also be grown: see 14.28.12a;
XSW measured Zd, the distance between Ni layer at interface and the average position of the first Si layer; reasonable geometrical models permit to exclude 5-, 8-fold coordination of the Ni atom at the interface

DATA COLLECTION
Technique: XSW
Dataset : Ni K fluorescence and reflectivity as a function of reflection angle Θ (32 angular intervals)

THEORY/DATA TREATMENT
1) fivefold Ni, 2) sevenfold Ni, 3) eightfold Ni

STRUCTURES EXAMINED
2) is preferred structure, 1) and 3) ruled out; Zd constant with thickness

2D UNIT CELLS (1 domain observed)

Cell	Ax (Å)	Ay (Å)	Bx (Å)	By (Å)	α (°)	Matrix	Pattern	Cell type
Bulk	3.838	0.000	-1.919	3.324	120.0	(1.000, 0.000) (0.000, 1.000)	(1x1)	b: bulk lattice
Surface 1	3.838	0.000	-1.919	3.324	120.0	(1.000, 0.000) (0.000, 1.000)	(1x1)	s1: commens. superlattice

3D COORDINATES

Si1-Ni2-Si3: periodically repeating bulk NiSi2 trilayer; Si4-Ni5-Si6: last NiSi2 trilayer;
Ni5-Si6-Si7: interface plane; 530Å sample used for data; Si11-Si12: periodically repeating bulk substrate bilayer

Dx/Dy in Å, or as a fraction of layer's unit cell vectors; atom 0 at the origin. Epir/subr are bulk repeat vectors.

No. of atoms: 12 Bulk z = 3.130 Å

Region	Chem el.	At. no.	Cell type	Site occ.	Rel. to	Dx ± εx		Dy ± εy		Dz ± εz		Dz/Bz(%) ± εz/Bz
epir		-2				1.919	Å	-1.108	Å	3.130	Å	
subr		-1				-1.919	Å	-1.108	Å	3.130	Å	
epil	Si	1	b	1.00	0	0.000	f	0.000	f	0.000	Å	0.0
epil	Ni	2	b	1.00	1	0.333	f	0.667	f	0.783	Å	25.0
epil	Si	3	b	1.00	2	0.333	f	-0.333	f	0.783	Å	25.0
intf	Si	4	b	1.00	3	-0.333	f	0.333	f	1.565	Å	50.0
intf	Ni	5	b	1.00	4	0.333	f	-0.333	f	0.783	Å	25.0
intf	Si	6	b	1.00	5	-0.667	f	-0.333	f	0.783	Å	25.0
intf	Si	7	b	1.00	6	0.000	f	0.000	f	2.130 ± .100	Å	68.7 ± 3.2
intf	Si	8	b	1.00	7	0.333	f	-0.333	f	0.783	Å	25.0
intf	Si	9	b	1.00	8	0.000	f	0.000	f	2.350	Å	75.1
intf	Si	10	b	1.00	9	0.333	f	0.667	f	0.783	Å	25.0
subl	Si	11	b	1.00	10	0.000	f	0.000	f	2.350	Å	75.1
subl	Si	12	b	1.00	11	-0.667	f	-0.333	f	0.783	Å	25.0

Si(111)-(1x1)-NiSi2(111) interface
14.28.12b BOND DISTANCES AND ANGLES

No. of distances/angles: 7

Interatomic dist. A-B (Å)	Atom A	Atom B	Atom C	Bond angle A-B-C (°)
2.350	Si1	Ni2	Si1(1,1)	109.5
2.350	Si1	Ni2	Si3(0,1)	180.0
2.350	Si1	Ni2	Si3	70.5
2.713	Si3	Si4(1,0)	Ni2(1,0)	54.8
2.713	Si3	Si4(1,0)	Si3(1,1)	90.0
2.350	Si4	Ni5	Si3	70.6
2.350	Si4	Ni5	Si4(1,0)	109.5

COMMON NAME : Si(111)-(1x1)-NiSi2(111) interface ILLUSTRATION: 114
CLASSIFICATION : 14.28.2
TECHNIQUE : HEIS
AUTHORS : E.J. van Loenen, J.W.M. Frenken, J.F. van der Veen and S. Valeri
REFERENCE : Phys. Rev. Lett., 54, 827 (1985)

SURFACE TYPE
Substrate : Si
Crystal face: 111
Temperature : RT*
Bulk lattice: diamond
2D bulk symm: p3m1
2D surf symm: p3m1

Adsorbate: NiSi2
Coverage : epilayer
Pattern : (1x1)
Matrix : (1.000, 0.000)
(0.000, 1.000)

STRUCTURE TYPE
Epitaxial (1x1) multilayer; ideal 3D fit of top Si bilayer against bottom NiSi2 trilayer with Si-Si bonds perpendicular to interface

SAMPLE PREPARATION (1 sample)
Treatment : 25Å epilayer from Ni deposition on (7x7), then annealed
Crystallinity:
Anal. methods:
Contamination: Si(111)(7x7) clean by AES and ISS

COMMENTS
Only the interfacial separation 0.75+2.31=3.06±0.08Å was measured; the sum is here decomposed in proportion to the ideal values of 0.77+2.35=3.12Å

DATA COLLECTION
Technique: HEIS; RBS of 100keV He ions with channeling
Dataset : energy- and angle-dependent yield around two off-normal channel directions

THEORY/DATA TREATMENT
Monte Carlo simulation

STRUCTURES EXAMINED
Interface between ideal bulk-like lattices with two possible registries and variable interfacial layer spacing

QUALITY OF EXPERIMENT-THEORY FIT
Visual

2D UNIT CELLS (1 domain observed)

Cell	Ax (Å)	Ay (Å)	Bx (Å)	By (Å)	α (°)	Matrix	Pattern	Cell type
Bulk	3.838	0.000	1.919	3.324	60.0	(1.000, 0.000) (0.000, 1.000)	(1x1)	b: bulk lattice
Surface 1	3.838	0.000	1.919	3.324	60.0	(1.000, 0.000) (0.000, 1.000)	(1x1)	s1: commens. superlattice

3D COORDINATES

Si1-Ni2-Si3: repeating bulk NiSi2 epilayer; Si4-Ni5-Si6: bottom NiSi2 trilayer;
Si7-Si8: top Si bilayer; Si9-Si10: repeating bulk Si bilayer

Dx/Dy in Å, or as a fraction of layer's unit cell vectors; atom 0 at the origin. Epir/subr are bulk repeat vectors.

No. of atoms: 10 Bulk z = 3.130 Å

Region	Chem el.	At. no.	Cell type	Site occ.	Rel. to	Dx ± εx	Dy ± εy	Dz ± εz	Dz/Bz(%) ± εz/Bz
epir		-2				3.838 Å	2.216 Å	3.080 Å	
subr		-1				1.919 Å	1.108 Å	3.130 Å	
epil	Si	1	b	1.00	0	0.000 f	0.000 f	0.000 Å	0.0
epil	Ni	2	b	1.00	1	0.667 f	0.667 f	0.770 Å	24.6
epil	Si	3	b	1.00	2	-0.333 f	-0.333 f	0.770 Å	24.6
intf	Si	4	b	1.00	3	0.333 f	0.333 f	1.540 Å	49.2
intf	Ni	5	b	1.00	4	-0.333 f	-0.333 f	0.770 Å	24.6
intf	Si	6	b	1.00	5	-0.333 f	-0.333 f	0.750 ± .080 Å	24.0 ± 2.6
intf	Si	7	b	1.00	6	0.000 f	0.000 f	2.310 ± .080 Å	73.8 ± 2.6
intf	Si	8	b	1.00	7	0.333 f	0.333 f	0.780 Å	24.9
subl	Si	9	b	1.00	8	0.000 f	0.000 f	2.350 Å	75.1
subl	Si	10	b	1.00	9	0.333 f	0.333 f	0.780 Å	24.9

Si(111)-(1x1)-NiSi2(111) interface
14.28.2 BOND DISTANCES AND ANGLES

Bond distances and angles are derived from coordinates

No. of distances/angles: 12

Interatomic dist. A-B (Å)	Atom A	Atom B	Atom C	Bond angle A-B-C (°)
2.346	Ni2	Si3(1,0)	Si4	54.0
2.310	Si3	Ni5	Si6	108.7
2.340	Ni5	Si6		
2.310	Si6	Si7		
2.346	Ni2	Si3(1,0)	Ni5(1,0)	109.2
2.310	Ni2	Si4	Si3	55.2
2.310	Ni2	Si4	Ni5	109.2
2.310	Ni2	Si4	Si6(1,0)	124.5
2.699	Si3	Si4	Si3(1,0)	90.7
2.699	Si3	Si4	Ni5	54.0
2.699	Si3	Si4	Si6(1,0)	89.1
2.310	Si3	Ni5	Si4	70.8

COMMON NAME : Si(111)-(1x1)-NiSi2(111) interface   ILLUSTRATION: 114
CLASSIFICATION : 14.28.8
TECHNIQUE : XSW
AUTHORS : E. Vlieg, A.E.M.J. Fischer, J.F. van der Veen, B.N. Dev and G. Materlik
REFERENCE : Surf. Sci., 178, 36 (1986)

SURFACE TYPE
Substrate : Si
Crystal face: 111
Temperature : RT*
Bulk lattice: diamond
2D bulk symm: p3m1
2D surf symm: p3m1
Adsorbate: NiSi2
Coverage : multilayers
Pattern : (1x1)
Matrix : (1.000, 0.000)
(0.000, 1.000)

STRUCTURE TYPE
Epitaxial bulk NiSi2 growth forming (1x1) interface; tabulated is B-type silicide orientation (same as in Van Loenen et al, PRL 54, 827 (1985), class. no. 14.28.2)

SAMPLE PREPARATION (1 sample)
Treatment : cycles of Ni evap. and 5 min 823 K anneals to req. thickness
Crystallinity:
Anal. methods: MEIS monitors thickness; HEIS determines number of Ni planes
Contamination: AES: <5% contamination before Ni depos.

COMMENTS
A-type growth also produced experimentally (for A-type swap fractional coordinates of the epilayer atoms); measured spacings between 1st Ni plane and (111) Bragg planes of Si: 3.41±0.03Å for B, 3.48±0.05Å for A; measured NiSi2(111) interplanar spacing: 3.108±0.004Å for B (0.4% contraction)

DATA COLLECTION
Technique: XSW; fluorescence measurement
Dataset : 12.4k eV x-rays for thin B-type sample, 14k eV otherwise; fluorescence measured about (111) Bragg reflection

THEORY/DATA TREATMENT
X-ray standing wave analysis

STRUCTURES EXAMINED
Interface with the Ni atoms either 7- or 5- fold coordinated

QUALITY OF EXPERIMENT-THEORY FIT
Visual

2D UNIT CELLS (1 domain observed)

Cell	Ax (Å)	Ay (Å)	Bx (Å)	By (Å)	α (°)	Matrix	Pattern	Cell type
Bulk	3.838	0.000	-1.919	3.324	120.0	(1.000, 0.000) (0.000, 1.000)	(1x1)	b: bulk lattice
Surface 1	3.838	0.000	-1.919	3.324	120.0	(1.000, 0.000) (0.000, 1.000)	(1x1)	s1: commens. superlattice

3D COORDINATES

Si1-Ni2-Si3: periodically repeating bulk silicide epilayer;
Si7-Si8: bulk-like bilayer; Si11-Si12: periodically repeating bulk substrate

Dx/Dy in Å, or as a fraction of layer's unit cell vectors; atom 0 at the origin. Epir/subr are bulk repeat vectors.

No. of atoms: 12   Bulk z = 3.170 Å

Reg ion	Chem el.	At. no.	Cell type	Site occ.	Rel. to	Dx ± εx		Dy ± εy		Dz ± εz		Dz/Bz(%) ± εz/Bz
epir		-2				1.919	Å	-1.108	Å	3.108	Å	
subr		-1				-1.919	Å	1.108	Å	3.170	Å	
epil	Si	1	b	1.00	0	0.000	f	0.000	f	0.000	Å	0.0
epil	Ni	2	b	1.00	1	0.333	f	0.667	f	0.778	Å	24.5
epil	Si	3	b	1.00	2	0.333	f	-0.333	f	0.778	Å	24.5
intf	Si	4	b	1.00	3	-0.333	f	0.333	f	1.552	Å	49.0
intf	Ni	5	b	1.00	4	0.333	f	-0.333	f	0.778	Å	24.5
intf	Si	6	b	1.00	5	-0.667	f	-0.333	f	0.754	Å	23.8
intf	Si	7	b	1.00	6	0.000	f	0.000	f	2.259	Å	71.3
intf	Si	8	b	1.00	7	0.667	f	0.333	f	0.807	Å	25.5
intf	Si	9	b	1.00	8	0.000	f	0.000	f	2.350	Å	74.1
intf	Si	10	b	1.00	9	-0.333	f	0.333	f	0.820	Å	25.9
subl	Si	11	b	1.00	10	0.000	f	0.000	f	2.350	Å	74.1
subl	Si	12	b	1.00	11	-0.333	f	-0.667	f	0.780	Å	24.6

COMMON NAME : Si(111)-(√3x√3)R30°-Pb
CLASSIFICATION : 14.82.1
TECHNIQUE : XSW
AUTHORS : B.N. Dev, G. Materlik, F. Grey and R.L. Johnson
REFERENCE : Springer Series in Surface Sciences, 11, 340 (1987)

ILLUSTRATION: 96

SURFACE TYPE
Substrate : Si
Crystal face: 111
Temperature : RT
Bulk lattice: diamond
2D bulk symm: p3m1
2D surf symm: p31m

Adsorbate: Pb
Coverage : 1/3 Pb/Si
Pattern : (√3x√3)R30°
Matrix : (1.000, 1.000)
(-2.000, 1.000)

STRUCTURE TYPE
Atomic adsorption in 1-fold coordinated top sites over unreconstructed, unrelaxed substrate terminated between bilayers

SAMPLE PREPARATION (1 sample)
Treatment : Pb evaporated from Knudsen cell at 0.4ML/min, 673 K anneal
Crystallinity: initially clear (7x7) LEED pattern
Anal. methods:
Contamination: no impurities detected by photoemission

COMMENTS
Multisite adsorption proposed if Pb evaporated onto sample at 523 K; the small coherent fraction noted by the authors may also indicate that more than one site have nearly equal binding energies

DATA COLLECTION
Technique: XSW
Dataset : the (111) and (220) reflections were measured

THEORY/DATA TREATMENT
X-ray standing wave analysis

STRUCTURES EXAMINED
Some surface relaxations and different sites could fit the (111) data, but could not simultaneously fit the (220) data

QUALITY OF EXPERIMENT-THEORY FIT
Visual

2D UNIT CELLS (1 domain observed)

Cell	Ax (Å)	Ay (Å)	Bx (Å)	By (Å)	α (°)	Matrix	Pattern	Cell type
Bulk	3.839	0.000	1.920	3.325	60.0	(1.000, 0.000) (0.000, 1.000)	(1x1)	b: bulk lattice
Surface 1	5.759	3.325	-5.759	3.325	120.0	(1.000, 1.000) (-2.000, 1.000)	(√3x√3)R30°	s1: commens. superlattice

3D COORDINATES

Pb1: overlayer, 1-fold coordinated to Si2; Si2-Si3: bulk-like bilayer;
Si4-Si5: periodically repeating bulk bilayer; 0.1Å error bar assumed for tabulation

Dx/Dy in Å, or as a fraction of layer's unit cell vectors; atom 0 at the origin. Epir/subr are bulk repeat vectors.

No. of atoms: 5 Bulk z = 3.130 Å

Region	Chem el.	At. no.	Cell type	Site occ.	Rel. to	Dx ± εx		Dy ± εy		Dz ± εz		Dz/Bz(%) ± εz/Bz
epir		-2					f		f		Å	
subr		-1				0.000	Å	2.217	Å	3.130	Å	
ovrl	Pb	1	s1	.33	0	0.000	f	0.000	f	0.000	Å	0.0
intf	Si	2	b	1.00	1	0.000	f	0.000	f	2.560 ± .100	Å	81.8 ± 3.2
intf	Si	3	b	1.00	2	-0.333	f	0.667	f	0.780	Å	24.9
subl	Si	4	b	1.00	3	0.000	f	0.000	f	2.350	Å	75.1
subl	Si	5	b	1.00	4	-0.333	f	0.667	f	0.780	Å	24.9

BOND DISTANCES AND ANGLES

Bond distances and angles are derived from coordinates

No. of distances/angles: 2

Interatomic dist. A-B (Å)	Atom A	Atom B	Atom C	Bond angle A-B-C (°)
2.560	Pb1	Si2	Si3	109.4
2.350	Si2	Si3	Si4	109.4

COMMON NAME : Si(111)-(√3x√3)R30°-Pb (ß phase) ILLUSTRATION: 96,97
CLASSIFICATION : 14.82.2
TECHNIQUE : LEED
AUTHORS : T.N. Doust and S.P. Tear
REFERENCE : Surf. Sci., 251/252, 568 (1991)

SURFACE TYPE
Substrate : Si
Crystal face: 111
Temperature : RT
Bulk lattice: diamond
2D bulk symm: p3m1
2D surf symm: p31m

Adsorbate: Pb
Coverage : 0.333 Pb/1x1
Pattern : (√3x√3)R30°
Matrix : (1.000, 1.000)
(-2.000, 1.000)

STRUCTURE TYPE
Atomic adsorption above T4 site; the three first-layer Si atoms are moved in as well as up; the Si right below Pb is moved down together with the Si atom below it; all the other Si's in the 2nd, 3rd, and 4th Si layers are moved upwards; the 5th Si layer and below are in bulk positions

SAMPLE PREPARATION (1 sample)
Treatment : annealed the Pb/Si above 350C
Crystallinity: good LEED pattern
Anal. methods: AES
Contamination:

COMMENTS
There exist two (√3x√3) phases of Pb on Si(111): the first (α) appears if the annealing temperature is below 350C; the other, irreversible, phase (ß) appears if annealed above 350C; the structure given here is for this ß phase

DATA COLLECTION
Technique: LEED; computer-controlled LEED diffractomet
Dataset : IV spectra measured for 223 beams; only 4 used in analysis; 40<E<200 eV

THEORY/DATA TREATMENT
Dynamical LEED (tensor LEED): 9 phase shifts; Vor=-5±1 eV

STRUCTURES EXAMINED
Four adatom adsorption models: T4 site with 0.33ML and 1ML coverages, and H3 site for the same two cases

QUALITY OF EXPERIMENT-THEORY FIT
RPE=0.32

2D UNIT CELLS (1 domain observed)

Cell	Ax (Å)	Ay (Å)	Bx (Å)	By (Å)	α (°)	Matrix	Pattern	Cell type
Bulk	3.839	0.000	1.920	3.325	60.0	(1.000, 0.000) (0.000, 1.000)	(1x1)	b: bulk lattice
Surface 1	5.759	3.325	-5.759	3.325	120.0	(1.000, 1.000) (-2.000, 1.000)	(√3x√3)R30°	s1: commens. superlattice

3D COORDINATES

Pb1: adatom Pb layer; Si2-Si4 : top Si layer (half bilayer); Si5-Si7: 2nd Si layer; Si14-Si15: repeating bulk layers

Dx/Dy in Å, or as a fraction of layer's unit cell vectors; atom 0 at the origin. Epir/subr are bulk repeat vectors.

No. of atoms: 15 Bulk z = 3.130 Å

Reg ion	Chem el.	At. no.	Cell type	Site occ.	Rel. to	Dx ± εx	Dy ± εy	Dz ± εz	Dz/Bz(%) ± εz/Bz
epir		-2				f	f	Å	
subr		-1				1.920 Å	1.108 Å	3.130 Å	
intf	Pb	1	s1	.33	0	0.000 ± .050 Å	0.000 ± .050 Å	0.000 ± .050 Å	0.0 ± 1.6
intf	Si	2	s1	.33	1	1.825 ± .050 Å	-1.054 ± .050 Å	1.430 ± .050 Å	45.7 ± 1.6
intf	Si	3	s1	.33	2	3.935 ± .050 Å	-0.165 ± .050 Å	0.000 ± .050 Å	0.0 ± 1.6
intf	Si	4	s1	.33	3	3.935 ± .050 Å	0.165 ± .050 Å	0.000 ± .050 Å	0.0 ± 1.6
intf	Si	5	s1	.33	1	0.000 ± .050 Å	0.000 ± .050 Å	2.400 ± .050 Å	76.7 ± 1.6
intf	Si	6	s1	.33	5	3.839 ± .050 Å	0.000 ± .050 Å	-0.380 ± .050 Å	-12.1 ± 1.6
intf	Si	7	s1	.33	6	3.839 ± .050 Å	0.000 ± .050 Å	0.000 ± .050 Å	0.0 ± 1.6
intf	Si	8	s1	.33	5	0.000 ± .050 Å	0.000 ± .050 Å	2.340 ± .050 Å	74.8 ± 1.6
intf	Si	9	s1	.33	8	3.839 ± .050 Å	0.000 ± .050 Å	-0.220 ± .050 Å	-7.0 ± 1.6
intf	Si	10	s1	.33	9	3.839 ± .050 Å	0.000 ± .050 Å	0.000 ± .050 Å	0.0 ± 1.6
intf	Si	11	s1	.33	8	1.920 ± .050 Å	1.108 ± .050 Å	0.730 ± .050 Å	23.3 ± 1.6
intf	Si	12	s1	.33	11	3.839 ± .050 Å	0.000 ± .050 Å	0.000 ± .050 Å	0.0 ± 1.6
intf	Si	13	s1	.33	12	3.839 ± .050 Å	0.000 ± .050 Å	0.000 ± .050 Å	0.0 ± 1.6
subl	Si	14	b	1.00	11	0.000 Å	0.000 Å	2.340 Å	74.8
subl	Si	15	b	1.00	14	0.000 Å	-2.217 Å	0.780 Å	24.9

Si(111)-(√3x√3)R30°-Pb (ß phase)
14.82.2 BOND DISTANCES AND ANGLES

Bond distances and angles are derived from coordinates

No. of distances/angles: 3

Interatomic dist. A-B (Å)	Atom A	Atom B	Atom C	Bond angle A-B-C (°)
2.400	Pb1	Si5	Si2	65.3
2.960	Si5	Si8	Si11	105.6
2.400	Si1	Si6	Si9	104.2

COMMON NAME : Si(111)-(√3x√3)R30°-Sn ILLUSTRATION: 96,97
CLASSIFICATION : 14.50.2
TECHNIQUE : XRD
AUTHORS : K.M. Conway, J.E. MacDonald, C. Norris, E. Vlieg and J.F. van der Veen
REFERENCE : Surf. Sci., 215, 555 (1989)

SURFACE TYPE
Substrate : Si
Crystal face: 111
Temperature : RT
Bulk lattice: diamond
2D bulk symm: p3m1
2D surf symm: p31m
Adsorbate: Sn
Coverage : 0.333 Sn/1x1
Pattern : (√3x√3)R30°
Matrix : (1.000, 1.000)
(-2.000, 1.000)

STRUCTURE TYPE
Atomic adsorption at T4 site; the three first-layer Si's are moved radially inwards; the Si below the T4 site is moved down, pushing the Si right below it downwards; other second-layer Si's and third-layer Si's below them are moved upwards; layers 4, 5, and 6 are laterally moved

SAMPLE PREPARATION (1 sample)
Treatment : Ar bombardment; annealing at 1080C; Sn from Knudsen cell
Crystallinity: Sharp 7x7 RHEED pattern
Anal. methods: RHEED
Contamination:

COMMENTS
RHEED used to check the crystallinity.

DATA COLLECTION
Technique: XRD; sync. rad., 5-circle diffractometer
Dataset : 32 in-plane reflections and 10 fractional order rods

THEORY/DATA TREATMENT
Least-square fitting together with the Keating model for strain energy

STRUCTURES EXAMINED
T4 and H3 sites; same with topmost Si bilayer rotated by 180°

QUALITY OF EXPERIMENT-THEORY FIT
Chi**2=1.26

2D UNIT CELLS (1 domain observed)

Cell	Ax (Å)	Ay (Å)	Bx (Å)	By (Å)	α (°)	Matrix	Pattern	Cell type
Bulk	3.839	0.000	1.920	3.325	60.0	(1.000, 0.000) (0.000, 1.000)	(1x1)	b: bulk lattice
Surface 1	5.759	3.325	-5.759	3.325	120.0	(1.000, 1.000) (-2.000, 1.000)	(√3x√3)R30°	s1: commens. superlattice

3D COORDINATES

Sn1: adsorbate layer at T4 site; Si2-Si7: 1st Si bilayer; Si8-Si13: 2nd Si bilayer; Si14-Si19: 3rd Si bilayer; Si20-Si21: periodic bulk layers

Dx/Dy in Å, or as a fraction of layer's unit cell vectors; atom 0 at the origin. Epir/subr are bulk repeat vectors.

No. of atoms: 21 Bulk z = 3.130 Å

Region	Chem el.	At. no.	Cell type	Site occ.	Rel. to	Dx ± εx	Dy ± εy	Dz ± εz	Dz/Bz(%) ± εz/Bz
epir		-2				f	f	Å	
subr		-1				1.920 Å	1.108 Å	3.130 Å	
intf	Sn	1	s1	.33	0	0.000 Å	0.000 Å	0.000 ± .300 Å	0.0 ± 9.6
intf	Si	2	s1	.33	1	1.737 Å	-1.003 Å	1.590 Å	50.8
intf	Si	3	s1	.33	2	4.022 Å	-0.317 Å	0.000 Å	0.0
intf	Si	4	s1	.33	3	4.022 Å	0.317 Å	0.000 Å	0.0
intf	Si	5	s1	.33	1	0.000 Å	0.000 Å	2.790 Å	88.1
intf	Si	6	s1	.33	5	3.839 Å	0.000 Å	-0.620 Å	-19.8
intf	Si	7	s1	.33	6	3.839 Å	0.000 Å	0.000 Å	0.0
intf	Si	8	s1	.33	1	0.000 Å	0.000 Å	5.070 Å	162.0
intf	Si	9	s1	.33	6	0.000 Å	0.000 Å	2.390 Å	76.4
intf	Si	10	s1	.33	7	0.000 Å	0.000 Å	2.390 Å	76.4
intf	Si	11	s1	.33	1	2.006 Å	1.158 Å	5.490 Å	175.4
intf	Si	12	s1	.33	11	3.753 Å	-0.150 Å	0.000 Å	0.0
intf	Si	13	s1	.33	12	3.753 Å	0.150 Å	0.000 Å	0.0
intf	Si	14	s1	.33	1	1.946 Å	1.123 Å	7.850 Å	250.8
intf	Si	15	s1	.33	14	3.813 Å	-0.045 Å	0.000 Å	0.0
intf	Si	16	s1	.33	15	3.813 Å	0.045 Å	0.000 Å	0.0
intf	Si	17	s1	.33	1	1.899 Å	-1.096 Å	8.640 Å	276.0
intf	Si	18	s1	.33	17	3.860 Å	-0.036 Å	0.000 Å	0.0
intf	Si	19	s1	.33	18	3.860 Å	0.036 Å	0.000 Å	0.0

Si(111)-($\sqrt{3}$x$\sqrt{3}$)R30°-Sn
14.50.2

BOND DISTANCES AND ANGLES

Bond distances and angles are derived from coordinates

No. of distances/angles: 3

Interatomic dist. A-B (Å)	Atom A	Atom B	Atom C	Bond angle A-B-C (°)
2.560	Si2	Sn1	Si5	51.6
2.181	Si2	Si6	Si9	105.4
2.355	Si5	Si8	Si11	100.3

COMMON NAME : Si(100)-(2x1)-2Sb
CLASSIFICATION : 14.51.7
TECHNIQUE : SEXAFS
AUTHORS : M. Richter, J.C. Woicik, P. Pianetta, K.E. Miyano, T.Kendelewicz, C.E. Bouldin, W.E. Spicer and I. Lindau
REFERENCE : J. Vac. Sci. Technol., A9, 1951 (1991)

ILLUSTRATION: 109

SURFACE TYPE
Substrate : Si
Crystal face: 100
Temperature : RT*
Bulk lattice: diamond
2D bulk symm: pmm
2D surf symm: pmm

Adsorbate: Sb
Coverage : 1 Sb/1x1
Pattern : (2x1)
Matrix : (2.000, 0.000)
(0.000, 1.000)

STRUCTURE TYPE
Sb dimer on unreconstructed substrate (dimer orientation perp. to that of the Si dimer formed after removal of the Sb layer, i.e. parallel to Si dimer formed by substitution of Si for Sb)

SAMPLE PREPARATION (1 sample)
Treatment : Sb depos. on clean Si at RT, then annealed to 375C and 550C
Crystallinity: LEED: (1x1) with diffuse (2x1) spots
Anal. methods: AES, LEED
Contamination: AES: no C, O

COMMENTS

DATA COLLECTION
Technique: SEXAFS; polariz.-dep. SEXAFS at SSRL
Dataset : SEXAFS above Sb L3M45M45 edge at 3 angles: k range 2.5-8Å-1

THEORY/DATA TREATMENT
Fourier transform, backtransform and 2-shell fit, using bulk AlSb and Sb as standards

STRUCTURES EXAMINED
3 models with 1ML coverage: 4-fold hollow, bridge, and 'modified bridge' (the preferred model); substrate unrelaxed

QUALITY OF EXPERIMENT-THEORY FIT
Visual

2D UNIT CELLS (1 domain observed)

Cell	Ax (Å)	Ay (Å)	Bx (Å)	By (Å)	α (°)	Matrix	Pattern	Cell type
Bulk	3.840	0.000	0.000	3.840	90.0	(1.000, 0.000) (0.000, 1.000)	(1x1)	b: bulk lattice
Surface 1	7.680	0.000	0.000	3.840	90.0	(2.000, 0.000) (0.000, 1.000)	(2x1)	s1: commens. superlattice

3D COORDINATES

Sb1-Sb2: dimer overlayer (similar to clean-surface dimer); Si3-Si4: periodically repeating bulk pair of layers; coordinates are derived from bond distances and angles, assuming unrelaxed bulk

Dx/Dy in Å, or as a fraction of layer's unit cell vectors; atom 0 at the origin. Epir/subr are bulk repeat vectors.

No. of atoms: 4

Bulk z = 1.358 Å

Region	Chem el.	At. no.	Cell type	Site occ.	Rel. to	Dx ± εx		Dy ± εy		Dz ± εz		Dz/Bz(%) ± εz/Bz
epir		-2					f		f		Å	
subr		-1				1.920	Å	1.920	Å	2.715	Å	
ovrl	Sb	1	s1	.50	0	0.000	f	0.000	f	0.000	Å	0.0
ovrl	Sb	2	s1	.50	1	-0.621 ± .005	f	0.000	f	0.000	Å	0.0
subl	Si	3	b	1.00	2	0.121 ± .001	f	-0.500	f	1.740 ± .060	Å	128.2 ± 4.4
subl	Si	4	b	1.00	3	-0.500	f	0.000	f	1.358	Å	100.0

BOND DISTANCES AND ANGLES

No. of distances/angles: 3

Interatomic dist. A-B (Å)	Atom A	Atom B	Atom C	Bond angle A-B-C (°)
2.910	Sb1	Sb2(1,0)	Si3(2,0)	100.2
2.633	Sb2	Si3	Sb2(0,-1)	93.7
2.633	Sb2	Si3	Si4	103.7

COMMON NAME : Si(111)-(7x7)-Te
CLASSIFICATION : 14.52.1
TECHNIQUE : SEXAFS
AUTHORS : P.H. Citrin, P. Eisenberger and J.E. Rowe
REFERENCE : Phys. Rev. Lett., 48, 802 (1982)

ILLUSTRATION: 96

SURFACE TYPE
Substrate : Si
Crystal face: 111
Temperature : RT*
Bulk lattice: diamond
2D bulk symm: p3m1
2D surf symm: cm

Adsorbate: Te
Coverage : 1 ML
Pattern : (1x1)
Matrix : (1.000, 0.000)
(0.000, 1.000)

STRUCTURE TYPE
Atomic on-bridge adsorption;
(7x7) structure inferred from LEED; modeled as (1x1) here
relationship with (7x7) structure unknown; only local
geometry near adsorbate determined

SAMPLE PREPARATION (1 sample)
Treatment : Ar+ bomb. and annealed, exposed to Te, annealed at 773 K
Crystallinity: LEED showed a (7x7) pattern
Anal. methods: AES determined 1 ML Te coverage
Contamination:

COMMENTS

DATA COLLECTION
Technique: SEXAFS; total yield SEXAFS
Dataset : SEXAFS from Te LIII edge vs polar angle and polarisation

THEORY/DATA TREATMENT
Fourier transform analysis of SEXAFS data with SiI(CH3)3 standard combined with analysis of amplitude ratios

STRUCTURES EXAMINED
Structure determined directly from first neighbor bond lengths and amplitude ratios

QUALITY OF EXPERIMENT-THEORY FIT
Visual

2D UNIT CELLS (3 domains observed)

Cell	Ax (Å)	Ay (Å)	Bx (Å)	By (Å)	α (°)	Matrix	Pattern	Cell type
Bulk	3.840	0.000	-1.920	3.326	120.0	(1.000, 0.000) (0.000, 1.000)	(1x1)	b: bulk lattice
Surface 1	3.840	0.000	-1.920	3.326	120.0	(1.000, 0.000) (0.000, 1.000)	(1x1)	s1: commens. superlattice

3D COORDINATES

Te1: bridge-site overlayer; Si2-Si3: repeating bulk bilayer (assumed structure);
coordinates are derived from bond lengths; 0.1Å error bar assumed for tabulation

Dx/Dy in Å, or as a fraction of layer's unit cell vectors; atom 0 at the origin. Epir/subr are bulk repeat vectors.

No. of atoms: 3 Bulk z = 3.130 Å

Region	Chem el.	At. no.	Cell type	Site occ.	Rel. to	Dx ± εx		Dy ± εy		Dz ± εz		Dz/Bz(%) ± εz/Bz
epir		-2					f		f		Å	
subr		-1				0.000	Å	2.217	Å	3.130	Å	
ovrl	Te	1	b	1.00	0	0.000	f	0.000	f	0.000	Å	0.0
subl	Si	2	b	1.00	1	0.500	f	0.000	f	1.506 ± .100	Å	48.1 ± 3.2
subl	Si	3	b	1.00	2	0.333	f	0.667	f	0.780	Å	24.9

BOND DISTANCES AND ANGLES

No. of distances/angles: 11

Interatomic dist. A-B (Å)	Atom A	Atom B	Atom C	Bond angle A-B-C (°)
2.440	Te1	Si2	Te1(1,0)	103.8
2.350	Si2	Si3(0,-1)	Si2(1,0)	109.6
2.541	Si3	Te1(1,1)	Si2(0,1)	56.3
2.440	Te1	Si2	Si3	101.8
2.440	Te1	Si2	Si3(0,-1)	148.0
2.440	Te1	Si2	Si3(-1,-1)	64.0
2.541	Te1	Si3(-1,-1)	Si2	59.7
2.541	Te1	Si3(-1,-1)	Si2(-1,-1)	96.5

Si(111)-(7x7)-Te
14.52.1

Bond Distances and Angles - Continued

Interatomic dist. A-B (Å)	Atom A	Atom B	Atom C	Bond angle A-B-C (°)
2.440	Si2	Te1	Si3(-1,-1)	56.3
2.350	Si2	Si3	Te1(1,1)	96.5
2.350	Si2	Si3(0,-1)	Te1(1,0)	59.7

COMMON NAME : SiC(100)-c(2x2) (C2H4 exposed) ILLUSTRATION: 118
CLASSIFICATION : 14.6.7a
TECHNIQUE : LEED
AUTHORS : J.M. Powers, A. Wander, P.J. Rous, M.A. Van Hove and G.A. Somorjai
REFERENCE : Phys. Rev., B44, 11159 (1991)

SURFACE TYPE
Substrate : SiC
Crystal face: 100
Temperature : RT
Bulk lattice: zincblende
2D bulk symm: pmm
2D surf symm: cmm
Adsorbate:
Coverage :
Pattern : c(2x2)
Matrix : (1.000,-1.000)
(1.000, 1.000)

STRUCTURE TYPE
C-rich C-terminated reconstruction with symmetric C pairs bridging Si pairs; slight relaxations in second layer; this structure possibly contains hydrogen (see structure 14.6.7b for a presumably H-free version)

SAMPLE PREPARATION (sample)
Treatment : heat cleaned (1175 K), exposed to 100L C2H4, further anneal
Crystallinity: slightly diffuse LEED pattern
Anal. methods: AES
Contamination: reduced by Si dep.; AES Si/C=1.0

COMMENTS
ß-SiC prepared as 4-6µ thick film on Si(100) wafer by chemical vapor deposition; single-domain SiC obtained by cutting Si wafer 0.5° off (100)

DATA COLLECTION
Technique: LEED; video camera
Dataset : IV curves for
(01),(10),(11),(20),(02),(0.5,0.5),(1.5,0.5
(01),(10),(11),(20),(02),(0.5,0.5),(1.5,0.5

THEORY/DATA TREATMENT
Dynamical LEED (autom. search, tensor LEED, lay dblg): pots from H-S w. fcts; Vor=-9±1 eV (fit), Voi=-6eV; ΘD=1430 K

STRUCTURES EXAMINED
7 different models with C pairs, dimers and monomers, Si dimers and monomers were examined with conventional dyn. LEED, optimizing only top layer; 2 best models were subject of automated optimization of 12 coordinates (for 4 atoms in 2 top layers)

QUALITY OF EXPERIMENT-THEORY FIT
RPE=0.24

2D UNIT CELLS (1 domain observed)

Cell	Ax (Å)	Ay (Å)	Bx (Å)	By (Å)	α (°)	Matrix	Pattern	Cell type
Bulk	3.090	0.000	0.000	3.090	90.0	(1.000, 0.000) (0.000, 1.000)	(1x1)	b: bulk lattice
Surface 1	3.090	-3.090	3.090	3.090	90.0	(1.000,-1.000) (1.000, 1.000)	c(2x2)	s1: commens. superlattice

3D COORDINATES

C1-C2: symmetric bonded C pair bridging Si3-Si4 pair; Si3-Si4: slightly relaxed substrate atoms;
C5-Si6: periodically repeating bulk pair of layers

Dx/Dy in Å, or as a fraction of layer's unit cell vectors; atom 0 at the origin. Epir/subr are bulk repeat vectors.

No. of atoms: 6 Bulk z = 1.093 Å

Region	Chem el.	At. no.	Cell type	Site occ.	Rel. to	Dx ± εx		Dy ± εy		Dz ± εz		Dz/Bz(%) ± εz/Bz
epir		-2					f		f		Å	
subr		-1				1.545	Å	-1.545	Å	2.185	Å	
intf	C	1	s1	.50	0	0.000	f	0.000	f	0.000	Å	0.0
intf	C	2	s1	.50	1	-0.202 ± .012	f	-0.202 ± .012	f	0.000 ± .025	Å	0.0 ± 2.3
intf	Si	3	s1	.50	2	0.851 ± .012	f	0.851 ± .012	f	1.620 ± .025	Å	148.3 ± 2.3
intf	Si	4	s1	.50	3	-0.500 ± .012	f	-0.500 ± .012	f	0.000 ± .025	Å	0.0 ± 2.3
subl	C	5	b	1.00	4	0.000 ± .016	f	0.500 ± .016	f	1.070 ± .025	Å	97.9 ± 2.3
subl	Si	6	b	1.00	5	0.500	f	0.000	f	1.093	Å	100.0

SiC(100)-c(2x2) (C2H4 exposed)
14.6.7a BOND DISTANCES AND ANGLES

Bond distances and angles are derived from coordinates

No. of distances/angles: 4

Interatomic dist. A-B (Å)	Atom A	Atom B	Atom C	Bond angle A-B-C (°)
1.250	C1	C2	Si3(-1,-1)	119.6
1.863	C1	Si4	C5	119.7
1.879	Si4	C5	Si6	109.2
1.879	Si4	C5	Si6(-1,0)	109.2

COMMON NAME : SiC(100)-c(2x2) (Si sublimation) ILLUSTRATION: 119
CLASSIFICATION : 14.6.7b
TECHNIQUE : LEED
AUTHORS : J.M. Powers, A. Wander, P.J. Rous, M.A. Van Hove and G.A. Somorjai
REFERENCE : Phys. Rev., B44, 11159 (1991)

SURFACE TYPE
Substrate : SiC
Crystal face: 100
Temperature : RT
Bulk lattice: zincblende
2D bulk symm: pmm
2D surf symm: cmm
Adsorbate:
Coverage :
Pattern : c(2x2)
Matrix : (1.000,-1.000)
(1.000, 1.000)

STRUCTURE TYPE
C-rich C-terminated reconstruction with symmetric C pairs bridging Si pairs; marked relaxations in second layer; this structure is presumably hydrogen-free (see structure 14.6.7a for a possibly H-containing version)

SAMPLE PREPARATION (sample)
Treatment : heat cleaned (1175 K), Si removed by 1300K anneal
Crystallinity: moderately sharp LEED pattern
Anal. methods: AES
Contamination: reduced by Si dep.; AES Si/C=1.0

COMMENTS
ß-SiC prepared as 4-6μ thick film on Si(100) wafer by chemical vapor deposition; single-domain SiC obtained by cutting Si wafer 0.5° off (100)

DATA COLLECTION
Technique: LEED; video camera
Dataset : IV curves for (01),(10),(11),(20),(0.5,0.5),(1.5,0.5), (0.5,1.5) beams at normal incidence; cumul.

THEORY/DATA TREATMENT
Dynamical LEED (autom. search, tensor LEED, lay dblg): pots from H-S w. fcts; Vor=-9±1 eV (fit), Voi=-6eV; ΘD=1430 K

STRUCTURES EXAMINED
7 different models with C pairs, dimers and monomers, Si dimers and monomers were examined with conventional dyn. LEED, optimizing only top layer; 2 best models were subject of automated optimization of 12 coordinates (for 4 atoms in 2 top layers)

QUALITY OF EXPERIMENT-THEORY FIT
RPE=0.22

2D UNIT CELLS (1 domain observed)

Cell	Ax (Å)	Ay (Å)	Bx (Å)	By (Å)	α (°)	Matrix	Pattern	Cell type
Bulk	3.090	0.000	0.000	3.090	90.0	(1.000, 0.000) (0.000, 1.000)	(1x1)	b: bulk lattice
Surface 1	3.090	-3.090	3.090	3.090	90.0	(1.000,-1.000) (1.000, 1.000)	c(2x2)	s1: commens. superlattice

3D COORDINATES

C1-C2: symmetric bonded C pair bridging Si3-Si4 pair; Si3-Si4: markedly relaxed substrate atoms;
C5-Si6: periodically repeating bulk pair of layers

Dx/Dy in Å, or as a fraction of layer's unit cell vectors; atom 0 at the origin. Epir/subr are bulk repeat vectors.

No. of atoms: 6 Bulk z = 1.093 Å

Region	Chem el.	At. no.	Cell type	Site occ.	Rel. to	Dx ± εx		Dy ± εy		Dz ± εz		Dz/Bz(%) ± εz/Bz
epir		-2					f		f		Å	
subr		-1				1.545	Å	-1.545	Å	2.185	Å	
intf	C	1	s1	.50	0	0.212 ± .023	f	0.212 ± .023	f	0.000 ± .100	Å	0.0 ± 9.2
intf	C	2	s1	.50	1	-0.212 ± .023	f	-0.212 ± .023	f	0.000 ± .100	Å	0.0 ± 9.2
intf	Si	3	s1	.50	2	0.825 ± .023	f	0.825 ± .023	f	1.600 ± .100	Å	146.5 ± 9.2
intf	Si	4	s1	.50	3	-0.439 ± .023	f	-0.439 ± .023	f	0.000 ± .100	Å	0.0 ± 9.2
subl	C	5	b	1.00	4	-0.062 ± .032	f	0.500 ± .032	f	1.070 ± .100	Å	97.9 ± 9.2
subl	Si	6	b	1.00	5	0.500 ± .032	f	0.000 ± .032	f	1.093 ± .100	Å	100.0 ± 9.2

SiC(100)-c(2x2) (Si sublimation)
14.6.7b

BOND DISTANCES AND ANGLES

Bond distances and angles are derived from coordinates

No. of distances/angles: 4

Interatomic dist. A-B (Å)	Atom A	Atom B	Atom C	Bond angle A-B-C (°)
1.310	C1	C2	Si3(-1,-1)	124.0
1.930	C1	Si4	C5	114.4
1.889	Si4	C5	Si6	104.2
1.889	Si4	C5	Si6(-1,0)	114.2

COMMON NAME : SiC(100)-p(2x1) ILLUSTRATION: 120
CLASSIFICATION : 14.6.8
TECHNIQUE : LEED
AUTHORS : J.M. Powers, A. Wander, M.A. Van Hove and G.A. Somorjai
REFERENCE : Surf. Sci., 260, L7 (1992)

SURFACE TYPE
Substrate : SiC
Crystal face: 100
Temperature : RT
Bulk lattice: zincblende
2D bulk symm: pmm
2D surf symm: pm
Adsorbate:
Coverage :
Pattern : (2x1)
Matrix : (2.000, 0.000)
(0.000, 1.000)

STRUCTURE TYPE
Si-rich Si-terminated reconstruction with asymmetric buckled Si dimers; small relaxations in deeper layers

SAMPLE PREPARATION (sample)
Treatment : heat cleaned at 1175 K, Si deposition and further annealing
Crystallinity: moderately sharp LEED spots
Anal. methods: AES
Contamination: reduced by Si dep.; AES Si/C=1.8

COMMENTS
ß-SiC prepared as 4-6μ thick film on Si(100) wafer by chemical vapor deposition

DATA COLLECTION
Technique: LEED; video camera
Dataset : IV curves for (01),(10),(11),(1,0.5),(0,0.5) beams at normal incidence; cumul. E range 920 eV

THEORY/DATA TREATMENT
Dynamical LEED (autom. search w. tensor LEED, RFS): pots from H-S w. fcts; Vor=-10±1 eV (fit), Voi=-6eV; ΘD=1430 K

STRUCTURES EXAMINED
3 different symmetric and buckled Si dimer models, with automated optimization of 18 coordinates (for 6 atoms in 3 top layers)

QUALITY OF EXPERIMENT-THEORY FIT
RPE=0.27

2D UNIT CELLS (1 domain observed)

Cell	Ax (Å)	Ay (Å)	Bx (Å)	By (Å)	α (°)	Matrix	Pattern	Cell type
Bulk	3.090	0.000	0.000	3.090	90.0	(1.000, 0.000) (0.000, 1.000)	(1x1)	b: bulk lattice
Surface 1	6.180	0.000	0.000	3.090	90.0	(2.000, 0.000) (0.000, 1.000)	(2x1)	s1: commens. superlattice

3D COORDINATES

Si1-Si2: buckled asymmetric dimer; C3-C4-Si5-Si6: slightly relaxed substrate atoms;
C7-Si8: periodically repeating bulk pair of layers

Dx/Dy in Å, or as a fraction of layer's unit cell vectors; atom 0 at the origin. Epir/subr are bulk repeat vectors.

No. of atoms: 8 Bulk z = 1.093 Å

Region	Chem el.	At. no.	Cell type	Site occ.	Rel. to	Dx ± εx		Dy ± εy		Dz ± εz		Dz/Bz(%) ± εz/Bz
epir		-2					f		f		Å	
subr		-1				-1.545	Å	1.545	Å	2.185	Å	
intf	Si	1	s1	.50	0	0.000	f	0.000	f	0.000	Å	0.0
intf	Si	2	s1	.50	1	-0.372 ± .008	f	0.000 ± .016	f	0.200 ± .025	Å	18.3 ± 2.3
intf	C	3	s1	.50	2	-0.070 ± .008	f	0.500 ± .016	f	0.990 ± .025	Å	90.6 ± 2.3
intf	C	4	s1	.50	3	0.479 ± .008	f	0.000 ± .016	f	0.020 ± .025	Å	1.8 ± 2.3
intf	Si	5	s1	.50	4	-0.740 ± .008	f	0.000 ± .016	f	1.030 ± .025	Å	94.3 ± 2.3
intf	Si	6	s1	.50	5	0.500 ± .008	f	0.000 ± .016	f	0.010 ± .025	Å	.9 ± 2.3
subl	C	7	b	1.00	6	0.000 ± .016	f	-0.500 ± .016	f	1.080 ± .025	Å	98.9 ± 2.3
subl	Si	8	b	1.00	7	-0.500	f	0.000	f	1.093	Å	100.0

SiC(100)-p(2x1)
14.6.8

BOND DISTANCES AND ANGLES

Bond distances and angles are derived from coordinates

No. of distances/angles: 14

Interatomic dist. A-B (Å)	Atom A	Atom B	Atom C	Bond angle A-B-C (°)
1.976	Si1	C4	Si1(0,1)	102.9
1.820	C3	Si6	C7	109.5
1.911	C4	Si5(1,0)	C3(1,0)	114.3
1.911	C4	Si5(1,0)	C7(1,0)	108.1
1.809	C4	Si6	C3	109.3
1.809	C4	Si6	C7	109.2
1.976	Si1	C4	Si5(1,0)	115.4
1.976	Si1	C4	Si6	104.9
1.885	Si2	C3	Si2(0,1)	110.1
1.885	Si2	C3	Si5	118.6
1.885	Si2	C3	Si6	96.9
1.922	C3	Si5	C4(-1,0)	114.3
1.922	C3	Si5	C7(-1,0)	108.4
1.820	C3	Si6	C4	109.3

COMMON NAME : SrTiO3(100)-(1x1) O-Ti-O termination
CLASSIFICATION : 38.22.8.1b
TECHNIQUE : LEED
AUTHORS : N. Bickel, G. Schmidt, K. Heinz and K. Mueller
REFERENCE : Phys. Rev. Lett., 62, 2009 (1989)

ILLUSTRATION: 154

SURFACE TYPE
Substrate : SrTiO3
Crystal face: 100
Temperature : 120 K
Bulk lattice: perovskite
2D bulk symm: p4m
2D surf symm: p4m

Adsorbate:
Coverage :
Pattern : (1x1)
Matrix : (1.000, 0.000)
(0.000, 1.000)

STRUCTURE TYPE
Bulk termination with 4.1% buckled top O-Ti-O compound layer (O outward), spacing between 1st O-Ti-O and 2nd Sr-O layers expanded 2%, spacing between 2nd Sr-O and 3rd O-Ti-O layers contracted 2%

SAMPLE PREPARATION (1 sample)
Treatment : Ar-ion sputtering at 550 K and annealing to 900K
Crystallinity:
Anal. methods:
Contamination:

COMMENTS
Best fit for the SrTiO3(100) surface was obtained with 2 domains with different layer terminations, coexisting with 1:1 ratio; RPE for either one termination was above 0.63, for a 1:1 mixture of the two terminations 0.529; O-Ti-O term. given here: see 38.22.8.1a for Sr-O term.

DATA COLLECTION
Technique: LEED; back-view LEED, CCD-based video camer
Dataset : I-V curves for (10), (11), (20), (12) and (22) beams

THEORY/DATA TREATMENT
Dynamical LEED (combined space method and renormalized forward scattering)

STRUCTURES EXAMINED
Varied were: spacing between Ti-subplane in top layer and Sr-O second layer, second interlayer spacing, buckling between the Ti- and O-subplanes in top layer

QUALITY OF EXPERIMENT-THEORY FIT
RPE=0.529

2D UNIT CELLS (1 domain observed)

Cell	Ax (Å)	Ay (Å)	Bx (Å)	By (Å)	α (°)	Matrix	Pattern	Cell type
Bulk	3.900	0.000	0.000	3.900	90.0	(1.000, 0.000) (0.000, 1.000)	(1x1)	b: bulk lattice
Surface 1	3.900	0.000	0.000	3.900	90.0	(1.000, 0.000) (0.000, 1.000)	(1x1)	s1: commens. superlattice

3D COORDINATES

Ti1-02-03: buckled top compound O-Ti-O layer, O outermost; Sr4-05: 2nd relaxed Sr-O compound layer; Ti6-07-08: 3rd relaxed O-Ti-O compound layer; Ti6-010: repeating set of substrate layers

Dx/Dy in Å, or as a fraction of layer's unit cell vectors; atom 0 at the origin. Epir/subr are bulk repeat vectors.

No. of atoms: 10

Bulk z = 1.950 Å

Reg ion	Chem el.	At. no.	Cell type	Site occ.	Rel. to	Dx ± εx		Dy ± εy		Dz ± εz		Dz/Bz(%) ± εz/Bz
epir		-2					f		f		Å	
subr		-1				3.900	Å	3.900	Å	3.900	Å	
intf	Ti	1	b	1.00	0	0.000	f	0.000	f	0.000	Å	0.0
intf	O	2	b	1.00	1	0.500	f	0.000	f	-0.080 ± .080	Å	-4.1 ± 4.1
intf	O	3	b	1.00	1	0.000	f	0.500	f	-0.080 ± .080	Å	-4.1 ± 4.1
intf	Sr	4	b	1.00	1	0.500	f	0.500	f	1.989 ± .039	Å	102.0 ± 2.0
intf	O	5	b	1.00	4	-0.500	f	-0.500	f	0.000	Å	0.0
subl	Ti	6	b	1.00	5	0.000	f	0.000	f	1.911 ± .039	Å	98.0 ± 2.0
subl	O	7	b	1.00	6	0.500	f	0.000	f	0.000	Å	0.0
subl	O	8	b	1.00	6	0.000	f	0.500	f	0.000	Å	0.0
subl	Sr	9	b	1.00	6	0.500	f	0.500	f	1.950	Å	100.0
subl	O	10	b	1.00	9	-0.500	f	-0.500	f	0.000	Å	0.0

SrTiO3(100)-(1x1) O-Ti-O termination
38.22.8.1b

BOND DISTANCES AND ANGLES

Bond distances and angles are derived from coordinates

No. of distances/angles: 10

Interatomic dist. A-B (Å)	Atom A	Atom B	Atom C	Bond angle A-B-C (°)
1.952	O1	Ti3(0,1)	O1(0,1)	175.3
2.843	O2	Sr4	O2(0,1)	86.6
1.952	O1	Ti3(0,1)	O2(0,1)	89.9
2.843	O1	Sr4	O1(1,0)	86.6
2.843	O1	Sr4	O2(0,1)	58.0
2.843	O1	Sr4	Ti3(0,1)	35.0
1.952	O2	Ti3(1,0)	O1(1,0)	89.9
1.952	O2	Ti3(1,0)	O2(1,0)	175.3
1.952	O2	Ti3(1,0)	Sr4(1,0)	126.7
2.843	O2	Sr4	O1(1,0)	58.0

COMMON NAME : SrTiO3(100)-(1x1) Sr-O termination
CLASSIFICATION : 38.22.8.1a
TECHNIQUE : LEED
AUTHORS : N. Bickel, G. Schmidt, K. Heinz and K. Mueller
REFERENCE : Phys. Rev. Lett., 62, 2009 (1989)

ILLUSTRATION: 155

SURFACE TYPE
Substrate : SrTiO3
Crystal face: 100
Temperature : 120 K
Bulk lattice: perovskite
2D bulk symm: p4m
2D surf symm: p4m

Adsorbate:
Coverage :
Pattern : (1x1)
Matrix : (1.000, 0.000)
(0.000, 1.000)

STRUCTURE TYPE
Bulk termination with 8.2% buckled top Sr-O compound layer (O outward), spacing between 1st Sr-O and 2nd O-Ti-O layers contracted 10%, spacing between 2nd O-Ti-O and 3rd Sr-O layers expanded 4%

SAMPLE PREPARATION (1 sample)
Treatment : Ar-ion sputtering at 550 K and annealing to 900K
Crystallinity:
Anal. methods:
Contamination:

COMMENTS
Best fit for the SrTiO3(100) surface was obtained with 2 domains with different layer terminations, coexisting with 1:1 ratio; RPE for either one termination was above 0.63, for a 1:1 mixture of the two terminations 0.529;
Sr-O term. given here: see 38.22.8.1b for O-Ti-O term.

DATA COLLECTION
Technique: LEED; back-view LEED, CCD-based video camer
Dataset : I-V curves for (10), (11), (20), (12) and (22) beams

THEORY/DATA TREATMENT
Dynamical LEED (combined space method and renormalized forward scattering)

STRUCTURES EXAMINED
Varied were: spacing between Sr-subplane in top layer and O-Ti-O second layer, second interlayer spacing, buckling between the Sr- and O-subplanes in top layer

QUALITY OF EXPERIMENT-THEORY FIT
RPE=0.529

2D UNIT CELLS (1 domain observed)

Cell	Ax (Å)	Ay (Å)	Bx (Å)	By (Å)	α (°)	Matrix	Pattern	Cell type
Bulk	3.900	0.000	0.000	3.900	90.0	(1.000, 0.000) (0.000, 1.000)	(1x1)	b: bulk lattice
Surface 1	3.900	0.000	0.000	3.900	90.0	(1.000, 0.000) (0.000, 1.000)	(1x1)	s1: commens. superlattice

3D COORDINATES

O1-Sr2: buckled top compound Sr-O layer, O outermost; Ti3-O4-O5: 2nd relaxed O-Ti-O compound layer;
Sr6-O7: 3rd relaxed Sr-O compound layer; Sr6-O10: repeating set of substrate layers

Dx/Dy in Å, or as a fraction of layer's unit cell vectors; atom 0 at the origin. Epir/subr are bulk repeat vectors.

No. of atoms: 10

Bulk z = 1.950 Å

Reg ion	Chem el.	At. no.	Cell type	Site occ.	Rel. to	Dx ± εx		Dy ± εy		Dz ± εz		Dz/Bz(%) ± εz/Bz
epir		-2					f		f		Å	
subr		-1				3.900	Å	3.900	Å	3.900	Å	
intf	Sr	1	b	1.00	0	0.000	f	0.000	f	0.000	Å	0.0
intf	O	2	b	1.00	1	0.500	f	0.500	f	-0.160 ± .080	Å	-8.2 ± 4.1
intf	Ti	3	b	1.00	1	0.500	f	0.500	f	1.755 ± .039	Å	90.0 ± 2.0
intf	O	4	b	1.00	3	0.500	f	0.000	f	0.000	Å	0.0
intf	O	5	b	1.00	3	0.000	f	0.500	f	0.000	Å	0.0
subl	Sr	6	b	1.00	3	-0.500	f	-0.500	f	2.028 ± .039	Å	104.0 ± 2.0
subl	O	7	b	1.00	6	0.500	f	0.500	f	0.000	Å	0.0
subl	Ti	8	b	1.00	7	0.000	f	0.000	f	1.950	Å	100.0
subl	O	9	b	1.00	8	0.500	f	0.000	f	0.000	Å	0.0
subl	O	10	b	1.00	8	0.000	f	0.500	f	0.000	Å	0.0

SrTiO3(100)-(1x1) Sr-O termination
38.22.8.1a

BOND DISTANCES AND ANGLES

Bond distances and angles are derived from coordinates

No. of distances/angles: 13

Interatomic dist. A-B (Å)	Atom A	Atom B	Atom C	Bond angle A-B-C (°)
2.762	O1	Sr2(1,1)	O1(1,1)	173.4
2.762	Sr2	O1	Sr2(1,1)	173.4
3.269	Sr2	Ti3	O1	57.5
3.269	Sr2	Ti3	O7	122.5
2.624	Sr2	O4(0,-1)	Sr6	88.1
2.762	O1	Sr2(1,1)	O1(1,0)	89.8
2.762	O1	Sr2(1,1)	Sr2(1,0)	45.1
2.762	O1	Sr2(1,1)	Ti3(1,1)	150.9
2.762	O1	Sr2(1,1)	Ti3(1,0)	91.8
2.762	O1	Sr2(1,1)	O4(1,1)	124.3
2.762	O1	Sr2(1,1)	O4(1,0)	60.9
1.915	O1	Ti3	Sr2(1,1)	57.5
1.915	O1	Ti3	O4(1,0)	90.0

COMMON NAME : Ta(100)-(1x1)
CLASSIFICATION : 73.1
TECHNIQUE : LEED
AUTHORS : A. Titov and W. Moritz
REFERENCE : Surf. Sci., 123, L709 (1982)

ILLUSTRATION: 12

SURFACE TYPE
Substrate : Ta
Crystal face: 100
Temperature : 260 K
Bulk lattice: bcc
2D bulk symm: p4m
2D surf symm: p4m

Adsorbate:
Coverage :
Pattern : (1x1)
Matrix : (1.000, 0.000)
(0.000, 1.000)

STRUCTURE TYPE
Bulk termination with multilayer relaxation

SAMPLE PREPARATION (1 sample)
Treatment : electropolished, flashed to 2800 K for 30-60 sec
Crystallinity:
Anal. methods:
Contamination: AES: clean, but H2 may be problem

COMMENTS

DATA COLLECTION
Technique: LEED
Dataset : I-V spectra: 5 non-degenerate beams at normal incidence, 50<E<280 eV

THEORY/DATA TREATMENT
Dynamical LEED (layer doubling): up to 12 phase shifts, α-optimized modified Slater exchange; VoiαE**1/3; ΘD=200 K

STRUCTURES EXAMINED
Varied top two interlaying spacings

QUALITY OF EXPERIMENT-THEORY FIT
RZJ=0.21

2D UNIT CELLS (1 domain observed)

Cell	Ax (Å)	Ay (Å)	Bx (Å)	By (Å)	α (°)	Matrix	Pattern	Cell type
Bulk	3.300	0.000	0.000	3.300	90.0	(1.000, 0.000) (0.000, 1.000)	(1x1)	b: bulk lattice
Surface 1	3.300	0.000	0.000	3.300	90.0	(1.000, 0.000) (0.000, 1.000)	(1x1)	s1: commens. superlattice

3D COORDINATES

Dx/Dy in Å, or as a fraction of layer's unit cell vectors; atom 0 at the origin. Epir/subr are bulk repeat vectors.

No. of atoms: 4

Bulk z = 1.650 Å

Reg ion	Chem el.	At. no.	Cell type	Site occ.	Rel. to	Dx ± εx	Dy ± εy	Dz ± εz	Dz/Bz(%) ± εz/Bz
epir		-2				f	f	Å	
subr		-1				1.650 Å	1.650 Å	1.650 Å	
intf	Ta	1	b	1.00	0	0.000 f	0.000 f	0.000 Å	0.0
intf	Ta	2	b	1.00	1	0.500 f	0.500 f	1.470 ± .030 Å	89.1 ± 1.8
intf	Ta	3	b	1.00	2	-0.500 f	-0.500 f	1.670 ± .030 Å	101.2 ± 1.8
subl	Ta	4	b	1.00	3	0.500 f	0.500 f	1.650 Å	100.0

BOND DISTANCES AND ANGLES

Bond distances and angles are derived from coordinates

No. of distances/angles: 3

Interatomic dist. A-B (Å)	Atom A	Atom B	Atom C	Bond angle A-B-C (°)
3.300	Ta1	Ta1(0,1)	Ta2	53.3
2.758	Ta1	Ta2	Ta3	67.8
2.870	Ta2	Ta3	Ta4	70.9

ILLUSTRATION: 12

COMMON NAME : Ta(100)-(1x1)
CLASSIFICATION : 73.4
TECHNIQUE : PED
AUTHORS : R.A. Bartynski, D. Heskett, K. Garrison, G.M. Watson, D.M. Zehner, W.N. Mei, S.Y. Tong and X. Pan
REFERENCE : Phys. Rev., B40, 5340 (1989)

SURFACE TYPE
Substrate : Ta
Crystal face: 100
Temperature : RT*
Bulk lattice: bcc
2D bulk symm: p4m
2D surf symm: p4m

Adsorbate:
Coverage :
Pattern : (1x1)
Matrix : (1.000, 0.000)
(0.000, 1.000)

STRUCTURE TYPE
Contraction of the first interlayer spacing

SAMPLE PREPARATION (1 sample)
Treatment : photoemission
Crystallinity:
Anal. methods:
Contamination:

COMMENTS
In agreement with LEED results

DATA COLLECTION
Technique: PED; hemispherical electron-energy analyzer
Dataset : 4f levels from 50-150 eV and at 65eV from -20 to 80 degrees

THEORY/DATA TREATMENT
Multiple scattering formalism

STRUCTURES EXAMINED
Variation of 1st interlayer spacing

QUALITY OF EXPERIMENT-THEORY FIT
Visual

2D UNIT CELLS (1 domain observed)

Cell	Ax (Å)	Ay (Å)	Bx (Å)	By (Å)	α (°)	Matrix	Pattern	Cell type
Bulk	3.298	0.000	0.000	3.298	90.0	(1.000, 0.000) (0.000, 1.000)	(1x1)	b: bulk lattice
Surface 1	3.298	0.000	0.000	3.298	90.0	(1.000, 0.000) (0.000, 1.000)	(1x1)	s1: commens. superlattice

3D COORDINATES

Dx/Dy in Å, or as a fraction of layer's unit cell vectors; atom 0 at the origin. Epir/subr are bulk repeat vectors.

No. of atoms: 2

Bulk z = 1.649 Å

Region	Chem el.	At. no.	Cell type	Site occ.	Rel. to	Dx ± εx	Dy ± εy	Dz ± εz	Dz/Bz(%) ± εz/Bz
epir		-2				f	f	Å	
subr		-1				1.649 Å	1.649 Å	1.649 Å	
intf	Ta	1	s1	1.00	0	0.000 Å	0.000 Å	0.000 Å	0.0
subl	Ta	2	b	1.00	0	1.649 Å	1.649 Å	1.484 ± .082 Å	90.0 ± 5.0

BOND DISTANCES AND ANGLES

Bond distances and angles are derived from coordinates

No. of distances/angles: 1

Interatomic dist. A-B (Å)	Atom A	Atom B	Atom C	Bond angle A-B-C (°)
2.856	Ta1	Ta2		

COMMON NAME : Ta(100)-(1x3)-O
CLASSIFICATION : 73.8.1
TECHNIQUE : LEED
AUTHORS : A. Titov and H. Jagodzinski
REFERENCE : Surf. Sci., 152/153, 409 (1985)

ILLUSTRATION: 53

SURFACE TYPE
Substrate : Ta
Crystal face: 100
Temperature : RT*
Bulk lattice: bcc
2D bulk symm: p4m
2D surf symm: pmm

Adsorbate: O
Coverage : 0.333 (O/Ta)
Pattern : (1x3)
Matrix : (1.000, 0.000)
(0.000, 3.000)

STRUCTURE TYPE
Atomic O adsorption under bridge in top Ta layer and above bridge in second Ta layer (4-fold coordinated interstitial site); top Ta layer buckled

SAMPLE PREPARATION (1 sample)
Treatment : exposure to >10E5 L O2, then annealing at 2550 K
Crystallinity:
Anal. methods:
Contamination: AES: clean

COMMENTS

DATA COLLECTION
Technique: LEED
Dataset : I-V curves for 11 non-degenerate beams, 40<E<160 eV

THEORY/DATA TREATMENT
Dynamical LEED: phase shifts for free neutral atoms with radii of 1.43Å (Ta) and 0.85Å (O)

STRUCTURES EXAMINED
1 or 2 O per unit cell allowed to induce Ta reconstruction; 9 basic (1x3) models examined for top Ta layer structure; assumed energetically equivalent O sites; bad agreement without Ta reconstruction; hence, first reconstruction optimized, then O added

QUALITY OF EXPERIMENT-THEORY FIT
RZJ=0.39

2D UNIT CELLS (2 domains observed)

Cell	Ax (Å)	Ay (Å)	Bx (Å)	By (Å)	α (°)	Matrix	Pattern	Cell type
Bulk	3.300	0.000	0.000	3.300	90.0	(1.000, 0.000) (0.000, 1.000)	(1x1)	b: bulk lattice
Surface 1	3.300	0.000	0.000	9.900	90.0	(1.000, 0.000) (0.000, 3.000)	(1x3)	s1: commens. superlattice

3D COORDINATES

Ta1-Ta2-Ta3: buckled, laterally relaxed top layer; O4: interstitial underlayer;
0.1Å error bars assumed for tabulation

Dx/Dy in Å, or as a fraction of layer's unit cell vectors; atom 0 at the origin. Epir/subr are bulk repeat vectors.

No. of atoms: 6

Bulk z = 1.650 Å

Region	Chem el.	At. no.	Cell type	Site occ.	Rel. to	Dx ± εx		Dy ± εy		Dz ± εz		Dz/Bz(%) ± εz/Bz
epir		-2					f		f		Å	
subr		-1				-1.650	Å	1.650	Å	1.650	Å	
intf	Ta	1	s1	.33	0	0.000	f	0.808 ± .010	f	0.000	Å	0.0
intf	Ta	2	s1	.33	0	0.000	f	0.192 ± .010	f	0.000	Å	0.0
intf	Ta	3	s1	.33	0	0.000	f	0.500	f	0.100 ± .100	Å	6.1 ± 6.1
intf	O	4	s1	.33	0	0.000	f	0.000 ± .010	f	0.430 ± .100	Å	26.1 ± 6.1
intf	Ta	5	b	1.00	0	0.500	f	0.000	f	1.650 ± .050	Å	100.0 ± 3.0
subl	Ta	6	b	1.00	5	-0.500	f	0.500	f	1.650	Å	100.0

Ta(100)-(1x3)-O
73.8.1 BOND DISTANCES AND ANGLES

Bond distances and angles are derived from coordinates

No. of distances/angles: 4

Interatomic dist. A-B (Å)	Atom A	Atom B	Atom C	Bond angle A-B-C (°)
3.051	Ta1	Ta3	Ta2(1,1)	85.6
1.949	Ta1	O4(0,1)	Ta1(0,1)	150.2
1.949	Ta1	O4(0,1)	Ta3(0,2)	155.9
3.010	Ta1	Ta5(0,1)	Ta1(1,1)	109.0

COMMON NAME : TaC(100)-(1x1) ILLUSTRATION: 149
CLASSIFICATION : 73.6.2
TECHNIQUE : LEED
AUTHORS : J.R. Noonan, H.L. Davis and G.R. Gruzalski
REFERENCE : J. Vac. Sci. Technol., 5, 787 (1987)

SURFACE TYPE
Substrate : TaC
Crystal face: 100
Temperature : RT*
Bulk lattice: NaCl
2D bulk symm: p4m
2D surf symm: p4m

Adsorbate:
Coverage :
Pattern : (1x1)
Matrix : (1.000, 0.000)
(0.000, 1.000)

STRUCTURE TYPE
Bulk termination with buckled top mixed layer (C outward, Ta inward)

SAMPLE PREPARATION (1 sample)
Treatment : single crystal cut, polished, annealed to 2273 K
Crystallinity: sharp (1x1) LEED pattern
Anal. methods:
Contamination: 0.01ML O

COMMENTS

DATA COLLECTION
Technique: LEED
Dataset : I-V spectra for 7 sets of non-degenerate beams at near- normal (≈0.1°) incidence; 20<E<350 eV

THEORY/DATA TREATMENT
Dynamical LEED: up to 10 phase shifts, bulk TaC potentials; Voi=-5.5 eV; ΘD=350 K(Ta), 1600K(C)

STRUCTURES EXAMINED
Variations in spacing between C and Ta in 1st and 2nd mixed layers

QUALITY OF EXPERIMENT-THEORY FIT
RZJ=0.068

2D UNIT CELLS (1 domain observed)

Cell	Ax (Å)	Ay (Å)	Bx (Å)	By (Å)	α (°)	Matrix	Pattern	Cell type
Bulk	3.150	0.000	0.000	3.150	90.0	(1.000, 0.000) (0.000, 1.000)	(1x1)	b: bulk lattice
Surface 1	3.150	0.000	0.000	3.150	90.0	(1.000, 0.000) (0.000, 1.000)	(1x1)	s1: commens. superlattice

3D COORDINATES

C1-Ta2: buckled top mixed layer; C3-Ta4: planar 2nd mixed layer;
C5-Ta6: periodically repeating mixed bulk layer

Dx/Dy in Å, or as a fraction of layer's unit cell vectors; atom 0 at the origin. Epir/subr are bulk repeat vectors.

No. of atoms: 6 Bulk z = 2.230 Å

Region	Chem el.	At. no.	Cell type	Site occ.	Rel. to	Dx ± εx		Dy ± εy		Dz ± εz		Dz/Bz(%) ± εz/Bz
epir		-2					f		f		Å	
subr		-1				1.575	Å	1.575	Å	2.230	Å	
intf	C	1	b	1.00	0	0.000	f	0.000	f	0.000	Å	0.0
intf	Ta	2	b	1.00	1	0.500	f	0.500	f	0.200 ± .080	Å	9.0 ± 3.6
intf	C	3	b	1.00	2	0.000	f	0.000	f	2.120 ± .020	Å	95.1 ± .9
intf	Ta	4	b	1.00	3	-0.500	f	-0.500	f	0.000	Å	0.0
subl	C	5	b	1.00	4	0.000	f	0.000	f	2.230	Å	100.0
subl	Ta	6	b	1.00	5	0.500	f	0.500	f	0.000	Å	0.0

BOND DISTANCES AND ANGLES

Bond distances and angles are derived from coordinates

No. of distances/angles: 6

Interatomic dist. A-B (Å)	Atom A	Atom B	Atom C	Bond angle A-B-C (°)
2.236	C1	Ta2	C1(1,1)	169.7
2.236	C1	Ta2	C1(1,0)	89.5
2.236	C1	Ta2	C3	95.1

TaC(100)-(1x1)
73.6.2

Bond Distances and Angles - Continued

Interatomic dist. A-B (Å)	Atom A	Atom B	Atom C	Bond angle A-B-C (°)
2.236	C1	Ta2	Ta4(1,0)	93.5
2.320	C1	Ta4	C3	90.0
2.120	Ta2	C3	Ta4	90.0

COMMON NAME : TaC(100)-(1x1)
CLASSIFICATION : 73.6.4
TECHNIQUE : LEED
AUTHORS : G.R. Gruzalski, D.M. Zehner, J.R. Noonan, H.L. Davis, R.A. DiDio and K. Mueller
REFERENCE : J. Vac. Sci. Technol., A7, 2054 (1989)

ILLUSTRATION: 149

SURFACE TYPE
Substrate : TaC
Crystal face: 100
Temperature : RT*
Bulk lattice: NaCl
2D bulk symm: p4m
2D surf symm: p4m

Adsorbate:
Coverage :
Pattern : (1x1)
Matrix : (1.000, 0.000)
(0.000, 1.000)

STRUCTURE TYPE
Bulk termination with buckled top two mixed layers (C moves outward, Ta inward) and contraction of first interlayer spacing

SAMPLE PREPARATION (1 sample)
Treatment : Ar-ion bombardment and annealing to 2900 K
Crystallinity: sharp (1x1) LEED pattern
Anal. methods: ARUPS, XPS and AES
Contamination:

COMMENTS

DATA COLLECTION
Technique: LEED
Dataset : I-V spectra for 7 non-degenerate beams at normal incidence; 20<E<350 eV

THEORY/DATA TREATMENT
Dynamical LEED (RFS): 10 phase shifts, different ΘD for Ta and C

STRUCTURES EXAMINED
Variations in spacing between C and Ta in 1st and 2nd mixed layers

QUALITY OF EXPERIMENT-THEORY FIT
RZJ=0.068

2D UNIT CELLS (1 domain observed)

Cell	Ax (Å)	Ay (Å)	Bx (Å)	By (Å)	α (°)	Matrix	Pattern	Cell type
Bulk	3.150	0.000	0.000	3.150	90.0	(1.000, 0.000) (0.000, 1.000)	(1x1)	b: bulk lattice
Surface 1	3.150	0.000	0.000	3.150	90.0	(1.000, 0.000) (0.000, 1.000)	(1x1)	s1: commens. superlattice

3D COORDINATES

C1-Ta2: buckled top mixed layer; C3-Ta4: buckled 2nd mixed layer;
C5-Ta6: periodically repeating mixed bulk layer

Dx/Dy in Å, or as a fraction of layer's unit cell vectors; atom 0 at the origin. Epir/subr are bulk repeat vectors.

No. of atoms: 6 Bulk z = 2.230 Å

Region	Chem el.	At. no.	Cell type	Site occ.	Rel. to	Dx ± εx		Dy ± εy		Dz ± εz		Dz/Bz(%) ± εz/Bz
epir		-2					f		f		Å	
subr		-1				1.575	Å	1.575	Å	2.230	Å	
intf	C	1	b	1.00	0	0.000	f	0.000	f	0.000	Å	0.0
intf	Ta	2	b	1.00	1	0.500	f	0.500	f	0.200 ± .100	Å	9.0 ± 4.5
intf	C	3	b	1.00	2	0.000	f	0.000	f	2.080 ± .100	Å	93.3 ± 4.5
intf	Ta	4	b	1.00	3	-0.500	f	-0.500	f	0.040 ± .100	Å	1.8 ± 4.5
subl	C	5	b	1.00	4	0.000	f	0.000	f	2.230	Å	100.0
subl	Ta	6	b	1.00	5	0.500	f	0.500	f	0.000	Å	0.0

BOND DISTANCES AND ANGLES

Bond distances and angles are derived from coordinates

No. of distances/angles: 6

Interatomic dist. A-B (Å)	Atom A	Atom B	Atom C	Bond angle A-B-C (°)
2.236	C1	Ta2	C1(1,0)	89.5
2.320	C1	Ta4	Ta2	46.4
2.080	Ta2	C3	Ta4(1,1)	91.0

TaC(100)-(1x1)
73.6.4

Bond Distances and Angles - Continued

Interatomic dist. A-B (Å)	Atom A	Atom B	Atom C	Bond angle A-B-C (°)
2.080	C3	Ta2	C1(1,1)	95.1
2.080	C3	Ta2	C1(1,0)	95.1
2.120	Ta2	C3	Ta4	90.0

COMMON NAME : Tb(0001)-(1x1)
CLASSIFICATION : 65.1
TECHNIQUE : LEED
AUTHORS : J. Quinn, J.S. Li, F. Jona and D. Fort
REFERENCE : Surf. Sci., 257, L647 (1991)

ILLUSTRATION: 19

SURFACE TYPE
Substrate : Tb
Crystal face: 0001
Temperature : RT*
Bulk lattice: hcp
2D bulk symm: p3m1
2D surf symm: p3m1

Adsorbate:
Coverage :
Pattern : (1x1)
Matrix : (1.000, 0.000)
(0.000, 1.000)

STRUCTURE TYPE
Multilayer relaxation

SAMPLE PREPARATION (1 sample)
Treatment : ion bombardment followed by annealing
Crystallinity: sharp LEED pattern
Anal. methods: AES
Contamination: AES C ratio of 0.1, Cl ratio of 0.03

COMMENTS

DATA COLLECTION
Technique: LEED; video LEED
Dataset : IV spectra for 3 non-equivalent beams at normal incidence and 7 non-equivalent beams at 8.5° off normal

THEORY/DATA TREATMENT
Dynamical LEED: RSS and layer doubling (VHT programs)

STRUCTURES EXAMINED
Variation of 1st and 2nd interlayer spacing

QUALITY OF EXPERIMENT-THEORY FIT
RPE=0.515, RVHT=0.248, RZJ=0.31

2D UNIT CELLS (1 domain observed)

Cell	Ax (Å)	Ay (Å)	Bx (Å)	By (Å)	α (°)	Matrix	Pattern	Cell type
Bulk	3.601	0.000	1.801	3.119	60.0	(1.000, 0.000) (0.000, 1.000)	(1x1)	b: bulk lattice
Surface 1	3.601	0.000	1.801	3.119	60.0	(1.000, 0.000) (0.000, 1.000)	(1x1)	s1: commens. superlattice

3D COORDINATES

Dx/Dy in Å, or as a fraction of layer's unit cell vectors; atom 0 at the origin. Epir/subr are bulk repeat vectors.

No. of atoms: 5

Bulk z = 2.847 Å

Region	Chem el.	At. no.	Cell type	Site occ.	Rel. to	Dx ± εx		Dy ± εy		Dz ± εz		Dz/Bz(%) ± εz/Bz
epir		-2					f		f		Å	
subr		-1				0.000	Å	0.000	Å	5.694	Å	
intf	Tb	1	s1	1.00	0	0.000	Å	0.000	Å	0.000	Å	0.0
intf	Tb	2	s1	1.00	0	1.801	Å	1.040	Å	2.736 ± .030	Å	96.1 ± 1.1
intf	Tb	3	s1	1.00	0	0.000	Å	0.000	Å	5.623 ± .030	Å	197.5 ± 1.1
subl	Tb	4	b	1.00	0	1.801	Å	1.040	Å	8.469	Å	297.5
subl	Tb	5	b	1.00	0	0.000	Å	0.000	Å	11.316	Å	397.5

BOND DISTANCES AND ANGLES

Bond distances and angles are derived from coordinates

No. of distances/angles: 2

Interatomic dist. A-B (Å)	Atom A	Atom B	Atom C	Bond angle A-B-C (°)
3.436	Tb1	Tb2		
3.557	Tb2	Tb3		

ILLUSTRATION: 158

COMMON NAME : Te(10-10)-(1x1)
CLASSIFICATION : 52.1
TECHNIQUE : LEED
AUTHORS : R.J. Meyer, W.R. Salaneck, C.B. Duke, A. Paton, C.H. Griffiths, L. Kovnat and L.E. Meyer
REFERENCE : Phys. Rev., B21, 4542 (1980)

SURFACE TYPE
Substrate : Te
Crystal face: 10-10
Temperature : 55 K
Bulk lattice: hexagonal
2D bulk symm: p1
2D surf symm: p1

Adsorbate:
Coverage :
Pattern : (1x1)
Matrix : (1.000, 0.000)
(0.000, 1.000)

STRUCTURE TYPE
Bulk termination with upper chain distortion, approximately conserving strong intrachain bonds (bulk consists of loosely-interconnected helical Te chains with 3-fold screw axis // c-axis, i.e. // (10-10) surface)

SAMPLE PREPARATION (1 sample)
Treatment : air cleavage, then Ar+ sputtering and annealing at 478 K
Crystallinity:
Anal. methods:
Contamination: monitored by LEED, AES and EELS

COMMENTS
'pseudo-dynamical' LEED = diffraction from topmost layers evaluated exactly, that from deeper layers kinematically

DATA COLLECTION
Technique: LEED
Dataset : I-V spectra: 9 inequivalent beams, 30<E<220 eV

THEORY/DATA TREATMENT
Pseudo-dynamical (see comm.), with kinematic initial search: 5 phase shifts, mfp=8Å; Vor=-13 eV; ΘD: bulk=160 K, surf=80K

STRUCTURES EXAMINED
Kinematic search for rigid movements of polymeric chains and intrachain relaxations, and rigid bond rotations of top-layer atoms; best-fit structures were refined with pseudo-dynamical search

QUALITY OF EXPERIMENT-THEORY FIT
Visual

2D UNIT CELLS (1 domain observed)

Cell	Ax (Å)	Ay (Å)	Bx (Å)	By (Å)	α (°)	Matrix	Pattern	Cell type
Bulk	4.457	0.000	0.000	5.927	90.0	(1.000, 0.000) (0.000, 1.000)	(1x1)	b: bulk lattice
Surface 1	4.457	0.000	0.000	5.927	90.0	(1.000, 0.000) (0.000, 1.000)	(1x1)	s1: commens. superlattice

3D COORDINATES

Te1-Te2-Te3: distorted outermost helical chain; Te4-Te5-Te6 and Te7-Te8-Te9: 2 bulk chains; 0.1Å lateral error bars assumed for tabulation

Dx/Dy in Å, or as a fraction of layer's unit cell vectors; atom 0 at the origin. Epir/subr are bulk repeat vectors.

No. of atoms: 9 Bulk z = 3.860 Å

Region	Chem el.	At. no.	Cell type	Site occ.	Rel. to	Dx ± εx		Dy ± εy		Dz ± εz		Dz/Bz(%) ± εz/Bz
epir		-2					f		f		Å	
subr		-1				2.228	Å	0.000	Å	3.860	Å	
intf	Te	1	b	1.00	0	0.000	f	0.000	f	0.000	Å	0.0
intf	Te	2	b	1.00	1	0.543 ± .022	f	0.665 ± .017	f	0.368 ± .100	Å	9.5 ± 2.6
intf	Te	3	b	1.00	2	0.368 ± .022	f	-0.305 ± .017	f	1.498 ± .100	Å	38.8 ± 2.6
intf	Te	4	b	1.00	3	-0.500	f	-0.333	f	1.783	Å	46.2
intf	Te	5	b	1.00	4	-0.404	f	0.667	f	1.038	Å	26.9
intf	Te	6	b	1.00	5	0.404	f	-0.333	f	1.038	Å	26.9
subl	Te	7	b	1.00	6	0.500	f	-0.333	f	1.783	Å	46.2
subl	Te	8	b	1.00	7	-0.404	f	0.667	f	1.038	Å	26.9
subl	Te	9	b	1.00	8	0.404	f	-0.333	f	1.038	Å	26.9

Te(10-10)-(1x1)
52.1

BOND DISTANCES AND ANGLES

Bond distances and angles are derived from coordinates

No. of distances/angles: 4

Interatomic dist. A-B (Å)	Atom A	Atom B	Atom C	Bond angle A-B-C (°)
2.869	Te1	Te2(-1,-1)	Te3(-1,-1)	96.0
2.864	Te1	Te3(-1,0)	Te2(-1,0)	102.1
2.862	Te2(-1,0)	Te3(-1,0)	Te1	102.1
2.866	Te4	Te6	Te5	102.3

ILLUSTRATION: 19

COMMON NAME : Ti(0001)-(1x1)
CLASSIFICATION : 22.1
TECHNIQUE : LEED
AUTHORS : H.D. Shih, F. Jona, D.W. Jepsen and P.M. Marcus
REFERENCE : J. Phys., C9, 1405 (1976)

SURFACE TYPE
Substrate : Ti
Crystal face: 0001
Temperature : 300 K
Bulk lattice: hcp
2D bulk symm: p3m1
2D surf symm: p3m1

Adsorbate:
Coverage :
Pattern : (1x1)
Matrix : (1.000, 0.000)
(0.000, 1.000)

STRUCTURE TYPE
Bulk-like termination with contraction of top interlayer spacing

SAMPLE PREPARATION (1 sample)
Treatment : cleaned with very long cycles of ion bomb. & anneal at 750 K
Crystallinity:
Anal. methods:
Contamination: AES: <2% monolayer of O

COMMENTS
Ti(0001) clean surface difficult to prepare because at T>700-800 K S and Cl tend to segregate to surface; because of its reactivity, the surface could only be made atomically clean at T>700 K;
best structure insensitive to nonstructural parameters

DATA COLLECTION
Technique: LEED
Dataset : 25 I-V spectra at 4 different angles of incidence; energy range 20-300 eV

THEORY/DATA TREATMENT
Dynamical LEED (layer KKR): 8 phase shifts (Moruzzi et al); domain-averaging over steps; Vor=-10 eV; Voi=-3eV; ΘD=342 K

STRUCTURES EXAMINED
ABAB hcp termination with the spacing between the first 2 planes varied in the range 2.39-2.31Å in steps of 0.053Å;
ABABC stacking fault

QUALITY OF EXPERIMENT-THEORY FIT
Visual

2D UNIT CELLS (1 domain observed)

Cell	Ax (Å)	Ay (Å)	Bx (Å)	By (Å)	α (°)	Matrix	Pattern	Cell type
Bulk	2.950	0.000	-1.475	2.555	120.0	(1.000, 0.000) (0.000, 1.000)	(1x1)	b: bulk lattice
Surface 1	2.950	0.000	-1.475	2.555	120.0	(1.000, 0.000) (0.000, 1.000)	(1x1)	s1: commens. superlattice

3D COORDINATES

Ti2-Ti3: repeating bulk pair of layers

Dx/Dy in Å, or as a fraction of layer's unit cell vectors; atom 0 at the origin. Epir/subr are bulk repeat vectors.

No. of atoms: 3

Bulk z = 2.340 Å

Region	Chem el.	At. no.	Cell type	Site occ.	Rel. to	Dx ± εx		Dy ± εy		Dz ± εz		Dz/Bz(%) ± εz/Bz
epir		-2					f		f		Å	
subr		-1				0.000	Å	0.000	Å	4.680	Å	
intf	Ti	1	b	1.00	0	0.000	f	0.000	f	0.000	Å	0.0
subl	Ti	2	b	1.00	1	0.333	f	0.667	f	2.290 ± .050	Å	97.9 ± 2.1
subl	Ti	3	b	1.00	2	-0.333	f	-0.667	f	2.340	Å	100.0

BOND DISTANCES AND ANGLES

Bond distances and angles are derived from coordinates

No. of distances/angles: 7

Interatomic dist. A-B (Å)	Atom A	Atom B	Atom C	Bond angle A-B-C (°)
2.950	Ti1	Ti1(1,1)	Ti2	58.9
2.854	Ti1	Ti2	Ti1(0,1)	62.2
2.854	Ti1	Ti2	Ti3(0,1)	145.5
2.854	Ti1	Ti2	Ti3	107.3
2.950	Ti2	Ti2(1,0)	Ti3(1,1)	59.4

Ti(0001)-(1x1)
22.1

Bond Distances and Angles - Continued

Interatomic dist. A-B (Å)	Atom A	Atom B	Atom C	Bond angle A-B-C (°)
2.894	Ti2	Ti3(1,1)	Ti2(1,0)	61.3
2.894	Ti2	Ti3(1,1)	Ti3	59.4

COMMON NAME : Ti(10-10)-(1x1)
CLASSIFICATION : 22.3
TECHNIQUE : LEED
AUTHORS : P.R. Watson and J.M. Mischenko III
REFERENCE : Phys. Rev., B42, 3415 (1990)

ILLUSTRATION: 20

SURFACE TYPE
Substrate : Ti
Crystal face: 10-10
Temperature : RT
Bulk lattice: hcp
2D bulk symm: pmm
2D surf symm: pmm

Adsorbate:
Coverage :
Pattern : (1x1)
Matrix : (1.000, 0.000)
(0.000, 1.000)

STRUCTURE TYPE
70% of surface bulk terminated with narrow top interlayer spacing (d1) contracted about 5%; second interlayer spacing (d2) expanded about 1-2%;
30% of surface bulk terminated with expanded (+6%) large (d2) interlayer spacing

SAMPLE PREPARATION (1 sample)
Treatment : extensive 650C Ar+ sputtering with 600C anneal (1972)
Crystallinity: high
Anal. methods: AES
Contamination: <5% C,O

COMMENTS
Bulk consists of pairs of narrowly-spaced (d1) layers separated by spacing d2 = 2 x d1;
non-unique surface; can terminate with domains of small or large interlayer spacing;
supercedes Surf. Sci. 220, L667 (1989)

DATA COLLECTION
Technique: LEED; photo/video
Dataset : I-V spectra for 11 beams at normal incidence and 13 at Θ=10°, π=270°; energy range 50-260 eV (total 3580)

THEORY/DATA TREATMENT
Dynamical LEED (layer doubling): 8 phase shifts from free atom potential; E-dependent Voi; ΘD=342 K

STRUCTURES EXAMINED
Mixtures of the two possible lattice terminations; variations of two topmost layer spacings

QUALITY OF EXPERIMENT-THEORY FIT
RPE=0.25, RZJ=0.12, R2=0.19

2D UNIT CELLS (2 domains observed)

Cell	Ax (Å)	Ay (Å)	Bx (Å)	By (Å)	α (°)	Matrix	Pattern	Cell type
Bulk	2.950	0.000	0.000	4.683	90.0	(1.000, 0.000) (0.000, 1.000)	(1x1)	b: bulk lattice
Surface 1	2.950	0.000	0.000	4.683	90.0	(1.000, 0.000) (0.000, 1.000)	(1x1)	s1: commens. superlattice

3D COORDINATES

Ti1-Ti2: narrowly spaced (contracted) pair of layers; Ti3-Ti4 and Ti5-Ti6: 2 pairs of narrowly spaced layers, together forming periodically repeating set of bulk layers; 0.1Å error bars assumed for tabulation

Dx/Dy in Å, or as a fraction of layer's unit cell vectors; atom 0 at the origin. Epir/subr are bulk repeat vectors.

No. of atoms: 6 Bulk z = 2.555 Å

Region	Chem el.	At. no.	Cell type	Site occ.	Rel. to	Dx ± εx		Dy ± εy		Dz ± εz		Dz/Bz(%) ± εz/Bz
epir		-2					f		f		Å	
subr		-1				1.475	Å	0.000	Å	2.555	Å	
intf	Ti	1	b	1.00	0	0.000	f	0.000	f	0.000	Å	0.0
intf	Ti	2	b	1.00	1	0.500	f	0.500	f	0.815 ± .050	Å	31.9 ± 2.0
subl	Ti	3	b	1.00	2	0.000	f	-0.500	f	1.784 ± .050	Å	69.8 ± 2.0
subl	Ti	4	b	1.00	3	-0.500	f	0.500	f	0.852	Å	33.3
subl	Ti	5	b	1.00	4	0.000	f	-0.500	f	1.703	Å	66.7
subl	Ti	6	b	1.00	5	0.500	f	0.500	f	0.852	Å	33.3

BOND DISTANCES AND ANGLES

Bond distances and angles are derived from coordinates

No. of distances/angles: 6

Interatomic dist. A-B (Å)	Atom A	Atom B	Atom C	Bond angle A-B-C (°)
2.480	Ti1	Ti2	Ti3	60.4
2.636	Ti2	Ti4	Ti5	126.0
2.896	Ti3	Ti4	Ti5	61.2

Ti(10-10)-(1x1)
22.3

Bond Distances and Angles - Continued

Interatomic dist. A-B (Å)	Atom A	Atom B	Atom C	Bond angle A-B-C (°)
2.950	Ti1	Ti1(-1,0)	Ti3	60.4
2.988	Ti1	Ti3	Ti4	90.3
3.293	Ti2	Ti3	Ti4	49.9

COMMON NAME : Ti(0001)-(1x1)-2Cd
CLASSIFICATION : 22.48.3a
TECHNIQUE : LEED
AUTHORS : H.D. Shih, F. Jona, D.W. Jepsen and P.M. Marcus
REFERENCE : Phys. Rev., B15, 5561 (1977)

ILLUSTRATION: 87

SURFACE TYPE
Substrate : Ti
Crystal face: 0001
Temperature : 300 K
Bulk lattice: hcp
2D bulk symm: p3m1
2D surf symm: p3m1

Adsorbate: Cd
Coverage : 2 (Cd/Ti)
Pattern : (1x1)
Matrix : (1.000, 0.000)
(0.000, 1.000)

STRUCTURE TYPE
Adsorbed Cd bilayer with hcp structure:
acABAB... layer stacking (ac=Cd, ABAB...=Ti)

SAMPLE PREPARATION (1 sample)
Treatment : see Shih, Jona, Jepsen and Marcus, Phys Rev. B15, 5550(1977)
Crystallinity: thickness monitored by LEED
Anal. methods:
Contamination: AES: no C, O

COMMENTS
Cd bilayer has lattice constants within 1% of bulk Cd parallel and perpendicular to surface

DATA COLLECTION
Technique: LEED
Dataset : I-V spectra for 7 beams at 2 angles of incidence, E<=145 eV

THEORY/DATA TREATMENT
Dynamical LEED: 8 phase shifts; 31 beams; self-consistent potentials for Cd, Ti; Vor=-6 eV, Voi=-3eV; vib amps=0.198Å

STRUCTURES EXAMINED
Truncated Ti bulk structure with 2 complete close-packed atomic layers of Cd; 4 stacking sequences examined: cbABAB..., abABAB..., bcABAB..., acABAB... (abc=Cd, ABC=Ti); top two interlayer spacings varied

QUALITY OF EXPERIMENT-THEORY FIT
Visual

2D UNIT CELLS (1 domain observed)

Cell	Ax (Å)	Ay (Å)	Bx (Å)	By (Å)	α (°)	Matrix	Pattern	Cell type
Bulk	2.950	0.000	-1.475	2.555	120.0	(1.000, 0.000) (0.000, 1.000)	(1x1)	b: bulk lattice
Surface 1	2.950	0.000	-1.475	2.555	120.0	(1.000, 0.000) (0.000, 1.000)	(1x1)	s1: commens. superlattice

3D COORDINATES

Cd1-Cd2: (1x1) bilayer; Ti4-Ti5: periodically repeating bulk layers;
0.1Å error bars assumed for tabulation

Dx/Dy in Å, or as a fraction of layer's unit cell vectors; atom 0 at the origin. Epir/subr are bulk repeat vectors.

No. of atoms: 5 Bulk z = 2.340 Å

Region	Chem el.	At. no.	Cell type	Site occ.	Rel. to	Dx ± εx		Dy ± εy		Dz ± εz		Dz/Bz(%) ± εz/Bz
epir		-2					f		f		Å	
subr		-1				0.000	Å	0.000	Å	4.680	Å	
ovrl	Cd	1	b	1.00	0	0.000	f	0.000	f	0.000	Å	0.0
ovrl	Cd	2	b	1.00	1	0.667	f	0.333	f	2.810 ± .100	Å	120.1 ± 4.3
intf	Ti	3	b	1.00	2	-0.667	f	-0.333	f	2.630 ± .100	Å	112.4 ± 4.3
subl	Ti	4	b	1.00	3	0.333	f	0.667	f	2.340	Å	100.0
subl	Ti	5	b	1.00	4	-0.333	f	-0.667	f	2.340	Å	100.0

BOND DISTANCES AND ANGLES

Bond distances and angles are derived from coordinates

No. of distances/angles: 3

Interatomic dist. A-B (Å)	Atom A	Atom B	Atom C	Bond angle A-B-C (°)
2.950	Cd1	Cd1(1,0)	Cd2	63.3
3.286	Cd1	Cd2	Ti3	115.9
2.950	Ti3	Ti3(1,0)	Ti4(0,-1)	59.4

COMMON NAME : Ti(0001)-(1x1)-4Cd ILLUSTRATION: 87
CLASSIFICATION : 22.48.3b
TECHNIQUE : LEED
AUTHORS : H.D. Shih, F. Jona, D.W. Jepsen and P.M. Marcus
REFERENCE : Phys. Rev., B15, 5561 (1977)

SURFACE TYPE
Substrate : Ti
Crystal face: 0001
Temperature : 300 K
Bulk lattice: hcp
2D bulk symm: p3m1
2D surf symm: p3m1
Adsorbate: Cd
Coverage : 4 (Cd/Ti)
Pattern : (1x1)
Matrix : (1.000, 0.000)
(0.000, 1.000)

STRUCTURE TYPE
4 adsorbed Cd layers with hcp structure: acac (registry wrt Ti substrate not determined, but here assumed equal to Cd bilayer case, cf. class. no. 22.48.3a)

SAMPLE PREPARATION (1 sample)
Treatment : see Shih, Jona, Jepsen and Marcus, Phys Rev. B15, 5550(1977)
Crystallinity: thickness monitored by LEED
Anal. methods:
Contamination: AES: no C, O

COMMENTS
Cd layer has lattice constants within 1% of bulk Cd parallel and perpendicular to surface

DATA COLLECTION
Technique: LEED
Dataset : I-V spectra for several beams at 2 angles of incidence, E<=145 eV

THEORY/DATA TREATMENT
Dynamical LEED: 8 phase shifts; 31 beams; self-consistent potential for Cd; Vor=-6 eV, Voi=-3eV; vib amps=0.198Å

STRUCTURES EXAMINED
Relaxation of top interlayer spacing of semi-infinite bulk-like Cd(0001) surface

QUALITY OF EXPERIMENT-THEORY FIT
Visual

2D UNIT CELLS (1 domain observed)

Cell	Ax (Å)	Ay (Å)	Bx (Å)	By (Å)	α (°)	Matrix	Pattern	Cell type
Bulk	2.950	0.000	-1.475	2.555	120.0	(1.000, 0.000) (0.000, 1.000)	(1x1)	b: bulk lattice
Surface 1	2.950	0.000	-1.475	2.555	120.0	(1.000, 0.000) (0.000, 1.000)	(1x1)	s1: commens. superlattice

3D COORDINATES

Cd1-Cd4: 4 layers with hcp structure; relationship to Ti5 assumed like that of Cd bilayer structure determination; Ti6-Ti7: periodically repeating bulk layers; 0.1Å error bar assumed for tabulation

Dx/Dy in Å, or as a fraction of layer's unit cell vectors; atom 0 at the origin. Epir/subr are bulk repeat vectors.

No. of atoms: 7 Bulk z = 2.340 Å

Reg ion	Chem el.	At. no.	Cell type	Site occ.	Rel. to	Dx ± εx		Dy ± εy		Dz ± εz		Dz/Bz(%) ± εz/Bz
epir		-2					f		f		Å	
subr		-1				0.000	Å	0.000	Å	4.680	Å	
ovrl	Cd	1	b	1.00	0	0.000	f	0.000	f	0.000	Å	0.0
ovrl	Cd	2	b	1.00	1	0.333	f	0.667	f	2.810 ± .100	Å	120.1 ± 4.3
ovrl	Cd	3	b	1.00	2	-0.333	f	-0.667	f	2.810	Å	120.1
ovrl	Cd	4	b	1.00	3	0.333	f	0.667	f	2.810	Å	120.1
intf	Ti	5	b	1.00	4	-0.333	f	-0.667	f	2.630	Å	112.4
subl	Ti	6	b	1.00	5	0.667	f	0.333	f	2.340	Å	100.0
subl	Ti	7	b	1.00	6	-0.667	f	-0.333	f	2.340	Å	100.0

BOND DISTANCES AND ANGLES

Bond distances and angles are derived from coordinates

No. of distances/angles: 2

Interatomic dist. A-B (Å)	Atom A	Atom B	Atom C	Bond angle A-B-C (°)
2.950	Cd1	Cd1(0,1)	Cd2	63.3
3.286	Cd1	Cd2	Cd3	117.6

COMMON NAME : Ti(0001)-(1x1)-Cd
CLASSIFICATION : 22.48.2
TECHNIQUE : LEED
AUTHORS : H.D. Shih, F. Jona, D.W. Jepsen and P.M. Marcus
REFERENCE : Phys. Rev., B15, 5550 (1977)

ILLUSTRATION: 87

SURFACE TYPE
Substrate : Ti
Crystal face: 0001
Temperature : 300 K
Bulk lattice: hcp
2D bulk symm: p3m1
2D surf symm: p3m1

Adsorbate: Cd
Coverage : 1.0 Cd/Ti
Pattern : (1x1)
Matrix : (1.000, 0.000)
(0.000, 1.000)

STRUCTURE TYPE
Atomic overlayer in 3-fold fcc hollow sites

SAMPLE PREPARATION (>2 sample)
Treatment : Cd exposure by heating 99.9999% pure Cd capsule
Crystallinity:
Anal. methods:
Contamination: by AES and LEED: no increase in C and O

COMMENTS
Growth of Cd layers reproduced more than twice with the same Ti sample and Cd source

DATA COLLECTION
Technique: LEED
Dataset : I-V curves for 7 beams and 2 angles of incidence; E<=145 eV

THEORY/DATA TREATMENT
Dynamical LEED: 8 phase shifts; 31 beams; self consistent potentials for Cd and Ti; Vor=-8 eV, Voi=-3 eV

STRUCTURES EXAMINED
Truncated Ti bulk with Cd at a) top sites (dz=3.0Å), b) 1 of 3 bridge sites (dz=2.9Å), c) hcp 3-fold hollow sites (dz=2.57 Å), d) fcc 3-fold hollow sites (1.4<dz<2.8Å); all calculations step-averaged (for 2 hcp terminations)

QUALITY OF EXPERIMENT-THEORY FIT
Visual

2D UNIT CELLS (1 domain observed)

Cell	Ax (Å)	Ay (Å)	Bx (Å)	By (Å)	α (°)	Matrix	Pattern	Cell type
Bulk	2.950	0.000	-1.475	2.555	120.0	(1.000, 0.000) (0.000, 1.000)	(1x1)	b: bulk lattice
Surface 1	2.950	0.000	-1.475	2.555	120.0	(1.000, 0.000) (0.000, 1.000)	(1x1)	s1: commens. superlattice

3D COORDINATES

Cd1: overlayer in fcc 3-fold hollow sites; Ti1-Ti2: repeating bulk pair of layers;

Dx/Dy in Å, or as a fraction of layer's unit cell vectors; atom 0 at the origin. Epir/subr are bulk repeat vectors.

No. of atoms: 3 Bulk z = 2.340 Å

Region	Chem el.	At. no.	Cell type	Site occ.	Rel. to	Dx ± εx		Dy ± εy		Dz ± εz		Dz/Bz(%) ± εz/Bz
epir		-2					f		f		Å	
subr		-1				0.000	Å	0.000	Å	4.680	Å	
ovrl	Cd	1	b	1.00	0	0.000	f	0.000	f	0.000	Å	0.0
subl	Ti	2	b	1.00	1	0.333	f	0.667	f	2.570 ± .100	Å	109.8 ± 4.3
subl	Ti	3	b	1.00	2	0.333	f	-0.333	f	2.340	Å	100.0

BOND DISTANCES AND ANGLES

Bond distances and angles are derived from coordinates

No. of distances/angles: 7

Interatomic dist. A-B (Å)	Atom A	Atom B	Atom C	Bond angle A-B-C (°)
2.950	Cd1	Cd1(1,1)	Ti2	61.4
3.083	Cd1	Ti2	Cd1(0,1)	57.2
3.083	Cd1	Ti2	Ti2(0,1)	118.6
3.083	Cd1	Ti2	Ti3	120.8
2.950	Ti2	Ti2(1,0)	Ti3	59.4

Ti(0001)-(1x1)-Cd
22.48.2

Bond Distances and Angles - Continued

Interatomic dist. A-B (Å)	Atom A	Atom B	Atom C	Bond angle A-B-C (°)
2.894	Ti2	Ti3	Ti2(1,0)	61.3
2.894	Ti2	Ti3	Ti3(1,0)	120.6

COMMON NAME : Ti(0001)-(1x1)-N
CLASSIFICATION : 22.7.2
TECHNIQUE : LEED
AUTHORS : H.D. Shih, F. Jona, D.W. Jepsen and P.M. Marcus
REFERENCE : Surf. Sci., 60, 445 (1976)

ILLUSTRATION: 58

SURFACE TYPE
Substrate : Ti
Crystal face: 0001
Temperature : 300 K
Bulk lattice: hcp
2D bulk symm: p3m1
2D surf symm: p3m1

Adsorbate: N
Coverage : 1.0 N/Ti
Pattern : (1x1)
Matrix : (1.000, 0.000)
(0.000, 1.000)

STRUCTURE TYPE
Atomic interstitial in octahedral sites between first and second Ti layers; slight expansion of Ti-Ti spacing; forms trilayer of TiN compound exposing (111) face

SAMPLE PREPARATION (2 sample)
Treatment : cleaned, then N introduced at 1.5E-8 torr
Crystallinity:
Anal. methods:
Contamination: monitored by AES and LEED

COMMENTS

DATA COLLECTION
Technique: LEED
Dataset : LEED I-V spectra: 3 beams at normal inc., 5 beams at Θ=8°, ϕ=-30°; (00) beam at Θ=20°, ϕ=-30°; 20<E<250 eV

THEORY/DATA TREATMENT
Dynamical LEED (layer KKR): 8 phase shifts (Moruzzi et al); domain-averaging over steps; Vor=-10 eV; Voi=-3eV; ΘD=342 K

STRUCTURES EXAMINED
1. atomic or molecular (N-N=1.098Å) adsorption in top, bridge and either one or both 3-fold hollow sites;
2. one of the 2 types of tetrahedral or octahedral interstitial sites between the 1st and 2nd Ti layers (Ti-N spacings varied in range 0.9-2.8Å); step-averaged

QUALITY OF EXPERIMENT-THEORY FIT
Visual

2D UNIT CELLS (1 domain observed)

Cell	Ax (Å)	Ay (Å)	Bx (Å)	By (Å)	α (°)	Matrix	Pattern	Cell type
Bulk	2.950	0.000	-1.475	2.555	120.0	(1.000, 0.000) (0.000, 1.000)	(1x1)	b: bulk lattice
Surface 1	2.950	0.000	-1.475	2.555	120.0	(1.000, 0.000) (0.000, 1.000)	(1x1)	s1: commens. superlattice

3D COORDINATES

N2: underlayer in octahedral site of slightly expanded Ti(0001) lattice;
Ti1-N2-Ti3: trilayer of TiN; Ti4-Ti5: repeating bulk pair of layers

Dx/Dy in Å, or as a fraction of layer's unit cell vectors; atom 0 at the origin. Epir/subr are bulk repeat vectors.

No. of atoms: 5 Bulk z = 2.340 Å

Region	Chem el.	At. no.	Cell type	Site occ.	Rel. to	Dx ± εx		Dy ± εy		Dz ± εz		Dz/Bz(%) ± εz/Bz
epir		-2					f		f		Å	
subr		-1				0.000	Å	0.000	Å	4.680	Å	
intf	Ti	1	b	1.00	0	0.000	f	0.000	f	0.000	Å	0.0
intf	N	2	b	1.00	1	0.667	f	0.333	f	1.220 ± .050	Å	52.1 ± 2.1
intf	Ti	3	b	1.00	2	-0.333	f	0.333	f	1.220 ± .050	Å	52.1 ± 2.1
subl	Ti	4	b	1.00	3	-0.333	f	-0.667	f	2.340	Å	100.0
subl	Ti	5	b	1.00	4	0.333	f	0.667	f	2.340	Å	100.0

Ti(0001)-(1x1)-N
22.7.2 BOND DISTANCES AND ANGLES

Bond distances and angles are derived from coordinates

No. of distances/angles: 8

Interatomic dist. A-B (Å)	Atom A	Atom B	Atom C	Bond angle A-B-C (°)
2.950	Ti1	Ti1(1,1)	N2	45.3
2.950	Ti1	Ti1(1,1)	Ti3	60.3
2.095	Ti1	N2	Ti1(1,0)	89.5
2.095	Ti1	N2	Ti3	90.5
2.976	Ti1	Ti3	Ti1(0,1)	59.4
2.976	Ti1	Ti3	N2	44.8
2.976	Ti1	Ti3	Ti3(0,1)	119.7
2.095	N2	Ti3	Ti3(0,1)	134.8

COMMON NAME : TiC(111)-(√3x√3)R30°-O
CLASSIFICATION : 22.6.8.3
TECHNIQUE : ICISS
AUTHORS : R. Souda, C. Oshima, S. Otani and Y. Ishizawa
REFERENCE : Surf. Sci., 199, 154 (1987)

ILLUSTRATION: 156

SURFACE TYPE
Substrate : TiC
Crystal face: 111
Temperature : RT*
Bulk lattice: NaCl
2D bulk symm: p3m1
2D surf symm: p31m

Adsorbate: O
Coverage : 1/3
Pattern : (√3x√3)R30°
Matrix : (1.000, 1.000)
(-2.000, 1.000)

STRUCTURE TYPE
Atomic adsorption in 'fcc' 3-fold hollow site on Ti-terminated unreconstructed, unrelaxed TiC substrate

SAMPLE PREPARATION (1 sample)
Treatment : TiC single xtal polished, annealed, and flashed at ≈2073 K
Crystallinity: clear LEED pattern
Anal. methods:
Contamination:

COMMENTS
A (1x1) structure occurring at higher temperatures was also studied: here all 'fcc' 3-fold hollows are occupied, but authors suggest that the Ti pinch towards or away from O in a random manner

DATA COLLECTION
Technique: ICISS; 1keV He+ beam
Dataset : scattering angle of 163°

THEORY/DATA TREATMENT
Impact collision ion scattering spectroscopy: intensity analyzed as function of angles of incidence

STRUCTURES EXAMINED
Suggestion that Ti atoms pinch towards O in 3-fold hollows

QUALITY OF EXPERIMENT-THEORY FIT
Visual

2D UNIT CELLS (1 domain observed)

Cell	Ax (Å)	Ay (Å)	Bx (Å)	By (Å)	α (°)	Matrix	Pattern	Cell type
Bulk	3.060	0.000	1.530	2.650	60.0	(1.000, 0.000) (0.000, 1.000)	(1x1)	b: bulk lattice
Surface 1	4.590	2.650	-4.590	2.650	120.0	(1.000, 1.000) (-2.000, 1.000)	(√3x√3)R30°	s1: commens. superlattice

3D COORDINATES

O1: overlayer in 'fcc' 3-fold hollow sites of bulk-like Ti2 layer;
C3-Ti4: periodically repeating bulk layers

Dx/Dy in Å, or as a fraction of layer's unit cell vectors; atom 0 at the origin. Epir/subr are bulk repeat vectors.

No. of atoms: 4 Bulk z = 2.500 Å

Reg ion	Chem el.	At. no.	Cell type	Site occ.	Rel. to	Dx ± εx	Dy ± εy	Dz ± εz	Dz/Bz(%) ± εz/Bz
epir		-2				f	f	Å	
subr		-1				1.530 Å	0.883 Å	2.500 Å	
ovrl	O	1	s1	.33	0	0.000 f	0.000 f	0.000 Å	0.0
intf	Ti	2	b	1.00	1	-0.333 f	0.667 f	1.000 Å	40.0
subl	C	3	b	1.00	2	-0.333 f	0.667 f	1.250 Å	50.0
subl	Ti	4	b	1.00	3	0.667 f	-0.333 f	1.250 Å	50.0

BOND DISTANCES AND ANGLES

Bond distances and angles are derived from coordinates

No. of distances/angles: 7

Interatomic dist. A-B (Å)	Atom A	Atom B	Atom C	Bond angle A-B-C (°)
2.030	O1	Ti2	Ti2(0,1)	138.9
2.030	O1	Ti2	C3	174.2
2.030	O1	Ti2	C3(0,-1)	86.0
2.030	O1	Ti2	Ti4	130.8
2.030	O1	Ti2	Ti4(0,-1)	84.3
3.060	Ti2	Ti2(1,0)	C3(1,0)	90.0

TiC(111)-(√3x√3)R30°-O
22.6.8.3

Bond Distances and Angles - Continued

Interatomic dist. A-B (Å)	Atom A	Atom B	Atom C	Bond angle A-B-C (°)
2.164	Ti2	C3	Ti4	90.0

COMMON NAME : TiO2(100)-(3x1)
CLASSIFICATION : 22.8.1
TECHNIQUE : LEED
AUTHORS : P. Zschack
REFERENCE : Springer Series in Surface Sciences, 24, 646 (1991)

ILLUSTRATION: 153

SURFACE TYPE
Substrate : TiO2
Crystal face: 100
Temperature : RT
Bulk lattice: rutile
2D bulk symm: pmm
2D surf symm: pmm

Adsorbate:
Coverage :
Pattern : (3x1)
Matrix : (3.000, 0.000)
(0.000, 1.000)

STRUCTURE TYPE
Strongly corrugated surface with alternating (110) and (-110) microfacets, due to missing rows along [001]:

1st three interlayer spacings expanded 5.4%, 5.4% and 10%, 4th contracted 21%

SAMPLE PREPARATION (1 sample)
Treatment : Ar-ion sputtering and annealing to 875 K
Crystallinity:
Anal. methods: GIXD
Contamination:

COMMENTS
There are also lateral displacements of the atoms, which were refined by grazing incidence x-ray diffraction; an alternate simple missing-row model, compatible with GIXD, was rejected by LEED analysis

DATA COLLECTION
Technique: LEED; LEED: pulse counting data acquisition
Dataset : I-V curves for 7 symm. inequivalent beams at normal incidence, E-range 20-100 eV

THEORY/DATA TREATMENT
Dynamical LEED (Tong/Van Hove package: composite layers; RFS)

STRUCTURES EXAMINED
GIXD Patterson fct. determined lateral Ti positions and integrated 2D electron density map, implying facetted missing row reconstruction; GIXD fit of lateral positions gives RX=0.079; in LEED, variation of top 5 interlayer spacings

QUALITY OF EXPERIMENT-THEORY FIT
RZJ=0.34

2D UNIT CELLS (1 domain observed)

Cell	Ax (Å)	Ay (Å)	Bx (Å)	By (Å)	α (°)	Matrix	Pattern	Cell type
Bulk	4.715	0.000	0.000	3.036	90.0	(1.000, 0.000) (0.000, 1.000)	(1x1)	b: bulk lattice
Surface 1	14.145	0.000	0.000	3.036	90.0	(3.000, 0.000) (0.000, 1.000)	(3x1)	s1: commens. superlattice

3D COORDINATES

O1 through O9: reconstructed region, forming (110) and (-110) microfacets;
Ti14-O19: set of layers forming bulk rutile lattice; 0.1Å error bars assumed for tabulation

Dx/Dy in Å, or as a fraction of layer's unit cell vectors; atom 0 at the origin. Epir/subr are bulk repeat vectors.

No. of atoms: 19 Bulk z = .918 Å

Region	Chem el.	At. no.	Cell type	Site occ.	Rel. to	Dx ± εx	Dy ± εy	Dz ± εz	Dz/Bz(%) ± εz/Bz
epir		-2				f	f	Å	
subr		-1				4.715 Å	3.036 Å	4.715 Å	
intf	O	1	s1	.33	0	0.239 ± .007 f	0.000 f	-3.275 ± .100 Å	-356.8 ± 10.8
intf	Ti	2	s1	.33	0	0.184 ± .002 f	0.500 f	-2.307 ± .100 Å	-251.4 ± 10.8
intf	O	3	s1	.33	0	0.144 ± .010 f	0.000 f	-1.339 Å	-145.9
intf	O	4	s1	.33	0	0.275 ± .008 f	0.500 f	-1.018 ± .100 Å	-110.8 ± 10.8
intf	O	5	s1	.33	0	-0.021 ± .012 f	0.500 f	-1.018 ± .100 Å	-110.8 ± 10.8
intf	Ti	6	s1	.67	0	0.000 f	0.000 f	0.000 Å	0.0
intf	Ti	7	s1	.67	6	0.331 ± .002 f	0.000 f	0.000 Å	0.0
intf	O	8	s1	1.00	6	0.065 f	0.500 f	0.718 ± .100 Å	78.2 ± 10.8
intf	O	9	s1	1.00	6	0.410 ± .010 f	0.500 f	0.718 ± .100 Å	78.2 ± 10.8
intf	O	10	b	1.00	6	0.695 f	0.000 f	1.239 Å	135.0
intf	Ti	11	b	1.00	6	0.500 f	0.500 f	2.157 Å	235.0
intf	O	12	b	1.00	6	0.305 f	0.000 f	3.075 Å	335.0
intf	O	13	b	1.00	6	0.805 f	0.500 f	3.597 Å	391.8
subl	Ti	14	b	1.00	6	0.000 f	0.000 f	4.514 Å	491.7
subl	O	15	b	1.00	14	0.195 f	0.500 f	0.918 Å	100.0
subl	O	16	b	1.00	14	0.695 f	0.000 f	1.439 Å	156.8
subl	Ti	17	b	1.00	14	0.500 f	0.500 f	2.357 Å	256.8
subl	O	18	b	1.00	14	0.305 f	0.000 f	3.275 Å	356.8
subl	O	19	b	1.00	14	0.805 f	0.500 f	3.797 Å	413.6

TiO2(100)-(3x1)
22.8.1 BOND DISTANCES AND ANGLES

Bond distances and angles are derived from coordinates

No. of distances/angles: 9

Interatomic dist. A-B (Å)	Atom A	Atom B	Atom C	Bond angle A-B-C (°)
1.961	O1	Ti2	O1(0,1)	101.4
1.961	O1	Ti2	Ti2(0,1)	140.7
1.961	O1	Ti2	Ti2(0,-1)	39.3
1.997	Ti2	O3(0,1)	Ti2(0,1)	98.9
1.997	Ti2	O3(0,1)	Ti6(0,1)	123.9
1.997	O3	Ti2	O1(0,1)	171.6
1.969	O4	Ti2	O1(0,1)	96.6
1.992	O4	Ti7(0,1)	Ti2(0,1)	74.7
1.852	O5	Ti6(1,1)	O3(1,1)	80.3

ILLUSTRATION: 161

COMMON NAME : TiSe2(0001)-(1x1)
CLASSIFICATION : 22.34.2
TECHNIQUE : LEED
AUTHORS : M. Kasch, E. Pehlke, W. Schattke, T. Kurberg, H.P. Barnscheidt, R. Manze and M. Skibowski
REFERENCE : Surf. Sci., 214, 436 (1989)

SURFACE TYPE
Substrate : TiSe2
Crystal face: 0001
Temperature : RT
Bulk lattice: CdI2
2D bulk symm: p3m1
2D surf symm: p3m1

Adsorbate:
Coverage :
Pattern : (1x1)
Matrix : (1.000, 0.000)
(0.000, 1.000)

STRUCTURE TYPE
Bulk termination with complete Se-Ti-Se sandwich and bulk layer stacking; slight interlayer spacing relaxations: top two Ti-Se spacings expanded 3.5%, and 1st sandwich-to-sandwich repeat distance c contracted by 1%

SAMPLE PREPARATION (2 sample)
Treatment : cleavage in ultra high vacuum
Crystallinity:
Anal. methods:
Contamination:

COMMENTS
Bulk structure consists of Se-Ti-Se sandwiches in which the spacing between the Ti and Se layers is 0.255c, where c is the spacing between two sandwich center planes (c=6.004Å)

DATA COLLECTION
Technique: LEED; movable electron energy analyzer
Dataset : I-V curves for 6 beams, E range 50-150 eV

THEORY/DATA TREATMENT
Dynamical LEED (Van Hove/Tong package): Vor=-9 eV, Voi=-2.7eV

STRUCTURES EXAMINED
Varied were: lattice constant c at the surface (distance between adjacent Ti-layers) and spacing between top and third Se layers

QUALITY OF EXPERIMENT-THEORY FIT
RZJ=0.49

2D UNIT CELLS (1 domain observed)

Cell	Ax (Å)	Ay (Å)	Bx (Å)	By (Å)	α (°)	Matrix	Pattern	Cell type
Bulk	3.535	0.000	1.768	3.061	60.0	(1.000, 0.000) (0.000, 1.000)	(1x1)	b: bulk lattice
Surface 1	3.535	0.000	1.768	3.061	60.0	(1.000, 0.000) (0.000, 1.000)	(1x1)	s1: commens. superlattice

3D COORDINATES

Se1-Ti2-Se3: top sandwich with 3.5% expanded spacings; Se4-Ti5-Se6: 2nd sandwich with bulk spacings between the Se4-Ti5 and Ti5-Se6 layers and with a 1% contracted distance to the top sandwich

Dx/Dy in Å, or as a fraction of layer's unit cell vectors; atom 0 at the origin. Epir/subr are bulk repeat vectors.

No. of atoms: 6 — Bulk z = 6.004 Å

Region	Chem el.	At. no.	Cell type	Site occ.	Rel. to	Dx ± εx		Dy ± εy		Dz ± εz		Dz/Bz(%) ± εz/Bz
epir		-2					f		f		Å	
subr		-1				0.000	Å	0.000	Å	6.004	Å	
intf	Se	1	b	1.00	0	0.000	f	0.000	f	0.000	Å	0.0
intf	Ti	2	b	1.00	1	-0.333	f	-0.333	f	1.577 ± .100	Å	26.3 ± 1.7
intf	Se	3	b	1.00	2	0.667	f	0.667	f	1.577 ± .100	Å	26.3 ± 1.7
subl	Se	4	b	1.00	3	-0.333	f	-0.333	f	2.836 ± .100	Å	47.2 ± 1.7
subl	Ti	5	b	1.00	4	-0.333	f	-0.333	f	1.531	Å	25.5
subl	Se	6	b	1.00	5	0.667	f	0.667	f	1.531	Å	25.5

BOND DISTANCES AND ANGLES

Bond distances and angles are derived from coordinates

No. of distances/angles: 10

Interatomic dist. A-B (Å)	Atom A	Atom B	Atom C	Bond angle A-B-C (°)
2.579	Se1	Ti2(0,-1)	Se1(1,0)	86.5
2.580	Se3	Ti2	Se1(1,0)	93.5
2.579	Se1	Ti2(0,-1)	Ti2(1,-1)	133.3

TiSe2(0001)-(1x1)
22.34.2

Bond Distances and Angles - Continued

Interatomic dist. A-B (Å)	Atom A	Atom B	Atom C	Bond angle A-B-C (°)
2.579	Se1	Ti2(0,-1)	Ti2	90.0
3.535	Ti2	Ti2(1,0)	Se1(1,1)	46.7
3.535	Ti2	Ti2(1,0)	Ti2(1,1)	120.0
3.535	Ti2	Ti2(1,0)	Ti2(1,-1)	60.0
2.580	Ti2	Se3	Ti2(0,-1)	86.5
2.580	Ti2	Se3	Ti2(-1,0)	86.5
2.580	Se3	Ti2	Se1(1,1)	180.0

ILLUSTRATION: 12

COMMON NAME : V(100)-(1x1)
CLASSIFICATION : 23.4
TECHNIQUE : LEED
AUTHORS : V. Jensen, J.N. Andersen, H.B. Nielsen and D.L. Adams
REFERENCE : Surf. Sci., 116, 66 (1982)

SURFACE TYPE
Substrate : V
Crystal face: 100
Temperature : RT*
Bulk lattice: bcc
2D bulk symm: p4m
2D surf symm: p4m

Adsorbate:
Coverage :
Pattern : (1x1)
Matrix : (1.000, 0.000)
(0.000, 1.000)

STRUCTURE TYPE
Bulk termination with 2-layer relaxation

SAMPLE PREPARATION (1 sample)
Treatment : very long anneals and sputtering to remove H, S, P, C and O
Crystallinity: sharp (1x1) LEED pattern
Anal. methods:
Contamination: AES: <0.01ML S, P, C and <0.05ML O

COMMENTS
R-factor defined by Adams et al, Phys. Rev. B20, 4789 (1979)
previously reported reconstructed V(100)-(5x1) is shown to result from 0.2 monolayer of O

DATA COLLECTION
Technique: LEED
Dataset : I-V spectra for 5 inequivalent beams at normal incidence, 50<E<350 eV

THEORY/DATA TREATMENT
Dynamical LEED: 10 phase shifts (Moruzzi et al);
Vor = -9.2±0.2 eV, Voi = -4.0±0.1eV, ΘD= 510±10 K

STRUCTURES EXAMINED
Truncated bulk structure with relaxations of first interlayer spacing from 1.3 to 1.7Å in 0.025 Å steps, and subsequent relaxation of second interlayer spacing

QUALITY OF EXPERIMENT-THEORY FIT
R2<0.074

2D UNIT CELLS (1 domain observed)

Cell	Ax (Å)	Ay (Å)	Bx (Å)	By (Å)	α (°)	Matrix	Pattern	Cell type
Bulk	3.028	0.000	0.000	3.028	90.0	(1.000, 0.000) (0.000, 1.000)	(1x1)	b: bulk lattice
Surface 1	3.028	0.000	0.000	3.028	90.0	(1.000, 0.000) (0.000, 1.000)	(1x1)	s1: commens. superlattice

3D COORDINATES

Dx/Dy in Å, or as a fraction of layer's unit cell vectors; atom 0 at the origin. Epir/subr are bulk repeat vectors.

No. of atoms: 3

Bulk z = 1.514 Å

Reg ion	Chem el.	At. no.	Cell type	Site occ.	Rel. to	Dx ± εx		Dy ± εy		Dz ± εz		Dz/Bz(%) ± εz/Bz
epir		-2					f		f		Å	
subr		-1				1.514	Å	1.514	Å	1.514	Å	
intf	V	1	b	1.00	0	0.000	f	0.000	f	0.000	Å	0.0
intf	V	2	b	1.00	1	0.500	f	0.500	f	1.410 ± .010	Å	93.1 ± .7
subl	V	3	b	1.00	2	-0.500	f	-0.500	f	1.530 ± .010	Å	101.1 ± .7

BOND DISTANCES AND ANGLES

Bond distances and angles are derived from coordinates

No. of distances/angles: 8

Interatomic dist. A-B (Å)	Atom A	Atom B	Atom C	Bond angle A-B-C (°)
2.564	V1	V2	V1(1,0)	72.4
2.564	V1	V2	V2(1,0)	126.2
2.564	V1	V2	V3	68.9
2.632	V2	V3(1,1)	V1(1,1)	54.5
2.632	V2	V3(1,1)	V2(1,0)	70.2
2.632	V2	V3(1,0)	V3	54.9

V(100)-(1x1)
23.4

Bond Distances and Angles - Continued

Interatomic dist. A-B (Å)	Atom A	Atom B	Atom C	Bond angle A-B-C (°)
2.632	V2	V3	V1	54.5
2.632	V2	V3	V2(0,-1)	70.2

COMMON NAME : V(110)-(1x1)
CLASSIFICATION : 23.2
TECHNIQUE : LEED
AUTHORS : D.L. Adams and H.B. Nielsen
REFERENCE : Surf. Sci., 107, 305 (1981)

ILLUSTRATION: 11

SURFACE TYPE
Substrate : V
Crystal face: 110
Temperature : RT*
Bulk lattice: bcc
2D bulk symm: cmm
2D surf symm: cmm

Adsorbate:
Coverage :
Pattern : (1x1)
Matrix : (1.000, 0.000)
(0.000, 1.000)

STRUCTURE TYPE
Slightly contracted bulk termination

SAMPLE PREPARATION (1 sample)
Treatment : 80 hrs of Ar+ bombardment at 673 K and annealing at 373-1073K
Crystallinity:
Anal. methods:
Contamination: AES: <0.02 monolayer of oxygen

COMMENTS
R2 = intensity R-factor (Adams et al, Phys. Rev. B20, 4789 (1979))

DATA COLLECTION
Technique: LEED
Dataset : normal incidence LEED I-V data for 6 beams in energy range 60-360 eV

THEORY/DATA TREATMENT
Dynamical LEED: RFS; 10 ph shs; 20 beams (symm.-reduced);
Vor=-9.2±0.3 eV, Voi=-4.7±0.3 eV, ΘD=512±36 K (all fit)

STRUCTURES EXAMINED
Variations in the first layer spacing only, in the range 1.90-2.40Å in 0.05Å steps (bulk value is 2.141Å)

QUALITY OF EXPERIMENT-THEORY FIT
R2=0.03 (see comment)

2D UNIT CELLS (1 domain observed)

Cell	Ax (Å)	Ay (Å)	Bx (Å)	By (Å)	α (°)	Matrix	Pattern	Cell type
Bulk	2.622	0.000	.874	2.472	70.5	(1.000, 0.000) (0.000, 1.000)	(1x1)	b: bulk lattice
Surface 1	2.622	0.000	.874	2.472	70.5	(1.000, 0.000) (0.000, 1.000)	(1x1)	s1: commens. superlattice

3D COORDINATES

Dx/Dy in Å, or as a fraction of layer's unit cell vectors; atom 0 at the origin. Epir/subr are bulk repeat vectors.

No. of atoms: 3

Bulk z = 2.141 Å

Region	Chem el.	At. no.	Cell type	Site occ.	Rel. to	Dx ± εx		Dy ± εy		Dz ± εz		Dz/Bz(%) ± εz/Bz
epir		-2					f		f		Å	
subr		-1				-1.748	Å	-1.236	Å	2.141	Å	
intf	V	1	b	1.00	0	0.000	f	0.000	f	0.000	Å	0.0
intf	V	2	b	1.00	1	0.500	f	0.500	f	2.130 ± .010	Å	99.5 ± .5
subl	V	3	b	1.00	2	-0.500	f	-0.500	f	2.141	Å	100.0

BOND DISTANCES AND ANGLES

Bond distances and angles are derived from coordinates

No. of distances/angles: 7

Interatomic dist. A-B (Å)	Atom A	Atom B	Atom C	Bond angle A-B-C (°)
2.622	V1	V1(1,0)	V1(1,1)	109.5
2.622	V1	V1(1,0)	V1(0,1)	54.7
2.622	V1	V1(1,0)	V2	70.5
2.613	V1	V2(-1,0)	V1(0,1)	54.9
2.613	V1	V2(-1,0)	V2	70.5
2.613	V1	V2(-1,0)	V3	109.3
2.622	V2	V2(1,0)	V3(1,0)	54.7

COMMON NAME : VN0.89(100)-(1x1)
CLASSIFICATION : 23.7.1
TECHNIQUE : LEED
AUTHORS : Y. Gauthier, Y. Joly, J. Rundgren, L.I. Johansson and P. Wincott
REFERENCE : Phys. Rev., B42, 9328 (1990)

ILLUSTRATION: 149

SURFACE TYPE
Substrate : VN0.89
Crystal face: 100
Temperature : RT
Bulk lattice: NaCl
2D bulk symm: none
2D surf symm: none

Adsorbate:
Coverage :
Pattern : (1x1)
Matrix : (1.000, 0.000)
(0.000, 1.000)

STRUCTURE TYPE
VN0.89 is a substoichiometric compound with N vacancies; bulk termination with a 8% buckled top VN compound layer, N moving outward, 7% contracted first interlayer spacing, planar second VN compound layer, 1% expanded 2nd interlayer spacing

SAMPLE PREPARATION (1 sample)
Treatment : cycles of Ar-ion bombardment and annealing to 1350 K
Crystallinity:
Anal. methods: AES
Contamination: AES: <1% ML O2 and other contaminants

COMMENTS
LEED analysis shows that there are no N vacancies in the surface region, so that the stoichiometry is VN there; here this structure is modeled as VN by neglecting N vacancies also in the bulk, where they in fact form disordered mixed layers with the bulk stoichiometry

DATA COLLECTION
Technique: LEED; movable spot photometer
Dataset : I-V spectra from 30 to 280 eV, 5 beams at normal incidence, 13 beams at 20° off-normal incidence, cumul. E range 2550 e

THEORY/DATA TREATMENT
Dynamical LEED: 10 phase shifts

STRUCTURES EXAMINED
Variation of top-layer buckling, first and second interlayer spacings; it was also checked that the vacancy concentration had no effect

QUALITY OF EXPERIMENT-THEORY FIT
D1=10.69

2D UNIT CELLS (1 domain observed)

Cell	Ax (Å)	Ay (Å)	Bx (Å)	By (Å)	α (°)	Matrix	Pattern	Cell type
Bulk	2.910	0.000	0.000	2.910	90.0	(1.000, 0.000) (0.000, 1.000)	(1x1)	b: bulk lattice
Surface 1	2.910	0.000	0.000	2.910	90.0	(1.000, 0.000) (0.000, 1.000)	(1x1)	s1: commens. superlattice

3D COORDINATES

N1-V2: buckled top compound VN layer, N outermost; V3-N4: bulk repeat compound VN layer

Dx/Dy in Å, or as a fraction of layer's unit cell vectors; atom 0 at the origin. Epir/subr are bulk repeat vectors.

No. of atoms: 4

Bulk z = 2.058 Å

Region	Chem el.	At. no.	Cell type	Site occ.	Rel. to	Dx ± εx		Dy ± εy		Dz ± εz		Dz/Bz(%) ± εz/Bz
epir		-2					f		f		Å	
subr		-1				1.455	Å	1.455	Å	2.058	Å	
intf	N	1	b	1.00	0	0.000	f	0.000	f	0.000	Å	0.0
intf	V	2	b	1.00	1	0.500	f	0.500	f	0.170 ± .008	Å	8.3 ± .4
subl	V	3	b	1.00	2	-0.500	f	-0.500	f	1.920 ± .006	Å	93.3 ± .3
subl	N	4	b	1.00	3	0.500	f	0.500	f	0.000	Å	0.0

BOND DISTANCES AND ANGLES

Bond distances and angles are derived from coordinates

No. of distances/angles: 5

Interatomic dist. A-B (Å)	Atom A	Atom B	Atom C	Bond angle A-B-C (°)
2.065	N1	V2	N1(1,1)	170.6
2.065	N1	V2	N1(1,0)	89.6
2.090	N1	V3	V2	47.0
2.065	V2	N1(1,1)	V2(1,1)	170.6

VN0.89(100)-(1x1)
23.7.1

Bond Distances and Angles - Continued

Interatomic dist. A-B (Å)	Atom A	Atom B	Atom C	Bond angle A-B-C (°)
2.090	V3	N1	V2	85.3

COMMON NAME : W(100)-(1x1)
CLASSIFICATION : 74.1.21a
TECHNIQUE : LEED
AUTHORS : J.B. Pendry, K. Heinz and W. Oed
REFERENCE : Phys. Rev. Lett., 61, 2953 (1988)

ILLUSTRATION: 12

SURFACE TYPE
Substrate : W
Crystal face: 100
Temperature : 400 K
Bulk lattice: bcc
2D bulk symm: p4m
2D surf symm: p4m

Adsorbate:
Coverage :
Pattern : (1x1)
Matrix : (1.000, 0.000)
(0.000, 1.000)

STRUCTURE TYPE
Multilayer relaxation

SAMPLE PREPARATION (1 sample)
Treatment : standard methods; indirect heating
Crystallinity:
Anal. methods:
Contamination:

COMMENTS

DATA COLLECTION
Technique: LEED; video LEED
Dataset : IV spectra for 4 symmetrical inequivalent beams, 20-500 eV

THEORY/DATA TREATMENT
Dynamical LEED (direct method)

STRUCTURES EXAMINED
Fitting of 1st and 2nd interlayer spacing

2D UNIT CELLS (1 domain observed)

Cell	Ax (Å)	Ay (Å)	Bx (Å)	By (Å)	α (°)	Matrix	Pattern	Cell type
Bulk	3.168	0.000	0.000	3.168	90.0	(1.000, 0.000) (0.000, 1.000)	(1x1)	b: bulk lattice
Surface 1	3.168	0.000	0.000	3.168	90.0	(1.000, 0.000) (0.000, 1.000)	(1x1)	s1: commens. superlattice

3D COORDINATES

Dx/Dy in Å, or as a fraction of layer's unit cell vectors; atom 0 at the origin. Epir/subr are bulk repeat vectors.

No. of atoms: 3

Bulk z = 1.580 Å

Region	Chem el.	At. no.	Cell type	Site occ.	Rel. to	Dx ± εx	Dy ± εy	Dz ± εz	Dz/Bz(%) ± εz/Bz
epir		-2				f	f	Å	
subr		-1				1.584 Å	1.584 Å	1.580 Å	
intf	W	1	s1	.10	0	0.000 Å	0.000 Å	0.000 Å	0.0
intf	W	2	s1	.10	0	1.584 Å	1.584 Å	1.481 Å	93.7
subl	W	3	b	.10	0	0.000 Å	0.000 Å	3.091 Å	195.6

BOND DISTANCES AND ANGLES

Bond distances and angles are derived from coordinates

No. of distances/angles: 2

Interatomic dist. A-B (Å)	Atom A	Atom B	Atom C	Bond angle A-B-C (°)
2.685	W1	W2		
2.759	W2	W3		

COMMON NAME : W(100)-(1x1)
CLASSIFICATION : 74.21
TECHNIQUE : LEED
AUTHORS : F.S. Marsh, M.K. Debe and D.A. King
REFERENCE : J. Phys., C13, 2799 (1980)

ILLUSTRATION: 12

SURFACE TYPE
Substrate : W
Crystal face: 100
Temperature : 470 K
Bulk lattice: bcc
2D bulk symm: p4m
2D surf symm: p4m

Adsorbate:
Coverage :
Pattern : (1x1)
Matrix : (1.000, 0.000)
(0.000, 1.000)

STRUCTURE TYPE
Bulk termination with top layer spacing contraction by 7.6%

SAMPLE PREPARATION (1 sample)
Treatment : flash to 2000 K in 1E-6 torr O2, then flash to 2500K in vacuo
Crystallinity:
Anal. methods:
Contamination: AES: no impurities

COMMENTS
This high-temperature structure later believed to be disordered version of the c(2x2) reconstruction (eds.)

DATA COLLECTION
Technique: LEED
Dataset : I-V curves for 10,1-1,20,21 beams at Θ=0° and for 00,10 beams at Θ=14° in [100] azimuth

THEORY/DATA TREATMENT
Dynamical LEED (layer doubling): 10 pseudo-rel. ph shs (also non-rel. tested); E-dep Vor, Voi; ΘD=318 K(surf), 450K(bulk)

STRUCTURES EXAMINED
Relaxation of top interlayer spacing

QUALITY OF EXPERIMENT-THEORY FIT
RZJ=0.30

2D UNIT CELLS (1 domain observed)

Cell	Ax (Å)	Ay (Å)	Bx (Å)	By (Å)	α (°)	Matrix	Pattern	Cell type
Bulk	3.160	0.000	0.000	3.160	90.0	(1.000, 0.000) (0.000, 1.000)	(1x1)	b: bulk lattice
Surface 1	3.160	0.000	0.000	3.160	90.0	(1.000, 0.000) (0.000, 1.000)	(1x1)	s1: commens. superlattice

3D COORDINATES

Dx/Dy in Å, or as a fraction of layer's unit cell vectors; atom 0 at the origin. Epir/subr are bulk repeat vectors.

No. of atoms: 3

Bulk z = 1.580 Å

Reg ion	Chem el.	At. no.	Cell type	Site occ.	Rel. to	Dx ± εx	Dy ± εy	Dz ± εz	Dz/Bz(%) ± εz/Bz
epir		-2				f	f	Å	
subr		-1				-1.580 Å	-1.580 Å	1.580 Å	
intf	W	1	b	1.00	0	0.000 f	0.000 f	0.000 Å	0.0
intf	W	2	b	1.00	1	0.500 f	0.500 f	1.460 ± .030 Å	92.4 ± 1.9
subl	W	3	b	1.00	2	-0.500 f	-0.500 f	1.580 Å	100.0

BOND DISTANCES AND ANGLES

Bond distances and angles are derived from coordinates

No. of distances/angles: 3

Interatomic dist. A-B (Å)	Atom A	Atom B	Atom C	Bond angle A-B-C (°)
2.669	W1	W2	W1(1,0)	72.6
2.669	W1	W2	W3	68.4
2.737	W2	W3		

COMMON NAME : W(100)-(1x1) ILLUSTRATION: 12
CLASSIFICATION : 74.2a
TECHNIQUE : LEED
AUTHORS : M.A. Van Hove and S.Y. Tong
REFERENCE : Surf. Sci., 54, 91 (1976)

SURFACE TYPE
Substrate : W
Crystal face: 100
Temperature : 300 K
Bulk lattice: bcc
2D bulk symm: p4m
2D surf symm: p4m

Adsorbate:
Coverage :
Pattern : (1x1)
Matrix : (1.000, 0.000)
(0.000, 1.000)

STRUCTURE TYPE
Bulk termination with top layer spacing contraction by 6.3%

SAMPLE PREPARATION (1 sample)
Treatment :
Crystallinity:
Anal. methods:
Contamination:

COMMENTS
This high-temperature structure later believed to be disordered version of the c(2x2) reconstruction (eds.)

DATA COLLECTION
Technique: LEED
Dataset : data of P.P. Wei, J. Chem. Phys. 53, 2939 (1970): examined (10),(20),(21) beams; E range 20-200 eV

THEORY/DATA TREATMENT
Dynamical LEED (layer doubling): 8 phase shifts, superpos. pot of charge densities; Voi=-5 eV; ΘD=380 K(bulk),550K(surf)

STRUCTURES EXAMINED
Relaxation of top interlayer spacing

QUALITY OF EXPERIMENT-THEORY FIT
Visual

2D UNIT CELLS (1 domain observed)

Cell	Ax (Å)	Ay (Å)	Bx (Å)	By (Å)	α (°)	Matrix	Pattern	Cell type
Bulk	3.160	0.000	0.000	3.160	90.0	(1.000, 0.000) (0.000, 1.000)	(1x1)	b: bulk lattice
Surface 1	3.160	0.000	0.000	3.160	90.0	(1.000, 0.000) (0.000, 1.000)	(1x1)	s1: commens. superlattice

3D COORDINATES

Dx/Dy in Å, or as a fraction of layer's unit cell vectors; atom 0 at the origin. Epir/subr are bulk repeat vectors.

No. of atoms: 3 Bulk z = 1.580 Å

Region	Chem el.	At. no.	Cell type	Site occ.	Rel. to	Dx ± εx		Dy ± εy		Dz ± εz		Dz/Bz(%) ± εz/Bz
epir		-2					f		f		Å	
subr		-1				-1.580	Å	-1.580	Å	1.580	Å	
intf	W	1	b	1.00	0	0.000	f	0.000	f	0.000	Å	0.0
intf	W	2	b	1.00	1	0.500	f	0.500	f	1.480 ± .100	Å	93.7 ± 6.3
subl	W	3	b	1.00	2	-0.500	f	-0.500	f	1.580	Å	100.0

BOND DISTANCES AND ANGLES

Bond distances and angles are derived from coordinates

No. of distances/angles: 3

Interatomic dist. A-B (Å)	Atom A	Atom B	Atom C	Bond angle A-B-C (°)
3.160	W1	W1(1,0)	W2(1,0)	126.1
2.680	W1	W2	W3	68.8
2.737	W2	W3		

COMMON NAME : W(100)-(1x1) disordered
CLASSIFICATION : 74.47
TECHNIQUE : LEED
AUTHORS : J.B. Pendry, K. Heinz, W. Oed, H. Landskron, K. Mueller and G. Schmidtlein
REFERENCE : Surf. Sci., 193, L1 (1988)

ILLUSTRATION: 14

SURFACE TYPE
Substrate : W
Crystal face: 100
Temperature : 450 K
Bulk lattice: bcc
2D bulk symm: p4m
2D surf symm: none

Adsorbate:
Coverage :
Pattern : disordered
Matrix : (1.000, 0.000) (0.000, 1.000)

STRUCTURE TYPE
Disordered version of W(100)-c(2x2) reconstruction, with top-layer W atoms randomly displaced laterally by 0.16Å in 4 equivalent [011] directions; disorder is here modeled as rigid shift of complete top layer in one direction

SAMPLE PREPARATION (1 sample)
Treatment : 'standard methods'
Crystallinity:
Anal. methods:
Contamination:

COMMENTS
The ordered surface with relaxation gives RPE=0.28 if the highest energy is restricted to 250eV;
variance in the R-factor=0.06, greater than the reduction achieved by including disorder

DATA COLLECTION
Technique: LEED
Dataset : normal incidence IV curves for 4 beams: 10, 11, 20, 21; E range 20-500 eV

THEORY/DATA TREATMENT
Dynamical diffuse tensor LEED as it influences integer order beams: 10 phase shifts; ΘD=400 K

STRUCTURES EXAMINED
Ordered and disordered surface with varying degrees of disorder, in conjunction with relaxed top layer; the displacement and relaxation were determined simultaneaously from the R-factor map

QUALITY OF EXPERIMENT-THEORY FIT
RPE=0.24 (see comment)

2D UNIT CELLS (1 domain observed)

Cell	Ax (Å)	Ay (Å)	Bx (Å)	By (Å)	α (°)	Matrix	Pattern	Cell type
Bulk	3.160	0.000	0.000	3.160	90.0	(1.000, 0.000) (0.000, 1.000)	(1x1)	b: bulk lattice
Surface 1	3.160	0.000	0.000	3.160	90.0	(1.000, 0.000) (0.000, 1.000)	disordered	s1: commens. superlattice

3D COORDINATES

W1: displaced top layer, here modeled as rigid (1x1); 0.1Å error bars assumed for tabulation

Dx/Dy in Å, or as a fraction of layer's unit cell vectors; atom 0 at the origin. Epir/subr are bulk repeat vectors.

No. of atoms: 3 Bulk z = 1.580 Å

Reg ion	Chem el.	At. no.	Cell type	Site occ.	Rel. to	Dx ± εx	Dy ± εy	Dz ± εz	Dz/Bz(%) ± εz/Bz
epir		-2				f	f	Å	
subr		-1				1.580 Å	1.580 Å	1.580 Å	
intf	W	1	b	1.00	0	0.000 f	0.000 f	0.000 Å	0.0
intf	W	2	b	1.00	1	0.464 ± .032 f	0.464 ± .032 f	1.450 ± .100 Å	91.8 ± 6.3
subl	W	3	b	1.00	2	0.500 f	0.500 f	1.580 Å	100.0

BOND DISTANCES AND ANGLES

Bond distances and angles are derived from coordinates

No. of distances/angles: 2

Interatomic dist. A-B (Å)	Atom A	Atom B	Atom C	Bond angle A-B-C (°)
2.531	W1	W2	W1(1,0)	74.8
2.737	W2	W3		

COMMON NAME : W(110)-(1x1)
CLASSIFICATION : 74.2b
TECHNIQUE : LEED
AUTHORS : M.A. Van Hove and S.Y. Tong
REFERENCE : Surf. Sci., 54, 91 (1976)

ILLUSTRATION: 11

SURFACE TYPE
Substrate : W
Crystal face: 110
Temperature : 300 K
Bulk lattice: bcc
2D bulk symm: cmm
2D surf symm: cmm

Adsorbate:
Coverage :
Pattern : (1x1)
Matrix : (1.000, 0.000)
(0.000, 1.000)

STRUCTURE TYPE
Unrelaxed bulk termination

SAMPLE PREPARATION (1 sample)
Treatment : see Buchholz and Lagally, Surf. Sci. 49, 508 (1975)
Crystallinity:
Anal. methods:
Contamination:

COMMENTS

DATA COLLECTION
Technique: LEED
Dataset : I-V curves for at least 11,20,02 beams at Θ=0, and 00,02 beams at Θ=7°,φ=90°

THEORY/DATA TREATMENT
Dynamical LEED (layer doubling): Moruzzi et al potential, 8 phase shifts; Vor=-10.0 eV, Voi=-5.0eV; ΘD=318 K

STRUCTURES EXAMINED
1. bulk termination with top layer spacing variation; 2. same with top layer atoms in 3-fold coordinated sites of the 2nd layer.

QUALITY OF EXPERIMENT-THEORY FIT
Visual

2D UNIT CELLS (1 domain observed)

Cell	Ax (Å)	Ay (Å)	Bx (Å)	By (Å)	α (°)	Matrix	Pattern	Cell type
Bulk	2.740	0.000	-.914	2.583	109.5	(1.000, 0.000) (0.000, 1.000)	(1x1)	b: bulk lattice
Surface 1	2.740	0.000	-.914	2.583	109.5	(1.000, 0.000) (0.000, 1.000)	(1x1)	s1: commens. superlattice

3D COORDINATES

Dx/Dy in Å, or as a fraction of layer's unit cell vectors; atom 0 at the origin. Epir/subr are bulk repeat vectors.

No. of atoms: 3

Bulk z = 2.230 Å

Region	Chem el.	At. no.	Cell type	Site occ.	Rel. to	Dx ± εx		Dy ± εy		Dz ± εz		Dz/Bz(%) ± εz/Bz
epir		-2					f		f		Å	
subr		-1				-0.913	Å	-1.292	Å	2.230	Å	
intf	W	1	b	1.00	0	0.000	f	0.000	f	0.000	Å	0.0
intf	W	2	b	1.00	1	0.500	f	0.500	f	2.230 ± .100	Å	100.0 ± 4.5
subl	W	3	b	1.00	2	-0.500	f	-0.500	f	2.230	Å	100.0

BOND DISTANCES AND ANGLES

Bond distances and angles are derived from coordinates

No. of distances/angles: 3

Interatomic dist. A-B (Å)	Atom A	Atom B	Atom C	Bond angle A-B-C (°)
2.740	W1	W1(1,0)	W1(1,1)	70.5
2.734	W1	W2	W1(1,1)	70.7
2.734	W1	W2	W3	109.3

ILLUSTRATION: 11

COMMON NAME : W(110)-(1x1)
CLASSIFICATION : 74.45
TECHNIQUE : HEIS
AUTHORS : R.J. Smith, C. Hennessy, M.W. Kim, C.N. Whang, M. Worthington and M. Xu
REFERENCE : Phys. Rev. Lett., 58, 702 (1987)

SURFACE TYPE
Substrate : W
Crystal face: 110
Temperature : 300 K
Bulk lattice: bcc
2D bulk symm: cmm
2D surf symm: cmm

Adsorbate:
Coverage :
Pattern : (1x1)
Matrix : (1.000, 0.000)
(0.000, 1.000)

STRUCTURE TYPE
Bulk-like termination

SAMPLE PREPARATION (1 sample)
Treatment : cycles of heating in O, then flashes to 2200 K
Crystallinity: sharp LEED pattern
Anal. methods:
Contamination:

COMMENTS
Normal component of the rms surface-atom vibration amplitude is 2.6 times larger than in the bulk (0.05Å), while parallel component is not significantly enhanced

DATA COLLECTION
Technique: HEIS; 0.5 to 2.0 MeV He+ ions
Dataset : E scan for scattering along [-1,-1,0], [0,-1,0], [-1,-1,1] directions, and angular scan in (001) plane about [0,-1,0]

THEORY/DATA TREATMENT
Monte Carlo computer simulations

STRUCTURES EXAMINED
Various model structures: 1. relaxations of first and second interlayer distances
2. variations of the bulk rms thermal vibration amplitude and anisotropic enhancement of the surface vibration ampl.

2D UNIT CELLS (1 domain observed)

Cell	Ax (Å)	Ay (Å)	Bx (Å)	By (Å)	α (°)	Matrix	Pattern	Cell type
Bulk	2.740	0.000	-.913	2.583	109.5	(1.000, 0.000) (0.000, 1.000)	(1x1)	b: bulk lattice
Surface 1	2.740	0.000	-.913	2.583	109.5	(1.000, 0.000) (0.000, 1.000)	(1x1)	s1: commens. superlattice

3D COORDINATES

Dx/Dy in Å, or as a fraction of layer's unit cell vectors; atom 0 at the origin. Epir/subr are bulk repeat vectors.

No. of atoms: 3

Bulk z = 2.230 Å

Region	Chem el.	At. no.	Cell type	Site occ.	Rel. to	Dx ± εx		Dy ± εy		Dz ± εz		Dz/Bz(%) ± εz/Bz
epir		-2					f		f		Å	
subr		-1				0.913	Å	1.292	Å	2.230	Å	
intf	W	1	b	1.00	0	0.000	f	0.000	f	0.000	Å	0.0
intf	W	2	b	1.00	1	0.500	f	0.500	f	2.230 ± .040	Å	100.0 ± 1.8
subl	W	3	b	1.00	2	-0.500	f	-0.500	f	2.230 ± .040	Å	100.0 ± 1.8

BOND DISTANCES AND ANGLES

Bond distances and angles are derived from coordinates

No. of distances/angles: 1

Interatomic dist. A-B (Å)	Atom A	Atom B	Atom C	Bond angle A-B-C (°)
2.740	W1	W1(1,0)		

COMMON NAME : W(100)-c(2x2)
CLASSIFICATION : 74.14
TECHNIQUE : LEED
AUTHORS : R.A. Barker, P.J. Estrup, F. Jona and P.M. Marcus
REFERENCE : Solid State Commun., 25, 375 (1978)

ILLUSTRATION: 13

SURFACE TYPE
Substrate : W
Crystal face: 100
Temperature : 120 K
Bulk lattice: bcc
2D bulk symm: p4m
2D surf symm: pmg

Adsorbate:
Coverage :
Pattern : c(2x2)
Matrix : (1.000, 1.000)
(-1.000, 1.000)

STRUCTURE TYPE
Zig-zag displacive reconstruction of top layer

SAMPLE PREPARATION (1 sample)
Treatment : see Felter et al, Phys. Rev. Lett. 38, 1138 (1977)
Crystallinity:
Anal. methods:
Contamination:

COMMENTS

DATA COLLECTION
Technique: LEED
Dataset : I-V spectra for (1,0), (1,1), (1/2,1/2), (3/2,1/2) beams at Θ=0°; (0,0) beam atΘ=5°,φ=0°; 30 eV<E<150eV

THEORY/DATA TREATMENT
Dynamical LEED (layer-KKR): 8 phase shifts (Kohn-Sham local density self-consistent potential); Voi=-4 eV; rms amps=0.7Å

STRUCTURES EXAMINED
1. perpendicular-shift buckled-top-layer model; 2. parallel-shift model with pmg symmetry; amount of shift and top layer spacings were varied

QUALITY OF EXPERIMENT-THEORY FIT
Visual

2D UNIT CELLS (2 domains observed)

Cell	Ax (Å)	Ay (Å)	Bx (Å)	By (Å)	α (°)	Matrix	Pattern	Cell type
Bulk	3.168	0.000	0.000	3.168	90.0	(1.000, 0.000) (0.000, 1.000)	(1x1)	b: bulk lattice
Surface 1	3.168	3.168	-3.168	3.168	90.0	(1.000, 1.000) (-1.000, 1.000)	c(2x2)	s1: commens. superlattice

3D COORDINATES

W1-W2: reconstructed top layer

Dx/Dy in Å, or as a fraction of layer's unit cell vectors; atom 0 at the origin. Epir/subr are bulk repeat vectors.

No. of atoms: 4

Bulk z = 1.580 Å

Region	Chem el.	At. no.	Cell type	Site occ.	Rel. to	Dx ± εx		Dy ± εy		Dz ± εz		Dz/Bz(%) ± εz/Bz
epir		-2					f		f		Å	
subr		-1				1.584	Å	-1.584	Å	1.580	Å	
intf	W	1	s1	.50	0	0.000	f	0.000	f	0.000	Å	0.0
intf	W	2	s1	.50	1	0.500	f	0.600 ± .004	f	0.000	Å	0.0
intf	W	3	b	1.00	2	-0.450 ± .004	f	-0.550 ± .004	f	1.530 ± .050	Å	96.8 ± 3.2
subl	W	4	b	1.00	3	0.500	f	-0.500	f	1.580	Å	100.0

BOND DISTANCES AND ANGLES

Bond distances and angles are derived from coordinates

No. of distances/angles: 4

Interatomic dist. A-B (Å)	Atom A	Atom B	Atom C	Bond angle A-B-C (°)
2.869	W1	W2(0,-1)	W1(1,1)	102.7
2.900	W1	W3	W1(0,1)	68.5
2.900	W1	W3	W2	76.9
2.900	W1	W3	W4	67.0

ILLUSTRATION: 13

COMMON NAME : W(100)-c(2x2)
CLASSIFICATION : 74.53
TECHNIQUE : XRD
AUTHORS : M.S. Altman, P.J. Estrup and I.K. Robinson
REFERENCE : Phys. Rev., B38, 5211 (1988)

SURFACE TYPE
Substrate : W
Crystal face: 100
Temperature : 175 K
Bulk lattice: bcc
2D bulk symm: p4m
2D surf symm: pmg

Adsorbate:
Coverage :
Pattern : c(2x2)
Matrix : (1.000, 1.000)
(-1.000, 1.000)

STRUCTURE TYPE
Zig-zag chain reconstruction with lateral relaxations in 1st and 2nd layers

SAMPLE PREPARATION (1 sample)
Treatment : flash to 2300 K and anneals at 1400K in 1E-7 torr O2
Crystallinity:
Anal. methods:
Contamination:

COMMENTS

DATA COLLECTION
Technique: XRD
Dataset : integrated intensities for 4 half-order Bragg spots

THEORY/DATA TREATMENT
Kinematic calculations; thermal vibrations fitted to measured intensities: rms ampls=0.05Å

STRUCTURES EXAMINED
Lateral distortions of top and second layers and buckling of top layer

2D UNIT CELLS (2 domains observed)

Cell	Ax (Å)	Ay (Å)	Bx (Å)	By (Å)	α (°)	Matrix	Pattern	Cell type
Bulk	3.168	0.000	0.000	3.168	90.0	(1.000, 0.000) (0.000, 1.000)	(1x1)	b: bulk lattice
Surface 1	3.168	3.168	-3.168	3.168	90.0	(1.000, 1.000) (-1.000, 1.000)	c(2x2)	s1: commens. superlattice

3D COORDINATES

W1-W2: planar top layer with zig-zag chains; W3-W4: planar second layer with smaller lateral relaxations; 0.1Å lateral error bars assumed for tabulation

Dx/Dy in Å, or as a fraction of layer's unit cell vectors; atom 0 at the origin. Epir/subr are bulk repeat vectors.

No. of atoms: 6

Bulk z = 1.580 Å

Reg ion	Chem el.	At. no.	Cell type	Site occ.	Rel. to	Dx ± εx		Dy ± εy		Dz ± εz		Dz/Bz(%) ± εz/Bz
epir		-2					f		f		Å	
subr		-1				1.584	Å	1.584	Å	1.580	Å	
intf	W	1	s1	.50	0	0.000	f	0.000	f	0.000	Å	0.0
intf	W	2	s1	.50	1	0.500	f	0.608 ± .022	f	0.000 ± .160	Å	0.0 ± 10.1
intf	W	3	s1	.50	2	-0.500	f	-0.044 ± .022	f	1.520 ± .160	Å	96.2 ± 10.1
intf	W	4	s1	.50	3	0.500	f	-0.520 ± .022	f	0.000	Å	0.0
intf	W	5	b	1.00	4	-0.510 ± .032	f	-0.490 ± .032	f	1.580 ± .160	Å	100.0 ± 10.1
subl	W	6	b	1.00	5	0.500	f	0.500	f	1.580	Å	100.0

BOND DISTANCES AND ANGLES

Bond distances and angles are derived from coordinates

No. of distances/angles: 5

Interatomic dist. A-B (Å)	Atom A	Atom B	Atom C	Bond angle A-B-C (°)
2.846	W1	W2(0,-1)	W1(1,0)	128.1
2.846	W1	W2(0,-1)	W3(1,0)	60.3
2.949	W1	W3	W4	51.9
2.475	W1	W3(0,-1)	W1(1,0)	74.9
2.714	W1	W4	W5	69.5

COMMON NAME : W(100)-c(2x2)
CLASSIFICATION : 74.59
TECHNIQUE : LEED
AUTHORS : H. Landskron, N. Bickel, K. Heinz, G. Schmidtlein and K. Mueller
REFERENCE : J. Phys. CM, 1, 1 (1989)

ILLUSTRATION: 13

SURFACE TYPE
Substrate : W
Crystal face: 100
Temperature : 140 K
Bulk lattice: bcc
2D bulk symm: p4m
2D surf symm: pmg

Adsorbate:
Coverage :
Pattern : c(2x2)
Matrix : (1.000, 1.000)
(-1.000, 1.000)

STRUCTURE TYPE
Reconstructed zigzag, reconstruction with lateral displacements; dimer model has only slightly worse fit; dimer model is not consistant with the glide-plane symmetry

SAMPLE PREPARATION (1 sample)
Treatment : heating in oxygen followed by flashing to 2200 K
Crystallinity: sharp LEED pattern
Anal. methods:
Contamination:

COMMENTS

DATA COLLECTION
Technique: LEED; video LEED
Dataset : IV spectra for 4 fract.-order beams at normal and 14° off normal incidence; cumulative E range 1079 eV

THEORY/DATA TREATMENT
Quasi-dynamical and full dynamical LEED

STRUCTURES EXAMINED
Zig-zag, dimer model and vertical shift model

QUALITY OF EXPERIMENT-THEORY FIT
RPE=0.26

2D UNIT CELLS (2 domains observed)

Cell	Ax (Å)	Ay (Å)	Bx (Å)	By (Å)	α (°)	Matrix	Pattern	Cell type
Bulk	3.168	0.000	0.000	3.168	90.0	(1.000, 0.000) (0.000, 1.000)	(1x1)	b: bulk lattice
Surface 1	3.168	3.168	-3.168	3.168	90.0	(1.000, 1.000) (-1.000, 1.000)	c(2x2)	s1: commens. superlattice

3D COORDINATES

W1-W2 form planer top layer with zigzag chains W3-W4 form planer 2nd layer with zigzag chains

Dx/Dy in Å, or as a fraction of layer's unit cell vectors; atom 0 at the origin. Epir/subr are bulk repeat vectors.

No. of atoms: 5

Bulk z = 1.580 Å

Region	Chem el.	At. no.	Cell type	Site occ.	Rel. to	Dx ± εx	Dy ± εy	Dz ± εz	Dz/Bz(%) ± εz/Bz
epir		-2				f	f	Å	
subr		-1				1.584 Å	1.584 Å	1.580 Å	
intf	W	1	s1	.50	0	0.000 ± .040 Å	0.000 ± .040 Å	0.000 Å	0.0
intf	W	2	s1	.50	0	0.339 ± .040 Å	3.507 ± .040 Å	0.000 Å	0.0
intf	W	3	s1	.50	0	1.740 ± .007 Å	1.740 ± .007 Å	1.470 ± .030 Å	93.0 ± 2.0
intf	W	4	s1	.50	0	1.768 ± .007 Å	4.936 ± .007 Å	1.470 ± .030 Å	93.0 ± 2.0
subl	W	5	b	1.00	0	0.170 Å	0.170 Å	3.070 ± .030 Å	194.3 ± 2.0

BOND DISTANCES AND ANGLES

Bond distances and angles are derived from coordinates

No. of distances/angles: 4

Interatomic dist. A-B (Å)	Atom A	Atom B	Atom C	Bond angle A-B-C (°)
3.524	W1	W2		
3.196	W3	W4		

W(100)-c(2x2)
74.59

Bond Distances and Angles - Continued

Interatomic dist. A-B (Å)	Atom A	Atom B	Atom C	Bond angle A-B-C (°)
2.866	W1	W3		
2.737	W3	W5		

COMMON NAME : W(211)-(1x1) ILLUSTRATION: 17
CLASSIFICATION : 74.55
TECHNIQUE : TOF-SARS
AUTHORS : J.W. Rabalais, O. Grizzi, M. Shi and H. Bu
REFERENCE : Phys. Rev. Lett., 63, 51 (1989)

SURFACE TYPE
Substrate : W
Crystal face: 211
Temperature : RT*
Bulk lattice: bcc
2D bulk symm: pm
2D surf symm: pm

Adsorbate:
Coverage :
Pattern : (1x1)
Matrix : (1.000, 0.000)
(0.000, 1.000)

STRUCTURE TYPE
Bulk termination with relaxation of top interlayer spacing by -9.3% and registry shift by 6.0%, relative to bulk

SAMPLE PREPARATION (1 sample)
Treatment : cleaned by annealing to 2300 K
Crystallinity:
Anal. methods:
Contamination:

COMMENTS

DATA COLLECTION
Technique: TOF-SARS; 2-5keV He+, Ne+ or Ar+ pulsed bea
Dataset : measurements made as a function of incident and azimuthal scans

THEORY/DATA TREATMENT
Classical trajectory simulations for back scattering, forward scattering, and direct recoiling

STRUCTURES EXAMINED
Relaxation of top interlayer spacing and registry

2D UNIT CELLS (1 domain observed)

Cell	Ax (Å)	Ay (Å)	Bx (Å)	By (Å)	α (°)	Matrix	Pattern	Cell type
Bulk	2.741	0.000	0.000	4.480	90.0	(1.000, 0.000) (0.000, 1.000)	(1x1)	b: bulk lattice
Surface 1	2.741	0.000	0.000	4.480	90.0	(1.000, 0.000) (0.000, 1.000)	(1x1)	s1: commens. superlattice

3D COORDINATES

Dx/Dy in Å, or as a fraction of layer's unit cell vectors; atom 0 at the origin. Epir/subr are bulk repeat vectors.

No. of atoms: 3 Bulk z = 1.292 Å

Region	Chem el.	At. no.	Cell type	Site occ.	Rel. to	Dx ± εx	Dy ± εy	Dz ± εz	Dz/Bz(%) ± εz/Bz
epir		-2				f	f	Å	
subr		-1				1.830 Å	2.238 Å	1.290 Å	
intf	W	1	b	1.00	0	0.000 f	0.000 f	0.000 Å	0.0
intf	W	2	b	1.00	1	0.628 ± .026 f	0.500 f	1.170 ± .070 Å	90.7 ± 5.4
subl	W	3	b	1.00	2	0.668 f	0.500 f	1.290 Å	100.0

BOND DISTANCES AND ANGLES

Bond distances and angles are derived from coordinates

No. of distances/angles: 3

Interatomic dist. A-B (Å)	Atom A	Atom B	Atom C	Bond angle A-B-C (°)
2.741	W1	W1(1,0)		
2.726	W1	W2(-1,0)		
2.590	W1	W3		

COMMON NAME : W(310)-(1x1) ILLUSTRATION: 18
CLASSIFICATION : 74.63
TECHNIQUE : LEED
AUTHORS : D.L. Adams and S.P. Andersen
REFERENCE : Springer Series in Surface Sciences, 24, 395 (1991)

SURFACE TYPE
Substrate : W
Crystal face: 310
Temperature : 300 K
Bulk lattice: bcc
2D bulk symm: pm
2D surf symm: pm

Adsorbate:
Coverage :
Pattern : (1x1)
Matrix : (1.000, 0.000)
(0.000, 1.000)

STRUCTURE TYPE
Unreconstructed surface with multilayer relaxations perpendicular to the surface by (-18.3,-0.7,-0.6,-1.7%) and registry shifts by (-1.8,-1.3,+2.3,-0.8%), relative to bulk

SAMPLE PREPARATION (1 sample)
Treatment :
Crystallinity:
Anal. methods:
Contamination:

COMMENTS

DATA COLLECTION
Technique: LEED; Omicron rear-view LEED
Dataset : 15 symmetry inequivalent beams; E range 50-375 eV

THEORY/DATA TREATMENT
Dynamical LEED (layer-doubling): 14 phase shifts from Mattheis potential

STRUCTURES EXAMINED
Varied top 5 interlayer spacings and registries
minimization carried out via customized R-factor

QUALITY OF EXPERIMENT-THEORY FIT
R=0.185

2D UNIT CELLS (1 domain observed)

Cell	Ax (Å)	Ay (Å)	Bx (Å)	By (Å)	α (°)	Matrix	Pattern	Cell type
Bulk	3.165	0.000	1.583	5.004	72.5	(1.000, 0.000) (0.000, 1.000)	(1x1)	b: bulk lattice
Surface 1	3.165	0.000	1.583	5.004	72.5	(1.000, 0.000) (0.000, 1.000)	(1x1)	s1: commens. superlattice

3D COORDINATES

0.1Å error bars assumed for tabulation

Dx/Dy in Å, or as a fraction of layer's unit cell vectors; atom 0 at the origin. Epir/subr are bulk repeat vectors.

No. of atoms: 5 Bulk z = 1.001 Å

Region	Chem el.	At. no.	Cell type	Site occ.	Rel. to	Dx ± εx	Dy ± εy	Dz ± εz	Dz/Bz(%) ± εz/Bz
epir		-2				f	f	Å	
subr		-1				1.583 Å	2.002 Å	1.001 Å	
intf	W	1	b	1.00	0	0.000 f	0.000 f	0.000 Å	0.0
intf	W	2	b	1.00	1	-0.709 ± .022 f	0.400 ± .020 f	0.818 ± .100 Å	81.7 ± 10.0
intf	W	3	b	1.00	2	0.294 ± .022 f	-0.600 ± .020 f	0.994 ± .100 Å	99.3 ± 10.0
intf	W	4	b	1.00	3	0.312 ± .022 f	0.400 ± .020 f	0.995 ± .100 Å	99.4 ± 10.0
subl	W	5	b	1.00	4	-0.704 f	-0.600 f	0.984 Å	98.3

BOND DISTANCES AND ANGLES

Bond distances and angles are derived from coordinates

No. of distances/angles: 4

Interatomic dist. A-B (Å)	Atom A	Atom B	Atom C	Bond angle A-B-C (°)
2.663	W1	W2(1,0)		
2.636	W1	W3		
2.576	W1	W3(0,-1)		
2.727	W2	W3(0,1)		

COMMON NAME : W(100)-(1x1)-2H
CLASSIFICATION : 74.1.10
TECHNIQUE : LEED
AUTHORS : M.A. Passler, B.W. Lee and A. Ignatiev
REFERENCE : Surf. Sci., 150, 263 (1985)

ILLUSTRATION: 48,50

SURFACE TYPE
Substrate : W
Crystal face: 100
Temperature : 300 K
Bulk lattice: bcc
2D bulk symm: p4m
2D surf symm: p4m

Adsorbate: H
Coverage : 2.0 (H/W)
Pattern : (1x1)
Matrix : (1.000, 0.000)
(0.000, 1.000)

STRUCTURE TYPE
Atomic adsorption in bridge sites of both azimuthal orientations (2H per unit cell)

SAMPLE PREPARATION (1 sample)
Treatment : high-T treatment in 10E-7 torr O2, then Kr+ ion bombardment
Crystallinity:
Anal. methods:
Contamination: monitored by AES

COMMENTS

DATA COLLECTION
Technique: LEED
Dataset : 4 symm.-inequivalent beams at Θ=0°, 40<E<260 eV; 6 symm.-inequivalent beams at Θ=6°, 90<E<260eV

THEORY/DATA TREATMENT
Dynamical LEED (RSP, RFS): 8 phase shifts (Mattheiss W potential, Moruzzi et al H potential)

STRUCTURES EXAMINED
Assumed 2 H atoms per unit cell, centered over bridge sites; H-W layer spacing and first W-W layer spacing were varied

QUALITY OF EXPERIMENT-THEORY FIT
RZJ=0.21

2D UNIT CELLS (1 domain observed)

Cell	Ax (Å)	Ay (Å)	Bx (Å)	By (Å)	α (°)	Matrix	Pattern	Cell type
Bulk	3.160	0.000	0.000	3.160	90.0	(1.000, 0.000) (0.000, 1.000)	(1x1)	b: bulk lattice
Surface 1	3.160	0.000	0.000	3.160	90.0	(1.000, 0.000) (0.000, 1.000)	(1x1)	s1: commens. superlattice

3D COORDINATES

H1-H2 form overlayer in bridge sites with 2 azimuthal orientations

Dx/Dy in Å, or as a fraction of layer's unit cell vectors; atom 0 at the origin. Epir/subr are bulk repeat vectors.

No. of atoms: 5

Bulk z = 1.580 Å

Region	Chem el.	At. no.	Cell type	Site occ.	Rel. to	Dx ± εx		Dy ± εy		Dz ± εz		Dz/Bz(%) ± εz/Bz
epir		-2					f		f		Å	
subr		-1				1.580	Å	-1.580	Å	1.580	Å	
ovrl	H	1	b	1.00	0	0.000	f	0.000	f	0.000	Å	0.0
ovrl	H	2	b	1.00	1	0.500	f	0.500	f	0.000	Å	0.0
intf	W	3	b	1.00	2	0.000	f	-0.500	f	1.170 ± .040	Å	74.1 ± 2.5
intf	W	4	b	1.00	3	-0.500	f	0.500	f	1.560 ± .200	Å	98.7 ± 12.7
subl	W	5	b	1.00	4	0.500	f	-0.500	f	1.580	Å	100.0

BOND DISTANCES AND ANGLES

Bond distances and angles are derived from coordinates

No. of distances/angles: 5

Interatomic dist. A-B (Å)	Atom A	Atom B	Atom C	Bond angle A-B-C (°)
1.966	H1	W3	H1(1,0)	107.0
1.966	H1	W3	H2	69.3
1.966	H1	W3	W4	82.8

W(100)-(1x1)-2H
74.1.10

Bond Distances and Angles - Continued

Interatomic dist. A-B (Å)	Atom A	Atom B	Atom C	Bond angle A-B-C (°)
3.160	W3	W3(1,0)	W4(1,0)	54.6
2.725	W3	W4	W5	70.2

COMMON NAME : W(100)-c(2x2)-N ILLUSTRATION: 48,50
CLASSIFICATION : 74.7.2
TECHNIQUE : LEED
AUTHORS : K. Griffiths, D.A. King, G.C. Aers and J.B. Pendry
REFERENCE : J. Phys., C15, 4921 (1982)

SURFACE TYPE
Substrate : W
Crystal face: 100
Temperature : 300 K
Bulk lattice: bcc
2D bulk symm: p4m
2D surf symm: p4m

Adsorbate: N
Coverage : 0.5 N/W
Pattern : c(2x2)
Matrix : (1.000, 1.000)
(-1.000, 1.000)

STRUCTURE TYPE
Atomic adsorption in hollow site, with shorter bond to 2nd-layer W atoms than to 1st-layer W atoms

SAMPLE PREPARATION (1 sample)
Treatment : 2000 K anneal in 1.0E-6 torr O2, then flash to 2500K in vacuo
Crystallinity:
Anal. methods: coverage determined by use of molecular
Contamination: monitored by AES

COMMENTS

DATA COLLECTION
Technique: LEED
Dataset : I-V spectra: 5 integer, 5 fractional order beams at Θ=0; 11 integer, 10 fractional order beams at Θ=9,φ=0°

THEORY/DATA TREATMENT
Dynamical LEED (CAVLEED): 37 integer, 36 fractional order beams; Voi=-5.0ev; Vor=-12.5 eV (fit)

STRUCTURES EXAMINED
Hollow adsorption site assumed; N-W interlayer spacing varied 0.46-0.64Å; spacing between 1st two W layers varied 1.44-1.63Å

QUALITY OF EXPERIMENT-THEORY FIT
RPE=0.55 (RR=0.2)

2D UNIT CELLS (1 domain observed)

Cell	Ax (Å)	Ay (Å)	Bx (Å)	By (Å)	α (°)	Matrix	Pattern	Cell type
Bulk	3.168	0.000	0.000	3.168	90.0	(1.000, 0.000) (0.000, 1.000)	(1x1)	b: bulk lattice
Surface 1	3.168	3.168	-3.168	3.168	90.0	(1.000, 1.000) (-1.000, 1.000)	c(2x2)	s1: commens. superlattice

3D COORDINATES

N1: overlayer in hollow sites

Dx/Dy in Å, or as a fraction of layer's unit cell vectors; atom 0 at the origin. Epir/subr are bulk repeat vectors.

No. of atoms: 4 Bulk z = 1.580 Å

Region	Chem el.	At. no.	Cell type	Site occ.	Rel. to	Dx ± εx		Dy ± εy		Dz ± εz		Dz/Bz(%) ± εz/Bz
epir		-2					f		f		Å	
subr		-1				1.584	Å	1.584	Å	1.580	Å	
ovrl	N	1	s1	.50	0	0.000	f	0.000	f	0.000	Å	0.0
intf	W	2	b	1.00	1	0.500	f	0.500	f	0.490 ± .060	Å	31.0 ± 3.8
intf	W	3	b	1.00	2	-0.500	f	-0.500	f	1.600 ± .060	Å	101.3 ± 3.8
subl	W	4	b	1.00	3	0.500	f	0.500	f	1.580	Å	100.0

BOND DISTANCES AND ANGLES

Bond distances and angles are derived from coordinates

No. of distances/angles: 7

Interatomic dist. A-B (Å)	Atom A	Atom B	Atom C	Bond angle A-B-C (°)
2.293	N1	W2	W3	47.9
2.090	N1	W3	W4	125.2
2.090	N1	W3	W4(0,-1)	125.2
2.090	N1	W3	W4(-1,0)	125.2
2.090	N1	W3	W4(-1,-1)	125.2

W(100)-c(2x2)-N
74.7.2

Bond Distances and Angles - Continued

Interatomic dist. A-B (Å)	Atom A	Atom B	Atom C	Bond angle A-B-C (°)
3.168	W2	W2(1,0)		
2.753	W2	W3	W4	70.7

COMMON NAME : W(100)-O disordered ILLUSTRATION: 51
CLASSIFICATION : 74.8.8
TECHNIQUE : DLEED
AUTHORS : P.J. Rous, J.B. Pendry, D.K. Saldin, K. Heinz, K. Mueller and N. Bickel
REFERENCE : Phys. Rev. Lett., 57, 2951 (1986)

SURFACE TYPE
Substrate : W
Crystal face: 100
Temperature : 120 K
Bulk lattice: bcc
2D bulk symm: p4m
2D surf symm: none

Adsorbate: O
Coverage : considerable
Pattern : disordered
Matrix : (2.000, 0.000)
(0.000, 2.000)

STRUCTURE TYPE
Atomic adsorption in hollow sites with lateral W relaxations towards O position; disordered structure modeled as (2x2) structure here

SAMPLE PREPARATION (1 sample)
Treatment : O added until low-T reconstruction of W(100) disappeared
Crystallinity:
Anal. methods:
Contamination: AES: no detectable impurities

COMMENTS
This structure is refinement of that reported by Heinz et al, Phys. Rev. Lett. 55, 2312 (1985) in which substrate atom displacements were not considered

DATA COLLECTION
Technique: DLEED
Dataset : diffuse LEED intensities at 400 k// points at 46 and 48 eV and normal incidence

THEORY/DATA TREATMENT
Dynamical diffuse LEED, with tensor LEED applied to substrate atom displacements

STRUCTURES EXAMINED
O in hollow sites (random occupation) 0.45-0.85Å above top W layer; relaxations of top W layer: buckling <=0.4Å; zig-zag disps as in W(100)-c(2x2); rotations around adsorption site; all cases tested for displacements of W atoms perpendicular to surface between ±0.2Å

QUALITY OF EXPERIMENT-THEORY FIT
RPE=0.05

2D UNIT CELLS (1 domain observed)

Cell	Ax (Å)	Ay (Å)	Bx (Å)	By (Å)	α (°)	Matrix	Pattern	Cell type
Bulk	3.160	0.000	0.000	3.160	90.0	(1.000, 0.000) (0.000, 1.000)	(1x1)	b: bulk lattice
Surface 1	6.320	0.000	0.000	6.320	90.0	(2.000, 0.000) (0.000, 2.000)	disordered	rd1: reconstr. lattice-gas dis

3D COORDINATES

O1: overlayer over relaxed hollows; W2-W3-W4-W5: relaxed top W layer, contracted toward O site

Dx/Dy in Å, or as a fraction of layer's unit cell vectors; atom 0 at the origin. Epir/subr are bulk repeat vectors.

No. of atoms: 7 Bulk z = 1.580 Å

Region	Chem el.	At. no.	Cell type	Site occ.	Rel. to	Dx ± εx	Dy ± εy	Dz ± εz	Dz/Bz(%) ± εz/Bz
epir		-2				f	f	Å	
subr		-1				1.580 Å	1.580 Å	1.580 Å	
ovrl	O	1	rd1	1.00	0	0.000 ± .016 f	0.000 ± .016 f	0.000 ± .100 Å	0.0 ± 6.3
intf	W	2	rd1	1.00	1	0.774 ± .016 f	0.774 ± .016 f	0.590 ± .100 Å	37.3 ± 6.3
intf	W	3	rd1	1.00	1	0.226 ± .016 f	0.774 ± .016 f	0.590 ± .100 Å	37.3 ± 6.3
intf	W	4	rd1	1.00	1	0.774 ± .016 f	0.226 ± .016 f	0.590 ± .100 Å	37.3 ± 6.3
intf	W	5	rd1	1.00	1	0.226 ± .016 f	0.226 ± .016 f	0.590 ± .100 Å	37.3 ± 6.3
intf	W	6	b	1.00	5	-0.452 ± .032 f	-0.452 ± .032 f	1.580 ± .100 Å	100.0 ± 6.3
subl	W	7	b	1.00	6	0.500 ± .032 f	0.500 ± .032 f	1.580 ± .100 Å	100.0 ± 6.3

W(100)-O disordered
74.8.8 BOND DISTANCES AND ANGLES

Bond distances and angles are derived from coordinates

No. of distances/angles: 3

Interatomic dist. A-B (Å)	Atom A	Atom B	Atom C	Bond angle A-B-C (°)
2.104	O1	W5		
2.170	O1	W6		
2.857	W2	W3(1,0)	W2(1,1)	95.5

COMMON NAME : W(100)-p(2x1)-disordered O
CLASSIFICATION : 74.8.12
TECHNIQUE : LEIS
AUTHORS : D.R. Mullins and S.H. Overbury
REFERENCE : Surf. Sci., 210, 481 (1989)

ILLUSTRATION: 52

SURFACE TYPE
Substrate : W
Crystal face: 100
Temperature : RT
Bulk lattice: bcc
2D bulk symm: p4m
2D surf symm: none

Adsorbate: O
Coverage : 0.5 ML
Pattern : disordered
Matrix : (1.000, 0.000)
(0.000, 1.000)

STRUCTURE TYPE
Missing-row reconstruction of substrate
oxygen is disordered in 2nd layer top sites

SAMPLE PREPARATION (1 sample)
Treatment : exposed to oxygen; annealed to 1300 K
Crystallinity:
Anal. methods:
Contamination:

COMMENTS

DATA COLLECTION
Technique: LEIS; rotatable spherical analyzer
Dataset : 0-70° scans along the [011] direction at 3 total scattering angles

THEORY/DATA TREATMENT
Shadow-cone geometry computation

2D UNIT CELLS (1 domain observed)

Cell	Ax (Å)	Ay (Å)	Bx (Å)	By (Å)	α (°)	Matrix	Pattern	Cell type
Bulk	3.160	0.000	0.000	3.160	90.0	(1.000, 0.000) (0.000, 1.000)	(1x1)	b: bulk lattice
Surface 1	6.320	0.000	0.000	3.160	90.0	(2.000, 0.000) (0.000, 1.000)	(2x1)	s1: commens. superlattice

3D COORDINATES

O1: disordered oxygen in top site on 2nd W layer W1: remaining row of missing-row reconstruction

Dx/Dy in Å, or as a fraction of layer's unit cell vectors; atom 0 at the origin. Epir/subr are bulk repeat vectors.

No. of atoms: 4 Bulk z = 1.580 Å

Reg ion	Chem el.	At. no.	Cell type	Site occ.	Rel. to	Dx ± εx	Dy ± εy	Dz ± εz	Dz/Bz(%) ± εz/Bz
epir		-2				f	f	Å	
subr		-1				1.580 Å	1.580 Å	1.580 Å	
ovrl	O	1	nd1	.50	0	0.000 Å	0.000 Å	0.000 Å	0.0
intf	W	2	s1	.50	0	1.580 Å	1.580 Å	0.060 ± .100 Å	3.8 ± 6.3
intf	W	3	b	1.00	0	0.000 Å	0.000 Å	2.000 Å	126.6
subl	W	4	b	1.00	0	1.580 Å	1.580 Å	3.580 Å	226.6

BOND DISTANCES AND ANGLES

Bond distances and angles are derived from coordinates

No. of distances/angles: 3

Interatomic dist. A-B (Å)	Atom A	Atom B	Atom C	Bond angle A-B-C (°)
2.000	O1	W3		
2.235	O1	W2		
2.959	W1	W2		

COMMON NAME : W(110)-(2x1)-O
CLASSIFICATION : 74.8.1
TECHNIQUE : LEED
AUTHORS : M.A. Van Hove and S.Y. Tong
REFERENCE : Phys. Rev. Lett., 35, 1092 (1975)

ILLUSTRATION: 44,45

SURFACE TYPE
Substrate : W
Crystal face: 110
Temperature : 300 K
Bulk lattice: bcc
2D bulk symm: cmm
2D surf symm: p2

Adsorbate: O
Coverage : 1/2 (O/W)
Pattern : (2x1)
Matrix : (2.000, 0.000)
(0.000, 1.000)

STRUCTURE TYPE
Atomic adsorption in 3-fold coordinated hollow sites

SAMPLE PREPARATION (1 sample)
Treatment : see Buchholz, Wang and Lagally, Surf. Sci. 49, 568 (1975)
Crystallinity:
Anal. methods:
Contamination:

COMMENTS
O in 2-fold coord. long-bridge ('center') site not excluded (with same O-W spacing);
later R-factor comparison (Van Hove, Tong and Elconin, Surf. Sci. 64, 75 (1979)) confirms this structure, with first spacing of 1.25±0.03Å and R2=0.17

DATA COLLECTION
Technique: LEED
Dataset : I-V curves for 5 non-degenerate beams at normal incidence

THEORY/DATA TREATMENT
Dynamical LEED (layer doubling): 8 ph shs (W: Moruzzi et al O: overl. at. ch. dens.; Vor=-10.0 eV,Voi=-5.0eV; ΘD=318 K(W)

STRUCTURES EXAMINED
1. top sites: (0,0) registry; 2. 2 types of short bridge sites: (0,1/2),(1/4,1/2);
3. center of diamond sites (=long bridge sites): (1/4,0); 4. 3-fold coordinated hollow sites: (0.188,0.354);
5. halfway between sites 3 and 4: (0.219,0.177)

QUALITY OF EXPERIMENT-THEORY FIT
Visual (see comments)

2D UNIT CELLS (4 domains observed)

Cell	Ax (Å)	Ay (Å)	Bx (Å)	By (Å)	α (°)	Matrix	Pattern	Cell type
Bulk	2.737	0.000	.912	2.580	70.5	(1.000, 0.000) (0.000, 1.000)	(1x1)	b: bulk lattice
Surface 1	5.473	0.000	.912	2.580	70.5	(2.000, 0.000) (0.000, 1.000)	(2x1)	s1: commens. superlattice

3D COORDINATES

O1: overlayer in 3-fold coord. hollow sites

Dx/Dy in Å, or as a fraction of layer's unit cell vectors; atom 0 at the origin. Epir/subr are bulk repeat vectors.

No. of atoms: 3 Bulk z = 2.235 Å

Reg ion	Chem el.	At. no.	Cell type	Site occ.	Rel. to	Dx ± εx		Dy ± εy		Dz ± εz		Dz/Bz(%) ± εz/Bz
epir		-2					f		f		Å	
subr		-1				1.824	Å	1.290	Å	2.235	Å	
ovrl	O	1	s1	.50	0	0.000	f	0.000	f	0.000	Å	0.0
intf	W	2	b	1.00	1	0.616	f	0.616	f	1.250 ± .100	Å	55.9 ± 4.5
subl	W	3	b	1.00	2	-0.500	f	-0.500	f	2.235	Å	100.0

BOND DISTANCES AND ANGLES

Bond distances and angles are derived from coordinates

No. of distances/angles: 3

Interatomic dist. A-B (Å)	Atom A	Atom B	Atom C	Bond angle A-B-C (°)
2.080	O1	W2(0,-1)	W2	76.4
2.080	O1	W2(0,-1)	W3	93.0
3.160	W2	W3		

COMMON NAME : Zn(0001)-(1x1)
CLASSIFICATION : 30.1
TECHNIQUE : LEED
AUTHORS : W.N. Unertl and H.V. Thapliyal
REFERENCE : J. Vac. Sci. Technol., 12, 263 (1975)

ILLUSTRATION: 19

SURFACE TYPE
Substrate : Zn
Crystal face: 0001
Temperature : 70 K
Bulk lattice: hcp
2D bulk symm: p3m1
2D surf symm: p3m1

Adsorbate:
Coverage :
Pattern : (1x1)
Matrix : (1.000, 0.000)
(0.000, 1.000)

STRUCTURE TYPE
Bulk termination with 2% top spacing contraction

SAMPLE PREPARATION (sample)
Treatment : see Baker and Blakely, Surf. Sci. 32, 45 (1972)
Crystallinity:
Anal. methods:
Contamination:

COMMENTS
Caution: result not accurate due to kinematic LEED analysis;
I-V data due to Baker and Blakely, Surf. Sci. 32, 45 (1972)

DATA COLLECTION
Technique: LEED
Dataset : I-V spectra: (00) beam at 70 K, 10<E<300 eV

THEORY/DATA TREATMENT
Kinematic LEED theory with constant-momentum transfer averaging: ΘD=220 K

STRUCTURES EXAMINED
Top layer spacing variations

QUALITY OF EXPERIMENT-THEORY FIT
Visual

2D UNIT CELLS (1 domain observed)

Cell	Ax (Å)	Ay (Å)	Bx (Å)	By (Å)	α (°)	Matrix	Pattern	Cell type
Bulk	2.660	0.000	-1.330	2.304	120.0	(1.000, 0.000) (0.000, 1.000)	(1x1)	b: bulk lattice
Surface 1	2.660	0.000	-1.330	2.304	120.0	(1.000, 0.000) (0.000, 1.000)	(1x1)	s1: commens. superlattice

3D COORDINATES

Dx/Dy in Å, or as a fraction of layer's unit cell vectors; atom 0 at the origin. Epir/subr are bulk repeat vectors.

No. of atoms: 4

Bulk z = 2.440 Å

Region	Chem el.	At. no.	Cell type	Site occ.	Rel. to	Dx ± εx		Dy ± εy		Dz ± εz		Dz/Bz(%) ± εz/Bz
epir		-2					f		f		Å	
subr		-1				0.000	Å	0.000	Å	4.880	Å	
intf	Zn	1	b	1.00	0	0.000	f	0.000	f	0.000	Å	0.0
intf	Zn	2	b	1.00	1	0.333	f	0.667	f	2.390 ± .050	Å	98.0 ± 2.1
subl	Zn	3	b	1.00	2	-0.333	f	-0.667	f	2.440	Å	100.0
subl	Zn	4	b	1.00	3	0.333	f	0.667	f	2.440	Å	100.0

BOND DISTANCES AND ANGLES

Bond distances and angles are derived from coordinates

No. of distances/angles: 3

Interatomic dist. A-B (Å)	Atom A	Atom B	Atom C	Bond angle A-B-C (°)
2.660	Zn1	Zn1(0,1)		
2.841	Zn1	Zn2	Zn1(0,1)	55.8
2.883	Zn2	Zn3	Zn2(0,-1)	54.9

COMMON NAME : ZnO(0001)-(1x1) ILLUSTRATION: 123
CLASSIFICATION : 30.8.2
TECHNIQUE : LEED
AUTHORS : A.R. Lubinsky, C.B. Duke, S.C. Chang, B.W. Lee and P. Mark
REFERENCE : J. Vac. Sci. Technol., 13, 189 (1976)

SURFACE TYPE
Substrate : ZnO
Crystal face: 0001
Temperature : RT
Bulk lattice: wurtzite
2D bulk symm: p3m1
2D surf symm: p3m1

Adsorbate:
Coverage :
Pattern : (1x1)
Matrix : (1.000, 0.000)
(0.000, 1.000)

STRUCTURE TYPE
Bulk termination (Zn at top) with contraction of top Zn-O spacing

SAMPLE PREPARATION (1 sample)
Treatment : Ar+ bombardment followed by thermal annealing below 873 K
Crystallinity:
Anal. methods:
Contamination:

COMMENTS
Data due to Chang and Mark, Surf. Sci. 46, 293 (1974); these results supercede those reported by Duke and Lubinsky, Surf. Sci. 50, 605 (1975); analysis of the non-specular beams was unsuccessful because of an unknown domain structure which made (10) and (01) spectra equal

DATA COLLECTION
Technique: LEED
Dataset : I-V spectra: (00) beam at Θ=7°, ϕ=30°, 20<E<200 eV

THEORY/DATA TREATMENT
Dynamical LEED: 4 phase shifts, both neutral-atom and singly ionic models; Vor=-10 eV; mfp=8-12Å; ΘD=920 K

STRUCTURES EXAMINED
Top layer spacing varied between 0 and 0.907Å

QUALITY OF EXPERIMENT-THEORY FIT
Visual

2D UNIT CELLS (1 domain observed)

Cell	Ax (Å)	Ay (Å)	Bx (Å)	By (Å)	α (°)	Matrix	Pattern	Cell type
Bulk	3.250	0.000	1.625	2.814	60.0	(1.000, 0.000) (0.000, 1.000)	(1x1)	b: bulk lattice
Surface 1	3.250	0.000	1.625	2.814	60.0	(1.000, 0.000) (0.000, 1.000)	(1x1)	s1: commens. superlattice

3D COORDINATES

Zn1-O2: topmost bilayer (contracted spacing); Zn3-O4 and Zn5-O6: 2 bulk bilayers, together forming repeating bulk set of layers

Dx/Dy in Å, or as a fraction of layer's unit cell vectors; atom 0 at the origin. Epir/subr are bulk repeat vectors.

No. of atoms: 6 Bulk z = 5.206 Å

Region	Chem el.	At. no.	Cell type	Site occ.	Rel. to	Dx ± εx	Dy ± εy	Dz ± εz	Dz/Bz(%) ± εz/Bz
epir		-2				f	f	Å	
subr		-1				0.000 Å	0.000 Å	5.206 Å	
intf	Zn	1	b	1.00	0	0.000 f	0.000 f	0.000 Å	0.0
intf	O	2	b	1.00	1	0.333 f	0.333 f	0.607 ± .100 Å	11.7 ± 1.9
subl	Zn	3	b	1.00	2	0.000 f	0.000 f	1.796 Å	34.5
subl	O	4	b	1.00	3	-0.333 f	-0.333 f	0.807 Å	15.5
subl	Zn	5	b	1.00	4	0.000 f	0.000 f	1.796 Å	34.5
subl	O	6	b	1.00	5	0.333 f	0.333 f	0.807 Å	15.5

BOND DISTANCES AND ANGLES

Bond distances and angles are derived from coordinates

No. of distances/angles: 3

Interatomic dist. A-B (Å)	Atom A	Atom B	Atom C	Bond angle A-B-C (°)
1.972	Zn1	O2	Zn1(1,0)	111.0
1.972	Zn1	O2	Zn3	107.9
1.796	O2	Zn3	O4	113.3

COMMON NAME : ZnO(10-10)-(1x1)
CLASSIFICATION : 30.8.2a
TECHNIQUE : LEED
AUTHORS : C.B. Duke, A.R. Lubinsky, B.W. Lee and P. Mark
REFERENCE : J. Vac. Sci. Technol., 13, 761 (1976)

ILLUSTRATION: 124

SURFACE TYPE
Substrate : ZnO
Crystal face: 10-10
Temperature : RT
Bulk lattice: wurtzite
2D bulk symm: pm
2D surf symm: pm

Adsorbate:
Coverage :
Pattern : (1x1)
Matrix : (1.000, 0.000)
(0.000, 1.000)

STRUCTURE TYPE
Relaxed bulk termination: top Zn-O layer buckles (O outward, Zn inward)

SAMPLE PREPARATION (1 sample)
Treatment : Ar+ bombardment and thermal annealing
Crystallinity:
Anal. methods:
Contamination: monitored by LEED and AES

COMMENTS
Lateral relaxations of top bilayer could not be determined by this analysis; top Zn and O were moved in 2 different directions: a) towards midpoint of their 2nd-layer neighbors; b) perpendicular to plane of their 3 neighbors

DATA COLLECTION
Technique: LEED
Dataset : I-V spectra: 30<E<112 eV

THEORY/DATA TREATMENT
Dynamical LEED: 4 phase shifts, both neutral-atom and singly ionic models; Vor=-10 eV; mfp=8-12Å; ΘD=920 K

STRUCTURES EXAMINED
1) covalent bond length conserving rotations of top layer ZnO bond between 0° and 21.1°, with 0° clearly preferred;
2) ionic reconstructions in which top Zn and O layers are expanded and contracted wrt other bulk layers; preferred structure has top Zn and O move in by 0.3 and 0.1 Å, resp.

QUALITY OF EXPERIMENT-THEORY FIT
Visual

2D UNIT CELLS (1 domain observed)

Cell	Ax (Å)	Ay (Å)	Bx (Å)	By (Å)	α (°)	Matrix	Pattern	Cell type
Bulk	3.250	0.000	0.000	5.207	90.0	(1.000, 0.000) (0.000, 1.000)	(1x1)	b: bulk lattice
Surface 1	3.250	0.000	0.000	5.207	90.0	(1.000, 0.000) (0.000, 1.000)	(1x1)	s1: commens. superlattice

3D COORDINATES

O1-Zn2: buckled top bilayer, O outward Zn5-O6 and Zn7-O8: 2 bulk bilayers, together forming repeating bulk set of layers; 0.1Å error bars assumed for tabulation

Dx/Dy in Å, or as a fraction of layer's unit cell vectors; atom 0 at the origin. Epir/subr are bulk repeat vectors.

No. of atoms: 8

Bulk z = 2.814 Å

Region	Chem el.	At. no.	Cell type	Site occ.	Rel. to	Dx ± εx		Dy ± εy		Dz ± εz		Dz/Bz(%) ± εz/Bz
epir		-2					f		f		Å	
subr		-1				-1.625	Å	0.000	Å	2.814	Å	
intf	O	1	b	1.00	0	0.000	f	0.000	f	0.000	Å	0.0
intf	Zn	2	b	1.00	1	0.000	f	0.655	f	0.840 ± .100	Å	29.9 ± 3.6
intf	Zn	3	b	1.00	2	0.500	f	-0.500	f	0.640 ± .100	Å	22.7 ± 3.6
intf	O	4	b	1.00	3	0.000	f	0.345	f	0.000	Å	0.0
subl	Zn	5	b	1.00	4	0.000	f	0.155	f	1.876	Å	66.7
subl	O	6	b	1.00	5	0.000	f	-0.655	f	0.000	Å	0.0
subl	Zn	7	b	1.00	6	-0.500	f	0.155	f	0.938	Å	33.3
subl	O	8	b	1.00	7	0.000	f	0.345	f	0.000	Å	0.0

ZnO(10-10)-(1x1)
30.8.2a

BOND DISTANCES AND ANGLES

Bond distances and angles are derived from coordinates

No. of distances/angles: 7

Interatomic dist. A-B (Å)	Atom A	Atom B	Atom C	Bond angle A-B-C (°)
1.983	O1	Zn2(0,-1)	O4(0,-1)	121.4
2.341	O1	Zn3	O4	110.2
2.341	O1	Zn3	O6	116.4
1.924	Zn2	O4	Zn2(1,0)	115.3
1.924	Zn2	O4	Zn3	114.8
1.796	Zn3	O4	Zn2	114.8
1.796	Zn3	O4	Zn5	113.3

COMMON NAME : ZnO(11-20)-(1x1)
CLASSIFICATION : 30.8.2b
TECHNIQUE : LEED
AUTHORS : C.B. Duke, A.R. Lubinsky, B.W. Lee and P. Mark
REFERENCE : J. Vac. Sci. Technol., 13, 761 (1976)

ILLUSTRATION: 125

SURFACE TYPE
Substrate : ZnO
Crystal face: 11-20
Temperature : RT
Bulk lattice: wurtzite
2D bulk symm: p1
2D surf symm: p1

Adsorbate:
Coverage :
Pattern : (1x1)
Matrix : (1.000, 0.000)
(0.000, 1.000)

STRUCTURE TYPE
Unrelaxed bulk termination

SAMPLE PREPARATION (1 sample)
Treatment : Ar+ bombardment and thermal annealing
Crystallinity:
Anal. methods:
Contamination: monitored by LEED and AES

COMMENTS

DATA COLLECTION
Technique: LEED
Dataset : I-V curves: (10) and (11) beams, 30<E<180 eV

THEORY/DATA TREATMENT
Dynamical LEED: 4 phase shifts, singly ionic bulk potential
Vor=-10 eV; mfp=8-12Å

STRUCTURES EXAMINED
Top layer spacing varied from 1.525 to 1.725Å

QUALITY OF EXPERIMENT-THEORY FIT
Visual

2D UNIT CELLS (1 domain observed)

Cell	Ax (Å)	Ay (Å)	Bx (Å)	By (Å)	α (°)	Matrix	Pattern	Cell type
Bulk	5.207	0.000	0.000	5.628	90.0	(1.000, 0.000) (0.000, 1.000)	(1x1)	b: bulk lattice
Surface 1	5.207	0.000	0.000	5.628	90.0	(1.000, 0.000) (0.000, 1.000)	(1x1)	s1: commens. superlattice

3D COORDINATES

Zn1-Zn2-O3-O4: planar top layer; Zn5-Zn6-O7-O8: periodically repeating planar bulk layer;
0.1Å error bar assumed for tabulation

Dx/Dy in Å, or as a fraction of layer's unit cell vectors; atom 0 at the origin. Epir/subr are bulk repeat vectors.

No. of atoms: 8

Bulk z = 1.625 Å

Region	Chem el.	At. no.	Cell type	Site occ.	Rel. to	Dx ± εx		Dy ± εy		Dz ± εz		Dz/Bz(%) ± εz/Bz
epir		-2					f		f		Å	
subr		-1				0.000	Å	2.814	Å	1.625	Å	
intf	Zn	1	b	1.00	0	0.000	f	0.000	f	0.000	Å	0.0
intf	Zn	2	b	1.00	1	0.500	f	0.333	f	0.000	Å	0.0
intf	O	3	b	1.00	2	-0.155	f	-0.333	f	0.000	Å	0.0
intf	O	4	b	1.00	3	0.500	f	0.333	f	0.000	Å	0.0
subl	Zn	5	b	1.00	4	-0.845	f	0.167	f	1.625 ± .100	Å	100.0 ± 6.2
subl	Zn	6	b	1.00	5	0.500	f	0.333	f	0.000	Å	0.0
subl	O	7	b	1.00	6	-0.155	f	-0.333	f	0.000	Å	0.0
subl	O	8	b	1.00	7	0.500	f	0.333	f	0.000	Å	0.0

BOND DISTANCES AND ANGLES

Bond distances and angles are derived from coordinates

No. of distances/angles: 12

Interatomic dist. A-B (Å)	Atom A	Atom B	Atom C	Bond angle A-B-C (°)
1.796	Zn1	O3	Zn2	113.3
1.796	Zn2	O4	Zn5(1,0)	113.3
2.042	Zn2	O7	Zn5	113.3

ZnO(11-20)-(1x1)
30.8.2b

Bond Distances and Angles - Continued

Interatomic dist. A-B (Å)	Atom A	Atom B	Atom C	Bond angle A-B-C (°)
2.042	Zn2	O7	Zn6	105.4
1.796	Zn1	O3	Zn6(0,-1)	113.3
2.042	Zn1	O4(-1,0)	Zn2(-1,0)	113.3
2.042	Zn1	O4(-1,0)	Zn5	105.4
2.042	Zn1	O8(-1,-1)	Zn5(0,-1)	105.4
2.042	Zn1	O8(-1,-1)	Zn6(-1,-1)	113.3
2.042	Zn2	O3	Zn1	113.3
2.042	Zn2	O3	Zn6(0,-1)	105.4
1.796	Zn2	O4	Zn1(1,0)	113.3

COMMON NAME : ZnS(110)-(1x1) ILLUSTRATION: 116
CLASSIFICATION : 30.16.2
TECHNIQUE : LEED
AUTHORS : C.B. Duke, A. Paton and A. Kahn
REFERENCE : J. Vac. Sci. Technol., A2, 515 (1984)

SURFACE TYPE
Substrate : ZnS
Crystal face: 110
Temperature : RT
Bulk lattice: zincblende
2D bulk symm: pm
2D surf symm: pm

Adsorbate:
Coverage :
Pattern : (1x1)
Matrix : (1.000, 0.000)
(0.000, 1.000)

STRUCTURE TYPE
Relaxed bulk termination with 26° tilt in top layer

SAMPLE PREPARATION (sample)
Treatment : see Duke et al, J. Vac. Sci. Technol. 18, 866 (1981)
Crystallinity:
Anal. methods:
Contamination:

COMMENTS
X-ray R-factor showed two minima corresponding to tilts of 2.5° and 26°; integrated R-factor RI clearly distinguishes in favor of 26°

DATA COLLECTION
Technique: LEED
Dataset : I-V curves for 14 beams:
10=-10,01=0-1,11=-11,1-1=-1-1,02,
20=-20,12=-12,1-2=-1-2,21=-21,2-1=-2-1,13=-

THEORY/DATA TREATMENT
Dynamical LEED: see Meyer et al, Phys. Rev. B19, 5194 (1979)
charge overlap potentials; mfp=6Å

STRUCTURES EXAMINED
Bond-length conserving rotation of top bilayer was used to determine vertical spacing between Zn and S in top layer; spacing between top 2 layers was varied; spacing between Zn and S in second layer was varied; registry of Zn and S in top layer was varied

QUALITY OF EXPERIMENT-THEORY FIT
RX=0.18, RI=0.09

2D UNIT CELLS (1 domain observed)

Cell	Ax (Å)	Ay (Å)	Bx (Å)	By (Å)	α (°)	Matrix	Pattern	Cell type
Bulk	3.825	0.000	0.000	5.409	90.0	(1.000, 0.000) (0.000, 1.000)	(1x1)	b: bulk lattice
Surface 1	3.825	0.000	0.000	5.409	90.0	(1.000, 0.000) (0.000, 1.000)	(1x1)	s1: commens. superlattice

3D COORDINATES

S1-Zn2: top bilayer with tilted Zn-S chain; Zn3-S4, Zn5-S6: bulk bilayers;
Zn7-S8: periodically repeating bulk bilayer; 0.1Å error bars assumed for tabulation

Dx/Dy in Å, or as a fraction of layer's unit cell vectors; atom 0 at the origin. Epir/subr are bulk repeat vectors.

No. of atoms: 8 Bulk z = 1.913 Å

Region	Chem el.	At. no.	Cell type	Site occ.	Rel. to	Dx ± εx		Dy ± εy		Dz ± εz		Dz/Bz(%) ± εz/Bz
epir		-2					f		f		Å	
subr		-1				1.913	Å	2.705	Å	1.913	Å	
intf	S	1	b	1.00	0	0.000	f	0.000	f	0.000	Å	0.0
intf	Zn	2	b	1.00	1	0.500 ± .026	f	0.206 ± .019	f	0.590 ± .100	Å	30.9 ± 5.2
intf	Zn	3	b	1.00	2	-0.500 ± .026	f	0.582 ± .019	f	1.403 ± .100	Å	73.4 ± 5.2
intf	S	4	b	1.00	3	0.500	f	-0.250	f	0.000 ± .100	Å	0.0 ± 5.2
intf	Zn	5	b	1.00	4	0.000	f	-0.250	f	1.913	Å	100.0
intf	S	6	b	1.00	5	-0.500	f	-0.250	f	0.000	Å	0.0
subl	Zn	7	b	1.00	6	0.000	f	0.750	f	1.913	Å	100.0
subl	S	8	b	1.00	7	0.500	f	-0.250	f	0.000	Å	0.0

ZnS(110)-(1x1)
30.16.2 BOND DISTANCES AND ANGLES

Bond distances and angles are derived from coordinates

No. of distances/angles: 13

Interatomic dist. A-B (Å)	Atom A	Atom B	Atom C	Bond angle A-B-C (°)
2.291	S1	Zn2	S1(1,0)	113.2
2.342	Zn3	S6(0,1)	Zn7	109.5
2.342	S4	Zn5	S6(1,0)	109.5
2.342	S4	Zn5	S6	109.5
2.342	S4	Zn5	S8	109.5
2.291	S1	Zn2	S4	122.8
2.299	S1	Zn3(0,-1)	S4(0,-1)	106.7
2.299	S1	Zn3(0,-1)	S6	114.8
2.342	Zn3	S4	Zn2	117.1
2.342	Zn3	S4	Zn3(1,0)	109.5
2.342	Zn3	S4	Zn5	109.5
2.342	Zn3	S6(0,1)	Zn5(0,1)	109.5
2.342	Zn3	S6(0,1)	Zn5(-1,1)	109.5

COMMON NAME : ZnSe(110)-(1x1)
CLASSIFICATION : 30.34.2a
TECHNIQUE : LEED
AUTHORS : C.B. Duke, A. Paton, A. Kahn and D.W. Tu
REFERENCE : J. Vac. Sci. Technol., B2, 366 (1984)

ILLUSTRATION: 116

SURFACE TYPE
Substrate : ZnSe
Crystal face: 110
Temperature : 200 K
Bulk lattice: zincblende
2D bulk symm: pm
2D surf symm: pm
Adsorbate:
Coverage :
Pattern : (1x1)
Matrix : (1.000, 0.000)
(0.000, 1.000)

STRUCTURE TYPE
Relaxed bulk termination with 4° tilt in top layer; analysis could not distinguish this result from 29°-tilt model (see class. no. 30.34.2b); 29° tilt is more likely

SAMPLE PREPARATION (2 sample)
Treatment : 1. single xtal: Ar+ and annealing; 2. thin film on GaAs(110)
Crystallinity: I-Vs identical for two samples
Anal. methods:
Contamination: AES: equal Zn and Se in thin film

COMMENTS
Scattering amplitudes of top 4 bilayers were calculated dynamically, but superposed kinematically ('quasi-dynamical LEED')

DATA COLLECTION
Technique: LEED
Dataset : I-V spectra for 14 inequivalent beams at normal incidence

THEORY/DATA TREATMENT
Quasi-dynamical LEED: mfp=8Å; Vor optimised

STRUCTURES EXAMINED
Bond length conserving rotations in top bilayer up to 35°; perpendicular and lateral displacements in first bilayer; smaller displacements in second layer

QUALITY OF EXPERIMENT-THEORY FIT
Visual

2D UNIT CELLS (1 domain observed)

Cell	Ax (Å)	Ay (Å)	Bx (Å)	By (Å)	α (°)	Matrix	Pattern	Cell type
Bulk	4.010	0.000	0.000	5.670	90.0	(1.000, 0.000) (0.000, 1.000)	(1x1)	b: bulk lattice
Surface 1	4.010	0.000	0.000	5.670	90.0	(1.000, 0.000) (0.000, 1.000)	(1x1)	s1: commens. superlattice

3D COORDINATES

Se1-Zn2, Zn3-Se4: 2 bilayers with tilted Zn-Se chains; Zn5-Se6: bulk bilayer;
Zn7-Se8: periodically repeating bulk bilayer; 0.1Å error bars assumed for tabulation

Dx/Dy in Å, or as a fraction of layer's unit cell vectors; atom 0 at the origin. Epir/subr are bulk repeat vectors.

No. of atoms: 8 Bulk z = 2.004 Å

Region	Chem el.	At. no.	Cell type	Site occ.	Rel. to	Dx ± εx		Dy ± εy		Dz ± εz		Dz/Bz(%) ± εz/Bz
epir		-2					f		f		Å	
subr		-1				2.005	Å	2.835	Å	2.004	Å	
intf	Se	1	b	1.00	0	0.000	f	0.000	f	0.000	Å	0.0
intf	Zn	2	b	1.00	1	0.500 ± .025	f	0.250 ± .018	f	0.098 ± .100	Å	4.9 ± 5.0
intf	Zn	3	b	1.00	2	-0.500 ± .025	f	0.512 ± .018	f	1.927 ± .100	Å	96.2 ± 5.0
intf	Se	4	b	1.00	3	0.500 ± .025	f	-0.250 ± .018	f	0.050 ± .100	Å	2.5 ± 5.0
intf	Zn	5	b	1.00	4	0.000 ± .025	f	-0.250 ± .018	f	1.979 ± .100	Å	98.8 ± 5.0
intf	Se	6	b	1.00	5	-0.500	f	-0.250	f	0.000	Å	0.0
subl	Zn	7	b	1.00	6	0.000	f	0.750	f	2.004	Å	100.0
subl	Se	8	b	1.00	7	0.500	f	-0.250	f	0.000	Å	0.0

ZnSe(110)-(1x1)
30.34.2a

BOND DISTANCES AND ANGLES

Bond distances and angles are derived from coordinates

No. of distances/angles: 8

Interatomic dist. A-B (Å)	Atom A	Atom B	Atom C	Bond angle A-B-C (°)
2.457	Se1	Zn2	Se1(1,0)	109.4
2.457	Se1	Zn2	Se4	112.2
2.433	Se1	Zn3(0,-1)	Se4(0,-1)	109.7
2.433	Se1	Zn3(0,-1)	Se6	111.4
2.457	Zn2	Se1	Zn2(-1,0)	109.4
2.457	Zn2	Se1	Zn3(0,-1)	106.7
2.473	Zn2	Se4	Zn3	109.3
2.473	Zn2	Se4	Zn5	107.5

COMMON NAME : ZnSe(110)-(1x1)
CLASSIFICATION : 30.34.2b
TECHNIQUE : LEED
AUTHORS : C.B. Duke, A. Paton, A. Kahn and D.W. Tu
REFERENCE : J. Vac. Sci. Technol., B2, 366 (1984)

ILLUSTRATION: 116

SURFACE TYPE
Substrate : ZnSe
Crystal face: 110
Temperature : 200 K
Bulk lattice: zincblende
2D bulk symm: pm
2D surf symm: pm
Adsorbate:
Coverage :
Pattern : (1x1)
Matrix : (1.000, 0.000)
(0.000, 1.000)

STRUCTURE TYPE
Relaxed bulk termination with 29° tilt in top layer; analysis could not distinguish this result from 4°-tilt model (see class. no. 30.34.2a); 29° tilt is more likely

SAMPLE PREPARATION (2 sample)
Treatment : 1. single xtal: Ar+ and annealing; 2. thin film on GaAs(110)
Crystallinity: I-Vs identical for two samples
Anal. methods:
Contamination: AES: equal Zn and Se in thin film

COMMENTS
Scattering amplitudes of top 4 bilayers were calculated dynamically, but superposed kinematically ('quasi-dynamical LEED')

DATA COLLECTION
Technique: LEED
Dataset : I-V spectra for 14 inequivalent beams at normal incidence

THEORY/DATA TREATMENT
Quasi-dynamical LEED: mfp=8Å; Vor optimised

STRUCTURES EXAMINED
Bondlength conserving rotations in top bilayer up to 35°; perpendicular and lateral displacements in first bilayer; smaller displacements in second layer

QUALITY OF EXPERIMENT-THEORY FIT
Visual

2D UNIT CELLS (1 domain observed)

Cell	Ax (Å)	Ay (Å)	Bx (Å)	By (Å)	α (°)	Matrix	Pattern	Cell type
Bulk	4.010	0.000	0.000	5.670	90.0	(1.000, 0.000) (0.000, 1.000)	(1x1)	b: bulk lattice
Surface 1	4.010	0.000	0.000	5.670	90.0	(1.000, 0.000) (0.000, 1.000)	(1x1)	s1: commens. superlattice

3D COORDINATES

Se1-Zn2: top bilayer with tilted Zn-Se chains; Zn3-Se4, Zn5-Se6: 2 bulk bilayers;
Zn7-Se8: periodically repeating bulk bilayer; 0.1Å error bars assumed for tabulation

Dx/Dy in Å, or as a fraction of layer's unit cell vectors; atom 0 at the origin. Epir/subr are bulk repeat vectors.

No. of atoms: 8 Bulk z = 2.004 Å

Reg ion	Chem el.	At. no.	Cell type	Site occ.	Rel. to	Dx ± εx		Dy ± εy		Dz ± εz		Dz/Bz(%) ± εz/Bz
epir		-2					f		f		Å	
subr		-1				2.005	Å	2.835	Å	2.004	Å	
intf	Se	1	b	1.00	0	0.000	f	0.000	f	0.000	Å	0.0
intf	Zn	2	b	1.00	1	0.500 ± .025	f	0.130 ± .018	f	0.687 ± .100	Å	34.3 ± 5.0
intf	Zn	3	b	1.00	2	-0.500 ± .025	f	0.626 ± .018	f	1.416 ± .100	Å	70.7 ± 5.0
intf	Se	4	b	1.00	3	0.500 ± .025	f	-0.250 ± .018	f	0.000 ± .100	Å	0.0 ± 5.0
intf	Zn	5	b	1.00	4	0.000 ± .025	f	-0.250 ± .018	f	2.004 ± .100	Å	100.0 ± 5.0
intf	Se	6	b	1.00	5	-0.500	f	-0.250	f	0.000	Å	0.0
subl	Zn	7	b	1.00	6	0.000	f	0.750	f	2.004	Å	100.0
subl	Se	8	b	1.00	7	0.500	f	-0.250	f	0.000	Å	0.0

ZnSe(110)-(1x1)
30.34.2b BOND DISTANCES AND ANGLES

Bond distances and angles are derived from coordinates

No. of distances/angles: 11

Interatomic dist. A-B (Å)	Atom A	Atom B	Atom C	Bond angle A-B-C (°)
2.244	Se1	Zn2	Se1(1,0)	126.6
2.455	Zn3	Se6(0,1)	Zn7	109.5
2.455	Se4	Zn5	Se6	109.5
2.244	Se1	Zn2	Se4	116.3
2.517	Se1	Zn3(0,-1)	Se4(0,-1)	108.5
2.517	Se1	Zn3(0,-1)	Se6	111.4
2.559	Zn2	Se4	Zn3	118.7
2.559	Zn2	Se4	Zn5	88.3
2.456	Zn3	Se4	Zn3(1,0)	109.5
2.456	Zn3	Se4	Zn5	109.5
2.455	Zn3	Se6(0,1)	Zn5(0,1)	109.5

COMMON NAME : ZnTe(110)-(1x1) ILLUSTRATION: 116
CLASSIFICATION : 30.52.2
TECHNIQUE : LEED
AUTHORS : C.B. Duke, A. Paton and A. Kahn
REFERENCE : J. Vac. Sci. Technol., A1, 672 (1983)

SURFACE TYPE
Substrate : ZnTe
Crystal face: 110
Temperature : 425 K
Bulk lattice: zincblende
2D bulk symm: pm
2D surf symm: pm
Adsorbate:
Coverage :
Pattern : (1x1)
Matrix : (1.000, 0.000)
(0.000, 1.000)

STRUCTURE TYPE
Relaxed bulk termination with 28° tilt in top layer

SAMPLE PREPARATION (1 sample)
Treatment : chemical polishing, ion bombardment and annealing
Crystallinity:
Anal. methods:
Contamination: monitored by LEED and AES

COMMENTS
This ZnTe structure was found to be remarkably similar to that of GaSb(110)-(1x1) reported in same paper (see class. no. 31.51.1a)

DATA COLLECTION
Technique: LEED
Dataset : I-V curves: 14 beams in range 30<E<210 eV

THEORY/DATA TREATMENT
Dynamical LEED: approximate multiple scattering model of Mayer et al, Phys. Rev. B19 5194 (1979); no thermal vibs

STRUCTURES EXAMINED
Following quantities were varied in order: 1) angle between plane of top ZnTe chains and surface; 2) spacings of two outermost layers; 3) displacements in second layer normal to surface; 4) reduced relaxations of anion parallel to surface

QUALITY OF EXPERIMENT-THEORY FIT
RX=0.26

2D UNIT CELLS (1 domain observed)

Cell	Ax (Å)	Ay (Å)	Bx (Å)	By (Å)	α (°)	Matrix	Pattern	Cell type
Bulk	4.306	0.000	0.000	6.089	90.0	(1.000, 0.000) (0.000, 1.000)	(1x1)	b: bulk lattice
Surface 1	4.306	0.000	0.000	6.089	90.0	(1.000, 0.000) (0.000, 1.000)	(1x1)	s1: commens. superlattice

3D COORDINATES

Te1-Zn2, Zn3-Te4: top 2 bilayers with tilted Zn-Te chains; Zn5-Te6: bulk bilayer;
Zn7-Te8: periodically repeating bulk bilayer; 0.1Å lateral error bars assumed for tabulation

Dx/Dy in Å, or as a fraction of layer's unit cell vectors; atom 0 at the origin. Epir/subr are bulk repeat vectors.

No. of atoms: 8 Bulk z = 2.153 Å

Region	Chem el.	At. no.	Cell type	Site occ.	Rel. to	Dx ± εx		Dy ± εy		Dz ± εz		Dz/Bz(%) ± εz/Bz
epir		-2					f		f		Å	
subr		-1				2.153	Å	3.045	Å	2.153	Å	
intf	Te	1	b	1.00	0	0.000	f	0.000	f	0.000	Å	0.0
intf	Zn	2	b	1.00	1	0.500 ± .023	f	0.221 ± .016	f	0.715 ± .050	Å	33.2 ± 2.3
intf	Zn	3	b	1.00	2	-0.500 ± .023	f	0.588 ± .016	f	1.575 ± .050	Å	73.2 ± 2.3
intf	Te	4	b	1.00	3	0.500	f	-0.250	f	0.050 ± .050	Å	2.3 ± 2.3
intf	Zn	5	b	1.00	4	0.000	f	-0.250	f	2.128 ± .050	Å	98.8 ± 2.3
intf	Te	6	b	1.00	5	-0.500	f	-0.250	f	0.000	Å	0.0
subl	Zn	7	b	1.00	6	0.000	f	0.750	f	2.153	Å	100.0
subl	Te	8	b	1.00	7	0.500	f	-0.250	f	0.000	Å	0.0

ZnTe(110)-(1x1)
30.52.2 BOND DISTANCES AND ANGLES

Bond distances and angles are derived from coordinates

No. of distances/angles: 14

Interatomic dist. A-B (Å)	Atom A	Atom B	Atom C	Bond angle A-B-C (°)
2.638	Te1	Zn2	Te1(1,0)	109.4
2.637	Zn3	Te4	Zn5	110.6
2.657	Zn3	Te6(0,1)	Zn5(0,1)	109.3
2.657	Zn3	Te6(0,1)	Zn7	109.8
2.616	Te4	Zn5	Te6	109.6
2.616	Te4	Zn5	Te8	109.2
2.638	Te1	Zn2	Te4	124.6
2.568	Te1	Zn3(0,-1)	Te4(0,-1)	106.2
2.568	Te1	Zn3(0,-1)	Te6	118.1
2.638	Zn2	Te1	Zn2(-1,0)	109.4
2.638	Zn2	Te1	Zn3(0,-1)	89.4
2.622	Zn2	Te4	Zn3	116.2
2.622	Zn2	Te4	Zn5	92.7
2.637	Zn3	Te4	Zn3(1,0)	109.5

COMMON NAME : Zr(0001)-(1x1) ILLUSTRATION: 19
CLASSIFICATION : 40.1
TECHNIQUE : LEED
AUTHORS : W.T. Moore, P.R. Watson, D.C. Frost and K.A.R. Mitchell
REFERENCE : J. Phys., C12, L887 (1979)

SURFACE TYPE
Substrate : Zr
Crystal face: 0001
Temperature : RT*
Bulk lattice: hcp
2D bulk symm: p3m1
2D surf symm: p3m1

Adsorbate:
Coverage :
Pattern : (1x1)
Matrix : (1.000, 0.000)
(0.000, 1.000)

STRUCTURE TYPE
Bulk hcp termination with 1%±2% top spacing contraction

SAMPLE PREPARATION (1 sample)
Treatment : Ar+ bombardment and annealing
Crystallinity:
Anal. methods:
Contamination: AES: only C detected

COMMENTS

DATA COLLECTION
Technique: LEED
Dataset : I-V spectra: 4 inequivalent beams for Θ=9°, φ=13° and Θ=φ=0°; 30<E<230 eV

THEORY/DATA TREATMENT
Dynamical LEED: 7 phase shifts, 61 beams; Vor=-10 eV, Voi=-1.32 E**1/3; ΘD=270 K

STRUCTURES EXAMINED
1. unreconstructed hcp with equal numbers of 2 domains; 2. hcp but with the last 3 layers in fcc configuration; 3. variations in top layer spacing up to ±7.5% for both models

QUALITY OF EXPERIMENT-THEORY FIT
RZJ=0.12

2D UNIT CELLS (1 domain observed)

Cell	Ax (Å)	Ay (Å)	Bx (Å)	By (Å)	α (°)	Matrix	Pattern	Cell type
Bulk	3.220	0.000	1.610	2.789	60.0	(1.000, 0.000) (0.000, 1.000)	(1x1)	b: bulk lattice
Surface 1	3.220	0.000	1.610	2.789	60.0	(1.000, 0.000) (0.000, 1.000)	(1x1)	s1: commens. superlattice

3D COORDINATES

Dx/Dy in Å, or as a fraction of layer's unit cell vectors; atom 0 at the origin. Epir/subr are bulk repeat vectors.

No. of atoms: 4 Bulk z = 2.570 Å

Region	Chem el.	At. no.	Cell type	Site occ.	Rel. to	Dx ± εx		Dy ± εy		Dz ± εz		Dz/Bz(%) ± εz/Bz
epir		-2					f		f		Å	
subr		-1				0.000	Å	0.000	Å	5.140	Å	
intf	Zr	1	b	1.00	0	0.000	f	0.000	f	0.000	Å	0.0
intf	Zr	2	b	1.00	1	0.333	f	0.333	f	2.540 ± .050	Å	98.8 ± 2.0
subl	Zr	3	b	1.00	2	-0.333	f	-0.333	f	2.570	Å	100.0
subl	Zr	4	b	1.00	3	0.333	f	0.333	f	2.570	Å	100.0

BOND DISTANCES AND ANGLES

Bond distances and angles are derived from coordinates

No. of distances/angles: 3

Interatomic dist. A-B (Å)	Atom A	Atom B	Atom C	Bond angle A-B-C (°)
3.220	Zr1	Zr1(1,0)	Zr2	59.2
3.148	Zr1	Zr2	Zr3	107.9
3.172	Zr2	Zr3	Zr4	108.2

ILLUSTRATION: 58

COMMON NAME : Zr(0001)-(1x1)-C
CLASSIFICATION : 40.6.1
TECHNIQUE : LEED
AUTHORS : P.C. Wong, J.R. Lou, and K.A.R. Mitchell
REFERENCE : Surf. Sci., 206, L913 (1988)

SURFACE TYPE
Substrate : Zr
Crystal face: 0001
Temperature : RT*
Bulk lattice: hcp
2D bulk symm: p3m1
2D surf symm: p3m1

Adsorbate: C
Coverage : 1 C/Zr
Pattern : (1x1)
Matrix : (1.000, 0.000)
(0.000, 1.000)

STRUCTURE TYPE
Atomic adsorption in octahedral interstitial sites between 1st and 2nd Zr layers, inducing slight expansion between these Zr layers

SAMPLE PREPARATION (1 sample)
Treatment : see W.T. Moore et al. J. Phys. C12, L887 (1979)
Crystallinity:
Anal. methods:
Contamination:

COMMENTS
Analysis could not distinguish whether C is adsorbed in deeper layers also

DATA COLLECTION
Technique: LEED
Dataset : I-V spectra for 12 independent beams: (10)(11)(20) at Θ=0° 9 beams at Θ=14°; 50<E<230 eV

THEORY/DATA TREATMENT
Dynamical LEED (RFS and layer doubling): 8 phase shifts

STRUCTURES EXAMINED
1) overlayer models: (B)ABAB, (C)AB, etc.; 2) underlayer in octahedral sites: A(C)AB.., C(B)AB, A(C)B(C)AB, C(B)AB, etc.; 3) underlayer in tetrahedral sites: A(A)BA, A(B)BA and many others (23 total); A,B,C denote Zr layers, () C layer; C-Zr layer distances were fit

QUALITY OF EXPERIMENT-THEORY FIT
RPE=0.63 for A(C)BA.., (BA repeated)

2D UNIT CELLS (1 domain observed)

Cell	Ax (Å)	Ay (Å)	Bx (Å)	By (Å)	α (°)	Matrix	Pattern	Cell type
Bulk	3.230	0.000	1.615	2.797	60.0	(1.000, 0.000) (0.000, 1.000)	(1x1)	b: bulk lattice
Surface 1	3.230	0.000	1.615	2.797	60.0	(1.000, 0.000) (0.000, 1.000)	(1x1)	s1: commens. superlattice

3D COORDINATES

C2: interstitial layer in octahedral sites between Zr1-Zr3; Zr4-Zr5: hcp substrate;
0.10Å error bars set for fitted coord.

Dx/Dy in Å, or as a fraction of layer's unit cell vectors; atom 0 at the origin. Epir/subr are bulk repeat vectors.

No. of atoms: 5 Bulk z = 5.280 Å

Reg ion	Chem el.	At. no.	Cell type	Site occ.	Rel. to	Dx ± εx		Dy ± εy		Dz ± εz		Dz/Bz(%) ± εz/Bz
epir		-2					f		f		Å	
subr		-1				0.000	Å	0.000	Å	5.280	Å	
intf	Zr	1	b	1.00	0	0.000	f	0.000	f	0.000	Å	0.0
intf	C	2	b	1.00	1	0.333	f	0.333	f	1.330 ± .100	Å	25.2 ± 1.9
intf	Zr	3	b	1.00	2	0.333	f	0.333	f	1.330 ± .100	Å	25.2 ± 1.9
subl	Zr	4	b	1.00	3	-0.667	f	-0.667	f	2.640	Å	50.0
subl	Zr	5	b	1.00	4	0.667	f	0.667	f	2.640	Å	50.0

Zr(0001)-(1x1)-C
40.6.1 BOND DISTANCES AND ANGLES

Bond distances and angles are derived from coordinates

No. of distances/angles: 8

Interatomic dist. A-B (Å)	Atom A	Atom B	Atom C	Bond angle A-B-C (°)
3.230	Zr1	Zr1(1,0)	Zr1(1,1)	120.0
3.230	Zr1	Zr1(1,0)	Zr1(1,-1)	60.0
3.230	Zr1	Zr1(1,0)	Zr1(0,1)	60.0
3.230	Zr1	Zr1(1,0)	C2	45.2
2.290	Zr1	C2	Zr1(1,0)	89.7
2.290	Zr1	C2	Zr3	180.0
2.290	C2	Zr3	Zr1(1,1)	89.5
2.290	C2	Zr3	Zr1(1,0)	44.8

COMMON NAME : Zr(0001)-(1x1)-N
CLASSIFICATION : 40.7.1
TECHNIQUE : LEED
AUTHORS : P.C. Wong and K.A.R. Mitchell
REFERENCE : Surf. Sci., 187, L599 (1987)

ILLUSTRATION: 58

SURFACE TYPE
Substrate : Zr
Crystal face: 0001
Temperature : RT*
Bulk lattice: hcp
2D bulk symm: p3m1
2D surf symm: p3m1

Adsorbate: N
Coverage : 1 N/Zr
Pattern : (1x1)
Matrix : (1.000, 0.000)
(0.000, 1.000)

STRUCTURE TYPE
Atomic adsorption in octahedral interstitial sites between 1st and 2nd Zr layers, inducing slight contraction between these Zr layers

SAMPLE PREPARATION (1 sample)
Treatment :
Crystallinity:
Anal. methods:
Contamination: checked by AES

COMMENTS
Analysis could not distinguish between bulk Zr stacking and other (non-hcp) stackings starting in 4th Zr layer

DATA COLLECTION
Technique: LEED
Dataset : I-V spectra for 9 independent beams:
(10)(11)(20) at Θ=0°,
(00)(10)(20)(2-1)(3-1)(3-2) at Θ=18°; 50<E<

THEORY/DATA TREATMENT
Dynamical LEED (RFS and layer doubling): 8 phase shifts

STRUCTURES EXAMINED
1) overlayer models; 2) underlayer structures, hcp substrate;
3) underlayer structures, non-hcp substrate; 4) overlayer plus single underlayer models

QUALITY OF EXPERIMENT-THEORY FIT
RPE=0.22

2D UNIT CELLS (1 domain observed)

Cell	Ax (Å)	Ay (Å)	Bx (Å)	By (Å)	α (°)	Matrix	Pattern	Cell type
Bulk	3.230	0.000	1.615	2.797	60.0	(1.000, 0.000) (0.000, 1.000)	(1x1)	b: bulk lattice
Surface 1	3.230	0.000	1.615	2.797	60.0	(1.000, 0.000) (0.000, 1.000)	(1x1)	s1: commens. superlattice

3D COORDINATES

Zr1-Zr3-Zr4-Zr5: hcp substrate; N2: interstitial layer in octahedral sites;
0.05Å error bars assumed for tabulation

Dx/Dy in Å, or as a fraction of layer's unit cell vectors; atom 0 at the origin. Epir/subr are bulk repeat vectors.

No. of atoms: 5 Bulk z = 5.280 Å

Region	Chem el.	At. no.	Cell type	Site occ.	Rel. to	Dx ± εx		Dy ± εy		Dz ± εz		Dz/Bz(%) ± εz/Bz
epir		-2					f		f		Å	
subr		-1				0.000	Å	0.000	Å	5.280	Å	
intf	Zr	1	b	1.00	0	0.000	f	0.000	f	0.000	Å	0.0
intf	N	2	b	1.00	1	0.333	f	0.333	f	1.300 ± .050	Å	24.6 ± 1.0
intf	Zr	3	b	1.00	2	0.333	f	0.333	f	1.300 ± .050	Å	24.6 ± 1.0
subl	Zr	4	b	1.00	3	-0.667	f	-0.667	f	2.640	Å	50.0
subl	Zr	5	b	1.00	4	0.667	f	0.667	f	2.640	Å	50.0

BOND DISTANCES AND ANGLES

Bond distances and angles are derived from coordinates

No. of distances/angles: 3

Interatomic dist. A-B (Å)	Atom A	Atom B	Atom C	Bond angle A-B-C (°)
3.230	Zr1	Zr1(1,0)	N2	44.7
2.273	Zr1	N2	Zr3	180.0
2.273	N2	Zr3	Zr4(1,0)	103.3

COMMON NAME : Zr(0001)-(2x2)-O ILLUSTRATION: 59
CLASSIFICATION : 40.8.1
TECHNIQUE : LEED
AUTHORS : K.C. Hui, R.H. Milne, K.A.R. Mitchell, W.T. Moore and M.Y. Zhou
REFERENCE : Solid State Commun., 56, 83 (1985)

SURFACE TYPE
Substrate : Zr
Crystal face: 0001
Temperature : RT*
Bulk lattice: hcp
2D bulk symm: p3m1
2D surf symm: p3m1

Adsorbate: O
Coverage :
Pattern : p(2x2)
Matrix : (2.000, 0.000)
(0.000, 2.000)

STRUCTURE TYPE
O atoms in octahedral holes within fcc reconstructed Zr, with layer stacking AcBaCb... (O atoms in lower case); O atoms form (2x2) 0.25ML interstitial layers

SAMPLE PREPARATION (1 sample)
Treatment : exposure to 1L oxygen and anneal at 250C
Crystallinity:
Anal. methods:
Contamination: AES: C(274)/Zr(170)=0.05-0.1

COMMENTS
O intercalation expands bulk Zr-Zr spacing from 2.57 to 2.74Å

DATA COLLECTION
Technique: LEED
Dataset : IV curves for 6 beams at normal incidence

THEORY/DATA TREATMENT
Dynamical LEED: 8 phase shifts (band structure potential for Zr, Demuth superposition potential for O); Voi=-5 eV

STRUCTURES EXAMINED
Zr-Zr spacing fixed at 2.57Å; Zr-O spacing varied; O atoms located above surface, between 1st and 2nd layers, and in octahedral interstitial sites; hcp and fcc Zr both tried

QUALITY OF EXPERIMENT-THEORY FIT
RPE=0.305

2D UNIT CELLS (1 domain observed)

Cell	Ax (Å)	Ay (Å)	Bx (Å)	By (Å)	α (°)	Matrix	Pattern	Cell type
Bulk	3.230	0.000	1.615	2.797	60.0	(1.000, 0.000) (0.000, 1.000)	(1x1)	b: bulk lattice
Surface 1	6.460	0.000	3.230	5.595	60.0	(2.000, 0.000) (0.000, 2.000)	p(2x2)	s1: commens. superlattice

3D COORDINATES

Zr1: (1x1) top layer; O2 forms (2x2) underlayer between Zr1 and Zr3 (also 1x1);
Zr3-O4: repeating bulk set of layers (Zr: 1x1; O: 2x2); 0.05Å error bars assumed for tabulation

Dx/Dy in Å, or as a fraction of layer's unit cell vectors; atom 0 at the origin. Epir/subr are bulk repeat vectors.

No. of atoms: 4 Bulk z = 2.740 Å

Reg ion	Chem el.	At. no.	Cell type	Site occ.	Rel. to	Dx ± εx		Dy ± εy		Dz ± εz		Dz/Bz(%) ± εz/Bz
epir		-2					f		f		Å	
subr		-1				-6.460	Å	-3.730	Å	2.740	Å	
intf	Zr	1	b	1.00	0	0.000	f	0.000	f	0.000	Å	0.0
intf	O	2	s1	.25	1	0.667	f	0.667	f	1.370 ± .050	Å	50.0 ± 1.8
subl	Zr	3	b	1.00	2	0.333	f	-0.667	f	1.370 ± .050	Å	50.0 ± 1.8
subl	O	4	s1	.25	3	-0.833	f	-0.333	f	1.370 ± .050	Å	50.0 ± 1.8

BOND DISTANCES AND ANGLES

Bond distances and angles are derived from coordinates

No. of distances/angles: 3

Interatomic dist. A-B (Å)	Atom A	Atom B	Atom C	Bond angle A-B-C (°)
3.230	Zr1	Zr1(1,0)	O2(0,-1)	134.3
2.314	Zr1	O2(-1,-1)	Zr3(-2,0)	91.5
2.314	O2	Zr3	O4(1,1)	91.5

3. Acknowledgements

The authors are grateful to the National Institute of Standards and Technology, particularly to the Standard Reference Data Program, for undertaking the support of the electronic Surface Structure Database and of this atlas, which is its printed version. We are also much indebted to our consultants, Prof. J.B. Pendry and Prof. G.A. Somorjai, for helpful advice and much needed organizational support.

Much of the hard work of collecting data and putting them into the required format was accomplished by the individuals shown as "major contributors".

A considerable amount of compiling was also accomplished by the following persons, whose contributions were absolutely essential to the success of this task:

At Imperial College (London):

James MacLaren, Philip Rous, Dilano Saldin and Dimitri Vvedensky;

At Lawrence Berkeley Laboratory:

Simon Bare, Brian Bent, Gregory Blackman, Istvan Böszörmenyi, Mark Bussell, José Carrazza, Peter Ditlevsen, Morgan Edwards, Sabrina Fu, David Godbey, Brian Hadden, Ian Harrison, Michael Hilton, David Jentz, Chi-Tzu Kao, David Kelly, Colette Knight, Mark Levin, Kenneth Lewis, Bruno Marchon, Mathew Mate, Peter McAnally, Brian Naasz, Pedro Nascente, Roger Nix, Frank Ogletree, Hiroko Ohtani, Pedro Pereira, Jim Powers, Philip Rous, Thomas Rucker, Erik Sowa, Daniel Strongin, Gil Vandentop, Gerard Vurens, Kevin Williams and Mu-Liang Xu.

4. Bibliography and Related Software

4.1. References

J.M. MacLaren, J.B. Pendry, P.J. Rous, D.K. Saldin, G.A. Somorjai, M.A. Van Hove and D.D. Vvedensky, "Surface Crystallographic Information Service, A Handbook of Surface Structures, Software and User's Manual" (D. Reidel, Dordrecht) 1987.

J.F. Nicholas, "An Atlas of Models of Crystal Surfaces" (Gordon and Breach, New York) 1965.

H. Ohtani, C.-T. Kao, M.A. Van Hove and G.A. Somorjai, "A Tabulation and Classification of Clean Solid Surfaces and of Adsorbed Atomic and Molecular Monolayers as Determined from Low-Energy Electron Diffraction", Progr. Surf. Sci. **23**, 155 (1986).

P.R. Watson, M.A. Van Hove and K. Hermann, SSD electronic database: "NIST Surface Structure Database – Ver. 1", NIST Standard Reference Data Program, Gaithersburg, Maryland (1993).

P.R. Watson, "Critical Compilation of Surface Structures Determined By LEED Crystallography" J. Phys. Chem. Ref. Data **16**, 953 (1987).

P.R. Watson, "Critical Compilation of Surface Structures Determined By Ion Scattering Compared with Those from LEED Crystallography" J. Phys. Chem. Ref. Data **19**, 85 (1990).

P.R. Watson, "Critical Compilation of Surface Structures Determined By SEXAFS Compared with Those from LEED and Ion Scattering" J. Phys. Chem. Ref. Data **21**, 123 (1992).

E.A. Wood, "The 80 Diperiodic Groups in Three Dimensions", Bell System Techn. Journ. **43** (Part 2), 541 (1964).

4.2. Software

SCIS (Surface Crystallographic Information Service)
marketed by Reidel Publishing (Dordrecht 1987, The Netherlands, or Kluwer Academic Publishers Group, USA).

SARCH/LATUSE/PLOT3D (Surface ARCHitect / LATtice USEr / PLOT 3 Dimensional systems)
to obtain copies contact either M.A. Van Hove or K. Hermann (see Appendix G).

BALSAC (Build and Analyze Lattices, Surfaces, and Clusters)
to obtain copies contact K. Hermann (see Appendix G).

5. Appendices

Appendix A: Entering New Structural Data into SSD

A.1. Submitting New Structures for Inclusion in SSD

Procedure

You may submit a new structure for inclusion in the next update of SSD. Any structure should satisfy the criteria spelled out in the Introduction (Sec. 1.2), but may have been published at any time.

The first step is to contact us, the database authors, for a decision on inclusion. You should send us a copy of a refereed and published paper that describes the structure determination and the result. Once accepted, the preferred form of submission is as an ASCII file, described in this Appendix. Such a structure should be sent in an electronic file, either on diskette or by electronic mail, to one of the authors' addresses shown in Appendix G.

Electronic submission

New structures for SSD are first entered into a file with the ASCII format. Then we, the database authors, translate this file into files with a non-ASCII Paradox format for general distribution and for publication in book form.

An ASCII file can be created and modified with any text editor or word processor. Thus, the data can be manually put in the desired form and entered into a file. The ASCII files used for SSD can also be written to (and read by) the independently marketed PC-based program SARCH (described in Sec. 1.5). This allows structures to be generated or modified interactively. Note that SARCH can only generate the structural data items (coordinates, etc.), but not the textual data items; these must be subsequently input manually with a text editor or word processor.

In Appendix A.2 we give a detailed definition of the ASCII file format used for SSD. Appendix A.3 shows a generic ASCII file for reference, while A.4 gives actual sample ASCII files. In A.5 we describe a utility program, called SSDCHK, available from us, with which you can test such an ASCII file for correct formatting and structural content.

A.2. Format of ASCII Files for SSD

This appendix specifies the detailed format of the ASCII files used to enter structures into SSD. The detailed rules for the file format are specified first. They are followed by a generic example in A.3, which serves as a guide when entering data items and which also indicates the length limits on each data item. Several actual examples are shown in A.4. The meanings of the individual data items are defined in 1.4.

Rules For ASCII File Format:

a. All data items must appear in the order given by the generic format given in A.3;
b. Each data item must appear either as a "free format" textual item or as a "fixed format" tabular item, as detailed in the generic format shown in A.3;
c. No line of information may exceed 80 characters in total, including all SSD-specific characters;
d. A blank line in the free format sections is ignored (it is not ignored in the fixed format tabular data sections);
e. A line can start with a command (##, #c, #tx or #fx, where x = a, b, c, d, e or f); all characters following such a command on the same line are ignored;
f. Lines without commands can contain either one free format item, or only fixed format items: no mixing of free and fixed format items is allowed on the same line;
g. The characters ## are mandatory and signal the beginning of a new structure: it must appear at the beginning of a new line;
h. The characters #ff are mandatory and signal the last line of a structure (see also #fx below). It must appear at the beginning of a new line;
i. The characters #c, at the beginning of a line, signal a comment line in a free format section (but a comment line may not appear between the blocks of data #tc − #fc and #td − #fd); the line is ignored. A comment line may NOT be embedded in tabular data;
j. The characters #tx (where x = a, b, c, d, e or f), at the beginning of a line, signal the beginning of a section of tabular data, using fixed formats; #tx must be paired with a subsequent #fx (with the same value of x: see next); characters following #tx on the same line are ignored, i.e. the actual data must start on the next line;
k. The characters #fx (where x = a, b, c, d, e or f), at the beginning of a line, signal the end of a section of tabular data, and the return to free format items. Characters following #fx on the same line are ignored, i.e. a free format item must start on a subsequent line;
l. Tabular data have a fixed format, i.e. each data item must fit within a specific range of columns, as displayed in the generic format in A.3:

   • the six blocks of tabular data (#ta-#fa through #tf-#ff) are mandatory; none may be left out;

   • numerical values in tabular data may be integer or real, but may NOT have exponential form; the number of digits after the decimal point is free;

   • in alphanumeric fields leading and trailing blanks will be ignored;

   • a blank numerical field means no number (rather than a zero number), while a blank alphanumeric field will be ignored;

   • tabs may not be used (since tab positions vary among word processors and text editors);

   • reminder: a group of lines containing tabular data must be preceded by a line starting with #tx (x = a, b,... or f), and must be terminated by a line starting with #fx (with the same value of x);

m. The entry ''xn:aaa'' in ''free format'' data is one item containing one field of data aaa; only one item is allowed per line; no item may be split over more than one line (except s5, see below):

aaa can be any string of characters (excepting the double quote '' and special ASCII characters, see below); the string may have any length up to the limit given below in the generic format; aaa can be null, i.e. ''xn:'' is allowed (or preferably omitted) and implies an empty field, meaning no character or no number (SSD will show a blank);

n. the labels xn must always be included in alphanumeric fields and must take on the exact values shown in the generic form below (e.g., s1 for the Name of the structure): the data will otherwise not be recognized; x in xn can only have the values: s (= Outline section), t (= Technique section), 2d (= 2D sections), 3d (= 3D section), b (= Bond distances and angles section); n in xn takes the values 1, 2, 3,...;

• the free format data ''xn:aaa'' must appear in the order shown in the generic format below: e.g., s5 cannot precede s4, and t2 cannot precede s7;

• items may be omitted: e.g., t6 may immediately follow t4, if t5 is omitted; if the item ''xn:aaa'' is absent, aaa is assumed null;

• in the generic format given in Appendix A.3, Ann and Nnn show the maximum allowed length of each field: up to nn characters are allowed; Ann allows alphanumeric characters and the special characters given below; Nnn represents a numeric field for real numbers or integers, i.e. digits with or without a decimal point (no exponent is allowed);

o. Item s5 (Authors) allows up to 120 characters by splitting it into two smaller items on successive lines, each limited to 60 characters: it is then composed of two items ''s5:first authors'' and ''s5:more authors'', with the same s5 label; the two fields will be joined by SSD and a blank space inserted between them (so a name should not be split among the two fields); if even more authors exist than will fit in 120 characters, they are inserted in the comments items [s28-s32]; however, these author names will not be searchable;

p. Authors' names may not include accents of any kind: accents should be omitted;

Note: the umlaut (double dotted vowel) in German names may also be replaced by an extra letter e: thus, Muller and Mueller are both acceptable (with the umlaut omitted in the first form): to search for such names in the SSD database, it is best to try both spellings;

q. Special characters: some special ASCII characters may be used in free-format items, but only as coded in the table below; special ASCII characters may NOT be entered as 3 digits while keeping the **ALT** key pressed simultaneously, as such characters may be corrupted or lost in file conversion and transmission, e.g., in electronic mail;

powers of 10 should be denoted as, for example: 1.5E-10;

the following special characters should be represented either by the corresponding fully-spelled-out 6-character sequence Altnnn, or preferably by the 2-character symbol of type \x (the latter requires less space):

$\approx$	Alt247	\= (approximate)
$\propto$	Alt224	\a (i.e. same as α)
$\pm$	Alt241	\+
∞	Alt236	\i
^{2}	Alt253	\2
$\surd$	Alt251	\R
$^\circ$	Alt248	\0 (0 = zero)
Å	Alt143	\A
α	Alt224	\a (l.c. = lower case)
β	Alt225	\b
Γ	Alt226	\G (u.c. = upper case = capital)
δ	Alt235	\d
ϵ	Alt238	\e
η	Alt252	\E
ϕ	Alt237	\f
Φ	Alt232	\F
μ	Alt230	\m
Π	Alt227	\P (l.c. is not available: use u.c. pi)
σ	Alt229	\s
Σ	Alt228	\S
τ	Alt231	\T
θ	Alt233	\t
Ω	Alt234	\O (O = u.c. letter O)

r. The minimum content for a meaningful structure is:
##
field ''s1:structure name''
all data blocks (#ta-#fa through #tf-#ff)
items [2d8], [2d14]

A.3. Generic ASCII File

The following pages list an ASCII file in which items have been inserted that describe the expected type of entries. These items also give the maximum field lengths in the forms Ann for alphanumeric items, and Nnn for numerical items. (Here, the numbers nn give the allowed number of characters for each item). The comment lines preceded by #c are used to provide additional information; this includes the last allowed column for the last character of subsequent textual items, marked by the symbol "=", and definitions of subsequent tabular items.

This user data submission file format can be downloaded from the NIST Standard Reference Data *anonymous ftp* site as follows:

ftp srd.nist.gov
login: anonymous
password: guest
cd forms
get format.SSD

Questions can be e-mailed to srdp@enh.nist.gov.

Generic ascii file:

```
#c-------------------------------------------------------------------
#c this is a comment line for file identification and date of
#c last change
## this command marks the beginning of a new structure
#c this bar shows allowed length 40:     >=
"s1:Name A40"
#c allowed length 20  >=
"s2:Class. No. A20"
#c                                        >=
"s3:Status A40"
#c                    >=
"s4:Technique A20"
#c                                                            >=
"s5:Authors A60"
#c                                                            >=
"s5:Authors A60"
#c                                        >=
"s6:Journal A40"
#c        >=
"s7:Vol A8"
#c     >=
"s8:Page A5"
#c          >=
"s9:Year N10"
#c                    >=
"s10:Substrate A20"
#c                    >=
"s11:Bulk lattice A20"
#c          >=
"s12:Cryst. face A10"
#c     >=
"s13:2D bulk symm. A4"
#c     >=
"s14:2D surf. symm. A4"
#c                    >=
"s15:Adsorbate A20"
#c                    >=
"s16:Coverage A20"
#c                    >=
"s17:Pattern A20"
#c s18:Matrix M11 N10, s19:M12 N10, s20:M21 N10, s21:M22 N10
#ta
        nn        nn        nn        nn
#fa
#c     >=
"s22:Temp. A5"
#c                                                            >=
"s23:Struct. type-1 A60"
"s24:-2 A60","s25:-3 A60","s26:-4 A60"
"s27:-5 A60"
#c                                                            >=
"s28:Comments-1 A60"
"s29:Comments-2 A60"
"s30:Comments-3 A60"
"s31:Comments-4 A60"
```

```
"s32:Comments-5 A60"
#c         >=
"s33:Illustration A8" not used

#c  >=
"t1:No. samples A2"
#c                                                              >=
"t2:Treatment A60"
#c                                            >=
"t3:Crystallinity A40"
#c                                                              >=
"t4:Analyt. methods A60"
#c                                            >=
"t5:Contamination A40"
#c                                            >=
"t6:Data colln. method A40"
#c                                                              >=
"t7:Data set-1 A60"
"t8:Data set-2 A60"
#c                                                              >=
"t9:Theory-1 A60"
#c                                                               >=
"t10:Theory-2 A60"
#c                                                               >=
"t11:Structs. examined-1 A60"
"t12:Structs. examined-2 A60"
"t13:Structs. examined-3 A60"
"t14:Structs. examined-4 A60"
"t15:Structs. examined-5 A60"
#c                                             >=
"t16:Fit A40"


#c 2d1:Bulk Ax N10, 2d2:Ay N10, 2d3:Bx N10, 2d4:By N10, 2d5:alpha N10
#c 2d6:No. domains N4, 2d7:No. cells (<=4) N4"
#tb
  nnn.nnnn  nnn.nnnn  nnn.nnnn  nnn.nnnn  nnn.nnnn
  nn  nn
#fb
#c                                             >=
"2d8:Cell-i (i=1, 2, ...) A40"
#c 2d9:C1 Ax N10, 2d10:C1 Ay N10, 2d11:C1 Bx N10, 2d12:C1 By N10"
#c 2d13:C1 alpha N10"
#tc
  nnn.nnnn  nnn.nnnn  nnn.nnnn  nnn.nnnn  nnn.nnnn
#fc
#c                  >=
"2d14:Ci pattern A20"
#c 2d15:Ci matrix M11 N10, 2d16:M12 N10, 2d17:M21 N10, 2d18:M22 N10
#td
        nn         nn         nn         nn
#fd

#c repeat 2d8-2d18 as many times as No. of cells for Cell-2 etc.
(see 2d7)

#c                                                              >=
```

```
"3d1:3D notes-1 A60"
"3d2:3D notes-2 A60"
"3d3:3D notes-3 A60"
"3d4:3D notes-4 A60"
#c 3d5: not used
#c 3d6:Bulk z N10
#c 3d7:No. of atoms (excl. epir and subr entries) N4
#c 3d8:Region A4, 3d9:Chem. el. A3, 3d10:Atom # N5, 3d11:Cell type A4
#c 3d12:Site occ. N6, 3d13:Rel. to N4, 3d14:Dx N10, 3d15:Units Dx A1
#c 3d17:Dy N10, 3d18:Units Dy A1, 3d20:Dz N10, 3d22:Dz/Bz N10
#c 3d16:Err. Dx N10, 3d19:Err. Dy N10
#c 3d21:Err. Dz N10, 3d23:Err. Dz/Bz N10
#te
  nnn.nnnn
  nn
epir      -2                 nnn.nnnnA  nnn.nnnnA  nnn.nnnn

subr      -1                 nnn.nnnnA  nnn.nnnnA  nnn.nnnn

regn el   nn cel n.nnn  nn   nnn.nnnna  nnn.nnnna  nnn.nnnn  nnn.nnnn
                             nnn.nnnna  nnn.nnnna  nnn.nnnn  nnn.nnnn
#c repeat 3d8-3d23 (last two lines) as many times as 3d7 indicates
#fe

#c                                                                >=
"b1:Dist/angles notes A60"
#c b2:No. of dist & angl N4
#c b3:Dist.A-B N10
#c b4:Atom-A A12, b5:-B A12, b6:-C A12
#c b7:Angle A-B-C N10
#tf
  nn
  nnn.nnnn  aaaaaaaaaa  aaaaaaaaaa  aaaaaaaaaa  nnn.nnnn
#c repeat b3-b7 (last line) as many times as b2 indicates
#ff this marks the end of a structure

## this marks the beginning of a new structure
"s1:Other name" starts next structure
#c etc.
#c----------------------------------------------------------------
```

A.4. Sample ASCII Files

The following five structures are put in the form of ASCII files for SSD and serve to illustrate the format and the style of data inclusion. They are included with the SSD softwae in file TEST.ASD.

```
Sample ASCII files

##---------------------------------------------------------------------
"s1:Al(311)-(1x1)"
"s2:13.30"
"s4:LEED"
"s5:J.R. Noonan, H.L. Davis, W. Erley"
"s6:Surf. Sci."
"s7:152/153"
"s8:142"
"s9:1985"
"s10:Al"
"s11:fcc"
"s12:(311)"
"s13:cm"
"s14:cm"
"s17:(1x1)"
#ta
    1.0000     0.0000     0.0000     1.0000
#fa
"s22:298K"
"s23:bulk termination with multilayer relaxation;"
"s24:no detectable lateral relaxation"

"t1:1"
"t2:electropolish in H2SO4/H3PO4; Ar+ sputter; 500C anneal"
"t5:AES: <0.05% ML Si"
"t6:Faraday cup"
"t7:34 LEED beams (21 symmetry-inequivalent); energy range"
"t8:50-300eV; normal incidence within 0.5\0"
"t9:dynamical LEED (Reverse Scattering Perturbation): Moruzzi-"
"t10:Janak-Williams potential; Voi=4.75eV; \tD=550K"
"t11:relaxation of top two interlayer spacings and lateral"
"t12:displacement of top layer"
"t16:RZJ=0.07, R2=0.083"

#tb
    2.8600     0.0000    -1.4319     4.7488   106.7800
   1   1
#fb
"2d8:s1"
#tc
    2.8600     0.0000    -1.4319     4.7488   106.7800
#fc
"2d14:(1x1)"
#td
    1.0000     0.0000     0.0000     1.0000
#fd
"3d1:0.05\A error bars assumed for tabulation of lateral"
"3d2:relaxation"
```

```
#te
    1.2270
   4
epir      -2

subr      -1                    0.0025A    -2.5881A     1.2270

intf Al    1 b  1.0000   0      0.0000f     0.0000f     0.0000        0.00

intf Al    2 b  1.0000   1      0.7270f     0.4540f     1.0680       87.04
                                0.0228f     0.0106f     0.0100        0.82
intf Al    3 b  1.0000   2     -0.2730f     0.4550f     1.3350      108.80
                                                        0.0200        1.63
subl Al    4 b  1.0000   3     -0.2720f    -0.5450f     1.2270      100.00

"b1:bond distances and angles are derived from coordinates"
#tf
  14
   2.8600 Al1          Al1(1,0)    Al2              59.26
   2.8600 Al1          Al1(1,0)    Al3(0,-1)        59.62
   2.7984 Al1          Al2         Al1(1,0)         61.45
   2.7984 Al1          Al2         Al3             119.65
   2.7993 Al1          Al2(-1,0)   Al1(0,1)        124.54
   2.7993 Al1          Al2(-1,0)   Al3             119.70
   2.8042 Al2          Al1(1,1)    Al3              62.33
   2.7993 Al2          Al1(1,0)    Al3(0,-1)        62.30
   2.9136 Al2          Al3(1,0)    Al2(1,0)         58.76
   2.9159 Al2          Al3         Al1(0,1)         88.42
   2.9159 Al2          Al3         Al3(1,0)         60.58
   2.8301 Al3          Al1(1,1)    Al3(1,0)         60.71
   2.9121 Al3          Al2(0,1)    Al3(0,1)        116.65
   2.9159 Al3          Al2         Al3(1,0)         58.76
#ff

##------------------------------------------------------------------------
"s1:Ti(0001)-(1x1)-N"
"s2:22.7.2"
"s4:LEED"
"s5:H.D. Shih, F. Jona, D.W. Jepsen and P.M. Marcus"
"s6:Surf. Sci."
"s7:60"
"s8:445"
"s9:1976"
"s10:Ti"
"s11:hcp"
"s12:(0001)"
"s13:p3m1"
"s14:p3m1"
"s15:N"
"s16:1.0 N/Ti"
"s17:(1x1)"
#ta
    1.0000     0.0000     0.0000     1.0000
#fa
"s22:300K"
"s23:atomic interstitial in octahedral sites between first and"
"s24:second Ti layers; slight expansion of Ti-Ti spacing;"
```

```
"s25:forms trilayer of TiN compound exposing (111) face"

"t1:2"
"t2:cleaned, then N introduced at 1.5E-8 torr"
"t5:monitored by AES and LEED"
"t7:LEED I-V spectra: 3 beams at normal inc., 5 beams at \t=8\0,"
"t8:\f=-30\0; (00) beam at \t=20\0, \f=-30\0; 20<E<250eV"
"t9:dynamical LEED (layer KKR): 8 phase shifts (Moruzzi et al);"
"t10:domain-averaging over steps; Vor=-10eV; Voi=-3eV; \tD=342K"
"t11:1. atomic or molecular (N-N=1.098\A) adsorption in top,"
"t12:bridge and either one or both 3-fold hollow sites;"
"t13:2. one of the 2 types of tetrahedral or octahedral"
"t14:interstitial sites between the 1st and 2nd Ti layers"
"t15:(Ti-N spacings varied in range 0.9-2.8\A); step-averaged"
"t16:visual"

#tb
    2.9500     0.0000    -1.4750     2.5548   120.0000
   1   1
#fb
"2d8:s1"
#tc
    2.9500     0.0000    -1.4750     2.5548   120.0000
#fc
"2d14:(1x1)"
#td
    1.0000     0.0000     0.0000     1.0000
#fd
"3d1:N2: underlayer in octahedral site of slightly expanded"
"3d2:    Ti(0001) lattice;"
"3d3:Ti1-N2-Ti3: trilayer of TiN;"
"3d4:Ti4-Ti5: repeating bulk pair of layers"
#te
    2.3400
   5
epir      -2

subr      -1                     0.0000A     0.0000A     4.6800

intf Ti    1 b  1.0000   0     0.0000f     0.0000f     0.0000       0.00

intf N     2 b  1.0000   1     0.6667f     0.3333f     1.2200      52.14
                                                       0.0500       2.13
intf Ti    3 b  1.0000   2    -0.3333f     0.3333f     1.2200      52.14
                                                       0.0500       2.13
subl Ti    4 b  1.0000   3    -0.3333f    -0.6667f     2.3400     100.00

subl Ti    5 b  1.0000   4     0.3333f     0.6667f     2.3400     100.00

#fe
"b1:bond distances and angles are derived from coordinates"
#tf
   8
   2.9500 Ti1          Ti1(1,1)    N2             45.25
   2.9500 Ti1          Ti1(1,1)    Ti3            60.28
   2.0951 Ti1          N2          Ti1(1,0)       89.50
   2.0951 Ti1          N2          Ti3            90.50
```

```
   2.9756 Ti1          Ti3          Ti1(0,1)         59.43
   2.9756 Ti1          Ti3          N2               44.75
   2.9756 Ti1          Ti3          Ti3(0,1)        119.72
   2.0951 N2           Ti3          Ti3(0,1)        134.75
#ff

##-------------------------------------------------------------------------
"s1:Ni(100)-c(2x2)-CO"
"s2:28.6.8.6"
"s4:PED"
"s5:L.G. Petersson, S. Kono, N.F.T. Hall, C.S. Fadley and"
"s5:J.B. Pendry"
"s6:Phys. Rev. Lett."
"s7:42"
"s8:1545"
"s9:1979"
"s10:Ni"
"s11:fcc"
"s12:(100)"
"s13:p4m"
"s14:p4m"
"s15:CO"
"s16:1/2 CO/Ni"
"s17:c(2x2)"
#ta
    1.0000    1.0000   -1.0000    1.0000
#fa
"s23:molecular on-top adsorption, C bonding to Ni"
"s28:average orientation of CO is determined to be within 12\0 of"
"s29:normal"

"t1:1"
"t2:CO adsorbed to 2.0L (see PRL 41, 117 & 41, 1831 (1978))"
"t5:checked by ARXPS: <3%ML C"
"t6:angle resolved x-ray photoemission"
"t7:polar-angle scans of C(1s) and O(1s) intensities"
"t9:single-scattering calculations for both a single CO molecule"
"t10:and a finite cluster"
"t11:C-O and Ni-C distance were assumed 1.15\A and 1.8\A, resp.;"
"t12:various C-O tilt angles tested;"
"t13:bulk Ni layer spacings assumed"
"t16:visual"

#tb
    2.4890    0.0000    0.0000    2.4890   90.0000
   1   1
#fb
"2d8:s1"
#tc
    2.4890    2.4890   -2.4890    2.4890   90.0000
#fc
"2d14:c(2x2)"
#td
    1.0000    1.0000   -1.0000    1.0000
#fd
"3d1:O1-C2: upright on-top molecule (C bonded to Ni3)"
```

```
#te
   1.7600
  4
epir      -2

subr      -1                    1.2445A    1.2445A    1.7600

ovrl O     1 s1 0.5000   0      0.0000f    0.0000f    0.0000     0.00

ovrl C     2 s1 0.5000   1      0.0000f    0.0000f    1.1500    65.34
                                0.0875f    0.0875f
intf Ni    3 b  1.0000   2      0.0000f    0.0000f    1.8000   102.27

subl Ni    4 b  1.0000   3      0.5000f    0.5000f    1.7600   100.00

#fe
"b1:bond distances and angles are derived from coordinates"
#tf
  4
  1.1500 O1          C2          Ni3           180.00
  1.8000 C2          Ni3
  2.4890 Ni3         Ni3(1,0)
  2.4890 Ni3         Ni4
#ff

##-------------------------------------------------------------------
"s1:Si(111)-(\R3x\R3)R30\0-Ga"
"s2:14.31.4"
"s4:LEED"
"s5:A. Kawazu and H. Sakama"
"s6:Phys. Rev."
"s7:B37"
"s8:2704"
"s9:1988"
"s10:Si"
"s11:diamond"
"s12:(111)"
"s13:p3m1"
"s14:p31m"
"s15:Ga"
"s16:0.3 Ga/Si"
"s17:(\R3x\R3)R30\0"
#ta
   1.0000    1.0000   -2.0000    1.0000
#fa
"s23:atomic adsorption in 4-fold coordinated T4 'top' site over"
"s24:top bilayer, with relaxations down into 2nd bilayer"
"s28:R-factors for the structures examined were, respectively:"
"s29:0.34, 0.45, 0.45, 0.25, 0.15"

"t1:1"
"t2:Ga molecular beam from Knudsen cell at 740K"
"t3:sharp LEED pattern"
"t7:I-V spectra: 9 beams at normal incidence, 30<E<190eV"
"t9:dynamical LEED"
"t11:Ga atoms in hollow site above the 4th layer Si; Ga atoms on"
"t12:3 atom Si clusters centered above 2nd layer Si; clusters"
```

```
"t13:centered above 4th layer Si; Ga substituted for 1/3 of top"
"t14:layer Si; T4 'top' site; in each case structural parameters"
"t15:were varied (see comments)"
"t16:RZJ=0.15"

#tb
    3.8394     0.0000     1.9197     3.3250    60.0000
   1   1
#fb
"2d8:s1"
#tc
    5.7591     3.3250    -5.7591     3.3250   120.0000
#fc
"2d14:(\R3x\R3)R30\0"
#td
    1.0000     1.0000    -2.0000     1.0000
#fd
"3d1:Ga1: in T4 'top' site over Si7;"
"3d2:Si2-Si7 and Si8-Si13: relaxed top 2 bilayers;"
"3d3:Si14-Si15: repeating bulk substrate layers;"
"3d4:0.1\A error bars assumed for tabulation"
#te
    3.1300
  15
epir       -2

subr       -1                     1.9197A    1.1083A    3.1300

ovrl Ga     1 s1 0.3333   0       0.0000f    0.0000f    0.0000         0.00
                                  0.0237f    0.0174f    0.1000         3.19
intf Si     2 s1 0.3333   1       0.3158f    0.3158f    1.3500        43.13
                                  0.0237f    0.0174f    0.1000         3.19
intf Si     3 s1 0.3333   2       0.3684f   -0.3158f    0.0000         0.00
                                  0.0237f    0.0174f    0.1000         3.19
intf Si     4 s1 0.3333   3      -0.6842f    0.6842f    0.0000         0.00
                                  0.0237f    0.0174f    0.1000         3.19
intf Si     5 s1 0.3333   4       0.6667f   -0.3509f    0.5800        18.53
                                  0.0237f    0.0174f    0.1000         3.19
intf Si     6 s1 0.3333   5      -0.3333f    0.3333f    0.0000         0.00
                                  0.0237f    0.0174f    0.1000         3.19
intf Si     7 s1 0.3333   6      -0.3333f   -0.6667f    0.6400        20.45
                                  0.0237f    0.0174f    0.1000         3.19
intf Si     8 s1 0.3333   7       0.6667f    0.3333f    1.8000        57.51
                                  0.0237f    0.0174f    0.1000         3.19
intf Si     9 s1 0.3333   8      -0.3333f    0.3333f    0.0000         0.00
                                  0.0237f    0.0174f    0.1000         3.19
intf Si    10 s1 0.3333   9      -0.3333f   -0.6667f    0.3400        10.86
                                  0.0237f    0.0174f    0.1000         3.19
intf Si    11 s1 0.3333  10       0.3414f    0.0000f    0.5100        16.29
                                  0.0237f    0.0174f    0.1000         3.19
intf Si    12 s1 0.3333  11      -0.3414f    0.3414f    0.0000         0.00
                                  0.0237f    0.0174f    0.1000         3.19
intf Si    13 s1 0.3333  12       0.6586f    0.3172f    0.0000         0.00
                                  0.0237f    0.0174f    0.1000         3.19
subl Si    14 b  1.0000  10       0.3333f    0.3333f    2.8600        91.37
                                  0.0110f    0.0301f    0.1000         3.19
```

```
subl Si    15 b  1.0000  14    0.3333f   -0.6667f     0.7800       24.92
                               0.0110f    0.0301f     0.1000        3.19
#fe
"b1:bond distances and angles are derived from coordinates"
#tf
   5
   2.4966 Ga1         Si2         Si5           124.89
   2.5700 Ga1         Si7         Si2            59.85
   2.3499 Si2         Si5         Si3           114.12
   2.3499 Si2         Si6         Si9           104.29
   2.4287 Si2         Si7         Si10          120.15
#ff

##---------------------------------------------------------------------
"s1:Si(111)-(1x1)-NiSi2(111) interface"
"s2:14.28.2"
"s4:HEIS"
"s5:E.J. van Loenen, J.W.M. Frenken, J.F. van der Veen and"
"s5:S. Valeri"
"s6:Phys. Rev. Lett."
"s7:54"
"s8:827"
"s9:1985"
"s10:Si"
"s11:diamond"
"s12:(111)"
"s13:p3m1"
"s14:p3m1"
"s15:NiSi2"
"s16:epilayer"
"s17:(1x1)"
#ta
    1.0000    0.0000    0.0000    1.0000
#fa
"s23:epitaxial (1x1) multilayer; ideal 3D fit of top Si bilayer"
"s24:against bottom NiSi2 trilayer with Si-Si bonds perpendicular"
"s25:to interface"
"s28:only the interfacial separation 0.75+2.31=3.06\+0.08\A was"
"s29:measured; the sum is here decomposed in proportion to the"
"s30:ideal values of 0.77+2.35=3.12\A"

"t1:1"
"t2:25\A epilayer from Ni deposition on (7x7), then annealed"
"t5:Si(111)(7x7) clean by AES and ISS"
"t6:RBS of 100keV He ions with channeling"
"t7:energy- and angle-dependent yield around two off-normal"
"t8:channel directions"
"t9:Monte Carlo simulation"
"t11:interface between ideal bulk-like lattices with two"
"t12:possible registries and variable interfacial layer"
"t13:spacing"
"t16:visual"

#tb
    3.8380    0.0000    1.9190    3.3238   60.0000
   1   1
#fb
```

```
"2d8:s1"
#tc
    3.8380     0.0000     1.9190     3.3238    60.0000
#fc
"2d14:(1x1)"
#td
    1.0000     0.0000     0.0000     1.0000
#fd
"3d1:Si1-Ni2-Si3: repeating bulk NiSi2 epilayer;"
"3d2:Si4-Ni5-Si6: bottom NiSi2 trilayer;"
"3d3:Si7-Si8: top Si bilayer;"
"3d4:Si9-Si10: repeating bulk Si bilayer"
#te
    3.1300
  10
epir      -2                        3.8380A    2.2159A    3.0800

subr      -1                        1.9190A    1.1079A    3.1300

epil Si    1 b  1.0000   0      0.0000f    0.0000f    0.0000        0.00

epil Ni    2 b  1.0000   1      0.6667f    0.6667f    0.7700       24.60

epil Si    3 b  1.0000   2     -0.3333f   -0.3333f    0.7700       24.60

intf Si    4 b  1.0000   3      0.3333f    0.3333f    1.5400       49.20

intf Ni    5 b  1.0000   4     -0.3333f   -0.3333f    0.7700       24.60

intf Si    6 b  1.0000   5     -0.3333f   -0.3333f    0.7500       23.96
                                                      0.0800        2.55
intf Si    7 b  1.0000   6      0.0000f    0.0000f    2.3100       73.80
                                                      0.0800        2.55
intf Si    8 b  1.0000   7      0.3333f    0.3333f    0.7800       24.92

subl Si    9 b  1.0000   8      0.0000f    0.0000f    2.3500       75.08

subl Si   10 b  1.0000   9      0.3333f    0.3333f    0.7800       24.92

#fe
"b1:bond distances and angles are derived from coordinates"
#tf
  12
    2.3458 Ni2         Si3(1,0)     Si4             53.96
    2.3458 Ni2         Si3(1,0)     Ni5(1,0)       109.16
    2.3100 Ni2         Si4          Si3             55.20
    2.3100 Ni2         Si4          Ni5            109.16
    2.3100 Ni2         Si4          Si6(1,0)       124.45
    2.6985 Si3         Si4          Si3(1,0)        90.66
    2.6985 Si3         Si4          Ni5             53.96
    2.6985 Si3         Si4          Si6(1,0)        89.10
    2.3100 Si3         Ni5          Si4             70.84
    2.3100 Si3         Ni5          Si6            108.70
    2.3400 Ni5         Si6
    2.3100 Si6         Si7
#ff
```

A.5. Utiility Program SSDCHK

Available from the authors is a utility program, called SSDCHK, with which you can test an ASCII file into which you have entered one or more new structures, using the format described earlier in this Appendix.

To run this program, make sure that the ASCII file to be checked is in the same subdirectory as SSDCHK and type:

SSDCHK

SSDCHK first asks for the name of the ASCII file containing the new structure(s). As an example you may use the TEST.ASD (TEST.ASD is supplied with the SSD software), even though it contains no errors to be detected. We recommend using file names ending in .ASD for such ASCII files. SSDCHK then tests the ASCII file for conformance with the file format rules given in Appendix A.2: item length limits, order of items, inclusion of mandatory items, presence and position of tabular blocks, etc. However, SSDCHK does not check the tabular formatting (like column positions). Nor does it check the contents of tabular blocks for geometrical consistency. The graphical presentation of the structures should be used for this purpose.

If SSDCHK finds errors, these are output to and described in file SSDCHK.ERR, which can be viewed interactively. (SSDCHK.ERR can also be printed after termination of SSDCHK to help find and correct the errors in the ASCII file).

If SSDCHK finds at least one satisfactory structure in the ASCII file, it asks you whether to visualize those structures on the screen. Visualizing allows you to use the various viewing and analysis features of the SSD graphics program SURVIS to check symmetry, atomic sites, bonding neighbors, bond lengths, bond angles, etc.

After SSDCHK is terminated, one should correct any errors in the ASCII file with a word processor or text editor. Then SSDCHK should be used again to check the corrected file.

Appendix B: List of Symbols and Abbreviations

General:

**	power (e.g., E**2 is E squared)
//	parallel (to surface)
α	exchange parameter (Xα Slater local exchange potential)
θ	polar incidence angle (of electron, photon)
θ_D	Debye temperature
λ	wavelength
ϕ	azimuthal incidence angle
Ω	ohm
at.	atomic
AV	(spin-polarized) asymmetry vs. E (voltage) curve
bcc	body-centered cubic
bct	body-centered tetragonal
class. no.	SSD structure classification number
cov.	coverage
cum.	cumulative
E	energy
(eds.)	editors' comment
Enn (nn=signed digits)	10 to the power nn
eV	electron volt
fcc	face-centered cubic
H	Hartree (atomic energy unit; 1 H = 27.21161 eV)
hcp	hexagonal close-packed
H-S	Herman-Skillman (wavefunction tabulation)
inv.	inversion
IV, I-V	intensity vs. energy (voltage) curve
KKR	Korringa-Kohn-Rostoker method
L	Langmuir
mfp	mean free path
ML	monolayer
m.s.	mean square, multiple scattering
para.	parallel (to surface)
perp.	perpendicular (to surface)
ph. shs.	phase shifts
pot.	potential
PRB	Phys. Rev. B journal
PRL	Phys. Rev. Lett. journal
PV	(spin-)polarization vs. E (voltage) curve
R	R-factor
R2	x-ray R-factor
rel., relat.	relativistic
rf	radio frequency
RI	x-ray R-factor
rms	root mean square

rms ampl, rms vibr. ampl.	root mean square vibration amplitude
RPE	Pendry R-factor
RR	Pendry RR-factor (relative R-factor)
RT	room temperature
RT*	assumed room temperature
RVH, RVHT	Van Hove/Tong R-factor
Ryd	Rydberg (atomic energy unit; 1 Ryd=13.59 eV)
RZJ	Zanazzi-Jona R-factor
SS	Surface Science journal
UHV	ultra-high vacuum
<u*u>	mean square vibration amplitude
vibr. ampl.	vibration amplitude
Voi	imaginary part of inner potential (optical potential)
Vor	real part of inner potential (muffin-tin zero)
w.f.	wave function
w.r.t.	with respect to

Techniques

AED	Auger Electron Diffraction
AFM	Atomic Force Microscopy
ALICISS	Alkali ICISS
At. diffr.	Atom Diffraction
At. scatt.	Atom Scattering
ARAES	Angular Resolved Auger Electron Spectroscopy
ARPES, ARUPS, ARXPD, ARPEFS	Angular Resolved (Ultraviolet / X-ray) Photoelectron Spectroscopy / Fine Structure
BSN	Beam-Set Neglect
CMA	Cylindrical Mirror Analyzer
CMTA	Constant-Momentum Transfer Averaging
CSM	Combined Space Method
CWMS	Curved-Wave Multiple Scattering
DLEED	Diffuse LEED
EAPFS	Electron Appearance Potential Fine Structure
EELS	Electron Energy Loss Spectroscopy
EH	Electron Holography
EM	Electron Microscopy
ESDIAD	Electron Stimulated Desorption Ion Angular Distribution
EXAFS	Extended X-ray Absorption Fine Structure
EXELFS	Extended Electron Energy Loss Fine Structure
EXFAS	Extended Fine Auger Structure
FIM	Field Ion Microscopy
FYNES	Fluorescence-Yield Near-Edge Structure
GIXD	Grazing-Incidence X-Ray Diffraction
GIXS	Grazing-Incidence X-Ray Scattering
He diffr.	Helium Diffraction
HEIS	High-Energy Ion Scattering
HREELS	High-Resolution Electron Energy Loss Spectroscopy

ICISS	Impact Collision Ion Scattering Spectroscopy
INS	Ion Neutralization Spectroscopy
Ion scatt.	Ion Scattering
IRS, IRAS	Infrared (Reflection-Absorption) Spectroscopy
ISS	Ion Scattering Spectroscopy
KSLA	Kinematic Sublayer Addition
LEED	Low-Energy Electron Diffraction
LEIS	Low-Energy Ion Scattering
LEPD	Low-Energy Positron Diffraction
MEED	Medium-Energy Electron Diffraction
MEIS	Medium-Energy Ion Scattering
Neutr. diffr.	Neutron Diffraction
NEXAFS	Near-Edge X-ray Absorption Fine Structure
NMR	Nuclear Magnetic Resonance
NPD	Normal Photoelectron Diffraction
OPD	Off-normal Photoelectron Diffraction
PED	Photoelectron Diffraction
PES	Photoelectron Spectroscopy
PEXAFS	Photoemission Extended X-ray Absorption Fine Structure
PLEED	(Spin-) Polarized LEED
RAIRS	Reflection-Absorption Infrared Spectroscopy
RBS	Rutherford Backscattering
RHEED	Reflection High-Energy Electron Diffraction
RFS	Renormalized Forward Scattering
RSP	Reverse Scattering Perturbation
SEELFS	Surface Electron Energy Loss Fine Structure
SEM	Scanning Electron Microscopy
SEXAFS	Surface Extended X-ray Absorption Fine Structure
SIMS	Secondary Ion Mass Spectroscopy
SPLEED	Spin-Polarized LEED
SSRL	Stanford Synchrotron Radiation Laboratory
STM	Scanning Tunneling Microscopy
TDS	Thermal Desorption Spectroscopy
TEAS	Thermal Energy Atomic Scattering
TED	Transmission Electron Diffraction
TEM	Transmission Electron Microscopy
TLEED	Tensor LEED
TOF-SARS	Time-of-Flight Scattering and Recoiling Spectroscopy
TPD	Temperature Programmed Desorption
UPS	Ultraviolet Photoelectron Spectroscopy
WF(C)	Work Function (Change)
XANES	X-ray Absorption Near-Edge Structure
XAS	X-ray Absorption Spectroscopy
XPS	X-ray Photoelectron Spectroscopy
XRD	X-Ray Diffraction
XSW	X-Ray Standing Wave

Journal Names

Acta Crys.	Acta Crystallographica
Appl. Phys.	Applied Physics
Appl. Surf. Sci.	Applications of Surface Science
Can. J. Chem.	Canadian Journal of Chemistry
Can. J. Phys.	Canadian Journal of Physics
Chem. Phys. Lett.	Chemical Physics Letters
Europhysics Lett.	Europhysics Letters
J. Am. Chem. Soc.	Journal of the American Chemical Society
Jap. J. Appl. Phys.	Japanese Journal of Applied Physics
J. Chem. Phys.	Journal of Chemical Physics
J. Phys. Chem.	Journal of Physical Chemistry
J. Phys.	Journal of Physics (London)
J. Vac. Sci. Technol.	Journal of Vacuum Science and Technology
Phys. Lett.	Physics Letters
Phys. Stat. Sol.	Physica Status Solidi
Phys. Rev.	Physical Review
Phys. Rev. Lett.	Physical Review Letters
S. Afr. J. Phys.	South African Journal of Physics
Surf. Sci.	Surface Science
Sol. St. Commun.	Solid State Communications
Springer Series in Surface Sciences (not abbreviated)	
Z. Phys.	Zeitschrift für Physik
Z. Naturf.	Zeitschrift für Naturforschung

Appendix C: Major Surface Structure Techniques

This database reports surface structural data from a wide variety of techniques. The most prominent techniques and their basic principles are the following, in alphabetical order (a more complete list is given in Appendix B):

a. angle-resolved photoelectron diffraction (ARPD):
also labelled ARPED, ARXPD, ARPEFS, NPD, OPD, etc., depending on author and mode of operation): photons cause the emission of photoelectrons in the energy range of 100–2000 eV, which are diffracted by the near-surface atoms before being detected as a function of emission angle and/or energy; the diffraction is simulated theoretically to obtain the structure that produces the best fit to the measured data; multiple scattering in the forward scattering direction must generally be included in such simulations; see C.S. Fadley, in ''Synchrotron Radiation Research: Advances in Surface Science,'' edited by R.Z. Bachrach (Plenum, New York, 1992), pp 421–517.

b. ion scattering:
at low, medium or high energies (called LEIS below about 2 keV: MEIS at intermediate energies and HEIS above about 500 keV; one also uses the name Rutherford backscattering at energies above about 2keV): ions are scattered by surface atoms, following classical trajectories that create shadow cones, thereby yielding atomic alignment information as a function of angular detection; theoretical simulations, which must include vibrational effects, are used to optimize corresponding structural models; see M. Aono, Y. Hou, C. Oshima and Y Ishizawa, Phys. Rev. Lett. **49**, 567 (1982), H. Niehus and G. Comsa, Surf. Sci. **140**, 18 (1984), J.F. van der Veen, Surf. Sci. Rep. **5**, 199 (1985).

c. low-energy electron diffraction (LEED):
electrons in the range 20–300 eV are elastically diffracted by a surface into sharp beams (for ordered surfaces) or diffuse angular distributions (for disordered layers); the intensities are generally measured as a function of energy and the resulting I-V curves simulated theoretically using a surface geometry that is made to vary until the fit with experiment is optimized; multiple scattering must generally be included in such simulations; see J.B. Pendry, ''Low-Energy Electron Diffraction'', Academic Press (London) 1974, M.A. Van Hove, W.H. Weinberg and C.-M. Chan, ''Low-Energy Electron Diffraction'', Springer-Verlag (Heidelberg) 1986.

d. surface extended x-ray absorption fine structure (SEXAFS):
photons are absorbed by surface atoms, which emit electrons that scatter back from nearby atoms and modulate the absorption probability as a function of the photon energy (and thereby of the electron energy in the range 100–1000 eV); surface sensitivity is obtained by detecting emitted electrons or ions; directional information can be obtained by using variable incident polarization; Fourier transformation, after correction for phase shifts, gives approximate interatomic distances, while a full theoretical simulation of the process allows more reliable fitting of the experimental data; see P.H. Citrin, J. Phys. (Paris), Colloque C8, 437 (1986), J.E. Rowe, in ''Synchrotron Radiation Research: Advances in Surface Science,'' edited by R.Z. Bachrach (Plenum, New York, 1992).

e. x-ray diffraction (XRD):
x-rays are diffracted from surfaces by choosing grazing incidence and/or grazing emergence and/or fractional-order diffraction conditions; intensities are measured as a function of scattering angles and/or energy; direct theoretical simulation, as in bulk x-ray diffraction, may be preceded by a Patterson function interpretation; see R. Feidenhans'l, Surf. Sci. Rep. **10**, 105 (1989).

f. x-ray standing waves (XSW):
bulk-reflected waves interfering with the incident x-ray wave create standing waves, which can be phase shifted across surface atoms, yielding their positions with respect to the bulk crystal structure; surface sensitivity is obtained by monitoring element-specific emission from foreign adsorbates, e.g. fluorescence; see P.L. Cowan, J.A. Golovchenko and M.F. Robbins, Phys. Rev. Lett. **44**, 1680 (1980).

Appendix D: List of Covalent Radii

This list of covalent radii is mainly provided as a guide for estimating bond lengths. The radii have been compiled from a variety of sources. All values are in Ångströms.

Element	Radius	Element	Radius	Element	Radius
1 (H)	0.4350	32 (Ge)	1.2248	63 (Eu)	1.9840
2 (He)	0.9300	33 (As)	1.2000	64 (Gd)	1.8180
3 (Li)	1.5199	34 (Se)	1.1600	65 (Tb)	1.8005
4 (Be)	1.1430	35 (Br)	1.1400	66 (Dy)	1.7951
5 (B)	0.9750	36 (Kr)	1.1200	67 (Ho)	1.7886
6 (C)	0.6550	37 (Rb)	2.4700	68 (Er)	1.7794
7 (N)	0.7500	38 (Sr)	2.1513	69 (Tm)	1.7687
8 (O)	0.7300	39 (Y)	1.8237	70 (Yb)	1.9396
9 (F)	0.7200	40 (Zr)	1.6156	71 (Lu)	1.7515
1 (H)	0.4350	32 (Ge)	1.2248	63 (Eu)	1.9840
2 (He)	0.9300	33 (As)	1.2000	64 (Gd)	1.8180
3 (Li)	1.5199	34 (Se)	1.1600	65 (Tb)	1.8005
4 (Be)	1.1430	35 (Br)	1.1400	66 (Dy)	1.7951
5 (B)	0.9750	36 (Kr)	1.1200	67 (Ho)	1.7886
6 (C)	0.6550	37 (Rb)	2.4700	68 (Er)	1.7794
7 (N)	0.7500	38 (Sr)	2.1513	69 (Tm)	1.7687
8 (O)	0.7300	39 (Y)	1.8237	70 (Yb)	1.9396
9 (F)	0.7200	40 (Zr)	1.6156	71 (Lu)	1.7515
10 (Ne)	0.7100	41 (Nb)	1.4318	72 (Hf)	1.5973
11 (Na)	1.8579	42 (Mo)	1.3626	73 (Ta)	1.4280
12 (Mg)	1.6047	43 (Tc)	1.3675	74 (W)	1.3705
13 (Al)	1.4318	44 (Ru)	1.3529	75 (Re)	1.3800
14 (Si)	1.1758	45 (Rh)	1.3450	76 (Os)	1.3676
15 (P)	1.0600	46 (Pd)	1.3755	77 (Ir)	1.3573
16 (S)	1.0200	47 (Ag)	1.4447	78 (Pt)	1.3873
17 (Cl)	0.9900	48 (Cd)	1.4894	79 (Au)	1.4419
18 (Ar)	0.9800	49 (In)	1.6662	80 (Hg)	1.5025
19 (K)	2.2620	50 (Sn)	1.5375	81 (Tl)	1.7283
20 (Ca)	1.9758	51 (Sb)	1.4000	82 (Pb)	1.7501
21 (Sc)	1.6545	52 (Te)	1.3600	83 (Bi)	1.4600
22 (Ti)	1.4755	53 (I)	1.3300	84 (Po)	1.4600
23 (V)	1.3090	54 (Xe)	1.3100	85 (At)	1.4500
24 (Cr)	1.2490	55 (Cs)	2.6325	86 (Rn)	1.4300
25 (Mn)	1.3500	56 (Ba)	2.1705	87 (Fr)	2.5000
26 (Fe)	1.2411	57 (La)	1.8725	88 (Ra)	2.1400
27 (Co)	1.2535	58 (Ce)	1.8243	89 (Ac)	1.8775
28 (Ni)	1.2460	59 (Pr)	1.8362	90 (Th)	1.7975
29 (Cu)	1.2780	60 (Nd)	1.8295	91 (Pa)	1.6086
30 (Zn)	1.3325	61 (Pm)	1.8090	92 (U)	1.5683
31 (Ga)	1.3501	62 (Sm)	1.8040		

Appendix E: The 17 2D Space Groups

This appendix illustrates the 17 two-dimensional space groups that exist for surfaces and interfaces in strictly two dimensions. SSD characterizes both ideal substrate terminations and actual surfaces by their respective space groups.

Additional space groups are possible if one allows symmetry operations across the interface, e.g., mirror symmetry about the interface, as may happen in twinning of two crystals. Then, 80 space groups exist; they have been tabulated by Wood (1964). However, situations where any of these additional space groups might be useful are probably rare in surface science. The 17 2D space groups always remain applicable, even if they do not convey all the useful information in those rare cases.

The following diagrams show, for each of the 17 2D space groups:

- one unit cell outlined in heavy lines (where not overdrawn by thin symmetry lines);
- all applicable symmetry elements: 2-, 3-, 4- and 6-fold rotation axes (perpendicular to the surface), drawn as black ovals, triangles, squares and hexagons; mirror lines (planes perpendicular to the surface) drawn as thin lines, glide lines (planes perpendicular to the surface) as dashed lines;
- a general point in the unit cell, symbolized by a yin-and-yang figure, together with all other points equivalent to it by symmetry; the left-handed and right-handed yin-and-yang symbols show how orientations are mirrored and rotated into each other.

The 17 two-dimensional space groups

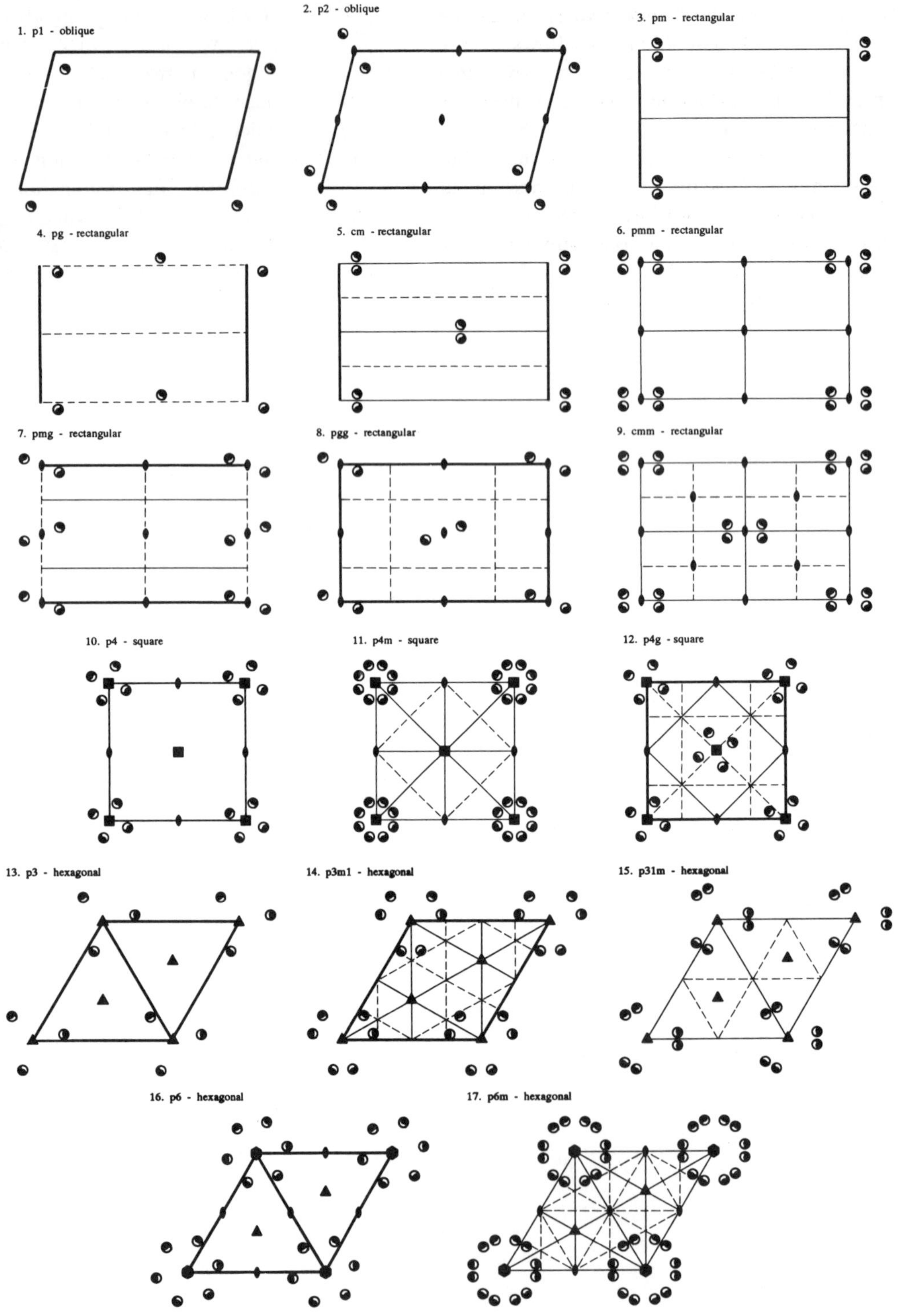

Appendix F: Examples of Superlattice Cells and Notations

This appendix illustrates many common 2-dimensional superlattices and their notations. In each case one common unit cell of the superlattice is drawn; occasionally, another equivalent unit cell is also included. The Wood, matrix and Bibérian (rect) notations are given wherever they apply. For centered unit cells, a non-centered cell is also shown outlined with thinner lines.

The illustrations assume either a hexagonal substrate lattice or a square substrate lattice. Lattice vectors of the (1×1) substrate unit cell are shown. The superlattice notation assumes this choice. This unit cell choice is non-unique, e.g., an angle of 60° could have been chosen instead of 120° for the hexagonal lattice; a different (1×1) unit cell can change the superlattice notation, especially the matrix notation. Therefore, the superlattice notation is also non-unique, i.e. other notations may also apply.

The Bibérian (rect) notation is only valid for a hexagonal substrate lattice and it is only used for rectangular or centered rectangular superlattice cells.

Other types of substrate lattices (rectangular or oblique) can be obtained by simple one-dimensional stretching or shearing of the illustrated lattices. The matrix notation remains valid under these modifications. However, some of the Wood notations are then no longer applicable. The Wood notation assumes that the two lattice vectors of the superlattice unit cell have the same angle between them as those of the substrate. This angle in general is not preserved in going to rectangular or oblique lattices.

Superlattices on hexagonal substrate lattice

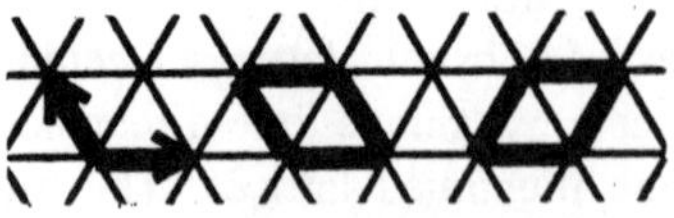

(1×1), $\begin{pmatrix}1 & 0\\0 & 1\end{pmatrix}$

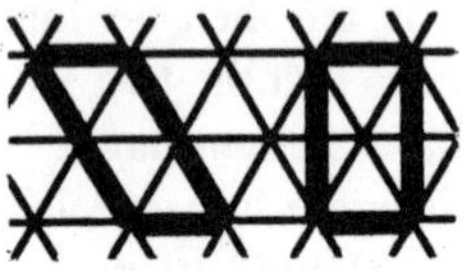

(1×2), $c(2\times2)$, $(\sqrt{3}\times1)rect$, $\begin{pmatrix}1 & 0\\0 & 2\end{pmatrix}$, $\begin{pmatrix}1 & 0\\1 & 2\end{pmatrix}$

(2×2), $c(2\sqrt{3}\times2)rect$, $\begin{pmatrix}2 & 0\\0 & 2\end{pmatrix}$

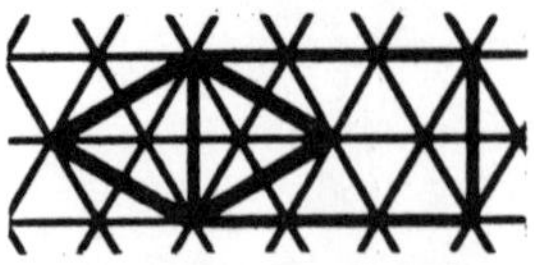

$(\sqrt{3}\times\sqrt{3})R30°$, $c(\sqrt{3}\times3)rect$, $\begin{pmatrix}2 & 1\\-1 & 1\end{pmatrix}$, $\begin{pmatrix}2 & 1\\1 & 2\end{pmatrix}$

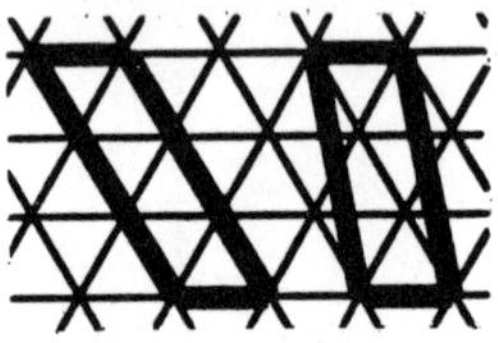

(1×3), $c(3\sqrt{3}\times1)rect$, $\begin{pmatrix}1 & 0\\0 & 3\end{pmatrix}$, $\begin{pmatrix}1 & 0\\1 & 3\end{pmatrix}$

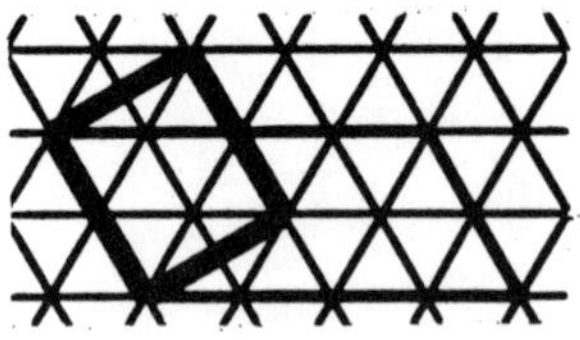

$c(4\times2)$, $(\sqrt{3}\times2)rect$, $c\begin{pmatrix}4 & 0\\0 & 2\end{pmatrix}$, $\begin{pmatrix}2 & 1\\0 & 2\end{pmatrix}$

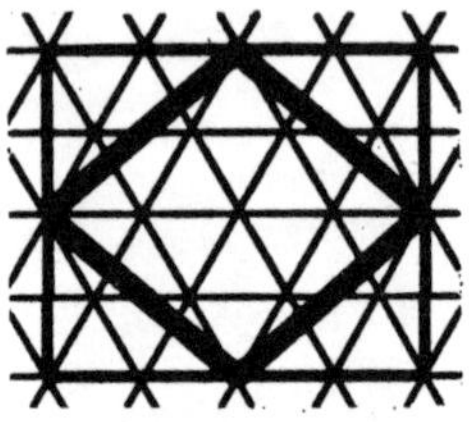

$c(2\sqrt{3}\times4)rect$, $\begin{pmatrix}3 & 1\\1 & 3\end{pmatrix}$, $\begin{pmatrix}3 & -2\\-1 & 2\end{pmatrix}$

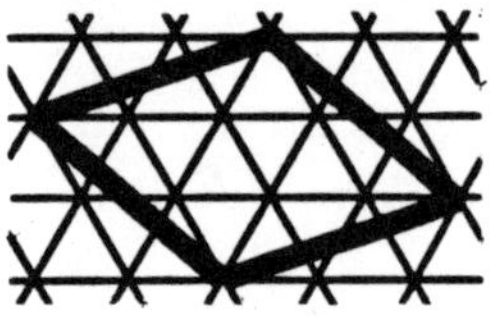

$(\sqrt{7}\times\sqrt{7})R\ arctg(\sqrt{3}/5)$, $\begin{pmatrix}3 & 1\\-1 & 2\end{pmatrix}$

Superlattices on square substrate lattice

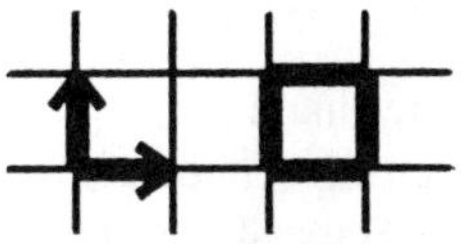

(1×1), $\begin{pmatrix}1 & 0\\0 & 1\end{pmatrix}$

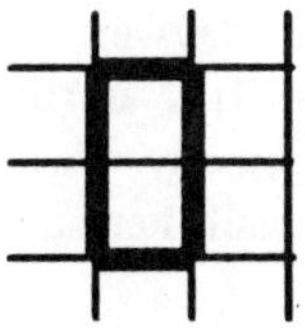

(1×2), $\begin{pmatrix}1 & 0\\0 & 2\end{pmatrix}$

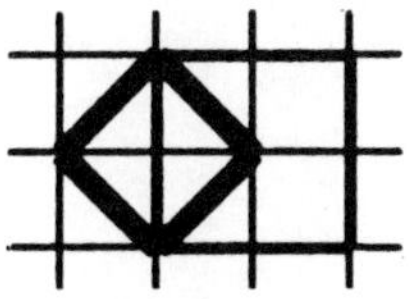

$c(2\times2)$, $(\sqrt{2}\times\sqrt{2})R\,45°$, $\begin{pmatrix}1 & 1\\-1 & 1\end{pmatrix}$

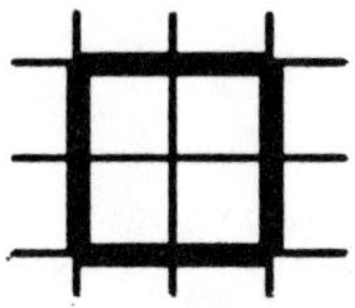

(2×2), $\begin{pmatrix}2 & 0\\0 & 2\end{pmatrix}$

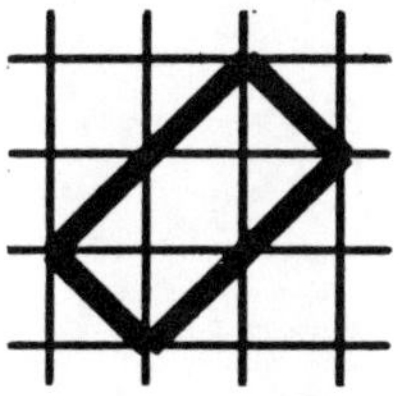

$(2\sqrt{2}\times\sqrt{2})R\,45°$, $\begin{pmatrix}2 & 2\\-1 & 1\end{pmatrix}$

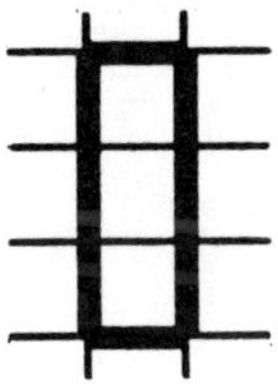

(1×3), $\begin{pmatrix}1 & 0\\0 & 3\end{pmatrix}$

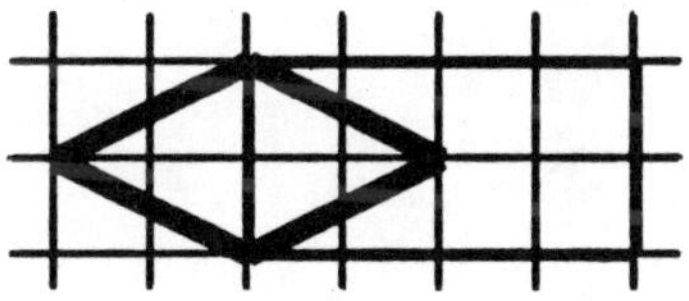

$c(4\times2)$, $\begin{pmatrix}2 & 1\\-2 & 1\end{pmatrix}$, $\begin{pmatrix}2 & 1\\0 & 2\end{pmatrix}$

Appendix G: Contacts

If you have questions or comments about the Surface Science Database, please contact:

Joan Sauerwein
National Institute of Standards and Technology
Standard Reference Data
Building 221, Room A320
Gaithersburg, MD 20899–0001
Internet: srdata@enh.nist.gov
Phone: (301) 975–2208
FAX: (301) 926–0416

If you have technical questions relating to the data, contact:

P.R. Watson
Department of Chemistry
Oregon State University
Corvallis, OR 97331, USA
e-mail: watsonp@ccmail.orst.edu

M.A. Van Hove
Materials Sciences Division
Lawrence Berkeley Laboratory
Berkeley, CA 94720, USA
e-mail: vanhove@lbl.bitnet or vanhove@lbl.gov

K. Hermann
Abteilung Theorie
Fritz-Haber-Institut
Faradayweg 4–6
D-14195 Berlin
Germany
e-mail: hermann@FHI-Berlin.MPG.DE

6. Index

This refers to section, subsection and appendix numbers.

2D Bulk Symmetry 1.4.2
2D Surface Symmetry 1.4.2

Abbreviations App. B
Adsorbate 1.4.2
Analytical Methods 1.4.5
ASCII Files for SSD App. A
Authors 1.4.1

BALSAC 1.5
Bibérian Notation App. F
Bibliography 4
Bond Distances and Angles 1.4.13
Bulk Lattice 1.4.2
  Repeat Vectors 1.4.12
  Substrate Unit Cell 1.4.10

Cell Type 1.4.12
Chemical Element 1.4.12
Classification Number 1.4.1
Coadsorbates 1.4.2
Commensurate Superlattice 1.4.11
Comments 1.4.4
Common Name 1.4.1
Contamination 1.4.5
Coordinates 1.4.12
Coverage 1.4.2
Criteria for Inclusion In SSD 1.2
Crystal Face 1.4.2
Crystallinity 1.4.5

Data Item
  Definition 1.4
  Format App. A
Data Set, Experimental 1.4.6
Disordered Lattice 1.4.11
Distance, Interatomic 1.4.13
Domain Orientations 1.4.10

Electronic Version of SSD 1.2
Epilayer 1.4.12
Error Bars 1.4.12

Files
  ASCII App. A
  Paradox App. A.1

Incommensurate Superlattice 1.4.11
Interatomic Distance 1.4.13
Interface 1.4.12
Interface Layer 1.4.12
Interlayer Spacing 1.4.12

Journal
  Abbreviations App. B
  Data Item 1.4.1

Lattice
  Bulk 1.4.10
  Super- 1.4.10

LATUSE 1.5
Matrix Notation 1.4.2, App. F
Method, Experimental 1.4.6

Non-reconstructive Lattice-gas Disorder 1.4.11
Notation, Superlattice App. F
Number of Samples 1.4.5

Occupancy 1.4.12
Overlayer 1.4.12

Pattern, LEED 1.4.2

Quality of Experiment-theory Fit 1.4.9

Radii, Covalent App. D
Randomly Mixed Layer 1.4.11
Reconstructive Lattice-gas Disorder 1.4.11
Rect Notation App. F
Region 1.4.12
Repeat Vectors, Bulk 1.4.12

SARCH 1.5
SCIS 1.5
Site
  Occupancy 1.4.12
  Orientations 1.4.10
Software, Related 1.5, 4
Space Groups 14.2, App.E
SSDCHK App. A.5
Structures Examined 1.4.8
Structure Type 1.4.3
Submission of New Structures App. A
Substrate 1.4.2
Superlattice
  Cells 14.10, App.F
  Commensurate 1.4.11
  Disordered 1.4.11
  Examples App. F
  Incommensurate 1.4.11
  Notations App. F
Surface 2D Unit Cells 1.4.10
Symbols, List App. B

Technique
  Data Item 1.4.1
  List of Abbreviations App. B
Temperature 1.4.2
Treatment
  Data 1.4.7
  Sample 1.4.5

Use of SSD 1.2
Utility SSDCHK App. A.5

Wood Notation App. F

American Institute of Physics

500 Sunnyside Boulevard
Woodbury, NY 11797-2999

Roland W. Schmitt	Chair, Governing Board
Marc H. Brodsky	Executive Director and CEO
Roderick M. Grant	Secretary
Arthur T. Bent	Treasurer and CFO
John S. Rigden	Director of Physics Programs
Darlene Carlin-Walters	Vice President, Publishing

The American Institute of Physics was founded in 1931; it is chartered as a membership corporation with leading societies in the fields of physics and astronomy as members. The Institute combines into one operating agency those functions which can best be done by the societies jointly. Its purpose is the advancement and diffusion of the knowledge of physics and its application to human welfare. About 100 000 physicists belong to its ten member societies: The American Physical Society, Optical Society of America, Acoustical Society of America, Society of Rheology, American Association of Physics Teachers, American Crystallographic Association, American Astronomical Society, American Association of Physicists in Medicine, American Vacuum Society, American Geophysical Union. In addition, there are 20 Affiliated Societies, 84 Corporate Associates, and 6500 members of the Society of Physics Students.

For itself and its member societies, AIP publishes 20 archival journals, 17 translated Soviet journals, a general interest journal, conference proceedings, handbooks, and bulletin-programs, as well as an integrated package of current awareness journals, magnetic tapes, and microfilm. AIP also serves the public by making available to the press and other channels of public information reliable communications on physics and astronomy and their progress; carries on extensive career planning and placement activities; encourages and assists in the documentation and study of the history and philosophy of recent physics; cooperates with local, national, and international organizations devoted to physics and the related sciences; and fosters the relations of the science of physics to other sciences and to the arts and industries.

American Chemical Society

1155 Sixteenth Street, NW
Washington, DC 20036-9976

Paul H. L. Walter	Chairman of the Board
Ned D. Heindel	President
John K. Crum	Executive Director
Brian A. Bernstein	Treasurer
D.H. Michael Bowen	Deputy Executive Director and Secretary
Robert J. Massie	Director, Chemical Abstracts Service
Robert H. Marks	Director, Publications Division*
Halley A. Merrell	Director, Membership Division
Sylvia A. Ware	Director, Education Division

The American Chemical Society was founded in 1876 as a scientific and educational society, and in 1937 was granted a national charter by an act of Congress. Its objectives include the advancement of chemistry, the promotion of research in science and industry, improvement of qualifications and usefulness of chemists, increased diffusion of chemical knowledge, and the promotion of scientific interest and inquiry.

ACS publishes 26 magazines, journals, and other scientific periodicals in addition to its participation in publication of the *Journal of Physical and Chemical Reference Data*. It also publishes books and other nonperiodical reports and operates the Chemical Abstracts Service, the key to the world's chemical literature. It pursues its objectives through its services to chemists and chemical engineers and to education, industry, and government.

*Inquiries regarding ACS participation in JPCRD should be directed to this division.

National Institute of Standards and Technology

U.S. Department of Commerce
Gaithersburg, MD 20899-0001

U.S. Department of Commerce	Ronald H. Brown, Secretary
Technology Administration	Mary L. Good, Under Secretary for Technology
National Institute of Standards and Technology	Arati Prabhakar, Director

The National Institute of Standards and Technology (NIST), formerly the National Bureau of Standards, was established by an act of Congress 3 March 1901. NIST works to strengthen U.S. industry's international competitiveness, advance science and engineering, and improve public health, safety, and the environment. With an annual budget of more than $350 million and about 3100 staff members, NIST conducts science and engineering research in commercially important fields such as advanced materials, information systems, biotechnology, optoelectronics, computer-integrated manufacturing, and sensor technology.

NIST's laboratory research is designed to support development of critical emerging technologies and the new measurement methods and standards necessary to make them commercially viable. The ability of U.S.-based industries to exploit these new technologies determines in large part the health of the U.S. economy.

Several new NIST programs are designed to help spur innovation at U.S. businesses through seed money for development of generic technologies; grants to states for support of technology transfer programs; and financial and technical assistance to help small and mid-sized companies adopt more efficient manufacturing methods.

Throughout its history, NIST has helped industry to develop high-quality products and processes by supplying standard reference materials and data; by continuously improving technologies for precision measurements of mass, time, length, voltage, radiation, and other units; and through active participation in voluntary standards-setting bodies.